W9-DCB-865

APPLICATION OF ACCELERATORS IN RESEARCH AND INDUSTRY

Application of Accelerators in Research and Industry
Twenty-first International Conference
CAARI 2010

University of North Texas
Denton, Texas 76203

Sandia National Laboratories
Albuquerque, New Mexico 87185

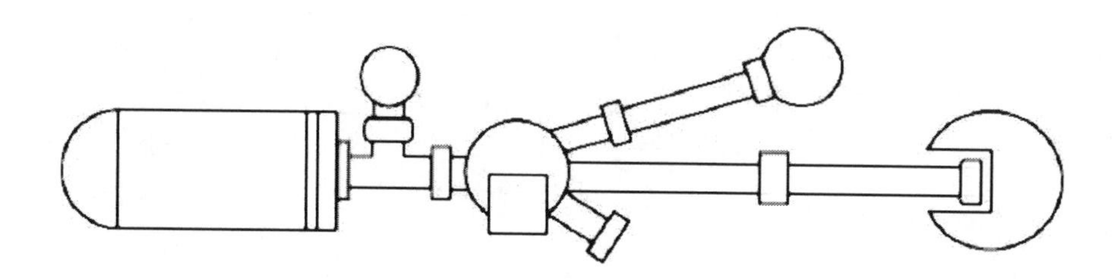

Previous Proceedings in the Series of conferences on the Application of Accelerators in Research and Industry

Year	Conference	Publisher	ISBN
2008	Twentieth	AIP Conf. Proceedings Vol. 1099	0-7354-0633-9
2006	Nineteenth	Elsevier, NIMB Vol 261	
2004	Eighteenth	Elsevier, NIMB Vol 241	
2002	Seventeenth	AIP Conf. Proceedings Vol. 680	0-7354-0149-7
2000	Sixteenth	AIP Conf. Proceedings Vol 576	0-7354-0015-6
1998	Fifteenth	AIP Conf. Proceedings Vol 475	1-56396-825-8
1996	Fourteenth	AIP Conf. Proceedings Vol 392	1-56396-652-2
1994	Thirteenth	Elsevier, NIMB Vol. 99	
1992	Twelfth	Elsevier, NIMB Vol. 79	

To learn more about these titles, or the AIP Conference Proceedings Series, please visit the webpage: **http://proceedings.aip.org/proceedings**.

APPLICATION OF ACCELERATORS IN RESEARCH AND INDUSTRY

Twenty-First International Conference

Fort Worth, Texas 8 – 13 August 2010

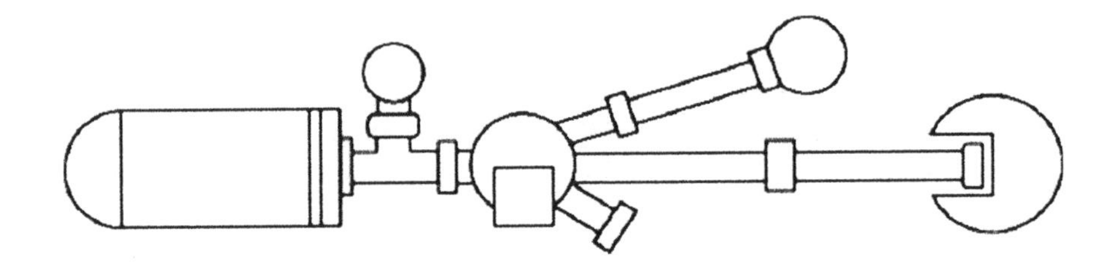

EDITORS

Floyd D. McDaniel
University of North Texas, Denton, Texas

Barney L. Doyle
Sandia National Laboratories, Albuquerque, New Mexico

SPONSORING ORGANIZATIONS
University of North Texas
Sandia National Laboratories

All papers have been peer reviewed

Melville, New York, 2011
AIP | CONFERENCE PROCEEDINGS ■ 1336

Editors

Floyd D. McDaniel
University of North Texas
1155 Union Circle #311427
Denton, Texas 76203

E-mail: mcdaniel@unt.edu

Barney L. Doyle
Sandia National Laboratories
PO Box 5800, MS 1056
Albuquerque, New Mexico 87185

E-mail: bldoyle@sandia.gov

L.C. Catalog Card No. 2011900630
ISBN 978-0-7354-0891-3
ISSN 0094-243X
Printed in the United States of America

AIP Conference Proceedings, Volume 1336
Application of Accelerators in Research and Industry
Twenty-First International Conference

Table of Contents

SECTION III - EARLY CAREER HOT TOPICS

SECTION VI - ION BEAM ANALYSIS

SECTION VII - ION BEAM MODIFICATIONS

SECTION VIII - MEDICAL APPLICATIONS AND RADIOISOTOPES

SECTION IX - NUCLEAR-BASED ANALYSIS

SECTION X - NUCLEAR PHYSICS

Editorial

This special volume of the American Institute of Physics Conference Series contains the "Proceedings of the 21st International Conference on the Application of Accelerators in Research and Industry (CAARI 2010)". This conference was held in Fort Worth, Texas, USA, August 8–13, 2010. The CAARI conference series is a biennial meeting that began as a "Conference on the Use of Small Accelerators for Teaching and Research" by Jerry Duggan in 1968, while he was a staff member at Oak Ridge Associated Universities. Jerry came to the University of North Texas in Denton, Texas, in 1973 and he continued the Conference series in 1974. Over the years, the number of participants in the Conference has grown, as well as the number of research topics presented, and the accelerated ion energies. In late 1974, Lon Morgan began helping Jerry with the industrial applications of accelerators and the Conference was renamed the *International Conference on the Application of Accelerators in Research and Industry.* Today, the Conference brings together scientists from all over the world who use particle accelerators in their research and industrial applications. The 21st CAARI conference had 513 attendees (100 of which were students and 10 were accompanying persons) from 38 countries.

CAARI 2010 was held in the Renaissance Worthington Hotel in the heart of downtown Fort Worth. The Hotel was within walking distance to 36 restaurants, 3 live comedy theatres, 2 movie theatres, museums, the Bass Concert Hall, nightclubs, bookstores and many beautiful walking areas. Areas of interest included historic downtown, the Stockyards National Historic District, the Cultural District, the Fort Worth Zoo and the Sundance Square for shopping.

The Conference Co-Chairs were Floyd Del McDaniel, University of North Texas, and Barney Doyle, Sandia National Laboratories. Registration began Sunday, August 8, and was followed by a welcoming reception with musical entertainment.

The main topical areas of the Conference were Accelerator Technology, Atomic/Nuclear Physics, Focused Ion Beams/Nanofabrication, Ion Beam Analysis/Modifications, Medical Applications/Radioisotopes, Nuclear-based Analysis, Radiation Effects, Safety, Security and Contraband Detection, and Teaching with Accelerators. Two new topics were introduced this year: Environmental/Earth Sciences and Advanced Energy, and Early Career Hot Topics. The Early Career area gave young researchers an opportunity to showcase their research and development through oral presentations.

The Conference began with welcoming remarks and the plenary session on Monday morning in which overviews were presented on aspects of the Conference topic areas. Plenary topics included the *ITER: a Fusion Energy Source* by Michael Ulrickson (Sandia National Labs), *A Rapier, Not a Broad Sword: Accelerator-based analyses of aerosols for global climate research* by Thomas Cahill (DELTA Group, UC-Davis), *Industrial Applications of Accelerators* by Robert Hamm (R&M Technical Enterprises), and *The Future of Accelerators: A DOE symposium and the stakeholders' view* by Walter Henning (Argonne National Labs).

The Industrial Exhibit Show began at 10:00 AM on Monday and ran through Wednesday at 3:30 PM. Monday afternoon and throughout the week, there were 6 parallel break-out oral sessions, which included 275 Review/Invited talks and 121 Contributed talks. Poster sessions were held Monday through Thursday showcasing 58 posters. The Monday through Wednesday poster sessions included a social hour with music from various local artists.

The Topic Editors meeting was held on Tuesday evening for assigning referees to all the submitted abstracts. A Student Appreciation Event was held Wednesday evening at the 8.0 Restaurant and Bar. Following the student event, the Topic Editors and Session Chairs got together for a topical "jam" session with Barney and Del. The Banquet was held on Thursday with entertainment provided by the Bonnie Norris Band.

Application of Accelerators in Research and Industry
AIP Conf. Proc. 1336, 1-2 (2011); doi: 10.1063/1.3586046

At the Banquet, awards were given out for the best student poster in each of the four poster sessions. The student posters were judged on scientific merit, presentation quality of the poster, and the quality of the student's oral presentation. The winners were: Mark Hanni (Colorado State University) for *Spectroscopy of High-L Rydberg Levels of Fr-like Th using the RESIS Technique*; John Weidner (Los Alamos National Laboratory) for *Accelerator Production of Ac-225 for Alpha-immunotherapy;* Ion Burducea (Horia Hulubei National Institute of Physics and Nuclear Engineering) for *Rutherford Backscattering Spectrometry of InN Thin Films;* and Mangal Dhoubhadel (University of North Texas) for *Synthesis and Characterization of Transitional Metal Nanosystems in Silicon.* Also during the banquet, we introduced the 31 student Conference Assistants and awarded them Certificates of Appreciation for their valuable assistance to their assigned session chairs and speakers.

Friday morning, we held a special class and plenary session. Konstantin Toms (University of New Mexico/CERN) gave a short class on *Brief Overview of the Standard Model for Experimentalists* and then a plenary talk about *The Developments and Experiments on the LHC.* Barney and Del wrapped up the week with an all-attendee discussion about *CAARI: Past, Present and Future.* We had a total of 454 presentations and this volume contains 161 manuscripts.

We would like to use this occasion to thank all of our referees and reviewers for their hard work. Without them these proceedings would have not been realized. This Conference was handled electronically through the CAARI web site, which was made possible through the work of Zdenek Nejedly of AnZ Solutions.

We thank our many sponsors for their support; this allowed us to support 30 students from around the world. We thank our Industrial Exhibitors for their continuing support. The exhibits allow those of us in the scientific community to keep up with the rapidly changing technologies. We also thank our Topic Editors and Session Chairs for their hard work and dedication to the field and to this Conference series. In many cases, the 73 Session Chairs each single-handedly put together an outstanding list of speakers for their technical session. There are several additional people whom we wish to recognize for their contribution to the success of this conference. First of all, we thank Margaret Hall, the Conference Secretary, who handled the non-scientific organization of the conference. Her negotiation skills and hard work were crucial for the success of this conference. We thank our extra helpers – Gina Atkinson, Lisa Clarke, Holly Decker, and Mistie Hagen - for their on-site registration and T-shirt booth assistance. We thank Scott Sifferd from Sandia who configured the computers for the registration desk, manuscript desk, and Cyber Café, and handled computer-related issues throughout the conference. A big "thank you" goes to our photographers, Mangal Dhoubhadel, Lucas Phinney, Sahil Naik and Ratish Patnaik; our banquet slideshow and CAARI 2010 Yearbook would not have been produced without them. We thank Karen Sloan, who handled all the technical aspects of the conference, and for her work with the Topic Editors, Session Chairs and authors.

The next conference in this series will be held August 5 - 10, 2012 at the Renaissance Worthington Hotel in downtown Fort Worth. Del, Barney and Karen look forward to seeing all of you in Fort Worth for the 22nd International Conference on the Application of Accelerators in Research and Industry. The web site for the CAARI conferences is www.caari.com. Check this site periodically for upcoming information about CAARI 2012.

Floyd Del McDaniel and Barney L. Doyle
CAARI 2010 Conference Co-chairs

Obituary

Carl Jerome Maggiore, Ph.D.
(1943 – 2011)

Carl was born in Grand Island, Nebraska. He graduated from Creighton University and received his PhD in physics from Michigan State University, where he completed his thesis work at the cyclotron. Carl's life in physics was characterized by a highly flexible intellect and an insatiable curiosity. He was a patient teacher to his summer students, and was always a gracious and generous colleague. Carl began his career as a researcher in the Environmental Sciences Laboratory of the Mt. Sinai School of Medicine in New York City and went on to design detectors for the Princeton Gamma Tech Company. He began to work at Los Alamos National Laboratory in April 1976. At Los Alamos, his main interest was in finding practical applications for accelerated particles. Initially, he worked with particles from the vertical accelerator at the Ion Beam Facility where he developed the first nuclear microprobe at a U.S. National Laboratory. He later founded the Ion Beam Materials Laboratory, which was specifically dedicated to using accelerated particles to modify or analyze materials, and for which he and his colleagues receive a Distinguished Performance Award. On his retirement from LANL, Carl worked for several years to test the feasibility of treating tumors using antiprotons, performing experimental work at CERN under the auspices of PBar Technology Corporation, and with collaborators both at CERN and the University of Aarhus, in Denmark.

In 2007, Carl began a second career as an artist, discovering a talent and passion for drawing and sculpture. Music was also an essential part of Carl's life. He loved the outdoors, and felt privileged to live in the mountains of New Mexico surrounded by so much beauty.

Carl Jerome Maggiore passed away on January 7, 2011 at his home in Los Alamos, New Mexico, following a 10-month battle with metastatic rectal cancer. Carl was a frequent attendee and session chair at CAARI and a great friend, and our community will sorely miss him.

Barney Doyle and Del McDaniel
Co-Chairs of CAARI 2010

Joe Tesmer
Los Alamos National Laboratory

CAARI 2010

Proceedings of the Twenty-First International Conference

APPLICATION OF ACCELERATORS IN RESEARCH AND INDUSTRY

Fort Worth, Texas

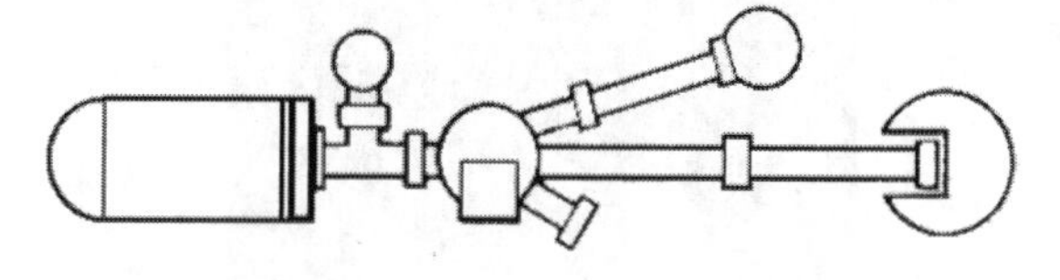

August 8-10, 2010

EDITORS

Floyd D. McDaniel — University of North Texas, Denton, TX, USA

Barney L. Doyle — Sandia National Laboratories, Albuquerque, NM, USA

TOPIC EDITORS

Itzik Ben-Itzhak — J.R. Macdonald Laboratory, KSU, Manhattan, KS, USA

David C. Cohen — IER ANSTO, Sydney, NSW, Australia

Philip L. Cole — Idaho State University, Pocatello, Idaho, USA

Alfredo Galindo-Urbarri — Oak Ridge National Laboratory, Oak Ridge, TN, USA

Richard Greco — Los Alamos National Laboratory, Los Alamos, NM, USA

Marianne E. Hamm — R&M Technical Enterprise, Pleasanton, CA, USA

Robert W. Hamm — R&M Technical Enterprise, Pleasanton, CA, USA

Richard P. Levy — ABC Foundation, Lake Arrowhead, CA, USA

Paolo Rossi — University of Padua, Padua, Italy

Justin M. Sanders — University of Southern Alabama, Mobile, AL, USA

Thomas Schenkel — Lawrence Berkeley National Laboratory, Berkeley, CA, USA

S. Thevuthasan — Pacific Northwest National Laboratory, Richland, WA, USA

Gyorgy Vizkelethy — Sandia National Laboratories, Albuquerque, NM, USA

George Vourvopoulos — Science Applications International Corporation, Orlando, FL, USA

Yongqiang Wang — Los Alamos National Laboratories, Los Alamos, NM, USA

GUEST EDITOR

Karen F. Sloan — Sandia National Laboratories, Albuquerque, NM, USA

Committees

Local Organizing Committee

F. Del McDaniel	Co-Chair	University of North Texas
Barney L. Doyle	Co-Chair	Sandia National Labs
Jerome L. Duggan	Chair Emeritus	University of North Texas
Margaret M. Hall	Conference Secretary	University of North Texas
Karen F. Sloan	Administrative Assistant	Sandia National Labs
Zdenek Nejedly	Webmaster	AnzSolutions

Session Chairs

John Baglin
IBM Almaden Research Center, USA

Nuno Barradas
Reactor Instituto Tecnologico e Nuclear, Portugal

Eleanor Blakely
Lawrence Berkeley National Laboratory, USA

Erik Blomquist
Uppsala University Hospital, Sweden

Stefano Cabrini
Lawrence Berkeley National Laboratory, USA

Kevin D. Carnes
Kansas State University, USA

David Chichester
Idaho National Laboratory, USA

Marshall Cleland
IBA Industrial Inc, USA

Eric Colby
SLAC National Accelerator Laboratory, USA

George B. Coutrakon
Northern Illinois University, USA

Giacomo Cuttone
LNS INFN, Italy

Daniel Dale
Idaho State University, USA

Morgan Dehnel
D-Pace, Inc, Canada

Oscar Dubon
University of California at Berkeley, USA

Ernst Esch
Los Alamos National Laboratory, USA

Engang Fu
Los Alamos National Laboratory, USA

Gary Glass
Louisiana Accelerator Center, USA

Mark Goldberg
Real Estate Hanassi, Israel

Siegbert Hagmann
GSI Darmstadt, Germany

Khalid Hattar
Sandia National Laboratories, USA

Charles Havener
Oak Ridge National Laboratory, USA

Adam Hecht
University of New Mexico, USA

Peter Hosemann
University of California at Berkeley, USA

Eugen Hug
Paul Scherrer Institute, Switzerland

Alan Hunt
Idaho Accelerator Center, USA

Jacek K. Jagielski
Inst. for Electronic Materials Technology, Poland

Milko Jaksic
Rudjer Boskovic Institute, Croatia

Qing Ji
Lawrence Berkeley National Laboratory, USA

Weilin Jiang
Pacific Northwest National Laboratory, USA

Brant Johnson
Brookhaven National Laboratory, USA

Tadashi Kamada
National Institute of Radiological Sciences, Japan

Gaylord King
G. King Consulting, USA

Thomas Kirchner
York University, Canada

Karen Kirkby
University of Surrey, United Kingdom

Naoki Kishimoto
National Institute for Materials Science, Japan

Patrick Kluth
The Australian National University, Australia

Robert D. Kolasinski
Sandia National Laboratories, USA

Sergey Kurennoy
Los Alamos National Laboratory, USA

Richard C. Lanza
Massachusetts Institute of Technology, USA

Stephen Lundeen
Colorado State University, USA

Willy Maenhaut
Ghent University, Belgium

Jan Meijer
Ruhr-Universitaet Bochum, Germany

Rahul Mehta
University of Central Arkansas, USA

Javier Miranda
Universidad Nacional Autonoma de Mexico

Gary E. Mitchell
North Carolina State University, USA

Ross E. Muenchausen
Los Alamos National Laboratory, USA

Tobin Munsat
University of Colorado, USA

Victor Orphan
SAIC, USA

Anand P. Pathak
University of Hyderabad, India

Arun Persaud
Lawrence Berkeley National Laboratory, USA

Carroll A. Quarles
Texas Christian University, USA

Bibhudutta Rout
University of North Texas, USA

Reinhold Schuch
Stockholm University, Sweden

David Schultz
Oak Ridge National Laboratory, USA

Christian Segebade
Germany

Lin Shao
Texas A&M University, USA

Richard Sheffield
Los Alamos National Laboratory, USA

Michelle Shinn
Jefferson Lab, USA

Jefferson L. Shinpaugh
East Carolina University, USA

V. Shutthanandan
Pacific Northwest National Laboratory, USA

Žiga Šmit
University of Ljubljana, Slovenia

S. G. Srivilliputhur
University of North Texas, USA

Martin Stockli
Oak Ridge National Laboratory, USA

Thomas Stoehlker
GSI Darmstadt, Germany

Daniel W. Stracener
Oak Ridge National Laboratory, USA

John A. Tanis
Western Michigan University, USA

Joseph Tesmer
Los Alamos National Laboratory, USA

James Tickner
CSIRO, Australia

Igor O. Usov
Los Alamos National Laboratory, USA

Alex Weiss
University of Texas at Arlington, USA

Doug P. Wells
Idaho Accelerator Center, USA

Stephen A. Wender
Los Alamos National Laboratory, USA

Zihua Zhu
Pacific Northwest National Laboratory, USA

SPONSORS

Financial support from sponsors, exhibitors and advertisers is an essential element of a successful conference. This support enables us to keep the registration fee to a minimum and to provide support for conference events, students and other attendees. We are very grateful to the sponsors, exhibitors and advertisers listed below for their support of CAARI 2010.

Research Sponsors

Los Alamos National Labs
Sandia National Laboratories

Industrial Sponsors

AccSys Technology, Inc.
High Voltage Engineering Europa B.V.
National Electrostatics Corp.
TDK-Lambda Americas

Exhibitors

AccelSoft Inc.
CAEN Technologies Inc.
Everson Tesla Inc.
GMW Associates
High Voltage Engineering Europa B.V.
HV Components/CKE (Div. of Dean Technology)
JP Accelerator Works
Kurt J. Lesker Company
National Electrostatics Corp.
ORTEC/Ametek
Pantechnik
RadiaBeam Technologies, Inc.
Raith USA Inc.
Saint-Gobain Crystals
TDK-Lambda Americas
Tech-X Corporation
Thermo Fisher Scientific
Varian Vacuum Technologies
Weiner, Plein & Baus Ltd.
XIA LLC
XScell Corporation

Advertisers

AccSys Technology Inc.
GMW Associates
Materials Research Society

Uniform Beam Distributions Of Charged Particle Beams

Nicholaos Tsoupas

CA, Brookhaven National Laboratory, 911B, Upton NY 11973, United States

Abstract. The use of octupole elements was originally suggested [1] to transform the transverse Gaussian distribution of a charged particle beam into a uniform beam distribution having rectangular cross section. The first experimental realization of this concept was materialized at the Radiation Effects Facility (REF) of the Brookhaven National Laboratory (BNL) [2] where the transverse Gaussian distribution of a 200MeV H^- beam was transformed into a rectangular uniform distribution. Later, the beam line of the NASA Space Radiation Laboratory (NSRL) facility [3,4] built at BNL, was specifically designed to generate uniform beam distributions with rectangular cross sections, in both, the horizontal and vertical directions, at the location of the target. The NSRL facility generates uniform beam distributions of various nuclear species, which may vary in energy, and also vary in size of the uniform beam distribution. We will present an overview of the method to generate uniform beam distributions, show some results from the NSRL facility, and suggest other possibilities for generating such beam distributions.

Keywords: Third order beam optics, uniform beam distribution

INTRODUCTION

Uniform beam distributions of charged particle beams over rectangular areas ranging from a few cm^2 to a few hundred cm^2 are desired to irradiate uniformly various materials, like biological samples or components of electronic circuits. Such uniform beam distributions are available in the NASA Space Radiation Laboratory (NSRL) of the Brookhaven National Laboratory (BNL) [3,4] where research in the medical, biological, and nuclear physics fields has been ongoing since the year 2000. Details on the NSRL facility, and on the beam optics of the NSRL beam line which produces uniform beam distributions , appear in Refs [3,4]. The main purpose of this paper is to provide the reader with the basic theory of generating uniform beam distributions, and also give an example of how to use this theory to transform a beam with Gaussian beam distribution, into a beam with the uniform beam distribution at the location of the target.

TRANSFORMATIONS OF A CHARGED PARTICLE BEAM DISTRIBUTION

A particle in a beam distribution is characterized by six coordinates $(x, x', y, y', \delta l, \delta p)$. The symbols (x, x') represent the horizontal displacement and angular divergence of the particle with respect to a six dimensional coordinate system whose origin is located on the central particle, and having the z-axis along the direction of the momentum p_0 of the central particle. The symbols (y, y') are the corresponding vertical coordinates of the particle, and the symbols δl, and δp are the displacement along the z-axis, and the momentum deviation respectively of a particle from the position and momentum of the central particle. Such a particle distribution can be described by a density distribution function $f(x, x', y, y', \delta l, \delta p)$ which is subject to the constraint: $\int f(x, x', y, y', \delta l, \delta p)\, dx\, dx'\, dy\, dy'\, d(\delta l)\, d(\delta p) = N_0$. N_0 is the total number of particles in the beam distribution. In this paper we will also use the notation $(x_1, x_2, x_3, x_4, x_5, x_6)$ for the six coordinates, and we assume that a charged particle beam as it emerges from an accelerator has a Gaussian (Normal) distribution in any of its six coordinates. In the next two subsections we discuss the linear and nonlinear transformation of such a Gaussian beam distribution as it is transported and focused on the target with linear and nonlinear magnetic elements.

Linear Transformation Of A Charged Particle Beam Distribution

A Gaussian beam distribution with two variables (x, x') which have a correlation σ_{12}, and standard deviations $(\sigma_{11})^{1/2}$, and $(\sigma_{22})^{1/2}$, respectively, can be expressed mathematically by the function (1) below.

Application of Accelerators in Research and Industry
AIP Conf. Proc. 1336, 11-15 (2011); doi: 10.1063/1.3586047

$$f(x,x') = \frac{1}{2\pi\sqrt{|\sigma|}} e^{-\frac{1}{2}(\frac{\sigma_{22}x^2}{|\sigma|} - \frac{2\sigma_{12}xx'}{|\sigma|} + \frac{\sigma_{11}x'^2}{|\sigma|})}$$

$$= \frac{1}{2\pi\sqrt{|\sigma|}} e^{-\frac{1}{2}(\tilde{X}\sigma^{-1}X)} \quad (1)$$

In Equation (1) the symbols X and $\tilde{X}$ are the

$$X = \begin{pmatrix} x \\ x' \end{pmatrix} \text{ and } \tilde{X} = (x, x')$$

position vector of a particle, and its transpose. The symbol $|\sigma|$ is the determinant of the matrix σ.

$$\sigma = \begin{Bmatrix} \sigma_{11} & \sigma_{12} \\ \sigma_{12} & \sigma_{22} \end{Bmatrix}$$

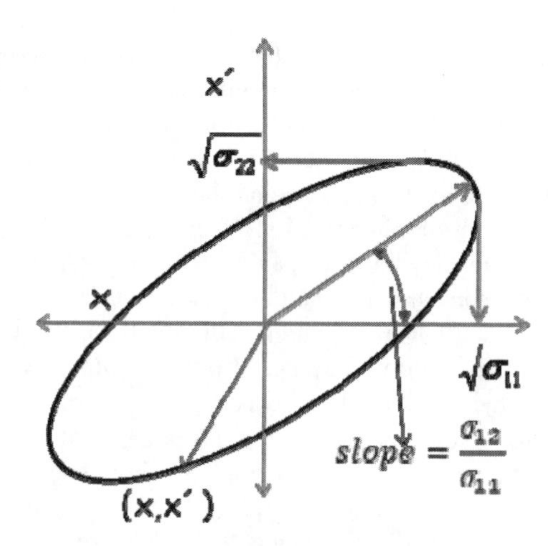

FIGURE 1. A schematic diagram showing the physical meaning of the matrix elements σ_{12} , σ_{11} , and σ_{22}. The projected particle distribution along the x-axis will have a Gaussian distribution with an rms value of $(\sigma_{11})^{1/2}$. The corresponding rms value for the x′ axis is $(\sigma_{22})^{1/2}$. Such an ellipse with specified values of σ_{12} , σ_{11} , and σ_{22} is uniquely defined and referred as the "beam ellipse".

The physical meaning of the correlation σ_{12} and the standard deviations $(\sigma_{11})^{1/2}$, and $(\sigma_{22})^{1/2}$ is shown in figure 1 where the ellipse encloses particles having one standard deviation in x and x'. Extending the notation to six dimensions the Gaussian distribution function can be expressed in terms of the σ–matrix by the function (2) below.

$$f(x_1, x_2, x_3, x_4, x_5, x_6) =$$

$$= \frac{1}{(2\pi)^2\sqrt{|\sigma|}} e^{-\frac{1}{2}(\tilde{X}\sigma^{-1}X)} \quad (2)$$

with $\int f(x_1,x_2,x_3,x_4,x_5,x_6)dx_1dx_2dx_3dx_4dx_5dx_6=1$

The transport of a charged particle beam from an initial coordinate system located at the point (I) to a final coordinate system located downstream at a point (F), can be expressed in terms of a 6x6 beam transport R matrix as: $X^{(F)} = RX^{(I)} = \sum_{j=1}^{6} R_{ij}x_j$. The R matrix satisfies the simplecticity condition: $\tilde{R}SR = S$, $R^{-1} = S^{-1}\tilde{R}S$, $|R| = 1$, where S is a 6x6 matrix with elements { S_{12}=1, S_{21}=-1, S_{34}=1, S_{43}=-1, S_{56}=1, S_{65}=-1, with the rest of the elements set to zero}(The symbol ~ indicates the transpose of a matrix). Using the properties of the R matrix mentioned above, it is easy to prove that the beam distribution at the coordinate system (F) can be expressed as:

$$f(x_1, x_2, x_3, x_4, x_5, x_6) =$$

$$= \frac{1}{(2\pi)^2\sqrt{|R\sigma_I\tilde{R}|}} e^{-\frac{1}{2}(\tilde{X}^F(R\sigma_I^{-1}\tilde{R})X^F)} =$$

$$= \frac{1}{(2\pi)^2\sqrt{|\sigma_F|}} e^{-\frac{1}{2}(\tilde{X}^F(\sigma_F^{-1}X^F)} \quad (3)$$

In equation (3) we made the substitutions

$$|R\sigma_I\tilde{R}| = |\sigma_F| \text{ and } R\sigma_I^{-1}\tilde{R} = \sigma_F^{-1}$$

Comparing the relations (3) and (2) we conclude that under a linear transformation R, the new particle distribution remains Gaussian, and the σ-matrix σ_F at the point (F) is related to the σ–matrix σ_I at the point (I) as $\sigma_F = R\sigma_I\tilde{R}$. In conclusion, a beam with a Gaussian distribution remains Gaussian under linear transformation. To modify a Gaussian beam distribution to a non-Gaussian one, higher order magnetic elements are required.

Simple Example Of Linear And Non Linear Transformation Of A Beam Distribution

Assume that the charged particle beam emanating from a single point O, shown in figure 2, is distributed In angle θ, as in the function (3a) below.

$$P(\theta) = \frac{1}{\sqrt{2\pi\sigma_\theta}} e^{-(\frac{\theta^2}{2\sigma_\theta})} \quad (3a)$$

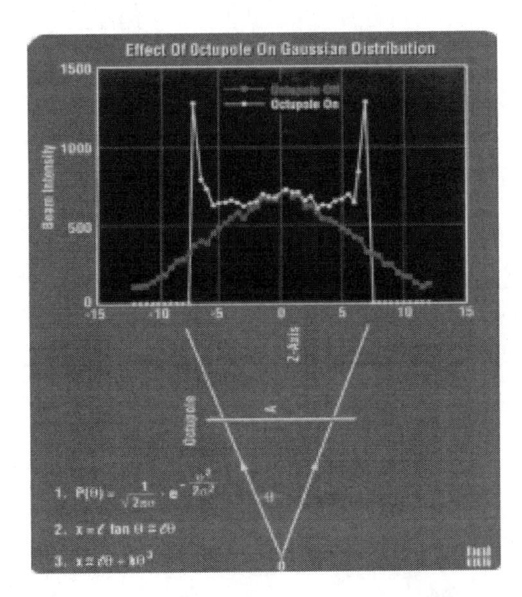

FIGURE 2. Particles emanating from point O subject to the density distribution function (3a) where σ_θ is the standard deviation (rms) value of the distribution, generated by a Gaussian distribution (red curve) along the x-direction. The insertion of an octupole element at the location A alters the distribution along the x-direction into a uniform distribution (yellow curve).

The beam transport from point O to a distance *l* (see fig.2) is equivalent to a linear transformation of the beam distribution by an R matrix which in this case represents a drift space of length *l*. It is easy to prove that, under the approximation $x \approx l\theta$, the beam is also distributed along the transverse coordinate x, normally as is shown by the red curve in figure 2. When the octupole, of strength k, at the location A is energized, the x coordinate of a particle at the location $z=l$ is $x \approx l\theta + k\theta^3$. Under this nonlinear relation, the Gaussian beam distribution of the beam, is transformed into the beam distribution shown by the yellow curve in figure 2, which exhibits a rather uniform distribution. The numerical values of the points of the yellow curve have been calculated, using the Monte Carlo method as applied to the algorithm, $x=l\theta + k\theta^3$. In the Monte Carlo calculations the beam coordinate θ was randomly selected from a set of coordinates having a Gaussian distribution.

GENERAL TRASFORMATION OF A GAUSSIAN BEAM DISTRIBUTION INTO A UNIFORM

Under a linear transformation $X^{(F)} = RX^{(I)} = \sum_{j=1}^{6} R_{ij} x_j^{(I)}$,

a Gaussian beam distribution is transformed always in a Gaussian beam distribution. Only the insertion of non-linear elements in the beam line, such as octupoles, may transform a Gaussian beam distribution into a non Gaussian one. The general equation which provides the coordinate transformation of a particle in a beam distribution, which is acted upon by quadrupole, sextupole and octupole, magnetic elements, is given by:

$$x_i^F = R_{ij} x_j^{(I)} + T_{ijk} x_j^{(I)} x_k^{(I)} + W_{ijkl} x_j^{(I)} x_k^{(I)} x_l^{(I)} \quad (4)$$

In this equation we assume summation over the repeated indices appearing in the variarables of each term. The R_{ij}, T_{ijk}, and W_{ijkl}, are the Taylor series expansion coefficients of the beam coordinates at the target $x_i^{(o)}$ in terms of the coordinates $x_i^{(I)}$ at the entrance of the beam line, and are refered as the first (R_{ij}) second (T_{ijk}) and third (W_{ijkl}) order aberration coefficients, respectively. In the rest of this section we provide the tasks required to transform a gaussian beam distribution into a uniform beam distribution at the location of the target. a) Establish the Gaussian beam parameteres (σ-matrix) at the beginning of the beam line. b) By exciting only the quadrupole elements of the beam line, generate, a highly correlated beam among the (x,x') beam coordinates at one location (O1) of the beam line (see fig.3) ; Also at this location (O1) the condition $(\sigma_{22})^{1/2} \ll (\sigma_{11})^{1/2}$ should be satisfied. Likewise at a different location (O2) of the beam line (see fig. 3), generate a highly correlated beam among the (y,y') coordinates with the condition $(\sigma_{11})^{1/2} \ll (\sigma_{22})^{1/2}$also satisfied. c) In conjuction with task b) above, use two quadrupoles, Q5 and Q6 in fig 3, placed downstream from the high beam correlation locations points (O1) and (O2), to adjust the beam size at the target location. d) Place octupole elements at the locations (O1) and (O2) and energize the octupoles. e) Compute the aberration coefficients R_{ij}, T_{ijk}, and W_{ijkl}, as defined by equation (4). The TRANSPORT code [5] may provide the values of these coefficents. f) Use Monte Carlo method to compute the particle coordinates at the target location as defined by equation (4). The random coordinates of each particle at the beginning of the beam line, are selectected from a specified Gaussian beam distribution defined by equation (3).

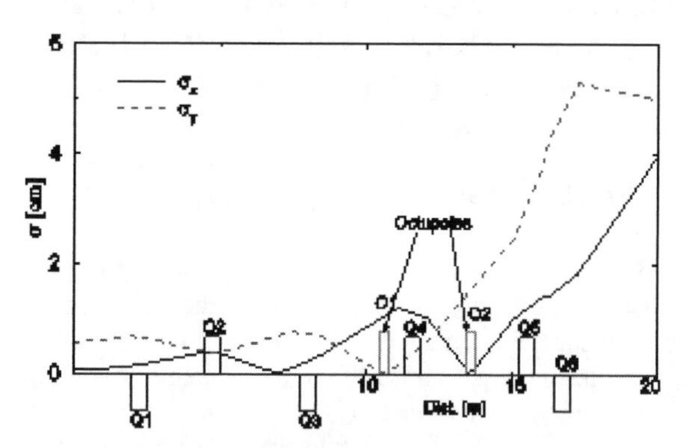

FIGURE 3. Horizontal (solid black) and vertical (dashed red) beam profiles generated by the first order elements (quadrupoles Q1 to Q8, shown as blue rectangles). A the location of the first/second green rectangles (octupoles) the first order beam optics constraints the (x,x′)/(y, y′) coordinates to be highly correlated. The last two quads (Q5,Q6) are needed to adjust the size of the beam profiles at the end of the line.

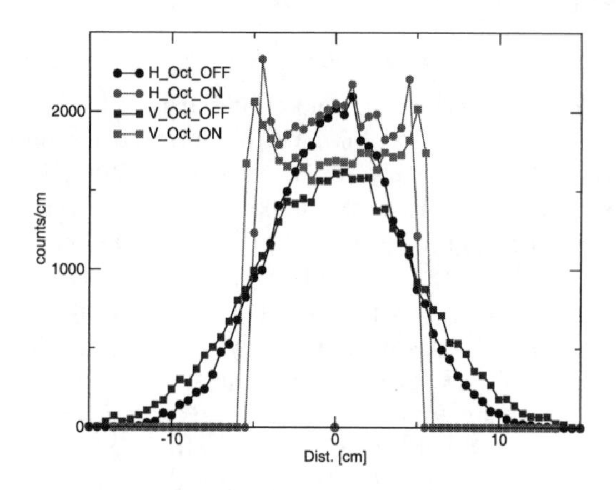

FIGURE 4. The horizontal (black circles/red circles) beam distribution at the target with the octupoles OFF/ON. The blue squares/green squares correspond to the vertical distribution with octupoles OFF/ON. The steepness of the beam distribution when the octupoles are ON depends on the correlation of the (x,x′) and (y, y′) coordinates at the location of the octupoles and the value of the horizontal and vertical beam emittance.

Applying the tasks (a) to (e) above to a beam line with six quadrupoles, we generated figure 3 which shows the first order horizontal (solid black line) and vertical (dashed red line) beam envelopes of a beam line which satisfies the beam constraints mention in tasks (a) through (c) above for generating a uniform beam.

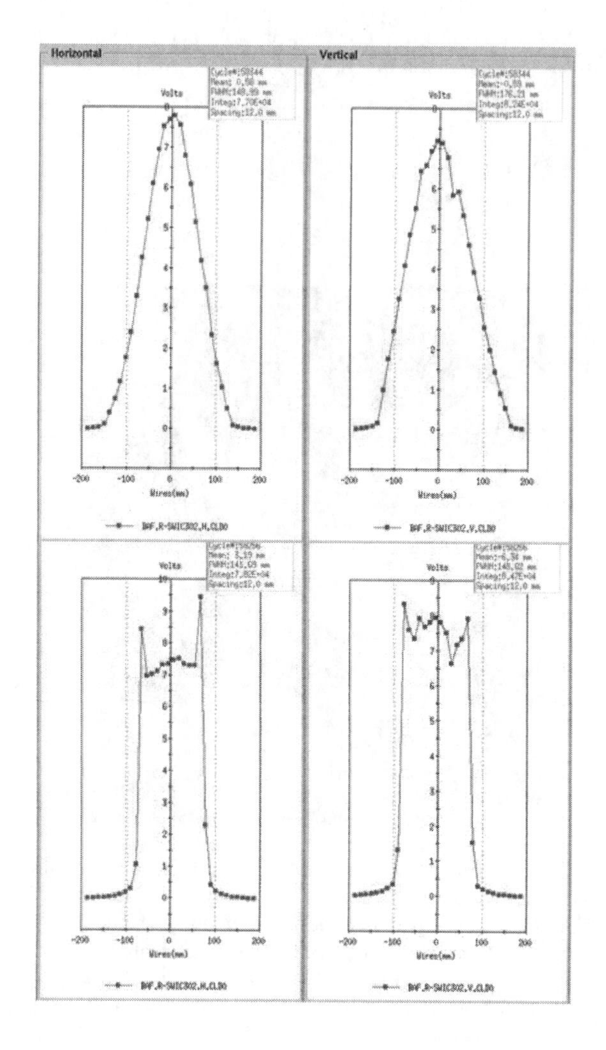

FIGURE 5. An experimentally measured horizontal and vertical beam distributions at the NSRL facility, at the target when the Octupoles are OFF (top). The same distributions as measured with Octupoles ON is shown in the bottom of the figure. The spacing of the harp wires is 8.5 mm. One of the vertical harp wires at y~25 mm is not calibrated correctly.

The same distributions as measured with Octupoles ON is shown in the bottom of the figure 5. The spacing of the harp wires which are used to measure the beam distribution, is 8.5 mm. The dip in the vertical beam profile at y~25 mm of figure 5 indicates that one of the vertical harp wires is not calibrated correctly. Figure 6 shows an isometric plot of the beam distribution at the target location of the NSRL facility. This distribution shown in figure 6 is measured by a two dimensional diode array placed at the target location of the NSRL facility.

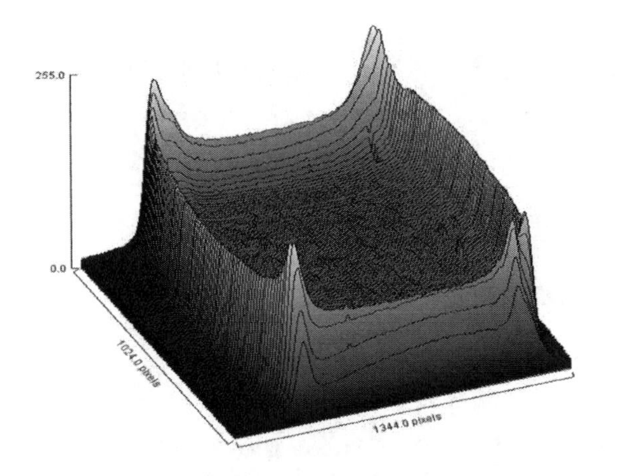

FIGURE 6. An isometric plot of the beam distribution at the target location of the NSRL facility as measured by a two dimensional diode array placed at the target location of the NSRL facility.

ACKNOWLEDGMENTS

The author would like to thank the CAD management, and the personnel of BNL for supporting and implementing the method of uniform beams to the NSRL facility.

REFERENCES

1. P. Meads "A Nonlinear Lens System to smooth the Intensity Distribution of a Gaussian Beam" IEEE Transactions on Nuclear Science, Vol. NS-30, No. 4, August 1983.
2. N. Tsoupas et al. "Uniform beam Distributions Using Octupoles" Proc. 14th Particle Accelerator Conf., San Francisco, California, May 6-9, 1991, p. 1695.
3. N. Tsoupas et al. "Uniform beam distributions at the NASA Space Radiation Laboratory" Phys. Rev. ST Accel. Beams 10, 024701 (2007).
4. K. Brown et al. The NASA Space Radiation Laboratory at Brookhaven National Laboratory: Preparation and delivery of ion beams for space radiation research Volume 618, Issues 1-3, 1 June 2010- 21 June 2010, Pages 97-107
5. K. L. BROWN et al. "TRANSPORT computer code" SLAC-91

Present Trends In The Configurations And Applications Of Electrostatic Accelerator Systems

Gregory A. Norton and George M. Klody

National Electrostatics Corp.
7540 Graber Road
Middleton, WI, United States

Abstract. Despite the worldwide economic meltdown during the past two years and preceding any stimulus program projects, the market for electrostatic accelerators has increased on three fronts: new applications developed in an expanding range of fields; technical enhancements that increase the range, precision, and sensitivity of existing systems; and new accelerator projects in a growing number of developing countries. From the single application of basic nuclear structure research from the 1930's into the 1970's, the continued expansion of new applications and the technical improvements in electrostatic accelerators have dramatically affected the configurations and capabilities of accelerator systems to meet new requirements. This paper describes examples of recent developments in cosmology, exotic materials, high resolution RBS, compact AMS, dust acceleration, ion implantation, etc.

Keywords: Pelletron, Electrostatic accelerator, RBS, PIXE, AMS, implantation
PACS: 29.20Ba, 81.70.-q, 87.80.Dj.

INTRODUCTION

Van de Graaff accelerators were first pressurized in 1933 [1]. The sole purpose of these first direct potential drop, electrostatic accelerators was to produce MeV energy protons to investigate nuclear structure. At that very early time, all equipment, including the accelerator itself, needed to be built in the laboratory.

The situation eased some what in the late 1940's when electrostatic accelerators became commercially available [2, 3]. However, basic nuclear physics research was still their only application. This drove the development of accelerator capabilities from approximately 4 MeV protons to heavy ion beams in excess of 300 MeV as higher energies were needed to overcome the Coulomb barrier. In the 1970's the world's highest voltage electrostatic accelerator [4] was built at just about the time new applications were being found for these machines and funding for basic nuclear research waned.

APPLICATIONS REDEFINE ACCELERATOR SYSTEM DESIGN

Up until the 1970's accelerators were commercially available, but laboratories still had to build additional equipment for targetry and detection. This changed in the late 1970's and early 1980's with the availability of complete turn key systems for implantation and materials analysis, which often needed to be compact to fit into existing lab spaces [5].

The use of ion implantation in the keV region was considered in the early 70's and by the early 1980's was becoming well established in the semiconductor industry [6]. Ever shrinking device sizes created a need for MeV ion implantation systems. These needed to be complete turn key systems that would work well in the production environment. The first systems were based on 1 MV tandem accelerators complete with cassette to cassette wafer handlers. Today, MeV implantation is commonplace [7], and research implanters in the MeV energy region continue to be built in a wide variety of configurations.

At this same time, techniques such as simple elastic scattering and atomic excitation, which were commonplace in physics labs, were being applied to elemental materials analysis in industry. The primary

Application of Accelerators in Research and Industry
AIP Conf. Proc. 1336, 16-20 (2011); doi: 10.1063/1.3586048

demand was in the semiconductor industry where the need for non-destructive elemental analysis was becoming the most apparent. This led to the development of the now commonly used techniques of RBS, Channelling RBS, Elastic Recoil Detection (ERD), Nuclear Reaction Analysis (NRA), and Particle Induced X-ray Emission (PIXE).

Although analysis systems were mainly sold to universities, demand was also growing in the private sector. In time, electrostatic accelerators became available with complete analysis end stations designed for true unattended operation. These systems use digital control of both the analysis end station and the complete accelerator system including the ion source.

In addition, severe demands were being placed on ion beam optics as the need for micro PIXE and micro RBS developed. This changed the overall footprint of the accelerator system dramatically. In the case of PIXE microprobes, the beam line after the energy analyzing magnet is often longer than the rest of the entire accelerator system. (figure 1.)

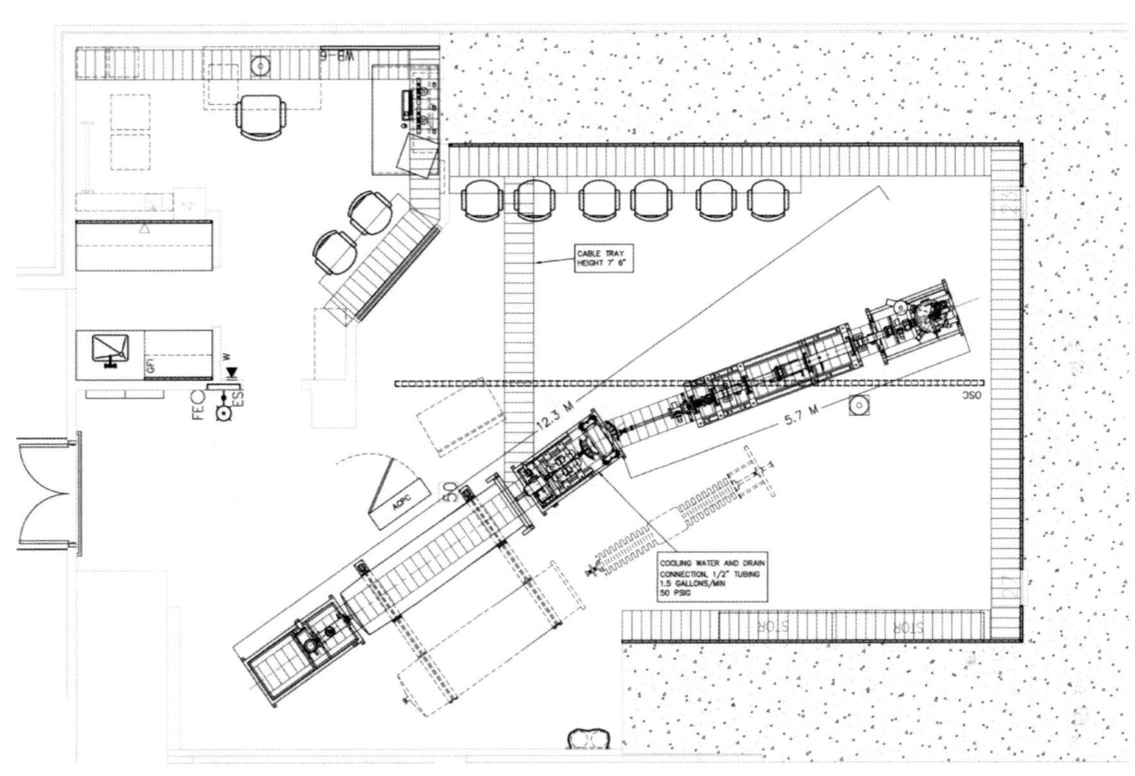

FIGURE 1. 1.7 MV Tandem Pelletron at the Department of Physics of the State University of New York at Geneseo equipped with a 10-20 micron He^+ microprobe. The microprobe beam line after the analyzing magnet is about the same length as the complete tandem assembly with injector and high energy analysis system.

All of these techniques are widely used and are continually being refined. High Resolution RBS systems are now commercially available to analyze thin films with Ãngström level resolution [8]. In addition, the digital control systems allow true remote operation. The person operating the accelerator does not need to be in a laboratory or even on the same continent.

Today, there are approximately 100 active laboratories using MeV accelerators for materials analysis. Approximately 50 of these are equipped with microprobes.

Accelerator Mass Spectrometry

In the late 1980's and early 1990's the use of electrostatic accelerators for radioisotope ratio measurements became established [9]. The requirements for this application are the most severe to date for the overall accelerator configuration. Ion sources have to be adapted to have multiple samples in the ion source at one time, the low energy injection systems are required to measure and inject both abundant and rare masses with very high mass resolution, the accelerator terminal must be equipped with the thick stripper for equilibrium stripping, and the high energy beam line must be equipped with an analysis system capable of very high precision measurement of abundant isotopes and the rare isotope. Fully automated digital control is needed to achieve the high sample throughput that is required in most cases. (figure 2.)

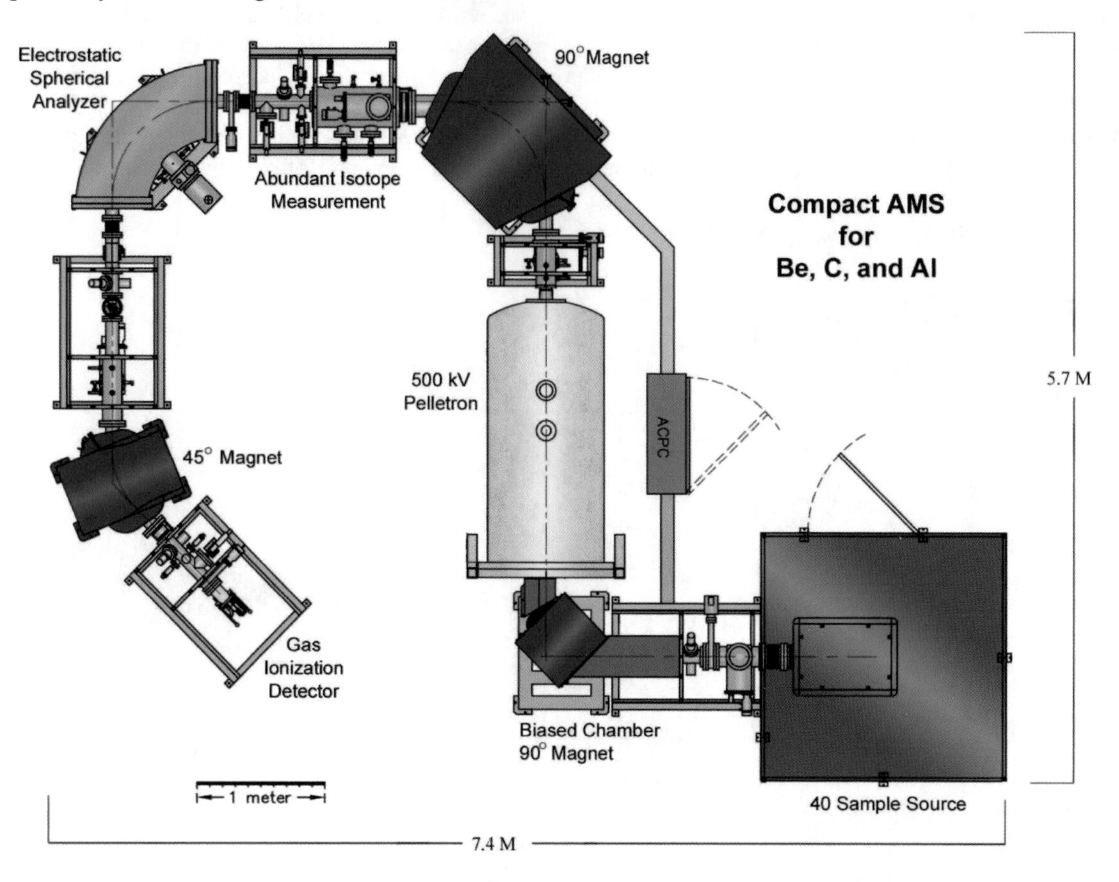

FIGURE 2. Model XCAMS is a complete AMS system based on a 500kV tandem Pelletron. This system is presently in use at the National Isotope Center, Institute of Geological and Nuclear Sciences Limited, Lower Hut, New Zealand. It is designed for beryllium, carbon and aluminium radioisotope ratio measurements.

The number of accelerators being used for AMS has been climbing steadily. It is the largest single market for electrostatic accelerators today. Originally, the accelerator systems were designed for carbon radioisotope measurements, i.e., carbon dating. Now the field has expanded to routine measurements for isotopes of beryllium, aluminium, chlorine, calcium, iodine, and the actinides. The MeV electrostatic accelerator based AMS system is the most sensitive spectrometer available with a background in the 1 out of 10^{16} region and sometimes better. The requirements for precisions are being pushed beyond 0.3% to the 0.1% region. As new applications within the AMS field develop, the demand on sample throughput continues to climb. It is now possible to purchase a commercially available system capable of running over 2000 samples of modern carbon in five days to better than 2% precision.

The varied uses of AMS have pushed the design of AMS systems to more compact machines with more novel configurations. Typically, AMS machines are based on tandem accelerators. This has changed with the advent of the Single Stage AMS system where the high energy acceleration tube has been eliminated and all final isotopic mass analysis and detection is done at voltage [11].

The approximately 70 active AMS laboratories around the world today support a highly varied range of applications, from the traditional carbon dating for archeology to earthquake studies, climatology, oceanography, and drug discovery in the field of pharmacokinetics [10].

ACCELERATORS BEHIND THE HEADLINES

The techniques mentioned above, developed in the field of nuclear physics, are finding increased applications. This directly results in the increased use of electrostatic accelerators in many high profile projects. However, the electrostatic accelerator itself is rarely mentioned. Table I lists a few of the areas where accelerators are playing a significant role in today's society. As this table shows, the range is extremely broad.

TABLE 1. Electrostatic Accelerators are occasionally used for very high profile experiments that are reported in the popular press. However, it is very rare for the accelerator or actual techniques used to be mentioned.

Story	Technique
Alzheimer's – Aluminum Scare (1990's)	PIXE
The Louvre – Art Authentication	RBS, PIXE
High Temperature Superconductors	RBS
Dating of the Shroud of Turin	AMS
Dating of the Dead Sea Scrolls	AMS
Dating of the "Ice Man" (Alps)	AMS
Study of Atherosclerosis	Micro PIXE
Solar Wind Studies	AMS
Psychology – Depression Studies	Isotopes - PET
Drug Development	AMS

RECENT APPLICATIONS

There is strong interest in the use of AMS for pharmacokinetics. Pharmacokinetics involves the bio availability of a compound, its distribution in the body, its resident time in the body or elimination time, and the formation of metabolites in the body. Metabolite formation is leading the interest in compound specific AMS, using a liquid chromatograph connected directly to an AMS source [12, 13]. Such a system is presently under development to allow the measurement of specific mass fractions through ^{14}C tagging. AMS is far more sensitive than any other technique, allowing extremely small amounts of ^{14}C to be added, which in turn allows earlier studies in humans in the drug development process [14].

There is renewed interest in radiation damage studies, particularly for the next generation of nuclear reactors. One way to investigate this type of damage is through the use of a triple beam facility where up to three ion beams impact the material simultaneously. This application requires ten's of MeV of heavy ions. In order to keep the accelerator size small, a single ended Pelletron® has been equipped with the Pantechnik Nanogan® ECR source. This source produces high charge states of nuclei from a wide variety of heavy ions including the noble gases which cannot be obtained from a tandem accelerator.

A single ended 3 MV Pelletron with ECR source is now in use at the Joint Accelerators for Nano-science and Nuclear Simulation (JANNuS) laboratory at Commissariat A l'Energie Atomique in Saclay. (figure 3) Table II shows beams that have been tested and some of the beam currents recorded. The triple beam facility itself went into operation on March 11, 2010, with beams from a 2.5 MV Van de Graaff, a 2 MV tandem Pelletron, and a 3 MV single ended ECR Pelletron [15] (Table 2).

FIGURE 3. 3 MV Single ended Pelletron equipped with a Pantechnik Nanogan Electron Cyclotron Resonance (ECR) source to provide high charge state positive ions. This system is presently in use as part of the triple beam facility at Joint Accelerators for Nano-Science and Nuclear Simulation (JANNuS) at the Commissariat A l'Energie Atomique in Saclay, France.

TABLE 2. Performance of the 3 MV single ended Pelletron with the Nanogan ECR source in the high voltage terminal. Both the source gas flow and the source emittance present significant challenges for use on a MV platform. [15]

ion	Analyzed Current	Beam Energy
H^{+}	144 μA	3 MeV
He^{++}	37 μA	6 MeV
Ne^{+}	35 μA	3 MeV
Ne^{+7}	8 nA	21 MeV
Ar^{+6}	3.1 μA	18 MeV
Ar^{+11}	8 nA	33 MeV
$^{132}Xe^{+12}$	220 nA	36 MeV

The technology developed for the triple beam facility also has applications in astrophysics. A 5 MV single ended Pelletron with an ECR source in the terminal is presently in manufacture for the nuclear science laboratory at the University of Notre Dame (USA). The accelerator will be located above the existing accelerator laboratory to provide high charge state heavy ions for reverse kinematics astrophysics reactions using hydrogen and helium targets. Shipment is scheduled for December, 2011 [16]. This "technology transfer" is not unusual. The design of an electron Pelletron for an infrared free electron laser has been used for antiproton cooling at Fermi National Laboratory. In addition, an accelerator system designed for carbon dating AMS is now also used to measure many other radioisotopes for a wide variety of applications as mentioned above.

In addition to nuclear astrophysics reactions, accelerators are also being used in the field of space science to investigate the behaviour of micrometeoroids. A 3 MV single ended Pelletron is presently in manufacture for the Center for Integrated Plasma Studies at the University of Colorado under the direction of Dr. Tobin Munsat. This group will install a dust source to produce micron size particles with speeds up to ten's of km/s into a simulated lunar environment [17].

CONCLUSION

In this time of economic turmoil, the demand for electrostatic accelerators has continued to be very strong. From August, 2008 to the present time, more than twenty new electrostatic systems have been sold. There is no single dominant design or application. In addition, existing accelerator laboratories have an extremely durable lifetime. Some electrostatic accelerators are still active after more than 40 years, and many labs continue to upgrade and add to their systems. We expect that this trend will continue as new applications demand new capabilities.

ACKNOWLEDGMENTS

The authors would like to express our appreciation to all those who have obtained and used electrostatic accelerators. It is the specifications and feedback from this group that drives the technology of all multi million Volt electrostatic accelerators.

REFERENCES

1. R.G. Herb, D.B. Parkinson and D.W. Kerst, Rev. Sci. Instr. 6, 261 (1935).
2. D. Allan Bromley, Nucl. Instr. and Meth. 122 (1974) 1-34.
3. G.A. Norton, J.A. Ferry, R.E. Daniel, and G.M. Klody, "A Retrospective of the Career of Ray herb", Proc. Of Heavy Ion Accelerator Technology: Eight International Conference, Argonne National Laboratory 1999, the American Institute of Physics, 3-23.
4. C.M. Jones, "The Oak Ridge 25 MV Tandem Accelerator", Proc. Of Third Int. on Electrostatic Accelerator Technology, Oak Ridge, Tennessee (April 13-16, 1981) 98.
5. G.A. Norton, G.M. Klody, Proc. Of the 14th Int. Conf. On the Application of Accelerators in Research and Industry, Denton, Texas, November, 1996, AIP Press 1997, 1109-1114.
6. P.H. Rose, Nucl. Instr. and Meth. B6 (1985) 1-8.
7. P.H. Rose, Solid State Technology, May 1997, 129-130.
8. T.J. Pollock, J.A. Haas, G.M. Klody, (these proceedings, abstract no. 151)
9. P.O Povinec, A.E. Litherland, K.F. von Redden, Radiocarbon Vol 51, Nrl, (2009) 45-78.
10. C. Tuniz, G. Norton, Nucl. Instr. and Meth. B 266 (2008) 1837-1845.
11. National Electrostatics Corp. patent no. 6,815,666.
12. Massachusetts Institute of Technology patent no. 6,867,415 and 6,707,035.
13. R.G. Liberman et al, Nucl. Instr. and Meth. B223-224 (2004) 82-86.
14. R.C. Garner, Drug Discovery Today, 10 9April, 2005) 449-451.
15. S. Miro, P. Trocellier, Y. Serruys, E. Bordas, H. Martin, N. Chaabane, S. Pellegrino, S. Vaubaillon, J.P. Galliern, (these proceedings abstract no. 57). Y. Serruys, private communication (2010).
16. URL: http://isnap.nd.edu/html/research_5mv.html
17. T. Munsat, (these proceedings, abstract no. 60)

Industrial Beamline Technologies And Approaches

M.P. Dehnel

D-Pace, Inc. P.O. Box 201, Nelson, B.C., Canada V1L 5P9

Abstract. This paper describes industrial charged particle beamline technologies and design approaches. Beamlines accommodate myriad constraints in the radioisotope production, electron beam processing, and ion implantation market segments, and some very strange yet interesting solutions result. In this paper, a detailed look at a particular injection beamline solution gives some sense of the complexity of research and development required, and the sophistication of the beamline solutions that are utilized. A brief review of beamline applications in each industrial segment follows.

Keywords: Beamline, Ion-Optics, PET, Cyclotron, Implantation, Electron, Injection, Magnets, Quadrupole, Steering, Diagnostic, Proton, Focus, Bombardment, Beams, Radioisotope
PACS: 81.40.Wx, 87.58.Ji, 41.75.Ak, 41.85.-p, 41.85.Lc, 87.58.Fg, 41.75.Ak, Cn, 28.41.Qb

INTRODUCTION

A beamline is a device that accomplishes charged particle transport (i.e. it is not the particle accelerator). Accelerator design has rightly garnered much attention. However, the vast array of strange injection and extraction beamline applications, constraints and innovative solutions is also impressive. In this paper, we define and discuss two types of beamlines: (1) injection lines – which are intermediate beamlines between the charged particle source, and the primary accelerator, and (2) extraction beamlines – which are used to transport charged particles at their final kinetic energy to a target. An injection line example is discussed in detail to demonstrate the wide range of constraints considered in beamline design, and to illustrate the elegant yet sophisticated solution that often must be developed. Extraction beamlines from the radioisotope production, electron beam processing, and ion implantation industries shall be briefly reviewed.

CONSTRAINTS

A beamline system is intended to provide the transport of a beam of charged particles from a source (i.e. ion or electron source, particle accelerator) to a target or another accelerator. The beamline system must accommodate a large selection of constraints, which include [1]:

- lack of space
- reliability of elements
- alignment & impact on other sub-systems
- adaptation of emittances between elements
- beam transmission
- vacuum quality
- space charge
- fringing fields
- technology of fabrication
- cost

as well as:

- charge, isotope & energy selection
- pulsing, bunching, scanning
- uniformity, purity
- current range, energy range
- target size, accelerator acceptances, beam matching
- source beam characteristics
- diagnostics
- trajectories, obstacles, rotation, up, down, left, right, spiral, phase space swap
- radiation
- etc.

INJECTION BEAMLINES

Injection beamlines are typically found in the radioisotope production industry. They are the beamlines that transport the source ion beam to the industrial linac, or cyclotron.

Consider the injection line for a commercial cyclotron such as those produced by IBA, EuroMeV, KAERI, ACSI, and Sumitomo. Such injection lines can be under 2 metres in length, refer to Figure 1, but accommodate a large range of constraints [2,3].

Application of Accelerators in Research and Industry
AIP Conf. Proc. 1336, 21-25 (2011); doi: 10.1063/1.3586049

TABLE 1. Focusing Element Aperture Selection

Device	Aperture Diameter (cm)	Beam Diameter (cm)	Emittance Growth (%)
Solenoid Magnet	6.0	3.1	14.0
Solenoid Magnet	8.0	3.1	4.3
Solenoid Magnet	10.0	3.1	1.8
Quadrupole Magnet	4.0	2.8	2.4
Quadrupole Magnet	5.0	2.8	1.3
Quadrupole Magnet	6.0	2.8	1.0

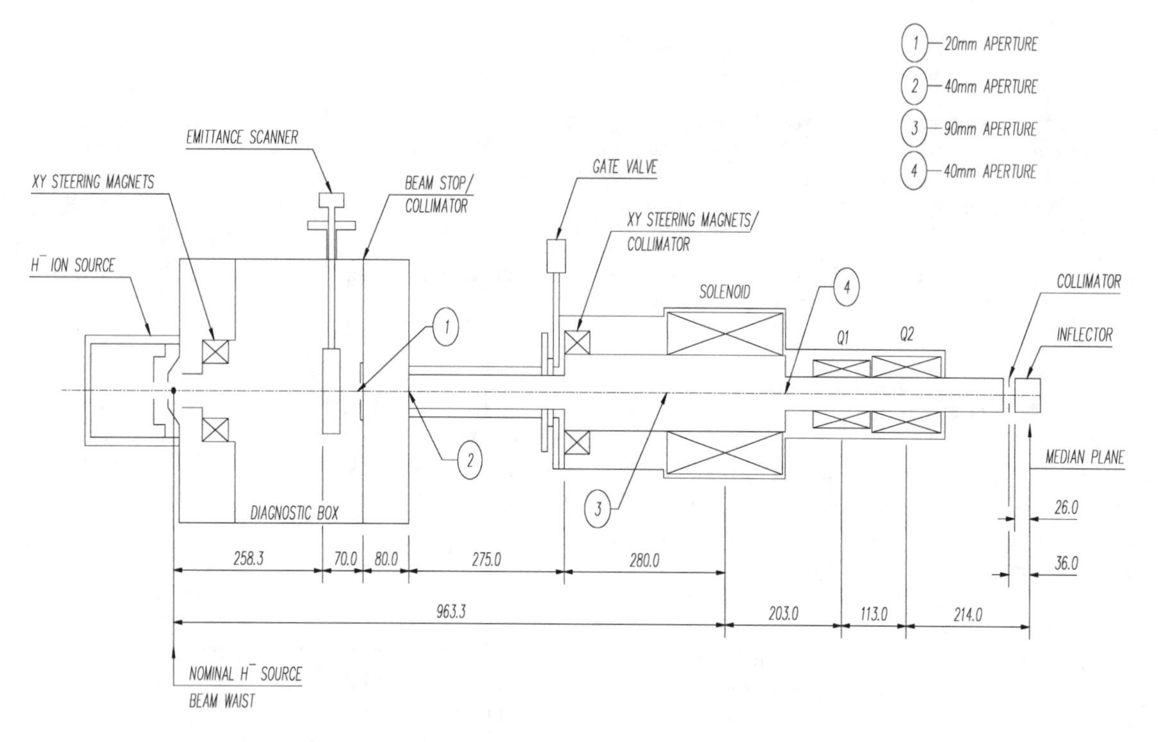

FIGURE 1. TRIUMF injection line for industrial cyclotrons. Dimensions in millimeters.

An external volume-cusp H^- ion source is attached to a vacuum box, Figure 2. The ion source and lenses yield a high brightness beam with electrons removed (minimizes space charge) [4]. The beam phase ellipses in xx' and yy', the dispersion, the kinetic energy, and the current are important input parameters to the beamline, and these can be measured by a Faraday Cup or Emittance Scanner mounted in the vacuum box, Figure 2. The vacuum box is differentially pumped. High capacity turbo-pumps, and special conductance features remove hydrogen gas (minimize H^- stripping, i.e. 5% beam loss/metre at 12 microTorr) in the upstream box [5,6]. Turbo-pumps maintain a vacuum level of ~1 microTorr, maximizing space-charge neutralization while minimizing stripping losses in the downstream box [5,6]. Figure 1 and 2 show XY steering is required to ensure a centred beam upon entry to the solenoid and quadrupole magnets. Focusing element type and aperture are determined by ion-optics studies, Table 1 [5]. High transmission, reliability, cost, reduced emittance growth, and beam matching to the cyclotron acceptance are optimized.

Rotate-able quadrupole magnets introduce coupling between xx' and yy' phase-planes to compensate for emittance growth due to cross-plane coupling introduced by the inflector [2,7]. Bunching the DC source beam to match the cyclotron RF frequency can yield significant accelerated current increases. A choice between electrostatic mirror, hyperboloid inflector and spiral inflector, Figure 3, and their electric, magnetic and aspect ratio parameters must be made to maximize acceptance [8]. The designer must also account for the axial bore field of the cyclotron, and at times an analyzing magnet is used for particle selection and beam purity, and for switching between ion sources. The designer ensures the beam has high transmission, is ion-optically matched for acceleration, and is directed on a prescribed orbit within the cyclotron.

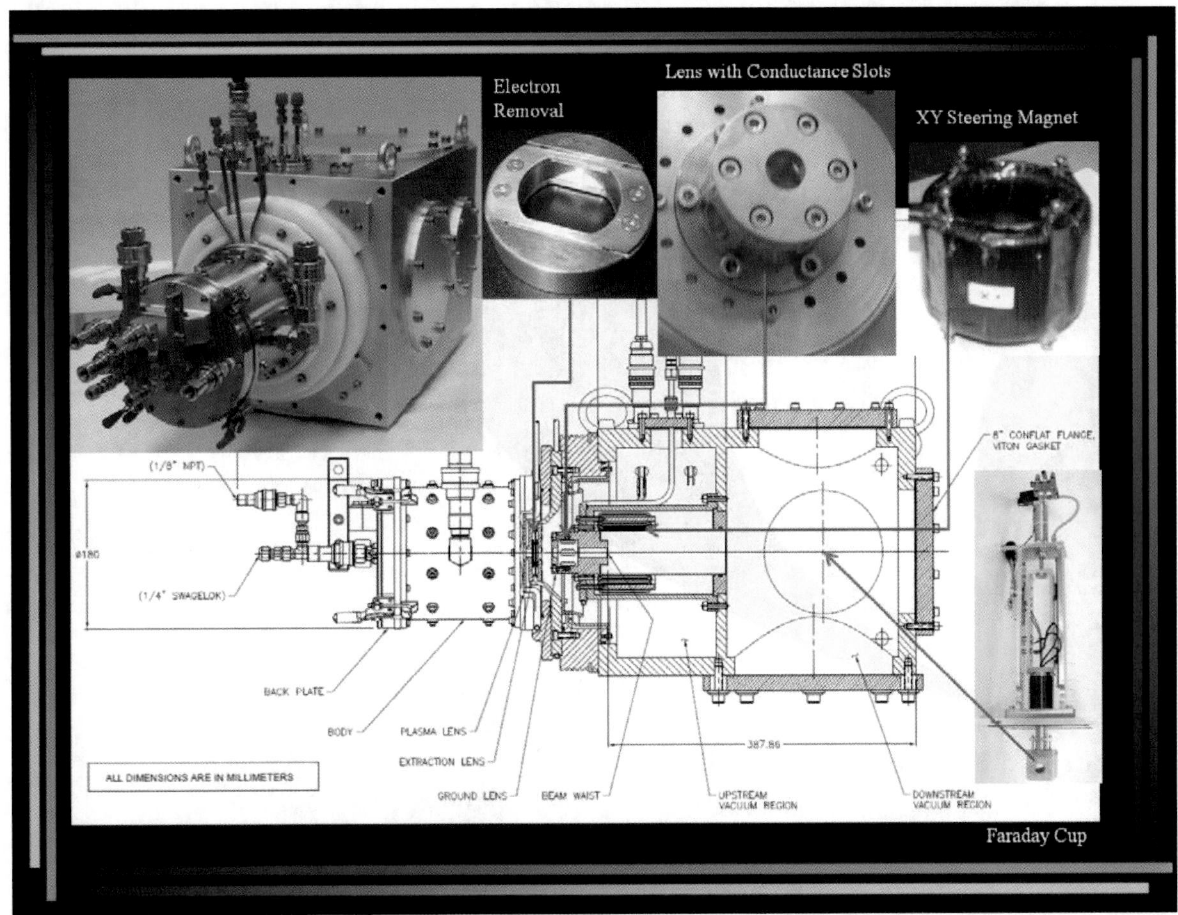

FIGURE 2. D-Pace's TRIUMF Licensed H^- Ion Source and Vacuum Box.

FIGURE 3. CYCLONE30 Spiral Inflector. Photo courtesy of IBA.

EXTRACTION BEAMLINES

Radioisotope Production

For Positron Emission Tomography (PET) applications (0-20 MeV, proton, Figure 4 Left Hand Side, LHS) , targets are often directly mounted to the cyclotron chamber [9], though, as beam currents on target increase, compact ~1 metre beamlines for PET with steering/focusing, diagnostics, target selection and shielding permit beamspot optimization on target, removes residual dose from the cyclotron, and shields maintenance personnel, Figure 4 – Right Hand Side, RHS [10]. Longer extraction beamlines for production of Single Photon Emission Computed Tomography (SPECT) radioisotopes by cyclotrons with proton energies in the 20 – 70 MeV range are used[11]. According to [12], as beam currents and production capacity have increased, dose to maintenance personnel has gone down by minimizing residual radiation at the cyclotron by transporting the beam to well-shielded target bunkers.

FIGURE 4. LHS – PET Targets on a Cyclotron, Photo Courtesy of Lewis Carroll. RHS – Compact PET Beamlines. Photo courtesy of D-Pace.

Electron Beam Processing

Industrial electron accelerators and associated beamlines are classified by kinetic energy and current. The electron energy defines depth of penetration, and the current determines material processing rate. Low energy accelerators range from 75 keV to 300 keV. These accelerators are direct current (DC) units providing up to 3 meter wide un-scanned beams. These machines cure inks, coatings and adhesives, crosslink thin plastic films. The beamline is a very short drift from source to target. Medium energy accelerators utilize electrons from 300 keV to 5 MeV. These multistage accelerators produce scanned beams up to 3 meters wide, Figure 5. They are used in cross-linking insulated wire and cable, heat shrinkable plastic tubing and films, plastic parts, and rubber tires. High energy accelerators produce electrons over 5 MeV. Microwave linear or radiofrequency resonant cavity devices are used as is beam scanning, Figure 6. The main application is the bulk sterilization of pre-packaged medical devices, and thick plastic piping and molded parts. This section is per [13].

FIGURE 5. LHS – DC, electrons, 550-1000 keV, 100+ mA. Photo courtesy of Wasik Associates. RHS – Dynamitron: DC, electrons, 550-5000 keV, 100+ mA. Photo courtesy of IBA.

Semiconductor Ion Implantation

This topic is well covered in these proceedings [14], so a brief discussion shall suffice. Ion implanters are extraction beamlines that have been developed with extremely thorough and careful design to not only yield machines which deposit ultra pure, uniform, parallel [15] and mono-energetic beams of ions in an incredibly reliable manner, but to also be well-suited to high volume manufacture. An ion implanter schematic is shown in Figure 7, as well as an analyzing magnet in which very careful attention to the fringe fields has been undertaken. An ion implanter nowadays may incorporate an analyzing magnet, mass resolving slit, booster, focusing elements, ExB filter for isomers, AC sweep magnet, and a beam collimating magnet.

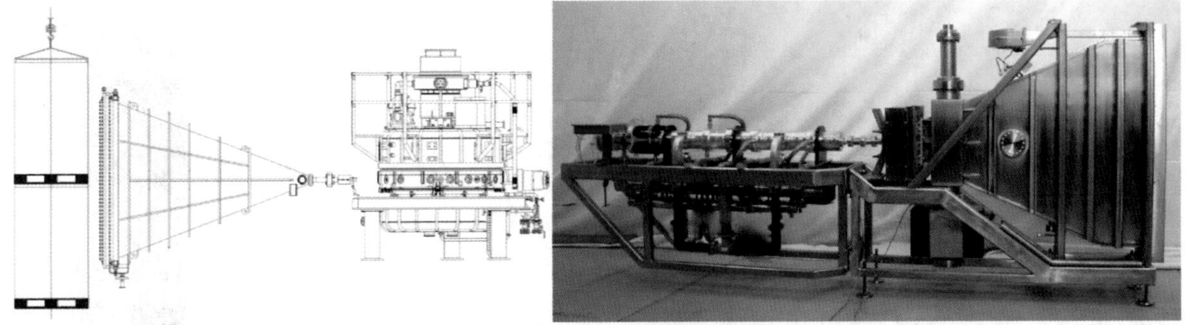

FIGURE 6. LHS – Rhodotron: AC, electrons, 2-10 MeV, 45-700 kW. Photo Courtesy of IBA. RHS – e-LINAC: AC, electrons, 3-25 MeV, 5-30 kW. Manufacturing photo courtesy of Mevex.

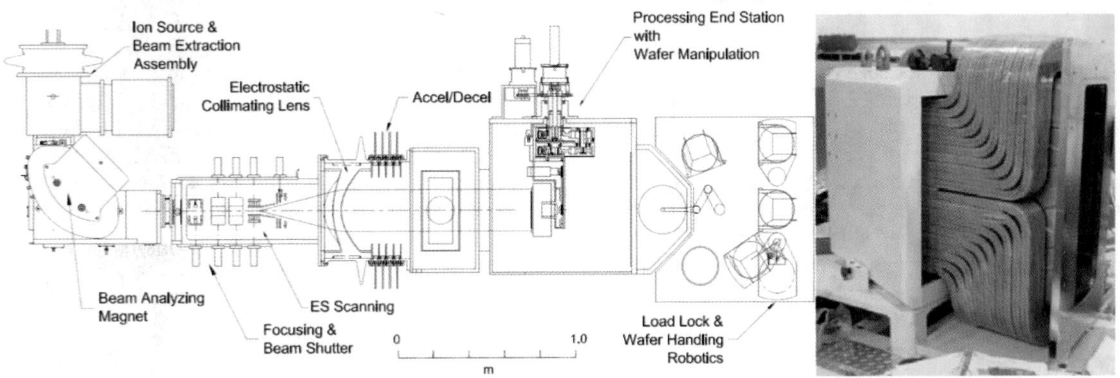

FIGURE 7. LHS – Parallel Implantation, Graphic Courtesy of Isys. RHS – Analyzing Magnet. Photo courtesy of Nick White.

ACKNOWLEDGMENTS

On behalf of D-Pace, the author gratefully acknowledges the financial contributions of the following Canadian government programmes: SRED, SICEAI, BDC, NRC and WEDC. In addition, the author wishes to express his appreciation and thankfulness to the following laboratory, companies, and individual for the contribution of graphic images for this paper: TRIUMF, Isys, Wasik Associates, Carroll & Ramsey, IBA, Mevex, and Nick White.

REFERENCES

1. J.L. Belmont, "Axial Injection and Central Region of the AVF Cyclotron", RNCP-Kikuchi Summer School on Accelerator Technology, 1986.
2. M.P. Dehnel, Doctoral Thesis, UBC, Vancouver, 1995.
3. R. Baartman, Proc. 14th Cyc. Conf., Cape Town, South Africa, (1995), p. 440.
4. T. Kuo et al, Proc. 14th Cyc. Conf., Cape Town, South Africa, (1995), p.1314.
5. R. Baartman, TR30-DN-23, TRIUMF Design Note, 1989.
6. R. Baartman & D. Yuan, EPAC, Vol. 2, Rome, (1988), p. 949.
7. W. Kleeven & R. Baartman, Particle Accelerators, Vol. 41, (1993), p. 55.
8. Balden et al, Proc. 12th Cyc. Conf., Berlin, (1989), p. 435.
9. J. Theroux et al, Proc. 18th Cyc. Conf., Catania, Italy, (2007), p. 361.
10. M.P. Dehnel et al, CAARI2008, AIP Vol. 1099, (2009), p. 504.
11. M.P. Dehnel et al, I.E.E.E. Trans. Ind. App's, Vol. 28, No. 6, pp. 1384-1391, Detroit, 1992.
12. W.J. Dickie et al, NIM B79, North Holland, (1993), p.929.
13. Private Communication, Kathleen Amm, DOE Industrial Accelerator Working Group Report, Draft 7, November 19, 2009.
14. B.A. MacKinnon & J.P. Ruffell, CAARI 2010, these proceedings, in press.
15. B.A. MacKinnon & M.L. King, NIM B74, pp. 469-473, 1993.

Dust Accelerators And Their Applications In High-Temperature Plasmas

Cătălin M. Ticoş[a] and Zhehui Wang[b]

[a]*National Institute for Lasers, Plasma and Radiation Physics, 409 Atomistilor Str., 077125 Bucharest, Romania*
[b]*Los Alamos National Laboratory, MS H846, Los Alamos, NM 87545 , USA*

Abstract. The perennial presence of dust in high-temperature plasma and fusion devices has been firmly established. Dust inventory must be controlled, in particular in the next-generation steady-state fusion machines like ITER, as it can pose significant safety hazards and potentially interfere with fusion energy production. Although much effort has been devoted to getting rid of the dust nuisance, there are instances where a controlled use of dust can be beneficial. We have recognized a number of dust-accelerators applications in magnetic fusion, including in plasma diagnostics, in studying dust-plasma interactions, and more recently in edge localized mode (ELM)'s pacing. With the applications in mind, we will compare various acceleration methods, including electrostatic, gas-drag, and plasma-drag acceleration. We will also describe laboratory experiments and results on dust acceleration.

Keywords: Dust, Accelerators, Plasma, Fusion
PACS: 52.25.Vy, 52.27.Lw, 52.40.Hf, 52.59.Dk

INTRODUCTION

The presence of dust in fusion machines is receiving increased attention from the plasma community [1]. Dust is produced in today's tokamaks and is a result of plasma-wall interactions, such as sputtering, arc melting of the machine wall surface, and condensation of impurities, or flaking of depositions, in particular, following a disruption or less violent but more frequent plasma instabilities, including Edge-Localized Modes (ELMs). The dust particles range in size from hundreds of nanometers to a few millimeters and have irregular shapes determined mainly by the formation process. Observation of dust is done "in situ" by high-speed cameras or at the end of a campaign, when it is collected directly from the walls of the machine. In the first case dust is self-illuminated as it is heated to high temperatures by the dense ion and electron fluxes and is transported by the cross-field flows. Dust is currently produced in small quantities due to the relative small size of the current tokamaks, but in the next generation fusion machine, ITER, it can become a problem for the plasma confinement and for the safety of operations. However, the prospects of using a limited quantity of dust particles to perform diagnostics inside the fusion plasma or to control some type of instabilities are encouraging [2,3]. In this respect, it is desirable to select the dust material that is compatible with magnetic fusion energy production, and control both the amount of the dust particles and dust speeds. Sometimes it might be necessary to have high speeds in order for the dust particles to reach further into the fusion plasma, near the core or at least way beyond the edge as the magnetic fluxes do not intersect with the fusion chamber wall.

DUST ACCELERATION TECHNIQUES AND APPLICATIONS

A few successful techniques have been employed to accelerate micron-size dust particles to speeds of the order of km/s. Electrostatic acceleration relies on the electrical charge deposited on a dust particle by contact charging. A typical spherical dust particle exposed to a strong electric field (up to 10^9 V/m) created by a high voltage needle can reach the limit of elementary charges beyond which self-breakup can occur [4]. The highly charged dust particle is then introduced in a Van de Graaf accelerator and exposed to high electric fields corresponding to voltages of the order of 1-5 MV [5]. This method is bulky and for higher speeds, multiple-stage acceleration is needed; however it is efficient and has been extensively employed to simulate micro-meteorites impacts with surfaces. The ultimate speed

Application of Accelerators in Research and Industry
AIP Conf. Proc. 1336, 26-28 (2011); doi: 10.1063/1.3586050

(u_d) of the electrostatic acceleration is related to the dust sphere radius (r_d) and the acceleration voltage (V) as $r_d u_d^2 = cV$, with c being a constant determined by the material properties of the dust grain. Therefore, high speed (> 1 km/s) is only feasible for submicron dust for a few MV of accelerating voltage. Still, this method is a "clean" method as no other material is mixed up with the dust particles at the end of the acceleration channel. Another technique is to accelerate the dust particles by puffing gas through a narrow nozzle in vacuum. The gas molecules or atoms flowing at the sound speed entrain in their motion the dust particles by transferring them their momentum through collisions. The drag force exerted by the gas is proportional with its density and the flow speed. The dust speed can in the best cases reach about 1 km/s since it cannot exceed the speed of sound in the gas. Thus, the use of light gases such as H or He is preferred. However, the terminal speed can be greatly increased by the use of a piston put in motion by the high gas pressure which subsequently pushes the dust particles. This type of technology based on compressed gas has been developed and refined over the years resulting in the development of gas gun injectors with multiple stages [6]. Millimeter-size pellets made of frozen ice have been launched at speeds up to 2 km/s for refueling tokamak plasmas. Heating of the compressed gas to about 450 °C has led to only a 10% increase in the terminal speed of the millimeter size pellets.

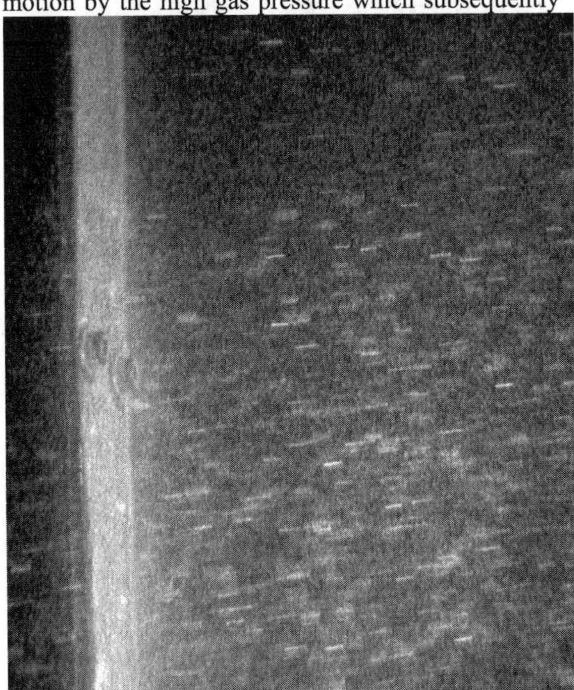

FIGURE 1. A dust shower produced by a plasma jet. Each needle-like segment corresponds a high-speed glowing dust grain of graphite. A probe is place on the left for reference. The exposure time is 4 microsec. The capacitor bank charging voltage was 8 kV. Further details about the accelerator operation can be found in [7].

An alternative technique which proved to be more effective in terms of attainable dust speed is based on dust acceleration by a dense plasma flow. A plasma jet is launched when currents of hundreds of kA flow between two coaxial electrodes for several hundred microseconds, a center rod and an outer conductive cylinder. The self-generated azimuthal magnetic field and the current density produce a **J**×**B** force which expels the plasma particles at about 10-60 km/s [7]. Dust particles released in the path the plasma jet are dragged in the flow direction, as show in Fig. 1.

The speed of dust particles measured *in situ* attained 3.7 km/s. The dust particles were made of carbon (graphite and diamond) compatible with fusion plasmas with a wide range of sizes, from a few microns to 60 μm. Compression of the plasma jet in a helical coil improved the performance of coaxial plasma accelerators [8]. Dust particles speed of up to 20 km/s have been measured indirectly, from the length of the particles' trace in aerogels or from the impact force detected by piezo-electric probes. The mass of particles was in the range of 10^{-10} to 10^{-1} g. The drag force exerted by the flowing ions colliding with the dust particles is by far the dominant influence on dust dynamics at relatively high plasma densities of 10^{17}-10^{22} m^{-3} [9], and is up to a factor of 10^3 larger than the force of gravity acting on a dust particle. High dust accelerations of 10^7 m/s^2 can be thus obtained on distances of the order of 1 m. One main drawback especially in fusion applications is the injection of plasma alongside the dust particles, which can mix with the fusion plasma and cool it down. Fast blocking valves can however reduce the effect of plasma flow leak into a fusion reactor. Table I gives the dust size range, the terminal dust speed, the acceleration distance and the main requirements of four dust acceleration techniques.

TABLE 1. Comparison of different dust acceleration techniques

Dust acceleration	Dust size	Terminal speed	Distance	Main requirements
Electrostatic	0.1- 2 μm	20 km/s	~10-20 m	Voltages 1-5 MVolts, ~0 current
Puffed gas	0.1-50 μm	100 m/s-1 km/s	~1 m	Gas pressure 10-100 bar
Multiple stage	0.1-5 mm	2.5 km/s	~ 5 m	Gas pressure 100-160 bar
Plasma Drag	0.1-60 μm	3.7-25 km/s	~ 1m	10-50 kV, 10-300 kA (pulsed)

There are several applications of highly accelerated dust particles. The study of surface resistance to impacts with microparticles flying at several km/s is particularly important for space shuttle and satellite development. In material science, the morphological properties of some surfaces can be modified by bombardment with high speed nano or microparticles. In fusion plasmas, dust particles can be used for local magnetic field diagnostics by injecting a controlled amount and by further monitoring the spectral emission or by real-time imaging with a high speed camera of the modifications suffered due to the interaction with the energetic plasma fluxes. It has been recently proposed to use highly accelerated dust particles for mitigating disruptions or local instabilities such as edge localized modes or ELM's in fusion plasmas [3]. Disruptions are large instabilities which terminate the discharge and put mechanical stress and heat on the machine walls. In the H confinement mode, several distinctive ELM types were observed early. They included *dithering, type III* and large amplitude or *type I ELMs*. More ELM types have been found since then. The Type I ELMy H-mode is a particular type of ELM which is observed in the high confinement regime (or H-mode) of the tokamak plasma. It is an intermittent mode during which the particle and energy losses to the wall are increased, yet allowing for the H-mode to be sustained. The Type I ELMy H-modes show experimental performances that could meet the necessary requirements of a fusion reactor when extrapolated to the next step devices such as ITER. A major drawback of the Type I ELMy H-mode is the periodic (commensurate with the externally applied heating power) large power loads on plasma facing components (PFCs), in particular the divertor. ELMs have been observed to be triggered by fuelling pellets on several tokamaks, no matter where the injection location is. This led to the idea to artificially induce more frequent ELMs by injecting pellets smaller than what is usually used for fueling, the concept now known as *ELM pacing* [10]. It is possible to use dust injection to induce more frequent ELMs than currently possible with the pellet techniques. Based on our estimates, single dust (ranging from 50 to several hundred microns in size) injection at a frequency of ~ 1 kHz and dust speed of up to several hundred m/s would be feasible and potentially useful.

REFERENCES

1. See for example, G. Federici et al., *Nucl. Fusion* **41**, 1967-2137 (2001); J. P. Sharpe, D. A. Petti and H. W. Bartels, *Fusion Eng. Design* **63-64**, 153-163 (2002); J. Winter, *Plasma Phys. Control. Fusion* **46**, B583-B592 (2004); and Z. Wang, C. H. Skinner, G. L. Delzanno et al., "Physics of Dust in Magnetic Fusion Devices" in *New Aspects of Plasma Physics*, edited by P. K. Shukla et al., World Scientific Singapore, 2008, pp. 394-475.
2. Z. Wang and G. A. Wurden, *Rev. Sci. Instrum.* **74**, 1887-1891 (2003).
3. Z. Wang, D. K. Mansfield, L. A. Roquemore, C. M. Ticos, and G. A Wurden, "Applications and Progress of Dust Injection to Fusion Technology" in *Multifacets of Dusty Plasmas*, edited by J. T. Mendonca et al., AIP Conference Proceedings 1041, American Institute of Physics, Melville, NY, 2008, pp. 135-138.
4. S. Hasegawa, Y. Hamabe, A. Fujiwara et al., *Int. J. Impact Eng.* **26**, 719-727 (2001).
5. G. L. Stradling, G. C. Idzorek, B. P. Shafer et al., Int. J. Impact Eng. **14**, 719-727 (1993).
6. S. K. Combs, *Rev. Sci. Instrum.* **74**, 1887-1891 (2003).
7. C. M. Ticos, Z. Wang, G. A. Wurden et al., *Phys. Rev. Lett.* **100**, 155002 (2008).
8. A. Hüdepohl, M. Rott, E. Igenbergs, *IEEE Trans. Magn.* **25**, 232-237 (1989).
9. C. M. Ticos, I. Jepu, C. P. Lungu, P. Chiru, V Zaroschi, A. M. Lungu, *Appl. Phys. Lett.* **97**, 011501 (2010).
10. P. T. Lang, G. D. Conway et al. *Nucl. Fusion* **44**, 665-677 (2004).

Features Of The J-PARC Linac

Tetsuya Kobayashi

Japan Atomic Energy Agency (JAEA)
2-4 Shirakata-Shirane, Tokai-mura, Naka-gun, Ibaraki-ken, 319-1195, Japan

Abstract. Japan Proton Accelerator Research Complex (J-PARC) will be one of the highest intensity proton accelerators in the world aiming to realize 1 MW class of the beam power. The accelerator consists of a 400-MeV linac, a 3-GeV rapid-cycling synchrotron (RCS) and a main ring synchrotron (MR), and the accelerated beam is applied to several experimental facilities. The linac, which is the injector for the RCS, has about 50 cavity modules to accelerate the beam up to 400 MeV. The acceleration field error in all of them should be within ±1% in amplitude and ±1 degree in phase because the momentum spread of the RCS injection beam is required to be within 0.1%. For the cavity field stabilization, a high-stable optical signal distribution system is used as the RF reference, and sophisticated digital feedback and feed-forward system is working well in the low level RF control system. Consequently the providing beam to the RCS is very stable, and the beam commissioning and the experiments of the application facilities have been progressed steadily.

Keywords: proton linac, RFQ, DTL, SDTL, ACS, RF Chopper, LLRF, klystron
PACS: 29.20.Ej

INTRODUCTION

Japan Proton Accelerator Research Complex (J-PARC) is the high-intensity proton accelerator facility in Japan, which is a joint project between KEK (High Energy Accelerator Research Organization) and Japan Atomic Energy Agency (JAEA) [1]. The J-PARC has several experimental facilities designed to pursue frontier science in particle physics, nuclear physics, materials science and life science.

The J-PARC accelerator, as shown in Fig. 1, is comprised of a 400-MeV linac, the 3-GeV rapid-cycling synchrotron (RCS), and the Main Ring (MR) of which the energy is 30-50 GeV. The 400-MeV beams from the linac are injected to the RCS with a repetition rate of 25 Hz, while the beams are further accelerated up to 600 MeV by the superconducting (SC) linac to be used for the basic study of the Accelerator-Driven Nuclear Waste Transmutation System (ADS). The maximum peak current is 50 mA for each of the RCS and ADS. The 1-MW beams from the RCS are mostly extracted to the Materials and Life Science Experimental Facility (MLF), where the muon-production and neutron-production targets are located in a series.

Since the J-PARC linac is the injector for the RCS, the output energy is not so high. However, the peak current is high. In addition, high quality and stability is required for the injection beam to the RCS ring. The momentum spread (Δp/p) of the RCS injection beam should be within ±0.1%. Therefore the cavity field stability is the one of the most important issues for the linac.

At the present stage of the Phase 1, the linac energy is reduced to 181-MeV acceleration, and RCS power is also reduced to 0.6 MW. The 400-MeV upgrade of the linac is in progress now.

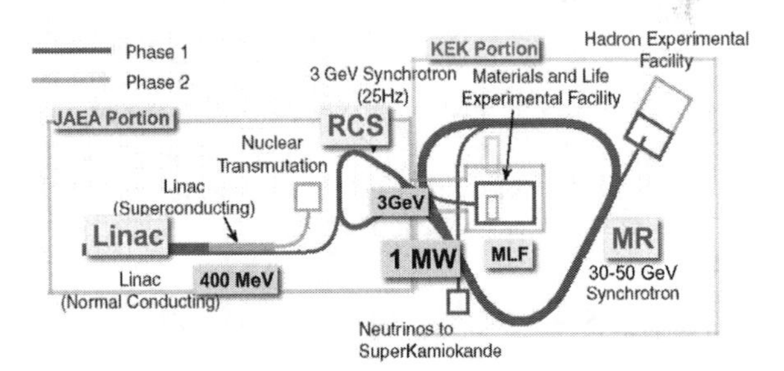

FIGURE 1. Layout of the J-PARC facility.

LINAC SCHEME

J-PARC linac scheme is summarized in Fig. 2 and its basic parameters are shown in Table 1.

The volume-production type of the negative hydrogen ion source [2], which is Cs free, is followed by a Radio-Frequency Quadrupole (RFQ) linac [3]. The RFQ linac [3] accelerates the 50-keV beams from

Application of Accelerators in Research and Industry
AIP Conf. Proc. 1336, 29-32 (2011); doi: 10.1063/1.3586051

the ion source up to 3 MeV. The Medium Energy Beam Transport (MEBT) is comprised of 8 quadrupole-magnets and two bunchers to match the RFQ beams to the following Drift Tube Linac (DTL). In addition, the RF deflector (RF chopper) [4] is installed in the MEBT to chops the 500-us long macro-pulse beam into the medium-bunches at the RCS revolution frequency of 1-MHz. TE11 like deflecting field in the chopper cavity kicks the beam. After the phase optimization of the chopper RF, the observed residual beam is well less than 10-3. It is consistent with the noise level.

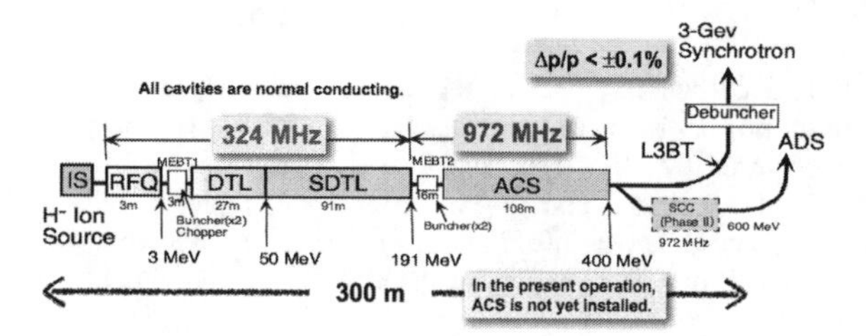

FIGURE 2. Scheme of the J-PARC Linac.

The three DTL cavities [5] accelerate the beams from 3 MeV up to 50 MeV. The DTL has compact electro-quadrupole magnets accommodated in the drift tube. The Separated-type DTL (SDTL) cavities [6], which have no quadrupole magnets inside, follow the DTL. The 32 SDTL cavities accelerate the beams up to 191 MeV. The RF frequency is 324 MHz in this low β section.

The Annular-ring Coupled Structure (ACS) [7] has been chosen for the high-energy structure to accelerate the beams up to 400 MeV. The ACS has coaxial symmetric structure, and its use for the first time in the world is a unique challenge. The RF frequency of the ACS is 972 MHz.

In the present phase, the ACS cavities are not installed yet, and the linac provides 181-MeV beam to the RCS. In this case, the last 2 cavities of the SDTL are applied as the debunchers.

RF SYSTEM AND PERFORMANCE

The overview of the RF system is shown in Fig. 3. Twenty 324 MHz and twenty-five 972MHz klystrons are needed. Totally 45 klystrons will be line up in the klystron gallery for the 400-MeV acceleration. In the SDTL section, one klystron drives a pair of 2 cavities. The high voltage DC power supply (HVDCPS) drives 4 klystrons. The electron gun of the klystron is a triode type, and each klystron has an anode modulator. The output peak power of the klystron is 3 MW. In the present operation of the 181-MeV acceleration, 20-unit of 324-MHz klystrons have been operated for 16,400 hours from 2006 Octorber to 2010 June. There are no serious troubles, such as a discharge in electron gun, vacuum degradation, or arcing in output window. The availability of the J-PARC accelerator averages 92.4% in recent three months.

TABLE 1. Basic Parameters of the J-PARC Linac

Acc.Ions Row	H^- (negative hydrogen)
Output Energy	400 MeV (present: 181 MeV)
Peak current	50 mA (present: 30 mA max)
Repetition	25 Hz (+ 25 Hz for ADS)
Beam Pulse width	500 μs (macro pulse)
RF Pulse width	650 μs (with build up time)
RF Frequency	324 MHz (~ 191 MeV) 972 MHz (~ 400 MeV)
Linac Length	300 m
Number of Klystrons	45

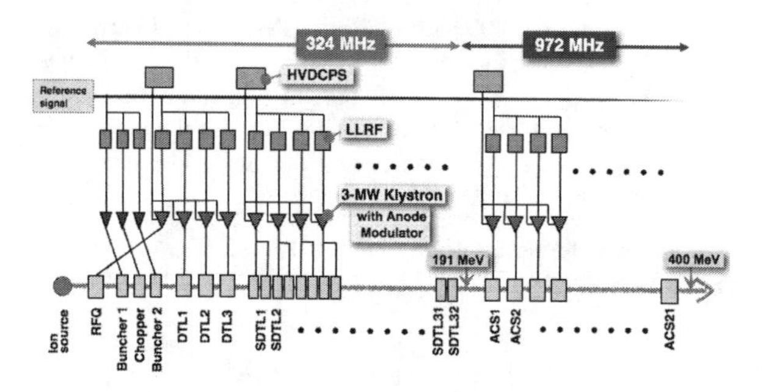

FIGURE 3. Overview of the RF system of the J-PARC linac.

LLRF Control System

Stability in the RF control is very critical for the high-intense beam acceleration to be achieved without unexpected beam loss. Since the momentum spread (Δp/p) at the RCS injection beam should be within 0.1%, the stability of the accelerating fields is required to be within ±1% in amplitude and ±1 degree in phase. To realize this stability, an FPGA-based digital feedback (FB) control is used in the Low Level RF (LLRF) control system, and a feed-forward (FF) technique is combined with the FB control for the beam loading compensation [8].

Figure 4 shows the block diagram of the LLRF control system for the SDTL cavities. The vector sum of the two cavity fields is controlled by the digital FB system installed in a compact PCI (cPCI) crate, which consists of the DSP with FPGA, Mixer & IQ modulator, and RF & CLK boards as shown in Fig. 5. The RF reference is 312-MHz or 960-MHz optical signal, which is received by an optical-electrical transceiver (O/E). This reference is directly used as the Local Oscillator (LO) signal. Then the cavity-monitor signals (324 MHz or 972 MHz) are down-converted by mixers into 12-MHz IF signals with the LO, and they are digitized by ADC's at 48-MHz CLK, then I/Q components of the two cavities are directly obtained. The I/Q values are compared with set-tables and proportional-integral FB control and FF control is made.

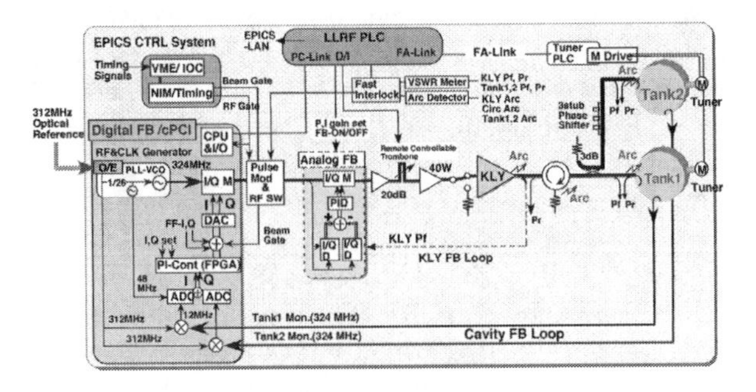

FIGURE 4. Block diagram of the LLRF system.

FIGURE 5. Digital FB/FF control system on cPCI crate.

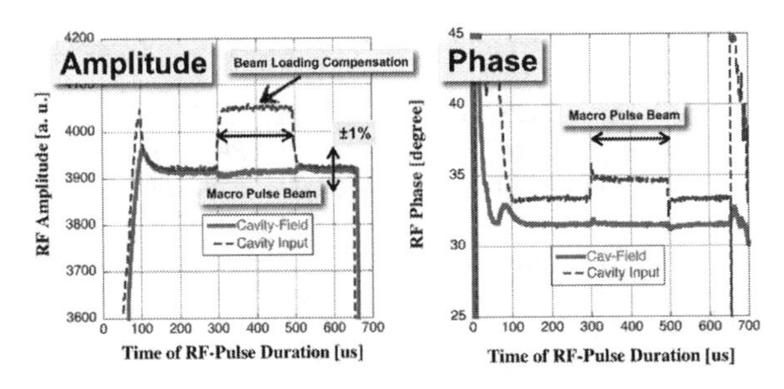

FIGURE 6. RF-pulse (amplitude and phase) waveforms. The solid lines shows the cavity-monitor. The dashed line shows the cavity-input as compensating the beam loading.

The result of the stabilization is shown in Fig. 6. Plotted are the RF-pulse waveforms (amplitude and phase, respectively). In the figure, the solid line shows the cavity-monitor, and the dashed line corresponds to the cavity-input as compensating the beam loading in the case of 200-μ s macro pulse width. As the result, the klystron amplitude/phase sag, which is due to the high voltage sag of the power supply, and the beam loading are successfully compensated. The stability of within ±0.2% in amplitude and within ±0.2 degrees in phase was accomplished.

The digital LLRF system has many other functions, for example, chopped beam loading compensation [9], Q-value monitoring of RF cavities and auto-tuning control of the cavity-feed RF frequency [10].

RF Reference System

About 50 cavity modules, including chopper, bunchers and debunchers, line up in the 300-m long linac. The synchronization of all these cavities with each other or the cavity-to-cavity phase stability is also very essential. Even though the FB control system functions perfectly for the each cavity, the stability of the mutual phase between the cavities depends on the reference signal. For this linac, a unique reference system, which is an optical signal multi-dividing transfer system with an optical amplifier, was originally invented independently.

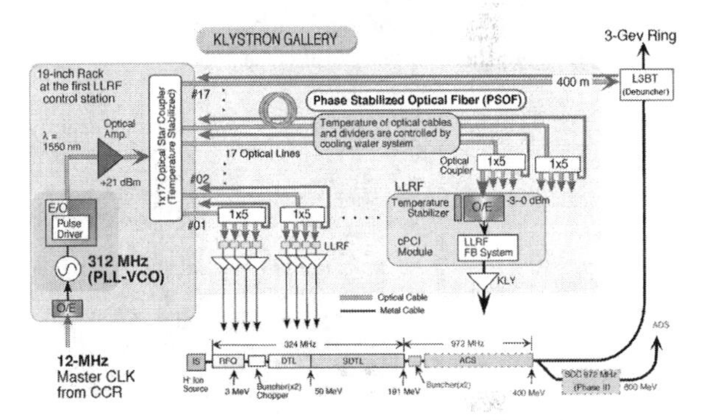

FIGURE 7. Schematic draw of the RF reference distribution system. For the 400 MeV upgrade, additionally, 960-MHz signal will be also generated and distributed from the master oscillator.

The schematic draw of the RF reference distribution system [11] is shown in Fig. 7. The RF reference signal is optically amplified and divided into all LLRF control systems. The Phase Stabilized Optical Fibers (PSOF), of which thermal coefficient is 0.4 ppm/°C, are used for the transfer lines. Temperature of the optical cables and the components (E/O, O/E and dividers) is controlled to be constant. The optical transceivers (E/O and O/E) were newly designed to reduce the transfer jitter of the optical link to less than 1 ps. For the 400-MeV upgrade, additionally, 960-MHz signal will be also generated and distributed from the master oscillator.

Figure 8 shows the trend graph of the RF reference phase (round-trip signal) from the oscillator to the linac end. It also shows the temperature in the klystron gallery. The phase stability was approximately ±0.06 degrees for a 4-day measurement period, while the room temperature change was approximately 3.5°C in the klystron gallery. Totally, a cavity field stability of within ±0.3% in amplitude and within ±0.3 degrees in phase including beam loading and temperature drift has been accomplished by the LLRF and Reference system [12].

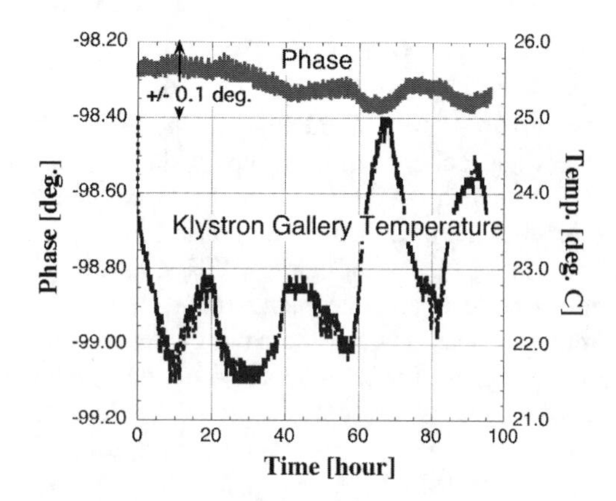

FIGURE 8. Trend graph of the RF reference phase (round-trip signal) from the oscillator to the linac end measured for 4 days.

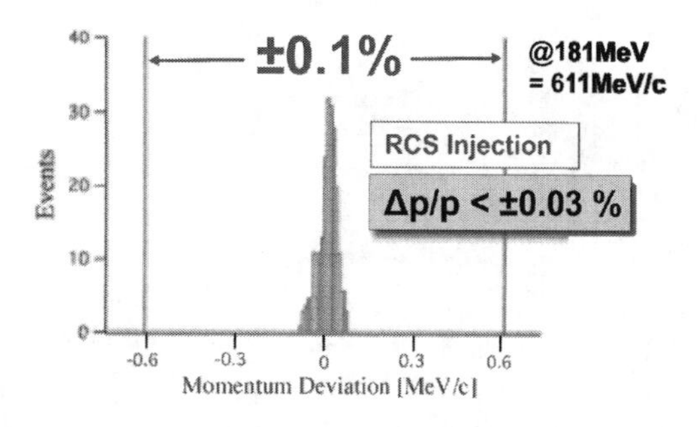

FIGURE 9. Measured momentum jitter of the RCS injection beam.

Because the most systems have been working successfully, the operation of the J-PARC accelerator has been progressing favorably with high reliability of the machines. As the result of the good performance of the cavity field stabilization including the beam loading compensation, the momentum jitter of less than ±0.03 % was obtained at the RCS injection in 181-MeV acceleration as shown in Fig. 9. Consequently, RCS injection beam is very stable and highly reproducible during long operation. Figure 9 indicates the pulse-to-pulse fluctuation of the beams. The momentum spread in the beam bunches has been not yet measured, but it was estimated to be within ±0.1% from the RCS operation parameters.

SUMMARY

The 400-MeV linac of the J-PARC, which is an injector for the RCS, is based on many newly developed technologies including the cavities, beam alignment methods, klystrons and LLRF control systems to accelerate high-intense beams. The developments of the linac system have been done successful, and the commissioning and the operation for the experiments are progressing well. The availability of the J-PARC accelerator is more than 90%. In particular, the good performance of the LLRF control system has the greatest contribution to the supply of highly-stable beam to the RCS.

The beam power of the RCS at the present stage is 0.6 MW, since the linac energy is 181 MeV. However it will increase to 1 MW in the near future because of the linac energy upgrade to 400 MeV now in progress.

REFERENCES

1. "The Joint Project for High-Intensity Proton Accelerators", KEK Report 99-4 and JAERI-Tech 99-056, 1999.
2. Oguri H, et al., "Development and operation of a Cs-free J-PARC H(-) ion source", Rev. Sci. Instrum. 79, 2008.
3. A. Ueno and Y. Yamazaki, "New field stabilization method of a four-vane type RFQ", Nucl. Instr. Meth. A300, 1991, pp. 15-24.
4. S. Wang, S. Fu and T. Kato, "The development and beam test of an RF chopper system for J-PARC", Nucl. Instr. Meth. A 547, 2005, pp. 302–312.
5. F. Naito, et al., "Development of the 50-MeV DTL for the JAERI/KEK Joint Project", Proc. of LINAC2000, 2000, pp. 563-565.
6. F. Naito, et al., "High-Power Test of The SDTL for The JAERI/KEK Joint Project", Proc. of LINAC2002, 2002, pp. 425-427.
7. H. Ao and N. Hayashizaki, "R&D Status of The Annular Coupled Structure Linac for The JAERI/KEK Joint Project", Proc. of LINAC2002, 2002, pp. 82-84.
8. S. Michizono, S. Anami, S. Yamaguchi and T. Kobayashi, "Digital Feedback System for J-PARC Linac RF Source", Proc. of the LINAC2004, 2004, pp. 742-744.
9. S. Michizono, et al., "Digital Feedback Control for 972-MHz RF System of J-PARC Linac", Proc of PAC09, WE5PFP082, 2009.
10. T. Kobayashi, et al., "Automatic Frequency Matching for Cavity Warming-up in J-PARC Linac Digital LLRF Control", Proc of PAC09, 2009, WE5PFP087.
11. T. Kobayashi, et al., "RF reference distribution system for J-PARC linac", Nucl. Instr. Meth. A585, 2008, p. 12.
12. T. Kobayashi, et al., "Performance of J-PARC Linac RF System", Proc. of PAC07, 2007, WE5PFP087.

Novel Linac Structures For Low-Beta Ions And For Muons

Sergey S. Kurennoy

Los Alamos National Laboratory, Los Alamos, NM 87545, USA

Abstract. Development of two innovative linacs is discussed. (1) High-efficiency normal-conducting accelerating structures for ions with beam velocities in the range of a few percent of the speed of light. Two existing accelerator technologies – the H-mode resonator cavities and transverse beam focusing by permanent-magnet quadrupoles (PMQ) – are merged to create efficient structures for light-ion beams of considerable currents. The inter-digital H-mode accelerator with PMQ focusing (IH-PMQ) has the shunt impedance 10-20 times higher than the standard drift-tube linac. Results of the combined 3-D modeling for an IH-PMQ accelerator tank – electromagnetic computations, beam-dynamics simulations, and thermal-stress analysis – are presented. H-PMQ structures following a short RFQ accelerator can be used in the front end of ion linacs or in stand-alone applications like a compact mobile deuteron-beam accelerator up to a few MeV. (2) A large-acceptance high-gradient linac for accelerating low-energy muons in a strong solenoidal magnetic field. When a proton beam hits a target, many low-energy pions are produced almost isotropically, in addition to a small number of high-energy pions in the forward direction. We propose to collect and accelerate copious muons created as the low-energy pions decay. The acceleration should bring muons to a kinetic energy of ~200 MeV in about 10 m, where both an ionization cooling of the muon beam and its further acceleration in a superconducting linac become feasible. One potential solution is a normal-conducting linac consisting of independently fed 0-mode RF cavities with wide apertures closed by thin metal windows or grids. The guiding magnetic field is provided by external superconducting solenoids. The cavity choice, overall linac design considerations, and simulation results of muon acceleration are presented. Potential applications range from basic research to homeland defense to industry and medicine.

Keywords: Accelerator; linac; ion; muon; beam
PACS: 29.20.Ej; 29.27.-a; 41.75.-i; 41.85.-p

H-MODE ACCELERATOR STRUCTURES WITH PMQ FOCUSING

Room-temperature accelerating structures based on inter-digital H-mode (IH) resonators are very efficient at low beam velocities, $\beta = v/c < 0.1$ [1]. Small sizes of the drift tubes (DTs), required for achieving high shunt impedances in the H-resonators, usually prevent placing conventional electromagnetic quadrupoles inside DTs. Inserting permanent-magnet quadrupoles (PMQs) inside small DTs of the H-structure, as we suggested in [2], promises both efficient beam acceleration and good transverse focusing. Further studies of this approach [3] based on EM 3-D modeling, beam-dynamics simulations, and engineering analysis, confirmed that IH-PMQ accelerating structures are feasible. Papers [3] studied only one or a few periods of the IH structures in the beam velocity range $\beta = 0.0325$-0.065, corresponding to the deuteron beam energies from 1 to 4 MeV. In [4] a complete short tank containing the IH-PMQ accelerating structures with vanes was designed. The beam velocity range was chosen to be $\beta = 0.0325$-0.05, so that the tank can serve as the first of two tanks in a 1-4 MeV compact deuteron (D^+) accelerator. We mostly concentrated on the end-cell design and the means to tune the electric field profile along the beam axis, as well as the frequency of the working mode to 201.25 MHz. This IH-PMQ tank was also used to prove feasibility of the thermal management with cooling channels in vanes, even at high duty factors.

The IH-PMQ tank [4] was designed with equal gaps between DTs, which leads to equal magnitudes of the on-axis electric field in (and voltages across) the gaps except the two end ones. The cell length L_c increases along the structure as $L_c = \beta_g\lambda/2$, where β_g is the design beam velocity and λ is the RF wavelength, so the average accelerating gradient per cell decreases along the tank. Beam dynamics simulations for the IH tank indicated no problems at low or moderate currents but for high currents (~50 mA) the beam particles can be lost at the DT walls. Due to small DT lengths in that design, from 1.8 cm, short PMQs inside DTs were not strong enough to focus high-current beams.

Application of Accelerators in Research and Industry
AIP Conf. Proc. 1336, 33-38 (2011); doi: 10.1063/1.3586052

Modified IH-PMQ Tank Design

The IH-PMQ tank was redesigned [5] with a higher injection energy (β_{in} = 0.04, corresponding to the D^+ energy of 1.5 MeV; β_{out} = 0.0543, energy ~2.8 MeV) so that its DTs are longer. The 3-D EM modeling of the tank was performed with the CST MicroWave Studio (MWS) [6] as in [2-4], but this time we used iterations of beam-dynamics and EM simulations to adjust the tank layout. The gap widths were tuned for the electric field strength to increase along the tank proportionally to the cell length, to keep the cell gradient nearly constant. The gap positions were adjusted to create a ramp of the synchronous phase along the tank from -45° to -35°, except -54° in the first gap and -41° in the last. The ramp maintains a constant RF-bucket height and provides better beam capture. The modified IH-PMQ tank is shown in Fig. 1. The cavity total length is 73.51 cm, and its radius is 11.92 cm; the shortest DT is DT3 (23 mm) and the longest one is DT20 (34.3 mm). The gap widths vary between 7.1 and 8.3 mm, except the very first and the last gap, which were reduced to 3.6 and 3.9 mm, respectively, to bring up the electric fields near the tank ends. All drift tubes up to DT11 have the bore radius of 5 mm, and all from DT12 have a larger bore radius of 5.5 mm. The outer radius of all DTs is the same, 14 mm. The field profile in the tank, as well as its frequency, can be tuned with two pairs of slug tuners in the side walls, one near the tank entrance and the other near its end (not shown in Fig. 1).

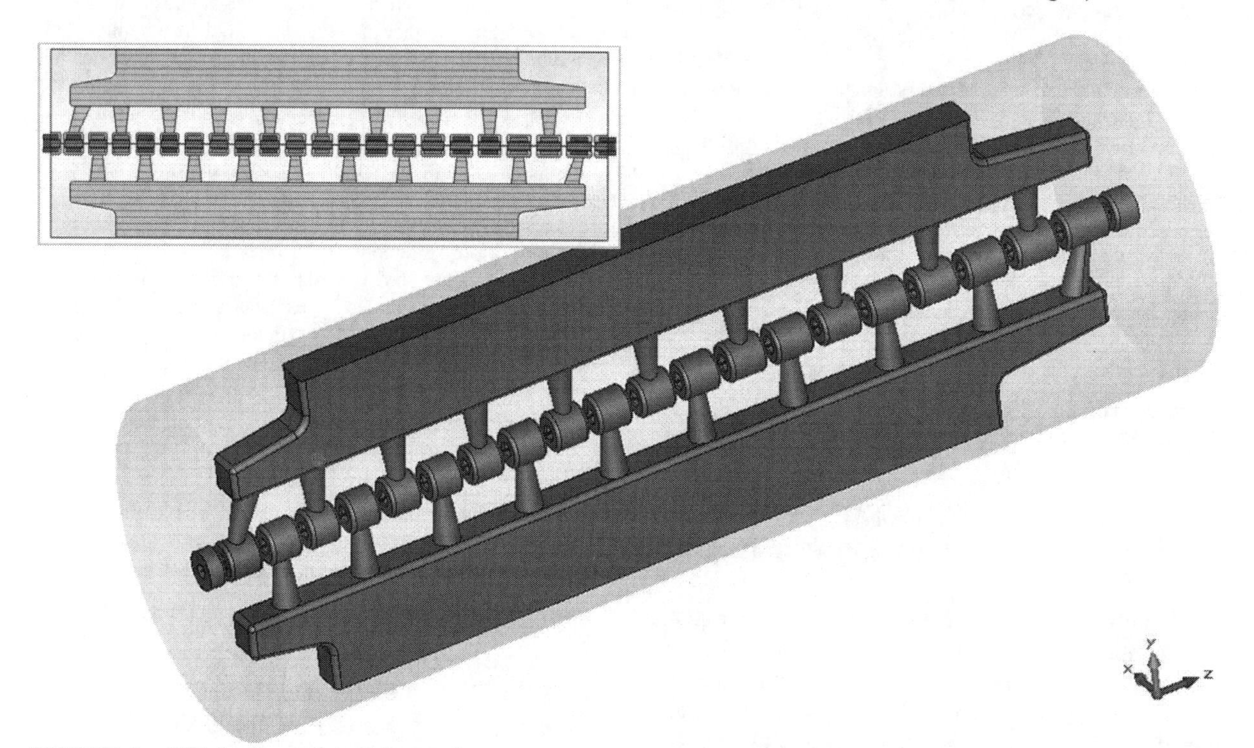

FIGURE 1. CST Studio model of the IH-PMQ tank for β = 0.04-0.0543. The tank outer wall is removed; the cavity inner volume is shown as a transparent cylinder. The inset shows PMQs inserted in DTs and arranged in FOFODODO lattice.

The EM parameters of the tank for the averaged accelerator gradient E_0 = 2.5 MV/m are listed in Tab. 1. Two families of the 16-segment PMQs are used for beam focusing in this tank: the short ones, 18.89 mm, in the first 12 DTs (0 to 11), and longer ones, 22.67 mm, from DT12 on (12-21). The remanent magnetic flux density for the PMQ SmCo segments is 1.0 T, a conservative value. The inner PMQ radius is 5.5 mm for the first PMQ family, and 6 mm for the second; the outer radius is 11 mm for both. The quadrupole gradients of the 16-segment PMQs are 170 and 142 T/m, but the integrated focusing strengths for both PMQ types are the same, 3.2 T.

TABLE 1. Electromagnetic Parameters of IH-PMQ Tank at the Averaged Tank Gradient of 2.5 MV/m.

Parameter (* = 100% duty factor)	Value
Quality factor Q	9973
Transit-time factors T, for 21 gaps	0.91-0.95
Effective shunt impedance ZT^2, MΩ/m	408
Surface (Cu) RF power loss P_{surf}, kW*	19.2
Maximal electric field E_{max}, MV/m ($/E_K$)	23.3 (1.58)
Max surface loss power density, W/cm^2*	113

The PMQ magnets are arranged in pairs to form an FOFODODO beam focusing lattice, as schematically illustrated in the inset of Fig. 1. To take into account the PMQ field overlaps, the static magnetic field for the whole array of 22 PMQs was calculated by the CST Electro-Magnetic Studio (EMS) [6].

Beam Dynamics In The IH-PMQ Tank

The design of the IH-PMQ tank in Fig. 1 was optimized for high-current beams using iterations of 3-D EM MWS calculations and a specialized linac design code that was developed based on the PARMILA [7] algorithms. It applies the transit-time factors calculated for individual cells from MWS results to fine-tune the beam velocity profile along the tank. After that we employed 3-D multi-particle beam dynamics simulations with PARMELA [7] and CST Particle Studio (PS) [6] to confirm the design. The emittances of the matched input beams to the IH tank were estimated using PARMTEQM [7] for a generic RFQ with the deuteron output energy 1.5 MeV and current 50 mA. The initial normalized transverse rms emittance in the RFQ was 0.13 π mm·mrad at 62 keV and 55 mA. Both PARMELA and PS multi-particle simulations have used the same exact 3-D fields from the CST Studio codes: RF fields from MWS and static magnetic field from EMS. We explored both "water-bag" and Gaussian initial distributions of the bunch particles generated by the PARMILA code, with up to 100K particles used in simulations. The same initial distributions were imported into both PARMELA and PS runs. Results of the beam dynamics simulations for the IH tank indicated no particle loss at the fractional level down to 10^{-5} even at 50 mA [5].

A cold model of the IH-PMQ tank – a simple aluminum model without cooling for low-power field measurements – has been recently manufactured. First measurements of the electric field for 201.25-MHz mode with bead pulls in the cold model as built showed that deviations from the calculated field profile are below 3% except for two gaps near the tank ends. The differences will be further reduced after the gap lengths in the cold model are adjusted to the design values.

CONCLUSIONS

We have demonstrated that normal-conducting IH-PMQ accelerating structures are feasible and efficient for beam velocities in the range of a few percent of the speed of light. Results of combined 3-D modeling for the IH-PMQ accelerator tank – electromagnetic computations, beam-dynamics simulations, and thermal-stress analysis – prove that H-mode structures (IH-, and the higher-mode CH-, etc) with PMQ focusing can work even at high currents. Due to the structure efficiency, the thermal management is simple and can be realized with cooling channels in vanes. The accelerating field profile in the IH-PMQ tank was tuned to provide the best beam propagation using coupled iterations of electromagnetic and beam-dynamics modeling.

One inherent limitation of all H-mode structures is a fixed velocity profile. Our study indicates also that some restrictions of H-PMQ structures at high currents can be caused by limited beam apertures. This conclusion is due to the fact that increasing the DT bore size leads to larger inner radii of PMQs, and increasing the inner radius of PMQ quickly weakens its focusing strength.

IH-PMQ accelerating structures following a short RFQ can be used in the front end of ion linacs or in stand-alone applications. In particular, we explored a compact efficient deuteron-beam accelerator to 4 MeV. Overall, H-PMQ ion linacs look especially promising for industrial and medical applications.

Large-Acceptance Linac For Low-Energy Muons

Beams of accelerated muons are of great interest for fundamental research as well as for applications in homeland defense and in industry. When a proton beam hits a target, pions are produced, and after the short 26-ns life time they decay into muons. Most of the created muons have low energies and are spread in all directions from the target. The 2.2-μs life time of muons is long enough to accelerate them; accelerated muons live much longer due to relativistic effects. Muon accelerators are being studied for the Neutrino Factory and Muon Collider (NF/MC) [8] but their huge size and cost prohibit other applications. In the NF/MC projects only a small fraction of all produced muons, those with relatively high energies and traveling in the forward direction, is used. This is justified by the need to have very low emittances for MC but requires a high-power proton driver – a $1B-class machine – to provide a sufficient muon flux.

We suggest an alternative approach of collecting and accelerating the copious low-energy (tens of MeVs) muons that can lead to smaller and cheaper systems. In homeland defense, muon beams can enable unique element analysis of special nuclear materials (SNM) via muonic X-rays or detection of high-Z materials by muon radiography even under heavily-shielded conditions. For industrial applications, muon radiography of large-scale machinery using cosmic-ray muons is becoming popular, but a high-intensity pencil beam of accelerated muons would be a much more valuable tool for dynamic investigation of machinery. A compact and inexpensive muon accelerator is also desired for medical use. These applications have different beam requirements: for muon radiography, a mono-energetic μ^+ beam with minimal divergence is ideal, but μ^- beams for SNM interrogation in cargo can benefit from a wide energy spread and large spot size. In all cases, however, the common task is to accelerate the low-energy muons to higher energies. Therefore, there is a significant need for a compact and efficient accelerator that can capture a large fraction of a divergent pion-muon beam from a production target and accelerate muons quickly. We propose a novel linear accelerator with large acceptance and high gradient that can provide an efficient capture and fast acceleration of low-energy muons.

Large-Acceptance Linac Design

Large transverse acceptances can be achieved with wide-aperture cavities. Keeping the "cloud" of pions and muons inside the aperture requires a strong continuous solenoidal magnetic field. The longitudinal acceptance can be increased by using high accelerating gradients. A potential solution that satisfies all these requirements is a normal-conducting (NC) linac with external solenoids. NC linac structures with large beam apertures typically operate in the π-mode, with adjacent cells coupled via open apertures, mainly at beam velocities $\beta = v/c \approx 1$. Our linac should start at much lower β-values: around $\beta = 0.5$ for μ^- produced by a proton beam (the peak of muon kinetic energy is ~16 MeV [9]), or even at $\beta = 0.25$ for μ^+ from a surface muon source (~4 MeV). Instead of π-mode, we suggest using the 0-mode – the TM_{010} mode of a pill-box resonator – in a chain of independently-fed cavities with wide apertures electrically closed by thin metal windows or grids. The two modes are compared in Fig. 2 for 805-MHz cavities with the design value $\beta_g = 0.8$. The cavity length $L_c = \beta_g\lambda/2 = 14.9$ cm; the aperture radius is $a = 6$ cm, about 40% of the cavity inner radii $R \approx 15$ cm. The highest electric field is near the axis in the 0-mode, while the maximum in the π-mode is at the septa, near the aperture edges. As the cavity lengths become shorter for smaller values of β, the π-mode cavity becomes more and more inefficient due to very large fields at the septa. The ratio E_{max}/E_K of the maximal electric field to the Kilpatrick field (for 805 MHz $E_K = 26.08$ MV/m) in 805-MHz cavities with $a = 6$ cm at the fixed gradient $E_0 = 35$ MV/m for β_g from 0.5 to 0.94 remains below 1.8 for the 0-mode but reaches above 6 for the π-mode at low β_g [10]. The maximum surface fields below $1.8E_K$ are usually considered safe with respect to RF breakdowns, but this limit can be reduced by a factor of 1.5-2 in solenoidal magnetic fields of a few T, e.g. [11]. The difference is even more drastic for the cavity surface-loss power: at $\beta_g = 0.5$ the π-mode cavity requires more than 8 times higher RF power compared to the 0-mode one with the same gradient [10]. The comparison becomes even more in favor of the 0-mode at larger values of a/R, i.e. wider apertures [12].

A linac based on isolated cavities with apertures closed by thin metal windows can work efficiently only for muons, due to their high penetrating ability. A fraction of the cavity RF power is deposited on the metal aperture windows, cf. Tab. 2, so their cooling must be addressed.

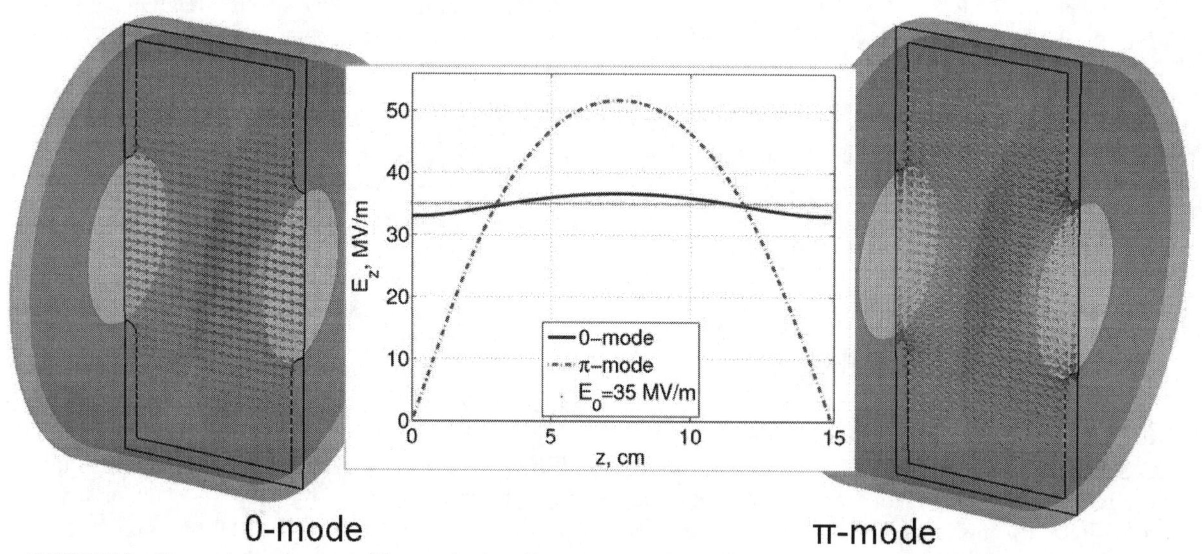

FIGURE 2. Comparison of two working modes in wide-aperture cavities: the electric field pattern and on-axis field profile.

The values in Tab. 2 are for the Cu cavities with a = 6 cm, the septum thickness 2 cm, at the fixed gradient E_0 = 35 MV/m. From the maximal values of the wall power-loss density in Tab. 2, these cavities can be operated at duty factors of a few percent. Still the power deposited on the aperture windows will be in the kW range. For a given 0-mode RF cavity, $P_a \propto a^4$ when $a \ll R$, and the window cooling for large apertures can become an engineering challenge. One possible approach is an edge-cooled thin-wall window, with all cooling channels located in the septum. Another option is a double-wall window with supporting ribs between two thin metal sheets. The channels between the ribs serve for cooling, e.g. by a cold gas.

TABLE 2. Parameters of 805-MHz 0-mode cavities at the gradient of 35 MV/m for various β_g

Parameter (* = at 100% duty factor)	$\beta_g = 0.5$	$\beta_g = 0.7$	$\beta_g = 0.94$
Length L, cm	9.31	11.17	17.50
Radius R, cm	14.86	14.67	14.55
Shunt impedance $Z_{sh}T^2$, MΩ/m	18.4	25.6	32.3
Surface-loss power P, MW*	2.56	2.60	2.80
Maximal surface-loss power flux, W/cm^2*	1595	1409	1292
Aperture surface loss P_a, %% / kW*	2.5 / 63	2.3 / 60	2.0 / 56

The linac frequency choice is important. Lower frequencies lead to larger cavities; the aperture size can also be larger, and the number of cavities is smaller. On the other hand, achieving high gradients is easier at higher frequencies, and high-frequency klystrons provide higher peak power. The cavity comparison in [12] between 402.5, 805, and 1300 MHz indicates that E_0 = 35 MV/m is easier in the latter case. There is some experimental evidence that in a pill-box cavity filled with high-pressure (HP, 15-30 atm) hydrogen gas the limiting value of the maximal electric surface field, even in high external magnetic fields, is restored to that without the external field independent of frequency [13]. A research program for HP cavities is being pursued in NF/MC R&D activities [13]. If this development succeeds, HP cavities would be uniquely suited for muon acceleration.

A NC muon linac consists of closely-packed 0-mode RF cavities, to maximize the real-estate accelerating gradient, with external solenoids surrounding the whole structure [10, 12]. We chose 805-MHz 0-mode RF cavities as a baseline for the μ^- linac starting at β_g = 0.5. To achieve the muon energy gain of 200 MeV in such a linac designed with continuously increasing cavity lengths (β_g = 0.5 to 0.94), gradient E_0 = 35 MV/m and synchronous phase φ = -30° in all cavities, the required number of cavities is 76 [12]. The final energy of the synchronous particle is 216 MeV; the total linac length is 10.19 m. Having a phase ramp (synchronous phase adjusted along the linac) is better for beam dynamics: starting with a larger phase, φ = -60°, increases the phase acceptance and provides beam bunching; gradually reducing it along the linac to φ = -20° restores the acceleration. Such a ramp leads to a higher muon flux from the linac as confirmed by simulations [9, 14]. It is customary in linacs to employ only a few types of cavities instead of making them all different: e.g. a few β_g = 0.5 cavities, then β_g = 0.6, etc. The resulting linac then consists of a smaller total number of RF cavities, N_{cav} = 67. The final energy is 217 MeV, and the total length is 10.00 m. The number of cavities of each type is 4, 6, 10, 22, and 25, for β_g = 0.5, 0.6, 0.7, 0.8, and 0.94 [10].

The 0-mode RF cavities in the linac form a chain of independent (uncoupled) cavities so the RF power should be fed into each of them separately: each cavity has an independent RF input. This provides flexibility in operating the linac (it can accommodate any phase ramp) but complicates the overall design since the RF feeds must fit between or inside the solenoid coils. To arrange the proper accelerating phases (~180°-degree phase shift per cavity) in a 0-mode chain, the structure can be fed inter-digitally from two RF inputs shifted by 180°, creating an accelerating field pattern in the linac similar to that in the π-mode. The number of RF feeds can be reduced by a factor of two by employing pairs of coupled 0-mode cavities, with one RF input per pair. The two cavities should then be coupled magnetically by slots in the common septum, while the apertures remain closed by windows. An estimate of the required total peak RF power for the above muon linac with 67 cavities of 5 types is 180 MW. Adding usual 15% for a realistic surface conductivity and RF losses in waveguides (~5%) increase that by 20%. The power estimate for a similar linac at 402.5 MHz with the same gradient is about 35% higher, and that at 1.3 GHz is 30% lower [12]. The minimal RF pulse length is defined by the cavity filling time of ~30 μs at 805 MHz. It may be advantageous to use klystrons with pulses of a few 100 μs that provide higher peak power.

Nevertheless, finding ways to reduce the required peak RF power is important.

The muon flux produced by the linac is estimated by simulations with a specialized Monte-Carlo tracking code [9, 14]. The code tracks particles in combined RF and magnetic fields taking into account pion and muon decays and particle interactions with window material. The 805-MHz linac described above with the magnetic field B = 5 T accelerates to 200 MeV about 10% of the muons at its entrance. Assuming a 50-μA 800-MeV proton beam on a 30-cm carbon target, the linac can deliver $3\cdot10^9$ μ^-/s at 200 MeV. Simulations show the muon-flux reduction by a factor of about 2 when either E_0 is reduced to 25 MV/m or B to 3 T. This reduction can be mitigated to some extent by optimizing the linac design for particular E_0 and B [14].

SUMMARY

We proposed an approach for collecting and accelerating the low-energy muons produced in pion decays when a proton beam hits a target. This method is complementary to the NF/MC projects and can lead to smaller and cheaper systems for various applications. Properties of 0-mode RF cavities and considerations for designing a high-gradient large-acceptance muon linac are presented. Our baseline design is a normal-conducting linac consisting of independently fed 0-mode RF cavities with wide apertures closed by thin metal windows. The guiding static axial magnetic field is provided by external super-conducting solenoids.

Accelerating low-energy muons presents unique challenges but the proposed approach – large-acceptance NC linac with external solenoidal field – looks feasible. The main engineering challenge is to combine the RF system, cooling, and external solenoids.

ACKNOWLEDGMENTS

The author would like to acknowledge contributions of his colleagues at LANL. The results for the IH-PMQ structures were obtained in collaboration with L. Rybarcyk, J. O'Hara, and E. Olivas, and this work is supported by the LANL Laboratory-Directed R&D (LDRD) program. The muon accelerator results were obtained in collaboration with H. Miyadera and A. Jason, and partially sponsored by the Defense Threat Reduction Agency.

REFERENCES

1. U. Ratzinger, *Nucl. Instr. Meth.* **A464**, 636 (2001); also Proceedings CAS 2000, CERN 2005-003, 2005, p. 351.
2. S. Kurennoy, L. Rybarcyk, and T. Wangler, "Efficient Accelerating Structures for Low-Energy Light Ions," Proceedings of Particle Accelerator Conference (PAC07), Albuquerque, NM, 2007, p. 3824.
3. S. Kurennoy et al, Proceed. EPAC08, Genoa, 2008, p. 3428; Proceed. Linac08, Victoria, BC, 2008, p. 954.
4. S. Kurennoy, J. O'Hara, E. Olivas, and L. Rybarcyk, "Development of IH Accelerating Structures with PMQ Focusing for Low-Beta Ion Beams," PAC09, Vancouver, BC, FR5REP070 (2009).
5. S. Kurennoy, J. O'Hara, E. Olivas, and L. Rybarcyk, "H-mode Accelerating Structures with PMQ Focusing for Low-Beta Ion Beams," Proceed. IPAC10, Kyoto, Japan, 2010, pp. 828.
6. CST Studio Suite, v. 2010, www.cst.com.
7. Los Alamos Accelerator Code Group, laacg.lanl.gov.
8. J.S. Berg et al, *Phys. Rev. ST-AB* **9**, 011001 (2006).
9. H. Miyadera, A. Jason, S. Kurennoy, "Simulations of a Large Acceptance Linac for Muons," PAC09, Vancouver, BC, FR5REP071 (2009).
10. S. Kurennoy, A. Jason and H. Miyadera, "Large-Acceptance Linac for Accelerating Low-Energy Muons," Proceed. IPAC10, Kyoto, Japan, 2010, pp. 3518.
11. R. Palmer et al, *Phys. Rev. ST-AB* **12**, 031002 (2009).
12. S. Kurennoy, "Baseline Design of High-Gradient Muon Linac for SNM Interrogation," Report LA-UR-10-00246, Los Alamos, 2010.
13. R. Sah et al, "Breakdown in Pressurized RF Cavities," Proceed. LINAC08, Victoria, BC, 2008, p. 945.
14. H. Miyadera, S. Kurennoy, and A. Jason, "Simulation of Large Acceptance Muon Linac," Proceed. IPAC10, Kyoto, Japan, 2010, pp. 3521.

Performance Evaluation Of An Irradiation Facility Using An Electron Accelerator

R. M. Uribe[(1)], E. Filppi[(2)], and K. Hullihen[(1)]

(1) Kent State University, Kent, Ohio, USA
(2) Case Western Reserve University, Cleveland OH, USA

Abstract. Irradiation parameters over a period of seven years have been evaluated for a radiation processing electron accelerator facility. The parameters monitored during this time were the electron beam energy, linearity of beam current, linearity of dose with the reciprocal value of the samples speed, and dose uniformity along the scanning area after a maintenance audit performed by the electron accelerator manufacturer. The electron energy was determined from the depth-dose curve by using a two piece aluminum wedge and measuring the practical range from the obtained curves. The linearity of dose with beam current, and reciprocal value of the speed and dose uniformity along the scanning area of the electron beam were determined by measuring the dose under different beam current and cart conveyor speed conditions using film dosimetry. The results of the experiments have shown that the energy in the range from 1 to 5 MeV has not changed by more than 15% from the High Voltage setting of the machine over the evaluation period, and dose linearity with beam current and cart conveyor speed has not changed. The dose uniformity along the scanning direction of the beam showed a dose uniformity of 90% or better for energies between 2 and 5 MeV, however for 1 MeV electrons this value was reduced to 80%. This parameter can be improved by changing the beam optics settings in the control console of the accelerator though.

Keywords: electron irradiation, electron accelerator performance, energy measurements, dosimetry
PACS: 41.75.Fr, 61.80.Fr, 87.53.Bn

INTRODUCTION

Electron accelerators are used for the treatment of various products or materials with the main purpose to modify their physical and chemical properties. Usually the process to achieve a determined change in the product will depend on the absorbed dose and this physical quantity on the operation parameters of the electron accelerator, namely the electron energy, the beam current, the extent of the radiation zone (scanning area), and the irradiation time[(1)]. Dose requirements will vary depending on the application, but the final result will depend on the control of the physical parameters of the accelerator mentioned above. There are some processes that require a documented measurement of the dose, like in the sterilization of medical devices or for food irradiation, and in those, a measurement of the dose during actual runs is performed by means of a dosimetric technique[(2)]. These processes are usually regulated by federal agencies like the FDA. There are other processes like the irradiation of polymers that do not require a measurement of the dose during actual runs, but for whom a quality control of the process is accomplished through the verification of the accelerator parameters during irradiation, once these parameters have been established through initial dosimetry methods. In this case it is important to have a good control on the physical parameters of the electron accelerator and to make sure that they do not drift away from their factory settings after routine maintenance and/or after accelerator shut down for maintenance or repairs. These quality assurance procedures will ensure that the process is being performed under the same irradiation conditions all the time.

The electron accelerator at the NEO Beam Facility in North East Ohio, has been in operation since May of 2000. The electron accelerator is a Dynamitron DPC-2000 with a maximum energy of 5.0 MeV and maximum beam power of 150 kW. The facility has been described in further detail elsewhere[(3)]. As part of the quality assurance procedures used to verify that the electron accelerator operates in a reliable way, a set of physical measurements has been conducted following scheduled shutdowns for maintenance for a period of seven years. These measurements include the determination of the electron beam energy, the linearity of the dose with electron beam current and with the inverse of cart conveyor speed, the uniformity of the

Application of Accelerators in Research and Industry
AIP Conf. Proc. 1336, 39-45 (2011); doi: 10.1063/1.3586053

radiation zone along the scanner of the accelerator at different high voltage settings, and the linearity of scan fraction setting with scan length.

The purpose of this work is to summarize the results of these measurements over the time period described above and present some recommendations based on these results.

EXPERIMENTAL

Some of the activities carried out follow the standard practices described by the ASTM regarding the characterization of electron beam facilities used in routine radiation processing to ensure that products have been treated with an acceptable range of absorbed doses(4) and include guidelines to measure the electron beam energy and dose uniformity along the scanned width of the electron beam. The rest of the measurements described herein have been developed in-house as more expertise was developed in the use of the electron accelerator.

Electron Energy Determinations

The energy of the electron beam at accelerating potentials of 1, 2, 3, 4, and 5 MV was determined using the depth-dose profile in an aluminum wedge method, according to the procedure outlined by the ASTM(4). The dosimeters used in these measurements were cellulose triacetate (CTA) films(5) supplied by Fuji Corp,[1] catalog number FTR-125. For each of the accelerating potential settings of the accelerator mentioned above, a CTA film was placed between the two sides of an aluminum wedge manufactured by Risø National Laboratory in Denmark and irradiated on the cart conveyor system using a beam current of 10 mA, a cart speed of 2.5 cm/s, and a scanning fraction of 1.0. After irradiation the UV absorbance along the film, produced by the irradiation, was measured using a Genesys 5 spectrophotometer fitted with a driving mechanism to move the film under the analyzing light beam of the spectrophotometer. These measurements produced a depth-dose profile along the film from which the practical range and the electron energy were obtained using the equation(4):

$$E(MeV) = 0.256 + 4.91R_p - 0.0248R_p^{\ 2} \qquad (1)$$

In this equation R_p is the practical range of the electrons in aluminum and E the initial energy of the electron beam.

Linearity Of The Beam Current

The linearity of the dose with beam current was measured by irradiating a set of dosimeters in the cart conveyor system at a series of beam currents, keeping the rest of the accelerator parameters constant. The dosimeters chosen for this experiment were Risø B3 radiochromic dye films(6) produced by Risø National Laboratory in Denmark. For each beam current setting, four dosimeters were placed inside paper envelopes and the envelopes taped to the upper surface of a polystyrene foam block 5 cm thick. This block was placed on a cart tray with the dosimeters facing the electron beam. The irradiation conditions were as follows: electron energy 3 MeV, cart speed 15.24 cm/s, scanning fraction 1.0, and the following beam currents: 10, 20, 30, 40, and 50 mA. At the end of the irradiation the dosimeters were heated in a laboratory oven to 60°C for 10-15 min and their absorbance measured at 604 nm in a Genesys 20 spectrophotometer. The dose was determined using the software package WINDose from GEX Corporation (Denver CO) with a dose calibration traceable to the National Institute of Standards and Technology (NIST).

Dose Linearity With The Reciprocal Value Of The Sample Speed

The reciprocal value of the speed is directly proportional to the time the samples spend under the electron beam. Since the dose is proportional to this time, then the dose should be proportional to the reciprocal value of the speed too. Any deviations from linearity will point to a problem in the speed setting in the control console of the accelerator. To check this, an experiment very similar to the one described in the previous section was performed for five different speed settings of the cart conveyor system. For each speed setting four radiochromic dye film dosimeters were placed inside paper envelopes and taped on top of a 5 cm thick polystyrene foam block. This block was then placed on the tray of a conveyor system cart and irradiated using the following accelerator conditions: electron beam energy 3 MeV, beam current 20 mA, scan fraction 1.0 and the following cart conveyor speeds: 5.08, 10.16, 15.24, 20.32, and 25.4 cm/s. The doses for each one of the conveyor speeds were determined after the irradiation using the same procedure as the one described in the section above.

[1] The mention of any commercial product or company is to clarify the procedure description only, and does not imply any endorsement on the part of the authors, or by Kent State University.

Dose Uniformity Along The Beam Width

The beam width is defined as the dimension of the irradiation zone produced by the electron beam perpendicular to the beam length and direction of the electron beam specified at a specific distance from where the beam exits the accelerator (see FIGURE 1). This dimension is noted at some defined fractional level "*f*" of the average maximum dose[4]. In this case "*f*" was taken at 90% of the maximum value of the dose. To determine the beam width and its uniformity, a 122 cm long strip of CTA film was taped to the upper edge of a cart conveyor system tray, along its width, for each electron beam energy selected for this experiment. Then, the strip was irradiated using the following conditions: electron beam current 10 mA, cart speed 2.54 cm/s, scan fraction 1.0, and electron beam energy of 1, 2, 3, 4, and 5 MeV. After the irradiations the absorbance along the CTA strip was measured at a wavelength of 280 nm using the same instrument and software as the one used for the determination of electron beam energy.

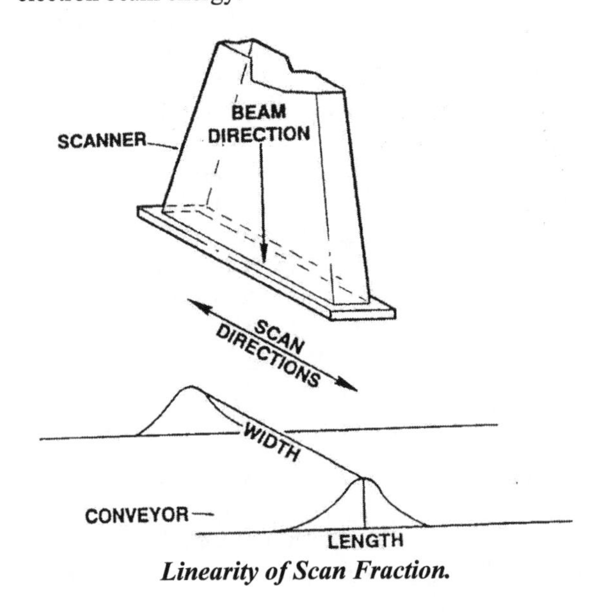

FIGURE 1. Diagram showing the scanner of the electron accelerator and the radiation zone produced by the electron beam used to define the beam length and the beam width. The electron beam is being scanned along the beam width in this diagram. Taken from reference 4.

Linearity of Scan Fraction.

The percent of scan width in terms of the scan fraction setting of the accelerator was verified in a similar way as the one for the experiment described in the previous section, the only differences were that the electron energy was fixed and the scan fraction setting changed for each different run. The experiment was run at the following scan fraction values: 0.25, 0.5, and 0.75

RESULTS AND DISCUSSION

Energy Determination

A typical depth dose curve consists of two sections[4]. In the first section called the build-up region, the dose increases with depth, due to the production of secondary electrons when the incident beam interacts with the irradiated material. This continues until a maximum value in the dose is observed. This corresponds to the maximum range of secondary electrons in the sample and marks the end of the first section. Then follows the second section, called the absorption section in which the dose decreases monotonically with depth. The practical range of the electrons in aluminum is determined from this section of the graph. To do so the middle section of the absorption part of the graph is approximated by a straight line and the point at which this line intersects the depth axis determines the practical range. Once this is done, the energy of the incident electrons can be obtained from equation (1). TABLE 1 shows how these measurements have changed over the time of this study. These results show that the energy of the electrons is usually within a few percent of the HV setting of the accelerator and that the maximum deviation between the energy setting of the accelerator and the electron beam energy determined from depth dose measurements has not been greater than 15% except for one time during the whole period of this study (the corresponding measurement at 1 MV in 2008). The estimated uncertainty in the technique is roughly 6.5%

TABLE 1. Changed of the electron energy as compared with the voltage setting of the electron accelerator over time.

HV Setting		Energy (MeV)			
(MV)	2004	2005	2007	2008	2010
1	1.04	1	1	1.15	1.06
2	2.02	1.96	2.01	2.037	2.02
3	3.09	3.022	2.98	2.993	3.02
4		4.1	3.96	3.994	4.04
5	4.96	5.06	5.16	5.018	

Linearity Of The Beam Current

FIGURE 2 shows the dose obtained from the analysis of a set of radiochromic films irradiated at five values of the electron accelerator beam current as described in the experimental section of this work. This current setting is determined from the difference between the current from the rectifier assembly circuit charging the high voltage terminal and the sum of the currents from the high voltage resistor and the beam tube resistor divider circuits of the accelerator. The graphs in the figure show that for the time period of this study all are very similar with the dose increasing linearly with the beam current setting of the accelerator; in all cases the experimental data can be fitted to a straight line with a correlation coefficient better than 0.99. It should be mentioned that at the present time there is no way to get an absolute value of the value of the current coming out of the accelerator through the titanium window. Doing that would require a system that would collect the electron beam in its entirety.

Linearity Of Dose With The Reciprocal Value Of The Cart Conveyor System Speed

The cart conveyor system of the NEO Beam facility consists of three sets of chains pulling the carts from the loading area into the beam room, then under the irradiation zone, and finally back to the unloading area. The chain moving the carts under the irradiation zone is the most important of the three, with a dual purpose: to keep the carts close to each other, with minimum space between them to use the beam more efficiently; and to move the carts under the beam with a well controlled speed[7]. The results obtained show that the dose is linear with this parameter with a slope of 300.6 kGy-s/cm and a correlation coefficient of 0.9963. Lack of linearity of the dose with this parameter would mean the need for a calibration of the speed control setting of the conveyor cart system in the control console of the accelerator.

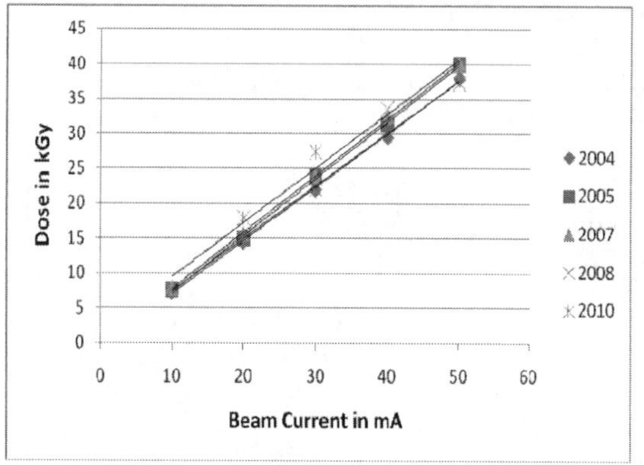

FIGURE 2. Dose as a function of electron beam current measured with radiochromic dye film (Risø B3) using the procedure described in the experimental part of this work. The data was collected for the different years listed in the legend of the figure.

Dose Uniformity Along The Beam Width

An example of the results of the dose uniformity measurements across the cart tray are presented in FIGURE 3 for 1, 3, and 5 MeV electron beam energies for measurements carried out in 2004. In this experiment the dose uniformity was determined along the full width of the cart conveyor system tray. The graphs show that the dose is fairly uniform across most of the tray, with a maximum variation in dose as represented by the standard deviations from the average value of the dose of 4.4% for 5 MeV, 2.7% for 3 MeV, and 4.5% for 1 MeV. A closer look at the graphs in FIGURE 3 shows an increase in the dose value towards the edge of the tray for 3 and 5 MeV electron energies. This effect is probably due to electron backscattering from the tray edge and not an effect due to the electron accelerator beam optics system. On the other hand the 1 MeV electron irradiation showed a decrease in the dose towards the edges of the tray. This effect has been observed in previous measurements[8] and is attributed to the large dispersion of the beam when it interacts with the air molecules between the titanium window and the position of the dosimeter at that energy. TABLE 2 shows the change of dose uniformity along the width of the cart conveyor tray over the time period for this study. The dose uniformity was characterized in terms of the slope and flatness of the dose measurements along the CTA film for dose values exceeding 90% of the maximum dose. The slope gives an indication of whether more electron beam energy is absorbed at one side of the conveyor cart tray as compared to the other and the flatness gives an indication of how much are the values of dose deviating from the average value within the extreme points at 90% of the dose. The slope was calculated by applying a linear regression analysis to the dose points included in this analysis. The flatness was calculated as the percent error of the dose measurements, that is, the percent standard deviation over the mean for the dose points included in this analysis. These results show that the dose uniformity along the width of the scanner of the electron accelerator has a slope close to zero, with a flatness value not exceeding 10%, so the dose delivered is essentially the same along the width of the cart conveyor system tray. However larger values in lack of flatness have been obtained for measurements performed in 2010, which will require the readjustment of the beam optics system of the accelerator.

Linearity Of Scan Fraction With Scan Fraction Setting

FIGURE 4 shows how the dose uniformity changes when the scan fraction setting of the electron accelerator is changed. The extent of the dose uniformity was determined in a similar way as for the case of dose uniformity along the scan width described above. From this figure it can be seen that the extent of the region where the dose is uniform is reduced when the scan fraction setting is reduced. TABLE 3 shows a comparison of the scan fraction setting with the measured scan length over the time period of this study. Even though the scan setting refers to the percentage of scan width with respect to the length of the scanner of the electron accelerator, the scan extent was measured over the tray of the cart conveyor system. As such it was expected that the percent measured scan width would be larger than the setting on the control console of the accelerator. That has been the case for the measurements taken up to 2008, however the last measurement taken on 2010 shows a larger reduction in dose uniformity when the scan fraction setting of the accelerator is reduced, again calling for some readjustment of the beam optics system of the electron accelerator.

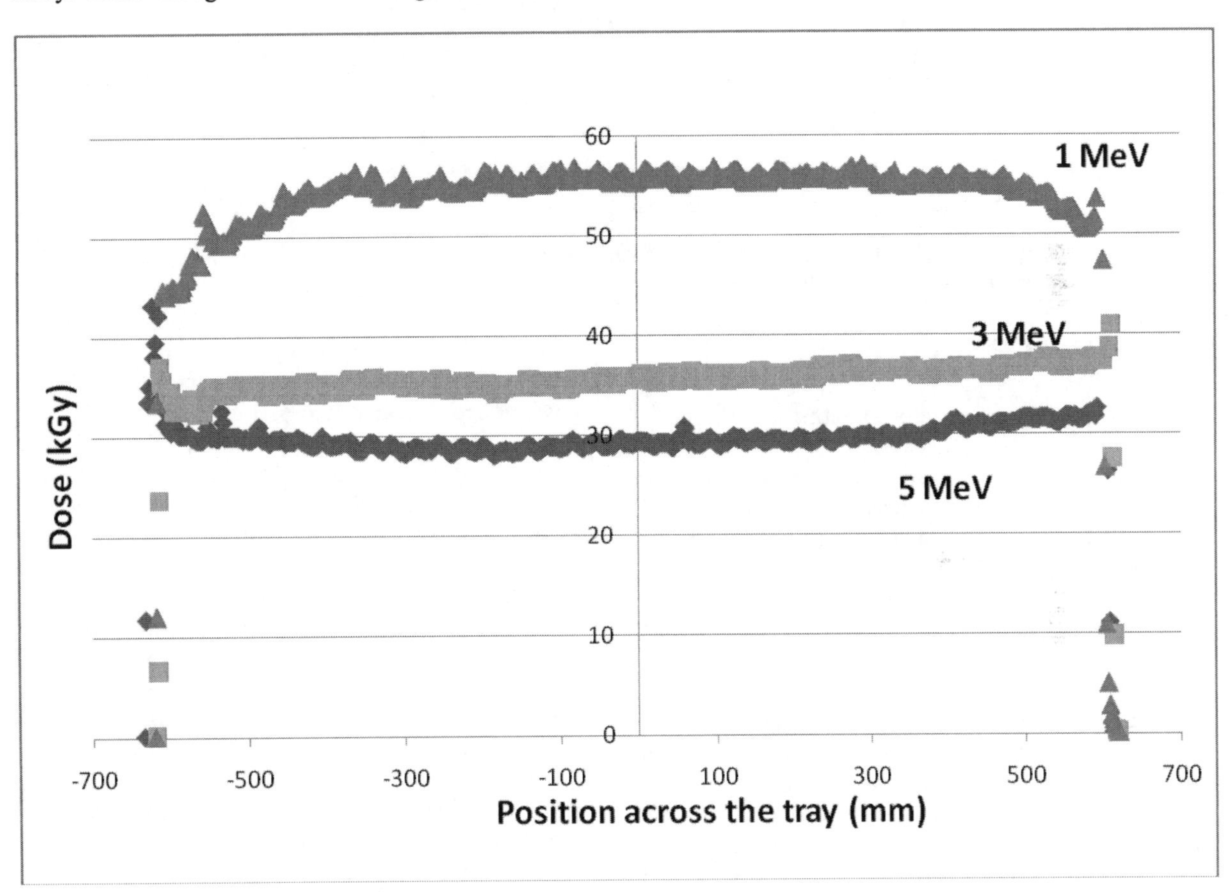

FIGURE 3. Dose uniformity (D), measured in kGy, across the cart conveyor tray for three different electron beam energies. The position across the tray was measured from the center of the tray, which corresponds to the center of the scan horn of the electron accelerator.

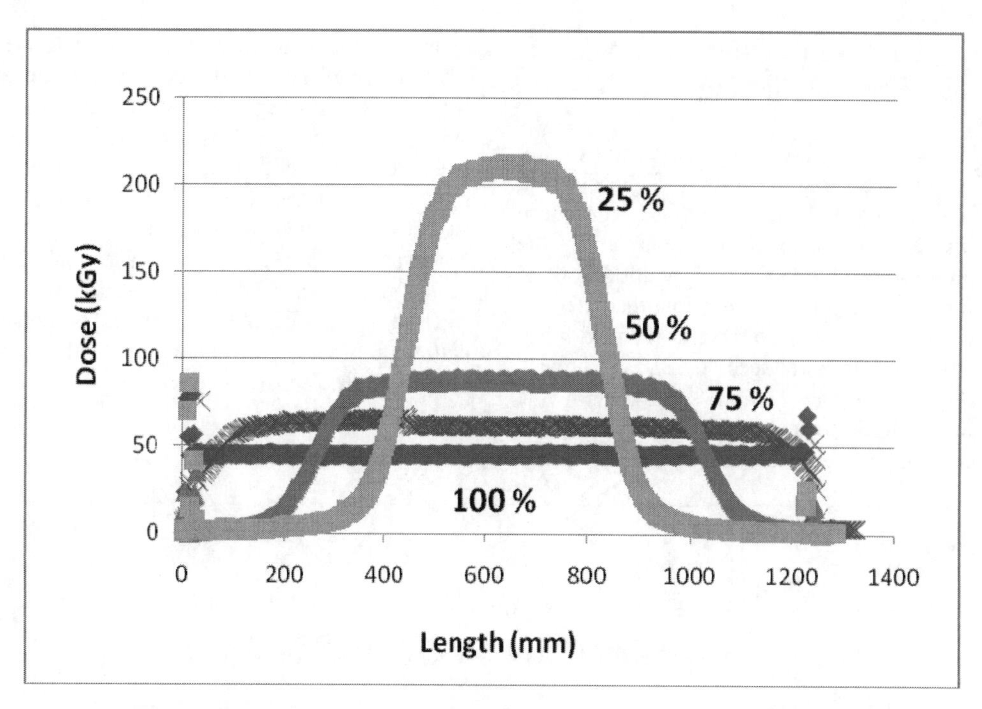

FIGURE 4. Change in dose uniformity with the scan fraction setting of the electron accelerator. The scan length was determined as the distance between points having 90% of the maximum dose at the selected scan fraction setting.

CONCLUSIONS

Benchmarks for accelerator performance after a shutdown have been established for an electron accelerator used for radiation processing. The measurements performed included the determination of energy at several high voltage settings, the verification of the linearity of the dose with the beam current setting of the accelerator, the verification of the linearity of the dose with the reciprocal value of the speed setting of the cart conveyor system, the uniformity of the dose along the scanning direction of the beam, and the verification of the scan dimension with the scan fraction setting on the accelerator control console. Validation of a particular process involving electron beam irradiation will depend on the dose delivered to the product and this on the accelerator parameters mentioned above[9]. All of them will affect the dose delivered to the product as well as the dose variations within it and as such it is important to keep them under controlled conditions. Some of these parameters like the scan dose uniformity and scan fraction can be modified without major problems by changing the beam optics conditions in the scanner circuit of the accelerator, but there are some other parameters, like the energy, beam current and cart conveyor speed, that would require a shutdown of the accelerator in order to bring them under the required specifications if there is a need for that. The results included above show that the energy and the linearity of dose with electron beam current have been stable over the years, however there has been some degradation in the beam optics of the electron accelerator causing some reduction in the dose uniformity along the scan width for some electron beam energies and when the scan fraction setting of the accelerator is reduced. Since the operation of the accelerator has shifted in recent years from a 50/50 time commitment on production and research to a 90/10 situation and usually production work is performed at an energy of 3.75 MeV and 100% scanning, those are the parameters for which dose uniformity is more critical to maintain. At the present time the dose is linear with the reciprocal value of the speed of the conveyor system. Measurements of this parameter in the future will give an indication of how often the system needs to be brought down for preventive maintenance.

TABLE 2. Dose uniformity along the beam width for energies between 1 and 5 MeV over a period of seven years. The slope (m) was measured in kGy/mm and the percent flatness (F) as the percent deviation from the average value for positions along the dosimeter film exceeding a value of 90% of the maximum dose.

Year>	**2004**		**2005**		**2007**		**2008**		**2010**	
Energy (MeV)	**m**	**F**	**m**	**F**	**m**	**F**	**m**	**F**	**m**	**F**
1	0	2.9	-0.0032	6.7	0.0170	7.9	0.0003	5.6	0.0100	9.1
2			0.0014	1.4	0.0007	3.1	0	1.4	0	3.9
3	0.0024	2.7	-0.0017	5.6	0.0019	1.8	0.0083	1.4	0.0129	7.6
4			-0.0009	5.5	0.0002	1.2	0.0019	1.0	0	8.0
5	0.0019	4.4	-0.0012	4.7	0.0012	1.9	0.0002	2.0		

TABLE 3. Scanned width measured with CTA film compared to the scan fraction setting of the electron accelerator over a period of seven years. Maximum uncertainty in these measurements was 10%.

	Year>>>	**2005**	**2007**	**2008**	**2010**
Scan	**100**	106	100	100	60
Length	**75**	92	90	86	48
(%)	**50**	57	57	55	28
	25	26	25	22	15

ACKNOWLEDGEMENTS

The authors would like to acknowledge the following individuals for their collaboration to carry out the irradiations: Mr. John Juras, Thomas Goodner, and Ben Wheeler.

REFERENCES

1. Korenev S., "Critical Analysis of Industrial Electron Accelerators", Rad. Phys. Chem. **71**, 535-537 (2004)
2. McLaughlin W. L., Boyd A. W., Chadwick K. H., McDonald J. C., and Miller A., "Radiation Processing Dosimetry", Taylor and Francis, London (1989), 251 pp.
3. Vargas-Aburto C. and Uribe R. M., "Electron Irradiation Facility for the Study of Radiation Damage in Large Solar Cell Arrays in the Energy Range $0.5 < E \leq 5$ MeV", Sol. En. Matls. Sol. Cells. **87**, 1-4, 629-636 (2005).
4. Practice for Dosimetry in an Electron-Beam Facility for Radiation Processing at Energies Between 300 keV and 25 MeV, ASTM standard ISO/ASTM 51649, ASTM International, 100 Barr Harbor Drive, PO Box C700, West Conshohocken, PA 19428-2959
5. Practice for Use of Cellulose Acetate Dosimetry Systems, ASTM standard ISO/ASTM 51650, ASTM International, 100 Barr Harbor Drive, PO Box C700, West Conshohocken, PA 19428-2959
6. Practice for the Use of a Radiochromic Film Dosimetry System, ASTM standard ISO/ASTM 51275, ASTM International, 100 Barr Harbor Drive, PO Box C700, West Conshohocken, PA 19428-2959
7. Hackett J. L., "A State of the Art Electron Beam Sterilization Facility", Rad. Phys. Chem., **52**, 1-6, 491-494 (1998)
8. Uribe R. M. and Vargas-Aburto C., "Dosimetry at the NEO Beam Facility II. Dose distribution along the electron beam scanning direction", Internal Progress Report pebt-01-4, Program on Electron Beam Technology, Kent State University, Oct-2001
9. Burns P., Drewell N. H., and McKeown J., "The measurement, control and validation of critical parameters in an electron beam sterilization facility", Nucl. Instrum. Meth. B **113**, 96-98 (1996).

Semiconductor Ion Implanters

Barry A. MacKinnon[a] and John P. Ruffell[b]

[a]Isys, 2727 Walsh Ave., Suite 103, Santa Clara, CA 95051, United States
[b]Group 3, LLC, Sunnyvale, CA 94086, United States

Abstract. In 1953 the Raytheon CK722 transistor was priced at $7.60. Based upon this, an Intel Xeon Quad Core processor containing 820,000,000 transistors should list at $6.2 billion! Particle accelerator technology plays an important part in the remarkable story of why that Intel product can be purchased today for a few hundred dollars. Most people of the mid twentieth century would be astonished at the ubiquity of semiconductors in the products we now buy and use every day. Though relatively expensive in the nineteen fifties they now exist in a wide range of items from high-end multicore microprocessors like the Intel product to disposable items containing 'only' hundreds or thousands like RFID chips and talking greeting cards. This historical development has been fueled by continuous advancement of the several individual technologies involved in the production of semiconductor devices including Ion Implantation and the charged particle beamlines at the heart of implant machines. In the course of its 40 year development, the worldwide implanter industry has reached annual sales levels around $2B, installed thousands of dedicated machines and directly employs thousands of workers. It represents in all these measures, as much and possibly more than any other industrial application of particle accelerator technology. This presentation discusses the history of implanter development. It touches on some of the people involved and on some of the developmental changes and challenges imposed as the requirements of the semiconductor industry evolved.

Keywords: semiconductor implantation plasma ion source wafer
PACS: 34.50.Bw 34.90.+q 81.05.-t 81.15.Jj 41.75.Ak 41.85.-p 41.85.Ja 41.85.Lc

HISTORY

Goldstein and Thompson identified the existence of rays of positive ions ("Kanalstrahlen")[1] and began to understand their nature in the period 1886 to 1921. Rutherford around 1906[2] actually observed the scattering of alpha particles ($^4He^+$) by gold foil providing strong evidence for the nuclear nature of the atom. Rutherford described his astonishment at the result, commenting that it was as if a careening cannon shell had been bounced back by a sheet of tissue paper! - It makes for reflection on the unimaginable concentration of mass in atomic nuclei. These events marked a start to the development of ion beam technology for use in the modification of materials.

It was realized that an ionized dopant atom, accelerated to an appropriate velocity could penetrate a target body to a somewhat well defined depth depending on its speed (energy). The idea that it suggested an alternative means of creating p-type and n-type semiconductors occurred to Moyer and separately to Shockley some time prior to 1955.

The technology became known as "Ion Implantation". In the semiconductor industry the ions are most commonly boron, arsenic, or phosphorus and the device substrate (usually Silicon) has its electrical properties modified by the presence of the dopant atoms within its crystal structure. A modern semiconductor device may be processed by an Ion Implanter more than 25 separate times. Today, manufacture of these small accelerators and associated wafer equipment is a $2 billion industry.

This presentation is a brief summary of a history which has been described in detail by Freeman[3], Wegmann[4], Rose[5], Moffatt[1], Armour[6], Pearton[7] among several others.

Doping germanium or silicon wafers to create p-type or n-type behavior can be done by introducing the dopant material in a gaseous form to the wafer in a "Diffusion Furnace", keeping the wafer temperature high enough to cause the dopant atoms to migrate into the wafer - a process called "Thermal Diffusion". In the early 1960s there were no alternatives to Thermal Diffusion but there are difficulties in controlling diffusion depths and uniformity.

Application of Accelerators in Research and Industry
AIP Conf. Proc. 1336, 46-51 (2011); doi: 10.1063/1.3586054

A Real Alternative - Ion Implantation Oversimplified:

- Make an ion beam.
- Accelerate it to the desired energy.
- Place a semiconductor wafer in the path of the beam.
- Scan the beam over the wafer area (since the beam is usually much smaller than the wafer).

Patents filed by Moyer[8] and Shockley[9] in 1954 first proposed bombarding a semiconductor with dopant ions to modify its bulk (as opposed to surface) electrical properties. The Shockley paper in particular identified many of the characteristics and associated processes that would be required to make Ion Implantation a commercially viable process in semiconductor production. Shockley identified mass separation, beam scanning and post implant annealing of the silicon crystal, all processes included in modern semiconductor implant recipes.

By the mid and late 1960's a flood of papers appeared in the scientific literature exploring the application of ion beam technology to semiconductor and other manufacture. Freeman in England[10] among others such as Bernas in Paris[11] had developed sources capable of providing beams of ions suitable for Ion Implantation. A slit extraction geometry producing a ribbon shaped beam allowed the transport of more intense beams without proportionate worsening of the beam control problem. As a side-benefit, such beams could be more efficiently 'painted' over the wafer surface.

In 1969 Lintott in England and in 1970 Accelerators, Inc. (later Veeco) in Texas shipped the first commercial implanters built specifically for semiconductor manufacture. Lintott was later acquired by Applied Materials of Santa Clara, CA. Rose[5] and Moffatt[1] have laid out family trees showing the genesis of the main early corporate players in Ion Implant.

In the United States in 1971, a group of scientists and engineers led by Peter Rose became interested in the commercial possibilities of Ion Implantation - specifically its application to semiconductor manufacture. These people had a background in low energy (2 MeV to 40 MeV) research accelerator technology at High Voltage Engineering Corp. of Burlington, Massachusetts where they and their colleagues had developed a range of pressurized electrostatic accelerators for research.

Rose[5] provides an entertaining and revealing description of that period of commercialization of the technology. In April 1971 Rose et al formed their company, Extrion in Gloucester, Massachusetts. Months later they sold their first implanter to National Semiconductor at the astonishingly low price of $65,000. Extrion's success was built partially on their early recognition of the need to provide a beam at the wafer that was as stable as possible in quality and shape over the required energy range (in their case 20 keV to 150 keV) and their achieving this by "pre-analysis". In this technique the beam is mass analyzed prior to rather than after acceleration to final energy and a beam 'waist' is formed at the entrance of the accelerating structure as shown in Fig. 1(b).

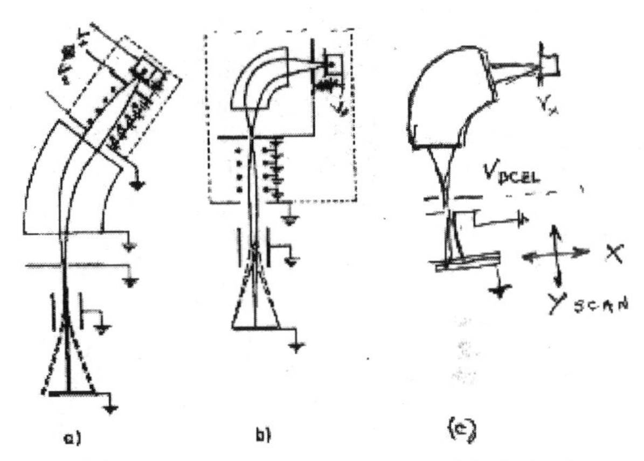

FIGURE 1. History in the making - The original sketch from Extrion showing three basic design concepts for a commercial Ion Implanter:
(a) Post-acceleration analysis.
(b) Analysis before acceleration.
(c) Extraction energy only with the option of post-analysis decel.

In 1975 Extrion was acquired by Varian Associates of Palo Alto, (now Varian Semiconductor Equipment Associates). The demand spread by the 1980s. Today the industry is populated by several large organizations.[12,13,14,15,16] Of these, Varian Semiconductor and Axcelis owe their genesis to Rose et al.

Differentiating performance requirements for ion implanters were continuously added as the processing demands of semiconductor devices evolved in the on-going quest to satisfy "Moore's Law"[17]. Some of these are:

- Implanter reliability.
- Increased wafer throughput.
- Improved implant uniformity over the wafer surface.
- Tighter implant angle control.
- Implant larger wafers - 25 mm diameter in 1960, 300 mm now, 450 mm when?
- Higher implant energy.

- Extremely low (eV) implant energy.

We now discuss developments aimed at these requirements.

RELIABILITY

In the 1960's sometimes only parts of broken wafers were loaded by hand with tweezers and up-time was hit-and-miss. Nowadays, Ion implanters have become highly automated production tools with guaranteed up-time in the high 90% range and MTBF in the hundreds and in many cases thousands of hours. An industry quip in the 1970's was that every implanter should ship with a living-breathing PhD. Today's production tools are operated 24/7 by production operators who visit each implanter in the fab from time to time during a shift, or if an alarm appears in the central control room. Chip yield, on a run of wafers is a vital measure of process accuracy and consistency - there are a few hundred to a few thousand chip die on a 300mm wafer so that the loss of even a single wafer due to process tool failure in the final stages of manufacture represents a very large financial loss. Also dose, uniformity, and implant angle spread (parallelism) must be tightly controlled to avoid significantly different transistor characteristics for chips from different locations on the wafer.

WAFER THROUGHPUT

Early "Serial" implanters processed one wafer at a time. As ion sources improved and beam current was increased throughput became limited by wafer heating. For high dose implants this led to "Batch" implanters where a number of wafers mounted near the outer circumference of a rapidly rotating disc repeatedly flash past the high current beam spot quickly enough that the thermal shock is limited, and the aggregate silicon area of the batch could tolerate much higher beam power. The spinning disc translates back and forth completing the required two dimensional scan. Several corrections are employed to compensate for dose and implant angle non-uniformities inherent in this geometry. Today's implanters are almost all Serial in order to meet much more stringent implant requirements. In a rare example of physics making our jobs easier, today's large wafers tolerate Serial high dose processing more easily since the beam power is distributed over a greater silicon area.

SOME NOTABLE DEVELOPMENTS

A few inflection points in implanter evolution are worth noting:

- Extrion (previously discussed)[5] (1971)
- The parallel beam implanter of Berrian et al at ASM Ion Implant[18] (1987)
- Fast magnetic scanning by Glavish[19] (1992)
- New techniques for ultra-low energy doping.
- Full wafer size ribbon beam by White et al at Diamond Semiconductor[20] (1993)

Parallel Implant

The ASM implanter was acquired by Varian and its architecture became the main-stay of their Medium Current product line. Other examples of approaches to parallel implantation are shown in Figures 2[21] and 3[22]. Varian also acquired the Diamond Semiconductor ribbon beam tool and updated versions have become important High Current implant tools.

Fast Magnetic Scan

This technique pioneered by Glavish[19] allays concerns about destruction of space charge neutralization by electrostatic scanning fields with the resulting worsening of beam control problems. An example is shown in Figure 2.

High And Low Energy Doping

Special variations have led to "High Energy" (up to about 4MeV) and "Ultra Low Energy" (down to a few hundred eV) implant tools. Axcelis[14], important in the High Energy market, departing from dc acceleration uses a novel compact Heavy-ion Linac.[23] The ion-beam challenges in delivering high currents at very low energies have led Varian and Applied Materials to a non-beamline method called "Plasma Doping" where the wafer is presented to a plasma of the required dopant and a momentary electrical bias on the wafer accelerates the dopant ions from the plasma directly into the wafer. There are competing beamline methods for low energy doping from SemEquip[16,21] using beams of Octadecaborane ($B_{18}H_{22}$) and Varian[13] using Carborane. When these "clusters" collide with the wafer, the ion energy is shared between the constituent atoms so that for example, a 1 mA, 20 keV Octadecaborane ion is the implant equivalent of about 18 mA of 1 keV boron ions.

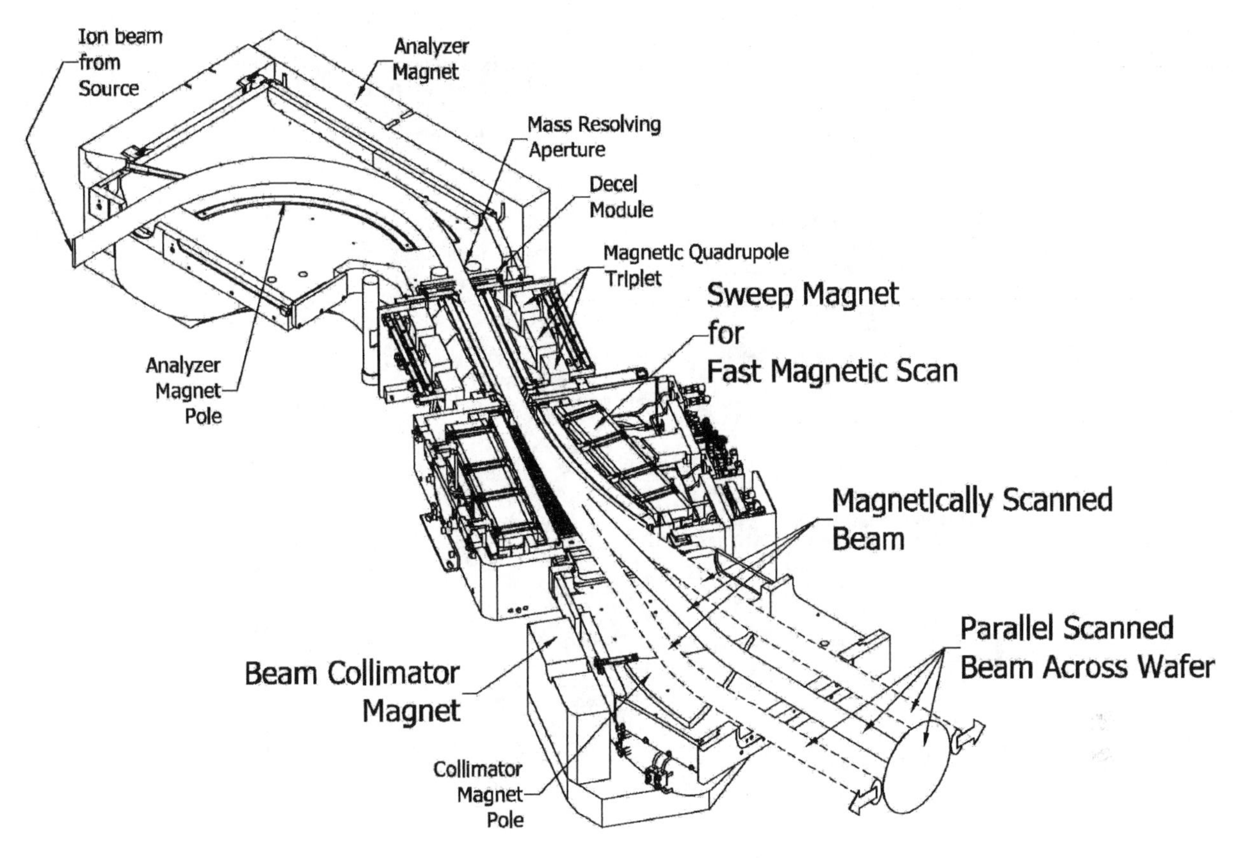

FIGURE 2. Beamline showing fast magnetic scanning and magnetic collimation.

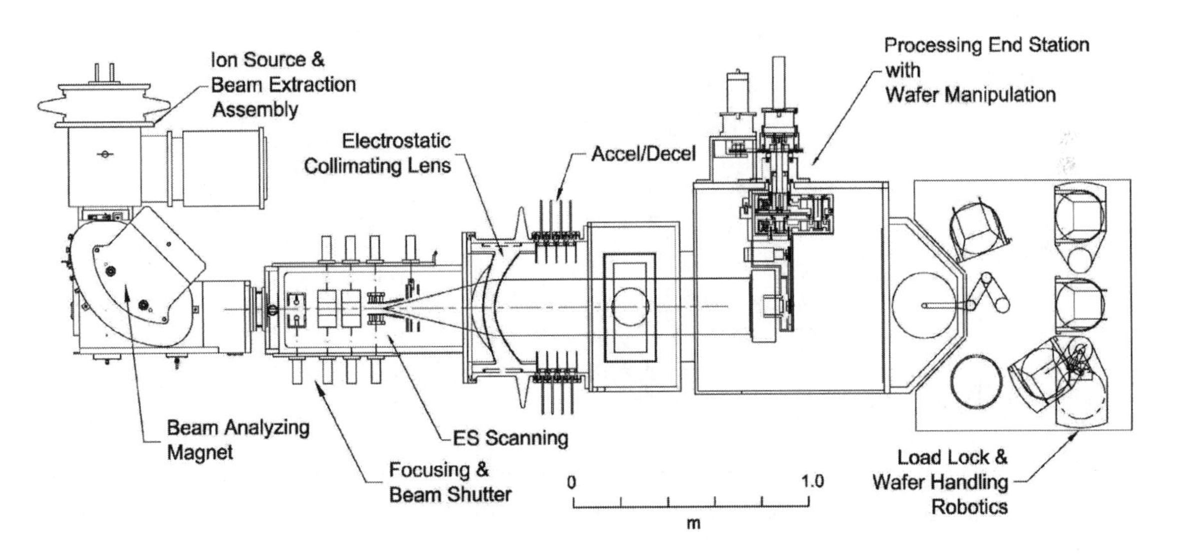

FIGURE 3. Beamline showing electrostatic scanning and collimation.

US005350926A

United States Patent [19]

White et al.

[11] **Patent Number: 5,350,926**

[45] **Date of Patent: Sep. 27, 1994**

[54] **COMPACT HIGH CURRENT BROAD BEAM ION IMPLANTER**

[75] Inventors: **Nicholas R. White**, Wenham, Mass.; **Manny Sieradzki**, Gloucester; **Anthony Renau**, West Newbury, all of Mass.

[73] Assignee: **Diamond Semiconductor Group, Inc.**, Gloucester, Mass.

[21] Appl. No.: **29,766**

[22] Filed: **Mar. 11, 1993**

[51] **Int. Cl.⁵** **H01J 37/317**

[52] **U.S. Cl.** **250/492.21**; 250/398; 250/251

[58] **Field of Search** .... 250/492.21, 492.3, 398, 250/397, 396 ML, 396 R, 251

[56] **References Cited**

U.S. PATENT DOCUMENTS

3,845,312	10/1974	Allison, Jr.	250/398
4,017,403	4/1977	Freeman	250/492
4,486,664	12/1984	Wollnik	250/396
4,661,712	4/1987	Mobley	250/492.2
4,745,281	5/1988	Enge	250/356
4,812,663	3/1989	Douglas-Hamilton et al.	250/492.2
4,914,305	4/1990	Benveniste et al.	250/492.3
5,091,655	2/1992	Dykstra et al.	250/492.2
5,126,575	6/1992	White	250/492.3
5,177,361	1/1993	Krahl et al.	250/396 ML
5,206,516	4/1993	Keller et al.	250/492.2

OTHER PUBLICATIONS

Instability Threshold for a Calutron (Isotope Separator) With Only One Isotope Species, Igor Alexeff, IEEE Transactions on Plasma Science, vol. P3-11, No. 3, Jun. 1983.

The Design of Magnets with Nondipole Field Components, White et al., Nuclear Instruments and Methods in Physics Research A258 (1987) 437-442 North-Holland, Amsterdam.

Primary Examiner—Jack I. Berman

Attorney, Agent, or Firm—Lahive & Cockfield

[57] **ABSTRACT**

A compact high current broad beam ion implanter capable of serial processing employs a high current density source, an analyzing magnet to direct a desired species through a resolving slit, and a second magnet to deflect the resultant beam while rendering it parallel and uniform along its width dimension. Both magnets have relatively large pole gaps, wide input and output faces, and deflect through a small radius of curvature to produce a beam free of instabilities. Multipole elements incorporated within at least one magnet allow higher order aberrations to be selectively varied to locally adjust beam current density and achieve the high degree of uniformity along the beam width dimension.

23 Claims, 3 Drawing Sheets

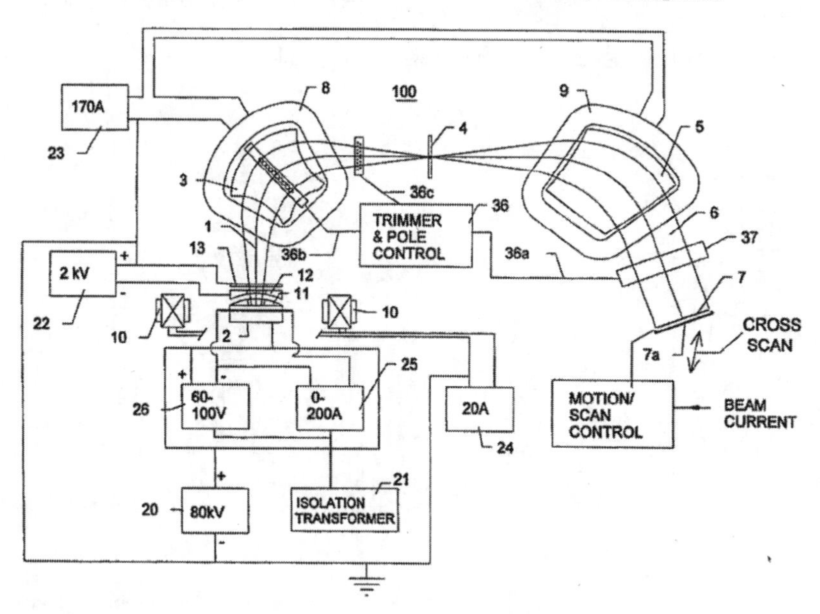

FIGURE 4. The patented White, full wafer width beam implanter architecture.

A FEW NOVEL IMPLANTERS

Flat panel displays which employ thin film transistors use ion implant steps in production. Both Nissin Ion and Mitsui have machines in this market. Accommodating the very large flat panels, meters in size requires spectacularly large implanters with ribbon beams one meter or more wide. The beam analyzing magnets sometimes have gaps into which a small child can walk without stooping.

High dose oxygen implanters were developed in the 1980s and later, to form a buried layer of SiO_2 in silicon wafers. In 1984 the authors where instrumental in the development of a commercial product with beam current up to 200mA and energy up to 200keV.[24] The beam was spread out into a "ribbon" which was wider than the wafer diameter thereby lowering the current density at the wafers to manage the 40kW beam power without wafer damage. Batches of wafers were deployed on the inside wall of a cone assembly rotating at high speed. This "Eaton NV-200" was probably the first commercial semiconductor ion implanter to use a wafer-size ribbon beam. The implant uniformity of about 5% required for the SIMOX process allowed simple beam spreading techniques using only quadrupole defocusing magnets. Building on the NV-200 work, White et al in 1993[20] developed techniques to deliver, in a wafer-size ribbon beam, the better-than-one-half-percent uniformity required for doping implants. This work, implemented by Diamond Semiconductor Group, Inc. [Fig. 4] became the basis of today's high current implanters sold by Varian and AIBT.

A new player in the field, AIBT[12] allows dual-mode operation using the White ribbon beam capability combined with 1D wafer scan or 2D wafer scanning depending on process requirements.

A large new application is "Layer Separation" using high dose hydrogen implants which cause lattice disruption at the implant depth, cleaving the surface layers and allowing the removal of gossamer thin silicon sheets. This is an established production process in the manufacture of Silicon On Insulator substrates by Soitec and others. Recently the technique has entered the Photovoltaic market being addressed by SiGen, Twin Creeks Technologies and others. These activities are characteristic of the commercial Ion Implantation industry where development occurs when radically new processes outside the range of available tools is required. Then, when large scale production creates sufficient demand, new products appear from both the existing players and new start-ups. An excellent summary of the current state of the art is given by Rose and Ryding.[25] This substantial industrial application of accelerators continues to grow.

ACKNOWLEDGEMENTS

For invaluable help in preparing this presentation, the authors wish to thank AIBT[12], Axcelis Technologies, Inc., Dr. Hilton Glavish of Zimec, Inc., Mike Racicot, Varian Semiconductor Equipment Associates, Dr. Nicholas White of Albion Systems and Liz Laughead of Isys.

REFERENCES

1. S. Moffatt, *Nucl. Instr. and Meth.* **B96,** 1995, pp. 1-6 (Proc 10th Int. Conf. Ion Implantation Technology, Catania, Italy (1994).
2. Rutherford, E., *Phil. Mag.* Series 6 **21**: 1911, pp. 669–688.
3. J. H. Freeman, *Radiation Effects and Defects in Solids* Vol. 100, London: Gordon and Breach, Science Publishers Inc., 1968, pp. 161-248.
4. L. Wegmann, *Nucl. Instr. and Meth.* **189,** 1981, pp. 1-6 (Proc. 3rd Int. Conf. Ion Implantation Equipment and Techniques, Kingston, Ontario, 1980).
5. Peter H. Rose, *Nucl. Instr. and Meth.* **B6,** 1985, pp. 1-8.
6. Dave G. Armour, in *Ion Implantation Technology*, ed. Seebauer et al, AIP Conference Proceedings 620, American Institute of Physics, Melville, NY, 2008, pp. 3-10.
7. S. J. Pearton, informal publication: http://pearton.mse.ufl.edu/rgw/IV/IV-1/a-History.doc.
8. J. W. Moyer, U.S. Patent No. 2,842,466 (8 July 1958).
9. W. Shockley, U.S. Patent No. 2,787,564 (2 April 1957).
10. J. H. Freeman, *Nucl. Instr. and Meth.* **22** (1963) 306; U.S. Patent No. 4,792,687 (20 Dec. 1988).
11. R. Bernas and A. O. Nier, *Rev. Sci. Instrum.* **19**,1947, pp. 895.
12. AIBT, San Jose, CA.
13. Varian Semiconductor Equipment Associates, Gloucester, MA.
14. Axcelis Technologies, Inc., Beverly, MA.
15. Nissin Ion Equipment Co. Ltd., Kyoto, Japan.
16. SemEquip, Inc., North Billerica, MA.
17. http://en.wikipedia.org/wiki/Moore's_law.
18. D. W. Berrian, R. E. Kaim, J. W. Vanderpot and J. F. M. Westendorp, *Nucl. Instr. and Meth.* **B37/38,** 1989, pp. 500; U.S. Patent No. 4,922,106 (1 May 1990)
19. Glavish, U.S. Patent 5,311,028 (10 May 1995).
20. White, Sieradzki and Reneau, U.S. Patent 5,350,926 (27 Sep. 1994).
21. Glavish et al, AIP Conference Proceedings 866, 2006, pp. 167-170 (Proc. 16th Int. Conf. Ion Implantation Technology, Marseille, France (2006).
22. Barry A. MacKinnon and Monty L. King, , *Nucl. Instr. and Meth.* **B74** (1993) 469-473 (Proc. 9th Int. Conf. Ion Implantation Technology, Gainesville, FL (1992).
23. H. F. Glavish, *Nucl. Instr. and Meth.* **B21,** 1987, pp. 218-223.
24. J. P. Ruffell, D. H. Douglas-Hamilton, R. E. Kaim and K. Izumi: *Nucl. Instrum. Meth.* **B21**, 1987, pp. 229-234.
25. Peter H. Rose and Geoffrey Ryding, *Rev. Sci. Instrum.* **77**, 2006, pp. 111101-01 - 111101-12.

The IBA Easy-E-Beam^{TM} Integrated Processing System

Marshall R. Cleland, Richard A. Galloway, Thomas F. Lisanti

IBA Industrial, Inc., 151 Heartland Blvd., Edgewood, NY 11717 USA

Abstract. IBA Industrial Inc., (formerly known as Radiation Dynamics, Inc.) has been making high-energy and medium-energy, direct-current proton and electron accelerators for research and industrial applications for many years. Some industrial applications of high-power electron accelerators are the crosslinking of polymeric materials and products, such as the insulation on electrical wires, multi-conductor cable jackets, heat-shrinkable plastic tubing and film, plastic pipe, foam and pellets, the partial curing of rubber sheet for automobile tire components, and the sterilization of disposable medical devices. The curing (polymerization and crosslinking) of carbon and glass fiber-reinforced composite plastic parts, the preservation of foods and the treatment of waste materials are attractive possibilities for future applications. With electron energies above 1.0 MeV, the radiation protection for operating personnel is usually provided by surrounding the accelerator facility with thick concrete walls. With lower energies, steel and lead panels can be used, which are substantially thinner and more compact than the equivalent concrete walls. IBA has developed a series of electron processing systems called Easy-e-Beam^{TM} for the medium energy range from 300 keV to 1000 keV. These systems include the shielding as an integral part of a complete radiation processing facility. The basic concepts of the electron accelerator, the product processing equipment, the programmable control system, the configuration of the radiation shielding and some performance characteristics are described in this paper.

Keywords:. industrial electron accelerators, high-power accelerators, direct-current accelerators, self-shielded accelerator systems, X-ray shielding, electron beam processing
PACS: Repl29.20.-c, 29.20. Ba

INTRODUCTION

An Easy-e-Beam^{TM} system is an all-in-one electron beam (EB) processing facility which can be used for the crosslinking of polymeric products, such as the insulation on high-performance electrical wires, the jackets on multi-conductor cables, heat-shrinkable plastic tubing for encapsulations, heat-shrinkable plastic film for wrapping fresh foods and for other packaging applications, plastic foam for padding furniture and the interiors of automobiles, and thin sheets of rubber compounds for making automobile tires. It includes an electron accelerator, a beam scanner and a product handling system, which are located within a self-shielded enclosure to protect operating personnel from the X-rays emitted when energetic electrons strike solid materials. Additional equipment outside but close to this enclosure are the Dynamiton® high-voltage generator, the radio-frequency oscillator, which energizes the high-voltage generator, the programmable logic controller (PLC), the insulating gas transfer and storage system and the product pay-off and take-up equipment. A drawing of a complete Easy-e-Beam^{TM} system for irradiating insulated wire and small diameter plastic tubing is shown in Figure 1. Some advantages of a compact, self-shielded facility in comparison to a large concrete vault are rapid and simple installation in an existing building, reduced floor space, and the ability to add an EB treatment process in line with an established manufacturing operation.

FACILITY DESCRIPTION

Electron Accelerator Assembly

The electron accelerator includes three major subassemblies, the high-voltage generator, the electron gun and electron acceleration tube, and the beam scanner. The Dynamitron® high-voltage generator embodies a multi-stage, cascaded-rectifier circuit that converts radio-frequency (RF) power to high-voltage, direct current (DC) power with an electrical efficiency greater than 60 percent. This type of generator can sustain load currents up to 100 milliamperes (mA). The output voltage depends on the number of rectifiers in the assembly, which can each produce 50 kilovolts (kV). The system shown in Figure 1 has been designed for an accelerating voltage of 800 kV, an electron

Application of Accelerators in Research and Industry
AIP Conf. Proc. 1336, 52-55 (2011); doi: 10.1063/1.3586055

beam current of 100 mA and a beam power of 80 kilowatts (kW). Easy-e-Beam™ systems can be designed for any voltage in the range from 300 kV to 1000 kV [1]. The high-voltage generator does not emit radiation and does not need to be shielded. The high voltage power is connected through a rigid, gas-insulated transmission line to the electron gun and acceleration tube, which are mounted in a right-angle extension of the pressure vessel inside the shielded enclosure.

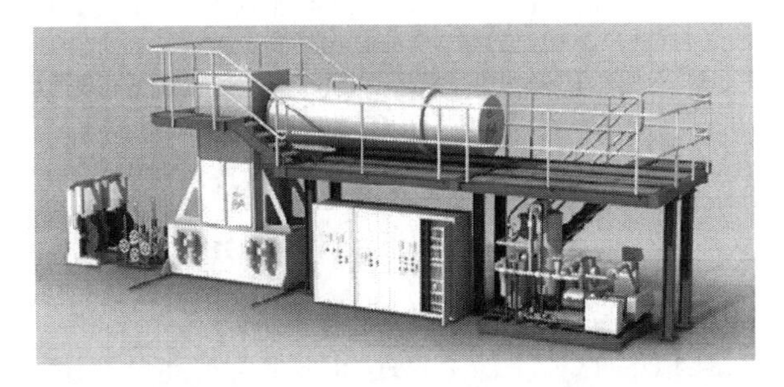

FIGURE 1. A drawing of a complete Easy-e-Beam™ system for irradiating insulated wire and plastic tubing. The horizontal cylindrical pressure vessel houses the high-voltage generator. The upper rectangular shield encloses the electron accelerator section. The lower rectangular shield contains the underbeam wire handling equipment. The cabinet next to the lower shield contains the radio-frequency oscillator. The equipment next to the oscillator is a gas pumping system for transferring the insulating SF6 gas into and out of the pressure vessel.

The electron gun uses a thermionic cathode without a control grid. The electrons are emitted from a directly-heated tungsten filament containing a small amount of rhenium. The beam current is determined by the temperature of the filament, which is controlled automatically to stabilize the emission. The electron beam is extracted and focused into the acceleration tube by a cylindrical anode whose accelerating voltage is a fixed fraction of the total acceleration voltage. This simple scheme allows the accelerating voltage to be changed over a range of two to one without affecting the optical properties of the beam.

The evacuated acceleration tube consists of a stack of metallic dynodes separated by glass rings. These dynodes are convoluted so that energetic electrons, which can be scattered by collisions with residual gases in the tube, are intercepted by the dynodes so they cannot strike the glass rings. The beam apertures in the dynodes are 3.0 inches (7.6 centimeters) in diameter. This provides sufficient pumping speed (vacuum conductance) so that a single vacuum pump at the grounded end of the tube can establish a satisfactory vacuum throughout the length of the tube. The dynodes are spaced 1.0 inch apart and the voltage between adjacent dynodes is only 33 kV when the accelerator is operating at full voltage. The length of the acceleration tube in an 800 kV system is 24 inches (61 cm). The intermediate potentials of the dynodes are obtained from a string of high-voltage resistors, called the beam tube divider, which is connected between the electron gun and the grounded end of the tube. This configuration establishes a uniform electric field within the tube, and prevents sparking within the tube.

The beam scanning system deflects the electrons through angles of plus and minus 30 degrees at a frequency of 200 hertz (Hz). The waveform of the magnetic field in the scanning magnet is essentially triangular to provide a uniform electron fluence (absorbed dose) across the width of the product handling fixture. However, abrupt steps are added to the magnet coil current at the peaks of the field to reduce the dwell time of the beam at the ends of the scan. This helps to compensate for the tendency of a triangular waveform to be rounded at the peaks due to attenuation of the high-frequency components. A secondary, small-amplitude scan is applied at right angles to the primary scan at a frequency of slightly less than 4000 Hz to avoid synchronizing both scan frequencies. This increases the effective width of the beam and increases the amount of current that can be transmitted through the beam window without overheating it. The width (long dimension) of the beam at the exit of the scanning horn is nominally 36 inches (91 cm). The beam exits through a thin metallic foil made with a titanium alloy. The beam window thickness is 40 micrometers (μm). It dissipates less than 30 kiloelectron volts (keV) of kinetic energy from 800 keV electrons, and it is cooled with a fast-flowing stream of air.

Product Handling System

A multiple-pass product handling (under-beam) system for irradiating small-diameter electrical wires or plastic tubing is shown in Figure 2. It fits snuggly inside the lower section of the shielded enclosure. This fixture and the heavy side panel of the enclosure can easily be rolled out to facilitate product loading. The pay-off and take-up reels and the apparatus (dancers) for controlling the tension in the wire or tubing are located outside of the enclosure as shown in Figure 1. The rotational speeds of the pay-off and take-up reels are adjusted automatically to maintain the proper tension in the products. The line speed of the wire or tubing is controlled by the rotational speed of the under-beam fixture, which is synchronized with the electron beam current to provide the desired dose in the products. The plan view in Figure 3 shows that the beam scanning horn is oriented at 45 degrees to the side panels of the radiation shield, and also at 45 degrees to the direction of the wire or tubing. This allows the beam window to be wider than the under-beam fixture to deliver the high beam current without overheating the beam window. It also reduces the width of the shielded enclosure.

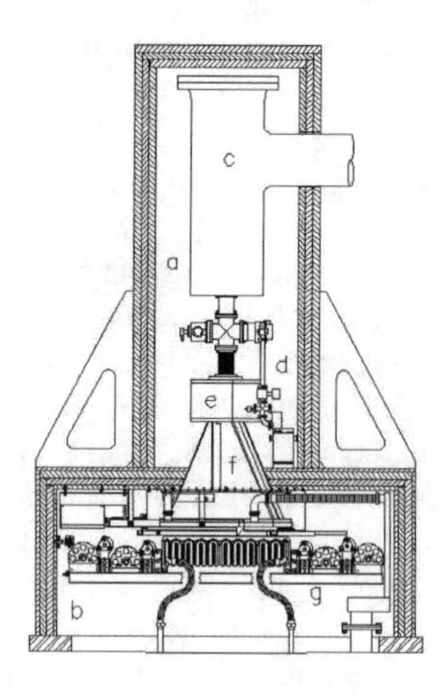

FIGURE 2. A drawing of an Easy-e-Beam™ system showing the upper (a) and lower (b) sections of the shielded enclosure, including the pressure vessel (c) housing the electron accelerator, the vacuum pump (d), the beam scanning magnet housing (e), the triangular beam scanning chamber (f) and the under-beam fixture (g) for wire and/or tubing.

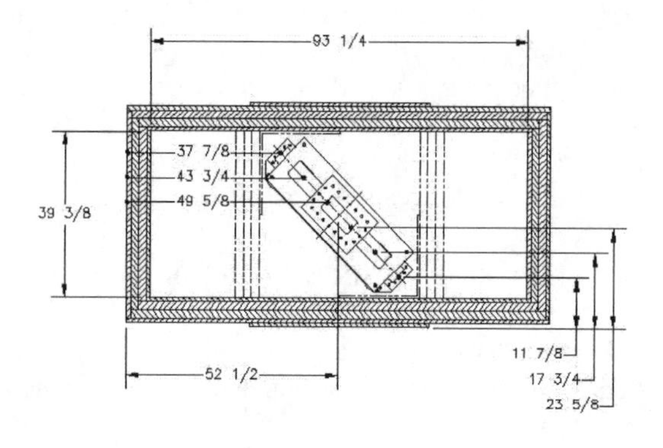

FIGURE 3. A plan view of the enclosure showing the 45 degree orientation of the beam scanning horn.

The under-beam fixture has four smooth, stainless-steel drums that are accurately machined and balanced for high-speed processing. The wire or tubing passes back and forth through the wide, scanning electron beam many times to make efficient use of the beam, to expose all sides of the products to the beam, and to reduce the temperature rise in the products. This is especially important for copper wire, which has a low heat capacity. The spacing of the wire or tubing on the drums is controlled by grooved roller guides that are free to rotate when in contact with the products. These guides can be changed to accommodate products with different diameters. Typical wire sizes irradiated with an 800 keV Easy-e-Beam™ range from American Wire Gauge (AWG) 8 to 24. The conductor diameter for AWG 8 is 0.1285 inch (3.264 millimeter) and the conductor diameter for AWG 24 is 0.0201 inch (0.0511 millimeter). The processing line speeds with AWG 8 wire can be as high as 820 feet (250 meters) per minute and with AWG 24 wire the line speeds can be as high as 3280 feet (1000 meters) per minute. The maximum line speed is usually limited by the pay-off and take-up equipment rather than by the electron beam current of the accelerator.

Additional Equipment

A Dynamitron® high-voltage generator converts radio-frequency (RF) power at a frequency of 100 kilohertz (kHz) to high-voltage, direct current (DC) power with a multi-stage, cascaded-rectifier circuit using parallel, capacitive coupling from a pair of semi-cylindrical electrodes surrounding the rectifier assembly to an array of semi-circular electrodes, which are connected to each rectifier junction. The RF power is produced with a self-tuning oscillator circuit energized with a high-power triode tube. The frequency is determined by the capacitance of the driving electrodes and the inductance of a resonant, iron-free transformer that increases the RF voltage from the triode tube. The self-tuning feature guarantees that this oscillator circuit always operates at its natural resonant frequency, so the inherent values and the stability of the capacitive and inductive components are not critical parameters. The maximum input DC power for the triode tube is 220 kW, which can generate up to 150 kW of RF power. This is substantially more than the power needed to produce an 80 kW electron beam. The oscillator components are enclosed in a separate cabinet and the RF power is transmitted to the high-voltage generator by a flexible coaxial cable. At the relatively low frequency of 100 kHz, this cable can be as long as needed for a convenient installation. In Figure 1, the oscillator cabinet is located beneath the elevated high-voltage generator.

All of this equipment is controlled and monitored by a programmable logic controller (PLC), which is shown next to the oscillator cabinet in Figure 1. The PLC enables the operator to set the critical process conditions, such as the accelerating voltage, the beam current, the scanning amplitude and the speed of the under-beam product handling system, to deliver the required absorbed dose to the products. These critical process conditions are regulated automatically. Other operating parameters are monitored and recorded for future reference. Information about the operating parameters can be accessed by the operator or maintenance personnel and presented on the PLC screen. A typical informational screen is shown in Figure 4.

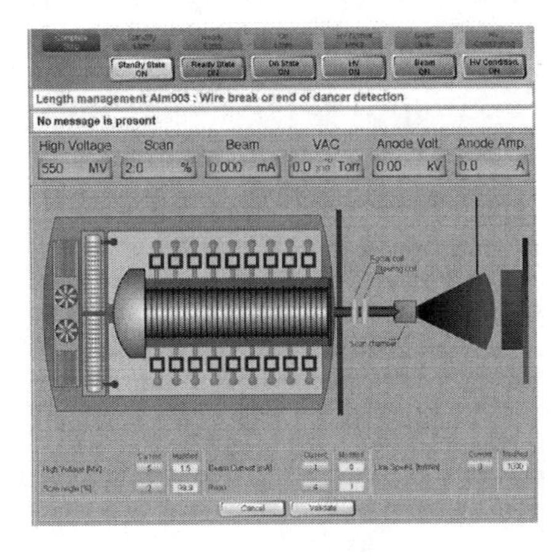

FIGURE 4. An example of the many screens that can be presented on the monitor of the PLC control system to inform the operator or maintenance engineer about the performance conditions of the Easy-e-Beam system.

The high-voltage generator and the electron accelerator section are insulated with compressed sulfur hexafluoride (SF6) gas at an overpressure of about 6 atmospheres. When this equipment needs to be opened for maintenance, the SF6 gas is transferred to a storage tank with a compressor. Before the gas is transferred back, the air is removed from the pressure vessel of the high-voltage generator and accelerator with a vacuum pump. During these procedures, the SF6 gas is passed through filters to remove moisture and preserve its insulating quality. This non-toxic gas is saved and reused because it is expensive and it also has a greenhouse effect in the atmosphere. The gas handling equipment is located under the high-voltage generator next to the oscillator cabinet in Figure 1. The gas storage tank can be placed in any convenient location, even outside of the factory building.

Self-Shielded Enclosure

The walls of the self-shielded enclosure are made with steel and lead panels to protect the operator and other people that might be near the Easy-e-Beam™ facility from X-rays, which are emitted when energetic electrons strike the products and the under-beam equipment within the enclosure. With electron energies less than 1.0 MeV, the use of steel and lead instead of concrete provides a more compact facility. Lead, because of its high atomic number, is especially effective for shielding X-rays with lower energies. For example, the side walls in the lower part of the enclosure nearest the stainless steel beam stop, which is the primary source of X-rays when there are no products in the beam, need to provide an attenuation factor of about 3.3×10^{-10}, assuming that the Easy-e-Beam is operating at its maximum ratings with 100 mA of beam current at 0.80 MeV, and assuming that the occupied area is 1.0 m from the beam centerline. This attenuation factor would require about 4.9 inches (12 cm) of lead, or 13 inches (33 cm) of steel or 47 inches (119 cm) of ordinary concrete.

The calculations for the thicknesses of the shielding panels have been done by following the methods given in the NCRP Report No. 51 [2]. As shown in Figure 2, the panels in the lower part of the enclosure are made with 4 inches of lead covered with 1 inch of steel on both sides of the lead. In the upper part of the enclosure, the panels are made with 6 inches of steel, which is sufficient because the X-rays coming upward from the beam stop through the scan horn are entering the side walls at steep angles. These thick panels will reduce the external radiation intensity to less than the average dose rate of 0.050 millirem/hour (0.50 microsievert/hour), which has been recommended by the U.S. Nuclear Regulatory Commission (NRC) for the general public in uncontrolled areas [3]. This low rate is only about 2 percent of the dose rate allowed for radiation workers who are aware of the safety requirements and whose absorbed dose is being monitored and recorded.

CONCLUSION

An all-in-one, self-shielded Easy-e-Beam™ facility can provide a practical, economical and reliable way to add electron beam processing to an existing manufacturing operation.

REFERENCES

1. M. R. Cleland, C. C. Thompson, H. Saito, T. F. Lisanti, R. G. Burgess, H. F. Malone, R. J. Loby and R. A. Galloway, "New high-current Dynamitron accelerators for electron beam processing," *Nuclear Instruments and Methods in Physics Research* **B79**, 861-864 (1993).
2. Anon, NCRP Report No. 51, Radiation Protection Design Guidelines for 0.1 – 100 MeV Particle Accelerator Facilities, Section 4.3, Methods of Calculating Shielding Thickness, National Council on Radiation Protection and Measurements, 7910 Woodmont Avenue, Bethesda, Maryland 20814, USA (1977).
3. Anon, Code of Federal Regulations, 10 CFR Part 20, Standards for Protection Against Radiation, Section 20.1301, Dose limits for individual members of the public (2002).

Design And Performance Of A 3 MV Tandetron™ Accelerator System For High-Current Applications

Nicolae C. Podaru, A. Gottdang, FRENA Group*, and D.J.W. Mous

High Voltage Engineering Europa B.V., P.O. Box 99, 3800 AB, Amersfoort, The Netherlands
**Saha Institute of Nuclear Physics, Sector-1, Bloc-AF, Bidhan Nagar, Kolkata 700064, India*

Abstract. The Saha Institute of Nuclear Physics, Kolkata, India will commission in 2011 a 3 MV Tandetron™ accelerator system. Hi-flux neutron production, nuclear reaction cross-section measurements and time of flight experiments are among the research activities to be performed with this system. Features such as high beam currents, high beam energy stability and low beam energy spread are necessary when conducting these types of experiments. At the same time, the beam energy must be known with high accuracy. This article reports the early results obtained during the in-house testing of the system. H beam currents of 500 μA have been transported through the system. The so-called "Q-snout" electrode lens ensures high particle transmission (~70%) through the accelerator even at 7% of the rated terminal voltage (TV). At present, the negative H ion beam current output of the SO120 multicusp ion source, rated at only 2 mA, combined with the Tandetron™ accelerators (with terminal voltage ranging from 1 to 6 MV) provides H beam powers of up to 10 kW.

Keywords: Tandem accelerators; High-current; High-power; Pulsed ion beams
PACS: 25.40.-h; 25.45.-z; 25.55.-e; 29.20.-Ba; 29.25.Ni; 29.27.Eg; 41.85.Lc; 87.10.Kn; 87.56.bd

INTRODUCTION

Medical, scientific and industrial applications such as boron neutron capture therapy (BNCT), positron emission tomography, nuclear resonance absorption, silicon cleavage for solar cell production, and isotope production require high-current MeV ion beams. In the recent years, HVE has extended the power range of Tandetron™ and Singletron™ accelerator systems to accommodate the demands of these specific applications. As an example, for BNCT, dedicated Singletron™ particle accelerators (3 MV) can deliver 20 mA H beam currents, yielding a total beam power of 60 kW.

With a rated beam power of 3 kW, the research activities to be performed include studies of nuclear reactions in stellar interiors, high-flux neutron production, and time of flight experiments. These research objectives require specific design criteria. The cross-section of the charged-particle-induced nuclear reaction, $\sigma(E)$, drops almost exponentially with decreasing energy E thus the absolute incident energy must be accurately known. For the same reason, the beam energy must have low energy spread. For example, in the *14N(p,γ)15O* (reaction in the CNO-cycle of a star) an error of 1.5 keV in beam energy at 100 keV leads to an error in the $\sigma(E)$ of about 20% [1]. In addition, the low reaction rate for various nuclear reactions of astrophysical interest requires high-current H and He ion beams, for both DC and pulsed operation modes. Nevertheless, long measurement times at constant beam energy are needed. The system has been designed for high-current ion beam transport (500 μA H), low beam energy spread (e.g. TV ripple 30 V_{RMS}), long term voltage energy stability (± 30V per hour), while also being able to determine the beam energy with high accuracy.

SYSTEM OVERVIEW

The schematic layout of the system is shown in Fig. 1. The light ion injection beam line contains a dual-source injector [2], a patented beam chopper – buncher system [3] and the necessary beam guiding and defining elements. The SO130 multicusp ion source with positive ion extraction is used for He^- production. A charge exchange canal (CEC) is used to produce the required He^- for injection into Tandetron™. Hydrogen is generated with the SO120 multicusp ion source by direct negative extraction. For both ion sources, a 30° magnet located at the exit of the dual-source injector is used to analyze the negative light ion beam. The third ion source of the system is the SO110 (cesium-sputter type), used for heavy

Application of Accelerators in Research and Industry
AIP Conf. Proc. 1336, 56-59 (2011); doi: 10.1063/1.3586056

negative ions extraction. This particular source configuration features a carousel that can contain up to 50 samples. Prior to the injection into the Tandetron™ accelerator the heavy ion beam is analyzed with a +45° magnet. As a future extension, an additional injector can be connected to the -45° magnet.

The all-solid-stated power supply of the accelerator is constructed around the high-energy acceleration tube, this being the first 3MV HVE tandem accelerator designed with such geometry (coaxial) and thus completing the already existing 2, 5 and 6 MV coaxial accelerators systems [4, 5]. The footprint of the accelerator is reduced with 35% when compared to the footprint of the previous type 3 MV "T-shaped" Tandetron™ accelerator. A more detailed description of the coaxial accelerator design was provided by Mous et al.[4].

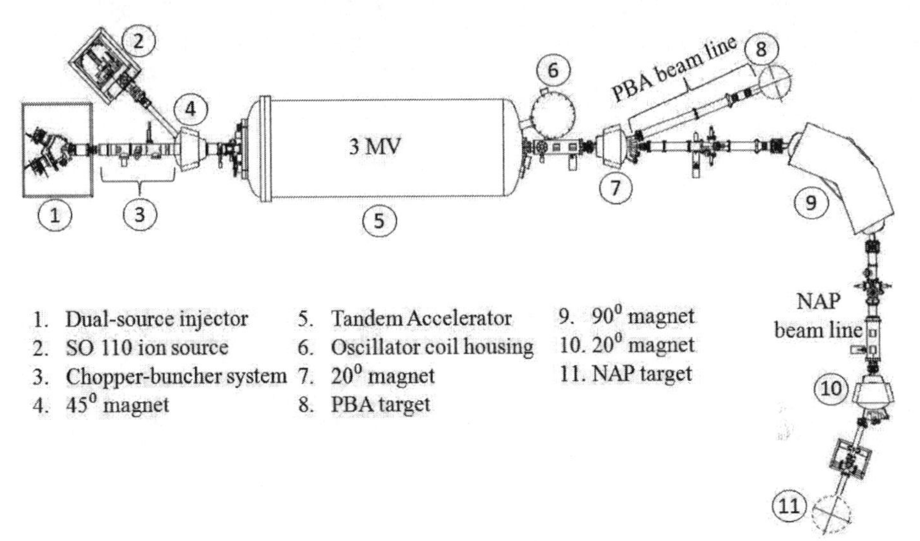

FIGURE 1. The principle layout of the 3 MV Tandetron™ particle accelerator system. The system footprint is 19.5 × 14.1 m^2

The high-energy side of system features two beam lines. One of the beam lines is dedicated to pulsed beam applications (PBA), while the other is designed for nuclear astrophysics experiments (NAP). Among ion optical beam guiding elements, the PBA beam line is equipped with a "fast" Faraday cup. It is capable of resolving beam pulses with nanosecond resolution. The NAP beam line contains components specifically designed to handle high power beam transport, including a 90° analyzing magnet with a dual-slit stabilization system.

DESIGN CONSIDERATIONS

Ion optical calculations indicate that beam power densities in excess of 1 kW/mm^2 could be achieved at several waist positions along the NAP beam line. At such power densities the ion beam can easily burn through any material within seconds. In order to transport 3 kW of beam power to the NAP target position, high power beam line components such as Faraday cups, protection apertures and slits have been designed. The 3 kW Faraday cup (FC) features active water cooling (10 l/min) and secondary electron suppression. The design of this FC has been determined using 3D thermal finite element analysis coupled with 3D fluid dynamics simulations. As a general rule, the high power components of this system expose to the ion beam a sheet of tantalum (reduction of radiation and high temperature melting point) while copper is used as a heat conductor between the cooling water channels and the tantalum.

When using high power beams, unintentional beam line component damage can be caused by operator or equipment malfunction. For high power systems, HVE has developed an in-house software extension (the so-called "BeamWizard"), completely integrated in the machine control software. The main purpose of the BeamWizard is to assist the operator when the beam power in the high-energy side exceeds a preset value. Routines included in the BeamWizard ensure that components which cannot take the full beam power (beam profile monitors, etc.) will not be exposed to the ion beam. If the beam power dumped on components (e.g. slits) exceeds the preset value, the beam is stopped by inserting the Faraday cup located at the entrance of the accelerator. Thus, BeamWizard automatically controls various parameters when high beam powers are used, taking system protective action if preset conditions are not fulfilled.

Beam width is mainly determined by the TV ripple and by the energy straggling (~100 eV) in the stripping process [5]. The beam energy spread of the ion sources (including the CEC contribution for He ion beams) is relatively small, being less than 10 eV. For experiments such as neutron production, both the absolute beam energy and the beam energy spread must be known with high accuracy, since the neutron field energy is dependent on both. The generating voltmeter, (GVM) [6], is commonly used to achieve long TV stability, in the 10^{-4} range. For higher long term TV stability, in the 10^{-5} range, a 90° analyzing magnet with a dual-slit stabilization system can be used, schematically depicted in Fig. 2. At the image / object of the analyzing magnet, two slits are positioned. Under the conditions that the magnetic field of the analyzing magnet is stable over time ($\Delta B/B = 10^{-5}$ per hour) and the object point of the magnet is kept constant by using the position feedback loop, then small variations in beam energy lead to beam position displacement in the dispersive plane of the magnet at the image point. The absolute beam current is measured per each jaw of the two slits. The relative difference of collected beam current on this slit is used as a feedback signal in order to control the TV of the accelerator. To precisely determine the magnetic field of the analyzing magnet, a set of 3 NMR probes covering a range from 0.09 T to 1.05 T is used. It is expected that the long term TV stability is increased by at least a factor of 10 when compared to the GVM stabilization method, achieving a TV stability per hour of ±30 V/ 3 MV. Furthermore, nuclear reactions can be used to determine in an absolute manner the relation between the beam energy and the magnetic field.

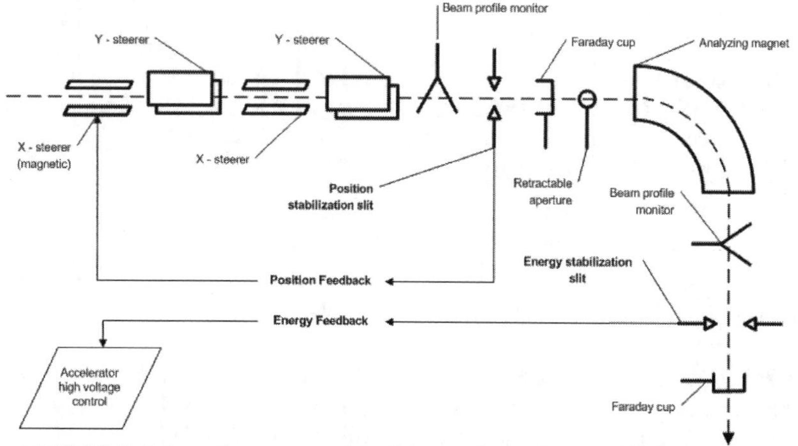

FIGURE 2. Schematic representation of the terminal voltage stabilization system

The design of the power supply (all-solid-state) of the Tandetron™ accelerators ensures low TV ripple (RMS), in the 10^{-5} regime.

Characteristic to the large diameter acceleration tubes incorporated in Tandetron™ accelerators, the electrodes feature large pumping apertures around the central aperture through which the beam is running, thus reducing the pressure levels in the acceleration tubes. The electrodes are also equipped with permanent magnets creating a spiraled magnetic field (50-80 G) along the beam axis. The magnetic field sweeps away the secondary electrons created from ion beam – rest gas interaction before gaining significant energy. This results in minimal radiation levels during operation and extended tube lifetime. To achieve good particle transmissions through the accelerator at any TV (200 kV to 3 MV for this particular system) the patented Q-snout electrode (electrostatic lens) matches the optics of the injector to that of the accelerator.

SYSTEM PERFORMANCE

The output of the three ion sources was measured. 20 keV He^- beam currents of ~50 μA were produced with the SO130 ion source after the CEC. For charge states 1 to 6 C^- ion beam currents of ~110 μA were measured from the SO110 ion source. Although rated at only 2 mA of H^-, the output current of the SO120 ion source sets no limits on system performance. The excellent particle transmission throughout the system requires only a maximum of 1 mA H^- ion source output.

The measured voltage ripple of the Tandetron™ accelerator is 27 V_{RMS} at the rated TV, in the 10^{-5} range. Particle transmissions through the accelerator were measured for three ionic species: H, He (both charge states) and C (for charge states 1 to 6). At 3 MV the particle transmission was 80% for H and 86% for C while at 7% of the TV the particle transmission was 67% for H and 71% for C.

6 MeV H beams with currents in excess of 400 μA were successfully transported through the NAP beam line with the assistance of the BeamWizard. All its protection features have been successfully tested. Visual inspection of the beam line components after transporting 3 kW beam powers to the NAP target revealed them as undamaged. At this point the dual-slit TV stabilization system has not been fully tested. However, past similar designs showed during in-house tests a TV stability of ±7 V / hour.

The chopper-buncher system has a pulse repetition frequency of up to 4 MHz. It allows injection of H^- and D^- at 30 keV, respectively He^- at 20 keV. Early experiments on pulsed H (4 MeV) and He^{2+} (6 MeV) ion beams were conducted. The aim was to measure the pulse width and intensity. The pulse signal was measured with the aforementioned "fast" Faraday cup and a fast sampling oscilloscope (bandwidth of 1GHz, sensitivity of 5 mV/div). Nanosecond pulses for H (1.66 ns FWHM) and He^{2+} (1.75 ns FWHM) were successfully produced. A single H pulse with a Gaussian shape and an intensity of 8 pC can be observed in Fig. 3.

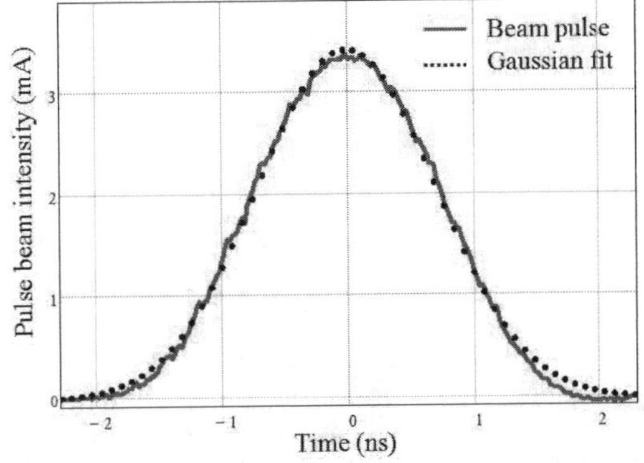

FIGURE 3. A single H beam pulse acquired with the "fast" Faraday cup. The FWHM of the beam pulse is 1.66 ns

CONCLUSIONS AND OUTLOOK

The first 3 MV coaxial type, high-current high-power, Tandetron™ accelerator system has been successfully tested in-house.

An excellent agreement between design values and test results has been found (particle transmission through the accelerator, high beam energy stability, nanosecond pulsing, etc). High-current (500 μA), high-power (3 kW) beams were safely transported to the NAP target position. The BeamWizard and the high power beam components (Faraday cups, protection apertures, etc) guarantee safe high-current high-power beam transport. The automatic component protection delivered by the BeamWizard provides a direction for the development of future high-power Tandetron™ and Singletron™ accelerator systems.

REFERENCES

1. A. Formicola, et al., Nucl. Instr. and Meth. B 266, 1828-1832 (2008).
2. D.J.W. Mous, U.S. Patent No. 7,244,952 (17 July 2007).
3. G.W.W. Quax, A. Gottdang, D.J.W. Mous, Rev. Sci. Instr. 81, 02A701 (2010).
4. D.J.W. Mous, J. Visser, R.G. Haitsma, Nucl. Instr. and Meth. B 219-220, 490-493 (2004).
5. B. Hartmann, S. Kalbitzer, and Ch. Klatt, Nucl. Instr. and Meth. B 124, 490-499 (1997).
6. D.J.W. Mous, R.G. Haitsma, T. Butz, R.-H. Flagmeyer, D. Lehmannm, J. Vogt, Nucl. Instr. and Meth. B 130, 31-36 (1997).

Can-AMS: The New Accelerator Mass Spectrometry Facility At The University Of Ottawa

W. E. Kieser[a], X-L Zhao[a], I. D. Clark[b], T. Kotzer[c] and A. E. Litherland[d]

a) IsoTrace Laboratory and Dept. of Physics, University of Ottawa
b) IsoTrace Laboratory and Dept. of Earth Sciences, University of Ottawa
c) Environmental Geochemistry, Cameco Corp. Saskatoon SK
d) IsoTrace Laboratory and Dept of Physics, University of Toronto

Abstract. The Canadian Centre for Accelerator Mass Spectrometry (AMS) at the University of Ottawa will be equipped with a new, 3 MV tandem accelerator with peripheral equipment for the analysis of elements ranging from tritium to the actinides. This facility, along with a wide array of support instrumentation recently funded by the Canada Foundation for Innovation, will be located in a new science building on the downtown campus of the University of Ottawa. In addition to providing the standard AMS measurements on ^{14}C, ^{10}Be, ^{26}Al, ^{36}Cl and ^{129}I for earth, environmental, cultural and biomedical sciences, this facility will incorporate the new technologies of anion isobar separation at low energies using RFQ chemical reaction cells for ^{36}Cl and new heavy element applications, integrated sample combustion and gas ion source for biomedical and environmental ^{14}C analysis and the use of novel target matrices for expanding the range of applicable elements and simplifying sample preparation, all currently being developed at IsoTrace. This paper will outline the design goals for the new facility, present some details of the new AMS technologies, in particular the Isobar Separator for Anions and discuss the design of the AMS system resulting from these requirements.

Keywords: Accelerator mass spectrometry, fluorinating matrices, gas ion sources, isobar separation, RF quadrupole reaction cells, ^{36}Cl analysis
PACS: 07.75.+h-; 07.77.; 34.70.+e-; 36.40.Qv; 37.10.-x; 82.80.Ms

INTRODUCTION: REQUIREMENTS AND DESIGN GOALS

In the 33 years since the initial development of Accelerator Mass Spectrometry (AMS), the range of applications of this technique has expanded greatly. During the same period, the commercially available equipment has evolved along two specialized lines: one in which increasingly smaller machines are used for ^{14}C analyses [1, 2, 3] and another in which larger machines are used for the isotopes of all other elements [4, 5]. More recently, much effort has gone into developing analytical strategies for the other elements with smaller accelerators [6, 7].

With the current level of specialization in the uses of AMS equipment, the need to establish a multi-purpose AMS facility does not occur frequently. However, in the case of the facility recently funded at the University of Ottawa, a number of research groups in a variety of fields, both within the university and across Canada, have supported this venture. A simplified overview of these research areas and the isotopic analyses that they require is given in Table 1. While the range of isotopes listed in this table would normally indicate the need for an accelerator with at least 5 MV terminal voltage, recent developments in the use of RF ion guided reaction cells for very low energy ions to suppress atomic isobars have shown sufficient promise that the expense and large space required for such a big machine may no longer be necessary. Another development in the choice of ion source matrices has provided a wider range of anions, often with larger currents, as well as additional atomic isobar selection capability. In this paper, we will review these new AMS technologies and then describe the equipment to be installed at the University of Ottawa, with reference to the analytical requirements and the integration of these new technologies.

Application of Accelerators in Research and Industry
AIP Conf. Proc. 1336, 60-66 (2011); doi: 10.1063/1.3586057

SEPARATION OF ATOMIC ISOBARS WITH LOW ENERGY REACTION CELLS

The use of ion – gas reactions to select one element and remove isobaric interferences is well known in the mass spectrometry of cations (positive ions); in particular they are associated with Inductively Coupled Plasma Mass Spectrometers [8]. Early attempts to use this technique with AMS ion sources and gas cells showed that at the acceleration voltages required to operate the ion source efficiently, the ion energy resulted in excessive scattering and loss of ions [9]. This work, however, led to the realization that ion – gas reactions at lower energies could very effectively discriminate between anion isobars [10] and resulted in the invention of a system [11,12] involving ion retardation, RF quadurpole ion guides and a reaction cell. A prototype system, the Isobar Separator for Anions (ISA) had been built and tests with a variety of ion – gas combinations have been carried out [13, 14, 15]. A schematic of this device, indicating how it fits into the low energy spectrometry side of an AMS system, is shown in Figure 2 of reference [12]. Essentially, an analyzing magnet[1] is used to separate the beams of the abundant isotopes from that of the rare one. The currents of the abundant isotopes are measured in off-axis Faraday cups, or are otherwise stopped and the beam with the mass of the rare isotope is collimated and focused onto the entrance of the ISA column. In order to reduce the energy of the beam to the appropriate level for the reaction [10], the ISA column is mounted on a high voltage deck held at a potential close to that of the total acceleration voltage of the ion source and the entrance lens of the column is a decelerating electric lens. A schematic of the ISA column is shown in Figure 1 of reference 13. In the version of the prototype used in references 12, 13 and 14, one gas cell was used both to cool the ions and for the isobar destroying reactions; later versions of this system will have separate cells for these functions so that different gases can be used for each. The initial decelerating lens is followed by two short quadrupole sections, separated by apertures, to provide a region for future differential pumping before the reaction cell and to control the final stages of deceleration. The reaction cell includes both the quadrupole guide and a set of four tapered bars located between the poles of the quadrupole. The quadrupole keeps the ions scattered off the reactant gas within the effective area by returning the ions to the axis and the tapered bars provide a weak axial electric field to keep the lower energy ions from moving backward in the cell. DC voltages can be applied to all apertures and quadrupole rods to adjust the beam energy in each section so they moved properly through the system. Following the reaction cell and one interconnecting quadrupole section, the ions are re-accelerated to their original ion source energy and injected into the accelerator.

Initial tests of this system were carried out for the case of ^{36}Cl analysis, in which the isobar, ^{36}S is difficult to remove by sample preparation chemistry and typically requires >5 MV accelerators so that it can be separated by a gas-filled magnet [16]. By using nanoampere beams of the more abundant stable chlorine and sulfur isotopes to simulate ^{36}Cl and ^{36}S and NO_2 as a reaction gas, an attenuation factor of $\sim 10^7$ was observed for the S^- ions [13]. At the optimum gas pressure, an attenuation of the Cl^- ions of ~0.75 also resulted; however, this is thought to be caused by excess reaction gas being present in the reacceleration section (which was not adequately pumped), as the destruction cross section for Cl^- increases rapidly with energy (Figure 1 in reference 12). Proper differential pumping of this section is expected to reduce this effect. Tests of a suite of ^{36}Cl samples prepared by PRIME Laboratory, Purdue University, were reported [14] and showed excellent linearity over two orders of magnitude in ^{36}Cl concentration.

The selection of NO_2 as a reaction gas was made on the basis of the cross section data presented in Figure 1 of reference 10, which shows a significant attenuation of S^- and a relatively weak attenuation of Cl^-. Recently, a test was carried out to determine the effectiveness of other oxidizing gases as reaction gases for S^- elimination. Figure 1 shows the relative attenuations for NO_2, N_2O and O_2. These results, in which the N_2O and O_2 show weak attenuations (similar to that for Cl^- in NO_2) indicate that the oxidizing nature of the gas in itself is not responsible for the S^- destruction.

FLUORIDE MATRICES FOR ENHANCED ANION PRODUCTION

The need to use anions for injection into the tandem accelerator has limited the range of elements which can be efficiently analyzed by AMS. The use of molecular anions containing the analyte element is a useful technique for extending this range [17], but can require additional sample preparation chemistry steps and / or introduce complications in the analysis with additional interferences involving mass, charge and energy in the high energy system [18]. The use of fluoride molecular anions for the analysis of ^{41}Ca [19]

[1] Normally an electric analyzer would also be used, but this was omitted for the preliminary tests.

is an example of the utility of this approach, and further studies of fluoride molecular anions have revealed considerable isobar discrimination capability [20]. These attributes led to a systematic study of fluoride anions by Zhao et al [21, 22, 23] in which much of the periodic table is surveyed and in which a systematic variation in the probability of forming fluoride anions from MF_n^- is observed. (M is any element and the integer n is a function of the available valence states of the element.) For each element, there is usually a distribution in n of the molecules formed; a striking feature is observed that in many cases, the n for the highest probability of formation for one element is different from that n for its isobar. A graphical display of these probabilities is provided in figure 1 of reference 21.

The use of fluoride anions, particularly for the heavier elements where n is often 4 or 5, places additional requirements on the low energy analyzing components of the AMS system. The bending power of the inflection magnet has to be sufficient to analyze up to 100 mass units more than the mass of the element and, as a result, this places additional requirements on the resolution of this magnet, so that tails in the energy distribution of neighbouring, more abundant isotopes can be kept sufficiently low to avoid generating too high a level of continuum background in the high energy system.

THE AMS EQUIPMENT FOR THE UNIVERSITY OF OTTAWA

The new AMS system for the University of Ottawa has been designed to accommodate both the users' need to analyze the wide range of isotopes and rare elements listed in table 1 as well as the technical developments outlined in the above sections. In the plan view of this system, provided in Figure 2, the accelerator, the closest injection line and the high energy analysis system are to be provided by High Voltage Engineering (Europa) B.V. The second low energy line on the left will be transferred from the IsoTrace Laboratory at the University of Toronto when the building in Ottawa is ready (expected completion date mid 2012).

First (Right Hand) Low Energy Analysis System

Despite the requirements to analyze many different elements on this system, the bulk of the analytical work is still expected to be on ^{14}C. However, the ^{14}C work is divided into two distinct areas: the conventional dating and low level (below ambient concentration) tracing measurements for cultural and planetary sciences and the higher level (above ambient concentration) tracing and survey measurements for bio-medical, pharmaceutical and environmental security measurements. Although thorough cleaning of the ion source can remove any cross contamination between these uses, as a general operational precaution one ion source is planned for each of these applications. As increasing use of gas source and integrated combustion or compound determination systems is expected for both levels, this protocol also provides a separation for these front-end components, including their connecting capillaries. Both these sources are connected to the AMS injection line through a rotatable 54° electric analyzer. Some degree of dispersion compensation is expected between this analyzer and the inflection magnet for one of the sources; as efficiency for the low level analyses is more critical, this work will be done on this source.

The 120° inflection magnet on this line will be able to bend up to mass 339 ($^{244}PuF_5$ anions) at full ion source extraction energy (35 kV). It will accept beams with a half angle divergence of up to 30 mr. In addition to the usual fast "bouncing" capability for routine analyses such as ^{14}C, in which the ^{12}C and ^{13}C beams can be switched in to the accelerator on a millisecond time scale, it will also be capable of a user controlled slower bouncing for the analysis of elements with more than one rare isotope of interest. This line is also equipped with an off-axis Faraday cup for measurement of the proton beam during tritium analyses.

The Accelerator

The accelerator is a standard 3 MV, medium current Tandetron. However, as one of the limiting factors in the analysis of heavier elements is the presence of a continuum, in energy, of ions which are transmitted through the high energy system to the final detector. As these originate from charge changing and scattering in the residual gas from the stripper canal in the high energy accelerator tube, the stripper canal will have a second stage of differential pumping, pumped by a separate turbopump. Also, to simplify the analysis of elements with more than one rare isotope, software controls for the "bouncing" of the terminal voltage will be provided for the user. This eliminates the need to change the setting of the high energy magnet on a short time scale.

The High Energy Analysis System

The high energy analysis components follow the now more common pattern of alternating the types of analyzers; this reduces the continuum background resulting from charge changing and scattering in the high energy accelerator tube and in the following vacuum lines. The principal analyzing magnet, with ρ = 2 m, bends the beam through 90° and has a bending power of 351 MeV-AMU, sufficient to bend $^{244}Pu^{+3}$ from $^{244}PuF_4^-$ with the accelerator terminal at 3.0 MV. This is followed by a Faraday cup / slit terminal voltage stabilizer assembly with adjustable position and a 65° electric analyzer. The second magnet is a "D" shaped switching magnet which in the 20° position has similar bending power to the primary 90° magnet. This will also permit the use of other analysis positions for future development work. The final detection system includes a removable Faraday cup and a two anode gas ionization detector.

The Isobar Separator Line

For installation in Ottawa, the current arrangement of the isobar separator line at IsoTrace will be augmented by the 45° rotatable spherical analyzer and the existing SIMS type source from the main injection line at IsoTrace. This will include a wide gap 90°, ρ = 50 cm magnet used to prepare the beam for injection into the ISA. Following the ISA, a second 90°, ρ = 1.5 m magnet will be used to inject the beam into the AMS system through a 0° port in the 120° injection magnet. This magnet, with Faraday cups and a Daly detector, will permit offline work to be carried out on the ISA while other analytical work proceeds on the rest of the system. For possible future work on isobar separation with positive ions, there is provision for installing a charge changing canal following this magnet.

CONCLUSIONS

The many applications required by a broadly based user community and several new technologies developed to extend the capabilities and increase the sensitivity of AMS are implemented in the design of this system for the University of Ottawa. In particular, this system will provide a platform for continuing research into promising new AMS technologies.

ACKNOWLEDGMENTS

The authors wish to thank Henri van Oosterhout and Dirk Mous for insightful discussions during the specification of the AMS system, John Eliades for his on-going work in obtaining data with the Isobar Separator, Colin Bray of the Geology Department at Toronto for providing the N_2O gas, and Reza Javahery, Lisa Cousins and the engineering group at Ionics Mass Spectrometry Group Inc. for the design of the RF components and the electronics of the Isobar Separator. Funding from the Canada Foundation for Innovation, the Ontario Research Fund and the Natural Sciences and Engineering Research Council of Canada is gratefully acknowledged.

TABLE 1. A sample of the research areas planning to use AMS procedures and the isotopes requiring analysis for the users of the University of Ottawa AMS facility

Research Area	Isotope Analysis Required
Geographical and antrhopological effects of climate change	^{14}C, ^{210}Pb
Discovery and handling of nuclear fuel materials; biological and geological monitoring of nuclear processes and waste materials	^{3}H, ^{36}Cl, ^{129}I, ^{90}Sr, ^{137}Cs, ^{236}U, ^{all}Pu, ^{241}Am, ^{244}Cm
Landscape evolution – exposure age dating	^{10}Be, ^{14}C, ^{26}Al, ^{36}Cl
Environmental toxicology – transport metabolism of toxins	^{10}Be, ^{14}C, ^{26}Al and others
Biomedical and pharmaceutical research	^{3}H, ^{14}C, ^{26}Al, ^{41}Ca,
Development of AMS technology	?
Geographical and antrhopological effects of climate change	^{14}C, ^{210}Pb

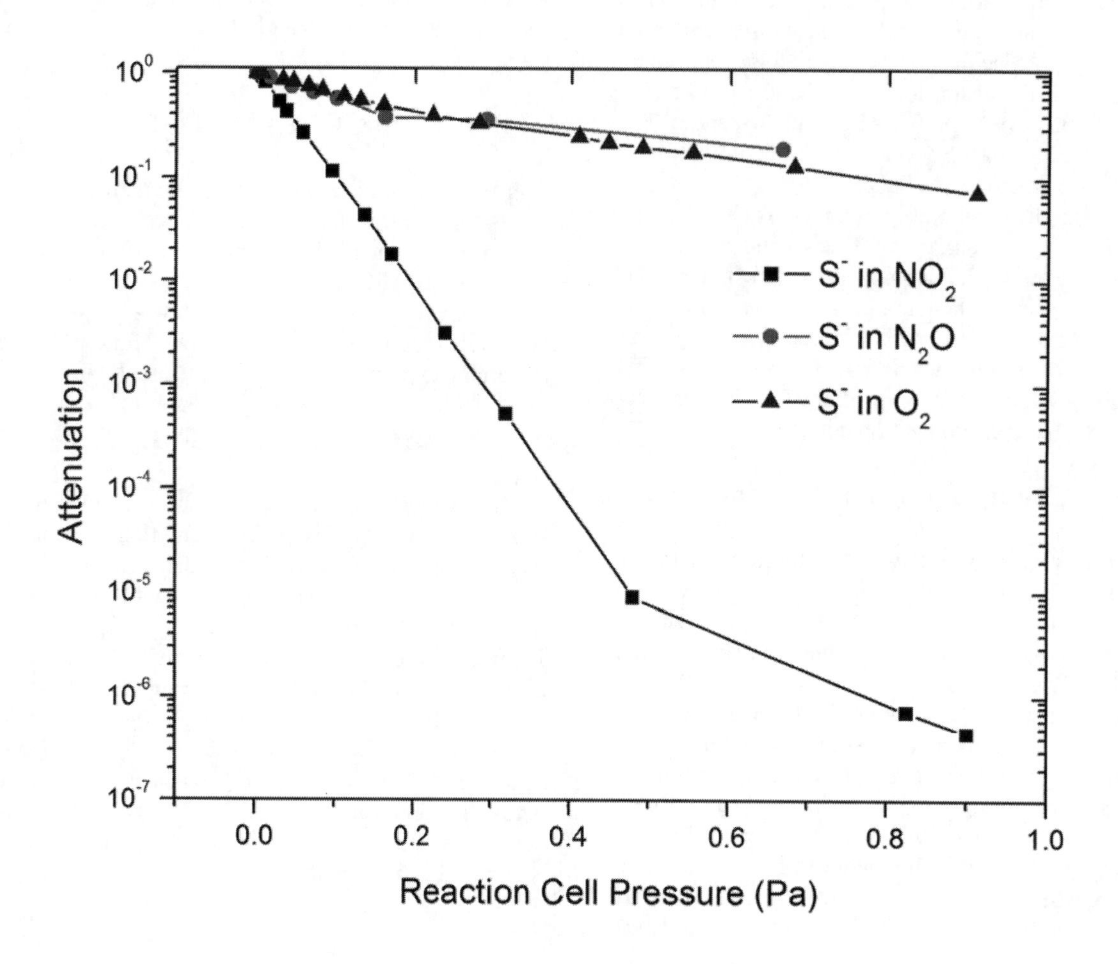

Figure 1. This Attenuation curves for S^- ions in three oxidizing gases in the reaction cell: NO_2, N_2O and O_2. The weak attenuation in the N_20 and O_2 shows a similar pattern to that of Cl^- in NO_2

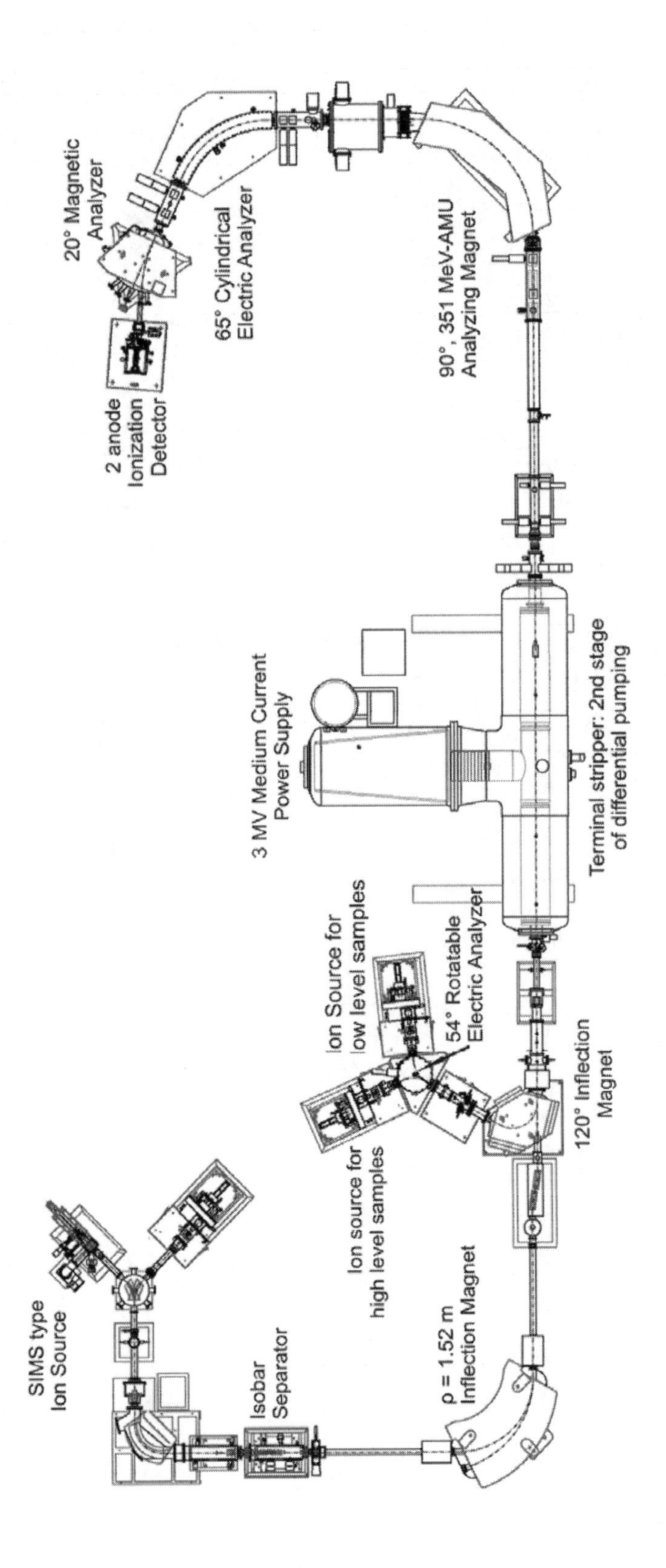

Figure 2. Plan view of the 3 MV AMS system to be installed at the University of Ottawa. The low energy spectrometry line closest to the accelerator, the accelerator and the high energy spectrometry line will be provided by High Voltage Engineering Europa, BV. The low energy line to the left is being set up for accelerator development work with the Isobar Separator and other experimental ion source techniques.

REFERENCES

1. H.A. Synal, S. Jacob and M. Suter, Nuclear Instruments and Methods B 172, 1-7 (2000)
2. H. A. Synal, M. Stocker and M. Suter, Nuclear Instruments and Methods B 223-224, 339-345 (2004)
3. G. Skog, Nuclear Instruments and Methods B 259, 1-6 (2007)
4. C. Maden, P.A.F. Anastasi, A. Dougans, S.P.H.T. Freeman, R. Kitchen, G. Klody, C. Schnabel, M. Sundquist, K. Vanner and S. Xu, Nuclear Instruments and Methods B 259, 131-139 (2007)
5. M. Klein, A. Gottdang, D.J.W. Mous, D.L. Bourlès, M. Arnold, B. Hamelin, G. Aumaître, R. Braucher, S. Merchel and F. Chauvet, Nuclear Instruments and Methods B 266, 1828-1832 (2008)
6. M.G. Klein, H.J. van Staveren, D.J.W. Mous and A. Gottdang, Nuclear Instruments and Methods B 259, 184-187 (2007)
7. H-A. Synal and L. Wacker, Nuclear Instruments and Methods B 268, 701-707 (2010)
8. S.D. Tanner, V.I. Baranov and D.M. Bandura, Spectrochimica Acta Part B 57, 1361-1452.(2002)
9. A.E. Litherland, I. Tomski and J.P. Doupé, Nuclear Instruments and Method B 204, 702-724 (2003)
10. D.B. Dunkin, F.C. Fehsenfeld and E.E. Ferguson, Chemical Physics Letters 15, 257-259 (1972)
11. A. E. Litherland, J. Doupé, W. E. Kieser, X-L Zhao, G. Javaheri, L. Cousins and I. Tomski, U.S. Patent No 7,436,498, (2008 October 21)
12. A.E. Litherland, I. Tomski, X.-L. Zhao, Lisa M. Cousins, J.P. Doupé, G. Javahery and W.E. Kieser, Nuclear Instruments and Methods B 259, 230–235 (2007)
13. J. Eliades, A.E. Litherland, W.E. Kieser, L. Cousins, S.J. Ye and X.-L. Zhao, Nuclear Instruments and Methods B 268, 839–842 (2010)
14. W.E. Kieser, J. Eliades, A.E. Litherland, X-L. Zhao, L. Cousins and S.J. Ye, Radiocarbon 52 (2010) in press
15. J. Eliades, X-L Zhao, W.E. Kieser and A.E. Litherland, Geostandards and Geochemical Research, in press
16. Henning W., Proc Trans Royal Philosophical Society of London A 323, 87–99 (1987)
17. Middleton R. 1990. A Negative Ion Cookbook. Revised version 1990. available at the BNL web site http://www.nd.edu/~nsl/nsl_docs/Making_Beam/SNICS/Negative%20Ion%20Cookbook.pdf
18. L.R. Kilius, X-L. Zhao, A.E. Litherland and W.E. Kieser, Nuclear Instruments and Methods B 123, 10-17 (1997)
19. S.P.H.T. Freeman, R.E. Serfass, J.C. King, J.R. Southon, Y. Fang, L.R. Woodhouse, G.S. Bench, J.E. McAninch Nuclear Instruments and Methods B 99, 557-561 (1995)
20. X-L. Zhao, A.E. Litherland, J.P. Doupé and W.E. Kieser, Nuclear Instruments and Methods B 223-224 199-204. (2004)
21. X.-L. Zhao, A.E. Litherland, J. Eliades, W.E. Kieser and Q. Liu, Nuclear Instruments and Methods B 268, 807–811 (2010)
22. X.-L. Zhao and A.E. Litherland, Nuclear Instruments and Methods B 268, 812-815 (2010)
23. X.-L. Zhao, J. Eliades, Q. Liu, W.E. Kieser, A.E. Litherland, S. Ye and L. Cousins, Nuclear Instruments and Methods B 268, 816–819 (2010)

Optimization Of A Mass Spectrometry Process

José Lopes [1,2,3,4], F. Corrêa Alegria [2,5], Luís Redondo [1,4], N. P. Barradas [3,4], E. Alves [3,4], Jorge Rocha [3]

(1) Instituto Superior de Engenharia de Lisboa, Rua Conselheiro Emídio Navarro, 1, 1959-007 Lisbon, Portugal.
(2) Instituto Superior Técnico, Av. Rovisco Pais, 1, 1049-001 Lisbon, Portugal
(3) Instituto Tecnológico e Nuclear, E.N. 10, Sacavém 2686-953, Portugal
(4) Centro de Física Nuclear da Universidade de Lisboa, Av. Prof. Gama Pinto 2, 1649-003 Lisboa, Portugal
(5) Instituto de Telecomunicações, Av. Rovisco Pais, 1, 1049-001 Lisbon, Portugal

Abstract. In this paper we present and discuss a system developed in order to optimize the mass spectrometry process of an ion implanter. The system uses a PC to control and display the mass spectrum. The operator interacts with the I/O board, that interfaces with the computer and the ion implanter by a LabVIEW code. Experimental results are shown and the capabilities of the system are discussed.

Keywords: Ion Implanter, LabVIEW, Mass Spectrum
PACS: 29.85.Ca; 07.05.Hd; 85.40.Ry

INTRODUCTION

Ion implantation is a materials engineering process by which ions can be introduced into any material, thereby changing their properties. One of the most relevant advantages of ion implantation is the selectivity of the implanted chemical species and in most of the cases isotope selection is easily achievable. In this sense mass spectrometry is fundamental to guarantee beam purity avoiding the contamination of the implanted samples [1-4]. The ion implanter installed at Instituto Tecnológico e Nuclear (ITN) is dedicated to materials research which makes isotope selection an important requirement. The system previously installed allowed printing the mass spectrum in a plotter, after deflecting the ion beam through a magnetic field controlled manually through the current source.

In this work we describe a new system developed to control the mass spectrometry through a PC application. The mass spectrum is given through the PC allowing an instantaneous analysis and correction of the experimental parameters in order to maintain the beam purity.

SYSTEM DESCRIPTION

The high current ion implanter installed at ITN, Sacavém, Portugal, represented in Figure 1, is the model 1090 of Danfysik [5-6]. The operating flexibility of the ion source (gas and sputter version), makes possible the production of ion beams from nearly all elements of the periodic table [7]. However, for safety reasons, radioactive or very hazardous elements are not handled.

The maximum ion beam acceleration voltage is 210 kV (160 kV post-acceleration plus 50 kV extraction). The acceleration tube is specially developed to minimize space charge effects and hence avoid expansion of the beam in an uncontrolled manner. Furthermore, by blocking back streaming electrons it minimizes X-ray emission from the acceleration section. With an electromagnetic two-dimensional beam scanning system the ion beams may be scanned for homogenous exposure over large areas. The maximum beam scanning area depends on the mass-energy product.

The global record of all the extracted ion species from the source (i.e. a mass spectrum) requires two signals, one proportional to the analyzing magnet magnetic field (x-axes) and another proportional to the beam current intensity (y-axes). The first signal is taken from a voltage proportional to the analyzing magnet current, available from the magnet power supply. The use of this signal comprises two difficulties: (i) the magnet current has hysteresis, thus it is advisable to continuously scan the current during a mass spectrum; (ii) the analyzing magnet is located on the post-acceleration potential. Therefore, an optic-

Application of Accelerators in Research and Industry
AIP Conf. Proc. 1336, 67-69 (2011); doi: 10.1063/1.3586058

fiber based circuit was developed to send the magnet current signal to the ground potential.

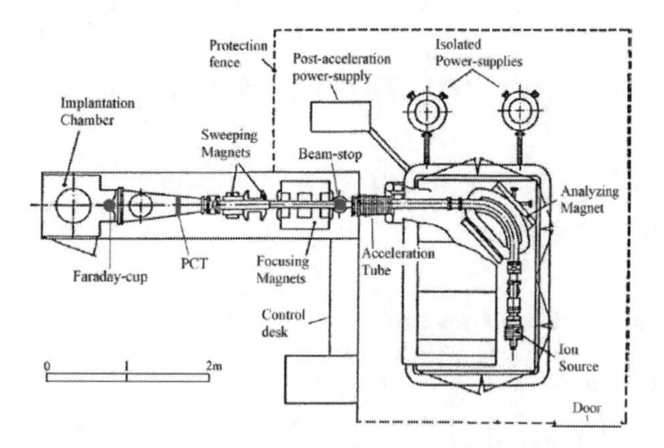

FIGURE 1. Layout of the high-current ion implanter installed at the Nuclear and Technological Institute, Sacavém, Portugal.

To carry out this control system a personal computer was equipped with multifunction input/output board from National Instruments, model USB-6251. Those boards have both analog inputs and outputs. The program that was chosen for this work was National Instruments LabVIEW, a graphical programming language specially created for instrumentation and measurement [8].

In FIGURE 2 we show the display of the application created. It has buttons to control the ion implanter and indicators (numerical and graphical). The application developed allows the control of the ion implanter through a PC and the exportation of the data obtained with the mass spectrometry.

As already mentioned, the mass spectrometry is essential step of ion implantation in order to make a precise selection of the implanted isotope. With this system, the mass spectrometry PC application displays the mass spectrum and offers the possibility to analyze online the results. The developed system still requires that the operator calculates the values for calibration. For this purpose, the user needs the values of two different elements. Usually the value obtained for Hydrogen (lowest calibration peak) is used, because it is the first deflected element by the analyzing magnet, together with the signal from the gas used in the ion source, usually Argon (highest calibration peak) which is the most intense ion current in the beam. The developed application allows for the user to save two files related to the obtained mass spectrometry: (1) a text format file with all the x-y data acquired by the DAQ; (2) a bmp format file with the mass spectrometry obtained. The txt file is extremely important because it provides the correct values to be used in the calibration calculation. The bmp file gives the user the possibility to study the mass spectrum and use it without printing, allowing a careful analysis with the help of a simple design program.

It is essential to mention that the files saved by the program will permit that the user gets a more accurate calibration reducing errors, and also reducing the time needed for this procedure.

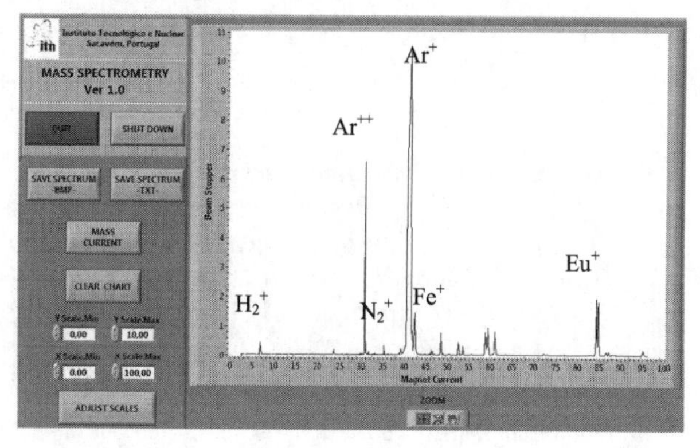

FIGURE 2. User interface of the control and monitoring application.

CONCLUSIONS

In this paper we presented the control system developed to improve the mass spectrometry of the ion implanter used by ITN for scientific research. The system used a personal computer, one multifunction input/output board, some custom made electronic interface modules and a LabVIEW application.

One of the goals was to acquire the signals of the ion implanter and give the information to the user through a PC. Currently, the user is able to: (i) observe the mass spectrometry through the PC; (ii) acquire x-y values with more accuracy; (iii) calculate the calibration more efficiently and quickly; (iv) print the mass spectrometry when desired, as it is saved automatically by the system.

The next step of this work is the full automation of the mass spectrometry. For that the system must be able to be self-calibrating. Another goal to achieve is the automatic identification of the elements in the mass spectrometry so any user can be able to know which element is present in the ion beam.

ACKNOWLEDGMENTS

JL thanks Fundação Calouste Gulbenkian for support. This work was supported by the Technological and Nuclear Institute, ITN, Sacavém, Portugal, and the Nuclear and Physics Center from the Lisbon University, CFNUL, Lisbon, Portugal.

REFERENCES

1. H.Ryssel, L. Ruge, "Ion Implantation", John Wiley & Sons Ltd., 1986.
2. E. Alves, C. Marques, R.C. da Silva, T. Monteiro, J. Soares, C. McHargue, L.C. Ononye, L.F. Allard, Nucl. Instr. and Meth. B 207, 55 (2003).
3. C. Marques, E. Alves, R.C. da Silva, M.R. Silva, A.L. Stepanov, Nucl. Instr. and Meth. B 218, 139 (2004).
4. G. Dearnaley, W. A. Grant, "Ion Implantation of Semiconductors", London, 1976.
5. P. A. Miranda, M. A. Chesta, S. A. Cancino, J. R. Morales, M. I. Dinator, J. A. Wachter and C. Tenreiro, Nucl. Instr. and Meth. B 248, 150 (2006).
6. B.R. Nielsen, P.Abrahamsen, S. Eriksen, Mater. Sci A 116, 193 (1989).
7. B. Torp, B.R. Nielsen, D.M. Rück, H. Emig, P. Spädtke, B.H. Wolf, Rev. Sci. Instr. 61, 595 (1990).
8. V. Havránek, V.Hnatowicz, A.Macková, V.Voseček, Nuclear Physics Institute ASCR v.v.i., 250 68 Rez, Czech Republic.

Compact AMS System At Yamagata University

Fuyuki Tokanai[a], Kazuhiro Kato[a], Minoru Anshita[a], Akihiro Izumi[a], Hirohisa Sakurai[a], Tsugio Saito[b]

[a] *Department of Physics, Faculty of Science, Yamagata University, Yamagata 990-8560, Japan*
[b] *Electronics equipment division, Hakuto Co., Ltd,. Tokyo 160-8910, Japan*

Abstract. A new compact accelerator mass spectrometry (AMS) system has been installed in the Kaminoyama research institute at Yamagata University. The AMS system is based on a 0.5 MV Pelletron accelerator developed by National Electrostatics Corp. The performance of the system was investigated using C series samples (C1-C8), standard samples (HOxII), and reagent graphite without any chemical treatment. The precision of ^{14}C measurements for the standard samples is typically higher than 0.3%. The ratio of ^{14}C to ^{12}C is less than 6×10^{-16} for the reagent graphite. In this paper, we present the performance of the new compact AMS system, as well as of the fully automated 20-reactor graphite lines equipped at the research institute.

Keywords: AMS, Microdosing, Radiocarbon dating
PACS: 82.80.Ms, 06.30.Gv, 81.05.U-

INTRODUCTION

Accelerator mass spectrometry (AMS) is the most sensitive technique for measuring extremely low ratios of isotopes in a very small sample of less than 1 mg. In particular, AMS measurements for ^{14}C, which has a half-life of 5730 years, have been widely utilized for radiocarbon dating in various fields such as archaeology, environmental science, geology, and space and earth sciences. In the medical and pharmaceutical areas, new interests in "microdosing study" using the AMS measurements for ^{14}C have been attracting much attention for drug development. A microdosing study is aimed at predicting the pharmacokinetics of new drugs by labeling them with ^{14}C and then administering them to human volunteers. Positron emission tomography (PET) and liquid chromatography–tandem mass spectrometry (LC–MS/MS) are also utilized in microdosing studies. They are expected to help accelerate drug development and reduce development cost. Since the microdose is defined as less than 1/100th of the dose calculated to yield a pharmacological effect of the test substance, and with a maximum dose of less than 100 μg [1], a very sensitive analytical method using AMS is required to measure the distributions of drugs labeled with ^{14}C in samples of blood, urine, feces and tissue. Following the guidance of the European Union (2003) [1] and United States (2006) [2], the Japanese Health Ministry issued a guidance on microdose clinical trials in 2008 [3]. However, there are few AMS facilities for conducting measurements of the samples obtained from microdosing studies.

To meet the requirement of ^{14}C AMS for microdosing and medical studies, Yamagata University installed an AMS (YU-AMS) system. An automated graphitization line was also installed at the same research institute for sample preparation. This AMS system is the first AMS system installed in a university in the Tohoku-Hokkaido region in Japan. The facility also provides radiocarbon dating for samples from other universities, institutes and public organizations. In this paper, we describe the status of the YU-AMS installation.

DESCRIPTION OF YU-AMS SYSTEM

Compact AMS System

The AMS system is a Pelletron accelerator (1.5SDH-1) developed by National Electrostatics Corp. Fig. 1 shows a schematic view of the AMS system. A multicathode negative ion source by cesium sputtering (MC-SNICS) source is equipped so as to convert the sample to negative carbon ions. The MC-SNICS source can hold 40 samples in one load. A positive cesium beam produced on a hot ionizer electrode is accelerated to about 6 keV and impinges on the graphitized sample in the cathode. It sputters the

Application of Accelerators in Research and Industry
AIP Conf. Proc. 1336, 70-74 (2011); doi: 10.1063/1.3586059

cathode and creates negative carbon ions from the samples. The negative carbon ions are accelerated by the cathode voltage, extractor electrode, and preacceleration tube. The acceleration energy is about 60 keV at the point of a low-energy (LE) magnet. A 45° electrostatic spherical analyzer (ESA) with a radius of 300 mm cuts the high-energy tail of the carbon beam. Since the 45° ESA is rotatable, it is possible to install a second ion source on a spare port at a position 90° from the first one. The beam is then accelerated by changing the bias voltage sequentially at the vacuum chamber of the LE magnet so that $^{12}C^{-}$, $^{13}C^{-}$, and $^{14}C^{-}$ are on the correct trajectory for transmission into the accelerator. The injection magnet at the LE part of the Pelletron accelerator has a bending angle of 90°, a radius of 305 mm, ME/Z^2 = 0.84, and $M/\Delta M$ = 29. The sequential injection times are about 300 μs for $^{12}C^{-}$, 1 ms for $^{13}C^{-}$, and the remainder of the time per cycle for $^{14}C^{-}$ microseconds per cycle, respectively, with a cycle repeat rate of 10 Hz.

The accelerator is 0.5 MV Pelletron (1.5SDH-1: NEC) with one pellet chain. It has a maximum terminal voltage of 0.5 MV and can accelerate a carbon ion with a 1+ charge state to 1 MeV. The accelerator terminal is housed in a steel tank filled with SF_6 gas at 80 psi to isolate the high-voltage terminal. Because argon gas is recirculated with two turbomolecular pumps at the center of the accelerator tube, two electrons of a C^- ion are stripped by the collision with argon gas, and a C^+ ion is produced. The C^+ ion is then accelerated and impinges the double-focusing magnet at the high-energy (HE) part of the accelerator. The HE magnet has a bending angle of 90°, a radius of 750 mm, ME/Z^2 = 15, and $M/\Delta M$ = 227. The ion currents of $^{12}C^{+}$ and $^{13}C^{+}$ are each measured with one of two Faraday cups installed at the offset beamline of $^{14}C^{+}$. After passing through a 90° ESA with a radius of 750 mm, the total energy of the $^{14}C^{+}$ ions is measured using a silicon solid-state detector (SSD). The ESA prevents such dissociated ions from reaching the final SSD detector.

The installation of the AMS was finished in the middle of March 2010. The performance tests were carried out at the end of March 2010, by measuring the C series standard samples (C1 – C8) and HOxII provided by IAEA and NIST, respectively, in accordance with the methods performed by previous works [4,5]. The graphite for each kind of standard sample was divided into three cathodes and loaded at different positions on the disk. The cathodes were each measured 8 times for 180 s. For the samples of HOxII standards, uncertainties of 0.186%, 0.133% and 0.351% were obtained for the ratios of $^{14}C/^{12}C$, $^{14}C/^{13}C$ and $^{13}C/^{12}C$. These values showed no significant differences during the measurement. Each measured value of C2 – C8 was subtracted by that of C1 for background estimation and normalized against HOxII standards. Finally, each value is combined into a single mean value. As listed in Tab. 1, the results are in very good agreement with the consensus values. Fig. 2 shows excellent relationship between the normalized against HOxII standards. Finally, each value is combined into a single mean value. As listed in Tab. 1, the results are in very good agreement with the consensus values. Fig. 2 shows excellent relationship between the measured and consensus values. The background property of only AMS was investigated by measuring the reagent graphite without any chemical treatment. The obtained ratio of $^{14}C/^{12}C$ is less than 5.4 $\times 10^{-16}$, which means that the background corresponds to 61000 years BP. Currently, we are investigating the

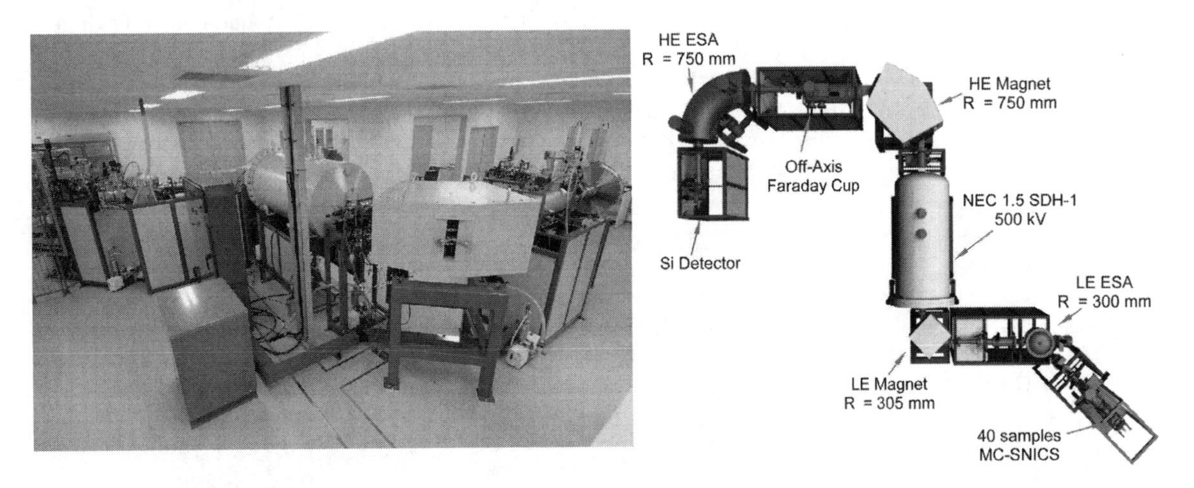

FIGURE 1. Photograph and schematic view of YU-AMS system. It consists of 40 samples of a multicathode negative ion source by cesium sputtering (MC-SNICS) source, a 45° electrostatic spherical analyzer (ESA), a 90° bending magnet with a biased chamber for sequential injection, an acceleration tube (NEC 1.5 SDH-1), a 90° analyzing magnet, off-set Faraday cups for measuring ^{12}C and ^{13}C beam current, 90° ESA and a solid-state detector for counting the ^{14}C beam.

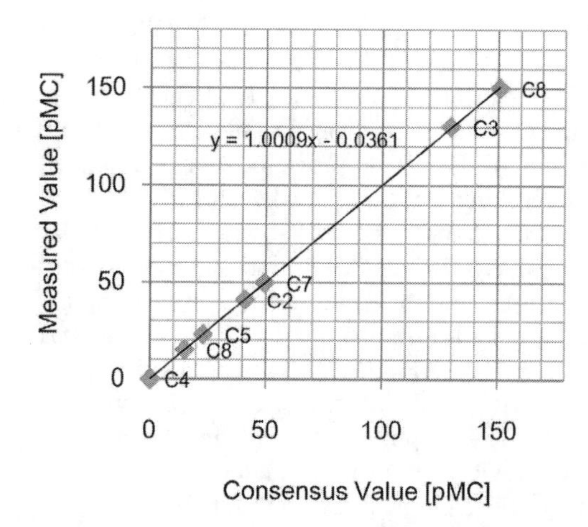

FIGURE 2. ^{14}C measurement results of YU-AMS for C-series provided by IAEA. Measured values of C2 – C8 were obtained by subtracting the C1value and normalizing against HOxII standards. The solid line is derived from the fit of a linear function to the measured values. The line is Y = 1.0009X–0.0361 (pMC), where X represents the consensus values and Y represents the measured ones.

long-term stability of the AMS system, and the dependence of the transmission and the machine background on stripper gas pressure [5,6], to realize stable and continuous measurements of both microdosing and carbon dating samples. normalized against HOxII standards. Finally, each value is combined into a single mean value. As listed in Tab. 1, the results are in very good agreement with the consensus values. Fig. 2 shows excellent relationship between the measured and consensus values. The background property of only AMS was investigated by measuring the reagent graphite without any chemical treatment. The obtained ratio of $^{14}C/^{12}C$ is less than 5.4 $\times 10^{-16}$, which means that the background corresponds to 61000 years BP. Currently, we are investigating the long-term stability of the AMS system, and the dependence of the transmission and the machine background on stripper gas pressure [5,6], to realize stable and continuous measurements of both microdosing and carbon dating samples.

Graphite Preparation System

In the ^{14}C measurement of an organic sample using the AMS system, the sample is physically and chemically treated to remove contaminants. After this treatment, it is combusted and chemically reduced to graphite, and compressed into an aluminum pellet that acts as a target for the cathode in the ion source.

We introduced an automated system for combusting the sample. Fig. 3 shows the schematic diagram of the graphite preparation system. The system consists of an elemental analyzer (EA Vario MICRO Cube, Elementar), a stable isotope ratio mass spectrometer (IRMS, IsoPrime), and an automatic cryogenic CO_2 trapping system (Koushin Rikagaku). Several institutes have already applied the EA to sample combustion and sending their gaseous products to successive lines [e.g., 7-10]. The sample is placed in a tin capsule and placed in the EA system. Although the system can analyze a total of 120 samples in continuous operation, the maximum number of treated samples is restricted to 20 because of limitation of the following cryogenic CO_2 trapping system. The sample in the capsule is combusted in a quartz reactor kept at 1150 °C, yielding CO_2, N_2 and SO_2 gases after purification and H_2O removal. By this method, the amounts of carbon, nitrogen and sulfur in the sample can be determined in addition to producing CO_2 needed to prepare the graphite. After carrying out the elemental analysis, the yield gases are transported to the IRMS and then to the cryogenic CO_2 trapping system by helium carrier gas. The rate of gas flow consisting of CO_2 after combustion plus the helium carrier gas is 200 ml/min. The flow splits in two pathways: one for the IRMS line with a flow rate of 20

TABLE 1. Percent modern carbon (pMC) data from consensus values of C-series provided by IAEA and those obtained using the YU-AMS system. Measured value of C2 – C8 was subtracted by that of C1 and normalized against HOxII standards.

Sample Type	Consensus Value IAEA standard (pMC)	Measured Value YU-AMS Data (pMC)
IAEA-C1	0 ± 0.02	0
IAEA-C2	41.14 ± 0.03	41.06 ± 0.25
IAEA-C3	129.41 ± 0.06	130.13 ± 0.24
IAEA-C4	0.20 - 0.44	0.04 ± 0.03
IAEA-C5	23.05 ± 0.02	22.98 ± 0.07
IAEA-C6	150.61 ± 0.11	150.09 ± 1.14
IAEA-C7	49.53 ± 0.12	49.77 ± 0.29
IAEA-C8	15.01 ± 0.17	15.09 ± 0.06

ml/min and the other for the cryogenic CO_2 trapping system with flow rate of 180 ml/min. The IRMS can precisely measure the $\delta^{13}C$, $\delta^{15}N$ and $\delta^{35}S$ of the sample. The isotope measurement is very useful for such tasks as the estimation of ancient diet from archaeological samples of bone collagen or burnt food remaining within cooking vessels excavated from archaeological sites.

The cryogenic CO_2 trapping system consists of 20 identical glass lines. Each glass line has a pneumatic valve and collects the CO_2 in a glass vessel, by means of a liquid nitrogen (LN_2) trap, from each corresponding sample combusted by the EA system. The CO_2 collecting system is operated fully automatically by a programmable logic controller (PLC). The trapped CO_2 is finally reduced to graphite by the following reaction at 630 °C with hydrogen and iron powder catalyst:

$$CO_2 + 2H_2 \rightarrow C + 2H_2O\ .$$

During the reaction, the produced H_2O is trapped using a Peltier cooling device. It takes about 6 h for the reaction to be completed.

The quantity of CO_2 gas collected from the EA is measured as the final pressure at the CO_2 trapping line. The collection efficiency of CO_2 is approximately 90%. The efficiency can be improved to 95% by optimizing the performance of the EA for the preparation of graphite. At present, the measurement of the background level and the evaluation of the memory effect for the system are under way. The produced graphite powder is finally pressed into an aluminum target container under 130 psi using a manual press system. The characteristics of the graphite were determined by scanning electron microscopy (SEM) and energy dispersive X-ray spectroscopy (EDS). As shown in Fig. 4, graphite resembling fluffy balls was produced. The EDS measurement of the graphite indicates that its production is accompanied with iron catalyst, but no other contaminant was found in the graphite sample. The details of our study on the production of such graphite will be published elsewhere.

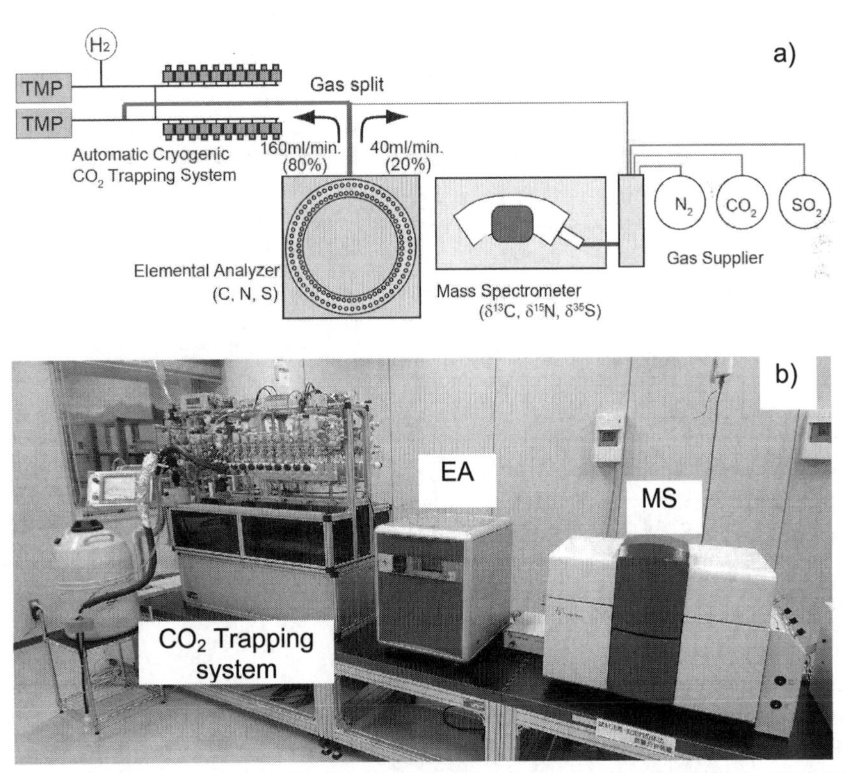

FIGURE 3. Schematic diagram and photograph of the graphite preparation system employed in the YU-AMS system. It consists of an elemental analyzer, a stable isotope ratio mass spectrometer, and an automatic cryogenic CO_2 trapping system. The line can produce 20 graphite samples per day.

FIGURE 4. Scanning electron microscopy (SEM) images of the graphite produced using the graphite preparation system employed at the YU-AMS system. Graphite resembling fluffy balls was produced.

CONCLUSION

A new compact accelerator mass spectrometry (AMS) system and fully automated 20-reactor graphite lines were installed at Yamagata University (YU) in the middle of March 2010. The performance tests of the YU-AMS system were carried out at the end of March 2010, by measuring the C series standard samples (C1 – C8) and HOxII provided by IAEA and NIST, respectively. The results show that the YU-AMS enables precise measurement of ^{14}C concentration. The long-term stability of the AMS system is currently being investigated with respect to precise and continuous measurements of both microdosing and carbon dating samples. The graphitization system has been automated employing an elemental analyzer (EA), a stable isotope ratio mass spectrometer (IRMS), and an automatic cryogenic CO_2 trapping system. The system is capable of producing 20 graphite samples per day. The collection efficiency of the CO_2 trapping system is approximately 90%. We are planning to improve the efficiency to 95% by optimizing the performance of the EA. Further evaluation of the YU-AMS system by comparing the ^{14}C ages of Japanese tree rings having dendrochronologically determined calendar ages with calibration data is under way. The final goal of this project is to apply the system in medical and pharmaceutical areas, particularly to microdosing studies to predict the pharmacokinetics of new drugs by labeling them with ^{14}C and administrating them to human volunteers.

ACKNOWLEDGMENT

We would like to thank Dr. K. Kobayashi, H. Fujine and S. Itoh for helpful suggestions about the compact-AMS system. We would like to thank Yoshimi Suzuki for helpful discussions about the SEM image of graphite.

REFERENCES

1. Position paper on non-clinical safety studies to support clinical trials for a single microdose, European Agency for the Evaluation of Medicinal Products (EMEA), 2003.
2. Guidance for Industry, Investigators, and Reviewers: Exploratory IND Studies, Center for Drug Evaluation Research (CDER), Food and Drug Administration, 2006.
3. Ministry of Health, Labour and Welfare Guidance. Tokyo, Japan: Microdose clinical studies, Ministry of Health, Labour and Welfare, Pharmaceutical and Medical Safety Bureau, 2008.
4. T. Nakamura et al., "High precision ^{14}C measurements with the HVEE Tandetron AMS system at Nagoya University", Nucl. Instrum. Methods B 223&224, 124–129 (2004).
5. K. Kobayashi et al., "The compact ^{14}C AMS facility of Paleo Labo Co., Ltd., Japan", Nucl. Instrum. Methods B 259, 31–35 (2007).
6. K. Liu et al., "A new compact AMS system at Peking University", Nucl. Instrum. Methods B 259, 23–26 (2007).
7. A. Th. Aerts-Bijma, H. A. J. Meijer, J. van der Plicht, "AMS sample handling in Groningen", Nucl. Instrum. Methods B 123, 221–225 (1997).
8. M. Yoneda et al., "AMS ^{14}C measurement and preparative techniques at NIESTERRA", Nucl. Instrum. Methods B 223&224, 116–123 (2004).
9. L. Wacker, M. Nemec, J. Bourquin, "A revolutionary graphitization system: Fully automated, compact and simple", Nucl. Instrum. Methods B 268, 931–934 (2010).
10. M Sakamoto et al., "Design and performance tests of an efficient sample preparation system for AMS-^{14}C dating", Nucl. Instrum. Methods B 268, 935–939 (2010).

Design Study Of Cyclotron Magnet With Permanent Magnet

Hyun Wook Kim[a], Jong Seo Chai[a,b]

[a] *Lab. Of Accelerator and Medical Engineering, SungKyunKwan University, Cheoncheon-dong, Jangan-gu, Suwon Gyeonggi-do 440-746, Korea*
[b] *Department of Energy Science, SungKyunKwan University, Cheoncheon-dong, Jangan-gu, Suwon Gyeonggi-do 440-746, Korea*

Abstract. Low energy cyclotrons for Positron emission tomography (PET) have been wanted for the production of radio-isotopes after 2002. In the low energy cyclotron magnet design, increase of magnetic field between the poles is needed to make a smaller size of magnet and decrease power consumption. The Permanent magnet can support this work without additional electric power consumption in the cyclotron. In this paper the study of cyclotron magnet design using permanent magnet is shown and also the comparison between normal magnet and the magnet which is designed with permanent magnet is shown. Maximum energy of proton is 8 MeV and RF frequency is 79.3 MHz. 3D CAD design was done by CATIA P3 V5 R18 [1] and the All field calculations had been performed by OPERA-3D TOSCA [2]. The self-made beam dynamics program OPTICY [3] is used for making isochronous field and other calculations.

Keywords: cyclotron; magnet, permanent magnet
PACS: 29.20.dg; 07.55.Db; 75.50.Ww

INTRODUCTION

The use of low energy cyclotron for PET is increasing in Korea. The first PET cyclotron of Korea- 'KIRAMS 13' was developed by Korea Institute of Radiological & Medical Sciences (KIRAMS) for that reason. After the development of 'KIRAMS 13', further research related with cyclotron design and cyclotron applications have been done in Korea. However the high power consumption and large size of cyclotron disturbed the wide use of cyclotron in small hospitals or clinics. The permanent magnet can be used in the cyclotron to solve these problems because this can generate additional magnetic field on mid-gap of poles without additional electric power consumption. Study of cyclotron magnet design with a permanent magnet is needed for this reason.

This paper describes considerations about the permanent magnet materials and the position of permanent magnet and design of isochronous magnet using the permanent magnet is shown. Only two strong permanent magnet materials are considered. 3D drawing of the design had been done by CATIA P3 V5 R18 [1] and the field calculation had been performed by OPERA-3D TOSCA [2]. Batch files were developed for the fast field calculation and it can generate node, mesh and field map automatically. The self-made beam dynamics program OPTICY [3] is used for isochronous field matching, calculation of tunes and phase error.

PERMANENT MAGNET CONDITIONS

Samarium-Cobalt and Neodymium-Iron-Boron permanent magnet is used in this step Fig. 1 shows the B-H curves of both magnet materials. A NdFeB magnet produces a stronger field than a SmCo magnet. Several simulations for location of permanent magnet with both magnet materials had been done and the result is shown in Table 1. The total volume of permanent magnet is a critical issue in this simulation. As the volume of permanent magnet increases, the flux density of magnetic field on mid-gap of poles also increases. However, if the permanent magnets exist in both positions of hill and return yoke; they work as resistance in magnetic circuit for each other. Therefore, only the return yoke is utilized for location of permanent magnet. The return yoke can provide large volume to permanent magnet so that can provide high flux between the poles. The material for the other part of magnet is low carbon steel ANSI 1008.

Application of Accelerators in Research and Industry
AIP Conf. Proc. 1336, 75-78 (2011); doi: 10.1063/1.3586060

TABLE 1. Permanent magnet conditions for simulation. 3D model, total volume of permanent magnet and maximum field value on mid-gap between the poles are shown. (Hill angle is 48° and hill gap and valley gap are 3 cm and 50 cm. These values are pre-design magnet parameters for simulation.)

Location	Hill		Return Yoke		Hill and Return Yoke	
3D CAD Model						
Total volume [cm^3]	13731.83		82938.05		96669.88	
	SmCo	NdFeB	SmCo	NdFeB	SmCo	NdFeB
Max. field value [T]	0.438	0.717	1.319	1.667	0.782	1.337

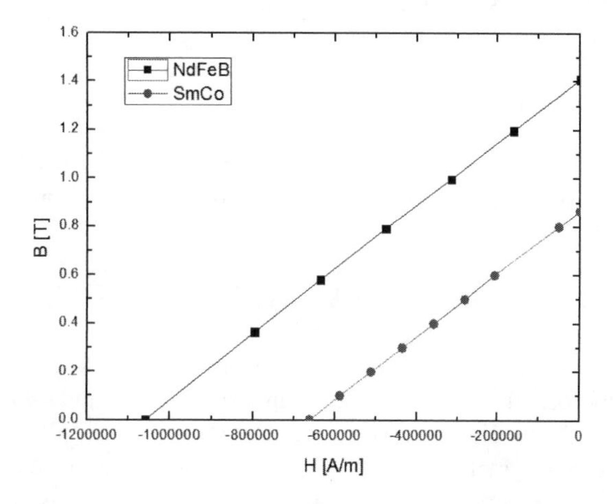

FIGURE 1. B-H Curves of NdFeB and Smco permanent magnet.

MAIN COIL CONDITIONS

To determine the proper field value for mid-gap between the two poles the current of coil is increased by 9.8 A for each step of simulation. The coil has 10 layers with 10 turns so a 980 Ampere-Turn increase is used for each step. Simulations had been done from 0 Ampere-Turn to 34300 Ampere-Turn. Height and width of one coil are 1cm and 0.98cm. The pole radius of pre-designed magnet is 35cm so the total length of coil is approximately 571m. Fig. 2 shows the maximum field value on mid-gap for each simulation and the power consumption for each coil current. The cases with the permanent magnet can make much higher flux than the cases without permanent magnet. The main coil current is set to 196 A and the electric power consumption of the coil is about 4.57 kW. With these results, the basic parameters for isochronous magnet design would be fixed in next step.

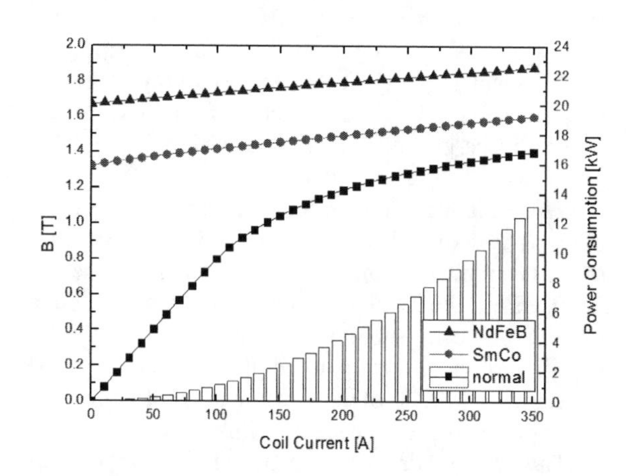

FIGURE 2. Maximum magnetic field value on mid-gap plane and power consumption of main coil. Left Y axis is for the three line graphs which indicate the magnetic field and right Y axis is for the bar graph that indicates the electric power consumption

ISOCHRONOUS MAGNET DESIGN

There are three steps for main magnet design. Basic calculations like magnet rigidity, RF frequency, gamma value and isochronous field had been done and 3D CAD drawing work is followed. Field simulation using OPERA-3D TOSCA [2] was done after that works. Shimming bars are designed to make an isochronous field. Table 2 shows the basic parameters of designed magnet. To accelerate protons to 8MeV, 0.41T-m magnet rigidity is needed at the extraction radius. The central field is set to 1.30 T so the extraction radius and RF frequency are set to 0.31 m and 79.3 MHz. The material of the magnet is low carbon steel ANSI 1008 and the return yoke is NdFeB permanent magnet.

To make an isochronous field, the average magnetic field of the mid-gap must increase gradually with the radius of the pole. The method whereby the hill-valley ratio is changed along the radius is a good way to make an isochronous field. Using the spline curve, 8 points were selected along the pole radius and those points have specified positions to make the field isochronous. To increase the magnetic flux at the outer radius part of the hill, a tapering method was used at the outer part of the hill [4]. The designed isochronous field and ideal isochronous field are shown in Fig. 3. The designed field is increased with respect to the ideal field by a 20 Gauss error boundary. Fig. 4 contains the tune diagrams for beam dynamics.

TABLE 2. Basic parameters of 8MeV H- beam cyclotron magnet.

Parameters	Values
Maximum Energy	8 MeV
Beam species	Negative hydrogen
Central field	1.30 T (79.3 MHz)
Pole radius	0.35 m
Extraction radius	0.30 m
Number of sectors	4
Hill / Valley gap	0.03 / 0.50 m
Hill angle	53°
B-field (min., max.)	0.37, 1.92 T

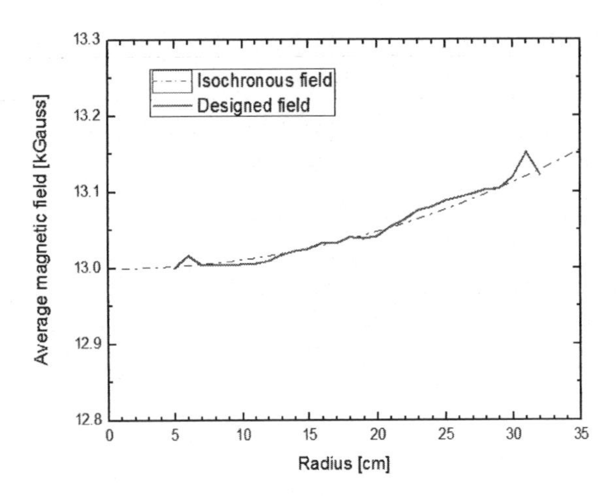

FIGURE 3. Average magnetic field graph of designed magnet with idle isochronous field.

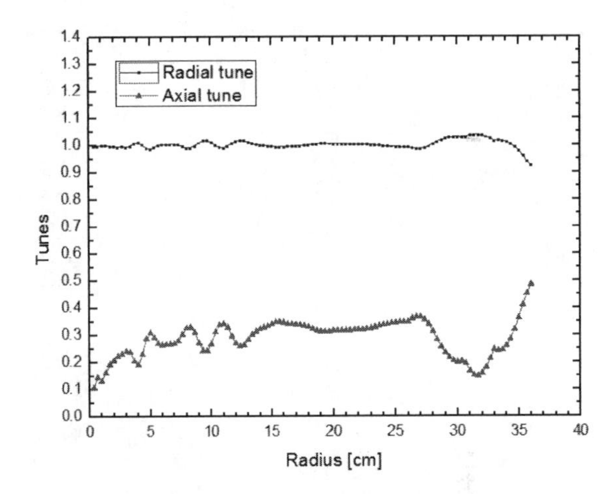

FIGURE 4. Radial and axial beam tunes.

CONCLUSION

The cyclotron magnet design using permanent magnet is finished. It can generate 1.92 T magnetic flux density on mid-gap with 19600 Ampere-Turn main coil and the power consumption of this magnet is only 4.57 kW. If that magnet does not have the permanent magnet on return yoke, it can generate only 1.33 T on the mid plane of two poles.

As the results of every step of design, it is shown that permanent magnet can be helpful for low energy cyclotron magnet design. It can reduce the power consumption of main coil and reduce the size of the magnet. In the case of very low energy cyclotron, even though there is no main coil, magnet design could be done with only permanent magnet. The next works of this paper are simulation of demagnetization of permanent magnet and heat distribution of whole magnet system.

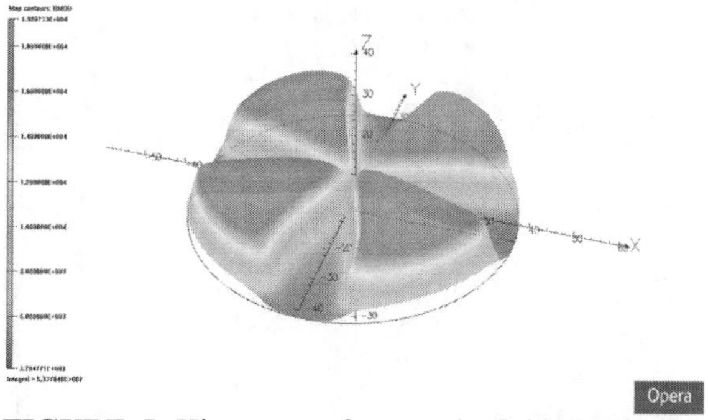

FIGURE 5. Histogram of magnetic field distribution on mid-gap.

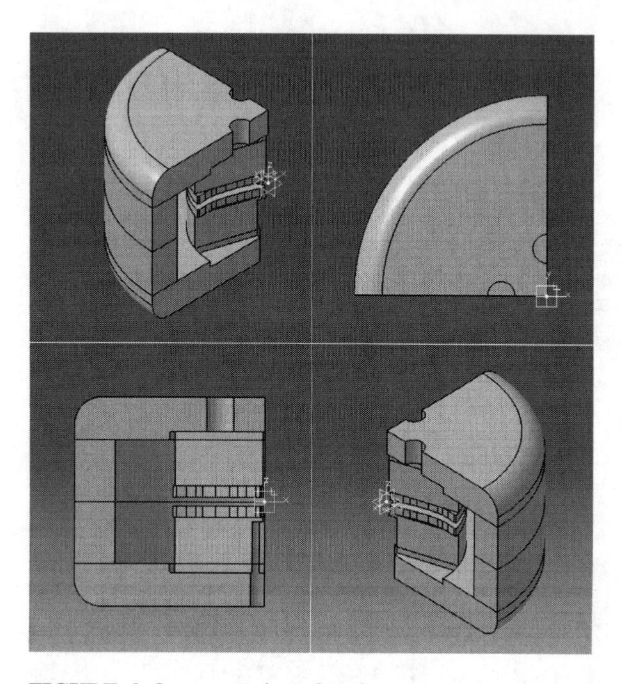

FIGURE 6. Quarter section of cyclotron magnet - 3D CAD design.

ACKNOWLEDGMENTS

This work was supported by Nuclear R&D program through the National Research Foundation of Korea funded by the Ministry of Education, Science and Technology. The project name is "Development of IT based power saving circular accelerator for positron emitter" and grant number is 0025953. Also Department of Energy Science and School of Information and Communication Engineering of SungKyunKwan University supported this project.

REFERENCES

1. Dassualt Systems, FR
2. Cobham, Vector Fields Ltd, UK.
3. S.H. Shin, Pohang Accelerator Laboratory, POSTECH, Pohang, Korea, 2007.
4. Jack T. Tanabe, "Iron Dominated Electromagnet Design", World Scientific, Chapter 9.

Versatile Low Level RF System For Linear Accelerators

James M. Potter

JP Accelerator Works, Inc., 2245 47th Street, Los Alamos, NM 87544

Abstract. The Low Level RF (LLRF) system is the source of all of the rf signals required for an rf linear accelerator. These signals are amplified to drive accelerator and buncher cavities. It can even provide the synchronizing signal for the rf power for a synchrotron. The use of Direct Digital Synthesis (DDS) techniques results in a versatile system that can provide multiple coherent signals at the same or different frequencies with adjustable amplitudes and phase relations. Pulsing the DDS allows rf switching with an essentially infinite on/off ratio. The LLRF system includes a versatile phase detector that allows phase-locking the rf frequency to a cavity at any phase angle over the full 360° range. With the use of stepper motor driven slug tuners multiple cavity resonant frequencies can be phase locked to the rf source frequency. No external phase shifters are required and there is no feedback loop phase setup required. All that is needed is to turn the frequency feedback on. The use of Digital Signal Processing (DSP) allows amplitude and phase control over the entire rf pulse. This paper describes the basic principles of a LLRF system that has been used for both proton accelerators and electron accelerators, including multiple tank accelerators, sub-harmonic and fundamental bunchers, and synchrotrons.

Keywords: RF Source, Linear Accelerator, Free Electron Laser, DDS, DSP
PACS: 29.20.Ej, 07.07.Tw, 07.57.Hm, 41.60.Cr, 84.30.Sk

INTRODUCTION

The low level rf (LLRF) circuit is the heart of the rf power system of a linear accelerator. The LLRF must provide a stable, low noise CW or pulsed rf signal which is amplified to power the accelerator cavities.

FREQUENCY FEEDBACK IN THE LOW LEVEL RF CIRCUIT

There is usually a form of frequency feedback to insure that the rf power is at the cavity resonant frequency. The cavity frequency can vary due to changing ambient conditions or variations in rf power level which may introduce thermal gradients in the structure. Either the rf source is phase locked to the cavity and the frequency changed to track the drifting cavity frequency or the frequency is fixed and the cavity is kept tuned to the drive frequency with a mechanical tuner or possibly a ferrite tuner. In some multi cavity systems the frequency is locked to one cavity and the other cavities are tuned to follow its frequency.

Modern Direct Digital Synthesis technology allows replacing many of the expensive and unstable components of a conventional rf system with an inexpensive DDS chip, a microcontroller and software algorithms.

TYPICAL NON-DDS LLRF

Figure 1 shows a typical non-DDS accelerator rf system. The signal source is a Voltage Controlled Crystal Oscillator (VCXO). The tuning range of a VCXO is typically limited to 0.1% of the center frequency. The tuning range can be more, but there is a trade-off between tuning range and oscillator stability that favors a narrow tuning range. If the frequency of the accelerator is not exactly known in advance and not easily tunable the VCXO may have to be purchased after the accelerator is built.

As shown in the figure the rf pulse is developed by inserting an analog rf switch in the output of the VCXO. The rf switch and rf amplifier performance must be balanced to get a good on-to-off amplitude ratio.

Application of Accelerators in Research and Industry
AIP Conf. Proc. 1336, 79-83 (2011); doi: 10.1063/1.3586061

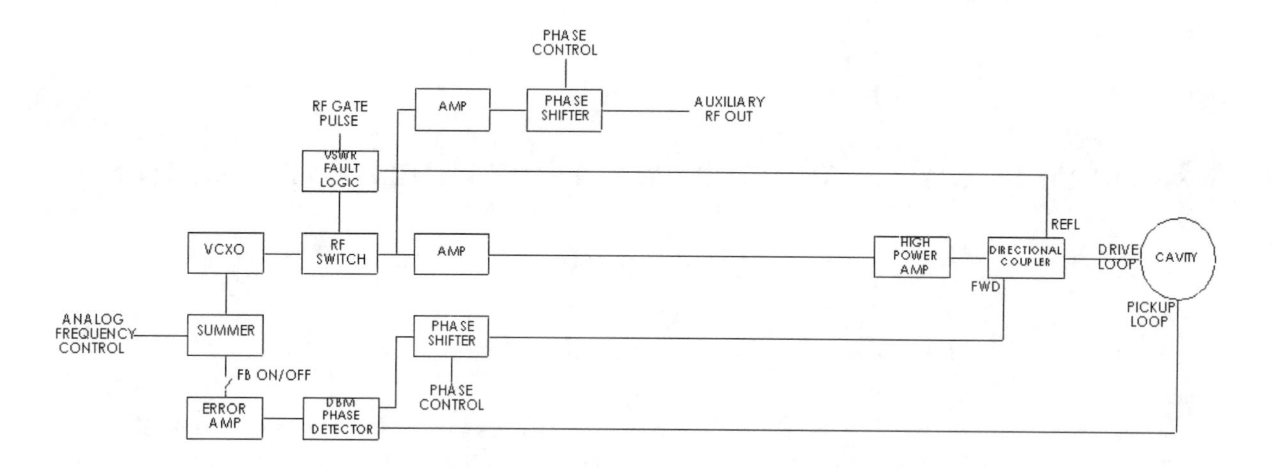

FIGURE 1. Typical non-DDS-based rf system.

The conventional frequency feedback loop uses a double balanced mixer (DBM) as a phase detector. There are other approaches that are more complex that will not be described here.1,2 To use the DBM in a frequency feedback loop there must be a phase shifter in one leg to set the DBM output to zero when the loop is open and the cavity is driven at resonance. With a 360° phase shifter range there are two nulls in the output. When the proper null is chosen and the loop is closed the frequency of the VCXO will track the resonance of the cavity.

If there is a second rf output it will have a phase shifter to set its phase relative to the first output.

BASIC DDS-BASED LLRF

Figure 2 shows a comparable rf accelerator with a DDS-based LLRF. The DDS is controlled by a microcontroller. Frequency and phase adjustments are made using digital commands over an Ethernet link. The DDS provides crystal controlled stability and low phase noise over a wide range of frequencies. The limitation is that the output frequency cannot be so near the Nyquist frequency that the desired signal and its image cannot be separated with rf filters. Details of the operation of a DDS chip can be found in the data sheet for the chip. The systems described here are based on the Analog Devices AD99593 chip which has four DDS cores and can have a reference clock up to 500 MHz. References to specific features of the DDS chip imply the AD9959.

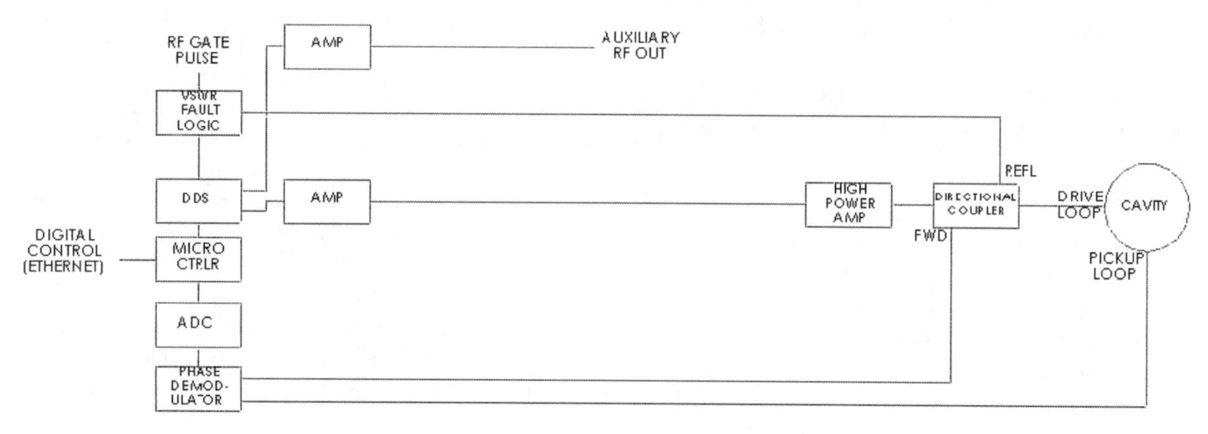

FIGURE 2. DDS-based rf system.

Phase information is obtained by comparing the forward amplitude rf signal with the cavity signal using a vector demodulator that has outputs proportional to the sine and cosine of the phase difference. These signals are read into the microcontroller on every rf pulse. When the frequency feedback loop is closed the last open loop readings become the reference. Subsequent readings are compared to the reference and an error signal is calculated in the microcontroller. When the error signal exceeds a predetermined threshold the frequency is stepped in small increments to reduce the error until it changes sign. A similar algorithm can step the motor on a slug tuner to correct the cavity resonant frequency instead of the rf source frequency.

The relative phase of a second, third or fourth channel can be set digitally through the phase offset registers in the corresponding DDS cores. No electronic or mechanical phase shifters are required.

AMPLITUDE AND PHASE CONTROL WITH A DDS-BASED LLRF

The DDS chip has inputs for modulating the amplitude but only with four bits of resolution which is not really useful for amplitude control. Additionally, the modulation ports are more valuable to gate the channels on and off, thereby replacing the analog rf switch and achieving a virtually infinite on/off ratio.

Amplitude Control

There are two routes to amplitude control depending on whether one has a linear rf system such as a klystron or a non-linear, Class C system. In the latter case the output rf amplitude is controlled by modulating the gain of the final power amplifier (FPA). A Digital Signal Processor (DSP) is used to rapidly digitize the rf envelope at frequent intervals, *e.g.* every microsecond. The weighted average of the previous N pulses is compared to the ideal amplitude.

The difference is used to make a small correction to the error output which is, in turn, used to make a change in the gain of the FPA. With the proper algorithm the amplitude converges to a stable value in a few hundred pulses. To quickly accommodate the large changes in power required from beam-on to beam-off two correction signals are maintained, one corresponding to beam-on the other to beam-off. A 200 MHz, 520 kW peak, pulsed, triode-based rf system has been built using this technique with a measured amplitude control accuracy of ±0.3%.

Phase Control

Direct phase control is available through the same four modulation bit inputs as the amplitude control but this feature is not used for the same reason it is not used for amplitude control. The phase can be controlled through the 14 bit programmable phase shift offset registers, however these must be programmed through the SPI ports. There are four SPI (clocked serial) ports that can be programmed at up to 200 Mb/s. Programming can be through one port or through all four ports giving an effective data rate of 800 Mb/s. Programming a single channel phase takes 40 bits plus some small delays between program words. At 800 Mb/s a channel can be updated in about 60 ns which is more than fast enough. It is practical to update two channels at 200 Mb/s in about 500 ns which is easily fast enough for a 1 μs sample rate.

Amplitude control can be obtained by combining two channels with identical frequencies and adjusting the phase between the channels. The DDS outputs are current sources so they can be combined simply by connecting them in parallel. The top graph in Figure 3 shows how two channels with amplitude V programmed with phases $+A$ and -A, respectively can be summed to yield an output with phase 0° and amplitude $2V\cos(A)$. By varying from 45° to 90° the amplitude can be varied from $\sqrt{2}*V$ to 0. The variation is approximately linear. If desired, the amplitude can be linearized by interpolating a simple lookup table.

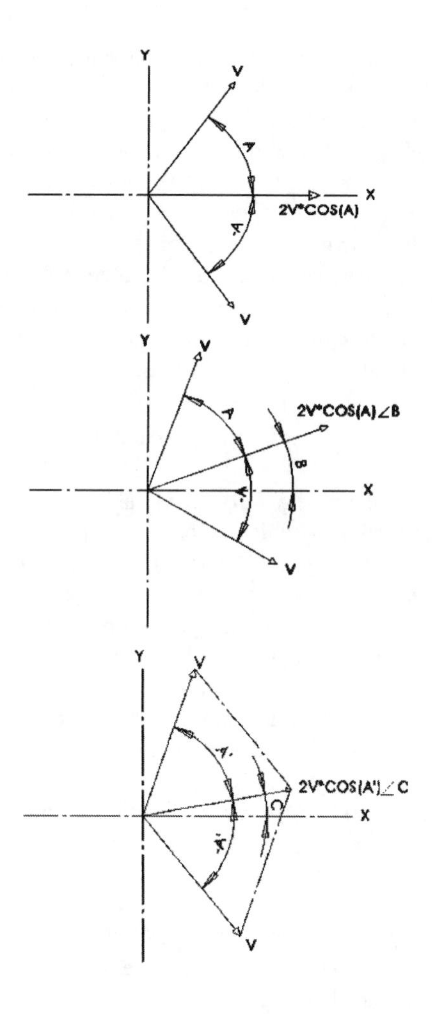

FIGURE 3. Vector sums of two DDS outputs

The phase of the output can be varied by shifting each output by the same phase, *B* as shown in the center graph of Figure 3. The output then has amplitude *2Vcos(A)* at phase angle *B*. By varying the angle of the two vectors independently the amplitude and phase can be adjusted simultaneously, see the bottom graph of Figure 3. There are more sophisticated algorithms but the one presented is adequate for control of amplitude and phase to better than 1% and 1°, respectively.

GENERATING HARMONICALLY RELATED OUTPUTS

The DDS is particularly useful for creating coherent signal with rational frequency ratios. In one case it was desired to have two outputs, one at 2856 MHz and one at exactly 2856/13 MHz, or ~219.7 Mhz. Since 2856 MHz is beyond the reach of current DDS technology the DDS output frequency was selected to be 2856/16 MHz, or 178.5 MHz. This output was then frequency doubled four times to reach 2856 MHz. Thus, two outputs are needed with a frequency ratio of exactly 13:16 are needed. For the AD9959 the output frequency is $N_0*2^{-32}*f_{ref}$, where N_0 is the frequency tuning word, f_{ref} is the reference clock frequency, typically 500 MHz, and 2^{32} is the capacity of the phase accumulator. To achieve frequencies f_1 and f_2, where f_1:f_2=13:16, the frequency tuning word ratio N_1:N_2 must have the ratio 13:16. This is achieved if N_1=13*M* and N_2=16*M*. There is a small penalty in frequency resolution but for a practical accelerator it is of no consequence. The basic frequency increment is $2^{-32} f_{ref}$ or ~0.116 Hz. With the constraint on the frequency ratio the resolution for f_1 is ~1.5 Hz and for f_2, ~1.8 Hz. After the frequency multipliers the output of channel 1 is 16 f_1 and the resolution is 24 Hz, which is much smaller than needed for room temperature 2856 MHz cavities with Qs of 15,000.

Figure 4 is a block diagram of the system described above. It has an additional feature that there is a gate signal that is synchronized with the 219.7 MHz signal. The synchronization was achieved by applying the gate to the D input of a D-flipflop and the 219.7 MHz to the trigger input. Both the leading and trailing edges of the output are synchronized to the rising edge of the 219. MHz input. The measured jitter between the 2856 MHz signal, the 219.7 MHz signal and the gate output leading edge is 3.5 ps, RMS.

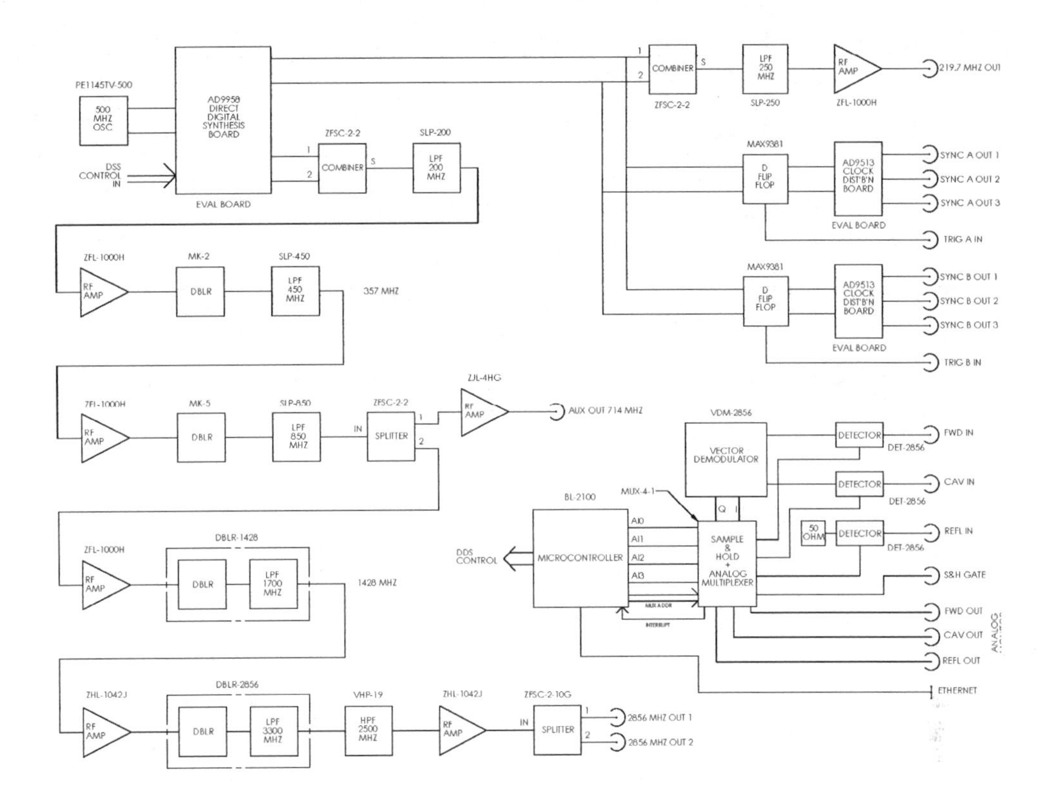

FIGURE 4. 2856 MHz/219.7 MHz DDS LLRF System.

ACKNOWLEDGMENTS

Thanks to Dave Schwellenbach and Nick Wilcox of National Security Technologies, LLC of Los Alamos, NM for assistance with the phase jitter measurements.

REFERENCES

1. A. Regan, S. Kwon, T.S. Rohlev, Y.M. Wang, M.S. Prokop, D.W. Thomson, "Design of the SNS Normal Conducting Linac RF Control System ," *20th Int'l Lin. Accel. Conf.*, Monterey, CA, USA, 2000, pp.e-proc. TUC17.
2. H. Ma, M. Champion, M. Crofford, K-U. Kasemir, M. Piller, L. Doolittle, A. Ratti, "SNS Low-Level RF Control System: Design and Performance," *Proc. 2005 Part. Accel. Conf.,* Knoxville, TN, USA, pp. 3479-3481.
3. "4-Channel 500 MSPS DDS with 10-Bit DACs AD9959", Analog Devices, Norwood, MA, USA (2005)

SECTION II – ATOMIC PHYSICS

Radiative Double Electron Capture Observed During O^{8+} + C Collisions At 38 Mev

A. Simon[a], A. Warczak[a], J. A. Tanis[b]

[a] *Institute of Physics, Jagiellonian University, Krakow, Poland*
[b] *Department of Physics, Western Michigan University, Kalamazoo, MI, USA*

Abstract. Radiative double electron capture (RDEC) is a one-step process during which two free (or quasifree) target electrons are captured into bound states of the projectile, e.g., into an empty K-shell, and the energy excess is released as a single photon. This process can be considered as a time inverse of double photoionization. As photon interacts with one electron only, RDEC is due to electron-electron interaction. Moreover, RDEC is investigated during bare ion-atom collisions and, unlike double photoionization, it does not have background from spectator electrons. Thus, RDEC may be the simplest, clean tool for exploration of the electron-electron interaction in the presence of electromagnetic fields generated during ion-atom collisions. Recently, an experiment dedicated to the RDEC process with 38 MeV O^{8+} ions colliding with a carbon foil was conducted at Western Michigan University using the tandem Van de Graaff accelerator. This experiment confirmed that the observation of such process is possible and allowed for estimation of the RDEC cross section of 3.2(1.9) b, which is significantly greater than the theoretically predicted value of 0.15 b.

Keywords: RDEC, REC, correlated double electron capture, oxygen, Van de Graaff accelerator
PACS: 32.30.Rj;32.80.Fb;34.70.+e

INTRODUCTION

Radiative double electron capture (RDEC) is a charge exchange process observed during ion-atom collisions that involves transfer of two target electrons into a bound state (usually $1s^2$) of the projectile with a simultaneous emission of a single photon [1, 2]. This process has to be compared with two-step double radiative electron capture (DREC) during which two electrons are captured independently and two single REC photons are emitted. RDEC may be described as a time inverse of double photoionization. As bare ions are used during the experiment, the observation of this process is not disrupted by interactions with other electrons as in case of double photoionization of more complex systems (high-Z atoms). Thus, RDEC can be considered as the simplest, clean tool for investigation of the electron-electron interaction in the presence of electromagnetic fields generated during ion-atom collision. The recent theoretical calculations of the RDEC cross section [3, 4, 5] suggest mid-Z ions and low energy collisions as the best systems for observation of the RDEC processes. The theoretical approach predicts an enhancement of the RDEC cross section during collisions with solid targets due to capture of electrons from the target valence band. Moreover, it points out that the capture to the excited state ($1s^12s^1$) of the projectile may be a significant contribution to the process. Based on this theory, bare oxygen ions at 38 MeV and a solid carbon target were chosen for the experiment.

EXPERIMENT

The measurements were carried out at Western Michigan University using the 6 MV tandem Van de Graaff accelerator. A schematic of the experimental setup is shown in Fig. 1. Emitted x-rays were registered by an ORTEC single crystal Si(Li) detector placed perpendicular to the beam direction. The target chamber was followed by a magnetic spectrometer with a set of surface barrier detectors that counted O^{7+} and O^{6+} ions, i.e., projectiles that captured one or two electrons. The data acquisition system allowed for registering x-rays in coincidence with particles that underwent single or double charge exchange with a time resolution of about 90 ns.

Application of Accelerators in Research and Industry
AIP Conf. Proc. 1336, 87-90 (2011); doi: 10.1063/1.3586062

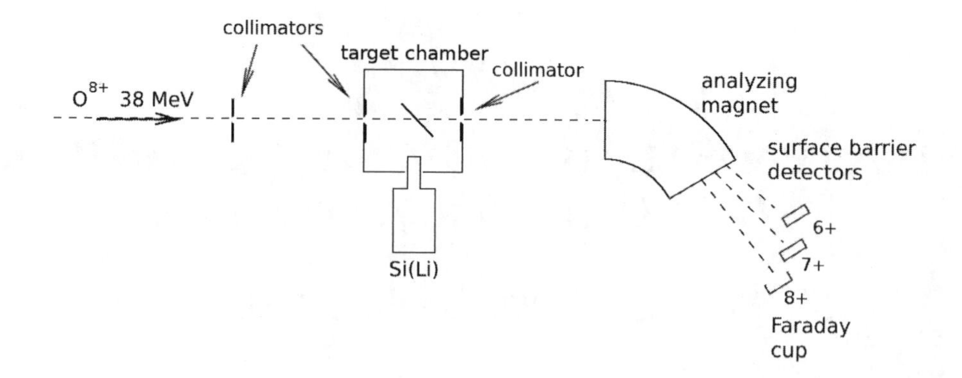

FIGURE 1. Schematic of the experimental setup.

RESULTS

The main goal of the experiment was observation of the x-rays generated by bare oxygen ions (O^{8+}) impinging on a thin (2 μg/cm2) carbon foil at an energy of 38 MeV. In Fig. 2, a singles spectrum obtained for bare oxygen ions is compared with PIXE analysis. As can be seen in

Fig. 2, no structure in the RDEC region of the PIXE spectrum was observed, nor was there evidence for REC around the photon energy of 2 keV. The spectra associated with double (O^{6+}) and single (O^{7+}) charge exchange are shown in Figs 3 (a) and (b), respectively. A structure in the RDEC region of the x-ray energy is visible in the double charge exchange channel. Additionally, a solid line representing a fit of background processes, i.e., REC (the Compton profile based on [6]), oxygen K-αline and bremsstrahlung, is shown in Fig. 3 (b). A structure in the RDEC region is clearly visible above the fitted curve in Fig. 3 (b) as well. In order to analyze the structure of the RDEC lines, the background fit was subtracted from both coincidence spectra. The resulting spectra are presented in Figs 4 (a) and (b). The structure of the RDEC line is due to various combinations of initial and final states of the electrons, as listed in Table 1. It can be noticed that the events in the RDEC range are more frequent in the spectra associated with single charge exchange rather than with double charge exchange channel (compare Figs 4 (a) and (b)). This is due to the fact that while passing through a solid foil it is very likely for the ion to undergo multiple collisions. Due to a high electron loss cross section, it is probable that after capture the projectile will be ionized. Moreover, the cross section for the removal of the L-shell electron is about an order of magnitude greater than that for the K-shell electron [7]. Thus, in case of double capture to the $1s^1 2s^1$ state, the 2s electron is most likely to be removed, while in case of double capture to $1s^2$ the final charge state of the ion will be less likely to change. This can be observed in Fig. 4, where the $1s^1 2s^1$ part of the RDEC peak is clearly visible in the single coincidence spectrum, while it is almost absent in the double charge exchange channel, compared with the structure associated with the capture to the $1s^2$ state which is still visible in the double charge exchange channel. As both the coincidence spectra contain RDEC events, the sum of them was further analyzed.

As the RDEC cross section was expected to be very low, the possibility of various pile-up effects was thoroughly investigated. It was shown in [7, 8] that the pile-up contributions are insufficient to describe the structure in the RDEC energy range of x-ray spectra. This means that the registered structure cannot be explained without incorporating the RDEC process.

The observed RDEC structure comprises at least two maxima which can be assigned to the RDEC

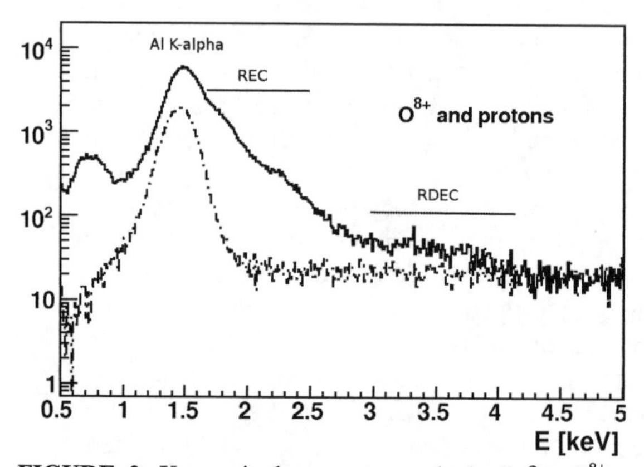

FIGURE 2. X-ray singles spectrum obtained for O^{8+} ions (solid line) compared with the results of PIXE analysis (dot-dashed line) of the carbon foil used during the experiment.

process. It is not only a result of capture to the ground ($1s^2$) and excited ($1s^1 2s^1$) states of the projectile, but can also be attributed to the capture of either K-shell or valence target electrons. The possible transitions and the associated x-ray energies are listed in Table 1. These transitions can be represented by a sum of Gaussian lines with widths equal to the FWHM of the individual line profiles. By fitting this sum to the sum of both coincidence spectra, the contribution of each of the transitions to the obtained data can be estimated. Here, as this is a preliminary analysis, a standard fitting procedure was used. The result is presented in Fig. 4 (c) and the fitting parameters are given in Table 1. Even with this rough estimate, it can be seen in Table 1 that the main contribution to the observed structure comes rather from the capture from the target K-shell, while transitions which involve at least one target valence electron are strongly suppressed. Moreover, the ratio of the areas under the fitted curves representing transitions from the target K-shell to the excited and ground projectile states (K-shell → $1s^1 2s^1$ and K-shell → $1s^2$) is equal 0.50(7). This value reflects the ratio of the RDEC cross sections for capture to the excited and ground states $\sigma_{RDEC}^{1s^1 2s^1} / \sigma_{RDEC}^{1s^2}$. The theoretical value of this ratio that can be estimated from [5] is about 0.7, which is close to the experimental result. This ratio will be further investigaded during the forthcoming continuation of this experiment, as a significant improvement of the data statistics is planned.

The numbers of counts in the coincidence spectra, $N_{RDEC}^{7+} = 326(27)$ and $N_{RDEC}^{6+} = 31(6)$ in the single and double charge exchange channel, respectively, gave the total number of RDEC counts $N_{RDEC} = N_{RDEC}^{7+} + N_{RDEC}^{6+} = 357(28)$. The ratio N_{RDEC} / N_{REC} obtained during the experiment was equal to 0.0092(6). When the angular distribution of the RDEC photons is assumed to be the same as for REC photons ($\sim \sin^2 \vartheta$, ϑ being the x-ray observation angle), the latter value gives the total RDEC cross section (σ_{RDEC}) of 5.5(3.2) b. When the contribution of the transition from the target K-shell to the projectile excited state is taken into account, the cross section for RDEC to the ground state, $\sigma_{RDEC}^{1s^2}$, and for capture to the excited state, $\sigma_{RDEC}^{1s^1 2s^1}$, are equal 3.2(1.9) b and 2.3(1.3) b, respectively. The $\sigma_{RDEC}^{1s^2}$ value is a factor of about 25 greater than the theoretical one.

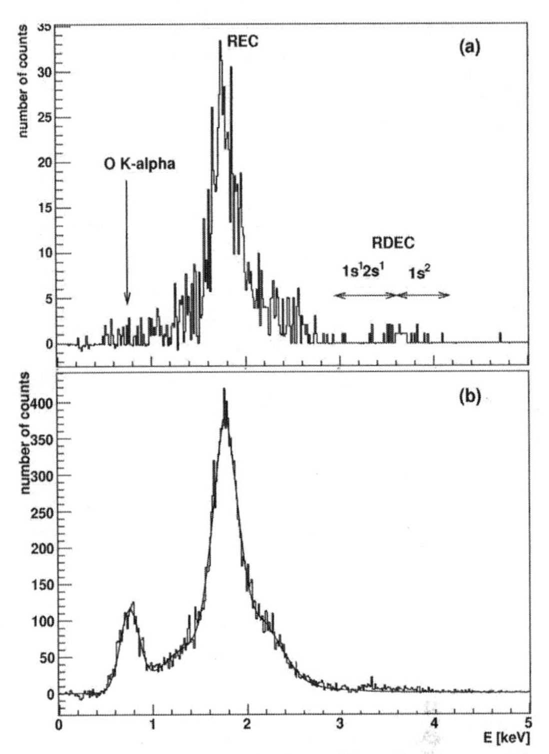

FIGURE 3. X-ray spectra registered in coincidence with double (a) and single (b) charge exchange. A structure in the RDEC region is visible in both spectra. Black line – background fit, i.e., combination of the Gaussian shape of the oxygen K-α line, the REC Compton profile and bremsstrahlung processes.

TABLE 1. Possible RDEC transitions and the corresponding x-ray energies compared with the results of the fit of the sum of the Gaussian lines to the obtained x-ray spectrum (see Fig. 4 (c)).

Transition	Peak position [keV]	Area under curve
2 · K-shell → $1s^1 2s^1$	3.04	118(11)
1 · K-shell + 1 · valence → $1s^1 2s^1$	3.31	0.0(1.9)
2 · valence → $1s^1 2s^1$ or 2 · K-shell → $1s^2$	3.62	237(11)
1 · K-shell + 1 · valence → $1s^2$	3.91	0.0(2.5)
2 · valence → $1s^2$	4.18	2.5(9.9)

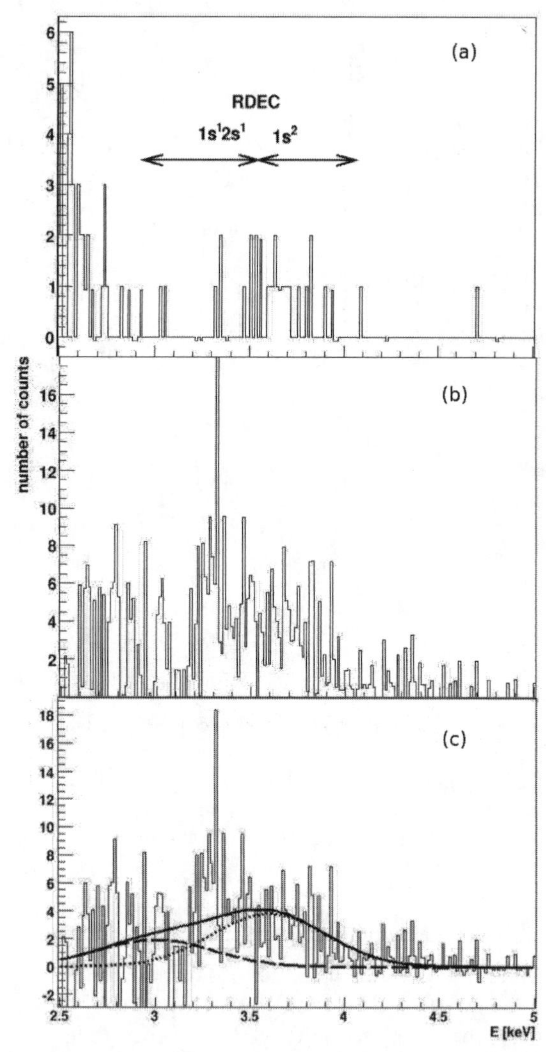

FIGURE 4. RDEC structure after subtraction of the background contribution from double (a) and single (b) coincidence channels. (c) is the sum of (a) and (b) with a fit of the sum of Gaussian lines representing various RDEC transitions (solid line). The non-zero contributions (see Table 1) are marked with dotted and dashed lines.

CONCLUSIONS

In this paper, an experimental observation of RDEC during collisions of bare oxygen ions with a carbon foil was presented. The results allowed for the experimental verification of the process and provided a test of the main theoretical predictions [3, 4, 5]. The obtained x-ray spectra revealed a complex structure of the RDEC line, which allowed for identification of capture to the projectile ground ($1s^2$) and excited ($1s^1 2s^2$) states. The ratio of the numbers of counts which could be associated with these two processes gave the ratio of the RDEC cross sections for capture to the excited and the ground states $\sigma_{RDEC}^{1s^1 2s^1} / \sigma_{RDEC}^{1s^2} = 0.50(7)$ which is close, considering the data statistics, to the theoretical value of 0.7. The capture from the target valence band, which, according to the theory, was supposed to significantly contribute to the RDEC process, was not confirmed. The obtained value of the cross section is a factor of 25 greater than the theoretical value estimated by Nefiodov [9]. This might be due to the assumption of the theoretical calculations that the captured electrons can be treated as quasifree, which is not the case in O^{8+} + C collisions.

ACKNOWLEDGMENTS

This work was supported by the Polish research Grant No. PB 1044/B/H03/2010/38.

REFERENCES

1. A. Warczak et al., *Nucl. Inst. Meth. B* **98**, 303 (1995)
2. G. Bednarz et al., *Nucl. Inst. Meth. B* **205**, 573 (2003)
3. A. I. Mikhailov, I. A. Mikhailov, A. Nefiodov, G. Plunien and G Soff, *Phys. Lett. A* **328**, 350 (2004)
4. I. Mikhailov, I. A. Mikhailov, A. N. Moskalev, A. V. Nefiodov, G. Plunien, and G. Soff, *Phys. Rev. A* **69**, 032703 (2004)
5. A. Nefiodov, A. I. Mikhailov and G. Soff, *Phys. Lett. A* **346**, 158 (2005)
6. F. Biggs, L. B. Mendelsohn and J. B. Mann, *At. Data and Nucl. Data Tables* **16**, 201 (1975)
7. A. Simon, A. Warczak, T. Elkafrawy and J. A. Tanis, *Phys. Rev. Lett.* **104**, 123001 (2010)
8. A. Simon, "Correlated Radiative Electron Capture in Ion-Atom Collisions", Ph.D. Thesis, Jagiellonian University, 2010, http://arxiv.org/abs/1008.5317
9. A. Nefiodov (private communication, 2009)

Transmission Of Fast Highly Charged Ions Through A Single Glass Macrocapillary

A. Ayyad, B. S. Dassanayake, A. Kayani, and J. A. Tanis

Department of Physics, Western Michigan University, Kalamazoo, MI 49008, USA

Abstract. The transmission of 3 MeV protons and 16 MeV O^{5+} ions through a single cylindrically-shaped insulating macrocapillary glass has been investigated in a preliminary study. The capillary had a diameter $d = 0.18$ mm and length $l = 14.4$ mm, giving an aspect ratio $l/d = 80$. The sample was mounted on a goniometer to permit precise positioning with respect to the incident beam direction. Results show that 3 MeV protons transmit through the capillary without energy loss at small tilt angles (near zero), and the 16 MeV O^{5+} ions show transmission through the sample also with little energy loss at the same small tilt angles and little change in their charge state (~ 1%). For a larger tilt angle (= 1.5°), appreciable losses in energy occurred for incident O^{5+} and the ions changed their charge state up to 7^{+}.

Keywords: fast ion guiding, insulating glass macrocapillaries, energy loss
PACS: 68.49. Sf, 34.50.Fa

INTRODUCTION

The interaction of charged particles with insulating capillaries has motivated the development of nanotechnology. Nanocapillaries having diameters of ~20 – 200 nm are embedded in a very thin foil, and the inner walls of these capillaries become the surface along which incident particles are transmitted if the incident beam charges up the inner wall of the capillaries. The electric field generated by charges on the inner walls of the capillaries can bend the trajectories of the incident beam, causing it to be guided through the capillary [1].

The interaction of slow (few keV) highly charged ions (HCI) with several kinds of insulating capillaries has been studied recently, experimentally and theoretically. These studies have shown that a considerable fraction of the ions are transmitted through the nanocapillary without close collisions with the inner wall [2]. Moreover, the transmission of slow negative ions (18 keV O^{-}) through insulating Al_2O_3 nanocapillary foils shows that these ions can also be guided through nanocapillaries [3]. Studies also include faster electrons transmitting through Al_2O_3 nanocapillaries [4], polyethylene terephthalate (PET) nanocapillaries [5], and a single glass macrocapillary [6]. Results for electrons (200 to 350 eV) through highly ordered insulating Al_2O_3 nanocapillaries with a large aspect ratio of 100 show evidence of guiding [3]. Even faster electrons (500 and 1000 eV) show guiding through PET capillaries with considerable energy loss increasing with tilt angle due to inelastic scattering with the inner surface of the capillary [5]. For the single glass macrocapillary, the transmitted electrons are found to lose energy for the same reason [4]. Guiding is also achieved with the single glass macrocapillary for slow highly-charged ions [7], but no energy loss is seen. Single glass capillaries have been applied with the intention of producing sub-micrometer sized beams that can be used for surface modification or to selectively damage the structure of biological cells [8].

In this preliminary study, fast highly charged ions of 3 MeV protons and 16 MeV O^{5+} ions transmitting through a single glass macrocapillary as a function of tilt angle were investigated. Analysis shows that 3 MeV protons transmit through the capillary without energy loss, while 16 MeV O^{5+} ions show transmission with some energy loss, especially for a larger tilt angle (= 1.5°), and changes in the charge state going from O^{5+} to O^{6+} and O^{7+}.

EXPERIMENT

As noted, transmission of 3 MeV protons and 16 MeV O^{5+} ions through a single glass macrocapillary was studied. A borosilicate, cylindrically-shaped single glass capillary prepared at the ATOMKI laboratory in Debrecen (Hungary) with a diameter $d = 0.18$ mm and length $l = 14.4$ mm for an aspect ratio l/d

Application of Accelerators in Research and Industry
AIP Conf. Proc. 1336, 91-93 (2011); doi: 10.1063/1.3586063

= 80 was used. The sample was covered with graphite on the front edge of the capillary and its holder to carry away excess charge deposited on it, and to enable reading the current on the sample. A goniometer was used to permit precise positioning with respect to the incident beam direction. The beam was obtained from the 6-MV tandem Van de Graaff accelerator at Western Michigan University. A collimated beam of about 1.5 mm diameter struck the sample for tilt angles ranging from -0.6° to 1.5°, and the transmitted ions were analyzed with a dipole magnet located about 2 m downstream to separate the emerging charge states. A movable silicon surface-barrier detector located about 1 m downstream from the magnet with a vertical aluminum slit about 6.6 mm wide was used to measure the intensities vs. magnetically analyzed position of the transmitted ions. The transmitted intensities were normalized to the current incident on the sample. A schematic of the experimental setup is shown in Fig. 1.

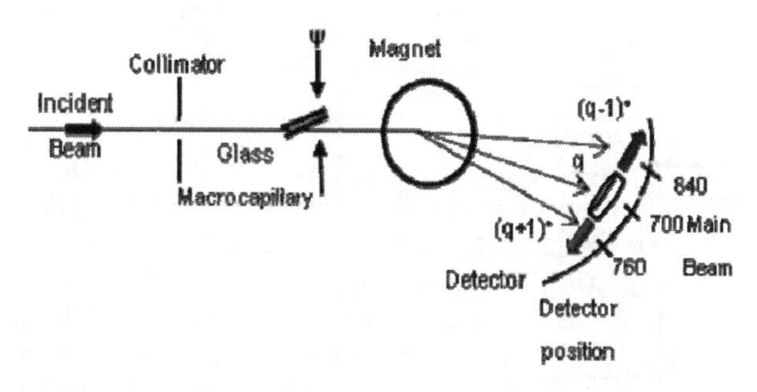

FIGURE 1. Schematic diagram of the experimental setup (top view).

RESULTS AND DISCUSSION

Transmitted intensities of 3 MeV protons and 16 MeV O^{5+} were measured for different angular positions of the surface barrier detector and for different tilt angles. The detector was set at magnet dial position of 700 (corresponding to a bend of about 5" for O^{5+}), and the magnetic field was then adjusted to a value such that the maximum beam intensity (1^+ for protons and 5^+ for oxygen ions) was incident on the detector. The detector was then moved, starting from low values of the detector position (see Fig. 1), over the full range where counts were expected. The major source of error in the measurements are the uncertainties in the number of counts recorded by the surface barrier detector, and are expected to be equal to the square root of the number of counts. For data off the main peaks there is also uncertainty in the background, which is expected to be larger relative to the peak heights. In future studies these uncertainties will be addressed more fully.

The transmitted spectra for 3 MeV protons are shown in Fig. 2. These data show that the beam traverses through the capillary with essentially no energy loss for all the angles measured between -0.6° to +1.4°, but the intensity falls off as the tilt angle goes away from zero on either side. For an angle of +0.4° (defined as zero angle) the maximum intensity is observed at dial setting 700. Tilting the sample to smaller or larger angles around zero shows that the beam intensity falls off with fewer counts (note that a log scale was used to plot the intensity).

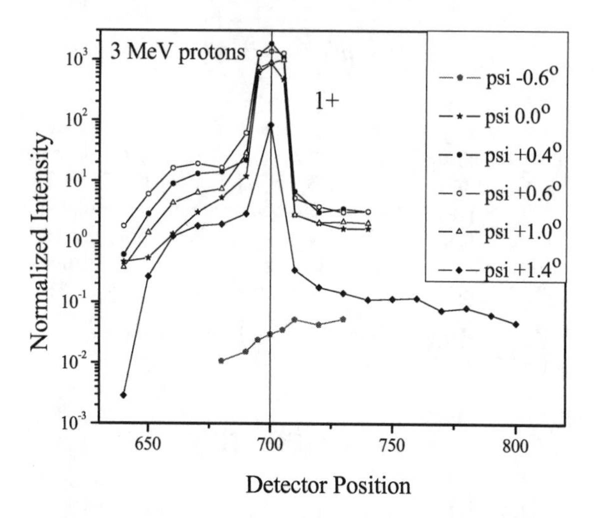

FIGURE 2. Normalized intensities vs. the detector position, varying from 640 to 800, for 3 MeV protons at several angles. The vertical line shows that the maximum intensity of charge state 1^+ is at dial position 700, as expected. Errors bars approximately equal to the statistical uncertainty in each point are expected.

In the same way the maximum beam intensity was observed for O^{5+} on the sample with the dial set at 700 for O^{5+}. For this beam, O^{6+} was found at an angular dial setting of 770 and O^{7+} at a setting of 840 for four different small tilt angles (-0.4° to +0.2°) as shown in Fig. 3(a) (note the log scale for the intensity). At a larger tilt angle of +1.5, Fig. 3(b) (note the linear scale), there was considerable energy loss for all three charge states and the transmitted beam was much lower. Furthermore, O^{6+} was the largest charge state, indicating that most of the incident O^{5+} changed charge while traversing through the macrocapillary. The reason for the shift in position is likely due to tilting the sample to +1.5° (the direction is as shown in Fig. 1), which redirects the beam trajectory through the magnet on a slightly different course. This means the beam will be shifted to the left slightly (see Fig. 1), and the analyzed beam will be shifted by 10 units downward on the dial setting in Fig. 3(b), compared to the incident beam at small tilt angles Fig. 3(a).

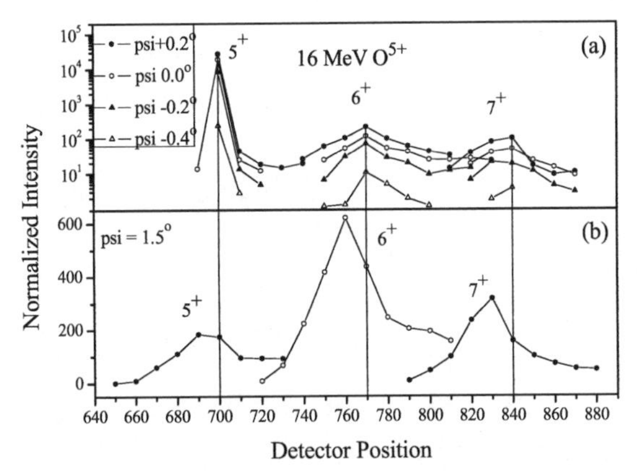

FIGURE 3. Normalized intensities vs. detector position varying from 650 to 880 for O^{q+} ions. (a) For four small tilt angles, (-0.4°, -0.2°, 0.0°, and +0.2°), the vertical lines show the maximum intensity at dial position 700 for O^{5+}, 770 for O^{6+}, and 840 for O^{7+}. (b) For a larger tilt angle (1.5°), the lines show that the maximum intensity is shifted by 10 units downward and found to be at dial settings of 690 for O^{5+}, 760 for O^{6+} and 830 for O^{7+} (see text for explanation). Errors bars approximately equal to the statistical uncertainty in each point are expected.

Figure 4 represents the maximum normalized intensity of transmitted ions found for each of the peaks shown in Figs. 2 and 3 plotted as a function of the tilt angle for both 3 MeV protons and 16 MeV O^{5+} ions. Tilting the sample around the zero position for 3 MeV protons shows that the maximum intensity appears at +0.4 (defined as zero) in Fig. 4(a) as expected. Fig. 4(b) shows the O^{5+} behavior, which is similar to the behavior seen for protons, except a log scale is used. The dashed line is drawn to represent the data where it was not taken. Since no data were taken for O^{5+} at +0.4°, it is possible that the maximum value occurs there. For the charge states O^{6+} and O^{7+} a different behavior is seen by the dashed line representing points where data was not taken. Here, the intensity of the O^{6+} and O^{7+} does not drop off with increasing tilt angle as it does for protons and O^{5+}, instead apparently continuing to rise. This can likely be attributed to nearly all the ions making close collisions with the inner walls of the capillaries for the tilt angle +1.5°, thereby causing a charge increase.

CONCLUSION

We have studied the transmission of 3 MeV protons and 16 MeV O^{5+} ions through an insulating single glass macrocapillary. It is found that the 3 MeV protons traverse through the glass macrocapillary without energy loss and without changing their charge state. On the other hand, 16 MeV O^{5+} ions traverse the capillary with energy losses that increase with tilt angle, and these ions also change their charge state as larger tilt angles are used. For the largest tilt angle studied (+1.5°), the majority of ions change their charge to the 6+ state, and each of the charge states observed (5+, 6+, and 7+) shift its position downward a little as seen by the detector, indicating redirection of the beam due to the capillary angle.

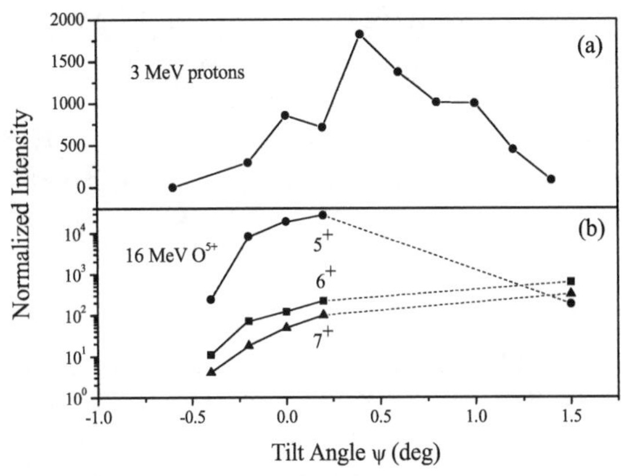

FIGURE 4. Maximum normalized intensity of the transmitted ions vs. tilt angle. (a) 3 MeV protons (dial position 700), (b) 16 MeV O^{5+} (main beam) (dial position 700), O^{6+} (dial position 770, except 760 for tilt angle of 1.5°), and O^{7+} (dial position 840, except 830 for tilt angle of 1.5°). (See Fig. 3 for the O^{q+} positions.)

ACKNOWLEDGMENTS

Preparation of the glass capillary sample by Ms. R. J. Bereczky of the ATOMKI laboratory in Debrecen (Hungary) is gratefully acknowledged. This work was supported by an award from Research Corporation.

REFERENCES

1. N. Stolterfoht, R. Hellhammer , R. Hellhammer, J. Bundesmann, D. Fink, Y. Kanai, M. Hoshino, T. Kambara, T. Ikeda , and Y. Yamazaki, Phys. Rev. A **76**, 022712 (2007); and references therein.
2. Y. Kanai, M. Hoshino, T. Kambara, T. Ikeda, R. Hellhammer, N. Stolterfoht , and Y. Yamazaki, J. Phys. Conf. Ser. **194**, 012068 (2009).
3. Guangzhi Sun *et al.*, Phys. Rev. A **79**, 052902 (2009).
4. A. R. Milosavljević, Gy. Vikor, Z. D. Pesic, P. Kolarz, D. Sevic, and B. P. Marinković, Phys. Rev. A **75**, 030901(R) (2007).
5. S. Das, B. S. Dassanayake , M. Winkworth, J. L. Baran, N. Stolterfoht, and J. A. Tanis, Phys. Rev. A **76**, 042716 (2007).
6. B. S. Dassanayake, S. Das, R. J. Bereczky, K. Tokesi, and J. A. Tanis, Phys. Rev. A **81**, 020701(R) (2010).
7. R. J. Bereczky, G. Kowarik, C. Lemaignan, F. Aumayr, and K. Tokesi, J. Phys. Conf. Ser. **194**, 132019 (2009).
8. T. Ikeda, T. M. Kojima, Y. Iwa, Y. Kanai, T. Kambara, T. Nebiki, T. Narusawa, and Y. Yamazaki, J. Phys. Conf. Ser. **58, 68** (2007).

Polarization Of The High-Energy End Of The Electron-Nucleus Bremsstrahlung In Electron-Atom Collisions

Renate Märtin[a,b], Roman Barday[c], Doris Jakubassa-Amundsen[d], Joachim Enders[c], Yuliya Poltoratska[c], Uwe Spillmann[b], Andrey Surzhykov[a,b], Günter Weber[a,b,e], Vladimir A. Yerokhin[a,b,f] and Thomas Stöhlker[a,b,e]

[a] *Physikalisches Institut, University of Heidelberg, Philosophenweg 12, Heidelberg 69120, Germany*
[b] *GSI Helmholtzzentrum für Schwerionenforschung, Planckstraße 1, Darmstadt 64291, Germany*
[c] *Institut für Kernphysik, Technische Universität Darmstadt, Schlossgartenstraße 9, Darmstadt 64289, Germany*
[d] *Mathematisches Institut, Universität München, Theresienstr. 39, München 80333, Germany*
[e] *Helmholtz-Institut Jena, Helmholtzweg 4, Jena 07743, Germany*
[f] *Saint-Petersburg State Polytechnical University, 29 Polytechnicheskaya st., St. Petersburg 195251, Russia*

Abstract. The linear polarization of bremsstrahlung radiation emitted in collisions of spin-polarized and unpolarized electrons with carbon and gold targets has been measured for an incident kinetic energy of 100 keV. We present preliminary results for the degree of linear polarization for incident unpolarized electrons as a function of bremsstrahlung photon energy.

Keywords: polarization transfer, x-ray polarimetry, bremsstrahlung, polarized electrons
PACS: 34.80.Nz, 07.60.Fs, 32.30.Rj

INTRODUCTION

Electron-nucleus bremsstrahlung occurring in electron-atom collisions is one of the basic photon-matter interaction processes and has been subject of extensive experimental and theoretical studies. Previous X-ray spectroscopy experiments mainly focused on the angular and energy differential cross section of bremsstrahlung as well as on its linear polarization properties [1, 2]. However, of special interest is the study of the bremsstrahlung polarization in regard to the polarization states of the involved collisions particles [3], namely the polarization transfer from the incident electron spin polarization to the bremsstrahlung photon polarization.

The linear polarization properties of radiation arising in collisions of electrons with ions or atoms can be conveniently described in terms of the Stokes parameters P_1 and P_2 [4], where P_1 denotes the linear polarization either in the reaction plane ($P_1 > 0$) or perpendicular to it ($P_1 < 0$). This parameter depends on the collision energy as well as on the target material. P_2 characterizes the linear polarization under 45° and 135° with respect to the reaction plane. A non-zero P_2 indicates a rotation of the polarization ellipse out of the reaction plane and it can be shown that this parameter is sensitive to the spin polarization properties of the collision partners [5]. The polarization ellipse is experimentally accessible by measuring the degree of linear polarization P_L and the orientation χ of the polarization ellipse with respect to the reaction plane. The experimental observables are related to the Stokes parameters by (compare Fig. 1):

$$P_L = \sqrt{P_1 + P_2} \quad \text{and} \quad \tan(2\chi) = \frac{P_2}{P_1}. \quad (1)$$

We report a measurement of the linear polarization of bremsstrahlung radiation performed at the polarized electron source SPIN of the TU Darmstadt [6] employing a novel position sensitive Si(Li) detector dedicated for Compton polarimetry in the photon energy range between 70 keV and a few hundred keV [7,8]. In this paper, we concentrate on the preliminary results obtained for P_L as a function of the emitted photon energy for the case of an unpolarized electron beam interacting with gold and carbon targets. Further measurements with polarized electrons are still the subject of detailed analysis and are just mentioned briefly.

Application of Accelerators in Research and Industry
AIP Conf. Proc. 1336, 94-96 (2011); doi: 10.1063/1.3586064

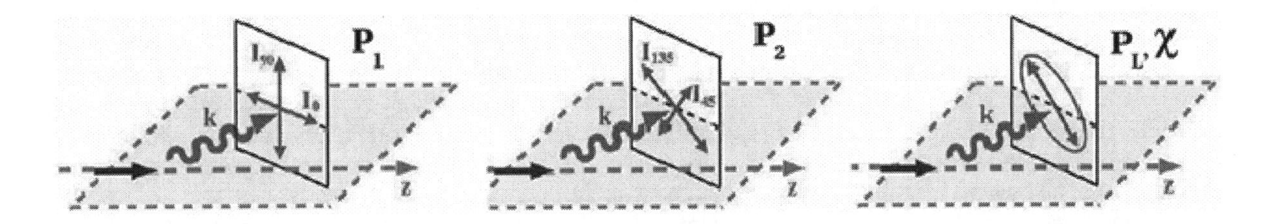

FIGURE 1. Illustration of the Stokes parameters P_1 and P_2 as defined with respect to the reaction plane given by the photon momentum k and the emitter momentum z [5]. The Stokes parameters are closely connected to the experimental observables, namely the degree of linear polarization P_L and the tilt χ of the polarization axis.

EXPERIMENTAL SETUP AND ENVIRONMENT

The experiment has been performed at the S-DALINAC Polarized INjector SPIN, which will serve as the future injector for the S-DALINAC electron accelerator at the TU Darmstadt. The SPIN electron source provides electrons with an energy of 100 keV and a high degree of electron spin polarization reaching up to 80 %. The spin orientation can be rotated by up to 360° by means of a Wien filter. In addition the source provides the opportunity to use an unpolarized electron beam too. During the experiment the degree of electron polarization averaged around 75% while the spin orientation was set to be parallel, anti-parallel and transverse with respect to the beam direction. A 47 nm thick gold foil and a 93 μm thick carbon foil served as bremsstrahlung production targets. For X-ray detection and polarization analysis a novel-type Si(Li) Compton polarimeter located at 130° or 60° with respect to the electron beam axis was used. The degree of linear polarization as well as the orientation of the polarization ellipse of the incident photons can be obtained from the two-dimensional Compton scattering distribution inside the detector, see [9] for details.

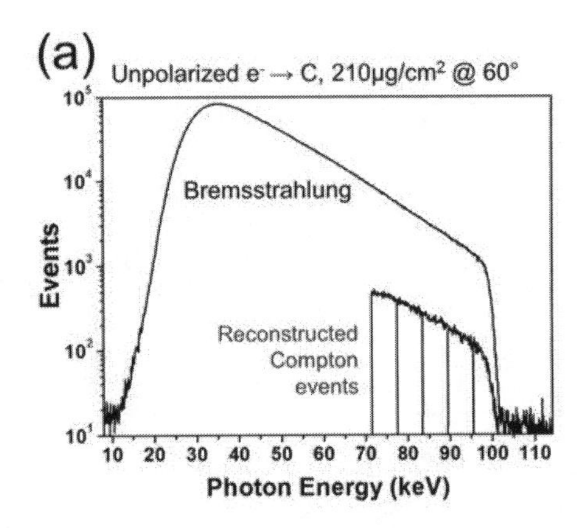

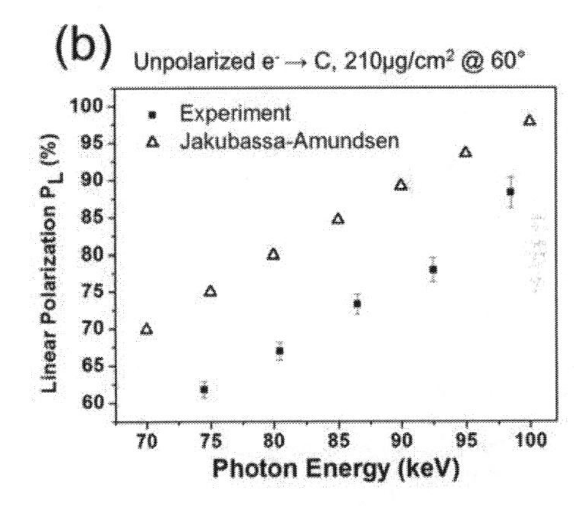

FIGURE 2. (a) Bremsstrahlung distribution of 100 keV electrons impinging on a carbon target recorded at 60°. The bottom curve shows the fraction of the reconstructed Compton events which can be used for the polarization analysis. (b) Preliminary: Degree of linear polarization for different energies of the emitted photon. Here only the statistical error is considered.

PRELIMINARY RESULTS

Fig. 2(a) shows the bremsstrahlung energy spectrum arising from an unpolarized electron beam impinging on the carbon target. The detector was located at 60° with respect to the beam direction and a 250 μm thick Fe absorber foil, mounted in front of the detector, was used to cut off the dominant low energy part of the spectrum. The underneath lying curve shows the corresponding energy spectrum as reconstructed from the Compton events which have been used for the polarization analysis. The degree of polarization has been evaluated for different photon energies reaching from the endpoint of the distribution down to 71.5 keV. Fig. 2(b) shows the preliminary degree of the linear polarization P_L versus the photon energy. Since P_2 is zero for the case of an unpolarized electron beam P_L is equal to P_1. Experimental data are compared to theoretical calculations by D. Jakubassa-Amundsen using semi-relativistic wave functions.

The bremsstrahlung distribution as recorded for the gold target under 130° is presented in Fig. 3(a). Fig .3(b) shows the corresponding linear polarization for observation angles of 60° and of 130°, respectively. The experimental data are compared to the theory by D. Jakubassa-Amundsen, as well as to complete relativistic calculations of an unscreened nucleus by A. Surzhykov and V. A. Yerokhin. Note that the data around 80 keV are affected by contributions from the K_β transition line which we assume to be unpolarized.

The discrepancy between experiment and theory can be ascribed to the not yet considered energy dependent correction for the polarimeter quality as well as to contributions by scattered and background radiation. In addition, target effects involving elastic scattering, energy loss, and straggling of electrons inside the target before bremsstrahlung emission might lead to a modification of the total linear polarization due to the superposition of different observation angles. This issue is currently subject of a detailed analysis.

For the final analysis, the comparison of the linear polarization of bremsstrahlung photons induced by unpolarized to transversely polarized electrons is of special interest. Since P_2 is non-zero for the polarized case a clear tilt of the polarization axis is visible as well as an enhancement of the linear polarization (see Eq. 1).

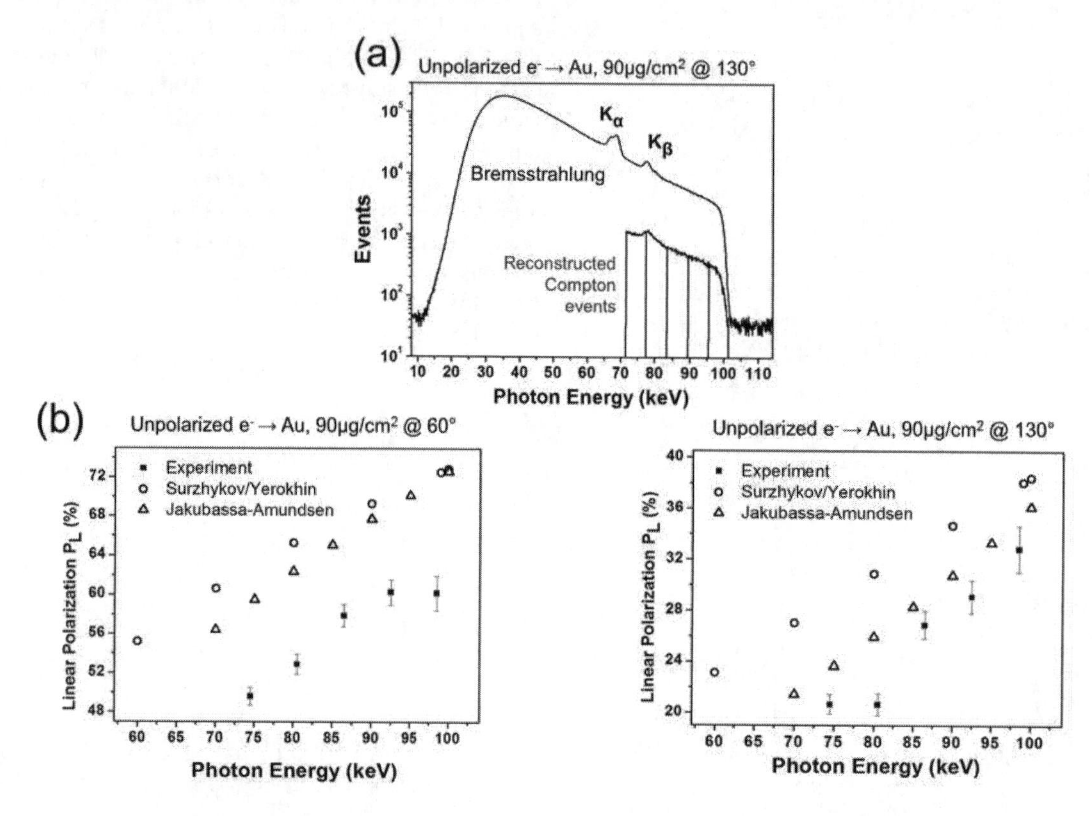

FIGURE 3. (a) Same as Fig. 2(a) but for a gold target and the polarimeter placed at 130° observation angle. (b) Preliminary results for the degree of linear polarization versus photon energy.

ACKNOWLEDGMENTS

This work was supported through SFB 634 of the Deutsche Forschungsgemeinschaft.

REFERENCES

1. H. W. Koch and J. W. Motz, Rev. Mod. Phys. 31, 920-955 (1959)
2. W. Nakel, Phys. Rep. 243, 317-353 (1994)
3. H. K. Tseng and R. H. Pratt, Phys. Rev. A 7, 1502-1515 (1973)
4. W. H. McMaster, Am. J. Phys. 22, 351 (1954)
5. A. Surzhykov et al., Phys. Rev. A 68, 022710 (2003)
6. C. Heßler, Conf. Proc. EPAC08, 1482-1484 (2008)
7. H. Bräuning et al., AIP Conference Proceedings 1099, 117 (2009)
8. G. Weber et al., JINST 5, C07010 (2010)
9. U. Spillmann et al., Rev. Sci Instr. 79, 083101 (2008)

Atomic-Orbital Close-Coupling Calculations Of Electron Capture From Hydrogen Atoms Into Highly Excited Rydberg States Of Multiply Charged Ions

Katharina Igenbergs[a], Markus Wallerberger[a], Josef Schweinzer[b] and Friedrich Aumayr[a]

[a]*Institute of Applied Physics, Vienna University of Technology, Association ÖAW-EURATOM, Wiedner Hauptstr.8-10/E134, 1040 Vienna, Austria*
[b]*Max-Planck-Institut für Plasmaphysik, EURATOM Association, Boltzmannstr.2, 85748 Garching, Germany*

Abstract. Collisions of neutral hydrogen atoms with multiply charged ions have been studied in the past using the semi-classical atomic-orbital close-coupling method. We present total and state-resolved cross sections for charge exchange as well as ionization. The advent of supercomputers and parallel programming facilities now allow treatment of collision systems that have been out of reach before, because much larger basis sets involving high quantum numbers are now feasible

Keywords: electron capture, charge transfer, cross section, ion-atom collision, atomic-orbital close-coupling.
PACS: 31.15xr, 52.20.Hv, 34.50.Fa, 52.75.-d, 52.25.Vy

MOTIVATION

Collisions between neutral hydrogen isotopes and multiply charged ions resulting in charge exchange (CX) and ionization (ION) have been the subject of a large number of studies in the past [1]. Major motivation arises from thermonuclear fusion research since these kinds of cross sections are needed for a variety of applications, in particular charge exchange recombination spectroscopy (CXRS) [2]. This diagnostic technique relies on the injection of fast neutral hydrogen (deuterium) atoms into the hot region of the plasma (e.g., by using the neutral heating beam). Through collisions with multiply charged impurity ions, the electron from the hydrogen can be captured into excited Rydberg states of the ion. The subsequent emission of photons with characteristic wavelengths and intensities, and Doppler broadening and shifts allows determination of the density and temperature of the plasma ions as well as the direction and the velocity of the plasma flow [2].

THEORETICAL METHOD – ATOMIC-ORBITAL CLOSE-COUPLING

We have applied the atomic-orbital close-coupling (AOCC) algorithm with both hydrogen-like states and pseudostates that represent the continuum [1,3]. The wavefunction of the active electron in the target and the projectile potentials is expanded in terms of basis states. Convergence of the solution of the close-coupling equations with increasing basis set size is achieved by systematically increasing the number of basis states on each center. The time-dependent Schrödinger equation is now solved in the truncated Hilbert space (i.e., the subspace spanned by the basis functions in matrix formulation) [4]. Therefore we need to calculate overlap and coupling matrix elements that can be subsumed under the term exchange matrix elements or exchange integrals.

Computational Challenges & Parallelization

The calculation of these matrix elements becomes more and more complex and time consuming when approaching high quantum numbers. Generally, we solve these exchange integrals using the Fourier-transform method [4]. We significantly optimized the respective numerical routines and are thus now able to calculate results using even larger basis sets up to much higher principal quantum numbers n, angular momentum quantum numbers l, and magnetic quantum numbers m [5].

Application of Accelerators in Research and Industry
AIP Conf. Proc. 1336, 97-100 (2011); doi: 10.1063/1.3586065

Additionally, we introduced 'non-coupling' basis states where the interaction with the other center is neglected. Such states can be used below ~10 keV/amu and above ~100 keV/amu. These non-coupling channels are centered on the hydrogen atom, where a large number of excitation and ionization channels can easily be included. In the intermediate energy region, however, these channels also interact significantly with the basis states on the ion center. Therefore, such an approximation cannot be made here. Nevertheless, this approximation is very useful when analyzing the behavior of AOCC calculations with different basis sets with respect to convergence of results. We focus our efforts mainly on impact energies that are of interest in neutral beam diagnostics of hot nuclear fusion plasmas (i.e., roughly between 10 and 100 keV/amu). In this region, all inelastic channels are in competition and therefore the AOCC approach faces great challenges especially when treating highly charged ions such as Ar^{q+} with q ≈ 15-18.

RESULTS

We present total CX and ION cross sections, as well as state-resolved CX cross sections. The total cross sections are compared to data from the literature and the *n*- and *nl*-resolved cross sections are used to calculate emission cross sections and effective emission coefficients. The latter are the quantities that are actually needed in the analysis of CXRS data for diagnosing thermonuclear plasmas. The effective emission coefficient is defined as the emission cross section averaged over a velocity distribution,

$$q^{eff}_{n_i \to n_f} = \iiint d^3\vec{v}\ \sigma^{emi}_{n_i \to n_f}\left(\left|\vec{v} - \vec{v}_{beam}\right|\right)\left|\vec{v} - \vec{v}_{beam}\right| f(|\vec{v}|)$$
$$f(\vec{v}) = f\left(|\vec{v}|\right) = f(v) = ce^{-v^2/w^2} \tag{1}$$
$$c \equiv \left(\frac{m}{2\pi k_B T}\right)^{3/2},\ w^2 \equiv \frac{2k_B T}{m}$$

where $f(v)$ is the Maxwellian speed distribution, k_B the Boltzmann constant, T the plasma temperature, and m the mass of the impurity ion. Using SI units throughout, the effective emission coefficients can be given in cm^3s^{-1}.

Be^{4+} + H(n=1,2)

Although a fairly light ion and by no means a new collisional system to be studied, the impact of fully stripped beryllium ions on neutral hydrogen in ground an excited states has recently received increased interest. With the decision to build ITER, fusion laboratories around the world have strongly focused their experimental activities on ITER-relevant research. In the course of these developments, the inner wall of the Joint European Torus (JET) is going to be changed to resemble the one planned for ITER. From 2011 onwards tungsten will be used in the divertor and beryllium for the first wall [6] of JET. Plasma wall interaction processes will lead to a high concentration of beryllium ions in the JET plasma. Therefore CXRS based on Be line emission will be an important and powerful diagnostic tool.

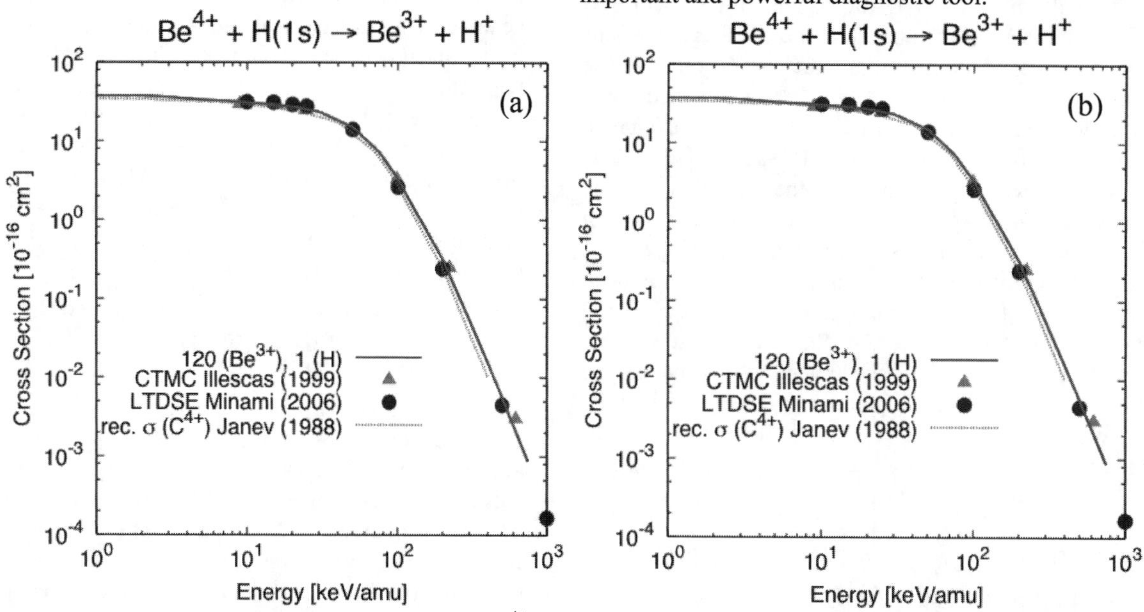

FIGURE 1. Total charge exchange cross sections for (a) Be^{4+} + H(1s) in comparison with data from [7], [8], [9] and for (b) Be^{4+} + H(n=2) in comparison with data from [10], [11], [12]. A more elaborate analysis of these results can be found in [3].

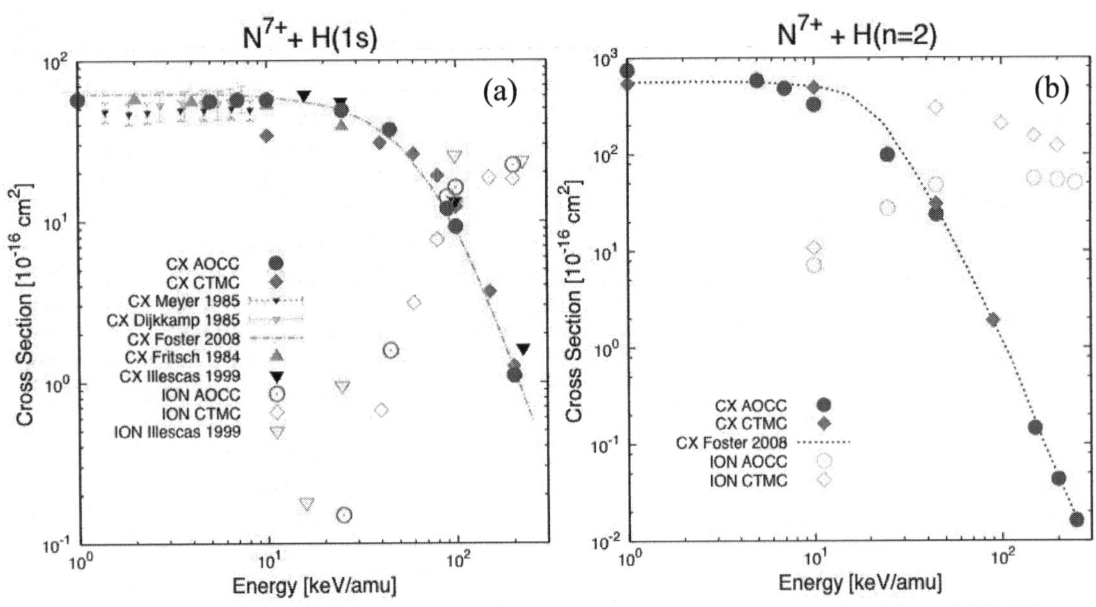

FIGURE 2. Total charge exchange cross sections and ionization cross sections calculated by AOCC (this work) and CTMC from [14] for (a) N^{7+} + H(1s) in comparison with data from [15], [8], and for (b) N^{7+} + H(n=2) in comparison with data from [16].

Fig. 1 shows total charge exchange cross sections for Be^{4+} impact on both H(1s) (Fig.1a) and excited H(n=2) (Fig.1b). The latter cross sections are statistically averaged cross sections for H(2s), H($2p_0$), H($2p_1$), and H($2p_{-1}$) [3]. Our data [3] for the H(1s) target show very good agreement with data calculated by Minami et al. in the lattice time-dependent Schrödinger equation (LTDSE) approach [7], classical trajectory Monte Carlo (CTMC) cross sections from Illescas and Riera [8], and recommended cross sections for C^{4+} ions from Janev et al. [9]. In the case of the H(n=2) target, CTMC cross section from Errea et al. [10], Landau-Zener data from Casaubon [11], and another set of CTMC calculations from Hoekstra et al. [12] are shown in Fig.1a for comparison. In this case the agreement is not as excellent as in the H(1s) case, but is nevertheless quite satisfactory.

N^{7+} + H(n=1,2)

Nitrogen is used as a seeding impurity for radiative plasma edge cooling at ASDEX Upgrade [13]. In order to limit the heat flux on the small wetted area of the divertor plates, radiating impurities can be puffed into the plasma chamber. When they get ionized and excited by electron (and ion) impact in the plasma edge, they distribute a fraction of this heat flux to larger areas by radiation, thus reducing the power load on divertor components and cooling the outermost plasma regions. At ASDEX Upgrade, the use of nitrogen seeding is accompanied by improved energy confinement due to higher plasma temperatures that more than compensates the negative effect of plasma dilution by nitrogen. Therefore, it recently became an interesting species for CXRS. In [14], we presented AOCC and CTMC charge exchange as well as ionization cross sections. For the latter, we had to include 34 unbound pseudostates on the hydrogen center. Fig.2 shows that these AOCC ION cross sections agree very well with CTMC [14]. The total CX cross sections are compared to earlier AOCC calculations including pseudostates (AO+) [15], CTMC data [8], as well as scaled cross sections calculated using the ADAS315 routine [16] included in the Atomic Data and Analysis Structure (ADAS) software and data package.

TOWARD HEAVY ION IMPACT CROSS SECTIONS

The major challenges in trying to expand the AOCC method beyond its present boundaries are twofold. On the one hand, computational times rise rapidly for larger basis sets involving high quantum numbers *n*, *l*, *m* required to treat capture to more highly charged and heavier ions and on the other, numerical instabilities when treating high *n* basis functions needs to be carefully addressed. For example, Ne^{10+} has a major capture channel for *n*=6 and the visible lines in the Ne^{9+} spectrum originate from even higher *n*-shells (*n*=10, 11, or 12). For fully stripped Ar^{18+} ions the major capture channel rises to *n*=10 and the visible lines come from transitions starting at *n*=14, 15, 16, and 17.

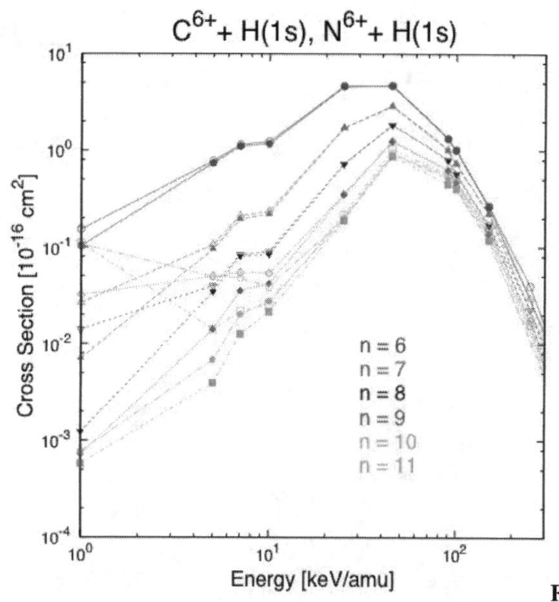

F

FIGURE 3. *n*-resolved cross sections of C^{6+} + H(1s) (closed symbols) in comparison to N^{6+} + H(1s) (open symbols).

For these noble gas ions, the fractional abundances of not fully-stripped ions at plasma temperatures is non-negligible. In single-electron transfer collisions, the AOCC method treats all passive electrons as perturbations to the potential of the active electron on the respective collision center. This means that elaborate potentials need to be found resulting in much more complex structures of the matrix elements. It is, nevertheless, a reasonable assumption that the influence of closely bound core electrons on the active electron that captures into very high *n*-shells is negligible. We therefore conducted a study of C^{6+} + H(1s) in comparison to N^{6+} + H(1s). The main capture channel of the active electron is $n = 4$, which is of course much lower than in the case of highly charged Ne or Ar ions. One would therefore expect differences in the cross sections as a result of the perturbed potential to be more pronounced. Fig.3 shows *n*-resolved CX cross sections of both collisional systems in comparison. At low energies, well below 10 keV/amu, there is a certain difference, but when looking at fusion relevant energy regions it becomes obvious that the cross sections do not deviate from each other anymore [14].

ACKNOWLEDGEMENTS

Katharina Igenbergs is a fellow of the Friedrich Schiedel Foundation for Energy Technology. Furthermore, her attendance of CAARI 2010 was also subsidized by a Mariette-Blau mobility scholarship of Vienna University of Technology. The results presented in this paper have been achieved in part using the Vienna Scientific Cluster (VSC).

This work, supported by the European Communities under the Contract of Association between EURATOM and the Austrian Academy of Sciences, was carried out within the framework of the European Fusion Development Agreement. The views and opinions expressed herein do not necessarily reflect those of the European Commission.

REFERENCES

1. W. Fritsch and C.D. Lin, *Phys. Reports* **202**, 1-97 (1991)
2. R. Isler, *Plasma Phys. Contr. Fusion* **36**, 171-208 (1994)
3. K. Igenbergs, J. Schweinzer, F. Aumayr, *J. Phys. B* **42**, 235206 (2009)
4. R. Shakeshaft, *J. Phys. B* **8**, L134-L136 (1975)
5. M. Wallerberger, K. Igenbergs, J. Schweinzer, F. Aumayr, *Computer Phys. Comm.* in print (2010)
6. J. Paméla, G.F. Matthews, V. Philipps, R. Kamendje, *J. Nucl. Mat.* **363-365**, 1-11 (2007)
7. T. Minami, M. Pindzola, D. Schultz, *J. Phys. B* **39**, 2877-2891 (2006)
8. C. Illescas, A. Riera, *Phys. Rev. A* **60**, 4546-4560 (1999)
9. R. Janev, R. Phaneuf, H. Hunter, *At. Data Nucl. Data Tables* **40**, 249-281 (1988)
10. L.F. Errea, C. Harel, H. Jouin, L. Mendez, B. Pons, Riera A., *J. Phys. B* **31**, 3527-3545
11. J.I. Casaubon, *Phys. Rev. A* **48**, 3680-3683 (1993)
12. R. Hoekstra, H. Anderson, F. Blieck, M. von Hellermann, C. Maggi, R.E. Olson, H. Summers, *Plasma Phys. Contr. Fusion* **40**, 1541-1550 (1998)
13. A. Kallenbach et al., *Plasma Phys. Contr. Fusion* **52**, 055002 (2010)
14. K. Igenbergs et al, submitted to *J. Phys. B* (2010)
15. W. Fritsch & C.D. Lin, *Phys. Rev. A* **29**, 3039-3051 (1984)
16. A. Foster, On the Behaviour and Radiating properties of Heavy Elements in Fusion Plasmas, PhD thesis, University of Strathclyde (2008)

Exploring Low Energy Molecular Ion Reactions With Merged Beams

C. C. Havener[a], I.N. Draganić[a] and V. M. Andrianarijaona[b]

[a]*Physics Division, Oak Ridge National Laboratory, Oak Ridge TN 37831, USA*
[b]*Department of Physics, Pacific Union College, Angwin CA 94508, USA*

Abstract. Charge transfer (CT) in molecular ion–neutral interactions can proceed through dynamically coupled electronic, vibrational, and rotational degrees of freedom. Using the upgraded Oak Ridge National Laboratory ion–atom merged–beams apparatus, absolute direct charge transfer is explored from keV/u collision energies where the collision is considered "ro–vibrationally frozen" to sub–eV/u collision energies where collision times are long enough to sample vibrational and rotational modes. Our first molecular ion measurement with the merged-beams apparatus has been performed for D_2^+ + H and is used to benchmark high energy sudden approximation theory and vibrational specific adiabatic theory for the $(H_2–H)^+$ complex. CT measurements have also been performed for D_3^+ + H from 2 eV/u to 2 keV/u and CO^+ + D from 20 eV/u to 2000 eV/u. With straightforward improvements to the apparatus, we plan to extend our measurements to key "destructive" rate coefficients for H_2^+ and CH^+ with H at temperatures relevant to the interstellar medium.

Keywords: Molecular ion, merged beams, low energy, charge transfer
PACS: 34.70.+e, 95.30.F

INTRODUCTION

Knowledge of charge transfer (CT) for molecular ion–neutral collisions is necessary for the modelling of high density–low temperature plasmas as found in cold divertor regions of a fusion tokamak. Accurate cross sections for the entire spectrum of excited molecules is necessary for understanding the formation of the detachment layer desired for the reduction of heat loads on divertor plates [1]. In our galaxy, the space between the stars is filled with a variety of particulate and gaseous matter with the main constituent hydrogen in atomic or molecular form. In the low density of the interstellar medium (ISM) including dense clouds, the majority of interstellar molecules are formed by ion–neutral reactions efficient at low temperatures as reflected in the UMIST database for astrochemistry [2]. Surprisingly, for several crucial ion–neutral reactions there is little or no experimental data. For the fundamental $(H_2–H)^+$ two electron system, while H^+ + H_2 collision processes have been extensively studied [3], the only measurement [4] for H_2^+ + H was performed at high energies between 10 keV/u and 50 keV/u and included dissociative and non-dissociative CT. These CT channels were found to dominate over target and projectile ionization at energies lower than 10 keV/u but are significantly below a sudden approximation calculation [5] at the high collision energies where ro-vibrational modes can be considered "frozen". Surprisingly for H_2^+ + H we know of no other CT measurements performed at these or lower collision energies. There is also limited theory at low energies; a fully quantal state–to–state calculation [6] is confined to energies 0.2 eV/u to 10 eV/u, while a state-to-state semi-classical calculation [1] is confined to energies 25 eV/u to 150 eV/u. Also, both experimental and theoretical investigations on isotopic variations of the H_2^+ + H reaction are completely missing and no calculations or cross section measurements extend to energies where rotational states are not "frozen".

The ion–atom merged–beams apparatus has been successful for many years in measuring CT for atomic ions, e.g., see the review [7]. With the intense high velocity molecular ion beams provided by the Multicharged Ion Research Facility (MIRF) High Voltage Platform, the upgraded ion-atom merged-beams apparatus [8] is now able to explore molecular ion-neutral reactions. Absolute direct charge transfer cross sections can be measured from keV/u collision energies where the collision is considered "ro-vibrationally frozen" to meV/u collision energies where collision times are long enough to sample vibrational and rotational modes. The first molecular

Application of Accelerators in Research and Industry
AIP Conf. Proc. 1336, 101-105 (2011); doi: 10.1063/1.3586066

ion measurement [9] with the merged–beams apparatus was performed for D_2^+ + H and is used to benchmark high–energy theory and vibrationally–specific adiabatic theory for the $(H_2\text{-}H)^+$ complex, the most fundamental ion-molecule two-electron system. Unlike CT with atomic ions, CT with molecular ions can have different reaction channels, e.g., direct CT, CT with dissociation, and CT with nuclear substitution where the atomic target interchanges with one of the atoms in the molecular ion. CT measurements are presented for D_3^+ + H from 2 eV/u to 2 keV/u and CO^+ + D from 20 eV/u to 2000 eV/u. The CO^+ measurements are compared to a vibrational state and orientation specific calculation [10]. Understanding such fundamental ion–molecular systems will form the basis for a better understanding of more complex systems.

APPARATUS

A description of the ion–atom merged–beams apparatus upgraded to accept beams from the CEA/Grenoble all-permanent magnet ECR ion source mounted on a 250 kV High Voltage (HV) platform has been described elsewhere [8]. Only a brief description is given here. In the merged-beam technique, fast (keV) molecular ion and atomic beams are merged producing center-of-mass collision energies from meV/u to keV/u. While designed for the production of highly charged ions, the ECR ion source has been found to produce intense molecular ion beams when configured to operate at high pressures and low microwave power. Figure 1 shows the mass scan of the ion beam extracted from the ECR ion source with injection of isotopically enriched D_2 gas at a pressure of 4.2×10^{-6} Torr, an extraction voltage of 16.4 kV, and direct microwave power of 3 W. A 6 mm x 6 mm slit is used in front of the analyzing Faraday cup. Note the intense beams of D_2^+ (172.3 μA) and D_3^+ (21.1 μA), ideal for a single pass merged-beams technique. In fact, the ECR ion source is found to produce and synthesize a wide variety of intense molecular ion beams [11]. The ECR ion source produces these intense molecular ions in vibrational and rotational excited states. In the case of D_2^+ with no dipole and with a relatively short flight time to the merge section, the molecular ions are expected to remain in their initial vibrational state distribution, which is most likely determined by Frank-Condon transitions [12] between ground state D_2 and D_2^+. Franck-Condon distributions [9] for a variety of H_2^+ isotopic systems are presented in Table 1.

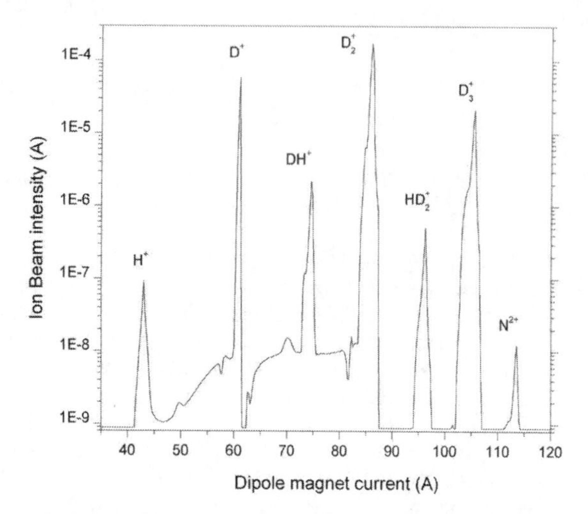

FIGURE 1. Ion beam spectra extracted from the ECR ion source during injection of D_2 gas.

The molecular ion beams with acceleration up to 150 kV can be electrostatically merged with a neutral ground state hydrogen (or deuterium) beam. The H beam is obtained by photodetachment of an 10 keV H^- beam as it crosses the optical cavity of a 1.06 μm 100 W CW Nd:YAG laser where kilowatts of continuous power circulate. The ion and neutral beams interact along a field free region, after which H^+ product ions are magnetically separated from the primary beams. The product signal H^+ ions are detected with a channel electron multiplier operated in pulse counting mode. The beam-beam signal rate (Hz) is extracted from (kHz) background with a two-beam modulation technique. The technique has been highly successful in providing benchmark charge transfer total cross sections for a variety of multiply charged atomic ions in collisions with H and D [7, 13]. The charge transfer measurements presented here correspond to our molecular ion studies to date.

TABLE 1. H_2 (ground state) to H_2^+ Franck-Condon distribution and the corresponding distributions for the isotopic systems [9]. The fits presented in Figure 2 use the distribution for H_2^+ + H.

v	0	1	2	3	4	5	6	7	8
H_2^+	0.119	0.190	0.188	0.152	0.125	0.075	0.052	0.037	0.024
HD^+	0.090	0.160	0.185	0.155	0.120	0.095	0.060	0.045	0.030
D_2^+	0.045	0.104	0.141	0.148	0.134	0.111	0.085	0.063	0.042

RESULTS

The merged-beam measurements for D_2^+ + H are presented in Figure 2 along with various theories for H_2^+ + H appropriate for the different collision energies. The wide range of collision energy, from 2,000 eV/u to 0.1 eV/u, corresponds to four orders of magnitude in collision time, from 1 a.u. to 1000 a.u., respectively. These collision times can be compared to a characteristic vibrational time of 50 a.u. and to the characteristic rotational time of 3000 a.u. calculated from a rotational energy < 0.01 eV and first vibrational excitation energy of ~0.5 eV for the D_2^+ molecule [9]. Above 200 eV/u, the collision can be considered ro-vibrationally frozen as the collision time is too short compared to the characteristic times of vibration and rotation. In this energy range our measurements (see Figure 2) are of the same order of magnitude as MacGraph's measurements [4] and are in good agreement with Errea's calculations [5] for H_2^+ + H. The internal energy of the molecular ions is too small compared to the collision energy, so no difference is expected between collisions with H_2^+ and D_2^+. CT through nuclear substitution is not expected as the relative motion is too fast.

At intermediate and lower energies, structures in the cross section as a function of energy are present both in the vibrationally–resolved semi-classical calculations [1] at intermediate energy (25 eV/u – 150 eV/u) and the quantal calculations at lower energy (0.2 eV/u – 10 eV/u) [6] summed over all final vibrational states. Calculations are shown with either an initial ground or the expected Franck-Condon distribution of initial vibrational states. Note that the predicted energy dependence at intermediate energies is not seen in the measurements. At low energies (0.2 eV/u - 10 eV/u), the Infinite Order Sudden Approximation (IOSA) prediction [6] is shown. The relatively slight difference in the structure suggested in our present measurements and the theoretical calculations might be expected as (D_2^+, D_2) and (H_2^+, H_2) are dynamically different in their vibrational modes. As direct CT and CT with nuclear substitution are not distinguishable in the H_2^+ + H calculations, the fact that the theoretical values (with substitution) are nearly a factor of four higher than the measured CT values (without substitution) suggest that substitutional CT could be an important process at these energies. Future measurements on H_2^+ + H compared to the present D_2^+ + H would directly assess this contribution.

Toward even lower energy (0.1 eV/u) where even rotational modes cannot be considered frozen anymore, our measurement seems to increase faster than the theoretical predictions weighted by the Franck-Condon distribution. This difference may be an

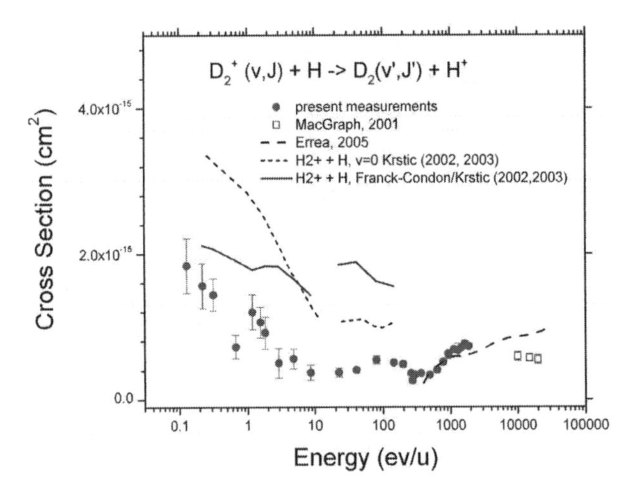

FIGURE 2. Ion-atom merged-beams total CT measurements [9] for D_2^+ + H compared to previous experiment [4] and existing theory [1,5,6]. The solid line is calculated from the state-to-state calculations of Krstic [1,6] weighted by a Franck-Condon distribution for H_2^+ + H of initial vibrational states (shown in Table 1).

indication that the rotationally hot D_2^+ produced by our ECR source results in an increased cross section. The sharp increase in the total cross section at low energies might also be understood as due to trajectory effects caused by the ion-induced dipole potential [13]. It is important to point out that the measurements below 1 eV/u were performed with a limited signal–to–noise which precluded extensive diagnostics on the beam-beam signal.

Preliminary measurements are shown in Figure 3 for CO^+ + D along with vibrational resolved state-to-state calculations [10] for CO^+ + H averaged within the IOSA approximation. Due to the current 150 kV ion beam acceleration limit of the merged–beams apparatus, measurements are performed with D instead of H to access the low center-of-mass collision energies. The following discussion assumes no difference between measurements with D and H for CO^+. The CT cross section for CO^+ + H using three initial vibrational states is shown; in the calculations, 9 and 11 channels are included for CO^+ and CO, respectively. At the high energies, where the vibrational motion could be considered frozen, the calculations for all possible initial vibrational states underestimate the observed CT cross section by at least a factor of 2.5. Our measured cross section drops sharply at around 40 eV/u, approaching the theoretical prediction at energies where collision times are on the order of vibration times. As for the D_2^+ + H system, the predicted CT cross section increases at lower energies due to the existence of exoergic channels. For CO^+ + H, exoergic channels exist even for CT between the v=v'=0 vibrational states.

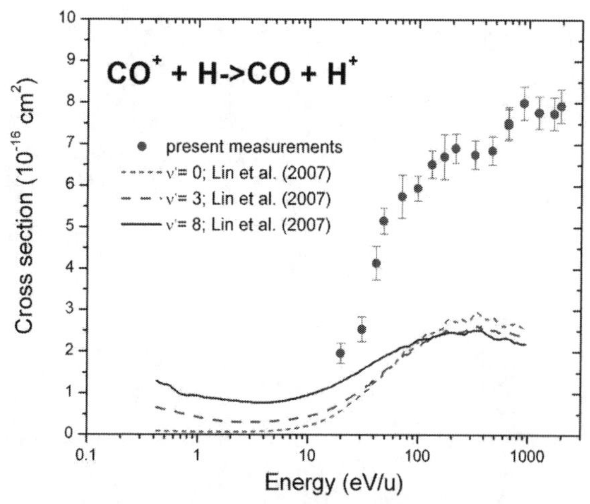

FIGURE 3. Ion-atom merged-beams total CT measurements for CO^+ + D compared to state-to-state calculations [10] for CO^+ + H using three initial vibrational states averaged over the incident orientation of the CO^+ (see text for details).

Figure 4 shows further details of these calculations: the cross section is shown for three different angles of approach between the H trajectory and the axis of the diatomic molecule along with an angle averaged result. The initial vibrational state distribution corresponds to a 300K Boltzman distribution. At high energies where the prediction is less dependent on vibrational states, the CT calculation is seen to be very sensitive to the collision angle. The calculation for zero angle shows the best agreement with our measurements and corresponds to the H approaching C along a trajectory aligned with the internuclear axis of the diatomic ion. At lower energies the discrepancy between theory and experiment decreases, probably due to the higher vibrational excitation present in our ions and the fact that these states have a predicted increased cross section.

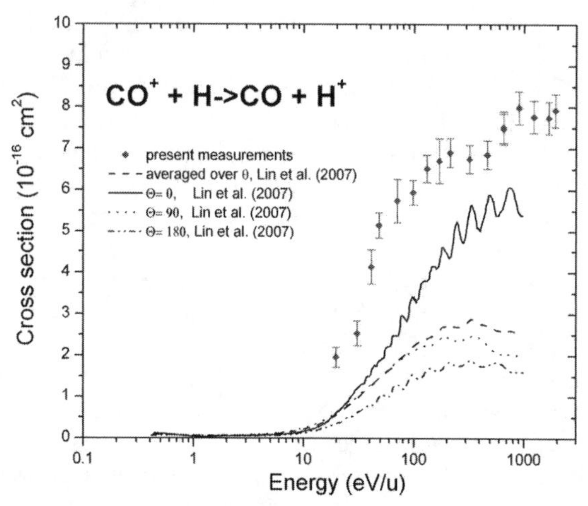

FIGURE 4. Ion-atom merged-beams total CT measurements for CO^+ + D compared to state-to-state calculations [10] for CO^+ + H using three initial and averaged orientations of the CO^+ for a 300K Boltzman distribution of the initial vibrational states (see text for details).

Figure 5 shows our CT measurements for D_3^+ + H. Unlike the diatomic systems discussed above, the triatomic molecular ion is not stable after charge transfer. The CT reaction is endoergic by ~4.4 eV when the D_3 dissociates into D_2 + D and ~9 eV when the products are D + D + D [14]. The measurements are consistent with a 4.4 eV/u threshold. At these energies, CT should be sensitive to the time scale of the vibrational states. While there is no known theory, the measurements are compared to a data compilation [15] for H_3^+ + Ar with the binding energy of the electron on Ar (15.76 eV) similar to H (13.6 eV).

At higher energies, the CT cross sections are seen to be similar while at eV/u energies both systems exhibit a threshold. The H_3^+ + Ar compilation shows a relatively flat somewhat stepwise increasing energy dependence at low energies with a sharper increase at higher energies where additional dissociation channels may become important. This dependence is also suggested in our D_3^+ + H measurements.

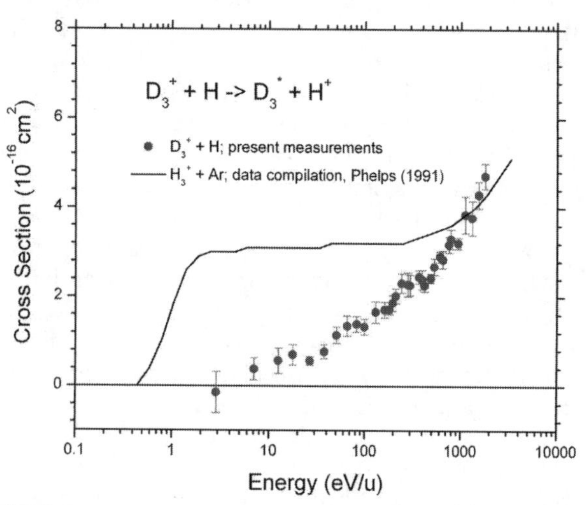

FIGURE 5. Ion-atom merged-beams CT measurements for D_3^+ + H compared to a data compilation for H_3^+ + Ar (see text for details).

SUMMARY AND FUTURE

The upgraded ion-atom merged-beams apparatus is now able to measure ion–neutral reactions for molecular ions with H from keV/u energies where the rotational and vibrational states can be considered frozen to eV/u energies where vibrational states are important. Measurements at meV/u also sample rotational states of the molecular ion. Our first measurement with D_2^+ + H is compared to H_2^+ + H theory where at high energies good agreement is found but at intermediate energies the predictions overestimate the cross section. At lower energies fully quantal calculations also overestimate the cross section but this discrepancy may be due to the importance of nuclear substitution or the rotationally hot ions from the ion source. Our measurements show an increasing trend at low energies. Measurements with CO^+ + D show a factor of 2.5 discrepancy with the predictions for CO^+ + H at the higher energies, improving slightly at eV/u energies. Our measurements with the D_3^+ + H system show some similarity with a H_3^+ + Ar compilation.

In the future, we will extend to lower energies our D_2^+ and CO^+ measurements to explore collision energies where rotational states are important. Comparison to planned H_2^+ + H measurements will directly assess the contribution of nuclear substitution in CT. Measurements will be made possible by some straightforward modifications to the apparatus. The main challenge will be to minimize and handle the higher background count rates which have been observed when using molecular ions. For measurements with H_2^+, there are significant H^+ backgrounds simply from collisional dissociation with background gas. This is in addition to the normal backgrounds of the primary H beam stripping on background gas producing H^+. Improvements in signal–to–noise will allow for more extensive signal diagnostics. Additional H_2 pumping will be supplied by the addition of a 2000 l/sec non-evaporative getter pump. While the additional pumping will help, replacement of the channel electron multiplier H^+ detector (50 nsec pulse width) with a discrete dynode detector (10 nsec pulse width) will increase the maximum count rate from 40 kHz to 400 kHz. Such an improvement was realized on the electron-ion crossed-beam experiment at MIRF which recently measured [16] electron-impact dissociation of the molecular ion CH_3^+. If the new measurements verify an increasing trend in the cross section for H_2^+ + H at thermal energies, this could have a significant impact on the H_3^+ formation in the ISM.

Measurements for the $CH^+ + H \rightarrow C^+ + H_2$ reaction are also planned. The above improvements are expected to aid the measurement of the C^+ signal by reducing the (C^+) background due to collisional dissociation of CH^+ on background gas. In addition, the demerge chamber must be redesigned to accommodate the proximity (in dispersion angle) of the primary CH^+ beam (~μA) from the signal C^+ channel. The current demerge chamber is built for a large dynamic range of ion energies needed to span the six orders of collision energy accessed by the apparatus. The new demerge chamber will be designed to maximize the dispersion between the primary CH^+ and signal C^+.

ACKNOWLEDGMENTS

Research supported by the Office of Fusion Energy Sciences and the Division of Chemical Sciences, Geosciences, and Biosciences, Office of Basic Energy Sciences of the U.S. Department of Energy. I.N.D is supported by the NASA Solar & Heliospheric Physics Program NNH07ZDA001N and acknowledges support from the ORNL Postdoctorial Research Associates Program administered jointly by the Oak Ridge Institute for Science and Education and Oak Ridge National Laboratory.

REFERENCES

1. P. S. Krstic and R. K. Janev, *Phys. Rev. A* **67**, 022708 (2003).
2. J. Woodall, M. Agundez, A. J. Markwick-Kemper and T. J. Miller. "The UMIST Database for Astrochemistry 2006." *A&A* **466**, 1197-1204 (2007).
3. D. W. Savin, P. S. Krstic, Z. Haiman and P. C. Stancil, *ApJ* **606**, L167 (2004).
4. C. MacGraph, *Phys. Rev. A* **64**, 062712 (2001).
5. L. F. Errea *et al.*, *Nucl. Instrum. Methods Phys. Res. B* **235**, 362 (2005).
6. P. S. Krstić, *Phys. Rev. A* **66**, 042717 (2002).
7. C. C. Havener, in *Accelerator-Based Atomic Physics Techniques and Applications,* eds. S M Shafroth and J C Austin (AIP, New York, 1997) pp. 117-135.
8. C. C. Havener, E. Galutschek, R. Rejoub, and D. G. Seely, *Nucl. Instrum. Methods Phys. Res. B* **261,** 129-32 (2007).
9. V. M. Andrianarijaona, J. J. Rada, R. Rejoub and C. C. Havener, J. *Phys. :Conf. Ser.* **194,** 012043 (2009).
10. C. Y. Lin *et al., Phys. Rev. A* **76**, 012702 (2007).
11. I. M. Draganic *et al.*, sub. to *Journal of Applied Physics*
12. V. M. Andrianarijaona, Ph.D. thesis, Université Catholique de Louvain, 2002.
13. C. C. Havener, D. G. Seely, J. D. Thomas and T. J. Kvale, *AIP Conf. Proc.* **1099**, 150 (2009).
14. R. Reichle, Ph. D. Thesis, University of Freiburg (2002).
15. A. V. Phelps, *J. Phys. Chem. Ref. Data* **21**, 883 (1992).
16. E. M. Bahati *et al., Phys. Rev. A* **79**, 052703 (2009).

Solar Wind Charge Exchange Studies Of Highly Charged Ions On Atomic Hydrogen

I. N. Draganić[a], D. G. Seely[b], D. McCammon[c], and C. C. Havener[a]

[a]*Physics Division, Oak Ridge National Laboratory, Oak Ridge, TN 37831, USA*
[b]*Department of Physics, Albion College, Albion, MI 49224, USA*
[c]*Department of Physics, University of Wisconsin, Madison, WI 53706, USA*

Abstract. Accurate studies of low-energy charge exchange (CX) are critical to understanding underlying soft X-ray radiation processes in the interaction of highly charged ions from the solar wind with the neutral atoms and molecules in the heliosphere, cometary comas, planetary atmospheres, interstellar winds, etc.. Particularly important are the CX cross sections for bare, H-like, and He-like ions of C, N, O and Ne, which are the dominant charge states for these heavier elements in the solar wind. Absolute total cross sections for single electron capture by H-like ions of C, N, O and fully-stripped O ions from atomic hydrogen have been measured in an expanded range of relative collision energies (5 eV/u - 20 keV/u) and compared to previous H-oven measurements. The present measurements are performed using a merged-beams technique with intense highly charged ion beams extracted from a 14.5 GHz ECR ion source installed on a high voltage platform at the Oak Ridge National Laboratory. For the collision energy range of 0.3 keV/u - 3.3 keV/u, which corresponds to typical ion velocities in the solar wind, the new measurements are in good agreement with previous H-oven measurements. The experimental results are discussed in detail and compared with theoretical calculations where available.

Keywords: Charge exchange, highly charged ions, atomic beams, merged beams, solar wind
PACS: 34.20+j; 41.75Ak; 37.20+j; 78.70 En.

SOLAR WIND CHARGE EXCHANGE PROCESSES IN HELIOSPHERE

The heliosphere is a plasma bubble created by the solar wind with a radius of 100 astronomic units around the sun. Neutral atoms and molecules penetrate into the heliosphere from the interstellar medium. The particle composition of the heliosphere has been derived from various space satellite missions such as the Röntgen Satellite (ROSAT), the Solar and Heliospheric Observer (SOHO), the Extreme Ultraviolet Explorer (EUVE), Chandra, etc.. The missions have used a combination of *in situ* particle measurements and spectroscopic observations. The most exciting and promising band passes for studying the heliosphere is the soft X-ray emission in the range of 0.3keV – 2 keV. Soft X-ray emission in this energy range was discovered from comet Hyakutake[1] by ROSAT, which was unexpected because comets are known to be very cold. Since then, soft X-rays have been observed from at least 13 other comets [2]. The most probable mechanism for the production of these X-rays involves solar wind charge exchange (SWCX) with neutrals in the cometary atmosphere [2, 3]. The SWCX process occurs when a highly charged solar wind ion captures an electron from a neutral in the cometary coma. As the electron radiatively relaxes, it has a large probability of emitting soft X-ray photons.

A second exciting breakthrough has been the development of flight-capable microcalorimeters which can be used to observe the diffuse soft X-ray background (SXRB) with high spectral resolution [4]. All-sky surveys have been conducted in the soft X-ray band for many years, and include measurements made by the Wisconsin sky survey [5], ROSAT [6], microcalorimeters on a sounding rocket [4], the Chandra X-ray Observatory (CXO) [7, 8], XMM-Newton [9] and recently with Suzaku [10, 11]. These observations have revealed a rich and dynamic diffuse background produced by a complex combination of X-ray sources. While much of this emission is due to extragalactic objects or hot gas in the local bubble, it was suggested by Cox [12] that the SWCX mechanism, originally proposed by Cravens [13] to explain cometary X-ray emission, might account for some of this soft X-ray background (SXRB). Cravens estimated that 25-50% of the SXRB could be attributed to heliospheric X-ray emission due to

Application of Accelerators in Research and Industry
AIP Conf. Proc. 1336, 106-110 (2011); doi: 10.1063/1.3586067

SWCX [13]. The later satellite observations with CXO, XMM-Newton, and Suzaku, as well as more detailed modeling [14], appear to confirm this conclusion. Therefore, studies of the SXRB thus offer another possible means of probing the properties of the heliosphere [15].

Using the SXRB to study the heliosphere hinges on our ability to understand the underlying atomic processes which produce the observed spectrum, namely SWCX [6]. The solar wind consists of protons, electrons, α particles and ~0.1% multiple charged ions of heavier species such as C, N, O, Ne, Mg, Si, S, and Fe. Collisions of multicharged ions with neutrals are believed to be responsible for a large fraction of the SXRB [2, 3, 14]. In the case of the SXRB, the dominant neutrals in the interstellar medium are H and He which stream through the heliopause [14, 15]. Typical ion velocities in the fast solar wind are 700 – 800 km s^{-1} corresponding to collision energies of 2.5−3.3 keV/u where u is the atomic mass unit. In the slow solar wind the velocities are 300 – 400 km s^{-1} or 460 – 830 eV/u [16, 17]. Central to our understanding of SWCX is an accurate knowledge of the CX process that produces the observed X-ray emission [6, 18]. Particularly important are data for bare and H-like ions of C, N, O, and Ne which are present in the solar wind and key ions in SWCX models of the SXRB [15, 18].

In the past few years three experimental groups, in particular, have performed laboratory simulations of SWCX [19, 20, 21]. Each group has made significant contributions to our understanding of the CX process. Our complimentary studies are performed with a ground state neutral beam target and a well defined collision energy. Experiments are performed using the Oak Ridge National Laboratory (ORNL) ion-atom merged-beams apparatus [23, 24]. This apparatus is unique and capable of studying CX collisions of highly charged ions with ground state atomic H over the entire range of collision energies from 100 eV/u to 10 keV/u relevant to SWCX studies. In addition to measuring absolute total CX cross sections, we plan to install the University of Wisconsin /Goddard Space Flight Center (UW/GSFC) microcalorimeter, or X-ray quantum calorimeter (XQC) [22], on the merged-path of the apparatus to measure the CX produced X-ray line emission with high spectral resolution. This is the same detector that has been flown on three sounding rocket flights [4].

ION ATOM MERGED BEAMS APPARATUS

The upgraded ORNL ion-atom merged-beams apparatus [23] is shown in Fig. 1. The success of the merged-beams technique is dependent on the availability of intense ion beam currents, and these can be produced by modern ECR technology. The intense multiply charged ion beams (1-50 μA) used here have energies up to $q\times150$ keV and are produced by a Grenoble all permanent-magnet ECR ion source on a high voltage platform at the ORNL Multicharged Ion Research Facility (MIRF) [25]. An example of the intense ion beams available is shown in Fig. 2 which corresponds to the charge state distribution of nitrogen optimized for extraction of H-like N^{6+}. Using typical gas mixing with helium, ion beams of H-like , He-like and Li-like nitrogen are routinely extracted with intensities that exceed 25 μA. The ECR ion source enables us to produce and extract similar intense ion beams for carbon and oxygen. With a typical ion beam transport efficiency of 50% between the high-voltage (HV) platform and the ion-atom merge path, the CX measurements can be performed with a highly charged beam intensity of between 1 and 10 μA. For O^{8+} ions, more than 2 μA has been analyzed on the platform by the use of isotopic enriched oxygen gas (A=18).

The ion beam is merged electrostatically with a ground state neutral hydrogen beam using a set of two spherical sector deflectors (see Fig. 1). A ground state H beam of up to 500 nA is produced by photodetachment of a 20 μA beam of H^- in a 100W CW Nd:YAG laser cavity (λ = 1.06 μm). Only about 50 nA of H is necessary for the total cross section measurements, while it is expected that the full 500 nA H beam will be used for the X-ray emission measurements. The 5-30 keV H^- beam is extracted from a duoplasmatron, analyzed by a 30° bending magnet, focused and steered horizontally and vertically so that subsequent to neutralization a nearly parallel neutral H beam will travel down the merge path, interact with the ion beam, and then be collected in the neutral detector.

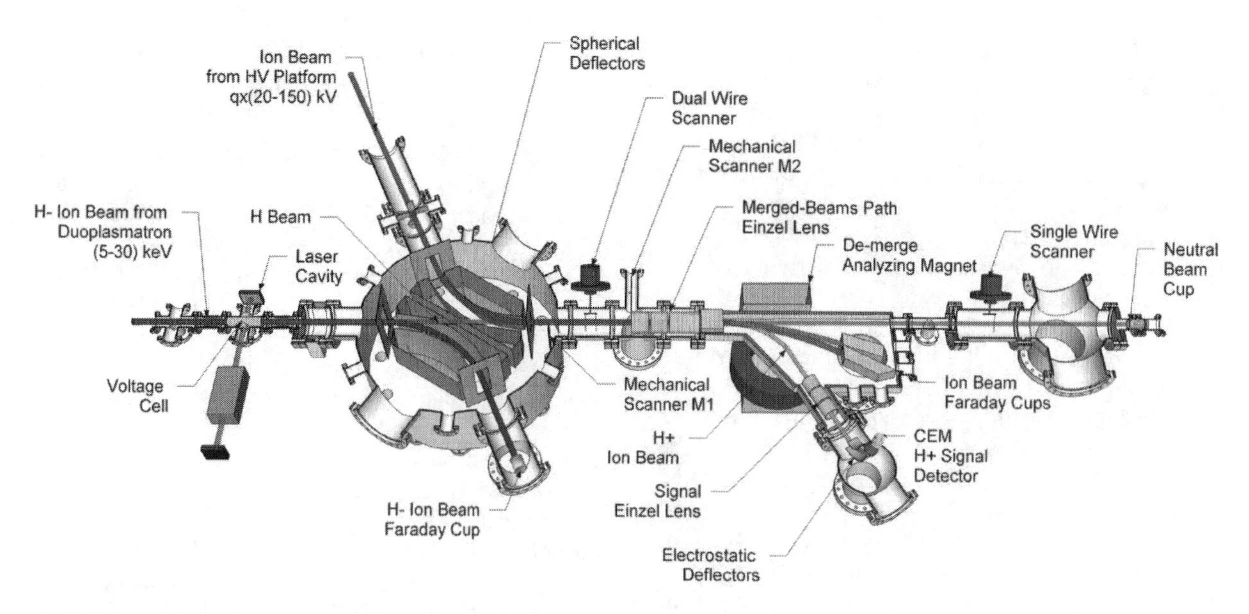

FIGURE 1. Schematic overview of the ORNL ion-atom merged-beams apparatus.

The merged beams interact in a field-free region over a distance of 37.5 cm, after which the primary beam is magnetically separated from the fast H beam and from the product or "signal" H^+ ions. The shortened merge-path of the upgraded apparatus [23] with the higher velocity primary beams from the new HV platform results in an improved angular acceptance of the apparatus allowing low energy measurements to be performed with H rather than just D (as was previously the case).

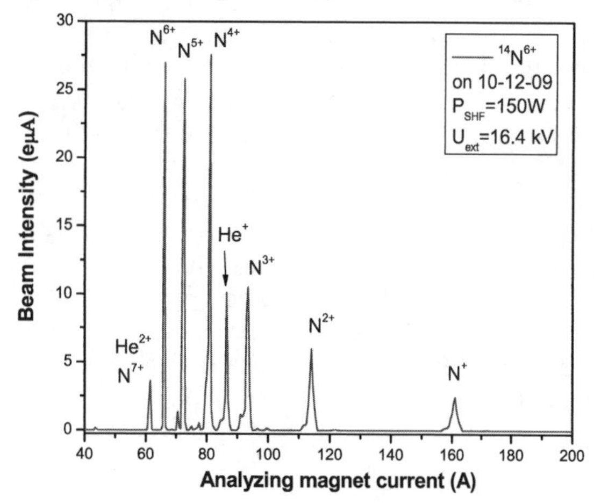

FIGURE 2. Charge state distribution obtained from the ORNL all-permanent ECR ion source with injection of nitrogen and helium gas. The charge state distribution was obtained using a direct microwave power of 150 W (P_{shf}) at frequency of 13.88 GHz and an extraction potential of 16.4 kV (U_{ext}). The N^{6+} ion beam current of 27 μA is recorded for an X-Y slit-defined beam size of 6x6 mm^2.

The merge path is kept at ultrahigh vacuum conditions (1×10^{-10} Torr) to minimize the H^+ background generated by stripping of the H beam as it travels through the background gas. To maximize the collection efficiency of H^+, a cylindrical Einzel lens is placed toward the end of the merge path. Due to a limited magnetic dispersion, the $X^{(q-1)+}$ product of the reaction is not measured separately but is collected with the X^{q+} beam in a special designed Faraday cup. The neutral-beam intensity is measured by secondary-electron emission from a stainless-steel plate, and the signal H^+ is recorded by a channel electron multiplier (CEM) operated in pulse counting mode. The H^+ undergo a second (vertical) dispersion out of the plane of magnetic analysis. This dispersion is designed to attenuate transmission into the CEM of photons from the ion Faraday cup. Cross sections are determined by measuring the signal produced by the beam-beam interaction over the merged length. For the ion-atom merged beams aproach, the CX cross section is determined from the H^+ signal rate [for details see Ref. 24]. In order to determine an absolute cross section from the measured signal, it is necessary to measure the overlap of the two beams. Two-dimensional profiles of the beam are measured at four positions along the merge path and used to calculate the overlap integral. A recently installed dual-wire scanner [27] developed at ORNL is optimized for merged-beams and facilitates the required considerable tuning of the beams. The CX signal is separated from background by a two-beam modulation technique [24].

The cross section for electron capture by a metastable ion can significantly differ from capture by the ground state ion. Ion beams from the ORNL ECR ion source can be diagnosed for metastables by observing the electron impact ionization signal below the ground state ionization threshold using the ORNL electron-crossed beams apparatus [28]. Based on our experience with diagnosing ion beams, neither H- or Li-like ion beams contain metastables. However, for He-like ion beams we can expect some metastable contaminants. For example, for our CX measurements of He-like C^{4+} + H, the metastable fraction of the C^{4+} beam was determined to be 0.05 ± 0.01 [29]. In that work Landau-Zener calculations were used to make small corrections of a few percent to the measured cross section accounting for the metastable contaminant. Note that the existence of metastables (with an inner shell vacancy) would clearly be evident in the X-ray emission measurements.

TOTAL CHARGE EXCHANGE CROSS SECTION MEASUREMENTS

We have performed merged-beam measurements of absolute total CX cross sections for the following highly charged ions:

a) N^{6+} on H in the collision energy range of 20 eV/u - 5 keV/u.

b) O^{7+} on H in the collision energy range of 10 eV/u - 20 keV/u.

c) C^{5+} on H in the collision energy range of 5 eV/u - 12 keV/u.

d) O^{8+} on H in the collision energy range of 100 eV/u - 2 keV/u.

Fig 3 shows the total charge exchange cross section of hydrogen-like nitrogen on atomic hydrogen. Our merged-beams measurement is in good agreement with previous measurements using a H-oven of Meyer et. al. [30] and Panov [31]. The Panov data, though, is shown to have some scatter (50% lower cross section at 3keV/u). Our measurements which extend to energies as low as 20 eV/u show a decreasing cross section which is somewhat surprising. Toward lower collision energies, ion-induced dipole trajectory effects can lead to an increasing cross section [26]. For this collision system no published theoretical calculation is known to exist but new theoretical studies are underway.

The total charge exchange cross section measurements of hydrogen-like oxygen on atomic hydrogen are presented in Fig 4. Our new measurements are lower than the obtained results of Meyer [30] and generally above the measurements of Panov [31] which again show scatter. The discrepancy with the previous measurements of Meyer et al. is outside of our estimated total systematic error of 15%. Again our extension to lower energies shows a decreasing cross section which is unexpected. No published theoretical calculation is known to exist but new theoretical studies are underway

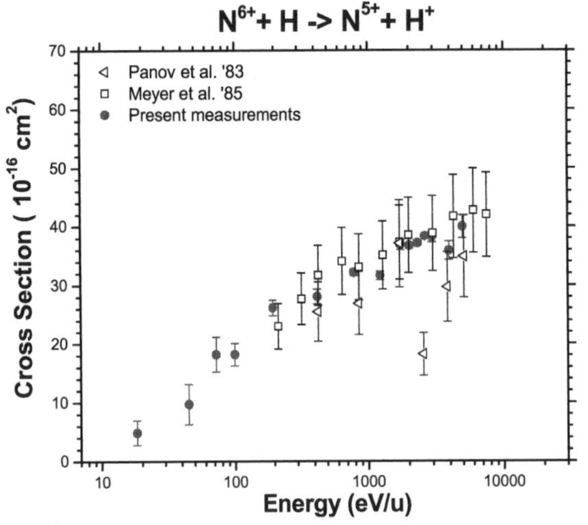

FIGURE 3. Merged-beams CX total cross section measurement for N^{6+} with atomic hydrogen compared to previous experimental data [30, 31].

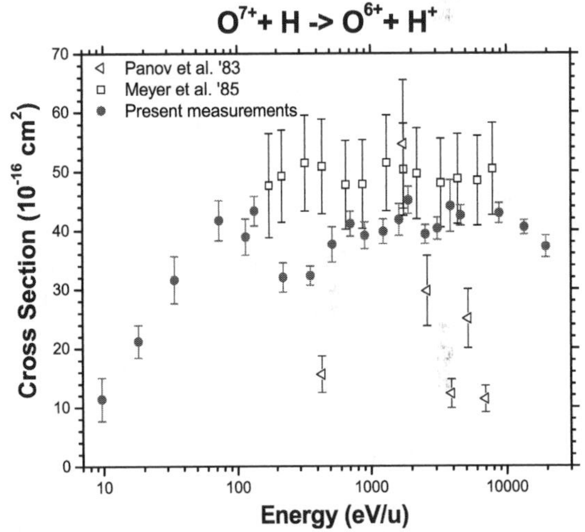

FIGURE 4 Merged-beams CX total cross section measurement of for O^{7+} with atomic hydrogen compared to previous experimental data [30, 31].

Measured CX cross sections (not shown) for C^{5+} on H and for O^{8+} on H [32, 33] are discussed and compared with the most recent theoretical calculations in Ref. 35, 36. Our merged-beams measurements show good agreement with the experimental data of Phaneuf et al. [34] for H-like carbon. For bare O^{8+} on H above 200 eV/u our merged-beams measurements show good agreement with previous hydrogen-oven measurements of Meyer et al.[30] which are below the more recent extensive theoretical study for this fundamental system [36]. The larger theoretical total cross section is attributed to the predicted increase of capture to the n=6 shell of O^{7+}. Below 200 eV/u our new measurements [33] do suggest a somewhat increasing trend which may be attributed to capture to the n=6 shell, but at a much lower energy than predicted. This discrepancy points to the need for state-selective information which will be provided by our planned X-ray emission measurements.

X-ray emission studies will be carried out using the X-ray Quantum Calorimeter (XQC), built and flown by the University of Wisconsin -Madison and NASA's Goddard Space Flight Center [4]. The XQC detector will be mounted above the merge-path and directly view EUV/Xray photons from the relaxation of electrons which have been captured to excited states. The X-ray detector is characterized by a high energy resolution (5-12 eV FWHM) along with high throughput (1000 times greater than dispersive X-ray detectors) [22, 37]. The redesign of the merged-beam path is in progress. Relative line intensities will be used to estimate the state selective CX cross sections for a detailed comparison with theory. We are not aware of any previous X-ray emission measurements performed for an atomic H target. Using this approach, X-ray spectra can be obtained over the whole SWCX collision energy range of interest.

ACKNOWLEDGMENTS

This research is supported by the NASA Solar & Heliospheric Physics Program NNH07ZDA001N, the Office of Fusion Energy Sciences and the Division of Chemical Sciences, Geosciences, and Biosciences, Office of Basic Energy Sciences of the U.S. Department of Energy. One of the authors (I.D.) gratefully acknowledges support from the ORNL Postdoctoral Research Associates Program administered jointly by the Oak Ridge Institute for Science and Education and Oak Ridge National Laboratory.

REFERENCES

1. C. M. Lisse et. al. *Science* **271**, 205-208 (1996).
2. T. E. Caverns, *Science* **296**, 1042-1045 (2002).
3. P. Beiersdorfer et. al. *Science* **300**, 1558-1559 (2002).
4. D. McCammon et al., *ApJ* **576**, 188-203 (2002).
5. D. McCammon and W. T. Sanders, *Annual Review of Astronomy and Astrophysics*, **28**, 657 (1990).
6. N. A. Schwadron, and T. E. Cravens, *ApJ* **544**, 558'566 (2000).
7. M. Markevitch, (2003), *Astr. J.* **583,** 70 (2003).
8. B. H. Wargelin, M. Markevitch, M. Juda, V. Karchenko, R. Edgar, and A. Dalgarno, *ApJ* **607**, 596 (2004).
9. S. L. Snowden, "*Soft X-ray Emission from Clusters of Galaxies and Related Phenomena*". Edited by Richard Lieu and Jonathan Mittaz, Published by Kluwer Academic Publishers, Dordrecht, The Netherlands, 2004, p.103.
10. R. Fujimoto, et al., *PASJ*, **59**, S133 - S140 (2007).
11. R. K. Smith, et al., *PASJ*, **59**, S141 - S150 (2007).
12. Cox, D. P. 1998, in "The Local Bubble and Beyond", edit by D. Breitschwerdt, M. J. Freybert, and J. Truemper (Berlin: Springer), 1998, pp 121-150.
13. T. E. Cravens, *GeoRL,* **24**, 105 (1997).
14. T. E. Cravens, ApJ, 532, L153-L156, (2000).
15. R. Pepino, V. Kharchenko, A. Dalgarno, and R. Lallement, *ApJ,* **617**, 1347-1352 (2004).
16. J. Geiss, et al., *Science*, **268**, 1033-1036 (1995).
17. M. Neugebauer, M., et al., *J. Geophys. Res.*, **103**, A14, 587 (1998).
18. V. Kharchenko, M. Rigazio, A Dalgarno, and V. A. Krasnopolosky, *ApJ,* **585**, L73-L75 (2003).
19. J. B. Greenwood, A. Chutjian, and S. J. Smith, *ApJ*, **529**, 605-6012 (2000); J. B. Greenwood, I. D. Williams, S. J. Smith, and A. Chutjian, ApJ, **533**, L175-L177 (2000).
20. P. Beiersdorfer, C. M. Lisse, R. E. Olson, G. V. Brown, and H. Chen, *ApJ,* **549**, L147-L149 (2001).
21. B. Bodewits, B., Z. Juhasz, R. Hoekstra, and A.G. G.M. Tielens, *ApJ*, **606**, L81 (2004).
22. McCammon, *J Low Temp Phys*, **151**, 715-720 (2008).
23. C. C. Havener, E. Galutschek, R. Rejoub and D. G. Seely, *Nucl. Instr. and Meth. B* **261**, 129-132 (2007).
24. C. C. Havener, M. S. Huq, H. F. Krause, P. A. Schultz and R. A. Phaneuf, *Phys. Rev. A*, **39**, 1725 (1989).
25. F. W. Meyer et.al, *Nucl. Instr. and Meth. B* **242** (2006) p. 71-78.
26. C. C. Havener, S. L. Hough, R. Rejoub, D. W. Savin, and M. Schnell. 2006, "Dipole Polarization Effects on Highly-Charged-Ion-Atom Electron Capture" in Proceedings of the XXIV International Conference on the Photonic, Electronic, and Atomic Collisions, edit by P. D. Fainstein et al., (World Scientific, 2006).
27. D. G. Seely, et al., *Nucl. Instr. and Meth. A* **585,** 69 - 75 (2008).
28. M. E. Bannister, *Phys. Rev. A*, **54**, 1435-1444 (1997).
29. F. W. Bliek, R. Hoekstra, M. E. Bannister, and C. C. Havener, *Phys. Rev. A*, **56**, 426-431 (1997).
30. F. M. Meyer, A. M. Howald, C. C. Havener, and R. A. Phaneuf, *Phys. Rev. A*, **32**, 3310- 3318 (1985).
31. M. N. Panov, A. A. Basalaev and K. O. Lozhkin, *Phys. Scrip.*, **T3**, 124-130 (1983).
32. I. N. Draganic et. al. (to be published).
33. C. C. Havener et. al. (in preparation).
34. R. A. Phaneuf, I Alvarez, F. W. Meyer and D. H. Crandall, *Phys. Rev. A*, **26**, 1892- 1906 (1982).
35. N. Shimakura, S. Koizumi, S Suzuki, and M Kimura, *Phys. Rev. A*, **45**, 7876- 7882 (1992).
36. T. G. Lee, m. Heese, A-T Le, and C. D. Lin, *Phys. Rev. A* **70**, 012702 (2004).
37. S. Otranto, R. E. Olson and P. Beiersdorfer, *Phys. Rev. A* **73**, 02273 (2006).

Magnetic Tunnel Junctions Fabricated Using Ion Neutralization Energy As A Tool

J.M. Pomeroy[a], H. Grube[a], P.L. Sun[b], R.E. Lake[c]

[a]*National Institute of Standards and Technology (NIST), 100 Bureau Dr., MS 8423, Gaithersburg, MD 20899-8423*
[b]*Dept. of Materials Science and Engineering, Feng Chia University, Taichung 40724, Taiwan*
[c]*Dept. of Physics and Astronomy, Clemson University, Clemson, SC 29634*

Abstract. The neutralization energy of highly charged ions (HCIs) is used during the fabrication of magnetic tunnel junctions (MTJs) to modify their electrical properties, providing additional flexibility in the devices' properties. While most ion species used in electronic fabrication (e.g., implantation or plasma cleaning/oxidation) have charge $q \approx 1$ and negligible neutralization energies, the HCIs utilized in this work are up to $q=44$ and carry as much as 52 keV per HCI of neutralization energy, ample to modify the chemical and electrical properties of the tunnel barriers at the impact site. Hundreds of MTJ devices have been fabricated, revealing some general characteristics of the HCI modified tunnel junctions: the electrical conductance increases linearly with the number of HCIs used; the conductance added per HCI depends on the initial tunnel barrier thickness and the barrier stoichiometry; the transport is usually tunneling (not Ohmic); and the conductance added by the HCI process can substantially increase the magneto-conductance. Recent experiments (not shown) have also revealed an enormous dependence on the charge state of the HCI used in tunnel barrier irradiation.

Keywords: Highly Charged Ions, Magnetic tunnel junctions, Ion-solid interactions
PACS: 85.70.Kh, 85.75.Dd, 34.35.+a, 79.20.Rf

INTRODUCTION

Ions are a fundamental part of mainstream technology, whether in halogen or fluorescent light bulbs, focused ion beam (FIB) milling and deposition, and semiconductor processing, e.g., sputter deposition, ion implantation, and plasma processing. In some of these cases, the ions enhance production of photons, and in others the ion charge provides a mechanism for applying force and creating momentum used to modify a target, but in none of them is the neutralization energy of the ion important to the function, since these ions are almost always near neutral, i.e., charge state q = 1, or 2. However, as *q* becomes larger, the neutralization energy increases rapidly and outpaces the kinetic stopping power, for example, Xe^{44+} releases ≈52 keV when it neutralizes. HCIs have become substantially more accessible due to the development of relatively small sources, like the electron beam ion trap (EBIT), that enable HCI production with relative ease and without the need for large accelerator facilities. The existence of these laboratory sources, coupled with scanning probes, secondary electron/ion, X-ray and other techniques has painted a picture of HCI-matter interactions as being especially violent at the surface[1-5], where surface energy densities reach many times higher than the most energetic of swift heavy ion collisions.

While most previous studies have examined exposed surfaces, both in real time and *post facto* with scanning probes, we have incorporated HCI-surface irradiation as a step in fabricating and tuning electronic devices[6-8]. Specifically, the neutralization energy of HCIs is used to modify the electrical integrity of thin insulating films used as a tunnel barrier in magnetic tunnel junctions (MTJs). MTJs are heavily utilized as read heads in computer hard drives and, recently, as memory elements in magnetoresistive random access memory (MRAM), but face scalability challenges in reducing the electrical resistance sufficiently to meet technical demands for future product generations. Since the electrical performance of the MTJ is exponentially sensitive to the integrity of the insulating barrier, HCI irradiation can tune the resistance over a dynamic range spanning orders of magnitude[7,8]. Conversely, the exponential sensitivity of the tunnel junction and the encapsulation of the irradiated surface also provides a novel scientific tool

Application of Accelerators in Research and Industry
AIP Conf. Proc. 1336, 111-114 (2011); doi: 10.1063/1.3586068

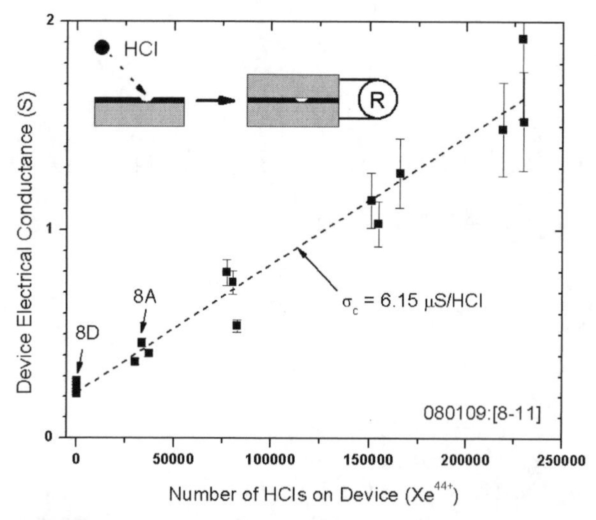

FIGURE 1. **(Inset - left)** A lower electrode structure is deposited and capped with an ultra-thin aluminum oxide tunnel barrier, which is then irradiated with a pre-selected dose of HCIs; **(Inset –right)** Immediately after irradiation, an upper electrode structure is deposited, encapsulating the irradiated tunnel barrier and enabling electrical measurements; **(Main panel)** Each HCI impact is found to contribute a conductance σ_c such that the total conductance of the MTJ increases linearly with the number of HCIs (Xe^{44+}) to which it was exposed; MTJs are [0.4 Co/30 Cu/1.0 Co/0.8 Al+Ox/2 Co/40 Cu] in nm, +Ox indicates plasma oxidation. Error bars represent the 1 sigma confidence interval as propagated from errors in the HCI density, the device size, the resistance measurement, and variations in the MTJ properties between the devices used in the data set.

for probing the electronic nature of nano-features formed by HCI-matter interactions. Here we present an overview of the device fabrication, characterization and general properties of these HCI modified MTJs.

DESCRIPTION OF SAMPLES

The MTJ samples described here are all fabricated completely *in situ* within a system of ultra-high vacuum (UHV) chambers attached to the NIST Electron Beam Ion Trap (EBIT) facility[9]; a more complete description of the apparatus and fabrication can be found elsewhere[10,11]. Briefly, thermally oxidized silicon chips (approximately 1 cm x 2 cm) are loaded into vacuum after 10+ hours of ultrasonic cleaning in distilled water, and then cleaned *in situ* with an oxygen plasma. The cleaned chips are transferred into a larger UHV chamber (< 3×10^{-8} Pa) for storage between processing steps. The chips are then individually transferred under vacuum to a UHV metals deposition chamber where the MTJ lower electrode structure (Co/Cu/Co or CoO/Co) is deposited and defined by a shadow mask that makes four individual MTJs per chip. Next, a uniform ultra-thin layer of aluminum is deposited over the entire chip, which is then immediately transferred to another chamber and exposed to oxygen plasma, completely oxidizing the aluminum. The sample is then stored in UHV for 12+ hours to allow the aluminum oxide to "relax." The chip is then transferred to a target chamber and exposed to HCIs (see left inset in Fig. 1). Only one MTJ on the chip is irradiated at a time (typically <5 min), and one of the four devices per chip is not exposed as a control. Immediately following irradiation, the chip returns to the deposition chamber where an electrode deposited (Co/Cu/Au) over the irradiated region defines the MTJ's size (≈3000 μm²). The upper and lower wires form a cross with the irradiated tunnel barrier between the upper and lower wires (right inset, Fig. 1). The completed chips are then removed from vacuum for measurement. A cross-sectional Transmission Electron Micrograph (TEM) image of a representative MTJ is shown in Fig. 2.

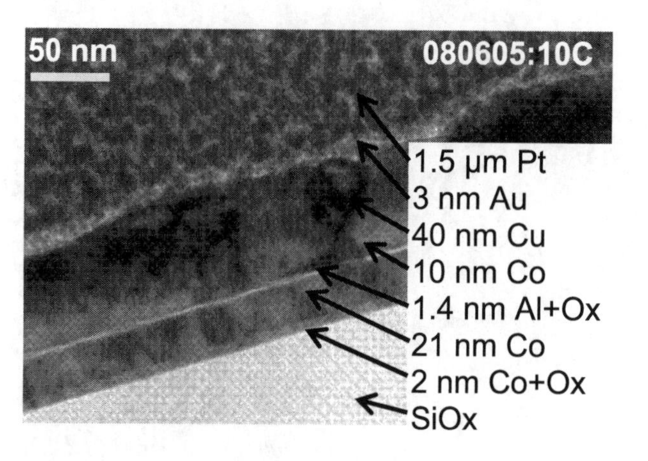

FIGURE 2. A TEM cross sectional image of a representative MTJ structure is shown. The aluminum oxide tunnel barrier is clearly seen as a bright, continuous line across the sample, other layers are as indicated. (Platinum was added for TEM only and is not present in any electrical measurements.)

CHARACTERIZATION OF DEVICES

Once removed from the vacuum system, the MTJs' electrical resistance (R) is measured using a four point geometry and corrected for a geometric artifact[12]. An example of the total electrical conductance g = 1/R vs. HCI dose is shown in Fig. 1, where each of the MTJs was exposed to a different fluence of Xe^{44+}. The Xe^{44+} is extracted from the EBIT with 8 keV x q of kinetic energy, accounting for a total stopping power in the device's tunnel barrier of approximately 6.89 keV (from SRIM) compared with 52 keV of neutralization energy. Fitting the electrical conductance

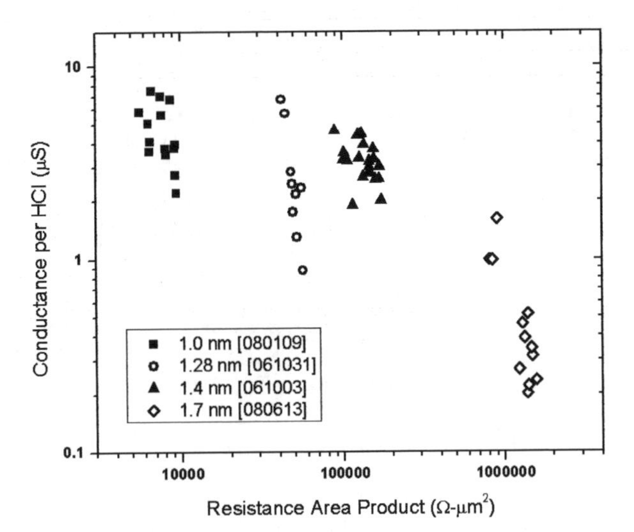

FIGURE 3. The conductance added per HCI (σ_c) is dependent on the thickness and stoichiometry of the tunnel barrier, as well as the ion charge state (not shown[13]). The σ_c values from MTJs with four different barrier thicknesses (see legend) and slightly different stoichiometries (variation within each thickness) reveals a σ_c dependence on the stoichiometry that is much stronger than the tunnel conductance dependence (~1/RA). All data utilize Xe^{44+}.

$g_T = \sigma_c N + g_{tun}$, where N is the number of HCIs and g_{tun} is the background (control) tunnel conductance, each HCI impact increases the conductance by $\sigma_c = 6.5$ μS, allowing large changes to be affected by very dilute HCI densities. For example, 150,000 HCIs in Fig. 1 (corresponding to 7.5 x 10^9 HCIs/cm^2) increased the conductance by more than a factor of five. For all materials and charge variations (from Xe^{26+} to Xe^{44+}) studied to date, the electrical conductance is always observed to increase linearly with the total number (N) of HCIs used (N ranging from 0 to 3 x 10^7 HCIs per device), so that σ_c characterizes the electrical response of the material system to the HCI dose. To illustrate the variation in σ_c accessible by small variations in the barrier thickness and stoichiometry, σ_c as a function of the MTJ's initial resistance-area (RA) products are shown in Fig. 3.

The RA product is directly proportional to electrical transparency of the unirradiated MTJ, but it is clear that HCI-induced change in transparency does not depend monotonically on the tunnel barrier's initial electrical transparency. In particular, variations in the barrier stoichiometry (potential energy barrier) for a single barrier thickness more strongly affect σ_c; as an example, the data for the barrier thickness s=1.28 nm has RA products ranging from 41.7 kΩ•μm^2 to 52 kΩ•μm^2, which corresponds to a range of initial potential barrier heights of 1.16 eV to 1.215 eV (from model) while σ_c varies from 0.88 μS to 6.7

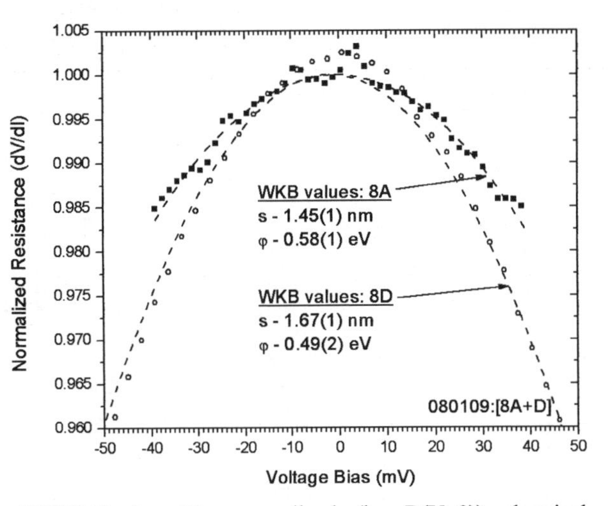

FIGURE 4. The normalized (by R(V=0)) electrical resistance as a function of applied voltage is tunneling (parabolic) for both irradiated (8A) and control (8D) MTJs as labeled in Fig. 1, (every other datum is hidden to reduce clutter). Fits using a WKB model for tunneling as shown as dashed lines with each data set and the parameters corresponding with each are inset.

μS. Experiments using a wide range of charge states (not shown, i.e., see[13], additional publications in preparation) show that σ_c can vary by more than three orders of magnitude between q = 26 and q = 44. Practically, for designing a device that can span many product generations without variation in initial deposition recipe, one would choose a large σ_c so that families of devices with decreasing RA products could be produced by a corresponding set of HCI fluences, all from the same fabrication recipe, or smaller to allow finer adjustments.

While the conductance per HCI impact σ_c is substantially less than a single quantum of conductance ($1/R_k$=38.8 μS, R_k = 25.8 kΩ) suggesting that the transport is non-Ohmic, the transport mechanism can be measured using the electrical resistance as a function of the bias voltage, as shown in Fig. 4. The resistance values for an irradiated (8A) and control (8D) MTJ are normalized by their respective zero voltage resistances to directly compare the relative resistance change with applied voltage, which has the same non-Ohmic trend for both. A modified WKB (Wentzel–Kramers–Brillouin) quantum tunneling model[7] fit is used to extract approximate tunnel barrier thickness s and potential barrier heights φ. For both the data sets shown, a dashed line represents the WKB fit to the data with the parameter values inset. The goodness of the fits affirms the hypothesis that the transport is tunneling. Other measurements and models of the MTJ resistance vs. temperature (not shown[8]) support the conclusion that the electrical transport remains tunneling after HCI irradiation.

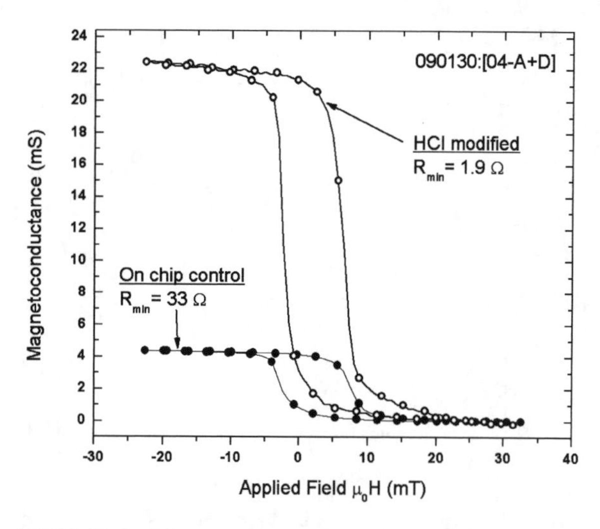

FIGURE 5. The magneto-conductance (MC) for a control and irradiated device (offset for each) shows an increase in the amount of MC in the irradiated device, indicating the HCI induced conductance channels retain magnetic sensitivity.

While these data demonstrate the versatility of the HCI irradiation method for modifying RA products over many orders of magnitudes (greater dynamic ranges shown elsewhere, e.g., see[7,8,13]), useful application to magnetic sensing devices or MRAM requires the maintenance of spin transfer, manifest as magneto-resistance (MR). The magneto-conductance vs. applied magnetic field (H) is presented in Fig. 5, i.e., each MTJ's conductance after subtracting the conductance when the magnetizations of the upper and lower Co layers are opposite. As is shown, the irradiated device has a larger magneto-conductance than the control, indicating that the conductance channels created by HCI must carry some spin information. Therefore, the HCI irradiation can adjust the RA product, while preserving the utility of the MTJs.

CONCLUSIONS

Magnetic tunnel junction devices fabricated using the neutralization energy of HCIs provides additional versatility to the numerous degrees of freedom provided by traditional electronic device fabrication, without requiring the device's fabrication recipe to be arduously re-adjusted. This method increases the conductance linearly, with a wide range of linear slopes possible (due to varying choices of tunnel barrier properties and ion charge state[13]), while preserving tunneling-type transport and magneto-conductance, essential for both magnetic sensor devices and MRAM.

REFERENCES

1. T. Schenkel, A. V. Hamza, A. V. Barnes and D. H. Schneider, *Prog. Surf. Sci.* **61** (2-4), 23-84 (1999).
2. F. Aumayr and H. Winter, *Phil. Trans. R. Soc. Lond.* **362**, 77-102 (2004).
3. M. Tona, H. Watanabe, S. Takahashi, N. Nakamura, N. Yoshiyasu, M. Sakurai, C. Yamada and S. Ohtani, *Nucl. Instrum. & Meth. B* **258** (1), 163-166 (2007).
4. J. P. Briand, S. Thuriez, G. Giardino, G. Borsoni, M. Froment, M. Eddrief and C. Sebenne, *Phys. Rev. Lett.* **77** (8), 1452-1455 (1996).
5. J. D. Gillaspy, *J. Phys. B: At. Mol. Opt. Phys.* **34**, R93-R130 (2001).
6. H. Grube, J. M. Pomeroy, A. C. Perrella and J. D. Gillaspy, *Mat. Res. Soc. Proc.* **960**, 0960-N0908-0902 (2007).
7. J. M. Pomeroy and H. Grube, *Nucl. Instrum. & Meth. B* **267** (4), 642-645 (2009).
8. J. M. Pomeroy, H. Grube, A. C. Perrella and J. D. Gillaspy, *Appl. Phys. Lett.* **91** (7), 073506 (2007).
9. J. D. Gillaspy, J. M. Pomeroy, A. C. Perrella and H. Grube, *J. Phys. Conf. Series* **58**, 451-456 (2007).
10. A. I. Pikin, C. A. Morgan, E. W. Bell, L. P. Ratliff, D. A. Church and J. D. Gillaspy, *Rev. of Sci. Instrum.* **67** (7), 2528-2533 (1996).
11. J. M. Pomeroy, H. Grube and A. C. Perrella, *Rad. Eff. and Def. in Sol.* **162** (7-8), 473-481 (2007).
12. J. M. Pomeroy and H. Grube, *J. of App. Phys.* **105** (9), 094503-094508 (2009).
13. R. E. Lake, J. M. Pomeroy and C. E. Sosolik, *J. Phys.-Condes. Matter* **22** (8), 084008 (2010).

Electron Spectroscopy In Heavy-Ion Storage Rings: Resonant and Non-Resonant Electron Transfer Processes

S. Hagmann[1,2*], Th. Stöhlker[2,3], Ch. Kozhuharov[2], V. Shabaev[4], I. Tupitsyn[4], Y. Kozhedub[4], H. Rothard[5], U. Spillmann[2], R. Reuschl[2,6], S. Trotsenko[2,3], F. Bosch[2], D. Liesen[2], D. Winters[2], J. Ullrich[7], R. Dörner[1], R. Moshammer[7], P.-M. Hillenbrand[2], D. Jakubassa-Amundsen[8], A. Voitkiv[7], A. Surzhykov[7], D. Fischer[7], E. de Filippo[9], X. Wang[7,10], B. Wei[10]

[1]Inst. f. Kernphysik Univ. Frankfurt, [2]GSI-Helmholtz-Zentrum, Darmstadt, [3]Helmholtz-Institut Jena, [4]Dept. of Physics, St. Petersburg State Univ., St Petersburg, Russia, [5]CIMAO-CIRIL-GANIL, Caen, [6]Univ. P. Marie Curie, ParisVI, [7]Max Planck Inst. f. Kernphysik, Heidelberg, [8]Mathem. Inst., LMU, München, [9]LNS, Catania, Italy, [10]Fudan Univ., Shanghai, China

Abstract. Whereas our understanding of total cross sections for ionization and capture processes in ion-atom collisions is widely viewed as having arrived at a state of adequate maturity, the same cannot be said at all about the dynamics of collisions, multi-electron processes or the electron continua (in target and projectile) which are at the origin of total cross sections. We depict how these processes can be studied favourably in storage ring environments. We present examples of resonant and non-resonant electron transfer processes, radiative and non-radiative. This is elucidated via the relation of the electron nucleus bremsstrahlung at the high energy tip of the bremsstrahlung spectrum to the radiative electron capture cusp (RECC) and a new approach to determining molecular orbital binding energies in superheavy quasi-molecules in resonant KK charge transfer.

Keywords: Ion-Atom Collisions, Electron Spectroscopy, Heavy Ion Storage Rings
PACS: 34.50.Fa, 34.70.+e, 29.20.db, 29.30.Dn

INTRODUCTION

Collisions of heavy ions with atoms encompass a supremely rich spectrum of mechanisms active in electron excitation into the target- as well the projectile continuum. The respective mechanisms range from weakly perturbing Coulomb excitation/ ionization e.g. for light targets ionized by swift ions all the way to strongly molecular promotion processes in adiabatic near symmetric collisions of the heaviest ions and to radiative electron transfer processes in their deep lying relation to electron-nucleus bremsstrahlung and photoionization. However, beyond total cross sections, where details of the electron emission pattern into the continua and their dependence on dynamical variables like momentum transfer are concerned the corresponding experiments still present a formidable challenge for the current most advanced a priori theories. For these classes of experiments, storage rings for heavy ions all the way up to Uranium, like the present ESR and the future NESR storage ring within the FAIR [1] project, now provide the user with a wide range of unique and unrivalled tools for spectroscopy as well as for the investigation of the dynamics of ion-atom and electron-ion collisions (see fig 1 future NESR with experimental installations).

At GSI, the ESR storage ring with a magnetic rigidity of 10Tm allows storing stable and exotic ions up to U^{92+} ions with specific energy ranging from 400AMeV down to well below 10AMeV (after deceleration). After electron cooling the momentum spread is reduced to values significantly below $1\ 10^{-4}$, this high luminosity beam thus permits a wide range of sensitive experiments.

The NESR storage ring [2] is the major workhorse for future atomic physics at FAIR. Energetic highly-charged heavy ions from SIS18 or exotic nuclei from the fragment separator SFRS will be injected into the NESR (see fig. 1). The high luminosity ring with a magnetic rigidity of 13Tm (compared to 10 Tm for the present ESR) allows storing U^{92+} ions and antiprotons

Application of Accelerators in Research and Industry
AIP Conf. Proc. 1336, 115-118 (2011); doi: 10.1063/1.3586069

with energy from 780AMeV down to 3AMeV and will in all aspects exceed the performance parameters of the current ESR.

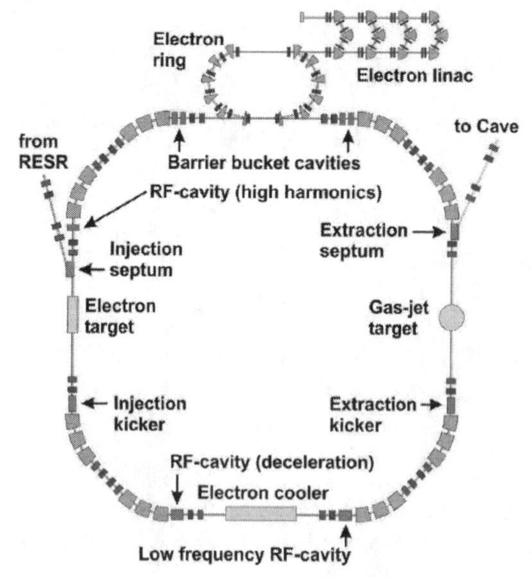

FIGURE 1. NESR storage ring depicting beam control and experimental installations.

ELECTRON SPECTROMETERS

For studies pertaining to electron spectroscopy we have installed a variety of imaging electron spectrometers. We have constructed one for forward Cusp electron spectroscopy ($v_e \approx v_{proj}$) which is an imaging 0^0-electron spectrometer in the ESR target zone. In combination with a reaction microscope this opens up many atomic fundamental processes for first studies in the realm of highly charged ions [3- 5].

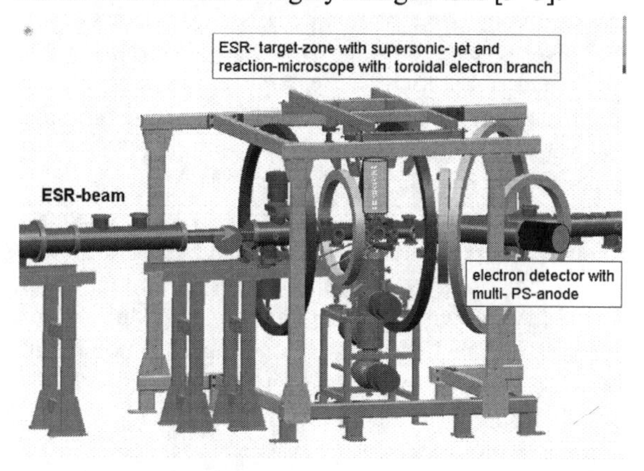

FIGURE 2. ESR target zone with reaction microscope and toroidal sector electron branch.

The imaging forward electron spectrometer is designed to reconstruct, after momentum analysis, on the 2D position sensitive detector the primordial vector momenta of electrons emitted in the target zone using telescopic imaging: a 60^0 dipole magnet is followed by a quadrupole triplett, another 60^0 dipole, momentum defining slits and a 2D position sensitive electron detector. The spectrometer has been successfully used in the first identification of electrons from the short wavelength limit of electron nucleus bremsstrahlung in the radiative electron capture into continuum cusp (RECC) [5]. Currently the quadrupole triplett is being replaced by a large-bore hysteresis-free unit in order to permit true telescopic operation and thus mapping of the primordial vector momenta of forward electrons onto the 2D position sensitive detector behind the image plane.

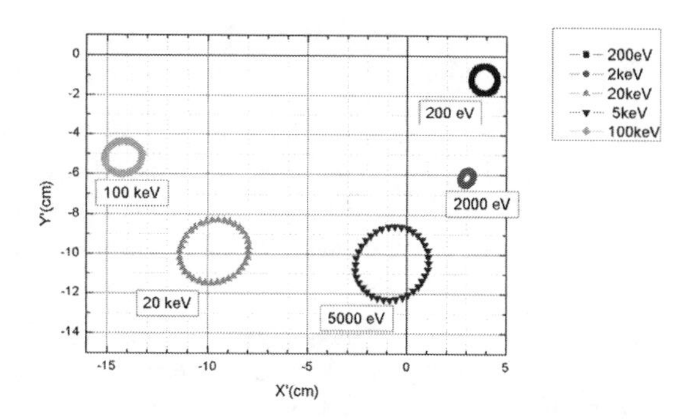

FIGURE 3. The positions of electrons on the detector plane. The position of the beam in this figure is X'= -15.4, Y'=0.

We also have designed for the longitudinal reaction microscope in the ESR a new magnetic toroidal electron sector (see figs. 2 and 3)[6] which permits simultaneously imaging electrons which are emitted with low to intermediate energies into the forward hemisphere, onto 2D position sensitive electron detectors.

ELECTRON TRANSFER PROCESSES

The recent observation that for electrons transferred into the projectile continuum the 00- cusp of electrons with $v_e \approx v_{projectile}$ does not emerge until a separation of around 5000 a.u. between recoiling target ion and projectile[7] elucidates the paramount importance of experiments on the dynamics of production of electron continua. Among these transfer processes the various radiative and non-radiative electron capture channels merit special consideration in comparison with electron loss to the projectile continuum appearing as a 0^0-cusp (ELC). It permits high resolution studies of the dynamics of the ionization for well defined shells of highly charged ionic projectiles with emphasis focusing on continuum states very close to threshold. In an experiment at the ESR of 90AMeV Be-like $U^{88+}(1s^2 2s^2)$ in collision with

N_2 molecules in the supersonic jet target[5] we have identified besides the radiative electron capture into the projectile cusp(RECC) which corresponds to the electron emission from the short wavelength limit of the electron nucleus bremsstrahlung, for the first

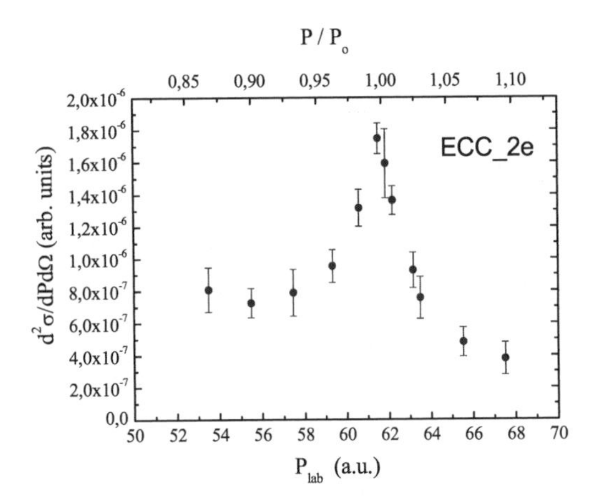

FIGURE 4. ECC-2e cusp for 90AMeV U^{88+} + N_2→U^{87+} + $\{N_2^{+*}\}$ + $e_{cusp}(0^0)$ from coincidences of the 0^0 cusp electron with the charge-changed projectile U^{87+}. The asymmetry of the cusp is clear evidence that the origin is not a double capture followed by autoionization in the projectile.

time at near-relativistic collision velocities a transfer ionization process where besides capture into a projectile bound state an electron is captured into a near threshold continuum state.

As a truly unique feature, the ESR routinely provides slow beams of bare and H-like very heavy ions, e.g. for high resolution spectroscopy. Upon completion of the toroidal sector arm for electrons in

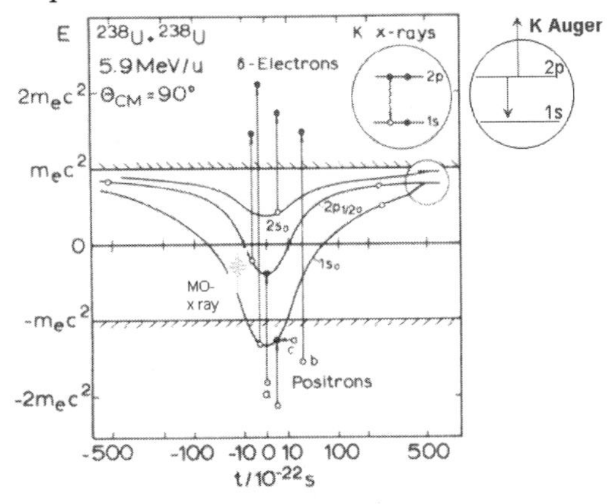

FIGURE 5. Energies of the innermost quasimolecular orbitals as a function of time, i.e. internuclear separation, for CM scattering angle 90^0. The steep dependence of the binding energies on the internuclear separation is clearly visible. Some dynamical processes filling inner shell vacancies are indicated (original of figure in ref 8). Note that binding energies are deduced from given total energies by subtracting m_ec^2 from the actual energies, the zero of the binding energy ordinate is thus at $+m_ec^2$, and diving occurs for a binding energy $<-2m_ec^2$. the reaction microscope, one will for the first time investigate simultaneously the impact parameter (b) dependence of the resonant 1s to 1s charge transfer probability P(b) in heavy symmetric systems like Xe^{53+} + Xe and the adiabatic ionization of the projectile.

In a symmetric collision system, as in Hydrogen-like $Xe^{53+}(1s)$+Xe, the projectile $Xe^{53+}(1s)$ carries a 1s vacancy into the collision. Due to the two possible indistinguishable path along the 1sσ and 2pσ MOs the vacancy may take during the collision [9, 10], the amplitudes of both must be added in order to calculate the probability of the transfer for a given trajectory of impact parameter b and distance of closest approach R_0.

The phase difference of the two amplitudes contains the area between both molecular orbitals $\Delta E(R)=E_{1s\sigma}(R)-E_{2p\sigma}(R)$ integrated over the path along the encounter(see illustration of molecular orbitals in fig. 5) and depends thus on the trajectories minimum internuclear distance R_0, i.e. the impact parameter b. This results - when using a simple 2-state calculation - for the collision velocity dependence of the transfer probability

$$P = \sin^2\left(\frac{1}{v}\int \Delta E(R)\frac{R}{\sqrt{(R^2 - R_0^2)}}dR\right) \quad (1)$$

in an oscillation of P between 0 and 1 as function of impact parameter b, v=collisions velocity.

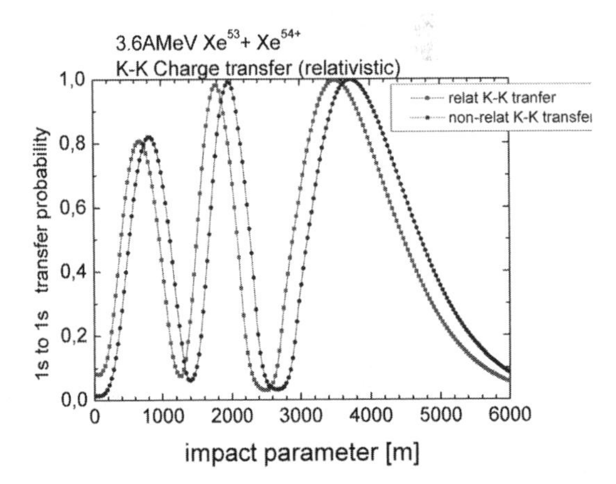

FIGURE 6. Comparison of relativistic and non-relativistic ab-initio calculations by V. Shabaev and I Tupitsyn for 1s to 1s charge transfer in 3.6 AMeV $Xe^{53+}(1s)$ + Xe^{54+}; note the transfer probability reaching unity even at impact parameters around 4000 fm. The smaller probability in the maximum near b ≈1000fm is attributed to the competing adiabatic ionization to the continuum. storage ring with experimental installations.

In fig. 6 a priori calculations from V. Shabaev et al. [11] for our planned experiment on K-K charge transfer in Xe^{53+}+Xe collisions show that transfer probabilities indeed oscillate and will reach for certain impact parameters values near unity. The location of the turning points as function of impact parameter b allows the determination of the energy difference in the integral in eq. 1.

SUMMARY

We have illustrated forward electron spectroscopy in storage rings with examples from recent electron cusp spectroscopy and have shown that it serves as a useful tool to unravel complex interaction dynamics.

For an experiment scheduled for the near future at the ESR we show how oscillations in P(b) of the 1s to 1s charge transfer very sensitively depend on the energy difference $E_{1s\sigma}$-$E_{2p\sigma}$ of the quasi-molecular orbitals as function of the internuclear separation and thus may eventually for very heavy collision systems, like U+U, serve as a truly unmistakable indicator of the $1s\sigma$ level's diving into the Dirac sea once it occurs in the course of a collisions.

*corresponding author

REFERENCES

1. FAIR Baseline Technical Report July, 2006, http://www.gsi.de/fair/reports/btr.html.
2. Ch. Dimopoulou et al. Phys. Rev ST10 020101 (2007)
3. D. Fischer et al. Ann. Rep. GSI 243 (2006)
4. K-U. Kühnel et al. Ann. Rep. GSI 276 (2007)
5. M. Nofal et al. Phys. Rev. Lett **99** 163201 (2007)
6. X. Wang, S. Hagmann, Meas. Sci. Technol. **18** (2007) 161
7. T. G. Lee et al. Phys Rev **A76** 050701 (2007)
8. H. Backe, Ch. Kozhuharov, in Prog Atom. Spect **C**, p459, ed H. Beyer, Plenum (1984)
9. S. Hagmann et al. Phys. Rev. **A36** 2603 (1987)
10. R. Schuch et al. Phys. Rev. **A37** 3313 (1988)
11. I. Tupitsyn et al. Phys. Rev **A82,** 042701 (2010)

Guiding Of Slow Highly Charged Ions Through A Single Mesoscopic Glass Capillary

R. J. Bereczky[a], G. Kowarik[b], C. Lemaignan[b], A. Macé[b], F. Ladinig[b], R. Raab[b], F. Aumayr[b], and K. Tőkési[a]

[a]*Institute of Nuclear Research of the Hungarian Academy of Sciences (ATOMKI) Debrecen 4001, Hungary, EU*
[b]*Institute of Applied Physics, TU Wien - Vienna University of Technology, 1040 Vienna, Austria, EU*

Abstract. We present experimental studies of the transmission of slow 4.5 keV Ar^{9+} ions through a single cylindrical-shaped glass capillary of macroscopic dimension with large aspect ratio. We find stable transmission of a micrometer-scale beam with considerable intensity after a charge-up phase, in which a self-organized process leads to the formation of a guiding electric field. We show the time evolution of the transmitted intensity through the sample for various capillary tilt angles.

Keywords: glass capillary, macroscopic capillary, ion guiding, temperature dependence
PACS: 34.50.Fa, 34.50.Dy

INTRODUCTION

In the past years research activities in the field of charged particle physics have turned to the investigation of charged particles interactions with cylindrical surfaces based on nanostructured materials and capillaries from nano- to macrometer size. These investigations with internal surfaces in microcapillaries introduced an alternative technique to study above-surface processes. The first measurements were performed using metallic microcapillaries, showing good agreement between theoretical predictions and experimental results [1]. Due to the discovery of the unexpected effect of charged particle guiding, the main interest later focused onto the investigation of the interaction between highly charged ions (HCI) and insulating nanocapillaries [2]. In contrast to the case of metallic capillaries, the experiments with insulating ones showed not only directional guiding of the ions, but also the remarkable fact that the ions keep their initial charge state as a consequence of a self-organized charge-up inside the capillary. The fact that most of the guided ions keep their initial charge state suggests that the ions do not touch the inner wall of the capillary during the transport process, i.e. a surprisingly well-tuned electric field is formed in each capillary. Ion guiding through the capillary ensues as soon as a dynamical equilibrium of self-organized charge-up by the ion beam, charge relaxation and reflection is established [3]. So the slow highly charged ions are able to pass through the capillary, keeping their initial charge states even though the capillary axis is tilted with large angles compared to the direction of the incident beam.

For the understanding the basic properties of the guiding for ions using several insulating materials like polyethylene terephthalate (PET), SiO_2, and Al_2O_3 [2-7] an intensive experimental investigation is started. Recently a great progress to explain guiding of positive ions through insulating, randomly distributed or ordered arrays of nanocapillaries has been achieved. During the latest experiments the guided transmission of low-energy electrons through insulator samples was also observed [8-10]. Another viewpoint of the experiments was how the guiding effect changes with the length or with the inner diameter of the capillary. In these investigations most of the experiments focused on the transmission ofcharged particles through tapered capillaries [11, 12].

Although during the past few years many research groups joined to this field of research and carried out various experiments with insulator capillaries many details of the interactions are still unknown.

Since the experiments so far have used almost thin insulating foils with randomly distributed capillaries (produced by swift heavy ion bombardment) or with ordered arrays of regular nanocapillaries, collective effects due to the presence of neighboring capillaries must be taken into account for an accurate simulation of the ion trajectories [3]. These collective effects make a full description of the interaction between

Application of Accelerators in Research and Industry
AIP Conf. Proc. 1336, 119-122 (2011); doi: 10.1063/1.3586070

charged particles and insulator capillary walls rather difficult.

Avoiding these difficulties, in this work we use a single, straight, cylindrical symmetric glass capillary with macroscopic dimension with large aspect ratio [13, 14]. We present experimental studies of the transmission of slow 4.5 keV Ar^{9+} ions through this capillary. We show the time evolution of the transmitted intensity through the sample for various capillary tilt angles.

EXPERIMENT

The measurements were carried out using beams of 4.5 keV Ar^{9+} ions produced by the Vienna 14.5 GHz electron cyclotron resonance ion source. We used a single, high aspect ratio glass capillary with length, L about 1cm, and with diameter, D=0.14 mm, so the aspect ratio, L/D is about 80. The capillary was produced in the Institute of Nuclear Research of the Hungarian Academy of Sciences (ATOMKI) by heating a straight glass tube of 1 cm diameter, made of Pyrex (borosilicate glass) and stretching it afterwards by pulling both ends with a constant force.

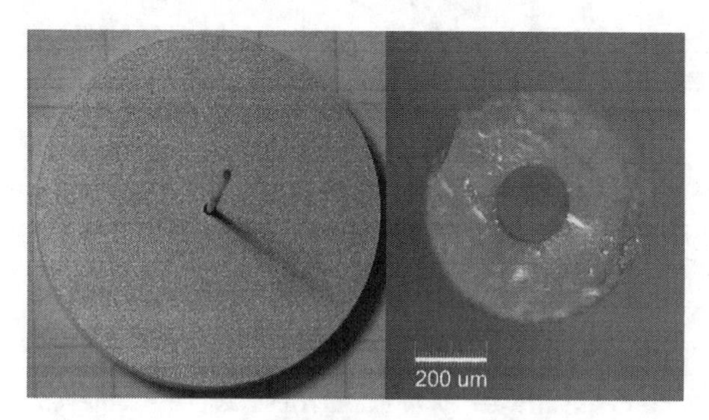

FIGURE 1. Top view of the fixed capillary (left), optical micrograph of the cross section of the capillary (right).

The capillary was mounted into an Al holder, as it is shown in Figure 1, and fixed by UHV suitable glue. In order to avoid strong fields and to assure rotational symmetry as well as proper charge transport, the outside of the capillary is covered by graphite. The acceptance angle of the glass tube is ±0.7°. The glass capillary was positioned in a differentially pumped UHV chamber, with a residual pressure better than 5 x 10^{-9} mbar.

Figure 2 shows a schematic view of the experimental setup. The beam diameter at the entrance of the capillary could be estimated to roughly 3 mm with an opening angle of full widths at half maximum (FWHM) of ~2.2°.

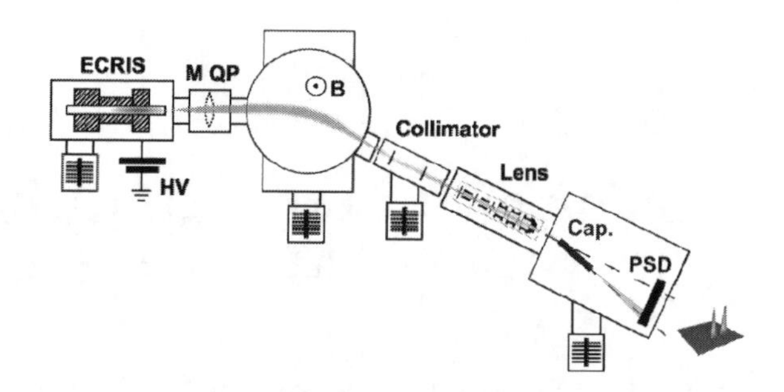

FIGURE 2. Schematic view of the experimental setup. The ion source (ECRIS) provides a variety of ion species, which are mass-charge-selected in a magnetic field and focused by magnetic quadrupol lenses (M QP) onto an entrance aperture. A system of electrostatic deflection plates and lens elements provides the final beam of 3 mm diameter at the capillaries entrance. The PSD is located excentric roughly 18 cm behind the capillary.

Approximately 18 cm behind the entrance of the capillary, a position sensitive detector (PSD) mainly consisting of a Chevron-type micro-channel-plate and a wedge-and-strip-anode is used to record the transmitted ions. The capillary holder itself could be moved in x, y and z-direction with high accuracy by stepper motors. Also the tilting against the beam axis in the horizontal plane (Φ-direction) was stepper-motor controlled, allowing step sizes as small as 0.02°. The location of the turning axis was chosen in a way to cross the entrance of the capillary in order to keep the beam facing side of the capillary in a fixed position when turning. When moving the capillary out of the vertical position of it, a reference intensity could be measured by allowing ion passage through a hole of known diameter (d=0.5mm) directly onto the channelplate detector. This direct beam measurement was also used to estimate the initial angular distribution entering the capillary. Tilting measurements were performed in a way, that the discharged capillary was positioned axial to the beam, from there the tilt angle was increased step-wise into one direction. The other tilt direction was either measured starting again from the straight position with a discharged capillary or by returning stepwise from the outmost deflection from the first tilt direction into the other one, passing the straight position in a charged-up condition.

RESULTS AND DISCUSSION

We investigated the time dependent transmission probability of highly charged ions passing through the single glass capillary (see Figure 3).

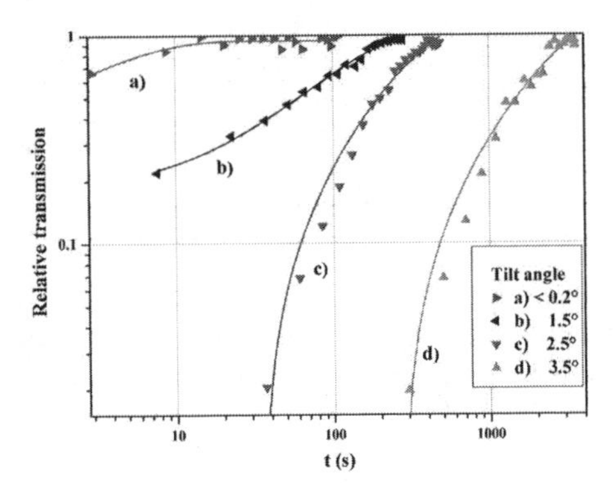

FIGURE 3. Time dependent relative transmission through the single glass capillary for various tilt angles. The values are normalized to 1 for the equilibrium case. The corresponding characteristic times of the charging up process are as follows: a) τ = 4s, b) τ = 143s, c) τ = 355s, d) τ = 3100s.

The time-dependence during the charging-up of the capillary was measured by tilting the capillary to 3.5° after having stable transmission in the straight direction.

We found that the experimentally observed values can be fitted with an exponential law $1\text{-}\exp^{(-t/\tau)}$, where τ is the time constant of the charging-up process). The results of the two parameter fit are shown in Figure 3 as solid lines. For tilt angles smaller than the capillary acceptance angle we immediately observed stable transmission because of the geometrical transmission. When the tilt angle compared to the beam axis is larger than the acceptance angle of the capillary, the injected ions hit the inner wall of the tube at least once. In agreement with the expectation, with increasing tilt angle the characteristic time of the charging-up increases.

We found that considerable stable transmission of guided ions could be observed for tilt angles of up to roughly 5°, after a charge-up period showing a smooth increase of ion transmission as well as deflection angle as a function of time (dose) until saturation is reached. The observations indicate that the initial charge-state is kept during the guiding.

We did systematic measurements to investigate the angular distribution of 4.5 keV Ar^{9+} projectiles passing through a single glass macrocapillary. As an example Figure 4 shows the angular distributions of the direct beam, collimated by a 0.5 mm aperture (a) in comparison with the guided distributions (b) for four different tilt angles as indicated. These angular distributions are calculated from the spatial ones at the PSD by a proportionality constant, which has been obtained by fitting the deflections of several measurements as function of tilt angle and assuming agreement between these values.

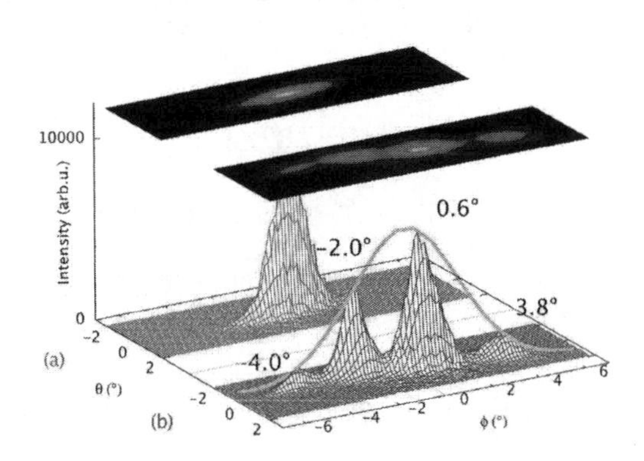

FIGURE 4. Image of the transmitted beam through a 0.5 mm of diameter reference aperture (a) and the macroscopic glass capillary for various tilt angles (b). The given angles correspond to azimuth (Φ) and elevation (Θ) angles.

The distribution of the direct beam ions is shown in order to get an impression of the initial distribution. The z-axis shows the number of ion impacts per unit area. The solid curve is the result of a Gaussian fit of the maxima of the distributions. Φ denotes the lateral deflection angle in the horizontal plane, the elevation angle is Θ. The elliptical shape of the beam results from imaging aberrations in the mass-charge-separation plane.

CONCLUSION

In this work, we have presented systematic measurements in collisions between slow HCIs and a macroscopic glass capillary with large aspect ratio. The observed data strongly support that the guiding effect known from nanocapillaries is also valid up to macroscopic dimensions of the order of millimeter. Possible tilt angles were as large as about 5° for still significant transmission. The charging-up of the insulating wall can be observed in the time-dependent transmission measurements. We found that the angular distributions of the transmitted ions have almost a similar width as the incident beam.

ACKNOWLEDGMENTS

The financial support received from the ITS-LEIF Project (RII3 026015) is gratefully acknowledged. This work was supported by the "Stiftung Aktion Österreich-Ungarn", the TéT Grant no. AT-7/2007, the Hungarian National Office for Research and Technology, as well as Austrian FWP (P17499), and the Hungarian Scientific Research Found OTKA (K72172). We also thank to Mr. Ferenc Turi for the technical support.

REFERENCES

1. K. Tőkési, L. Wirtz, C. Lemell and J. Burgdörfer, *Phys. Rev. A* **61** (2001), 020901 (R).
2. N. Stolterfoht, J.-H Bremer, V. Hoffmann, R. Hellhammer, D. Fink, A. Petrov, and B. Sulik, *Phys. Rev. Lett.* **88**, 133201 (2002)
3. K. Schiessl, W. Palfinger, K. Tőkési, H. Nowotny, C. Lemell, J.Burgdörfer, *Phys. Rev. A* **72**, 062902 (2005)
4. R. Hellhammer, Z.D. Pesic, P. Sobocinski, D. Fink, J. Bundesmann and N. Stolterfoht, *Nucl. Instr. and Meth. Phys. Res. B* 233 (2005)
5. M.B. Sahana, P. Skog, Gy. Vikor, R.T. Rajendar Kumar and R. Schuch, *Phys. Rev. A* **73** (2006)
6. M. Fürsatz, W. Meissl, S. Pleschko, I .C. Gebeshuber, N. Stolterfoht, HP. Winter and F. Aumayr, *J. Phys. Conf. Proc. HCI* 58 (2007)
7. P. Skog, I.L. Soroka, A. Johansson and R. Schuch, *Nucl. Instr. and Meth. Phys. Res. B* 258 (2007)
8. S. Das, B.S. Dassanayake, M. Winkworth, J.L. Baran, N. Stolterfoht and J.A. Tanis, *Phys. Rev. A* **76** (2007)
9. A.R. Milosavljević, G. Vikor. Z.D. Pešić, P. Kolarž, D. Šević, B.P. Marinković, S. Mátéfi-Tempfli, M. Mátéfi - Tempfli, and L. Piraux, *Phys. Rev. A* **75** (2007) 030901 (R).
10. B.S. Dassanayake, S. Das, R.J. Bereczky, K. Tőkési and J.A. Tanis, *Phys. Rev. A* **81** (2010) 020701(R).
11. T. Ikeda, Y. Kanai, T.M. Kojima, Y. Iwai, T. Kambara, Y. Yamazaki, M. Hoshino, T. Nebiki and T. Narusawa, *Appl. Phys. Lett.* 89 (2006)
12. A. Cassimi, et al., *Int. J. Nanotechnol.* 5 (2 0 0 8)
13. R.J. Bereczky, G. Kowarik, F. Aumayr, K. Tőkési, *Nucl. Instr. and Meth. Phys. B* 267, 317 (2009)
14. G. Kowarik, R.J. Bereczky, F. Aumayr, K. Tőkési, *Nucl. Instr. and Meth. Phys. B* 267, 2277 (2009)

Recent Applications Of The Lattice, Time-Dependent Schrödinger Equation Approach For Ion-Atom Collisions

D.R. Schultz[a], S.Y. Ovchinnikov[b], J.B. Sternberg[b], and J.H. Macek[a,b]

[a]*Physics Division, Oak Ridge National Laboratory, Oak Ridge, TN 37831*
[b]*Department of Physics and Astronomy, University of Tennessee, Knoxville, TN 37496*

Abstract. Contemporary computational methods, such as the lattice, time-dependent Schrödinger equation (LTDSE) approach, have opened opportunities to study ion-atom collisions at a new level of detail and to uncover unexpected phenomena. Such interactions within gaseous, plasma, and material environments are fundamental to diverse applications such as low temperature plasma processing of materials, magnetic confinement fusion, and astrophysics. Results are briefly summarized here stemming from recent use of the LTDSE approach, with particular emphasis on elucidation of unexpected vortices in the ejected electron spectrum in ion-atom collisions and for an atom subject to an electric field pulse.

Keywords: ion-atom collisions.
PACS: 34.50.Fa, 34.70.+e, 34.80.-t.

INTRODUCTION

By about fifteen year ago, computational power and techniques had advanced to the point that the time-dependent Schrödinger equation (TDSE) could be solved directly on a 3-dimensional grid to provide a novel description of ion-atom collisions and related phenomena. This approach, the lattice TDSE (LTDSE) method [1], was aimed at treating situations not readily amenable or not sufficiently accurately by other theoretical methods such as perturbation theory or atomic and molecular orbital close coupling approaches. Along with other contemporary methods (e.g., [2]), LTDSE has become not only a vehicle to explore such interactions but also, as computational power has continued to advance, a tool for calculations needed in applications such as astrophysics and fusion energy research. Particularly for the later application, we have recently applied LTDSE, augmented by and tested against results from other methods such as the atomic orbital close coupling (AOCC) and classical trajectory Monte Carlo (CTMC) methods, to calculation of charge transfer in ion-atom collisions needed to understand plasma behavior and diagnostics based on neutral beam injection in fusion devices. Such studies have included detailed treatment of state-selective charge transfer in collisions of Be^{4+} with atomic hydrogen [3] owing to the use of beryllium in fusion device walls, and for He^{2+} + H [4] because of the importance of plasma diagnostics based alpha particles. We have also used these complementary methods, LTDSE, AOCC, and CTMC, to treat plasma-relevant excitation and charge transfer involving proton collisions with H(2s) [5] because of the importance in some instances of processes involving the metastable state of hydrogen created through electron-impact excitation.

We have also applied the LTDSE method to treat the spectrum of electrons ejected in collisions that are promoted into the two-center continuum [6], motivated by the pioneering measurements of such spectra via the "reaction microscope" or COLTRIMS method [7-11]. That work led to new insight into the formation of the σ-π distribution effects [12] observed experimentally. However, limits on the size of the numerical lattice prohibited reaching an asymptotic internuclear distance using LTDSE or similar methods [13]. To go beyond this limitation, we have very recently developed a new approach, the regularized LTDSE (RLTDSE) method [14], in order to propagate the solution to distances where we are assured that the electronic probability distribution is comparable to that which would be observed experimentally. In so doing, we have unexpectedly observed vortices in the electronic wavefunction [15]. We have also found that certain of these vortices, being born in the collision, surprisingly survive to asymptotic distances in the electronic continuum and may thus be observable in experiments. In particular, we have shown how

Application of Accelerators in Research and Industry
AIP Conf. Proc. 1336, 123-126 (2011); doi: 10.1063/1.3586071

vortices appear in proton-hydrogen collisions, rotate around the nuclei, and interact, thereby transferring angular momentum from nuclear to electronic motions.

To illustrate recent progress using these methods in elucidating the origin and evolution of these vortices in an unexpectedly wide range of situations, we describe in the following Section our treatment of a two-electron ion-atom system and of an atom subject to a short electric field pulse.

VORTICES IN ATOMIC WAVEFUNCTIONS

As described previously [15,16], enabled by RLTDSE calculations, we examined the time-dependence of the electronic wavefunction (for proton-hydrogen collisions at 5 keV) throughout the evolution from near the collision to large internuclear separations, and noticed, unexpectedly, that deep minima existed when the nuclei were close, some of which persisted to the largest final distances considered. After examining the electronic probability current, we confirmed that these deep minima were vortices and were in fact quantized. Since that first study revealing them, we have found them in electron-impact ionization [17] also surprisingly since their formation comes about through a different topology of the potential experienced by the ejected electron (a repulsion in the electron-impact case rather than an attraction for proton-impact), and found that the vortex resulted in what had been an unexplained feature of the spectrum of ejected electrons.

Having found these vortices in both electron- and proton-impact of atomic hydrogen, we were motivated to determine if they were also present in a multi-electron target atom in order to treat a system that would be more readily amenable to experimental study. To that end, we have extended our methods to a higher number of dimensions in order to treat two-electron systems, and applied them to H^+ + He collisions. Full treatment of this two-electron collision in the lattice approach requires a 6-dimensional grid, still a daunting challenge to implement even on the largest supercomputers. Therefore, in order to explore a wide range of impact parameters and energies, we have begun with a 4-dimensional, planar model so that a number of calculations can be performed as pilot studies before performing 6-dimensional calculations. This follows our previous treatment in 4-dimensions of ionization in antiproton-helium collisions using LTDSE [18]. In the present case, using RLTDSE, the outgoing flux of electrons can be propagated to asymptotic distances in order to obtain not only total cross sections as in earlier work but the spectrum of ejected electrons.

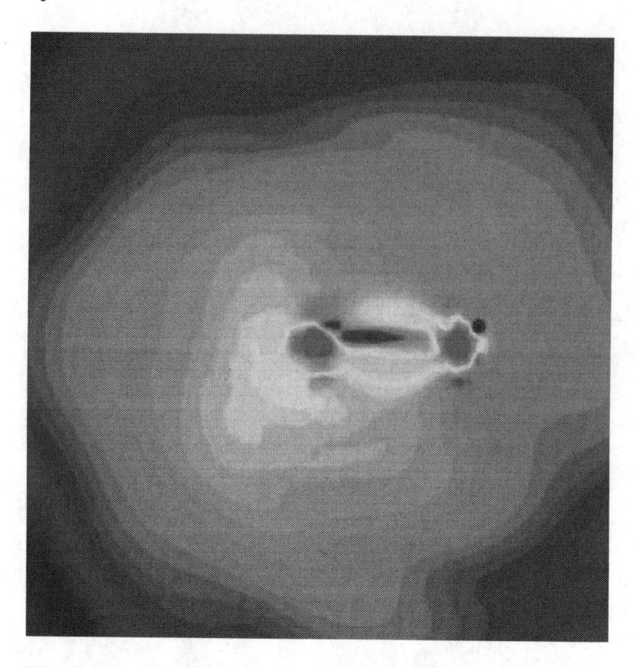

Figure 1. The electronic probability density for transfer ionization in a 4-dimensional model of 5 keV H^+ + He collisions at an impact parameter of 1 a.u. and an asymptotic distance following the collision. The horizontal and vertical axes are the electronic velocity scaled by the incoming proton velocity and range from -2 to 2 on each axis, yielding an image of the ejected electron momentum distribution. The transfer ionization probability density is obtained by integrating over the captured electron and the spectrum of the other ejected electron. Electrons near the target He atom are centered at the origin (high density at the center of the figure) and those near the projectile are to the right (high density concentration near coordinates (1,0)). Owing to the scaled coordinates, the bound state probability is strongly localized very near the two nuclei at the asymptotic internuclear separation shown here. The deep minima near the two-centers are vortices formed in the near collision that persist to asymptotic distances.

We show in Figure 1 the first result of this 4-dimensional calculation of the spectrum of ejected electrons (electronic velocity distribution) for transfer ionization in 5 keV H^+ + He at an impact parameter of 1 atomic unit (a.u.) at an asymptotic distance following the collision. The two densest regions show the high electron probability density associated with either the target or the projectile. The dark oval or circular "holes" near the projectile and target velocities are vortices that we can trace from their origin in the collision (within a few atomic units of projectile-target separation). This result confirms the existence and the similar origin of vortices as found for the case of atomic hydrogen targets in a two-electron target, and will hopefully motivate experiments.

Noting that the vortices originate from a particular topological effect of the potential experienced by the electron in ion-impact, and another stemming from the repulsion in the electron-impact case, we have also recently investigated [19] their formation resulting from a short electric field pulse on atomic hydrogen. Consideration of the electric pulse has simplified the topology and still led to vortex formation and persistence to asymptotic distances where they could be detected in experiments. Formation of the vortices by an electric field pulse also opens up the possibility of purposefully manipulating the vortices, which we have also shown [19] behave as quasiparticles (i.e., they persist, are driven by the field, and follow a trajectory like a particle). Figure 2 shows an elementary manipulation of the vortices created by applying an oppositely directed, second electric pulse immediately following the first.

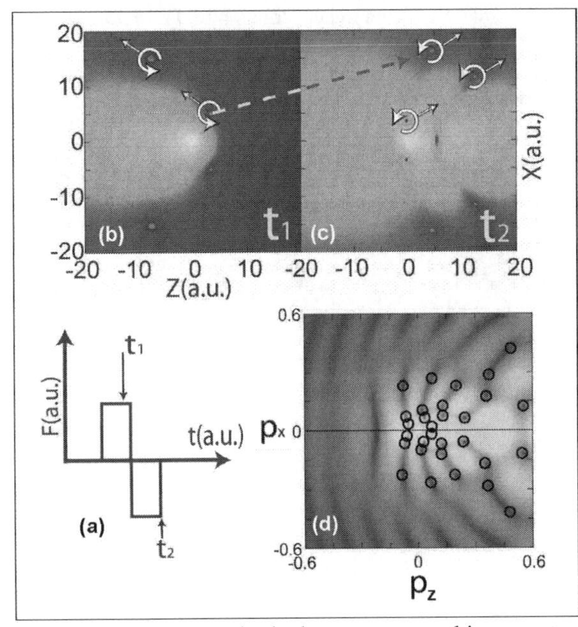

FIGURE 2. An atomic hydrogen atom subject to two successive short electric field pulses (a) of opposite sign (the first directed to the left, negative values of z), each of 16 a.u. duration, and 0.1 a.u. in magnitude. Ring vortices are formed on the righthand side of the atom, expand around the atom, and move off downstream shown (b) at $t=t_1=12$ a.u. As the field direction is reversed, the next vortices to form do so on the lefthand side of the atom, expand around the atom, and move off to the right. Note that the field turns the previously formed vortices around and drives them to the right as well. The vortex positions and direction of motion at the end of the second pulse, $t=t_2=32$ a.u. are shown in (c). The dashed arrow identifies the position and direction of the second vortex formed during the first pulse at the end of the second pulse. The spectrum of ejected electrons resulting from the two electric field pulses at asymptotic distances is shown in (d) showing the positions of vortices, primarily those formed latest in time and being associated with the lowest electronic momentum.

We have also found a correspondence of this phenomenon, that is, the onset of the vortices in solution of the TDSE, to its analog in fluid dynamics. In particular, closed line, ring vortices spontaneously appear associated with points in potential flows where density and pressure vanish. In the hydrodynamic interpretation, when the ground state hydrogen atom is subjected to the electric field step, a roughly spherical, thin layer of low pressure forms to completely surround the hydrogen atom. A ring vortex then develops on this surface at an area of low density and pressure along the direction of the electric field at a radius of about 5 a.u. The vortex trajectory follows the rules of hydrodynamics, propagating along the path of low pressure. As a result, the ring vortex follows this roughly spherical layer, forming on the side facing into the pulse, and being pushed by the force of the pulse to expand around the high density and high pressure core of the atom. The first ring vortex forms at about 4.3 a.u. after the turn-on of the electric field, associated with transitions to the lowest excited state, followed by additional vortices corresponding to the subsequent onset of excitation to higher states including the continuum.

CONCLUSIONS

The past decade has seen the growing utility in applying the LTDSE and other contemporary methods to treat ion-atom collisions because of the growing capacity of available computer systems and improving computational techniques. This has allowed their use in complementing and going beyond traditional theoretical methods such as perturbation theory and atomic and molecular orbital close coupling approaches for use in a wide range of applications such as astrophysics, low temperature plasma science, and fusion energy research. In addition, these contemporary methods have lead to uncovering and elucidating new phenomena, an example of which is the origin, evolution, and asymptotic persistence of vortices in a wide range of atomic-scale interactions. Further study of vortex formation will likely lead to new insights into the fundamental dynamics of ion-, electron-, and photon-impact of atoms, such as the transfer of angular momentum, and could lead to a new ability to manipulate vortex quasiparticles with tailored trains of pulses.

ACKNOWLEDGMENTS

Research was sponsored by the Office of Basic Energy Sciences and Office of Fusion Energy Sciences, U.S. Department of Energy, under contract No. DE-AC05-96-OR22464 with Oak Ridge National Laboratory, which is managed by UT-Battelle, LLC, and by Grant No. DE-FG02-02ER15283 from the Office of Basic Energy Sciences to the University of Tennessee.

REFERENCES

1. D.R. Schultz, M.R. Strayer, and J.C. Wells, *Phys. Rev. Lett.* **82**, 3976 (1999).
2. H.J. Lüdde, S. Henne, T. Kirchner, and R.M. Dreizler, *J. Phys. B* **29**, 4423 (1996); O.J. Kroneisen, H.J. Lüdde, T. Kirchner, and R.M. Dreizler, *J. Phys. B* **32**, 2141 (1999).
3. T. Minami, M.S. Pindzola, T-G. Lee, and D.R. Schultz, *J. Phys. B* **39**, 2877 (2006).
4. T. Minami, T-G. Lee, M.S. Pindzola, and D.R. Schultz, *J. Phys. B* **41**, 135201 (2008).
5. M.S. Pindzola, T-G. Lee, T. Minami, and D.R. Schultz, *Phys. Rev. A* **72**, 062703 (2005).
6. D.R. Schultz, C.O. Reinhold, P.S. Krstic, and M.R. Strayer, *Phys. Rev. A* **65**, 052722 (2002).
7. R. Dörner, H. Khemliche, M. H. Prior, C. L. Cocke, J. A. Gary, R. E. Olson, V. Mergel, J. Ullrich, and H. Schmidt-Böcking, *Phys. Rev. Lett.* **77**, 4520 (1996).
8. M. A. Abdallah, C. L. Cocke, W. Wolff, H. Wolf, S. D. Kravis, M. Stöckli, and E. Kamber, *Phys. Rev. Lett.* **81**, 3627 (1998).
9. R. Moshammer, A. Perumal, M. Schulz, V. D. Rodríguez, H. Kollmus, R. Mann, S. Hagmann, and J. Ullrich, *Phys. Rev. Lett.* **87**, 223201 (2001).
10. M. Schulz, R. Moshammer, D. Fischer, H. Kollmus, D. H. Madison, S. Jones and J. Ullrich, *Nature* (London) **433**, 48 (2003).
11. J. Ullrich, R. Moshammer, A. Dorn, R. Dörner, L. Ph. H. Schmidt, and H. Schmidt-Böcking, *Rep. Prog. Phys.* **66**, 1463 (2003).
12. J.H. Macek and S.Y. Ovchinnikov, *Phys. Rev. Lett.* **80**, 2298 (1998).
13 E.Y. Sidky and C.D. Lin, *Phys. Rev. A* **60**, 377 (1999).
14. T-G. Lee, S.Y. Ovchinnikov, J.B. Sternberg, V. Chupryna, D.R. Schultz, and J.H. Macek, *Phys. Rev. A* **76**, 050701(R) (2007).
15. J.H. Macek, J.B. Sternberg, S.Y. Ovchinnikov, T-G. Lee, and D.R. Schultz, *Phys. Rev. Lett.* **102**, 143201 (2009).
16. S.Y. Ovchinnikov, J.H. Macek, J.B. Sternberg, T-G. Lee, and D.R. Schultz, CP1099, Applications of Accelerators in Research and Industry: 20th International Conference, edited by F.D. McDaniel and B.L. Doyle, AIP Conference Series (2009), p. 164.
17. J.H. Macek, J.B. Sternberg, S.Y. Ovchinnikov, and J.S. Briggs, *Phys. Rev. Lett.* **104**, 033201 (2010).
18. D.R. Schultz and P.S. Krstić, *Phys. Rev. A* **67**, 022712 (2003).
19. S.Y. Ovchinnikov, J.B. Sternberg, J.H. Macek, T-G. Lee, and D.R. Schultz, *Phys. Rev. Lett.* (2010) in press.

Dipole Excitation With A Paul Ion Trap Mass Spectrometer

J. A. MacAskill, S. M. Madzunkov and A. Chutjian

Atomic and Molecular Physics Group, Jet Propulsion Laboratory, California Institute of Technology, Pasadena, CA 91109 USA

Abstract. Preliminary results are presented for the use of an auxiliary radiofrequency (*rf*) excitation voltage in combination with a high purity, high voltage *rf* generator to perform dipole excitation within a high precision Paul ion trap. These results show the effects of the excitation frequency over a continuous frequency range on the resultant mass spectra from the Paul trap with particular emphasis on ion ejection times, ion signal intensity, and peak shapes. Ion ejection times are found to decrease continuously with variations in dipole frequency about several resonant values and show remarkable symmetries. Signal intensities vary in a complex fashion with numerous resonant features and are driven to zero at specific frequency values. Observed intensity variations depict dipole excitations that target ions of all masses as well as individual masses. Substantial increases in mass resolution are obtained with resolving powers for nitrogen increasing from 114 to 325.

Keywords: Paul ion trap, dipole excitation, mass spectrometry
PACS: 37.10.Ty, 07.75.+h

INTRODUCTION

The development of linear quadrupole and Paul ion trap mass spectrometers as a technology is a major research effort with the immediate application of space-flight mass spectrometry. Ongoing efforts continue to increase one's fundamental understanding of the ion kinematics and ultimately the science return obtainable with these devices. In addition, this research effort provides numerous paths for technical improvements in a wide range of areas including design, radiofrequency (*rf*) generation, sample introduction, and analysis.

The basic operation of a Paul ion trap has been widely studied and reviews of its operation can be found in numerous sources [1, 2]. Recent work [3] has focused on the improvement of the spectral quality of the *rf* drive frequency for the simplest mode of operation of the Paul ion trap. This mode is such that both end caps are grounded, with the high voltage *rf* connected to the ring electrode providing a pure quadrupolar trapping potential. The trap operation in this configuration provides excellent performance for nominal mass range 2 – 400 amu with drive waveforms in the range 400 – 2000 MHz, and 30 – 2000 V. Described herein are first studies based on the addition of resonant excitation to the operation of the trap to expand its overall functionality through extended mass range, increased mass resolution, and mass selective ejection as described in Refs. [1, 2].

EXPERIMENT

The Paul ion trap used in this work is similar to that presented earlier [3], but with several improvements. The original ion trap was replaced with a new high mechanical precision (r_0 = 10.000 ± 0.005 mm), chemically pure titanium ion trap with a single one mm diameter hole in each end cap. The wire-wound ionizer was replaced with a tantalum disc emitter. An added einzel lens provided improved beam uniformity and focusing capability. Finally, the original mounting configuration [3] was replaced with one providing electrical isolation of the end caps of the Paul ion trap.

The quadrupole drive waveform to the ring electrode is provided by one of the JPL direct digital synthesis (DDS) *rf* generators [3]. For the work presented here, the quadrupole frequency is Ω_0 = 1245.14 kHz, with a voltage program of V_Q = 105 – 1215 V. This provides an effective mass range of ~ 7 – 83 amu in the absence of the auxiliary drive. The DDS *rf* generator is operated in an open-loop configuration to prevent spectral distortions arising from the feedback. The *rf* used for the auxiliary drive to the end caps is provided by an Agilent 33220A programmable waveform generator. The Agilent 33220A drives the

Application of Accelerators in Research and Industry
AIP Conf. Proc. 1336, 127-131 (2011); doi: 10.1063/1.3586072

primary of a 1:2 toroidal transformer. The secondary is center-tapped with each end of the secondary connected to an end cap and the center tap tied to ground. This gives an approximate gain of unity, relative to the input, for each of the end cap waveforms and maintains a 180° relative phase shift between them. In this configuration, the *rf* auxiliary drive provides a dipole potential with frequency ω_D that is variable from 0 – 5 MHz, and of voltage amplitude V_D = 5 V. Isolating the end caps and making connections to any external devices can modify the electrical characteristics of the tuned circuit providing the *rf* drive. The ability of the DDS *rf* generator to self-resonate alleviates this problem by automatically selecting a new operating frequency for the drive *rf* corresponding to changes in inductance or capacitance brought on by the connection of an external device.

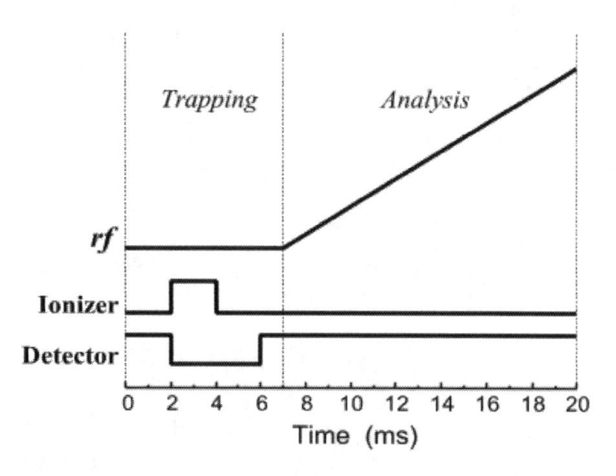

FIGURE 1 Timing sequence of the radiofrequency (*rf*) trapping and analysis potentials, the electron ionization pulse, and the detector *on* period. The dipole excitation is continuous and at fixed amplitude and frequency during this time.

Mass spectra from the trap are recorded as two-dimensional data structures. The ion signal is measured with a counter operating at 500 kHz. The counter stores the results in a buffer spanning one second. Individual mass spectra contained in that one second buffer, dictated by the modulation frequency of the *rf* drive, are summed together and stored as one frame. The results presented here use a modulation frequency of either 25 Hz or 50 Hz. The process is repeated for the next second, and results stored in an additional frame. This two-dimensional data structure allows a time evolution of the mass spectra to be recorded and analyzed. It allows for chromatographic measurements on ion traps with a coupled gas chromatograph [4], for monitoring *rf* stability through mass spectral peak drift, or in this case programmable scans of resultant mass spectra for varying mass spectrometer parameters. As such, it is an effective tool for diagnosing mass spectrometer performance as well as providing additional analytical capabilities.

The timing information for the ionizer, ion detector, and *rf* mass analysis is shown in Fig. 1. For the data presented here the counter frequency is 500 kHz. Operating the ion trap at 50 Hz then provides 10^4 counter channels spanning 20 msec. The ionizer is on for 2 msec, or 1000 counter channels. During this time the ion detector is gated off and remains off for an additional 2 msec. At 7 msec, or channel 3500, V_Q is ramped from its trapping level of 105 V to 1215 V in the following 13 msec or 6500 channels to provide mass analysis of the trapped ions.

RESULTS

Shown in Fig. 2 is an intensity plot of data taken during a frequency scan of $\omega_D = 0 \rightarrow 2.5$ MHz to span 2 Ω_0, in 1.25 kHz increments. This intensity plot contains 2000 mass spectra of the rest gas, at a pressure of P = 3.8 × 10^{-8} Torr. The dipole excitation frequency is shown on the abscissa and ion detection time is shown on the ordinate.

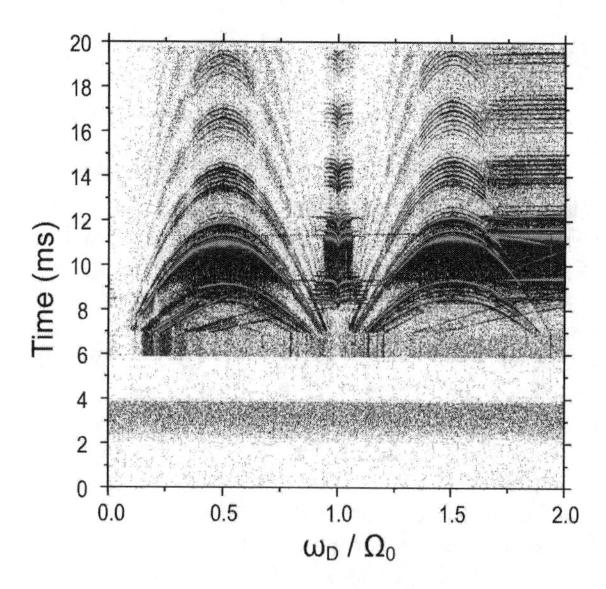

FIGURE 2 Intensity plot of the background-gas mass spectra through dipole frequency ω_D scans in the range 0-2 Ω_0. The intensity plot shows the transitions of the ionizer and detector during the trapping phase, and the variation in ion ejection time and intensity throughout the analysis phase.

The effect of the dipole excitation on ejection times, for all of the trapped ions, is precisely measured in this fashion for both the trapping (0 – 7 ms) and analysis (7 – 20 ms) portions of the quadrupole drive. While the detector is gated off (2 – 6 ms), the ion signal is severely attenuated with slightly elevated count rates during the ionization (2 - 4 ms). These events are identified as the darker, horizontal bands in Fig. 2. The ion signal rates increase after the detector is gated on and ion signal is measured for the remaining 1 ms of the trapping portion. During this time (6 – 7 ms), an interesting feature appears in Fig. 2 resembling a line spectrum for varying dipole excitation frequency. The total ion population produced during the ionization pulse undergoes destabilization for the remaining portion of the trapping-only phase at discrete values of the dipole excitation frequency. The analysis phase begins and varied effects on the ion ejections are observed with increasing dipole excitation frequency. The ion ejections depicted in Fig. 2 for the analysis phase show strong symmetries about several resonant frequencies.

The ion ejections depicted in Fig. 2 for the analysis phase show strong symmetries about several resonant frequencies. Data in Fig. 2 are reduced to emphasize these symmetries and resonances for the ion ejection times and signal intensities. This reduction is accomplished with the aid of software that performs a peak search on each frame and returns the ejection times and heights of all peaks. Each peak is then sorted according to ion mass. The results of this data reduction are shown in Fig. 3 for the mass lines m/z = 14, 16, 17, 18, 19, 28, 29, 32, 40, and 44 amu.

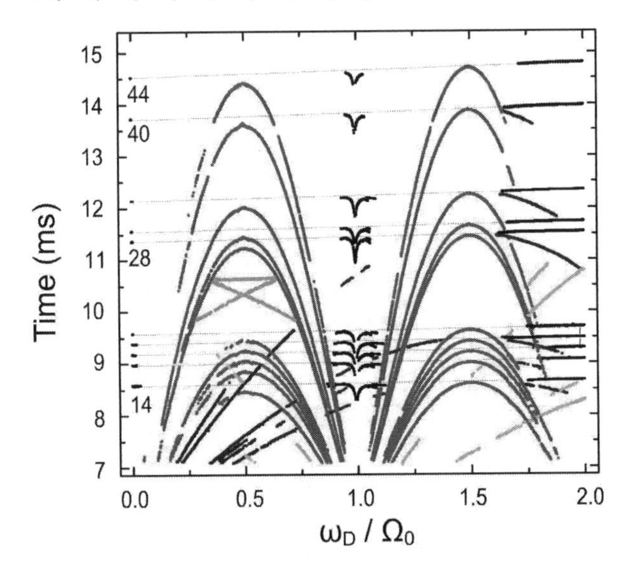

FIGURE 3 . Ion ejection times plotted as a function of dipole frequency ω_D for the masses 14, 16, 17, 18, 19, 28, 29, 32, 40, 44. One notes that the frequencies $\omega_D \sim 0$ and 2 Ω_0 , as well as $\omega_D = 0.5\ \Omega_0$ and 1.5 Ω_0 have relatively little affect on the mass spectra. All masses are annihilated at $\omega_D = \Omega_0$

Dashed lines in Fig. 3 correspond to ejection times dictated by the quadrupole potential only for the ions listed above. The slight slope of these lines is attributed to the open-loop configuration of the quadrupole *rf* that results in a slight decrease in amplitude over the course of the ~35 minute frequency scan. The structures shown in Figs. 2 and 3 are skewed slightly by this voltage variation and applying a correction results in near-perfect symmetry. In the vicinities of $\omega_D = 0$ and $\omega_D = 2\ \Omega_0$ the resultant mass spectra are largely unaffected by the presence of the dipole potential. The spectra are annihilated with the application of a dipole field at $\omega_D = \Omega_0$ but quickly revert to that of a quadrupole-only condition on both sides. At $\omega_D = 0.5\ \Omega_0$ and $\omega_D = 1.5\ \Omega_0$, the spectra are also largely unaffected by the application of the dipole potential. The ion ejection times are reduced in a continuous fashion as the dipole frequency moves away from these half-resonant points.

The points in Fig. 3 are color-coded according by symmetry. The red points correspond to ion ejections symmetric about the half resonant values $\omega_D = 0.5\ \Omega_0$ and $\omega_D = 1.5\ \Omega_0$. The ejection times for these points are only dependent on frequency. Additional dipole excitation scans (not shown here), suggest that the amplitude of the dipole potential dictates the point intensities. Portions of the $\omega_D = 0.5\ \Omega_0$ set of ejections (left-hand maximum) are observed with V_D as low as 0.1 V. The $\omega_D = 1.5\ \Omega_0$ set of ejections (right-hand maximum) only start to appear with $V_D > 1$ V. As the dipole amplitude is increased further to bring in more of the $\omega_D = 1.5\ \Omega_0$ structure, the $\omega_D = 0.5\ \Omega_0$ structure becomes attenuated and distorted.

The blue points in Fig. 3 are symmetric about $\omega_D = 1.5\ \Omega_0$ and exhibit a different dipole frequency dependence. A curve tracing through any set of blue points scales with a factor of three relative to its red counterpart. A similar structure about $\omega_D = 0.5\ \Omega_0$ is absent in all of the scans preformed to date. It is unclear whether these blue points arise from harmonic interactions or from a higher-order moment of the combined electric field.

The cyan points in Fig. 3 are similar to the blue in terms of their frequency dependence. These ejections appear to be coupled to harmonics or a higher-order moment. As these groups correspond to resonant values of $\omega_D = -0.5\ \Omega_0$ (fictitious) and $\omega_D = 2.5\ \Omega_0$, it more likely that they arise from a harmonic coupling. Attempts at scanning the dipole frequency beyond $\omega_D = 2\ \Omega_0$ to reveal additional structures have been unsuccessful to date. Assuming an amplitude dependence for the activation of these higher frequency resonant structures in line with those observed for the $\omega_D = 0.5\ \Omega_0$ and $\omega_D = 1.5\ \Omega_0$ activations, the activation energies required would far exceed those capable with the Agilent 33220A waveform generator directly. The use of a ferrite core transformer is also limited in voltage amplification due to the higher frequency operation. As such, the higher frequency resonant structure may exist but is so far unobserved.

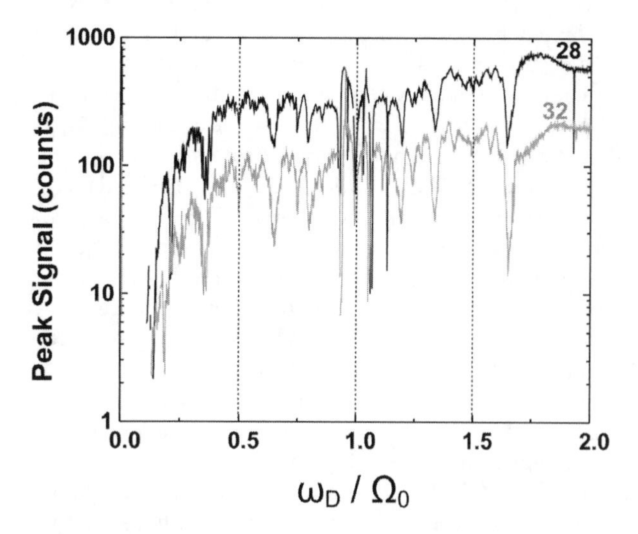

FIGURE 4 Ion ejection times plotted as a function of dipole frequency ω_D for the masses 14, 16, 17, 18, 19, 28, 29, 32, 40, 44. One notes that the frequencies $\omega_D \sim 0$ and $2\ \Omega_0$, as well as $\omega_D = 0.5\ \Omega_0$ and $1.5\ \Omega_0$ have relatively little affect on the mass spectra. All masses are annihilated at $\omega_D = \Omega_0$.

The green points in Fig. 3 represent observed structures within the spectral scan that are so far unexplained. These structures do not share the frequency dependence of the groups mentioned above. The top horizontal portion of this structure does not correspond to an integer m/z value. Its persistence suggests broadband frequency independence – possibly a ghost peak attributed to a higher-order moment of the electric field displacing m/z = 28. If this structure is related to ghosting of the quadrupole-only mass spectra, all of the masses would be affected but only the most intense species, m/z = 28 in this case, would be visible. The remaining cross-like portion of this structure is only apparent within the dominant dipole structure (in red) centered about $\omega_D = 0.5\ \Omega_0$. It may in part be dependent on the horizontal structure. There is also the possibility that the cross structure is associated with a real mass but exhibiting weaker frequency dependences than those of the blue or cyan groups in Fig. 3.

The ion ejection times reveal one aspect of the effect of the dipole excitation. The intensity of the ion signal along any of these curves reveals additional effects of the dipole excitations. Data are given in Fig. 4 for the ion signal intensity for ejections denoted by the red and black points in Fig. 3 for the m/z = 28 and 32 mass lines. Apparent are drops in intensities at particular dipole frequencies that affect all masses. These are seen in Fig. 2 as the dark vertical bands extending through the intensity plot. Interestingly, several of these bands correspond to those present in the trapping phase of the quadrupole drive. The resonant behavior of the ion ejection times and their respective symmetries is also reflected in the intensity profiles. However, the symmetries present in the intensity profiles extend much further, exhibiting both odd and even symmetries at numerous fractional resonant frequencies.

The variations in ion intensities about the $\omega_D = 0.5\ \Omega_0$ set of ejections are reduced significantly from those of the $\omega_D = 1.5\ \Omega_0$ set, highlighting the influence of the dipole excitation amplitude mentioned earlier. The lower frequency portion of the intensity spectrum for both m/z 28 and 32 shows distortion and attenuation relative to those at the higher frequency. The present data afford more symmetry and higher-frequency structure than those of previous studies of motional resonances in ion traps [5].

The resonant structures in the ion signal intensities for both m/z 28 and 32 occur at the same frequencies over most of the frequency scan. This suggests that these effects are also largely mass independent. The only exceptions are the prominent dips occurring at the transition points between quadrupole-only and quadrupole + dipole excitations. These transition points occur at frequencies varying according to ion mass.

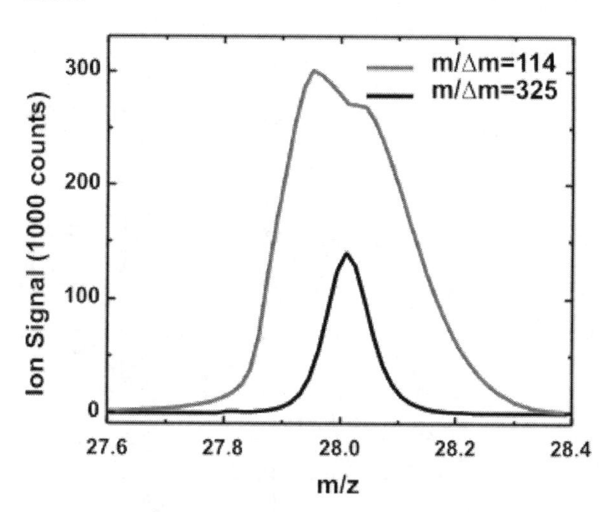

FIGURE 5 Mass spectral peak shapes for m/z = 28 ($^{28}N_2^+$). The lower resolution peak corresponds to the quadrupole-only configuration described in the text with $\Omega_0 = 1250.14$ kHz and the end caps grounded. The higher resolution peak is with the same quadrupole drive frequency and a dipole excitation of $\omega_D = 1687.50$ kHz and $V_D = 5V$.

Fig. 5 shows mass spectral data for m/z = 28 with two different *rf* configurations. The first configuration is with the end caps grounded, $\Omega_0 = 1250.14$ kHz, and a modulation frequency of 50 Hz. The resolving power (m/Δm) for m/z = 28 in this configuration is 114.

Adding only a dipole excitation, ω_D = 1687.50 kHz and V_D = 5V, to this configuration increases the resolving power to 325 and halves the ion signal. In both cases, the peak width is taken as full width half maximum. The addition of the dipole excitation generally provides improved spectral peak shapes through suppression of the wings, particularly on the low-mass side, and narrowing the peak width. Other configurations, not shown here, have been tested with varying modulation, quadrupole, and dipole frequencies, as well as varying quadrupole and dipole amplitudes. The results of these tests have shown resolving powers as high as 680 for m/z = 28. However, the improvements to peak shape come at the cost of diminished intensity. The selection of an operational dipole excitation frequency for optimum peak width remains, so far, unrelated to either the dipole excitation frequency or amplitude. Optimal operational frequencies are selected to reside at frequencies distant from those with extrema in the intensity profile of Fig. 4, and also lacking multiple ejection times for single m/z values as identified in Fig. 3.

SUMMMARY

The addition of a pure dipole potential to a high spectral purity quadrupole trapping potential can have profound impacts on the mass spectra obtained with a Paul ion trap. High mass range extensions are possible through compression of ion ejection times; and the ejected ion signal varies remarkably with dipole frequency. However, the mass spectral resolution and quality of signal appear to have no direct relation to the applied dipole field. The present results also depict a richness that is not fully conveyed in previous work on motional resonances in an ion trap [5, 6]. It is apparent that a study of the ion ejection times or net ion signal alone is inadequate for describing the full effect of dipole excitations on resultant mass spectra obtained with a Paul ion trap. Furthermore, a study of ejection times or signal intensity only provides details of the ion destabilization and cannot provide direct insight into the ion kinematics during trapping.

ACKNOWLEDGMENTS

We thank R. Schaefer (JPL) for technical assistance and development of the DDS *rf* generators. This work was supported by the Advanced Environmental Monitoring and Control Project of NASA's Exploration Technology Development Program. Copyright 2010, California Institute of Technology.

REFERENCES

1 R. E. March, *J. Mass Spectrom.* **32**, 351 - 369 (1997).
2 F. G. Major, V. N. Gheorghe, G. Werth, *Charged Particle Traps,* Berlin: Springer - Verlag, 2005.
3 R. T. Schaefer, J. A. MacAskill, M. Mojarradi, A. Chutjian, M. R. Darrach, S. M. Madzunkov, and B. J. Shortt, *Rev. Sci. Instrum.* **79**, 095107 (2008).
4 A. Chutjian, B. J. et al: *SAE Technical Paper Series* (paper 07ICES-331, available at http://www.sae.org as a refereed conference proceeding).
5 X. Z. Chu, M. Holzki, R. Alheit, and G. Werth, *Int. J. Mass Spectrom. Ion Processes* **173**, 107 - 112 (1998).
6 R. Alheit, S. Kleineidam, F. Vedel, M. Vedel, and G. Werth, *Int. J. Mass Spectrom. Ion Processes* **154**, 155 - 169 (1996).

SPARC: The Stored Particle Atomic Research Collaboration At FAIR

Th. Stöhlker[1,2,3,4*], H. F. Beyer[1], A. Bräuning-Demian[1], C. Brandau[1], A. Gumberidze[4,5], R. E. Grisenti[6], S. Hagmann[1, 6], F. Herfurth[1], Ch. Kozhuharov[1], Th. Kühl[1], D. Liesen[1], Yu. Litvinov[1], R. Maertin[1,3], W. Nörtershäuser[1], O. Kester[1], N. Petridis[6], W. Quint [1], U. Schramm[7], R. Schuch[8], U. Spillmann[1],S. Trotsenko[1,2], G. Weber[1,2], D. Winters[1], on behalf of the SPARC Collaboration

[1] *Helmholtz-Zentrum für Schwerionenforschung GSI, Darmstadt, Germany*
[2] *Helmholtz-Institut Jena, Jena, Germany*
[3]*Physikalisches Institut, Ruprecht-Karls Universität Heidelberg, Germany*
[4]*ExtreMe Matter Institute EMMI and Research Division, Helmholtz-Zentrum für Schwerionenforschung GSI, Darmstadt, Germany*
[5] *Frankfurt Institute for Advanced Studies FIAS, Johann Wolfgang Goethe Universität Frankfurt, Germany*
[6]*Institut für Kernphysik, Johann Wolfgang Goethe Universität Frankfurt, Germany*
[7] *FZD Forschungszentrum Dresden- Rossendorf, Germany*
[8] *Dept. of Physics, Stockholm University, Stockholm, Sweden*

Abstract. The future international accelerator Facility for Antiproton and Ion Research (FAIR) encompasses 4 scientific pillars containing at this time 14 approved technical proposals worked out by more than 2000 scientists from all over the world. They offer a wide range of new and challenging opportunities for atomic physics research in the realm of highly-charged heavy ions and exotic nuclei. As one of the backbones of the Atomic, Plasma Physics and Applications (APPA) pillar, the Stored Particle Atomic Physics Research Collaboration (SPARC) has organized tasks and activities in various working groups for which we will present a concise survey on their current status.

Keywords: Ion-Atom Collisions, Heavy-Ion Storage Rings, QED, Fundamental Interactions, High-Resolution Spectroscopy.
PACS: 91.52.+r; 13.40._f; 34; 31.30.Jv

INTRODUCTION

The International Facility for Antiproton and Ion Research, FAIR [1, 2] - whose legal body has been formally established in an international ceremonial contract signing of all founding countries in October 2010 - is promising the highest intensities for relativistic beams of stable and unstable heavy nuclei, combined with the strongest available electromagnetic fields, for a broad range of experiments. This will then allow the extension of atomic physics research across virtually the full range of atomic matter, i.e. concerning the accessible ionic charge states as well as beam energies. For atomic physics (AP), one of the major research fields served by FAIR, the scientific program will concentrate on two central research areas, correlated electron dynamics including production of electron-positron pairs in strong ultra-short electromagnetic fields and fundamental interactions between electrons and heavy nuclei – in particular the interactions described by Quantum Electrodynamics, QED [3,4]. It is further considered to use atomic physics techniques to determine properties of stable and unstable nuclei and to perform tests of predictions of fundamental theories besides QED.

In order to reach the desired physics objectives, a large number of diverse experimental installations will be available, each equipped with novel and sophisticated instrumentation. Within the Stored Particle Atomic Physics Research Collaboration (SPARC) the atomic physics community has formed twelve experimental and one theoretical working groups to advance the development and construction of the project [4].

Application of Accelerators in Research and Industry
AIP Conf. Proc. 1336, 132-137 (2011); doi: 10.1063/1.3586073

THE SPARC WORKING GROUPS AND THEIR ACTIVITIES

In the following selected current activities of the various working groups within SPARC are summarily presented. For an exhaustive description of the individual tasks the reader is referred to the SPARC technical report [4]. Many tasks are shared between the various working groups in correspondence to the four main experimental areas for atomic physics at the new facility. The R&D activities of the various working groups also emphasize essential precursor experiments at the present experimental storage ring ESR of GSI and focus as well on the development of new prototype set-ups for FAIR. Main activities in atomic physics at FAIR will begin at a high-energy experimental cave served by the SIS100 synchrotron and at a new experimental storage ring NESR (the "SchwerIonenSynchrotron"s are characterized by the maximum magnetic rigidity, given in Teslameter, for beams in the ring, for SIS100 and SIS300 this means 100Tm and 300Tm, respectively, the latter corresponding to a maximum projectile specific energy of approximately 35AGeV for U^{92+}). The main experimental areas of the future New Experimental Storage Ring NESR at FAIR are depicted in Fig. 1.

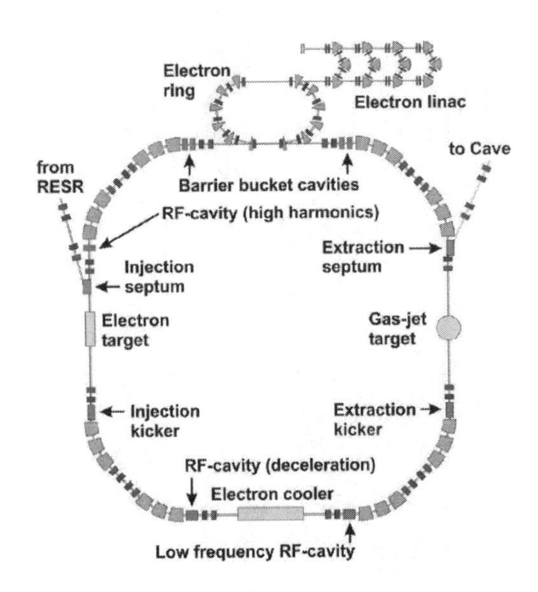

FIGURE 1. Experimental Storage Ring NESR at future FAIR depicting beam-control and experimental installations.

The installations for laser cooling and laser spectroscopy, e.g. at the future synchrotrons SIS100/SIS300, will mostly be dedicated to experiments with Li-like very heavy ions. They exploit the Doppler boost of optical laser photons into the XUV regime as seen in the rest frame of the counter-propagating ions; an ensuing frequency shift of up to a factor $2\gamma\approx70$ at the maximum beam energy available at SIS300 [4, 5] enables high precision fluorescence laser spectroscopy and laser cooling for arbitrary Li-like ions up to the highest nuclear charges on the basis of conventional laser systems.

A laser cooling experiment conducted for 122 AMeV C^{3+} ions at the ESR [8] enabled for the very first time cooling of a highly-charged ion down to an unsurpassed longitudinal momentum spread $\Delta p_{\|}/p<4\cdot10^{-7}$ by exploration of the Doppler boost, similar as it is finally planned for the experiments at SIS100/300. It is anticipated that after implementation of mode-hop free frequency scan with a new laser system no electron-precooling may be necessary. In addition, at the ESR, Time Dilation in Special Relativity has been successfully tested using $^7Li^+$ at an velocity of β =0.34 [9] in accordance with the most stringent previous test [10].

In the high-energy cave for Atomic Physics, biophysics/materials research and also plasma physics, which is situated near the SIS100/300 double ring facility, experiments in atomic physics and applications in radiobiology, space and materials research, warm and hot dense matter with extracted beams from SIS100 will be performed. The investigation will concentrate on atomic structure, e.g. new concepts for QED in extreme fields, collision studies at moderate and highly relativistic energies (ionization, capture and pair production) to generate insight into the correlated many-body dynamics via ultra-short and super-intense field pulses ($<10^{-18}$ s), and avenues towards the discovery of new electromagnetic interaction processes. Besides being a topic central to our understanding of the general atomic Coulomb excitation processes in the far relativistic regime [28-34] pair production has surfaced as a topic deserving imminent attention as dominant source of luminosity limiting beam losses due to capture by pair production[30, 32 and references therein]. As well irradiation schemes of individual samples for biological or solid material research will be developed.

A detailed layout for the high-energy cave including a magnetic spectrometer for the charge state analysis of the ions behind the target has been finished; calculations for various spectrometers, e.g. for photons and electrons and positrons, have been worked out [4].

As a feasibility test towards future experiments planned a new spectrometer containing a goniometer for high resolution x ray spectroscopy of Li-like Uranium via resonant coherent excitation (RCE) has been installed in Cave A and has been successfully commissioned [12] with a cooled 191.5 AMeV U^{89+}

beam from the ESR to study the $1s^22s_{1/2}$-$1s^22p_{3/2}$ transition energy.

The future NESR storage ring can be regarded as the next-generation ESR with significantly optimized features, as the significantly enhanced acceptance, and expanded novel experimental installations, e.g. concerning the analysis of the then accessible greater range of projectile final charge states after the collisions and a separate electron target. The NESR will serve in combination with further auxiliary storage rings also as an accumulator and storage/cooler ring both for stable and exotic ions and as well for antiprotons [2]. Compared to all the other heavy ion storage rings currently operating or under construction, the NESR will acquire a leading position by being the most versatile and flexible among its peers, providing brilliant beams up to bare uranium over a wide range of energies, starting at 740 MeV/u down to 4 MeV/u. We wish to emphasize that at the future NESR the segmented dipole sectors, each equipped with additional position sensitive particle detectors, will permit to cover a much wider dynamic range of charge changing collisions. A large variety of experimental installations for atomic physics experiments will be implemented here [4]. For atomic structure investigations and QED tests a new electron target, internal target and electron- and X-ray spectrometers will be built. High resolution recombination studies will be done with a specially designed dense and cold electron target. Collision experiments for ultra-high resolution spectroscopy and for comprehensive studies of collision systems beyond the range of validity of perturbation theory are planned at the internal target with electron-, recoil-ion, various X-ray spectrometers and optical/UV spectrometers, respectively.

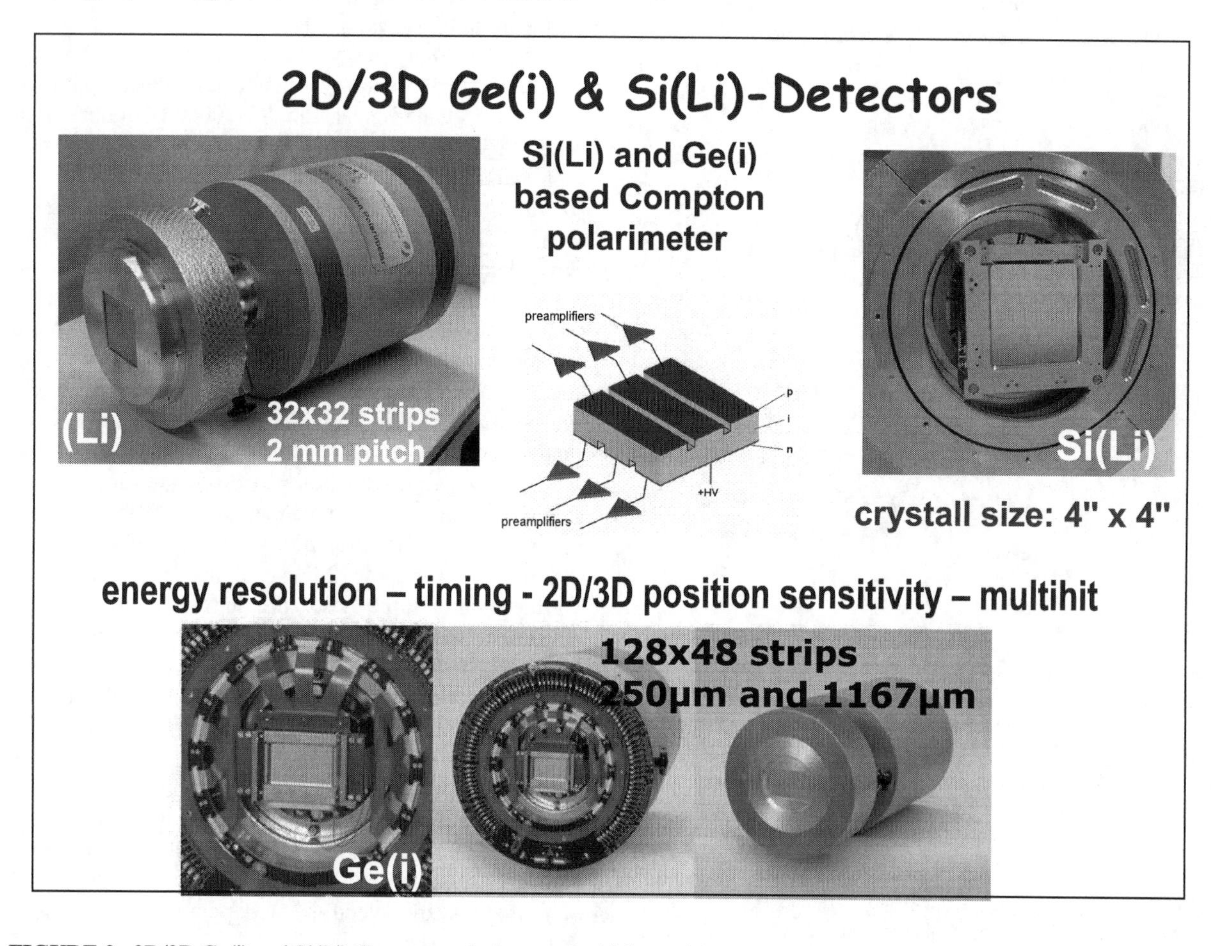

FIGURE 2. 2D/3D Ge(i) and Si(Li) Compton polarimeters for high precision spectroscopy from 50 to 500 keV photon energy.

An experiment on photo-recombination (PR) with the three uranium isotopes A=236, 237, 238 at the ESR measured isotope shifts and hyperfine effects in the resonant contribution of PR, i.e., in dielectronic recombination (DR) [14]. This is a powerful example for future applications of atomic spectroscopy techniques for model-independent tests of nuclear properties - as they are proposed for the short lived, radioactive nuclei profusely produced at the future super-fragment separator SFRS serving the NESR. Isotope effects on DR resonances were previously studied with stable isotopes, e.g., hyperfine quenching of metastable states, hyperfine effects and nuclear size contributions; a dielectronic recombination (DR) experiment has successfully measured the isotope shift between $^{142}Nd^{57+}$ and $^{150}Nd^{57+}$ at the ESR [11]. With the new DR experiments with unstable nuclei at the ESR the spectrum of physics issues that can be addressed employing this atomic physics tool has been broadened.

The x-ray polarimetry and spectroscopy program of the SPARC collaboration relies strongly on the availability of 2D and 3D position-, time and energy sensitive Si(Li) and Ge(i) μ-strip detector systems with their excellent capabilities concerning spectroscopy and imaging as well as polarization sensitivity. Here elementary radiation-matter interaction processes such as REC [40,41] and Bremsstrahlung [20] as well as fundamental physics e.g. parity-nonconservation effects [39,42] are in the center of interest. Newly developed position-sensitive Ge(i) and Si(Li) μ-strip detectors were taken into operation [15] (see fig.2). The detector performance of the Ge(i) was tested in a beam time at the ESRF synchrotron facility. Here, the position response was tested and as well its suitability as a Compton polarimeter was established using monochromatic radiation at energies of 60, 115 and 210 keV, respectively [16]. In order to accomplish this and guarantee a spatial resolution below 100 μm as required by use in the FOCAL spectrometer (FOcussing Compensated Asymmetric Laue geometry) system it was found to be necessary to as well measure the interaction depth and thus get a fully 3D spatial readout of a 2D μ-strip detector system .

A twin crystal-spectrometer assembly, operated in the **fo**cusing **c**ompensated **a**symmetric **L**aue (FOCAL) geometry has been developed for accurate spectroscopy of fast highly charged heavy ions in the hard-X-ray region. Coupled to the focusing crystal optics is a specially developed two-dimensional position-sensitive X-ray detector which is necessary for retaining spectral resolution also for fast moving sources.

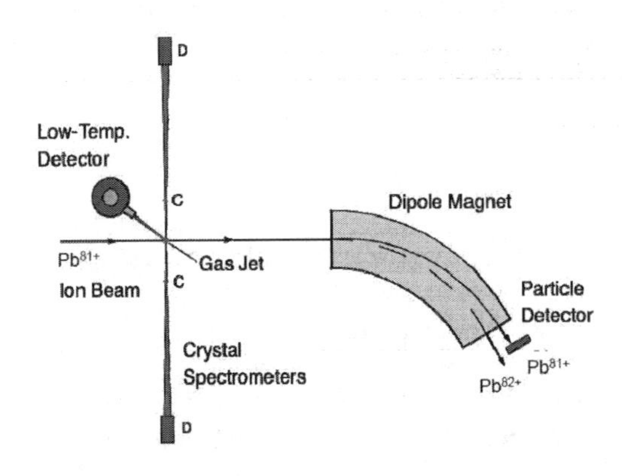

FIGURE 3 Experimental arrangement at the ESR target zone for measuring the ground-state Lamb shift of H-like Pb with simultaneous installation of the high resolution FOCAL spectrometer and a micro-calorimeter.

A commissioning experiment of the two-arm transmission crystal X-ray spectrometer FOCAL [17] was performed using Pb^{81+} X-rays produced in collisions with Ar atoms at about 210 MeV/u, β=0.59 (fig. 3). The spectrometer, in combination with the high-performance position-sensitive μ-strip Germanium detector allowed one to observe the spatial distribution of the reflection (close to 60 keV) in two dimensions. For this purpose, the energy dispersion as well as the time resolution of the detectors turned out to be indispensable for the X-ray spectroscopy program of the SPARC collaboration [18].

Using a novel Johann type spectrometer in the soft X-ray regime the excited levels of He-like U were studied. The focusing properties of the spherical crystal used maximize the number of X-rays diffracted towards the detector and lead to an increasing efficiency without any significant drawback for the achievable resolution [19]. For the energy of the observed transition $1s^2p\,^3P_2 \rightarrow 1s^2s\,^3S_1$ an accuracy at the edge of the sensitivity for two-electron QED contributions, 0.76 eV, was achieved.

Extensive investigations on the production of cryogenically cooled liquid hydrogen and helium droplet beams at the experimental storage ring ESR were carried [23] out with the goal to achieve high area densities for these low-Z internal targets, which are of immense relevance for spectroscopic studies. The results show that an area density of up to10^{14} cm^{-2} is achieved for both light gases by expanding the liquid through sub-10μm diameter nozzles [22]. The achieved area density is comparable with the previous results for the hydrogen internal target and represents an improvement by about four orders of magnitude for the helium target. The droplet target provides in addition a very well localized target (below 1 mm) which is of utmost importance for accurate very high resolution X-ray spectroscopy as well as for nuclear reaction studies planned within the EXL (Exotic nuclei studied in light-ion induced reactions at the NESR storage ring) collaboration [21].

The 0^0-forward electron spectrometer installed at the internal jet target of the ESR storage ring[20] has been used for a wide range of forward emission electron studies. By using beams of Be-like uranium at 90 MeV/u, four distinct channels of forward electron emission produced in collisions with nitrogen molecules were identified. By utilizing projectile X-ray/electron coincidences this technique enabled us to disentangle radiative and non-radiative (direct) capture into the continuum states of the projectile, for the first time electrons in coincidence with bremsstrahlung photons from the short wavelength limit of electron nucleus bremsstrahlung were measured. The spectrometer is currently upgraded with a large hysteresis-free quadrupole triplet to enable true telescopic operation.

Recoil ion spectroscopy will be important also as part of the envisaged experimental program at the NESR. First recoil ion spectroscopy experiments have been performed at the ESR at the internal target section making use of a longitudinal configuration for the detection of recoil ions and electrons. With 230 AMeV U^{90+} +Ar and decelerated 13AMeV U^{92+} +Ne [13] studies of the dynamics of pure target ionization and multiple ionization accompanied by electron capture, respectively, were successfully completed. In addition, modifications of the standard configuration of the reaction microscope [24] where the extraction E-field and the guiding B-field are parallel but configured at an angle of typically 10^0 with respect to the projectile axis are currently under investigation for the NESR. In order to assure optimal background free performance and an unambiguous detection of electrons and recoil ions magnetically, a design of a dispersive toroidal configuration has been worked out [25]. This is intended to be employed, for example, for studies of relativistic and QED effects in kinematically complete electron impact ionization of H-like U^{91+}.

Using the feature of the NESR to decelerate beams of very heavy bare and H-like ions with one or two 1s vacancies, like U, to adiabatic collision velocities near and below the Coulomb barrier the door to precision spectroscopy of the innermost molecular orbitals (MO) near the DIRAC–sea for superheavy quasi-molecules with an asymptotic united atom charge >173 via combined x-ray and electron/positron spectroscopy is opened [35-37]. The decisive progress over previous experiments [38 and references therein] is the ability of the NESR to provide decelerated heavy ions with a 1s vacancy in the incoming trajectory of the collisions and thus an ensuing enhancement of the pertaining MO x ray and positron production cross sections by many orders of magnitude over the cross sections found for beams without incoming 1s vacancies.

The HITRAP facility, one of the cornerstones of the future FLAIR facility, to be used by the SPARC (highly charged ions) and the FLAIR (low-energy antiprotons) collaborations has been installed at the ESR [6]. The trap facility HITRAP will employ deceleration of heavy highly-charged ions and antiprotons from 4 MeV/u down to cryogenic temperatures. First commissioning experiments with 4AMeV Ni^{28+} beams from the ESR have been successfully performed and beams were successfully decelerated to 0.5 AMeV [26], (for technical details of the HITRAP project, see the HITRAP Technical Design Report [6]).

SUMMARY

The SPARC collaboration has been formed to exploit the multitude of new and challenging opportunities for atomic physics research provided by the future international accelerator facility FAIR; this collaboration will be in charge of the development and construction of the set-ups and instrumentations needed for the planned experiments. The working groups are currently developing a broad range of activities and are strongly involved in design studies, development of prototype set-ups as well as in precursor experiments.

* corresponding author

REFERENCES

1. W.F. Henning (Ed.), Internal Accelerator Facility for Beams of Ions and Antiprotons, GSI-Darmstadt, November, 2001, http://www.gsi.de/GSI-Future/cdr/.
2. FAIR Baseline Technical Report July, 2006, http://www.gsi.de/fair/reports/btr.html. and FAIR, Technical Design Report NESR, GSI-Darmstadt (2008)
3. Th. Stöhlker et al. Hyp. Int. 172 135 (2006)
4. Technical Report of the SPARC Collaboration, http://www.gsi.de/sparc.
5. H. Backe, GSI Workshop on its Future Facility, October 18–20, 2000.
6. Th. Beier et al. (Eds.), HITRAP Technical Design Report, GSI,Darmstadt, 2003 http://www.gsi.de/documents/DOC-2003-Dec-69-2.pdf;
7. Technical Report of the FLAIR Collaboration, http://www.gsi.de/flair.
8. M. Bussmann et al COOL09 ;http://www.JACoE.org
9. C. Novotny et al., Phys. Rev. A80 022107 (2009).
10. S. Reinhardt et al. Nature Physics 3 861 (2007)
11. C. Brandau et al., Phys. Rev. Lett 100 073201 (2008)
12. Y. Nakano, Ann. Rep. GSI 272 (2009)
13. D. Fischer et al. Ann. Rep. GSI 276 (2007), 243 (2006)
14. C. Brandau et al. Ann Rep GSI 354 (2009)
15. U. Spillmann et al. Rev SCi Inst 79 083101 (2009) .

16. U. Spillmann et al., GSI Scientific Report 2009, p. 371
17. H.F. Beyer et al., Spectrochim. Acta B 64 (2009) 736.
18. U. Spillmann et al., Ann Rep GSI 291 (2008) R. Reuschl, PhD Dissertation, Frankfurt(2009)
19. M. Trassinelli et al., Ann. Rep GSI 353 (2009)
20. M. Nofal et al. Phys. Rev. Lett 99 163201 (2007)
21. Technical Proposal of the EXL coll. http://www.gsi.de/~wwwnusta/tech_report/05-exl.pdf.
22. M. Kühnel et al. Nucl Inst Meth A602 311 (2009)
23. N Petridis et al . Ann. Rep GSI 372 (2009)
24. J. Ullrich et al.,Rep. Prog Phys 66 1463 (2003)
25 X. Wang, S. Hagmann, Meas. Sci. Technol. 18 (2007) 161.
26. F. Herfurth et al. Ann Rep GSI 374 (2009)
27. A. Artemyev et al. Phys. Rev. A79 032713 (2009)
28. D. Ionescu et al. Phys. Rev. A59 3527 (1999)
29. O. Busic et al. Phys. Rev. A70 062707 (2004)
30. H. Meier et al. Phys. Rev. A63 032713 (2001)
31. M. Lee et al. Phys. Rev. A63 062712 (2001)
32. A. Belkacem et al. Phys. Rev. A56 2806 (1997)
33. G. Deco et al. J. Phys. B22 1043 (1989)
34. D. Ionescu et al. Eur. Physics Journal D18 301 (2002)
35. R. Schuch, 5th Sparc workshop, Predeal, Romania (2008)
36. R. Schuch, Intern. Symposium on Heavy Ion Physics, GSI, Nov 2008
37. W. Greiner et al, QED of supercritical fields, Springer, Heidelberg 1985
38. H. Backe, Ch. Kozhuharov, in Prog Atomic Spectr. C, p 459, ed H. Beyer, Plenum(1984)
39. F. Ferro, PRA **81**, 062503 (2010).
40. Th. Stöhlker, European Physical Journal-Special Topics 169: 5-14 (2009)
41 J. Eichler and Th. Stöhlker (2007)., Physics Reports **439**, 1 (2007)
42. V.M. Shabaev, Physical Review A **81**, 052102 (2010)

Origin, Evolution And Imaging Of Vortices In Atomic Processes

J. H. Macek

Department of Physics and Astronomy, University of Tennessee, Knoxville, TN-37996
and Oak Ridge National Laboratory, Oak Ridge, TN

Abstract. Calculations of the time-dependent wave function for proton impact on atomic hydrogen using the Lattice-Time-Dependent-Schrödinger-Equation (LTDSE) method [1] follow the wave function from microscopic to macroscopic times. Isolated zeros, now identified as vortices, appear when the target and projectile nuclei are separated by a few atomic units [2]. Such structure has apparently been observed frequently in *ab. initio.* calculations. Our work shows that such structures persist to macroscopic distances and appear as "holes" in electron momentum distributions. They can, in principle, be observed in reaction microscope studies. Such observation is formally justified by the "imaging theorem", which can be derived from first principles. Similarly, two-electron momentum distributions as observed, e.g., in (e,2e) coincidence measurements may show isolated zeros [3]. Calculations using correlated wave functions support the interpretation that a minima in (e,2e) for helium targets corresponds to a vortex. In this case the "imaging theorem" can be applied to argue that there must be vortices in the two-electron wave function[4]. A general discussion of vortices in quantum mechanics will be illustrated using exact LTDSE calculations, and, for interpretive purposes, the time-dependent first Born approximation. These calculations show that the plane wave B1 amplitudes have no vortices, but the time-dependent B1 amplitudes do. It will be further shown that angular momentum transfer is the key to forming vortices in the time-dependent theory
Keywords: Atomic collisions, vortex, Born.
PACS: 30.50.Fa

THE IMAGING THEOREM

The LTDSE method for calculating atomic wave functions has proven to be quite accurate for dynamics of excitation and electron transfer reactions in one-electron species. The total ionization cross sections agree with that obtained by other time-dependent methods to within a few percent. Computation of the momentum distribution is somewhat more problematical, since most methods project onto asymptotic continuum eigenstates [5]. In the case of ion-atom collisions such states are not known unless the separation of the atomic nuclei becomes extremely large.

An alternative to projecting onto asymptotic states is simply to integrate to times that are essentially asymptotic using that the electron propagator $G(r,t;r',t')$ given in atomic units (a.u.) goes over to the explosion factor $\exp[(ir^2)/2t]$ times a plane wave factor $\exp[i\boldsymbol{k}\cdot\boldsymbol{r}]$ and an unimportant phase factor in the limit that t becomes infinite and t' is set to zero. Then the time-dependent wave function yields the one-electron imaging theorem

$$\lim_{r\to\infty}\left[\left|\Psi(r,t)\right|^2 d^3r\right]_{r=kt} = |\underline{\Psi}(k,t_0)|^2\, d^3k,$$

where the underline denotes the Fourier transform.

This result shows that one may extract momentum distributions by taking the Fourier transform at sufficiently large times or by integrating the time-dependent Schrödinger equation to effectively infinite times. Integrating to infinite times is not normally employed since the wave function is singular in r due to the explosion factor. Also any infinite limit cannot be directly treated numerically and some sort of analytic approximation is needed for large times, for example fitting onto asymptotic wave functions.

One may integrate to infinite times by using a new variable $\tau = 1/t$ and integrating to $\tau = 0$. To do this one must remove a divergent oscillating factor, namely the explosion factor given above. This is easily done but one numerical problem still remains in that the dimensions of the ionized part of the atomic wave

Application of Accelerators in Research and Industry
AIP Conf. Proc. 1336, 138-141 (2011); doi: 10.1063/1.3586074

function becomes infinitely large as τ goes to zero. One treats this increase in size of the continuum part by using coordinates scaled to this increase with increasing time. We introduce the scaled coordinate $q = r / vt$ where v is a parameter with the dimensions of velocity so q is dimensionless. The exact value for v is of no consequence since it simply affects the size of the time step in the integration routine. To avoid complications at the point $t = 0$, we actually scale according to a time-like variable $R_s^2 = b_s^2 + v_s^2 t^2$, where b_s and v_s are constants chosen to optimize numerical calculations in the region $t = 0$. In much of the work reported here we use a scaling impact parameter and scaling velocity equal to the true impact parameters and relative velocities of the ions. With this choice of the scaled variables $q = r / R(t)$ the transformation from t to τ, r to q, and removal of the explosion factor gives a τ-dependent Schrödinger equation that is identical to the equation given by Solvev [5] in the hidden crossing theory of atomic collisions.

RESULTS

Low Energy Ion-Atom Collisions

Figure 1a). shows a contour plot of the magnitude of time dependent wave function for proton impact on atomic hydrogen computed with RLTDSE method when the projectile and target are separated by a distance of 4.0 a.u.. Also shown on the plot are grey arrows giving the direction of the probability current at each point in space. It is seen that the probability current circulates around some point not located at the target and projectile nuclei. This circulation indicates a vortex [6] and one can verify that the time-dependent wave function has an exact zero at the center of the circulating current. As time becomes large, the vortex evolves in a complicated way and its effects show up as vortices in the final wave function. According to the imaging theorem, the electron momentum distribution has holes corresponding to the vortices in the final state. This shows that vortices are observable structures in electron momentum distributions as measured by imaging techniques such as the COLTRIMS technique [7].

Low Energy Electron-Atom Ionization (E,2e)

Electron momentum distributions in low and intermediate energy electron-atom ionization are usually expressed in terms of (e,2e) electron-electron correlation measurements. The theory of such processes has involved various approximate descriptions of two outgoing electrons in the field of a positively charged ion. One approximate description is the 3C final state meaning that three Coulomb distortion factors, one for each particle-particle interaction, multiply a standard plane wave for two electrons [4]. The associated transition matrix element $T(k_i, k_a, k_b)$, where k_i, k_a, k_b stand for the initial electron momentum, the scattered electron momentum and the ejected electron momentum respective, is computed without further approximation. In the special case of symmetric geometry where k_a and k_b have the same magnitude and are symmetrically oriented relative to a plane through k_i, Barackder and Briggs found an exact zero of the singlet amplitude. We have repeated their calculation and do indeed find an exact, isolated zero at $k_a + k_b = (1.04, 0.0, 0.62)$. We now know that such isolated exact zeros must be vortices, however, to verify the vortex nature of the zero we have plotted the direction of the velocity field in the neighborhood of the zero. Figure 1b) is a plot of arrows pointing in the direction of the velocity field in the neighborhood of the exact zero. One can see that the velocity field circulates around the exact zero of the T-matrix element consistent with vortex structure.

High Energy Ion-Atom Ionization

At high impact energies the ionization of atoms is typically described in terms of the plane wave Born , B1, approximation. In this approximation the mean value <L> of the angular momentum vector for excited states vanishes owing to rotational symmetry about the momentum transfer axis. Because <L> vanishes the electron density does not circulate and there can be no vortices. In the time-dependent first Born approximation, B1T, the mean value <L> does not vanish and there can be vortices. Figure 2a) shows a contour plot in the scattering plane of the electron momentum distribution in the B1T approximation for the ionization of atomic hydrogen by proton impact at an impact parameter of 1 a.u., and v=4 a.u.. The binary encounter ridge and the direct ionization peak are visible as lighter features on a dark background.

Also visible is a dark spot, which is circled for clarity, on the lower left hand side of the direct ionization peak. This feature is a vortex as can be verified from the circulation of the velocity field.

The example of the Born approximation shows that even at moderately high impact energies where electron momentum distributions are thought to be well understood, unanticipated additional features may appear due to the transfer of angular momentum from relative motion to internal motion. Such transfer is not represented in the B1 approximation, however the time-dependent first order approximation B1T does describe angular momentum transfer, even at high velocity. The time-dependent approximation is essential because it incorporates the velocity v and impact parameter b so that there is angular momentum of relative motion bv. Even though bv is constant in the calculation, it represents a source of angular momentum that may be transferred to the active electron. If that transfer goes to ionized electrons it can only show up as circulation of electron probability about some point, i. e. a vortex. The surprising feature of Figure 2a) is that the circulation is about some point not centered on the target nuclei.

High Energy Electron-Atom Ionization

The Born approximation is known to give the high energy limit of transition amplitudes when the momentum transfer is held fixed and the relative velocity becomes infinite. Alternatively, if the scattering angle is held fixed, then the collision of the incident projectile with the target nuclei cannot be neglected. For electron impact on atoms the high energy limit at fixed scattering angle should be the Coulomb Born CB1 approximation. This approximation has been used to study (e,2e) processes for the K-shell of carbon atoms [8]. There it was found that the angular distribution of ejected electrons were very much different for the B1 and CB1 approximations. The difference was never explained, but can now be traced to the effect of vortices.

To see this consider a polar plot of the number of electrons ejected by fast electrons incident along the z-axis and scattered upwards in the scattering plane as shown in Figure 2b. The dashed curve is the secondary distribution in the B1 approximation. It has the expected shape that is rotationally symmetric about the momentum transfer axis. In this plot the momentum transfer points downward at an angle of 250° to the vertical. The CB1 distribution looks completely different. Its most prominent feature is the near zero number of electrons at an angle where the B1 distribution has a local minima. In the opposite direction to the minima there is a prominent maxima. These minima and maxima lie on a line that is perpendicular to the momentum transfer axis. A second difference is and equalization of the intensity if direction parallel and anti-parallel to the momentum transfer.

To understand the zero in the distribution a thin solid curve that shows the squared magnitude of sum of the B1 amplitude and the p-wave, m=1 CB1 amplitude. The CB1 part was chosen because it corresponds to a vortex contribution. As is immediately apparent the vortex contribution exactly reproduces the zero and, in the opposite direction, adds coherently to increase the cross section by about a factor of 4. Clearly, the vortex accounts for the most prominent effect of the Coulomb deflection. The second effect, namely the change in the magnitude of the cross section in directions parallel to the momentum transfer axis are due to a change of the relative phases of the m=0 zero components in CB1 as compared with B1. While we have found no simple way to compute these phases, it is apparent that two effects, namely, a relative phase change, and the addition of a vortex term account for the ejected electron distribution due to high energy electron impact when the scattering angle is held fixed as the incident energy is increased without limit. In contrast, if the momentum transfer is held fixed the distribution approaches the B1 distribution. This result shows that vortices are needed to understand electron momentum distributions produced by fast charged particles.

SUMMARY

Vortices are a feature of time-dependent wave functions for atomic systems. Electron momentum distributions show the vortices in coordinate space wave functions as holes in electron momentum distributions. Such holes can, in principle, be observed in coincidence measurements. They are shown to occur in collisions in both the low and the high energy regimes. In the latter case they account for structures seen in the distribution of ejected electrons in the limit that the energy of the incident particle becomes large while its scattering angle is fixed. No vortices appear when the high energy limit is taken holding momentum transfer fixed.

ACKNOWLEDGMENTS

This research is sponsored by the Office of Basic Energy Sciences, U.S. Department of Energy under grant number DE-FG02-02ER15283..

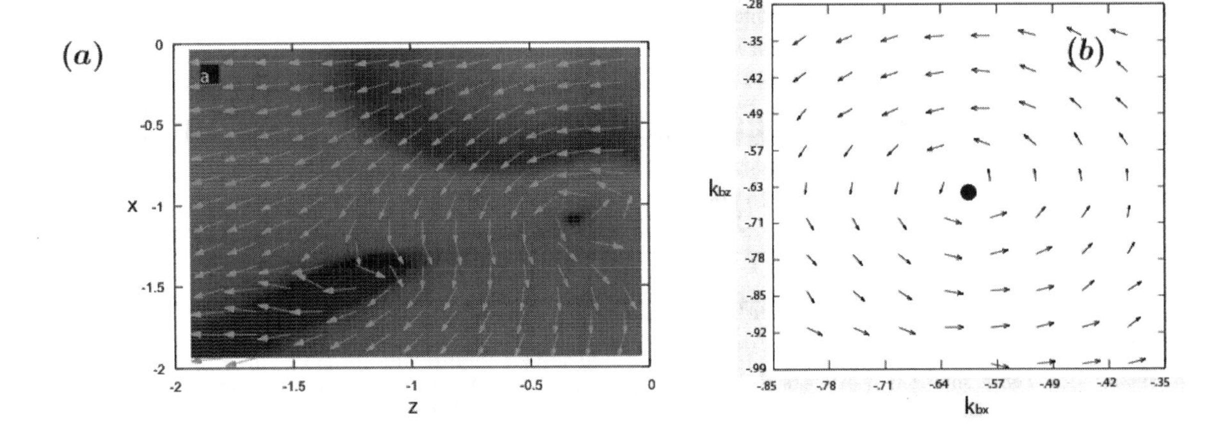

FIGURE 1. (a) is a contour plot of the electron wave function for proton impact on atomic hydrogen at 5 keV impact energy, an impact parameter of 1 a.u., and 4.0 a.u. separation of the atomic nuclei showing a vortex in the wave function at z=-0.42 and x=-1.0. (b) shows a vortex in the (e,2e) T-matrix element at $k_b = (-0.61, 0.82, -0.65)$ for (e,2e) with helium targets at an impact energy of 67.5 eV. The arrows show the direction of the velocity field. The labels are in a.u.

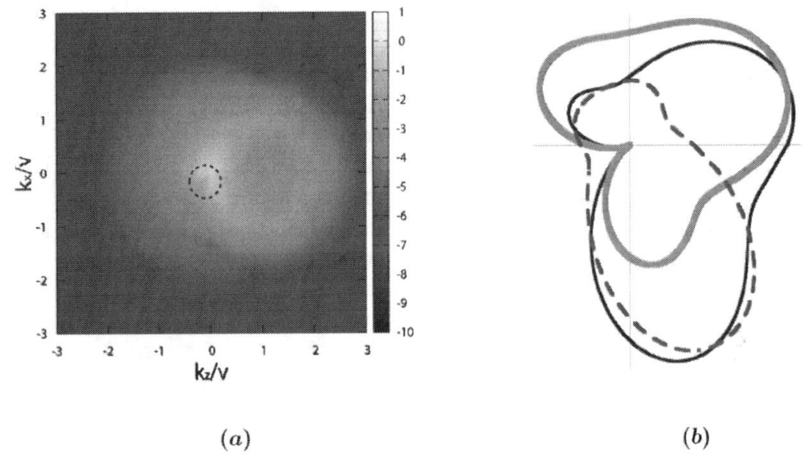

(a) (b)

FIGURE 2. (a) is a contour plot of the momentum distribution of electrons ejected by proton impact on atomic hydrogen. The circle marks the location of a vortex. (b) is a polar plot of the cross section for ejection of K-shell electrons in carbon by 1801.2 eV incident electrons scattered at 20o. The energy of the ejected electron is 9.6 eV. The dashed curve is the plane wave Born approximation, the thick grey curve is the Coulomb Born approximation and the thin black curve is the plane wave Born approximation plus the 2p, m=1 Coulomb Born amplitude.

REFERENCES

1. D. R. Schultz, M. R. Strayer, J. C. Wells, Phys. Rev. Lett. 82, 3976-3979(1999).
2. J. H. Macek et al., Phys. Rev. Lett, 102, 143201 (2009).
3. J.H. Macek et al., Phys. Rev. Lett 104, 033201 (2010).
4. J. Berakdar, J. S. Briggs, J.Phys.B 27, 4271-4280 (1994).
5. E. A. Solov'ev, Zh. Eksp. Teor. Fiz. 70, 872-882 (1976) [Sov. Phys. JETP 43, 453-458 (1976)].
6. I. Bialynicki-Birula, Z Bialynicka-Birula, C. Sliwa, Phys. Rev. A 61, 032110 (2000).
7. J. Ullrich et al., Rep. Prog. Phys. 66, 1463-1538 (2003).
8. J. Botero and J. H. Phys. Rev. A 45, 154-165 (1992).

as prominent nor as separated as they were in the ZRP model, any interference effects between these rings are not easily seen in these static pictures. However, examination of the wave function in time throughout propagation shows that the entire picture shown in Figure 4a is created by the interference of waves scattered from the target and the projectile.

Unfortunately, although a proton on hydrogen collision is the most efficient and least noisy calculation that can currently be done using the RTDSE method, it is also one of the worst systems for producing the Fermi shuttle. A screened potential representing a higher N atom should produce a more intense Fermi shuttle effect [7].

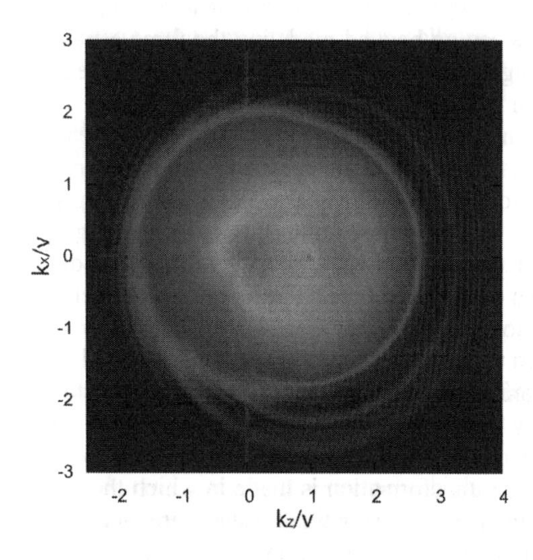

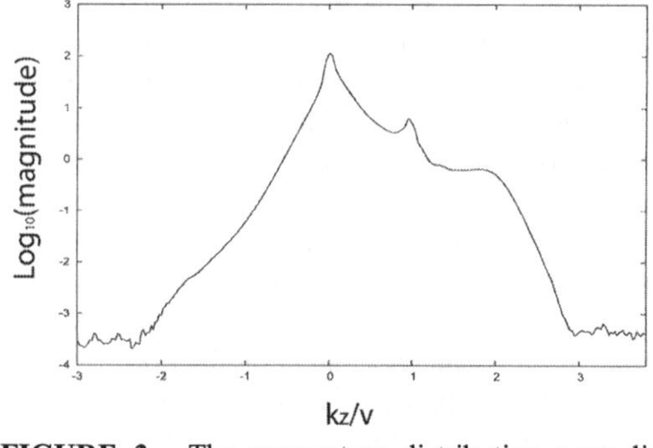

FIGURE 3. The momentum distribution normalized by velocity for RTDSE model for v = 2.8 a.u. protons on hydrogen at an impact parameter of b=1 a.u. is shown here. a) shows a slice of the full three-dimensional wave function along the ky =0 axis with the colors representing the log of the amplitude of the wave function. b) is a two-dimensional slice of a) along kx=0.

THE RTDSE METHOD FOR C+ + XE

In order to model the collision of higher N atoms, the RTDSE method described in the previous section is adapted to use screened Hatree-Fock based potentials as described by Green [8]. Unfortunately, because the target atom is in an excited state, rather than the ground state the previous method of settling the wave function on the grid by propagating in imaginary time can no longer be used. This results in a noisier and less accurate calculation, but the hope is that the enhancement of the Fermi shuttle mechanism more than makes up for this. Once again, the wave function is propagated in a non-expanding space until the united atom limit, and then is transformed to an expanding space as was described above. Since the target is not spherically symmetric, the calculation must be performed for different orthogonal orientations and averaged.

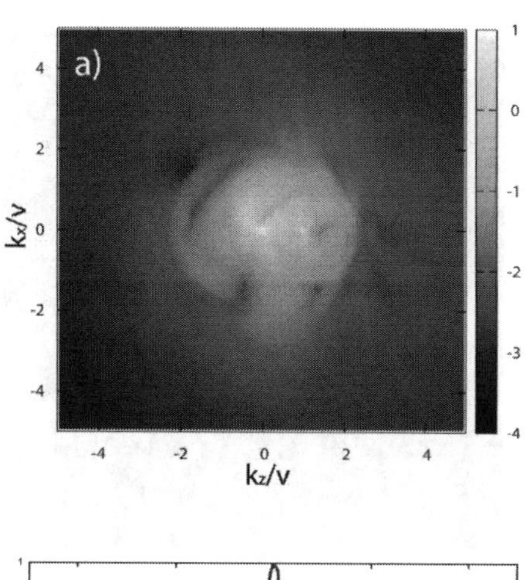

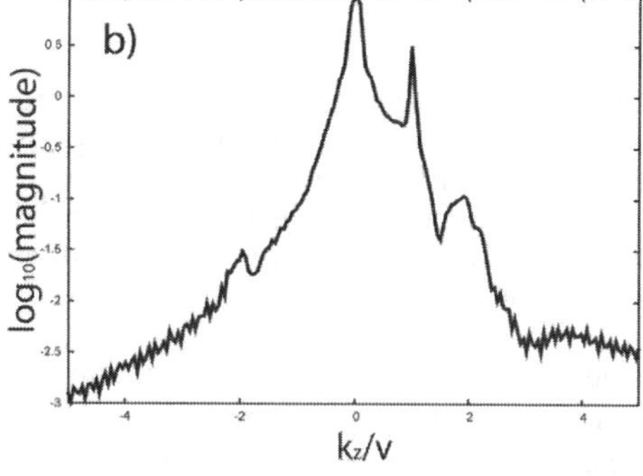

FIGURE 4. The momentum distribution normalized by velocity for RTDSE model for v = 4 a.u C+ incident on Xe at an impact parameter of b=1 a.u. is shown here. a) shows a slice of the full three-dimensional wave function along the ky =0 axis with the colors representing the log of the amplitude of the wave function. b) is a two-dimensional slice of a) along kx=0.

The results at 2000 a.u. C^+ incident upon Xe at v = 4 a.u. are seen in Figure 5. The picture that arises is more complicated than the previous calculations since the potentials are much more complicated and support more structure. The first few Fermi shuttle rings are very pronounced, however. In fact the binary encounter ridge and the backscatter ridge look quite similar to the classical prediction shown in Figure 1. There are places where the rings almost completely disappear due to destructive interference as was found in the ZRP model. A two dimensional slice along k_x=0 of Figure 5a is shown in Figure 5b, where prominent peaks are seen at v=-2 and v=2, as well as the peaks representing the locations of the target and the projectile. Fermi shuttle rings beyond these are below the noise floor of the calculation.

CONCLUSION

Although the standard explanation for the Fermi shuttle mechanism in ion-atom collisions is purely classical and depends on the definition of classical orbits, we have shown several purely quantum mechanical models for these collisions which also produce the Fermi shuttle. Because these are purely quantum mechanical models the concept of a classical orbit must be disposed of and the Fermi shuttle features must be explained by interference effects in the wave function. This alternate explanation for the formation of the Fermi shuttle can, and indeed does lead to differences in these features between classical expectations and what one would expect to actually measure in such a collision. Interference effects can enhance the ring structure in some directions and eliminate them in other directions. Understanding these interference effects is essential to understanding the Fermi shuttle mechanism.

ACKNOWLEDGMENTS

This research is sponsored by the Division of Chemical Sciences, Office of Basic Energy Sciences, U.S. Department of Energy under grant DE-FG02-02ER15283.

REFERENCES

1. E. Fermi, *Physical Review* **75**, 1169. (1949)
2. B. Sulik et al., *Radiation Physics and Chemistry* **76** (2007) 483
3. Jianyi Wang, Joachim Burgdörfer and Anders Bárányi, *Phys. Rev. A* **43** (1991) 4036
4. J. H. Macek, S. Yu. Ovchinnikov, and E. A. Solov'ev, *Phys. Rev. A* **60** (1999) 1140
5. D.R. Schultz, M.R. Strayer, and J.C. Wells, *Phys. Rev. Lett.* **82** (1999) 3976
6. Teck-Ghee Lee et al., *Phys. Rev. A* **76**, 050701 (2007)
7. H. Rothard et al. *Nucl. Inst. And Meth. B* **230** (2005) 419
8. Green A. E. S. *Adv. Quantum Chem.* **7** (1973) 221

Work Towards Experimental Evidence Of Hard X-Ray Photoionization In Highly Charged Krypton*

E. Silver[1], J. D. Gillaspy[2], P. Gokhale[2], E. P. Kanter[3], N. S. Brickhouse[1], R. W. Dunford[3], K. Kirby[1], T. Lin[1], J. McDonald[4], D. Schneider[5], S. Seifert[3], and L. Young[3]

[1] *Harvard-Smithsonian Center For Astrophysics, 60 Garden Street, Cambridge, MA 02138*
[2] *National Institute of Standards and Technology, 100 Bureau Drive, Gaithersburg, MD 20899*
[3] *Argonne National Laboratory, 9700 South Cass Avenue, Argonne, IL 60439*
[4] *George E. Wahlen Department of Veterans Affairs Medical Center, Salt Lake City, UT 84148*
[5] *Lawrence Livermore National Laboratory, P.O. Box 808, Livermore, CA 94551*

Abstract. Ions of almost any charge state can be produced through electron-impact ionization. Here we describe our first experiments designed to photoionize these highly charged ions with hard x-rays by pairing an electron and photon beam. A spectral line at 12.7(1) keV with an intensity corroborated by theory may be the first evidence of hard x-ray photoionization of a highly charged ion.

Keywords: x-rays, photoionization, spectroscopy, EBIT, synchrotron
PACS: 32.30.R, 52.70.La,78.70.En, 32.80.Aa, 37.10.Ty, 32.80.Fb, 29.20.dk

INTRODUCTION

The Electron Beam Ion Trap (EBIT) has facilitated the production of charge states of any naturally occurring element. Much of our motivation for studying these highly charged ions (HCIs) stems from their abundance amongst matter in the universe [1]. Whereas HCIs in space are created by both electron impact and photo-ionization, the ionization mechanism in an EBIT is driven solely by electron impact. Therefore, studying photoionization of HCIs is an important step toward improving our understanding of several astrophysical phenomena.

Although we have made tremendous progress in collecting atomic data for electron-impact ionization for most charge states, our analogous databases for photoionization such as the National Institute of Standards and Technology's XCOM [2] span only neutral atoms. Empirical data for photoionization would expand such databases and serve as a benchmark for theoretical predictions of photoionization cross sections for HCIs. Although groups in Germany [3] and France [4] have had success in extreme ultraviolet and soft x-ray photoionization, hard x-ray photoionization has not yet been observed.

EXPERIMENTAL SETUP

The experimental setup at the Argonne National Laboratory combined the Advanced Photon Source (APS), a synchrotron radiation-based x-ray beam, and a portable EBIT cooled to cryogenic temperatures. The electron and photon beams were aimed to perpendicularly intersect at the center of a cylindrical copper drift tube. The electron beam was centered on the drift tube axis by optical alignment of the electron gun cathode and by fine tuning to minimize the reflected electron beam current. The photon beam was centered on the drift tube axis by passing it through narrow slits cut into the sides of the drift tube and adjusting the height, pitch angle, and roll angle (with respect to the photon beam direction) using three stepper motors that supported the EBIT. The rate of photons entering and exiting the EBIT was monitored with ionization chambers, allowing us to observe when the photon beam began to be occluded by the slits. An energy dispersive lithium-drifted silicon (SiLi) detector with a resolution of 131 eV at a photon energy of 6.4 keV was used to determine the x-ray spectra of emitting ions.

The EBIT was cooled to 4K and mounted on an adjustable cradle. For the data presented here, the

Application of Accelerators in Research and Industry
AIP Conf. Proc. 1336, 146-149 (2011); doi: 10.1063/1.3586076

energy of the electron beam was 2.9 keV (enough to produce Kr XXVII), the current 32 mA, and the trap depth was 40 eV. A photon beam at 18.00 keV was extracted from the APS using a monochromator. A background data set was collected with the photon beam turned off, and a control data set was collected with a photon beam energy of 14.97 keV and the electron beam turned off.

THEORY

Charge State Balance

We ran a simulation of the charge state evolution inside the trap, using a code developed by D. Knapp at the Lawrence Livermore National Laboratory, based on the known electron beam energy and current, as well as estimates for the pressure, ion temperature, and effective beam radius. The code contains most of the physics described by Penetrante et al. [5], including electron impact ionization and collisional charge exchange. The simulation, shown in Figure 1, indicated that Kr XXVII (neon-like) accounts for 73% of the final charge state distribution and that Kr XXVI (sodium-like) accounts for another 23%. Therefore, the majority of photons emitted by krypton is from highly charged, neon-like krypton.

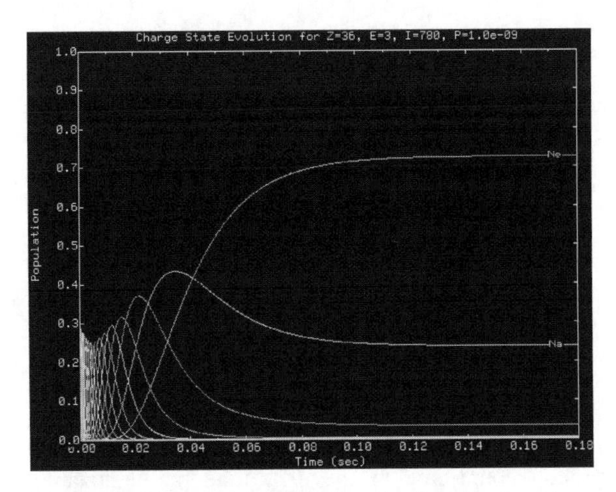

Figure 1. Charge state evolution for Krypton based on approximate trap conditions.

Expected Line Center Position

Neutral krypton has a $K\alpha_1$ emission line at 12.649 keV and a $K\alpha_2$ emission line at 12.598 keV [6]. The lines for Kr XXVII are shifted to 12.67(13) keV for $K\alpha_2$ and 12.72(13) for $K\alpha_1$ as calculated [7] using the Flexible Atomic Code [8]. Since these lines are separated by less than the detector resolution of 185 eV at 12.7 keV, they appear blended into a single peak.

Expected Intensity

The expected number of photons from $K\alpha$ photoionization for Kr XXVII can be approximated as

$$N = \frac{N_\gamma N_i F_{ill} \sigma}{A}$$

where N_γ is the number of photons entering the trap over the run period, N_i is the number of trapped Kr XXVII ions, F_{ill} is the fraction of these ions illuminated by the photon beam, σ is the photoionization cross section and A is the cross sectional area of the photon beam. This formula assumes that the photon beam's cross sectional area is less than or equal to the area of the ion cloud in the plane orthogonal to the photon beam—otherwise, not all photons pass through the ion cloud.

The number of trapped Krypton ions was estimated to be 4×10^6 (half of the calculated trap capacity for ions of charge 16 for the electron beam current and energy used), of which 73% were Kr XXVII according to the charge state evolution simulation. The photon beam's cross sectional area had rectangular dimensions of 0.004m × 0.00015m. Approximately 20% of the ions were illuminated by the photon beam. The cross section for K-shell photoionization of Ne-like krypton at 18 keV is predicted to be 0.86×10^{-20} cm^2/atom [9], approximately 16% lower than that of neutral krypton [2].

The photon beam flux was determined by a 10 cm gas ionization chamber filled with air which has an electron-ion pair production energy of 34.4 eV. The detector had an efficiency of 0.82(02)% at the 18 keV photon beam energy. A chamber current of 4.5 μA implied that 3.72×10^{17} photons entered the EBIT over the 15.8 hour run time. Correcting for the detector's solid angle of collection of 3.08×10^{-4} sr yielded a predicted value for photon counts equal to 8.

EXPERIMENTAL RESULTS AND DISCUSSION

Figure 2 is a plot of photon counts vs energy with the electron beam turned off and the photon beam operating at 14.97 keV. The first three lines indicate photoionization of neutral copper in the drift tube slits. The cluster of lines near 14.97 keV represents photons from the x-ray source, scattered both elastically and inelastically into the detector. Figure 3 shows a histogram in the energy range of 0 keV to 4.5 keV for the data set with the photon beam turned off and the electron beam operating at 2.9 keV. The first line at 0.47 keV is a blend of several lines from hydrogen- and helium-like carbon, nitrogen, and oxygen. The two most prominent lines are of Kr XXVII followed by two lines for Kr XXVI. The last two lines above the energy of the electron beam indicate radiative recombination (RR).

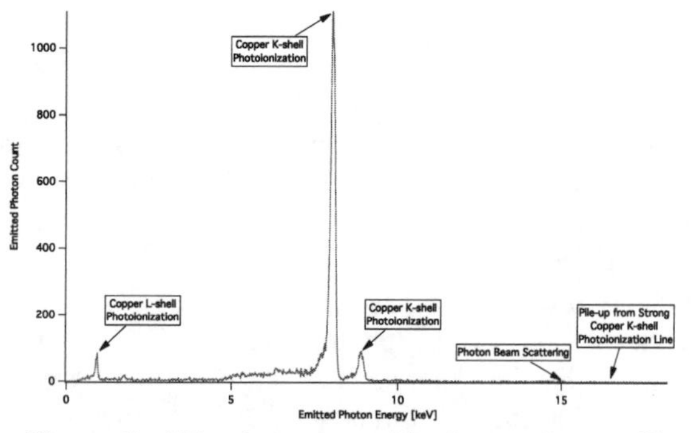

Figure 2. 445 minute run with electron beam off and photon energy of 14.97 keV (19.2 eV bin width)

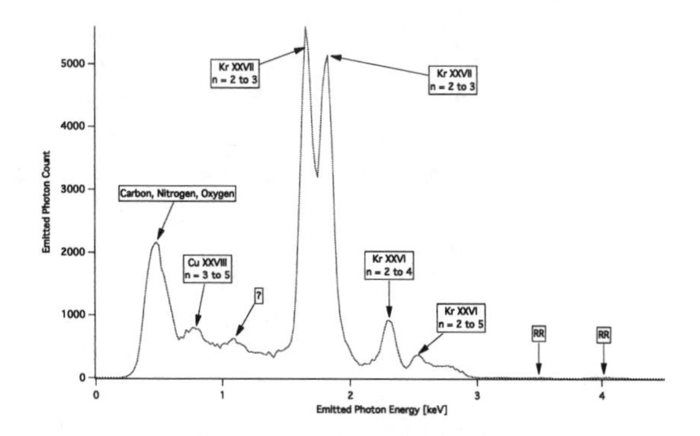

Figure 3. 11 minute run with photon beam off and EBIT electron beam operating at 2.9 keV (19.2 eV bin width).

Figure 4 shows another histogram for the experimental data set with the photon beam operating at 18 keV and the electron beam at 2.9 keV. Between 0 keV and 4.5 keV the resolvable lines are identical to those in Figure 3, which also had the electron beam on.

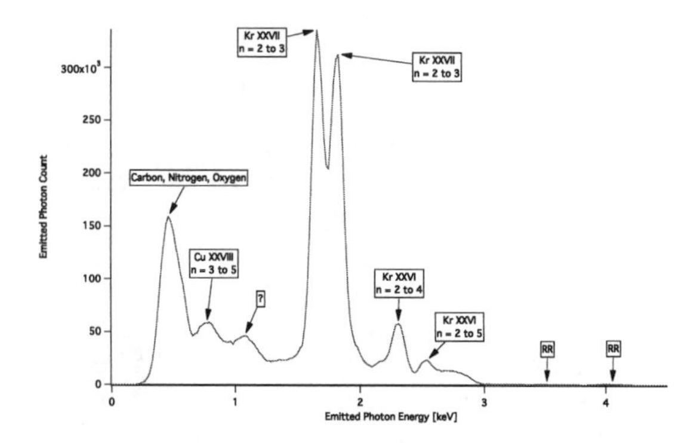

Figure 4. 948 minute run with EBIT electron beam energy of 2.9 keV and incident photon energy of 18 keV (bin width of 19.2 eV).

Figures 5, 6, and 7 are histograms of the spectra obtained for runs with the electron-beam only, photon-beam only, and photon plus electron beam, respectively. The histograms were binned starting at 0.32 keV with a bin width of 377 eV, twice the detector resolution at 12.7 keV. Each plot spans the region of interest at photon energies greater than 8 keV. As expected, the control run in Figure 5 has few counts which can be attributed to background.

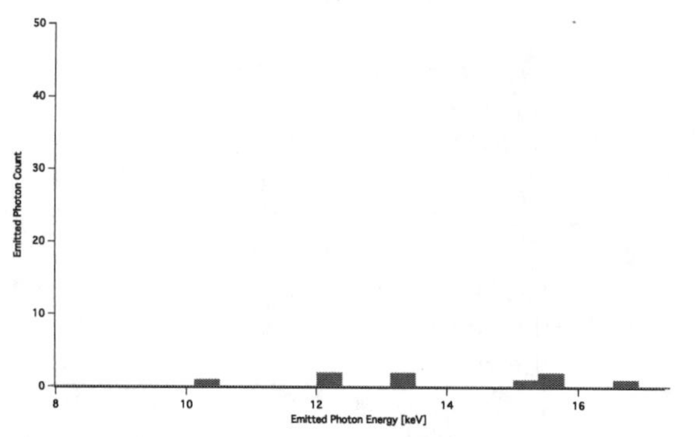

Figure 5. Histogram in region of interest for electron-beam only run(see Figure 3).

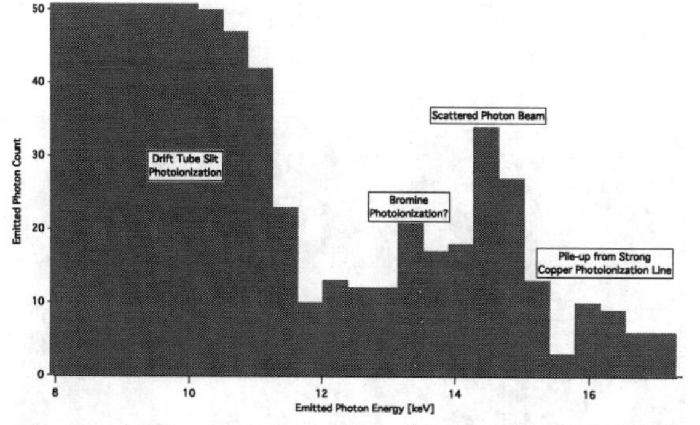

Figure 6. Histogram in region of interest for "photon-beam" only run

In the experimental run (see Figure 7), the photon beam was precisely positioned to avoid photoionizing the drift tube slits. This explains the greater width and photon count rate in the drift tube photoionization range for Figure 6. Nonetheless, copper alone does not account for all of the lines between 8 keV and 12.38 keV. Bromine, a component of the Br-128 anticorrosive material used in the Beryllium window coating, has Kβ lines in the bin spanning 13.138 keV and 13.515 keV. However, its stronger $K\alpha_1$ and $K\alpha_2$ lines, which should be in the 11.630 keV to 12.007 keV bin, are absent. Both plots show evidence of photon beam scattering at the respective photon beam energies.

In Figure 7, the bin containing Kr XXVII's Kα radiation shows a distinct peak of 32 counts. Although this result is only marginally statistically significant, the approximate predicted intensity of 8 counts is roughly in agreement with the number of counts measured over the background.

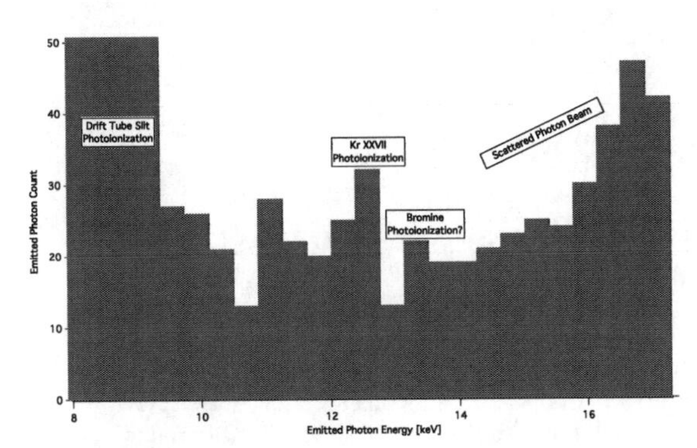

Figure 7. Histogram in region of interest for run with both electron and photon beams on.

CONCLUSION

In conclusion, we have paired an electron beam ion trap with a beam of hard x-rays from a synchrotron and searched for K-shell emission driven by state-specific core photoionization of neon-like iron. However, the photoionization signal was only marginally statistically significant, and we believe that repeating this experiment with a microcalorimeter detector, higher beam current, and more photon flux will yield a more conclusive result.

ACKNOWLEDGMENTS

The authors wish to thank D. Caldwell and G. Austin for their expertise in designing and setting up the movable support structure for the EBIT.

*This work was supported by NASA Grant NNX08AK33G and an Argonne National Laboratory LDRD grant.

REFERENCES

1. T.T. Fang and C.R. Canizares, *Astrophys. J.* **539**, 532 (2000).
2. M.J. Berger and J.H. Hubbell, *XCOM: Photon Cross Sections Database*, National Institute of Standards and Technology, 1998.
3. M C Simon et al, *J. Phys. B: At. Mol. Opt. Phys.* **43** 065003 (2010).
4. J.-M. Bizau et al, *Phys. Rev. Lett.* **84**, 435 (2000). 5. B. M. Penetrante et al., *Phys. Rev. A* **43** 4861 (1991).
6. J. A. Bearden, *Rev. Mod. Phys.* **39**, 78 (1967).
7. Y. Ralchenko (private communication).
8. M. F. Gu, *Can. J. Phys.* **86**, 675 (2008).
9. M.B. Trzhaskovskaya, V.K. Nikulin, and R.E.H. Clark, *At. Data Nucl. Data Tables* **94**, 71 (2008).

Control Of Screening Of A Charged Particle In Electrolytic Aqueous Paul Trap

Jae Hyun Park and Predrag S. Krstić

Physics Division, Oak Ridge National Laboratory, P.O. Box 2008, Oak Ridge, TN 37831-6372, USA

Abstract. Individual charged particles could be trapped and confined by the combined radio-frequency and DC quadrupole electric field of an aqueous Paul trap. Viscosity of water improves confinement and extends the range of the trap parameters which characterize the stability of the trap. Electrolyte, if present in aqueous solution, may screen the charged particle and thus partially or fully suppress electrophoretic interaction with the applied filed, possibly reducing it to a generally much weaker dielectrophoretic interaction with an induced dipole. Applying molecular dynamics simulation we show that the quadrupole field has a different effect at the electrolyte ions and at much heavier charged particle, effectively eliminating the screening by electrolyte ions and reinstating the electrophoretic confinement.

Keywords: Aqueous Paul trap, electrolyte screening, electrophoretic, molecular dynamics.
PACS: 37.10.Ty, 72.20.Jv, 82.45.Tv, 82.45.Gj, 82.45.-h

INTRODUCTION

Standard, conventional applications of a quadrupole Paul trap [1], in 2D and 3D options, are in the fields of mass spectrometry and ion trapping, respectively. More recent applications include the reduced dimension micro/nano traps and are in the area of quantum information processing, for coherent quantum-state manipulation of trapped atomic ions, for functional studies with fluorescent proteins, for laser sideband cooling of the motion, for formation of ordered structures of trapped ions, etc. By default, the trap confinement functions are used in vacuum, and are characterized by the applied DC (U) and AC (V) electric fields of frequency Ω, building a quadrupole potential

$$\varphi(x,y,t) = [U - V\cos(\Omega t)]\frac{x^2 - y^2}{2R_0^2} \qquad (1)$$

applied to a charged particle of mass M and charge Q in a 2D trap of lateral dimension R_0. Application for controlling of motion and position of biomolecules, for example in a nanopore for DNA sequencing [2-5], has raised interest in an aqueous Paul trap. We have previously shown by molecular dynamics simulation that nanoscale quadrupole Paul trap can confine charged particles in an aqueous environment [5]. If the trap is filled with a medium of viscosity ξ (for water $\xi=8.9\times10^{-4}$ Pa•s) then the classical, Langevin equation of motion of a charged particle is of Mathieu type with damping which takes the form

$$\frac{d^2x}{d\tau^2} + b\frac{dx}{d\tau} + (a - 2q\cos 2\tau)x = g(\tau) \qquad (2)$$

where g(τ) is the stochastic Gaussian fluctuating term, arising from the Brownian motion,

$$\tau = \frac{\Omega t}{2}, a = \frac{4QU}{MR_0^2\Omega^2}, q = \frac{2QV}{MR_0^2\Omega^2}, b = \frac{2\xi}{M\Omega} \qquad (3)$$

are dimensionless parameters which define the functions of the trap, Q and M are the trapped particle charge and mass, respectively, and R_0 is the characteristic dimension of the trap. In particular, with neglecting Brownian fluctuations we derive the stability regions of the solutions of Eq. (2), shown in Figure 1 for various viscosity parameters b. The trap will preserve its confinement functions even at zero parameter a (proportional to the DC bias U), the limits of stability for parameter q (proportional to the AC bias V) will strongly depend on the viscosity parameter b. Thus, the q parameter in vacuum (b=0) is limited to values less than approximately 0.9, while for b=92, it is limited to values less than 4500.

However, if the water in the trap contains electrolytes, as is often a case with trapped charged biomolecules (for example DNA and KCl electrolyte), it is expected that counter-charged ions in electrolyte

Application of Accelerators in Research and Industry
AIP Conf. Proc. 1336, 150-153 (2011); doi: 10.1063/1.3586077

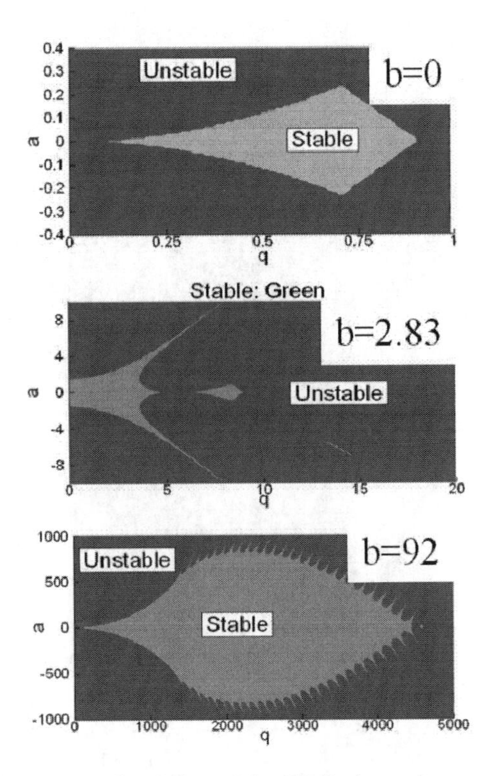

FIGURE 1. of stability of the 2D Paul trap in vacuum (b=0) and in damping environments of viscosity factor b≠0.

may screen the trapped particle. This might have a negative effect on the trapping stability of the molecule because the molecule charge Q in parameter q decreases (see Eq. (3)) by screening.

The main goal of this work is to demonstrate, using a series of molecular dynamics simulations, that the quadrupole electrophoretic field of a 2D aqueous Paul trap removes and prevent the screening, preserving the confining functions in presence of an electrolyte. Direct inspiration for this work has come from the recent publication of Liu et al. [6], who investigated reduction of the electrostatic screening with applied electrophoretic field of a model of charged biomolecule in a cylindrical micro-pore filled with an electrolyte. They used continuum approach to the problem, by solving Poisson-Nernst-Planck equations. and showed that, due to the imbalance between electrophoretic drift and diffusion of ions, the effective Debye length in electrolyte around the molecule increases with increase of external bias (reflecting decrease of the concentration of the counter-ions around the molecule, i.e. descreening).

In the next section, using a charged buckyball as a model of a biomolecule, we apply atomistic, molecular dynamic approach in a aqueous quadrupole 2D Paul nanotrap filled with solvated KCl, and show the suppression of the screening effect. While the screening suppression shown by Liu et al. [6] might be present in the Paul trap too, a more efficient effect, emerging from difference in the stability ranges of the buckyball and electrolyte ions, cause the total de-screening.

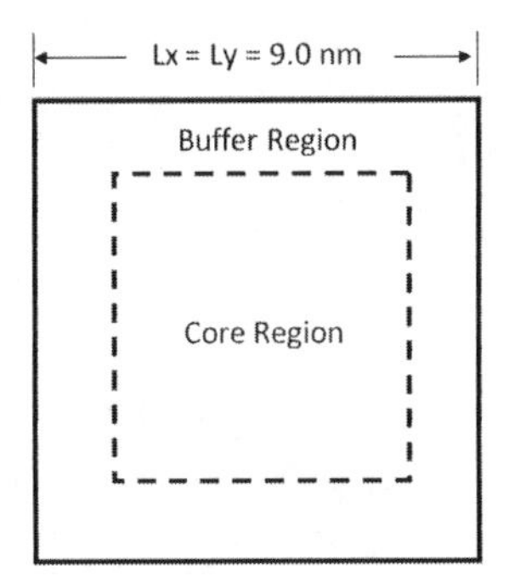

FIGURE 2. Computational space for the molecular dynamics calculation.

COMPUTATION AND RESULTS

The molecular dynamics simulations were performed with the modified GROMACS 4.0 [7] in canonical ensemble (NVT) at T=300 K. The simulation box consisted of core region and the surrounding buffer region, as shown in Figure 2. The $6\times6\times8$ nm^3 core region was placed in the center of simulation box and 1.5 nm-thickness buffer regions were attached to the core region in x- and y-direction, resulting in the total simulation box of $9\times9\times8$ nm^3, containing 11286 water molecules in the core region and 11001 water molecules in the buffer. A buckyball of diameter 0.8 nm and of negative charge -12e uniformly distributed over 60 carbon atoms. The buckyball, 20 dissociated KCl molecules, and water molecules were initially located in the core region while buffer region is filled with water only. The periodic boundary conditions were applied along the axis of the 2D Paul trap (z-axis) to all particles in the trap, while in the x and y directions the periodic boundary conditions are applied only to water molecules. As a consequence, water molecules could freely move in and out the buffer region in x-y directions. However, once the ions or buckyball reached border of the buffer region they could not return back to the core box and their charges were reset to zero. The system temperature was regulated using Nosé-Hoover thermostat with time constant of 0.1 ps, while the simulation time-step was 1.0 fs. The cut-off scheme was introduced to describe the electrostatic interactions with the cut-off radius of 4 nm. In these simulations we used the MD potentials optimized as in Refs. [8-10].

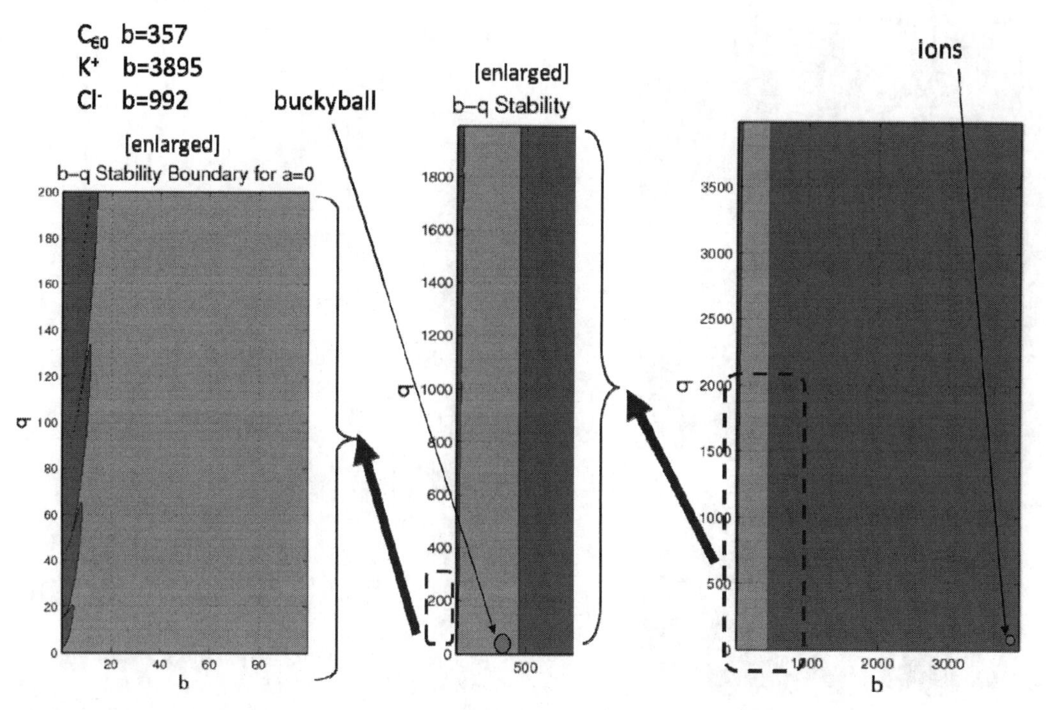

FIGURE 3. Stability of buckyball and electrolytic ions in the aqueous 50 nm 2D Paul trap with V=1 V and Ω=5 GHz.

We apply AC voltage V=1 V of frequency Ω=1 GHz in the Paul trap of R_0=50 nm. Since the DC component of voltage (U in Eq. (1)) is easily screened at electrodes by the electrolyte ions, we apply only the AC field. This leads to the buckyball q-factor of 1.30 in Eq. (3), and 2.0 and 1.32 for potassium and chlorine ions (of charge ±e), respectively. Due to large difference in masses, the buckyball b-factor in Eq. (3) is close to 357, while for K^+ and Cl^- it reaches values of ~3895 and 3992, respectively. The consequence of these large differences is shown in Fig. 3. While buckyball is well in the stable region of the solution of Eq. (2), the electrolytic ions are not stable, and might be expelled from the core region of the Fig. 2 (i.e. from the trap confinement region).

Figure 4 shows molecular visualization of the ion removal process in aqueous Paul trap. Only electrolytic ions and buckyball are shown in the figure. With here applied high frequency of the AC quadrupole field, it is enough 10 ns to fully expel the electrolytic ions from the "core" part of the simulation cell (Fig. 2). The process of descreening shows an exponential decrease with time, and 75% of ions are expelled in the first 2 ns. Negative chlorine ions are leaving the simulation volume somewhat faster, being helped by the Coulomb repulsion of the negatively charged buckyball. The molecular dynamics simulations take into account both inter-ion Coulomb interactions and Brownian motion and the MD simulation shows that the Brownian motion does not influence the stability behavior of this particular system The control MD calculation, performed under identical conditions, but without applied qudrupole AC electrical field, shows a complete screening of the buckyball by the K^+ ions.

CONCLUSIONS

Molecular dynamics simulations were performed on a charged buckyball in a 2D Paul trap, immersed in a water solution of potassium chloride. Without loss of generality, an electrophoretic quadrupole field of high frequency in GHz range was applied to speed up MD calculations. The conditions of confinement stability were altered between buckyball and the electrolytic ions. A simple but remarkable result was obtained: A charged particle was not screened (and can be easily de-screened) by the electrolytic ions in an aqueous 2D Paul trap. This opens possibilities for efficient use of Paul trap for confinement and control of biomolecules and other charged particles in electrolytic aqueous environments.

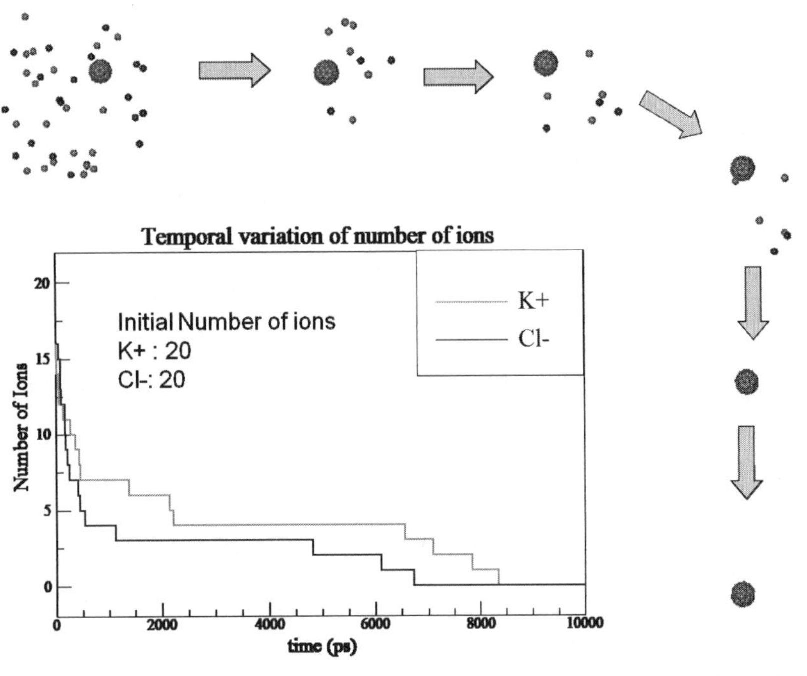

FIGURE 4. Result of the MD simulation. Numbers of potassium and chlorine ions as functions of time in the aqueous Paul trap, residual in the confinement computational volume of the charged (-12e) buckyball, from initial 20 KCl molecules.

ACKNOWLEDGMENTS

This work was supported by the DNA Sequencing Technology Program of the National Human Genome Research Institute of the National Institute of Health (1R21HG004764-01, 1RC2HG005625-01). The computations were performed on Kraken (a Cray XT5) at the National Institute for Computational Sciences (http://www.nics.tennessee.edu). PSK acknowledge partial support of the U.S. Department of Energy, through Oak Ridge National Laboratory, managed by UT-Battelle, LLC, under contract DE-AC05-00OR22725. JHP acknowledge support through ORNL Postdoctoral Program, administered by ORISE.

REFERENCES

1. W. Paul, *Rev. Mod. Phys.* **62**, 531-540 (1990).
2. D. Branton, et al, *Nature Biotechnol.* **26**, 146–153 (2008).
3. H. Liu, J. He, J. Tang, Hao Liu, P. Pang, Di Cao, P. S. Krstić, S. Joseph, S. Lindsay, and C. Nuckolls, *Science* **327**, 64-67 (2009).
4. S. Joseph, P. S. Krstić, W. Guan, and M. A. Reed, *Nanotechnology* **21**, 015103 (2010).
5. X. Zhao and P. S. Krstić, *Nanotechnology* **19**, 195702 (2008).
6. Y. Liu, J. Sauer, and R. W. Dutton, *J. App. Phys.* **103**, 084701 (2008).
7. D. B. Hess, C. Kutzner, D. van der Spoel, E. Lindahl, *J. Chem. Theory Comput.* **4**, 435-447 (2008).
8. H. J. C. Berendsen, J. R. Grigera and T. P. Straatsma, *J. Phys. Chem.* **91**, 6269-6271 (1987).
9. S. Koneshan, J. C. Rasaiah, R. M. Lynden-Bell, and S. H. Lee, *J. Phys. Chem. B* **102**, 4193-4204 (1998).
10. Y. Guo, N. Karasawa, W. A. Goddard, *Nature* **351**, 464-467 (1991).

Inelastic Transmission Of Electrons Through A Single Macro-Glass Capillary And Secondary Electron Emission

B.S. Dassanayake[1], S. Das[1,2] and J.A. Tanis[1]

[1]*Department of Physics, Western Michigan University, Kalamazoo, Michigan, 49008, USA*
[2]*Manne Siegbahn Laboratory, Stockholm University, Frescativägen 26, 11418 Stockholm, Sweden*

Abstract. The transmission of electrons with incident energies 300-1000 eV through single glass capillary samples of macroscopic dimensions has been investigated. The transmitted spectra for all energies indicate dominance of inelastic electrons as the tilt angle ψ increases, with the ratio of inelastically to elastically transmitted intensity for tilt angles > 2.5° (the indirect region of transmission) reaching a constant value of about 2:1. When the secondary electron emission was analyzed, higher yield was observed with decreasing primary beam energy and with increasing sample tilt angle.

Keywords: Glass capillary, macro-capillary, electron transmission
PACS: 34.50.Fa, 34.80.Dp

INTRODUCTION

The interaction of slow highly charged ions (HCIs) and electrons with insulator surfaces has been vigorously studied in the last few decades [1,2]. In recent years focus has been shifted to the more challenging and difficult transmission of charged particle through linear structures such as capillaries of nano- and macro-meter scales [3-7]. These studies are not only from the fundamental understanding of ion/electron insulator interactions [3-5] view point, but also due to its potential applications in various fields [6,7].

The ability of HCIs to traverse through the capillaries with a large fraction of them retaining their initial charge state with negligible energy loss is called "guiding" [8]. Studies on ion guiding have revealed that the inner walls of the capillaries collect charges in a self-arranging manner such that electrostatic repulsion inhibits close collisions with the surface and, in turn, prevent electron capture to the projectile, allowing ions to be guided through [8]. The guiding effect of slow HCIs has been observed for many insulating materials of different densities and aspect ratios such as PET [8], SiO_2 [9], Al_2O_3 [4] and polycarbonate [3] in the last few years. On the other hand, there have only been a few studies reported on electron transmission through capillary arrays, either experimentally or theoretically [4,5].

A recent study on fast electrons through PET [5] has reported that transmitted electrons suffered significant energy losses due to the inelastic interactions with the capillary walls, unlike in the case of slower electrons through Al_2O_3 [4], yet still are guided through the capillaries. Because of these inelastic processes it has been suggested that electron guiding is fundamentally different than ion guiding [5].

In this work, we report the transmission of fast electrons (300-1000 eV) through a single glass capillary of macroscopic dimensions. Since the collective effect of many capillaries do not need to be taken into account, the analysis and interpretation of the outcome is much simpler compared to transmission through capillary foils. The resulting spectra for all the energies have been found to exhibit significant energy losses which increase with sample tilt angle ψ. The characteristics of the energy loss mechanism are discussed as well as the possible processes giving rise to the inelasticity of the transmitted electrons. Furthermore, secondary electron (energies < 50 eV) emission from the capillary is discussed in a qualitative manner for different beam energies.

Experimental Setup

The experiment was conducted in the tandem Van de Graaff accelerator laboratory at Western Michigan University. The glass capillary, hereafter referred as sample A (B) of diameter (D) 0.18 (0.23) mm, length (L) 14.4 (16.8) mm, aspect ratio L/D = 88 (73), and half-width due to aspect ratio (geometrical capillary

Application of Accelerators in Research and Industry
AIP Conf. Proc. 1336, 154-157 (2011); doi: 10.1063/1.3586078

opening), Γ_{aspect} = 0.36° (0.39°) used for the measurements, were prepared at the ATOMKI laboratory in Debrecen, Hungary. The glass capillary samples, which were coated with graphite on the front side to carry away incident excess electrons, were mounted in a goniometer having two degrees of rotational freedom: tilt angle (ψ) rotation about a vertical axis and azimuthal rotation (ϕ) about a horizontal axis with respect to the incident beam. The sample was mounted in an aluminum holder connected to ground through a current meter so that the transmitted intensity could be normalized to the incident beam. A commercially available filament was used as the source of electrons. The beam was collimated with a double collimator of diameters 1.5 and 2.0 mm, respectively, and transmitted electrons were analyzed using an electrostatic parallel-plate analyzer. Electron events were counted using a channel electron multiplier. The background pressure in the scattering chamber was about 10^{-6} Torr during the experiment. A schematic of the experimental setup can be found elsewhere [10].

Results and Discussion

Transmission of 300, 500, 800, and 1000 eV electrons through sample A and 300 and 500 eV through sample B were investigated. The transmitted intensities were normalized with respect to the incident beam. All the data were taken after the transmission intensity presumably reached a steady state at every tilt angle ψ, about 1-2 hours after putting the beam on the sample. Measurements were made by changing the spectrometer angle θ in small steps while keeping tilt angle ψ constant to acquire profiles for electrons transmitted through the glass capillary for different tilt angles and energies.

Significant intensities of the electrons transmitted through the capillary were observed up to $\psi \sim 6°$ for 500 and 1000 eV, whereas intensities were observed only up to 3° for 300 eV. These angles clearly exceed the angle for electrons to travel in a straight line without touching the inner walls of the capillary (~2.5°), indicating that the electrons interact with inner capillary surface at least once before being transmitted through.

Measured electron energy spectra at tilt angles 0°, 0.5°, 1.0°, 1.5° 2.0° and 3.0° for $\psi \approx \theta$, where the maximum transmitted intensity was observed, for 500 and 1000 eV are shown in Fig. 1. (Spectra for both energies exhibit slightly higher energies than 500 and 1000 eV due to imprecise calibration of the power supply used for electron acceleration). It can be seen that the overall transmitted intensity decreases strongly as the tilt angle increases. Notably, the spectra show energy losses when the sample is tilted to larger angles, suggesting that a fraction of the electrons undergo inelastic scattering with the inner surface of the capillary while traversing the sample. This is most prominent when the tilt angle exceeds 2.5°, where the direct beam (primary electrons go through without touching the inner wall of the capillary) dominance is lost. The spectra in Fig. 1 also indicate the existence of two different regions of transmission due to the dominance of inelastically transmitted electrons at higher tilt angles compared to elastically transmitted electrons at lower angles.

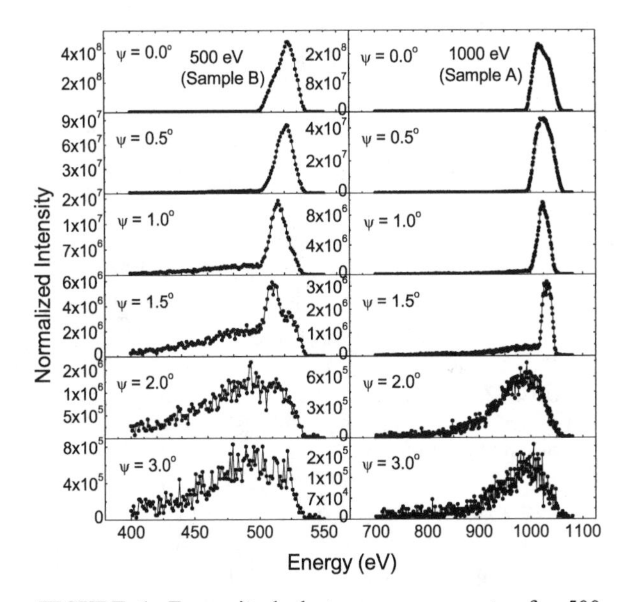

FIGURE 1, Transmitted electron energy spectra for 500 (sample B) and 1000 eV (sample A). All spectra were taken for $\psi \approx \theta$, where maximum transmitted intensity was observed.

To analyze elastic and inelastic characteristics of the transmitted spectra further, the variation of the ratio of inelastically to elastically transmitted electron intensities with sample tilt angle ψ was analyzed for $\psi \approx \theta$ spectra for all energies for both samples. The inelastically transmitted intensity was obtained by integrating the inelastic portions of the spectra, energies of 200-300, 400-500, 700-800 and 700-990 for 300, 500, 800 and 1000 eV, respectively, whereas energies of 300-330, 500-550, 800-860 and 990-1080 for 300, 500, 800 and 1000 eV, correspondingly, were considered as the elastically transmitted range. To determine these regions the elastic contribution was chosen based on the bare beam (without the sample) electron transmission spectrum, and then the inelastic region was taken to be the spectrum for all energies lower than this. The obtained result for $I_{inelastic}/I_{elastic}$ is illustrated in Fig 2.

As the results indicate, all ratios follow the same trend, i.e., an increase in inelasticity with increasing tilt angle. For tilt angles beyond 2.5° (indirect region) the inelastic transmission reaches a constant value of about 2:1 to the elastic contribution for all incident energies. This result also demonstrates the difference in transmission characteristics in the direct and indirect regions.

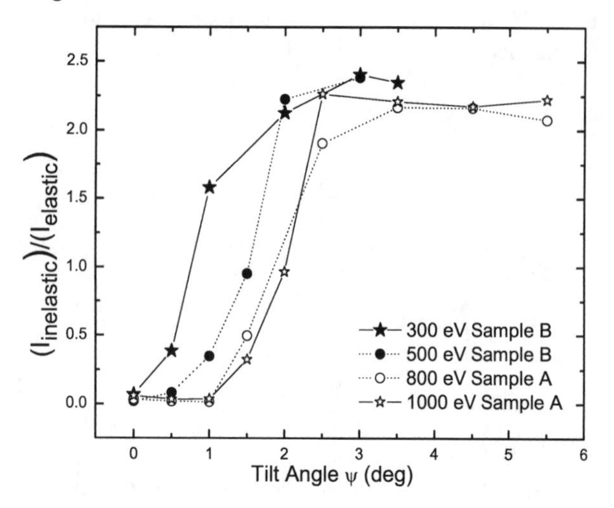

FIGURE 2, Variation of inelastic to elastic ratio of transmitted electrons vs. sample tilt angle for samples A and B at different electron energies.

Interaction of energetic beams with solid surfaces depends on several factors. For beams with glancing angles of incidence, the type of interaction can be divided into three categories in terms of impact parameter values [11]: (1) charged particles which approach close to the target atoms can undergo Rutherford scattering and lose energy, being diverted from the initial direction of trajectory as a result, (2) charges can be reflected by the surface potential losing less energy than the previous case, (3) charges which have intermediate impact parameter values compared to the previous two cases can lose their energy and be diverted from the initial direction due to interaction with inner shell electrons of the atoms. Which type of interaction is dominant in a given instance depends on factors such as incident angle, material topology, and most importantly incident energy of the charged particles.

If the incident angle of the primary incoming beam is < 2.5° (direct transmission), the majority of incoming electrons transmit through the capillary without making any interactions with the inner wall, consequently not losing much initial energy as seen in Fig. 1. The transmission in this region is dominated by elastically transmitted electrons as seen in Fig. 2. But even in this region, some component of the beam can interact with the capillary walls as a result of beam divergence effects (at $\psi = 0°$, ~20% of the incident beam interacts with the inner walls). When the incident beam approaches the indirect region (tilt angles > 2.5°), electrons start interacting with sample inner surface. During the inner wall interaction phase, some incident electrons are repelled electrostatically due to Coulombic forces from already accumulated charges on the surface, giving rise to the elastic intensity contribution in Fig. 1. Other electrons penetrate into the material and closer to atomic nuclei to scatter off (Rutherford scattering), diverting from the initial trajectory and even losing a fraction of their energy. If lower energy electrons interact with the surface, they can scatter (quasi-) elastically from atoms close to the surface and be transmitted toward the exit, whereas for higher energies, due to higher penetration depths, inelastic processes such as ionization and excitation of inner shell electrons of the material atoms can cause incident electrons to lose energy. This results in significant energy losses as seen in Fig. 1, and, if not lost within the bulk of the material, lowers the transmitted intensity.

Fig. 2, where lower energies show a higher inelastic contribution with tilt angle, suggests the small percentage of the beam which makes inner wall interactions in the direct region can still make a considerable contribution to the total transmitted fraction even after losing some of its initial energy. Furthermore, the transmitted lower primary beam energy electrons have been seen to agree with Rutherford predictions in a recent publication [10], whereas the elastic "Coulombic deflection" process was found to dominate at higher energies, justifying the observation in Fig. 2.

When a solid is irradiated by a beam of electrons, the bombarded surface emits electrons which can be classified into two categories: true secondary electrons which have energies ranging from 0 to 50 eV, and elastically and inelastically scattered electrons which have energies that go downward from that of the primary incident electron beam [1]. In order to understand the secondary emission process from the glass capillary sample, a study was conducted at 300, 500 and 1000 eV for both sample A and B using spectra at $\psi \approx \theta$. The data were acquired by obtaining full spectra of the transmission for a few selected sample tilt angles. The transmitted electron intensity within the secondary emission region was found by integrating the spectral region below 50 eV, and the fraction of secondary electrons transmitted was obtained by dividing it by the total transmitted intensity. Acquired results are given in Fig. 3.

At low primary beam energies, though the internal secondary electrons produced can escape efficiently (due to low escape depths), only a few secondary electrons are generated as a result of the low energy of the primary electrons. As the energy increases to 500 eV, the secondary electron emission rises in accordance with the results of Ref. [1], which show this to be the maximum energy for secondary electron production. At higher energies of 1000 eV, even though the secondary electron generation is bigger, the nature of the escape process causes a rapid decrease in the number of internal secondary electrons that escape compared to the increase in generation of internal secondary electrons [12]. This effect can be seen playing a role in the results shown in Fig. 3, where the fraction of secondary electrons produced seems to decrease with increasing energy. Since the measurements were conducted primarily in the direct region of the transmission of electrons, the increase in the secondary electron fraction with increasing sample tilt angle can be due to the increasing interaction of primary electrons with the inner sample surface.

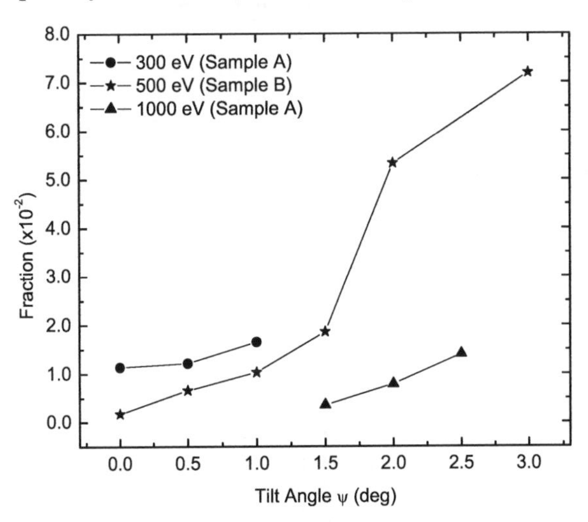

FIGURE 3, Transmitted secondary electron fraction vs. tilt angle for 300, 500 and 1000 eV. The secondary electron fraction was obtained by normalizing to the total transmitted electron intensity.

It is also evident from Fig. 3 that for deeper penetration into the bulk (higher energies), the probability of secondary electrons to escape the sample is smaller. This can cause the sample to charge more negatively and further inhibit close collisions with the capillary wall. So, the majority of electron transmission through the capillary at higher energies is driven by Coulombic reflection of the primary beam. However, due to the inability of electrons to make strong charge patches, surface interactions of electrons are possible at any given energy [10].

In summary, we have demonstrated that when electrons transmit through a macro-meter scale glass capillary a large fraction tends to lose energy. This is likely due to scattering from the inner surface while traversing the capillary tube, giving rise to an inelastic transmission of electrons. The inelastic contribution can be due to insufficient deposition of electrons on the inner walls of the capillary, allowing traversing electrons to interact with the walls. A fraction of the incident electrons in the direct region of transmission was found to show energy losses, probably due to divergence effects of the primary beam. For lower incident energies the inelastic fraction in the direct region was found to dominate the total transmission compared to higher energies. The transmitted secondary electron intensity was found to decrease with increasing energy, probably as a result of increasing penetration and escape depths.

ACKNOWLEDGEMENT

Preparation of capillary samples by Ms. R.J. Bereczky of ATOMKI and assistance from Ms. A. Ayyad of WMU with experimental work is gratefully acknowledged. This work was supported by an award from Research Corporation.

REFERENCES

1. H. Seiler, *J. Appl. Phys.* **54** R1 (1983).
2. Z. G. Song, C. K. Ong and H. Gong, *Appl. Surf. Sci.* **119** 169 (1997).
3. N. Stolterfoht, R. Hellhammer, Z. Juhász, B. Sulik, V. Bayer, C. Trautmann, E. Bodewits, A. J. de Nijs, H. M. Dang, and R. Hoekstra, *Phys. Rev. A* **79** 042902 (2009) and references there in.
4. A.R. Milosavljević, G. Vikor. Z.D. Pešić, P. Kolarž, D. Šević, B.P. Marinković, S. Mátéfi-Tempfli, M. Mátéfi - Tempfli, and L. Piraux, *Phys. Rev. A* **75** 030901(R) (2007).
5. S. Das, B.S. Dassanayake, M. Winkworth, J.L. Baran, N. Stolterfoht, and J.A. Tanis, *Phys. Rev. A* **76** 042716 (2007).
6. T. Nebiki, M.H. Kabir, and T. Narusawa, *Nucl. Instrum. Meth. Phys. Res. B* **249** 226 (2006).
7. C. Lemell, A.S. El-Said, W. Meissl, I.C. Gebeshuber, C. Trautmann, M. Toulemonde, J. Burgdorfer and F. Aumayr, *Solid-State Electron.* **51** 1398 (2007).
8. N. Stolterfoht, J.H. Bremer, V. Hoffmann, R. Hellhammer, D. Fink, A. Petrov, and B. Sulik, *Phys. Rev. Lett.* **88** 133201 (2002).
9. M.B. Sahana, P. Skog, Gy. Víkor, R.T. Rajendra Kumar, and R. Schuch, *Phys. Rev. A* **73** 040901(R) (2006).
10. B.S. Dassanayake, S. Das, R.J. Bereczky, K. Tőkési and J.A. Tanis, *Phys. Rev. A* **81** 020701(R) (2010).
11. A. Lagutin, *Proceedings of RuPAC*, Zvenigorod, Russia (2008).
12. A. Shih, J. Yater, C. Hor, and R. Abrams, *Appl. Surf. Sci.* **111** 251 (1997).

n-Selective Single Capture Following Xe^{18+} And Xe^{54+} Impact On Na(3s) And Na*(3p)

S. Otranto[1], R. E. Olson[2], V. G. Hasan[3] and R. Hoekstra[3]

[1]*CONICET and Dto. de Física, Universidad Nacional del Sur, 8000 Bahía Blanca, Argentina.*
[2] *Physics Department, Missouri University of Science and Technology, Rolla, MO 65401 USA*
[3]*KVI-Atomic and Molecular Physics, University of Gröningen, The Netherlands*

Abstract. State selective single charge exchange n-level cross sections are calculated for collisions of Xe18+ and Xe54+ ions with Na(3s) and Na*(3p) over the energy range of 0.1 to 10.0 keV/amu. The CTMC method is used which includes all two-body interactions. Experimental state-selective cross sections and their corresponding transverse momentum spectra for Xe18+ are found to be in reasonable accord with the calculations.

Keywords: highly charged ions, charge exchange.
PACS: 34.70.+e

INTRODUCTION

The planned construction of a large, international, high temperature tokamak fusion reactor (ITER) in Cadarache, France [1] has renewed the interest in studying multiply-charged ion electron capture collisions. Starting in the 1970s, the tokamak fusion reactor program stimulated the studies of charge exchange processes involving collisions of multiply charged ions of C, N and O with atomic hydrogen (see [2] for a review). The collision energies of interest were in the range 1-80 keV/amu. These studies were prompted by the use of the photon emission cascade following the charge exchange process to estimate the impurity ion concentrations, the plasma temperature and its rotation.

Charge exchange processes between heavy rare gases and multiply charged projectiles are expected to play a major role in the new, large ITER reactor. In order to extract the heat from the reactor, it is proposed that charge exchange collisions in the scrape off layer and divertor will give rise to photon emission that will uniformly heat the plasma facing walls, thereby removing the possibility of the hot plasma touching the wall and causing burn-through. Heavy rare gas ions will be used as the charge exchange medium that interacts with the H (D) atoms.

Although it is well known that the H*(n=2) target provides a high fraction of the photon flux in these collisions, it has been experimentally infeasible up to now to provide experimental cross sections to test theoretical predictions. However, ground state alkali atoms have very similar cross sections due to their nearness in ionization potentials and provide substitutes that can be experimentally explored. The KVI group has recently succeeded in obtaining state-selective data for Xe18+ + Na(3s) collisions using the MOTRIMS (Magneto Optical Trap Recoil Ion Momentum Spectroscopy) method [3]. The same group has already optically pumped Na(3s) to obtain Na*(3p) targets together with He2+ and C4+ projectiles in the past [4-5].

In this work, we use the 3-body CTMC method to obtain state selective electron capture data for the reactions:

$$Xe^{q+} + Na(3s) \rightarrow Xe^{(q-1)+}(n) + Na^{+},$$
$$Xe^{q+} + Na^{*}(3p) \rightarrow Xe^{(q-1)+}(n) + Na^{+}. \quad (1)$$

Application of Accelerators in Research and Industry
AIP Conf. Proc. 1336, 158-161 (2011); doi: 10.1063/1.3586079

The projectiles under study will be Xe18+ and Xe54+. The Na*(3p) target is included in our study to provide guidance for future experimental studies. The energy range investigated is 0.1 – 10 keV/amu, termed intermediate since the collision speeds surround those of the target electron which corresponds to 9.4 keV/amu for Na(3s) and 5.6 keV/u for Na*(3p).

THEORETICAL METHOD

The classical trajectory Monte-Carlo (CTMC) method has been used to calculate cross sections for single electron capture [6]. Hamilton's equations were solved for a mutually interacting three-body system. The center-of-mass of the Na target is frozen at the beginning of each simulation. The active electron evolves under the central potential model developed by Green et al. from Hartree-Fock calculations [7], and later generalized by Garvey et al. [8]. The CTMC method directly includes the ionization channel and is not limited by basis set size for the prediction of capture to very high-lying excited states. The Na(3s) and Na*(3p) states are distinguished through their respective ionization potentials and the classical angular momentum restriction $l^2<1$, and $1<l^2<4$ respectively [9].

Since the electron tends to be captured to high n-values with minimal contributions from the s-, p-, and d-states, quantum defects play a minor role and the orbital energies for the captured electron are similar to those obtained with bare projectiles. We then represent the electron-projectile interaction by a Coulomb potential of charge 18+ and 54+ respectively.

RESULTS

In Figs. 1 and 2 are presented the n-selective capture cross sections for Na(3s) and Na(3p) targets as a function of collision energy for the Xe18+ and Xe54+ systems, respectively. We note that at 1 keV/amu the total charge exchange cross sections (not shown here) for both targets under consideration scale linearly with the projectile charge as previously predicted [10]. Furthermore, the Na*(3p) total charge exchange cross sections are approximately a factor of three larger than those for the ground state. One can see that the capture process is sharply peaked in n-value at the lowest energy leading to large cross sections for only a few n-values. However, at the highest energy the n-distribution is considerably broadened due to strong mixing of the capture channels in the increasingly important small impact parameter collisions.

The theoretical cross sections presented for Xe18+ on Na(3s), Fig. 1, can be compared to experimental data obtained by the KVI group. The measurements were obtained using the newly developed MOTRIMS method [3]. In Fig. 3 are shown the data over the experimental energy range. Overall, there is good agreement between theory and experiment, indicating that theory is able to represent these strong perturbation collisions.

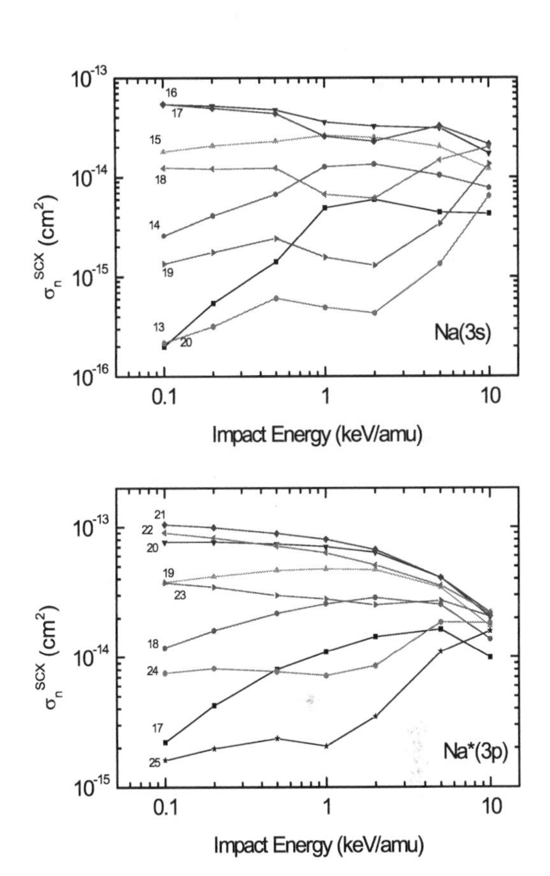

FIGURE 1. n-state selective capture cross sections for Na(3s) and Na*(3p) targets as a function of collision energy for Xe^{18+}. The final n-levels are labeled next to each curve.

A more strenuous test of theory is to compare the n-level cross sections differential in their transverse momenta. By so doing, we are testing the impact parameter dependence of the n-level transition probabilities since the transverse momentum magnitude is inversely proportional to impact parameter. In Fig. 4 are shown the transverse momentum comparisons for the n = 14, 16 and 18 levels in 2.23 keV/amu collisions of Xe18+ on Na(3s).

The data collection and analyses are described in the Ph.D. thesis of V. G. Hasan [11]. Theory reproduces the overall trends in the data. As expected, the lower n-values are formed in the hard collision, low impact parameter region, resulting in large transverse momentum. However, there are systematic differences. The main one is that theory tends to overestimate the transverse momentum peaks. From this, one can infer that theory is underestimating the range of the collisions, or overestimating the strength of the repulsive Coulomb interaction during the collision. We must also mention that in Fig. 3 and 4 we have normalized the experimental data to our cross sections in the peak region because the experimental data are relative, not absolute in magnitude.

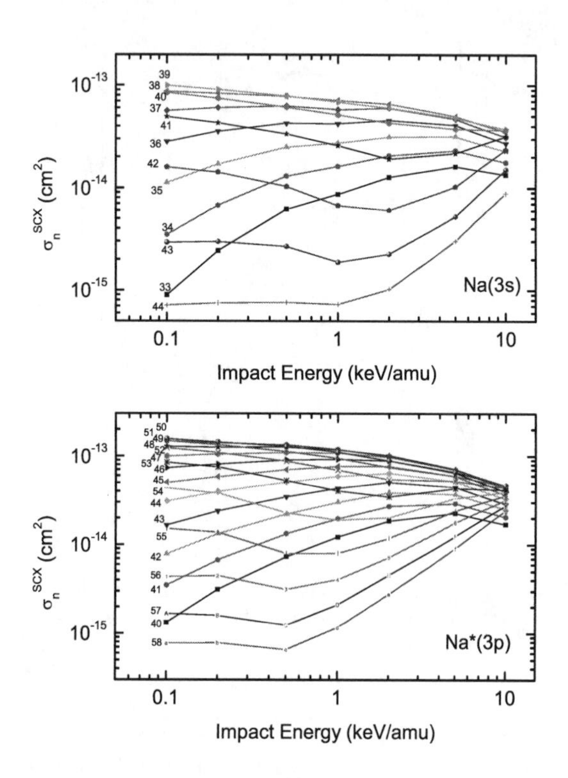

FIGURE 2. n-state selective capture cross sections for Na(3s) and Na*(3p) targets as a function of collision energy for Xe^{54+}.

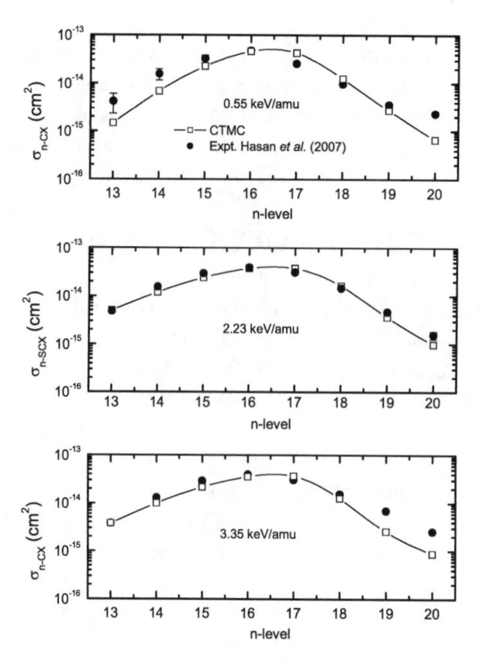

FIGURE 3. n-sate selective capture cross sections for 0.55 keV/amu, 2.23 keV/amu and 3.35 keV/amu Xe18+ collisions on Na(3s). The experimental data shown are from ref. [3].

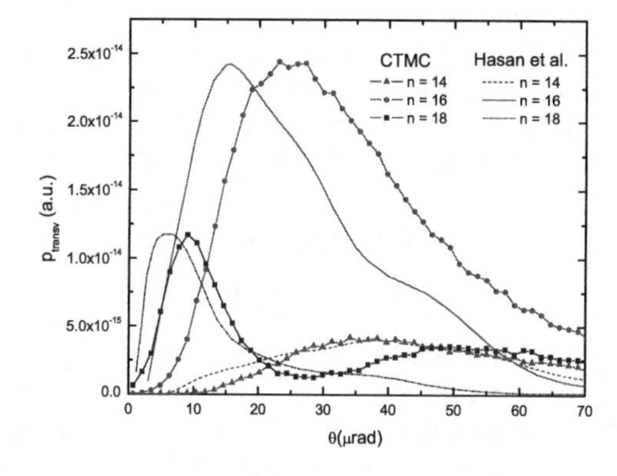

FIGURE 4. Recoil ion transverse momentum distribution at 2.23 keV/amu Xe18+ collisions on Na(3s) target (capture to n=14, 16 and 18). The experimental data shown are from Hasan (ref.[11]).

CONCLUSIONS

In this work we have studied state selective charge exchange processes for collisions of Xe18+ and Xe54+ ions with Na(3s) and Na*(3p) over the energy range of 0.1 to 10.0 keV/amu. As expected, the calculated capture cross sections from Na*(3p) are larger than those from Na(3s) and they populate larger n-values of the projectile at the same collision energy.

We have compared our theoretical predictions for the n-state selective capture cross sections and their corresponding transverse momentum spectra with available experimental data for Xe18+ +Na(3s) from the KVI group. Within the impact energy range experimentally explored, we found that our calculations are in reasonable accord with the data. These results show that the CTMC method provides a fast and confident platform for these studies.

ACKNOWLEDGMENTS

Work at UNS supported by PGI 24/F049, PICT-2007-00887 of the ANPCyT and PIP 112-200801-02760 of CONICET (Argentina).

REFERENCES

1. www.iter.org
2. R.C. Isler, Plasma Phys. Control. Fusion **36**, 171 (1994).
3. V.G. Hasan, S. Knoop, R. Morgenstern and R. Hoekstra, J. Phys: Conf. Series **58**, 199 (2007).
4. S. Schippers, P. Boduch, J. van Buchem, F. W. Bliek, R. Hoekstra, R. Morgenstern and R. E. Olson, J. Phys. B: Atom. Molec. Opt. Phys. **28**, 3271 (1995).
5. S. Schippers, R. Hoekstra, R. Morgenstern and R. E. Olson, J. Phys. B: Atom. Molec. Opt. Phys. **29**, 2819 (1996).
6. R. E. Olson and A. Salop, Phys. Rev. A **16**, 531 (1977).
7. A. E. S. Green, D. L. Sellin and A. S. Zachor, Phys. Rev. **184**, 1 (1969).
8. R. H. Garvey, C. H. Jackman and A. E. S. Green, Phys. Rev. A **12**, 1144 (1975).
9. R. E. Olson, Phys. Rev. A **24**, 1726 (1981).
10. R. E. Olson, K. H. Berkner, W. G. Graham, R. V. Pyle, A. S. Schlachter and J. W. Stearns J W, Phys. Rev Lett **41**, 163 (1978).
11. V.G. Hasan, "MOTRIMS investigations of electron removal from Na by highly charged ions", Ph.D. Thesis, University of Gröningen, The Netherlands, (2008).

M Sub-Shell Cross Sections For 75-300 keV Proton Impact On W, Pt And Pb

Sam J. Cipolla

Physics Department, Creighton University, Omaha, NE 68178 USA
Email: samcip@creighton.edu

Abstract. M sub-shell x-ray production cross sections from 75-300 keV proton bombardment of thick elemental targets of W, Pt, and Pb were measured and compared with ECPSSR and relativistic RPWBA-BC cross sections using different data bases of fluorescence yields, Coster-Kronig factors, and x-ray transition rates. With a few exceptions, the differences between the various data base comparisons were not significant. For different sub-shells, either ECPSSR or RPWBA-BC compared better with the measurements. In all cases, agreement with theory improved as the collision energy increased.

Keywords: M sub-shell cross sections, Ion-atom collisions, ECPSSR
PACS: 32.80Hd, 32.70Fw

INTRODUCTION

Much research has been devoted to K and L x-ray production cross section measurements from proton bombardment for the purpose of testing theory and in PIXE applications[1]. In comparison, partly because of the complexity and compactness of the x-ray energy spectra, there has been less M-shell research, especially M sub-shell measurements. Furthermore, there are relatively few M-shell measurements at low energies where theories are pushed to their limit. Presented here are measurements of M sub-shell x-ray production cross sections for proton impact on thick targets of W, Pt and Pb in the 75-300 keV energy range. Previous measurements by others[2-5] in this energy range have mostly reported total M-shell cross sections for these elements.

The ECPSSR theory[6] has been found to give a good description of measured K- and L-shell ionization cross sections in intermediate and high-energy for proton collisions. For the M shell, the ECPSSR theory, as calculated through ISICS[7], uses non-relativistic hydrogenic wave functions without any relativistic corrections and employs the united atom binding energy perturbation. The RPWBA-BC theory[8] is similar to ECPSSR, except that relativistic hydrogenic wave functions are employed. It is expected that RPWBA-BC will differ slightly from ECPSSR at high energies, but could differ significantly at lower collision energies. Different atomic data bases of fluorescence yields, Coster-Kronig yields, and x-ray transitions rates can have an effect on the theory comparisons. Use of recent data bases in the theory calculations are compared with using ones commonly employed in this work.

EXPERIMENTAL PROCEDURE

The experimental system has been fully described in previous work[9], so only a brief description will be given here. A mass-analyzed and collimated proton beam produced in a Cockcroft-Walton accelerator was made to impact thick elemental target foils. A biased secondary-electron suppression cage surrounded the target holder which was connected to a calibrated current integrator to measure the proton charge, and therefore the number of protons, N_p, on the targets. The targets were oriented at 45° to the beam direction and to a high-resolution Si(Li) x-ray detector equipped with an ultra-thin window. An additional 6-μm doubly-aluminized Mylar absorber was used to reduce the count rate from softer x-rays. The detector efficiencies were determined from a combination of K-shell cross section measurements and radioactive standards using the Gallagher-Cipolla model[10].

Application of Accelerators in Research and Industry
AIP Conf. Proc. 1336, 162-165 (2011); doi: 10.1063/1.3586080

DATA ANALYSIS

Peak x-ray counts, N_x, were obtained by fitting Gaussians plus an estimation of the underlying background to the cluster of expected peaks. The width of each Gaussian was not freely fitted, but was actively determined during a fit according to the methods described in ref. 11. All the major M x-ray peaks plus several minor peaks were extracted from the fits.

The x-ray production cross section of a particular x-ray transition, denoted by i, for protons of incident energy E was determined from the thick-target equation,

$$\sigma(E)_i^x = \frac{Y(E)_i S(E)}{N\varepsilon_i}\left[\frac{(\mathrm{d}Y/\mathrm{d}E)_i}{Y(E)_i} + \frac{(\mu/\rho)_i}{S(E)}\right] \tag{1}$$

where $Y(E)_i = (N_x)_i/N_p$ is the x-ray yield, N is the target atom density, ε_i is the detector efficiency, S(E) is the proton stopping power in the target (from SRIM[12]), $(\mu/\rho)_i$ is the mass absorption coefficient for x-ray self absorption in the target (from XCOM[13]), and $(dY/dE)_i$ is the slope of the measured Y(E) vs. E curve evaluated at the incident proton energy. The yield slope was obtained by fitting the yield data to $Y(E) = A(E-C)^B$ and differentiating; A, B and C are the fitted parameters.

The sub-shell x-ray production cross sections, σ_s^x (E), were obtained from a representative x-ray production cross section, $\sigma(E)_i^x$, according to,

$$\sigma(E)_s^x = \sigma(E)_i^x \frac{\Gamma_s}{\Gamma_i} \tag{2}$$

where Γ_i is the transition rate for the particular x-ray transition, denoted by i, and Γ_s is the total radiative transition rate for the corresponding sub-shell, s (s=1 - 5). The M_5 shell ionization cross section was determined from the M_α ($M_5N_{6,7}$) x-ray production cross section, M_4 was determined from M_β (M_4N_6) , M_3 from M_γ (M_3N_5), M_2 from M_2N_4, and M_1 was determined from $M_1N_{,3}$ for all the targets.

Theoretical sub-shell x-ray production cross sections, $\sigma_x^x(E)$, are obtained from sub-shell ionization cross sections, $\sigma_s^I(E)$, as

$$\sigma_s^x = \omega_s F_s \sigma_s^I \tag{3}$$

where ω_s is the fluorescence yield and F_s is the Coster-Kronig vacancy enhancement factor due to intra-shell vacancy-filling transitions.

RESULTS AND DISCUSSION

Figures 1 through 4 show the experiment/theory M sub-shell x-ray cross section ratios using the McGuire[14] or Chauhan-Puri[15] atomic databases of fluorescence yields and Coster-Kronig factors in calculating the theoretical ECPSSR x-ray production cross sections, and using the relative x-ray transition rates from Bhalla[16] or Chen-Crasemann[17]. The measured cross sections are also compared with the relativistic RPWBA-BC cross section calculations from Chen and Crasemann using their transition rates and McGuire's atomic parameters. ECPSSR M sub-shell x-ray production cross sections were calculated from the most recent version of ISICS. For the M shell, ISICS employs only the united atom calculation of electron binding energies, as is done in RPWBA-BC, and uses no relativistic correction. It should also be noted that the Coulomb deflection factor used by Chen and Crasemann is different from the calculation of the ECPSSR theory for the M shells. The difference between the two calculations decreases in going from the inner to outer M shells, and significantly increases as the proton energy decreases. The differences also increase as Z of the target atom increases.

Overall, Figures 1-5 show that use of the McGuire atomic parameter database tends to produce better agreement between theory and experiment, except for the M_4 and M_5 comparisons for W where Chauhan-Puri gives better agreement. There is practically no difference between using either data base of x-ray transition rates. For the M_1 shell, ECPSSR and RPWBA-BC theories both trend the measurements similarly as seen in Fig. 1. For the M_2 shell, the measurements are in slightly better agreement with ECPSSR, as shown in Fig. 2. Fig. 3 shows that the measurements are in better agreement with ECPSSR throughout the entire energy range for the M_3 shell, with RPWBA-BC/experiment ratio being consistently about 50% higher that the ECPSSR ratio. For the M_4 (and similarly for M_5 which is not shown) shells, Fig. 4 shows that both theories agree well with the measurements at the higher collision energies, but the agreement with RPWBA-BC continues down to the lower energies for all three target systems whereas ECPSSR under-predicts at the lower energies. A general trend noted in all the comparisons is increasing agreement as the impact energy increases for all elements and for each sub-shell, a trend found in K- and L-shell comparisons as well.

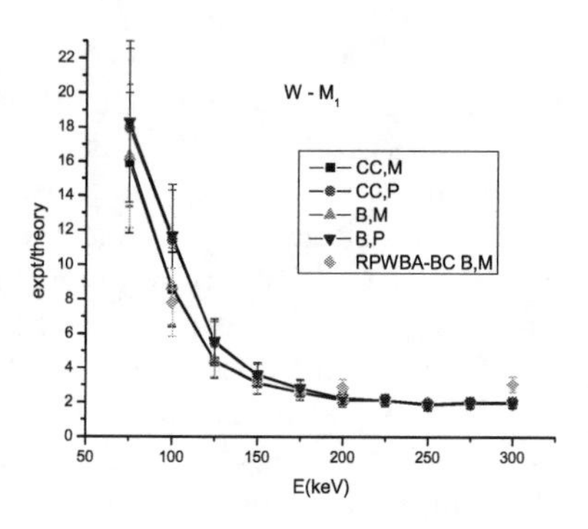

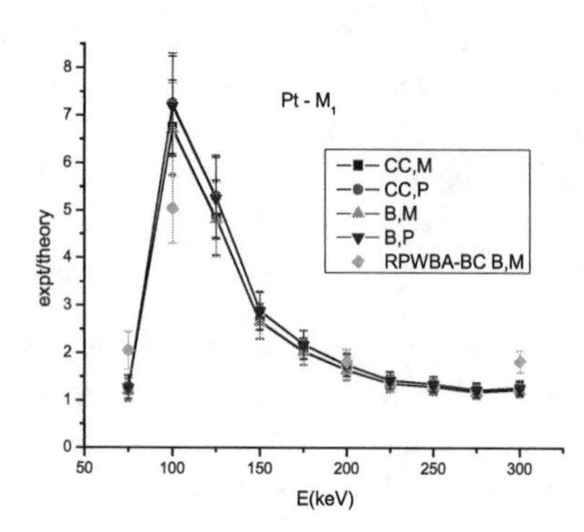

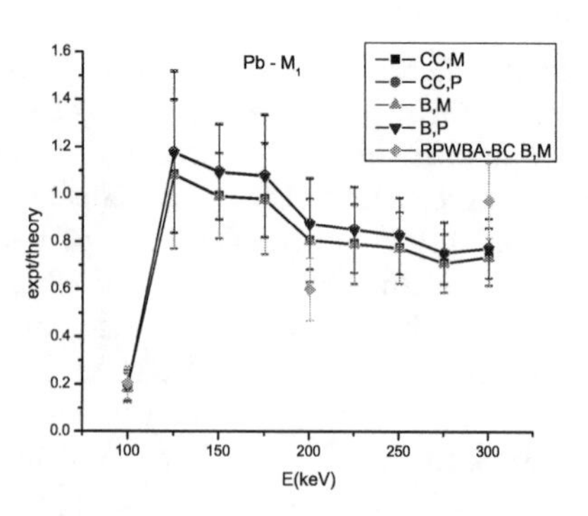

FIGURE 1. Comparison of measured W, Pt, and Pb M_1 x-ray production cross section with ECPSSR theory (ISICS) using x-ray transition rates from ref. 17 (CC) or ref. 16 (B) and fluorescence/Coster-Kronig yields from ref. 14 (M) or ref. 15 (P). The measurements are also compared with the relativistic RPWBA-BC theory of ref. 8.

CONCLUSION

There is essentially no difference between the relativistic Hartree-Slater x-ray transition rates of Bhalla and relativistic Dirac-Fock calculations of Chen-Crasemann for these collision systems. The Chauhan-Puri data base of fluorescence and Coster-Kronig yields, which is based on the limited calculations of Chen and Crasemann[17], gave slightly better agreement between theory and experiment over using the non-relativistic McGuire database. For the M_1 and M_2 x-ray production cross sections for these collision systems, both ECPSSR and RPWBA-BC compared similarly with experiment. For M_3, the ECPSSR theory compared better than the RPWBA-BC theory. On the other hand, for M_4 and M_5, the measurements were in good agreement with RPWBA-BC throughout the entire energy range. Overall, agreement between experiment and theory for all sub-shells improved as the collision energy increased.

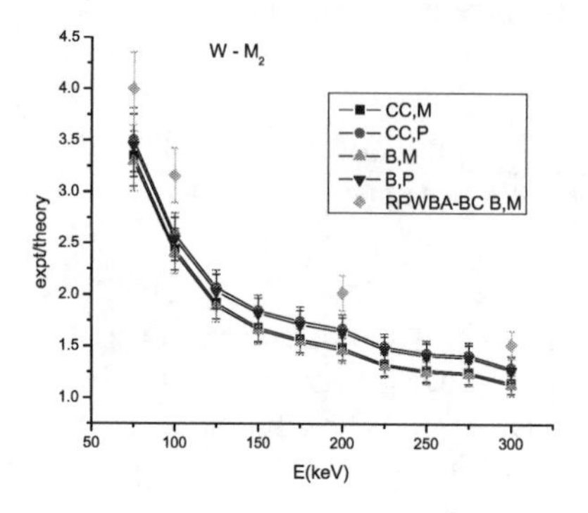

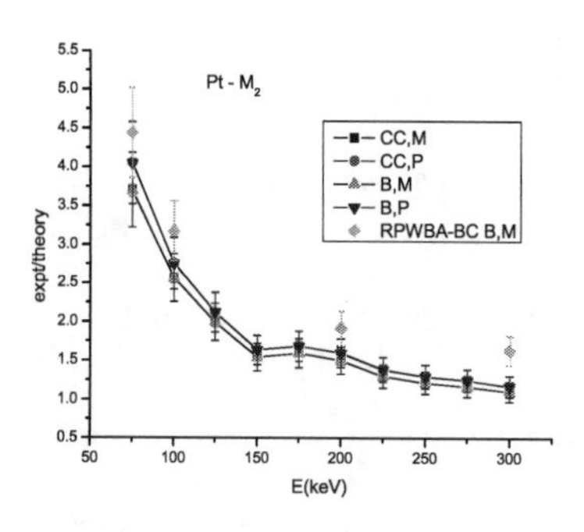

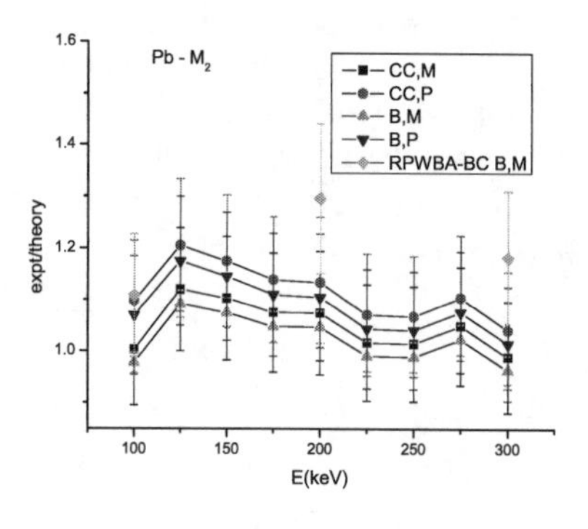

FIGURE 2. Same as fig. 1 for the M_2 x-ray production cross sections.

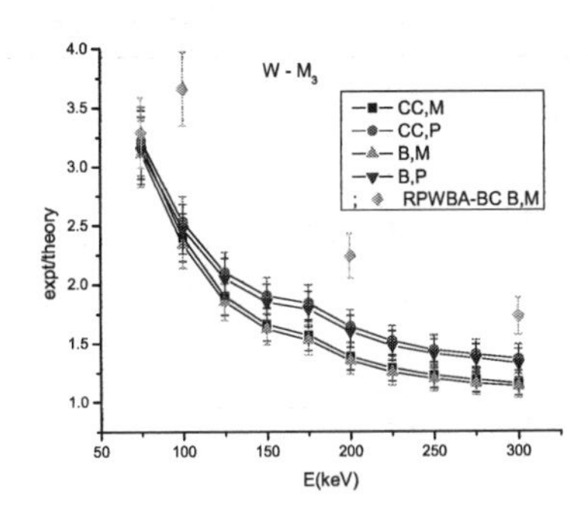

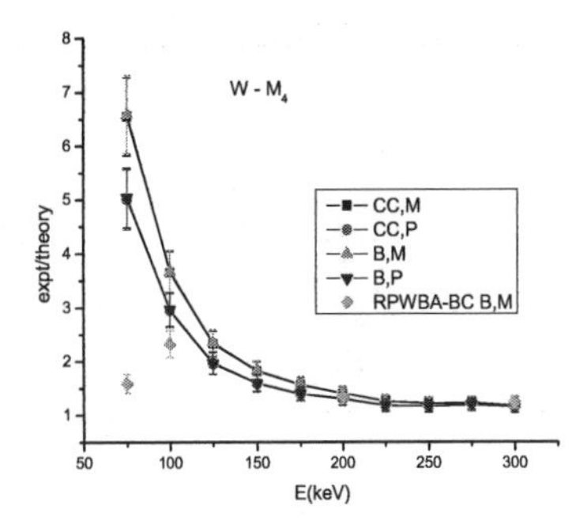

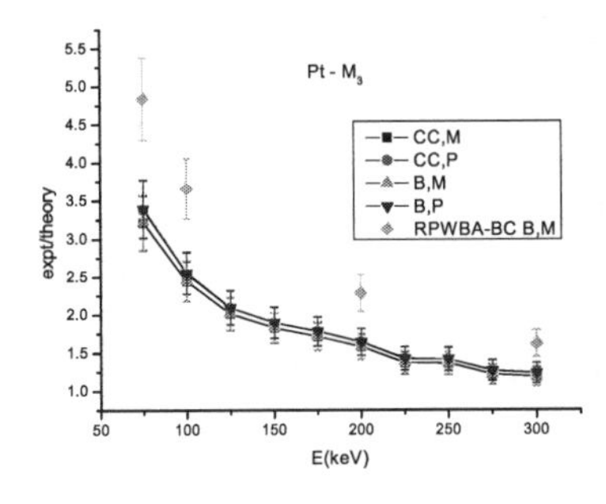

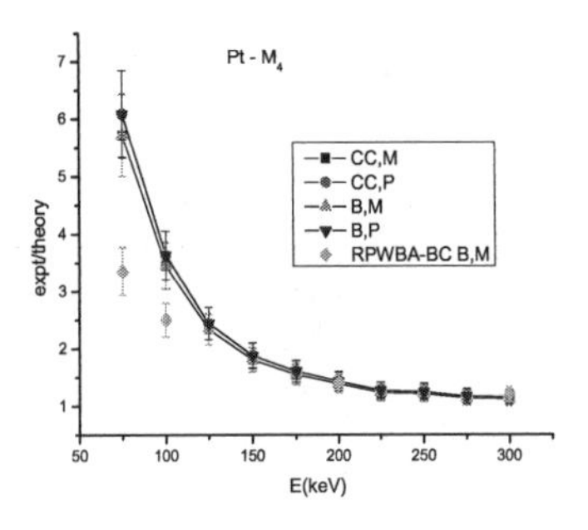

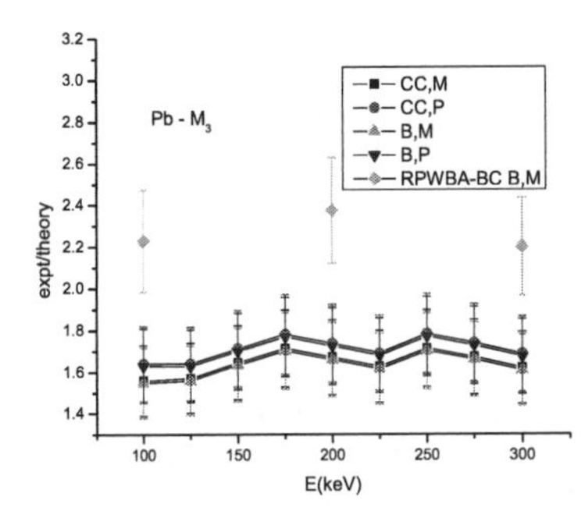

FIGURE 3. Same as fig. 1 for the M_3 x-ray production cross sections.

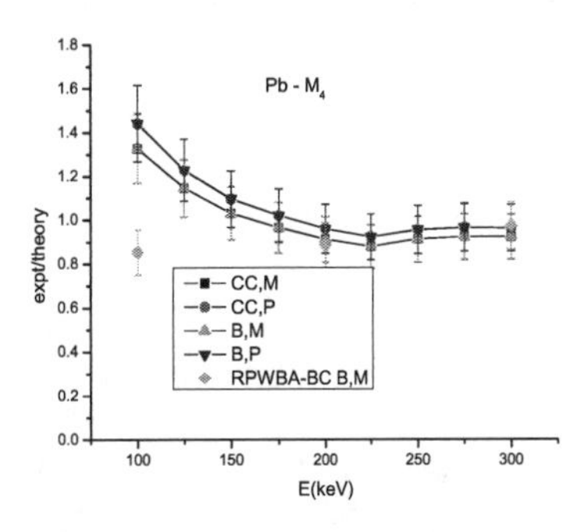

FIGURE 4. Same as fig. 1 for the M_4 x-ray production cross sections.

REFERENCES

1. G. Lapicki, *Nucl. Instr. Methods B*, **189**, 8 (2002).
2. L. Rodríguez-Fernandez, J. Miranda, J.L. Ruvalcaba-Sil E. Segundo, and A. Oliver, *Nucl. Instr. Methods B* **189**, 27 (2002).
3. R. Mehta, J.L. Duggan, J.L. Price, F.D. McDaniel, G. Lapicki, *Phys. Rev. A*, **26**, 1883 (1982).
4. N.V. de Castro Faria, F.L. Freire, Jr., A.G. de Pinho, and E.F. da Silveira, *Phys. Rev. A*, **28**, 2770 (1983) 2770.
5. M. Sarkar, H. Mommsen, W. Sarter and P. Schürkes, *J. Phys. B: At. Mol. Phys.* **14**, 3163 (1981).
6. W. Brandt and G. Lapicki, *Phys. Rev. A* **23**, 1717 (1981).
7. Z. Liu and S. Cipolla, *Comput. Phys. Comm.* **97**, 315 (1996); S. Cipolla, *Comput. Phys. Comm.* **176**, 157 (2007); S. Cipolla, *Comput. Phys. Comm.* **180**, 1716 (2009)
8. M.H. Chen and B. Crasemann, *At. Data & Nucl. Data Tables*, **30**, 257 (1989).
9. S.J. Cipolla and Brian P. Hill, *Nucl. Instru. & Meth. B* **241**, 129 (2005).
10. W. Gallagher and S. Cipolla, *Nucl. Instru. & Meth.* **122**, 405 (1974); S. Cipolla and S. Watson *Nucl. Instru. & Meth. B* **10/11**, 946 (1985).
11. S. Cipolla, *J. Phys. B:At. Mol. Opt. Phys.* **42**, 205201 (2009).
12. J.F. Ziegler, SRIM-2003, *Nucl. Instrum. Methods B* **219-220**, 1027 (2004).
13. M. Berger and H. Hubbell, XCOM: Photon Cross-Section on a Personal Computer, NBS, Gaithersburg, USA, Report NBSIR 87-3595 (July 1987).
14. E. McGuire, *Phys. Rev. A*, **5**, 1043 (1972).
15 Y. Chauhan an S. Puri, *At. Data & Nucl. Data Tables*, **94**, 38 (2008).
16. C. Bhalla, *J. Phys. B: At. Mol. Phys.*, **3**, 916 (1970).
17. M.H. Chen and B. Crasemann, *At.. Data Nucl. Data Tables* **30**, 170 (1984).

Artificially Structured Boundary For Antihydrogen Studies

C. A. Ordonez

Department of Physics, University of North Texas, Denton, Texas

Abstract. It may be possible to confine antiprotons using an artificially structured boundary, as part of a process for synthesizing antihydrogen. An artificially structured boundary is defined at present as one that produces a spatially periodic static field, such that the spatial period and range of the field is much smaller than the dimensions of a cloud, plasma or beam of charged particles that is confined by the boundary. A modified Kingdon trap could employ an artificially structured boundary at the location of inner electrodes. The artificially structured boundary would produce a multipole magnetic field that keeps confined particles from reaching the inner electrodes. The magnetic field would be sufficiently short in range to affect the particle trajectories only in close proximity to the inner electrodes. The conditions for producing such a magnetic field have been assessed. The results indicate that the magnetic field must be an octupole or higher order field.

Keywords: Antihydrogen, Antiprotons, Ion Traps, CERN AD.
PACS: 37.10.Ty, 52.27.Jt, 52.65.Cc, 52.65.Pp, 52.58.Qv

INTRODUCTION

Antihydrogen has been produced at the CERN Antiproton Decelerator by mixing antiprotons and positrons within nested Penning traps [1–3]. Much current research at the CERN Antiproton Decelerator is now aimed at trapping antihydrogen. However, the antiproton kinetic energy associated with various drifts within the nested Penning traps may be too large for trapping antihydrogen that is synthesized from the antiprotons [4]. Antihydrogen atoms that have sufficiently low kinetic energy may be useful for conducting fundamental studies such as CPT and gravity tests [5]. The work presented here concerns the possibility of trapping non-drifting antiprotons using a modified Kingdon trap. A non-rotating cloud or plasma of antiprotons that is confined within a modified Kingdon trap may be ideally suited for application of a recently developed evaporative cooling technique [5]. A preliminary discussion of antihydrogen studies that may be possible with an artificially structured boundary is provided in Ref. [6]. Various other possibilities are described in Refs. [7-10] regarding the possible utility of artificially structured boundaries.

A Kingdon trap nominally employs a combination of electrostatic and centrifugal forces for confining charged particles [11-13]. The trap has been used for a number of different measurements [14-16], and it is closely related to a number of other systems [17-21]. Prior theoretical studies regarding charged-particle confinement in Kindgon traps include Refs. [22-25]. Figure 1 illustrates the electrode configuration of a simple Kingdon trap. The two electrodes shown in Fig. 1 consist of a straight wire and a coaxial cylindrical electrode with end caps. Charged particles that are confined within a Kingdon trap typically experience an orbital drift between the pair of cylindrically symmetric coaxial electrodes. The orbital drift provides a radially outward centrifugal force that keeps particles from reaching the inner electrode. The electric field produced by the electrodes provides a radially inward force that keeps the particles from reaching the outer electrode. Axial confinement is also provided electrostatically. The term "cylindrical trap" is used to refer to a trap that produces a radial electric field that has an inverse dependence on the distance from the trap's axis of symmetry. A recent variation of the Kingdon trap produces a spherically symmetric Coulomb-like field [26]. The term "spherical trap" is used to refer to a trap that produces a radial electric field that has an inverse square law dependence.

In Ref. [25], criteria for confining charged particles that have a drifting Maxwellian velocity distribution were obtained for both cylindrical and spherical traps. It was found that the average kinetic energy associated with the drift must be much larger than the average thermal energy, if the confinement criteria are satisfied. In the work presented here, the possibility of reducing or eliminating the need for an orbital drift in a Kingdon trap is explored. Particles are kept from

Application of Accelerators in Research and Industry
AIP Conf. Proc. 1336, 166-169 (2011); doi: 10.1063/1.3586081

reaching the location of inner electrodes by employing an artificially structured boundary for introducing a multipole magnetic field. The magnetic field is sufficiently short in range to affect the particle trajectories only in close proximity to the location of the inner electrodes. The conditions for producing such a magnetic field are assessed. To simplify the analysis, the magnetic field is assumed to be produced by a number of straight wires.

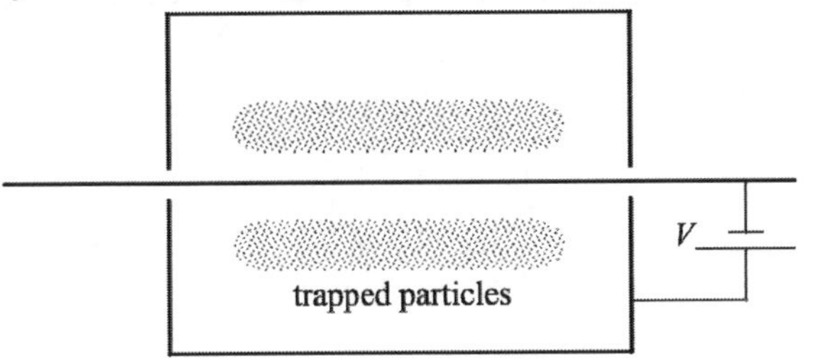

FIGURE 1. A simple Kingdon trap configuration.

CONDITIONS FOR PRODUCING A SUITABLE MAGNETIC FIELD

Figure 2 illustrates cylindrical and spherical trap configurations. Also shown are Cartesian coordinates (x,y,z) and polar coordinates (r,θ), with the polar coordinate system located on the z = 0 plane. The cylindrical trap configuration has a number of outer cylindrical electrodes of inner radius r_1, and the spherical trap configuration has an outer spherical electrode of inner radius also denoted r_1. The rest of the structure consists of a number of small-radius electrodes and current-carrying wires. Each wire is straight and is parallel to the z axis of the Cartesian coordinate system. There are five wires shown in Fig. 2(c) that have locations labeled 1 - 5 and that have (x,y) coordinates (0,0), (r_0,0), ($-r_0$,0), (0,r_0), and (0,$-r_0$), respectively. Each wire passes through a set of small-radius electrodes. It is assumed that the small-radius electrodes have applied voltages that produce the spatial variation of the electric potential associated with the cylindrical trap or spherical trap at distances sufficiently far from the electrodes. The assumption should be valid provided that the small-radius electrodes are sufficiently small and closely spaced.

The radial range of the magnetic field is determined on the z = 0 plane, considering different numbers of wires to be present. Table 1 provides a summary of the results. The first column lists the number N of wires that are considered to be present. The second column lists the location number and the current of each wire that is present. The wire at the location labeled 1 has a radial coordinate r = 0. The other wire locations have the same radial coordinate r = r_0 and have angle coordinates θ with values 0, π, π/2, or 3π/2. The magnetic field **B** is calculated as the superposition of fields produced by wires that are present, $\mathbf{B} = \sum^{N}_{i=1} \mathbf{B}_i$. Each wire is approximated as having negligible thickness and being infinitely long. The wire at the location labeled 1 (on the z axis) carries current I and produces a magnetic field given by

$$B_1(x,y) = \frac{\mu_0 I}{2\pi}\left(-\frac{y}{x^2+y^2}\hat{i} + \frac{x}{x^2+y^2}\hat{j}\right), \quad (1)$$

where μ_0 is the permeability of free space, and SI units are used. The other wires carry oppositely directed current given by $-I/(N-1)$, such that the net axial current is zero except when only one wire is present. The magnetic field produced by each of the other wires is as follows: $\mathbf{B}_2(x,y) = -\mathbf{B}_1(x-r_0,y)/(N-1)$, $\mathbf{B}_3(x,y) = -\mathbf{B}_1(x+r_0,y)/(N-1)$, $\mathbf{B}_4(x,y) = -\mathbf{B}_1(x,y-r_0)/(N-1)$, $\mathbf{B}_5(x,y) = -\mathbf{B}_1(x,y+r_0)/(N-1)$. The Cartesian coordinates are written in terms of the polar coordinates as x = r cosθ and y = r sinθ.

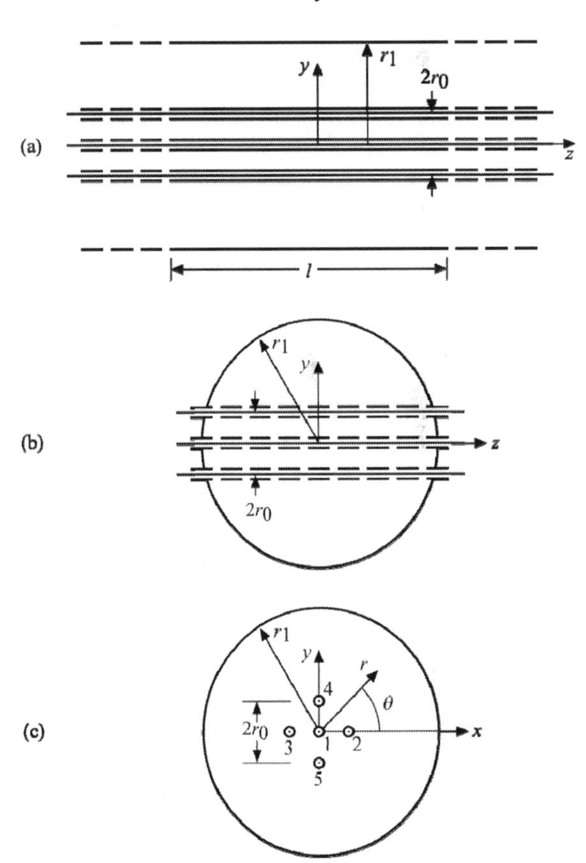

FIGURE 2. Cross-sectional views at the x = 0 plane of cylindrical (a) and spherical (b) trap configurations. Both configurations have the same cross-sectional view at the z = 0 plane (c), with five current-carrying wire locations that are labeled. The illustrations are not to scale. The relationships r0 << r1 << l are assumed.

TABLE 1. Summary of results. If more than one value or a range of values for θ_{min} (or θ_{max}) is specified, then $\rho(r > r_0,\theta)$ has the same minimum (or maximum) value at all specified values of θ for a given r. No solution exists for r_{sn} with $N \leq 3$.

N	Location (current)	θ_{min}	θ_{max}	$\alpha = 2$ r_{sn}	$\alpha = 3$ r_{sn}
1	1 (I)	$0 \leq \theta_{min} < 2\pi$	$0 \leq \theta_{max} < 2\pi$	-	-
2	1 (I), 2 (-I)	0	π	-	-
3	1 (I), 2 (-I/2), 3 (-I/2)	0, π	$\pi/2$, $3\pi/2$	-	-
5	1 (I), 2-5 (-I/4 each)	0, $\pi/2$, π, $3\pi/2$	$\pi/4$, $3\pi/4$, $5\pi/4$, $7\pi/4$	4.1	4.7

Suppose that the confined particles have a non-drifting Maxwellian velocity distribution associated with temperature T in energy units at a location $r = r_s$, where $r_0 < r_s < r_1$. For $r_0 < r < r_s$, a cyclotron radius scale length $\rho(r,\theta) = [m(T+2\Delta U)/(q^2B^2)]^{1/2}$ is defined to be approximately equal to the average cyclotron radius of a particle of mass m and charge q within a magnetic field of magnitude $B(r,\theta)$. Here, $\Delta U(r) = U(r_s)-U(r)$, where U(r) is the potential energy of the particle. For $r_0 < r < r_s$, the cyclotron radius $\rho(r,\theta)$ varies with θ for a given value of r (i.e., if r is held fixed) such that the same maxima and minima occur at different r. The values of θ at which minima occur in $\rho(r,\theta)$ for a given value of r with $r > r_0$ are denoted θ_{min}, while the values of θ at which maxima occur in $\rho(r,\theta)$ for a given value of r with $r > r_0$ are denoted θ_{max}. The values of θ_{min} and θ_{max} are listed in Table 1.

Let r_s be defined such that, at $r = r_s$, the minimum cyclotron radius is equal to r_1. Thus, r_s is defined such that $\rho(r_s,\theta_{min}) = r_1$. The cyclotron radius in the region $r \geq r_s$ has global minima at (r_s,θ_{min}), and particles located at $r > r_s$ may be considered to be effectively unmagnetized. The maximum cyclotron radius at $r = r_0$ is given by $\rho_0 = \rho(r_0,\theta_{max}) = (m[T+2\Delta U(r_0)])^{1/2}(4\pi r_0)/(\mu_0|I||q|)$ for the configurations considered with $N > 1$ (hereafter). The following relation is found between r_s and ρ_0:

$$\frac{r_{1n}}{\rho_{0n}} = \frac{r_{sn}(r_{sn}^{N-1}[\Theta(1-N)+1]-1)}{2\sqrt{1+2u(r_{sn})}}, \tag{2}$$

Here, normalized parameters are used that are defined as follows: $r_{1n} = r_1/r_0$, $\rho_{0n} = \rho_0/r_0$, $r_{sn} = r_s/r_0$, and $u(r_{sn}) = [U(r_s)-U(r_0)]/T$. Also, $\Theta(x)$ is the Heaviside step function: $\Theta(x < 0) = 0$, $\Theta(x \geq 0) = 1$. An expression for the difference in normalized potential energy is [25]

$$u(r_n) = \frac{qV}{T}\left(\frac{r_{1n}}{r_n}\right)^{\alpha-2}\frac{\beta_\alpha(r_n)}{\beta_\alpha(r_{1n})}, \tag{3}$$

where $r_n = r/r_0$. Here, V is the difference between the electric potentials at $r = r_1$ and $r = r_0$. (Also, $qV > 0$.) For the cylindrical trap, $\alpha = 2$ and $\beta_2(r_{1n}) = \ln(r_{1n})$. For the spherical trap, $\alpha = 3$ and $\beta_3(r_{1n}) = r_{1n}-1$.

Let $\omega_s = u(r_{1n})-u(r_{sn})$ denote the normalized depth of the electric potential energy well at the normalized radial coordinate r_{sn}. The condition $\omega_s >> 1$ is considered to be a necessary (but not necessarily sufficient) condition for good confinement of an unmagnetized, non-rotating plasma. Evaluating an expression for ω_s and rearranging gives

$$\frac{qV}{T} = \frac{\omega_s}{1-\left(\frac{r_{1n}}{r_{sn}}\right)^{\alpha-2}\frac{\beta_\alpha(r_{sn})}{\beta_\alpha(r_{1n})}}. \tag{4}$$

Equations (2) - (4) are solved numerically for r_{sn}. The condition $r_{1n} >> 1$ is assumed in the present study, and $r_{1n} = 10$ is used. A condition for the magnetic field to keep the particles from reaching the inner electrodes is considered to be $\rho_{0n} << 1$, and $\rho_{0n} = 0.1$ is used. A condition for the magnetic field to be sufficiently short in range to affect the particle trajectories only in close proximity to the inner electrodes is considered to be $r_{sn} << r_{1n}$. The value of r_{sn} increases with ω_s, and no solution exists if ω_s is too large. For the parameter values considered, and assuming the condition $\omega_s = 10$, numerical solutions yield $r_{sn} > 4$ for $N \leq 5$ (see Table 1). A value $r_{sn} > 4$ is too large to satisfy the condition $r_{sn} << r_{1n}$. It is possible to consider higher N values, with the assumption that Eq. (2) is applicable for larger N. It is found that for $N = 8$ and $\alpha = 2$, a numerical solution yields $r_{sn} = 2.3$, which marginally satisfies the condition $r_{sn} << r_{1n}$. The results indicate that a magnetic field that is sufficiently short in range to affect the particle trajectories only in close proximity to the inner electrodes must be an octupole or higher order field.

CLASSICAL TRAJECTORY MONTE CARLO STUDY

The results of the preceding section indicate that the configuration illustrated Fig. 2 with N = 5 is not suitable for producing a magnetic field that is sufficiently short in range to affect the particle trajectories only in close proximity to the inner electrodes. Nevertheless, a classical trajectory Monte Carlo study was carried out to assess whether the inner electrodes would be shielded by the magnetic field. The methodology used is similar to that described in Ref. [6]. The value of ω_s was chosen small enough for the condition $r_{sn} << r_{1n}$ to be marginally satisfied. However, as might be expected, many trajectories reached the outer electrode. The results were somewhat mixed. It was found that particles could reach the region $r < r_0$. For the cylindrical trap, shielding of the electrode locations was observable. In particular, no trajectories approached the z axis, thereby indicating that shielding would be possible for an electrode located on the z axis. However, for the spherical trap, many trajectories approached very close to the z axis. (Keep in mind that for these simulations, the wires are treated as having negligible thickness.) Some simulations were done in which the electric field was turned off for trajectories that approached within a distance of r_0 from the z axis for the cylindrical trap or within a distance of r_0 from the coordinate origin for the spherical trap. Under such conditions, the results indicated that shielding should be possible for all electrodes.

CONCLUDING REMARKS

The possibility of developing an alternative to the nested Penning trap for producing antihydrogen from trapped antiprotons provides an underlying motivation for the research reported here. Such an alternative might allow antihydrogen atoms to be formed with lower kinetic energy than is presently possible. Research on a modified Kingdon trap has been reported here. The modified Kingdon trap would employ an artificially structured boundary that produces a multipole magnetic field to keep antiprotons (or other charged particles) from reaching inner electrodes. The magnetic field would be sufficiently short in range to affect the trajectories of antiprotons only in close proximity to the inner electrodes. The conditions for producing such a magnetic field were assessed. The results of the present work indicate that a configuration similar to that illustrated in Fig. 2 may be promising, with N > 5, without a wire on the z axis, and with adjacent wires having oppositely directed currents.

ACKNOWLEDGMENTS

The author would like to thank J. R. Correa for helpful comments and suggestions. This material is based upon work supported by the Department of Energy under Grant No. DE-FG02-06ER54883.

REFERENCES

1. M. Amoretti et al., *Nature* (London) **419**, 456 (2002).
2. G. Gabrielse et al., *Phys. Rev. Lett.* **89**, 213401 (2002).
3. G.B. Andresen et al., *J. Phys. B: At. Mol. Opt. Phys.* **41**, 011001 (2008)
4. C.A. Ordonez and D.L. Weathers, *Phys. Plasmas* **15**, 083504 (2008).
5. G.B. Andresen et al., *Phys. Rev. Lett.* **105**, 013003 (2010).
6. C.A. Ordonez, *J. Appl. Phys.* **106**, 024905 (2009).
7. C.A. Ordonez, *IEEE Trans. Plasma Sci.* **38**, 388 (2010).
8. C.A. Ordonez, *J. Appl. Phys.* **104**, 054903 (2008).
9. C.A. Ordonez, *Phys. Plasmas* **15**, 114507 (2008).
10. J.R. Correa, C.A. Ordonez, and D.L. Weathers, *Nucl. Instrum. Methods in Phys. Res.* **B241**, 909 (2005).
11. K.H. Kingdon, *Phys. Rev.* **21**, 408 (1923).
12. D.P. Moehs, D.A. Church, and R.A. Phaneuf, *Rev. Sci. Instrum.* **69**, 1991 (1998).
13. S. Robertson and D. Alexander, *Phys. Plasmas* **2**, 3 (1995).
14. D.A. Church, *Phys. Rep.* **228**, 253 (1993); and references therein.
15. D.P. Moehs, D.A. Church, M.I. Bhatti, and W.F. Perger, *Phys. Rev. Lett.* **85**, 38 (2000).
16. S.J. Smith, A. Chutjian, and J.A. Lozano, *Phys. Rev. A* **72**, 062504 (2005).
17. J.M. Burke, W.M. Manheimer, and E. Ott, *Phys. Rev. Lett.* **56**, 2625 (1986).
18. W.G. Mourad, T. Pauly, and R.G. Herb, *Rev. Sci. Instrum.* **35**, 661 (1964).
19. R.A. Douglas, J. Zabritski, and R.G. Herb, *Rev. Sci. Instrum.* **36**, 1 (1965).
20. Q. Hu, R.J. Noll, H. Li, A. Makarov, M. Hardman, and R.G. Cooks, *J. Mass Spectrom.* **40**, 430 (2005).
21. R. Blumel and I.Garrick-Bethell, *Phys. Rev. A* **73**, 023411 (2006); and references therein.
22. R.H. Hooverman, *J. Appl. Phys.* **34**, 3505 (1963).
23. R.R. Lewis, *J. Appl. Phys.* **53**, 3975 (1982).
24. C.E. Johnson, *J. Appl. Phys.* **55**, 3207 (1984).
25. C.A. Ordonez, *Phys. Plasmas* **15**, 072508 (2008).
26. S. Robertson, *Phys. Plasmas* **2**, 2200 (1995).

SECTION III – EARLY CAREER HOT TOPICS

Statistical Requirements For Pass-Fail Testing Of Contraband Detection Systems

David M. Gilliam

National Institute of Standards and Technology, 100 Bureau Dr., Gaithersburg, MD 20899-8463
david.gilliam@nist.gov

Abstract: Contraband detection systems for homeland security applications are typically tested for probability of detection (PD) and probability of false alarm (PFA) using pass-fail testing protocols. Test protocols usually require specified values for PD and PFA to be demonstrated at a specified level of statistical confidence CL. Based on a recent more theoretical treatment of this subject [1], this summary reviews the definition of CL and provides formulas and spreadsheet functions for constructing tables of general test requirements and for determining the minimum number of tests required. The formulas and tables in this article may be generally applied to many other applications of pass-fail testing, in addition to testing of contraband detection systems.

Keywords: contraband, detection, statistics, confidence, probability, false alarm
PACS: 89.20.Dd

INTRODUCTION

It is the intention of this paper to give a very brief tutorial on statistical inferences from pass-fail testing of detection systems for contraband detection in homeland security applications, beginning with some understanding of what is meant by probability of detection (PD), probability of false alarm (PFA), and confidence level (CL). (The terms "confidence level" and "level of statistical confidence" are used interchangeably in this paper.) A more extensive theoretical treatment of these issues is given in an earlier paper [1].

In subsequent sections, formal definitions of these quantities and some quick ways of finding relationships between them will be given, but it may be helpful to first consider an example illustrating some of these relationships: For even the most effective detection system, the PD will go from near unity to near zero as the amount of contraband substance decreases toward its limit of detectability. The exact determination of PD and PFA for a given system would require an infinite number of tests, but as a practical matter one wants to know what can be inferred about PD and PFA with a finite number of tests. Consider the case in which 29 correct detection results are found in a set of 30 trials. These results could be expressed statistically in a range of different ways, including the following inferences: PD > 0.95 with CL of 44.6 %, or PD > 0.90 with Cl of 81.6 %, or PD > 0.85 with CL of 95.2 %. All of these are correct and equivalent inferences.

DEFINITIONS

The definitions of PD and PFA are easily understood. It is assumed that PD is a fixed property of the detection system for a particular amount of substance, and that a very good estimate of the true value of PD could be determined by carrying out a very large number of tests on the system. It is assumed that as the number of tests is increased, a limiting value for PD would be approached as the ratio of the number of correct detections to the total number of trials. Analogous assumptions are made regarding PFA for tests where no contraband is present.

The definition of confidence limit (CL) is not so easily understood. The formal definition of CL is stated in equation (1) below, but this equation does not give one much feeling for the meaning of statistical confidence. It may be helpful to know that the quantity CL can be loosely interpreted as an approximation of the likelihood that any system that gives m successes in a single set of n trials will have a true PD value greater than or equal to a chosen value, PD_c. Even this statement requires some careful thought, but it is somewhat more tractable than equation (1).

Application of Accelerators in Research and Industry
AIP Conf. Proc. 1336, 173-175 (2011); doi: 10.1063/1.3586082
2011 American Institute of Physics 978-0-7354-0891-3/$30.00

The formal definition of the confidence level, CL, for the probability of detection (PD) can be stated as follows: If *m* successes are found in a single set of *n* pass-fail trials, then for any chosen value of PD, designated PDC, there is a corresponding number called the confidence level CL(m, n, PDC), defined by the equation

$$CL(m,n,PD_c) = \sum_{j=0}^{m-1} b(j;n,PD_c), \qquad (1)$$

where $b(j;n,PD_c)$ is the binomial discrete density function,

$$b(m;n,p) = \frac{n!}{m!(n-m)!} p^m (1-p)^{n-m},$$

the variable *p* can take on either PD or PFA values.

To find the minimum value of *m* establishing the PD_c of interest with a preselected, fixed level of confidence, CL, one must invert the inequality

$$\sum_{j=0}^{m-1} b(j;n,PD_c) \geq CL$$

to obtain

$$m = INVBINCDF(CL,n,PD_c) + 1, \qquad (2)$$

where INVBINCDF(CL, n, PD_c) is the inverse cumulative binomial distribution function, which is available in many statistical software packages and in general spreadsheet applications, such as EXCEL*, as the function CRITBINOM(n, PD_c, CL). Although equation (2) is not expressed in closed form using common functions, it is still easy to find the required number of successful tests *m* using a single function call in a commonly available spreadsheet. It is also useful to know that the EXCEL spreadsheet function BINOMDIST (m-1, n, PD_c, TRUE) gives $CL(m, n, PD_c)$ while BINOMDIST(m ,n, p, FALSE) gives $b(j; n, p)$. In these spreadsheet functions, PD_c, CL, and p must be entered as a number between zero and one, not as a percentage (e.g., 0.9 not 90 %).

An analogous definition of CL applies to testing for the probability of false alarms (PFAs) in systems where no more than a negligible amount of a threat agent is present.

The formal definition of the confidence level, CL, for the probability of false alarm (PFA) can be stated as follows: If *m* false alarms occur in a single set of *n* pass-fail trials, then for any chosen value of PFA, designated PFA_c, there is a corresponding number called the confidence level CL(m, n, PFA_c), defined by the equation

$$CL(m,n,PFA_c) = \sum_{j=m+1}^{n} b(j;n,PFA_c)$$

To find the maximum value of *m* establishing the PFA_c of interest with a preselected, fixed level of confidence, CL, one must invert the inequality

$$\sum_{j=m+1}^{n} b(j;n,PFA_c) \geq CL$$

to obtain

$$m = n - 1 - INVBINCDF(CL,n,1-PFA_c) \qquad (3)$$

where INVBINCDF(C, n, 1-PFA_c) is the same function mentioned above in the discussion of relationships for PD. The EXCEL spreadsheet expression corresponding to CL(m, n, PFA_c) is [1-BINOMDIST(m, n, 1 – PFA_c,TRUE)].

With equations (2) and (3), one may construct tables of the permissible number of errors in the testing, while still establishing a given PD or PFA at the desired level of confidence. By tabulating *n-m* in tests for PD, one can combine the results of equations (2) and (3) in a single table, such as Table I. If equation (2) returns a value of *n+1* or if equation (3) returns a value of *-1*, then the value of *n* is too small to establish PD or PFA at the desired level of confidence, even when all of the tests gave a correct result; this condition is indicated by the entry "*" in Table I.

MINIMUM NUMBER OF TESTS

It is seen in Table I that there is a minimum number of tests that are required to establish any PD or PFA at a specified level of confidence; this minimum number corresponds to the smallest value of *n* with zero as the permissible number of errors. Equation (4) gives a simple formula for calculating this minimum number of required tests:

n_{min} = Smallest Integer Exceeding a, where

$$a = \frac{\ln(1-CL)}{\ln(PD)} \quad or \quad a = \frac{\ln(1-CL)}{\ln(1-PFA)}. \qquad (4)$$

* Certain commercial software is identified in this paper in order to specify the calculational procedure adequately. Such identification is not intended to imply recommendation or endorsement by the National Institute of Standards and Technology, nor is it intended to imply that the materials or equipment identified are necessarily the best available for the purpose.

Table 1. Numbers of Permissible Errors for 68% Confidence Level

PD→	0.95	0.9	0.8	0.7	0.6	0.5
PFA→	0.05	0.1	0.2	0.3	0.4	0.5
n=2	*	*	*	*	*	0
n=3	*	*	*	*	0	0
n=4	*	*	*	0	0	1
n=5	*	*	*	0	0	1
n=6	*	*	0	0	1	1
n=7	*	*	0	0	1	2
n=8	*	*	0	1	2	2
n=9	*	*	0	1	2	3
n=10	*	*	0	1	2	3
n=11	*	0	0	2	3	4
n=12	*	0	1	2	3	4
n=13	*	0	1	2	3	5
n=14	*	0	1	2	4	5
n=15	*	0	1	3	4	6
n=16	*	0	1	3	4	6
n=17	*	0	2	3	5	7
n=18	*	0	2	3	5	7
n=19	*	0	2	4	6	7
n=20	*	0	2	4	6	8
n=21	*	0	2	4	6	8
n=22	*	0	2	5	7	9
n=23	0	1	3	5	7	9
n=24	0	1	3	5	7	10
n=25	0	1	3	5	8	10

The number n_{min} increases more rapidly as PD approaches 1 (or PFA approaches 0), than as CL approaches 1. Figure 1 shows n_{min} as a function of PD for various choices of CL, corresponding to 1, 2, or 3 standard deviation fractions in normal distributions. When planning tests with preconceived criteria for PD, PFA, and CL, equation (4) gives a simple way of calculating the minimum number of pass-fail tests required to establish the desired probabilities and confidence criteria.

DISCUSSION

In Table I, PD values are listed in the same columns as PFA values corresponding to 1-PD, but there is no implication that PD and PFA must be correlated in this way for any given system; this is just a convenient way to show both PFA and PD results in a single table.

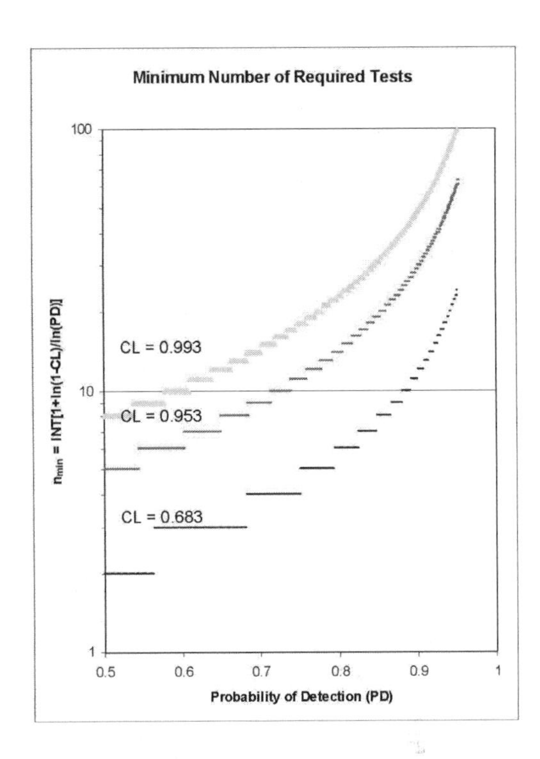

Figure 1. Minimum number of required tests, as a function of PD and Confidence Level, CL

Equation (4) for n_{min} shows that requiring either PD or CL to be extremely near unity (or PFA to be extremely near zero) can necessitate impractically large numbers of pass-fail tests. If such demanding criteria are required, then one should try to find some other method of qualification, rather than the use of pass-fail testing.

The results presented in equations (2), (3), and (4) make it possible to design pass-fail testing protocols based on routine functions calls in readily available statistical software packages or general spreadsheet applications.

It is intended that the statistical definitions and requirements in this paper supersede those in the original publication of the ANSI Standard N42.41 [2] on active interrogation systems (in the pending revision of that standard).

REFERENCES

1. D. Gilliam, S. Leigh, A. Rukhin, W. Strawderman, J. Res. Natl. Inst. Stand. Technol. **114** (2009) 195-199.
2. American National Standards Institute, ANSI N42.41-2007, Appendix B, Statistical Considerations, IEEE, 3 Park Avenue, New York, NY 10016-5007, USA, 15 February 2008.

Theoretical Approach Of The Reduction Of Chromatic And Spherical Aberrations In An Acceleration Lens System For Hundreds Of keV Gaseous Ion Nanobeam

Takeru Ohkubo, Yasuyuki Ishii and Tomihiro Kamiya

Takasaki Advanced Radiation Research Institute, Japan Atomic Energy Agency (JAEA)
1233 Watanuki-machi, Takasaki, Gunma, 370-1292, Japan

Abstract. The focused gaseous ion beam (gas-FIB) system composed of a series of electrostatic lenses, called "acceleration lens system", has been developed to form nanobeams using gaseous ions generated from a plasma ion source. Ion beams are accelerated and focused simultaneously by a pair of electrodes. A new all-in-one compact acceleration system including an acceleration tube is now under development to form 300 keV ion nanobeam. Chromatic and spherical aberrations are, however, hindrance to form nanobeams with their smaller sizes in diameter. A deceleration lens, which performs like a defocusing lens, was theoretically introduced to downstream of the present acceleration lens system to reduce the aberrations. Ion beam optics simulations were carried out to show that this aberration reduction technique is effective to reduce chromatic and spherical aberrations. As a result, we reduced the chromatic aberration coefficient by 26%, the spherical aberration coefficient by 17% and a beam diameter by 17%, with the deceleration energy of 15 keV. In case of using an electrostatic acceleration tube with 100 mm length, the final beam diameter of 103 nm at 300 keV is obtained by the all-in-one acceleration lens system with the total acceleration length of only 650 mm.

Keywords: ion beam, microbeam, nanobeam, acceleration lens, aberration
PACS: 41.85Lc, 41.85.Ne

INTRODUCTION

Progresses in formation techniques of ion microbeam from tens of keV to several MeV range have brought us microscopic analyses or micro-fabrications performed in micrometer or submicrometer scale. In the tens of keV range, there is commercially available compact Focused Ion Beam (FIB) systems using liquid metal ion sources with a sharp generation point (around 10 nm) [1-3]. FIBs have been also used for various micromachining and analyses in short depths of materials. Since the demagnifications of the lens system of FIBs are at most 100, they are insufficient to use a plasma ion source with an extracted beam size of at least 100 μm in order to form nanobeam of less than 1μm in diameter. Developments of plasma ion sources to obtain higher brightness and smaller source size are so far reported in ref.[4-6], the brightness is not high enough to make a bright nanometer ion source. On the other hand, in the several MeV range, the microbeams of various ion species are formed by a long microbeam system with an object distance around 10 m in a large facility [7-10]. Proton microbeams in MeV range are used especially for the material micro-fabrication by Proton Beam Writing (PBW) and/or the microscopic element analysis by Particle Induced X-ray Emission in micrometer scale (micro-PIXE) [11-14]. Though these techniques are expected to be great tools for medical diagnoses or nano-fabrications in various laboratories, the large microbeam system is an obstacle to expand its application. The shortening of the long MeV microbeam system is however difficult as long as using the conventional technique with magnetic or electrostatic lenses, because the long object distance is required to obtain large demagnification. In order to use the MeV proton microbeam widely for those applications, compact system to form microbeam or nanobeam is necessary.

A rather compact system has been developed at Japan Atomic Energy Agency (JAEA) for various gaseous ion nanobeams in keV range [15]. This focused gaseous ion beam (gas-FIB) system using a plasma ion source is composed of a series of two electrostatic lenses, called "acceleration lens system", each of which is a pair of disk-shaped electrode with a small center hole (hereinafter referred to as an acceleration lens). Since an ion beam is accelerated

Application of Accelerators in Research and Industry
AIP Conf. Proc. 1336, 176-180 (2011); doi: 10.1063/1.3586083

and focused simultaneously by the acceleration lens system, a distance from an object plane to an image plane becomes very short. Our acceleration lens system has an extraction stage and two acceleration stages. In each acceleration lens, the ion beam is strongly focused and weakly defocused around the object-side electrode and the image-side electrode, respectively [16]. A typical trajectory of an ion focused by a rotationally symmetric electrode in a cylindrical coordinates is shown in Fig.1, where ε_{ob} and ε_{im} are the kinetic energies of an ion beam at an object and an image sides, E_{ob} and E_{im} the electric fields in the object side and the image side, respectively; fields are represented at positive sign in the direction of ion acceleration. Demagnification of an acceleration lens M is described as follows,

$$M = \frac{\theta_{im}}{\theta_{ob}} \cdot \sqrt{\frac{\varepsilon_{im}}{\varepsilon_{ob}}} \quad (1)$$

where θ_{ob} and θ_{im} are angles of an ion beam respect to z axis. The demagnification of an acceleration lens equals to the demagnification of a general lens (θ_{im}/θ_{ob}) multiplied by the square root of the ratio of ion beam energy at an image plane to that at an object plane. Using the acceleration lens for focusing ion beams, much higher demagnification can be therefore obtained by a much shorter object distance comparing to the other large ion microbeam system using the general lenses. A beam diameter of 160 nm has been so far obtained with 46 keV hydrogen molecule ion beam focused by the compact acceleration lens system of 300 mm length and the demagnification has become more than 1000 [15].

A high energy gas-FIB to form gaseous ion nanobeam of hundreds of keV and with a beam diameter of 100 nm is presently desirable to be applied to PBW for nanofabrication or nano-scale analyses. As the next step, a new all-in-one acceleration lens system is under development as shown in Fig.2. The present acceleration lens system is connected to a following acceleration tube for 300 kV. An acceleration tube also works as an electrostatic lens and the ion beam from the present acceleration lens system is focused again

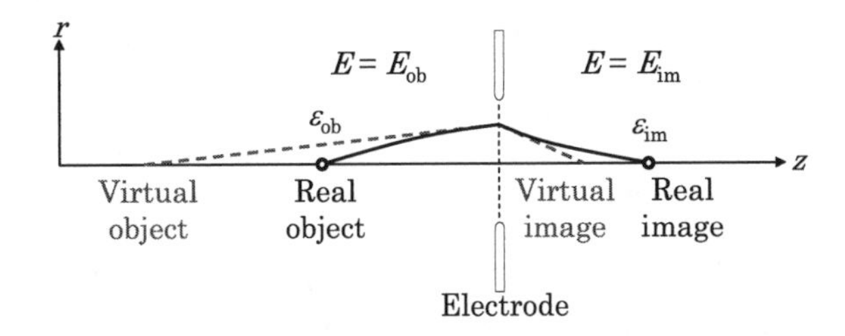

FIGURE 1. Typical trajectory of an ion focused by a rotationally symmetric electrode in the cylindrical coordinates.

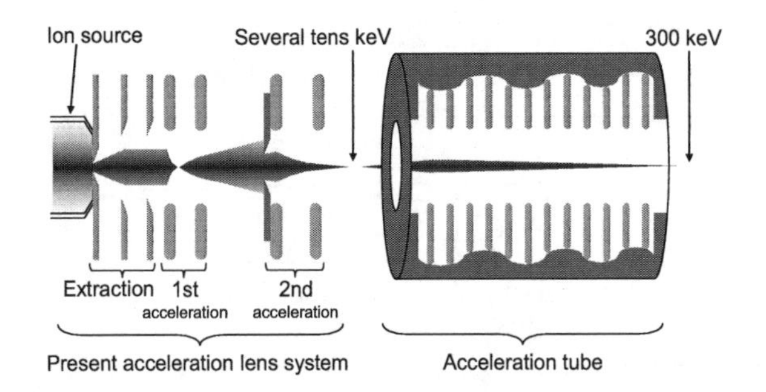

FIGURE 2. Concept of compact all-in-one system to form 300 keV ion nanobeam with 100 nm in diameter.

at the downstream of the acceleration tube, directly, though demagnification of the acceleration tube become at most 2 calculated by eq.(1). Since the beam diameter at the present acceleration lens system becomes the object size for the acceleration tube, the beam diameter has to be formed as small as possible. The obtained beam diameter of 160 nm by the present acceleration lens system is however comparable to chromatic and spherical aberrations. Reductions of the aberrations are necessary to make the beam diameter smaller.

In this study, a deceleration lens which performs like a defocusing lens is introduced to downstream of the present acceleration lens system in order to reduce chromatic and spherical aberrations. Schematic views of the second acceleration stage in the acceleration lens system are shown in Fig.3: (a) with a deceleration lens, (b) without a deceleration lens. Each dotted line is an envelope of the ion beam with slightly lower energy (by an energy spread) comparing to the central energy of the ion beam described as each solid line, respectively. At the second acceleration stage, the lower energy ion beam is both focused and defocused more strongly than the central energy ion beam. Aberrations generated by the focusing effect are corrected by this defocusing effect as shown in Fig.3. Ion beam optics simulations are carried out to show how much chromatic and spherical aberrations can be reduced in the new acceleration lens system and how small a nanobeam can be formed by the 300 kV all-in-one acceleration lens system.

OPTICS SIMULATION SETUP

Some optics simulation codes for charged particle acceleration are sometimes very complex, difficult in modification, and time-consuming. In the case of the acceleration lens system, a simulation code can be simplified by the following three factors: (1) an ion is accelerated by rotationally symmetric electrostatic lens, (2) an ion moves in the vicinity of the z axis in the cylindrical coordinates, (3) there are no magnetic fields. The ion accordingly behaves in the paraxial approximation described as following equation, called as paraxial ray equation,

$$\frac{d^2r}{dz^2}+\frac{1}{2V}\cdot\frac{dV}{dz}\cdot\frac{dr}{dz}+\frac{1}{4V}\cdot\frac{d^2V}{dz^2}\cdot r=0 \quad (2)$$

where V is the electric potential. The simulation code to solve the paraxial ray equation was developed to optimize the design of acceleration lens system to form smaller ion nanobeam. An ion trajectory in an acceleration lens system (so-called tracking) and chromatic and spherical aberration coefficients at an image plane were calculated using the code. The beam diameter at the image plane D is calculated as follows,

$$D=\sqrt{\left(M^{-1}D_{\mathrm{ob}}\right)^2+\left(C_{\mathrm{ch}}\frac{\Delta\varepsilon}{\varepsilon_{\mathrm{im}}}\theta_{\mathrm{im}}\right)^2+\left(C_{\mathrm{sp}}\theta_{\mathrm{im}}{}^3\right)^2} \quad (3)$$

where, D_{ob} is the diameter of an object point, $\Delta\varepsilon$ the energy spread of an ion beam, C_{ch} and C_{sp} the chromatic and the spherical aberration coefficients, respectively. C_{ch} and C_{sp} are calculated by the long equations shown in ref.[17].

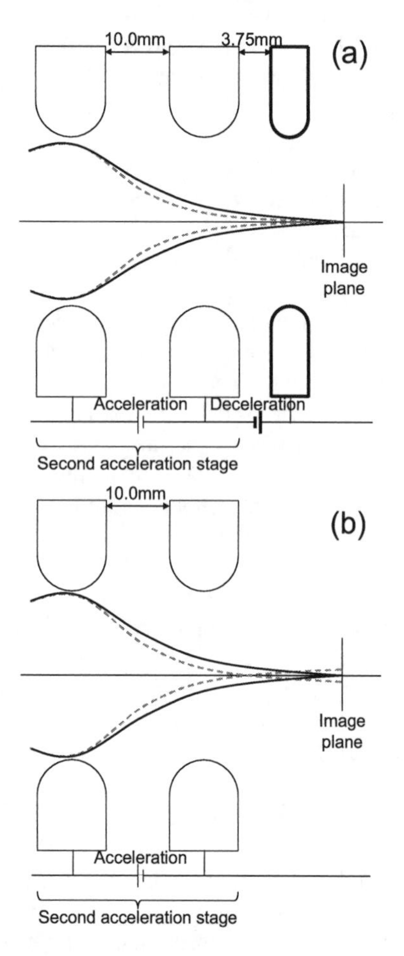

FIGURE 3. Schematic view of the aberration correction technique using a deceleration lens: (a) with a deceleration lens, (b) without a deceleration lens. Each dotted line is an envelope of the ion beam with slightly lower energy (by an energy spread) comparing to the central energy of the ion beam described as each solid line, respectively.

In the present acceleration lens system, hydrogen molecule ion (H_2^+) beam is extracted from a duoplasmatron-type ion source having an anode with a hole of 200 μm in diameter by applying the voltage of 0.6 kV to an extraction electrode in the extraction stage [18]. An object distance for the first acceleration lens becomes 300 mm by the effect of acceleration in the extraction stage, while the actual distance from the object aperture to the first acceleration lens is only 30 mm. This effect is a key part to realize high demagnification with short distance. The extracted ions are accelerated and focused with the voltages of 6.4 kV and 39 kV at the first and second acceleration stages, respectively. The first focal point of the ion beam is formed inside the first acceleration lens with the demagnification of about 100. The beam diverges from the first focal point at the first lens and is focused again at the second lens with the demagnification of about 10: accordingly total demagnification becomes about 1000.

SIMULATION RESULT

Ion beam optics simulation was carried out to show how much chromatic and spherical aberrations could be reduced in the new acceleration lens system by the introduced additional deceleration lens for the 46 keV hydrogen molecule ion nanobeam. A deceleration lens was introduced to downstream of the present acceleration lens system to reduce the aberrations as shown in Fig.3. The position and the thickness of the additional electrode were optimized in order to keep away from space dielectric breakdown and to form the focal point outside the acceleration lens system considering with the case of a use for 50 kV gas-FIB system without an acceleration tube. Even if the deceleration voltage was given to the added deceleration lens, beam energy of 46 keV and the position of the focal point could be fixed by adjustments of the acceleration voltages at the first and second acceleration stage. The simulation results are shown in Fig.4: (a) chromatic and spherical aberration coefficients, C_{ch} and C_{sp} respectively, (b) beam diameter with different energy spread at the image plane, as a function of deceleration voltage. The aberration coefficients are gradually reduced by increasing the deceleration voltage. The deceleration voltage has however a limit for dielectric breakdown in the actual acceleration lens system. The distance of 3.75 mm between electrodes of deceleration lens cannot keep stability at the voltage over 15 kV because of a space dielectric breakdown endurance; the

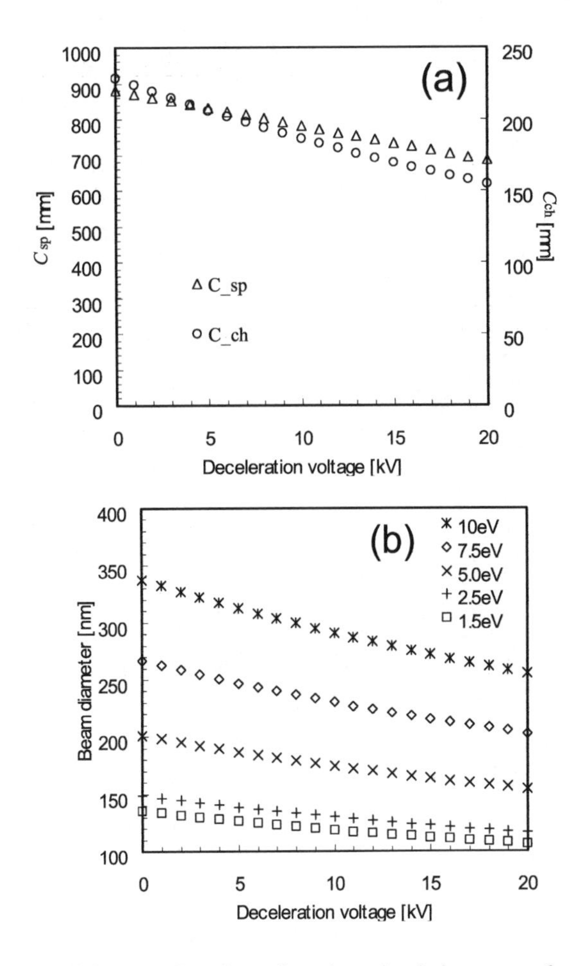

FIGURE 4. Results of optics simulation to reduce aberrations. (a) chromatic and spherical aberration coefficients, C_{ch} and C_{sp} respectively, (b) beam diameter with different energy spread at the image plane, as a function of deceleration voltage.

threshold for the breakdown was assumed as at most 4 kV/mm. In addition, the beam current became maximal at the deceleration voltage of 12 kV by the effect of reductions of the aberrations and was decreased by increasing the deceleration voltage over 12 kV, because the beam diverged wider and wider and the loss of beam current increased more and more at the aperture inside the acceleration lens system. Therefore, the deceleration voltage of over 15 kV is useless for the realistic gas-FIB. As a result, the deceleration energy of 15 keV for the 46 keV hydrogen molecule ion beam reduced the chromatic aberration coefficient by 26%, the spherical aberration coefficient by 17% and the beam diameter by 17% to be 109nm with the energy spread of 1.5 eV in the new acceleration lens system.

An acceleration tube for 300 kV was furthermore introduced to downstream of the new acceleration lens system. The acceleration tube was designed so that the chromatic and spherical aberrations became at

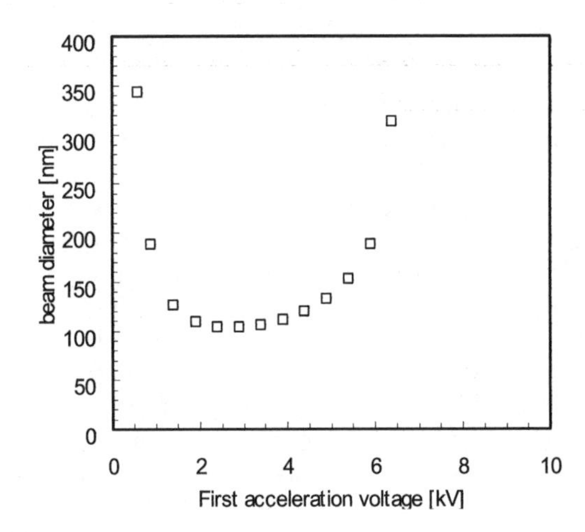

FIGURE 5. The final beam diameter with an energy spread of 1.5 eV at the image plane, as a function of the first acceleration voltage.

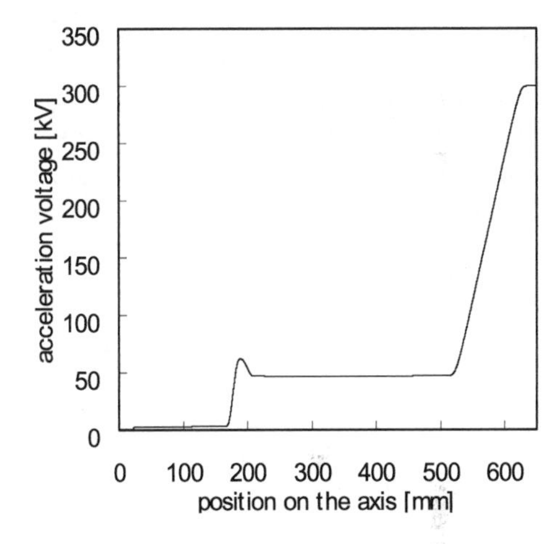

FIGURE 6. Optimal distribution of acceleration voltage along the axis coordinate.

minimum [19]. As the final focal point was formed outside the acceleration tube, the distance from the front of the acceleration tube to the end of the new acceleration lens system was determined to be 300 mm. Total acceleration length was limited within only 650 mm. The final beam diameter of the all-in-one acceleration lens system was calculated by the ion beam optics simulation including the results of the aberration reductions. There were several optimizations for voltages of the acceleration lenses. A result of an optimization for the first acceleration voltage to obtain the minimal final beam diameter is shown in Fig.5. Optimal distribution of acceleration voltage along the axis coordinate is also shown in Fig.6. The final beam diameter of 103 nm was obtained at the first acceleration voltage of 3.0 kV, which was the best result among those optimizations.

SUMMARY

A deceleration lens was theoretically introduced to downstream of the present acceleration lens system in order to reduce chromatic and spherical aberrations. The deceleration energy of 15 keV for the 46 keV hydrogen molecule ion beam reduced the chromatic aberration coefficient by 26%, the spherical aberration coefficient by 17%. The beam diameter was also decreased by 17% because of the aberration reductions, which meant the beam size of this nanobeam was determined by those aberrations rather than an image size calculated from demagnification, in short, those aberrations were much larger than the image size. Therefore, more reductions of aberrations must enable us to form smaller beam size. The final beam diameter of 103 nm was obtained by optimizations in the all-in-one acceleration lens system with an acceleration tube for 300 kV.

The all-in-one acceleration lens system is expected to become the lens system inside a high energy gas-FIB system, which forms, for instance, 300 keV proton nanobeam with 100 nm in diameter. Total acceleration length of 650 mm is short enough to make the high energy gas-FIB compact and it will be a great tool for applications in nanofabrication and/or "nano" analyses.

ACKNOWLEDGMENTS

This work was partly supported by MEXT Grant-in-Aid forYoung Scientists (B) 21760709.

REFERENCES

1. V. E. Krohn and G. R. Ringo, Appl. Phys. Lett. **27** (1975) 479.
2. R. L. Seliger, J. W. Ward, V. Wang and R. L. Kubena, Appl. Phys. Lett. **34** (1979) 310.
3. J. S. Huh, M. I. Shepard and J. Melngailis, J. Vac. Sci. Technol. B **9** (1991) 173.
4. Y. Lee, R. A. Gough, T. J. King, Q. Ji and K. N. Leung, Microelectron. Eng. **46** (1999) 469.
5. Q. Ji, T. J. King, K. N. Leung and S. B. Wilde, Rev. Sci. Instrum. **73** (2002) 822.
6. B. W. Ward, John A. Notte and N. P. Economou, J. Vac. Sci. Technol. B **24** (2006) 2871.
7. B.E. Fischer, R. Spohr, Nucl. Instr. and Meth. **168** (1980) 241.
8. L. A. Braby and W. D. Reece, Radiat. Prot. Dosimetry **31** (1990) 311.
9. T. Kamiya, T. Suda and R. Tanaka, Nucl. Instr. and Meth. B **118** (1996) 447.
10. M. Oikawa, T. Satoh, T. Sakai, N. Miyawaki, H. Kashiwagi, S. Kurashima, M. Fukuda, W. Yokota and T. Kamiya, Nucl. Instr. and Meth. B **260** (2007) 85.
11. J. A. van Kan, A. A. Bettiol and F. Watt, Nano Lett. **6** (2006) 579.
12. F. Watt, M. B. H. Breese, A. A. Bettiol and J. A. van Kan, Materials Today **10** (2007) 20.
13. N. Uchiya, T. Harada, M. Murai, H. Nishikawa, J. Haga, T. Sato, M. Oikawa, Y.Ishii and T. Kamiya, Nucl. Instr. and Meth. B **260** (2007) 405.
14. Y. Shimizu, K. Dobashi, T. Kusakabe, T. Nagamine, M. Oikawa, T. Satoh, J. Haga, Y. Ishii, T. Ohkubo, T. Kamiya, K. Arakawa, T. Sano, S. Tanaka, K. Shimizu, S. Matsuzaki, M. Utsugi and M. Mori, Int. J. Immunopathol. Pharmacol. **21** (2008) 567.
15. Y. Ishii, A. Isoya and T. Kojima, Nucl. Instr. and Meth. B **210** (2003) 70.
16. M. M. Elkind, Rev. Sci. Instr. **24** (1953) 129.
17. E. Munro, "A Set of Computer Programs for Calculating the Properties of Electron Lenses" Cambridge University Report, (1975).
18. Y. Ishii, A. Isoya, K. Arakawa, T. Kojima and T. Tanaka, Nucl. Instr. and Meth. B **181** (2001) 71.
19. Y. Ishii, T. Ohkubo and T. Kamiya, Proceedings of ICNMTA2010.

A New Facility For Non-Destructive Assay With A Time-Tagged ^{252}Cf Source

L.Stevanato [a], M. Caldogno [a], R. Dima[b], D. Fabris[b], Xin Hao[a] , M. Lunardon[a,b], S. Moretto[a,b], G. Nebbia[b], S. Pesente[a,b], L. Sajo-Bohus[c], G. Viesti[a,b]

[a] *Dipartimento di Fisica, Università di Padova, Via Marzolo 8, I-35131Padova, Italy*
[b] *INFN Sezione di Padova, Via Marzolo 8, I-35131Padova, Italy*
[c]*Laboratorio de Fisica Nuclear, Universidad Simon Bolivar, Apartado 89000, 1080 A Caracas, Venezuela*

Abstract. A new facility for Non-Destructive Assay based on a time-tagged ^{252}Cf spontaneous fission source is now in operation at the Padova University. The system is designed to analyze samples with dimensions on the order of 20 × 20 cm^2, the material recognition being obtained by measuring simultaneously transmission of neutrons and gamma rays as a function of energy.

Keywords: Non-destructive assay, neutron and gamma-ray transmission, time of flight measurement, ^{252}Cf source.
PACS: 81.70.-q, 25.85.Ca, 28.20.Fc

INTRODUCTION

A new laboratory system for non-destructive assays of small samples is now in operation at the University of Padova. The system employs a time-tagged fission source so that the transmitted neutrons and gamma rays are detected by a time-of-flight system. In this way not only the energy integrated transmission is derived, but also the neutron transmission is measured as a function of energy. Moreover, the transmitted gamma-ray spectrum is also recorded in a specialized detector, allowing one to study the attenuation of gamma-rays as a function of energy.

The characterization of a given material is first obtained using the ratio R between the energy integrated absorption coefficients of neutrons and gamma rays which depend on the atomic number Z [1,2,3]. In addition, direct signatures to identify light elements such as C,N,O in the sample are obtained using the measured transmission versus neutron time of flight through the energy dependence of the neutron cross section [4]. This allows one to determine the relevant elemental ratios (C/O and C/N) that are normally used to identify threat organic materials [3].

On the heavy element side, the discrimination is optimized by comparing each transmitted gamma-ray spectrum with a library of reference spectra for different elements. For samples with know thickness, this method allows the identification of the material atomic number Z with an uncertainty typically lower than 3 Z units [5].

The present system employs a linear array of 8 fast plastic scintillators for time-of-flight measurements. Vertical sliding of the object provides a first 2D reconstruction of the average Z inside the sample by measuring the R ratio. Furthermore, for well defined positions on the 2D image, neutron and gamma ray attenuation as a function of energy are measured, yielding more precise information on the nature of the inspected object. The description of the system and results obtained so far are presented in this paper.

THE NON-DESTRUCTIVE ASSAY SYSTEM

The system, shown in Fig.1, employs a 10^6 fissions/s ^{252}Cf source which is resident in its biological shield when not used. All functionalities of the system are controlled using the LabView software. In the automated scanning mode the software controls positioning of the sample and data taking for each view using a CAEN VME DAQ system. Data sorting and analysis are performed by means of the ROOT package.

The measurement of the neutron/gamma attenuation ratio R is based on a linear array of 8 scintillation detectors each made by a 2" × 2" right

Application of Accelerators in Research and Industry
AIP Conf. Proc. 1336, 181-183 (2011); doi: 10.1063/1.3586084

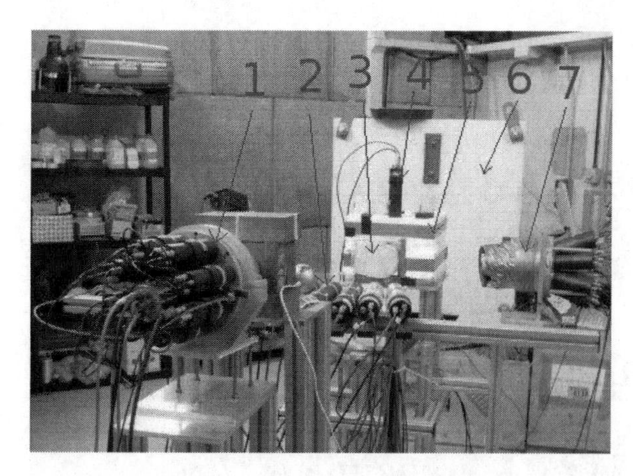

FIGURE 1. The Non-Destructive Assay system. The ^{252}Cf source is resident in the biological shield (6) and is driven in the Teflon collimator (5) housing the tagging detector (4) when the system is in operation. The object (3) is placed on a the computer controlled platform. The linear array of fast plastic (2), the transmission gamma ray detector (1) and the E=2.2 MeV gamma detector (7) are also indicated.

cylinder of EJ228 fast plastic coupled to a PHOTONIS XP2020 or ETL9214 photomultiplier. An additional detector of the same type, placed inside a Teflon collimator, is used to tag the fission source when extracted from its biological shield.

Such detectors improve substantially the resolution in the time of flight measurement compared to our previous experiments [3]. The time resolution was studied by ^{60}Co γ–γ coincidences with two of the above detectors, as a function of the energy threshold. It is found that the time resolution is well below δt=0.5 ns [FWHM] (for thresholds above 150 keV) to be compared with δt=1.5 ns [FWHM] obtained in our previous work. This reflects in an overall typical resolution better than δt=0.8 ns [FWHM] in routine operation with the ^{252}Cf source and thresholds as low as 100 keV. This improvement in time resolution allows the use of a shorter (100 cm) flight path respect to that previously used (160 cm).

The sample positioning platform is centered at about 30 cm from the source. In such conditions the size of the pixel at the sample position is about 1.6 cm.

Finally, two additional detectors are used in the system. A standard NaI(Tl) 3" x 3" scintillator with an anti-coincidence shield is placed behind the scintillator array at a distance of about 160 cm from the sample. It is used to perform energy-dependent gamma-ray attenuation measurements as those described in ref. 5. An additional Compton-suppressed gamma-ray detector is placed sideways, close to the sample position. It is used to measure the average hydrogen content of the sample by looking at the 2.2 MeV capture gamma-ray.

MATERIAL RECOGNITION

The R value is defined as the ratio between the neutron (μ_n) and gamma-rays (μ_γ) absorption coefficient. They are calculated from the logarithmic ratio between the intensity of the neutron and gamma-rays with (I_n and I_γ) and without ($I_{n,0}$ and $I_{\gamma,0}$) the sample.

$$R=\mu_n / \mu_\gamma = \ln (I_n / I_{n,0}) / \ln (I_\gamma / I_{\gamma,0})$$

The experimental calibration of R, obtained with the ^{252}Cf source, is reported in Fig. 2 as obtained by using 14 samples of elements with Z=6-83 and 10 compound samples (mainly plastics) spanning the range Z_{ave}=2.7-8. The latter compounds were included to study in details the response of the system for threat organic material as explosive and drugs. Looking globally at the measured R values and the interpolation functions, it is clear that the data exhibit a rather nice monotonic behaviour with much larger incremental ratio for light samples (Z<13) with respect to the heavier ones. The Z resolving power achievable with this method is a function of the statistical accuracy in the measure of R but in any case it is clear that this technique offers the best material discrimination for relatively light elements.

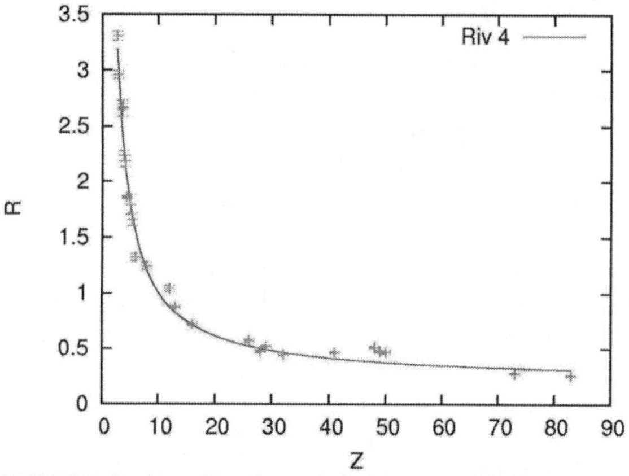

FIGURE 2. R calibration of the detector #4 of the linear array, the interpolation function R=R(Z) is also reported.

A typical R 2D-image, reported in Fig.3, has been obtained with a phantom including melamine, graphite and sulfur parts. Moreover, in order to verify the possibility of determining the elemental composition of samples made of few slices of different materials, we have stacked together three 2 cm thick slabs of plexiglas, aluminum and graphite. The stack was realized so that the attenuation due to the single plexiglas, to the sum of plexiglass and aluminum and then to all three materials could be determined. From

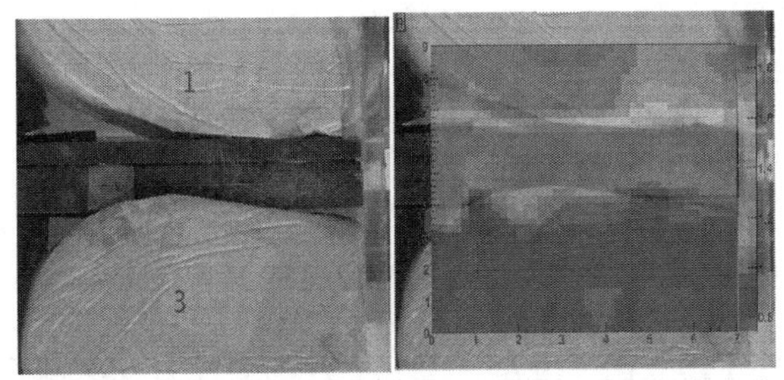

FIGURE 3. Example of phantom for R ratio 2D imaging containing melamine (1), graphite (2) and sulphur (3) (left) and measured distribution (right).

the measured data, the R value of the aluminum and graphite is obtained by a subtraction procedure. The results confirm the possibility of identifying an object "hidden" inside a background material when the background is measured directly.

In addition, the measured neutron transmission as a function of the time of flight is used to identify elements typical of organic materials like C,N,O that exhibit distinctive features in the neutron cross section within the energy range accessible to the present set-up. Recently, a new analysis has been introduced as compared to our previous work [3] resulting in the determination, from the measured transmission, of the effective total macroscopic cross section. This task has been tested first with samples of known thickness and composition. The reconstructed data compare well with expected values obtained by a gaussian smearing of the neutron cross section available from the Evaluated Nuclear Data File [6]. In the case of composite materials (such as water, paper or plastics) of known thickness, the density of the material is determined independently by using the transmission of gamma-rays as discussed in [5]. It is thus possible to reconstruct also in this case the macroscopic experimental cross section and compare it with a library of predicted spectra derived from the elemental cross sections assuming different compositions. A χ^2 analysis is used to determine the most probable elemental composition of the sample. As an example, results relative to a water sample are reported in Fig.4. The predicted spectrum is calculated using a weighted sum of the smoothed spectra of neutron cross section on oxygen and hydrogen. The ratio 2H:O close to 1 is obtained by the chi-square search, shown in the inset of Fig.4.

This procedure was repeated for other material like paper, polyethylene and melamine finding always a good agreements between extracted and expected elemental ratios. However the sensitivity of the χ^2 analysis decreases in increasing the complexity of the search.

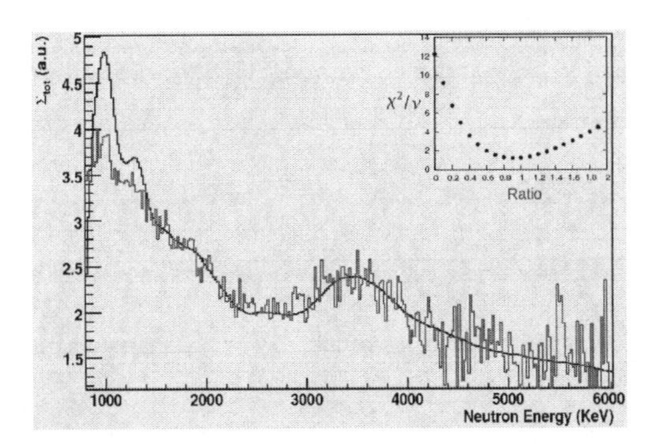

FIGURE 4. Measured macroscopic cross section from a sample of water (dotted line) compared with the one derived properly summing H and O cross sections with empirical weights that minimize the chi-square (solid line). Results from the χ^2 analysis are shown in the inset. For details see the text.

CONCLUSIONS

The new laboratory system for non-destructive assay of small samples is in operation at the University of Padova. It allows us to determine 2-D image of the atomic number Z of samples. In case of samples containing layers of different materials, the single layer atomic number is determined by a subtraction procedure. The system allows also to obtain multiple views by rotating the sample.

Additional measurements can be performed at specific positions in the 2D image. This gives us more precise information about the composition of the sample by looking at the energy dependent transmission of neutrons and gamma rays. Finally, the determination of the average hydrogen content in the sample is in progress by measuring in an ancillary detector the yield of the E_γ= 2.2 MeV from neutron capture reactions.

ACKNOWLEDGMENTS

This work is supported by the "Fondazione Cassa di Risparmio di Padova e Rovigo" under the "2007 Progetti di Eccellenza" scheme.

REFERENCES

1. R. J. Rasmussen et al., *Nucl. Instr. Meth.* B124, 611-614 (1997)
2. J.B. Eberhardt et al, *Appl. Radiat. Isot.* 63 , 179-188 (2005)
3. G. Viesti et al, *Nucl. Instr. Meth.* A593 , 592-596 (2008)
4. R. Lanza et al, IEEE Trans. Nucl. Sci., 49, 1919-1924 (2002)
5. G. Viesti et al, Nucl. Instr. Meth, A606, 816-820 (2009)
6. Available at http://www.nndc.bnl.gov/

950 keV X-Band Linac For Material Recognition Using Two-Fold Scintillator Detector As A Concept Of Dual-Energy X-Ray System

Kiwoo Lee[a], Takuya Natsui[a], Shunsuke Hirai[a], Mitsuru Uesaka[a], Eiko Hashimoto[b]

[a]The University of Tokyo, 2-22 Tokai-mura, Ibaraki-ken 319-1188, Japan
[b] Japan atomic energy agency (JAEA), 4-49 Muramatsu Tokai-mura, Ibaraki-ken, Japan

Abstract. One of the advantages of applying X-band linear accelerator (Linac) is the compact size of the whole system. That shows us the possibility of on-site system such as the custom inspection system in an airport. As X-ray source, we have developed X-band Linac and achieved maximum X-ray energy 950 keV using the low power magnetron (250 kW) in 2 μs pulse length. The whole size of the Linac system is $1\times1\times1$ m^3. That is realized by introducing X-band system. In addition, we have designed two-fold scintillator detector in dual energy X-ray concept. Monte carlo N-particle transport (MCNP) code was used to make up sensor part of the design with two scintillators, CsI and CdWO4. The custom inspection system is composed of two equipments: 950 keV X-band Linac and two-fold scintillator and they are operated simulating real situation such as baggage check in an airport. We will show you the results of experiment which was performed with metal samples: iron and lead as targets in several conditions.

Keywords: X-band Linac, material recognition, two-fold scintillator, dual energy X-ray.
PACS: 29.20.Ej; 87.59.B-

INTRODUCTION

X-ray tube has been normally used as X-ray source in the custom inspection system in an airport. It generates X-ray less than 500 keV ordinarily. 500 keV is almost the maximum energy X-ray tube can make at this writing. The reasons to apply X-ray tube to the custom inspection system are its compact size and its X-ray energy. The X-ray energy below 500 keV is available in a public place without heavy shielding around it.

On the other hand, the Linac appears in the field of inspection system as a new X-ray source, because it generates higher energy X-ray than X-ray tube. Higher X-ray can go through thicker or denser target readily. In many cases, the thicker or denser target such as radioactive material, gold and weapons is composed of high atomic number material. Therefore we can have more chance to find out the suspicious items, because the suspicious items are made of denser, in other word, high atomic number material often. Additionally, they are needed the concealment in most cases. That means they become thicker than other non-suspicious items. According to the development of Linac, the accelerating energy of electron beam becomes higher and the energy of bremsstrahlung X-ray becomes also higher when the electron beam is collided into a metal target. For example, the container inspection system applies 4 ~ 8, 6 ~ 9 MeV S-band Linac to generate high energy X-ray which can penetrate thick object readily [1, 2]. In addition, the whole size of accelerating system becomes smaller than S-band system introducing the high frequency acceleration such as X-band Linac. Based on these advantages, the accelerator is applied to custom inspection or medical application as alternative X-ray source instead of X-ray tube.

We have developed 950 keV X-band Linac as X-ray source for material recognition in the custom inspection system. That produces X-ray less than 1 MeV which is permissible in the public place without the particular restriction under the law for the prevention of radiological damage in Japan. 9.4 GHz X-band frequency provides small size of accelerating tube, magnetron and power supply that induce the compact size of entire system. The Linac is available in an airport in the custom inspection system expecting to widen the range of material recognition up to high atomic number material that was impossible before using X-ray tube because of its low X-ray energy.

Application of Accelerators in Research and Industry
AIP Conf. Proc. 1336, 184-188 (2011); doi: 10.1063/1.3586085

Two-fold scintillator is introduced in the part of detection in dual energy X-ray concept. It is composed of two scintillators: CsI and CdWO4 in the two-fold configuration such as Fig. 1. CsI scintillator placed in front of CdWO4 has thin thickness to absorb the low energy X-ray. That means we can expect low energy X-ray image from CsI scintillator. CdWO4 is thicker than CsI and placed after CsI. Therefore it absorbs high energy X-ray that goes through CsI and high energy X-ray image is obtained from it. Under the two-fold scintillator configuration, we can get two images: low energy X-ray image and high energy X-ray image from two scintillators during one-time of X-ray irradiation from Linac. These images are used to make reconstructed image which inform of the atomic number of target material through numerical calculation.

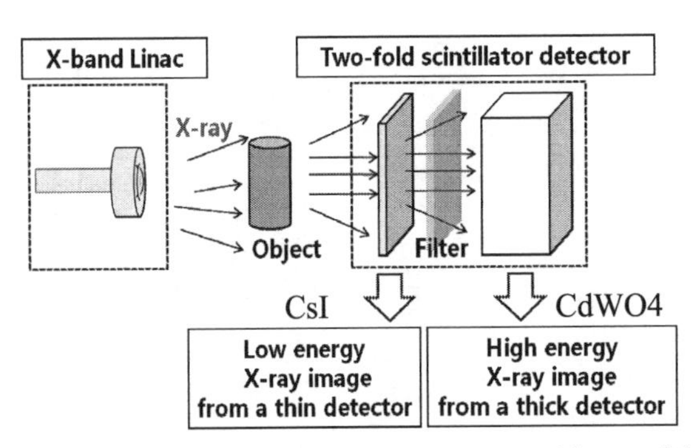

FIGURE 1. Dual energy X-ray concept with two-fold scintillator detector

In the research, we suggested iron and lead for the target materials to prove the ability of material recognition in our system. High atomic number material such as lead cannot be distinguished using X-ray tube, because low energy X-ray from X-ray tube cannot penetrate lead easily in certain thickness range and it does not give the information to the detector behind lead target. In our system, we want to show the possibility of the material recognition in an atomic number range from 13, iron to 82, lead.

EXPERIMENTAL DETAILS

950 keV X-Band Linac

Fig. 2 is the configuration of 950 keV X-band Linac that we have developed for the on-site custom inspection system. Length of the accelerating tube is 30 cm (Table 1). Short length of tube is one of the features of X-band Linac. X-band Linac system is supplied small RF power, 250 kW, from magnetron. The size of magnetron is $0.1{\times}0.2{\times}0.2$ m^3. As a result, many parts of X-band system are small. That is why this system can be downsized to $1 \times 1 \times 1$ m^3 [3]. The Linac generates 950 keV X-ray from tungsten target at the end of accelerating tube and the dose rate of X-ray is designed 0.2 Gy/min at 1-m distance.

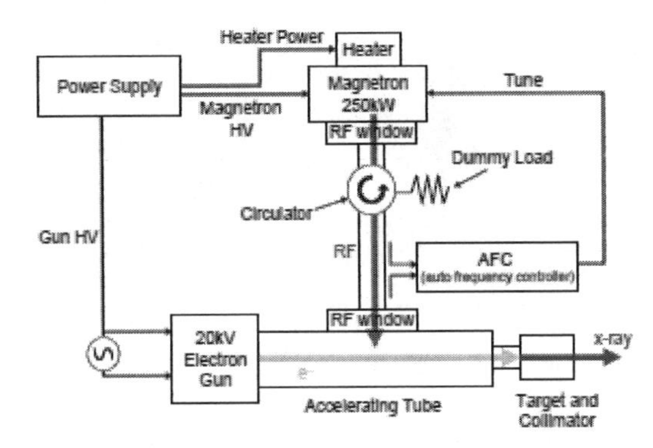

FIGURE 2. Schematic view of 950 keV X-band Linac

TABLE 1. Specifications of 950 keV X-ray source

Resonant frequency	X-band 9.4 GHz
RF source	250 kW magnetron
Cavity type	On-axis coupling
Shunt impedance	~ 70 MW/m
Gun type	Thermionic, Diode, 20 keV
Tube length	~ 30 cm

Two-fold Scintillator Detector

The thickness of scintillator is significant aspect to determine the energy of absorbed X-ray. It is very important, because the difference of absorbed X-ray energies between two scintillators determines the ability of material recognition. For example, the bigger difference of absorbed X-ray between two scintillators, the better ability of material recognition. MCNP monte carlo simulation was performed to determine each scintillator thickness; CsI and CdWO4 comparing the absorbed X-ray energy in each scintillator.

Target

The reason we provide iron and lead samples is that our goal in the research is recognition low atomic number material from high atomic number material (Fig. 3 a). In many cases, high atomic number material is used in illicit way such as gold and radioactive source. Lead represents the high atomic number material. If we can distinguish lead from iron, that means we can distinguish high atomic number material around 82 from low atomic number material around 26. They are placed in the cylinder concealments made of polyethylene and polyvinyl chloride. Small size of concealment is put inside of bigger one in five steps of the thickness. Every concealment has 5 mm thickness

(Fig. 3 b). Therefore it is increased in 10 mm when X-ray goes through the concealment; front and back parts of it. When the step goes up, X-ray has to go through 10 mm thicker concealment.

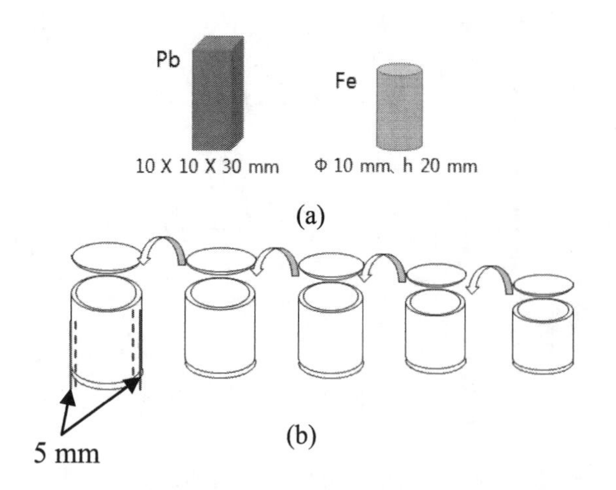

FIGURE 3. Target samples (a) and concealments (b)

Line sensor controller generates 10 Hz pulses which go to line sensor and 950 keV X-band Linac (Fig. 4). While the target goes by in front of line sensor in a proper velocity, X-ray from Linac goes through the target and reaches the two-fold scintillator in 10 Hz.

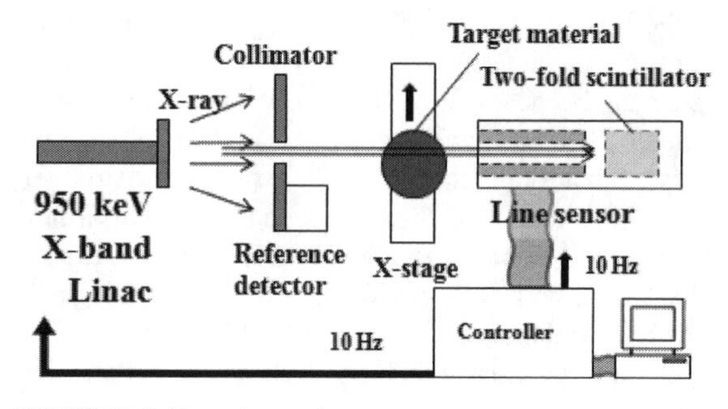

FIGURE 4. Experimental setup

CALCULATIONS

$$T(t,Z) = \frac{\int_0^E N(E)\exp(-\mu_Z t) E P_d(E) dE}{\int_0^E N(E) E P_d(E) dE} \quad (1)$$

where N(E) is X-ray energy spectrum, $P_d(E)$: the energy response function, μ_Z: the mass attenuation coefficient, t: the mass thickness of the irradiated material with atomic number Z. Definition of transparency is the ratio of X-ray intensity after a target to the incident X-ray intensity (see Eq. 1). We can calculate the transparency with the pixel value of the image experimentally. We define I_1 to the transparency of low energy X-ray for CsI and I_2 to the transparency of high energy X-ray for CdWO4. R is the ratio of the natural log of transparency of low energy X-ray for CsI to the natural log of transparency of high energy X-ray for CdWO4: R= $\ln I_2/\ln I_1$. The relation between R and I_1 shows certain pattern according to the atomic number (Fig. 5). The theoretical values of R and I_1 become the reference data to assign the experimental values the atomic number [4].

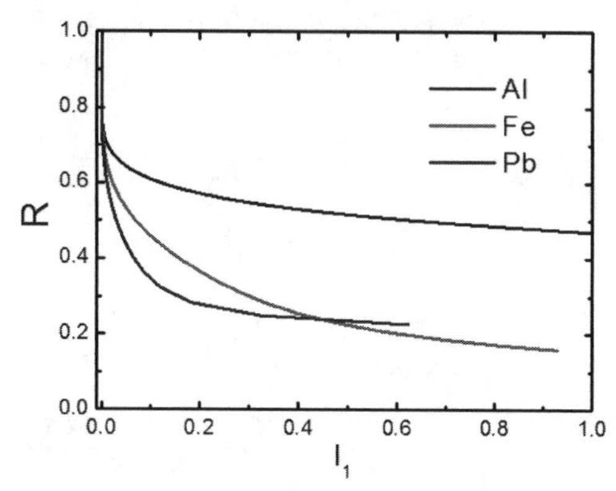

FIGURE 5. R and I_1 graph based on the attenuation calculation result

RESULTS AND DISCUSSION

We obtain the reconstructed images for iron and lead inside of the concealments according to the every step of thickness of the concealment from 10 to 50 mm (Fig. 6). In the polyethylene container, Fe and Pb show the atomic number difference between experimental and theoretical value from 3 to 10 for Fe, 2 to 5 for Pb. In the polyvinyl chloride container, Fe and Pb have the difference from 9 to 12 for Fe, 1 to 3 for Pb (Table 2). Fig. 7 (a) shows the comparison of atomic number of Fe to experimental results in every thickness of concealment. In polyvinyl chloride case, there is bigger difference between real atomic number of Fe: 26 and experimental value than polyethylene case. Because polyvinyl chloride has higher density: 1.38 – 1.41 g/cm^3 than polyethylene: 0.926 – 0.940 g/cm^3, low energy X-ray cannot penetrate the polyvinyl chloride easily. Therefore the transparency of CsI becomes higher by high energy X-ray that is left behind target. In addition, when the thickness of concealment is thicker, the atomic number difference between two concealments cases: red and blue line on the graph does not become bigger (Fig. 7 (a)). And atomic number in polyvinyl chloride case becomes even lower than polyethylene case in Fig. 7 (b).

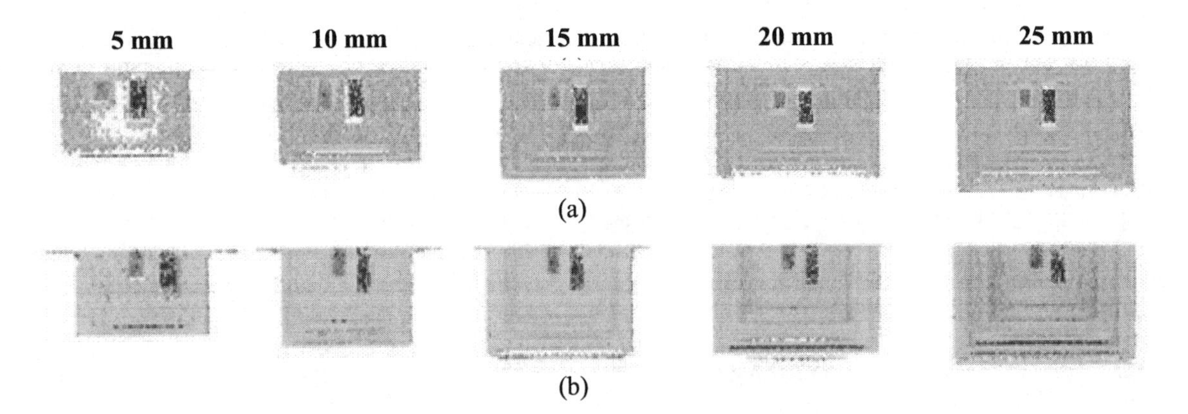

FIGURE 6. Image reconstruction of Fe (light green) and Pb (purple) in polyethylene (a) and polyvinyl chloride (b) container

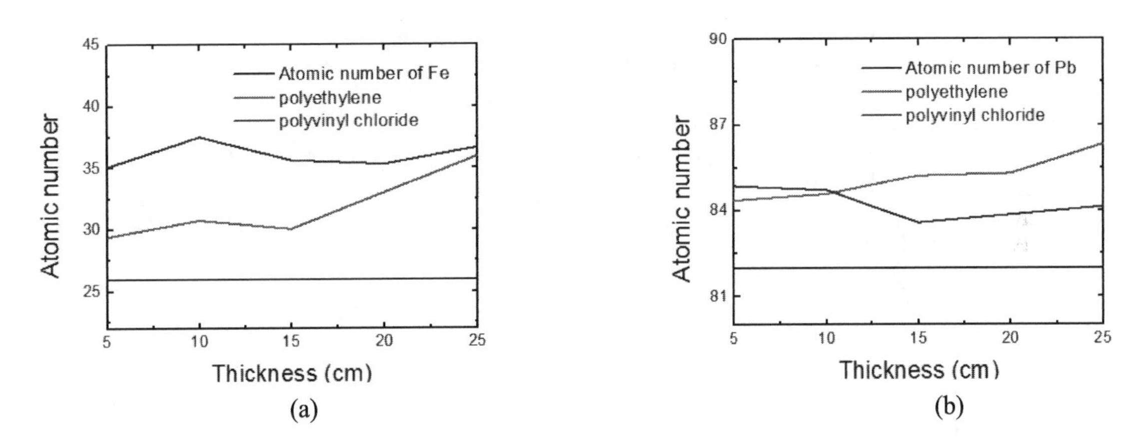

FIGURE 7. Atomic number of Fe and comparison to the experimental results in each concealment according to the thickness of concealment (a) and same comparison with Pb (b)

TABLE 2. Atomic number difference between experimental and theoretical values in the case of polyethylene concealment.

Thickness of container	Fe (26)	Pb (82)
5	3.41	2.35
10	4.76	2.58
15	4.06	3.22
20	7.02	3.31
25	9.99	4.34

TABLE 3. Atomic number difference between experimental and theoretical values in the case of polyvinyl chloride concealment

Thickness of container	Fe (26)	Pb (82)
5	9.09	2.86
10	11.53	2.71
15	9.64	1.58
20	9.33	1.86
25	10.7	2.14

This is from the fact that thicker concealment blocks more low energy X-ray in it than thin one. In particular, Pb in polyvinyl chloride concealment shows irregular pattern in graph, because most of low energy X-ray is blocked by Pb and concealment simultaneously. As a result, the difference between experimental value and theoretical value is getting smaller as the thickness of container is thicker and density of composed material of concealment is higher (Table 2, 3). That is not a good result even though the difference between experimental value and theoretical value is smaller, because it shows the irregular pattern. In order to improve the two-fold scintillator, we decide to increase the CsI scintillator thickness because when the target is thick, low energy X-ray that is mainly absorbed in CsI scintillator is reduced dramatically in

the target and just few low energy X-ray reach CsI scintillator. Therefore we can get more low energy X-ray in that case with thicker scintillator. In addition, there is another aspect we have to concern such as the size of photodiode and the scattered X-ray. The size of photodiode is $15\times15\times2$ mm^3. One photodiode expresses 2 mm part of the target, which is not available for the baggage inspection system. We plan to reduce pixel size to be smaller than 1 mm. Scattered X-ray increased by high energy X-ray induces incorrect transparency and make the spatial resolution down. The improved one will be equipped with the new collimator in front of two-fold scintillator, which can reduce its collimating gap until 2 mm and reduce scattered X-ray as a result

SUMMARY

We have performed the experiment of the material recognition in the custom inspection system using 950 keV X-band Linac. 950 keV X-band Linac is operated in 9.4 GHz RF frequency. X-band system realizes the compact system for practical uses. As the detector part, two-fold scintillaotr is proposed to get 2 images during one-time irradiation, which promises simpler and faster inspection. Numerical calculation to find out the atomic number of a target shows the results that iron and lead can be distinguished in the concealment. The difference between experimental value and theoretical value for Fe is from 3 to 10 and for Pb 2 to 5 in the polyethylene container. In the polyvinyl chloride container, Fe and Pb have the difference of atomic number from 9 to 12 for Fe, 1 to 3 for Pb. These differences are getting bigger as the thickness of container is thicker and density of composed material of concealment is higher. In severe situation such as target is lead and the concealment is polyvinyl chloride in thickness 10 to 50 mm, the experimental result of atomic number does not follow the pattern that we expect to become bigger difference. Because low energy X-ray hardly go through the thicker and dense material, the transparency of CsI that is mainly caused by low energy X-ray becomes higher by high energy X-ray. That makes big difference between real atomic number and experimental value and even causes irregular pattern in the extreme case such as lead target or thick concealment. We suggest increasing the thickness of CsI to 2 mm to absorb more low energy X-ray in the next experiment.

The custom inspection system we have developed shows the ability of material recognition. In particular, the light metal: iron and heavy metal: lead can be distinguished even in the concealment. However there are several limits such as thicker concealment and target. To solve the problem, we plan to make upgrade of two-fold scintillator: increase of CsI thickness, downsizing the channel size to 1 mm, and narrower collimator to suppress the scattered X-ray. We expect that our system can be available in the real situation in an airport to distinguish any suspicious items in baggage, because it has several advantages such as the compact size of the whole system, higher X-ray energy than X-ray tube, the controlling of X-ray dose in the Linac system and so on.

REFERENCES

1. Ogorodnikov S, Petrunin V., "Processing of interlaced images in 4-10 MeV dual energy customs system for material recognition [J]" Physical Review Special Topics-Accelerators and Beams, 2002, 5(10): pp1-11.
2. Ch. Tang, et al., *Proc. of LINAC 2006*, Knoxville, Tennessee USA, TUP007.
3. T. Yamamoto, et al., Proc. of European Particle Accelerator Conference '06, June 26 - 30, 2006, Edinburgh, Scotland, WEPCH182.
4. K. Lee, et al., Design and experiment of dual-energy X-ray material recognition using a 950 keV X-band Linac, Nuclear Instruments and Methods in Physics Research A, in press.

Measurement Of Hydrogen Capacities And Stability In Thin Films Of AlH Deposited By Magnetron Sputtering

A. Dissanayake[1], S. AlFaify[1], E. Garratt[1], M. I. Nandasiri[1], R. Taibu[1], G. Tecos[1], N. M. Hamdan[2] and A. Kayani[1]

[1] *Department of Physics, Western Michigan University, Kalamazoo MI 49008, United States*
[2] *Department of Physics, American University of Sharjah, Sharjah, United Arab Emirates*

Abstract. Thin, hydrogenated aluminum hydride films were deposited on silicon substrates using unbalanced magnetron (UBM) sputtering of a high purity aluminum target under electrically grounded conditions. Argon was used as sputtering gas and hydrogenation was carried out by diluting the growth plasma with hydrogen. The effect of hydrogen partial pressure on the final concentration of trapped elements including hydrogen has been studied using ion beam analysis (IBA) techniques. Moreover, in-situ thermal stability of trapped hydrogen in the film was carried out using Rutherford Backscattering Spectrometry (RBS), Non-Rutherford Backscattering Spectrometry (NRBS) and Elastic Recoil Detection Analysis (ERDA). Microstructure of the film was investigated by SEM analysis. Hydrogen content in the thin films was found decreasing as the films were heated above 110° C in vacuum.

Keywords: Aluminum hydride, hydrogen storage materials, Ion beam analysis, hydrogen thermal stability, RBS, ERDA
PACS: 60

INTRODUCTION

Hydrogen is the most abundant element in the universe and has the potential to be used as a fuel. If used as fuel, unlike fossil fuel, hydrogen will be nonpolluting and forms water as a byproduct. Moreover, specific energy content of hydrogen is 33.2 kWh/kg, which is three times larger than gasoline and diesel[1]. There is growing interest in the use of hydrogen as the main fuel for stationary and mobile applications, especially using fuel cells. Hydrogen fuel and fuel cells could help significantly reduce pollution and greenhouse gas emissions[2]. The conventional method of hydrogen storage is based on either increasing the pressure or by liquefying it. This is done to decrease the volume required to store a given weight of hydrogen. These options are not practical and have disadvantages. The main disadvantage of conventional storage methods is that energy is required to pressurize or liquefy hydrogen that decreases the overall efficiency of the energy producing device. Moreover, the conventional methods are not safe for on-board storage of hydrogen for mobile applications. The storage of hydrogen in metal hydride form offers a possible solution to these problems. This requires the development of sturdy alloys that without compromising their own structure absorb hydrogen by bonding with elements thus forming hydrides and on demand release the stored hydrogen. Increasing on-board gravimetric hydrogen storage capacities of solid storage materials has been identified as one of the most-challenging problems for moving towards hydrogen based economy[3].

Metal hydrides are the most promising materials for hydrogen storage. Hydrogen can be stored in metal hydrides under moderate temperature and pressure and they have enormous safety advantage over other forms of hydrogen storage. Al hydride is becoming increasingly attractive due to their high hydrogen storage capacity of 10 wt% [4]. Moreover, Al is cheap and abundantly available on earth.

Extensive work have been done on bulk materials for hydrogen storage application, but recently, thin hydride films are the emerging field of research[5,6], especially, to understand the basic science of hydrogen storage and diffusion process. Investigations of materials using thin films offer many advantages, such as (1) novel electron and electromagnetic spectroscopic techniques can be used to characterize the material, (2) ion beam analysis techniques can be used to quantify the uptake and the transport of hydrogen and resulting data can be used to accurately model the diffusion process. Unbalance Magnetron Sputtering (UBM) deposition is a technique to deposit

Application of Accelerators in Research and Industry
AIP Conf. Proc. 1336, 189-192 (2011); doi: 10.1063/1.3586086

TABLE 1. Deposition parameters and elemental concentrations of trapped elements in UBM deposited AlH thin films. Ar flow rate of 140 sccm and 30 W magnetron power was used for deposition

Sample Name	H_2 Flow Rates (SCCM)	H at %	Al at %	O at %
S-8	10	4.5	85.5	10
S-9	15	4.7	82.3	13
S-10	20	6.9	82.1	11
S-11	25	7.6	80.4	12

thin films of metal hydrides. By varying the deposition conditions, high performance hydrogen storage materials can be designed[7,8].

In this paper, we have deposited hydrogenated thin films of AlH on silicon substrates and studied the concentration of hydrogen with respect to varying partial pressure of hydrogen during the growth process. Ion beam analysis techniques that include Rutherford Backscattering Spectrometry (RBS), Non-Rutherford Backscattering Spectrometry (NRBS), and Elastic Recoil Detection Analysis (ERDA) techniques were used to determine the elemental concentration of deposited films. Moreover, thermal stability of hydrogen was carried out in-situ using a UHV compatible non gassy button heater and pre and post heated AlH samples were analyzed by IBA measurements. Deuterium was used to study the adsorption process for heat treated AlH film. The aims of this work are (1) to determine the effect of partial pressure of hydrogen on the final concentration of hydrogen, (2) to determine the temperature at which hydrogen is released from the material and (3) to determine whether deuterium can be adsorbed at the onset temperature of hydrogen release.

EXPERIMENT

The schematic diagram of the Ultra High Vacuum (UHV) stainless steel Unbalanced Magnetron Sputtering (UBM) chamber used for the thin films deposition of AlH is shown in figure 1. The sputtering target of 2.54 cm in diameter, which in our case was high purity aluminum, goes on the magnetron sputter head. Silicon was used as substrate which was mounted on a 5.6 cm in diameter circular stainless steel substrate holder. The distance between the sputter head and the substrate holder was held fixed at 20 cm. Ultra high vacuum was produced inside the chamber by cryogenic pump. The pressure inside the chamber was monitored by an ion gauge. A capacitance monometer was also attached with chamber to monitor gas pressure during the deposition process. The chamber was pumped for 24 hours before carrying out the deposition. Argon was used as sputtering gas and to get the hydrogenated samples, hydrogen was added

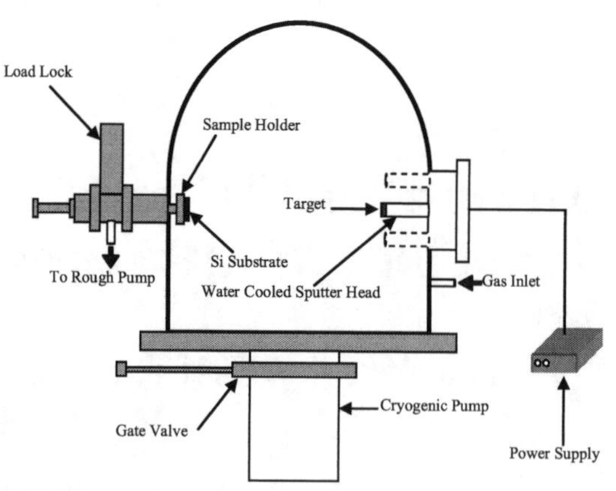

FIGURE 1. The schematic diagram of UHV magnetron sputtering deposition chamber.

along with argon during the growth process. For samples, hydrogen flow rates of 10, 15, 20 and 25 sccm were used for samples S-8, S-9, S-10, and S-11 respectively. Flow rate of argon was fixed at 140 sccm and samples were deposited for 3 hrs. Plasma inside the chamber was stimulated by a DC power supply and was used in constant power mode. The base pressures for the deposition was in lower 10^{-8} Torr range and all the films were deposited at 30 mTorr of working pressure. Deposition parameters and elemental concentration of trapped elements is given in table 1.

Characterization of the samples was carried out using ion beam analysis (IBA) techniques. Ion beam analysis of these samples was carried out using 6 MV tandem Van de Graaff accelerator facility of the department of physics, Western Michigan University, Kalamazoo Michigan. RBS and NRBS spectra were recorded using 2.5 MeV beams of H^+ ions at incident angle of 67.5 ° and exit angle of 68.9 ° using surface barrier silicon detector at 160^0 scattering angle. This configuration was chosen so that the position of the sample remains the same during the back and forward scattering measurements. Proton beam was used mainly to determine the Al and O concentration in the sample. At this energy enhanced O cross section

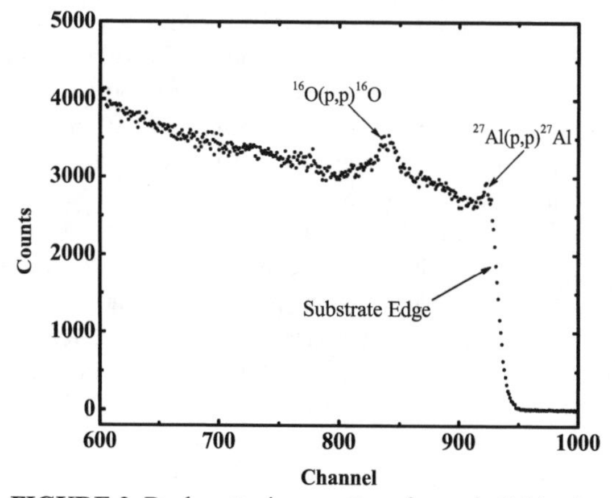

FIGURE 2. Backscattering spectra of sample S-11 taken at the scattering angle 160° with 2.5 MeV H^+ beam.

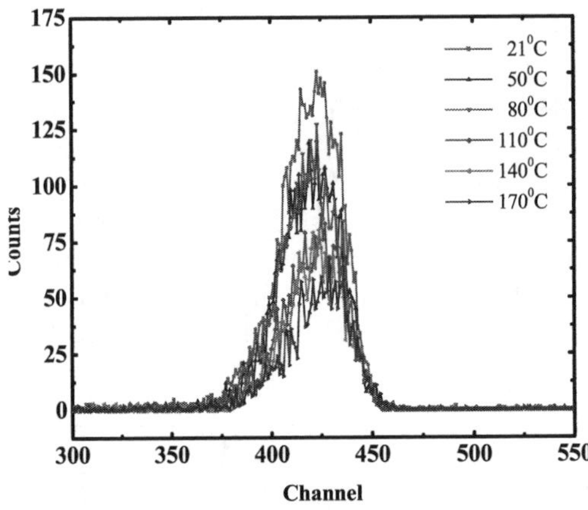

FIGURE 3. Overlapped forward recoil spectra obtained using 16 MeV O^{4+} beams for the sample S-11.

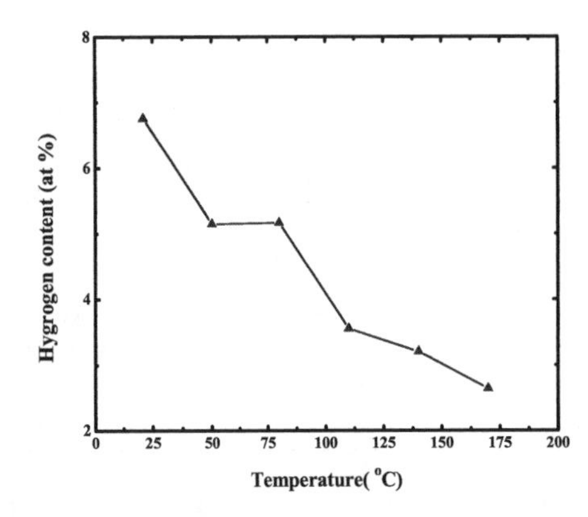

FIGURE 4. Hydrogen content (at %) with annealing temperature for the sample S-11.

makes it easy to quantify. Typical backscattering spectrum of sample S-11 is shown in figure 2.ERDA spectra of the samples were obtained using 16 MeV O^{4+} beam, incident at 67.5^0 from normal to the sample surface. Forward scattered ions were detected using a silicon surface barrier detector at a scattering angle of 45°, with an exit angle of 67.5°. A mylar foil of 13 µm was placed in front of the detector to prevent other forward scattered species other than hydrogen to enter the detector. Using the SIMNRA program[9], the set of target composition profile parameters was adjusted until a single set of parameters could be used to accurately simulate both, back and forward scattered spectra.

To investigate the thermal stability of hydrogen in AlH thin film, an experiment was designed to determine the concentration of elements in situ after going through a thermal cycle in vacuum. The sample was heated for 10 min in an evacuated stainless steel scattering chamber from 21 °C to 170 °C at an interval of 30 °C, using an UHV compatible non-gassy button heater. Target temperature was achieved at the rate of 3°C/min and temperature was determined by a K-type thermocouple attached to the sample holder. After each heating cycle, forward scattering spectra were taken, which is shown in figure 3. Change in the hydrogen concentration was plotted with respect to the temperature, which is shown in figure 4. Backscattering spectra were taken at the beginning and at the end of the experiment.

To study the adsorption of hydrogen at the temperature were desorption of hydrogen was observed, we used deuterium. Since the electronic properties of isotopes are the same, the adsorption process of deuterium in AlH will be the same as for hydrogen. The advantage of using deuterium is distinguishing the uptake from the existing hydrogen in the sample. Vacuum heated sample S-11 was chosen for this study. Using button heater, S-11 sample was successively heated at 110 ^{0}C and 140 ^{0}C for 10min, while heating, pressure inside the chamber was increased from 9.0×10^{-7} Torr to 7.0×10^{-6} Torr by bleeding deuterium through a precision leak valve. After the heating process forward recoil spectra were taken with 16 MeV O^{4+} beam. The forward spectrum of this sample is shown in figure 5.

Finally, to study the surface topography of the sample, scanning electron microscopy (SEM) image of the sample was taken. The SEM image of the sample is also shown in the figure 6.

RESULTS AND DISCUSSION

An important issue relating to AlH film properties is the temperature regime at which hydrogen loss from the film occurs, especially the temperature at which hydrogen evolution begins. The thermal stability not only determines the material candidacy for a particular application, such as hydrogen storage, but also helps

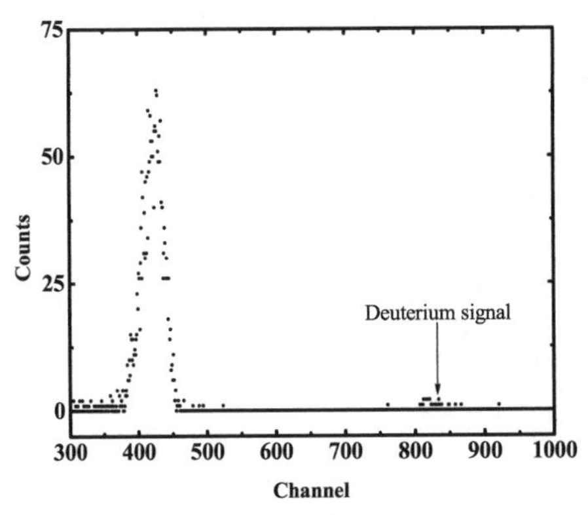

FIGURE 5. Forward scattering spectrum of sample S-11 with Deuterium adsorbed on the surface.

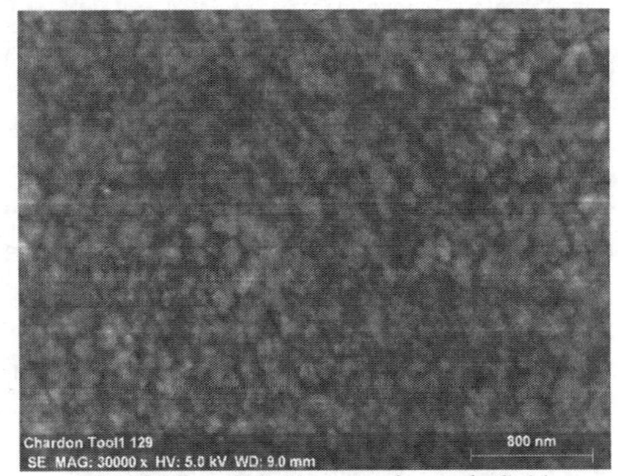

FIGURE 6. SEM image of sample S-11.

understand the structure and properties of the film.

The backscattering spectra of sample S-11 with 2.5 MeV proton beam is shown in figure 2. Aluminum peak is clearly seen in the spectrum. A small concentration of trapped oxygen is also seen in the film. The origin of oxygen in the film is not exactly known. However, we believe that when we take the sample out of the chamber, it might have reacted with atmospheric oxygen or the partial pressure of water might have contributed to the oxygen concentration during the growth process. How oxygen is bonded in the film is still unknown and is a subject of our future work. Figure 3 is a forward recoil spectra of sample S-11. The area under the individual peaks is the number of O^{4+} encounters with hydrogen and is proportional to the concentration of hydrogen in the film.

Figure 3 shows the overlapped spectra of the sample S-11, taken after each heating cycle. Area of the hydrogen peak is clearly seen decreasing with increasing temperature. We believe that slightly higher area of the virgin sample is because of the water vapors on the surface. After the first heating, water vapors on the surface are degassed to the vacuum that resulted in the drop of hydrogen concentration after first heating cycle. After each heating cycle forward spectra of sample was simulated with SIMNRA and hydrogen content with respect to temperature of the sample S-11 was plotted which is shown in figure 4. The onset temperature for hydrogen release for sample S-11 was found to be in between 80-110 °C.

Figure 5 shows forward scattering spectrum of sample S-11 that was subjected to adsorption process using deuterium. Small amount of adsorbed deuterium on the surface of the film is visible in figure 5. However, deuterium was not absorbed by the film. The surface of the metal plays the most important part in the absorption process. The surface makes an interface boundary with deuterium and offers a barrier for deuterium diffusion. Deuterium needs to dissociate at the surface to diffuse into the bulk of the metal. Diffusion into the bulk is the rate limiting process. Deuterium starts diffusing into the bulk in the beginning and occupies the sites available in the near surface region. When the active sites in the near surface region of the metal are occupied by deuterium[10], it forms a hydride layer that decreases further diffusion. Adsorbed deuterium needs to diffuse into the bulk via chemisorption process to make room for further diffusion of the deuterium to take place. For Al, chemisorption is a very slow process and this hydride layer acts as a barrier for further diffusion of deuterium. Moreover, Al is also highly reactive to oxygen and presence of oxygen also decreases the absorption rates due to formation of oxide layer on the surface, this increases the activation energies for hydrogen/deuterium absorption.

Figure 6 shows SEM image of sample S-11. Image was taken using S-4500 Hitachi machine which has minimum resolution at 60 nm. Close examination of these SEM images reveal that thin film consists of grains which are in order of hundred nano meters with grain boundaries and voids. The crystal structure of this material is not known. It is a subject of future study.

CONCLUSION

We have shown a mean of comparing the thermal stability of thin films of $AlH_,$ deposited on silicon substrates using ion beam analysis techniques. The results from ion beam analysis are both qualitative and quantitative and do not depend on the stoichiometry. From our results, we conclude that under the deposition parameters chosen, maximum trapping of 7.6 at% at 25 sccm of hydrogen can be obtained and hydrogen evolution from AlH film take place around 110 °C. For future measurements we will use CVD process and substrate heating to obtain high hydrogen content phase of Al hydride. Better depth resolution will be obtained using He^+ beam.

ACKNOWLEDGMENT

We acknowledge the technical assistance of Mr. Benjamin Gaudio, Mr. Allen Kern and Mr. Rick Welch of the department of physics, Western Michigan University.

REFERENCES

1. I.K. Kapdan, F. Kargi, *Enzyme Microb Technol* **38**, 569 (2006).
2. Mustafa Balat. Int. J. Hydrogen Energy, 33, (2008) 4013.
3. R.H. Jones, G. Thomas, *J. Miner met Mater Soc* **59**, 50 (2007).
4. W. Grochala, P.P. Edwards, *Chem Rev* **104**:1283, (2004) 315.
5. L.Z. Ouyang, H. Wang, C.Y. Chung, J.H. Ahnc, M. Zhu, *J Alloys Compd*, **422**, 58 (2006).
6. H. Akyildiz, M. Özenba¸s, T. Öztürk, *Int J Hydrogen Energy* **31(10),** 1379 (2006).
7. K. Higuchi, K. Yamamoto, H. Kajioka, K. Toiyama, M. Honda, S. Orimo and H. Fuji, *J. Alloys and Compounds* **330-332,** 526 (2002).
8. K. Higuchi, H. Kajioka, K. Toiyama, H. Fujii and Y. Kikuchi, *J. Alloys and Compounds* **293/295**, 484 (1999).
9. M. Mayer, SIMNRA User's Guide, Technical Report IPP 9/113, Max-Planck-Institut fur Plasmaphysik, Garching, Germany, (1997).
10. Holtz RL. Basic user's guide for NRL 6323 hydrogen storage system, 1996.

Measurement Of Gas Electron Multiplier (GEM) Detector Characteristics

Seongtae Park[a], Edwin Baldelomar[a], Kwangjune Park[b], Mark Sosebee[a], Andy White[a], Jaehoon Yu[a]

[a]University of Texas at Arlington, Arlington, TX 76019
[b]KAERI, Daejeon, Korea 305-600

Abstract. The High Energy Physics group of the University of Texas at Arlington has been developing gas electron multiplier detectors to use them as sensitive gap detectors in digital hadron calorimeters for the International Linear Collider, a future high energy particle accelerator. For this purpose, we constructed numerous GEM detectors that employ double GEM layers. In this study, two kinds of prototype GEM detectors were tested; one with 28x28 cm^2 active area double GEM structure with a 3 mm drift gap, a 1 mm transfer gap and a 1 mm induction gap and the other with two 3x3 cm^2 GEM foils in the amplifier stage with a 5 mm drift gap, a 2 mm transfer gap and a 1 mm induction gap. The detectors' characteristics from exposure to high-energy charged particles and other radiations were measured using cosmic rays and ^{55}Fe radioactive source. From the ^{55}Fe tests, we observed two well separated characteristic X-ray emission peaks and confirmed the detectors' functionality. We also measured chamber gains to be over 6000 at a high voltage of 395 V across each GEM electrode. The responses to cosmic rays show the spectra that fit well to Landau distributions as expected from minimum ionizing particles.

Keywords: Gas Electron Multiplier, Digital Hadron Calorimeter, KPIX, Cosmic Ray, Fe-55
PACS: 29.40.Vj, 29.40.Cs, 29.40.Gx

INTRODUCTION

The Gas Electron Multiplier (GEM) is a charge amplification gas detector developed at CERN in 1997[1]. The invention was motivated in order to overcome the drawbacks that traditional gas detectors (MWPC or MSGC) have, such as discharge or spark problems at the operational high voltages typically at 10s of kV. Since the first invention of the GEM, it has been used in various fields of application, e.g., X-ray radiography[2], X-ray polarimeter[3], GEM photomultiplier[4].

Future high energy physics experiments will place severe demands on the jet energy resolution of hadron calorimetry. A notable example of jet energy measurement requirements is found in the specification of the calorimetry for experiments at the International Linear Collider (ILC). The ILC is being planned in anticipation of the need for precision studies and measurements of new physics expected to emerge from experiments at the CERN Large Hadron Collider (LHC). Much of the possible new physics program requires high-precision measurements of jet energies and jet-jet invariant masses. The best past detectors have managed to achieve $50\%/\sqrt{E}$ jet energy resolution. However, for future experiments, such as those at the ILC, we require $30\%/\sqrt{E}$ or better.

The High Energy Physics (HEP) group at University of Texas at Arlington (UTA) has been developing GEM detectors to use as sensitive gap detectors in Digital Hadron Calorimeter (DHCAL). A DHCAL is a device which is used in a high energy particle collider to measure the energy of the particles resulting from collisions. It counts the number of particles to calculate the total energy of the incident particles that produce a shower while passing through the medium surrounding the collision point. It is composed of passive medium (absorber) and multiple active detectors. Steel slabs are used as the absorber and thin active detectors are sandwiched in-between the absorbers to measure the energy of the particles passing through the detectors.

PROTOTYPE GEM CHAMBER CONSTRUCTION

In order to study the basic operation characteristics of GEM detectors, we constructed two different types

Application of Accelerators in Research and Industry
AIP Conf. Proc. 1336, 193-196 (2011); doi: 10.1063/1.3586087

of prototype double GEM chambers (see Fig.1). The chamber (GEM33) shown in Fig.1 (a) was constructed with GEM foils made in the Gas Detector Development (GDD) workshop at CERN, and the foils have dimension of 3x3 cm^2. And the other GEM chamber (GEM3030) shown in Fig.1 (b) has 30x30 cm^2 GEM foils UTA HEP group jointly developed with 3M Inc.

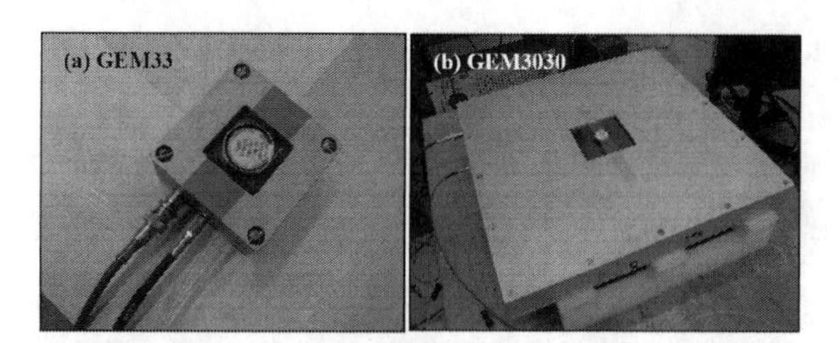

FIGURE 1. The GEM33 was installed with 3x3 cm^2 GEM foils from CERN and the GEM3030 with 30x30 cm^2 GEM foils from 3M. Both chamber were ran with the same gas mixture, Ar:CO_2=80:20.

Foils in GEM33 were produced through the standard chemical etching technique and have holes with smallest diameter 55 μm arranged in 140 μm pitch. The drift gap, transfer gap and induction gap are 5 mm, 2 mm and 1 mm, respectively. A 25 μm thick copper clad Kapton film is used as the cathode. A single resistor network is used to distribute high voltage to each of the five electrodes in the detector. To collect the avalanched electron charge, a 2cmx2cm square copper pad from ordinary printed circuit board (PCB) is used as the single channel anode board.

GEM3030 is constructed in a similar way to GEM33 with different dimensions and structure. In the sampling calorimeters, a limited number of layers are used and the thickness of each layer is also limited to control the overall cost of the detector. To minimize the total thickness of the detector, the three gaps (drift gap, transfer gap, induction gap) in the amplifier are optimized to 3 mm, 1 mm, 1 mm, respectively. As a prototype detector, we constructed 30x30 cm^2 GEM chamber with a 28x28 cm^2 active area. Again, two GEM foils are used in the detector amplification gap.

For the safe operation of the detector at high voltage, it is very important to keep the gap to avoid discharges or sparks which could destroy whole readout electronics during the data taking. To maintain the gap size, we designed special G-10 spacers for each gap. Each spacer has carefully laid out channels and notches on the frame for an effective gas distribution. The spacer segmented the 28x28 cm^2 active detector area into 5x5=25 grids and 1 mm thick grid barriers divided the squares. The readout board has 64 copper pixlated electrodes with dimension 1x1 cm^2 arranged in 8cmx8cm grid to cover the active area. Each anode pad was connected to the input channel of the frontend readout electronics which was attached on the opposite side of the anode board. The 64 channel KPiX [5] readout electronics chip being developed at the Stanford Linear Accelerator Center are used for data acquisition.

Since the GEM detector utilizes electron avalanche in a gas medium, the gas and high voltage are essential part of the GEM detector. The gas and high voltage directly affect the detector gain and operations. In this study, we used Ar:CO_2=80:20 mixture at 99.999% purity as the gas medium, and the high voltages were distributed through a single resistive network in both detectors GEM33 and GEM3030.

Readout Electronics

We constructed single channel readout electronics (SCR) for the test of GEM33. Since the signals from the GEM detector are charged pulses, we need to convert the charge signal to voltage for the further processing in the subsequent electronic circuits. We used a charge sensitive preamplifier (A225, AmpTek) for the conversion of the charge signal. A225 has sensitivity of 0.83 μV/electron and includes a shaping amplifier with peaking time of 2.4 μs. Another amplifier (A206, AmpTek) following the A225 is used in this circuit to further amplify the A225 output by a factor of 10.

In order to record the signal height in the computer, the analog signal has to be converted into a digital value. The A206 output, however, is a voltage pulse with 4 μs width, thus it must be held for a sufficient time period for an external analog to digital converter (ADC) can perform the conversion. We used a peak holder (PH300, AmpTek) for this purpose. The PH300 holds the maximum height of the A206 output pulse for 10 μs in the current setting of the circuits. The output of the PH300 is routed to an external ADC, and the converted digital values are transferred to the DAQ computer. Since, A206 has a discriminator in it, we used the output of the discriminator to generate a trigger signal for the PH300 and the external ADC. We utilized the LabView (ver. 8.0, National Instruments) for DAQ software.

The 13 bit KPiX readout system is used for the data acquisition of GEM3030. The KPiX system is composed of an FPGA (field-programmable gate array) board, an interface board and an anode board. It is designed for future ILC experiments. It has 4 terminals to accept external triggers such as beam line triggers or external scintillators. To minimize the environmental noise during the data acquisition, the FPGA board and the interface board are connected by optical cables which electronically isolate the interface board and the anode board from the FPGA board. While the ultimate goal of the KPiX chip is to handle 1024 channels, an available intermediate 64-channel version is used for our prototype detector. The data taken with the KPiX electronics is stored in a file which is packed using CERN ROOT [6] framework. The data can be analyzed using various programming languages such as C++ or Java.

Radioactive Source And Cosmic Ray Test Results

To characterize the two prototype GEM detectors, we have conducted many tests using radioactive sources and cosmic rays. ^{55}Fe is a well known standard X-ray source and often used to test gas detectors. From the measurement with ^{55}Fe, we observed two well separated ^{55}Fe characteristic emission peaks, 5.9 keV main peak and 3 keV Ar-escape peak in both chambers GEM33 and GEM3030, which meant the detectors were functioning correctly. Figure 2 (a) shows typical ^{55}Fe spectrum measured with GEM33 with high voltage V_{gem}=400 V across each GEM foil. The solid circles represent the data and the solid line the Gaussian fit to the data. The energy resolution measured at FWHM is about 21%. We obtained similar results from 64 individual channels in GEM3030.

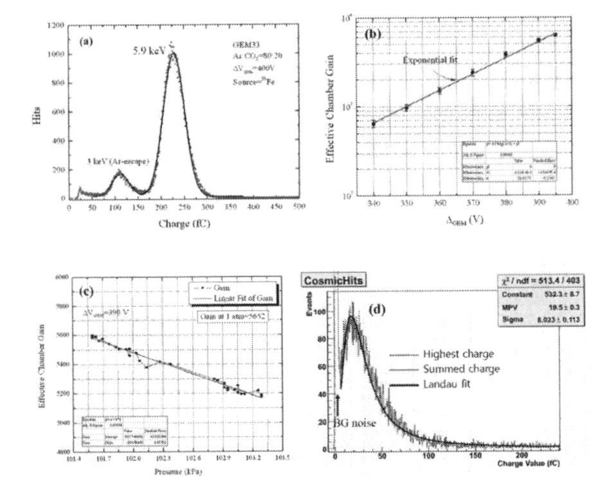

FIGURE 2. (a) The spectrum of ^{55}Fe was measured with GEM33 at high voltage V_{gem}=400 V. (b) High voltage dependence of the effective chamber gain. (c) Atmospheric pressure dependence of the effective chamber gain. (d) Cosmic run result with GEM3030. The measurement was carried out at high voltage of V_{gem}=396 V.

We measured the effective detector gain of GEM3030 as a function of the high voltage (HV) applied across each GEM electrode. As shown in Fig.2 (b), the detector gain depends on the HV exponentially and is measured to be about 6000 at the operational voltage 395 V across each GEM foil. There, however, is an apparent limitation in increasing the gain because of the HV limitation due to discharges. The rate of discharge is closely related to the GEM foil condition, gas mixture and HV.

We also measured the pressure dependence of the chamber gain. As shown in Fig.2 (c) the effective chamber gain has negative dependency on atmospheric pressure. The gain decreased with increasing pressure with a slope of about -244/kPa. From the result, we were able to correct the chamber gains to that at 1 atm. This result shows that it is important to control the pressure to maintain a stable gain in gas detectors.

When charged particles traverse through the material, they lose energy through ionization. The fluctuation of the energy deposit is well described by Landau distribution [7,8]. Cosmic ray muons are minimum ionizing particles (MIPs), and their energy deposit conforms to the Landau distribution. We took cosmic ray data to measure the response of GEM3030 prototype detector to MIPs. Coincidences of two 19x19 cm^2 scintillators are used to generate trigger signals for cosmic ray muons. Two counters were overlapped to form a 8x8 cm^2 coincidence area which covers the whole 64 active anode pad area. The separation of the two trigger counters was 30 cm and sandwiched the prototype GEM detector in between them. The final event count from a logic unit was about 40/min, and the coincidence logic signal was sent to FPGA board of the KPiX readout system to be used as triggers. The pulse heights from the recorded data are then analyzed.

Figure 2 (d) shows the MIP charge distributions from these cosmic ray triggered events. The thin dotted line represents the charge of the cosmic ray muon. Since we used external triggers from two scintillators that cover the entire 8cmx8cm active area, the KPiX system recorded data from all 64 channels whenever the trigger came in. Since cosmic ray muon most likely passes through one channel, only the highest charge value among 64 channel data is picked as the charge of the muon. One, however, must consider the possibility of charge sharing with the neighboring pads. In this case one has to add those shared charge together to correctly obtain total charge deposit by the cosmic ray. In order to obtain the correct total charge, first we went through the 64 channel data and identified the pad with the highest charge and the value of the charge, and then searched for the next highest charge. Once we found the next highest charge, examined if the channel is one of the nearest neighbors of the highest charge channel. If the channel belongs one of the nearest neighbors of the highest charge channel, the charge is added to the highest charge value and logged into the histogram. The MIP distribution in Fig.2 (d) shows the highest charge in dashed histogram while the two channels summed charge in the thin solid line. As can be seen in the figure, very little difference between the highest charge and summed charge histogram is observed. This means that the charge sharing among the neighbor pads are not significant in this setup. Finally, the thick solid line in Fig.2 (d) is the result of Landau fit with most probable value of 19.5 fC. Though we could not see full range of the Landau spectrum due to insufficient electronics dynamic range, the fit result showed that the data was well described by the Landau distribution.

CONCLUSIONS

As an effort to develop the Digital Hadron Calorimeter for future accelerator experiment, we have been working on GEM detectors. In this study, two prototype GEM detectors were constructed and characterized. For the analog signal test, we developed single channel readout electronics and DAQ program using LabView. Using the single channel GEM33 detector with the homemade DAQ system, we observed well separated two ^{55}Fe characteristic peaks, 5.9 keV main peak and 3 keV Ar-escape peak, with energy resolution of 21% at FWHM, which confirmed these prototype GEM detectors' functionality. We measured the effective chamber gain as a function of HV using a ^{55}Fe radioactive source. The chamber gain depends on the high voltage across GEM electrodes exponentially and measured to be about 6000 at an operational voltage of 395 V applied across each GEM foil. The chamber gain also depends on the atmospheric pressure. The chamber gain decreases with increasing pressure with a slope of -244/kPa. We measured GEM chamber responses to MIP using cosmic ray muons. The result showed a good Landau shaped spectrum with most probable value of 19.5 fC. The team will take these prototype chambers to high intensity charged particle beams for further characterization and is moving toward constructing large scale GEM chambers of dimension 33cmx100cm in the near future.

ACKNOWLEDGMENTS

We would like to acknowledge the support from the U.S. Department of Energy (DE-FG02-96ER40943, LCRD), and from the University of Texas at Arlington. We also would like to thank M. Breidenbach, D. Freytag and R. Herbst from SLAC for collaboration.

REFERENCES

1. F. Sauli, *Nucl. Instr. and Meth.* **A 386**, 531-534 (1997).
2. A. Bressan, R. De Oliveira, A. Gandi, J. C. Labbé, L. Ropelewski, F. Sauli, D. Mörmann, T. Müller, H.J. Simonis, *Nucl. Instr. and Meth.* **A 425,** 254-261 (1999).
3. E. Costa, P. Soffitta, R. Bellazzini, A. Brez, N. Lumb and G. Spandre, *Nature* **411**, 662-665 (2001).
4. A. Breskin, A. Buzulutskov, R. Chechik, B.K. Singh, A. Bondar and L. Shekhtman, *Nucl. Instr. and Meth.* **A 478**, 225-229 (2002).
5. D. Freytag, R. Herbst, J. Brau, M. Breidenbach, R. Frey, G. Haller, B. Holbrook, R. Lander, T. Nelson, V. Radeka, D. Strom and M. Tripathi, *IEEE Nuclear Science Symposium Conference Record*, 3447-3450 (2008).
6. Refer to http://root.cern.ch/.
7. L.D. Laudau. *J. Exp. Phys. (USSR)* **8**, 201-205 (1944).
8. D.H. Wilkinson, *Nucl. Instr. and Meth.* **A 383**, 513-515 (1996).

Tomographic Back-Projection Algorithm for "Incomplete" Compton X-Rays Detectors

Cristiano Lino Fontana[1], Giuseppe Baldazzi[2], Andrea Battistella[3], Michele Bello[3], Dante Bollini[4], Marcello Galli[5], Giuliano Moschini[1,3], Gianluigi Zampa[6], Nicola Zampa[6], Paolo Rossi[1]

(1)Department of Physics of the University of Padua and INFN, Via Marzolo 8, Padua 35131, Italy
(2)Department of Physics, University and INFN, Bologna, Italy
(3)National Laboratories of Legnaro, INFN, Legnaro (Padua), Italy
(4)INFN, Bologna, Italy
(5)ENEA, Bologna, Italy
(6)INFN, Trieste, ItalyContinue Here

Abstract. A "Compton" detector finds the direction of an X-ray by letting it interact with a gaseous, liquid or thin solid material (Tracker) and employing no collimators. This paper takes into account the case of an "incomplete" solid Tracker where the recoiling electron travels only a few dozen microns and cannot be followed. However, impact positions and incoming and outgoing energies are measured. In this situation, exploiting the Compton Scattering formula, one is only able to identify a cone whose surface the X-ray belongs to. On the other hand, Compton tomography luckily requires only a few views (for example rotating the apparatus in just four positions around the subject), as the "electronic collimation" that takes place in each position already extracts X-rays coming from many directions. A back-projection algorithm that combines the reconstructed "cones" in space, weighed according to the Klein-Nishina formula, has been applied to the special case of small animal SPECT (Single Photon Emission Computed Tomography). Here the source is close to the detector, every imaged point in space is calculated from many rays that are emitted at different angles, and the algorithm totally differs from that of the Compton imaging in Astronomy. Tomography reconstruction has been validated by employing simulated data, generated using GEANT4 that includes the Doppler Broadening in the energies of scattered photons, which is due to moving non-free electrons. Space resolution has been assessed.

Keywords: Small animals SPET, Compton Camera, Molecular Imaging, Gamma Detectors, Radio-Isotope markers, Pharmaceuticals distribution.
PACS: 87.58.-b, 87.58.Ce, 87.58.Pm

INTRODUCTION

The necessity of studying the targeting capability of pharmaceuticals in oncology brought to the development of different imaging techniques. Some examples are Positron Emission Tomography (PET) and Single Photon Emission Computed Tomography (SPECT), which involve the detection of gamma-rays emitted from radio-isotope markers [1-6]. Here we propose a Gamma Camera with an approach significantly different from the more conventional Collimated Anger Cameras (CAC). CACs employ lead collimators consisting of a matrix of millimeter-holes separated by small septa, which produces an image with the down side of rejecting a large fraction of gamma-rays. Our approach is based instead on the "Compton Concept" [7], in which gamma-rays elastically interact with a detector (Tracker) and then are absorbed by a second downstream detector (Calorimeter). To get the incident gamma-ray direction it is always necessary to measure at least the interaction point and the released energy on both detectors. A further measurement of the recoiling electron would lead to a full reconstruction of the event kinematics. However, in this paper we shall consider the case of a Tracker that is unable to measure the electron's direction, since this latter travels only a few dozens of microns inside a solid medium.

The lack of bulky collimators leads to many advantages like the compactness of the apparatuses

Application of Accelerators in Research and Industry
AIP Conf. Proc. 1336, 197-200 (2011); doi: 10.1063/1.3586088

that can be used on "small" subjects like prostate, thyroid and of course small animals like mice. Moreover gamma-rays with higher energies like I-131 (364 keV), beta+ emitters (511 keV) or Co-58 (810 keV) can now be profitably exploited, because the collimators were unable to efficiently filter such energies and they interact less inside the body of the subject.

DESIGN AND SIMULATION

The Compton Camera we are modeling consists of two detectors [7]: the Tracker is a Silicon Drift Detector (SDD) developed for the LHC-Alice experiment [8,9] and the calorimeter is based on a scintillating crystal optically coupled to a position sensitive multi-anode photomultiplier, a Hamamatsu H8500.

The design has been tested with a simulation developed with the GEANT4 library. Gamma-rays of 140 and 511 keV have been tracked through a mouse, air, and detectors according to the geometry of Fig. 1. The Tracker is a 1 mm wafer of Silicon with dimensions 7 cm x 7cm. The calorimeter is a 5 cm x 5 cm gamma-rays total absorber. The detectors were assumed to be ideal, i.e. with no experimental errors. We employed the GEANT4 "Livermore low-energy package", a model of electromagnetic processes down to 250 eV and up to 100 GeV, which also describes the recoiling electrons.

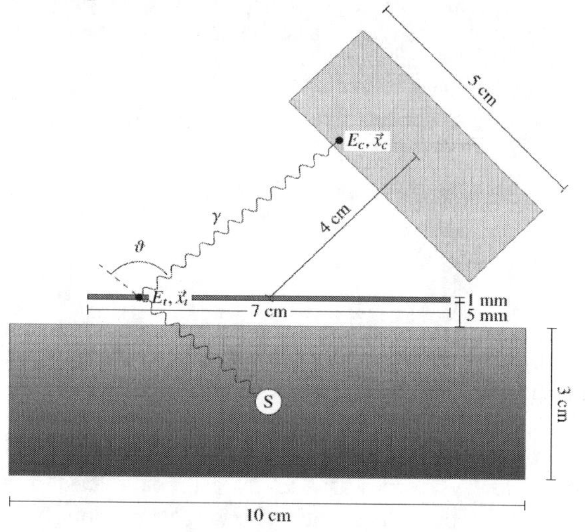

FIGURE 1. Sketch of the Compton Camera prototype. Formula and dimensions are those used in the GEANT4 simulations.

Recoiling Electrons Path

A simulation of the recoiling electrons, scattered by 140 keV photons, showed that their mean path inside the Silicon is a few dozens of micrometers, and that an electron could hardly escape from 1 mm Si Tracker with an energy sufficient to be detected. Therefore all the information comes from the photon.

Doppler Broadening

The GEANT4 Livermore low-energy package is able to accurately simulate the Compton Scattering in low-energy interactions, when electrons inside the atoms have to be considered non-free and moving, being their energy not negligible compared to that of the incoming photon. In such situation ("Doppler Broadening") the Compton formula is not totally valid [10]

$$\cos(\theta) = 1 - m_e c^2 \left[\frac{1}{E_c} - \frac{1}{E_t + E_c} \right]$$

where E_t is the energy released on the Tracker, E_c the energy released on the Calorimeter, and m_e is the electron mass. The "Doppler Broadening" entails an unavoidable "physical" limit in the achievable resolution, which will be investigated here. To this aim we have simulated events with no experimental errors.

IMAGE RECONSTRUCTION

The Image reconstruction from "incomplete" Compton Scattered data, with no information on the recoiling electron, requires special algorithms. In fact we merely obtain the gamma-ray's outgoing direction and scattering angle and with them a cone of possible solutions of the incoming direction [11]. At least three different cones are needed to have a unique solution. In this work, a Filtered Back-Projection algorithm has been developed for this kind of image reconstruction [12].

The reconstruction follows these steps:

1) Determination of intersections of reconstructed cones and a unitary sphere centered on the detector;
2) Weighing of the cones with the Klein-Nishina formula;
3) Fourier Transformation of the data distribution;
4) Deconvolution with the Point Spread Function;
5) Inverse Transformation of the distribution.

Cone Determinations And Weighing

The intersection curves of the cones with the unitary sphere are the roots of the equation:

$$S(\vec{u}) = \vec{d} \cdot \vec{u} - \cos(\theta)$$

where $\vec{d}$ is the direction of the scattered photon and θ is the scattering angle. The domain of $S(\vec{u})$ is restricted to the unitary sphere surface (bi-dimensional) on which we are searching the intersections. To this aim we have employed the "Newton" algorithm, which is among the fastest converging methods and keeps the elaboration time to a reasonable amount. The cones of different angular apertures are weighed with the Klein-Nishina formula. Lines in Figure 3 are an example of such intersections.

The distributions are then plotted on a four-dimensional histogram $g(\varphi,\theta,x_t,y_t)$, where (φ,θ) are the coordinates on the sphere and (x_t,y_t) are those on the Tracker.

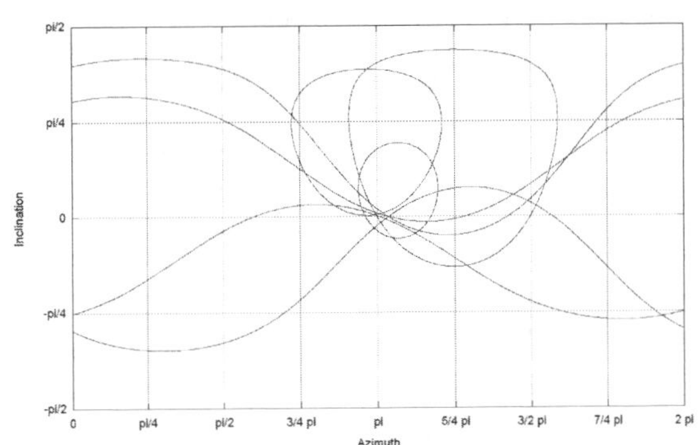

FIGURE 2. Example of cone intersections on the unitary sphere.

Fourier Transforms And Deconvolution With PSF

The distributions $g(\varphi,\theta,x_t,y_t)$ have been corrected in the sphere domain (φ,θ) by "deconvolving" them with the Point Spread Function (PSF). This latter was evaluated by simulating (with GEANT4) a parallel beam coming from the mouse to the Tracker, and then calculating the incoming direction with the Compton Formula. Fig. 3 shows the PSF on the sphere domain projected on the azimuth coordinate φ. The corresponding PSF, in the 2D space domain, projected on the Tracker plane is 4 mm (FWHM).

In particular the "deconvolution" is calculated in the frequency space by dividing the Fourier-transformed spectrum $F\{g(\varphi,\theta,x_t,y_t)\}$ by the PSF spectrum F{PSF}[12]. This is an unstable step because F{PSF} could vanish on some points. To solve this issue, the F{PSF} was modified by setting all the points below a certain threshold to the value of that threshold. This latter was determined by trials: i.e. by lowering it until no appreciable difference on the results was found.

No further low-pass filter was applied to the data because it reduced the sharpness of the final image.

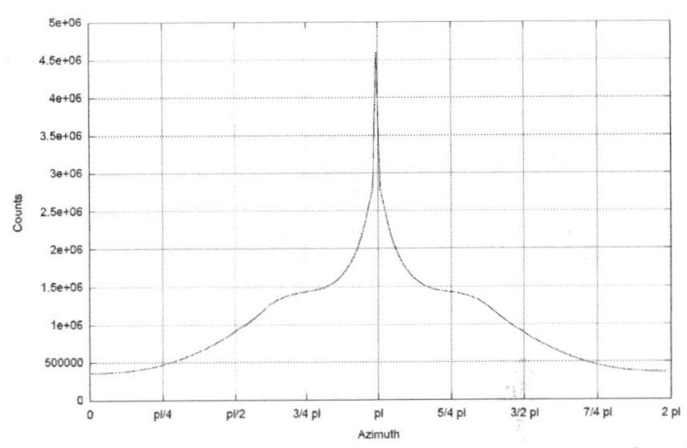

FIGURE 3. Point Spread Function projected on the azimuth coordinate, obtained by the simulation of a collimated 140 keV beam directed toward the center of the Tracker.

Projection On Tracker

The filtered histogram

$$g'(\varphi,\theta,x_t,y_t)$$

gives the projections of the data on the spheres centered on the coordinates of the tracker (x_t,y_t). Furthermore, by selecting a specific direction, i.e. a spherical point (φ,θ), one obtains the projection on the tracker plane along that direction. By selecting a direction orthogonal to the Tracker plane, an orthogonal projection of the source is obtained. Figures 4 through 6 show different images of sources on the Tracker plane.

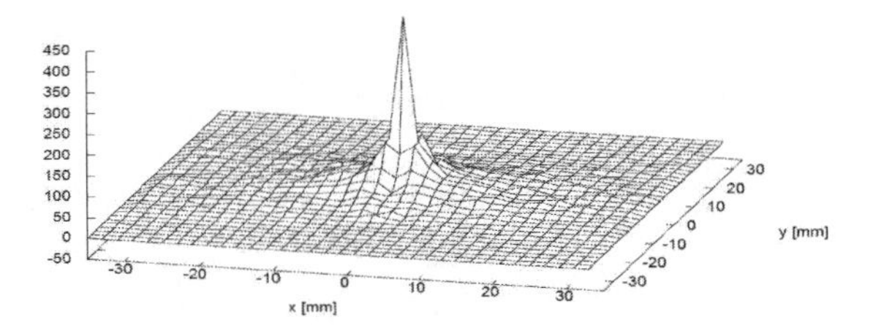

FIGURE 4. Orthogonal projection, on the Tracker plane, of a point-like source, inside the body of a simulated mouse.

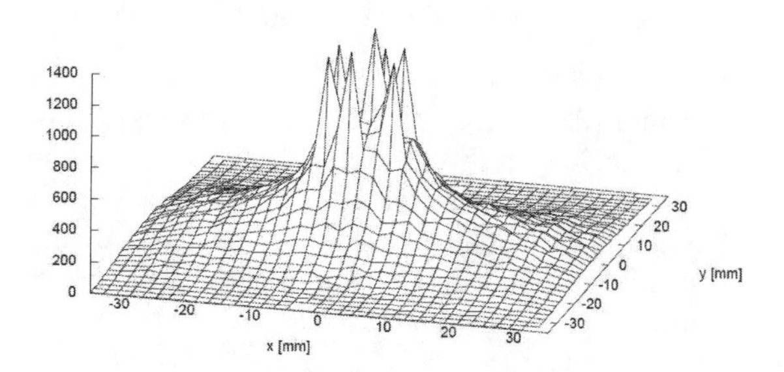

FIGURE 5. Orthogonal projection, on the Tracker plane, of a set of seven point-like sources 6 mm far from each other, inside the body of a simulated mouse.

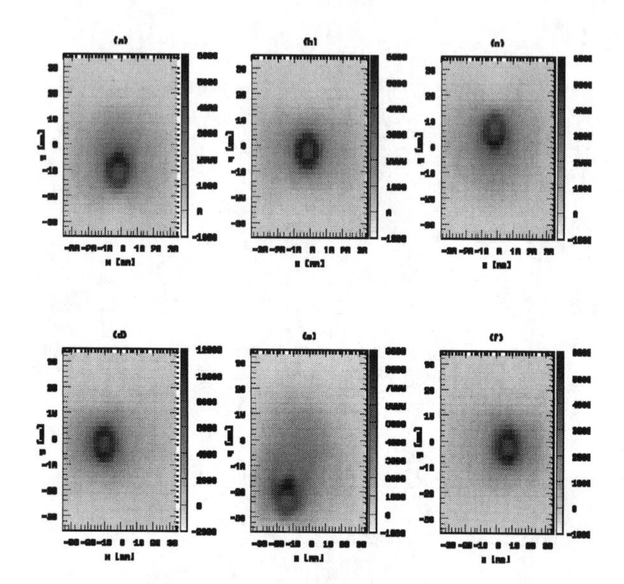

FIGURE 6. Projection along different directions, on the Tracker plane, of a circumference-like source of 12 mm diameter, inside the body of a simulated mouse. Fig 6 (b) shows the orthogonal projection of the source. The center of the detector has a better resolution as expected, but even on the border the circumference is still well recognizable.

RESULTS AND CONCLUSIONS

With this algorithm we were able to obtain a resolution of 6 mm FWHM on the Tracker plane. This resolution represents a lower limit for a CC, since it is just due to the Doppler Broadening effect and the scattering of gamma-rays inside the subject.

There are still margins of improvement in the root finding algorithm, taking into account and solving the following issue: the continuous nature of the Newton integration method is not compatible with the "quantized" nature of the data "binning", and brings to non-uniform distributions of the interaction points along the line of zeroes of each event.

At this point of the analysis, one is able to generate images of the radio-active sources inside the body of a mouse from different projections, and with these images define a "sinogram" of the source and perform a 3D tomographic reconstruction of the same. These further developments are not dealt with here.

REFERENCES

1. R. Pani, et al., *Nucl, Instr. And Meth. A* 571 (2007) 187.
2. R. Pani, et al., *Nucl, Instr. And Meth. A* 571 (2007) 268.
3. R. Pani, et al., *Nucl, Instr. And Meth. A* 571A (2007) 475.
4. R. Pani, et al., *AIP, CP* 1099 (2009) 488.
5. L. Melendez-Alafort, et al., *Nucl. Instr. And Meth. A* (2007) 484.
6. G, Moschini, et al., "The Scintirad Project". National Laboratories of Legnaro, Italy, INFN-LNL-221, 2007.
7. P. Rossi, et al., *Nucl Instr. And Meth. A* (2010), doi:10.1016/j.nima.1010.07.018
8. A. Vacchi, et al., *Nucl Instr. And Meth. A* 306 (1991) 187.
9. A. Rashecsky, et al., *Nucl Instr. And Meth. A* 485 (2002) 54.
10. C. E. Odonez, et al., Nuclear Science Symposium IEEE, 2 (1997) 1361-1365.
11. L. C. Parra, *Nuclear Science, IEEE Transactions on*, 47 (2000) 1543-1550.
12. D. Xu, et al., *Nuclear Science, IEEE Transactions on*, 53 (2006) 2787-2796.
13. GNU Scientific Library, Reference Manual. http://www.gnu.org/software/gsl/

Ion Beam Analysis Of Silicon-Based Surfaces And Correlation With Surface Energy Measurements

Qian Xing[a], N. Herbots[a], M. Hart[a], J. D. Bradley[a], B. J. Wilkens[a], D. A. Sell[a], Clive H. Sell[b], Henry Mark Kwong Jr[b]., R. J. Culbertson[a], S. D. Whaley[a]

[a]*Department of Physics, Arizona State University, Tempe, AZ 85287-1504*
[b]*Associated Retina Consultants, 7600 N 15th Street, Suite 155, Phoenix, AZ 85020*

Abstract. The water affinity of Si-based surfaces is quantified by contact angle measurement and surface free energy to explain hydrophobic or hydrophilic behavior of silicone, silicates, and silicon surfaces. Surface defects such as dangling bonds, surface free energy including Lewis acid-base and Lifshitz-van der Waals components are discussed. Water nucleation and condensation is further explained by surface topography. Tapping mode atomic force microscopy (TMAFM) provides statistical analysis of the topography of these Si-based surfaces. The correlation of the above two characteristics describes the behavior of water condensation at Si-based surfaces. Surface root mean square roughness increasing from several Å to several nm is found to provide nucleation sites that expedite water condensation visibly for silica and silicone. Hydrophilic surfaces have a condensation pattern that forms puddles of water while hydrophobic surfaces form water beads. Polymer adsorption on these surfaces alters the water affinity as well as the surface topography, and therefore controls condensation on Si-based surfaces including silicone intraocular lens (IOL). The polymer film is characterized by Rutherford backscattering spectrometry (RBS) in conjunction with 4.265 MeV $^{12}C(\alpha, \alpha)^{12}C$, 3.045 MeV $^{16}O(\alpha, \alpha)^{16}O$ nuclear resonance scattering (NRS), and 2.8 MeV elastic recoil detection (ERD) of hydrogen for high resolution composition and areal density measurements. The areal density of hydroxypropyl methylcellulose (HPMC) film ranges from 10^{18} atom/cm^2 to 10^{19} atom/cm^2 gives the silica or silicone surface a roughness of several Å and a wavelength of 0.16 ± 0.02 μm, and prevents fogging by forming a complete wetting layer during water condensation.

Keywords: ERD, NRS, RBS, AFM, polymer film, silicone, silica, Si(100), contact angle, surface free energy, surface topography, intraocular implant.
PACS: 61.05.Np, 82.80.Yc, 68.37.Ps, 87.64.Dz, 68.55.am, 68.35.Md, 68.55.jd, 68.55.Nq, 82.30.Rs

INTRODUCTION

Water affinity and condensation behavior of surfaces are of interdisciplinary interest across physics, material science, engineering, and medical applications[1-4]. Defects such as dangling bonds and contaminants render the Si(100) surface hydrophilic. One can attribute this behavior to the surface free energy and intermolecular interaction between water molecules and solid surfaces[5,6]. Cleaning, etching, and passivation of crystalline Si result in a lower surface energy, leading to a hydrophobic behavior of the surface.

Contact angle and surface tension are used to quantify water affinity utilizing Young's equation combined with Van Oss theory,

$$\gamma_{SL} = \gamma_S - \gamma_L \cos\theta \qquad (1)$$

$$\gamma = \gamma^{LW} + 2\sqrt{\gamma^+\gamma^-} \qquad (2)$$

the modified Young-Dupré equation thus becomes

$$(1+\cos\theta)\gamma_L = 2(\sqrt{\gamma_S^{LW}\gamma_L^{LW}} + \sqrt{\gamma_S^+\gamma_L^-} + \sqrt{\gamma_S^-\gamma_L^+}) \qquad (3)$$

with γ_L = liquid's surface tension, γ_S = solid's surface tension, γ_{SL} = surface tension at solid-liquid interface, γ^+ and γ^- = Lewis acid-base component of surface tension, γ^{LW} = Lifshitz-van der Waals component of surface tension, θ = contact angle made at the liquid/solid interface[7]. Large surface tension components or surface free energy result in the contact angle being small and surface being hydrophilic, while small surface tension results in the surface being hydrophobic. Although quartz silica and silicone surface free energy measurements have been conducted elsewhere[6], the fused quartz and medical silicone samples used in this paper need to be quantified and compared.

Application of Accelerators in Research and Industry
AIP Conf. Proc. 1336, 201-207 (2011); doi: 10.1063/1.3586089

The surface topography is also known to affect the condensation/nucleation behavior of water. The Kelvin equation,

$$\ln\left(\frac{p}{p_e}\right) = \frac{2\gamma V_M}{rRT} \qquad (4)$$

where p = actual vapor pressure, p_e = equilibrium vapor pressure, γ = surface tension of liquid, V_M = molar volume of liquid, r = radius of the droplet, R = gas constant, and T = temperature shows the fact that the curvature of the water droplet affects condensation behavior. With the surface being rough due to capillary type features and the water droplet curvature $\frac{1}{r}$ being small (implying "large" r), water nucleation and condensation becomes easier. Conversely, water droplets would be more difficult to nucleate and condense as the surface becomes smoother and water droplet curvature becomes larger. Thus a change in surface topography results in changing of the water condensation behavior. However, experimentally applying the above concept to the surface of silica, silicone, and HPMC polymer coated surfaces given surface roughness parameters in the nm to µm range needs to be investigated.

Medical issues such as water condensation at the liquid/air interface of intraocular implants during vitro-retinal surgery after cataract extraction and intraocular lens (IOL) implantation[1-3] provided the motivation for this study of Si-based surface water condensation behavior.

METERIALS AND METHODS

Materials And Preparations

Numerous experiments were conducted to carry out this research. De-ionized (DI) water is of 2 MΩ cm resistivity unless otherwise noted. Fused silica wafers are from Medtronic, and were cleaned via ultrasound agitation for 10 minutes in DI water of 18 MΩ cm resistivity. The silicone lenses used were Bausch & Lomb HD-500 intraocular lens (IOL). HPMC ($C_{32}H_{60}O_{19}$) CAS-9004-25-3, was 86 kDa molecular weight from Sigma-Aldrich and is hydrated using DI water.

Both the silica wafers and the silicone lenses are coated with HPMC film at room temperature, soaked in water-hydrated HPMC for 2 hours at various concentrations (from 0.20% w.t. to 1.00% w.t), then air dried under class 10K ventilation hood for a minimum of 24 hours.

Contact Angle And Roughness Measurements

The Sessile drop method[8] is used to conduct the contact angle measurement. The liquid droplets range in sizes from 0.4 µL ~ 5.0 µL and are delivered using a syringe with a 23 gauge cannula. The contact angle is measured using computer fitting techniques on the digital images. Both droplet size and contact angle are computed via computer processing.

Roughness and length-scale of roughness of surfaces were determined by tapping mode atomic force microscopy (TMAFM, Agilent). The AFM was operated in air with a silicon tip in AC mode. Gwyddion was the software used to produce images and roughness parameters: R_q = root mean square roughness, λ_a = average wavelength of the profile, and Δ_a = average absolute slope.

Rutherford Backscattering Spectroscopy

Areal density was used to analyze the HPMC film using He^{++} Rutherford backscattering spectrometry (RBS). RBS has been used to analyze polymer and substrate profiles including diffusion profiles, the aging process of organic photovoltaic cells, and polymers used for medical purposes[9-11]. Energy loss of ions can be used to determine the film thickness[12]. Both the energy loss due to the in-path of incident α particles and the out-path of the scattered α particles need to be considered. Since Si exists in the substrate only and not in the film, one can measure the energy shift of the Si edge to determine the energy loss,

$$\Delta E = kE_0 - E \qquad (5)$$

with ΔE = energy loss, k = kinematic factor, E_0 = incident ion energy, E = ion energy at the detector. Therefore to determine the areal density of the film,

$$Nt = \frac{kE_0 - E}{k\frac{1}{|\cos\theta_1|}(\varepsilon_{E_0}) + \frac{1}{|\cos\theta_2|}(\varepsilon_{kE_0})} \qquad (6)$$

where using Bragg's rule to compute the energy stopping factor,

$$\varepsilon_{C_{32}H_{60}O_{19}} \approx \frac{32\varepsilon_C + 60\varepsilon_H + 19\varepsilon_O}{32+60+19} \qquad (7)$$

with θ_1 = incident angle to sample normal and θ_2 = detector angle to sample normal. ε = the stopping cross section of alpha particles in the HPMC. An incident angle of 0° (normal to the surface) was used for films prepared with of 0.33% - 1.00% w.t. HPMC in water, while 65° incident angle was chosen for films prepared with < 0.33% w.t. HPMC in water. The incident angles were chosen so that the silicon edge of the RBS spectra would be exposed for thicker films, while being able to enhance the energy loss resolution for thinner films. Detector angle is 170° and incident energy is 2 MeV with the beam current ~20 nA.

HPMC consists of lighter C and O atoms which are typically difficult to detect on heavier substrates with a high degree of resolution. By implementing 4.265 MeV $^{12}C(\alpha, \alpha)^{12}C$ and 3.045 MeV $^{16}O(\alpha, \alpha)^{16}O$ nuclear resonance scattering (NRS), one can determine the carbon versus oxygen composition at high resolution near the surface. Again in these measurements, the incident beam is normal to the sample; detector angle is 170°, and beam current is ~20 nA.

The sample is separately measured for the height of the carbon and oxygen signals at 4.265 MeV and 3.045 MeV respectively. To maintain consistency, the Si signal is used to normalize all spectra,

$$\frac{N_C t}{N_O t} = \frac{\dfrac{\sigma_{Si}|_{E=C-Resonance}}{\sigma_C|_{E=C-Resonance}} \cdot \dfrac{H_C}{H_{Si}}\Bigg|_{E=C-Resonance}}{\dfrac{\sigma_{Si}|_{E=O-Resonance}}{\sigma_O|_{E=O-Resonance}} \cdot \dfrac{H_O}{H_{Si}}\Bigg|_{E=O-Resonance}} \quad (8)$$

with the subscripts of Si, O, and C representing silicon, oxygen, and carbon respectively, $\frac{H_C}{H_{Si}}$ and $\frac{H_O}{H_{Si}}$ = normalized yield over Si signal, σ_C and σ_O = scattering cross section at resonance, $N_C t$ and $N_O t$ = areal density of carbon and oxygen in HPMC film. The carbon and oxygen signal heights, respectively, were extracted from the RBS spectra by performing a linear fit to the background silicon signal and then subtracting the fitted background.

Elastic recoil detection (ERD) of the hydrogen atom is conducted to provide a comparative method to the above areal density measurement, as well as determining the hydrogen composition ratio. Incident α particles at 2.8 MeV with a 10.6 μm mylar filter at the detector are used. Projectile angle is 75° in reference to the sample normal and elastic recoil detector angle is 30°. Once again the beam current is kept at ~20 nA. RUMP simulation is then implemented to determine the areal density (Nt) of the HPMC film and the relative hydrogen composition compared to oxygen and carbon.

Ion beam damage in polymer films is a well known and impeding issue[13]. Therefore it was necessary to construct damage curves, which can then be used to extrapolate the effective non-damaged yield in the subsequent areal density measurement using regression modeling[14].

Sample charging due to the ion beam incidence is observed for both silica wafer and silicone lens samples. Grounding with an aluminum foil wrapping with a 4 mm ~ 10 mm diameter hole minimizes the charging effect; enabling consistent, reproducible data collection. However, the charging effect still was significant enough to require an offset in terminal voltage to compensate for the required resonance energy; and hence this offset was introduced after adding the conductive wrapping. Resonance yield was not stable at the fixed carbon resonance of 4.265 MeV and oxygen resonance of 3.045 MeV. Therefore, the scattering yield was maximized around 4.265 MeV and 3.045 MeV by adjusting the offset due to the charging effect to obtain the resonance energy for collection of the spectra. The experiment was conducted in vacuum of 10^{-7} torr ~ 10^{-6} torr.

RESULTS AND DISCUSSION

Condensation Behavior Versus Surface Free Energy And Surface Topography

The contact angle and related surface free energy components for silica and silicone samples are shown in Table 1. Surface free energy of silica is greater than that of silicone, particularly the Lewis basic component γ^-. This results in silica's hydrophilic behavior versus silicone's hydrophobic behavior, which is shown by their respective contact angles.

Surface topography images by TMAFM are shown in Figures 1a and 1b. Table 2 lists the roughness parameters of the respective samples. Notably, the roughness of the silicone is 9 times greater than that of the silica and the slope of features in silicone is about 7 times greater. However, the topographical wavelength of both samples is in the same range. Therefore, silicone provides more nano-scale capillary features.

TABLE 1. Surface Energies of Silicone and Silica Correspond to the Hydrophobic and Hydrophilic Behavior of the Respective Surfaces.

Type	γ^{LW} (mJ/m^2)	γ^+ (mJ/m^2)	γ^- (mJ/m^2)	Contact Angle with Water θ°
PDMS Silicone	19.4[15]	0.8[15] (γ^{polar})		107.2[15] 104.7 ± 1.4 (from this work)
Silica	41.3[16]	2.21[16]	35.68[16]	30.77 (calculated using Equation (3)) 32.2 ± 1.9 (from this work)

TABLE 2. Roughness Parameters via AFM.

Surface Type	R_q (nm)	λ_a (μm)	Δ_a
Silicone	3.7(2)	0.28(3)	0.067(4)
Quarts	0.42(5)	0.22(3)	0.0094(8)
HPMC 1% on Silicone	0.31(1)	0.16(1)	0.0098(4)
HPMC 1% on Silica	0.192(9)	0.16(2)	0.0064(7)
HPMC 0.2% on Silica	0.34(4)	0.16(2)	0.0110(7)

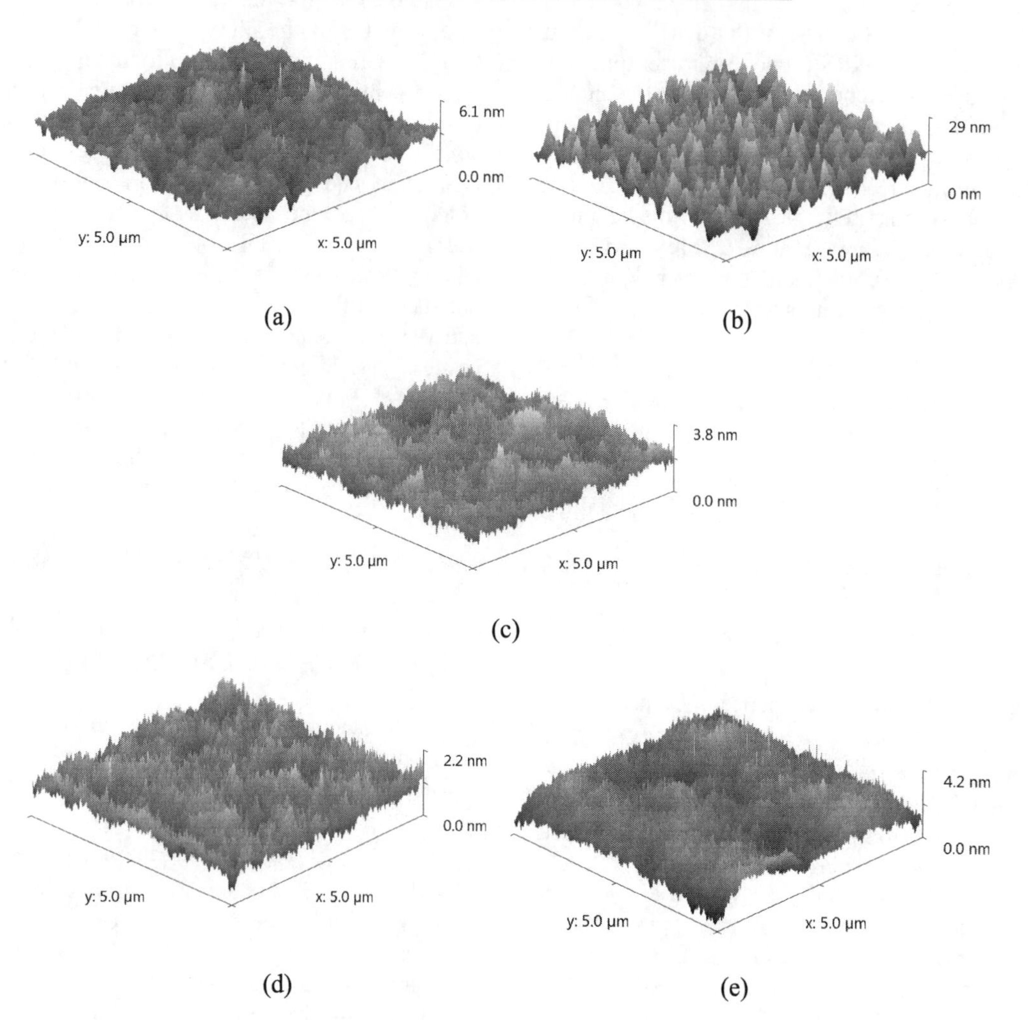

FIGURE 1. AFM 5 x 5μm images of (a) silica, (b) silicone, (c) HPMC 1% on silicone (d) HPMC 1% on silica, and (e) HPMC 0.2% on silica.

From Tables 1 and 2 we can explain the differences in the condensation behavior that silicone and silica experience which further show how the morphology and water affinity affect condensation behavior. Silicone is much rougher than silica, and hence one can imagine that each peak-valley-peak as a capillary-like structure. This allows for almost immediate condensation as seen in Figure 2c. However, because the roughness effectively isolates each condensed water drop, the total condensation area remains finely fogged from a macroscopic view, as seen in Figure 2d. Since the surface free energy is relatively low compared to the silica, the condensed water drops tend to form water "beads".

The condensation behavior on the silica is quite different, as seen in Figures 2a and 2b. Because of its smoothness, it takes much longer than the silicone for water droplets to nucleate on its surface. However, once these droplets are nucleated, due to the high surface free energy, the resulting condensation then tends to coalesce faster than that on the silicone. As a result, instead of a fine mist like fog as seen on the silicone, the condensation rapidly takes on the form of "puddles" of water spaced over the silica.

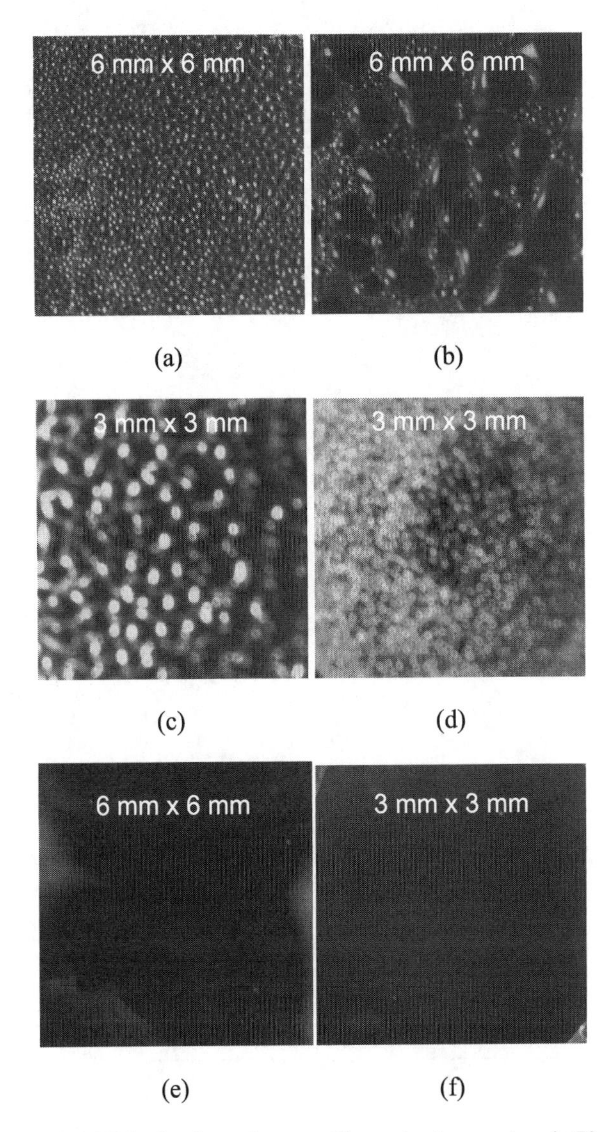

FIGURE 2. Condensation test. The water temperature is 70 °C. (a) 30 seconds after the silica is placed over the hot water. (b) 90 seconds after silica placement. The nucleated drops quickly coalesce into larger droplets. (c) 3 seconds after silicone is placed over the hot water. (d) 90 seconds after silicone placement. The nucleated drops tend to remain in place and coalescence is much slower than silica. (e) HPMC 1% coated silica over the hot water. (f) HPMC 1% coated silicone over the hot water. Complete wetting layer forms evenly.

HPMC polymer coating was used to alter the condensation behavior of silica and silicone. TMAFM was used to determine how the HPMC coating modified the surface topography of the respective silica and silicone substrate as shown in Figures 1c, 1d, and 1e. Roughness parameters in Table 2 demonstrate the wavelength was reduced significantly. HPMC is a water hydratable polymer which forms a water cage around its polymer chain via hydrogen bonding and has a high level of water affinity[17]. The condensation behavior rapidly forms a complete wetting layer as shown in Figures 2e and 2f.

Polymer Film Composition And Areal Density Measurement

HPMC polymer film was characterized by areal density measurements via ion beam energy loss in conjunction with the respective damage curve extrapolation as shown in Figure 3. The HPMC areal density was obtained by measuring the energy shift as defined in Equation (6) above, with various incident α particle flux at 2 MeV to obtain the damage curve. Then the damage curve is fitted using exponential regression modeling. The concept and implementation of the damage curve was necessary since even small flux of α particle causes a reduction in polymer film areal density measurement. Using damage curves allows for the use of IBA as an analysis tool on films that are significantly damaged in the measurement process. More details regarding the damage curve will be reported elsewhere[18]. These results have demonstrated that HPMC film with an areal density ranging from 10^{18} atom/cm^2 ~ 10^{19} atom/cm^2 can effectively prevent fogging on the silica and silicone surfaces by forming a complete wetting layer to improve visual clarity. A film that is too thick will cause visual distortion, while too thin of a film may have a wide range of effects, from merely experiencing initial heavy fogging and then clearing, to not being effective at all due to complete fogging, or a combination of these effects.

The HPMC polymer film carbon and oxygen ratios near the film's surface is determined by 4.265 MeV $^{12}C(\alpha, \alpha)^{12}C$ and 3.045 MeV $^{16}O(\alpha, \alpha)^{16}O$ nuclear resonance scattering (NRS). Damage curves were used to compensate for the ion beam damage. Using the method specified in Equation (8) above, the resulting measured ratio gives,

$$\frac{N_C^{HPMC}}{N_O^{HPMC}} \approx 1.64 \pm 0.08 \qquad (9)$$

which is in excellent agreement with the stoichiometric ratio of $\frac{C_{32}}{O_{19}} \approx 1.68$.

2.8 MeV elastic recoil detection (ERD) of hydrogen with RUMP simulation fitting is shown in Figure 4a. The areal density is 11600 (10^{15} atom/cm^2), with increments of 100 (10^{15} atom/cm^2) for each simulation step. The areal density measured by ERD is in excellent agreement with the areal density as extracted by the Si signal energy loss method, which gave a range of 11500 ~ 12500 (10^{15} atom/cm^2) as an areal density in Figure 4b. The hydrogen composition to oxygen and carbon ratio was also determined to be

$N_C : N_O : N_H = 1.64 : 1.00 : 2.70$ with N_H step of 0.05, compared to the HPMC stoichiometric ratio of $1.68 : 1.00 : 3.16$. The hydrogen content is 15% less than the bulk HPMC stoichiometric ratio would suggest. This discrepancy is suspected to be from the loss of hydrogen atom due to the ion beam damage; however, it is not fully understood at this time and bears further investigation.

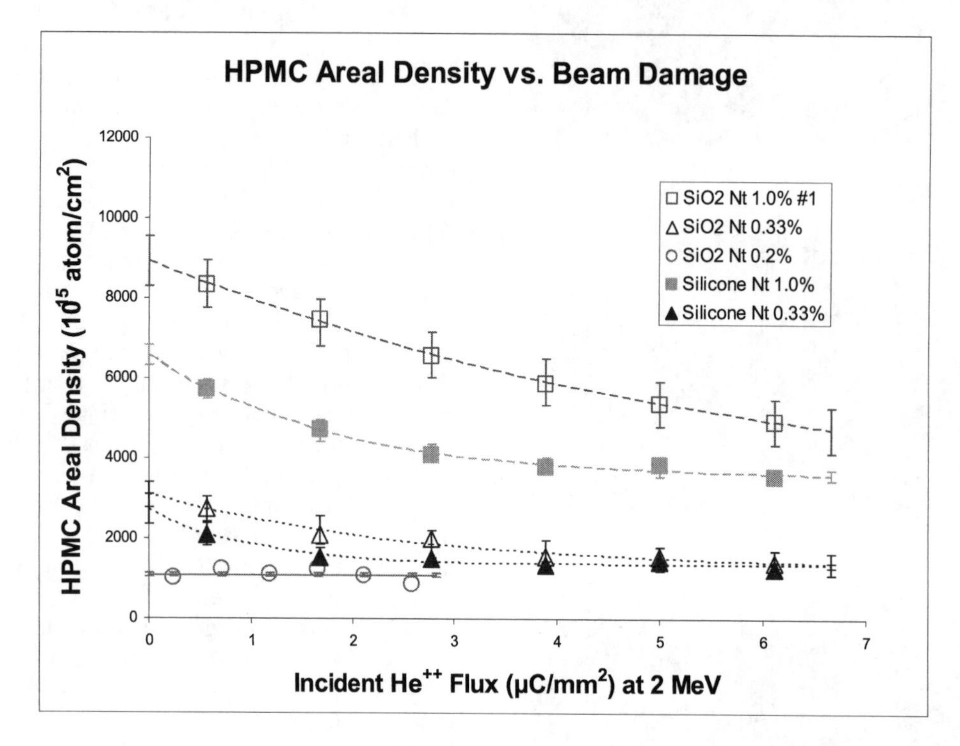

FIGURE 3. Damage curves conducted on SiO_2 (silica) and silicone substrates shown with polymer concentrations from 0.2% to 1.0%.

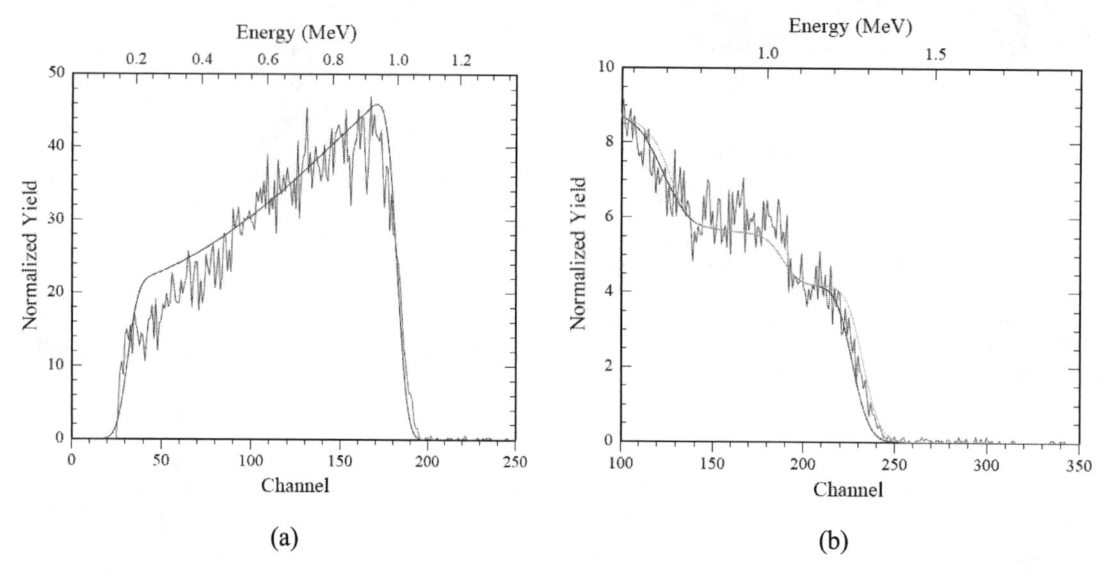

FIGURE 4. (a) ERD of HPMC film on silica substrate, with RUMP simulation. Simulation gives an areal density of 1.16×10^{19} atom/cm^2. (b) RUMP simulation executed twice on RBS spectrum of same HPMC film as in Figure 4a; areal density of simulation 1 (upper curve) = 1.15×10^{19} atom/cm^2, areal density of simulation 2 = 1.25×10^{19} atom/cm^2.

CONCLUSIONS

Characterization of a solid surface's water affinity via contact angle measurement using the Sessile Drop method, coupled with surface analysis using TMAFM, help explain the behavior of water condensation during liquid nucleation and condensation. Polymer adsorption on solid surfaces alters the water affinity, as well as the topography of the surface. Characterization of the polymer modified surface via IBA and TMAFM is correlated to the resulting change in behavior of water condensation and formation of wetting layer on the HPMC polymer surface.

The development of the damage curve was essential in enabling the use of IBA as a tool to characterize HPMC polymer, and allows for the compensation of the damage which the HPMC experiences during the IBA process. NRS and ERD, in combination with standard RBS, enable for the high resolution areal density measurement and composition determination of the HPMC film. TMAFM, along with the determination of surface free energy via the Sessile Drop method open up a new methodology to characterize and explain surface interactions in terms of surface free energy and surface topography.

Characteristics of the surface and the polymer film have been successfully quantified in this work. However, the condensation pattern is still qualitative in nature. Future work will involve quantifying this condensation pattern and relating it to surface characteristics, thus creating a truly predictive model for water condensation.

REFERENCES

1. R. Porter, Ophthalmology, 107 (4), 778-782 (2000).
2. Y. Yuriko, Japanese Journal of Ophthalmic Surgery, 18 (3), 383-386 (2005).
3. T. H. Levin, A New, Simple Technique to Prevent Water Condensation on Intraocular Lenses During Vitrectomy. NASA Technical Reports Server, Jet Propulsion Laboratory. NASA (2001).
4. M. Bjorkqvist, J. Paski, J. Solonen, and V. P. Lehto, IEEE Sensors Journal, 6 (3), 542-547 (2006).
5. H. Gouin, J. Phys. Chem. B, 102 (7), 1212-1218 (1998).
6. A. Carre, J. Adhesion Sci. Technol., 21 (10), 961-981 (2007).
7. R. S. Faibish, J. Colloid and Interface Sci., 256, 341-350 (2002).
8. E. Rame, J. Colloid and Interface Sci., 185 (1), 245-251 (1997).
9. P. J. Mills, P. F. Green, C. J. Palmstrom, J. W. Mayer, and E. J. Kramer, Appl. Phys. Lett., 45 (9), 957-959 (1984).
10. R. de Bettignies, J. Leroy, M. Firon, C. Sentein, S. Bailly and S. Guillerez, "Ageing Process in Organic Photovoltaic Solar Cell: Accelerated Lifetime and RBS Measurements", European Conference on Hybrid and Organic Solar Cells (2006).
11. M. Parizek, N. Kasalkova, L. Bacakova, P. Slepicka, V. Lisa, M. Blazkova, and V. Svorcik, Int. J. Mol. Sci., 10, 4352-4374 (2009).
12. J. W. Mayer and E. Rimini, Ion Beam Handbook for Material Analysis, Academic Press, New York (1977).
13. M. P. de Jong, L. J. van IJzendoorn, and M. J. A. de Voigt, Nucl. Instr. And Meth. In Phys. Research B.,161, 207-210 (2000).
14. J. M. Shaw, N. Herbots, Q. B. Hurst, D. Bradley, R. J. Culbertson, V. Atluri, and K. T. Queeney, J. Appl. Phys., 100, 104109 (2006).
15. Diversified Enterprises, Critical Surface Tension and Contact Angle with Water for Various Polymers (2009).
16. A. Zdziennicka, J. of Colloid and Interface Science, 340, 243-248 (2009).
17. M. T. Levitt, Proceedings of the National Academy of Sciences, 104 (30), 12336-12340 (2007).
18. Q. Xing, M. Hart, N. X. Herbots, J. D. Bradley, R. J. Culbertson, D. A. Sell, C. H. Sell, H. M. Kwong, Proceedings from the 17th International Conference on IBMM, (to be published).

The Characterization Of Secondary Electron Emitters For Use In Large Area Photo-Detectors

Slade J. Jokela†, Igor V. Veryovkin†, Alexander V. Zinovev†, Jeffrey W. Elam††, Qing Peng††, and Anil U. Mane††

†Materials Science Division, Argonne National Laboratory, 9700 S. Cass Ave., Building 200, Argonne Illinois 60439-4831

††Energy Systems, Argonne National Laboratory, 9700 S. Cass Ave., Argonne Illinois 60439

Abstract. The Large-Area Picosecond Photo-Detector Project is focused on the development of large-area systems to measure the time-of-arrival of relativistic particles with, ultimately, 1 pico-second resolution, and for signals typical of Positron-Emission Tomography (PET), a resolution of about 30 pico-seconds. Our contribution to this project is to help with identification and efficient fabrication of novel electron emitting materials with properties optimized for use in such detectors. We have assembled several techniques into a single ultra-high vacuum apparatus in order to enable characterization of both photocathode and secondary electron emission (SEE) materials. This apparatus will examine how photocathode quantum efficiency and SEE material electron yield correlate to surface chemical composition, state, and band structure. The techniques employed in this undertaking are X-ray photoelectron spectroscopy (XPS) for surface chemical composition, ultraviolet photoelectron spectroscopy (UPS) for the determination of band structure and surface work function, as well surface cleaning techniques such as argon-ion sputtering. To determine secondary electron emission yields and quantum efficiencies of detector materials, we use electron optics from a low energy electron diffraction (LEED) system whose set of hemispherical electrodes allows for efficient collection of secondary and photo electrons. As we gain a stronger insight into the details of mechanisms of electron emission from photocathodes and SEE materials, we will be able to lay a foundation for the larger collaborative effort to design the next generation of large-area photo-detectors. We present our preliminary results on the SEE materials from our as-yet completed characterization system.

Keywords: Secondary electron emission, Atomic Layer Deposition, Magnesium Oxide, Aluminum Oxide, XPS
PACS: 79.20.Hx, 79.20.La, 82.80.Pv

INTRODUCTION

The Large-Area Picosecond Photodetectors (LAPPD)[1] Project is a collaborative effort by several national labs, universities and small companies. Our three-year goal is to develop commercializable, lower-cost large-area detector systems capable of measuring the time of arrival of relativistic particles with 1 picosecond resolution and capable of measuring signals typical of positron-emission tomography (PET) with 30 picosecond resolution.

The Surface Chemistry group at Argonne National Laboratory (ANL) has been tasked with characterizing the thin film candidate materials for electron amplification (MgO and Al_2O_3) in the micro-channel plates that are being constructed for the LAPPD project, determining optimal film thickness and providing feedback on techniques intended to maximize secondary electron emission.

Secondary electron emission has been studied on a broad range of materials for many decades. The data from these experiments have been gathered in databases and reviewed extensively, showing that different studies on the same material rarely produce agreeing results. It has been known for quite some time that various differences in experimental instrumentation and conditions, as well as surface composition and morphology all play a role in a material's emission of secondary electrons.

EXPERIMENTAL SETUP

In order to further our understanding of the major contributing factors we are building at ANL an experimental apparatus for all-round characterization of emissive properties that includes Ultraviolet Photoelectron Spectroscopy (UPS), X-ray Photoelectron Spectroscopy (XPS), Ar-ion sputtering

Application of Accelerators in Research and Industry
AIP Conf. Proc. 1336, 208-212 (2011); doi: 10.1063/1.3586090

for cleaning and film damage studies, and Secondary Electron Yield (SEY) measurements all in one Ultra-High Vacuum (UHV) chamber. While each of these techniques have been used before in conjunction with SEY characterization, to our knowledge, none have included them all in one UHV system, a necessity if one has to prevent changes in surface composition between measurements.

The surface composition of our samples is studied using XPS. Our XPS system implements a Mg K-alpha X-ray source (1253 eV) and a hemispherical electron energy analyzer (HA100 from VSW Scientific Instruments). The X-ray beam is neither collimated nor passed through a monochromator. However, the X-ray emission is narrow enough to obtain elemental composition as well as some chemical information from the samples. We intend to determine what correlations exist between surface composition and changes in band structure and surface work function using UPS. Our UPS system is comprised of a helium UV source and the aforementioned hemispherical analyzer. The primary operational mode of the UPS uses the He-I emission at 21.22eV (He-II emission is also possible). 5 keV Argon-ion sputtering, incident at approximately 45° to the sample surface, is used to remove surface contaminants and precursor molecules from the ALD process. While the removal of this material can be detected using XPS, it may also produce a change in the band structure, detectable by UPS. The SEY is measured using the electron gun from a Low Energy Electron Diffraction system (LEED, manufactured by Vacuum Generators) for a continuous-beam of fixed energy electrons, usually 950 eV for these experiments. Pulsed electron gun operation will be implemented in future studies. The electron beam had a diameter of about 1 mm with currents between 0.03 and 2.0 μA. The kinetic energy of the electrons is varied by applying a negative potential to the sample using a Keithley Source Meter instrument (Model 2410), which also samples the electrical current flow. The initial beam current, I_{beam}, is sampled by applying a positive 1100 Volt bias to the sample, preventing all secondary electrons from escaping the sample. We then vary the sample voltage from -950V to 0 V in one Volt increments, measuring the current flow at every point. The gain, γ, is then calculated using the following equation

$$\gamma = \frac{I_{collector}}{I_{beam}} = 1 - \frac{I_{sample}}{I_{beam}}, \qquad (1)$$

where instead of using an external collector to collect the secondary electrons, we simply measure the current flowing from the voltage source used to bias the sample, I_{sample}.

SAMPLE PREPARATION

This report details the findings on MgO and Al_2O_3 films fabricated byAtomic Layer Deposition (ALD)[2]. These samples were deposited on boron-doped conductive-Si substrates and were prepared by the Energy Systems Division of ANL. ALD is ideally suited for the deposition of a conformal secondary emissive layer in the pores of micro-channel plates. More detail on the ALD technique can be obtained at the LAPPD Project website.[1] The samples used in the study of the effects of surface composition consisted of an Al_2O_3 film of thickness 113Å and an MgO film of thickness 290 Å. For our initial experiment studying how film thickness affects SE emission, a series of MgO films were created with thicknesses of 20, 30, 40, 55, 77, 110, and 200 Å. These samples were created together in the same growth run, limiting the chances of differences in sample composition and surface contamination due to ALD precursor molecules.

RESULTS

The Electron Dose Effect

Monitoring of the secondary electron yield as a function of primary electron energy showed a previously known effect that has been called the electron-dose effect.[3] This effect is characterized by what is usually a decrease in secondary electron yield as a function of electron dose. We have monitored this effect in both Al_2O_3 and MgO. However, the results are drastically different for each material, showing the expected decrease in emission with dose for Al_2O_3 (Fig. 1) but showing an increase in emission with dose for MgO (Fig. 2). Exponential decay fits show that the SEY approaches 3.2 for Al_2O_3 and 9.0 for MgO.

During these experiments, the electron-beam current steadily increased from 0.05 to 0.15 μA. This change in current was slow enough that it did not affect the gathering of individual SEY curves, but over the course of 1.5 hours that yield data was collected, the current increase prevents us from analyzing current density effects. Future improvements to the system hardware as well as startup procedures will be made to stabilize beam current.

Prior experiments have shown that surface contaminants, such as carbon,[4] can decrease electron yield. While these samples have been exposed to air between growth and characterization, allowing for CO_2 to chemisorb and dissociate on the surface,[5] it has also been suggested that in some materials, carbon can be stimulated to move to the surface from deeper within the sample.[6]

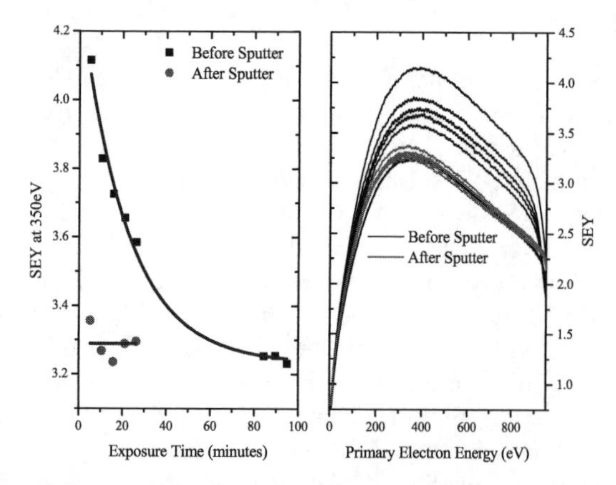

FIGURE 1. Secondary electron yield curves (right) for Al_2O_3 showing the electron-dose effect. The plot on the left shows emission at 350 eV as a function of exposure time. After Ar-ion sputtering, the emission stabilizes, presumably from the removal of a carbonaceous layer, observed in XPS spectra in Fig. 3.

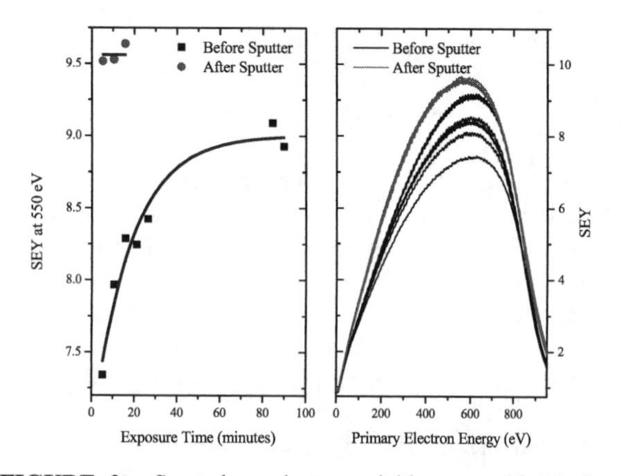

FIGURE 2. Secondary electron yield curves (right) for MgO showing the electron-dose effect (black). The plot on the left shows emission at 550 eV as a function of exposure time. After Ar-ion sputtering (red), the emission stabilizes, presumably from the removal of a carbonaceous layer, observed in XPS spectra in Fig. 4. Sample charging clearly present for higher primary electron energies, where the applied sample bias approaches 0V.

In our experiments, both Al_2O_3 and MgO films show carbon contamination in their XPS spectra (Figs. 3 and 4, respectively). This carbon is most likely from ALD precursor molecules and atmospheric contamination. Additionally, MgO shows a second carbon and oxygen peak. To determine whether these species are responsible for the dose effect observed on these samples, we sputter cleaned them using the 5keV Ar-ion source. After sputter cleaning, the carbon peaks, as well as the extra oxygen peak for the MgO sample, were virtually eliminated. The simultaneous removal of the second carbon and oxygen peak on the MgO sample would indicate the presence of a C-O bond.

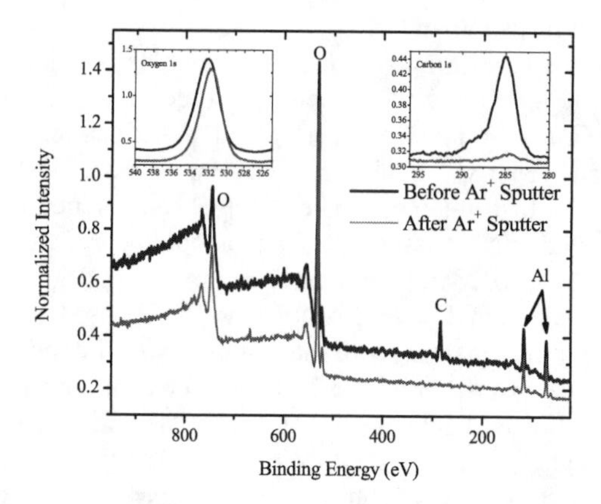

FIGURE 3. XPS spectra for Al_2O_3 before (blue) and after (red) Ar-ion sputtering. Spectra are normalized to the 532eV oxygen 1s peak height. The near-complete elimination of carbon appears to stabilize secondary electron emission.

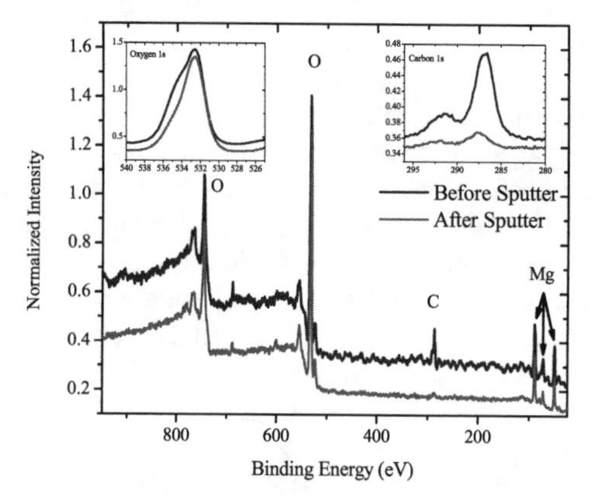

FIGURE 4. XPS spectra for MgO before (blue) and after (red) Ar-ion sputtering. Spectra are normalized to the 532eV oxygen 1s peak height. Of note is a second carbon and oxygen peak and their corresponding decrease in intensity after sputtering, indicating a C-O bond. This may be responsible for the difference in electron-dose effect between Al_2O_3 and MgO.

Observing the secondary electron emission in a different point on the sample after sputter cleaning revealed that in both cases, the emission was stabilized at their large-dose values, approximately 3.3 for Al_2O_3 and 9.6 for MgO.

The results indicate that the carbon surface contamination is responsible for the dose effect. Our results from samples prior to sputter cleaning seem to indicate that carbon was being removed via electron-stimulated desorption, confirmed by the stabilization seen after Ar-ion sputter cleaning. However, what still remains unclear is why the carbon compounds affect the samples differently, increasing emission in Al_2O_3 and decreasing the emission in MgO prior to their removal. It's unlikely that this can simply be explained by the emission characteristics of the carbon itself. It's more likely that the carbon somehow changes the effective work function of the surface, lowering it in the case of Al_2O_3 and increasing it in the case of MgO. This assertion is backed up by the presence of a C-O bond on the MgO surface.

Secondary Electron Emission Vs. Film Thickness

We are also interested in determining the optimal film thickness for our secondary electron emission materials. So far, we have only tested MgO samples for this experiment. Prior work has been done on samples thicker than 100Å.[7, 8] Their results show that electron-induced conduction from the excitation of electrons into the conduction band during secondary emission results in high emission for MgO. However, once the films reach thicknesses greater than the penetration depth of the primary electrons, the emission begins to decrease due to the resistance of the additional MgO film. In this case, sample charging occurs due to the lack of adequate compensation for the secondary electrons emitted from the material.

In our experiment we probe the range from 20 to 200 Å. Each sample's secondary electron emission was monitored for 1.5 hours allowing the sample's emission to stabilize, accounting for the electron-dose effect. In the final comparison, only the final, stabilized SEY curve was used in the comparison between samples of different thickness. The results show quite clearly that sample charging is affecting the emission curves for higher electron energies and is more apparent for thicker samples. The charging is only noticeable for higher energy electrons due to the configuration of our system; at these energies the sample's bias potential is close to 0V, allowing the charge to pull the secondary electrons back into the sample.

From this, we have concluded that the optimal film thickness for maximum emission resides between 110 and 200 Å (Fig. 5). This is in agreement with other findings showing the maximum escape depth of secondary electrons in MgO to be approximately 180 Å.[9] Additionally, XPS spectra show that at 55 Å, the Si emission peak is no longer visible (Fig. 6). Any additional material will simply result in increased charging due to the increased resistance and decreased compensation for the emitted secondary electrons.

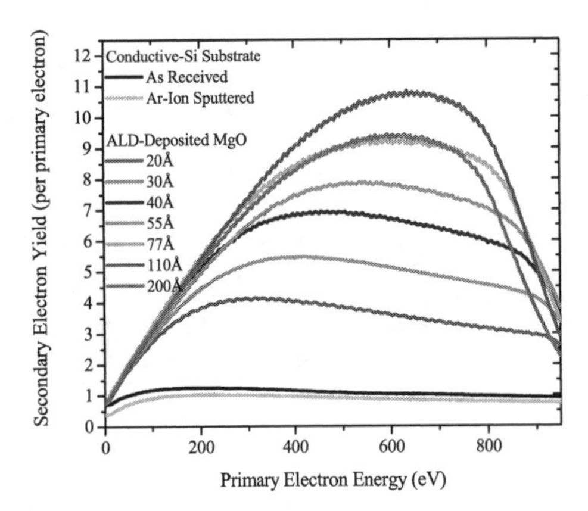

FIGURE 5. Secondary electron yield curves vs. MgO film thickness. The SEY increases until about 110Å where the emission begins to decrease.

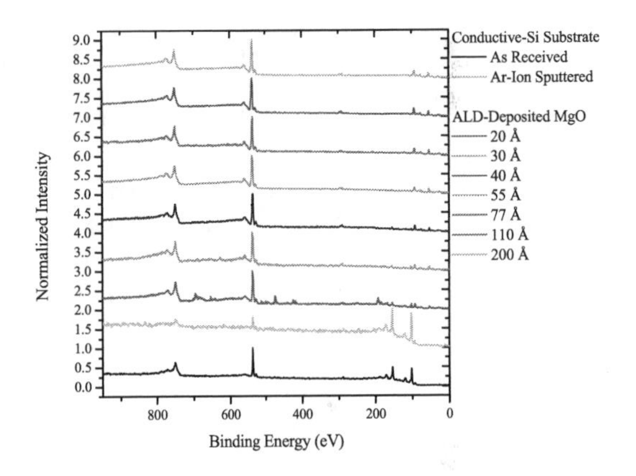

FIGURE 6. XPS Spectra corresponding to Fig. 5. The XPS spectra are normalized to the largest peak intensity (oxygen 1s, 532 eV, in all but the sputter cleaned Si sample). Of note are the Si peaks near 102 and 153 eV, which disappear with an MgO thickness of 55 Å.

CONCLUSIONS

The SEY's dependence on electron dose for both alumina and magnesia films, known as the electron-dose effect, appears to be due to surface contamination. The difference in trend, decreasing SEY for Al_2O_3 and increasing SEY for MgO, could be a result of changes in the effective work function or band structure of the sample surface. This dose effect is eliminated once the sample is sputter cleaned.

Of the film-thicknesses tested, the sample of 110 Å showed maximum secondary electron emission, in agreement with another study showing the maximum escape depth of secondary electrons to be approximately 180 Å,[9] others report it to be between 60 and 100 Å.[10]

Our future work will involve the continued enhancement of capabilities of our characterization system. We have plans to include pulsing operation of the electron beam to mitigate charging of the sample surface, providing us with more accurate SEY curves for insulating materials. Additional charge compensation techniques, such as increasing conductivity through sample heating, will also be explored. Furthermore, we have yet to fully implement the study of band structure using UPS, a crucial step in determining how surface contaminants affect the work function of the material. Finally, secondary neutral mass spectrometry (SNMS) is being considered for examination of molecules removed from the sample surfaces from electron stimulated desorption (ESD) as well as temperature programmed desorption (TPD). This will be used to confirm our findings that carbon is responsible for the electron-dose effect in our samples.

ACKNOWLEDGEMENTS

The submitted manuscript has been created by UChicago Argonne, LLC, Operator of Argonne National Laboratory ("Argonne"). Argonne, a U.S. Department of Energy Office of Science laboratory, is operated under Contract No. DE-AC02-06CH11357. The U.S. Government retains for itself, and others acting on its behalf, a paid-up nonexclusive, irrevocable worldwide license in said article to reproduce, prepare derivative works, distribute copies to the public, and perform publicly and display publicly, by or on behalf of the Government. This work was supported in part by American Recover and Relief Act (ARRA) funds.

REFERENCES

1. The Large-Area Photodetector Project website http://psec.uchicago.edu.
2. George, S.M., Ott, A.W., and Klaus, J.W., *J. Phys. Chem* **100**, 13121 (1996).
3. Kumar, P., Watts, C., Gilmore, M., and Schamiloglu, E., *IEEE Trans. Plasma Sci.* **37**, 1537 (2009).
4. Kirby, R., "Instrumental effects in secondary electron yield and energy distribution measurements" at *31st ICFA Advanced Beam Dynamics Workshop on Electron Cloud Effects*, 2004.
5. Ochs, D., Brause, M., Braun, B., Maus-Friedrichs, W., and Kempter, V., *Surf. Sci* **397**, 101 (1998).
6. Henrist, B., Hilleret, N., Scheuerlein, C., Taborelli, M., and Vorlaufer, G., in *The Proceedings of EPAC 2002, Paris, France*, 2002, pp. 2553.
7. Lee, J., Jeong, T., Yu, S., Jin, S., Heo, J., Yi, W., Jeon, D., and Kim, J.M., *Appl. Surf. Sci.* **174**, 62 (2001).
8. Meyza, X., Goeuriot, D., Guerret-Piécourt, Tréheux, D., and Fitting, H.-J., *J. Appl. Phys.* **94**, 5384 (2003).
9. Goldstein, B. and Dresner, J., *Surf. Sci* **71**, 15 (1978).
10. See, A. and Kim, W.S., *Appl. Phys. Letters* **81**, 1098 (2002).

Exploration Of Activity Measurements And Equilibrium Checks For Sediment Dating Using Thick-Window Germanium Detectors

Jacob A. Warner[a], Kathryn E. Fitzsimmons[b], Eva M. Reynolds[c], Laura G. Gladkis[a], Heiko Timmers[a]

[a]*School of Physical, Environmental and Mathematical Sciences, University of New South Wales at the Australian Defence Force Academy, Canberra ACT 2602, Australia*
[b]*Department of Human Evolution, Max Planck Institute for Evolutionary Anthropology, Deutscher Platz 6, D-04103 Leipzig, Germany*
[c]*Research School of Earth Sciences, Australian National University, Canberra ACT 0200, Australia*

Abstract. Activity measurements on sediment samples for trapped-charge geological dating using gamma-ray spectroscopy are an important verification of the field-site dose rate determination. Furthermore gamma-ray spectroscopy can check if the natural decay series are in secular equilibrium which is a crucial assumption in such dating. Typically the activities of leading members of the Thorium and Uranium decay series are measured, which requires Germanium detectors with thin windows and good energy resolution in order to effectively detect the associated low energy gamma-rays. Such equipment is not always readily available. The potential of conventional Germanium detectors with thick entrance window has been explored towards routine gamma-ray spectroscopy of sediment samples using higher energy gamma-rays. Alternative isotopes, such as Ac-228 and Pb-212 for the Thorium series, and Pa-234m, Ra-226 and Bi-214 for the Uranium series, have been measured in order to determine the mass-specific activity for the respective series and possibly provide a check of secular equilibrium. In addition to measurements of the K-40 activity, with the alternative approach, the activities of both decay series can be accurately determined. The secular equilibrium condition may be tested for the Thorium series. Measurement accuracy for Pa-234m is, however, not sufficient to permit also a reliable check of equilibrium for the Uranium series.

Keywords: Natural Decay Series, Gamma-Ray Detection, Trapped-Charge Dating, Luminescence Dating.
PACS: 07.85.Nc, 93.30.Fd, 93.85.Np

INTRODUCTION

A commonly used group of techniques for determining the age of geological sediment samples is trapped-charge dating, which includes the methods of optically stimulated luminescence (OSL), thermoluminescence (TL) and electron spin resonance (ESR). These techniques measure the total dose of ionizing radiation received by a sediment sample since burial. Combining this result with a measurement of the dose rate of ionizing radiation in the sediment then gives the age according to

$$Age = \frac{Dose}{Dose\ Rate} \qquad (1)$$

For optimal dating, the accuracy achievable with the TL, OSL and ESR techniques and that of the dose rate determination need to be similar. This is aided, if field recordings of total gamma-ray dose rates from portable sodium iodide detectors are verified by laboratory gamma-ray spectroscopy with Germanium detectors [1]. Such spectroscopy is particularly important if field recordings are impossible. The main contributions to the dose rate are from radioactive potassium (K-40; ~46%), see Figure 1(a), with a typical mass-specific activity of the order of 310 Bq/kg, and from the Thorium, Uranium and Actinium natural decay series (~53%) with mass-specific activities of the order of 41 Bq/kg, 37.3 Bq/kg and 1.7 Bq/kg, respectively [2].

The K-40 activity can be measured via the characteristic 1461 keV gamma-ray line. Since the half-life of the mother of each of the three decay series

Application of Accelerators in Research and Industry
AIP Conf. Proc. 1336, 213-218 (2011); doi: 10.1063/1.3586091

is very much larger than the half-lives of all other members in the series, the activities in each series are in secular equillibrium, i.e. the activities are effectively identical. The measurement of the mass-specific activity of one radioisotope in the series can thus determine the total activity of the series. This assumes, however, that the decay equilibrium is not disturbed by physio-chemical transport processes which may change the radioisotope concentrations. Such a disturbance can significantly distort the dating if the long-lived isotopes of Uranium and Thorium heading the two respective series, and detailed in Figure 1(b,c), are mobile. Due to the long half-lives of these isotopes disequilibrium may have then persisted over geologically relevant time-scales. Furthermore, element-specific transport processes may be ongoing at varying transport rates. In such open geological systems secular equillibrium may never be achieved.

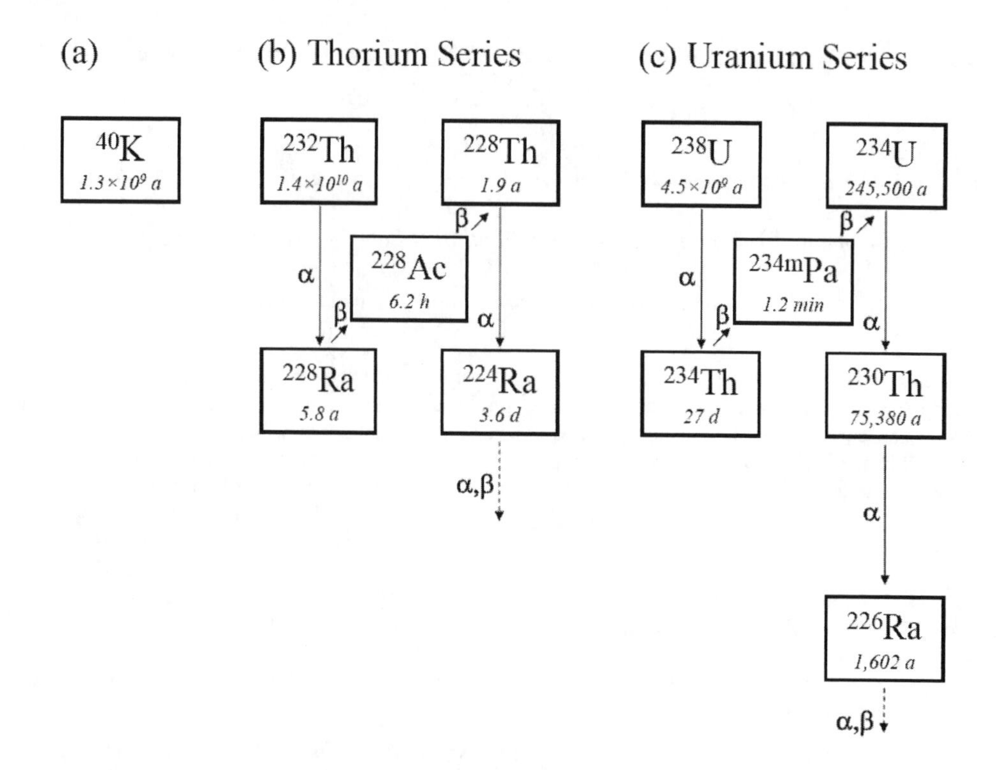

FIGURE 1. Overview of radioactive potassium and the leading members of the Thorium and Uranium decay series. Half-lives and decay modes are indicated. (a) Radioactive potassium. (b) Beginning of the Thorium series. (c) Beginning of the Uranium series.

Long-term disequilibria may be searched for by comparing the activities of isotopes at the beginnig of a decay series with the activities of isotopes towards its end. Olley *et al.* [3] showed for many Australian samples that deviations from the secular equillibrium for the long-lived isotopes at the beginning of the Uranium series are common for fluvial sediments. In contrast for the same samples disequilibria are uncommon for the Thorium series. This is partly because of the long half-lives of U-234 and Th-230 in the Uranium series and because the element Uranium can form water-soluble complexes. In many fluvial sediment samples activity differences of up to 50% have been observed for the U-238, U-234, Th-230, and Ra-226 isotopes heading this series [3].

For the reliable dating of sediments, routine checks for disequilibria at the beginning of the Uranium and Thorium series are thus advantageous, since affected samples may be deselected. Such tests can be achieved with laboratory gamma-ray spectroscopy. Several important gamma-ray lines have, however, very low energies. E.g. in the Uranium series the U-238 activity can be measured via a 16 keV line and the Th-230 activity is accessible via a 68 keV line. Lines at such low energies require a Germanium detector with a thin entrance window and good energy resolution. Thin window detectors, because they are designed for low energy gamma-ray spectroscopy, often have poor efficiency for detecting the high energy lines best suited for the determination of the activity of an equilibrated series. E.g. in the Uranium series Bi-214 may be effectively measured via a 1764 keV line and the K-40 activity is accessible via a 1461 keV line. Furthermore, specialized Germanium detectors with thin entrance windows are not readily available, whereas conventional Germanium detectors without entrance window and suitable lead shielding exist in many research labs.

In order to faciliate routine measurements of mass-specific activities in sediment samples and also to search for disequilibria, the potential of a conventional Germanium detector for such measurements has been explored. In order to widen the applicability of the results to non-ideal detection conditions, a Germanium detector with somewhat deteriorated energy resolution has been used in comparison with state-of-the-art gamma-ray spectroscopy.

Some of the samples studied in this work are from Pilot Creek, a small ephemeral channel located in the southeastern Australian highlands. This site is of interest since since the remains of terrestrial megafauna have been found, the timing and causes of their extinction being uncertain.

EXPERIMENTAL DETAILS

Two sets of sediment samples were investigated. Five samples were from aeolian desert dunes from the Strzelecki and Tirari Deserts in Australia [4]. Due to the type of sedimentation and its location, deviations from secular equilibrium in these samples are unlikely. Six further samples were collected from the fluvial sediments at Pilot Creek near Canberra, Australia. Since these samples have experienced water through-flow during their burial history, Uranium mobility, and consequently disequilibrium, is of potential concern for the reliability of dating at this site.

Following retrieval on site samples were dried at 121°C and then milled to a fine, consistent powder using a titanium mill. Approximately 80 g of the material was weighed, then compressed into standard containers forming approximately 10-15 mm thin disks to ensure reproducible detection geometry. The containers were sealed with an O-ring to avoid Radon escape. Sample geometry and thickness were chosen to maximize sample mass and minimize self-absorption within the sample. For gamma-ray spectroscopy the sample container was placed on top of the Aluminum can of a conventional, upright, cylindrical high purity Germanium detector (Ortec). Both, detector and sample were inside the bore of a cylindrical lead castle. The thickness of the low radiation lead shielding around and above the sample and detector was in excess of 100 mm. Underneath, detector and sample were shielded by the filled liquid nitrogen dewar of the Germanium detector. On the inside the lead bore was lined with a Cd/Cu shield to suppress neutrons from the lead and associated x-rays.

Due to previous applications in high radiation environments the energy resolution of the detector had deteriorated to a FWHM of about 10 keV for the Co-60 lines, thus simulating non-ideal measurement conditions, which may prevail in a routine analysis situation. Measurement times of 144 h per sample were deliberately chosen to be relatively long to maximize counting statistics. Detector operation was stable and detection dead time was insignificant.

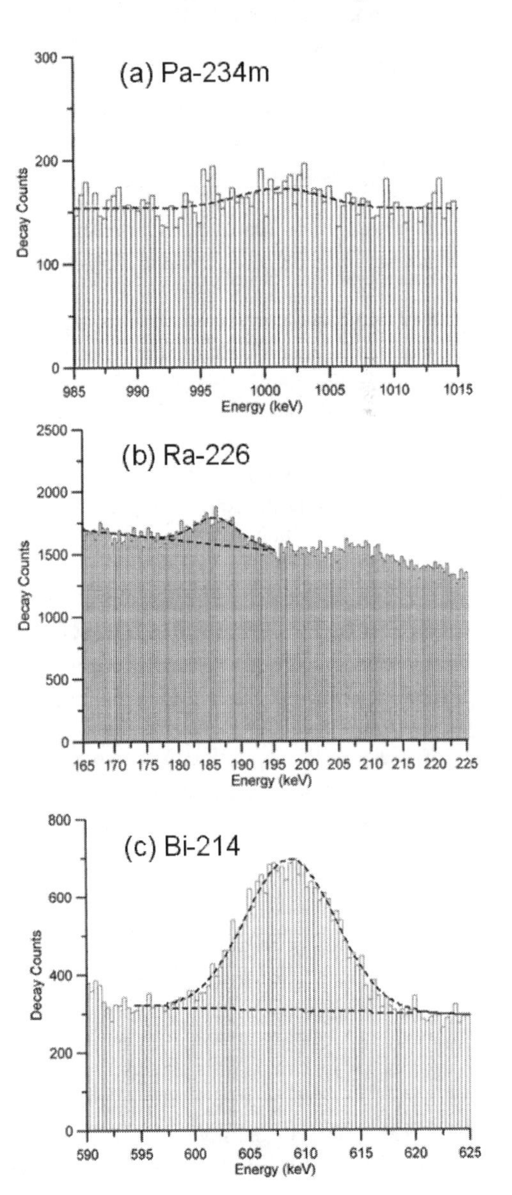

FIGURE 2. Gamma-ray lines in the Uranium series measured with a gamma-ray detector with a 10 KeV FWHM energy resolution for Co-60. Fits are also shown. (a) 1001 keV line associated with Pa-234m. (b) 186 keV line associated with Ra-226. (c) 609 keV line associated with Bi-214.

Peaks in the gamma-ray spectrum were fitted with Gaussians. K-40 was measured via the 1461 keV line. Due to its minor significance the Actinium series was not considered. The activity of the Thorium series was determined using the 239 keV (Pb-212) and the 911 keV and 969 keV lines (Ac-228), respectively. The activity of the Uranium series was quantified by measuring the lines at 1001 keV (Pa-234m), 186 keV (Ra-226), and 609 keV (Bi-214). The quality of data and the fitting of these three lines is shown in Fig. 2. In all cases the background activity for each line, measured without a sample, was subtracted. For the Pilot Creek samples the contribution of environmental background photons to the lines was typically 10% (K-40), 6% (Pb-212), 14% (Ac-228), 20% (Bi-214), 26% (Ra-226), 40% (Pa-234m), respectively. This emphasizes the importance of effective lead shielding as described above.

All eleven samples were independently measured by a commercial laboratory using a thin window detector, their own standard containers and measurement protocol with typical measurement times of 48 h. These reference data were used to calibrate the count rates measured with the resolution-deteriorated detector in terms of mass-specific activity (Bq/kg). In order to isolate the effect of detector resolution, the Ra-226 activity was also measured with a state-of-the-art conventional Germanium detector with thick entrance window which had an FWHM energy-resolution for Co-60 of better than 3 keV.

RESULTS AND DISCUSSION

The results obtained from the seven high energy gamma-ray lines considered have been compiled in Fig. 3. They are discussed below in comparison to the state-of-the-art reference measurements with a thin window detector.

K-40

The 1461 keV line characteristic of K-40 is well separated from any neighboring lines. The detected mass-specific count rate was calibrated by best matching the reference results using a multiplicative factor, see Figure 2 (a). Statistical and fitting uncertainties contribute about equally and the accuracy is better than 2%. The samples 'a-d' from aeolian sedimentation have significantly lower K-40 activities than the samples from the Pilot Creek fluvial sediment. This can be accounted for by the mineralogy of the aeolian samples which are dominantly quartz sand; compared with mica-bearing schistose sources for the fluvial Pilot Creek sediments.

Thorium Series

Figure 1(b) illustrates that any mobility affecting the two Thorium isotopes leading the Thorium series may be reflected in the activity of the Ac-228 isotope. Comparing the Ac-228 activity with activities further along the series can thus provide a test of secular equilibrium. Using a Germanium detector without thin window the Ac-228 activity may be measured via two associated gamma-ray lines at 911 keV and 969 keV. The detected mass-specific count rate was calibrated by best matching the reference results using a multiplicative factor. Figure 2(b) shows that the two Ac-228 measurements with the set-up used here reflect the reference data for Ra-228 and Th-228. The spread of the two Ac-228 data points indicates that the measurement accuracy compares well with the somewhat better accuracy of the Ra-228 and Th-228 reference measurements.

The activity towards the end of the Thorium series is readily obtained with a conventional Germanium detector by measuring the Pb-212 line at 239 keV, which is typically the strongest line in the spectrum. Superb energy resolution is not critical for this measurement. Measurement results are shown in Fig.2(c). Since a reference measurement was not available, the absolute activities for Pb-212 have been calibrated using a multiplicative factor against the Ra-228 and Th-228 reference results in Fig. 2(b). Since the aeolian deposits studied in this work (Samples 'a-e') are understood to be in secular equilibrium [4], this is justified. Indeed, following this calibration, all activities in panel (c) for Pb-212, representing the end of the Thorium series, agree with those shown in panel (b), representing the beginning of the series. Disequilibrium at the beginning of the Thorium series is thus not present for these samples. Combining activity measurements for Ac-228 and Pb-212 is thus well suited to determine the activity of the series, as well as reassuring the validity of the assumption of a constant dose-rate.

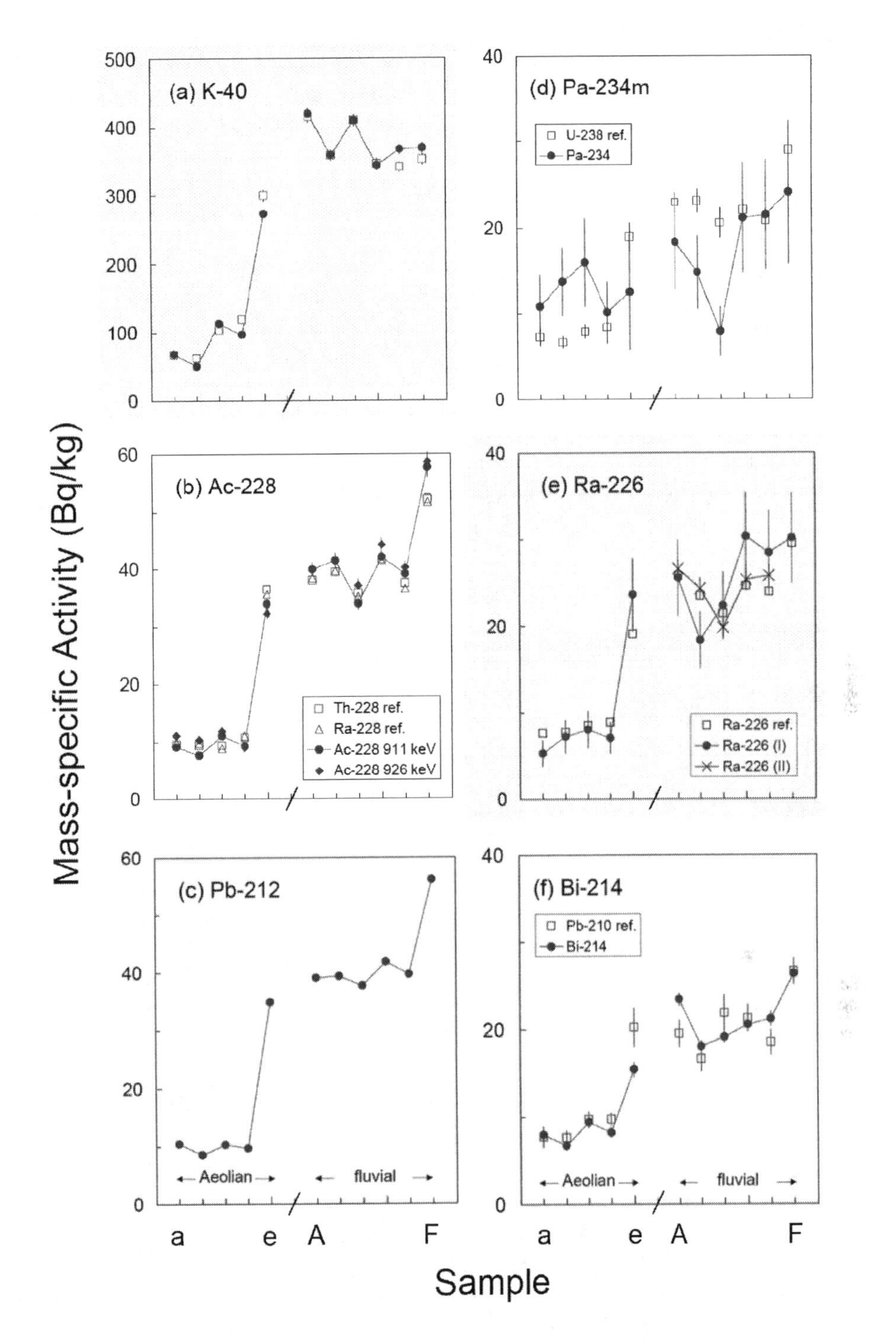

FIGURE 3. Mass-specific activities measured in this work with a resolution-deteriorated Germanium detector without thin entrance window (filled symbols). The data are compared with independent measurements by a commercial laboratory, which have been used as reference (ref.) and for calibration (open symbols). Five aeolian sediment samples (a-e) and six fluvial sediment samples (A-F) have been studied. Details are discussed in the text. (a) The results for K-40 agree within uncertainties. (b) Measurements for the 911 keV and the 926 keV lines associated with Ac-228. (c) Results for Pb-212. (d) Measurements for Pa-234m together with reference data for U-238. (e) The Ra-226 results are consistent with the reference data for the same isotope. The crosses indicate results from a re-measurement with improved energy resolution (f) Bi-214 measurements compared with Pb-210 reference results.

Uranium Series

Without using the conventional lines below 150 keV, the activity of the members of the Uranium series above Ra-226 may only be measured effectively via two gamma-ray lines which are associated with Pa-234m. Both, the 760 keV and 1001 keV lines have, however, very weak intensities [1]. The stronger one of the two lines at 1001 keV is typically 100 times weaker than the 1460 keV line associated with K-40. Figure 2 (a) shows that due to the long counting time of 144 h the 1001 keV line could be detected with typically 400 total counts above Compton background. The associated statistical uncertainty of about 5% is, however, outweighed by fitting uncertainties in excess of 20%. The experimental accuracy of the Pa-234m measurements shown in Fig. 3(d) is therefore only sufficient to hint at the pronounced difference in activity between samples 'a-d' and 'A-F'. The measured mass-specific count-rate could only be matched to the reference results for U-238, the grandmother of Pa-234m, by using in addition to a multiplicative factor the additive constant of 7 Bq/kg, implying that the detection sensitivity is of that order.

The Ra-226 activity is known to be often in disequilibrium with the activities of its supporting isotopes U-238 and U-234 [2]. The line at 186 keV is sufficiently high in energy to be detected without thin entrance window. However, the peak shown in Fig. 2(b) is contaminated by a 185 keV line associated with the Actinium series. Without separating the contamination this measurement is thus only representative of the Ra-226 activity, if the activity ratio of the Actinium and Uranium series is relatively constant, which is often the case. The peak was integrated as shown in Fig. 2(b) and mass-specific count rates were calibrated using in addition to a multiplicative factor an additive constant of 4 Bq/kg indicating that this is a lower sensitivity limit. The data in Fig. 3(e) agree within the relatively large error bars dominated by the fitting uncertainty with the reference results for the same isotope. Using an improvised lead castle and a Germanium detector with better energy resolution some of the samples were measured again, which is shown in Fig. 2(e) as Ra-226 (II). The better energy resolution reduced the fitting uncertainty and improved the agreement with the reference data.

Finally, the 609 keV line representative of Bi-214 was measured, see Figure 2(c). In Fig. 3(f) the Bi-214 data were calibrated with a multiplicative factor and an additive constant of 3 Bq/kg using the Pb-210 reference result also shown. As indicated by the error bars, the Pb-210 measurement, via the 47 keV line, has limited accuracy, so that the 609 keV line provides a good alternative for the determination of the mass-specific activity of the Uranium series. The results suggest that using the set-up explored the activity of the Uranium series can be determined with acceptable accuracy, however, equilibrium checks are only of limited assistance.

CONCLUSIONS

A constant dose rate is assumed for trapped-charge dating. Field measurements of that dose rate can be supported by laboratory gamma-ray spectroscopy with Germanium detectors. Furthermore, this allows for checks of the assumption of a constant dose rate, which may be affected by disequilibria in the natural decay series. This is particularly relevant for samples from water-lain sediments.

The use of higher energy gamma-ray lines for the determination of the mass-specific activities of the Thorium and Uranium series, as well as that of radioactive potassium, has been explored. A resolution-limited Germanium detector was used to simulate a less-than-ideal routine measurement situation. Furthermore the efficacy of such a set-up for providing routine tests of activity equilibration at the beginning of the Thorium and Uranium series has been evaluated.

With the gamma-ray lines selected and the detector used, the mass-specific activities of both decay series can be accurately determined in addition to that of K-40. The secular equilibrium condition may be tested for the Thorium series. Measurement accuracy for the Pa-234m and Ra-226 activities are, however, not sufficient to permit also reliable checks for the Uranium series, where disequilibrium is common in water-lain environments.

REFERENCES

1. H. L. Oczkowski, *Physica Scripta*, **64**, pp. 518-528, 2001.
2. M. J. Aitken, *Thermoluminescence Dating*, Academic Press, London, 1985.
3. J. M. Olley, A. Murray, R. G. Roberts, *Quarternary Science Reviews,* **15**, 1996, pp. 751-760.
4. K. E. Fitzsimmons *et al.*, *Quaternary Science Reviews* **26**, 2007, pp. 2598–2616.

Structural Modifications And Mechanical Degradation Of Ion Irradiated Glassy Polymer Carbon

Malek Abunaemeh[a,b], Mohamed Seif[c], Abdalla Elsamadicy[d], Claudiu Muntele[a] and Daryush Ila[a,b]

[a]Center for Irradiation of Materials, Alabama A&M University, Normal, AL 35762
[b]Physics Department, Alabama A&M University, Normal, AL 35762
[c]Mechanical Engineering Department, Alabama A&M University, Normal, AL 35762
[d]Physics Department, University of Alabama in Huntsville, Huntsville, AL 35899

Abstract. The TRISO fuel has been used in some of the Generation IV nuclear reactor designs. It consists of a fuel kernel of UOx coated with several layers of materials with different functions. Pyrolytic carbon (PyC) is one of the materials in the layers. In this study we investigate the possibility of using Glassy Polymeric Carbon (GPC) as an alternative to PyC. GPC is used for artificial heart valves, heat-exchangers, and other high-tech products developed for the space and medical industries. This lightweight material can maintain dimensional and chemical stability in adverse environment and very high temperatures (up to 3000°C). In this work, we are comparing the changes in physical and microstructure properties of GPC after exposure to irradiation fluence of 5 MeV Ag equivalent to a 1 displacement per atom (dpa) at samples prepared at 1000, 1500 and 2000°C. The GPC material is manufactured and tested at the Center for Irradiation Materials (CIM) at Alabama A&M University. Transmission electron microscopy (TEM) and Raman spectroscopy were used for analysis.

Keywords: Raman Spectroscopy, Transmission electron microscopy, Glassy polymeric carbon.
PACS: 68.37.Lp, 68.37Yz, 68.30-j

INTRODUCTION

The Tristructural-isotropic (TRISO) fuel particles were originally developed in Germany for high-temperature gas-cooled reactors [1]. TRISO is a type of micro fuel particle [2]. It consists of a fuel kernel composed of uranium oxide (UOX) in the center [2, 3], coated with four layers of three isotropic materials. These four layers are a porous buffer layer made of carbon, followed by a dense inner layer of pyrolytic carbon (PyC) [3], followed by a ceramic layer of SiC to retain fusion products at elevated temperatures and to give the TRISO particle more structural integrity, followed by a dense outer layer of PyC [2,3]. The TRISO fuel particle is designed not to crack due to the stresses from different processes (such as differential thermal expansion, gas pressure, or fission fragments entrapment) at temperatures beyond 1600°C. It can contain the fuel in the worst accident scenarios in a properly designed reactor. The pebbles or the graphite blocks containing the TRISO fuel is directly immersed in the cooling fluid that extracts the heat outside of the reactor core while keeping the inside within the operational temperature limits [4]. The purpose of this study is to understand the changes in fundamental properties (chemical and mechanical stability) of GPC prepared at various temperatures (1000, 1500 and 2000ºC). The samples were bombarded with 5MeV Ag to 1 displacement per atom (dpa). This study will help establish GPC eligibility for future irradiation testing and experimentation for a specific application in an extreme radiation environment in the nuclear reactor.

EXPERIMENTAL DETAILS AND DISCUSSION

The GPC samples were prepared at three different temperatures (1000, 1500, and 2000ºC)[5] and were then irradiated with 5MeV Ag with a fluence of 1.8×1016 ions/cm2, (equivalent to 1 dpa) at room temperature.

TEM was used for characterizing the microstructures of the pre irradiated GPC [5]. Figure 1(A-C) are typical HRTEM images of GPC prepared at 1000, 1500 and 2000°C. All of three samples show the graphite-like layered structure. The basal layer of

Application of Accelerators in Research and Industry
AIP Conf. Proc. 1336, 219-221 (2011); doi: 10.1063/1.3586092

the pre irradiated GPC sample that was prepared at 1000ºC (figure 1A) are not uniform aligned with each other. At 1500ºC, figure 1B, the layers appear to be more uniform and are getting aligned more neatly. As noticed in figure 1C, the layers are even more uniform and they are aligned in a nice structure, more uniformly when compared with figures 1A and 1B. Based on these results, high processing temperatures would be more desirable for obtaining more isotropic GPC material.

Raman Spectroscopy was used to monitor the changes in the chemical bonding for the GPC samples. Figure 2 shows the Raman spectroscopy spectra for the pre irradiated GPC samples, while Figure 3 shows the post irradiated samples. The D (distorted) lines and the G (graphitic) lines show that the sample is destroyed after irradiation. Table 1 shows the D and G calculation before and after Ag irradiation

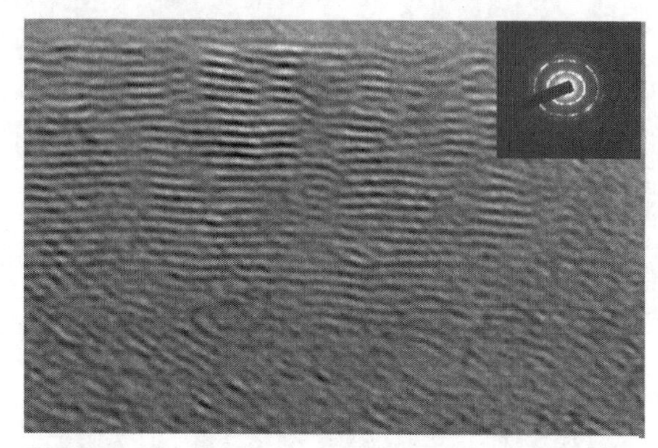

FIGURE 1A. TEM of pre-Irradiated GPC prepared at 1000˚C

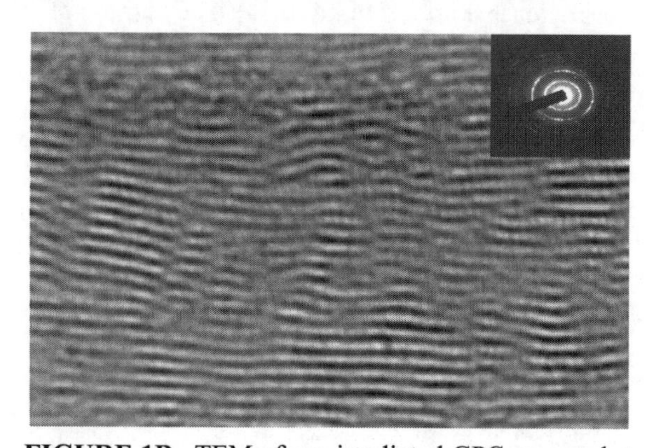

FIGURE 1B. TEM of pre-irradiated GPC prepared at 1500ºC

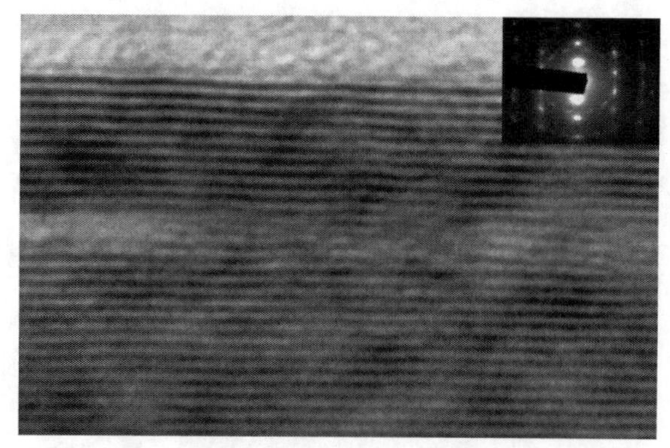

FIGURE 1C. TEM of pre-irradiated GPC prepared at 2000ºC

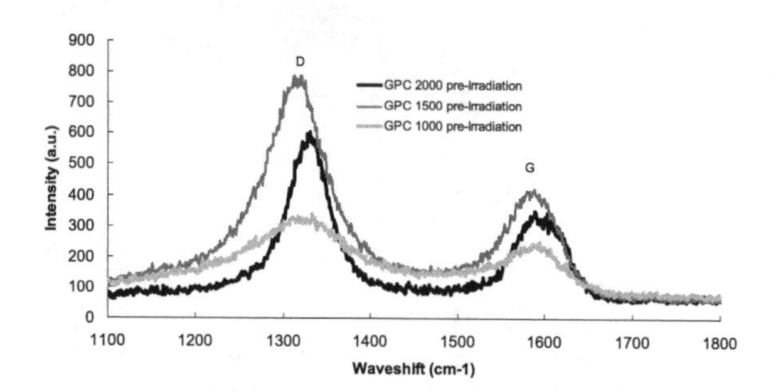

FIGURE 2. Raman spectroscopy of pre irradiated GPC samples prepared at various temperatures.

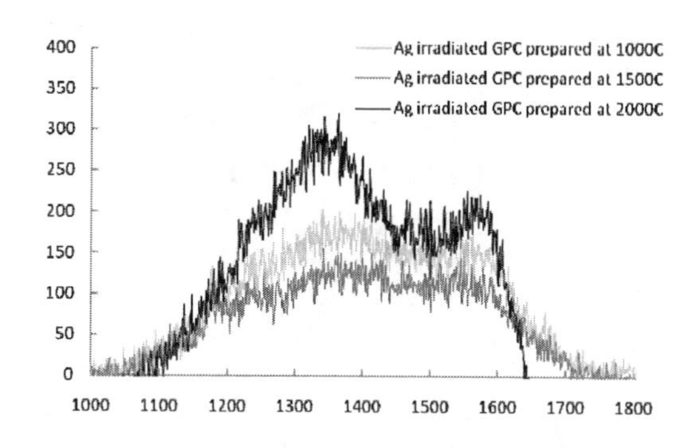

FIGURE 3. Raman spectroscopy of Ag irradiated GPC samples prepared at various temperatures

TABLE 1. D/G ratio for pre-irradiated and Ag irradiated GPC that was prepared at various temperatures

D/G of sample	GPC prepared at 1000ºC	GPC prepared at 1500ºC	GPC prepared at 2000ºC
D/G of pre-irradiated GPC	1.7	1.65	1.75
D/G of Ag irradiated GPC	2.94	1.63	3.9

CONCLUSIONS

TEM results showed that the higher the temperature the more ordered, graphite-like, layered structure appears in the pre irradiated GPC samples that are prepared at three different temperatures (1000, 1500 and 2000°C). These layers indicate that GPC can stand high temperatures. Raman showed that the samples were graphitized before irradiation took place and completely damaged and distorted after irradiation took place as table 1 indicated. The next step is to do TEM on the Ag irradiated GPC samples prepared at 1000, 1500 and 2000°C with distances of 2μm, 5μm and 10μm. The final TEM would be matched with SRIM simulations. We will also compare the results of the Ag in GPC with PyC. This result will help on further testing to determine if GPC was a better barrier for fission products than PyC to be used in the TRISO fuel for the next generation of nuclear reactors.

ACKNOWLEDGMENTS

This research was supported and funded by the AAMURI Center for Irradiation of Materials, NSF Alabama GRSP EPSCoR, and DoE NERI-C project number DE-FG07-07ID14894. We would like to also acknowledge the help from the department of nuclear engineering and radiological sciences at the University of Michigan, the center for irradiation of materials and the department of Physics at Alabama A&M University.

REFERENCES

1. Association of German Engineers (VDI), the Society for Energy Technologies (publ.) (1990). AVR - Experimental High-Temperature Reactor, 21 Years of Successful Operation for A Future Energy Technology. Association of German Engineers (VDI), The Society for Energy Technologies. pp. 9–23. ISBN 3-18-401015-5 http://www.nea.fr/abs/html/nea-1739.html
2. M D. A. Petti, J. Buongiorno, J. T. Maki and G. K. Miller, "Key differences in the fabrication, irradiation and high temperature accident testing of US and German TRISO-coated particle fuel, and their implications on fuel performance". Nuclear Engineering and Design, Volume 222, Issues 2-3, June 2003, Pages 281-297
3. L. Tan, T.R. Allen, J.D. Hunn, J.H. Miller, "EBSD for microstructure and property characterization of the SiC-coating in TRISO fuel particles", Journal of Nuclear Materials 372 (2008) 400–404
4. D. Olander J. Nucl. Mater. 389 (2009) 1-22.
5. M. Abunaemeh, M. Seif, Y. Yang, L. Wang, Y. Chen, I. Ojo, C. Muntele and D. Ila, "Characterization of Changes in Properties and Microstructure of Glassy Polymeric Carbon Following Au Ion Irradiation", Material Research Society Fall 2009 Proc. Vol. 1215-V16-26

SECTION IV – ENVIRONMENTAL/EARTH SCIENCES AND ADVANCED ENERGY

Ion Beam Analyses Of Bark And Wood In Environmental Studies

J.-O. Lill[a], K.-E. Saarela[b], L. Harju[b], J. Rajander[a], A. Lindroos[c], S.-J. Heselius[a]

[a]*Accelerator Laboratory, Turku PET Centre, Åbo Akademi University, Porthansg. 3, FI-20500 Turku, Finland*
[b]*Department of Chemical Engineering, Åbo Akademi University, Biskopsg. 8, FI-20500 Turku, Finland*
[c]*Department of Natural Sciences, Åbo Akademi University, Domkyrkotorget 1, FI-20500 Turku, Finland*

Abstract. A large number of wood and bark samples have been analysed utilizing particle-induced X-ray emission (PIXE) and particle-induced gamma-ray emission (PIGE) techniques. Samples of common tree species like Scots Pine, Norway Spruce and birch were collected from a large number of sites in Southern and Southwestern Finland. Some of the samples were from a heavily polluted area in the vicinity of a copper-nickel smelter. The samples were dry ashed at 550°C for the removal of the organic matrix in order to increase the analytical sensitivity of the method. The sensitivity was enhanced by a factor of 50 for wood and slightly less for bark. The ashed samples were pressed into pellets and irradiated as thick targets with a millimetre-sized proton beam. By including the ashing procedure in the method, the statistical dispersion due to elemental heterogeneities in wood material could be reduced. As a by-product, information about the elemental composition of ashes was obtained. By comparing the concentration of an element in bark ash to the concentration in wood ash of the same tree useful information from environmental point of view was obtained. The obtained ratio of the ashes was used to distinguish between elemental contributions from anthropogenic atmospheric sources and natural geochemical sources, like soil and bedrock.

Keywords: PIXE, elemental analyses, wood, bark, dry-ashing, environmental studies.
PACS: 29.30Kv; 89.60.-k.

INTRODUCTION

Long term elemental monitoring of metal emission to environment is a tedious and expensive task. The use of biological indicators with a long time span is therefore tempting. Many tree species can reach ages of more than a hundred years and are therefore of special interest as potential biological indicators. The radial elemental profiles of stem wood have been used in aim to construct historical records of the chemical exposure during the growth of the tree. The interpretation of these results is, however, complicated. A more direct approach is to measure the elemental concentrations in the bark of the trees. Pöykiö et al. have used pine bark for monitoring of environmental pollution in the Kemi-Tornio area in northern Finland [1]. From samples taken 1.2 km to 30 km from a chromium opencast mine and steel works respectively, pollution factors for chromium of above 100 were obtained.

From analytical chemical point of view tree based materials are problematic due to their heterogenic structures. Already within annual growth rings there are large seasonal variations in the elemental concentrations [2]. When going in the radial direction from the pith to the outmost sapwood there are large variations in the chemical composition [2]. Different layers of tree bark show even larger variations.

Main elements like Ca, K, Mn, P, S and Cl occur in wood materials on µg/g level [3]. The concentrations of heavy metals like Ni, Cu, Zn and Pb are in the range 0.1 µg/g – 10 µg/g, which means that the analytical method used for the analysis needs to be sensitive.

Mostly wet-chemical methods like atomic absorption spectrometry (AAS), atomic emission spectrometry (AES) and inductively coupled plasma mass spectrometry (ICP-MS) have been used for analysis of inorganic constituents in wood [3]. With these methods quite small amounts (100 mg – 200 mg) of the samples are dissolved before the chemical analysis. To overcome the problems with heterogeneity and in order to increase the sensitivity of the method we have combined dry ashing with PIXE in our work.

The main objective of the work is to find out how PIXE analyses combined with dry ashing of pine wood and bark can be used for environmental monitoring and for distinguishing between elemental contributions

Application of Accelerators in Research and Industry
AIP Conf. Proc. 1336, 225-229 (2011); doi: 10.1063/1.3586093

from anthropogenic atmospheric sources and natural geogenic sources.

EXPERIMENTAL

Sampling And Sample Preparation

Stem wood and bark samples of Scots pine (*Pinus sylvestris*), Norway spruce (*Picea abies*) and Silver birch (*Betula pendula*) were analysed for their elemental concentration in this study. Tree discs were cut at a height of ca 0.5 m above the ground usually during forest felling. For the elemental analysis of stem wood the laboratory samples were taken in a sectorial pattern from the discs in order to ensure a representative sampling of all wood layers [5,6]. Due to large concentration variations in different bark layers, whole bark samples were taken from the discs to be analysed separately from the wood samples. Totally several hundred samples were studied in this project but in this paper only the results for pine wood and bark are presented.

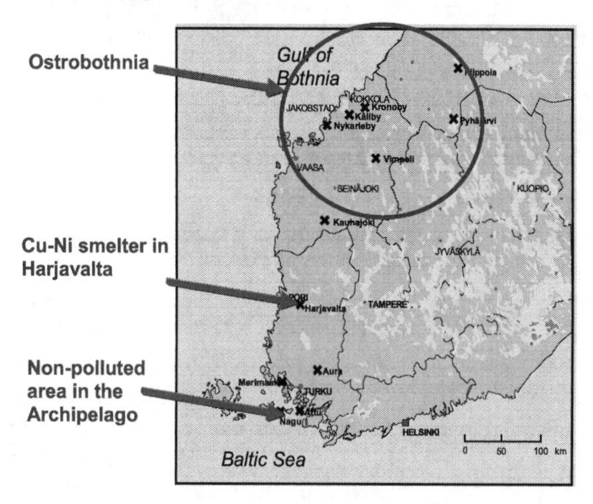

FIGURE 1. Sampling sites in Finland.

Figure 1 shows the main sampling sites of our study in Western and Southwestern Finland. The majority of samples were taken from non-industrialised and thus relatively clean areas. Harjavalta with a copper-nickel smelter represents a site with heavy metal pollution [7]. Here the samples were taken ca 6 km northeast of the smelter complex which is one of the main point sources of metal emission in Finland [8].

All categories of tree related materials were dry ashed in the same manner for the PIXE analysis. About 10 g to 40 g of the material was weighed into porcelain crucibles and dried at 105°C. The dry ashing was performed in a programmable oven (Vulcan 3-130) by slowly increasing the temperature to 550°C [5,9,10]. The ashes were pressed to pellets and stored in desiccators until irradiation. The ash percentages and elemental concentrations are given on dry weight basis.

The following bio-fuel based CRMs from Swedish University of Agricultural Sciences, Umeå, Sweden were used for the evaluation of the PIXE analyses: Wood Fuel (Pine), NJV 94-5, Energy Forest (Salix), NJV 94-3 and Energy Grass (*Phalaris arundinaceae L.*), NJV 94-4. Other CRMs used were Pine Needles (1575) from NIST (National Institute of Science and Technology, Gaithersburg, MA, USA) and Beech Leaves (CRM 100) from BCR (Community Bureau of Reference, Brussels, Belgium).

Experimental Set-Up For External Beam PIXE

The samples were irradiated outside the cyclotron vacuum system using a proton beam collimated to diameter of 1 mm and an energy of 3 MeV incident on the target. The current of the particle beam was measured by monitoring the light emission in the beam path [11]. The light emission is due to molecular transitions in N_2 and induced by the particle beam.

The emitted X-rays were measured using a 25 mm^2 intrinsic germanium planar detector in 90° with respect to the beam path. The spectrum acquisition time was about 500 s. The GUPIX software was employed to fit the recorded spectra and to calculate the elemental concentrations. The calibration was checked at each run by analyzing a pressed pellet of the granite standard G2 (USGS). A pellet of spectroscopically pure graphite was also irradiated at the beginning of each run to assure the quality.

RESULTS AND DISCUSSION

Evalution Of Dry Ashing And PIXE By Analysing Pine Needles

Test results of the combined dry ashing and PIXE method is given in Table 1. Elemental concentrations obtained for both dry ashed and non-ashed samples of the biological CRM Pine Needles (1575) are listed as well as consensus values given in a compilation by Roelandts and Gladney [12]. Limits of detection (LOD) for the two types of materials are also included in Table 1. As can be seen from the table the accuracy and reliability obtained by the PIXE method combined with dry ashing is good for most elements having concentrations clearly above the LOD values.

TABLE 1. Particle Induced X-ray Emission analyses of pellets of Pine Needles (NIST 1575). Concentrations and limits of detection (LOD) in µg/g are calculated per dry weight of the biological material. The Y indicates that the concentrations are above the limit of quantification (3.3•LOD). The consensus values from the compilation of Roelandts and Gladney [12] are included in the last column.

	Biological			Dry ashed (2.52%)			Consensus
	Conc.	LOD		Conc.	LOD		
Si	0.0	9371.5	N	931.0	1326.48	?	1290
P	1071.6	618.6	?	1163.3	167.43	Y	1210
S	1600.1	158.4	Y	683.4	47.27	Y	1250
Cl	429.4	80	Y	43.2	23.77	?	296
K	4496.4	36.3	Y	2905.9	9.35	Y	3690
Ca	4330.6	34.7	Y	3497.8	17.34	Y	4200
Ti	28.3	13.9	?	23.1	3.79	Y	14
Mn	899.0	6	Y	832.3	1.46	Y	660
Fe	250.2	15.8	Y	220.8	6.17	Y	193
Ni	1.7	2	?	2.5	0.31	Y	2.7
Cu	3.0	1.5	?	2.9	0.22	Y	2.9
Zn	64.5	1.3	Y	68.4	0.20	Y	65
Ga	0.1	1	N	0.3	0.21	?	0.09
Br	6.7	1.5	Y	1.1	0.18	Y	7.5
Rb	16.8	1.7	Y	10.7	0.15	Y	11.2
Sr	7.4	1.8	Y	4.2	0.17	Y	4.7
Zr	0.0	4.6	N	0.4	0.26	?	0.18
Nb	3.3	2.1	?	0.2	0.20	?	
Mo	0.7	3.8	N	0.2	0.22	?	0.14
Ag	0.0	12.2	N	0.4	0.60	?	0.1
Sn	6.6	24.6	N	1.6	1.10	?	
Ba	0.0	115.5	N	3.5	3.92	?	7.1
Pb	14.7	2.9	Y	10.7	0.53	Y	10.6

For Pine Needles with 2.52 % ash the LOD values for heavy metals are lowered by almost 1 logarithmic concentration unit by dry ashing. The inorganic constituents of Pine Needles are almost entirely retained in the ash after combustion of the organic matter. It is interesting to note that also lead is quantitatively retained to the ash. A drawback of the dry ashing at 550°C is that most of chlorine and bromine are lost. Almost 50 % of the sulfur is also lost during the ashing procedure. The losses have been discussed more in detail elsewhere [10].

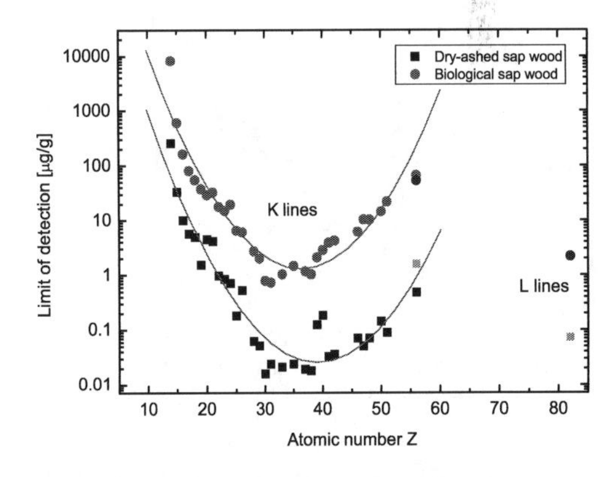

FIGURE 2. Limits of detection for elements in sap wood of spruce. The continuous curves are fitted to the values to guide the eye. The lower values (black squares) represent limits for elements in wood enriched by dry ashing at 550°C. The higher values (red circles) are limits of detection for untreated material pressed to pellets and analysed by PIXE.

Improved Limits of Detection By Dry Ashing

Figure 2 shows the limit of detection (LOD) when analysing sapwood of spruce with and without dry ashing. Direct PIXE analysis of untreated wood gave e.g., for Zn (Z =30) a LOD value of ca 1 µg/g. After dry ashing of the wood sample (ash-% = 0.29) the LOD value of Zn are improved to ca 0.02 µg/g.

The limits of detection are thus lowered by almost 2 logarithmic concentration units by dry ashing. For tree bark samples the improvement is slightly lower as the ash-% is in the range 1 – 3 %. The precision of the determination of the ash content by dry ashing expressed as relative standard deviation (RSD) obtained under repeatability conditions is ca 2 % [9] and of the PIXE method 1 % – 2 % [13].

Comparison of Elemental Concentrations in Pine Wood from Different Areas

In figure 3 the mean elemental concentrations in pine wood samples from Harjavalta and Ostrobothnia are compared to the concentrations of the corresponding elements in pine wood samples from the island Nagu, a relatively unpolluted site in the Turku archipelago. The highest concentration ratio, also called pollution factor [1], was obtained for Mn (2.6) in wood samples from Ostrobothnia. The anomalies for the heavy metals Fe, Cu, Ni, Pb and Zn are quite small for wood samples from Harjavalta. The copper smelter has been active since 1945 and nickel smelting started at 1959. Usually the precipitation of pollutants also leads to some degree of soil contamination. However, it can be concluded that airborne heavy metal emission only very weakly affects the stem wood concentration data.

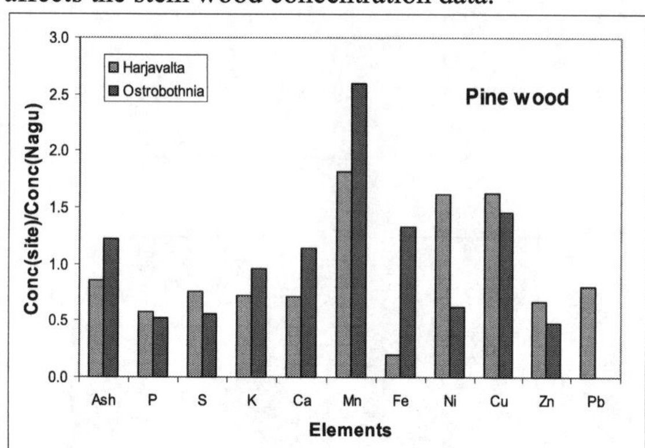

FIGURE 3. Mean elemental concentrations in pine wood from the sites Harjavalta and Ostrobothnia are compared to concentrations in pine wood from an unpolluted area (Nagu) in the Turku archipelago.

Comparison of Elemental Concentrations in Pine Bark from Different Areas

Also in case of pine bark the mean elemental concentrations in samples from Harjavalta and Ostrobothnia are compared to the concentrations in bark samples from the unpolluted reference site Nagu (Fig. 4). As can be expected high concentration ratios (range 25 -30) were found for Cu and Ni in pine bark samples from Harjavalta. A large number of pine bark samples from a large area in Ostrobothnia showed a high mean concentration ratio (ca 13) for Mn. This anomaly will be discussed more in detail in the next section. The results given in figure 4 show that direct elemental analyses of pine bark can be used quite successfully for monitoring of airborne heavy metal emission after comparison of chemical data for bark samples from an unpolluted area.

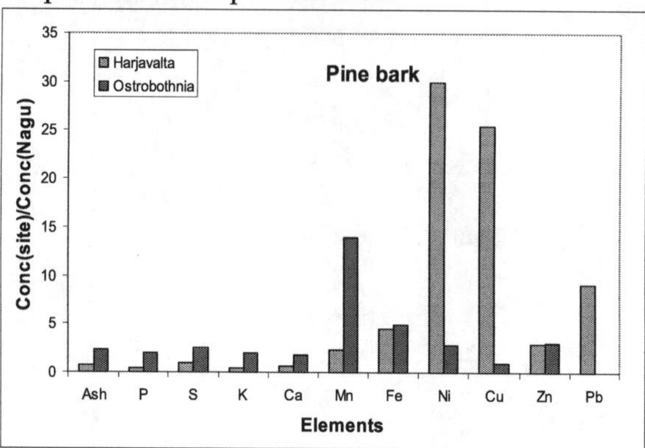

FIGURE 4. Mean elemental concentrations in pine bark from the sites Harjavalta and Ostrobothnia are compared to concentrations in pine bark from an unpolluted area (Nagu) in the Turku archipelago.

Bark Ash to Wood Ash Ratios – A Way of Normalisation

By comparison of elemental contents in ashes of bark and wood of individual trees, a normalisation is obtained [14, 3]. In figure 5 the mean concentration ratios for pine bark to pine wood for the areas Harjavalta and Ostrobothnia are presented.

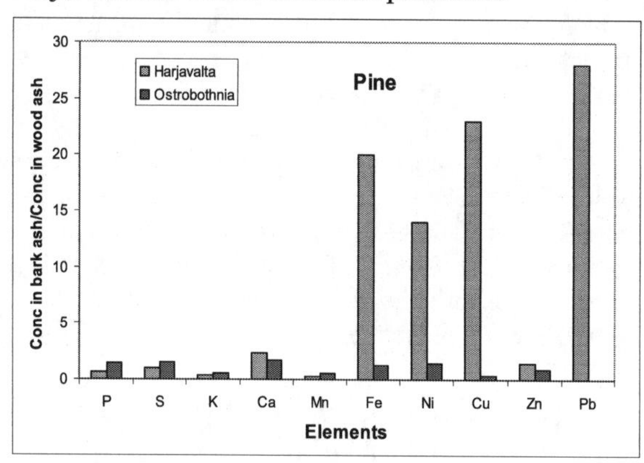

FIGURE 5. Mean elemental concentrations in pine bark ash from the sites Harjavalta and Ostrobothnia are normalized to concentrations in pine wood ash from the same sites. By using the obtained ratios of the ashes one can distinguish between elemental contributions from anthropogenic atmospheric sources and natural geochemical sources, like manganese in the soil in Ostrobothnia.

For relatively clean areas like Ostrobothnia, the mean ratios of the concentration in bark ash to the concentration in wood ash for different elements were close to 1. This means that ashes of bark and stem wood have quite similar chemical composition. For Ca the ratios were around 2 for all sampling sites. For the pine samples from the polluted area Harjavalta, the mean concentration ratios of Cu, Fe, Ni and Pb are clearly elevated and in the range 14 – 28 (Fig. 5), indicating the effect of airborne pollution. Similar ratios were also obtained for spruce samples from Ostrobothnia and Harjavalta [3].

It is interesting to note that by the normalization procedure a quite normal ratio for Mn is obtained for the pine bark samples from Ostrobothnia. This means that the soil and soil water in the Ostrobothnia area have a high content of Mn due to natural geochemical sources, which in turn increases the Mn concentration both in stem wood and bark.

CONCLUSION

Tree related materials are heterogeneous and large amounts of samples are needed to obtain statistically representative results. This problem was overcome by dry ashing of 10 g to 40 g of materials prior to the PIXE analysis. The ash percentages of wood (0.2 % – 0.4 %) and bark (1 % – 3 %) are quite low and thus a clear improvement of limit of detection (LOD) of the analytical method was also achieved. In studies of the effects of anthropogenic atmospheric pollution using bark, no samples from reference areas area are needed, when using the method of normalisation presented in this work.

ACKNOWLEDGMENTS

The help from Leif Österholm with the grinding of the wood sample is acknowledged. The authors thank Erkki Stenvall, Per-Olof Eriksson and Stefan Johansson for technical assistance and Eero Kuokkanen (Harjavalta) for help with the sampling work.

REFERENCES

1. P. Pöykiö, P. Perämäki and M. Niemelä, Intern. J. Environ. *Anal. Chem.* **85:2**, 127-139 (2005).
2. L. Harju, J-O. Lill, K-E. Saarela, S-J. Heselius, F. J. Hernberg and A. Lindroos, *Nucl. Instr. and Meth.* **B109/110**, 536-541 (1996).
3. K-E. Saarela, "Elemental analysis of wood materials by external millibeam thick target PIXE", Ph.D. Thesis, Åbo Akademi University, 2009.
4. A. Ivaska and L. Harju, "Analysis of Inorganic Constituents" in *Analytical Methods in Wood Chemistry, Pulping and Papermaking*, edited by E. Sjöström and R. Alén , Heidelberg, Springer Verlag, 1999, pp. 287–304.
5. L. Harju, J-O. Lill, K-E. Saarela, S-J. Heselius, F. J. Hernberg and A. Lindroos, *Fresenius J. Anal. Chem.* **358**, 523-528 (1997).
6. K-E. Saarela, L. Harju, J-O. Lill, J. Rajander, A. Lindroos, S-J. Heselius and K. Saari, *Holzforschung* **56**, 380-387 (2002).
7. T. M. Nieminen, "Response of Scots pine (Pinus sylvestris L.) to a long-term Cu and Ni exposure", Finnish Forest Research Institute, Research Papers 942, 2005.
8. N.A. McEnroe and H.-S.Helmisaari, *Environ. Pollution* **113**,11-18 (2001).
9. K-E. Saarela, J-O. Lill, F. J. Hernberg, L. Harju, A. Lindroos and S-J. Heselius, *Nucl. Instr. and Meth.* **B103,** 466-472 (1995).
10. L. Harju, J. Rajander, K-E. Saarela, J-O. Lill, S-J. Heselius, A. Lindroos, "Losses of elements during dry ashing of biological materials", Proc. 10th Int. Conf. on Particle Induced X-ray Emission and its Analytical Applications, Portorož, Slovenia, June 4-8, 2004 (Eds M. Budnar and M. Kavčič) Electronic edition, Ljubljana, 945.1-3 (2004).
11. J-O. Lill, *Nucl. Instr. and Meth.* **B150,** 114-117 (1999).
12. I. Roelandts and E. S. Gladney, *Fresenius J. Anal. Chem* **360,** 327-338 (1998).
13. J-O. Lill, K-E. Saarela,, F. J. Hernberg, S-J. Heselius and L. Harju,, *Nucl. Instr. and Meth.* **B83**, 387-393 (1993).
14. K-E. Saarela, L. Harju, J. Rajander, J-O. Lill, S-J. Heselius, A. Lindroos, K. Mattsson, *Sci. of the Total Environment* **343**, 231-241 (2005).

SECTION V – FOCUSED ION BEAMS AND NANOFABRICATION

Ion Beam Synthesis Of Silicon Nano-Crystals For Electronics And Photonics

B. Schmidt, K.-H. Heinig

Research Center Dresden-Rossendorf, POB 51 01 19, 01314 Dresden, Germany

Abstract. The present contribution addresses Si nanocrystal (NC) formation by conventional ion implantation into the gate oxide and by a non-conventional IBS approach of ion beam mixing of SiO_2/Si interfaces in thin gate oxides, with special emphasis on well-controlled size and position tailoring. The two approaches will be compared and related technological challenges discussed. Compared to conventional Si NC synthesis by Si^+ ion implantation into the gate oxide, we take advantage of the self-alignment ion beam mixing process, i.e., the Si NCs are formed in SiO_2 at a well-controlled small distance of ~2 nm from the Si/SiO_2 interfaces. The technical applications in non-volatile nanocrystal memories and in light emitting field-effect transistors (LEFET) are demonstrated.

Keywords: Ion beam synthesis, silicon nanocrcrystals, charge storage, electroluminesence
PACS: 81.07.-b, 81.16.Dn, 61.72.up, 61.80.Jh, 73.63.Bd, 78.60.Fi

INTRODUCTION

At conventional low- and medium-fluence ion implantation (<10^{16} ions/cm^2) in microelectronics, the concentration of introduced dopants is usually below the solubility limit of them in silicon. At certain annealing temperatures the impurity atoms are dissolved in Si and located on crystal lattice sites or, as a small part, remain as soluted interstitial atoms. In high-fluence ion implantation ($\geq 10^{16}$ ions/cm^2), or so-called ion beam synthesis (IBS), a far-from-equilibrium state (supersaturated solid) is achieved which relaxes towards thermodynamic equilibrium during subsequent annealing by phase separation through precipitation and ripening (Ostwald-Ripening) of nanoclusters (NC) [1] (see Fig1).

Although there are possibilities to tailor the mean NC size in conventional IBS (mainly by variation of ion fluence and annealing temperature and time), the possibilities for tailoring the NC size distribution by the variation of ion implantation parameters are rather limited.

Ion irradiation not only causes supersaturation and damage to the substrate but also affects the interface between a thin layer and the substrate or interfaces between thin stacked layers. As schematically shown in Fig. 2, primary energetic ions produce collisional cascades, which cause substantial interface mixing leading to non-stoichiometric and non-stable phases near the interface. Subsequent annealing restores the interface region rapidly via spinodal decomposition. However, the tails of the mixing profiles do not reach the recovered interfaces by diffusion; thus, phase separation proceeds via nucleation and growth of NCs

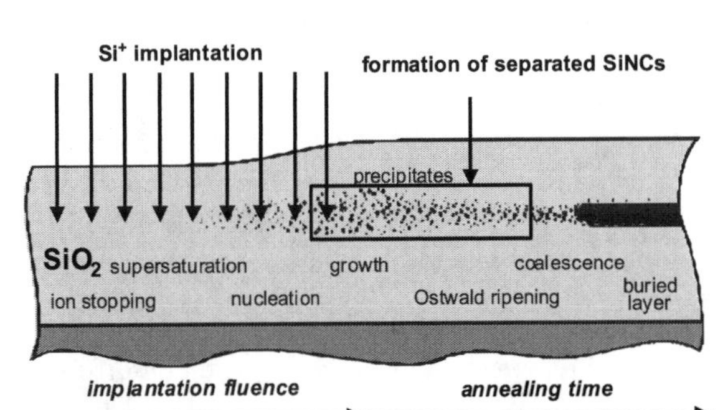

FIGURE 1. Scheme of ion beam synthesis of nanostructures. High dose ion implantation into a solid leads to supersaturation of impurity atoms. NCs nucleate and grow during ion implantation or, for impurity atoms immobile during implantation, during subsequent thermal treatment. The mean NC size as well as their spatial and size distribution changes during Ostwald ripening. At very high fluences, buried layers can form by coalesence of NCs.

Application of Accelerators in Research and Industry
AIP Conf. Proc. 1336, 233-238 (2011); doi: 10.1063/1.3586094

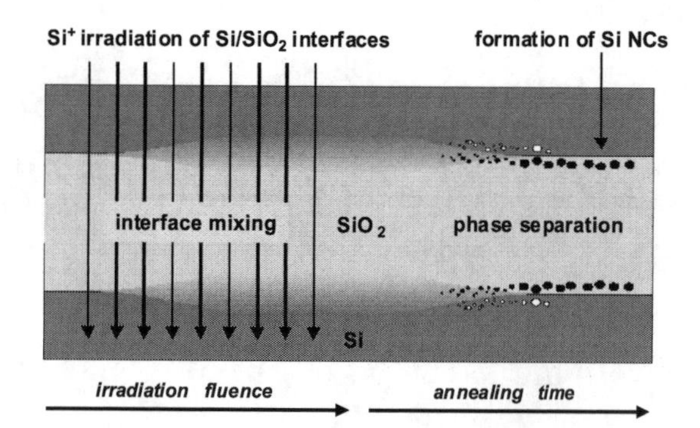

FIGURE 2. Scheme of ion beam induced self-alignment of a NC δ-layer near a layer/substrate interface by ion irradiation through the interface and subsequent thermal treatment leading to the phase separation in the ion mixed non-stoichiometric near-interface layer through precipitation and Ostwald ripening of NCs.

near the interface. The competition between interface restoration and nucleation self-aligns nearly monodispersive δ-layers of NCs in proximity to the interface. Recently, self-organization of a δ-layer of Ge NCs in SiO_2 has been found [2,3] close to the Si/SiO_2 interface after appropriate Ge^+ implantation into a SiO_2 layer on a Si substrate. The NC δ-layers was found when the following conditions were fulfilled : (i) a negligibly small amount of ions is implanted in the Si/SiO_2 interface region, and (ii) the cascade of energetic O and Si recoils created by primary collisions produces a few displacements per atom (dpa) in the Si/SiO_2 region. Different contributions to the self-organization and self- alignment of NC δ-layers have been described and investigated and predicted by kinetic MC simulations and reaction-diffusion equations [4]. Very recently, this prediction has been verified experimentally by transmission electron microscopy [5,6].

The IBS of semiconducting NCs has attracted much interest due to their compatibility with CMOS technology and due to unique physical properties of NCs as zero-dimensional objects. A great effort is currently devoted to their applications in new microelectronic devices, e.g. charge storing Si-NCs in non-volatile memory circuits [7-10] and light emitting field-effect transistors (MOSFET). However, the fabrication of monodispersive and small Si NCs (≤ 3 nm) separated from the transistor channel by a thin (2-3 nm) oxide represents a strong challenge. At the common IBS of Si NCs, low energy ion implantation into SiO_2 with ion energies < 5 keV and relatively high ion fluences in the order of 1-2x10^{16} cm^{-2} is applied [8,9]. Alternatively, the approach for ion irradiation induced self–alignment of Si NCs near SiO_2/Si interfaces can also be applied to the fabrication of non-volatile multi-dot floating gate memory devices [11].

The present contribution addresses to self-organization processes of silicon nanoparticles during ion irradiation of flat Si/SiO_2 interfaces during Si^+ ion irradiation and to the fabrication of Si-NC memory and light emitting MOSFET devices using this approach.

EXPERIMENTAL

A MOS-like $Si/SiO_2/Si$ structure was selected and subjected to Si^+ ion irradiation to mix the interfaces of this layer stack. The ion irradiation experiments were carried out on 6" (100)Si wafers covered with a 15 nm thermally grown SiO_2 layer and a 50 nm poly-Si layer deposited by LPCVD on top of the SiO_2 layer. The upper poly-Si layer was used as an capping layer to prevent any influence of contaminants from ambient and annealing atmosphere (mainly humidity) on the NC formation process in the thin gate oxide [12]. The poly-$Si/SiO_2/Si$ stack was irradiated with 50 keV Si^+ ions at fluences in the range of 3x10^{15} – 1x10^{16} cm^{-2}. After ion irradiation a highly n^+-doped and 250 nm thick poly-Si gate was deposited onto the irradiated stack. Subsequently, the samples were annealed (RTA) at different temperatures (950–1100 °C) and times (5–180 s) and further processed for the fabrication of nMOSFETs in the standard 0.6 μm CMOS process line. Transistors with long-channel (20 μm) and short-channel (0.6 μm) gate length and 20 μm gate width have been fabricated and electrically tested with respect to programming window, endurance and retention. On the same integrated MOSFET devices alternating (AC) square-wave voltage signals were applied to the gate with the p-Si substrate as the reference electrode and an amplitude between ±7 and ±11V (peak-to-peak voltage V_{pp} of 14 to 22 V) by a 20 MHz NF1930 arbitrary function generator. The source and drain voltages were set to V_s =V_d =0 V. The EL spectra were recorded using an Andor DU401A-UV-BR-DD CCD camera working at T = -70 °C (thermoelectrical cooling), a Shamrock 303i grating spectrometer, and an optical microscope.

RESULTS

The competition between interface restoration and nucleation self-aligns δ-layers of Si NCs in SiO_2 along the two interface. This self-alignment of δ-layers of Si NCs with the SiO_2/Si interfaces has been predicted by atomistic computer simulations [13] and is demonstrated in Fig. 3. The Si precipitates in SiO_2 have developed to Si NC δ-layers, which are aligned with the SiO_2/Si interfaces. The mean Si NC diameter is 2 nm and the mean distance from the interfaces is 2 nm, too. In each δ-layers, the Si NC areal density is in

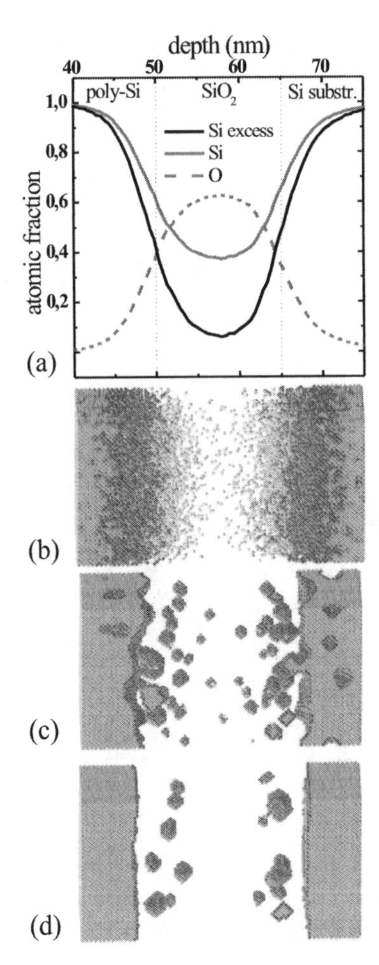

FIGURE 3. Self-aligned Si NC formation by ion irradiation of SiO_2/Si interfaces: (a) TRIDYN results of Si ion irradiation (E = 50 keV, D = $1x10^{16}$ cm^{-2}) through a layer stack of 50 nm poly-Si, 15 nm SiO_2, into the Si substrate. KLMC simulation snapshots referring (b) to the as-irradiated state, (c) to the early state of phase separation , and (d) to a later stage.

the order of 10^{12} cm^{-2}. This prediction has been verified experimentally by energy filtered transmission electron microscopy (EFTEM) [6]. As can be seen in Fig. 4, bright and dark areas are visible in dark and white regions of the SiO_2 layer, respectively. These spherical regions refer to Si NCs in the oxide. Accordingly, the Si NCs are separated from the SiO_2/Si interfaces by a mean distance of 3 nm, i.e., they are aligned in a δ-layer at each interface. The mean diameter of the NCs was estimated to be 3 nm. These morphological features are in accordance with the KLMC simulation predictions from [13].

Electrical measurements of memory properties [11] show that devices exhibit significant memory windows at low gate voltages. Devices with a memory window of about 0.1V for write/erase voltages of – 7V/+7V

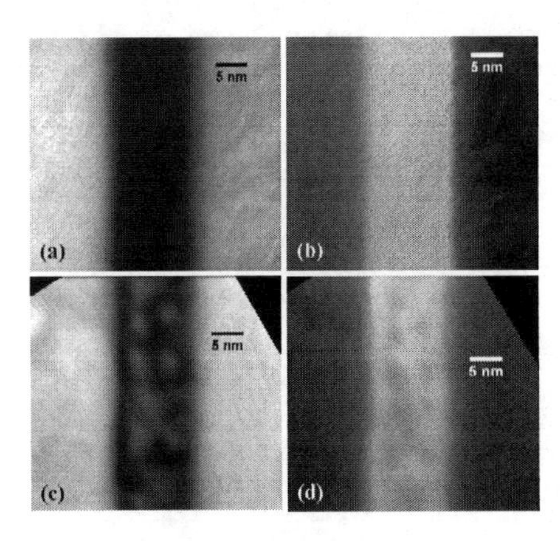

FIGURE 4. Energy-filtered XTEM images of the poly-Si/SiO_2/Si structure: (a,b) referring to the as-deposited state and (c,d) after Si^+ ion irradiation (E = 50 keV, D = $7x10^{15}$ cm^{-2}) and post-irradiation annealing (T = 1050 °C, t = 120 s).

and a programming time of t_{pp} = 10 μs have been achieved. In terms of memory window and tansistor characteristics, an implanted fluence of 5-$7x10^{15}$ Si^+ cm^{-2} and annealing at T_A =1050 °C for t_A = 30 s appear as promising conditions for device fabrication (see Fig. 5a). The memory window can be increased by annealing at higher temperatures (T = 1100 °C) and/or for longer times (> 120 s) proceeding to further phase separation Si NC formation. As shown in Fig. 5b, in this case a large stable memory window of ΔV_{th} ~ 3 V can be achieved.

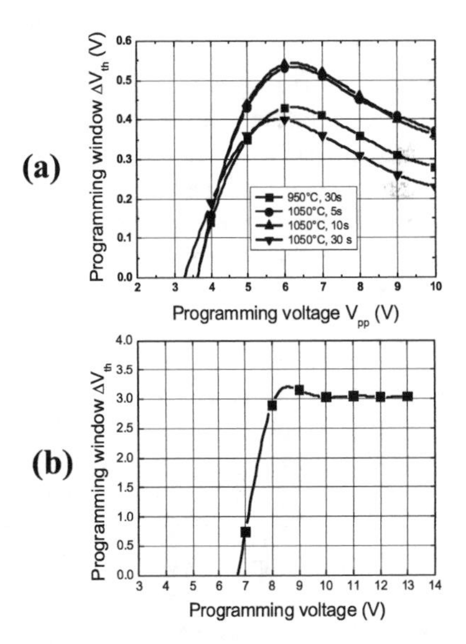

FIGURE 5. Memory window ΔV_{th} versus programming voltage V_{pp} (pulse length t_{pp} = 10ms) of n-channel MOSFETs with self-aligned Si NCs in the gate oxide. Si^+ irradiation was performed at an ion energy of 50 keV and a fluence of $7x10^{15}$ cm^{-2}: (a) for low thermal budget at different RTA annealing parameters; (b) for high thermal budget and annealing at T = 1050 °C for 120 s.

No degradation in memory windows was observed for devices after 10^7 write/erase cycles with V_{pp} = +7V/-7V, t_{pp} = 1ms programming conditions (Fig. 6), which means that the fabricated memory devices exhibit a superior endurance (very limited degradation up to 10^7 cycles). It was found that data retention time tested at 85 °C is too low for EEPROM application (100 days at room temperature and 8 h at 85 °C).

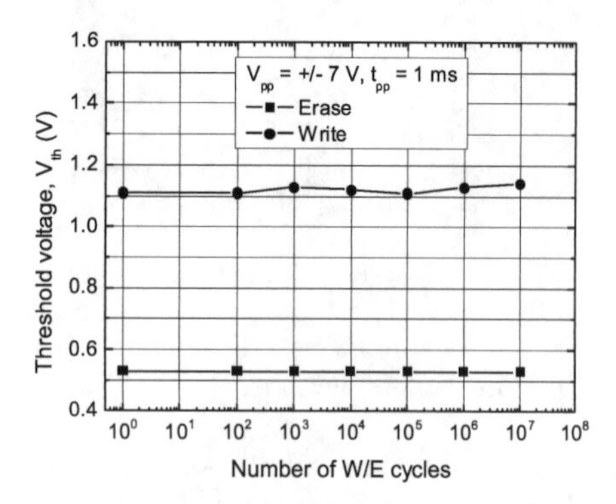

FIGURE 6. Endurance characteristics of nMOSFETs obtained with write/erase voltages of V_{pp} = ±7 V and pulses of t_{pp} = 1 ms. The sample was irradiated with 50 keV Si^+ ions at a fluence of $7x10^{15}$ cm^{-2} and annealed at 1050 °C for 30 s

The fabricated nMOS devices exhibit maximum memory windows at low gate voltages indicating that the charge storage nodes are located near to the Si/SiO_2 interface. This implies the possibility of direct charge carrier tunneling into Si NCs during low programming voltages. The lowering of the programming window at programming voltages V_{pp} > 6 V (Fig. 5a) can be explained by trap-assisted tunneling through the gate insulator. Besides retention, all electric parameters of the devices fulfill current requirements to non-volatile memory devices. The low data retention might be explained by direct re-tunneling of charge carriers to the channel which could be a common problem of multi-dot floating-gate memories.

With the application of an AC voltage to the gate electrode of the LEFET device - following the model of Walters *et al.* [14] - excitons are formed within the Si NCs due to alternating electron and hole injection from both electrodes. Since the emitted wavelength depends on the NC bandgap energy, i.e., the NC size, the luminescence with a center wavelength of about 750 nm as shown in Fig. 3 is attributed to radiative recombination of excitons within Si NCs of 2-3 nm in diameter which is consistent with the TEM result (see Fig. 4) [14,15]. The EL spectra in Fig. 7(a) and (c) were obtained from integrated devices of 20x20 μm^2 size [16]. As confirmed by light transmission

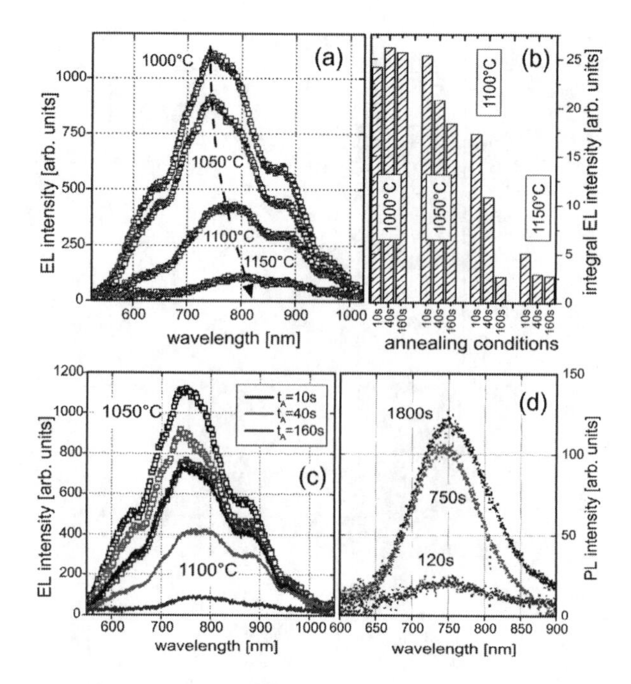

FIGURE 7. (a) EL spectra after 40 s annealing at different temperatures V_{pp}=22 V and f =10 kHz in (a)–(c). (b) Bar graph plot of the integral EL intensities according to the total sample set (annealing for 10, 40, and 160 s at each temperature). (c) EL spectra after isothermal annealing at 1050 °C (squares) and 1100 °C (lines). The different annealing times are indicated by different colors. (d) PL spectra obtained at simplified structures after 120, 750 (RTA) and 1800 s (furnace) annealing at 1050 °C.

calculations (not shown here), the undulating modulations of the typically Gaussian-shaped luminescence profiles are related to multiple internal reflections within the Si_3N_4/SiO_2 multilayer stack covering the LEFET devices. These deviations are clearly not correlated to the NCs fabrication method since the PL spectra in Fig. 7(d), which are obtained at simple 50 nm poly-Si/14.5 nm SiO_2/Si stacks, reveal a typical behaviour of tiny Si NCs. With the electro- and photoluminescence results in Fig. 7, different stages of NC evolution can be traced as a function of the annealing temperature and/or time. During the annealing at 1000 °C the overall EL intensity is close to its maximum value and still slightly increasing [see the bar graph plot of Fig. 7(b)]. With a higher thermal budget (1050 °C) the intensity is decreasing especially during the 1100 °C annealing down to a very low but stable level such as for 1150 °C thermal treatment. Whereas an increasing intensity indicates that the process of precipitation and Si NC formation has not been finished up to this state (1000 °C), the dissolution of NCs is beginning at 1050 °C: the evolution of the NC size follows Gibbs-Thomson's relation, i.e., a diffusion controlled ripening process [4,6]. NCs adjacent to the Si substrate and poly-Si gate dissolve faster, the closer they are located to the Si/SiO_2 interfaces. This accelerates the dissolution of small, close NCs (1100 °C) and stabilizes bigger more distant ones (related to the Si substrate and gate, respectively) like for 1100 °C, 160 s and 1150 °C annealings in Fig. 7(b). During Ostwald ripening the Si NCs compete with each other in plane whereas some Si NCs grow at the expense of smaller ones. This corresponds at the same time to (i) the red-shift of the EL spectra in Fig. 7(a), i.e., a radiative exciton recombination in larger NCs (which are characterized by a smaller bandgap), and (ii) the quenched luminescence due to a considerably reduced number of light emitting sites. It has to be mentioned that the denoted LEFET annealing temperatures represent only a part of the total thermal budget necessary for complete CMOS device fabrication. This explains why the overall PL intensity in Fig. 7(d) is still increasing during isothermal annealing indicating that the process of Si formation has not been finished after 1050 °C, 750 s, while seemingly the EL intensity already decreases with 1050 °C, 40 s thermal treatment (Fig. 7(c)).

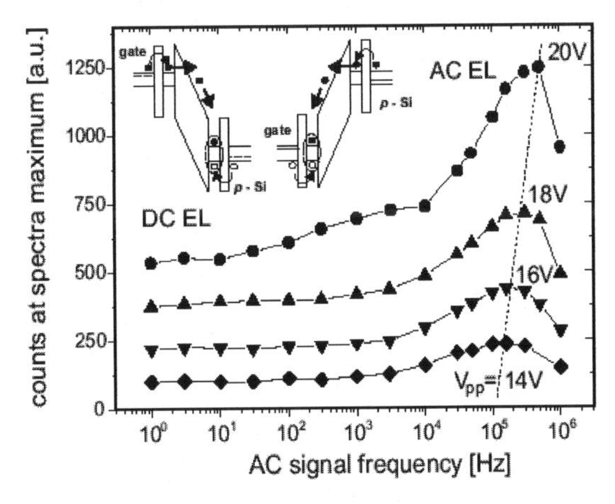

FIGURE 8. Dependence of EL intensity on ac gate voltage amplitude and frequency for a LEFET annealed at 1050 °C for 40 s. A schematic band diagram illustrates the quasi DC mode EL mechanism in the inset (f ≤ 10 kHz) for negative and positive ac square-wave gate voltage periods.This is the Style for Figure Captions. Center this if it doesn't run for more than one line.

The LEFET has two luminescence options depending on the applied gate signal frequency (Fig. 8). With similar EL spectra we refer both modes to radiative recombination of excitons in the Si NCs. At low frequencies, where both NC layers are saturated with charges of opposite sign, a constant (DC) current related EL is dominating (see the inset in Fig. 8). The applied gate voltage drops mainly across the remaining 5 nm thin SiO_2 layer sandwiched between the NC arrays. This enables a DC leakage current due to F-N tunneling of electrons which recombine radiatively with stored holes at the NCs of the opposite side. The DC current can be easily suppressed increasing the gate oxide thickness to quench the tunneling probability (with positive consequences for the device long time reliability). In Fig. 8 for frequencies f ≥ 10 kHz, an AC related EL emerges out of the DC luminescence background. This type of EL has been explained by Walters *et al.* [14] as a field-effect driven mechanism. The luminescence intensity increases with increasing number of switching cycles up to 100 - 500 kHz and collapses at 1 MHz AC signal frequency. The highest EL intensity occurs at much higher AC frequencies than obtained by Walters *at al.* [14] or predicted by Carreras *et al.* [17] (f ≤ 50 kHz) . Here, the gate capacitive charging time constant is clearly reduced due to our very small transistor. In addition the 2 nm thin direct tunneling oxide enables higher charging currents with shorter charging times as confirmed by their memory device characteristics [11,18]. The dependence of the EL intensity maximum on the AC voltage amplitude reveals the significance of the charge transfer rate. The EL quenches at about f = 1 MHz which corresponds supposably to the physical limit of radiative recombination of excitons in tiny Si NCs, i.e., their radiative lifetime [14,17].

ACKNOWLEDGMENTS

The authors especially thanks M. Klimenkov and T. Gemming for carrying out EF-TEM investigations. The authors also would like to thank P. Dimitrakis and E. Votintseva for characterization of nMOSFETs by electrical measurements and S. Probst for the luminescence measurements.

REFERENCES

1. S.Reiss, K.H. Heinig, *Nucl. Instr. Meth. Phys. Res. B* **102**, 256 (1995).
2. K.H. Heinig, B. Schmidt, A. Markwitz, R. Grötzschel, M. Strobel, S. Oswald, *Nucl. Instr. Meth. Phys. Res. B* **148**, 969 (1999).
3. J. von Borany,R.Grötzschel,K.H.Heinig, A.Markwitz,B. Schmidt,W. Skorupa, Solid-State Electron. **43**, 1159 (1999).
4. K.H. Heinig, T. Müller, B. Schmidt,M. Strobel, M. Möller, Appl. Phys. A **77**, 17 (2003).
5. L. Röntzsch,K.H. Heinig,B. Schmidt,A. Mücklich, Nucl. Instr. Meth. B **242,** 149 (2006).
6. L. Röntzsch, K.H. Heinig, B. Schmidt, A. Mücklich, W. Möller, J. Thomas, T. Gemming, phys. stat. sol. (a) **202,** R170 (2006).
7. V. Beyer, J. von Borany, in: Materials for Information Technologies, edited by E. Zschech and H. Mikolajick, Springer Verlag, Berlin, 2005, 139.

8. P. Normand, E. Kapetanakis, P. Dimitrakis, D. Tsoukalas, K. Beltsios, N. Cherkashin, C. Bonafos, G. Benassayag, H. Coffin, A. Claverie, V. Soncini, A. Agarwal, M. Ameen, Appl. Phys. Lett. **83**, 168 (2003).
9. P. Normand, E. Kapetanakis, P. Dimitrakis, D. Skarlatos, K. Beltsios, D. Tsoukala, C. Bonafos, G. Benassayag, N. Cherkashin, A. Claverie, J.A. van den Berg, V. Soncini, A. Agarwal, M. Ameen, M. Perego, M. Fraciulli, Nucl. Instr. Meth. Phys. Res. B **216**, 228 (2004).
10. T. Müller, K.H. Heinig, W. Möller, C. Bonafos, H. Coffin, N. Cherkashin, G. Benassayag, S. Schramm, G. Zanchi, A. Claverie, M. Tence, C. Colliex, Appl. Phys. Lett. **85**, 2373 (2004).
11. B. Schmidt, K.H. Heinig, L. Röntzsch, T. Müller, K.H. Stegemann, E. Votintseva, Nucl. Instr. Meth. Phys. Res. B **242**, 146 (2006).
12. B. Schmidt, D. Grambole, F. Herrmann, Nucl. Instr. Meth. B **191**. 482 (2002).
13. L. Röntzsch, K.H. Heinig, B. Schmidt, Mater. Sci. Semicond. Proc. **7**, 357 (2004).
14. R. J. Walters, G. I. Bourianoff, and H. A. Atwater, Nature Mater. **4**, 143 (2005).
15. A. Puzder, A. J. Williamson, J. C. Grossman, and G. Galli, J. Chem. Phys. **117**, 6721 (2002).
16. V. Beyer, B. Schmidt, K.-H. Heinig, and K.-H. Stegemann, Appl. Phys. Lett. **95**, 193501 (2009).
17. J. Carreras, J. Arbiol, B. Garrido, C. Bonafos, J. Montserrat, Appl. Phys. Lett. **92**, 091103 (2008).
18. P. Dimitrakis, P. Normand, E. Votintseva, K.-H. Stegemann, K.-H. Heinig, B. Schmidt, J. Phys.: Conf. Ser. **10**, 7 (2005).

Fabrication Of Nanomechanical Devices Integrated In CMOS Circuits By Ion Beam Exposure Of Silicon

G.Rius [a,c], J. Llobet [a], X. Borrisé [a,b] and F. Pérez-Murano [a]

[a]*Instituto de Microelectronica de Barcelona (IMB-CNM, CSIC). Campus de la Universitat Autonoma de Barcelona 08193-Bellaterra. Spain*

[b]*Institut Català de Nanotecnologia (ICN). Campus de la Universitat Autonoma de Barcelona 08193-Bellaterra. Spain*

[c]*Surface Science Laboratory, Toyota Technological Institute, 2-12-1 Hisakata, Tempa-ku, Nagoya 468-8511, Japan*

Abstract. We present a novel approach for the fabrication of nanomechanical devices integrated in CMOS circuits. It is based on ion beam patterning by direct exposure of silicon or poly-silicon surfaces and standard microfabrication silicon etching. We have studied the optimal processing conditions in terms of: i) selectivity, and ii) resolution (ion beam fluence and patterning mode, reactive ion etching parameters), as well as, iii) the procedure to ensure that electronic devices in CMOS circuits are not damaged by the ion beam processing. As a result, nanomechanical devices have been successfully patterned

Keywords: Focused Ion Beam, Nanomechanics, NEMS, CMOS integration.
PACS: 85.85.+j, 85.40.Hp, 81.16.-c, 84.40.Lj

INTRODUCTION

Since its early development, the semiconductor microelectronics industry has been using photolithography as a way to pattern the diverse layers of integrated circuits. A photolithography process flow requires (see figure 1) coating the substrate with a photosensitive resist by spinning or spraying, irradiating it with UV light through a mask containing opaque patterns (selective exposure) and finally developing the resist [1]. Along the years, the resolution of photolithography has continuously been improved through the use of (i) shorter wavelength in projection exposure tools, and (ii) optimized resists and masks. The most modern implementation of photolithography is deep UV (DUV) immersion lithography, which achieves 22 nm half pitch resolution [2]. However, DUV lithography requires extremely high investment costs for the exposure system and related tools, resists and masks, only accessible for industrial applications where massive fabrication can compensate the investment costs. Even for microelectronics, the cost of lithography may limit or slow down the miniaturization trend.

This situation encourages the development of alternative nanopatterning methods that could combine high resolution and high throughput, together with lower costs assumable for research activities, prototyping and production of small series micro/nano integrated systems. Examples of such methods are nanoimprint [3], nanostencil lithography [4], scanning probe lithography [5] and several forms of charged beam mask-less lithography, like e-beam [6] or ion-beam lithography [7].

We present our recent results of the use of ion beam exposure of silicon in combination with selective silicon etching as a new approach to integrate nanomechanical devices in CMOS circuits.

PATTERNING OF SILICON BY ION BEAM EXPOSURE AND SILICON ETCHING

The approach that we present for fast patterning of silicon at the nanometer scale relies on the fact that the incident ion beam particles locally modify the silicon. A robust mask for reactive ion etching (RIE) can be obtained by the simple irradiation of the silicon surface

Application of Accelerators in Research and Industry
AIP Conf. Proc. 1336, 239-242 (2011); doi: 10.1063/1.3586095

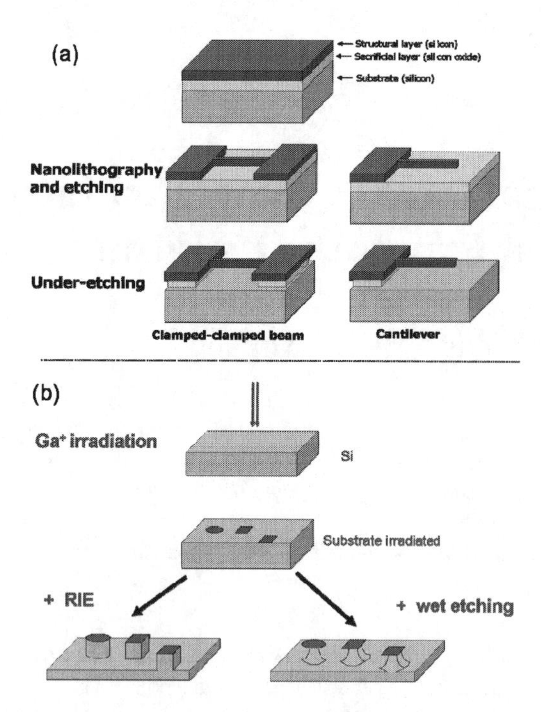

FIGURE 1. (a) Standard photolithography process flow. (b) A robust mask for RIE and wet etching can be obtained by local irradiation of the silicon surface by the Ga+ beam.

by the Ga^+ beam (see figure 1(b)). Previous works have reported the use of Ga^+ FIB exposure as a mask for wet etching of silicon [8-10], as well as a mask for reactive ion etching [11-12]. The SEM image of figure 2 (a) shows a line array pattern realized by irradiating the silicon surface with a fluence of $1.6{\cdot}10^{11}$ ions/cm and transferring the pattern into the silicon by RIE.

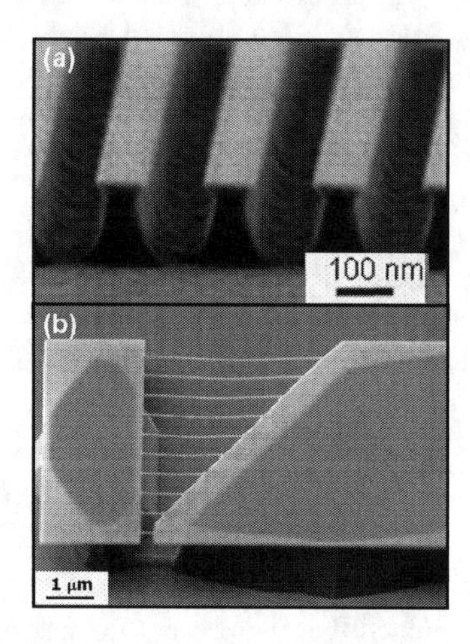

FIGURE 2. (a) SEM image that shows a line array pattern realized by irradiating the silicon surface with a fluence of $1.6{\cdot}10^{11}$ ions/cm and reactive ion etching. (b) SEM image showing an exposed pattern consisting of a series of thin parallel line and two squared areas. After wet etching in TMAH, the lines are under-etched, so that they become released but laterally anchored to the substrate.

The RIE process is performed by an inductive coupled plasma process using a Plasmalab80+ system from Oxford Instruments Plasma Technology, with the following etching conditions: simultaneous introduction of SF6 (20 sccm) and C4F8 (30 sccm), pressure of 15 mTorr, electrode radiofrequency power of 20 W and coil radiofrequency power of 220 W.

It is extremely relevant the robustness of the Ga^+ irradiated silicon as a RIE mask. This fact is later confirmed when performing a wet etching (i.e. higher isotropy), as it is shown in the SEM images of figure 2(b). In this case, the exposed pattern consists of a series of thin parallel lines and two squared areas. After wet etching in TMAH, the lines are under-etched, so that they become released and anchored to the substrate through the lateral areas. The thickness of the lines is around 25 nm, which fits with the penetration depth of 30 keV Ga^+ implanted in silicon.

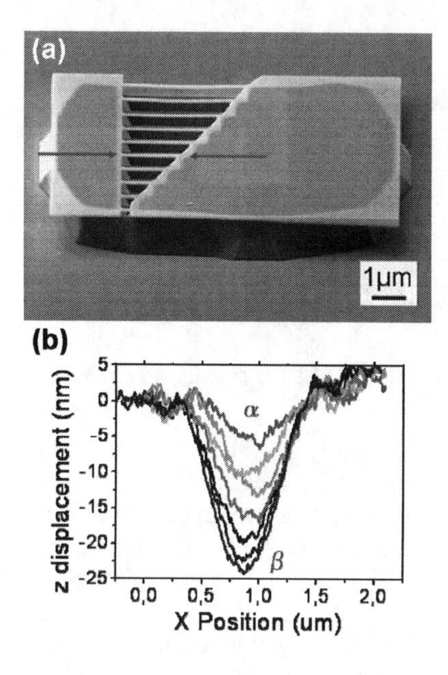

FIGURE 3. Mechanical properties of the resulting beams studied by AFM. (a) The AFM tip is scanned along the beam marked with an arrow in contact mode for different values of the contact force. (b) The deflection of the beam is recorded for each loading force.

The resulting beams are interesting for building-up nano-electromechanical systems (NEMS) because of its small size. Analogous example of NEMS fabrication using ion beam exposure and etching has been present by Bischoff et al. [13]. We have studied the mechanical properties of the resulting beams by atomic force microscopy (AFM) (figure 3), following a similar procedure as describe in [14]: the AFM tip is scanned along the beam in contact mode for different values of the contact force. Then, the maximum deflection of the beam is recorded for each applied force. From these data, one can extrapolate the elastic constant of the beam. As an example, for the beam marked with an arrow in figure 3(a), we have derived an elastic constant of 23 +/- 2 N/m. These preliminary results indicate that the beam has a Young Modulus at least as high as of silicon.

CMOS INTEGRATION

Nanomechanical resonators offer good performance to be used as high resolution mass sensors [15-19]: their change of resonance frequency is directly related with the added mass on top of the resonator. The sensitivity increases when the resonator is made smaller, since its relative change of mass becomes larger. However, when submicron resonators are made, detection of their oscillation becomes increasingly challenging.

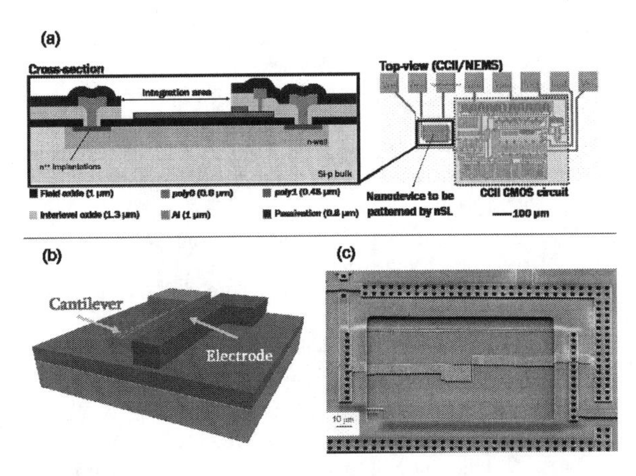

FIGURE 4. (a) Sketch showing the area dedicated to the nanopatterning (the so-called integration area). Integration area is directly connected to the close integrated circuit. (b) Sketch of the nanomechanical device: an oscillating cantilever driven by the electrode that also enables for capacitive detection. (c) SEM image of a nanomechanical device monolithically integrated into a CMOS circuit fabricated by a combination of Ga^+ beam exposure and RIE.

Following an all-electric actuation and detection scheme (based on capacitive transduction, figure 4(b)), a monolithic integration provides an optimized readout of the output signal through a decrease of parasitic capacitances at NEMS output [20]. In comparison with heterogeneous integration or direct external connections, monolithic integration allows the signal to be subsequently amplified and conditioned 'on chip'. This approach opens perspectives of closed-loop operation as stand-alone electromechanical oscillators. NEMS/CMOS devices also offer unique advantages in terms of compactness and packaging for portable applications like sensors or electronic devices (mobile phones, PDAs, etc).

Our integrative fabrication approach consists of using standard layers of a CMOS technology as the structural and sacrificial layers of the nanomechanical device, respectively. Device patterning is defined by means of a dedicated post-process module (figure 4 (a)). The fabrication of mechanical structures with critical dimensions approaching to 100 nm requires the use, integration and compatibility with advanced patterning methods.

Figure 4 (c) shows a nanomechanical device monolithically integrated into a CMOS circuit, in particular, fabricated by a combination of Ga^+ beam exposure and RIE. In this case, the structural layer is poly-silicon, which is the same material layer used for defining the gate of the MOS transistors in the CMOS circuits. Again, Ga^+ exposure of this material resulted in an efficient mask for RIE. Specifically, the device consists of two devices composed of submicron size cantilevers and the actuating/read-out electrode [21]. As the NEMS integration area is larger than the field size of the FIB system, the device is patterned by several consecutive exposures with some overlapping to compensate for stitching errors. As it can be observed, the resolution is good and, apparently, limited by the intrinsic granular morphology of the poly-silicon layer.

It is known that the radiation caused by charged beams can induce damage to CMOS circuits [22]. In order to evaluate the collateral effects of present fabrication approach, we have analyzed 'on-line' the effect that electron and ion beam irradiation of CMOS substrates produces on electronic devices. We have used a test set-up that mimics the conditions derived from our patterning strategy. The NEMS fabrication area is identical to the area used for integrating the nanomechanical devices into the CMOS circuit. A NMOS transistor is located very close to the NEMS fabrication area. Electrical contacts to the transistor are performed through the metal pads and lines (the same that are used for the standard probe station electrical characterization of the CMOS devices) by means of the micropositioners needles installed within the chamber of the dual beam system. Therefore, this system allows us to polarize on line and *in situ* the NMOS transistor, simultaneous to ion beam exposure.

A typical behavior observed when submitting the CMOS/NEMS to consecutive different exposures is shown in figure 5. More details can be found in [23]. During the exposure, the NMOS transistor is polarized with a gate-source voltage of 5 V and a drain-source voltage of 5 V (on state). The drain source current is recorded as a function of time during the exposure experiments to monitor the eventual change of electronic device operation. First, the NEMS fabrication area is exposed to the ion beam using the

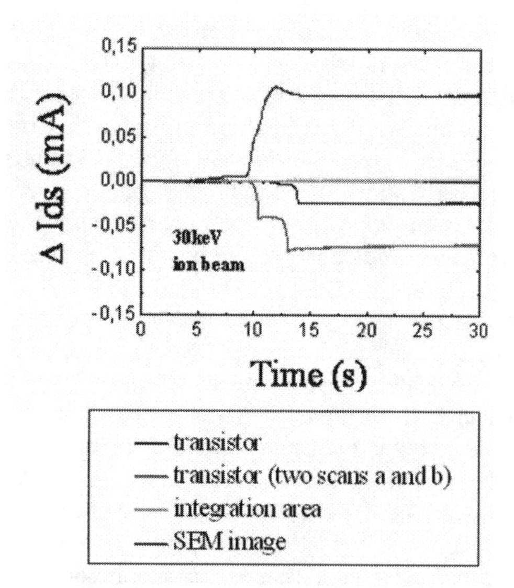

FIGURE 5. On line detection of the change of the drain source current when the NMOS transistor is polarized at 5 V for four different types of exposures. A NMOS transistor is located very close to the NEMS fabrication area.

same fluence as that used during patterning. No significant change in the current level is detected. The second and third exposures consists on a single (second) and double (third) exposures over the transistor. They are performed by fast scanning one time a square area of 20 x 35 μm^2 covering the transistor location. It is found that only when the ions are incidental to the active area of the transistor, the current decreases. As a final experiment, we have repeated the second exposure using the same fluence but with 5 keV electrons. In this case, the current is recovered to practically the initial level, indicating effective charge neutralization.

The response of the NMOS transistor to the exposure experiments indicates that the post-processing of CMOS circuits with ion-beams does not produce damage to the CMOS devices if the exposure is restricted to the NEMS fabrication area and exposure of the transistor is avoided. In addition, as dual-beam systems are often used for the realization of ion-beam induced patterning, it is important to consider and minimize the irradiation of the CMOS circuits with electrons. It is worth to mention that, as the penetration range for ions is much shorter than for electrons, it results even safer to perform the pattern with ions rather than with electrons. Being necessary to take into account the ion beam related processes involved, as hereby has been presented, this can represent an opportunity to establish simpler and submicron patterning methodology for the fabrication of NEMS.

ACKNOWLEDGMENTS

This work has been partially supported by projects CHARPAN, NMP-CT-2005-515803, and NANOFUN, TEC2009-14517-C02-01.

REFERENCES

1. M. J. Madou. Fundamentals of microfabrication. CRC Press. 2002
2. M. P. Brown, and K. Austin. Appl. Phys. Lett. 85, 2503 (2004).
3. H. Schit. J.Vac Sci Technol. B 26, 458-480 (2008)
4. J. Arcamone, A. Sánchez-Amores, J. Montserrat, M. A. F. van den Boogaart, J. Brugger, F. Pérez-Murano. J. Micro/Nanolith. MEMS MOEMS 6, 013005 (2007)
5. A. Knoll, D. Pires, O. Coulembier, Ph. Dubois , J. Hedrick , J. Frommer, U. Duerig. Adv.Mat. 22, 3361 (2010)
6. E. Platzgummer, H. Loeschner, G. Gross, Photomask Technology 6730, 73033 (2007)
7. A. A. Tseng, Small 1, 594 (2005)
8. B. Schmidt, L. Bischoff, J. Teichert, Sensors and Actuators A, 61, 369 (1997)
9. L. Bischoff, B. Schmidt, H. Lange, D. Donzev, Nuclear Instruments and Methods in Physics Research B, 267, 1372 (2009)
10. J. Brugger, G. Beljakovic, M. Despont, N. F. De Rooij, P. Vettiger. Microelec.Eng. 35, 401 (1997)
11. H. X. Qian, W. Zhou, J. Miao, L. E. N. Lim, X. R. Zeng. J. Micromech. Microeng. 18, 035003 (2008)
12. N. Cherukov, K. Grigoras, A. Peltonen, S. Franssila, I. Tottonen. Nanotechnology, 20, 065307 (2009)
13. B. Schmidt, S. Oswald, L. Bischoff. Journal of the Electrochemical Society, 152, G875 (2005)
14. S. G. Nilsson, X. Borrisé, and L. Montelius. Applied Physics Letters 85, 3555 (2004)
15. K. L. Ekinci, Y. T. Yang, M. L. Roukes. J. Appl. Phys. 95, 2682 (2004)
16. J. Verd, A. Uranga, G. Abadal, J. Teva, F. Torres, F. Pérez-Murano. Appl. Phys. Lett. 91, 013501 (2007)
17. J. Arcamone, M. Sansa, J. Verd, A. Uranga, G. Abadal, N. Barniol, M. van den Boogaart, J. Brugger, F. Perez-Murano. Small, 5, 176-180 (2009)
18. B. Lassagne, D. Garcia-Sanchez, A. Aguasca, A. Bachtold. Nanoletters 8, 3735 (2008)
19. K. Jensen, K. Kim, A. Zettl, Nature Nanotechnology 3, 5333 (2008)
20. G. Abadal, Z. J. Davis, B. Helbo, X. Borrisé, R. Ruiz, A. Boisen, F. Campabadal, J. Esteve, E. Figueras, F. Pérez-Murano , N. Barniol. Nanotechnology, 12, 100 (2001)
21. J. Arcamone , G. Rius , J. Llobet , X. Borrisé and F. Pérez-Murano. J. Phys.: Conf. Ser. 100 052031 (2008)
22. F. Campabadal, S. Ghatnekar, G. Rius, C. Fleta, J. Rafí, E. Figueras, J. Esteve. Proc. SPIE, 5836, 667 (2005)
23. G. Rius, J. Llobet, M. J. Esplandiu, L. Solé, X. Borrisé, F. Perez-Murano. Microelec.Eng. 86, 892 (2009)

First Results From A Multi-Ion Beam Lithography And Processing System At The University Of Florida

Brent Gila[1], Bill R. Appleton[1], Joel Fridmann[2] ,Paul Mazarov[3], Jason E. Sanabia[2], S. Bauerdick[3], Lars Bruchhaus[3], Ryo Mimura[3] and Ralf Jede[3]

[1]*NIMET Nanoscale Research Facility, University of Florida, Gainesville, FL. 32611*
[2]*Raith USA, Inc., Ronkonkoma, NY, 11779*
[3]*Raith GmbH, 44263 Dortmund, Germany*

Abstract. The University of Florida (UF) have collaborated with Raith to develop a version of the Raith ionLiNE IBL system that has the capability to deliver multi-ion species in addition to the Ga ions normally available. The UF system is currently equipped with a AuSi liquid metal alloy ion source (LMAIS) and ExB filter making it capable of delivering Au and Si ions and ion clusters for ion beam processing. Other LMAIS systems could be developed in the future to deliver other ion species. This system is capable of high performance ion beam lithography, sputter profiling, maskless ion implantation, ion beam mixing, and spatial and temporal ion beam assisted writing and processing over large areas (100 mm2) – all with selected ion species at voltages from 15 – 40 kV and nanometer precision. We discuss the performance of the system with the AuSi LMAIS source and ExB mass separator. We report on initial results from the basic system characterization, ion beam lithography, as well as for basic ion-solid interactions.

Keywords: FIB, LMAIS, Au ion, Si ion, GaAs
PACS: 81.16.Rf, 81.05.Ea, 81.15.Jj, 81.16.Nd

INTRODUCTION

The abilities and applications of focused ion beam systems appear to be limitless. Manufacturers and users alike are discovering more ways to create unique 3 dimensional structures from ion beam removal of materials, chemical vapor assisted deposition and etching and create uniquely modified surfaces, all scaled on the nanometer dimensions. Throughout the literature for over 20 years(1-3 and references within), FIB has be mostly employed with a gallium beam due to: its liquid metal state just above room temperature, low volatility, low reactivity with the needle material, low vapor pressure, and excellent vacuum and electrical stabilities. This seems to meet the needs for the majority of the FIB users and applications. While there are several focused ion beam tool s on the market meeting many of the conventional users needs (TEM sample preparation, slice and view 3D characterization, single process step applications), many of these are not feasible for step and repeat processing over large areas and offer other options for ion sources. For this reason, we have collaborated in developing an ion beam lithography (IBL) system that is flexible for user specific applications with multi-ion species sources. The use of non gallium ions sources has also endured for over 20 years (4-7), but the number of these tools is small compared to the gallium ion source tools operating today and most are home built by the laboratories operating them.

IBL SYSTEM OVERVIEW

The Raith ionLiNE system is a lithography tool capable of direct write focused ion beam patterning at nanometer level resolution and accuracy over large areas. The ion column is normal to the stage allowing for an ion beam centric system for high resolution, small working distances and compact optics. The laser interferometer stage is designed for maximum stability for nanometer scale reproducibility and enables this system to excel at lithography over more conventional FIB systems. This stage enables to ionLiNE user to perform "blind writes" onto sample areas that are up to 100mm away with nanometer precision. Stitching and overlay specifications for the system are both 60nm and we achieved 16nm and 37nm respectively (mean + 2σ) after installation at our facility at the University of Florida. The computer control software has a user friendly interface, advanced GDSII editor and pattern

Application of Accelerators in Research and Industry
AIP Conf. Proc. 1336, 243-247 (2011); doi: 10.1063/1.3586096

generator interface for imaging, exposures and beam control. These features allow for a system that provides: automated patterning tool, multiple process step and repeat, advanced beam control and GDSII editor for complex designs, large area process by automated stitching or continuous writing, automated mark recognition and alignment for overlay patterning. In addition to this list, the stage/pattern generator combination allows for fixed beam moving stage (FBMS) exposures. This exposure mode fixes the beam in a position and moves the stage, similar to a machinist endmill, to allow for designs to be created, without conventional stitching and associated stitching error, that are over a meter long. A gas injection system was added to further expand capabilities.

Since the primary application for the Raith ionLiNE (as for any lithography tool) is patterning and not imaging the area of interest, the Raith ionLiNE does not employ an electron column as is done for the conventional SEM/FIB systems. Non-destructive sample navigation is accomplished with an optical macroscope (coarse positioning) and by the laser interferometer stage (fine positioning). High resolution imaging is performed with the ionLiNE by detection of ion induced secondary electrons emitted from the sample surface. For lithography, the important function of imaging is mark detection and registration, which allows for nanometer accuracy in placement of the focused ion beam pattern onto the area of interest.

The Raith Ga ion column is the standard column shipped with the ionLiNE systems. It is capable of beam energies of 15-40kV and provides a beam current stability of less than 1% change per hour. This column is designed in a way that allows for an additional aperture and mass filter to be inserted below the beam limiting aperture. This provides the possibility of using different gallium isotopes and charged species for greater control of surface modifications, implantations, and milling. This also allows for the traditional gallium source to be removed and other liquid metal alloy ion sources (LMAIS) to be employed. This opens the door for a world of possible ions and energies to be used for nanofabrication and modification with only slight modifications to the ion column and software.

Chamber vacuum is achieved via a turbo molecular pump and the ion column is differentially pumped with an ion gettering pump. Typical vacuum levels are low to mid E-7 mbar for the column and low to mid E-8 mbar for the ion column. The loadlock shares the pumping system with the chamber while the ion column is isolated from the chamber via a gate valve during loadlock routines.

Au-Si LMAIS

The initial system arrived with a traditional gallium source and we used this to familiarize ourselves with the tool and the techniques. This Ga ion source was used to determine stage operating parameters and run stage accuracy tests and drift tests. We did employ the Ga ion beam in resist exposures tests and processed a number of devices in GaAs based materials relying on the FBMS speed and PMMA's higher sensitivity to ion beam exposure, typically 3 to 4 times higher sensitivity. We created an isolation pattern around a pair of ohmic contacts on GaAs and isolation was provided by inductively couple plasma (ICP) etching into the GaAs based epi layers. The total length of the isolation structure was 1.2mm and required approximately 7 seconds to expose, this work will be published elsewhere.

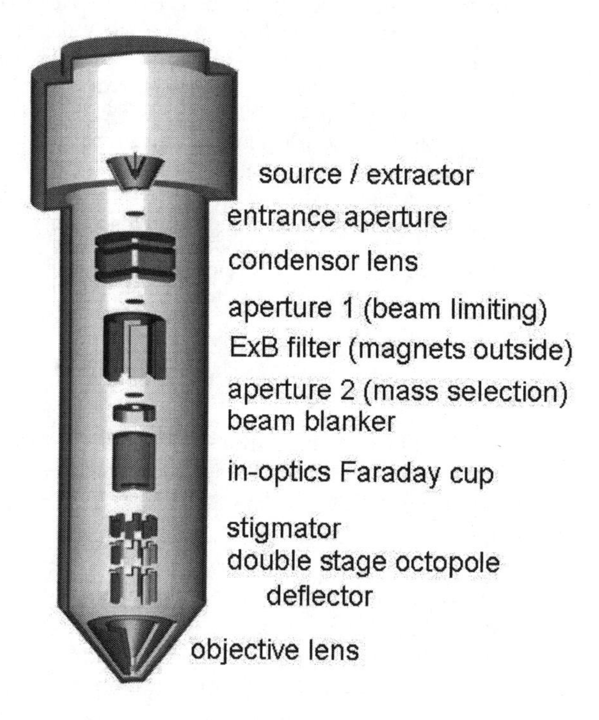

Figure 1. Illustration of the Raith ionLiNE column with mass filter (ExB) installed.

The column was then modified by adding the mass filter (also known as a Wien filter) and an additional aperture to provide for mass resolution. An illustration of the column and its components are shown in figure 1. The apertures located in position 1 (beam limiting) and 2 (mass limiting) range from 1000μm down to 5μm which allow for beam currents from 1nA for high rate milling to 1 pA for imaging and resist exposures. Additional software modules were added to the basic system software to control the second aperture and the electric field for the mass filter. The addition of the mass filter assembly creates more modes to operate the column beyond the selectable beam current. This mass filter is a mass trajectory filter separating the paths of the different ion masses based upon the applied electric field, the magnetic field and the velocity of the ions. The additional lower aperture is used to further block unwanted ions and isotope ions. The beam crossover position within the ExB filter affects the beam spot and the mass resolution. This provides two resolution modes to operate the column and the mode depends on the desired feature or process. Best imaging resolution mode, which supplies the smallest spot size, occurs when the beam cross-over is located within the center of the ExB filter. Best mass resolution mode, which supplies the larges mass separation, occurs when the beam is at cross-over at the second aperture. The position of the beam cross-over is controlled by the condenser lens for a given magnetic field. Several magnetic field strengths are available and we selected 1050 Gauss.

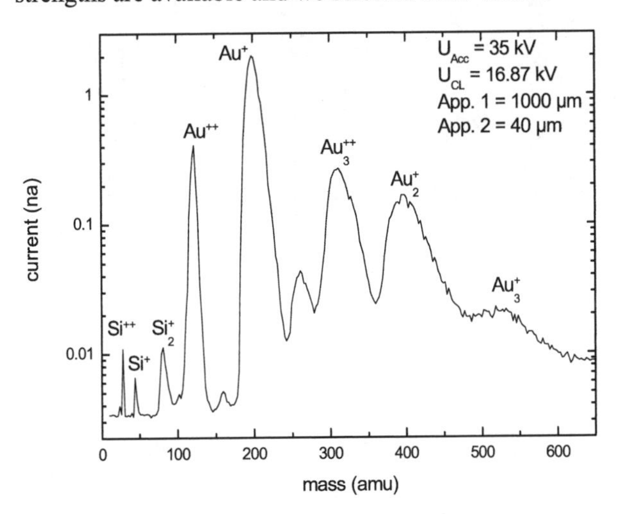

Figure 2. Mass spectrum for a typical Au-Si LMAIS in the Raith ionLiNE column.

The alloy source we selected to use first was an Au-Si eutectic (~30 at% Si). This source provides a wide range of ions for milling, doping, surface modification, chemical vapor assisted deposition and imaging. The mass spectrum from our Au-Si LMAIS is shown in figure 2. The condenser lens setting for this data collection was set for best mass resolution. As shown, there are a large variety of ions and ion clusters available for processing. Doubly charged ions and ion clusters expand the capabilities of the system by allowing higher and lower ion energies, respectively. If a doubly charged species is selected (such as Si++ or Au++) the effective acceleration voltage is double the column setting. This allows for a larger range of effective beam energies without the need for higher column voltages. If an ion cluster such as Au^+_2 is selected, then the landing voltage when the ion cluster breaks on the surface is half of the column voltage. The effects of these different ions and clusters on milling and deposition are being currently investigated.

Processing GaAs

Gallium Arsenide is a widely used semiconductor material and its roll and application has benefitted from FIB processing and modification (8-10). Several other semiconductor in the III-V family (III-P, III-Sb, III-As) have also benefitted from FIB technology(11-13) however these tend to be based more on surface modification and less on milled feature dimensions. In nearly all accounts of Ga ion beam processing of III-V materials (except for the III-Nitride materials) the results of the milling process lead to a roughened surface with a high density of Ga droplets forming on the milled surface. Techniques to remove these Ga droplets include in-situ gas assisted etching with iodine (14) or post FIB etching. We investigated processing III-V semiconductors with an Au ion beam and a Si ion beam to determine if a cleaner milling surface was feasible without requiring extra processing. According to TRIM calculations, the sputter yield of GaAs for a 30kV ion beam at an angle of 0° (in terms of Ga:As atom yield, modeled at 1000 incident ions) is 1.87:3.77 for Si+ ions, 2.92:6.35 for Ga+ ions, and 3.42:7.61 for Au+ ions, an obvious trend of heavier mass and shorter projected range results in higher material removal. A sample of undoped, semi-insulating (001) GaAs was milled with a 30kV Ga+ ion beam. Milling parameters such as beam current, dwell time and scan loops were altered to produce the lowest number of Ga droplets and smoothest sidewall and bottom features.

For the Ga ion beam, the beam current was changed from 100pA to 30pA to 6pA and the scan loops were varied from 5000 to 50000, while fluence was kept in the range of 30,000 µC/cm2 to 70,000 µC/cm^2. The lowest Ga droplet formation was found with a low beam current and lower scan loop counts. For scan loop counts over 20,000, the dwell time was too short for the pattern generator to manage, causing roughening of the sidewall and bottom. The effect of fluence seems to only affect the depth of the milled feature (within the range we studied). The images of these features, Figures 3 a-c, are similar to those found in the literature. The depth of the 70,000 µC/cm^2 milled feature is approximately 550nm.

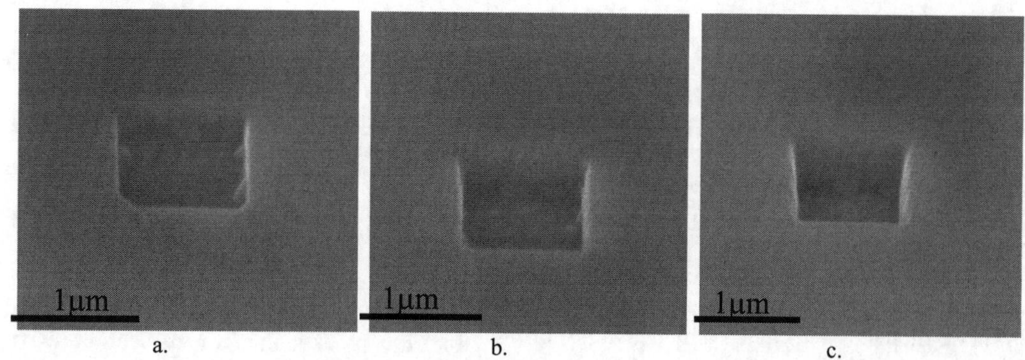

Figure 3. GaAs milled with a Ga+ ion beam at a 6pA beam current, 5000 scan loops, and a) 30,000μC/cm^2, b) 50,000 μC/cm^{2}and c) 70,000μC/cm^2. Ga droplets are visible at the bottom of each feature. Boxes are 1μm square tilted at 45° for imaging.

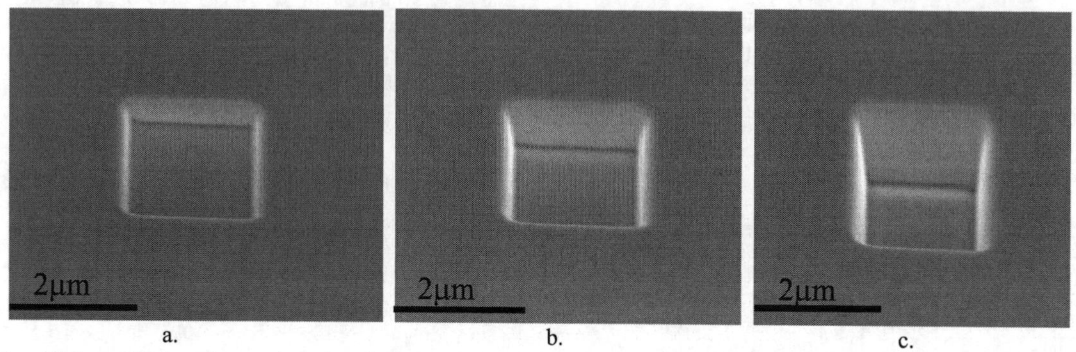

Figure 4. GaAs milled with an Au+ ion beam at a 30pA beam current, 100 scan loops, and a) 30,000μC/cm^2, b) 50,000μC/cm^{2}and c) 70,000μC/cm^2. Ga droplets are not visible at the any of the milled features. Boxes are 2μm square tilted at 35° for imaging.

For the Au+ ion beam, the beam current was maintained at 30pA and the scan loops were maintained at 100, while the fluence was varied in the range of 30,000 to 70,000 μC/cm^2. These were the initial conditions attempted and as seen from the images, Figure 4 a-c, the only conditions required. There was no evidence of droplet formation from the Au+ milling. The effect of fluence is similar to the Ga ion milling, seems to only affect the depth of the feature milled. The depth of the 70,000 μC/cm^2 milled feature is approximately 1562nm, a surprisingly almost 3 times deeper than the Ga ion beam milling. The lack of Ga droplets seems to indicate that the droplet formation is more dependent on the ion species impinging on the surface and less on the previously published conclusion regarding As desorption from local heating (15). This may be related to the implantation of Au in the subsurface at a much lower range, 12nm for Au and 17nm for Ga in GaAs for a 30kV beam condition from TRIM simulations. An additional factor is that an Au ion has a higher sputter yield for pure gallium (8.02) than a Ga ion has for pure gallium (6.77), leading to more efficient removal of surface gallium. This cause and effect is a current on-going investigation.

In addition to milling features into the surface of GaAs, the Si+ and Si++ ion beams were employed to ion implant n-type regions in GaAs. The idea of employing a focused ion beam to dope a semiconductor material is not new or unique (16-18). With the ion column control and reproducibility, we were able to mill alignment marks into the GaAs surface with an Au+ ion beam, then within a few minutes, have the Si ion beam ready for implantation without removing the sample from the FIB system. This saved both time and resources in producing prototype devices. The results of the Si ion implantation and device performance is to be published elsewhere.

SUMMARY

In summary, the collaboration efforts of the University of Florida and Raith have produced a unique FIB system that provides a user friendly interface, highly flexible beam and stage control, and a flexible column for multi ion sources. This system is capable of delivering ions from any practical LMAIS, and dozens of elements are available for these sources. A process that may currently seem unfeasible using the conventional Ga ion beam may become feasible with a different ion, like the case of the droplet free GaAs milling. Tools and concepts like these can only continue to expand the roll of FIB in the research and manufacturing fields.

ACKNOWLEDGEMENTS

This work was partially supported by grants from the Office of Naval Research and the Army Research Office.

REFERENCES

1. J. Gierak; Semicond. Sci. Technol. 24 (2009) 043001
2. J. Orloff; Rev. Sci. Instruments 64(5) p.1105 (1993)
3. L.A. Giannuzzi and F.A. Stevie, Introduction to Focused Ion Beams, Springer, 2005
4. L.W. Swanson, Nuclear Instruments and Methods in Physics Research 218, p.347 (1983)
5. J. Teichert, M.A. Tuinov; Meas. Sci. Technol. 4, p.754 (1993)
6. T. Ishitani, K. Umemura, Y. Kawanami; JVST B 6(3), p.931 (1988)
7. J. Gierak, A. Madouri, E. Bourhis, L. Travers, D. Lucot, J.C. Harmand; Microelectronic Eng. 87, p.1386 (2010)
8. A Lugstein, B. Basnar, E Bertagnolli; JVST B 20(6), p.2238 (2002)
9. J.G. Pellerin, D.P. Griffis, P.E. Russell; JVST B 8(6), p.1945 (1990)
10. Q. Wei, J. Lian, W. Lu, L. Wang; Phys Rev Lett 100, p.076103-1 (2008)
11. M. Taniwaki, S. Morita, N. Nitta: Applications of Accelerators in Res. and Ind. 20th Conf., p.524 (2009)
12. P. Schmuki, U. Schlierf, T. Herrmann, G. Champion; Electrochimica Acta 48, p. 1301 (2003)
13. A. Lugstein, M. Weil, B. Basnar, C. Tomastik, E. Bertagnolli; Nuclear Instr. And Methods in Phys. Res B 222, p. 91 (2004)
14. Y.L.D. Ho, R. Gibson, C.Y. Hu, M.J. Cryan, J.G. Rarity, P.J. Heard, J.A. Timpson, A.M. Fox, M.S. Skolnick, M. Hopkinson, A. Tahraoul; JVST B 25(4), p.1197 (2007)
15. J. H. Wu, W. Ye, B. L. Cardozo, D. Saltzman, K. Sun, H. Sun, J. F. Mansfield, R. S. Goldman; APL 95, p.153107 (2009)
16. D. Reuter, C. Riedesel, P. Schafmeister, C. Meier, A.D. Wieck; APL 82(3), p.481 (2003)
17. P.J.A. Sazio, J.H. Thompson, G.A.C. Jones, E.H. Linfield, D.A. Ritchie, M. Houlton, G.W. Smith; JVST B 14(6), p. 3933 (1996)
18. M.M. Hashemi, Y. Li, K. Kiziloglu, M. Wassermeier, P.M. Petroff, U.K. Mishra; JVST B 10(6), p.2945 (1992)

Comparison Of Electromagnetic, Electrostatic And Permanent Magnet Quadrupole Lens Probe-Forming Systems For High Energy Ions

Alexander D. Dymnikov and Gary A. Glass

Louisiana Accelerator Center/Physics Dept/, Louisiana University at Lafayette
320 Cajundome Blvd., Lafayette Louisiana 70506, United States

Abstract. The focusing system is an essential part of any ion microbeam system and focusing of MeV ion beams is generally accomplished using quadrupole lenses. There are two types of quadrupole lenses requiring the application of either voltage or current to provide the excitation, but there is also the possibility of utilizing lenses constructed from permanent magnets. All of these lens types have different advantages and disadvantages. Most microprobes employ electromagnetic quadrupoles for focusing, however electrostatic lenses have several advantages with respect to electromagnetic lenses, including significantly smaller size, no hysteresis effects, no heating, the utilization of highly stable voltage supplies, focusing which is independent of ion mass, and construction from industrial grade materials. The main advantage of the permanent magnetic lens is that it does not require the application of external power which can significantly reduce the overall lifetime cost. In this presentation, the short probe-forming systems comprised from all these types of quadrupole lenses are compared and the smallest beam spot size and appropriate optimal parameters of these probe-forming systems are determined.

Keywords: Ion microprobe; Quadrupole lens; Spherical aberration; Chromatic aberration; Doublet.
PACS: 41.75.-i; 41-85.-p; 41-85.Gy; 41-85.Lc; 41.90.+e

INTRODUCTION

As it is known the strong focusing quadrupole lenses were used as an alternative to the axial symmetrical lenses for high energy ions, unfortunately, the quadrupole lens is more complex than the axial symmetrical, in construction, in calculations and in operation. Also, the focusing quadrupole systems usually do not produce a regular image of the specimen because in two mutually perpendicular planes *xoz* and *yoz* they have different positions of the focal points and the main focal planes and different magnifications and aberrations. Quadrupole lenses have been used widely in accelerator applications where it is only necessary to focus ions into a very small spot, such as the high energy microprobe and the scanning ion microscope.

There are three types of quadrupole lenses: electromagnetic, electrostatic and permanent magnetic and all these lenses have different advantages and disadvantages. Most microprobes employ electromagnetic quadrupoles for focusing, however electrostatic lenses have several advantages with respect to electromagnetic lenses: (a) they can be made much smaller than electromagnetic lenses, (b) electrostatic lenses do not suffer from hysteresis effects, (c) negligible current is drawn from the power supply, and (d) stable voltage is more readily achieved than stable current in electromagnetic systems. In particular, the field strength required for focusing the beam onto the target is independent of ion mass, thus ideally suited for heavy-ion beams. Another advantage is that electrostatic lenses can be constructed from industrial-grade material that facilitates reproductions. The main advantage of the permanent magnetic lens is that it does not require the power supply so that the cost of operation of this lens can be minimum. The electromagnetic lens is the most expensive quadrupole lens due to the requirement for precision machining.

In this paper quadrupole doublets comprised from electromagnetic, electrostatic and permanent magnet quadrupole lenses are compared. The smallest beam spot size and appropriate optimal parameters of these probe-forming systems have been found.

OPTIMIZATION

We use our special program of optimizing analytical and numerical calculations to obtain the

Application of Accelerators in Research and Industry
AIP Conf. Proc. 1336, 248-252 (2011); doi: 10.1063/1.3586097

best design for our system. Beam focusing is understood as the result of non-linear motion of a set of particles through the system and the beam spot produced on the target is the result of this motion. The set of particles has a four-dimensional volume (the phase volume, or four-dimensional emittance $em_{xy}=em_x\, em_y$) and for a given brightness, the phase volume is proportional to the beam current and *vice versa.* The beam has an envelope surface inside which all particles of the beam are located. For a given phase volume (or beam current) the shape of the beam envelope can vary and the beam envelope is said to be *optimal* if the spot size on the target has a *minimum* value for a given emittance. Figure 1 schematically illustrates the configuration of a microprobe doublet focusing system. The beam of a given emittance is defined by a set of two matching slits: objective and divergence slits. For a given emittance em_{xy}, the shape of the beam envelope is the function of the half-width (or radius) r_1(r_{1x} and r_{1y}) of the objective slit and of the distance l_{12} between two slits. The size r_2 (r_{2x} and r_{2y}) of the second (divergence) slit is determined by the expressions: $r_{2x} = em_x \times l_{12}/r_{1x}$, $r_{2y} = em_y \times l_{12}/r_{1y}$. The optimal parameters r_{1x}, r_{1y1}, r_{2x}, r_{2y} and l_{12} determine the *optimal matching slits*. The *optimal probe-forming system* comprises the optimal excitations, optimal matching slits and optimal geometry. For each emittance we find the parameters of the *optimal matching slits*. We consider the non-linear motion of the beam accurate to terms of 3rd order for systems with the quadrupole symmetry.

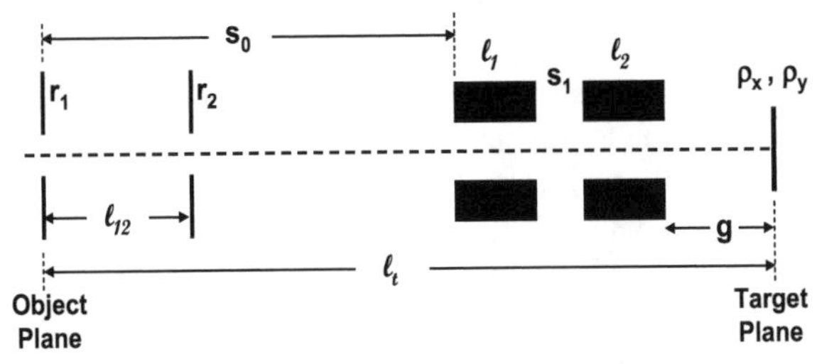

Fig.1 Doublet focusing system configuration

Matrix Approach

The essential feature of our optimization is the matrix approach for non-linear beam motion. In this approach we obtain and use analytical expressions for the matrizant (or transfer matrix) and for the envelope matrix using a technique known as the matrizant method [1, 2]. We use this technique for solving the nonlinear equations of motion in quadrupole lenses accurate to terms of the third order. These equations are replaced by two vector linear equations (for *x*- and *y*- planes) in the 12-dimensional phase moment space or by one equation in the 24-dimensional phase moment space. Writing the non-linear equations in a linearized form allows us to construct the solution using a 12×12 (or 24×24) third order matrizant. As a result of this linearization it becomes possible to use all the advantages of linear differential equations over non-linear ones, including the independence of the matrizant of the choice of the initial point of the phase space.

Assuming a uniform density of particles in the plane of the two slits, knowing the third order matrizant and choosing by a random method *N* particle we can obtain the position of these particles in the image plane or in the target (or specimen) plane.

From the nonlinear matrizants $R_x(z/0)$ and $R_y(z/0)$, where *z* is the longitudinal axis, we find the coefficients of spherical and chromatic aberrations. We use the following notations: d_x, d_y are the demagnifications, $C_{sx}, C_{sy}, C_{sxy}, C_{syx}$ are the spherical aberration coefficients in the object space and $c_{sx}, c_{sy}, c_{sxy}, c_{syx}$ are the spherical aberration coefficients in the image space, C_{px} and C_{py} are the chromatic aberration coefficients in the object space and c_{px}, c_{py} are the chromatic aberration coefficients in the image space. We suppose that the first slit is located at $z = 0$ and $z = z_{im}$ is the position of the target:

$$x(z_{im}) = R_{x1,1}(z_{im}/0)\, x_0 + \frac{dR_{x1,2}(z_{im}/0)}{dE} \Delta E\, x_0' + R_{x1,6}(z_{im}/0)\, (x_0')^3 + R_{x1,12}(z_{im}/0)\, x_0' \,(y_0^1)^2 \quad (1)$$

$$y(z_{im}) = R_{y1,1}(z_{im}/0)\, y_0 + \frac{dR_{y1,2}(z_{im}/0)}{dE} \Delta E\, y_0' + R_{y1,6}(z_{im}/0)\, (y_0')^3 + R_{y1,12}(z_{im}/0)\, y_0' \,(x_0^1)^2 \quad (2)$$

$$d_x = \frac{1}{R_{x1,1}(z_{im}/0)}, \quad d_y = \frac{1}{R_{y1,1}(z_{im}/0)} \tag{3}$$

$$\begin{aligned} C_{sx} &= R_{x1,6}(z_{im}/0), \quad C_{sy} = R_{y1,6}(z_{im}/0), \\ C_{sxy} &= C_{syx} = R_{x1,12}(z_{im}/0) = R_{y1,12}(z_{im}/0) \end{aligned}, \tag{4}$$

$$C_{px} = \frac{dR_{x1,2}(z_{im}/0)}{dE}E, \quad C_{py} = \frac{dR_{y1,2}(z_{im}/0)}{dE}E, \tag{5}$$

$$c_{sx} = \frac{C_{sx}}{d_x^3}, \quad c_{sy} = \frac{C_{sy}}{d}, \quad c_{sxy} = \frac{C_{sxy}}{d_x d_y^2}, \quad c_{syx} = \frac{C_{syx}}{d_y d_x^2}. \tag{6}$$

$$c_{px} = C_{px}/d_x, \quad c_{py} = C_{py}/d_y. \tag{7}$$

All considered systems have approximately the same geometry: approximately the same lengths of lenses and drift spaces, the same total length $l_t = 6.25$ *m* and the same set of working distances $g = 0.09$ *m*, 0.18 *m* and 0.27 *m*. All systems were considered for two four dimensional emittances: $em_{xy}=em_x\ em_y=10^{-18}\ m^2$ and $em_{xy}=em_x\ em_y=10^{-19}\ m^2$.

The Electrostatic Doublet As The Four Rod Design

We use the four rod design for the electrostatic doublet by using four ceramic rods 6-10 mm in diameter and 100-200 mm long for the entire set of two quadrupoles. Evaporating a thin layer of gold onto the cylindrical surface in bands creates the 8 positive and negative electrodes.

The Results of Calculations

The following results of calculations are shown in Tables 1-3.

1. Electrostatic systems have the biggest chromatic and spherical aberrations. For g = (0.09 m, 0.18 m, 0.27 m) we have c_{px}=(13.55 m, 12.66 m, 12.71 m), c_{py}=(11.39 m, 12.26 m, 12.22 m), c_{sx}=(44.50 m, 78.34 m, 171.9 m), c_{sy}=(0.7706 m, 6.712 m,17.69 m), $c_{sxy}=c_{syx}$=(-0.9139 m, -5.419 m, -15.07 m). For electromagnetic doublet for the same working distance these coefficients are equal: c_{px}=(6.48 m, 6.35 m, 6.31 m), c_{py}=(6.01 m, 6.15 m, 6.19 m), c_{sx}=(8.91 m, 22.30 m, 49.02 m), c_{sy}=(0.325 m, 1.957 m, 6.208 m), $c_{sxy}=c_{syx}$=(3.40 m, 15.5 m, 42.9 m).
2. Every optimal system for a given working distance and for a given emittance has approximately the same spot size ρ_x and approximately the same spot size ρ_y. For example, for the emittance $em_{xy}=10^{-18}\ m^2$ and for g = 0.09 m we have: for the electromagnetic doublet (ρ_x, ρ_y) = (0.3, 0.5); for the electrostatic doublet (ρ_x, ρ_y) = (0.35, 0.55).
3. For the electrostatic doublet the first optimal slit is approximately 10-20 percent bigger than for both magnetic systems. For example, for the emittance $em_{xy}=10^{-18}\ m^2$ and for g = 0.09 m we have: for the electromagnetic doublet ($r_{1,x}$, $r_{1,y}$) = (5.7μm, 26.2μm), for the electrostatic doublet ($r_{1,x}$, $r_{1,y}$) = (6.4μm, 32.8μm).
4. For the electrostatic doublet the second optimal slit is approximately 10-25 percent less than for both magnetic systems. For example, for the emittance $em_{xy}=10^{-18}\ m^2$ and for g = 0.09 m we have: for the electromagnetic doublet ($r_{2,x}$, $r_{2,y}$) = (604.8μm, 393.1μm), for the electrostatic doublet ($r_{2,x}$, $r_{2,y}$) = (475.8μm, 359.5μm).

SUMMARY

The probe-forming electromagnetic, electrostatic and permanent magnet quadrupole lens systems are compared. All considered systems have approximately the same geometry: approximately the same lengths of lenses and drift spaces, the same total length $l_t = 6.25$ *m* and the same set of working distances $g = 0.09$ *m*, 0.18 *m* and 0.27 *m*. Demagnifications and the coefficients of chromatic and spherical aberrations were found. All systems are considered for two four dimensional emittances:

em_{xy}=em_x em_y=10^{-18} m^2 and em_{xy}=em_x em_y=10^{-19} m^2. For these two emittances and for the set of working distances the size of optimal slits and the corresponding spot size were determined.

ACKNOWLEDGEMENTS

This work was supported in part by The National Science Foundation, Grant Numbers 0821693 and 0960222.

REFERENCES

1. A.D. Dymnikov and G.M. Osetinskij, Sov. J. Part. Nucl. (USA), **20**, 293-310 (1989).
2. A.D. Dymnikov and R. Hellborg, *Nucl. Instr. and Meth.* **A330**, 323-342 (1993).

Table 1. Electromagnetic Doublet

g[m]	d_x	d_y	$c_{sx}[m]$	$c_{sy}[m]$	$c_{sxy}=c_{syx}[m]$	$c_{px}[m]$	$c_{py}[m]$
0.09	-17.64	-90.21	8.905	0.3249	3.400	6.479	6.008
0.18	-14.92	-48.30	23.30	1.957	15.53	6.351	6.149
0.27	-11.07	-30.89	49.02	6.208	42.89	6.312	6.194

	$em_{xy}=em_x\ em_y=10^{-18}\ m^2$						$em_{xy}=em_x\ em_y=10^{-19}\ m^2$					
g[m]	$r_{1x}[\mu m]$	$r_{1y}[\mu m]$	$r_{2x}[\mu m]$	$r_{2y}[\mu m]$	$\rho_x[\mu m]$	$\rho_y[\mu m]$	$r_{1x}[\mu m]$	$r_{1y}[\mu m]$	$r_{2x}[\mu m]$	$r_{2y}[\mu m]$	$\rho_x[\mu m]$	$\rho_y[\mu m]$
0.09	5.656	26.24	604.8	393.1	0.3	0.5	2.385	11.07	453.6	294.8	0.2	0.12
0.18	5.629	18.22	698.3	492.5	0.5	0.6	2.374	7.685	523.7	369.4	0.24	0.24
0.27	5.494	14.82	754.6	556.9	0.55	0.8	2.317	6.351	565.9	417.6	0.24	0.34

Table 2. Electrostatic Doublet

g[m]	d_x	d_y	$c_{sx}[m]$	$c_{sy}[m]$	$c_{sxy}=cs_{yx}[m]$	$c_{px}[m]$	$c_{py}[m]$
0.09	-17.64	-90.21	44.50	0.7706	-0.9139	13.55	11.39
0.18	-14.92	-48.30	78.34	6.712	-5.419	12.66	12.26
0.27	-11.07	-30.89	171.9	17.69	-15.07	12.71	12.22

	$em_{xy}=em_x\ em_y=10^{-18}\ m^2$						$em_{xy}=em_x\ em_y=10^{-19}\ m^2$					
g[m]	$r_{1x}[\mu m]$	$r_{1y}[\mu m]$	$r_{2x}[\mu m]$	$r_{2y}[\mu m]$	$\rho_x[\mu m]$	$\rho_y[\mu m]$	$r_{1x}[\mu m]$	$r_{1y}[\mu m]$	$r_{2x}[\mu m]$	$r_{2y}[\mu m]$	$\rho_x[\mu m]$	$\rho_y[\mu m]$
0.09	6.421	32.84	475.8	359.5	0.38	0.4	2.708	13.85	356.8	269.6	0.15	0.17
0.18	7.641	24.74	516.1	361.6	0.5	0.55	3.222	10.43	387.0	271.2	0.22	0.24
0.27	7.057	19.7	754.6	566.7	0.6	0.65	2.976	8.308	424.9	324.8	0.28	0.3

Table 3. Permanent Magnetic Doublet

$g[m]$	d_x	d_y	$c_{sx}[m]$	$c_{sy}[m]$	$c_{sxy}=c_{syx}[m]$	$c_{px}[m]$	$c_{py}[m]$
0.09	-16.89	-88.99	13.97	0.2198	3.468	6.830	5.680
0.18	-14.21	-47.60	26.30	1.618	14.945	6.457	6.105
0.27	-11.62	-30.70	55.36	6.470	46.95	6.356	6.151

	$em_{xy}=em_x\ em_y=10^{-18}\ m^2$						$em_{xy}=em_x\ em_y=10^{-19}\ m^2$					
$g[m]$	$r_{1x}[\mu m]$	$r_{1y}[\mu m]$	$r_{2x}[\mu m]$	$r_{2y}[\mu m]$	$\rho_x[\mu m]$	$\rho_y[\mu m]$	$r_{1x}[\mu m]$	$r_{1y}[\mu m]$	$r_{2x}[\mu m]$	$r_{2y}[\mu m]$	$\rho_x[\mu m]$	$\rho_y[\mu m]$
0.09	4.547	23.96	660.4	500.1	0.35	0.55	1.917	10.10	495.3	375.1	0.12	0.22
0.18	5.314	17.80	701.4	530.3	0.5	0.6	2.241	7.508	526.0	397.7	0.2	0.26
0.27	5.670	14.99	721.9	558.7	0.6	0.8	2.391	6.319	541.3	418.9	0.24	0.32

Investigation Of Multi-Resolution Support For MeV Ion Microscopy Imaging

Harry J. Whitlow[a], Rattanaporn Norarat[a], Timo Sajavaara[a], Mikko Laitinen[a], Kimmo Ranttila[a], Pauli Heikkinen[a], Väinö Hänninen[a], Mikko Rossi[a], Pete Jones[a], Jussi Timonen[a], Leona K. Gilbert[b], Varpu Marjomäki[b], Minqin Ren[c], Jeroen A. van Kan[c], Thomas Osipowicz[c] and Frank Watt[c]

[a] Department of Physics, PO BOX (YFL), FI-40014 University of Jyväskylä, Finland
[b] Department of Biological and Environmental Sciences, PO BOX 35, FI-40014 University of Jyväskylä, Finland
[c] Center for Ion Beam Applications, Department of Physics, National University of Singapore, Singapore

Abstract. To minimize the dose applied to the specimens during imaging in a MeV ion microbeam we have investigated new concepts that allow collection of images with multi-resolution support. To test the concept, a set of reference PIXE microbeam images with well-characterised noise were segmented using both a direct down-sampling technique and wavelet decomposition. The results show both techniques could be used to select fields of view with <2% of the fluence required to collect a normal image, however, there is no compelling reason to select one technique over the other.

Keywords: Nuclear microbeam, ion microprobe, PIXE, biomedical, wavelet, image-decomposition
PACS: 42.30.Va, 42.66.Si, 87.85.Pq, 29.30.Kv

INTRODUCTION

Over the past 20 years, or so, wavelet-based image processing has attracted considerable attention. Images of natural objects such as images produced by ion microbeams are made up of features on a range of size scales. Image analysis using wavelets is based on image decomposition into a hierarchal tree of sub-images. These techniques facilitate image-analysis and manipulation on different scales simultaneously. Filter banks and down-sampling are used to give a multi-resolution representation with localisation in both spatial and frequency domains[1]. Wavelet approaches are powerful tools for image de-noising while preserving edge information and discontinuities, cryptography, image merging and "invisible" watermarking[1]. The industry-standard JPEG2000 image compression standard[1,2] uses a wavelet-based coding and decoding to facilitate both lossless and lossy compression that can allow progressive decoding of an image with progressively improving fidelity[1,3].

From an image-processing viewpoint an ion microprobe can be considered to be an instrument that sequentially samples an analogue image (2D-sample) by probing with an ion beam and quantises the information received (detector - data acquisition) to produce a digital image. The counting of quanta gives a linear mid-tread[1] quantisation (e.g. a Particle Induced X-ray Emission (PIXE) intensity map). Unlike normal images, such as a photographs and X-ray shadowgraphs, ion microprobe images are generally speckled. The intensity is generally Poisson distributed[4] where the probability p of detecting n quanta per pixel is $p(n) = \lambda^n ne^{-\lambda}/n!$. The mean number of counts λ give the variance, $\lambda = \sigma^2$. When λ is small, there is a significant probability, $p(0) \xrightarrow{\lambda \to 0} 1 - \lambda$ that there are no counts in a pixel. The significance of the zero-count pixels is that they carry no information and we cannot differentiate between pixels corresponding to zero intensity in the image function and zero counts from Poissonian noise[5]. The rapid transition over one pixel width from white to black implies the high-frequency information in these speckled images is significant.

The objective of this work was to investigate if a multi-resolution supported imaging strategy can be used to reduce the fluences needed for rapidly finding the field of view when using ion microprobes. This is particularly important when imaging biomedical

Application of Accelerators in Research and Industry
AIP Conf. Proc. 1336, 253-256 (2011); doi: 10.1063/1.3586098

specimens, where low–fluence techniques[6,7] are necessary to avoid detrimental ion induced changes in the specimen[8]. The work was motivated by the development of a new MeV ion microbeam (DREAM[9]) at the Pelletron accelerator in Jyväskylä that will feature high speed post-focus scanning coupled with a new generation of data collection system that is based on time-stamping to facilitate on-the-fly image transformation in lateral as well as time dimensions.

BASIS OF THE METHOD

The underlying idea is based on the different reaction of the human visual system to low- and high-frequency information in an image. For localisation of fields of interest it is generally sufficient to present only the low frequency information. The approach used is very akin to the classic filter-bank used to decompose images signals in wavelet analysis[1,3]. Fig. 1 a) illustrates the wavelet decomposition filter bank.

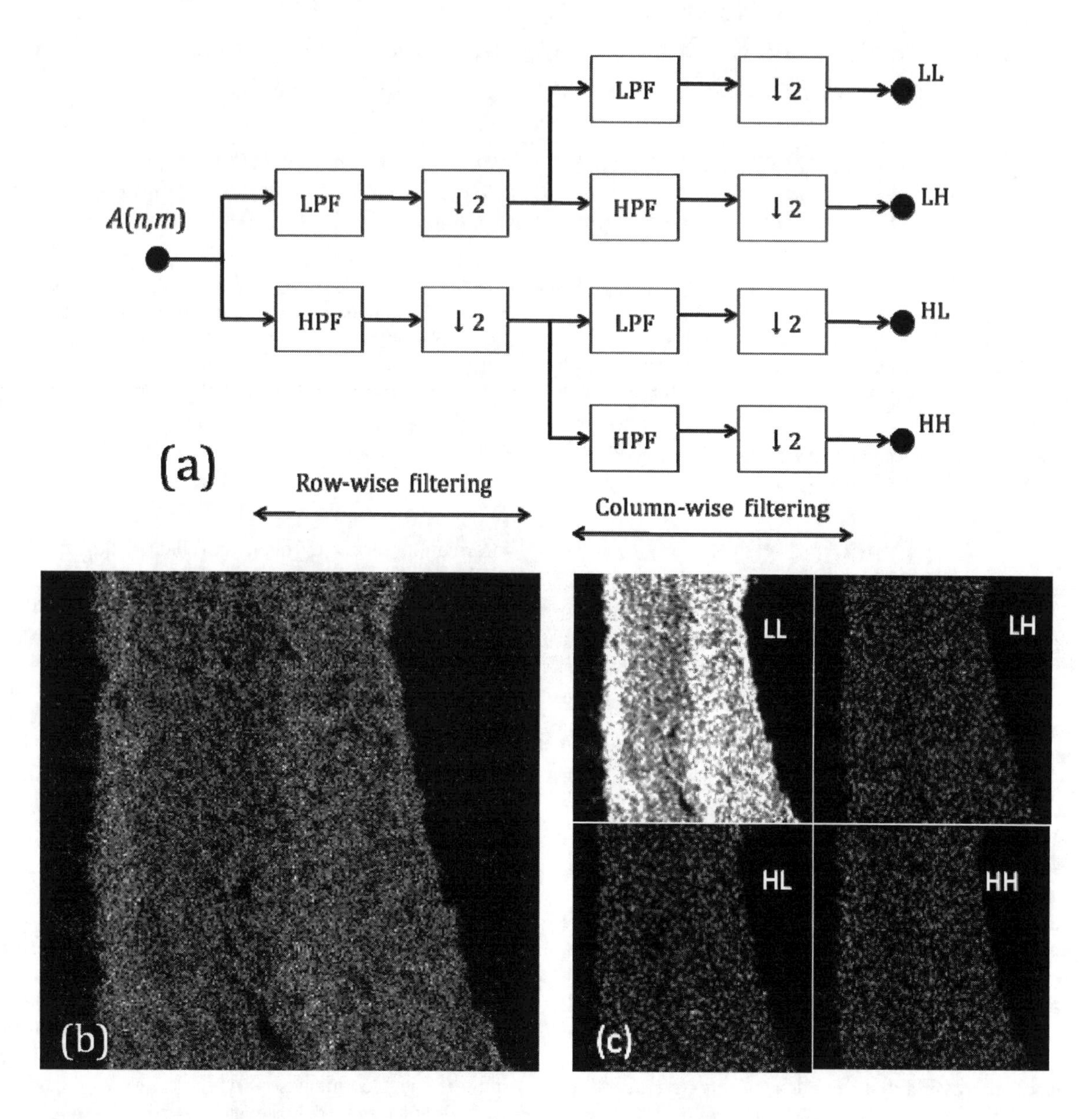

FIGURE 1. (a) Filter bank for wavelet decomposition. The input image *A(n,m)* is fed to a set of low-pass (LPF) and high-pass filters (HPF) followed by a down-sampling step. The filters act first row–wise and then column-wise to decompose the image into different bands LL, LH, HL and HH in the horizontal and vertical directions. (b) K Kα image *A*(*n*,*m*) (see text) used as input to the filter bank. The size of the field is 1100 ×1100 μm (c) decomposed images from the wavelet filter bank. (The images have been adjusted to give optimal printed contrast).

Fig. 1 b) is the ion microbeam image. Fig 1 c) the first level decomposition. It is seen the LL band image is a lower resolution image of the original image, where the low-frequency information is preserved. The high-frequency information in LH, HL and HH (Fig. 1 (c)) can be used for de-noising and signal compression[1,3]. Unlike normal wavelet decimation, which takes place in one shot, we decimate the image on-the-fly to concentrate the counts into a series of lower resolution images. In this way, zero pixels associated with Poissonian noise are suppressed making the image easier to visually interpret.

EXPERIMENTAL

A PIXE image of K Kα X-rays from a thin section of a rabbit aorta measured using 2 MeV protons measured at the National University of Singapore was used as a test. The image was chosen as it represents a typical natural image with a region of pixels with zero counts and an area with fairly, but not completely, uniform distribution of K. In the uniform contrast region, the counts per pixel were close to Poisson distributed over more than three decades with a mean of 1.85, a maximum of 7 counts per pixel, and ~33% of pixels had zero counts.

A data file was generated of x and y pixel coordinate pairs for each K Kα X-ray photon that makes up the image. By randomising the pixel coordinate pair sequence in the file, images corresponding to different number of detected photons (i.e. ion fluence) could be simulated by reading the corresponding number of pixel coordinates.

The procedure to form multi-resolution images is as follows:

- A series of 256×256 pixel images containing 1000, 4000, 16000 and 64000 counts were created.
- The wavelet decomposition was calculated using the Daubechies db4 wavelet[3,10] and the set of different resolution images extracted from the decomposition vector[3]. To overcome artefacts associated with boundary distortions[3], symmetric boundaries with half-point boundary values were used.
- The extra pixels associated with periodic boundaries[3] were trimmed from the images.
- The contrast of each image was expanded to fill a 256-level greyscale.

The possibility of improving the 1000 count 32×32 pixel approximation image using wavelet de-noising was carried out by selecting the threshold settings[3] on a cut-and-try basis. The decimation by direct down-sampling was carried directly by re-binning counts into pixels according to the decomposition level.

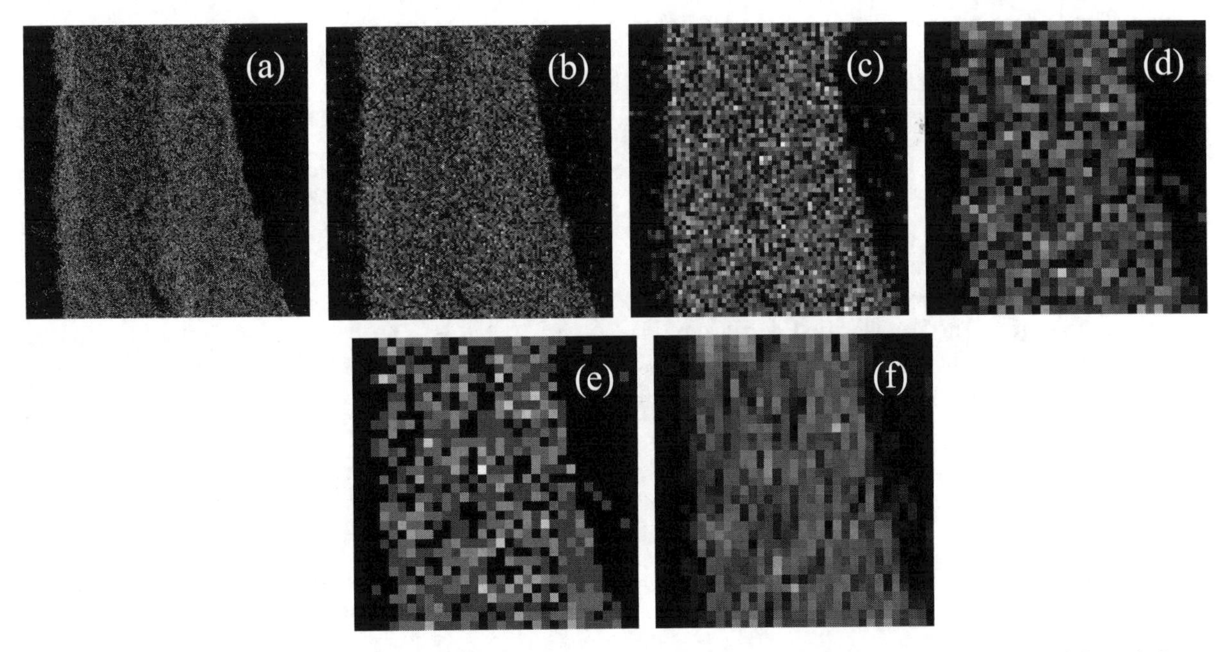

FIGURE 2. Multi-resolution representation of the image data using Daubechies wavelet db4. a) 64000 counts in full resolution 256×256 pixel image with field size 1100 ×1100 μm. (b) 16000 counts in 128×128 pixel level-1 image approximation. (c) 4000 counts in 64×64 pixel level-2 image approximation. (d) 1000 counts in 32×32 pixel level-3 image approximation. (e) 1000 counts in direct down-sampled 32×32 pixel image. (f) De-noised image of (d) using a soft threshold set at 0 and 12 for Level 1 and 2 wavelet coefficients, respectively. (The images have been adjusted to give optimal printed contrast.)

RESULTS AND DISCUSSION

Fig. 2 (a-d) presents the contrast expanded and level-3, level-2, level-1 and level-0 (full resolution) wavelet decomposed images for 1000, 4000, 16000 and 64000 counts, respectively. Comparison of Fig. 2 (a) and (d) reveals that even for 1000 counts (<2% of the fluence required for the full resolution image) the salient (low frequency) image components required for location of regions of interest are well reproduced. Comparison of the wavelet-decimated and direct down-sampled images (Fig. 2 (d) and (e)) shows they are very similar, but not identical. The difference can presumably be associated with choice of wavelet[3,10]. Whereas direct down sampling is inherently lossy, wavelet decimation is loss-less. The extra information in the decomposition vectors of the latter can be used for de-noising. Figure.2 (f) shows such wavelet de-noising smoothes the image in Fig 2 (d) and removes outlying non-zero pixels. The processing time for the wavelet decomposition and filtering using a 2.33 GHz dual core processor is less than 100 ms. The method above can hence be carried out on-the-fly to give processed and de-noised images in real time.

Use of the multi-resolution support for locating fields of interest requires that at each level the fluence is uniform over the field of view. This is not well-suited where conventional slow magnetic[11] or mechanical rastering of the beam is used. To facilitate straightforward multi-resolution support in DREAM[9], a fast electrostatic beam scanning system is being constructed that when used in conjunction with time-stamped data collection will facilitate different scan modes by reading pixel coordinates from a sequence file.

A further interesting possibility is to use the pattern writing capabilities of modern microprobes to collect images with different resolutions in different regions of the image in order to minimise the fluence in high resolution images. This would find application in imaging of low confluences of cells on a substrate. Rendering of region-dependent resolution images is already implemented in the JPEG2000 standard[1,2].

CONCLUSIONS

It is shown that by using wavelet decomposition or direct mean-filtering combined with down-sampling techniques to create multi-resolution ion microbeam images, low-fluence scans suitable for locating fields of interest can be obtained with about 2% of the fluence required for a normal quality image. The reduced resolution image quality can be further visually improved by de-noising using a wavelet method. There is no clear advantage of one technique over the other.

ACKNOWLEDGMENTS

This project was supported under the auspices of Academy of Finland (Centre of Excellence in Nuclear and Accelerator Based Physics, Ref. 213503, and senior researcher grant 129999. Additional support is acknowledged from the OSKE MIMMA project. HJW is also grateful for travel support from the Magnus Erhnrooth Foundation.

REFERENCES

1. S. Jayaraman, S. Esakkiajan and T. Veerakumar, in *Digital Image Processing* (Tata McGraw Hill Education Private Ltd., New Delhi, 2009)
2. *JPEG Image Coding System, JPEG 2000 Final Committee Draft V. 1.0, 16 March 2000*, M. Bolick (ed.)
3. M. Misiti, Y. Misiti, G. Openheim, J. M. Poggi, *Wavelet Toolbox 4, Users guide* (Mathworks, 2010) http://www.mathworks.com/access/helpdesk/help/pdf_doc/wavelet/wavelet_ug.pdf
4. P. Sigmund, *Particle penetration and Radiation Effects*, Springer, Hiedelberg, 2006, pp. 380-382.
5. I. Rodrigues, J. Sanches, J. Bioucas-Dias, *Denoising of medical images corrupted by Poisson noise*, Proc. IEEE Int. Conf. on Image Processing ICIP 2008, (IEEE, 2008) 1756, DOI: 10.1109/ICIP.2008.4712115
6. M. Ren, H. J. Whitlow, A. Sagari A. R., J. A. van Kan, T. Osipowicz and F. Watt, *J .Appl. Phys.* **103**, 014902 (2008).
7. H. J. Whitlow, M. Ren, J. A. van Kan, T. Osipowicz and F. Watt , *Nucl. Instr. and Meth. B* **267**, 2149 (2009)
8. F. Watt, P. S. P. Thong, A. H. M. Tan, S. M. Tang, *Nucl. Instr. and Meth. B* **130,** 216 (1997)
9. R. Norarat, et al., *Nucl. Instr. and Meth. B* (In submission).
10. I. Duabechies, *Ten lectures on wavelets*, CBMS-NSF Regional Conf. Series in Applied Mathematics, SIAM (1992).
11. G. W. Grime and F. Watt, *Beam Optics of Quadrupole Probe-Forming Systems*, (Adam Hilger, Bristol, 1984) pp. 111.

Thermoelectric Properties Of SiO_2/SiO_2+Au Nano-Layered Superlattices Modified By MeV Si Ions Beam

J. Chacha[1], S. Budak[1], C. Smith[2], D. McElhaney[1], M. Pugh[1], K. Ogbara[3], K. Heidary[1], R. B. Johnson[3], C. Muntele[2], D. Ila[2]

[1]*Department of Electrical Engineering, Alabama A&M University (4900 Meridian St, Normal, AL 35762, USA)*
[2]*Center for Irradiation of Materials, Alabama A&M University Research Institute, (P.O. Box 1447, Normal, AL35762, USA)*
[3]*Department of Physics, Alabama A&M University, Normal, AL USA (4900 Meridian St, Normal, AL 35762, USA)*

Abstract. The efficiency of thermoelectric devices and materials is limited by the properties of n- and p-type (semi)conductors. Effective thermoelectric materials have a low thermal conductivity and a high electrical conductivity. The performance of thermoelectric materials and devices is shown by a dimensionless figure of merit, The performance of thermoelectric materials and devices is shown by a dimensionless figure of merit, ZT = $S^2\sigma T/K$, where S is the Seebeck coefficient, σ is the electrical conductivity, T is the absolute temperature and K is the thermal conductivity. ZT can be increased by both increasing S and σ, or decreasing K. In this study we have prepared a thermoelectric generator from 100 alternating layers of SiO_2/SiO_2+Au superlattice films using ion beam assisted deposition (IBAD). In order to determine the stoichiometry of SiO_2 and Au in the grown multilayer films, Rutherford Backscattering Spectrometry (RBS) was used. The 5 MeV Si ions bombardment was performed using the CIM Pelletron ion beam accelerator to make nanodots and/or nanoclusters in the multi-layer superlattice thin films to decrease the cross plane thermal conductivity, increase the cross plane Seebeck coefficient and increase the cross plane electrical conductivity.

Keywords: Ion bombardment, thermoelectric properties, multi-nanolayers, Rutherford backscattering, Figure of merit.
PACS: 84.60.Rb; 81.15.Jj; 82.80.Yc; 62.23.Eg

INTRODUCTION

Thermoelectric devices have gained importance in recent years as viable solutions for applications such as spot cooling of electronic components, remote power generation in space stations and satellites etc. These solid-state devices have long been known for their reliability rather than their efficiency; they contain no moving parts, and their performance relies primarily on material selection, which has not generated many excellent candidates [1-3]. A thermoelectric device produces electrical power from heat flow across a temperature gradient [4]. Thermal to electrical energy conversion, through thermoelectric and thermionic materials, has been proposed to be much more efficient in lower dimensional materials at the nanoscale [5]. A growth in the commercial applications of thermoelectric devices depends primarily on increasing the figure of merit *ZT* of the materials used in the devices [6-7]. The figure of merit is defined by $ZT = S^2\sigma T / \kappa$, where **S** is the Seebeck coefficient, **σ** is the electrical conductivity, **T** is the absolute temperature, and **κ** is the thermal conductivity [8-9]. The figure of merit can be increased by increasing **S,** by increasing **σ,** or by decreasing **κ.** In this study we report on the growth of 100 alternating multi-layer super-lattice of SiO_2/SiO_2+Au films using the ion beam assisted deposition (IBAD), and high energy Si ions bombardments of the films for reducing thermal conductivity and increasing electrical conductivity.

EXPERIMENTAL

We have deposited the 100 alternating layers of SiO_2/SiO_2+Au nano-layers films on silicon and fused silica (suprasil) substrates using Ion Beam Assisted Deposition (IBAD). The multilayer films were sequentially deposited to have a periodic structure consisting of alternating SiO_2 and SiO_2+Au layers. The two electron-gun evaporators in the chamber were turned on and off alternately to deposit the SiO_2 and SiO_2+Au films. The base pressure obtained in IBAD chamber was about $4x10^{-6}$ Torr during the deposition

Application of Accelerators in Research and Industry
AIP Conf. Proc. 1336, 257-259 (2011); doi: 10.1063/1.3586099

process. The growth rate for the deposited film thickness was monitored by a gold coated Inficon Quartz Crystal Monitor (QCM). The film geometries used for deposition of SiO_2/SiO_2+Au nano-layers films are shown in Fig.1a and 1b for the thermal conductivity and for Seebeck coefficient measurements, respectively. The gold layers shown in Fig.1b have been used for the electrical and thermal contacts during the cross plane Seebeck coefficient measurements. For the cross plane electrical conductivity, the similar film geometry was used as in the geometry of Seebeck coefficient measurement. The cross plane electrical conductivity was measured by the 4-probe contact system and the cross plane thermal conductivity was measured by the 3ω (3rd harmonic) technique. The electrical conductivity, thermal conductivity and Seebeck coefficient measurements have been performed at room temperature (300 K). Detailed information on the 3ω (3rd harmonic) technique could be reached in refs [10-11].

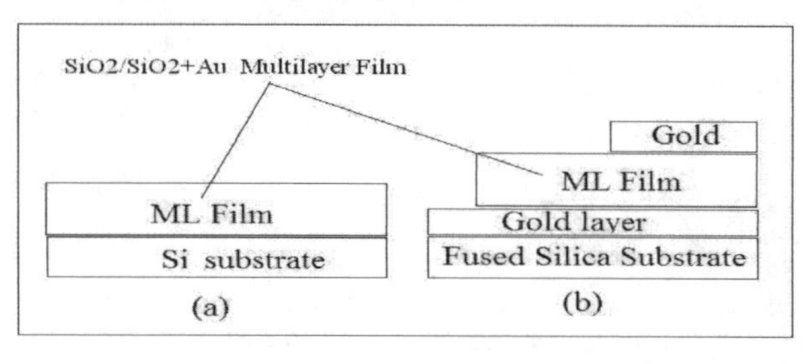

FIGURE 1. Used film geometries from the cross-section, (a) For thermal conductivity, (b) for Seebeck measurements

In order to make nano dots and/or nano clusters in the layers of the superlattice film systems, 5 MeV Si ions bombardments were performed with the Pelletron ion beam accelerator at the Alabama A&M University Center for Irradiation of Materials (AAMU-CIM). The energy of the bombarding Si ions was chosen based on SRIM simulations. The fluences used for bombardment were $1x10^{12}\,ions/cm^2$, $5x10^{12}\,ions/cm^2$, $1x10^{13}\,ions/cm^2$, $5x10^{13}\,ions/cm^2$, and $1x10^{14}\,ions/cm^2$. We used Rutherford Backscattering Spectrometry (RBS) to determine the stoichiometry of SiO_2 and Au in the films.

RESULTS AND DISCUSSION

After deposition, the films thickness was measured using a Fabry-Perot optical interferometer. The thickness of the 100 alternating layers SiO_2/SiO_2+Au thin film was found to be 243 (±10) nm. Fig.2 shows the He RBS spectrum of SiO_2/SiO_2+Au multi-layer films on GPC substrate. As seen from this graph, the amount of Au is quite less with respect to the deposited amount of the Si. Future works will include detailed studies of alternating composition of Au on our nanolayers to see the effects of the Au rates in the efficiency of the multilayer systems. Fig. 3 shows the thermoelectric properties of 100 alternating layers of SiO_2/SiO_2+Au virgin (unbombarded) and 5 MeV Si ions bombarded thin films at five different fluences. Fig. 3a. shows the square of the Seebeck coefficient of the thin films. Since the original Seebeck values are negative this indicates that electrons are the main charge carriers in the systems. The unbombarded thin film has a Seebeck coefficient of –123.74 μV/K. This value decreased as the ion bombardment fluence increased. The Seebeck coefficient decreased with the first irradiation of fluence $1x10^{12}\,ions/cm^2$. Subsequent irradiation showed an incremental increase in the Seebeck coefficient but not as high of a figure as observed in the unbombarded sample. For the coming studies, we will be focusing on the optimization of the suitable ion energy and fluences to achieve higher efficiencies of the fabricated thermoelectric devices. The MeV Si ions bombardments effects on the electrical and the thermal conductivity values as function of varying fluence could be seen in the Fig. 3b and 3c respectively. As seen from fig. 3b, a remarkable increase in the electrical conductivity was observed at $1x10^{13}\,ions/cm^2$ and $1x10^{14}\,ions/cm^2$. As seen from fig. 3c, the thermal conductivity value decreased at $1x10^{12}\,ions/cm^2$. After this fluence, the thermal conductivity showed an increase until $5x10^{12}\,ions/cm^2$. The lowest value for the thermal conductivity is at $5x10^{12}\,ions/cm^2$. From $5x10^{12}\,ions/cm^2$ on, the thermal conductivity started to increase. Fig. 3d. shows the calculated dimensionless figure of merit values by applying the equation given above. The ZT decreased after irradiating with $1x10^{12}\,ions/cm^2$ and there was no significant improvement after further irradiation. Therefore, for this film system, we see a negative effect on the figure of merit.

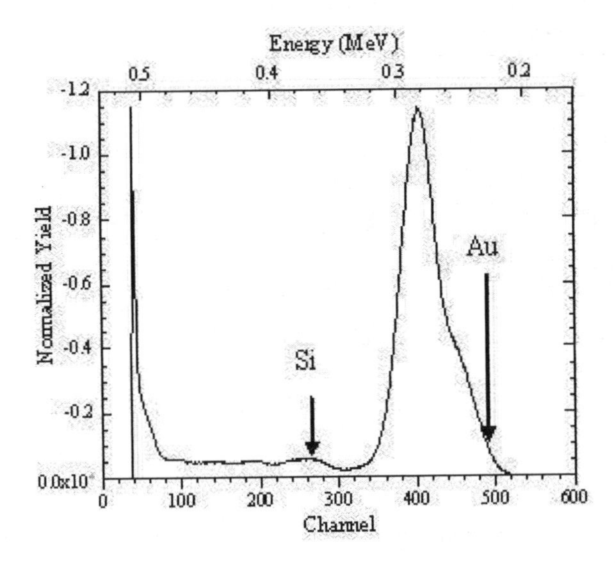

FIGURE 2. He RBS spectrum of SiO_2/SiO_2+Au multi-layer films.on GPC substrate.

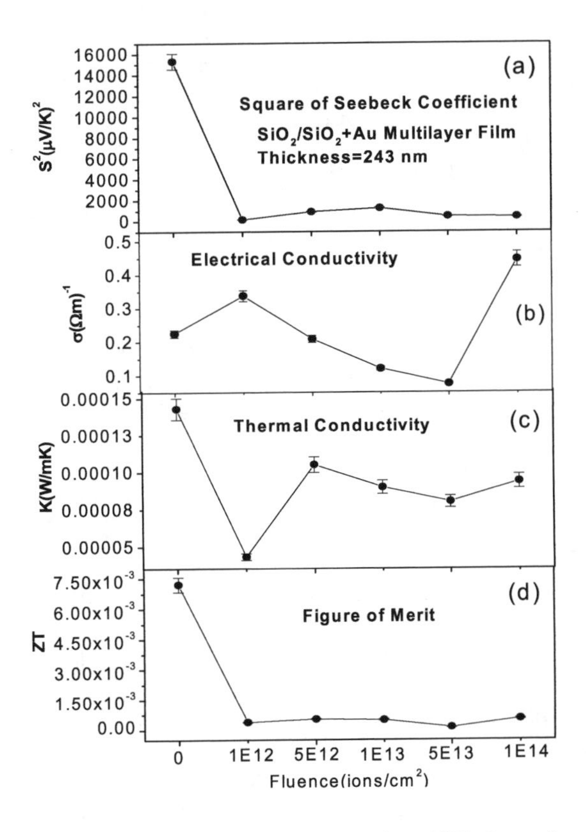

FIGURE 3. Thermoelectric Properties of 100 alternating nanolayers of SiO_2/SiO_2+Au multi-layer films.

CONCLUSION

We studied the effects of the MeV Si ions bombardments on the thermoelectric properties of the thermoelectric generators from the 100 alternating multilayer of SiO_2/SiO_2+Au thin films. We would like to observe an increase of the Seebeck coefficient and electrical conductivity, and a decrease of the thermal conductivity. We observed such modifications in previously studied systems. The important part here is to optimize the suitable fluence, applied bombardment energies, and film thickness. However, as these results show, the system described here did not show such modifications, except for the $1x10^{12}\, ions/cm^2$ region. Future work will involve irradiations with fluences around this value, to see if theis is really the optimal value or not for this system. Also, different film thicknesses, different energies and fluencies will be tried, in an attempt to reach higher conversion efficiency. We would also like to do TEM and optical absorption to show the formation of nanodots and/or nanoclusters.

ACKNOWLEDGMENTS

Research sponsored by the Center for Irradiation of Materials (CIM), National Science Foundation under NSF-EPSCOR R-II-3 Grant No. EPS-0814103, DOD under Nanotechnology Infrastructure Development for Education and Research through the Army Research Office # W911 NF-08-1-0425.

REFERENCES

1. A. Bulusu, D. G. Walker, Superlattices and Microstructures 44 (2008) 1-36.
2. S. Budak, C. Muntele, B. Zheng, D. Ila, Nuc. Instr. and Meth. B 261 (2007) 1167.
3. S. Budak, S. Guner, C. Muntele, D. Ila, Nuc. Instr. and Meth. B 267 (2009) 1592-1595.
4. G. J. Snyder and E. S. Toberer, Nature Materials 7 (2008) 105.
5. Paothep Pichanusakorn, Prabhakar Bandaru, Material Science and Engineering R 67 (2010) 19-63.
6. M. Chitroub, F. Besse, H. Scherrer, Journal of Alloys and Compounds 467 (2009) 31-34.
7. S. Guner, S. Budak, R. A. Minamisawa, C. Muntele, D. Ila, Nuc. Instr. and Meth. B 266 (2008) 1261.
8. B C. -K. Huang, J. R. Lim, J. Herman, M. A. Ryan, J. -P. Fleural, N. V. Myung, Electrochemical Acta 50 (2005) 4371.
9. T.M. Tritt, ed., Recent Trends in Thermoelectrics, in Semiconductors and Semimetals, 71, (2001).
10. L. R. Holland, R. C. Smith, J. Apl. Phys. 37 (1966) 4528.
11. D. G. Cahill, M. Katiyar, J. R. Abelson, Phys. Rev.B 50 (1994) 6077.

Ion Beam Analysis Of Nitrogen Incorporated Ultrananocrystalline Diamond (UNCD) Thin Films

S. AlFaify[a], E. Garratt[a], A. Dissanayake[a], D. C. Mancini[b], A. Kayani[a]

[a] *Department of Physics, Western Michigan University, Kalamazoo, MI 49008, USA*
[b] *Center for Nanoscale Materials, Argonne National Laboratory, 9700 S. Cass Avenue, Building 440, Argonne, IL 60439-4812, USA*

Abstract. Determination of the elemental composition is important to correlate the properties of nitrogen incorporated Ultrananocrystalline Diamond (UNCD) thin films with their growth conditions. Films were deposited by CVD deposition technology and nitrogen incorporation was introduced by diluting the growth Ar/CH_4 plasma with N_2 gas. Deposition of UNCD thin films was carried out on tungsten (~ 15nm) coated Si substrates with varying concentrations of N_2 diluted to the growth plasma. Scanning electron microscopy (SEM) and Raman spectroscopy (RS) were used to confirm the characteristic morphology of the UNCD film and its dominant sp^3 bonding respectively. The deposited films were smooth on the submicron scale with the RMS roughness value of 2.9-5.1 nm. Reflectometry spectroscopy analysis (RES) technique was used to measure the films thicknesses. To obtain the elemental composition of the UNCD thin films, Rutherford Backscattering Spectrometry (RBS), Non-Rutherford Backscattering Spectrometry (NRBS), Elastic Recoil Detection Analysis (ERDA) and Nuclear Reaction Analysis (NRA) were performed. Deposited UNCD films contained less than 5 at.% of H while N content incorporated in the films was estimated to be lower than 1 at.%. The intermixing region between the substrate and the film was found to be negligible. Moreover, amorphous phase as determined by Raman analysis was found to be increasing for the sample deposited with N_2.

Keywords: UNCD Film, MPCVD, Nanostructure, Nanocrystalline Diamond, RBS, NRA.
PACS: 61.46.-W

INTRODUCTION

Ultrananocrystalline diamond (UNCD) is best described as comprising diamond grains of 3-5nm in size bound together by abrupt boundaries estimated to be 0.5nm in width[1]. When deposited in microwave enhanced plasma chemical vapor deposition (MPCVD) system[2] using hydrogen-poor Ar / CH_4 gas mixture, UNCD film has unique structural properties that allow its physical characteristics such as optical transparency, electrical and thermal conductivity, film stress, and surface smoothness to be controlled in ways not possible with other diamond materials. The structure of UNCD is due to the high rate of renucleation that occurs as a result of the special plasma chemistry[3]. Surface roughness is an important characteristic of this material which is required for a range of applications including novel integrated electronic devices, micro and nano electromechanical systems (MEMS/NEMS)[4].

Another key characteristic of all UNCD films is their phase purity. UNCD film consists of a mixture of diamond phase (~ 90%) and amorphous phase within the volume fraction of the grain boundaries (GBs) that amounts for less than 10% and made of carbon in several different bonding states including sp^2 (graphitic)[5]. Moreover, UNCD films exhibit a unique set of complementary properties, such as chemical inertness, low friction and high wear resistance, high hardness, and good electrical transport properties when appropriately doped[6]. If H_2 is added to the UNCD film, it will incorporate into the atomically abrupt grain boundaries satisfying sp^2 carbon dangling bonds, resulting in increased resistivity[7] whereas, N incorporation in the UNCD by diluting the plasma mixture with N_2 gives rise to the electrical conductivity of the film by five orders of magnitude to be around 250Ω^{-1} cm^{-1} [8]. Due to the fact that properties of UNCD film are dependent on its elemental concentrations, the present work will

Application of Accelerators in Research and Industry
AIP Conf. Proc. 1336, 260-263 (2011); doi: 10.1063/1.3586100

introduce the use of the Ion beam analysis (IBA) to quantify H, N and other elemental contents in the film.

EXPERIMENTS

Samples deposition was carried out in the clean room of the Center for Nanoscale Materials at Argonne National Laboratories using state of art 915 MHz microwave plasma CVD Reactor from Lambda Technology . Tungsten (W) coated Si wafers were used as substrates. Thin W interlayer (~ 15nm) has been known to improve the adhesion of UNCD layer to the Si substrate and also to increase the diamond nucleation density[9]. Ar, CH_4, H_2, and N_2 were used as source gases and their flow rates were controlled by precision mass flow controllers. When the chamber pressure approximately reaches to 12 Torr, microwave power of 800 W is applied to ignite the plasma. Then processing pressure of 60-90 Torr are maintained. Absorbed microwave power is adjusted between 3-4.5 kW in order to cover the 8 inches diameter substrate with the plasma discharge. Substrate temperature is kept around 700°C. Details of the CVD reactor and standard deposition process of UNCD films are given elsewhere[10]. Each sample was deposited with different H_2 and N_2 gas concentrations in its growth plasma to investigate the effect of H_2 and N_2 flow rates on the elemental concentrations and the nanostructure of the film. Growth plasma of the first sample, UNCD deposited with 0%N_2, was diluted with 5% of H_2while the second one was deposited with 10%N_2 and 0% H_2.

After the depositing process Raman spectroscopy (RS), scanning electron microscopy (SEM), atomic force microscopy (AFM), and Reflectometry spectroscopy (RES) were carried out to confirm the signature of UNCD structure and thickness of the deposited films.

Elemental concentrations of the deposited UNCD films were obtained by ion beam analysis (IBA) measurements. Western Michigan University 6.0 MV tandem Van de Graaf accelerator was used to perform Rutherford and Non-Rutherford back scattering spectrometry (RBS & NRBS), Elastic Recoil Detection Analysis (ERDA), and Nuclear Reaction Analysis (NRA). RBS using 3 MeV He^{++} as incident ions was used to compare the thickness of the two deposited samples and also to investigate the interface region of the thin film and the substrate. Thicknesses of the samples were also obtained by the RES measurements which were approximately 114 nm and 140 nm for UNCD deposited with 0% N_2 and UNCD deposited with 10%N_2 respectively. Moreover, the He^{++} beam was also used to carry out ERDA at incident angle of 67.5° and 45° scattering angle to obtain concentration of H in the film. For the NRA measurements, beam of 1.1 MeV deuterium ions (D^+) is directed on the samples at an angle of 60° from the surface normal. Reaction products were detected at a scattering angle of 170°. The detected particles for these measurements were protons from the $^{14}N(d,p)^{15}N$, $^{16}O(d,p)^{17}O$ and $^{12}C(d,p)^{13}C$ nuclear reactions. Peak areas from the measured yields were converted to N or O and C concentrations. Since the depth resolution when using NRA is not nearly as good as that for the RBS measurements with He, our primary purpose for using ERDA and NRA was to determine the concentration of H, N and/or O, and C in the films. Using SIMNRA program, the set of target composition profile parameters was adjusted until a single set of parameters could be used to accurately simulate RBS, NRBS, ERDA and NRA spectra.

RESULTS AND DISCUSSION

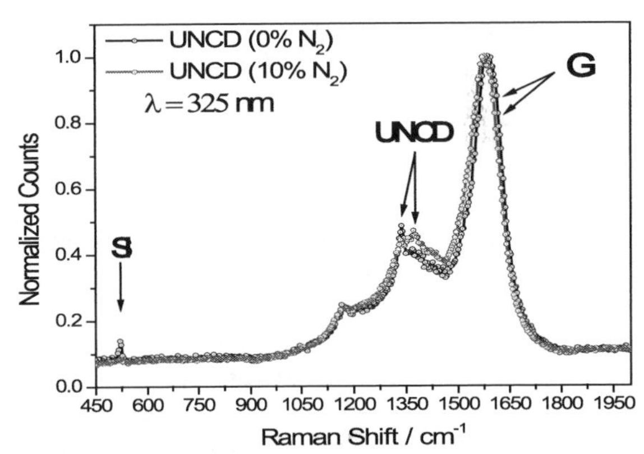

FIGURE 1. UV Raman scans of UNCD samples UNCD (0% N_2) and UNCD (10% N_2).

For the two deposited UNCD samples i.e. UNCD (0% N_2) and UNCD (10%N_2), basic characterizations results confirmed the signature of UNCD layer on the substrates. RS is widely used to check the bonding structures of UNCD films. Fig.1 shows Raman spectra of UNCD samples with and without N incorporation. It consists of three dominant peaks: the substrate peak usually appears around 520cm^{-1} for silicon, diamond peak around 1333 cm^{-1} which is the signature of UNCD nanostructure which represents sp^3 bonding. The peak around 1580 cm^{-1} is the assigned *G* band of the disordered carbons. *G* peak is due to the sp^2 bonded carbons, associated with the graphite like structure at the GBs[11], which arises from the in-plane stretching modes. A couple of differences can be noted between Raman spectra of the two UNCD films. The spectra show a slight increase in the ratio of the *G*-band to the diamond peak intensities as the N_2 content in the plasma is increased, which is an indication of increased amorphous phase in the film. Moreover, the area of the peak around 1333cm^{-1} increases with the increase of N_2 in the plasma as well, which indicates the increased of grains sizes[12].

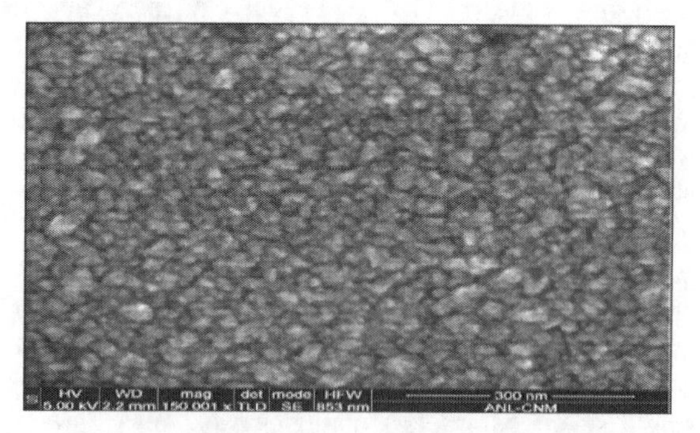

FIGURE 2. SEM micrograph of UNCD (0% N_2)

SEM analysis of the UNCD sample UNCD (0% N_2) is shown in Fig.2. Granular structure that defines UNCD layers can be seen clearly. The morphology of UNCD is made of grains separated by highly energetic twisted GBs[2]. From the experimental work, it is noticeable that obtaining SEM images with high resolution for the N incorporated UNCD film is more difficult than for the undoped UNCD sample. This might be because of increased electron scattering from the structure of the nitrogen incorporated UNCD, which also indicates increased in the grain size and GBs of the film. This result is in direct correlation with the RS measurements which show increased disorder as N_2 content is increased in the film plasma.

Low surface roughness is one of the characteristics of UNCD films which was obtained by AFM measurements. Films were found smooth on the submicron scale. The root mean square (RMS) roughnesses were 2.9 nm and 5.1 nm for UNCD (0% N_2) and UNCD (10% N_2) respectively.

The ERDA spectrum is shown in Fig. 3, and the H contents of UNCD (0% N_2) and UNCD (10% N_2) samples were 3 and 5 at% respectively. During the growth process, UNCD (0% N_2) film was deposited with 5% of H_2 in the plasma mixture, whereas, sample UNCD (10% N_2) was deposited with 0% of H_2 dilution. However, H was found trapped in the N incorporated sample. Owing to the fact that N_2 introduction to the growth plasma increases the volume of GBs[13] where H can passivate the dangling bonds[14], it is possible for the N incorporated sample to trap more hydrogen than the undoped one. This H incorporation is probably from CH_4 used in the growth process or it might have come from the water vapors in the chamber. In general, UNCD films are grown in H_2 poor or even H_2 free plasma conditions; however, literature has shown the H contents in the UNCD films in range of 3 to15 at.% [15]. The plasma chemistry of UNCDs growth is still an active field of research and not yet fully understood[16].

RBS results obtained by 3 MeV He^{++} ion beam at scattering angle of 170^0 are given in Fig. 4. Interface region as shown in the Fig. 4, for doped and undoped samples, is clean and shows little to no intermixing between the film and the substrate. From Fig. 4, it is also clear that there is a thickness difference between UNCD (0% N_2) and UNCD (10% N_2) films which agrees with the data obtain from the RES measurements.

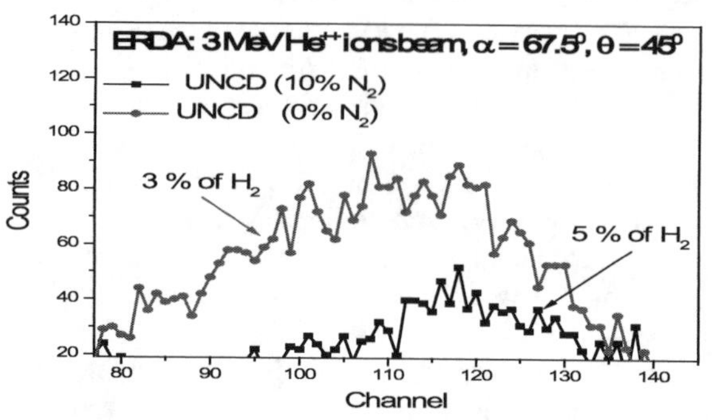

FIGURE 3. ERDA of UNCD films deposited with different N_2 concentrations in the growth plasma.

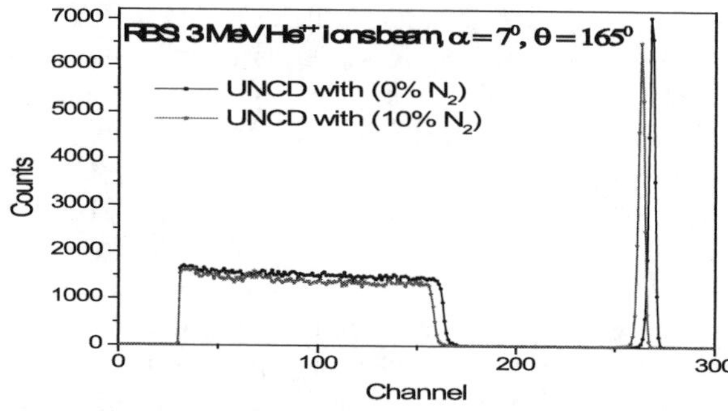

FIGURE 4. RBS of UNCD films deposited with different N_2 contents in the plasma mixture.

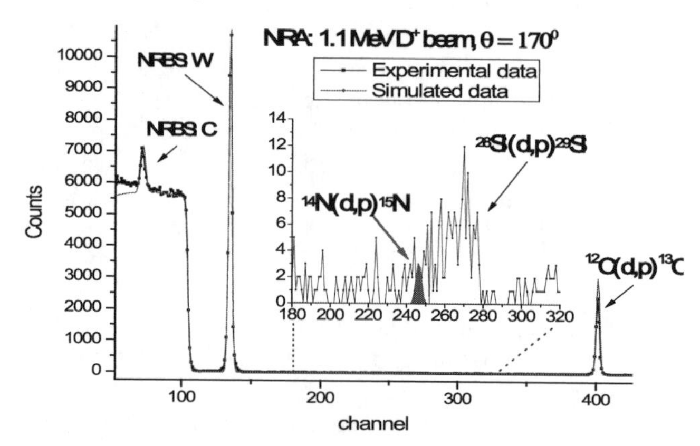

FIGURE 5. NRA of UNCD film deposited with 10% of N_2 concentration in the plasma mixture. Inset figure is zoomed section of the NRA spectra where ^{14}N (d,p)^{15}N cross section peaks should appear, the simulated N peak accounts for about 1at.% of N content.

NRA with deuterium (D^+) beam was used to find the concentration of N incorporated in the film. In addition, NRBS part of the D^+ spectra were used to quantify C concentration. Fig.5 shows spectra of the UNCD (10% N_2) sample with 1.1 MeV D^+ beam measured at 170° scattering angle. N content in the UNCD sample deposited with N_2 using NRA was difficult to quantify. However, we were able to estimate that N concentration in UNCD is less than 1 at.%. In Fig.5, zoomed section of the NRA spectra where $^{14}N(d,p)^{15}N$ cross section peaks should appear is shown. Difficulty to find out the exact nitrogen content in N incorporated UNCD is due to the existence of the $^{28}Si(d,p)^{29}Si$ nuclear reaction that almost overlaps with the nitrogen nuclear reaction. Careful examinations, simulations, and comparison between NRA data of UNCD (0% N_2) and UNCD (10% N_2) samples show a small peak for the $^{14}N(d,p)^{15}N$ reaction, therefore, we estimate that less than 1 at. % of N content is actually incorporated in the UNCD (10% N_2) film. This conclusion agreed with pervious result obtained by secondary ion mass spectroscopy (SIMS) [17]. Nevertheless, SIMS method has its limitation as well. Ongoing research is underway to accurately quantify nitrogen content in N incorporated UNCD films.

CONCLUSIONS

Investigation of the elemental composition and nanostructure of UNCD films have been carried out. Films were deposited by CVD deposition system and N incorporation was accomplished by diluting N_2 into the Ar/CH_4 gas mixture of the growth plasma. The quality of UNCD films including their unique bonding structures, surface roughnesses, and morphologies have been obtained by RS, AFM, and SEM respectively. ERDA data showed hydrogen content in the UNCD samples which are higher in the case of UNCD (10% N_2) film, likely due to the fact that N_2 introduction to the film growth plasma increases GBs volume where H can reside. RBS spectra obtained by 3 MeV He^{++} ion beam confirmed the thickness differences between UNCD (0% N_2) and UNCD (10% N_2) samples. That agrees with estimations of the samples thickness preformed by the RES. Only C (97 at.%) and H (3 at.%) were seen in 0% N_2 incorporated sample and C (94 at.%), H (5 at.%) and N(~1 at.%) were seen in 10% N_2 incorporated sample. This indicates that deposited films are of high quality and where free of impurities. The intermixing region between the substrate and the layer of UNCD was also found to be negligible.

ACKNOWLEDGMENTS

Use of the Center for Nanoscale Materials was supported by the U. S. Department of Energy, Office of Science, Office of Basic Energy Sciences, under Contract No. DE AC02-06CH11357.

REFERENCES

1. D. M. Gruen, Annu. Rev. Mater. Sci. 29 (1999) 211.
2. J. Phillip. P. Hess. T. Feygeison. J.E. Butler. S. Chattopadhyay. K.H. Chen. L.c. Chen, J Appl. Phys .93 (2003) 2164.
3. L. C. Qin, D. Zhou, A. R. Krauss, D. M. Gruen, Nanostruct. Mater. 10 (1998) 6491.
4. Auciello, O., Birrell, J., Carlisle, J.A., Gerbi, J.E., Xiao, X., Peng, B., Espinosa, H.D, *Journal of Physics: Condensed Matter* 16, 2004, R539–R552.
5. J.E. Gerbi, J. Birrell, M. Sardela and J.A. Carlisle, *Thin Solid Films* 473, 2005, p. 41.
6. S. Bhattacharyya, O. Auciello, J. Birrell, J.A. Carlisle, L.A. Curtiss, A.N. Goyette, D.M. Gruen, A.R. Krauss, J. Schlueter, A. Sumant and P. Zapol, *App. Phy. Let.* 79 (2001) 1441.
7. C. Liu, X. Xiao, J. Wang, B. Shi, V.P. Adiga, R.W. Carpick, J.A. Carlisle, O. Auciello, *J. Appl. Phys.* 102, 2007, p. 074115.
8. J. E. Butler and A. V. Sumant. *Chemical Vapor Deposition*, 14 (2008) 145.
9. Nevin N. Naguib, Jeffrey W. Elam, James Birrell, Jian Wang, David S. Grierson, Bernd Kabius, Jon M. Hiller, Anirudha V. Sumant, Robert W. Carpick, Orlando Auciello and John A. Carlisle, *Chemical Physics Letters* 430, Issues 4-6, 2006, P. 345.
10. J. Asmussen, T. A. Grotjohn, T. Schuelke, M. F. Becker, M. K. Yaran, D. J. King, S. Wicklein, and D. K. Reinhard, *Appl. Phys. Lett.* 93 (2008) 031502.
11. P. Achatz, J. Garrido, M. Stutzmann, O.A. Williams, D.M. Gruen, A.Kromka, D. Steinmueller, *Appl. Phys. Lett.* 88 (2006) 101908.
12. S. Bhattacharyya, O. Auciello, J. Birell, J. A. Carlisle, L. A. Curtiss, A. N. Goyette, D. M. Gruen, A. R. Krauss, J. Schlueter,A. Sumant, and P. Zapol, *Appl. Phys. Lett.* 79 (2001) 1441.
13. Birrell J, Gerbi JE, Auciello O, Gibson JM, Gruen DM, Carlisle JA , *J Appl Phys*. 93 (2003) 5606.
14. P. Reichart, G. Datzmann, A. Hauptner, R. Hertenberger, C. Wild and G. Dollingeer, *Science* 306 (2004) 1537.
15. Sh. Michaelson, O. Ternyak, and A. Hoffman O. A. Williams. D. M. Gruen, *Appl. Phys. Lett.* 91 (2007) 103104.
16. Luming Shen, Zhen Chen, *International Journal of Solids and Structures,* V. 44, Issue 10, 2007, P. 3379.
17. T.D. Corrigan, D.M. Gruen, A.R. Krauss, P. Zapol and R.P.Y. Chang, *Diamond Relat. Mater.* **11,** 2002, p. 43.

Characterization Of GeO_2 Nanocrystals Prepared By Microwave Annealing

V. Saikiran, N. Srinivasa Rao, G. Devaraju and A. P. Pathak*

School of Physics, University of Hyderabad, Central University (P.O), Hyderabad 500 046, India

Abstract. GeO_x films have been deposited on silicon substrate using RF magnetron sputtering. The as-deposited samples were annealed at 900^0C using microwave annealing. All the samples were subsequently characterized by X-ray diffraction (XRD) to observe the GeO_2 nanocrystal formation. Raman spectroscopy and Transmission electron microscopy (TEM) measurements were also carried out to confirm the presence of the nanocrystals. The film topography was studied by atomic force microscopy (AFM).

Keywords: RF sputtering, GeO_2 nanocrystals, XRD, and Microwave annealing.
PACS: 81.15.Cd; 61.46.Hk; 61.10.Nz; 81.40.Ef;

INTRODUCTION

Nanocrystals have novel physical and chemical properties which are different from those of the corresponding bulk materials. Semiconductor nanocrystals exhibit interesting electronic and optical properties that depend on the particle size, making them potential candidates for important applications. In addition to elemental semiconductor nanocrystals, Germanium dioxide (GeO_2) also attracts great attention due to its wide ranging applications. It is a dielectric oxide which exhibits visible light photoluminescence (PL) and has a refractive index and linear coefficient of thermal expansion higher than those for SiO_2 [1–3]. Besides, it is also a promising material for optical waveguides and nano-connections in optoelectronic communication and vacuum technology. Due to these fascinating properties, great attention has been paid to the preparation of GeO_2 nano-structures, including nano-particles, nanofibers and nano-sheets [4–8]. On the other hand, Germanium dioxide has been the subject of many studies due to its importance as a mineral, and its commercial importance as a piezoelectric and glass material. The crystal structure and properties of GeO_2 have been studied using various techniques [9-11]. Germanium dioxide is also used as a catalyst in production of polyethylene terephthalate resin, and for production of other germanium compounds [12].

Researchers are looking for various synthesis methods in order to optimize structural and optical properties of nanocrtystals for diverse applications. The properties of nanocrystals are not only determined by its size but also by its shape, composition and structure. An appropriate method helps to make use of nanocrystals properties for a wide range of applications. Physical properties of GeO_2 films grown by using DC reactive sputtering have already been reported [13]. In this paper, we present a simple synthesis method for GeO_2 nanocrystals by RF magnetron sputtering and subsequent microwave annealing. Importance of microwave annealing in this context has also been discussed.

EXPERIMENTAL DETAILS

The GeO_2 Thin films were prepared by the RF-magnetron sputtering technique by using a target of Ge of purity 99.999% (2 inch in diameter). Prior to deposition, the substrates were cleaned in alchohol and acetone. Eventually, substrates were rinsed with a large amount of de-ionized water, dried in hot air and immediately placed inside the chamber. In situ Ar pre-sputtering of the target was performed in order to clean the target material. After vacuum pumping, the reactive sputtering was performed with an Ar pressure of 20 SCCM (standard cubic centimeter per minute) and oxygen pressure of 20 SCCM, was introduced through a mass flow controller. The RF power was maintained at 30 W. The films were deposited on (1 0 0) oriented p-type Si substrates. These as deposited films were subjected to microwave annealing at 900^0C.

Raman scattering spectra of the films were obtained before and after annealing in backscattering

Application of Accelerators in Research and Industry
AIP Conf. Proc. 1336, 264-268 (2011); doi: 10.1063/1.3586101

configuration with a Raman microscope using a 514.5 nm Ar+ laser excitation source. X-ray diffraction measurements were carried out with CuKα X-rays with λ=0.154nm in a glancing angle incidence geometry. The atomic force microscopy (AFM) measurements were carried out using SPA400, Seiko Instruments Inc. TEM images were obtained with a FEI Tecnai TEM instrument operating at 200 kV. All the measurements were carried out at room temperature.

RESULTS AND DISCUSSION

Fig. 1 shows the XRD spectrum of the GeO_x as deposited and annealed film at 900^0C using microwave annealing. All the diffraction peaks including (100), (011), (110), (102), (111), (200), (021), (003), (112), (013), (202), (210) and (211) can be clearly indexed to the hexagonal structure of GeO_2. In the present XRD measurements, the angle of the incident beam to the substrate surface was approximately 2^0 and the detector was rotated to scan the samples. The calculated lattice constants of GeO_2 from experimental data are a = 4.9821 and c = 5.6444 Å which are close the values from JCPDS tables where a = 4.9858 Å and c = 5.6473 Å. The average size of the nanoparticles is calculated using Scherrer's formula given by

$$D = \frac{0.9\,\lambda}{\beta\ Cos\theta_B}$$

Where λ is the wavelength of the X-ray source (*i.e,* 0.15406 nm for CuKα), β = FWHM (in radians) and θ_B is Bragg angle and D is the diameter of the nanocrystal. The theoretical and experimental 2θ values of all planes have been listed in TABLE 1. The size of the nanocrystal was estimated to be around 40 nm. All the peak positions are in good agreement with the expected values.

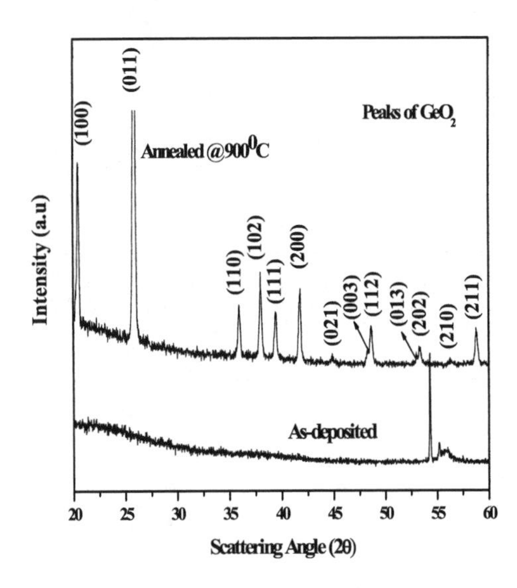

FIGURE 1. GIXRD spectra of as-deposited and annealed samples at 900^0C

TABLE 1. Different planes and XRD peak positions observed for GeO_2.

Peak (hkl)	Experimental (2θ)	Theoretical value (2θ)
100	20.56	20.55
011	25.96	25.95
110	36.00	35.99
102	38.05	38.04
111	39.53	39.48
200	41.87	41.80
021	44.93	44.91
003	48.39	48.31
112	48.71	48.68
013	53.07	53.02
202	53.42	53.37
210	56.36	56.32
211	58.87	58.85

Raman spectra of GeO_x pristine and annealed films have been shown in Fig.2. The peak at 521 cm^{-1} is attributed to the optical phonon mode of the Si substrate [14]. The peak at 442 cm^{-1} in the as deposited sample shows a broad feature which may be due to amorphous nature. But, the sharp peak at 443 cm^{-1} in annealed films is related to the crystalline GeO_2 [15]. Besides, we observe a peak at 301 cm^{-1}, which may be due to Ge optical phonons.

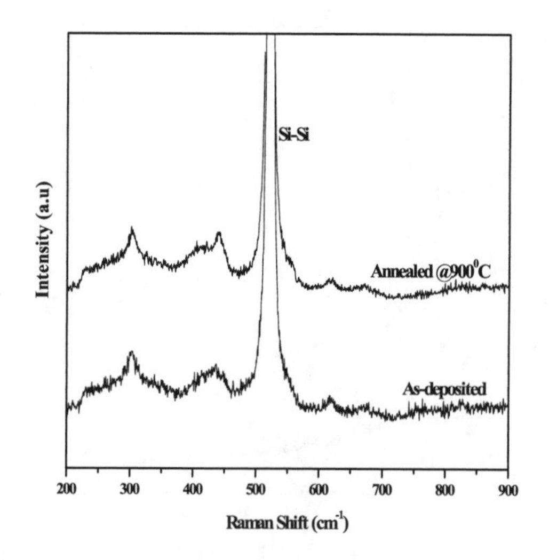

FIGURE 2. Raman spectra of as-deposited and annealed samples at 900^{0}C

Fig. 3a Shows an image of Fourier transform (FT) pattern representing the (100) and (011) zone axis's pattern of the GeO_2 nanocrystal structure. Diffraction spots of this FT pattern were used to measure lattice plane distances in these NCs from Fig 3b. The lattice spacing of the GeO_2 nanocrystals was determined using FTs of TEM images. The calculated lattice values from diffraction pattern are 4.3 Å and 3.4 Å for (100) and (011) planes respectively.

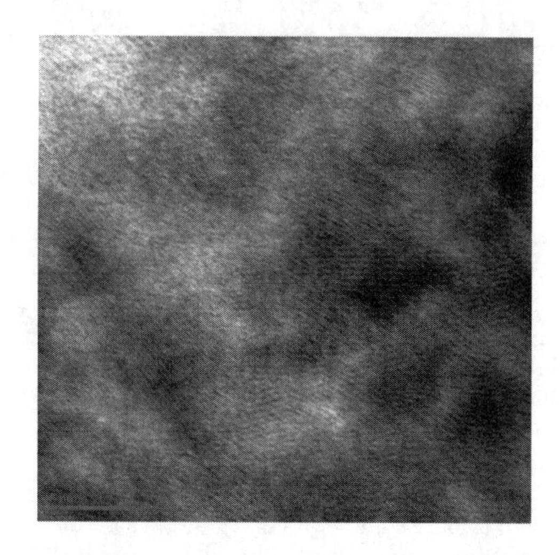

FIGURE 3a: TEM image of annealed sample at 900^{0}C

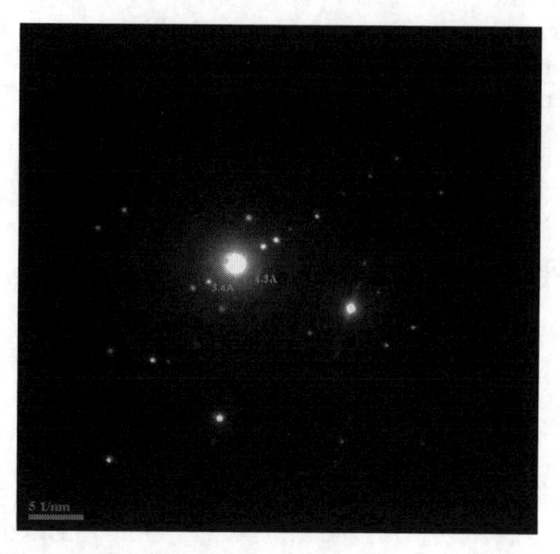

FIGURE 3b: Diffraction image of annealed sample at 900^{0}C

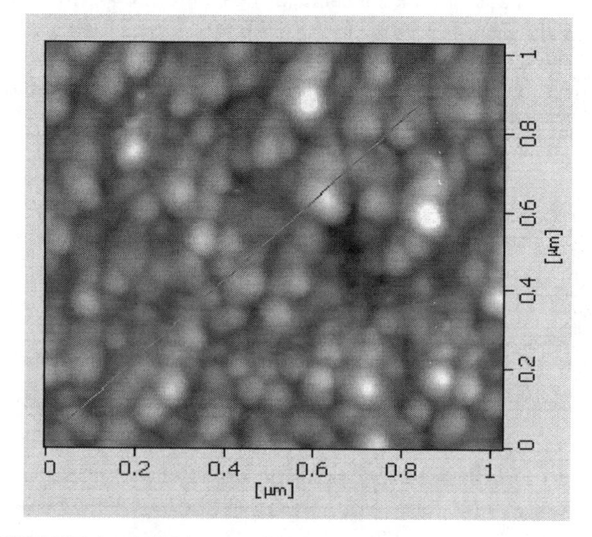

FIGURE 4a: AFM image of as deposited sample

Surface morphology of the sample is observed by AFM and image has been shown in Fig 4a. The as deposited film shows a smooth surface with rms surface roughness of about 1.3 nm. Sectional view of line is also plotted and shown in fig 4b.

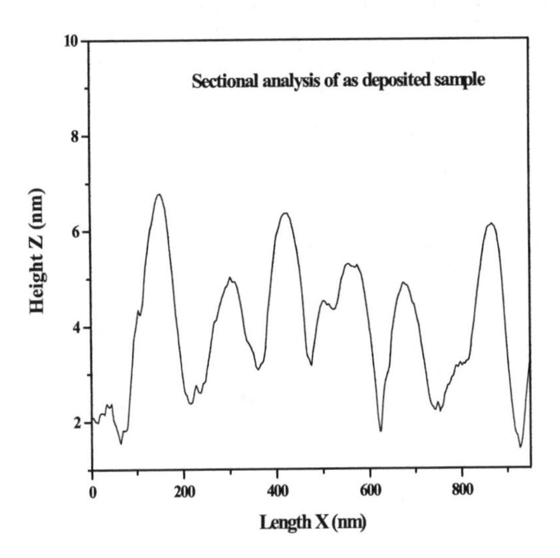

FIGURE 4b: AFM Sectional plot of as deposited sample

Most of the times preparation of GeO_x films using reactive sputtering method yields amorphous films with a composition close to GeO_2. Formation of GeO_2 nanocrystals in the present system is due to microwave annealing which is carried in air at 900^0C. XRD spectra also show the amorphous nature of as deposited films but it clearly indicates the formation of nanocrystals as a result of microwave annealing. Eventually, it is understood that GeO_x to GeO_2 transformation takes place due to annealing which is evident from the results discussed above. It is possible to tune the properties of nanocrystal by changing the sputtering deposition parameters like RF power, Ar, O_2 flow rates and distance between target to substrate. One can also change the microwave annealing parameters to alter the structural and optical properties of the nanocrystals.

Microwave energy is a form of electromagnetic energy with the frequency range of 300MHz to 300 GHz. Microwave heating is a process in which the materials couple with microwaves, absorb the electromagnetic energy volumetrically, and transform into heat. This method is different from other conventional annealing methods where heat is transferred between objects by the mechanisms of conduction, radiation and convection. In conventional heating, the surface of the material is heated first and is then followed by the heat propagating inward. This process results in a temperature gradient from the surface to the bulk. However, microwave heating generates heat uniformly within the material first and then heats the entire volume [16]. This heating mechanism is advantageous due to enhanced diffusion processes, reduced energy consumption, rapid heating rates and hence considerably reduced processing times, decreased sintering temperatures, improved physical and mechanical properties, simplicity, unique properties, and lower environmental hazards. These are features that have not been observed in conventional processes [17-20]. Microwave energy has been in use for variety of applications for few decades. These applications include communication, food processing, wood drying, rubber vulcanization, medical therapy, polymers, etc. It is also useful in process control, drying of ceramic sanitary wares, calcinations, and decomposition of gaseous species by microwave plasma, powder synthesis, and sintering [21, 22]. The current application of microwave annealing to synthesize nanocrystals adds to the novelty of microwave energy.

CONCLUSIONS

We have synthesized GeO_2 nanocrystals on silicon substrate by RF magnetron sputtering and subsequent microwave annealing. XRD indicates that as-deposited film shows amorphous nature whereas annealed films show crystalline nature. Raman spectrum also shows the presence GeO_2 formation after performing microwave annealing. AFM shows the smooth surface with small roughness. TEM also confirms the presence of GeO_2 nanocrystals from lattice image and diffraction pattern. The results obtained here vis-à-vis the importance of microwave annealing has been discussed in detail.

ACKNOWLEDGMENTS

N.S.R. and G.D. thank CSIR, New Delhi for award of SRF. V.S.K thanks CSIR, New Delhi for award of JRF. A.P.P thanks Centre for Nanotechnology for project funding. We greatly acknowledge the help of Dr K C James Raju and Mr A Rambabu in microwave annealing. We thank Mr P K Kulriya for his help in XRD measurements. We also acknowledge the help of Mr. Manohar Reddy and Mr. Durga Prasad for their kind support during TEM measurements, performed at the Centre for Nanotechnology established by DST New Delhi at University of Hyderabad.

*Corresponding author e-mail: appsp@uohyd.ernet.in
Tel: +91-40-23010181/23134316
Fax: +91-40-23010181 / 23010227

REFERENCES

1. K.P. Kalyanikutty, G. Gundiah, A. Govindaraj, C.N.R. Rao, 2005 J. Nanosci. Nanotechnol. **5** (2005) 425.
2. Y.J. Zhang, J. Zhu, Q. Zhang, Y.J. Yan, N.L. Wang, X.Z. Zhang, Chem. Phys. Lett. **317** (2000) 504.
3. J.Q. Hu, Q. Li, X.M. Meng, C.S. Lee, S.T. Lee, Adv. Mater. **14** (2002)1396.
4. M. Adachi, Nakagawa, K. Sago, K. Murata, Y. Nishikawa, Chem. Commun. **18** (2005) 2381.
5. M.J. Werner, S.R. Fribery, Phys. Rev. Lett. **79** (1997) 4143.
6. P. Viswanathamurthi, N. Bhattarai, H.Y. Kim, M.S. Khil, D.R. Lee, E.K. Suh, J. Chem. Phys. **121** (2004) 441.
7. T. Kawai, Y. Usui, K. Kon-No, Colloid Surface A **149** (1999) 39.
8. H.P. Wu, J.F. Liu, M.Y. Ge, L. Niu, Y.W. Zeng, Y.W. Wang, G.L. Lv, L.N. Wang, G.Q. Zhang, J.Z. Jiang, Chem. Mater. **18** (2006) 1817.
9. M. Grimsditch, A. Polian, V. Brazhkin and D. Balitskii, J. Appl. Phys., **83** (1998) 3018.
10. B. Houser, N. Alberding, R. Ingalls and E. D. Corzier, Phys. Rev. B, **7**(11) (1988) 6513.
11. N. Suresh, G. Jyoti, S. C. Gupta, S. K. Sikka, Sangeeta and S. C. Sabharwal, J. Appl. Phys., **76**(3) (1994) 1530.
12. Thiele, Ulrich K. International Journal of Polymeric Materials **50** (3) (2001) 387
13. Thomas Lange, Walter Njoroge, HansjoÈrg Weis, Manfred Beckers, Matthias Wuttig, Thin Solid Films **82** (2000) 365
14. R.A. Asmar, J.P. Atnas, M. Ajaka, Y. Zaatar, G. Ferblantier, J.L. Sauvajol, J. Jabbour, S. Juillaget and A. Foucaran, J. Cryst. Growth **279 (2005)** 394.
15. G. Kartopu, S.C. Bayliss, V.A. Karavanskii, R.J. Curry, R. Turan and A.V. Sapelkin, J. Lumin. **101 (2003)** 275.
16. P. Yadoji, R. Peelamedu, D. Agrawal, R. Roy, Materials Science and Engineering B 98 (2003) 269.
17. D. Agrawal, Journal of Materials Education **19** (1999) 49.
18. D.E. Clark, D.C. Folz, J.K. West, Materials Science and Engineering A **287** (2000) 153
19. C. Leonelli, P. Veronesi, L. Denti, A. Gatto, L. Iuliano, Journal of Materials Processing of Technology **205** (2008) 489
20. R.R. Menezes, P.M. Souto, R.H.G.A. Kiminami, Journal of Materials Processing Technology **190** (2007) 223
21. D. Agrawal, Sohn, International Symposium Advanced Processing of Metals and Materials, **4** (2006) 183.
22. D.E. Clark, W.H. Sutton, Annual Review of Materials Science **26** (1996) 299

SECTION VI – ION BEAM ANALYSIS

Ion Beam Irradiation Studies Of Ultrananocrystalline Diamond (UNCD)

A. Kayani *, E. Garratt*, S. AlFaify *, A. Dissanayake*, G. Tecos*, D. C. Mancini† and M. Syed±

*Department of Physics, Western Michigan University, Kalamazoo, MI 49008, USA
† Center for Nanoscale Materials, Argonne National Lab, Argonne, IL 60439-4812, USA
±Rose-Hulman Institute of Technology, Terre Haute, IN 47803, USA

Abstract. Investigations into the effects of high-energy ion bombardment of ultrananocrystalline diamond (UNCD) thin films was performed using 3 and 6 MeV protons and 24 MeV F^{4+}, with the fluence of 2.1 x 10^{17} ions/cm^2, 2.9 x 10^{17} ions/cm^2, and 6.7 x 10^{15} ions/cm^2 respectively. Objective of the research is to investigate the effect of structural damage on the physical properties of the material and compare it with the structure of unirradiated and N doped UNCD. Pre- and post-irradiated samples were analyzed by ion beam analysis (IBA) measurements, Raman spectroscopy, atomic force microscopy (AFM) and scanning electron microscopy (SEM). IBA measurements including Rutherford backscattering spectrometry (RBS), non-Rutherford backscattering spectrometry (NRBS) and elastic recoil detection analysis (ERDA) were used to determine elemental concentration of pre- and post-irradiated samples. Visible Raman spectra corresponding to samples irradiated at 3 and 6 MeV protons did not show much variation. For 24 MeV F^{4+} irradiated sample, significant changes were observed, particularly the loss of a shoulder at 1179 cm^{-1} and sharpening of the G peak at around 1532 cm^{-1}, indicating possible significant changes at the grain boundary and increase in sp^2 phase. AFM measurements show a reduction in RMS roughness after bombardment possibly due to the graphitization of the UNCD surface. The results of IBA measurements did not show any change in the elemental concentration or interface region between film and substrate.

Keywords: Ion beam analysis, UNCD, nanocrystalline diamond, polycrystalline diamond, ion beam irradiation
PACS: 60

INTRODUCTION

Ultrananocrystalline diamond (UNCD) films are thin films of diamond with a very fine-grain structure including grain size of 2-5 nm and abrupt grain boundaries. When deposited in microwave enhanced plasma CVD system (MPCVD) using a H-poor Ar / CH_4 gas mixture, UNCD gets a unique structure that allows controllable optical transparency, electrical and thermal conductivity, and film stress. The structure of UNCD is due to the high rate of renucleation that occurs as a result of the special plasma gas chemistry. Low surface roughness is an important characteristic of UNCD which is required for range of applications including novel integrated electronic devices, micro and nano electromechanical systems (MEMS/NEMS). UNCD films consists of a mixture of predominantly sp^3 diamond grains with sp^3 and sp^2 phases that are less than 10% of the film by volume residing in the grain boundaries. These films exhibit a unique set of properties comparable to that of bulk diamond that includes chemical inertness, low friction and high wear resistance, high hardness, and good electrical transport properties when appropriately doped[1].

The conductivity of UNCD increases by as much as five orders of magnitude when nitrogen gas (N_2) is added to the plasma during the growth. The resulting incorporation of N into the UNCD films also has a profound impact on the field emission and electrochemical behavior of the resulting material[2]. The exact role of N in UNCD is still not well understood. UNCD films with and without N doping have recently been shown to have mild n- and p-type semiconductor characteristics[3]. Investigation of the bonding structure of UNCD suggest that the observed increase in conductivity for N-doped UNCD films may be a result of the increase in overall grain boundary volume, while the grains themselves remain pure diamond[4]. Density-functional-based tight-binding MD simulations suggest that the increase in the conductivity is due to the increase of states in the band gap attributed to the bonding disorder and impurities

Application of Accelerators in Research and Industry
AIP Conf. Proc. 1336, 271-275 (2011); doi: 10.1063/1.3586102

in the grain boundaries[5]. Moreover, vibrational and optical studies on N doped UNCD films have shown a narrowed energy gap with a graphite-like ordered structure, which may be responsible for the high electrical conductivity[6]. Hence, evidence points towards the change in nanostructure of N-doped UNCD relative to undoped UNCD resulting in the increase in conductivity.

Highly dosed ion-implanted UNCD films have shown an amorphous phase that resulted in enhanced electron field emission. Ion irradiation induced enhancement in the field emission properties of diamond and carbon fibers[7] have been reported, which are attributed to the production of defects. A graphitic phase has been observed in ion irradiated diamond structures that increases with the increase in ion dosage. Thus, controlled ion beam irradiation of UNCD films may prove useful in creating disorder that can be compared to the effects on the physical properties of the material, brought about when N doping in UNCD is carried out.

For our initial studies, we have carried out ion beam irradiation studies of undoped UNCD films using high energy protons and F^{4+} and compared it with unirradiated UNCD, both N doped and undoped. Pre- and post-irradiated samples were analyzed by ion beam analysis (IBA) measurements, Raman Spectroscopy, AFM and SEM. IBA measurements including Rutherford backscattering spectrometry (RBS), non-Rutherford backscattering spectrometry (NRBS) and elastic recoil detection analysis (ERDA) were used to determine elemental concentration of pre- and post-irradiated samples and to investigate the effect of ion beam irradiation on the interfacial region of the samples. Our goal is to understand the effect of structural change on the physical properties of UNCD films which is brought upon to the samples via ion bombardment and compare that to structural differences between N-doped and undoped UNCD.

EXPERIMENT

Thin films of UNCD for our experiments were deposited on Si substrates in the Center of Nanoscale Materials (CNM), Argonne National Lab (ANL) using microwave plasma enhanced (MPE) chemical vapor deposition (CVD) system, the MPR-915-1800 CVD reactor from Lambda Technology, dedicated to UNCD growth. Before the deposition process, a thin layer (~10 nm) of W was deposited on the Si substrates by sputter deposition. The W interlayer has been shown to improve the adhesion of UNCD to the Si substrate and also increase the diamond nucleation density. This reduces the overall roughness of the UNCD films by eliminating interfacial voids and allows thinner UNCD

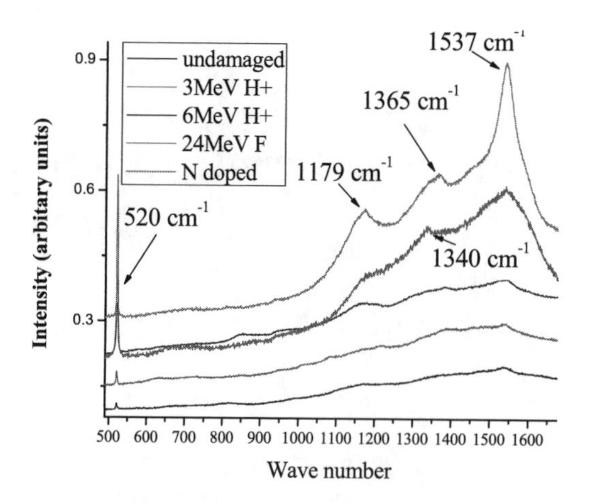

FIGURE 1. Raman spectra of N doped and undoped, and post irradiated samples

to be deposited. The W-coated Si substrate was placed on the substrate holder and microwave power of 800 watts was used to ignite the discharge. Ar, CH_4 and H with flow rates of 600, 2, and 10 sccm respectively were used as source gases and their flow rates are controlled by a precision multiple channel mass flow controller. Processing pressures of 120 mbr are achieved by controlling the throttle valve. Absorbed microwave power of 4 kW was used to cover the 8 inch diameter substrate wafer with the plasma discharge. Substrate temperature was fixed at 550 °C. N doping of UNCD was carried out by diluting the growth plasma with 10 % N_2. Samples were deposited for 5 hours. Film thicknesses for undoped and doped UNCD measured using reflectometry spectroscopy were 114 nm and 140 nm respectively.

Ion beam irradiation of undoped UNCD samples was carried at Western Michigan University's 6 MV Van de Graaff accelerator lab. Smaller wafer sections of roughly 1 cm^2 size were exposed at room temperature, separately, to 3 and 6 MeV protons and 24 MeV F^{4+}, with the fluence of 2.1 x 10^{17} ions/cm^2, 2.9 x 10^{17} ions/cm^2, and 6.7 x 10^{15} ions/cm^2, respectively. High energy was used so that beam specie won't stop in the film. Computer modeling of the irradiation effects of the two different ion beam species (H^+ and F^{4+}) were performed using the Stopping Range of Ion in Materials (SRIM) program[8]. The ion beam was allowed to pass through a gold foil to diverge the beam followed by an 8mm diameter circular collimator which defines the area of the sample to be exposed to the ion beam. Prior to the beam irradiation, an image of the beam on the sample was obtained using photographic film. Charge collected on each sample was measured directly from the current on the aluminum target holder itself and

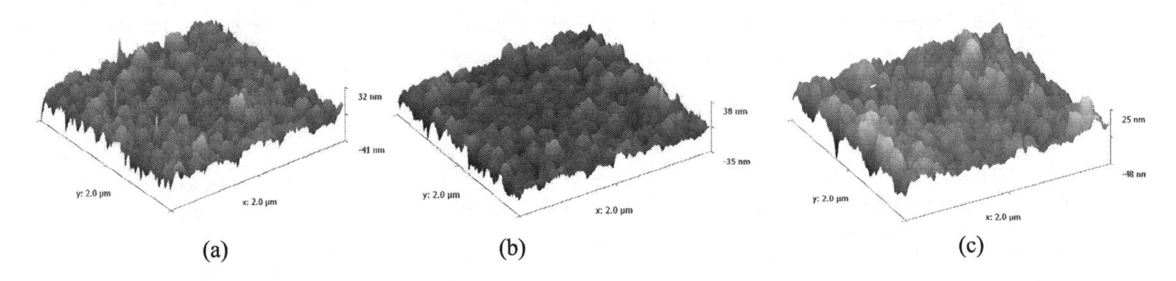

FIGURE 2. AFM image (a) unirradiated sample, (b) 24 MeV F^{4+} irradiated sample. (c) 6 MeV proton irradiated sample.

integrated to derive the total charge collected. Secondary electron emission was suppressed by a +200 V bias placed on the sample holder. After exposure the samples were left in the scattering chamber at ~10^{-6} Torr for 24 hours.

All UNCD samples were analyzed by Raman spectroscopy, AFM, and SEM. For Raman spectroscopy, 633 nm wavelength was used. Raman spectra of unirradiated N-doped and pre- and post-irradiated UNCD samples are shown in figure 1. AFM measurements were carried out in tapping mode, resulting in $2\mu m$ x $2\mu m$ images for unirradiated, and 24 MeV F^{4+} and 6 MeV irradiated undoped samples are shown in figures 2(a), 2(b) and 2(c) respectively. AFM measurement of N-doped sample was also carried out in tapping mode but the image obtained is not shown. An SEM image of unirradiated undoped UNCD sample is shown in figure 3(a). The SEM image clearly shows fine grain granular structure of UNCD film.

To obtain a complete picture of trapped impurities and elemental composition of N doped and pre and post irradiated samples, IBA measurements were carried out using a 6 MV tandem Van de Graaff accelerator of the department of physics at Western Michigan University. For the measurements, a beam of 1.5 MeV proton ions was used for RBS/NRBS measurements. For ERDA measurements to determine H concentration, a 16 MeV O^{4+} beam was used. To obtain the concentration of elements in the samples, SIMNRA computer simulation of the spectra with the original data[9] was carried out. RBS/NRBS experimental and simulated spectra of unirradiated and 24 MeV F^{4+} irradiated samples are shown in the figure 6. IBA measurements were also used to investigate the interfaces regions, UNCD/W and W/Si, of pre- and post-irradiated undoped samples, which were found to be abrupt without any change associated with the ion beam irradiation.

RESULTS AND DISCUSSION

Raman spectroscopy is widely used to confirm the UNCD structure by determining the presence of sp^3 and sp^2 carbon. The peak that appears around 1150 cm^{-1} in the visible Raman spectra is considered a finger print of UNCD[10]. Moreover, a single sharp peak around 1332cm^{-1} is used to determine crystalline quality[11]. Figure 1 shows Raman spectra of N-doped, undoped and irradiated UNCD samples. In all the cases, four dominant peaks are seen: the substrate peak that appears around 520 cm^{-1} for Si, diamond peak around 1140-1170cm^{-1} which is signature of UNCD structure and the peaks around 1350 and 1550 cm^{-1} that are the respective *D* and *G* modes of sp^2 carbon. These peaks can be due to the in-plane stretching modes of the sp^2-bonded carbon at the grain boundaries of UNCD[12]. Visible Raman spectra corresponding to samples irradiated at 3 and 6 MeV protons did not show much variation with respect to the unirradiated sample. For the 24 MeV F^{4+} irradiated sample, however, significant changes are observed, particularly the loss of a shoulder at 1179 cm^{-1} and sharpening of G peak at around 1532 cm^{-1} and an increase in the $I(G)/I(D)$ ratio, all indicating significant increase in the sp^2 bonded carbon.. The nanodiamond crystallinity decreases for the sample irradiated with 24 MeV F^{4+}. This is evident from the observation that D peak in the sample shifts from 1365 cm^{-1} to 1340 cm^{-1}. The increase in G peak intensity with respect to D peak for N-doped and 24 MeV F^{4+} irradiated spectra also points to the increased disorder because of the high scattering cross-section of sp^2 sites for visible Raman transitions.

Similar differences were observed between the Raman spectra of the N doped and undoped UNCD samples. The spectra also show an increase in the ratio of the G-band to the D-band peak intensities for nitrogen doped sample, which is an indication of increased sp^2 bonded carbon. Moreover, the sharpness of the peak around 1350cm^{-1} increases for N doped sample with respect to the unirradiated sample, which indicates increased grain size.

AFM measurements of pre and post irradiated samples indicate a smoothening effect associated with ion beam irradiation. Figures 2 (a), (b) and (c) are images of unirradiated, 24 MeV F^{4+}, and 6 MeV H^+ irradiated undoped samples respectively. The overall RMS roughness for each sample in the area surveyed is 3.9, 3.4, and 3.1 nm for the unirradiated, 24 MeV F^{4+}, and 6 MeV H^+ irradiated undoped samples, respectively. Although all the samples are smooth on a submicron scale, the ion beam irradiated samples are smoother than the unirradiated samples. Reduction in RMS roughness after bombardment may be due to the graphitization of the UNCD surface. The RMS roughness of N-doped UNCD sample was 5.24 nm. Higher RMS roughness of N-doped sample can be attributed to increased grain sizes when grown. Figures 3(a) show the SEM image of the unirradiated sample. The morphology of the sample indicates that

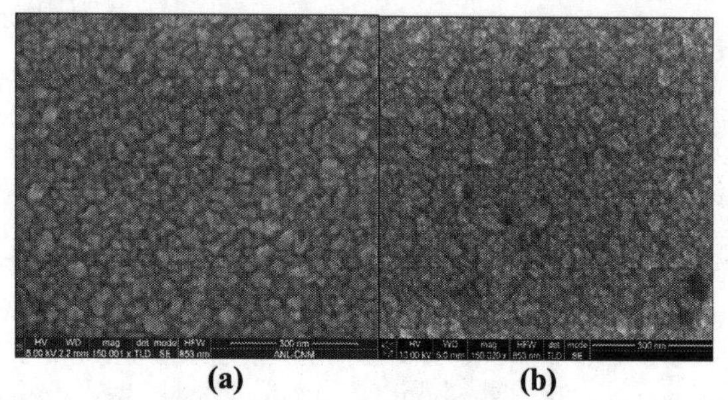

(a) (b)

FIGURE 3. SEM image of unirradiated (a) undoped (b) N-doped sample.

UNCD film is made of highly energetic twisted grains separated by grain boundaries. SEM images of the sample deposited with 10% of N_2 in the plasma, shown in figure 3(b), indicates that there is an increase in the average grain size with the increase in the N content in plasma as well as increase in the width of the grain boundaries. This result is in direct correlation with the Raman spectroscopy measurements which suggests the increase in the grain size on N-doped UNCD.

RBS/NRBS data for the unirradiated and 24 MeV F^{4+} irradiated samples are shown in figure 4. Simulated backscattering data using SIMNRA very closely matched experimental measurements for the unirradiated sample while peaks for C and W matched very closely for 24 MeV F^{4+} irradiated sample data, but a drop in the background height was observed at the Si/W edge which is attributed to the single crystal nature of Si which produces channeling effect. Substrate edge in the data obtained for unirradiated sample closely matched with that of irradiated sample indicating negligible intermixing of substrate and the film. Elemental concentration of 97 at% C and 3 at% H was obtained by simulating the back and forward scattering data using SIMNRA computer program. N concentration in N-doped UNCD sample could not be obtained. It has been reported that N content in N-doped UNCD is estimated to be less than 1 at%. Since the N peak in the RBS/NRBS spectra sits on a huge background of Si substrate, it is therefore not possible to find the N content using this technique.

CONCLUSIONS

Investigations into the damage associated with the ion beam bombardment of UNCD films have been carried out and compared with the unirradiated undoped and N-doped UNCD film. Films were deposited using MPECVD process on the Si substrates and N doping in UNCD was carried out by introducing N_2 into the growth plasma. High energy protons and F^{4+} beam were used to irradiate in the undoped UNCD. High energy was used so that beam specie won't stop in the film. Raman spectroscopy data for 24 MeV F^{4+} irradiated sample indicates uniform increase in fraction of sp^2 bonded carbon. For N-doped UNCD sample, however, localized increase in the sp^2 fraction with increased grain size is sp^3 bonded carbon is observed. Decreased roughness in AFM studies of pre and post irradiated samples also indicated the increase in graphitic phase in the irradiated samples. However, increased roughness in N-doped UNCD may be due to increased grain size. Overall all the samples were found smooth on submicron scale. No change in the elemental concentration and interface region of the sample was observed.

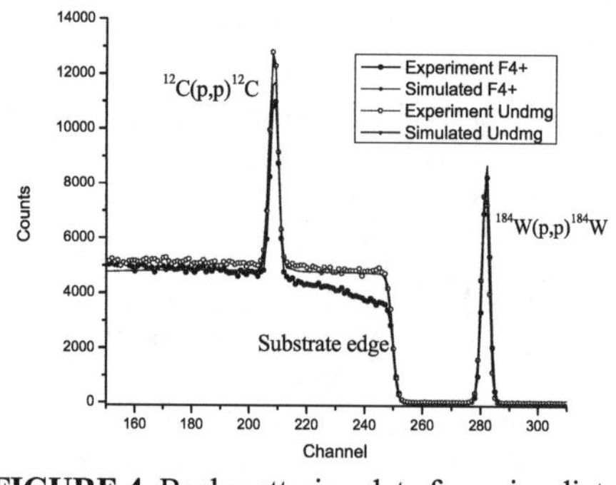

FIGURE 4. Backscattering data for unirradiated and F^{4+}.

ACKNOWLEDGEMENTS

Use of the Center for Nanoscale Materials (CNM) was supported by the U. S. Department of Energy, Office of Science, Office of Basic Energy Sciences, under Contract No. DE AC02-06CH11357. The authors would like to thank Mr. Benjamin Gaudio, Mr. Allan Kern and Mr. Rick Welch for providing technical assistance for this work.

REFERENCES

1. M. Hajra, C.E. Hunt, M. Ding, O. Auciello, J. Carlisle and D.M. Gruen, *J. of App. Phy*, **94**, 4079 (2003).
2. Q. Chen, D. M. Gruen, A. R. Krauss, T. D. Corrigan, M. Witek, and G. M. Swain, *J. Electrochem. Soc.* **148**, E44 (2001).
3. D. M. Gruen, P. C. Redfern, D. A. Horner, P. Zapol, and L. A. Curtiss, J. Phys. Chem B, **103**, 5459 (1999).
4. J. Birrell, J. E. Gerbi, O. Auciello, J. M. Gibson, D. M. Gruen, and J. A. Carlisle, *J. Appl .Phys*, 93, (2003) 5606.
5. P. Zapol, M. Sternberg, L. A. Curtis, T. Frauenheim, and D. M. Gruen, Phys Rev B, vol. 65, (2002) 045403.
6. Chien-Chung Teng, Shin-Min Song, Chien-Min Sung, and Chhiu-Tsu Lin, *J. Nano. mat,* Article ID 621208, (2009)
7. S. Talapatra, J. Y. Cheng, N. Chakrapani, S. Trasobares, A. Cao, R. Vajtai, M. B. Huang, and P. M. Ajayan, *Nanotechnology* 17, 305 (2006).
8. J. F. Ziegler, M.D. Ziegler and J.P. Biersack, SRIM-The Stopping and Range of Ions in Matter. 2008
9. M. Mayer, *SIMNRA User's Guide*, Technical Report IPP 9/113, Max-Planck-Institut fur Plasmaphysik, Garching, Germany, (1997).
10. R. E. Shroder and R. J. Nemanich. *Phys. Rev.* B 41, 3738 (1990).
11. D. S. Knight and W. B. White, *J. Mater. Res.* **4**, 385 (1989)
12. P. Achatz, J. Garrido, M. Stutzmann, O.A. Williams, D.M. Gruen, A.Kromka, D. Steinmueller, *Appl. Phys. Lett.* **88**, 101908 (2006).

IBIXFIT: A Tool For The Analysis Of Microcalorimeter PIXE Spectra

A. Taborda[a,b], L. C. Alves[a,c], N. P. Barradas[a,c], P. C. Chaves[a,b,d], M. A. Reis[a,b]

[a] *Instituto Tecnológico e Nuclear, Estrada Nacional 10, Sacavém, Apartado 21, 2686-953 Sacavém, Portugal*
[b] *Centro de Física Atómica da Universidade de Lisboa, Av. Prof. Gama Pinto, 2, 1649-003 Lisboa, Portugal*
[c] *Centro de Física Nuclear da Universidade de Lisboa, Av. Prof. Gama Pinto, 2, 1649-003 Lisboa, Portugal*
[d] *Instituto Superior Técnico da Universidade Técnica de Lisboa, Av. Rovisco Pais, 1049-001 Lisboa, Portugal*

Abstract. PIXE analysis software has been for long mainly tuned to the needs of Si(Li) detector based spectra analysis and quantification methods based on K_α or L_α X-ray lines. Still, recent evidences related to the study of relative line intensities and new developments in detection equipment, namely the emergence of commercial microcalorimeter based X-ray detectors, have brought up the possibility that in the near future PIXE will become more than just major lines quantification. A main issue that became evident as a consequence of this was the need to be able to fit PIXE spectra without prior knowledge of relative line intensities. Considering new developments it may be necessary to generalize PIXE to a wider notion of ion beam induced X-ray (IBIX) emission, to include the quantification of processes such as Radiative Auger Emission. In order to answer to this need, the IBIXFIT code was created based much on the Bayesian Inference and Simulated Annealing routines implemented in the Datafurnace code [1]. In this presentation, the IBIXFIT is used to fit a microcalorimeter spectrum of a $Ba_xSr_{(1-x)}TiO_3$ thin film sample and the specific possibility of selecting between fixed and free line ratios combined with other specificities of the IBIXFIT algorithm are shown to be essential to overcome the problems faced.

Keywords: Spectra fitting, PIXE, Microcalorimeter X-ray detector
PACS: 07.05.Tp, 29.40.Vj, 33.20.Rm

INTRODUCTION

Particle Induced X-ray Emission, or PIXE for short, has seen its conversion into an analytical technique as a consequence of the development of solid state detectors and their associated electronics in the decade of 1960 [2]. During the next decade (the 1970's) PIXE, proposed by Johansson [3,4], would experience an impressive development, such that by the 1980's [5] it was established much in its present shape, and ready for applications.

During the 30 years that have passed since, quantification of K_α and L_α lines were the main issue and little attention was given to other lines and/or second order phenomena. Still, as applications start to request more and more from the PIXE technique this picture starts to change, in particular due to the need to untangle close overlapping lines with high precision. Presently, as will be shown in this paper, new software tools are required to be able to properly explore new technological developments such as microcalorimeter EDS high resolution X-ray detectors [6,7,8,9], which bring to light second order effects so clearly [10,11,12] that they can no longer be disregarded from sample analysis without a clear notion that information present in the spectra is not being used.

MATERIALS AND METHODS

In this work, the PIXE analysis of a $Ba_xSr_{(1-x)}TiO_3$ thin film deposited on top of several nanometer thick commercial substrate of Pt, Ti and SiO2 layers deposited on top of a Si bulk ($Ba_xSr_{(1-x)}TiO_3$ / Pt / Ti / SiO_2 / Si) is used to show the versatility of the IBIXFit code as well as the capacity of microcalorimeter based EDS detectors to resolve close laying lines and second order contributions. RBS measurements were performed using a 2.0 MeV 4He ion beam. The experimental data were analysed with the IBA Datafurnace NDF v9.3b [1] confirming the structure of the substrate and the thicknesses of the Pt, Ti and SiO_2 layers were determined as being of 55 nm, 18 nm and 100 nm, respectively. The standard characterization of the sample by PIXE and RBS was carried out at the ITN 2.5 MV Van de Graaff set-ups [13]. A Gresham Sirius Si(Li) of 150 eV resolution detector positioned

Application of Accelerators in Research and Industry
AIP Conf. Proc. 1336, 276-280 (2011); doi: 10.1063/1.3586103

at a 70° angle to the beam direction was used to collect a first spectrum upon irradiation of the sample with a 1.0 MeV H^+ beam and using a My50 absorber in front of the detector window. In this work, the studied spectrum was collected at the ITN 3 MV Tandetron HRHE-PIXE set-up [14] using the Vericold Technologies GmbH POLARIS microcalorimeter detector having a normal operation resolution of 21 eV at 1.486 keV (Al-K_α) and 30 eV at 10.550 keV (Pb-$L_{\alpha 1}$). The sample was irradiated with a 1.0 MeV H^+ ion beam and the spectrum was collected using a grazing incidence angle of 70°. The beam spot was made approximately circular by using a set of two off centred collimators and thus assuring that the beam was kept within the target limits.

The collected spectrum was deconvoluted using the IBIXFit_3v8 computer code, which initial version was presented at the 2007 PIXE conference in Mexico [15].

The fitting model was based on the X-ray line energies from Zschornack [16] and the initial line intensity ratios were based on Scofield's work [17]. In the fitting model construction, the elements known to be present in the sample (Ba, Sr, Ti, Pt, Si) were included first. These were assumed to contribute to the spectrum with both main lines as well as multi-ionisation and radiative Auger lines. Afterwards, unknown impurity elements clearly seen on the microcalorimeter PIXE spectrum were added to the fitting model to improve the results obtained.

RESULTS AND DISCUSSION

In Figure 1, the spectrum collected with the microcalorimeter detector upon the irradiation of the thin film sample with a 1.0 MeV proton beam is presented. The number of counts in the spectrum is presented on logarithmic scale to allow for the observation of all the lines. A first observation and identification of the several X-ray lines of the elements known to be present in the sample can be made. In the low energy region of the spectrum, the Si K_α and K_β X-ray lines (1.739 keV and 1.835 keV, respectively) are clearly separated and are easily identified, as well as the Pt M_α (2.050 keV) and M_β (2.127 keV) X-ray lines. The Sr L_α and $L_{\beta 1}$ X-ray lines, with energies 1.805 keV and 1.871 keV, are also resolved, but not seen clearly in this Figure due to the x-axis scale, since these energies are very close to that of the K_β X-ray lines of Si. In the higher energy region of the spectrum, it is possible to identify the Ti K_α and K_β X-ray lines perfectly resolved from the L group lines of Ba, which is not at all the case in the spectrum collected using the Si(Li) detector (as can also be seen from Figure 1).

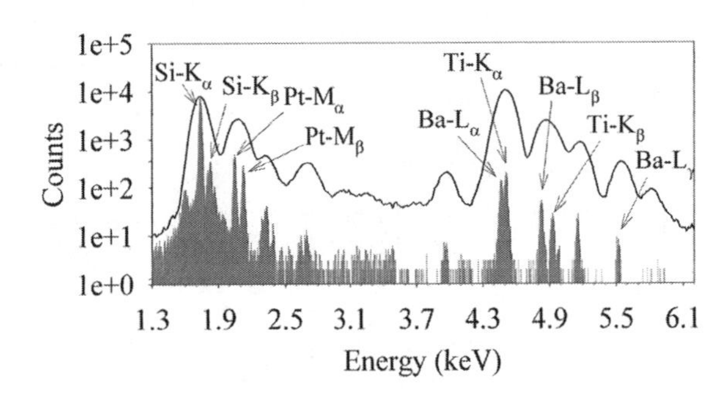

FIGURE 1. Spectra of the thin film sample, $Ba_xSr_{(1-x)}TiO_3$ / Pt / Ti / SiO_2 / Si, irradiated with a 1.0 MeV H^+ beam. The spectrum collected using the POLARIS microcalorimeter detector is represented as the gray area and is displayed in the energies range of 1.3 keV and 6.3 keV. The spectrum of the same sample collected with a Si(Li) detector is also plotted (solid black line) for comparison.

Such a high-resolution in a large range of X-ray energies implies that, in dealing with microcalorimeter based spectra analysis, exhaustive care must be taken, starting as soon as the energy calibration. In fact, a "small" error of 1% in energy vs channel calibration implies that the calculated centroid peak position is completely out of place, no longer representing just a slight misfit.

The identification of all the X-ray lines from the elements in the sample is crucial in the process of fitting a spectrum. In the fitting of microcalorimeter spectra such as the one in this work, this is of even greater importance since many of the resolved X-ray lines might have never been noticed to be present in non-microcalorimeter solid state detectors based PIXE spectra. Being further that, in IBIXFit, the fitting model is completely defined by the user essentially without making use of a third party defined database, extreme care is required.

The first step in the construction of the model to be fitted using IBIXFit is the simulation of the Bremsstrahlung background. The background simulation is presented in Figure 2 and both components of the background model used are represented by the dark gray solid line. The black solid thick line is the resulting simulated Bremsstrahlung background model.

In the case of PIXE spectra collected using Si(Li) detectors, it is usually sufficient to include the K_α and K_β X-ray lines in the fitting models in order to describe the Si X-ray lines. Figure 3 shows that the Si K_α and K_β X-ray lines alone (black solid thin lines), on top of the background, are not sufficient to describe the Si contribution to the spectrum.

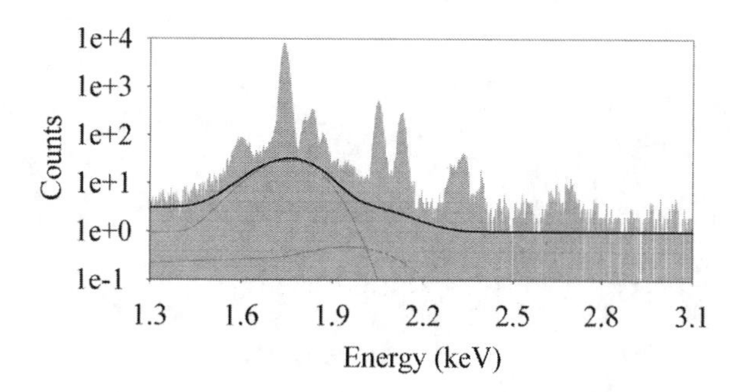

FIGURE 2. Bremsstrahlung background simulation by IBIXFit (black solid thick line) using a fitting model based only on the Bremsstrahlung background. The components of the background model are represented by the dark gray solid line.

A careful look at the energies of close by lines shows that these lines are not commonly observed due to both the lack of resolution as well as low peak to background ratios. An example of such is the lines group seen to the left of the Si K_α X-ray line, corresponding to Si Radiative Auger Emission (RAE) contributions. The high resolution also implies that multi-ionisation satellites must be included to properly fit the larger peaks.

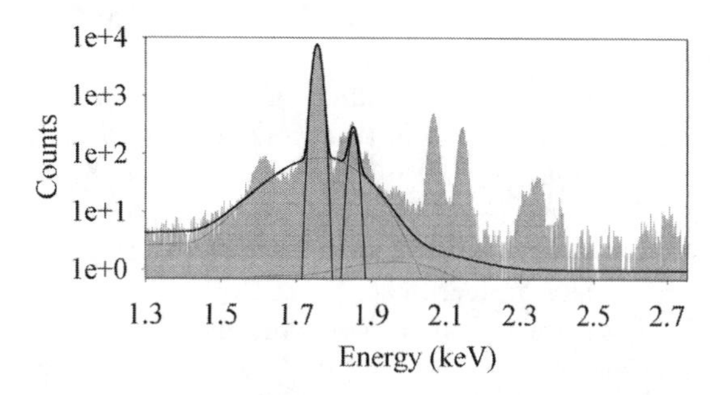

FIGURE 3. Fit of the low energy region of the spectrum, represented by the black solid thick line, using a model based only on the Si K_α and K_β X-ray lines (black solid thin lines) and on the background lines (dark gray solid lines).

The fit of the Si energy region of the spectrum when adding the Si RAE and multi-ionisation lines to the IBIXFit fitting model is seen in Figure 4. In Figure 4, Si K_α and K_β X-ray lines are represented with a black solid thin line as well as the multi-ionisation lines and the fitted RAE lines are represented by a dark gray solid line. From Figure 4 and onwards the background lines are not represented to keep the Figures as easy to understand as possible. Figure 4 also makes evident the usefulness of the possibility of selecting between fixed and free line intensity ratios. The evaluation of the fitting results leads to the observation that IBIXFit tries to numerically fit the Sr L_α and L_β X-ray lines with Si RAE and multi-ionisation lines.

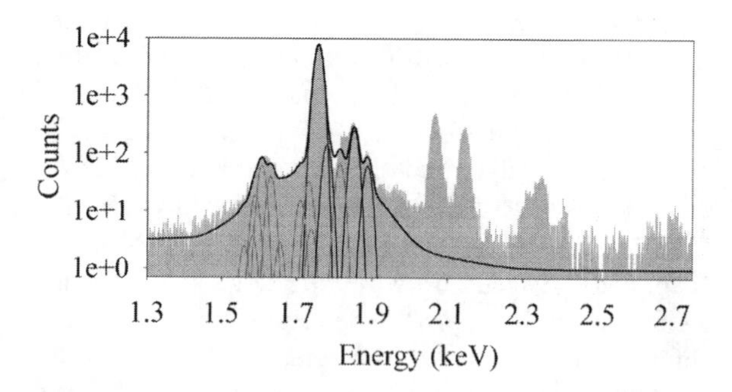

FIGURE 4. Fit of the low energy region of the spectrum with free line intensity ratios. The final fit is represented by the black solid thick line. Fitting model includes the Si RAE and multi-ionisation lines. The Si K_α and K_β X-ray lines are represented with a black solid thin line as well as the multi-ionisation lines. RAE lines are represented by a dark gray solid line.

Fixing the intensity ratios of these lines is therefore essential to obtain a fit with physical meaning. In this case, the RAE line intensity ratios were fixed with respect to the most intense RAE line, and the multi-ionisation lines were fixed with respect to the K_α X-ray line intensity. The fit obtained using the fixed RAE and multi-ionisation intensity ratios is shown in Figure 5.

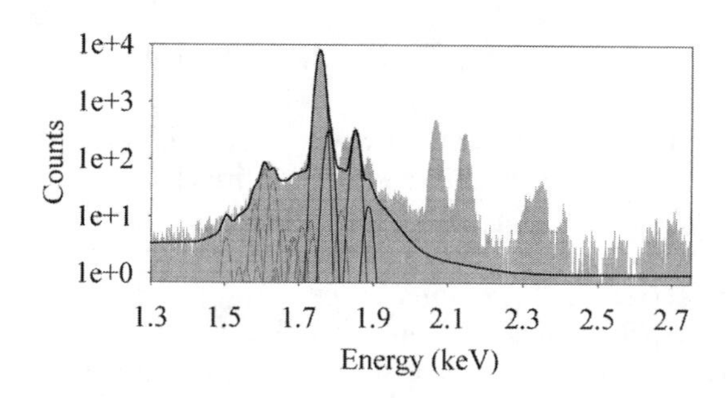

FIGURE 5. Fit of the low energy region of the spectrum with free and fixed line intensity ratios. The final fit is represented by the black solid thick line. Fitting model includes the Si RAE and multi-ionisation lines. The Si K_α and K_β X-ray lines are represented with a black solid thin line as well as the multi-ionisation lines. RAE lines are represented by a dark gray solid line.

The IBIXFit spectrum fitting using an exhaustive model including all the relevant X-ray lines as well as multi-ionisation and the more intense RAE lines for each of the known elements in the sample are shown in the graphs of Figure 6. The fitted spectrum is presented in Figure 6 (a) graph, where the experimental data is represented as the light gray area

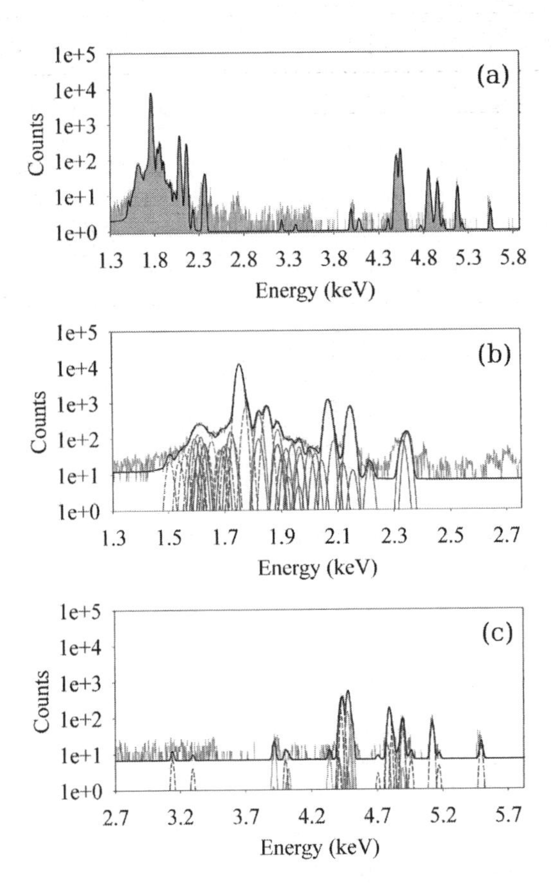

FIGURE 6. IBIXFit fitting results of the spectrum using an exhaustive model that includes the known elements in the sample. (a) fitted spectrum, the experimental data is represented as the light gray area and the fit is represented by the black solid thick line. (b) fitted lines in the lower energy region of the spectrum, where Si is represented by the red short dashed lines, Sr by the pink solid line and Pt by the dark green solid line. Fitted curve represented by the blue solid line. (c) fitted lines in the higher energy region of the spectrum, where the light green solid line represents the Ti fitted lines and the Ba lines are represented by the purple dashed line. Fitted curve represented by the blue solid line.

and the fit is represented by the black solid thick line. Figures 6 (b) and (c) show all the fitted lines, in the lower energy (b) and higher energy (c) regions of the spectrum. In these cases, the experimental data corresponds to the light gray line and the fitted curve is the blue solid thick line. In Figure 6 (b), the fitted X-ray lines for Si, Sr and Pt are individually plotted and represented by the red short dashed lines for Si, the pink solid line for the Sr and the dark green solid line for the Pt lines. The fitted lines for Ti and Ba are represented in Figure 6 (c) where Ti is represented by the light green solid lines and Ba by the purple dashed lines.

In the region between 3.1 keV and 3.5 keV a continuum of probably Ba originated RAE lines is observed. To partially confirm this, the most intense

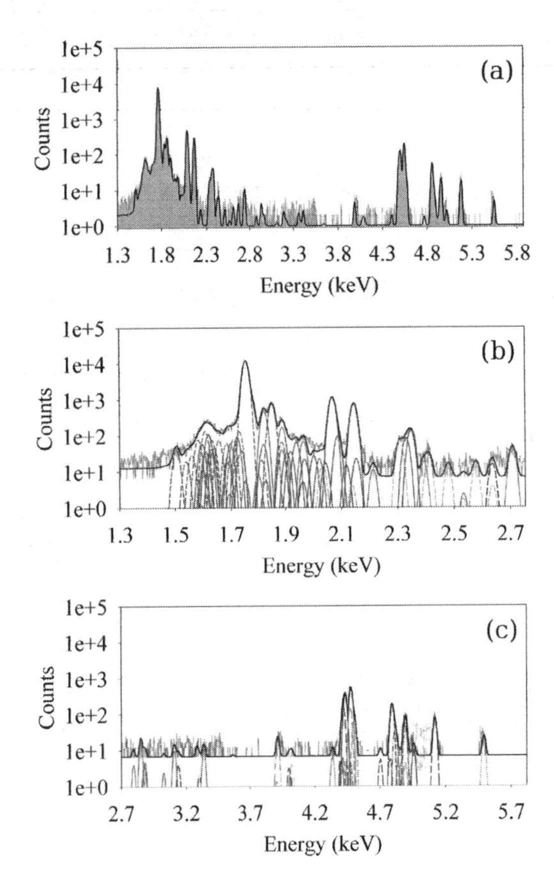

FIGURE 7. IBIXFit fitting results of the spectrum using an exhaustive model incorporating the known elements in the sample as well as the impurities found. (a) final fit of the spectrum, the experimental data is represented as the light gray area and the fit is represented by the black solid thick line. (b) fitted lines in the lower energy region of the spectrum, where Si is represented by the red short dashed lines, Sr by the pink solid line, Pt by the dark green solid line, Mo is represented by the cyan dashed line, Cl by the black dashed line and Rh by the gray solid line . Fitted curve represented by the blue solid line. (c) fitted lines in the higher energy region of the spectrum, where the light green solid line represents the Ti fitted lines, the Ba lines are represented by the purple dashed line and Rh is represented by the gray solid line. Fitted curve represented by the blue solid line.

lines were chosen deprecating the other RAE lines. Apart from this, the observation of the fitted spectrum in Figure 6 strongly suggests the presence of impurities in the sample, since there are significant peaks that were not fitted with the model constructed based on the known elements and after including all second order transitions expected to have highest intensity.

Figure 7 shows a set of 3 graphs as the ones in Figure 6, representing the final fit of the spectrum after adding the X-ray lines assumed to be from impurities into the fitting model. These impurities were found to be Mo, Rh and Cl and are represented by the cyan dashed line (Mo), the dark gray solid line (Rh) and the black dashed line (Cl).

Since IBIXFit is a Bayesian inference [15] numerical fitting program and over 100 lines were included in the model, the ouput results of the IBIXFit code were carefully evaluated, considering several aspects, namely the areas and errors reported for each of the lines in fitting model. In Table 1, the fitted areas and corresponding Bayesian inference errors for the more intense X-ray lines of the elements in the thin film sample are presented. The errors are displayed as percentage of the area value to allow a clear perception of the precision of the fitting results. It can be seen that in this list, more than half of the errors are below 5%, Mo and Rh L_α being the only α lines presenting error above 5%.

TABLE 1. Fitted areas and respective errors of some of the more intense X-ray lines of the elements in the sample.

Line	Energy (keV) [16]	Fitted Area	Error (%)
Si K_α	1.739	9.5883E+04	< 1.0
Si K_β	1.835	3.4790E+03	3.9
Ti $K_{\alpha 1}$	4.504	2.0137E+03	3.7
Ti $K_{\alpha 2}$	4.510	5.8831E+02	6.0
Sr L_α	1.805	2.1399E+03	2.3
Sr L_β	1.871	2.2344E+01	3.2
Ba $L_{\alpha 1}$	4.466	1.3337E+03	4.7
Ba $L_{\beta 1}$	4.827	6.2900E+02	6.5
Ba L_γ	5.531	1.5193E+02	5.5
Pt M_α	2.050	5.7203E+03	2.0
Pt M_β	2.127	3.4673E+03	1.7
Mo L_α	2.293	1.6394E+02	9.9
Mo L_β	2.394	6.1671E+01	10.0
Rh L_α	2.694	9.1553E+01	6.9
Rh L_β	2.834	4.9626E+00	6.9
Cl K_α	2.621	3.1801E+01	< 1.0

CONCLUSIONS

A few of the potentialities of microcalorimeter based X-ray spectra were pointed out and the need for the capabilities of the IBIXFit code under these conditions has been clearly shown. In fact it becomes clear that IBIXFit it is not only successful at fitting microcalorimeter spectra of complex samples, but also that it is of crucial importance. The need to select between fixed and free line ratios is met by IBIXFit allowing for the correct fitting of spectra, including the fit of lines newly seen in PIXE spectra, such as Radiative Auger Emission lines. The ability to observe RAE lines in PIXE spectra and, furthermore, being able to fit them using IBIXFit is remarkable.

ACKNOWLEDGMENTS

This work was partially supported by the Portuguese Foundation for Science and Technology, FCT, fellowships SFRH/BD/27557/2006 and SFRH/BD/43379/2008. This work is also funded in the framework of project REEQ/377/FIS/2005 of the Portuguese Foundation for Science and Technology, FCT. A. Taborda and N. P. Barradas would also like to thank the International Atomic Energy Agency for support under the Research Contract No. 14365.

REFERENCES

1. N. P. Barradas, C. Jeynes, R. P. Webb, *Appl. Phys. Letters* **71**, 291 (1997).
2. E. Elad, *Nucl. Instr. Meth.* **37**, 327 (1965).
3. S. A. E. Johansson, T. B. Johansson, *Nucl. Instr. Meth.* **137**, 473 (1976).
4. T. B. Johansson, R. Akselsson, S. A. E. Johansson, *Nucl. Instr. Meth.* **84**, 141 (1970)
5. S. A. E. Johansson, J. L. Campbell, *Proton-induced X-ray Emission: A novel technique for elemental analysis*, John Wiley and Sons, 1988.
6. T. Jach, J. N. Ullom, W. T. Elam, *Eur. Phys. J. Special Topics* **169**, 237 (2009).
7. C. Isaila, F. v. Feilitzsch, J. Hohne, C. Hollerith, K. Phelan, B. Simmnacher, R. Weiland, D. Wernicke, Nucl. *Instr. in Phys. Res. A* **559**, 734 (2006).
8. C. Hollerith, B. Simmnacher, R. Weiland, F. v. Feilitzsch, C. Isaila, J. Jochum, W. Potzel, J. Hohne, K. Phelan, D. Wernicke, T. May, *Rev. Sci. Instrum.* **77**, 053105 (2006).
9. J. N. Ullom, J. A. Beall, W. B. Doriese, W. D. Duncan, L. Ferreira, G. C. Hilton, K. D. Irwin, G. C. O'Neil, C. D. Reintsema, L. R. Vale, B. L. Zink, *Nucl. Instr. Meth. in Phys. Res. A* **559**, 422 (2006).
10. M. A. Reis, L. C. Alves, N. P. Barradas, P. C. Chaves, B. Nunes, A. Taborda, K. P. Surendran, A. Wu, P. M. Vilarinho, E. Alves, *Nucl. Instr. Meth. in Phys. Res. B* **268**, 1980 (2010).
11. M. A. Reis, N. P. Barradas, P. C. Chaves, A. Taborda, *X-Ray Spectrom.*, PIXE analysis of multilayered targets, (2011) In press.
12. M. A. Reis, P. C. Chaves, A. Taborda, *X-Ray Spectrom.* Radiative Auger studies with EDS High Resolution PIXE (2011) In press.
13. M. A. Reis, P. C. Chaves, V. Corregidor, N. P. Barradas, E. Alves, F. Dimroth, A. W. Bett, *X-Ray Spectrom.* **34**, 372 (2005).
14. P. C. Chaves, A. Taborda, M. A. Reis, *Nucl. Instr. Meth. in Phys. Res. B* **268**, 2010 (2010).
15. M. A. Reis, P. C. Chaves, L. C. Alves, N. P. Barradas, *X-Ray Spectrom.* **37**, 100 (2008).
16. G. Zschornack, *Handbook of X-Ray Data*, Springer, 2007.
17. J. H. Scofield, *At. Data Nucl. Data Tables* **14**, 121 (1974).

Integration Of SIMS Into A General Purpose IBA Data Analysis Code

N. P. Barradas [1,2], J. Likonen [3], E. Alves [1,2], L.C. Alves [1,2], P. Coad[4], A. Hakola[3] and A. Widdowson [4]

(1) Instituto Tecnológico e Nuclear, E.N. 10, Sacavém 2686-953, Portugal
(2) Centro de Física Nuclear da Universidade de Lisboa, Av. Prof. Gama Pinto 2, 1649-003 Lisboa, Portugal
(3) Association Euratom-TEKES, VTT, PO box 1000, 02044 VTT, Espoo, Finland
(4) Euratom/CCFE Fusion Association, Culham Science Centre, OX14 3DB, UK

Abstract. IBA techniques such as RBS, ERDA, NRA, or PIXE are highly complementary, and are often combined to maximize the extracted information. In particular, they have different sensitivities to various elements and probe different depth scales. The same is true for secondary ion mass spectrometry (SIMS), that can have much better detection limits for many species. Quantification of SIMS data normally requires careful calibration of the exact system being studied, and often the results are only semi-quantitative. Nevertheless, when SIMS is used together with other IBA techniques, it would be highly desirable to integrate the data analysis. We developed a routine to analyse SIMS data, and implemented it in NDF, a standard IBA data analysis code, that already supported RBS, ERDA, resonant and non-resonant NRA, and PIXE. Details of this new routine are presented in this work.

Keywords: SIMS; Ion Beam Analysis; Rutherford backscattering; NDF
PACS: 34.50.Bw; 82.80.Yc; 81.05.Ea

INTRODUCTION

Ion beam analysis (IBA) techniques such as Rutherford backscattering spectrometry (RBS), elastic recoil detection analysis (ERDA), nuclear reaction analysis (NRA), or particle-induced X-ray emission (PIXE) are highly complementary, because they have different sensitivities for various elements and probe different depth scales. On the other hand ion mass spectrometry (SIMS), can, for many species, probe a much wider range of concentrations, with much better sensitivity and lower detection limits. SIMS is, however, very different from the other IBA techniques, both experimentally and in the methods used to analyse the data. Nevertheless, in cases where SIMS is used in conjunction with other IBA techniques, it would be highly desirable to integrate the data analysis, ensuring self-consistency between all results. A routine to analyse SIMS data was developed and implemented in NDF [1,2], a standard code for IBA data analysis. Details of this new routine are presented in this work.

The new code was tested by studying deuterium retention in carbon samples subject to plasma irradiation in the JET tokamak [3,4]. SIMS and IBA techniques were extensively used for the analysis of those first wall components removed from JET. Lateral microbeam PIXE was performed on some samples, that is, cross sections of the samples were measured, obtaining in this way depth profiles with PIXE that probe as deep as SIMS did. RBS, ERDA, and NRA data were also obtained from the same samples. All the data are fitted simultaneously, with the same depth profile.

EXPERIMENTAL DETAILS

A large number of samples from different regions of tiles removed from the JET after the 2005–2007 campaign [5] were analysed with multiple techniques. The carbon-based samples also contain hydrogen and deuterium, beryllium, boron, oxygen, tungsten, nickel contaminants and some other trace impurities such as iron and cobalt, that were not measured with SIMS. Also the carbon isotopic ratio can change near the surface which is due to the $^{13}CH_4$ puffing experiment done in 2007. Rutherford backscattering was made with a 2 MeV ^4He beam detected at a 140° scattering angle and with different angles of incidence, with a ^1H beam at energies up to 2.4 MeV, detected at a 140°

Application of Accelerators in Research and Industry
AIP Conf. Proc. 1336, 281-285 (2011); doi: 10.1063/1.3586104

scattering angle and normal incidence. ERDA was made with a 2 MeV ^{4}He beam at a 24° recoil detection angle; in this case, RBS spectra were simultaneously collected at an 160° scattering angle. NRA was made with a ^{3}He beam with energies between 0.8 and 2.3 MeV, taking advantage of the ^{2}H(^{3}He,p)α reaction with ^{2}H, with Q=18352 keV. A stopping foil was used to stop the α particles from hitting the detector, so only the p were detected.. Lateral microbeam PIXE was done on some samples: a 2 MeV proton beam, focused down to a 3x4 μm^2 beam spot size, was used to measure a cross section of the samples. The SIMS analysis was made with a double focussing magnetic sector instrument using a 5 keV O_2^+ primary ion beam, at currents typically 500 nA.

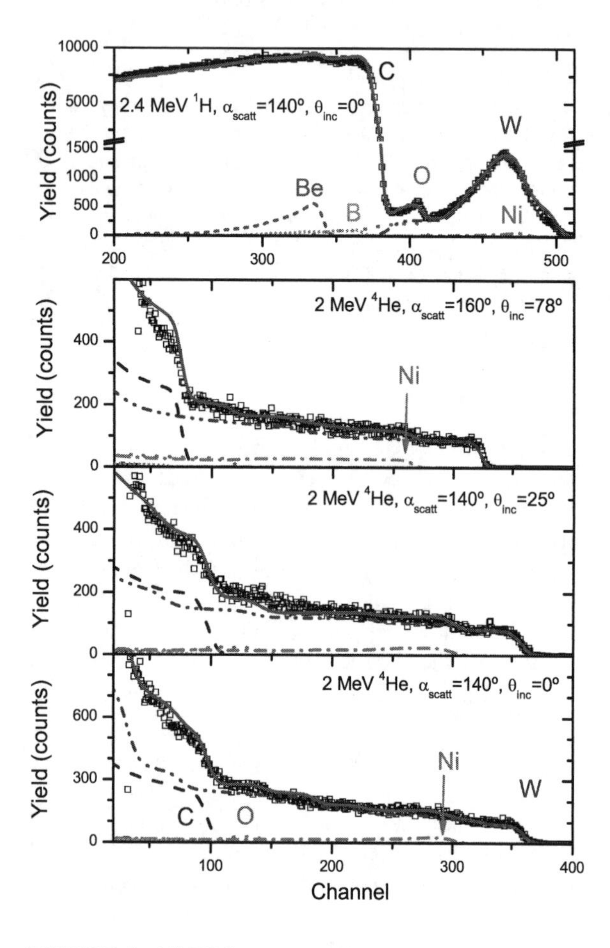

FIGURE 1. All RBS spectra collected for a given sample, including ^{4}He and ^{1}H beams.

CALCULATIONS

The NDF code already had implemented the techniques RBS, ERDA, both resonant and non-resonant NRA, and PIXE [6]. In all these techniques, NDF takes as input a depth profile, calculates the corresponding theoretical spectrum, and compares the calculation to the experimental data. This can be done both interactively, with the user changing manually the depth profile until a satisfactory fit is achieved, or in an automated way with different fitting algorithms. The crucial point is that any number of spectra from any combination of the techniques implemented can be analysed simultaneously: that is, from the assumed depth profile one theoretical spectrum is calculated for each of the experiments made. Optimisation is made globally on all data together. This is the ideal way of extracting the complementary information that each technique provides. For instance, the depth profile of the major heavy elements is easily extracted with ^{4}He-RBS (particularly if different angles of incidence and/or different beam energies were used), ^{1}H-RBS can be used for enhanced sensitivity to light elements, ^{4}He-ERDA for hydrogen isotopes, PIXE for trace elements, and NRA for specific isotopes. This complementarity is not only related to the elements present, given that the techniques probe very different depth scales. ERDA at the usual beam energies probes a few hundreds of nm; ^{1}H-RBS can probe several μm, and standard PIXE even deeper.

In all these techniques, translating a depth profile into the calculated spectrum involves a long series of complex calculations, particularly if advanced physics, such as plural and multiple scattering, is included [6]. In SIMS, in principle this could also be the case. One could, from the depth profile, and including chemical information, use data bases of experimental sputter rates and secondary ion yields, calculate neutralisation for the set-up used, and finally calculate measured yields [7]. However, these processes are very complex and there is as yet no theory that leads to satisfactory quantitative results: the sputter rate for one given ion depends strongly on the exact overall sample composition, on the incident ion species, angle of incidence and energy, and the relationship between the detected ion yield and the total sputter yield depends on many parameters. In practice, careful calibration of the exact system being studied is necessary (and IBA techniques are often used for that), and SIMS is often a semi-quantitative technique.

In NDF, the depth profile is translated into a yield vs. depth in two steps. In the first step, the concentration of one given element or isotope at a given depth is translated into a yield using a response function that must be input by the user. This response function will normally incorporate the results of the calibration of the system. However, in many cases the same element observed with SIMS is also seen, albeit in a narrower surface layer, in a measurement of the same sample made with RBS, NRA, or ERDA. In this case, there is an intrinsic self-calibration, as the calibration of the SIMS yield is made by a measurement of the exact same sample, and is obtained during the data analysis process. In fact, the response function in NDF can be fixed as defined by the user, but it can also be a fit parameter. NDF will find the optimal solution that constrains the elemental concentration to the value that it must have in the depth range accessed by the IBA technique, and will adjust the SIMS response function accordingly to obtain good agreement in that depth range. Outside that depth range, the response function thus derived is used as well.

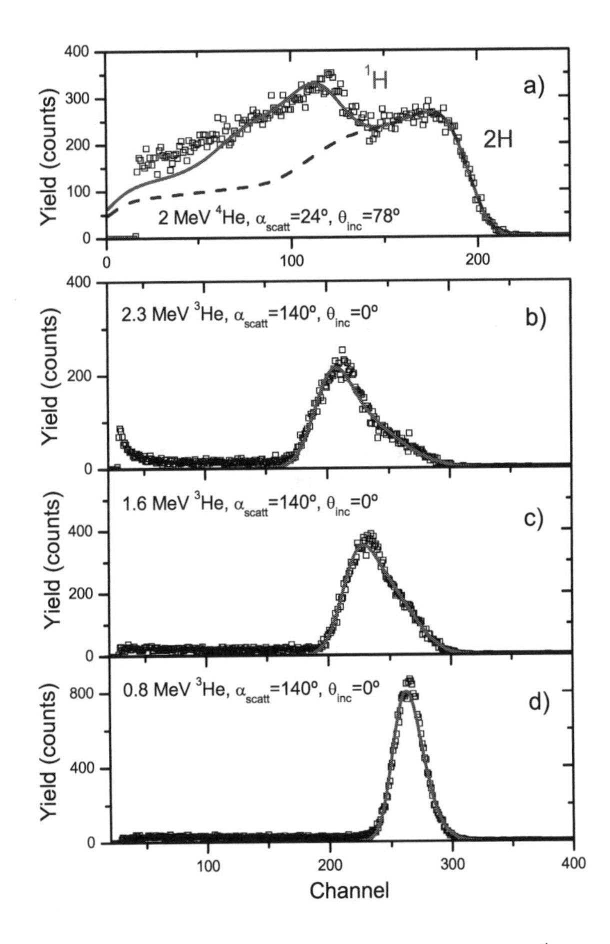

FIGURE 2. Data (squares) and fits (solid lines) for a) ^{4}He-ERDA spectrum, and b), c) and d) ^{3}He NRA spectra, that probe D only. Protons, which are a reaction product, are detected.

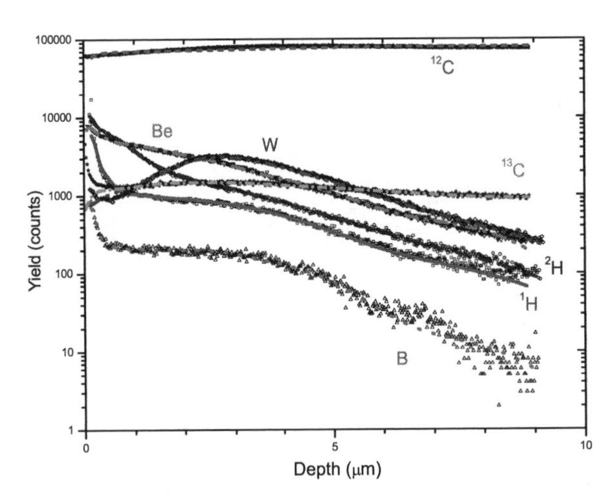

FIGURE 3. SIMS data (symbols) and fits (lines) for all elements but Ni.

In the second step, the depth, which is an areal density in NDF, is translated into a real depth scale. NDF uses areal density (with unit 10^{15} at./cm^2) in all its internal calculations, given that this is the natural quantity for techniques such as RBS, ERDA, or even PIXE, that are not sensitive to density and only depend on the amount of matter crossed by the beam: SIMS, on the contrary, is a real depth profiling technique (at least if one accepts the approximation usually made that the sputter rate is constant and known). To calculate a thickness corresponding to a given areal density, the density of the layer in question must be known. NDF can use a weighted average of the bulk elemental densities in that layer, but this will often lead to a large error. If, however, the density is known, then it can be input by the user. Finally, if a given element has a well defined depth profile visible in SIMS and in a different IBA technique, then the density is determined by the data being analysed. In this case, NDF can treat the density as a free parameter, which is effectively determined from the self-consistent analysis made.

RESULTS AND DISCUSSION

We will illustrate the possibilities of the new approach using one given sample, from the JET outer poloidal limiter. The four RBS spectra collected from this sample are shown in Fig. 1. The ^{4}He-RBS show that there is W and Ni at the surface and their concentrations are derived. From the proton spectrum most of the W profile is clearly seen, although it is superimposed to the Ni signal. Also, a peak corresponding to surface O is observed. The Be signal is observed as a small peak on top of the C signal, but whether its concentration changes with depth cannot be derived. Finally, the depth resolution is not enough to distinguish ^{12}C from ^{13}C (due to the $^{13}CH_4$ puffing), and the B signal is submerged in the much larger C signal.

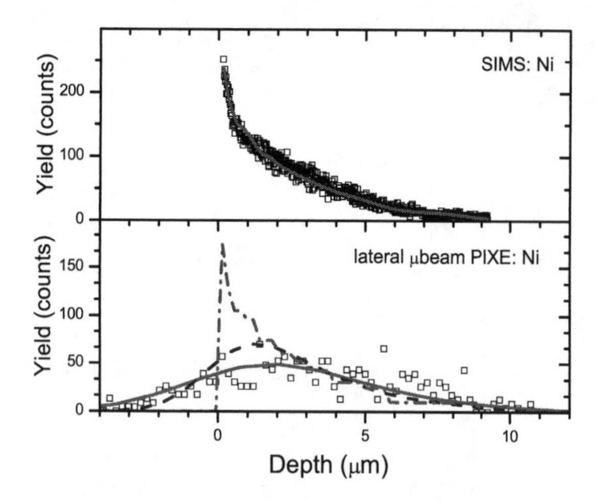

FIGURE 4. SIMS and lateral microbeam PIXE data for Ni. For the lateral microbeam PIXE, simulations for the final profile obtained, but with 10 nm (dash-dotted line) and the nominal 3 μm (dashed line) depth resolution are also shown.

The ^1H and ^2H signals are clearly seen in the ERDA spectrum shown in Fig. 2, and their surface concentrations can be determined, even if the superposition of the two signals leads to some uncertainty. The NRA data are sensitive only to ^2H, and the three different beam energies, when analysed simultaneously, impose strong constraints on the ^2H profile, taking advantage of the fact that the cross section [8] has a broad but well defined peak around 700 keV.

The SIMS data for all the elements except Ni are shown in Fig. 3. First of all, the ^1H and ^2H profiles are quantitatively constrained by the ERDA and NRA data, that is, response functions could be derived by the simultaneous fitting procedure. The same is true for ^{12}C, Ni, W, and Be. The B is not really seen in the RBS, and the quantification achieved with SIMS is a maximum limit only, given that a large B content would be observed in the proton RBS spectrum. For ^{13}C no internal calibration is obtained from the data, and we simply assumed the natural isotopic distribution for the deepest depth probed with SIMS, when the concentration of the other elements is close to zero.

The W SIMS signal can be directly compared to the W signal observed in the proton RBS spectrum. This effectively determines the density and thus the depth scale. The fit indicates that density is about 16% smaller than bulk graphite, assuming that sputter rate in SIMS was fairly constant.

The data obtained for Ni both with SIMS and lateral microbeam PIXE are shown in Fig. 4. It is clear that, while SIMS shows a Ni content constantly decreasing with depth, only a very broad signal is seen with PIXE. This is due, first of all, to the different depth resolution of the two techniques. SIMS has a nominal depth resolution around 10 nm at the surface. The microbeam PIXE beam spot was 3 μm. Simulations made for PIXE both with 10 nm and 3 μm are also shown in the figure. It is clear that, for 10 nm depth resolution, the calculated signal approaches the SIMS data. For 3 μm, however, one would expect that a good simulation is reached; however, the calculation is still not as broad as the observed signal, and the best fit was achieved for a resolution of 5.5 μm, which is unrealistically high. This is due to surface roughness that is extensive in irradiated samples. Its effect will be totally different in a technique such as SIMS that probes the sample from the surface in the usual way, and in cross- sectional microbeam PIXE, where the actual surface is probed from the side. In the first case, the effect of roughness is initially small (or non-existent in the limit case of normal incidence) and increases gradually. In cross section analysis, the roughness is immediately fully included in the signal observed. This means that the difference (taken in quadrature) between the apparent and the real depth resolution in PIXE can be taken as due to roughness, which we thus estimate to be around 4.6 μm in this sample.

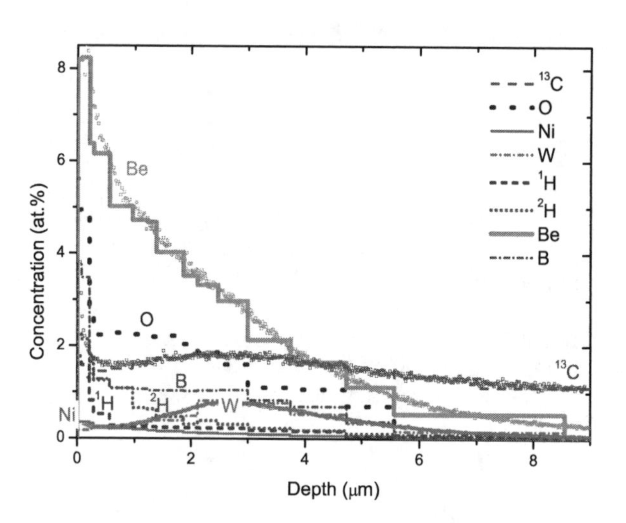

FIGURE 5. Final depth profile reached. The ^{12}C is not shown in this figure. The fitted depth profile is shown as different curves for each element. The point-by-point calculation (symbols) is also shown for selected elements.

The final depth profile determined is shown in Fig. 5. All the information derived from the 5 techniques used is included. We note that all the spectra from all the techniques, shown in Figs 1 to 4, were fitted with this same depth profile; one single fit was made, not separate fits to each spectrum. Note that NDF makes also an inverse calculation, deriving from each data point a concentration point [9]. This is also implemented for SIMS, using the response function input by the user (or derived during the fit). The results of this point-by-point calculation are shown for some of the elements only.

SUMMARY

We developed a simple routine to handle SIMS data, as well as any other real depth profiling techniques where a calibration of the depth and concentration scale can be obtained either from calculation or from comparison with other data, and included it in the code NDF, that already implemented RBS, ERDA, resonant and non-resonant NRA, and PIXE. We demonstrate the power of combining multiple techniques when analysing complex samples, by giving the example of a carbon based sample irradiated in the JET tokamak. The new routine is open source, and is available in the NDF web site http://www.itn.pt/facilities/lfi/ndf/uk_lfi_ndf.htm

REFERENCES

1. N.P. Barradas, C. Jeynes, and M.A. Harry, *Nucl. Instrum. Methods B* **136-138**, 1163 (1998).
2. N.P. Barradas, C. Jeynes, K.P. Homewood, B.J. Sealy, and M. Milosavljevic, *Nucl. Instrum. Methods B* **139**, 235 (1998).
3. J. Roth, E. Tsitrone, A. Loarte, Th. Loarer, G. Counsell, R. Neu, V. Philipps, S. Brezinsek, M. Lehnen, P. Coad, Ch. Grisolia, K. Schmid, K. Krieger, A. Kallenbach, B. Lipschultz, R. Doerner, R. Causey, V. Alimov, W. Shu, O. Ogorodnikova, A. Kirschner, G. Federici, A. Kukushkin, *J. Nucl. Mater.* **390–391**, 1 (2009).
4. G.J. van Rooij, V.P. Veremiyenko, W.J. Goedheer, B. de Groot, A. W. Kleyn, P. H. M. Smeets, T. W. Versloot, D. G. Whyte, R. Engeln, D. C. Schram, and N. J. Lopes Cardozo, *Appl. Phys. Lett.* **90**, 121501 (2007).
5. L.C. Alves, E. Alves, N.P. Barradas, R. Mateus, P. Carvalho, J.P. Coad, A.M. Widdowson, J. Likonen, S. Koivuranta, *Nucl. Instrum. Methods B* **268**, 1991 (2010) .
6. N.P. Barradas, K. Arstila, G. Battistig, M. Bianconi, N. Dytlewski, C. Jeynes, E. Kótai, G. Lulli, M. Mayer, E. Rauhala, E. Szilágyi, M. Thompson, *Nucl. Instrum. Methods B* **262**, 281 (2007)
7. D.P. Woodruff and T.A. Delchar, Modern Techniques of Surface Science - Second Edition, Cambridge: Cambridge University Press, 1994, pp. 338-355.
8. W.Moller and F.Besenbacher, *Nucl. Instrum. Methods* **168**, 11 (1980).
9. P. Børgesen, R. Behrisch and B.M.U. Scherzer, *Appl. Phys. A* **27**, 183 (1982).

Classical And Quantum Rainbow Scattering From Surfaces

H. Winter, A. Schüller, M. Busch, J. Seifert and S. Wethekam

Institut für Physik, Humboldt Universität zu Berlin, Newtonstrasse 15, D-12489 Berlin, Germany

Abstract. The structure of clean and adsorbate covered surfaces as well as of ultrathin films can be investigated by grazing scattering of fast atoms. We present two recent experimental techniques which allow one to study the structure of ordered arrangements of surface atoms in detail. (1) Rainbow scattering under axial surface channeling conditions, and (2) fast atom diffraction. Our examples demonstrate the attractive features of grazing fast atom scattering as a powerful analytical tool in studies on the structure of surfaces. We will concentrate our discussion on the structure of ultrathin silica films on a Mo(112) surface and of adsorbed oxygen atoms on a Fe(110) surface.

Keywords: Ion surface scattering, surface structure, rainbow scattering, fast atom diffraction
PACS: 79.20.Rf, 79.60.Bm, 34.20.Cf, 68.49.Bc

RAINBOW SCATTERING UNDER AXIAL SURFACE CHANNELING

Whenever in a collision an extreme in angular deflection is achieved, this gives rise to an enhancement in the flux of scattered projectiles, the so called "rainbow" in close analogy to the atmospheric phenomenon for scattering of sun light by drops of water. In surface scattering, collisional rainbows are present owing to the corrugated atomic interaction potential in front of the well ordered surface of a crystal. For a detailed discussion on this topic, we refer to a reviewing paper by Kleyn and Horn [1]. A specific collision regime is met for scattering of fast atoms and ions from surfaces under a grazing angle of incidence where the collision with the solid proceeds in the regime of surface channeling [2,3], i.e. a steering of projectiles in collisions with target atoms of the topmost surface layer. For impact along low indexed azimuthal crystal directions, the corrugation of the interaction potential across axial channels formed by strings of surface atoms leads to an extreme in the azimuthal deflection and consequently to rainbow structures in the angular distribution for scattered projectiles [4,5].

In this paper we will discuss recent work on rainbow scattering under axial surface channeling to study the structure of surfaces. Based on recent observations of quantum scattering for suitably chosen scattering conditions, rainbow scattering under axial surface channeling shows quantum mechanical interference effects, giving rise to a rich diffraction pattern in the angular distributions of scattered particles [6-8]. However, for incidence angles above typically some degrees and collision energies above some keV, excitation phenomena at the surface result in substantial decoherence and a classical scattering process. We will present studies for both collision regimes, i.e. classical as well as quantum rainbow scattering, and have selected as representative examples two studies on the structure of metal surfaces which are covered with a monolayer oxide film or with an ordered sub-monolayer arrangement of oxygen atoms. These studies demonstrate that rainbow scattering in the regime of axial surface channeling has indeed the potential as a future surface analytical tool.

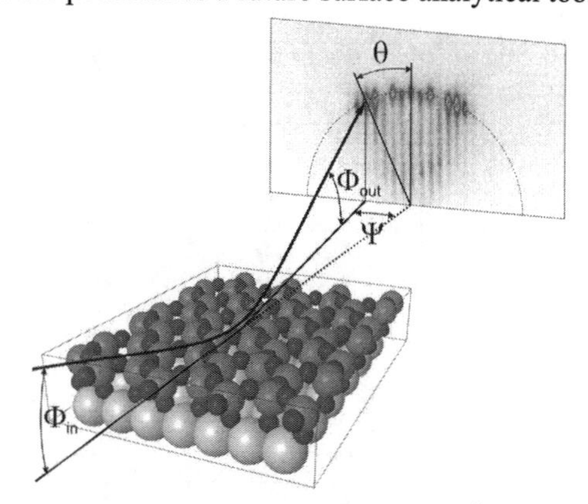

FIGURE 1. Sketch of collision geometry for axial surface channeling.

A sketch of the scattering geometry is shown in Figure 1 where fast atoms with energies of typically keV are scattered under a grazing polar angle of incidence $\Phi_{in} \approx 1°$ along a low indexed axial channel of a well ordered crystal surface. In the regime of

Application of Accelerators in Research and Industry
AIP Conf. Proc. 1336, 286-290 (2011); doi: 10.1063/1.3586105

surface channeling, the energy of motion normal with respect to atomic strings is conserved so that $\Phi_{in}^2 = \Phi_{out}^2 + \Psi^2$ for the angle of exit Φ_{out} and the azimuthal angle Ψ. The deflection angle in the detection plane θ is related to Ψ via $\theta \approx \Psi/\tan\Phi_{in}$ and for grazing scattering about two orders of magnitude larger than Ψ. Key feature of the experimental setup is a position sensitive micro-channelplate detector which allows one to record with high efficiency angular distributions for scattered projectiles. In the regime of classical or quantum rainbow scattering, defined peaks are identified which can be used to obtain information on the interaction potential and on details on the structure of the surface.

Classical Rainbow Sattering Under Axial Surface Channeling: Structure Of Monolayer Silica Film Of Mo(112)

Thin oxide films are widely used in a variety of applications ranging from electronic devices to catalysis [9]. Grown as ultrathin films on metal substrates, the problem of low conductivity of bulk oxides is avoided so that established surface science tools are applicable. An important oxide in advanced technology is silica (SiO_2) for which the preparation of ultrathin well-ordered films has been demonstrated on a Mo(112) surface [10]. Despite considerable efforts in order to clear up the geometrical structure of this film by a number of different experimental and theoretical methods, the detailed structure of the silica film on Mo(112) was a matter of debate [11]. Two conflicting models for the film structure were discussed: (1) A honeycomb-like 2D-network model formed by corner-sharing SiO_4 tetrahedra [12,13], (2) a cluster model consisting of isolated SiO_4 tetrahedra where all oxygen atoms are bound to the molybdenum surface [11,14,15]. In the left panels of Figure 2 we present top views of the arrangement of atoms for the two structural models.

For scattering along a low indexed axial channel, the resulting corrugation of the potential leads to an out-of plane scattering where the maximum for the azimuthal angular deflection shows an intensity enhancement for scattered atoms, the so called collisional "rainbow" [1]. This is illustrated in the right panel of Figure 2 by curves in a plane normal to $[\bar{1}\bar{1}1]$ which represent equi-potential planes averaged along atomic strings and projections of trajectories for a spatially constant flux of incoming projectiles. Note the substantial differences in the shape of the potential planes and in the trajectories for the two different structural models.

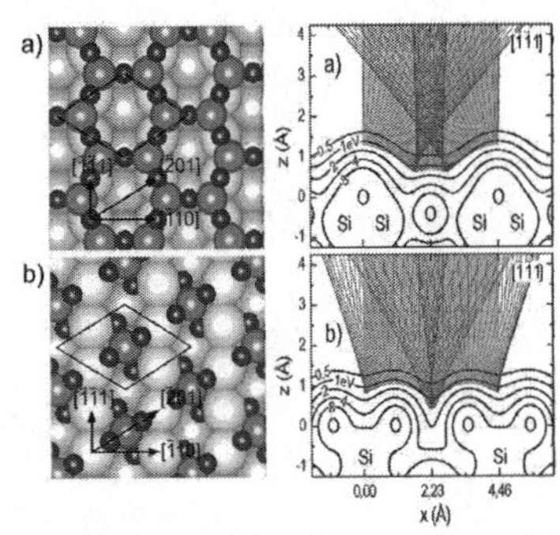

FIGURE 2. Panel a: sketch of network model, panel b: sketch of cluster model (large circles: Si, small circles: O). Right panels: calculated trajectories in plane normal to $[\bar{1}\bar{1}1]$ axial channel for 2 keV He at $\Phi_{in} = 1.5°$.

The preparation of the silica films and the scattering experiments were performed in an UHV chamber at a base pressure in the upper 10^{-11} mbar range. The Mo(112) surface was initially prepared by cycles of grazing sputtering with 25 keV Ar^+ ions and annealing at temperatures of about 1900 K for several minutes. Deposition of silicon atoms was performed using an electron beam evaporator from a high purity silicon rod. Following the recipe given in [16], the monolayer silica films were prepared by deposition of 1.2 ML silicon at 900 K at a partial oxygen pressure of $5 \cdot 10^{-8}$ mbar on an oxygen covered Mo(112) surface and after subsequent annealing at about 1200 K well ordered silica films showing sharp c(2x2) LEED patterns are obtained.

In the upper panel of Figure 3 we display a 2D intensity plot for scattering of 2 keV He atoms from the ultrathin silica film on Mo(112) along the $[\bar{1}\bar{1}1]$ axial channel. The circular shape of the intensity pattern reflects the symmetry for axial channeling where the two outer peaks at the maximum of azimuthal angular deflection stem from rainbow scattering. The large rainbow angle indicates a considerable corrugation of the interaction potential across this axial channel. The intense peak in the middle stems from scattering within the axial channels giving rise to a focusing in forward direction. The angular distribution is analyzed by projecting the intensity within an annulus (see figure) onto the scale of the deflection angle Θ and is plotted as solid curve in the lower panel of Figure 3 revealing an intense peak at $\Theta = 0$ and two rainbow peaks at $\Theta_{RB} \approx 50°$.

For a test on the two conflicting structural models, we performed computer simulations of projectile trajectories based on classical mechanics where we have chosen a superposition of inter-atomic pair potentials in the Thomas-Fermi approximation proposed by O'Connor and Biersack [17] resulting in a two-dimensional continuum potential by averaging over atomic strings. The intensity as function of deflection angle Θ was derived from the number of trajectories within a defined interval of deflection angles where broadening of the beam via thermal lattice vibrations is taken into account by convolution of with a Gaussian line shape. The cluster model (dotted curve) results in several rainbow peaks at smaller angles which are not in accord with the experiment, whereas the calculations based on the 2D-network model (dotted curve) reproduces the rainbow peaks and the overall shape of the angular distributions including the intensity for smaller deflection angles. Thus the analysis of these experiments is in clear-cut favour of the 2D-network model and discards the alternative approach.

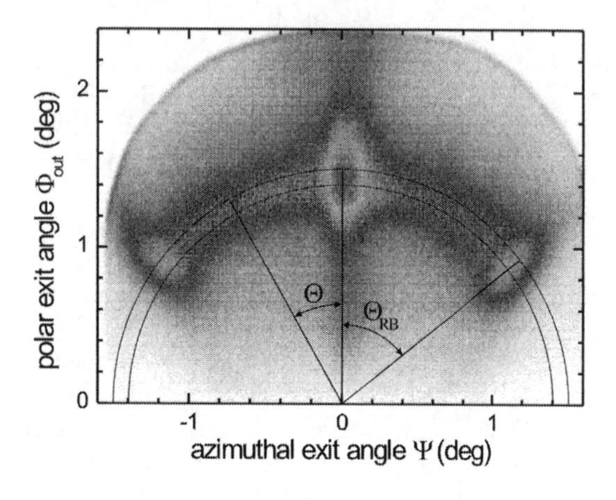

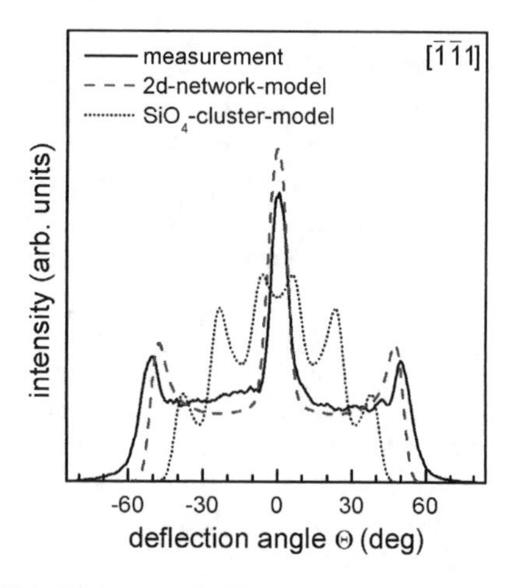

FIGURE 3. Upper panel: 2D intensity pattern for scattering of 2 keV He atoms from ultrathin silica film on Mo(112) along $[\bar{1}\bar{1}1]$. Lower panel: Intensity for scattered atoms projected on deflection angle. Solid curve: measurement, dashed curve: simulations for 2D-network model, dotted curve: simulations for SiO_4-cluster model.

Quantum Rainbow Sattering Under Axial Surface Channeling: Positions Of Oxygen Atoms On Fe(110)

For sufficiently small angles of incidence and projectile energies, the momentum transfer to surface atoms and electronic excitations of a solid with a band gap are on such a low level that in combination with the high angular resolution, diffraction phenomena can be observed for the scattered particles [6-8,18]. Striking feature of Fast Atom Diffraction (FAD) under axial surface channeling is the presence of quantum scattering for matter waves with de Broglie wavelengths which are about two to three orders of magnitude smaller than typical periodicity lengths for ordered structures of surfaces. Already first studies have shown that FAD can be considered as a powerful tool to study surface structures with fast atomic beams which extends the regime of the established method of He atom scattering (HAS) [19] from thermal energies to keV energies. Attractive features of FAD are an easy scanning of the de Broglie wavelength over a wide range by the acceleration voltage and the efficient detection by means of direct imaging using position sensitive detectors so that complete diffraction patterns can be recorded in time scales of seconds up to minutes. As a representative example of the method, we discuss here studies on the positions of adatoms on a Fe(110) surface.

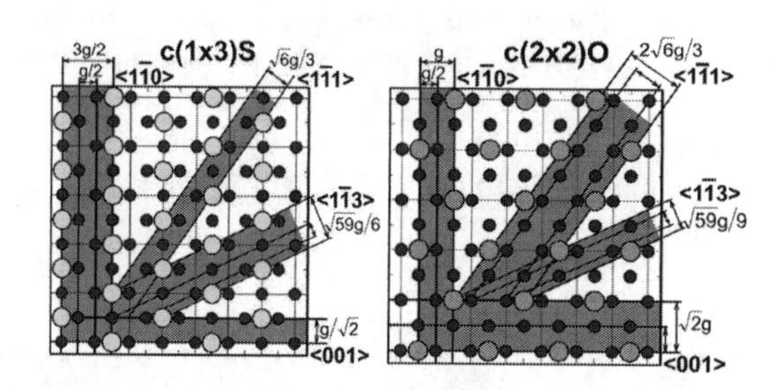

FIGURE 4. Sketch of structural model for c(1x3)S (left panel) and c(2x2)O (right panel) on Fe(110) surface.

For an O_2 dose of 2 Langmuir (1 L = 1 Langmuir = 1.33 10^{-6} mbar sec) and subsequent annealing at 770 K for about 10 minutes, one observes via LEED a well defined c(2x2)O/Fe(110) phase. For prolonged heating at 1000 K segregation of S impurities to the surface leads to a well ordered c(1x3)S/Fe(110) superstructure. A sketch of the structural models of the two phases is shown in Figure 4 where the adatoms are represented by larger circles and the substrate atoms by smaller full circles. Furthermore we have highlighted some prominent low indexed axial channels formed by the adatoms for these superstructures. In the left panels of Figure 5, we show angular distributions for scattering of 0.6 and 0.8 keV ^{3}He atoms along some low indexed channels which are commensurate with the substrate lattice, however, differ in width for the two structures and for different directions.

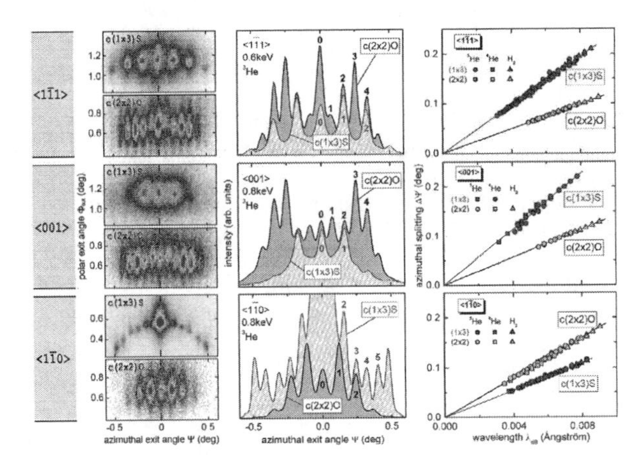

FIGURE 5. Left panels: angular distributions for scattering of 0.6 keV (top) and 0.8 keV (middle and bottom) ^{3}He from c(1x3)S and c(2x2)O on Fe(110). Middle panels: projections of intensities on azimuthal angle. Right panels: azimuthal splittings ΔΨ as function of de Broglie wavelength λ_{dB}.

The data shows well defined diffraction patterns where from the Bragg condition for constructive interference n λ_{dB} = d sinΨ (n = diffraction order) the splittings with respect to the azimuthal angle ΔΨ = λ_{dB}/d scale with the inverse of the channel width d ($\lambda_{dB} = h/(2\ M\ E)^{1/2}$ is the de Broglie wavelength of a particle of mass M and kinetic energy E and h is Plancks constant). Note the high angular resolution in the experiments where the overall azimuthal deflection is less than 1°. The projections of distributions onto the azimuthal exit angle Ψ are shown in panels in the middle and reveal diffraction orders up to n = 6. Plots of ΔΨ vs. λ_{dB} in the right panels give the expected linear dependences of slope 1/d. Within the experimental resolution the extracted widths d are in good agreement with the structural model displayed in Figure 4. With the lattice constant g = 2.86 Å of the Fe crystal, we obtain, e.g., for the $[1\bar{1}1]$ channel of c(1x3)S/Fe(110) d_{exp} = (2.36 ± 0.07) Å in full accord with g $6^{1/2}$ / 3 = 2.34 Å for the channel width for the unreconstructed substrate. Thus FAD provides detailed information on the lateral arrangements of atoms in the topmost surface layer.

In Figure 6 we show projections of intensities with respect to the deflection angle Θ for scattering of 0.5 keV to 1 keV ^{3}He atoms from c(2x2)O/Fe(110) under Φ_{in} = 0.85° along $[1\bar{1}0]$. A closer inspection of the diffraction patterns reveals a change of relative intensities for different diffraction orders with projectile energy. This variation of intensities depends strongly on the corrugation of the He surface interaction potential. At 1 keV the zeroth order diffraction peak is prominent, but is almost missing at 0.45 keV, whereas for the diffraction order n = 1 the opposite behaviour is found. Based on the approximation for scattering from a hard wall potential [20,21], one can deduce the full corrugation Δz of the potential across the channel as function of the energy of motion normal to the axial strings $E_\zeta = E\ \sin^2\Phi_{in}$.

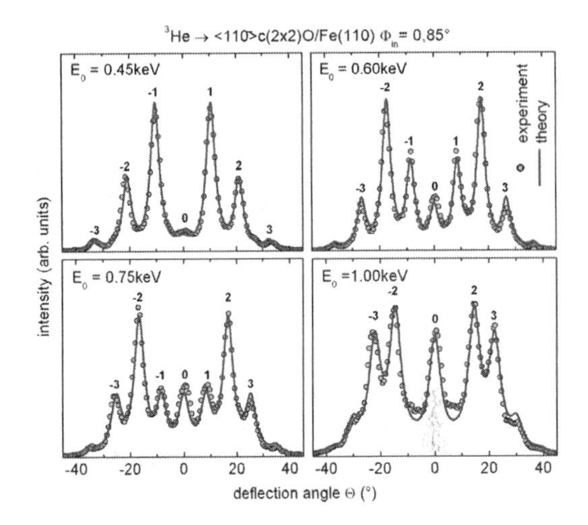

FIGURE 6. Projection of intensity with respect to deflection angle Θ for scattering of 0.45 keV (upper left panel), 0.60 keV (upper right panel), 0.75 keV (lower left panel), and 1.00 keV ^{3}He atoms (lower right panel) from c(2x2)O/Fe(110) under Φ_{in} = 0.85° along $[1\bar{1}0]$. Solid curves: best fits to Lorentzian lineshapes (for details see text).

In semiclassical theory for scattering from sinusoidal hard wall [20,21] the intensity of a diffraction spot of order n is $I_n \sim [J_n(\pi\ \Delta z\ (1+\cos\Theta_n)/\lambda_{dB\zeta})]^2$ with J_n being the Bessel function of order n, Δz the full corrugation, and $\Theta_n = \arccos[1-(n\lambda_{dB\zeta}/d)^2)]^{1/2}$ with the Broglie wavelength for the motion of atoms normal with respect to the surface plane $\lambda_{dB\zeta} = h\ /\ (2\ M\ E_\zeta)^{1/2}$. The solid curves in Figure 6 are best fits to the data by a sum of peaks with Lorentzian lineshape where the resulting peak heights are obtained from I_n with Δz as fit parameter. We have taken into account corrections cause by using the hard wall approximation [22] and have plotted in Figure 7 the resulting Δz as function of E_ζ.

The corrugation Δz is very sensitive to the positions of O atoms which are located for the $[1\bar{1}0]$ azimuth at a height h atop axial strings of Fe substrate atoms as sketched in the insert. Based on individual He – O and He – Fe interatomic potentials derived from HF wave functions [22], we have calculated the corrugation Δz for this channel as function of potential energies E_ζ for different heights h (curves in Fig. 7). The analysis for all accessible orders of diffraction patterns for four low indexed channels finally results in h = (0.9 ± 0.2) Å which is consistent with recent DFT calculations for this system [23]. Similar work for the SiO_2/Mo(112) system was performed recently [24].

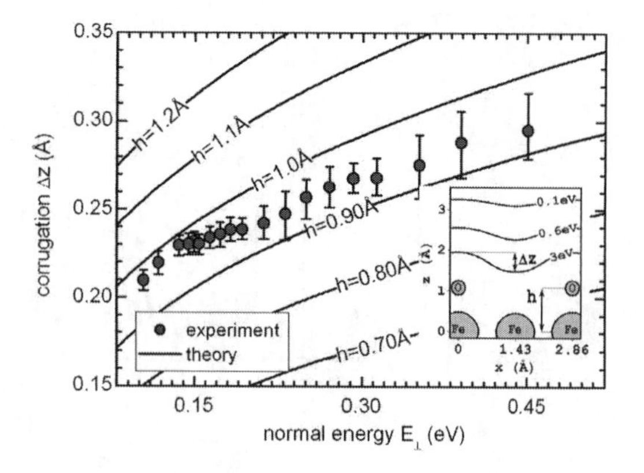

FIGURE 7. Full corrugation Δz (full circles with error bars) as function of normal energy E_ζ. Solid curves: corrugation of potential for different distances h of oxygen atoms to Fe substrate. Insert: sketch of positions for Fe and O atoms along $[1\bar{1}0]$ channel and contours of equi-potential planes.

ACKNOWLEDGMENTS

The assistance of K. Maass, G. Lindenberg, in performing the experiments is gratefully acknowledged. This work was supported by the DFG in Sonderforschungsbereich 546 "Struktur, Dynamik und Reaktivität von Übergangsmetalloxid-Aggregaten".

REFERENCES

1. A. W. Kleyn and T. C. M. Horn, *Phys. Rep.* **199** (1991) 191.
2. D.S. Gemmell, *Rev. Mod. Phys.* **46** (1974) 129.
3. H. Winter, *Phys. Rep.* **367** (2002) 387.
4. D. Danailov, K. Gärtner, and A. Caro, *Nucl. Instrum. Meth.* **B 153** (1999) 191.
5. A. Schüller, G. Adamov, S. Wethekam, K. Maass, A. Mertens, and H. Winter, *Phys. Rev.* **A 69** (2004) 050901(R).
6. A. Schüller, S. Wethekam, and H. Winter, *Phys. Rev. Lett.* **98** (2007) 016103.
7. P. Rousseau, H. Khemliche, A. G. Borisov, and P. Roncin, *Phys. Rev. Lett.* **98** (2007) 016104.
8. A. Schüller and H. Winter, *Phys. Rev. Lett.* **100** (2008) 097602.
9. H. J. Freund, *Faraday Discussions* 114 (1999) 1.
10. T. Schroeder, M. Adelt, B. Richter, M. Naschitzki, M. Bäumer and H.-J. Freund, *Surf. Rev. Lett.* **7** (2000) 7.
11. M. S. Chen and D. W. Goodman, *J. Phys.: Cond. Matter* **20** (2008) 264013.
12. J. Weissenrieder, S. Kaya, J.-L. Lu, H.-J. Gao, S. Shaikhutdinov, H.-J. Freund, M.M. Sierka, T.K. Todorova and J. Sauer, *Phys. Rev. Lett.* **95** (2005) 076103.
13. S. Kaya, M. Baron, D. Stacchiola, J. Weissenrieder, S. Shaikhutdinov, T.K. Todorova, M. Sierka, J. Sauer, H.-J. Freund, *Surf. Sci.* **601** (2007) 4849.
14. M. Chen and D.W. Goodman, *Surf. Sci.* **601** (2007) 591.
15. I.N. Yakovkin, *Surf. Rev. Lett.* **12** (2005) 449.
16. T.K. Todorova, M. Sierka, J. Sauer, S. Kaya, J. Weissenrieder, J.-L. Lu, H.-J. Gao, S. Shaikhutdinov and H.-J. Freund, *Phys. Rev.* **B 73** (2006) 165414.
17. D. J. O'Connor and J. P. Biersack, *Nucl. Instrum. Meth.* **B 15** (1986) 14.
18. J. R. Manson, H. Khemliche, and P. Roncin, *Phys. Rev.* **B 78** (2008) 155408.
19. D. Farias and K. H. Rieder, *Rep. Prog. Phys.* **61** (1998) 1575.
20. U. Garibaldi, A.C. Levi, R. Spadacini, and G.E. Tommei, *Surf. Sci.* **48** (1975) 649.
21. R. Masel, R. Merill, and W. Miller, *J. Chem. Phys.* **65** (1976) 2690.
22. A. Schüller, M. Busch, J. Seifert, S. Wethekam, H. Winter, and K. Gärtner, *Phys. Rev.* **B 79** (2009) 235425.
23. P. Blonski, A. Kienja, and J. Hafner, *Surf. Sci.* **590** (2005) 88.
24. J. Seifert, A. Schüller, H. Winter, R. Wlodarczyk, J. Sauer, and M. Sierka, *Phys. Rev.* **B 82** (2010) 035436.

Enhanced High Resolution RBS System

Thomas J. Pollock, James A. Hass and George M. Klody

National Electrostatics Corp., Middleton, Wisconsin, U. S. A. 53562-0310

Abstract. Improvements in full spectrum resolution with the second NEC high resolution RBS system are summarized. Results for 50Å TiN/HfO films on Si yielding energy resolution on the order of 1keV are also presented. Detector enhancements include improved pulse processing electronics, upgraded shielding for the MCP/RAE detector, and reduced noise generated from pumping. Energy resolution measurements on spectra front edge coupled with calculations using 0.4mStr solid angle show that beam energy spread at 400KeV from the Pelletron® accelerator is less than 100eV. To improve user throughput, magnet control has been added to the automatic data collection. Depth profiles derived from experimental data are discussed. For the thin films profiled, depth resolutions were on the Angstrom level with the non-linear energy/channel conversions ranging from 100 to 200eV.

Keywords: high resolution, nanotechnology, RBS, thin films, channeling, depth profile, channel plates, magnetic spectrometer
PACS: 68.49.-h; 82.80.Yc; 7.81.+a

INTRODUCTION

To meet the needs of the nanotechnology field National Electrostatics Corp. (NEC) has continued to improve the capabilities of their High Resolution Rutherford Backscattering (HR RBS) system [1]. Rutherford Backscattering is primarily used to measure elemental depth profiles. For standard RBS, depth resolution is on the order of 100Å using a solid-state detector. By employing a magnet for energy separation of the backscattered particles and a micro channel plate/resistive anode encoder detector in place of the solid-state detector, depth resolution is improved to the monolayer level for the near surface region [2,3].

EQUIPMENT

The NEC HR RBS system works in conjunction with the standard RC43 materials analysis system (Fig. 1). Equipment modifications that have improved signal quality include using a NEG vacuum pump. This eliminates the noise on the detector signal caused by mechanical vibration and the high frequency electronics of a turbo pump. New detector circuitry and detection electronics have also been implemented. Adding resistive elements to the circuit improves matching of the delay for the signal processing electronics (Fig. 2). The new electronics improve the peak detection of the analogue signal from the MCP/RAE.

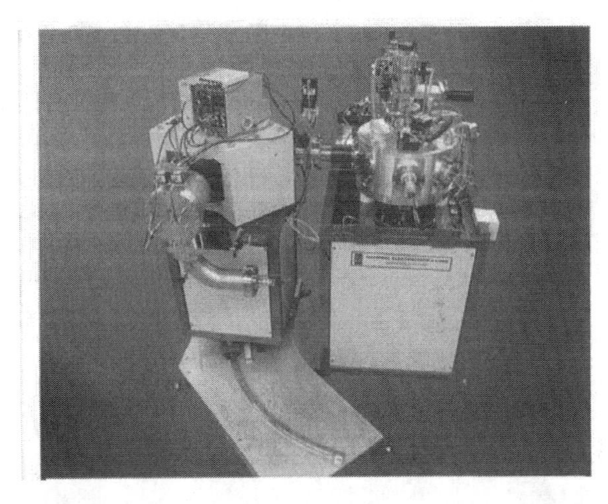

FIGURE 1. The high resolution RBS detector (left) on the 90° port of the standard version of the NEC RC43 analytical endstation. The HR RBS detector is mounted on a track for easy movement between ports. The bellows allows in situ +-5° scattering angle adjustments via a lead screw at the base at an accuracy of 0.05°.

Using a 90° 305mm bending radius magnet ($ME/q^2 = 2.2$) the energy window of the detector $\Delta E/E0$ is about 0.3 (Fig. 3). The detector is a 100mm X 15mm chevron type Micro Channel Plate with a maximum gain of $1x10^7$ followed by a Resistive Anode Encoder. Variable apertures placed before the magnet set the

Application of Accelerators in Research and Industry
AIP Conf. Proc. 1336, 291-294 (2011); doi: 10.1063/1.3586106

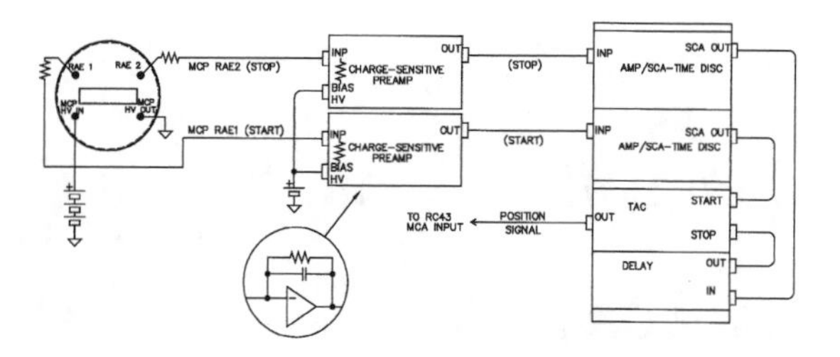

FIGURE 2. Schematic of signal processing electronics with resistive elements added to the RAE pickoffs.

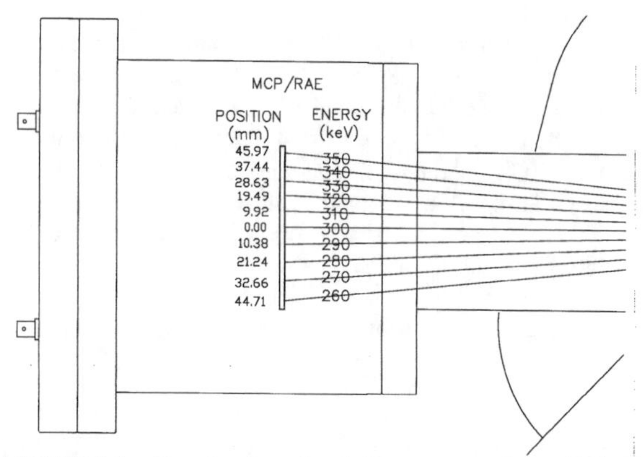

FIGURE 3. Ray tracing simulation generated for 400keV alpha beam scattering @ 90° from Si. Shows detector energy window $\Delta E/E0$ = 90/300 = 0.3 for 300keV central ray.

solid angle of the detector at a maximum of 0.4mStr. To collect a full energy spectrum, a minimum of 2 magnet settings is required. To automate this, the magnetic field control has been added to the sample position table to facilitate automatic data collection (Fig. 4).

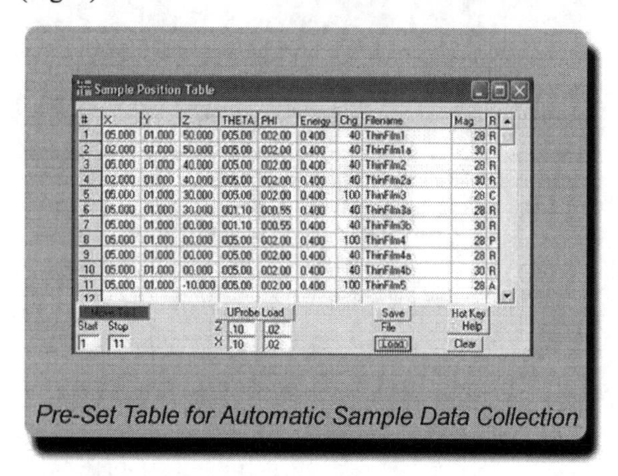

FIGURE 4. Sample position table for qued data collection showing HR RBS magnet control column.

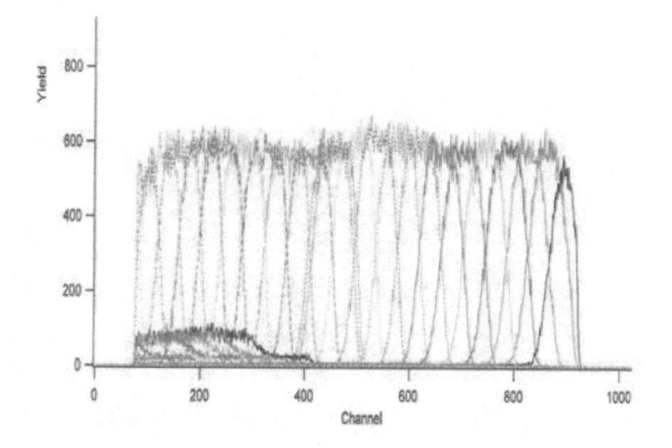

FIGURE 5. Hf thin film peaks for several magnet settings with central peak edge at 30.2Amps.

Particle energy vs detector position is non linear [2]. A series of thin film spectra (Fig. 5) can be used to determine the position to energy conversion. Rather than varying E with a fixed magnetic field, we varied B with a fixed ion energy, since the magnetic field is easier to measure. From these spectra, the position on the detector of the front edge of an Hf film is plotted vs $(B0/B)^2$ (Fig 6) which equals (E/E0). B0 is the magnetic field for a particle trajectory at the center of the detector. By applying a least squares fit [4] to the data in Fig. 6 a detector position to energy conversion was determined. Comparison of this experimental data with the ray tracing simulation of Fig. 3 shows good agreement. Channel number can now be substituted for detector position using a linear conversion, finally arriving at the desired channel to energy conversion for the data.

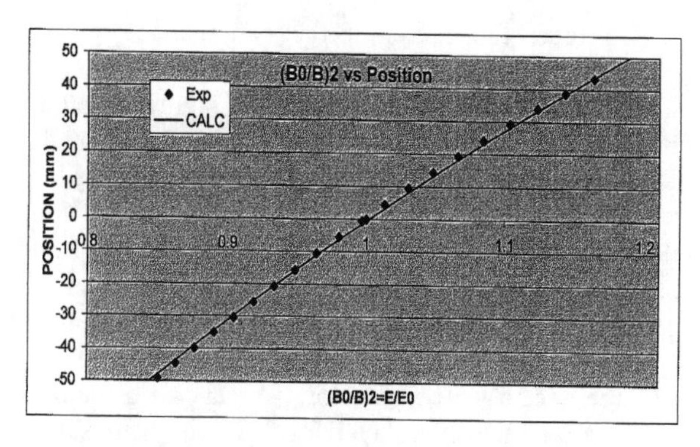

FIGURE 6. Energy calibration plot of magnetic field ratio vs detected ion position for experimental (triangles) and calculated (line) data.

To take full advantage of the energy resolution of the detector, a liner/driver is used inside the Pelletron® accelerator to minimize energy spread of the incident alpha beam from terminal voltage ripple. Voltage ripple at 400kV was 21Vrms (Fig. 7).

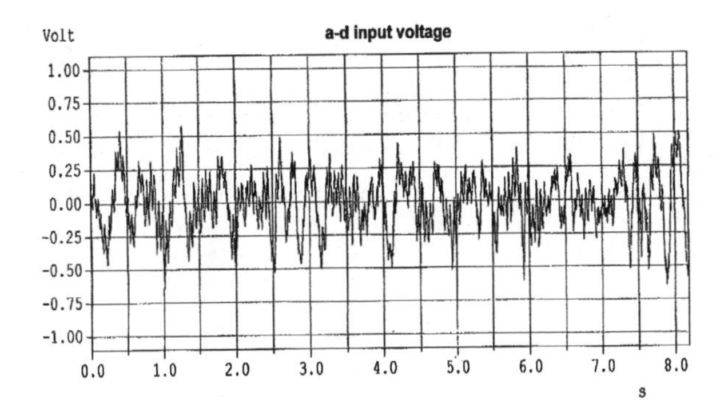

FIGURE 7. Plot of 3SDH terminal ripple showing 21Vrms for 0.4MV with voltage stabilizing liner at 80% gain. Vertical axis is calibrated at 100V/V.

RESULTS

The energy calibration was applied to data from a sample of 100ÅSi/50ÅTiN/20ÅHfO2 (Fig. 8). The incident 400 keV alpha particle beam scattered at an angle of 90° with a beam to sample normal angle of 20° and an exit angle of 70°. The calculated energy resolution, based on the slope of the Si front edge, is 1.6keV.

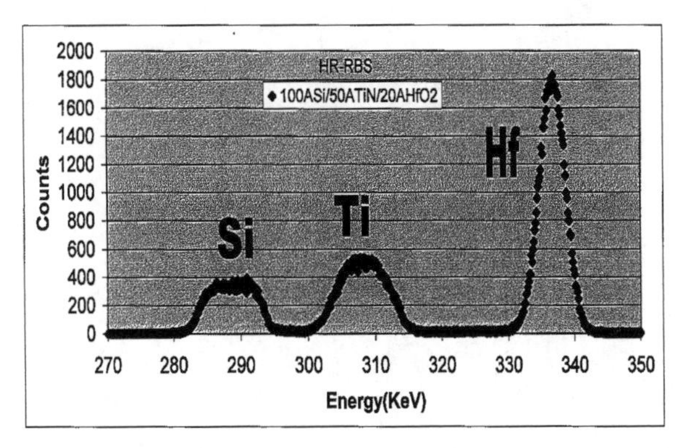

FIGURE 8. Energy calibrated spectra of 100ÅSi/50ÅTiN/20ÅHfO2 at magnet setting of 28Amps.

From the full energy spectra, all the near surface elements are visible (Fig. 9) and (Fig. 10).

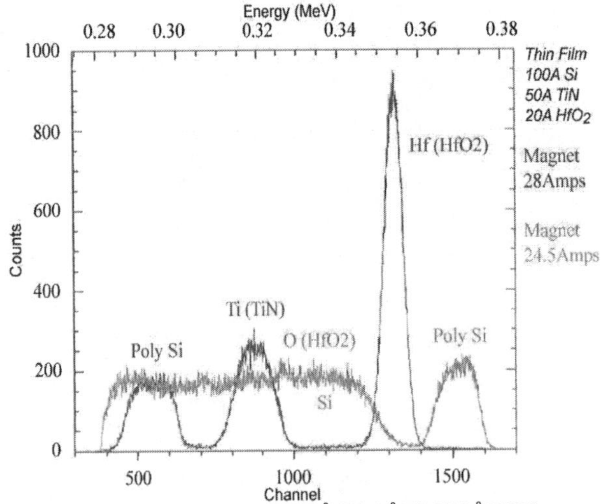

FIGURE 9. Two spectra of 100ÅSi/50ÅTiN/20ÅHfO2 at magnet setting of 24.5Amps and 28Amps.

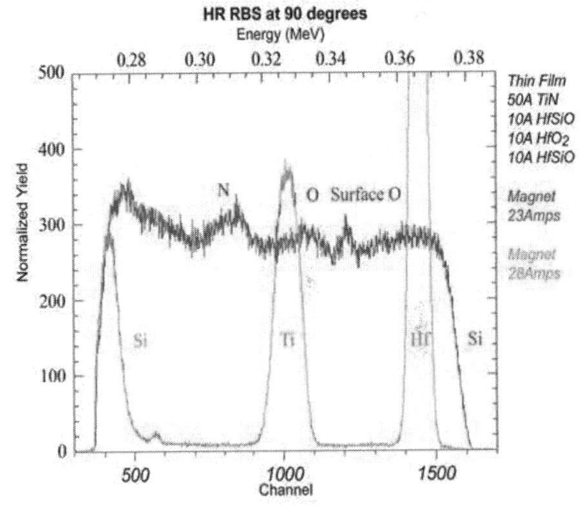

FIGURE 10. Two spectra of 50ÅTiN/10ÅHfSiO/10ÅHfO2/10ÅHfSiO at magnet setting of 23Amps and 28Amps.

Fig. 11 shows the depth profiles for Ti, Hf and Si from the spectra in Fig. 10. An approximate sample description must be entered so the correct energy loss/Å is used by the depth profile program.

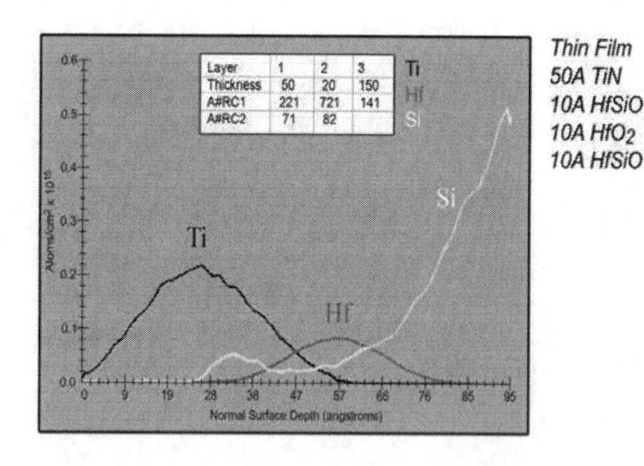

FIGURE 11. Depth profile for spectra in Fig. 10 with sample description, which is used to get correct energy loss/Å.

SUMMARY

Considering the 0.9keV kinematic broadening for a 400keV beam scattered from Si in our experimental setup, and the condition of the surface due to repeated use, a front edge energy resolution of 1.6keV is reasonable. Reducing the solid angle to 0.25mStr or 0.1mStr would reduce the kinematic broadening, but the data will then suffer from poorer statistics due to reduction in counts. Efforts continue to optimize energy resolution thru magnet pole design, etc. Improvements in software to automate energy calibration and depth profile generation are being implemented.

REFERENCES

1. T. Pollock, J. Haas, G. Klody, CP1099 Applications of Accelerators in Research and Industry 20th International Conference pp 323-325, 2009.
2. K. Kimura, M. Kimura, Y. Mori, M. Maehara, H. Fukuyama, AIP Conf Proc. 475 (1999) 500.
3. T. Osipowicz, H. Seng, T. Chan, B. Ho, Nucl. Instr. And Meth. B 249 (2006) 915.
4. A. C. Melissinos, Experiments in Modern Physics, Academic Press, New York, 1966, p. 462.

Development Of PIXE Measurement Of Ca Changes Resulting From Viral Transduction In Cells

Harry J. Whitlow[a], Orapin Chienthavorn[b], Hannele Eronen[a], Timo Sajavaara[a], Mikko Laitinen[a], Rattanaporn Norarat[a] and Leona K. Gilbert[c]

[a] *Department of Physics, PO BOX 35 (YFL), FI-40014 University of Jyväskylä, Finland,*
[b] *Department of Chemistry, Faculty of Science, Kasetsart University, Thailand,*
[c] *Department of Biological and Environmental Sciences, PO BOX 35 (Ambiotica), FI-4001 University of Jyväskylä, Finland*

Abstract. Ca is a life-element of particular interest because it is both bound to proteins, and as Ca^{2+} which functions as a signal molecule in apoptosis. Here we report development of chemical-matrix blind assaying the Ca fluxes from transduced HepG2 cells using particle induced X-ray emission. The cells were transduced with recombinant baculoviruses hosting the DNA for non-structural protein 1 (NS1) of the human pavovirus B19. Different recombinant baculoviruses were used that carried different DNA payloads of this NS1. Two different approaches have been developed to assay Ca in cells. The first is where the cells were directly cultured using a self-supporting pioloform as a substrate. In the second approach the cells are permeabilized, and bound-Ca content in the debris, and unbound-Ca in the wash solutions were measured using an internal V reference standard. The results support a difference in the Ca contents depending on the payload of the infecting virus, however the PIXE signals were too close to the minimum detection limit to draw reliable conclusions.

Keywords: Particle Induced X-ray Emission (PIXE), Ca concentrations, HepG2 cells, parvovirus, B19, baculovirus
PACS: 87.64.kd, 87.85.Xd, 87.17.Uv, 87.80.Dj, 87.19.xd, 87.17.Pq, 82.80.Yc, 82.80.Ej

INTRODUCTION

It is remarkable that all living cells have a very similar basic chemistry despite the wide diversity of organisms, which they constitute. Ca, which is the focus of this study, is of particular interest as a life element. [1,2]. Ca plays many roles; it is incorporated in calcium hydroxyapatite (the mineral content of bone) and Ca^{2+} ions are an important signal substance that influence processes in cells (e.g. contraction, embryonic development, neuro-signal transmission etc.) by phosphorylating of a wide range of proteins with specific CaM kinases that are in-turn activated by Ca^{2+}-responsive calmodulin [2]. There are large differences in Ca^{2+} concentration in cells which typically range from mM in the extra–cellular fluids, endoplasmic reticulum (ER) and mitochondria down to sub-μM levels in the cytosol. Cells that are attacked by disease agents such as human natural killer cells [3] $CD3^-$ and $CD16^+$ and B19 virus, release Ca^{2+} from the ER to the cytosol. This Ca^{2+} signal triggers apoptosis, autophagy and necrosis [3-5]. The role of different proteins causing cell death processes can be studied by mutating the DNA of the cell-death agent. This requires quantitative assaying of Ca bound to proteins and free Ca^{2+} in the cytosol. Under pathological conditions, (e.g. viral infections) the mitochondria can become compromised and the Ca^{2+} within may elicit pathological states leading to cell death [6,7].

Here we report on development of Ca assay methods based on Particle Induced X-ray Emission (PIXE). The advantage compared to competing methods is its blindness to chemical-bonding and matrix effects and high sensitivity. The challenge is to adapt cell culture techniques to obtain biomedical relevant information from PIXE.

MEASUREMENT OF Ca IN CELLS CULTURED ON PIOLOFORM

We investigated cell growth directly on a substrate for thin-PIXE analysis. Pioloform (polyvinyl butyral) [8] was used because it contains C, H and O and has low concentrations of heavy elements that would interfere with the PIXE measurements. The pioloform films were prepared on sterilized Al and stainless steel supporting rings [1]. (Plastic rings floated in the culture media.) To promote cell adhesion and

Application of Accelerators in Research and Industry
AIP Conf. Proc. 1336, 295-298 (2011); doi: 10.1063/1.3586107

proliferation on the pioloform substrate, the coated rings were subjected to a "soft" O_2 plasma treatment to render the pioloform surface more hydrophilic [9]. HepG2 cells (an immortalised liver carcinoma cell line [10]) were cultured by placing the pioloform film on its support on the bottom of a Petri dish containing 7.5×10^5 cells in 5 ml cell culture media. The cells were incubated (24 h 37 °C in 5% CO_2:95% air.) The cells were then either fixed, or transduced with recombinant baculoviruses at 50% transduction efficiency, (% of enhanced green fluorescent protein (EGFP) expression 24 h post-transduction in target cells). The viruses used carried different DNA payloads. These were EGFP (E), EGFP fused with B19 NS1 (ENS1) and NS1 that is mutated in its metal binding domain (ΔM). After culture the cells were fixed in 5% paraformaldehyde in phosphate buffer solution and dehydrated in successive 50%, 75% and 98% ethanol: water baths for 10 min, swabbed to remove excess liquid and dried in a laminar flow cabinet. The live-cell culture and virus handling were carried out in a clean-room (biosafety level 2) at the Jyväskylä Nanoscience Center. After fixing the cell samples do not pose a risk and could be handled in the normal laboratory environment.

Fig. 1 shows an optical microscope image of cells growing on pioloform that had been exposed to a "soft" oxygen plasma treatment using a parallel plate reactive ion etch (REI) using 20 W of 14 MHz rf power in 24 mbar of O_2 for 15 s. It is seen that the cells grow together in cohorts. If the plasma step was omitted, the pioloform surface is hydrophobic and the cells did not proliferate well. (Only isolated cells were observed which is atypical for cancer cells.) The effect of pioloform and ring on cell viability was tested by adding a drop of 0.4 % typan blue on the cells grown on the pioloform rings. The dead (stained) and living cells (unstained) were counted. The viability of cells cultivated on pioloform was 82% compared to 78.4% for cells cultivated in the plate. The difference in viabilities is not statistically significant at the 95% confidence level.

PIXE essentially measures the coverage of elements. To turn this into a biochemically pertinent quantity (Ca mass per cell) the areal density of cells must be determined. This was done by measuring the number of cell nuclei per unit area by staining the cells with Hoechst stain. Adding Miowol with Dabco (photobleaching inhibitor) improved the sharpness of the nuclei images. The Hoechst stain binds to DNA (nuclei) and by counting the number of fluorescent blue regions in ten 300×300 μm fields, the number of cells per unit area were determined. To avoid unwanted perturbation of the Ca levels by the stain this procedure was carried out on a separate sample that was cultured at the same time.

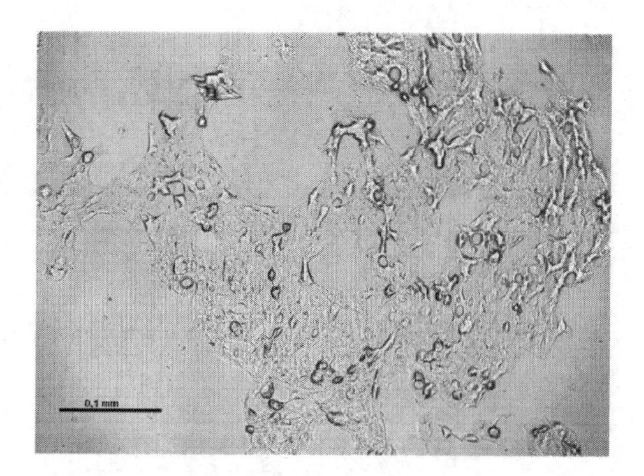

FIGURE 1. HepG2 cells cultured on pioloform treated with "soft" O_2 plasma bombardment. The scale bar is 100 μm.

3 MeV $^4He^{2+}$ ions from the Pelletron accelerator in Jyväskylä were used for PIXE analysis. X-rays were detected using Si:Li X-ray detector with ~190 eV fwhm resolution for Mn Kα X-rays with a 40 μm Mylar filter at 135° to the beam direction. Normalisation was done by counting ions backscattered from a rotating Au plated vane. The beam cross-sectional area was determined to be 2.9×3.5 mm^2 by exposing a PMMA coated Si wafer. The drawback of this method was the low sensitivity, which made reliable measurement of the Ca Kβ signal from K Kβ difficult and that the measurements are in terms of a coverage rather than concentration.

MEASUREMENT OF CA IN DROPLETS OF CELLS

It was established that the PIXE signals from cells cultivated directly on pioloform were very low which limited the sensitivity. Moreover, cytotoxic effects from the rings may perturb cell growth behaviour. A different approach was therefore developed were a known amount of cells was dropped on the pioloform film and dried. This is more preferable from a bioassay viewpoint because it allows treatments in different ways. However, it has the drawback that the droplet may dry heterogeneously and the sample must be spiked with an internal elemental concentration standard.

The elemental standard compound must satisfy a number of criteria. These include: no overlap of X-ray lines with life elements; large excitation cross-sections; readily available as a non-volatile water soluble compound such as halide, nitrate, acetate etc. that is anhydrous, or have a well-defined amount of water of crystallization; the compound should be stable and not be strongly oxidizing or reducing or have an aggressive nature. A number of candidate standards were tested. It turns out there is no ideal reference compound for spiking samples where life-elements are to be analyzed. Ammonium monovanadate, $NH_4(VO_3)$, was selected as it is water-soluble, the N, H and O content do not contribute interfering X-rays and V is a rarely occurring life-element. It is found in some V excreting bacteria [11] and bound in vanabin proteins in the blood of *Ascidiidae* (sea squirts) [12]. The Ca concentration was taken to be the product of a calibration coefficient and the ratio of the Ca Kβ to the V Kα peak areas

To quantify the Ca molar concentration the calibration coefficient was determined. 20 μl of 10 mM $NH_4(VO_3)$ solution was added to 10, 40 and 100 μl of 10 mM, and 10 and 40 μl of 1 mM $Ca(NO_3)_2$ solutions. These solutions were dropped onto separate pioloform films on stainless steel rings and allowed to dry under a filtered-air laminar flow hood. Fig. 2 presents the calibration data obtained using 2 MeV $^1H^+$ ions together with a least squares straight-line fit to the data. It is clear from Fig. 2 that sub-mM Ca levels can be analysed. The fitted calibration constant corresponded to a ratio of 0.279 Ca Kβ counts per V Kα count for equal molar concentrations of Ca and V.

For droplet analysis for Ca fluxes in transduced cells, 25 million cells were pelleted down (2500 rpm for 3 min) and permeabilized in 40 ug/mL digitonin in transport buffer (2mM Mg-acetate, 20 mM Hepes, 110 mM k-acetate, 1 mM EGTA, 5 mM Na-acetate, 1 mM DTT, pH 7.3) for 5 min on ice. Disrupted cells were then washed 3 times for 5 min with 100 μl transport buffer. After each wash step the disrupted cells were pelleted down (13 000 rpm for 1 min) and the wash solutions pooled. At the final wash the pelleted cell debris was re-suspended in 300 μl of transport buffer and 5 μl of 10 mM $NH_4(VO_3)$ standard solution was mixed together and then placed on the pioloform ring to dry. Similarly the pooled washes were added to the pioloform ring and left to air dry. The Ca concentration could then be determined from the Ca Kβ and reference V Kα yields because the dilution factors and number of cells were defined.

Comparing the two methods the spiked droplet approach has the advantages over the direct cultivation approach that the sensitivity is increased because it directly measures concentration rather than a coverage and higher sensitivity because a greater number of cells contribute to the analytical signal.

It was found that the very high K concentration in the samples, (about 10^4 times that for Ca) gave considerable problems because the tail from the K Kβ

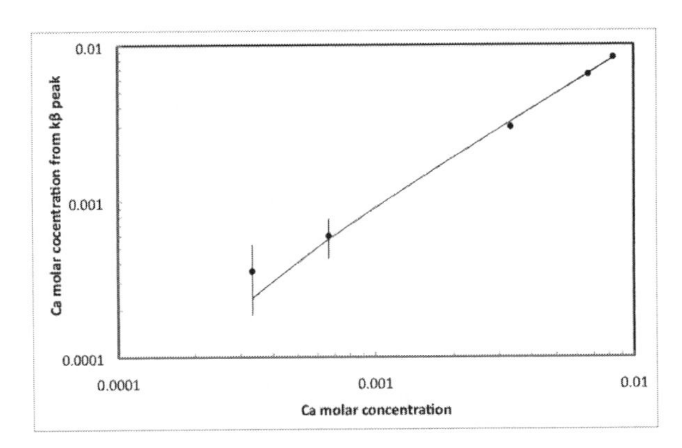

FIGURE 2. Molar concentration curve for Ca in $CaCl_2$ solutions from the Ca Kβ using a $NH_4(VO_3)$ reference. The line represents a linear least squares fit.

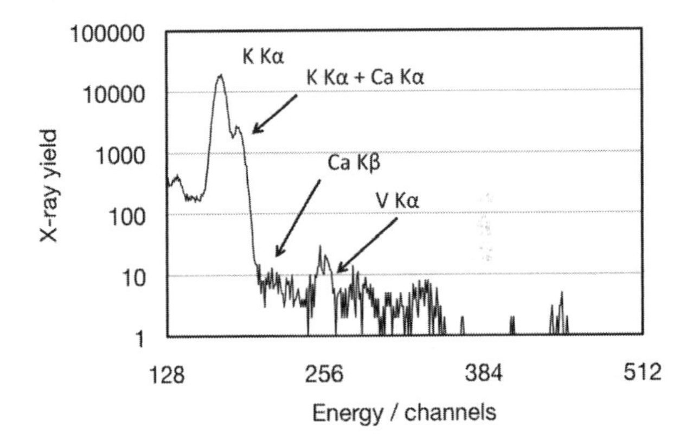

FIGURE 3. 2 MeV proton PIXE spectrum for cell debris from ENS1 virus infected HepG2 cells that has been spiked with a $NH_4(VO_3)$ reference. The Ca concentration of 13±5 mM is statistically significant.

line extends almost into the energy interval corresponding to Ca Kβ peak. Mammalian intracellular and extracellular K concentrations are ~140 mM and ~5 mM, respectively [10]. The corresponding levels for Ca^{2+} levels in normal cells are <1 μM and 1-2 mM, respectively [13]. Analysis of spectra where no Ca was present revealed the Ca Kβ peak appears as a small, weakly significant peak just separated from the K Kβ tail. (Fig. 3) Careful fitting of the background revealed the Ca level was either below or similar to the MDL level (97% significance level). An improvement of the MDL by a factor 30 is needed to assay Ca at deep sub-mM levels. To achieve this: Firstly, we have installed an X-ray detector with larger solid angle and better energy resolution at faster count rates, giving more precise background assignment between the K and Ca Kβ. Moreover, the improved resolution will further reduce the MDL by concentrating the Ca Kβ signal into fewer channels, enhancing signal to background ratio. Secondly, the background from pile-up of low-energy X-rays in the

Ca Kβ region can be reduced by using a more aggressive X-ray filtering. Thirdly, the MDL can be improved by increasing the areal density of cells. The Ca Kβ peak is clearly seen and easily quantifiable and mapped [14,15] in microbeam analysis of tissue sections with higher e.g. areal density, e.g. Fig. 1 of ref 12. This can either be done by using larger number of cells per culture, or by using a MeV ion microprobe to analyze single- or small cell cohorts. Large solid angle detectors on the other hand will degrade the MDL by reducing the precision for signals of only a few counts.

The data was close to the minimum detection limit no statistically reliable conclusions can be drawn. However, the results suggest there is a difference between the Ca levels in the cells, and disrupted cells and wash that depends on the DNA payload carried by the virus.

CONCLUSIONS

Two methods for PIXE-based Ca-flux analysis for transduced cells have been developed. Direct cultivation on pioloform gave low sensitivity because of low cell coverage. Droplets of cells spiked with ammonium mono-vanadate reference solution was more quantitative. Thus the droplet method is better suited than the direct cultivation approach for assaying the Ca content of cells using PIXE. Measured Ca levels in normal and B19 virus infected cells were comparable with the minimum detection limit (MDL) (10-20 mM Ca) because of swamping by K Kβ X-rays. The MDL needs to be reduced to allow more reliable conclusions to be drawn.

ACKNOWLEDGMENTS

Dr. Minqin Ren and Professor Frank Watt gave useful advice. This work was supported by the Academy of Finland; Centre of Excellence in Nuclear and Accelerator Based Physics, Ref. 213503, contracts 122061 and 129999, and an ASEM-DUO stipend for OC and HJW from the Office of the Higher Education Commission, Ministry of Education Thailand.

REFERENCES

1. M. Q. Ren, "Nuclear Microscopy : Development and Applications in Atherosclerosis, Parkinson's disease and Materials physics", Ph.D. Thesis, University of Jyväskylä, 2007.
2. B. Alberts, D. Bray, K. Hopkin, A. Johnston, J. Lewis, M. Raff, K. Roberts, P. Walter, *Essential cell biology,* 2nd Ed. (Garland Science, New-York, 2004) pp. 554.
3. Y. Osimi, K. Orchim and S. Miyasaki, *J. Physiology* **495.2,** 319 (1996).
4. P. Pinton, C. Giorgi, R. Siviero, E. Zecchini and R. Rizzotom, *Oncogene* **27,** 6407 (2008).
5. P. Pinton, D. Ferrari, E. Rappizzi, F. Di Virgillo, T. Pozzon and R. Rizzoto, *The EMBO Journal* **20,** 2690 (2001).
6. J. Nykky, J. E. Tuusa, S. Kirjavainen, M. Vuento and L. Gilbert, Int. J. of Nanomedicine. (In press)
7. M. R. Duchen, *J. Physiology* **529.1,** 57 (2000).
8. Pioloform product overview, Wacker Polymers, Burghausen, 2005.
9. S. Gorelick, P. Rahkila, A. Sagari A. R, T. Sajavaara, S. Cheng, L. B. Karlsson, J. A. van Kan, H. J. Whitlow, *Nucl. Instr. and Meth. B* **260,** 130 (2007).
10. HB-8065 Data sheet (ATCC /LGC standards, 2009) (http://www.lgcstandards; http://www.atcc.org/ATCCAdvancedCatalogSearch/ProductDetails/tabid/452/Default.aspx?ATCCNum=HB-8065&Template=cellBiology
11. A. N. Antipov, N. N. Lyalikova, N. P. L'vov, *IUBMB Life* **49,** 137 (2000).
12. T. Ueki, T. Adachi, S. Kawano, M. Aoshima, N. Yamaguchi, K. Kanamori, H. Michibata. *Biochim Biophys Acta* **1626,** 43 (2003).
13. ibid [2] Table 12-1 pp. 390.
14. M. Q. Ren, W.-Y Ong, X.S. Wang and F. Watt, *Experimental Neurology* **184,** 947 (2003).
15. F. Watt, R. Raajendran, M. Q. Ren, B. K. H. Tan, B. Halliwell, *Nucl. Instr. and Meth. B* **249,** 646 (2006).

Fabrication Of Buried Self-Organized Stripes In The Ni/C_{60} Composite

J. Vacik[1,2], V. Lavrentiev[1], P. Horak[1], K. Narumi[3]

[1]Nuclear Physics Institute (NPI) of AS CR, CZ-25068 Husinec - Rez, Czech Republic
[2]Research Center Rez, 250 68 Husinec - Rez, Czech Republic
[3]Japan Atomic Energy Agency (JAEA), 1233 Watanuki, Takasaki, 370-1292 Gunma, Japan

Abstract. A periodic array of striped nanoscopic domains, embedded in the epitaxially grown Ni layer, was formed after thermal annealing of the Ni/Ni+C_{60}/C_{60}/Ni thin multilayer sequence deposited on the MgO(001) monocrystal. The composite was annealed in 100°C/1 hr increments in the range of 200 – 600°C, and the structural evolution of the composite was analyzed mainly by Rutherford Backscattering and Scanning Electron Microscopy. The periodic system of the stripes was revealed after annealing at 600°C. The possible mechanism of the domain formation is suggested as follows: temperatures below 500°C incite diffusion (and consecutive disintegration) of the C_{60} molecules; temperatures above 500°C induce forceful intermixing of the upper Ni/Ni+C_{60}/C_{60} layers; due to the pronounced immiscibility of the Ni and C (C_{60}) components the process of the phase separation is triggered. In the confined crystalline matrix of the buffer Ni layer the separation proceeds coordinately according to the Ni lattice template - the stripe domains are formed in a direction parallel to the main crystallographic orientation (001) of the Ni epilayer.

Keywords: Fullerene; Nickel; Composite materials; Self-organization; Periodic stripes; RBS; SEM
PACS: 61.48.+c; 81.05.Tp; 81.16.Dn

INTRODUCTION

The preparation of well-defined structures on sub-microscopic scales is of the interest to material scientists and other specialists in order to perform basic studies on the sub-micro-structural systems and to develop functional materials with designed properties. Several techniques have been acknowledged to achieve these goals, with one of the most interesting approaches using the process of self-organization. Though the self-organization of inorganic matter is a common phenomenon often observed in nature [1], the essential physics of the regular pattern formation is still far from a complete understanding. The periodic patterns or structural ornaments (e.g., fractals) are usually formed far away from the thermodynamic equilibrium as a result of conflicting processes evolving in the systems. Though the mechanism of spontaneous order of matter is not well elucidated, it has been acknowledged that the process of self-organization has a potential to become an interesting method for the fabrication of periodic structures with specific parameters and properties. Because of the complexity and stochastic nature of the phenomena, the full 'control' of such a technique might, however, be difficult to approach.

Here we report on the fabrication of the nanoscopic stripe architecture spontaneously formed in a thin Ni/Ni+C_{60}/C_{60}/Ni multilayer after gradual thermal annealing up to 600°C. As the metal/fullerene interfaces exhibit unusual (charge) transport and other properties [2,3], they have become (during the last two decades) a subject of an intensive experimental investigation, which has led to several practical applications, e.g., in electronics [4,5], and elsewhere [6]. Despite the long-term research on the metal/fullerene systems, a number of open issues (e.g., their thermal stability, spontaneous phase separation, aging, etc.) still remains. Our main attempt was to inspect one of the issues, i.e., the thermal response of a hybrid metal/fullerene superstructure in a broad temperature range (200 - 600°C). We have analyzed a complex sandwich system Ni/Ni+C_{60}/C_{60}/Ni combining mixture (Ni+C_{60}) and multilayer (Ni/C_{60}/Ni) structures that have already shown an interesting thermal response (i.e., self-organization [7,8] and micro-objects formation [9] at room and elevated temperatures). A similar combination of the mixture and multilayer of Ni and C_{60} has been investigated in our previous work [10], where an interesting structural evolution of the Ni+C_{60}/C_{60}/Ni composite was observed: i.e., unusual growth of the

Application of Accelerators in Research and Industry
AIP Conf. Proc. 1336, 299-302 (2011); doi: 10.1063/1.3586108

buried C_{60} and Ni epilayers (at lower temperatures 100 - 350°C), and a collapse of the sandwich construction (due to intermixing and peeling of the upper adlayers), and a creation of an intriguing nanostructure in the remaining (Ni) film (at higher temperatures ~ 500°C). The nanostructure was formed in the buffer Ni layer, but due to the lack of relevant instrumentation, it remained unrevealed. The aim of our new inspection was to clarify the hidden nanostructure in a similar system that has been processed in an identical way.

In general, thermal annealing may lead to disintegration of the thermodynamically unstable (metal/fullerene) hybrid composites. It is therefore of (practical as well as basic) interest to know, how the structural transformation of the inspected hybrids proceeds. The observed formation of the periodic stripes was an unexpected 'side effect' of the thermal processing, pointing out the unusual spontaneous creativity of the transition metal - fullerene interactions.

EXPERIMENTAL

The $Ni/Ni+C_{60}/C_{60}/Ni$ superlattice was prepared using the MBE method with a background pressure of about 10^{-8} Torr. Electron beam bombardment of the 99.9 % Ni pellets and resistive heating of the 99.9% C_{60} powder was used for Ni and C_{60} evaporation. As a substrate MgO(001) single crystal was prepared. The deposition kinetics was as follows: a) deposition rate of Ni and C_{60} ~ 0.5 Å/s, b) thickness of the layers ~ 100 - 250 nm, c) temperature of the substrate during the deposition: 500°C for the buffer Ni layer, and 120°C for other adlayers. The $Ni+C_{60}$ mixture part was prepared by co-deposition of both Ni and C_{60} components with the same deposition rates. The final $Ni/Ni+C_{60}/C_{60}/Ni/MgO(001)$ thin multilayer system was gradually annealed in an Ar (+3% H_2) flow atmosphere between 200 and 600°C and the structural evolution of the sample was analyzed after each 100°C/1 hr annealing step using Rutherford Backscattering / Channeling (RBS / Channeling: 2.5 MeV α, 10 nA, 5 μC; NPI AS CR Rez, [11]) and Scanning Electron Microscopy (SEM – JEOL JSM 5600). Several supportive measurements were also performed using Atomic Force Microscopy (AFM - NT-NDT) and High Resolution Transmission Electron Microscopy (HRTEM - JEOL 2010) systems.

RESULTS AND DISCUSSION

Fig. 1 shows the evolution of the RBS patterns of the $Ni/Ni+C_{60}/C_{60}/Ni/MgO(001)$ multilayer sequence as-deposited and annealed isochronally for 1 hr at the temperatures of 200, 300, 400 and 500°C. The change of the RBS spectra (above 300°C) reflects the process of (thermally incited) degradation of the system. The pronounced dip and plateau in the as-deposited multilayer (spectrum range 320-420 channels), corresponding to the C_{60} and $Ni+C_{60}$ layers, are gradually (with the growing temperature) diminishing and merging (the layers merge together at 600°C). Similarly, the buffer Ni layer and also the substrate's Mg and O RBS edges evidence gradual broadening and flattening. All these changes reflect a dramatic restructuring of the multilayer composite: (i) a forceful intermixing of the Ni, C_{60} and $Ni+C_{60}$ layers, and (ii) a possible in-diffusion of the Ni and C phases into the MgO substrate (this phenomenon was also observed in [10]). Due to immiscibility of the Ni and C_{60} components, the intermixing of the hybrid layers results in a high internal stress that is rapidly increasing with the temperature. Above 500°C it leads to an entire destruction of the multilayer construction with a fully intermixed upper $Ni/Ni+C_{60}/C_{60}$ layers that are at 600°C peeled of the remaining part, which consists of the buffer Ni layer and indiffused C (C_{60} fragments).

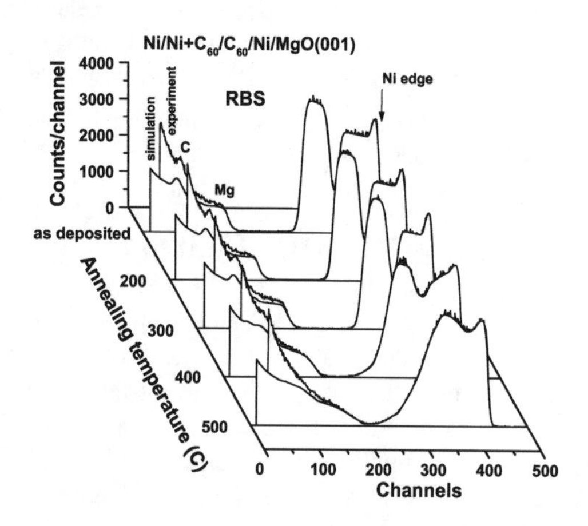

FIGURE 1. RBS spectra of the thin film superstructure $Ni/Ni+C_{60}/C_{60}/Ni/MgO(001)$ before and after isochronal annealing for 1 hour at 200, 300, 400 and 500°C.

The process of the rapid structural degradation of the multilayer system is well demonstrated in Fig. 2 representing evolution of the Ni relative concentration depth profiles (evaluated from the RBS data by the SIMNRA 6.05 code [12]) for the as-deposited hybrid system, after annealing at 500°C (when the multilayer was still compact) and at 600°C (when the three intermixed upper layers peeled of). In the SIMNRA evaluation, surface roughness (RMS) was taken into account: for the as-deposited system and the remaining buffer Ni layer the surface roughness was 1.5 nm and 3.7 nm, respectively, for the system annealed at 500°C the surface roughness achieved ~ 30 nm.

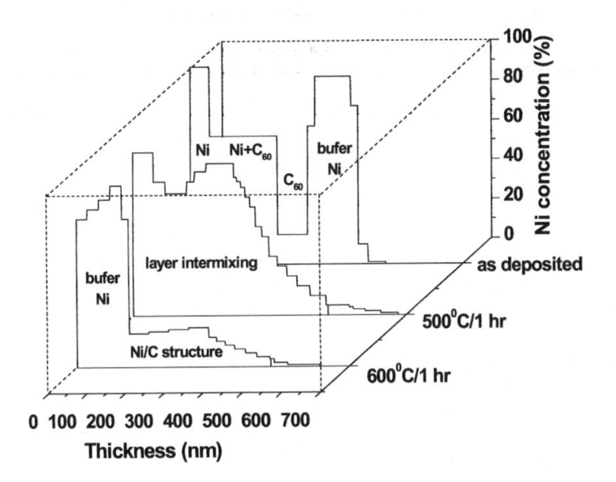

FIGURE 2. Ni relative concentration depth profiles (evaluated by the SIMNRA 6.05 code) for the as-deposited hybrid multilayer, and after 1 hour annealing at 500°C and 600°C.

Fig. 3 shows the additional RBS / Channeling study of the buffer Ni epilayer (after annealing at 600°C) that evidences (similarly as in [10]) the existence of a certain Ni/C - based structure buried in the Ni film (at the interface with the MgO substrate). The main part (estimated to ~ 85%) of the hidden (buried) Ni/C system is based on the C phase; the Ni phase is evidently growing epitaxially. As follows from Fig. 2 and 3, Ni (and also C) indifused into the MgO substrate, which is consistent with the observation in [10], where a mosaic of dense Ni particles was found in the subsurface zone of the MgO monocrystal.

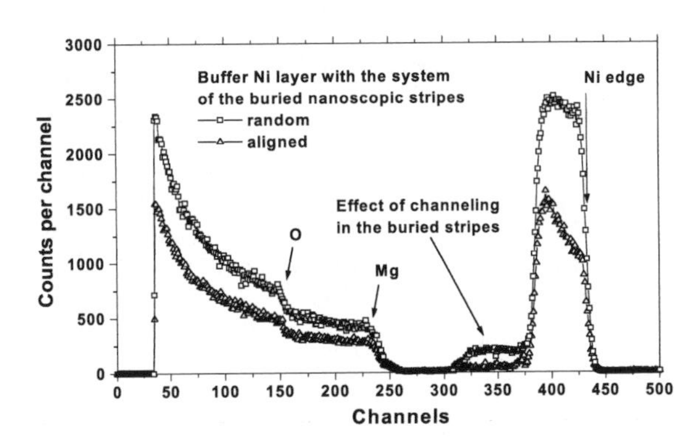

FIGURE 3. RBS / channeling study of the buffer Ni layer after peeling of the intermixed Ni/Ni+C_{60}/C_{60} adlayers.

To reveal the buried Ni/C structure the surface of the Ni layer was sputtered by a Ga focused ion beam. The sputtering removed the Ni film at several localities and revealed, surprisingly, a well-developed periodic system of the nanoscopic stripe domains. In Fig. 4, a part of the stripe system is demonstrated. It was found that the stripes are growing in parallel with the (001) crystallographic orientation of the MgO(001) substrate in a depth of about 200 nm, with the width and spacing of around 80 nm, and height estimated to 50 nm.

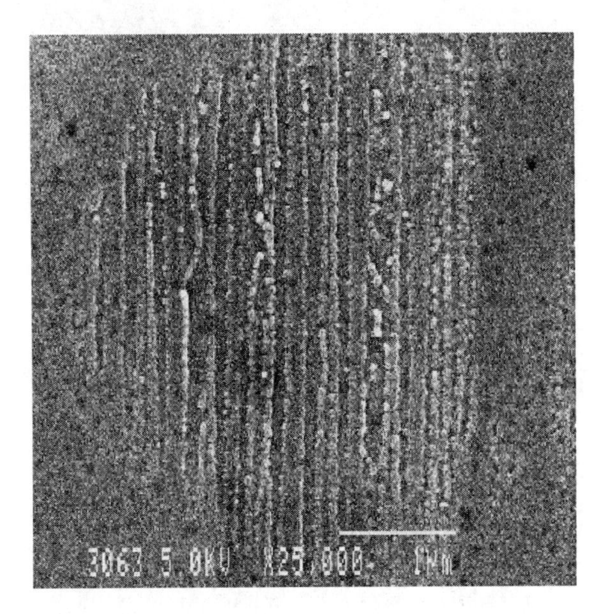

FIGURE 4. SEM micrograph of the buried nanoscopic domains revealed by sputtering of the buffer Ni layer using the Ga focus ion beam.

Additionally, the spatial position of the stripe domains was inspected by the cross-section HRTEM. Using the Ga FIB and a surface reinforcement by W (~ 100 nm), a thin cut (~ 50 nm) of the sample was prepared. Fig. 5 demonstrates the results of the analysis: the stripe system (a white band with a thickness of 50-80 nm) is grown on the MgO substrate in the (001) crystallographic orientation. Besides the pronounced C band, several other bands (less pronounced) can also be seen in the MgO subsurface layer. This again is in accordance with previous observation found in [10] and mentioned above.

The self-organized buried nanoscopic stripes is a phenomenon that resembles the formation of the surface stripe system developed in a mixture of Ni and C_{60} phases after their co-deposition on MgO(001) at certain deposition kinetics [7]. The mechanism of this effect was assigned to the drift-and-release propagation of the stress arisen in the hybrid mixture of Ni+C_{60}. The stress was due to the local accumulation of the C_{60} fragments (amorphous carbon, a-C) that are immiscible with the Ni matrix. Similarly, in the case of the buried stripe system, it is supposed that the pronounced Ni/a-C immiscibility is a main driving force of the buried stripes formation.

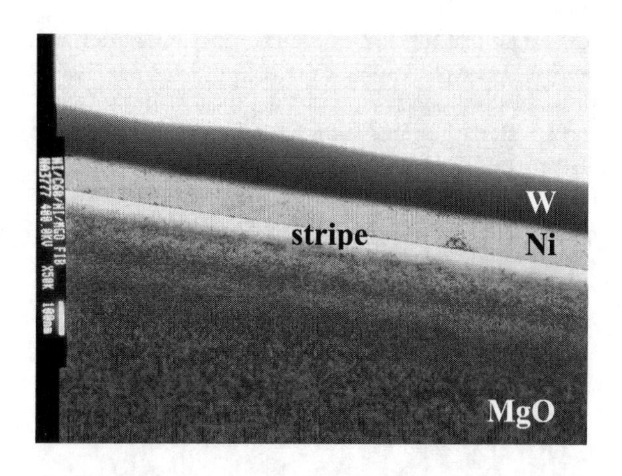

FIGURE 5. Cross section HRTEM micrograph of the interface between the MgO substrate and remaining part of the multilayer system after annealing at and 600ºC.

Obviously, during the thermal annealing, the C_{60} and C_{60} fragments (from the fullerite overlayer) in-diffuse in a massive way into the buffer Ni (towards the Ni/MgO interface) and trigger (at elevated temperatures ≥ 500ºC), the forceful restructuring of the hybrid multilayer system. The composite is (at this temperature) in a dynamic state seeking a structural configuration with the lowest surface energy. The nanoscopic stripe system is a result of the system's tendency (governed by the Ni/a-C immiscibility) to adapt its microstructure to the energetically most favorable form. As the minimal energy way to break symmetry of the uniform Ni crystal is a periodic variation in just one direction, a system of parallel stripe domains, as observed, is created. In the confined crystalline matrix of the buffer Ni layer the phase separation proceeds coordinately according to the template of the Ni lattice - the domains are formed in direction parallel with the main crystallographic orientation of the Ni epilayer.

CONCLUSIONS

In conclusion, we have reported on the thermal response of the Ni/Ni+C_{60}/C_{60}/Ni/MgO(001) super-multilayer system annealed in a broad temperature range from 200 - 600ºC. It has been shown that the metal/fullerite superlattice is susceptible to thermal processing and exhibits unusual structural alterations. Though at low temperature regimes constructive phenomena might occur (e.g., homo-epitaxial growth of the buried fullerite, as was observed in a similar multilayer system in [10]), at elevated temperatures a series of disruptive processes (fullerene molecule fragmentation, intermixing of layers, phase separation, etc.) determines the structural evolution. The collapse of the hybrid system is surprisingly accompanied with a constructive phenomenon – a spontaneous formation of the nanoscopic stripes, buried within the Ni epilayer. The pattern formation is due to the phase separation, which proceeds coordinately in the confined periodic template of the buffer Ni matrix.

ACKNOWLEDGMENTS

The authors acknowledge the financial support from the Academy of Sciences of the Czech Republic (Research projects No. KAN400480701 and IAA400320901), Grant Agency of the Czech Republic (Research project No. 106-09-1264) and Ministry of Education of the Czech Republic (Research program Nos. LC 06041).

REFERENCES

1. P. Ball, The Self-Made Tapestry, *Pattern Formation in Nature*, Oxford University Press, New York, 1999.
2. M.S. Dresselhaus, G. Dresselhaus and P.C. Eklund, *Science of Fullerenes and Carbon Nanotubes*, Academic Press, London, 1996.
3. H. Yonehara and C. Pac, *Appl. Phys. Lett.* **61**, 575-576 (1992).
4. K. Pichler, M. G. Harrison, R. H. Friend, and S. Pekker, *Synth. Metals* **56**, 3229-3234 (1993).
5. R. C. Haddon, A.S. Perel, R. C. Morris, T. T. M. Palstra, A. F. Hebard, and R. M. Fleming, *Appl. Phys. Lett.* **67**, 121-123 (1995).
6. A. C. Mayer, M. T. Lloyd, D. J. Herman, T.G. Kasen, G. G. Malliaras, *App. Phys. Lett.* **85**, 6272-6274 (2004).
7. J. Vacik, H. Naramoto, S. Yamamoto, K. Narumi, K. Myiashita, *J. Chem. Phys.* **114**, 9115-9119 (2001).
8. J. Vacik, V. Lavrentiev, V. Hnatowicz, V. Vorlicek, S. Yamamoto, H. Stadler, *J. Alloys and Comp.* **483**, 374-377 (2009).
9. J. Vacik, H. Naramoto, K. Narumi, S. Yamamoto, H. Abe, *Nucl. Instr. and Meth. B* **206**, 395-398 (2003).
10. J. Vacik, H. Naramoto, K. Narumi, S. Yamamoto, H. Abe, *Nucl. Instr. and Meth. B* **219-220**, 862-866 (2004).
11. http://neutron.ujf.cas.cz/vdg/index.html.
12. http://www.rzg.mpg.de/~mam/.

Particle-Induced X-Ray Emission (PIXE) Of Silicate Coatings On High Impact Resistance Polycarbonates

Qian Xing, M. A. Hart, R. J. Culbertson, J. D. Bradley, N. Herbots, Barry J. Wilkens, David A. Sell, Clarizza Fiel Watson

Department of Physics, Arizona State University, Tempe, AZ 85287-1504

Abstract. Particle-Induced X-ray Emission (PIXE) analysis was employed to characterize hydroxypropyl methylcellulose (HPMC) $C_{32}H_{60}O_{19}$ polymer film via areal density measurement on silicon-based substrates utilizing the differential PIXE concept, and compared with Rutherford backscattering spectrometry (RBS) results. It is demonstrated in this paper that PIXE and RBS measurements both yield comparable results for areal densities ranging from 10^{18} atom/cm^2 to several 10^{19} atom/cm^2. A collection of techniques including PIXE, RBS, tapping mode atomic force microscopy (TMAFM), and contact angle analysis were used to compute surface free energy, analyze surface topography and roughness parameters, determine surface composition and areal density, and to predict the water affinity and condensation behaviors of silicates and other compounds used for high impact resistance vision ware coatings. The visor surface under study is slightly hydrophilic, with root mean square of surface roughness on the order of one nm, and surface wavelength between 200 nm and 300 nm. Water condensation can be controlled on such surfaces via polymers adsorption. HPMC polymer areal density measurement supports the analysis of the surface water affinity and topography and the subsequent control of condensation behavior. HPMC film between 10^{18} atom/cm^2 and 10^{19} atom/cm^2 was found to effectively alter the water condensation pattern and prevents fogging by forming a wetting layer during condensation.

Keywords: RBS, PIXE, contact angle, TMAFM, surface free energy, silicate, SiO_2, ERD, NRS, polymer
PACS: 61.05.Np, 82.80.Yc, 82.80.Ej, 68.37.Ps, 87.64.Dz, 68.55.am, 68.35.Md, 68.55.jd, 68.55.Nq, 82.30.Rs

INTRODUCTION

Water condensation is a concern for many types of vision ware such as sports visors, eye glasses, goggles etc. Developing products with anti-fogging properties has been of interest for both science and engineering, based on the plethora of products on the market[1]. The recent success of utilizing HPMC polymer film on silica and silicone to enhance visibility during medical applications is based on the concept of forming a complete wetting layer on the substrate during water condensation, which therefore enhances vision clarity[2]. In order to achieve this complete wetting layer, both the surface topography and the surface free energy play critical roles. To extend the above concept to vision ware, one has to understand the complexity of the eyewear surfaces, commonly coated with high impact resistance coating such as silicate, and the underlying substrate, typically a form of polycarbonate.

Contact angle measurements have been used to quantify water affinity and surface free energy utilizing Young's Equation combined with Van Oss theory[3]. Large surface tension components result in the contact angle being small and surface being hydrophilic due to the strong intermolecular forces between the surface and water. Small surface tension components result in the surface being hydrophobic. The three surface tension components of solid surfaces from Van Oss theory can be calculated by measuring the contact angles using three distinct types of test liquids to the solid surface[4]. One can then calculate surface tension through the following,

$$\begin{pmatrix}\sqrt{\gamma_S^{LW}}\\ \sqrt{\gamma_S^{+}}\\ \sqrt{\gamma_S^{-}}\end{pmatrix}=\frac{1}{2}\cdot\begin{pmatrix}\sqrt{\gamma_{L1}^{LW}} & \sqrt{\gamma_{L1}^{-}} & \sqrt{\gamma_{L1}^{+}}\\ \sqrt{\gamma_{L2}^{LW}} & \sqrt{\gamma_{L2}^{-}} & \sqrt{\gamma_{L2}^{+}}\\ \sqrt{\gamma_{L3}^{LW}} & \sqrt{\gamma_{L3}^{-}} & \sqrt{\gamma_{L3}^{+}}\end{pmatrix}^{-1}\cdot\begin{pmatrix}(1+\cos\theta_{L1})\gamma_{L1}\\ (1+\cos\theta_{L2})\gamma_{L2}\\ (1+\cos\theta_{L3})\gamma_{L3}\end{pmatrix} \quad (1)$$

with $\theta=$ contact angle made at the liquid/solid interface, $\gamma=$ surface tension, subscript of $S=$ solid, $L1-L3$ represent the three types of liquid respectively, the superscript of γ^{+} and $\gamma^{-}=$ Lewis

Application of Accelerators in Research and Industry
AIP Conf. Proc. 1336, 303-309 (2011); doi: 10.1063/1.3586109

acid-base component of surface tension, and γ^{LW} = Lifshitz-van der Waals component of surface tension. This approach to analyzing surface free energy will be used to quantify the water affinity of the vision ware samples used in this paper.

PIXE is used to help characterize and quantify the substrates (silica and factory treated polycarbonate) and the HPMC film. PIXE has been used by previous labs and organizations to analyze such diverse samples[5-7]. Differential PIXE was used to determine the depth profiling and depth depletion assuming a homogeneous or layered profile structure[8]. In this work, differential PIXE is explored as a method to identify the areal density of the film by using the substrate's (on which the film is applied) Si K X-ray emission. Note that the film has no Si, and hence the substrate and film are distinguishable. We then explore the possibility of using differential PIXE to identify the polymer, and hence apply it to the possible identification of the HPMC layer. The method and theory of using the differential energy of PIXE (differential PIXE) has been used elsewhere[9-11]. The motivation is to control condensation on such surfaces via polymers adsorption in sport eyewear applications.

MATERIALS AND METHODS

Materials And Preparations

Numerous materials were required for this experiment. De-ionized (DI) water is of 2 MΩ cm resistivity unless otherwise noted. Fused silica wafers are from Medtronic, and were sonicated 10 minutes in DI water of 18 MΩ cm resistivity. Hydroxypropyl methylcellulose (HPMC), $C_{32}H_{60}O_{19}$ CAS-9004-25-3, was 86 kDa molecular weight from Sigma-Aldrich and is hydrated using DI water. Glycerin was ordered from Sigma-Aldrich (G2289, CAS number 56-81-5), as was α-bromonaphthalene (17640, CAS number 90-11-9). The visors are Oakley brand and were selected randomly.

Both silica and the Oakley visors were treated with HPMC film at room temperature, soaked in water-hydrated HPMC for 2 hours at various concentrations (from 0.33% w.t. to 1.00% w.t.), and then air dried under a class 10K ventilation hood for a minimum of 24 hours.

Techniques

The Sessile drop method[12] was used to conduct the contact angle measurements. The liquid droplets' size ranged from 0.3 μL ~ 10 μL and was delivered using a syringe with a 23 gauge cannula. The contact angle was measured using computer fitting techniques for the digital images. Both droplet size and contact angle were computed via computer processing. The three components including Lewis acid-base and Lifshitz-van der Waals components of surface tension γ_S^+, γ_S^-, and γ_S^{LW} from Equation (1) above were calculated using three distinct types of test liquids: water, glycerin, and α-bromonaphthalene[4]. The energy calculation tool was provided by SurfTen 4.3[13-14].

Roughness and length-scale of the roughness of surfaces were determined by tapping mode atomic force microscopy (TMAFM, Agilent). The AFM was operated in air with a silicon tip in AC mode. Gwyddion was the software used to produce images and roughness parameters; R_q = root mean square roughness, λ_a = average wavelength of the profile, and Δ_a = average absolute slope.

Rutherford backscattering spectrometry (RBS) combined with RUMP software provided composition and areal density measurement for the visor substrate as well as the polymer film areal density as a comparison to the PIXE method described below. 2 MeV He^{++} with incident angle of 8° and detector angle of 170° together with RUMP simulation were used to identify the composition; incident He^{++} at 2.8 MeV with incident angle of 8° and detector angle of 170° together with RUMP simulation was used to determine the areal density of the Si layer. HPMC stoichiometry near the surface was previously determined via 4.265 MeV $^{12}C(\alpha, \alpha)^{12}C$, 3.045 MeV $^{16}O(\alpha, \alpha)^{16}O$ nuclear resonance scattering (NRS), and 2.8 MeV elastic recoil detection (ERD) of hydrogen[2].

Particle-Induced X-ray Emission

Areal density measurements via incident α particle energy loss was used to analyze the HPMC film areal density on both the silica and visor substrates (see Figure 1). Since silicon exists only in the substrate, the ratio of the number of silicon K X-rays detected and the amount of incident α particles is related to the areal density of silicon in the substrate accessible by the α particles through the following relation,

$$(\rho t)_Z = \frac{N_X}{N_i} F_{X,Z} \quad (2)$$

with $(\rho t)_Z$ in g/cm^2 = areal density of element Z, N_X = number of X-rays, N_i = number of incident ions, and $F_{X,Z}$ is a constant for a given element, transition, primary ion beam, detector, and geometry[15]. At certain incident energies, α particle can penetrate through a limited film depth into the Si-based substrate due to the energy loss via the in-path of the HPMC film and

Si-based substrate. The amount of incident α particles is then compared to the amount of emitted silicon K X-rays with and without the HPMC film. Hence one can calculate the energy loss due solely to HPMC in-path where,

$$\Delta E_{in-path} = E_0 - E \approx [\frac{1}{|\cos\theta_{in}|}(\varepsilon)_{E_{in-path}}]Nt(HPMC) \quad (3)$$

where E_0 is the incident α particle energy just before striking the HPMC film, E is the energy of the α particle at the substrate interface $Nt(HPMC)$ is equal to the areal density of the HPMC film, θ_{in} the incoming angle of the α particles to the sample normal, and $\varepsilon_{E_{in-path}}$ the energy loss cross section of HPMC. Incident energy used is between 1.4 MeV ~ 2.2 MeV. Since the number of incident in-path α particles backscattered by the HPMC layer at the given energy level is negligible (< $10^{-5} N_i$), the silicon K X-ray attenuation due to out-path of HPMC layer attenuation is between 0.5% ~ 5.0% for 10^{18} atom/cm^2 ~ 10^{19} atom/cm^2 of the HPMC film if the detector is at the sample normal. Thus, one should expect with the systematic error, yields of the actual areal density are lower than this measurement. The experiment is conducted in a vacuum of 10^{-7} torr ~ 10^{-6} torr without a filter. The incident angle is 8° and detector angle is 37° to the sample normal for silica samples; and an incident angle of 30° and detector angle of 75° to sample normal for visor samples since the large incident angle is necessary due to the thinner silicate layer on visor samples. Controlling the beam current is critical to minimize detector dead time as well as pileup[16]. Detection of X-rays was most consistent when the beam current < 0.2 nA, therefore data collection was taken at 0.1 nA. Each spectrum received a charge of 0.5 μC. Sample charging due to the incident particles was also noticed for both the silica wafer and visor samples. Using an aluminum foil wrapping with a 4 mm ~ 10 mm diameter hole minimized the charging effect and stabilized the spectra; enabling consistent, repetitive data collection. GUPIX[17] was used to calculate the area count of the silicon K X-rays.

Composition analysis of the Oakley visors utilized proton incident particles. Incident energy is at 1.8 MeV, incident angle is 40° and detector angle is 40° to sample normal in a vacuum of ~10^{-1} torr without filter, and a beam current < 0.3 nA.

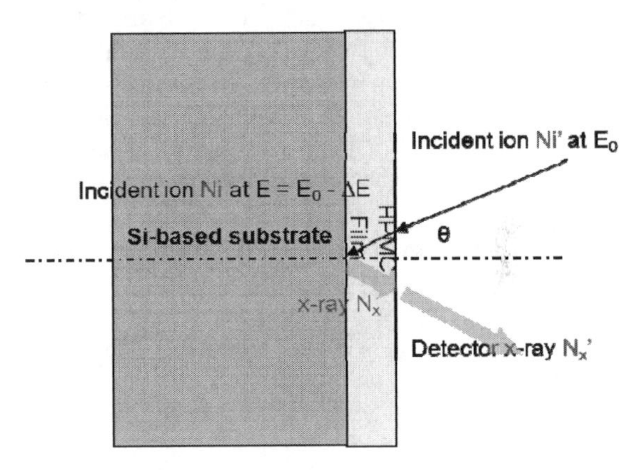

FIGURE 1. Pictorial representation and geometry of Si-based substrate with HPMC film layer placed on top.

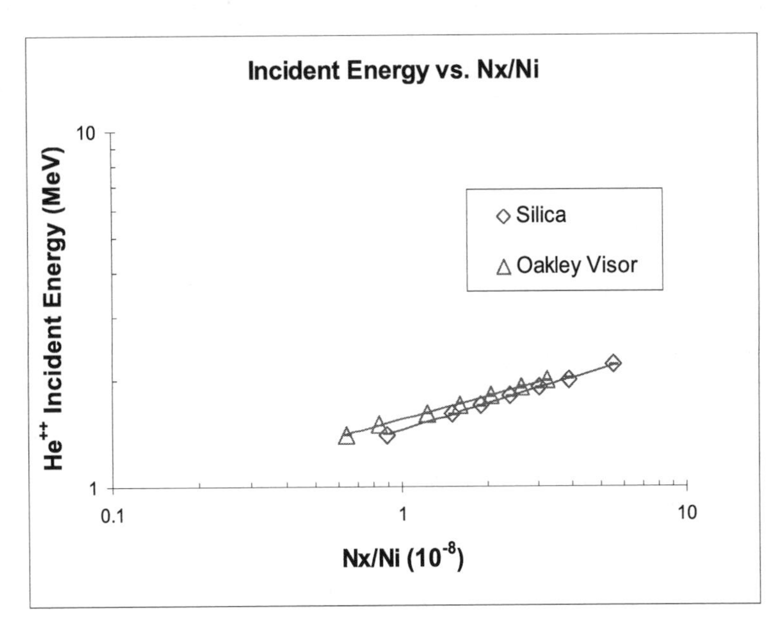

FIGURE 2. Incident energy of α particle plotted against Nx/Ni ratio. Note that the above data is based on silica and Oakley visor substrates without HPMC film by varying the incident energy.

RESULTS AND DISCUSSION

Polymer Film Areal Density On Silica

The relationship of incident α particle energy E vs. the ratio of detected silicon K X-ray count over the incident particle count $\frac{N_X}{N_i}$ is shown in Figure 2, a silica substrate without HPMC film, and is used as the baseline for the relative yield graph. The log-log regression modeling uncertainty is analyzed to serve as the basis to obtain the error of the areal density calculation utilizing the differential PIXE method. Areal density via the RBS method is obtained from RUMP simulation of the spectra. The uncertainty is analyzed from the roughness of determined by analyzing the Si leading edge of which indicated the roughness of the film is observed. Areal density of the HPMC film and comparison to its RBS measurement are shown in Table 1. The measurements of the areal density by PIXE and RBS are in agreement.

TABLE 1. Area density results of HPMC polymer film on silica substrate.

HPMC gel concentration	Nt (10^{15} atom/cm^2) He^{++} PIXE	Nt (10^{15} atom/cm^2) He^{++} RBS
1.00% w.t.	7500 ± 500	-
0.50% w.t.	4270 ± 560	5000 ± 1000
0.40% w.t.	3040 ± 580	2700 ± 700
0.33% w.t.	2100 ± 600	-

Once the substrate relative yield graph is set up, areal density measurement is straight forward using this method. Another advantage of this method, when compared to using RBS, is that thicker films can be handled without the elevation of the incident particle energy. However, the thicker the film, the X-ray attenuation due to the out-path of HPMC becomes more significant. Results from both methods are in-line with similar order of resolution.

Visor High Impact Resistance Coating

The proton PIXE spectra in Figure 3 demonstrated that there is significant Si and S on both the convex and concave sides of the visor. RUMP simulation of the RBS spectra, Figure 4, confirms the presence of the silicon and sulfur in the estimated stoichiometric formula of $SiO_2C_{2.5}$ with an areal density of 1.3×10^{19} atom/cm^2. Therefore, in order to utilize the energy loss method for areal density measurement, the incident α particle angle is increased to 75° to make the in-path longer, which will ensure a large enough silicon thickness.

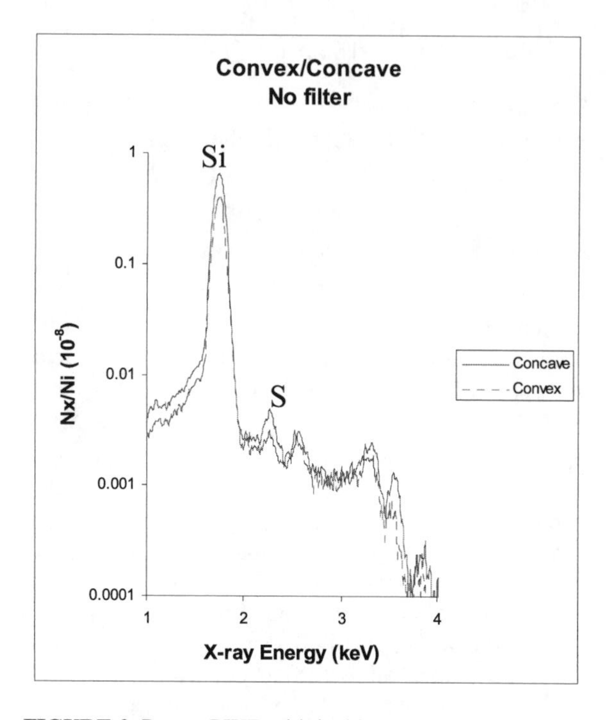

FIGURE 3. Proton PIXE with incident energy of 1.88 MeV, incident angle 40°, detector 40° to the sample normal showing significant silicon signal.

Analysis of the surface topography on the convex and concave sides of the visors was done using TMAFM. From Figures 5a, 5b, and Table 2, both concave and convex sides have similar root mean square roughness of ~1 nm, which provides nucleation sites for water condensation. Contact angle and surface free energy information is shown in Table 3 and 4, and leans slightly hydrophilic. The roughness parameter of a typical feature was ~1 nm and average wavelength of 0.2 μm ~ 0.3 μm, provides significant nucleation sites for rapid condensation as shown in Figure 6, on the area without the HPMC film. A slightly hydrophilic surface will indicate that the surface condensation pattern will be puddle-like upon ripening. Thus the combination of surface topography and surface free energy predicts the type of water condensation.

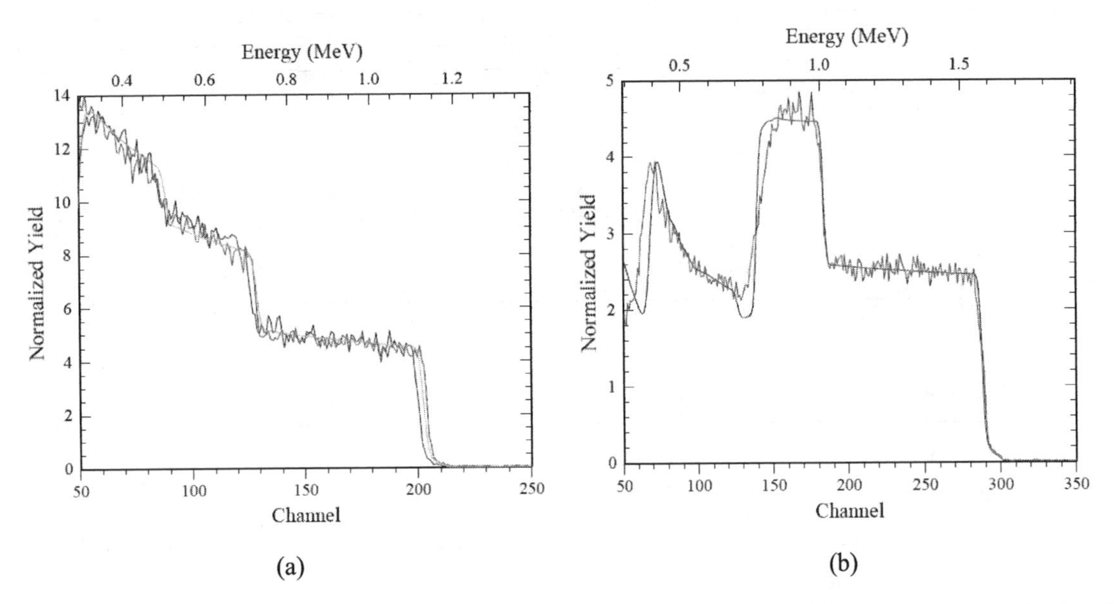

FIGURE 4. He^{++} RBS spectra, with incident energy of (a) 2.0 MeV, with convex and concave sides of visor having similar spectra, and (b) 2.8 MeV, incident angle 8°, detector 170° to the sample normal. Simulation fit via RUMP to the stoichiometric formula $SiO_2C_{2.5}$, concave side areal density ~$1.3x10^{19}$ atom/cm^2.

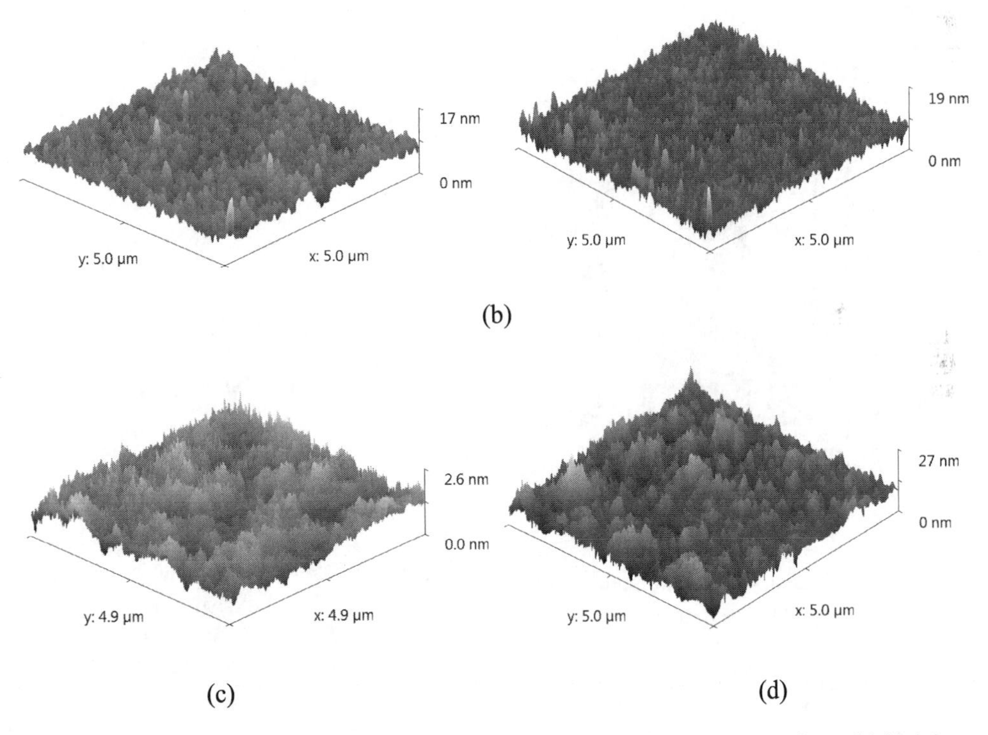

FIGURE 5. (a) is the concave surface with high impact resistance coating; (b) is the convex surface with high impact resistance coating; (c) is the concave side with 1% HPMC film; (d) is the concave side with 0.2% HPMC film.

TABLE 2. Summary of the convex and concave roughness parameters.

Roughness Parameters	R_q (nm)	λ_a (µm)	Δ_a
Concave	1.14(9)	0.27(1)	0.0201(8)
Convex	1.49(15)	0.22(2)	0.032(2)
Concave with 1% HPMC film	0.25(2)	0.19(2)	0.0069(7)
Concave with 0.2% HPMC film	1.77(1)	0.21(2)	0.055(5)

TABLE 3. Contact angle measurements for the Oakley visor.

Surface	$\theta_{water}{}^{o}$	$\theta_{glycerin}{}^{o}$	$\theta_{\alpha\text{-bromonaphthalene}}{}^{o}$
Concave side	61.5 ± 1.4	54.7 ± 0.4	32.2 ± 1.6
Convex side	74.5 ± 1.0	78.4 ± 1.7	36.7 ± 1.5

TABLE 4. Surface free energy measurements for the Oakley visor.

Surface	Surface tension (mJ/m^2)	γ^{LW} (mJ/m^2)	γ^- (mJ/m^2)	γ^+ (mJ/m^2)
Concave side	40.6 ~ 49.7	37.8 ± 1.7	16.8 ± 5.3	0.7 ± 0.4
Convex side	34.2 ~ 47.8	36.0 ± 1.8	18.0 ± 4.8	0.4 ± 0.5

Polymer Film Areal Density On Visor

The HPMC polymer film on the visor alters the surface topography, as well as water affinity and thus alters the condensation behavior, as shown in Figure 6. Surface topography and roughness parameters are shown in Figures 5c, 5d, and Table 2 respectively. The hydratable polymer mesh provides high water affinity which will enhance the coalescence of the nucleated droplets. Roughness scaling from several angstroms to several nm will provide ample nucleation sites, which will induce rapid condensation. The combined topography of the aforementioned characteristics leads to a rapid condensation which is immediately followed by coalescence, and the subsequent elimination of the fogging.

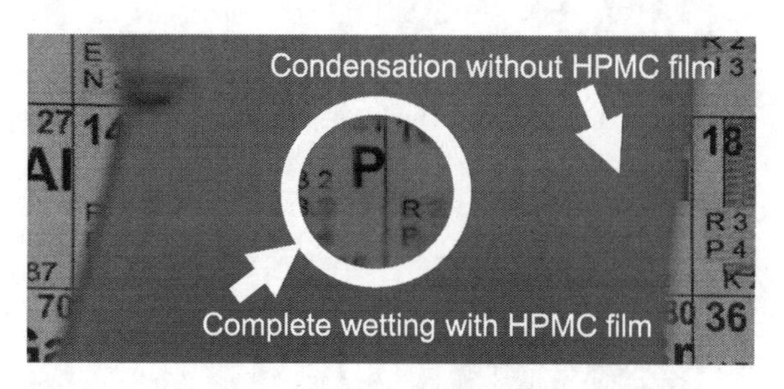

FIGURE 6. The condensation behavior on the HPMC polymer film differs when compared to the non-polymer film area.

HPMC film areal density measurement and comparison to RBS using the energy loss method described above, is shown in Figure 2 and Table 5. The results show that the HPMC film with areal density ranging from 10^{18} atom/cm^2 ~ 10^{19} atom/cm^2 can effectively prevent fogging on the visor surfaces by forming a complete wetting layer to improve visual clarity. However, initial fogging is noticeable while the wetting layer is formed. It was also observed that the HPMC film on the visor surface has a larger wavelength than the HPMC film on the silica wafer. After the initial fogging, rehydration of the polymer forms a complete wetting layer and no further fogging is observed afterwards. Finally, from Table 5, the areal density derived from PIXE is of the same order as that derived from RBS; however, the result differences of the visor substrate were more significant than that of the silica substrate. We noticed that the stoichiometric variations in the RBS samples ranging from $SiO_2C_{2.5}$ to SiO_3C_3, indicating slight compositional fluctuations at different sample locations of the visor coating may be the cause of error from the PIXE method, while the RBS result uncertainty indicates roughness of the HPMC. Hence, the respective areal density differences may be due to the observed variations and needs to be explored further.

TABLE 5. Area density results of HPMC polymer film on visor substrates.

HPMC gel concentration	Nt (10^{15} atom/cm^2) He^{++} PIXE	Nt (10^{15} atom/cm^2) He^{++} RBS
1.00% w.t.	16560 ± 170	-
0.50% w.t.	9430 ± 150	6500 ± 500
0.33% w.t.	1430 ± 170	2400 ± 400
0.20% w.t.	1640 ± 160	2000 ± 300

CONCLUSIONS

Areal density measurements using differential He^{++} PIXE were used to successfully characterize the HPMC film on Si based surfaces. The substrate allowed for the generation of a reference graph based on incident particle energy, which was then applied to the areal density measurement. The α-PIXE was necessary due to the thickness and composition of both the HPMC film and the Si based substrate involved.

The high impact resistance silicate coating, with Si and S, were characterized using proton PIXE, He^{++} RBS, TMAFM, and contact angle measurement. The results established the relation between the coating's condensation behavior, its surface free energy, topography, and hydrophilic/hydrophobic behaviors.

Polymer adsorption on the visor coating alters the surface water affinity, as well as the topography of the surface, and hence alters the water condensation behavior. The addition of an HPMC film on the Oakley visors helps to minimize the fogging problem.

Control of fogging on sports visors is effective via HPMC polymer adsorption. However, it was observed that the HPMC film wavelength was larger for sports visors than the film wavelength for the silica wafer. Thus, further investigation is needed to help explain the origin of this wavelength difference, whether the film application methodology, substrate type, or other undetermined factor is influencing this parameter.

REFERENCES

1. D. Lee, M. F. Rubner and R. E. Cohen, *Nano Letters,* 6 (10), 2305–2312 (2006).
2. Q. Xing, N. Herbots, M. Hart, J. D. Bradley, B. J. Wilkens, D. A. Sell, C. H. Sell, H. M. Kwong, R. J. Culbertson, and S. D. Whaley, *Proceeds of the 21st International Conference on the Application of Accelerators in Research and Industry,* to be published (August 8, 2010).
3. R. S. Faibish, *J. Colloid and Interface Sci.,* 256, 341-350 (2002).
4. A. Carre, *J. Adhesion Sci. Technol.,* 21 (10), 961-981 (2007).
5. B. Nsouli, M. Roumie, K. Zahraman, J. P. Thomas, M. Jaksic, Z. Pastuovic, P. Dole and M. Nasredine, *Nucl. Instr. And Meth. In Phys. Research B,* 198, 201-207 (2002).
6. H. Kabir, "Particle Induced X-ray Emission (PIXE) Setup and Quantitative Elemental Analysis", Ph.D. Thesis, Kochi University of Technology, 2007.
7. E. A. Preoteasa, C. Ciortea, B. Constantinescu, D. Fluerasu, S. E. Enescu, D. Pantelica, F. Negoita and E. Preoteasa, *Nucl. Instr. And Meth. In Phys. Research B,* 189, 426-430 (2002).
8. G. Demortier and J. L. Ruvalcaba-Sil, *Nucl. Instr. And Meth. In Phys. Research B,* 118, 352-358 (1996).
9. G. Demortier, S. Mathot and B. Van Oystaeyen, *Nucl. Instr. And Meth. In Phys. Research B,* 49, 46-51 (1990).
10. I. Brissaud, J. P. Frontier and P. Regnier, *Nucl. Instr. And Meth. In Phys. Research B*, 12, 235-244 (1985).
11. J. Miranda, *Nucl. Instr. And Meth. In Phys. Research B,* 118, 346-351 (1996).
12. E. Rame, *J. Colloid and Interface Sci.*, 185 (1), 245-251 (1997).
13. C. D. Volpe and S. Siboni, *Calculations of Acid-Base Surface Tension Components*: *SurfTen 4.3*, University of Italy, 2004: Retrieved from the Internet on July 28, 2010 from http://devolmac.ing.unitn.it:8080/mathpad4.html
14. C. Della Volpe, D. Maniglio, M. Brugnara, S. Siboni, and M. Morra, *Nucl. Instr. And Meth. In Phys. Research B,* 271, 434-453 (2004).
15. R. J. Culbertson and B. J. Wilkens, *Short Course on Ion Beam Analysis,* Tempe: Arizona State University, 2009, pp. 166.
16. W. J. Teesdale, J. A. Maxwell, A. Perujo, J. L. Campbell, L. Van der Zwan and T. E. Jackman, *Nucl. Instr. and Meth. In Phys. Research B,* 35, 57-66 (1988).
17. J. A. Maxwell, J. L. Campbell and W. J. Teesdale, *Nucl. Instr. and Meth. In Phys. Research B,* 43 218-230 (1989).

Hydrogen In Group-III Nitrides: An Ion Beam Analysis Study

K. Lorenz [1,2], S. M. C. Miranda [1], N. P. Barradas [1,2], E. Alves [1,2], Y. Nanishi [4], W. J. Schaff [5], L.W. Tu [6], V. Darakchieva [1,2,3]

(1) Instituto Tecnológico e Nuclear, E.N. 10, Sacavém 2686-953, Portugal
(2) Centro de Física Nuclear da Universidade de Lisboa, Av. Prof. Gama Pinto 2, 1649-003 Lisboa, Portugal
(3) IFM, Linköping University, SE-581 83 Linköping, Sweden
(4) Department of Photonics, Ritsumeikan University, Shiga 525-8577, Japan
(5) Department of Electrical and Computer Engineering, Cornell University, Ithaca, NY 14853, USA
(6) Department of Physics, National Sun Yat-Sen University, Kaohsiung 80424, Taiwan, ROC

Abstract. The doping mechanisms of InN, a promising material for novel optoelectronic and electronic devices, are still not well understood. Unintentional hydrogen doping is one possibility that could explain the unintentional n-type conductivity in high-quality nominally undoped InN films. We measured a series of state-of-the-art InN samples grown by molecular beam epitaxy with 2 MeV 4He-ERDA and RBS, showing the presence of relatively high amounts of hydrogen not only at the surface, but also in a deeper layer. Strong depletion of hydrogen due to the analysing beam was observed and taken into account in the analysis. Here, we report on the details of the analysis and show how the results correlate with the free-electron concentrations of the samples.

Keywords: ERDA; RBS; indium nitride; hydrogen
PACS: 82.80.Yc; 81.05.Ea; 02.60.C

INTRODUCTION

InN and related group-III-nitride alloys are key materials for contemporary and future optoelectronic and electronic devices such as high-brightness blue and white LEDs, multi-junction solar cells, high-frequency transistors, THz emitters and chemical sensors [1,2]. The doping mechanisms of InN are however still not well understood, and the origin of the unintentional n-type conductivity in InN is strongly debated. Unintentionally introduced hydrogen is a possible source of doping but until very recently it was believed that the amount of H in high-quality undoped InN films is insufficient to explain the observed free electron concentrations.

We measured a series of state-of-the-art InN samples grown by molecular beam epitaxy (MBE) with elastic recoil detection (ERDA) and Rutherford backscattering (RBS). Strong depletion of hydrogen due to the analysing beam was observed, and taken into account in the analysis. For that purpose, consecutive spectra were collected, and the results were interpreted with the bulk molecular recombination model (BMR) of Adel et al. [3], allowing us to determine the initial and final concentration of hydrogen. The presence of relatively high amounts of hydrogen not only at the surface, but also in a deeper layer, is demonstrated and quantified. Here, we report on the details of the analysis, and show how the results correlate with the free-electron concentrations of the samples.

EXPERIMENTAL DETAILS

State of the art unintentionally doped InN films with (0001) orientation (c-plane) and different thicknesses, from 350 nm to 1600 nm, were grown by MBE on sapphire substrates. Different buffers were used, including GaN buffer layers [4], GaN templates [5], and low temperature InN buffer layers [6].

ERDA experiments were made using a 2 MeV ^{4}He beam at 12° incidence with the surface of the sample. The Si surface barrier detector was located at a 24° scattering angle, so the outgoing angle is also 12°. Full experimental details are given elsewhere [7]. The depth resolution was around 35 nm at the surface, allowing us to clearly distinguish different layers within the depth probed.

The set-up had initially a 10 μm Kapton stopping foil. The large amount of ^{4}He beam particles that was forward scattered and stopped on the foil created hydrogen recoils in the foil, that were detected and led

Application of Accelerators in Research and Industry
AIP Conf. Proc. 1336, 310-313 (2011); doi: 10.1063/1.3586110

to a fairly large low energy background which reduced drastically the sensitivity to hydrogen in layers below the surface [8]. Therefore we replaced the Kapton by an 8 μm aluminium foil. RBS spectra were collected simultaneously with the ERDA, using a Si surface barrier detector located at a 160° scattering angle in the Cornell geometry. The beam was defined via a slit system. The beam height is fixed to be 0.6 mm, while its width can be adjusted between 0.2 and 1 mm. The area of the beam spot, necessary to calculate the beam fluence, was measured before each experiment by observing the beam spot on a Mylar foil.

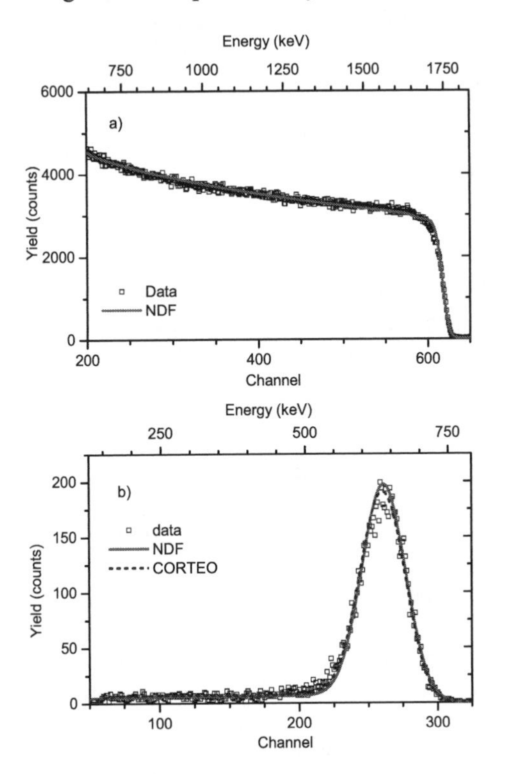

FIGURE 1. a) RBS and b) ERDA data for one InN sample irradiated with a fluence of 13.1×10^{15} $^{4}He/cm^{2}$. For the ERDA, the simulations calculated with NDF and the Monte Carlo code are nearly indistinguishable.

To study the depletion of hydrogen due to the analysing beam, spectra were collected consecutively in the same spot. While one spectrum for a given collected charge was stored, the beam was stopped by means of a shutter, so all the damage occurred during the actual measurements. From each sample, several spots were normally measured in this way. In total, 90 spots from 16 samples were measured, leading to many hundreds of spectra. We measured samples with different growth directions and present here the data for five c-plane samples. While H-incorporation in non-polar samples seems to show slightly different behaviour, the experimental aspects of the ERDA measurements including out-diffusion are comparable for differently oriented InN.

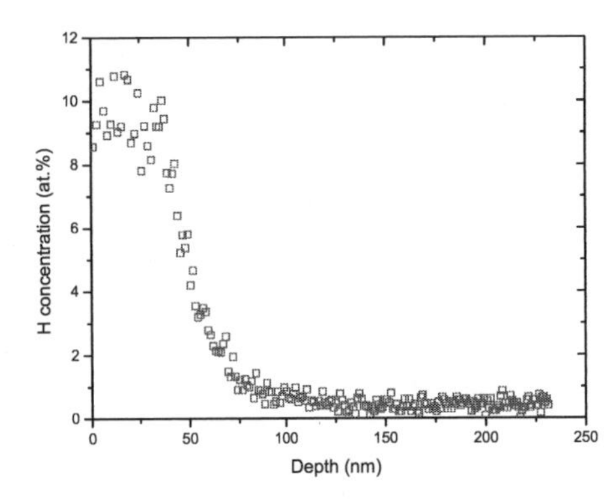

FIGURE 2. Hydrogen depth profile obtained by a direct conversion from the data using the known cross sections.

RESULTS AND DISCUSSION

We show in Fig. 1 the RBS and ERDA spectra collected for a c-plane sample for a fluence of 13.1×10^{15} $^{4}He/cm^{2}$, the highest value for this given sample. Two separate regions are clearly distinguished in the ERDA data: a surface layer around 50 nm thick, containing high amounts of hydrogen, and a deeper sub-surface layer extending to the largest depth probed, with much less hydrogen.

The RBS data is used to determine the collected charge and thus normalise the ERDA data. This process relies on the known dead time corrections of the RBS and ERDA spectra; on an accurate calculation of pile-up [9], which changes the observed yield; and on the stopping power of ^{4}He in InN. Given that using SRIM with the Bragg rule may lead to fairly large inaccuracies, we used experimental values for the stopping power that we measured for this purpose [10].

The data were analysed with the code NDF, which is capable of performing automated fits [11,12]. The maximum depth accessed by the experiment is ≈230 nm. For such depths the spectra collected for the initial low beam fluences have low statistics, with less than 1 count per channel on average. In this case, a χ^2 minimisation of the distance between the simulated curve and the data is very sensitive to the exact number of counts in each channel, and can easily lead to differences between the integral simulated and observed yield around 10%. Therefore, to analyse the data, first we made an automated fit that led to a good first approximation. Then, we manually adjusted the hydrogen content of the surface and of the deeper layer in order to obtain the same simulated and experimental total yield in the surface peak and in the low energy region, respectively.

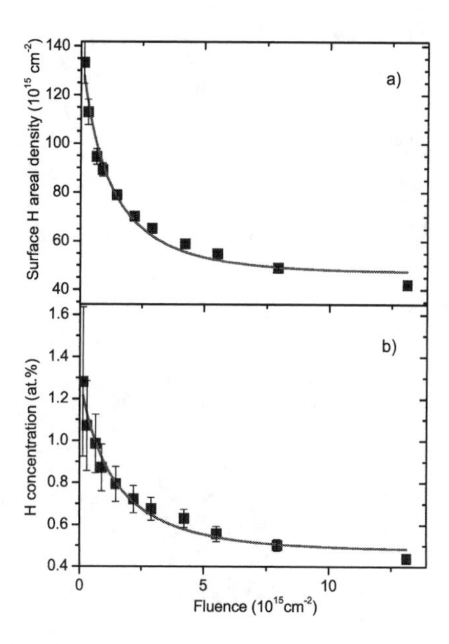

FIGURE 3. a) Surface hydrogen areal density and b) hydrogen concentration in the sub-surface region for one given InN sample as a function of the ^{4}He fluence. The solid lines are the fits using the bulk molecular recombination model.

The simulations included the effect of double scattering events [8], which lead to a low energy background that must be considered in order to determine quantitatively the low hydrogen concentrations below the surface. To check the accuracy of the simulations made, we used the Monte Carlo code CORTEO [13], that intrinsically includes effects such as higher order plural scattering and multiple scattering, that can have an important influence on the result, at least in some experimental conditions such as ERDA with heavy ion beams [14]. The simulation obtained for the exact same depth profile as obtained with NDF is also shown in Fig. 1b). It is extremely close to the NDF simulation, which is about 3 orders of magnitude faster even including the dual scattering calculation, validating its results.

The hydrogen profile derived from the ERDA data in Fig. 1 is shown in Fig. 2. In this case, each data point was directly converted into a concentration value simply by using the known scattering cross section [15], after subtraction of the background due to double scattering and pile-up.

The hydrogen surface areal density and the hydrogen concentration in the ≈230 nm thick layer below the surface are shown in Fig. 3 as a function of the beam fluence. Initially, there is a rapid decrease of the hydrogen content, which then stabilises. We analysed the data with the BMR model [3], which is based on the assumption that hydrogen effusion is due to breaking up hydrogen bonds due to the energy deposited by the beam, leading to recombination into H_2 molecules that easily escape the sample. According to this model the data points in Fig. 3 were fitted using equation

$$\rho(\Phi) = \left[\frac{1}{1/\rho_f + (1/\rho_0 - 1/\rho_f)\exp(-K\Phi)}\right],$$

where ρ_0 and ρ_f are the initial and final hydrogen concentrations and K is an effective molecular release cross section.

The BMR model has been previously applied to depletion of nitrogen in InN films irradiated with heavy ions [16]. In the same work, it was shown that nitrogen depletion only occurred for irradiation with ions as heavy as Ag or Au, and was not observed for irradiation with F and S beams. Therefore, we do not expect any loss of nitrogen due to irradiation with ^{4}He. Furthermore, ^{35}Cl-ERDA measurements on some samples have shown a N to In ratio very close to 1 within errors [17].

From the BMR model analysis we determine, for this sample, the release cross section, which is the fundamental parameter in the BMR model, to be K=3.5 Å^2 for depletion from the surface, and K=5.2 Å^2 for depletion from the deeper layer. These values have a large error due to the uncertainty in the beam cross section. Also, there is a fairly large variability, within a factor of 2, in the K values determined for different c-plane InN films, and even for different spots of the same sample. Nevertheless, the range of values obtained is of the order of magnitude expected [18,19].

We also determine the initial and final hydrogen concentration in the deeper sub-surface layer, to be 1.8 and 0.45 at.%, respectively. Large variations from sample to sample were observed, depending on the crystal orientation. Typically, c-plane films show final bulk concentrations from 0.2 to 0.45 at%. We note that, in different spots of the same sample, we have measured variations of up to 30%, which can be due to slightly different conditions of the sample surface.

We show in Fig. 4 the dependence of the bulk free electron concentration on the hydrogen concentration of the sub-surface layer in c-plane InN, showing scaling between the bulk electron concentration and the H concentration measured at depths extending to 230 nm. In Fig. 4 we used the hydrogen concentration with the highest charge corresponding to a fluence where out-diffusion already saturated since here the uncertainty is easier to estimate than for the initial point which relies on the validity of the BMR model and is subject of high statistical error. In fact, according to the BMR model hydrogen release ceases if H atoms are separated by more than a characteristic recombination distance. However, this would mean

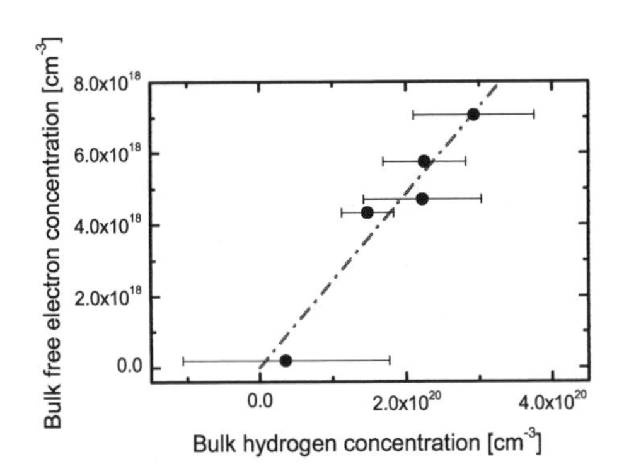

FIGURE 4. Bulk free electron concentrations vs. hydrogen concentration in the sub-surface region. The value for the film with the lowest free electron concentration represents the difference between hydrogen and carbon concentrations in the film [17].

derived from the last measured point that the saturation level should be the same for all InN samples, which is not observed, indicating that the BMR model may not be applicable to the present case despite the good fit to the experimental data and the reasonable values for the effective molecular release cross section K. The varying final hydrogen concentration for different samples suggests different incorporation sites for H in the InN lattice and out-diffusion takes place preferentially for weakly bonded H, possibly H decorating dislocations or grain boundaries. Also the presence of H_2 molecular hydrogen in the samples cannot be excluded. Assuming that higher out-diffusion takes place for samples with higher H concentrations the correlation of H vs. free electron concentration would be even more pronounced if the initial H-concentrations could be used. Our findings thus support the important role of unintentional hydrogen doping in the n-type conductivity of InN films [20].

SUMMARY

We studied the hydrogen content of MBE grown InN with ion beam analysis techniques, namely RBS and ERDA with a ^{4}He beam. Strong depletion of the hydrogen content under the analysing beam was observed and a careful analysis of the data was necessary to obtain the hydrogen concentration. Two regions are clearly observed, one at the surface (up to 50 nm depth) with very high amounts of hydrogen, that are likely to be due to surface contamination, and a sub-surface region extending at least to the observable depth of around 230 nm, with smaller hydrogen concentrations, around typically the 1 at.% order of magnitude.

ACKNOWLEDGMENTS

This work is partially supported by FCT Portugal under contract PTDC/FIS/100448/2008 and program Ciência 2007. We acknowledge support from the Swedish Research Council (VR) under grant No.2005-5054. We thank Filomena Baptista for technical support. NPB would like to thank the International Atomic Energy Agency for support under Research Contract No. 14365.

REFERENCES

1. E. F. Schubert, *Light-emitting Diodes*, Cambridge, UK, Cambridge University Press, 2003.
2. J. Wu, *J. Appl. Phys.* **106**, 011101 (2009).
3. M. E. Adel, O. Amir, R. Kalish, L. C. Feldman, *J. Appl. Phys.* **66**, 3248 (1989).
4. H. Lu, W. J. Schaff, J. Hwang, H. Wu, G. Koley, and L. F. Eastman, *Appl. Phys. Lett.* **79**, 1489 (2001).
5. C.-H. Liang, et al, *Appl. Phys. Lett.* **90**, 172101 (2007).
6. Y. Nanishi, Y. Saito, T. Yamaguchi, M. Hori, F. Matsuda, T. Araki, A. Suzuki, and T. Miyajima, *Phys. Status Solidi A* **200**, 202 (2003).
7. V. Darakchieva, et al, *Appl. Phys. Lett.* **96**, 081907 (2010).
8. N. P. Barradas, K. Lorenz, V. Darakchieva, E. Alves, these proceedings.
9. S. Molodtsov, A. F. Gurbich, *Nucl. Instrum. Methods B* **267**, 484 (2010).
10. N. P. Barradas, E. Alves, Z. Siketić, I. Bogdanović Radović, these proceedings.
11. N. P. Barradas, C. Jeynes, and M. A. Harry, *Nucl. Instrum. Methods B* **136-138**, 1163 (1998).
12. N. P. Barradas, C. Jeynes, K. P. Homewood, B. J. Sealy, and M. Milosavljevic, *Nucl. Instrum. Methods B* **139**, 235 (1998).
13. F. Schiettekatte, *Nucl. Instrum. Methods. B* **266**, 1880 (2008).
14. N. P. Barradas, et al, *Nucl. Instrum. Methods B* **262**, 281 (2007).
15. A. F. Gurbich, *Nucl. Instrum. Methods B* **268**, 1703 (2010).
16. S. K. Shrestha, H. Timmers, *Nucl. Instrum. Methods B* **249**, 257 (2006).
17. V. Darakchieva, N. et al, *Physica B* **404**, 4476 (2009).
18. S. K. Shrestha, K. S. A. Butcher, M. Wintrebert-Fouquet, H. Timmers, *Nucl. Instrum. Methods B* **219-220**, 686 (2004).
19. M. Abdesselam, J. P. Stoquert, S. Chami, M. Djebara, A. C. Chami, M. Siad, *Nucl. Instrum. Methods B* **267**, 108 (2009).
20. C. S. Gallinat, G. Koblmüller, J.S. Speck, *Appl. Phys. Lett.* **95**, 022103 (2009).

A Double Scattering Analytical Model For Elastic Recoil Detection Analysis

N. P. Barradas [1,2], K. Lorenz [1,2], V. Darakchieva [1,2,3], E. Alves [1,2]

[(1)] *Instituto Tecnológico e Nuclear, E.N. 10, Sacavém 2686-953, Portugal*
[(2)] *Centro de Física Nuclear da Universidade de Lisboa, Av. Prof. Gama Pinto 2, 1649-003 Lisboa, Portugal*
[(3)] *IFM, Linköping University, SE-581 83 Linköping, Sweden*

Abstract. We present an analytical model for calculation of double scattering in elastic recoil detection measurements. Only events involving the beam particle and the recoil are considered, i.e. 1) an ion scatters off a target element and then produces a recoil, and 2) an ion produces a recoil which then scatters off a target element. Events involving intermediate recoils are not considered, i.e. when the primary ion produces a recoil which then produces a second recoil. If the recoil element is also present in the stopping foil, recoil events in the stopping foil are also calculated. We included the model in the standard code for IBA data analysis NDF, and applied it to the measurement of hydrogen in Si.

Keywords: double scattering; ERDA; indium nitride
PACS: 82.80.Yc; 81.05.Ea; 02.60.Cb

INTRODUCTION

Data analysis of techniques such as Rutherford backscattering (RBS) or elastic recoil detection (ERDA) relies on accurate simulations of the spectra expected for a given sample structure. In complex cases, effects such as plural and multiple scattering have to be taken into account [1,2], and analytical calculations to simulate those effects are often not available.

In Monte Carlo simulations (MC), in principle, all effects can be included, including the exact geometrical configuration of the detection system [3,4]. In practice, the calculations can be very slow, and so-called acceleration techniques [3,5,6] have been developed to increase the efficiency of the calculations. These include using in the MC simulations virtual (larger) detectors, restricting the possible scattering angles, and artificially increasing the mean free paths of the beam particles. This means that what one calculates is not necessarily equivalent to what one measures, particularly at low energies where the very high cross sections make MC unpractical, and at very grazing angles where the acceleration techniques can lead to strong distortions in the calculations. Nevertheless, MC methods have proven very successful in making detailed calculation of plural and multiple scattering effects [7]. However, even if the codes are in principle available, practically all the publications involving MC have been made by the authors of the codes. The recent development of a user friendly Windows interface for the MC code Corteo [6] is however bound to increase its usage.

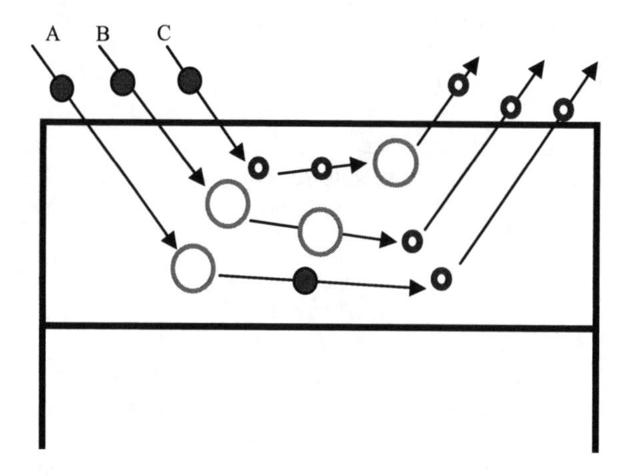

FIGURE 1. Double scattering trajectories in ERDA. The figure represents the target film..Full circles represent the primary ion. Small open circles represent the detected recoil. Large open circles represent a target atom other than the recoil.

Nevertheless, MC simulation is still not appropriate for widespread routine data analysis, for which analytical codes continue to be the first choice. Improving the quality of the analytical simulations is

Application of Accelerators in Research and Industry
AIP Conf. Proc. 1336, 314-318 (2011); doi: 10.1063/1.3586111

thus essential. Double scattering (DS), which is the simplest particular case of plural scattering, leads to a low energy background which, in RBS, decreases the sensitivity to low Z elements present in the sample. In ERDA, it reduces the sensitivity to the presence of the recoil atom in deeper layers. We have previously developed an analytical model of double scattering for RBS [8]. We now extended it to ERDA, and included it in the code NDF [9,10]. Here we give details of the calculation method, and show its relevance in the measurement of hydrogen in Si and in indium nitride with 4He-ERDA.

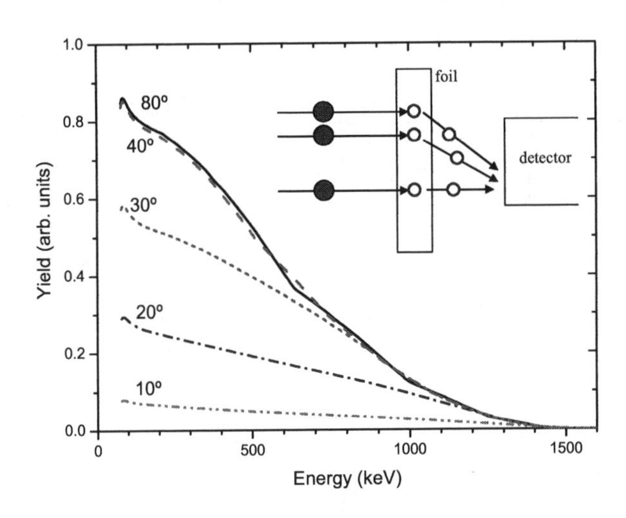

FIGURE 2. Contribution of production of recoils in the stopping foil for a 165 nm thick SiH0.20 target film on Si. A 1.25×1020 at./cm2 Mylar stopping foil is considered. Each curve is for one maximum value of the recoil angle in the stopping foil. The insert shows the geometry and different possibilities. Clearly, the result will depend on the exact dimensions of stopping foil and detector and distance between theml. Large open circles represent a target atom other than the recoil.

DOUBLE SCATTERING MODEL

The first analytical calculations of DS in RBS were presented by Weber et al. [11]. They considered only normal incidence, and imposed a minimum scattering angle α_{min}=15°. The Rutherford cross section is not defined for a 0° scattering angle, and it has only been experimentally verified for scattering angles down to 15° [12]. Eckstein and Mayer generalised the algorithm to any geometry [13], using 120 solid angle intervals (i.e. for the direction of the beam after the first scatter event), and α_{min}=20°, obtaining good agreement with experiments made close to normal incidence. Barradas [8] showed that the concept of a minimum scattering angle is not appropriate in grazing angle geometry, because small changes in angle can lead to large changes in the actual trajectory of the ions. Instead, one must consider the deviation of each trajectory from the corresponding single scattering trajectory. The resulting algorithm leads to excellent results in grazing angle RBS, where DS can be very large. If one imposes instead α_{min} around 20°, a calculated low energy background with approximately the correct shape is obtained, but one order of magnitude too low. This is relevant for ERDA, which is usually made in grazing angle geometry.

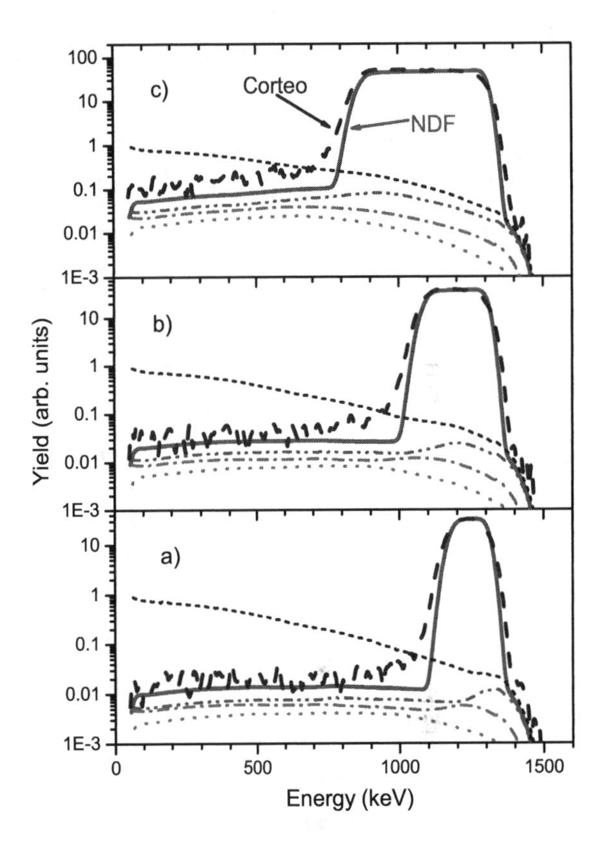

FIGURE 3. Calculations for target films on Si: a) 165 nm $SiH_{0.20}$, b) 275 nm $SiH_{0.23}$, c) 550 nm $SiH_{0.30}$. Thick solid and dashed lines are the NDF and Corteo calculations. Dash-dotted and dash-dot-dotted are the contributions of trajectories of type A and C, as calculated with NDF. The short dashed lines are the contribution of production of recoils in the Mylar stopping foil. The dotted lines are the NDF calculations imposing a hard 20° cut-off angle.

Repplinger et al. presented an analytical calculation for ERDA, where α_{min} was calculated for each incident ion/target pair on the grounds of an analogy between nuclear and electronic energy loss [14]. It is not stated in the paper how many different solid angle intervals were calculated. They included in the calculations recoils produced by forward scattered primary beam ions in the stopping foil used to prevent the primary ^4He beam to reach the detector, which also contained hydrogen. The low energy background thus calculated had approximately the correct shape, but it was 6 to 20 times smaller than the observed background, for SiH_x target films of thickness between 165 and 550 nm on Si and x between 0.2 and 0.3.

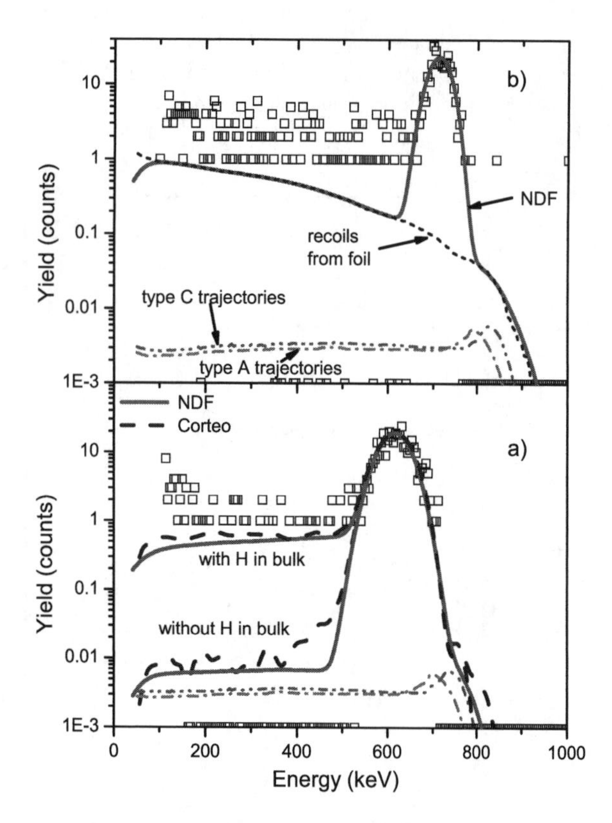

FIGURE 4. Data (squares) collected from the same Si sample, with a) aluminium stopping foil and b) Kapton stopping foil. The lower and upper thick solid and dashed lines in a) are the NDF and Corteo calculations without H in the bulk and with 0.06 at.% H in the bulk. The thick solid line in b) is the NDF calculation assuming no H in the bulk. Dash-dotted and dash-dot-dotted are the contributions of trajectories of type A and C, as calculated with NDF. The short dashed line in b) is the contribution of production of recoils in the stopping foil calculated with NDF.

We extended the DS algorithm previously developed for RBS [8] to ERDA. The algorithm is general, without any restrictions on incident beam, detected particle, number of elements and of layers in the target, or detection geometry. The three main changes towards the RBS algorithm are the following:

1. In RBS, it is the same ion that undergoes the two scattering events. In ERDA, as shown in Fig. 1, there are different possibilities. In trajectory A, a primary beam ion is scattered off a target atom, and then proceeds to create the recoil which is detected. In trajectory B, the primary ion creates an intermediate recoil, different from the detected particle, and the intermediate recoil then creates the recoil which is detected. In C, the primary ion creates the recoil, which then scatters off a target atom before being detected. We consider only cases A and C, i.e., we ignore case B where two consecutive recoils are produced. The ERDA cross section is only high for large recoil angles, for which the kinematic factor is low; for small recoil angles, the kinematic factor is high but the cross section is low. That is, two consecutive recoil events lead with high probability to a final recoil with energy too small to be detected; conversely, they only lead to a final recoil with energy high enough to be detected with very small probability.

2. The recoil cross section is strongly dependent on the scattering angle. For ^4He on hydrogen, the cross section deviates strongly from Rutherford. Furthermore, in events of type C, where the recoil is scattered off target atoms, if the recoil is hydrogen, the cross section for scattering on light elements is almost always very strongly non-Rutherford, with a strong angular dependence. The algorithm developed considers all this, by introduction of cross section tabular data on angle and energy for each reaction. Many of the most used cross sections are included in NDF, and it is easy to introduce new ones as needed, particularly with SigmaCalc and IBANDL [15].

3. Production of recoils in the stopping foil by forward scattered ions is also calculated. However, consider the insert of Fig. 2. It is clear that, depending on the exact geometrical configuration of the stopping foil and of the detector, and distance between them, different angles of scattering in the stopping foil can lead to the detection of particles. An integration on the recoil angle must be made, which depends on the set-up used, and on the exact configuration of any slits that may exist. Fig. 2 shows the calculated contribution to DS of recoils produced in the stopping foil, for a 165 nm thick $SiH_{0.2}$ target film on Si measured with 2.9 MeV ^4He and a 1.25×10^{20} at./cm^2 Mylar ($C_{10}O_4H_8$) stopping foil, for different maximum recoil angles, assuming scattering in a cone around the nominal beam direction, which is an ideal situation that real set-ups often do not match. Up to 40°, there is a fast increase of the calculated DS yield, which then stabilises because recoils produced in the stopping foil at a larger angle are stopped within the stopping foil itself.

RESULTS AND DISCUSSION

We show in Fig. 3 calculations corresponding to three SiH_x target films: 165 nm with x=0.2, 275 nm with x=0.23, and 550 nm with x=0.3, measured with 2.9 MeV ^4He and a 1.25×10^{20} at./cm^2 Mylar stopping foil, that is, corresponding to what Repplinger et al. showed in Fig. 2 of [14]. Angle-dependent non-Rutherford cross sections were used for ^{4}He on ^{1}H [16], and ^{1}H on Si [17] and on hydrogen [18].

Some of the NDF calculations presented in Fig. 3 do not include the recoils produced in the stopping foil, in order to compare with simulations made with the MC code Corteo [6] (we note that Corteo, in its current version, does not calculate recoils produced in the stopping foil), where 1×10^8 trajectories were calculated without acceleration techniques. In general terms, the NDF and Corteo simulations are very close; the width of the signal calculated with Corteo is slightly larger, and there are some extra counts on the low energy side of the large hydrogen signal. These effects are due to multiple scattering, as shown previously for RBS [1]. More importantly, the shape of the DS background calculated with NDF is very close to Corteo. Quantitatively, the yield calculated with NDF is too small, as the MC simulation is larger by a factor up to 2. This is due to events that involve more than two large angle collisions, that is, plural scattering with 3, 4, 5, or more scattering events [1], that are not included in the NDF dual scattering calculation.

We also show in Fig. 3 the contribution to DS of type A and type C trajectories, as well as the contribution due to recoils produced in the stopping foil with a maximum angle of 40°. We note that, even if we included all the contributions due to DS, by adding the Corteo result with the NDF stopping foil results, the simulated yield (not shown here) would still be much smaller than the data reported in ref. [14]. This can be due to some effect not included in the simulations shown here, but it can also be due to roughness of the samples analysed or even to actual diffusion of hydrogen to the Si substrate.

A Si sample was measured using a 2 MeV ^{4}He beam in consecutive days, in different spots in order to ensure that hydrogen loss, if it occurs, during the experiment was not affecting the measurement. The only difference between the two measurements is that on one day the stopping foil was Kapton, and on the next day it was Al. The data are shown in Fig. 4. We first consider the simulations made with NDF assuming that hydrogen is only present in the surface; for both stopping foils the calculated yield at energies lower than the surface peak is below the observed data. Believing that the simulations are actually correct, this would mean that a sub-surface layer around 300 nm thick has a small hydrogen content, with 0.06 at.% leading to a calculated integrated low energy yield equal to the experimental data collected with the Al stopping foil (see Fig. 4). A concentration of hydrogen in the stopping foil around 8 at.% would be needed to explain the data, which is not realistic. We note however that we cannot exclude effects such as scattering in the chamber, in the slits, in the detector aperture, or even electronics noise in the system, that might lead to the observed background. The 0.06 at.% must then be taken as the maximum hydrogen concentration below the surface. The sensitivity to hydrogen in the bulk is much better with the Al stopping foil, on the one hand because of the smaller background, and on the other hand due to the difficulties in the calculation of the hydrogen recoils produced in the stopping foil. It is clear that, in ^{4}He-ERDA, a hydrogen-free stopping foil is essential to measure small quantities of hydrogen [19]. The Corteo simulations, made with exactly the same hydrogen concentrations, are very close to the simulations made with NDF. As in the previous example, they are slightly larger.

Finally, data collected from one InN sample using the Al stopping foil is shown in Fig. 5. Full experimental details are given elsewhere [20,21]. The angle dependent cross section for 1H on N was taken from [22]. Again, the simulations made with both NDF and Corteo that assume no hydrogen is present in the sub-surface layer are one order of magnitude smaller than the observed data. Adjusting the hydrogen concentration in the bulk until the background calculated with NDF has the same integrated area as the data, we can determine that the hydrogen concentration is 0.17 at.%. The determination of such small values is relevant, because the free electron carrier concentration in unintentionally doped InN scales with the bulk hydrogen concentration [20].

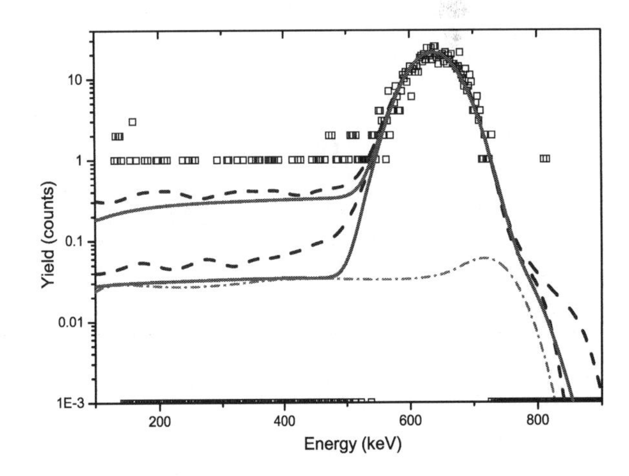

FIGURE 5. Data collected from an InN sample. The solid lines are calculated with NDF, and the dashed lines with Corteo. The lower lines assume that hydrogen is only present in the surface. The upper lines include 0.17 at.% hydrogen in the bulk.

SUMMARY

We developed an analytical algorithm to calculate the contribution of double scattering events in ERDA spectra. The model includes events where the primary ion first scatters off any of the target atoms, then produces a recoil, and events where the recoil is produced in the first scattering event, and then is scattered off any of the target atoms before being detected. The angular dependence of non-Rutherford cross sections is taken into account whenever the necessary data are available. Comparisons with Monte Carlo calculations show that the analytical calculations underestimate the low energy background by a factor up to 2, which is due to multiple scattering events not included in the analytical calculation.

We applied the code to measurements of hydrogen in Si made with an Al and a Kapton stopping foil. For the Kapton stopping foil the sensitivity to the hydrogen content of sub-surface layers is poor. For the Al stopping foil the sensitivity is much improved, and strong limits on the maximum hydrogen bulk content can be established. Finally, we show that small concentrations of hydrogen in InN can be determined with good accuracy using a standard ERDA set-up with a 2 MeV ^{4}He beam.

ACKNOWLEDGMENTS

NPB would like to thank the International Atomic Energy Agency for support under Research Contract No. 14365, Dr François Schiettekatte for useful discussions on Monte Carlo calculations, and the four at Praia da Foz for strong encouragement to continue to work.

REFERENCES

1. P. Bauer, E. Steinbauer, J.P. Biersack, *Nucl. Instrum. Methods B* **64**, 711 (1992).
2. N.P. Barradas, et al., *Nucl. Instrum. Methods B* **262**, 281 (2007)
3. K. Arstila, T. Sajavaara, J. Keinonen, *Nucl. Instrum. Methods B* **174**, 163 (2001).
4. R.D. Franich, P.N. Johnston, I.F. Bubb, *Nucl. Instrum. Methods B* **219-220**, 87 (2004).
5. R.D. Franich, P.N. Johnston, I.F. Bubb, N. Dytlewski, D. D. Cohen *Nucl. Instrum. Methods B* **190**, 252 (2002).
6. F. Schiettekatte, *Nucl. Instrum. Methods B* **266**, 1880 (2008) .
7. S. Giangrandi, K. Arstila, B. Brijs, T. Sajavaara, A. Vantomme, W. Vandervorst, *Nucl. Instrum. Methods B* **257**, 1936 (2009).
8. N.P. Barradas, *Nucl. Instrum. Methods B* **225**, 318 (2004).
9. N.P. Barradas, C. Jeynes, and M.A. Harry, *Nucl. Instrum. Methods B* **136-138**, 1163 (1998).
10. N.P. Barradas, C. Jeynes, K.P. Homewood, B.J. Sealy, and M. Milosavljevic, *Nucl. Instrum. Methods B* **139**, 235 (1998).
11. A. Weber, H. Mommsen, W. Sarter, A. Weller, *Nucl. Instrum. Methods* **198**, 527 (1982).
12. H. H. Andersen, F. Besenbacher, P. Loftager, W. Möller, *Phys. Rev. A* **21**, 1891 (1980).
13. W. Eckstein, M. Mayer, *Nucl. Instrum. Methods B* **153**, 337 (1999).
14. F. Repplinger, J. P. Stoquert, P. Siffert, *Nucl. Intrum. Methods B* **80/81**, 24 (1993).
15. Ion Beam Analysis Nuclear Data Library, which also hosts the SigmaCalc code by Prof. A. F. Gurbich. http://www-nds.iaea.org/ibandl/
16. A.F. Gurbich, *Nucl. Instrum. Methods B* **268**, 1703 (2010).
17. A.F. Gurbich, *Nucl. Instrum. Methods B* **145**, 578 (1998).
18. H.R. Worthington, J.N. Mcgruer, D.E. Findley, *Physical Review* **90**, 899 (1953).
19. A. Tripathi, O. Kruse, H.D. Carstanjen, *Nucl. Instrum. Methods B* **219-220**, 435 (2004).
20. V. Darakchieva, et al., *Appl. Phys. Lett.* **96**, 081907 (2010).
21. K. Lorenz, N. P. Barradas, E. Alves, F. Munnik, Y. Nanishi, W. J. Schaff, L.W. Tu, V. Darakchieva, these proceedings.
22. I. Bogdanović Radović, Z. Siketić, M. Jaksic, A.F. Gurbich, *J. Appl. Phys.* **104**, 074905 (2008).

Stopping Power Of He, C And O In InN

N. P. Barradas [1,2], E. Alves [1,2], Z. Siketić [3], I. Bogdanović Radović [3]

[(1)] Instituto Tecnológico e Nuclear, E.N. 10, Sacavém 2686-953, Portugal
[(2)] Centro de Física Nuclear da Universidade de Lisboa, Av. Prof. Gama Pinto 2, 1649-003 Lisboa, Portugal
[(3)] Ruđer Bošković Institute, P.O. Box 180, 10002 Zagreb, Croatia

Abstract. Group III nitrides such as InN, GaN, and their alloys are increasingly important in a host of optoelectronic and electronic devices. The presence of unintentional impurities is one of the factors that can strongly affect the electronic properties of these materials, and thus ion beam analysis techniques can play a fundamental role, in particular heavy ion elastic recoil detection analysis tracing and quantifying these contaminations. However, stopping powers in InN and GaN have not yet been measured, and data analysis relies on using the Bragg rule, which is often inaccurate. We have used a bulk method, previously developed by us and applied successfully to other systems, to determine experimentally the stopping power of several ions in InN. The results of our measurements and bulk method analysis are presented.

Keywords: stopping power; indium nitride, heavy ions
PACS: 34.50.Bw, 82.80.Yc, 81.05.Ea

INTRODUCTION

Group III nitrides such as InN and GaN and their alloys are very important materials in the field of optoelectronic and electronic devices such as high-brightness blue and white LEDs, multi-junction solar cells, high-frequency transistors, THz emitters and chemical sensors [1]. Unintentional impurities are often present, strongly affecting the electronic properties of these materials [2]. Heavy ion elastic recoil detection analysis (HI-ERDA) is ideally suited to determine the profile of light impurities in a heavy matrix such as InN [3]. The accuracy of the depth profiles obtained depends critically on the stopping powers used. However, to our knowledge, no experimental values of stopping power of any ion in InN have been reported so far, and popular interpolative schemes such as SRIM [4] used together with the Bragg rule may lead to large deviations around 10% or more, particularly for oxides, nitrides, and for heavy ions where data are very sparse or inexistent.

We have measured the stopping power of ^{4}He, ^{12}C and ^{16}O in InN thin films grown on sapphire. He is often used in Rutherford backscattering (RBS) studies, while C and O are common impurities. We applied a method previously developed [5,6], where the stopping power curve is taken as a fit parameter, using the effect it has on the height and width of the film signal, together with a well known marker layer to determine the solid-angle.charge product. A Bayesian inference [7] Markov chain Monte Carlo (BI/MCMC) algorithm, included in the code NDF [8,9] is used, , leading to determination of the uncertainty of the stopping curve. NDF is a standard code for IBA data analysis, that participated in the International Atomic Energy Agency intercomparison of IBA software [10].

EXPERIMENTAL DETAILS

An InN film with nominal thickness 400 nm grown by molecular beam epitaxy on a r-plane sapphire substrate was used. A 10 nm thick Au marker layer was deposited on top by electron beam evaporation, in order to serve as an internal reference between different experiments. ^{4}He RBS experiments were done at Instituto Tecnológico e Nuclear at a 160° scattering angle with beam energies 1, 2 and 2.3 MeV. The ^{12}C and ^{16}O RBS experiments at 3, 6, 10 and 15 MeV were done at the Ruđer Bošković Institute using three detectors simultaneously, at 118°, 150° and 165° scattering angles [6].

CALCULATIONS

We performed a Bayesian inference analysis with the Markov chain Monte Carlo method (BI/MCMC). The details of the method have been given previously

Application of Accelerators in Research and Industry
AIP Conf. Proc. 1336, 319-322 (2011); doi: 10.1063/1.3586112

[5,6] and are beyond the scope of this paper. Basically, the method relies on simulating the collected spectra. The stopping power curve (of InN as a compound, not of In and N individually) is a free parameter, constrained to reasonable shapes by using the ZBL85 parameterisation [11]. We note that this parameterisation is used only because it provides a convenient way of obtaining curves that have fairly realistic shapes (for instance, one single maximum), based on a few (a1-a8) parameters that can be automatically changed. No fitting is done; instead, very many simulations are produced (50000 for each ion in this work), each based on a different set of a1-a8 ZBL85 parameters and thus with a different stopping power curve, all of them consistent with the data within the errors. Besides the stopping power, other experimental parameters such as the energy calibration and the solid-angle.charge product are varied within their uncertainties. In the end, the average of all the stopping power curves that lead to simulations consistent with the data is taken as final result, and their standard deviation is the associated uncertainty.

This process relies on the quality of the simulations produced. NDF includes the effect of double scattering [12], pile-up [13], and we analysed pulse-height spectra instead of energy spectra, i.e. we took into account the pulse height defect coming from the detector dead layer and non-ionising energy losses in the detector [14].

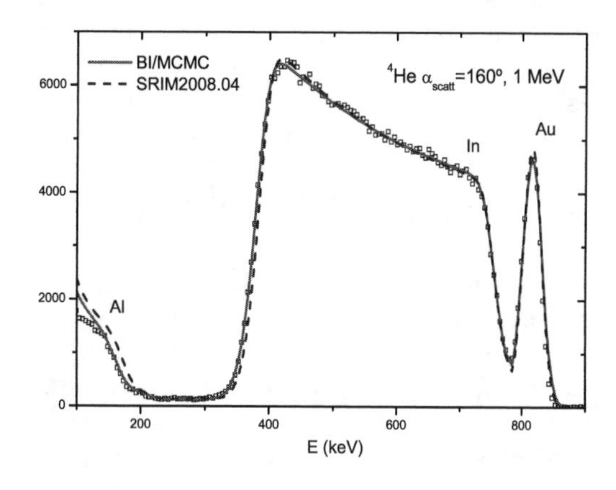

FIGURE 1. Spectrum collected with 1 MeV 4He. Simulations using SRIM stopping (dashed) and the stopping curve determined here (solid) are shown.

RESULTS AND DISCUSSION

The ^{4}He experiment has the double purpose of determining the stopping power as well as the InN areal density. To do that, we need to rely on the Al_2O_3 substrate signal, which depends on the stopping of ^{4}He on Al2O3. In the analysis, we used the experimental values for the molecular stopping of ^{4}He on Al_2O_3 [5]. The associated uncertainty is propagated to the uncertainty in the InN areal density, and thus to the uncertainty in the stopping of not only ^{4}He, but also ^{12}C and ^{16}O, in InN. Another source of uncertainty is the InN stoichiometry. In low quality MBE InN samples, N:In ratios around 1.05 have been reported [3]. This leads to changes in the stopping power around 1%. The results shown below include these sources of uncertainty.

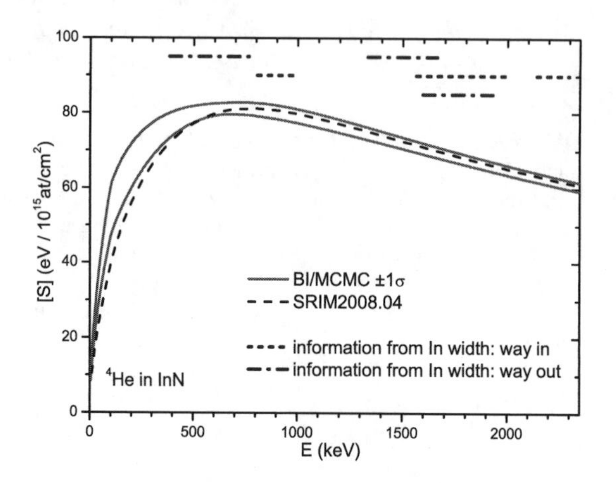

FIGURE 2. Limits of confidence of the stopping power of ^{4}He in InN (average plus minus one standard deviation, solid lines). Stopping calculated with SRIM and the Bragg rule is also shown (dashed). The regions of energy covered by the In signal (short dashed for the way in and dash-dotted for the way out) are shown. These results are valid in the 380-2300 keV range accessed by the beam inside the InN film.

We show in Fig. 2 the best fit of the 1 MeV spectrum of ^{4}He on InN, using SRIM stopping and the values now determined. First of all, the Au marker is clearly visible, while being separated from the In signal. The In signal is not superimposed to the N or to the background signal from the Al_2O_3. The information about the stopping power is extracted from the width of the In signal, and, in this case, also from the position of the Al edge. A fundamental point is that the energy calibration is the same as for the 2 and 2.3 MeV spectra, given that the settings were not changed and the experiments were made in one run, and therefore cannot be adjusted independently to give a better fit in this spectrum only. For this purpose, the pulse height deffect must be calculated, because it is a non-linear effct that depends on the energy of the detected ions [13]. If it is ignored, the energy calibration is not linear, and it is not possible to fit the low and high energy spectra simultaneously. It is clear that the Al edge cannot be well fitted using SRIM stopping, as the simulated edge is at a energy higher than the data. On the other hand, the low energy In signal as simulated with SRIM is also at an higher energy than measured. These two facts together show that, at the fairly low energies involved, SRIM underestimates the stopping power.

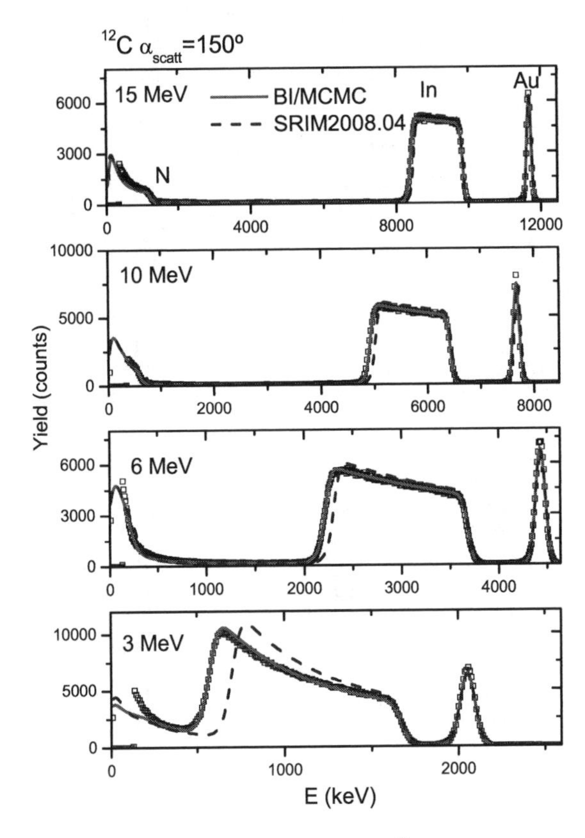

FIGURE 3. Spectra collected with ^{12}C with the 150° detector. Simulations using SRIM stopping (dashed) and the stopping curve determined here (solid) are shown.

We show in Fig. 2 the stopping power of ^{4}He on InN; what we show is the average value of the 50000 stopping power curves calculated, plus one and minus one standard deviation. We also show the values calculated with SRIM 2008.04 and the Bragg rule. The newly determined values are in excellent agreement with the SRIM prediction in the high energy range, and slightly above SRIM below the Bragg peak. The information on stopping power is mostly derived from the In signal, and the lowest energy of the ^{4}He beam, after scattering off In, is around 380 keV. This leads to a high uncertainty below that value, where the only signal that provides any information is the position of the Al leading edge in the 1 MeV spectrum, which is at around 150 keV. We therefore state that the measurement presented is valid between 380 and 2300 keV.

We show in Fig. 3 the spectra collected at 150° with the ^{12}C beam, together with simulations made using SRIM stopping. Clearly, SRIM stopping is too low to explain the data: The width of the In signal as calculated with SRIM stopping is much too small, which means that SRIM calculates an energy loss in the InN that is much smaller than observed.

The stopping curve determined with BI/MCMC is shown in Fig. 4, where the uncertainty includes the uncertainty in the InN areal density and sotichiometry, and in the Al_2O_3 stopping power. The determined stopping is about 9% larger than SRIM values. These results are valid in the 650-15000 keV range that is in fact accessed by the beam inside the InN film, taking into account the beam on the way in before scattering, and on the way out after being backscattered. The simulations using the BI/MCMC stopping are also shown in Fig. 3.

Finally, we show in Fig. 5 the stopping power determined for ^{16}O. Again, a large deviations from SRIM stopping is observed. The BI/MCMC and SRIM values are similar around 5 MeV, but the measured stopping values are higher than SRIM both below and above that value. Again, we stress that these results are valid in the 310-15000 keV range that is probed by the beam inside the InN film,

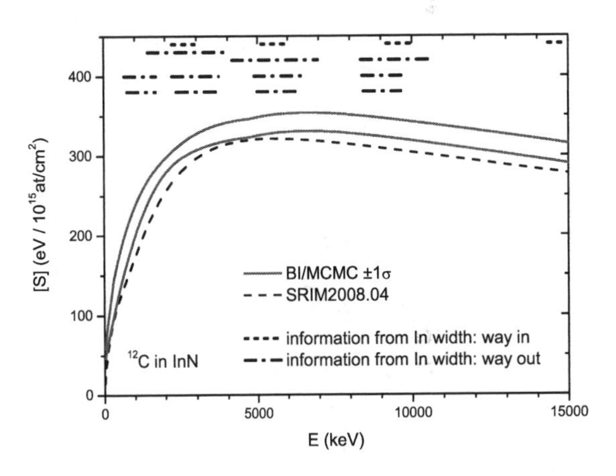

FIGURE 4. Limits of confidence of the stopping power of ^{12}C in InN (average plus minus one standard deviation, solid lines). Stopping calculated with SRIM and the Bragg rule is also shown (dashed). The regions of energy covered by the In signal (short dashed for the way in and dash-dotted for the way out) are shown. These results are valid in the 650-15000 keV range accessed by the beam inside the InN film..

The deviations found between the newly determined stopping power curves and SRIM calculations should not be surprising. The SRIM calculations presented used the Bragg rule, making a weighted average of the In and N elemental stopping powers, which is long known to be inaccurate [15] in the energy range of interest here, i.e. close to 1 MeV/nucleon, particularly in insulating and compounds with light elements. A discussion of the underlying physics is beyond the scope of this work, but we note that measurement of the stopping power of relevant compound materials is often essential to perform accurate IBA experiments [16].

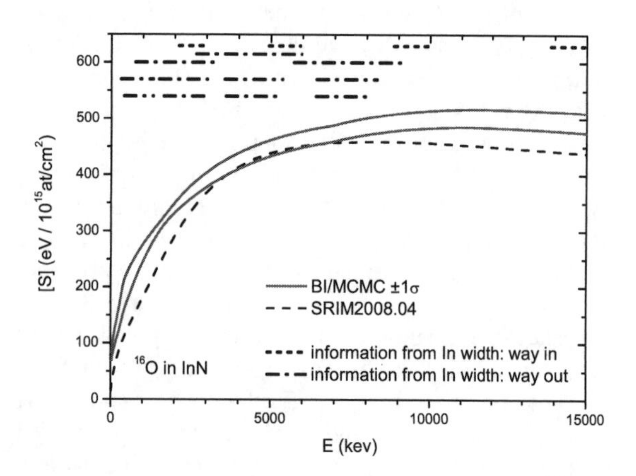

FIGURE 5 Limits of confidence of the stopping power of ^{16}O in InN (average plus minus one standard deviation, solid lines). Stopping calculated with SRIM and the Bragg rule is also shown (dashed). The regions of energy covered by the In signal (short dashed for the way in and dash-dotted for the way out) are shown. These results are valid in the 310-15000 keV range accessed by the beam inside the InN film.

SUMMARY

We determined the stopping of ^4He up to 2.3 MeV, ^{12}C and ^{16}O up to 15 MeV, in indium nitride, using a Bayesian inference method that relies on simple RBS spectra of thin film or bulk samples. The method leads not only to a stopping curve, but also to confidence limits on the results derived. These are, to our knowledge, the first measurements of any ion in this important system for optoelectronic applications. While for ^4He the results agree well with SRIM stopping (using SRIM for the elements In and N together with the Bragg rule) at energies above the stopping power maximum, for ^{12}C and ^{16}O the values determined are above SRIM in almost all the energy range accessed.

ACKNOWLEDGMENTS

We would like to thank the International Atomic Energy Agency for support under Research Contract No. 14365, Dr. Vanya Darakchieva for providing the InN sample, Micaela Fonseca for the Au deposition, and Filomena Baptista for technical assistance.

REFERENCES

1. L. Hsu and W. Walukiewicz, *J. Appl. Phys.* **104**, 024507 (2008).
2. V. Darakchieva, et al, *Physica B* **404**, 4476 (2009).
3. K.S.A. Butcher, A.J. Fernandes, P.P-T. Chen, M. Wintrebert-Fouquet, H. Timmers, S.K. Shrestha, H. Hirshy, R.M. Perks, B.F. Usher, *J. Appl. Phys.* **181**, 123702 (2007).
4. J.F. Ziegler, J.P.Biersack, and M.D.Ziegler, *SRIM - The Stopping and Range of Ions in matter*, Maryland, SRIM Co., 2008.
5. C. Pascual-Izarra, N. P. Barradas, G. García, A. Climent.Font, *Nucl. Instrum. Methods B* **239**, 135 (2005).
6. Z. Siketić, I. Bogdanović Radović, E. Alves N. P. Barradas, *Nucl. Instrum. Methods B* **268**, 1768 (2010).
7. T. Bayes, *Philosophical Transactions of the Royal Society of London* **53**, 370–418 (1763).
8. N.P. Barradas, C. Jeynes, and M.A. Harry, *Nucl. Instrum. Methods B* **136-138**, 1163 (1998).
9. N.P. Barradas, C. Jeynes, K.P. Homewood, B.J. Sealy, and M. Milosavljevic, *Nucl. Instrum. Methods B* **139**, 235 (1998).
10. N.P. Barradas, et al, *Nucl. Instrum. Methods B* **262**, 281 (2007).
11. J. F. Ziegler, J. P. Biersack, and U. Littmark, *Stopping and Ranges of Ions in Solids* (Pergamon, New York, 1985).
12. N.P. Barradas, *Nucl. Instrum. Methods B* **225**, 318 (2004).
13. S. Molodtsov, A. F. Gurbich, *Nucl. Instrum. Methods B* **267**, 484 (2010).
14. C. Pascual-Izarra, N. P. Barradas, *Nucl. Instrum. Methods B* **266**, 1866 (2008).
15. A. S. Lodhi and D. Powers, *Phys. Rev. A* **10**, 2131 (1974).
16. Y. Zhang and W. J. Weber, *Phys. Rev. B* **68**, 235317 (2003).

SECTION VII – ION BEAM MODIFICATIONS

Conducting Well-Controlled Ion Irradiations To Understand Neutron Irradiation Effects In Materials

F. U. Naab, E. A. West, O. F. Toader, G. S. Was

Department of Engineering and Radiological Sciences, College of Engineering, University of Michigan 2355 Bonisteel Boulevard, Ann Arbor, MI 48109-2104, USA

Abstract. A firm understanding of the effect of radiation on materials is required to develop predictive models of materials behavior in-reactor and provide a foundation for creating new, more radiation-tolerant materials. Ion irradiation can serve this purpose for nuclear reactor components and is becoming a key element of materials development for advanced nuclear reactors. Ion irradiations can be conducted quickly, at low cost, and with precise control over irradiation temperature, temperature uniformity, dose rate, dose uniformity and total dose. During proton irradiations the 2σ (twice the standard deviation) of the sample temperature is generally below ~7 °C, the dose rate variation ~3 %, the dose uncertainty ~3 %, and there is an excellent temperature and dose uniformity across the irradiated area. In this article, we describe the experimental setup and irradiation procedure used to conduct well-controlled ion irradiations at the University of Michigan.

Keywords: ion irradiation, radiation effect on materials, metal and alloys, nuclear energy
PACS: 61.82.Bg

INTRODUCTION

Ion irradiations were first conducted in the Michigan Ion Beam Laboratory at the University of Michigan in 1990. Since then about 100 articles were published on the topic of understanding the effects of radiation on materials using ions as the irradiating particle. One of these articles [1] validates the use of proton irradiation to emulate the full extent of neutron-irradiation effects in light water reactors. An article published in 1995 [2] described the experimental setup to conduct ion irradiations. The facility has continued to evolve over the years and it now has the capability to conduct higher and more uniform damage rates using proton irradiation, very high damage rates using heavy ions, and excellent sample temperature monitoring and control using a two dimensional thermal imager. The subsequent sections of this paper describe the current experimental setup used to perform ion irradiations.

ION BEAM PRODUCTION

The ion beam is produced using the 1.7-MV Tandetron accelerator in the Ion Beam Laboratory at the University of Michigan. Figure 1 shows a schematic drawing of the Tandetron accelerator used to do ion irradiations. Two ion sources are attached to the accelerator: a Torvis source [3,4] from National Electrostatics Corporation to produce proton beams and a sputter source [3,5] from Peabody Scientific to produce heavy ion beams (e.g. Fe and Ni). Typical ion currents at the target are ~60 μA for protons and ~1-10 μA for heavy ions, resulting in damage rates of ~1 dpa/day and ~100-1000 dpa/day, respectively. The experiments are conducted on the 15° beamline using a specially designed end station (stage) shown in Fig. 2. The beamline vacuum pressure in the vicinity of the irradiated samples is maintained at approximately 5-10 × 10^{-8} torr for the duration of the irradiation.

STAGE DESIGN AND ASSEMBLY

The execution of a successful irradiation requires, (1) secure specimen mounting, (2) proper beam alignment, (3) control of dose and dose uniformity, and (4) control of temperature and temperature uniformity. The ion irradiation stage designed and built at the University of Michigan satisfies all of these requirements. This section describes the unique stage design.

The base of the stage is a solid copper block with a cylindrical base and a rectangular extension for sample mounting (Fig. 2).

Application of Accelerators in Research and Industry
AIP Conf. Proc. 1336, 325-331 (2011); doi: 10.1063/1.3586113

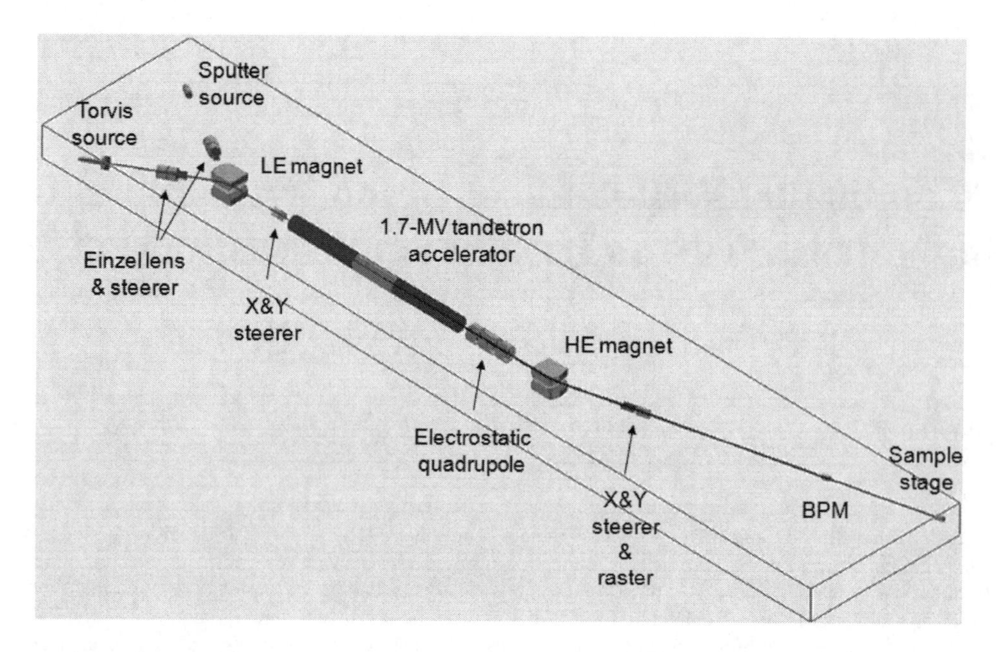

FIGURE 1. Drawing of the accelerator components used to transport the ion beam from each of the ion sources to the samples to be irradiated. The line from the Torvis source to the samples represents the trajectories of hydrogen ions.

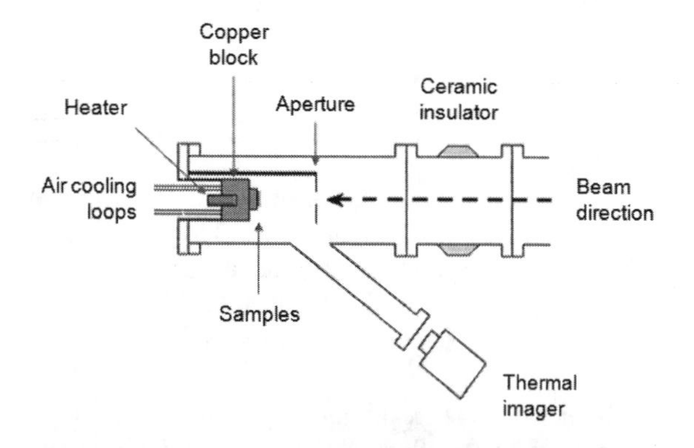

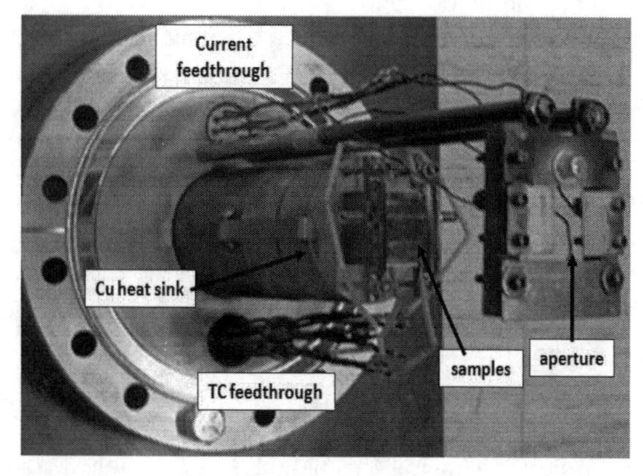

FIGURE 2. Top: simplified schematic of the irradiation stage mounted in the beamline. Bottom: photography of the assembled irradiation stage.

The Cu block acts as a heat sink via air cooling loops (one on each side of the heater), and a heat source via a resistive cartridge heater that is inserted into a cylindrical hole (Fig. 3), providing optimum control of temperature. On top of the copper block is a shim that contains a reservoir of liquid indium or tin during irradiation (Fig. 4). This solid/liquid/solid interface between the copper block and samples ensures excellent contact between all three layers and maintains temperature uniformity across the samples. The specimens are secured to the stage with hold-down bars (Fig. 4). Specimens used for either tensile testing or transmission electron microscopy analysis may be irradiated. The number of specimens that can be uniformly irradiated depends on the specimen width, and is typically 8 when 2 mm wide specimens are used. The hold-down bars are held in place using four screws. Thermocouples are located around the heater and cooling loops in the back of the Cu block (Fig. 3) to measure the temperature and ensure that the temperature of the indium or tin remains above the melting point. As shown in Fig. 4, J-type thermocouples are spot welded onto the samples. The placement of the thermocouples is critical, as they must not be in contact with either the irradiated region of the specimens or the hold-down bars. The iron and constantan wires are threaded through insulating sheaths to prevent shorts.

In front of the samples is a set of four rectangular tantalum sheets forming an aperture. The aperture is used to select the 10 mm × 18 mm area of the specimens to be irradiated. Each Ta sheet (electrode) is electrically insulated. They are used to set the ion beam raster amplitudes in the horizontal and vertical directions and align the rastered beam with the samples by evenly balancing the beam current on each electrode with the XY steerers (Fig. 5). The aperture is aligned with the samples using a bench-top laser. The laser beam is diffused using a plastic film to illuminate the whole aperture. The stage is electrically insulated from the beamline by the ceramic insulator shown in Fig. 2.

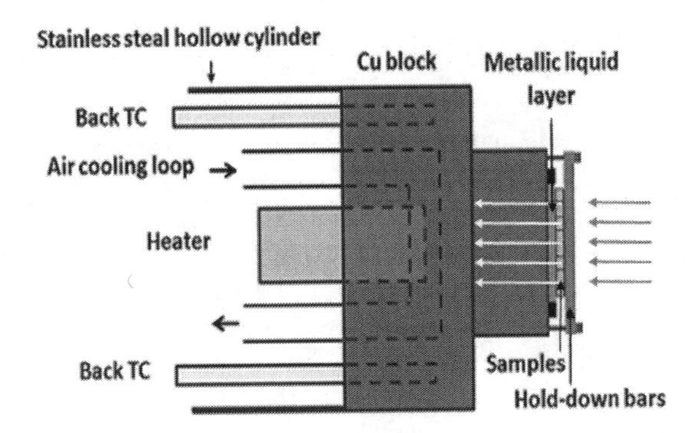

FIGURE 3. Cross-section of the copper block assembly on the irradiation stage. The gray arrows represent the rastered beam on the samples and the white arrows the heat transferred from the sample surface to the cylindrical copper base. The heat is finally dissipated by air cooling and the heater provides the appropriate temperature gradient to dissipate the beam power.

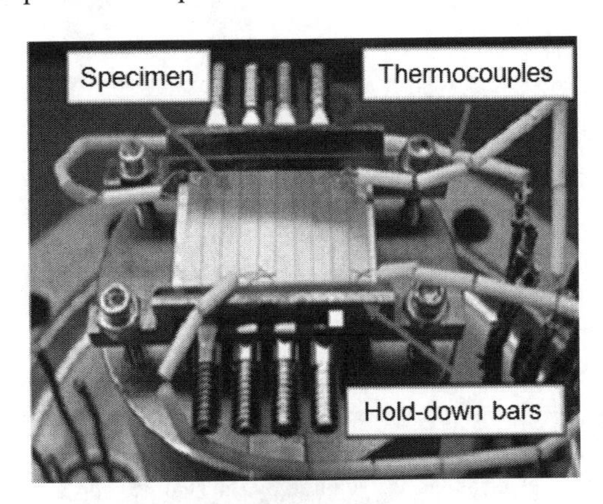

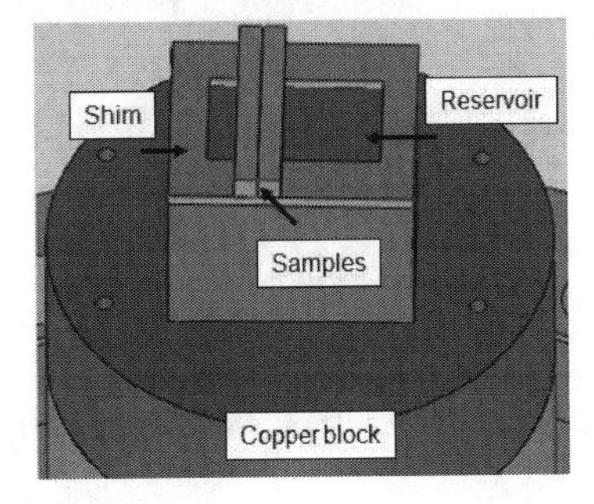

FIGURE 4. Top: picture of the samples mounted on the stage. Bottom: schematic showing the shim underneath the samples. The volume of the reservoir that is filled with In or Sn metal is 10 mm × 18 mm × 0.5 mm.

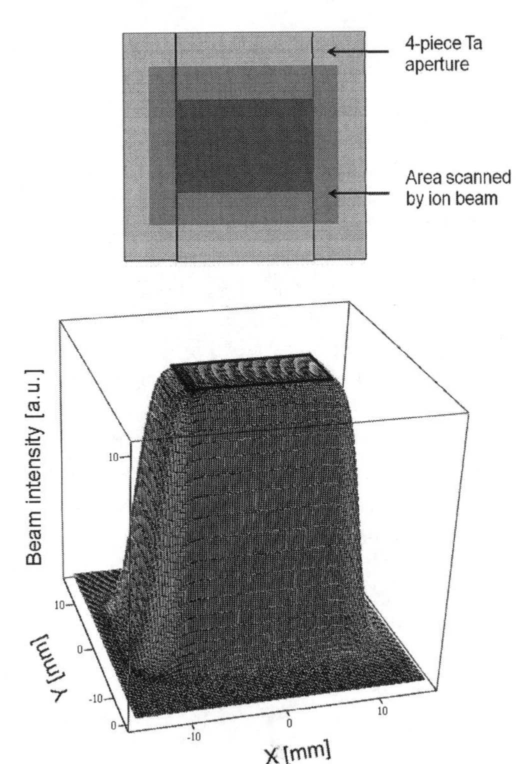

FIGURE 5. Top: front-view schematic of the four-electrode aperture and the area rastered by the ion beam. Bottom: plot of the rastered beam intensity (in arbitrary units) assuming that the focused beam has a Gaussian intensity distribution with a FWHM of 3 mm. The rectangle represents the limits of the aperture (18 mm × 10 mm). The integrated beam intensity in the area limited by the aperture is 2/3 of the total beam intensity integration.

BEAM ALIGNMENT AND CURRENT MEASUREMENT

The ion beam is focused in the experimental beamline with an electrostatic quadrupole and imaged using a beam profile monitor (BPM) located at a distance of 1.4 m from the samples. The focused ion beam has approximately a Gaussian intensity with a FWHM (full width at half maximum) of 3 mm as measured with the BPM. During the irradiation the beam is rastered across the aperture and Ta electrodes to produce an irradiated region of uniform intensity over the samples (Fig. 5).

The positioning of the rastered area is evaluated by reading the ion beam current on each of the Ta electrodes and samples using LabVIEW©. In order to uniformly irradiate all of the specimens, the size of the rastered area is adjusted until the samples receive 2/3 of the total beam current and the rest of the current is balanced on the Ta electrodes. This ratio is equivalent to rastering the center of the ion beam about 0.45 × FWHM or 1.35 mm beyond the edges of the aperture.

Figure 5 shows a plot of the rastered beam intensity on the aperture and Ta electrodes. To scan the ion beam the same distance beyond the horizontal and vertical edges of the aperture (e.g. 0.45 × FWHM), the ion beam current on the left and right electrodes should be about 72 % of the ion beam current on the top and bottom electrodes. LabVIEW© continuously updates and displays every ~1 second the current on the samples and Ta electrodes and records these values in an output file every ~1 minute.

The end station shown in Fig. 2 is insulated from the rest of the beamline and acts as a Faraday cup to measure the ion beam current on the samples. In this way, the system accurately measures the total beam current (sample current + aperture electrodes current). Since there is no electron suppression voltage between the aperture electrodes and the end station there is an exchange of electrons between them (Fig. 6). The relation between the measured (I_{ms}) and real beam current (I_{rs}) on the samples is described by:

$$I_{ms} = I_{rs} + \frac{I_{rs}}{q} eE_s \Omega - \frac{I_{ra}}{q} eE_a \quad (1)$$

where the second term represents the electron current emitted from the samples that is collected by the aperture electrodes and the third term the electron current emitted from the aperture electrodes that is fully collected by the end station. The rest of the quantities in Eq. 1 are: q the charge state of the incident ion, eEs the number of electrons emitted from the samples per incident ion, Ω the fraction of electrons emitted from the samples that are collected by the aperture electrodes, I_{ra} the real beam current on the aperture electrodes, and eEa the number of electrons emitted from the aperture electrodes per incident ion. Similarly, Eq. 2 describes the relation between the measured (I_{ma}) real beam current on the aperture electrodes:

$$I_{ma} = I_{ra} + \frac{I_{ra}}{q} eE_a - \frac{I_{rs}}{q} eE_s \Omega \quad (2)$$

From our measurements, Ω = 0.23±0.03 and the value of electron emission for proton beams between 2 and 3 MeV is eEs = eEa = 0.30±0.05 electrons/ion, which produce a difference between I_{ms} and I_{rs} of 5 %, where $I_{ms} < I_{rs}$.

In the case of heavy ions (e.g. Fe^{++} at 5 MeV) eEs = eEa = 12±2 electrons/ion and the electrons emitted from the Ta electrodes completely cancel out the beam current on the samples. To avoid this effect for heavy ions, the Ta electrodes forming the aperture should be placed before the ceramic insulator along the ion beam trajectory (see Fig. 2). This capability is not yet implemented in our system and an alternative method is currently used to set the beam raster amplitude. A white ceramic plate is placed inside a chamber between the BPM and the end station. When the beam strikes the ceramic a blue light is emitted allowing for visual determination of the focused and rastered beam. The focused beam on the ceramic is observed as a circle of about 3 mm in diameter. The ceramic has a rectangle drawn on its surface with the dimensions of the desired target rastered area. Each side of the rectangle is equal to the corresponding side of the aperture plus twice the beam diameter. Using the blue light emitted from the ceramic, the raster amplitudes are set to match the limits of the rectangle. In this way, the center of the beam is rastered beyond the edges of the aperture a distance of about 1.5 mm and the current on the samples is assumed to be 2/3 of the total beam current. After setting the raster amplitudes, the beam is directed onto the samples and aligned by balancing the current on each of the aperture electrodes. During the irradiation, the total beam current is monitored every 15 minutes by insertion of a Faraday cup in the 15° beamline located just before the BPM.

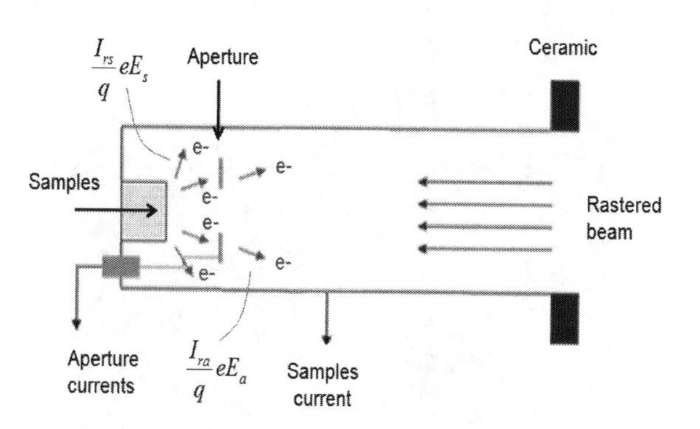

FIGURE 6. Schematic of the end station housing the stage and the electron emission effect. The two terms represent the total electron current emitted by the samples and aperture electrodes.

TEMPERATURE CALIBRATION AND CONTROL

Before starting an irradiation the copper block is heated with the cartridge heater to the irradiation temperature (e.g. 400 °C) as measured by the spot-welded J-type thermocouples. The temperature spread of the four thermocouples is typically ~3 °C. Then the IRCON® Stinger thermal imager shown in Fig. 2 is calibrated by using the thermocouple temperature output to set the sample emissivity. The thermal image of the sample is shown in Fig. 7. Using the

Stinger software three areas of interest (AOIs) are created on each sample. The emissivity value of each AOI is determined by adjusting the value until the AOI temperature matches the average temperature of the thermocouples.

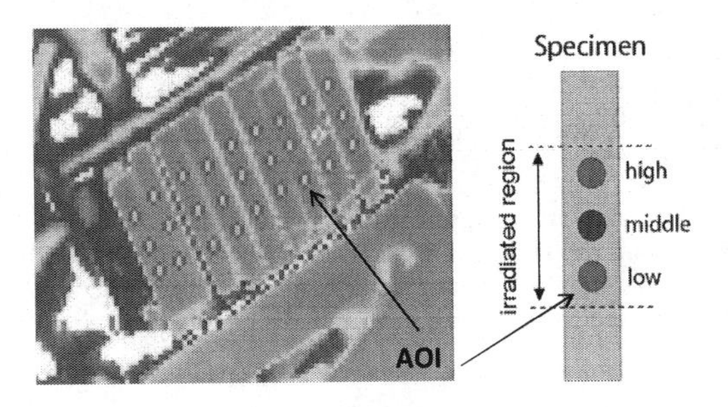

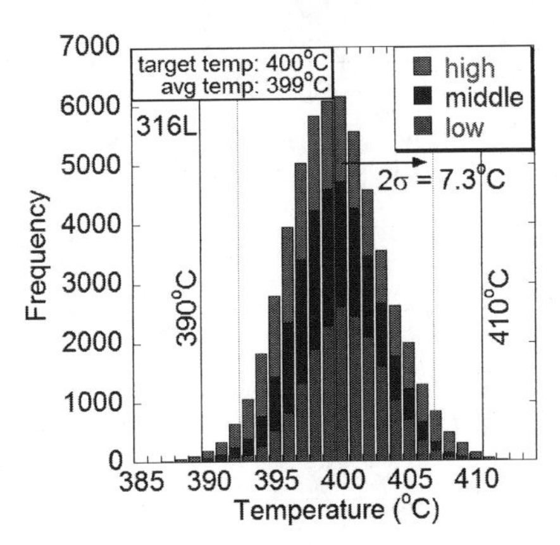

FIGURE 7. Top: on the left is a typical thermal image of the austenitic samples during irradiation with 2 MeV protons to a dose of 7 dpa at a target temperature of 400 °C and on the right a sample schematic with three AOIs. Bottom: AOI temperature profiles for one sample during irradiation.

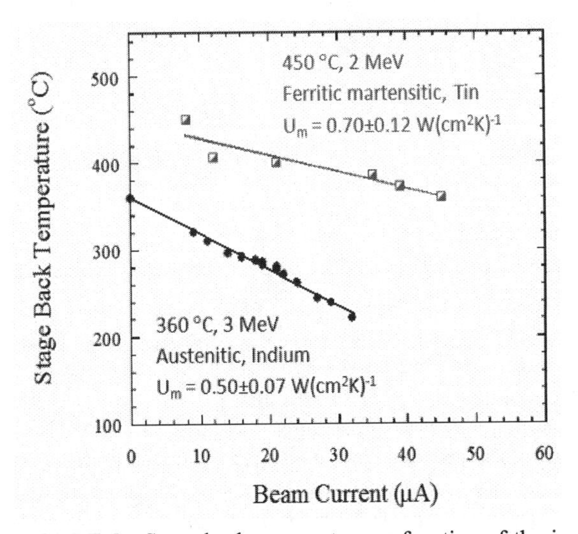

FIGURE 8. Stage back temperature as function of the ion beam current on the samples. Each set of data has its own irradiation temperature, proton beam energy, stainless steel alloy and metallic liquid layer.

During the irradiation, each AOI temperature is collected through the Stinger software with a frequency of 15 Hz and averaged over a period of one second. The averaged values are transferred to LabVIEW© which is interfaced with the Stinger software and reads and averages them at intervals of 15 to 60 seconds. LabVIEW© continuously updates and displays the temperature of each AOI, generates online plots to visualize the AOI temperature trends, and records every AOI temperature in an output file. By adjusting the heater voltage and air flow, the AOI temperatures are maintained within ±10 °C of the irradiation temperature. Alarms are set to alert the experimenter if any recorded temperature is more than 8 °C from the target temperature.

The purpose of the backside thermocouples (TCs) shown in Fig. 3 is to ensure that the Cu block temperature never approaches the melting temperature of the metallic liquid layer which would cause the liquid metal to solidify and temperature uniformity would be compromised.

The ion beam power per unit area incident on the samples is:

$$P_b = \frac{E}{A}\frac{I}{q} \tag{3}$$

where E is the ion energy, A the irradiated area, I the ion beam current on the samples and q the charge of each ion. The heat flow (ϕ) from the sample surface to the Cu block base (see Fig. 3) can be approximated as a one dimensional problem and expressed as:

$$\phi = U(T - T_b) \tag{4}$$

where T is the temperature of the sample surface (irradiation temperature), T_b the temperature at the junction between the squared extension and cylindrical base of the copper block, and U is the heat transfer coefficient given by:

$$U^{-1} = (L_S / k_S + L_{ML} / k_{ML} + L_{Cu} / k_{Cu}) \tag{5}$$

where L is the layer thickness (S: sample, ML: metallic liquid and Cu: copper block) and k the thermal conductivity for each of the materials. Figure 8 shows the dependence of the stage back temperature as function of the ion beam current on the samples for two cases with different irradiation temperatures, proton beam energies, irradiated stainless steel alloys and metallic liquid layers. The slope of the linear fit allows measuring the heat transfer coefficient, U_m (Fig. 8). The calculated values (U_c) using Eq. 5 are 0.65 W/(cm^{2}K) and 0.80 W/(cm^{2}K) for the austenitic and ferritic martensitic alloys, respectively. There is a good agreement between measured and calculated values. The difference is attributed to the positioning of the back TC in the Cu block.

TABLE 1. Possible (p,n) reactions for 2 MeV and 3 MeV protons incident on 316L stainless steel and resulting isotopes.

Reaction				Cross section (mb)	
Parent isotope	(p,n) Energy threshold (MeV)	Daughter	Half-life	2 MeV	3 MeV
Ni-61	2.49	Cu-61	3.35 h	N/A	9.5
Ni-64	2.52	Cu-64	12.7 h	N/A	13.9
Cr-53	1.39	Mn-53	3.7×10^6 yr	5	42
Cr-54	2.21	Mn-54	312.1 d	N/A	30
Mn-55	1.03	Fe-55	2.73 yr	4.5	60.5
Fe-57	1.65	Co-57	271.8 d	0.9	12.4

VERIFICATION OF IRRADIATION CONDITIONS

Temperature Uniformity

After the irradiation, the AOI temperatures of each specimen are plotted to verify that the temperature was maintained within the ±10 °C bounds. An example of such a profile is shown in Fig. 7, where the average temperature of the three AOIs for the sample agrees within one degree with the irradiation temperature and almost 100 % of the values are within the ±10 °C range.

Dose And Dose Rate

Post-irradiation β counting measurements are conducted to determine the residual sample activity following proton irradiation. The sample activity is low, but is useful in ensuring that the samples are irradiated uniformly. When stainless steel (SS) samples are irradiated with high energy protons many nuclear reactions occur. Table 1 lists the (p,n) interactions in SS when irradiated with 2 MeV and 3 MeV protons along with the threshold energies and cross sections for the reactions. The β counting measurements are performed using an Alpha/Beta Scalar (Model # 2929) produced by Ludlum Measurements Inc. The measurements for each alloy are normalized to time and the irradiated surface area of the specimen. The specimen to specimen variation in β activity for a given alloy is typically less than 5 % of the average (Fig. 9). These measurements verify the dose uniformity across the samples. In the case of heavy ion irradiations this test is not possible since there is not residual activity.

The uniformity of the dose rate along the irradiation can be verified by plotting the ion beam current on the sample. Due to the very good stability of the Torvis source it is possible to keep the proton beam current on the samples within ±1 μA for irradiations lasting up to 10 days (longest proton irradiation done in our facility). In the case of heavy ion irradiations, the sputter source has current fluctuations on the samples of about ±0.2 μA out of a total current of 1 μA.

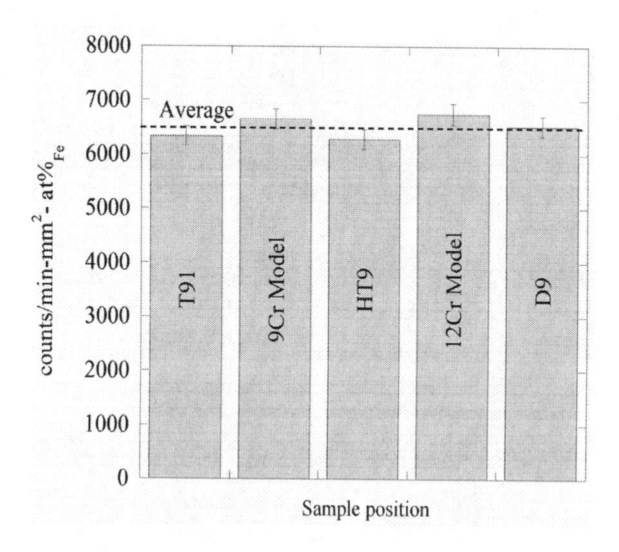

FIGURE 9. Beta activity of the ferritic-martensitic alloy specimens irradiated with protons at 2 MeV to a dose of 7 dpa at 500 °C as function of the position on the stage.

The alloy hardness is a function of the irradiation dose, and so microhardness measurements are made over the extent of the irradiated region to check the uniformity of the dose delivered to each sample. The pre-irradiation and post-irradiation hardness of the specimens is measured using a Micromet®-II Vickers Microhardness tester with a load of 25 g to ensure that the indent is confined to the irradiated depth of the specimen in the case of proton irradiations. Figure 10 shows the hardness measurements comparing the values inside and outside the irradiated area. The hardness uniformity inside the irradiated area verifies the dose uniformity along the length of the sample.

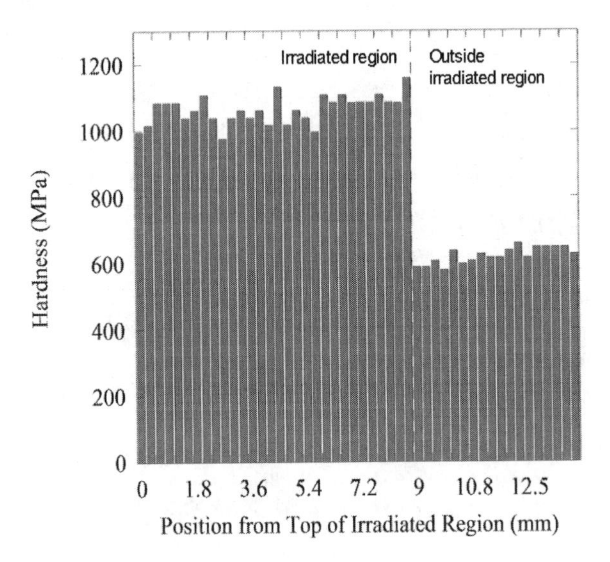

FIGURE 10. Hardness measurements inside and outside the irradiated area to check the dose uniformity of a ferritic-martensitic sample irradiated with 2 MeV protons to a dose of 7 dpa at 400 °C.

SUMMARY

The experimental setup to perform well-controlled ion irradiations was described in detail. The system provides excellent temperature monitoring and control through a combination of a 2-dimensional thermal imager and the supply of both heating and cooling to a stage with a liquid metal interface to the samples being irradiated. Accurate tracking of dose, dose rate and dose uniformity is achieved by strict definition of the irradiated area and the measurement of current on that area, and use of residual activity to assure uniform irradiation. During proton irradiations the 2σ of the sample temperature is generally less than ~7 °C, the dose rate variation ~3 %, the dose uncertainty ~3 %, and there is an excellent temperature and dose uniformity across the irradiated area. The degree of control of the key irradiation parameters makes ion irradiation the ideal tool for studying radiation effects processes in materials.

REFERENCES

1. G. S. Was, J. T. Busby, T. Allen, E. A. Kenik, A. Jenssen, S. M. Bruemmer, J. Gan, A. D. Edwards, P. M. Scott, P. L. Andresen, *J. Nucl. Mater.* **300** (2002) 198-216.
2. D. L. Damcott, J. M. Cookson, V. H. Rotberg, G. S. Was, *Nucl. Instr. and Meth. B* **99**, 780-783 (1995).
3. F. Naab and O. Toader, *Proceedings of the 11th International Conference on Heavy Ion Accelerator Technology* (2009) 147-150. http://accelconf.web.cern.ch/AccelConf/HIAT2009/index.htm
4. M. L. Sundquist, J. R. Adney, R. C. Schmidt, *Nucl. Instr. and Meth. B* **99** (1995) 684-687.
5. R. Middleton, *Nucl. Instr. and Meth.* **214** (1983) 139-150.

Ion Beam Studies Of Semiconductor Nanoparticles For The Integration Of Optoelectronic Devices

S.V.S. Nageswara Rao[1], V.S. Vendamani[1], Sandeep K. Satrasala[1], Santanu K. Padhe[1], Srinadha Rao K.[1], S. Dhamodaran[2] and Anand P. Pathak[3]

(1)Department of Physics, Pondicherry University, Kalapet, Pondicherry Pondicherry 605014, India
(2)Department of Physics, Indian Institute of Technology (IIT Kanpur),Kanpur, India
(3) School of Physics, University of Hyderabad, Central University P.O, Hyderabad Andhra Pradesh 500046, India

Abstract. Radiation hardened microelectronics compatible and tunable light emitting devices are essential for the integration of electronic and optoelectronic devices. Porous or nano-crystalline silicon seems to be a promising material for accomplishing this task. Ion beam patterned structures of light emitting silicon nanoparticles can be used for achieving multiple wavelength emission. Here we present a study on the influence of oxidation, pre and post anodization irradiation on emission parameters and on the stability of passivating bonds of porous silicon.

Keywords: Porous silicon, nano-crystalline silicon, radadiation damage and irradiation
PACS: 78.55 Hx, 81.35+k, 81.60Cp, 61.80-x

INTRODUCTION

Radiation hardened microelectronic compatible and tunable light emitting devices are essential for the integration of electronic and optoelectronic devices. Integration of optoelectronic devices requires band-gap tuning because different optoelectronic device demands different band-gap. Needless to mention the role of silicon in today's modern integrated circuit technology. Most of the semiconductor and photovoltaic industry are based on silicon technology. However silicon is not suitable for optoelectronics due to its poor efficiency to emit visible radiation or to detect infrared radiation. Now essentially there are two major research aspects in this research area: 1) to search for various tunable light emitting materials and 2) to attain light from silicon itself or at least from silicon compatible materials. The second aspect has got its own significance because silicon is the best understood element and the present microelectronic industry is totally based on silicon technology. Porous / nanocrystalline silicon seems to be a promising material for accomplishing this task. Ion beam modification of emission parameters is critical for achieving the required spatial band gap tuning. Radiation damage studies are essential because these devices are expected to be used in radiation prone areas (like space and nuclear laboratories). The other important aspect is to understand the influence of energetic ions and Gamma radiation on the stability of surface passivation bonds (Si-H and Si-O). In fact this information is essential for understanding the global nature of Si-H bonds in semiconductors and for improving long-term reliability of semiconducting devices.

The origin of porous silicon dates back to 1956 when A. Uhilr at Bell labs observed some black / brownish particles while electropolishing silicon wafers[1]. Later it was understood that the incomplete etching of silicon leads to a sponge like material, named as porous silicon. In 1972, Theunissen showed for the first time that the porous silicon is crystalline in nature[2]. The interesting and unexpected observation took place in 1990 by Canham[3] that is the emission of intense visible light from porous silicon. He reported the first photoluminescence observation from silicon based material and attributed it to the quantum confinement effects. Following this important observation, there are innumerable reports (for example ref. 1-17) in this field ranging from material aspects to device possibilities. However the origin of light emission phenomenon is still under debate[3-11].

Application of Accelerators in Research and Industry
AIP Conf. Proc. 1336, 332-336 (2011); doi: 10.1063/1.3586114

Quantum confinement and as wells the surface states are attributed to be prime reasons for the observed visible emission.

The pore size, porosity and hence the emission wavelength are functions of various growth parameters among which silicon doping concentration (hole concentration) is one important parameter. The free carrier concentration can be altered by creating defects via ion implantation to achieve spatial selectivity[17]. Here we present a study on the influence of pre and post anodization irradiation on emission parameters of porous silicon.

EXPERIMENTAL

Silicon samples cut from P- type boron doped (1 - 10 Ω cm, 1 - 50 Ω cm and 100 - 2000 Ω cm) single crystal silicon wafers (100) were anodically etched in aqueous HF solution for preparing light emitting porous silicon. Silicon samples were initially cleaned in 10% HF prior to anodization. Then these samples were etched for required time at a constant current density. These etched samples were dried with ethyl alcohol for 2 minutes and followed by pentane for 3 minutes. Formation of porous silicon was confirmed by the intense visible (red – orange) emission under UV lamp. Samples were further characterized by SEM (Hitachi S 3400N), FTIR (Thermo Nicolet Model: 6700) and PL (Flourolog FL3-11 or a 405 nm laser excitation) measurements. Porosity is estimated by gravimetric method. Gamma irradiation was performed at Pondicherry University at a constant dose rate of 8.33 KGy/h.

RESULTS AND DISCUSSION

Preparation Of Porous Silicon

An electrochemical etching setup has been designed and developed for preparing light emitting porous silicon samples. As shown in Figure 1a, bright visible emission is observed under UV excitation. Fig. 1b shows the PL spectra of four porous silicon samples prepared under same conditions. The emission wavelength is found to be 682 $\pm$ 3.5 nm. SEM (Fig. 1c) study shows a uniform sponge like surface confirming the presence of porous silicon layer. The average diameter of the pores is measured to be about 1.04 $\pm$ 0.17 μm. EDX and FTIR (Fig. 1d & e) measurements show the presence of various essential passivating Si-O and Si-H complexes.

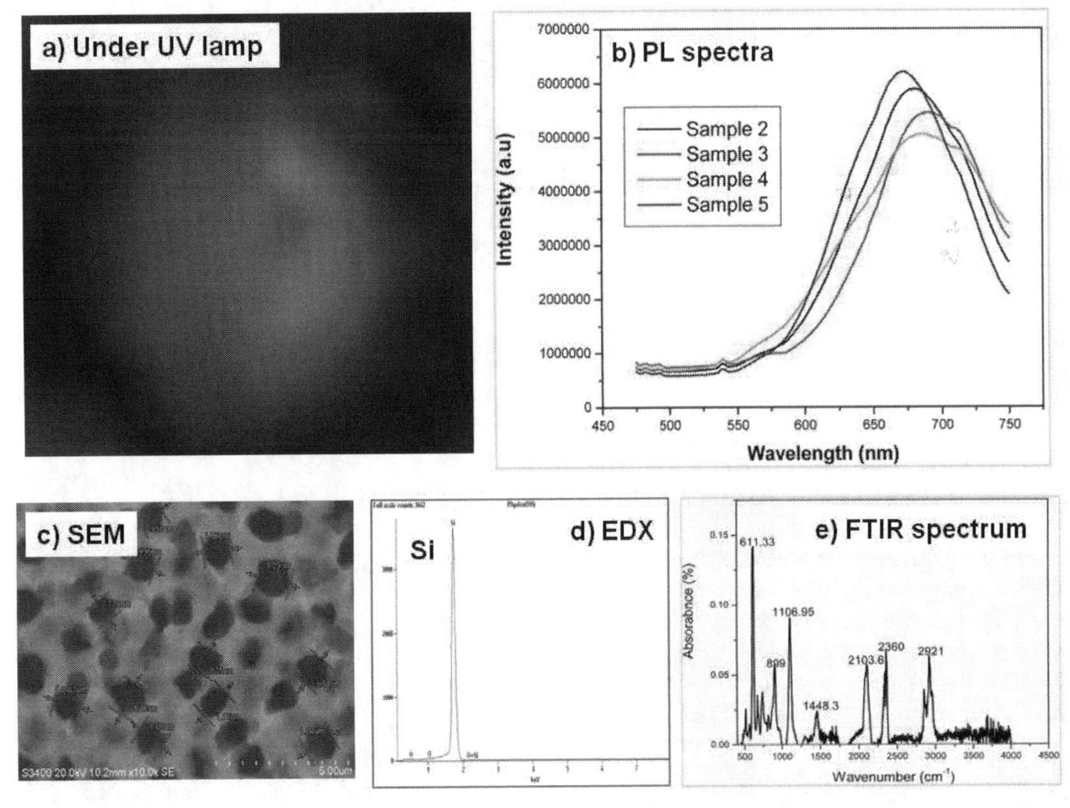

FIGURE 1: Formation of porous silicon: a) Bright visible luminance (red to orange) under UV illumination, b) PL spectra of four porous silicon samples prepared under similar conditions, c) SEM image showing the sponge like porous layer, d) EDX and e) FTIR spectra of a freshly anodized porous silicon.

Aging Effects

The emission wavelength of porous silicon is found to decrease with time. Fig. 2a shows the emission wavelenght of porous silicon as a function of time after synthesis. Freshly anodized porous silicon is expected to be fully passivated by Si-H bonds as it is etched in aqueous HF solution. FTIR measurements (Fig. 2b and 2c) show reduction of Si-H_x bonds and increase of Si-O-Si bonds with time. The observed oxidation of silicon nanoparticles can improve the confinement effects by preventing the loss of excited carriers. This is in agreement with the observed blue shift in emission wavelength. As shown in Fig. 2, the observed aging effects in our samples are saturated within few days.

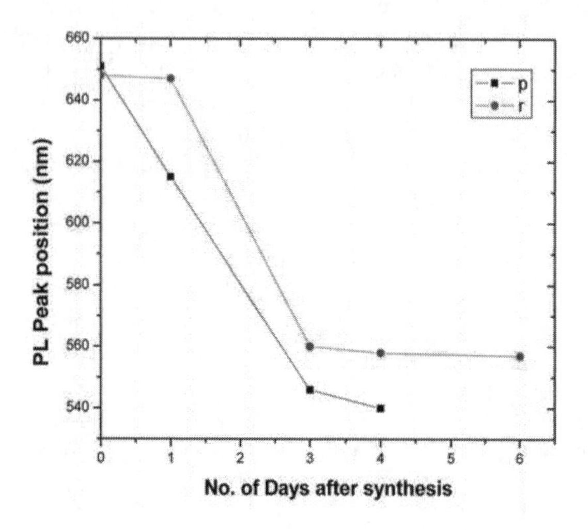

FIGURE. 2a: Aging effect: PL peak position (emission wavelength) as a function of time after anodization.

Effects Of Pre-Anodization Irradiation

Silicon samples were first subjected to gamma radiation and then anodized to study the influence of radiation damage of silicon on the formation of porous silicon. Porosity of the samples was measured by gravimetric method as a function of pre-anodization irradiation dose. Fig. 3 shows an important observation, that the porosity decreases linearly with increase in dose. Gamma irradiation is expected to reduce the sample resistivity by reducing effective carrier concentration of silicon. This in turn will reduce the efficiency of hole transfer mechanism (from bulk to the silicon surface) needed for promoting electrochemical etching. This will result in the observed reduction of porosity. Hence, porosity of porous silicon can be tuned by the proposed pre-irradiation process. We also observe an increase in the oxygen and hydrogen intake during the anodization of pre-irradiated silicon as compared to that of un-irradiated silicon.

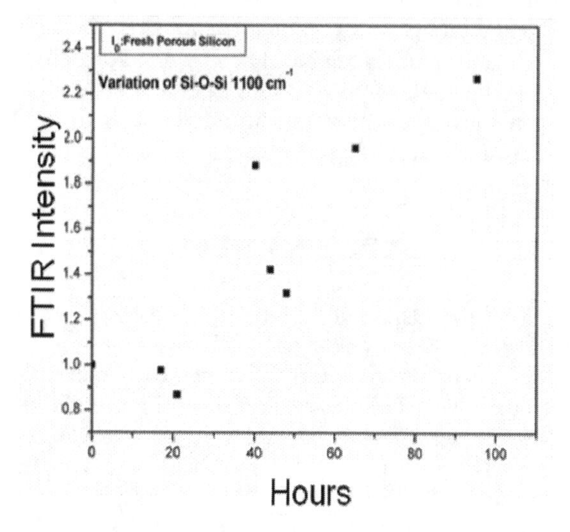

FIGURE 2b: Aging effect (FTIR): Intensity of Si-O-Si peak as a function of time after anodization.

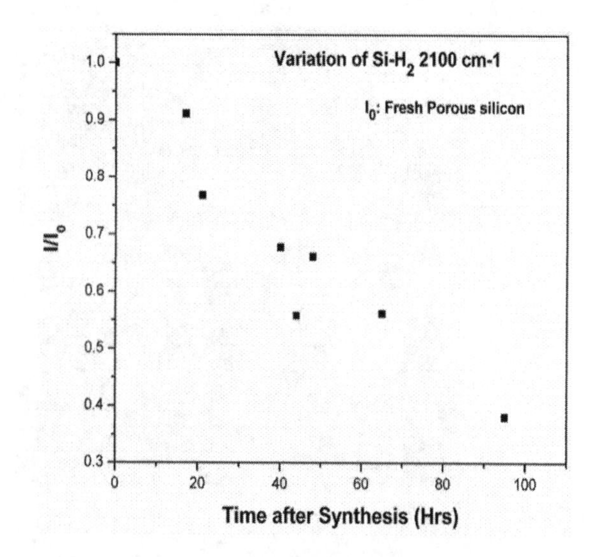

FIGURE. 2c: Aging effect (FTIR): Intensity of SiH_x peaks as a function of time after anodization.

Effects Of Post-Anodization Irradiation

Fig. 4 shows the SEM images recorded on several irradiated and un-irradiated samples at different magnification levels. As mentioned earlier all these samples show visible luminescence under UV excitation. SEM analysis confirms the presence of porous network on all irradiated samples though with different geometry. There seems to be a thick layer covering over the sponge like porous silicon network. Such a thick layer is not observed over any of the pristine samples. Hence this it is reasonable to assume that this layer is formed only during the gamma irradiation. The most possible candidate would be the silicon dioxide (SiO_2) as some of the Si-H and strained Si-Si bonds are expected to be dissociated during the irradiation[18]. Further the EDX and FTIR analysis show good amount of oxygen in this sample.

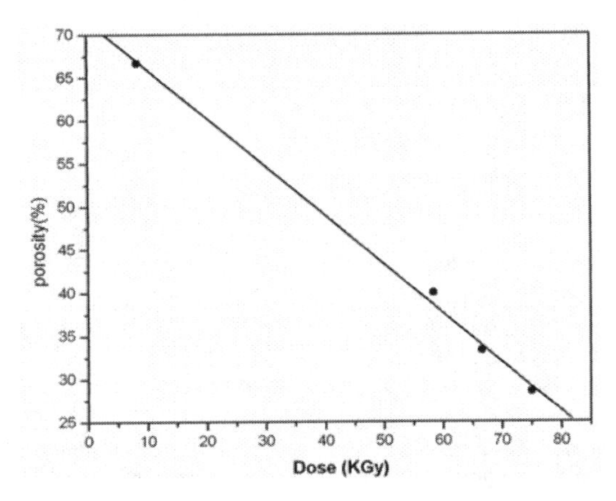

FIGURE. 3: Porosity as a function of pre-anodization irradiation dose.

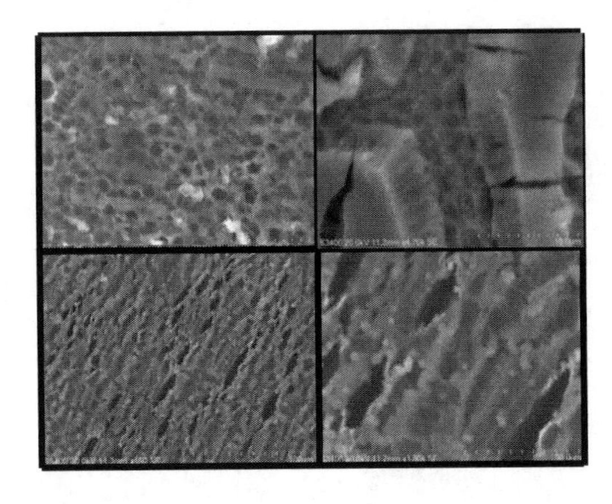

FIGURE. 4: SEM images of porous silicon with different doses (a: before irradiation). b) Formation of thick oxide after irradiation, c and d) elongation pores after irradiation.

Another important observation is that there is a significant change in the pore geometry and pore size. It seems that the pore size increases with increase in gamma dose. This is possible because lattice defects and strained Si-Si bonds are mobile in the presence of e-h pairs[18]]. This increase in pore size leads to the merging of nearby pores and eventually leads to the formation of columnar pores. This change in geometry and the increase in the pore size are expected to alter the size of the nanosized silicon crystallites situated on pore walls. As per the quantum confinement model this change in size should lead to change in emission wavelength. This expected change in wavelength is observed in our PL measurements shown in Fig. 5. A significant blue shift (~ 100 nm) in the emission wavelength can be observed as a result of gamma irradiation (Dose: 5.8 kGy). The reason for the observed blue shift could be the possible reduction of Si particle size due to irradiation enhanced oxidation of porous silicon. Hence the gamma irradiation promotes aging effects of porous silicon and stabilizes its emission parameters. Porosity measurements could not be performed because the new oxide layer grown on the surface prevented the proper dissolution of porous silicon in KOH solution.

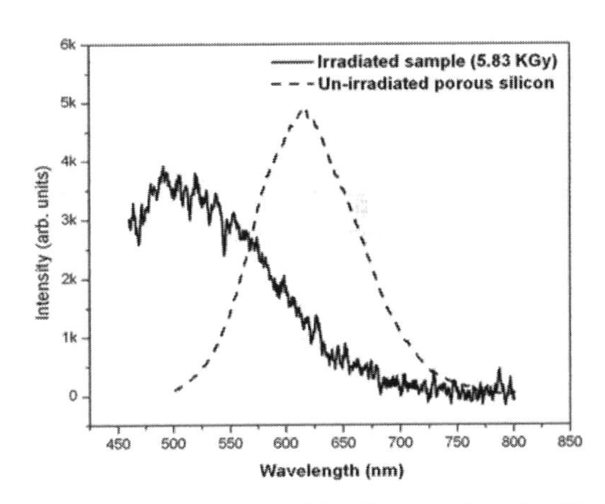

FIGURE. 5: PL spectra of irradiated and un-irradiated porous silicon.

CONCLUSION

Good quality bright light emitting porous silicon samples were prepared using electrochemical etching. It is shown that the oxidation of porous silicon causes blue shift in the emission wavelength. The porosity is found to decrease linearly with increase in pre-anodization irradiation dose. It is also shown that the post-anodization irradiation of porous silicon can alter and stabilize the emission wavelength by promoting aging effects. SEM analysis shows a significant change in the geometry and size of pores which in turn may alter the size of silicon nanocrystallites sitting on the pore walls. We conclude by saying that the irradiation techniques can be used to tune and stabilize emission parameters of porous silicon for the monolithic integration of electronic and optoelectronic devices.

ACKNOWLEDGMENTS

Most of the sample characterization and gamma irradiation experiments were performed in the Central Instrumentation Facility (CIF) of our University and we thank CIF and its staff members for their kind cooperation. We would like to thank Inter University Accelerator Centre (IUAC), New Delhi for the financial support in the form of an UFR project. SVSN would like to thank the Department of Science and Technology (DST), India and Pondicherry University, India for International Travel Support for presenting these results at CAARI21.

REFERENCES

1. A. Uhlir, *Bell Syst. Tech. J.* **35**, 333 (1956) (A technical report of Bell laboratories).
2. M.J.J. Theunissen, *J. Electronchem. Soc.*, **119,** 351 (1972).
3. L.T. Canham, *Appl. Phys. Lett.,* **57**, 1047 (1990).
4. L.T. Canham, Phys. World 5, 41 (1992).
5. O. Bisi et al. , *Surface Science Reports* **38,** 1 (2000).
6. A.G. Cullis *et. al., J. Appl. Phys.* **82**, 909 (1997).
7. K.D. Hischman *et. al, Nature (London)* **384,** 338 (1996).
8. H.Z. Song and X.M. Bao, *Phys. Rev.* ***B*** **55**, 6988 (1997).
9. E.F. Steigmeier *et. al, J. Luminance* **57**, 9 (1993).
10. V. Lehmann *et. al., Appl. Phys. Lett. 58 (1991) 586 (1991).*
11. M.S. Brandt *et. al., Soild. State. Commu.* ***81,*** *1046 (1990).*
12. Tanenbaum et al., Appl. Phys. Lett. 68, 1705 (1996).
13. Sabar D. Hutagalung et. al, *Solid State Science and Tech.,* ***16,*** *100 (2008).*
14. Nayfeh et al., *Appl. Phys. Lett.* **77**, 4086 (2000).
15. Xi-Mao *et. al., J. Appl. Phys.* **79**, 1320 (1996).
16. Pavesi *et. al., Appl. Phys. Lett.,* **65**, 2182 (1994).
17. M.B.H. Breese *et. al, Phys. Rev.* ***B*** **73**, 035428 (2006).
18. S.V.S. Nageswara Rao *et. al. Phys. Rev.* B, **75**, 235202 (2007) and refs. therein.

Method For Silicon Surface Texturing Using Ion Implantation

Nirag Kadakia[a], Sebastian Naczas[a], Hassaram Bakhru[a] and Mengbing Huang[a]

[a]*College of Nanoscale Science and Engineering, State University of New York at Albany*
255 Fuller Road, Albany NY 12203

Abstract. As the semiconductor industry continues to show more interest in the photovoltaic market, cheaper and readily integrable methods of silicon solar cell production are desired. One of these methods—ion implantation—is well-developed and optimized in all commercial semiconductor fabrication facilities. Here we have developed a silicon surface texturing technique predicated upon the phenomenon of surface blistering of H-implanted silicon, using only ion implantation and thermal annealing. We find that following the H implant with a second, heavier implant markedly enhances the surface blistering, causing large trenches that act as a surface texturing of c-Si. We have found that this method reduces total broadband Si reflectance from 35% to below 5%. In addition, we have used Rutherford backscattering/channeling measurements investigate the effect of ion implantation on the crystallinity of the sample. The data suggests that implantation-induced lattice damage is recovered upon annealing, reproducing the original monocrystalline structure in the previously amorphized region, while at the same time retaining the textured surface.

Keywords: solar cells, ion implantation, texturing, hydrogen, blistering, silicon
PACS: 88.40.jj

INTRODUCTION

Silicon remains the dominant material in commercial solar cells, despite lacking many characteristics of "ideal" single-junction semiconductor photovoltaics. Among these is its high refractive index—above 3.5 for wavelengths of 300-900 nm—which leads to power losses of up to 30% on cells without added absorption mechanisms[1,2]. Accordingly, silicon-based solar cells are often coated with an antireflective coating of index ~2 and whose thickness is chosen to be a quarter of the wavelength of the incident light, so the reflected beam within the layer interferes deconstructively with the incident beam and no photons are reflected. Clearly, this condition is only met for a particular wavelength and at a single incident angle; a single layer coating can reduce overall reflectance to about 20%[3].

Alternatively, the substrate surface can be roughened to reflect photons among themselves and increase the absorption of varying wavelengths and incident angles. Such a texturing process is often done through chemical etching , though recent work has shown that various other textures and low reflectivities (<5%) can be achieved through combinations of chemical etching, photolithography, e-beam evaporation, reactive ion etching and/or femtosecond lasers[4-6].

However, the integration of such lab-tested processes into commercial markets is precluded by either high costs or low throughput. Recently, the semiconductor industry is showing more interest in the photovoltaic market, and as such readily integrable and scaleable methods of solar cell production are needed. Accordingly, we have developed a method of texturing silicon that uses no chemicals, dry etching, or UHV deposition processes, relying rather on a combination of ion implantation followed by conventional thermal annealing. These processes are standard and highly optimized in large scale fabrication facilities, allowing for easy incorporation of the process in an assembly line fashion.

EXPERIMENTAL

Our work is predicated on the phenomenon of silicon blistering by implanted hydrogen. When large fluences (>3e16) of H+ ions are implanted into Si substrates, thermal annealing causes the atomic hydrogen to agglomerate at vacancy sites created by the implantation damage, forming gas phase molecular hydrogen. Small bubbles merge into larger ones and eventually push against the Si surface, causing blisters

Application of Accelerators in Research and Industry
AIP Conf. Proc. 1336, 337-340 (2011); doi: 10.1063/1.3586115

~ 3-5 μm in width and ~200 nm in height. With further annealing, these blisters may "pop" as subsurface stress overcomes the bonds at the surface and causes the Si to exfoliate, forming craters[7]. We initially tested this phenomenon by implanting Si (100) samples of thicknesses ~300 μm with H_2+ ions at room temperature at energies of 40 keV per molecule and fluences of 1e17/cm^2. Following implantation, the samples were annealed in an inert Ar ambient at various temperatures ranging 400° – 1100° C for 75 minutes. Optical microscope photos show that up to 1000° C, shallow blisters of several microns in diameter form, and some of them have popped into craters (Figure 1). At 1100° C, more exfoliation occurs.

Since the hydrogen bubbles nucleate at vacancy sites and grow through ripening processes, it seems likely that they may form at far higher densities if there were more nucleation sites. Accordingly, we repeated the experiment, but this time following the H implantation with an Ar implantation to a fluence of 5e15/cm^2 at 90 keV, corresponding to a depth of 100 nm. Since Ar is inert, we do not expect it to radically affect the electrical properties of silicon, while its larger mass allows it to more effectively damage the lattice and provide nucleation sites for the H bubbles.

RESULTS AND DISCUSSION

We find that with the Ar implant situation partway between the surface and the H implant (~270 nm) , the bubbles are densely packed (Figure 2). Owing to the far greater number of nucleation sites available, the bubbles, while denser, are also somewhat smaller, about ~1 um in diameter, though AFM data reveals that the height is the same as in the H-only implant, about 200 nm. Since the bubbles are not as close to the surface, the pressure they exert at T < 900° C is too small to exfoliate the Si. At 1100° C, an entirely different morphology characterized by an interconnected trench-like structure forms. AFM data reveals that these structures have depths of up to 1 μm and aspect ratios up to 100%.

While the implanted Ar distribution peaks at around 100 nm below the surface, the implantation creates an amorphous Si (a-Si) layer extending from the surface to a depth of approximately 300 nm. Crystal regrowth through the solid phase epitaxy process then occurs as the samples are annealed at temperatures above 500-600 °C. This can be seen in the Rutherford backscattering/ion channeling (RBS/C) data, by comparing the channeled spectrum of the virgin sample to that of the annealed samples (Figure 3). Previous work done on the implantation of H into 6H-SiC has attributed high dechanneling yields to

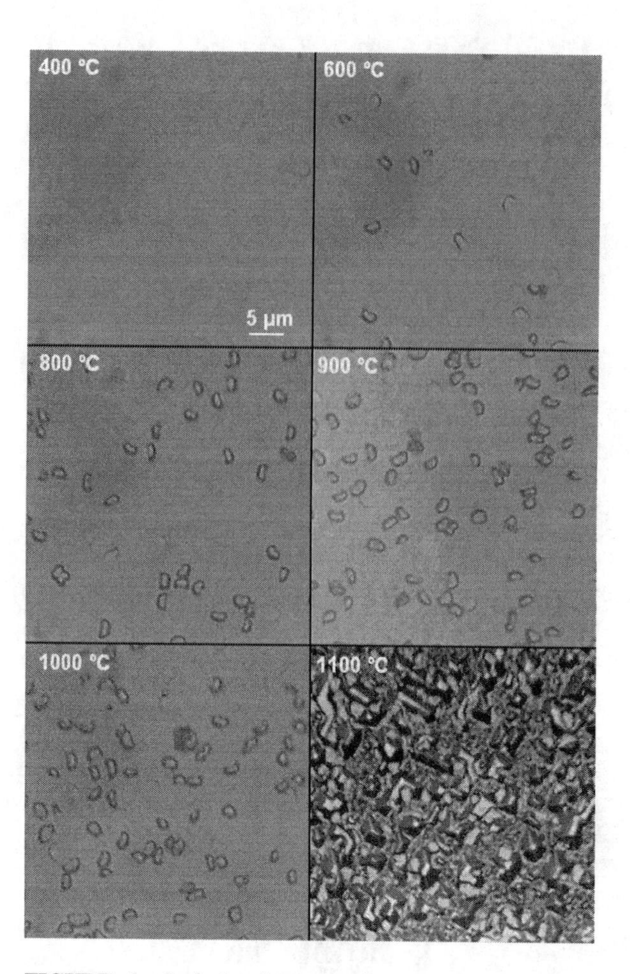

FIGURE 1. Optical microscope images of H-implanted samples annealed at various temperatures for 75 minutes.

beam misalignment caused by the slight surface curvature and the changes in the local lattice caused by the tension from the bubbles [8]. This suggests that dechanneling yields will decrease once exfoliation occurs and the strained layer is removed. However, in our data, it can be seen that while virtually no exfoliation occurs below 900 °C, the dechanneling yield nevertheless continues to decrease up to this temperature, indicating crystal changes independent of surface morphology. This suggests that high dechanneling yields may be caused by factors attributed not only to surface curvature, but continued grain growth as well. In any case, it is clear from the RBS/C spectra that following annealing at 1100 °C for 75 minutes the crystal quality of ion-implanted Si completely recovers that of the virgin Si.

It is interesting to note that in the 1100° C annealed samples, exfoliation can occur at depths greater than both the H and Ar concentration peaks. This difference may be understood by considering the effects of crystal stresses on the formation of H platelets. In the case with the H implant only, the location of maximum stresses coincides with the highest damage

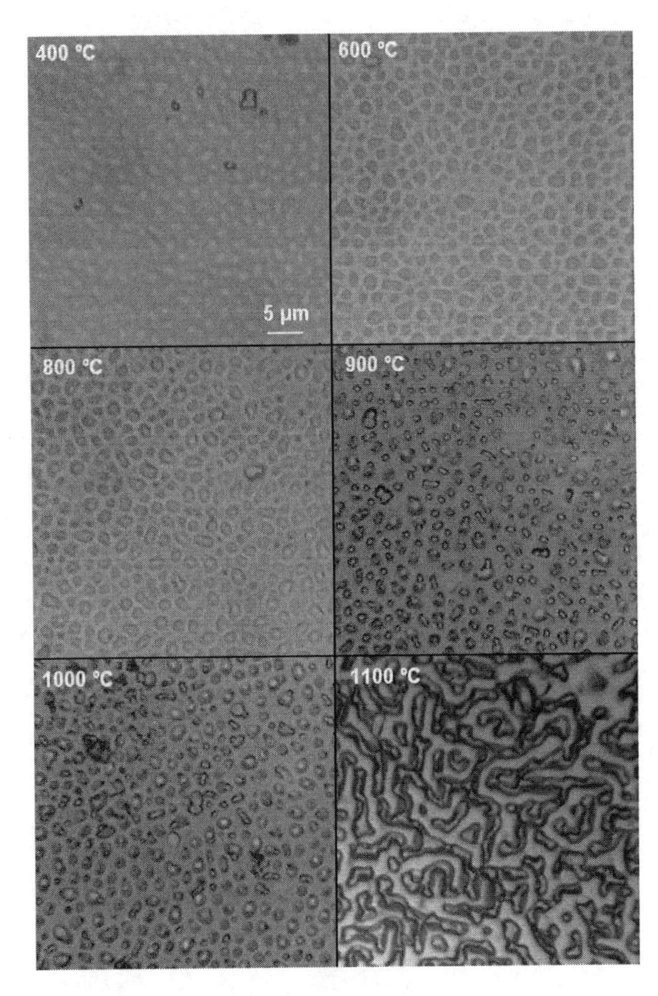

FIGURE 2. Optical microscope images of H- and Ar-implanted samples annealed at various temperatures for 75 minutes.

distribution, being actually shallower than the implanted H distribution in most cases. Prior work has shown that as the damage level becomes too high (even with the Si remaining in the crystalline phase), the location of maximum stresses can shift from the highly damaged region[9]. In our highly damaged samples, an amorphous region extends far below the surface. This layer is critical to the modification of stress distributions in the Si crystal: plastic flow can occur in a-Si, causing the relaxation of strains and thus a significant reduction of stresses in the implanted region[10]. Point defects initially produced by the argon implantation could outdiffuse from the implanted region and accumulate in the underlying crystalline Si, yielding the maximum stress at a location below the amorphous-crystalline Si interface, much deeper than the implanted H range[11]. It would be energetically favorable for H atoms to trap at the location of maximum stresses upon annealing, resulting in Si eruption at a depth ~ 1 μm below the surface in our samples.

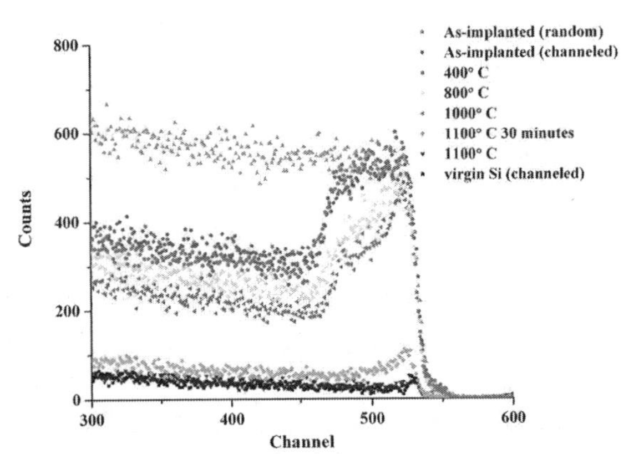

FIGURE 3. Rutherford backscattering/ion channeling of the H and Ar co-implanted and annealed samples. All samples are annealed at 75 minutes unless otherwise indicated.

Hydrogen profiling using the $^{15}N(p,\alpha\gamma)^{12}C$ reaction at 6.385 MeV was performed., and the role of stresses in exfoliation can be further elucidated by considering the H profiles following annealing. We find most of the implanted H fluence remains in the sample following the 400° C anneal; however at temperatures 600° C and higher, nearly all the H has outdiffused from the sample: the surface morphology remains irrespective of the presence of hydrogen. The exfoliation which occurs only at very high temperatures is therefore related to the H-induced strains which cause the eruption at high thermal budgets, rather than the internal gas pressure of the H_2 itself. These strains remain after the H has left the sample, and play a key role in the evolution of the surface morphology, particularly exfoliation.

For the samples with and Ar implant at 90 keV, reflection data has been taken at normal incidence (Figure 4). At annealing temperatures less than 900 °C, the data displays oscillatory features resulting from optical interference due to the a-Si layer above the crystalline substrate. As suggested by RBS/C data, the amorphous Si layer begins to recrystallize and forms grains after annealing at 900 °C, reducing this interference. Here, reflection decreases by ~ 50% compared to virgin Si, consistent with the appearance of a high density of craters on the surface. A dramatic reduction in reflectance occurs for the sample annealed at 1100 °C, with a reflectivity below 5% over a wide range of the solar spectrum. We have also evaluated the diffuse reflection loss of this sample and found that its value is greater than that of the polished virgin Si (1-2%), but remains below 5%. Therefore the actual light transmittance is expected to exceed 90%.

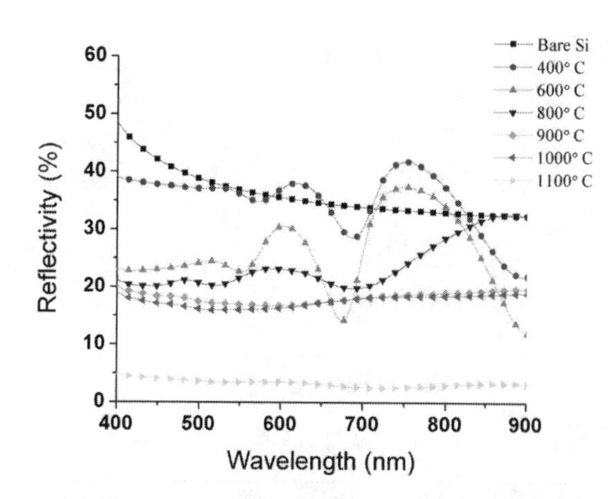

FIGURE 4. Rutherford backscattering/ion channeling of the H and Ar co-implanted and annealed samples. All samples are annealed at 75 minutes unless otherwise indicated.

CONCLUSION

To summarize, we have developed an approach to fabricating surface textures suited for silicon antireflection, based on ion implantation of H and Ar combined with thermal annealing. Using only a two-step process, we can produce highly textured silicon while recovering the crystallinity of the virgin sample, and whose normal incidence specular reflection is below 5%, and total diffuse + specular reflection is below 10%. Being fully compatible with current semiconductor technologies, the approach would offer a viable solution to large-volume and low-cost production of Si photovoltaic devices.

ACKNOWLEDGMENTS

We would like to thank Art Haberl and Wayne Skala in the University at Albany Ion Beam Lab for help with ensuring that the implantation and characterization ran smoothly.

REFERENCES

1. A. Goetzberger, J. Luther, and G. Willeke, Solar Energy Materials & Solar Cells 74, 1 (2002).
2. Handbook of Photovoltaic Science and Engineering, edited by A. Luque and S. Hegedus (Wiley, New York, 2003).
3. W. Zhou, M. Tao, L. Chen, and H. Yang, J. Appl. Phys. 102, 103105 (2007).
4. C.-H. Sun, P. Jiang and B. Jiang, Appl. Phys. Lett. 92, 061112 (2008).
5. C.-H. Sun, W.-L. Min, N.C. Linn, P. Jiang and B. Jing, Appl. Phys. Lett. 91, 231105 (2007).
6. M. Shen, J.E. Carey, C.H. Crouch, M. Kandyla, H.A. Stone, and E. Mazur, Nano Lett. 8, 2087 (2008).
7. B. Terreault - Phys. Status Solidi. A 204, 2129 (2007).
8. W. Jiang, W.J. Weber, S. Thevuthasan, and R. Grotzschel, Nucl. Instrum. Methods Phys. Res. B 166-167, 374 (2000)
9. T. Höchbauer, A. Misra, M. Nastasi, and J.W. Mayer, J. Appl. Phys. 92, 2335 (2002).
10. C.A. Volkert, J. Appl. Phys. 70, 3521 (1991).
11. A.I. Titov and S.O. Kucheyev, Nucl. Instrum. Methods Phys. Res. B 168, 375 (2000).

SHI Effects On Ge+SiO_2 Composite Films Prepared By RF Sputtering

N. Srinivasa Rao, A.P. Pathak*, N. Sathish, G. Devaraju and V. Saikiran

School of Physics, University of Hyderabad, Central University (P.O), Hyderabad 500 046, India

Abstract. Ge+SiO_2 composite films were deposited on Silicon substrate using RF magnetron sputtering. The as-deposited samples were irradiated with 150 MeV Ag^{+12} ions at a fixed fluence of $3x10^{13}$ ions/cm^2. These samples were subsequently characterized by X-ray diffraction (XRD) and Raman spectroscopy to understand the crystallization behavior. Formation of Ge nanocrystal in amorphous silicon dioxide film was studied using transmission electron microscopy (TEM). We also studied the surface morphology of these high energy irradiated samples by Atomic Force Microscopy (AFM). The basic mechanism for ion beam induced crystallization in these films has been discussed.

Keywords: Swift heavy ions, crystallization, RF sputtering, XRD, TEM and AFM
PACS: 61.80.Lj; 81.40.Ef; 81.15.Cd; 61.10.Nz; 68.37.Lp; 68.37.Ps

INTRODUCTION

Nanocomposite materials consisting of small nanocrystals in a host matrix are of current interest for their potential applications in electronic and photonic materials. Particularly, semiconductor nanoparticles exhibit new quantum phenomena which have potential application in a variety of photonic devices. Synthesis of such semiconductor nanocrystals, has been very promising field of research over the last decade. A large number of experimental investigations have been carried out in order to explore the interesting physical properties of nanocrystals or quantum dots. The properties of such nanostructured materials are quite different compared to the corresponding bulk materials. Several works have been reported on the synthesis of Ge nanocrystals prepared by various methods, such as RF co-sputtering [1], dc sputtering [2], ion implantation [3,4], evaporation–condensation [5], electron beam evaporation [6], chemical vapour deposition [7], and pulsed laser deposition [8]. Ion beam induced crystallization of materials has attracted enough interest in the recent years. Several such studies have been carried out on amorphized Si and implanted samples [9-12]. This important phenomenon of ion beam induced crystallization has been demonstrated using both, low energies as well as swift heavy ions [13]. Already Si nanocrystals embedded in SiO_2 [14-16] and Ge nanocrystals embedded in GeO_2 [17] matrix have been prepared by ion irradiation. Atom beam sputtering (ABS) is used to prepare Ge nanocrystals [18-19]. ABS has also been used to synthesis of metal–polymer nanocomposite for optical applications and to study the controlled growth of gold nanoparticles by ion irradiation [20-21]. Crystallization of Ge implanted SiO_2 films by swift heavy ion irradiation was also reported [22]. Effect of ion irradiation on already prepared Ge nanocrystals has also been studied [23].

The aim of the present work is to synthesize Ge nanocrystals by irradiating Ge+SiO_2 composite films with swift heavy ion bombardment. Specifically, we demonstrate the effects of swift heavy ion irradiation on Ge+SiO_2 films prepared by RF magnetron sputtering. The samples were characterized by transmission electron microscopy (TEM), X-ray diffraction, Raman spectroscopy and atomic force microscopy (AFM). Here we present TEM image of the formed nanocrystals that provides direct evidence of the nanocrystal formation. The characterization results obtained by using various techniques are consistent and have been discussed in detail.

EXPERIMENTAL DETAILS

The composite films were deposited at ambient temperature by co-sputtering the Ge and SiO_2 target using RF magnetron sputtering. Prior to deposition, substrates were cleaned using acetone, isopropyl alcohol and deionized water. The sputtering was carried out in a high pure Ar gas (99.999%) atmosphere. Pre-sputtering of the target was done in

Application of Accelerators in Research and Industry
AIP Conf. Proc. 1336, 341-344 (2011); doi: 10.1063/1.3586116

order to clean and remove any surface impurities. After vacuum pumping, the sputtering was performed with an Ar pressure of 20 SCCM (standard cubic centimeter per minute) was introduced through a mass flow controller. The RF power was maintained at 20 W. The chamber was evacuated upto 5 x 10^{-6} torr before the sputtering and the working pressure during the sputtering was 5 x 10^{-3} torr. The films were deposited on (1 0 0) oriented p-type Si substrates of 0.5 mm thickness purchased commercially. The films were also deposited on Carbon coated Nickel grids for TEM measurements and the same grids were subjected to swift heavy ion irradiation. Subsequently, these composite thin films of Ge and silica were irradiated with 150 MeV Ag^{+12} ions. The mean projected range of these ions is about 20 microns, which is much larger than the film thickness. The electronic energy loss and nuclear energy loss of 150 MeV Ag ions are 13 keV/nm and 0.048 keV/nm respectively. The irradiation was performed at fixed fluence of $3x10^{13}$ ions/ cm^2 using the 15UD Pelletron accelerator at Inter University Accelerator Centre (IUAC), New Delhi India. The ion beam current was maintained low (0.5–2 pnA) to avoid heating of the samples. The samples were kept at an angle of 5^0 with respect to the beam axis to avoid the channeling effect. Grazing incidence x-ray diffraction (GIXRD) spectra were recorded by Bruker D8 Advance diffractometer at an incidence angle of 2^0, using a CuK_α (1.5406 Å) at IUAC, New Delhi. The Raman measurements were carried out at room temperature using the 514.5 nm line of an Ar ion laser as an excitation source. Size of Ge nanocrystals embedded in Silica matrix has been obtained using advanced transmission electron microscopy (TEM), namely, FEI TECNAI, G^2 F-TWIN D2083 TEM, with electron-accelerating voltage of 200 kV. The atomic force microscopy (AFM) measurements were carried out using SPA400, Seiko Instruments Inc. All the measurements were carried out at room temperature.

RESULTS AND DISCUSSION

The thickness of the composite films used in this study is around ~150nm. XRD spectra of as-deposited and irradiated samples have been shown in Fig.1.
It is clear from the spectra that the pristine sample shows amorphous nature where as irradiation leads to crystallization of these films. The average size of the nanoparticles is calculated using Scherrer's formula given by

$$D = \frac{0.9\,\lambda}{\beta\ Cos\theta_B}$$

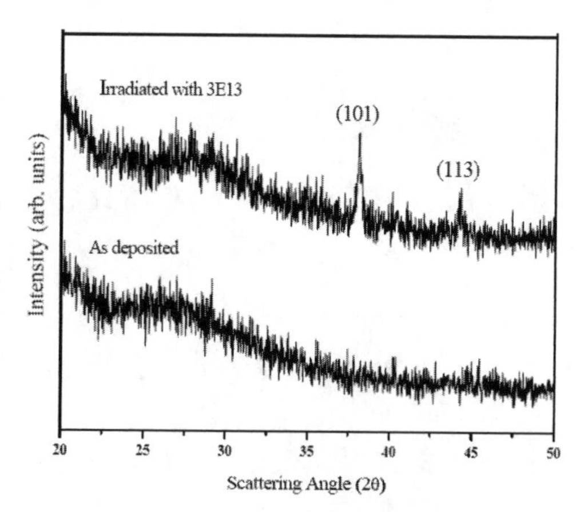

FIGURE 1. GIXRD spectra of as deposited and irradiated samples.

where λ is the wavelength of the X-ray source (i.e, 1.5406 Å for CuKα), β = FWHM (in radians) and θ_B is Bragg angle and D is the diameter of the nanocrystal. The size of the nanocrystal was estimated to be around 31 nm.

The Raman spectra of as deposited and irradiated sample have been shown in Fig.2. The as-deposited sample shows a broad Raman peak centered at 270 cm^{-1} which corresponds to amorphous Ge [24]. The appearance of peak around 290 cm^{-1} for the irradiated sample indicates the crystallization of the film. The broad feature in the irradiated Raman spectrum along with the peak at ~290 cm^{-1} indicates partial amorphous nature of the film. It is known that mechanical compressions can lead to an increase in the frequency of Ge–Ge bond vibrations. Due to the 4% mismatch in the lattice constants, the shift for transverse optical modes can reach 10 cm^{-1} [25].

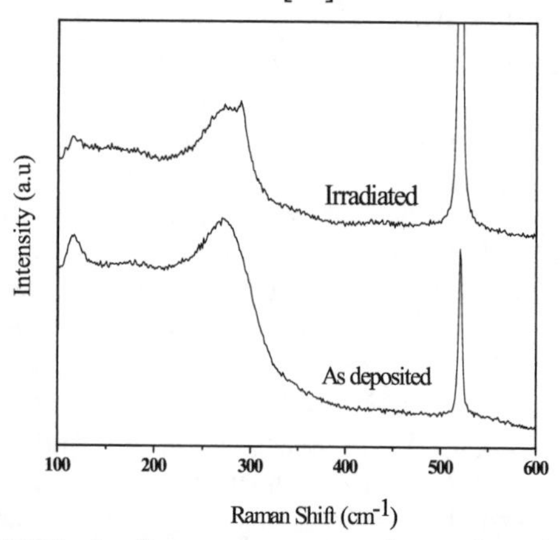

FIGURE 2. Raman spectra of as deposited and irradiated samples

The effect of mechanical stresses would lead to a shift of peaks toward the high-frequency region. It is also possible that the structural change of the Ge nanocrystals or stress in the high Ge concentration films will also result the peak position to be lower than the bulk Ge value. Raman spectra of the samples show a shift in the peak position and an asymmetrical broadening indicates the existence of crystalline Ge.

Figure 3 shows TEM image of typical Ge nanocrystals obtained from the irradiated sample. Here the formation of nanocrystallites is clearly seen in the TEM micrographs of the Ag ion irradiated sample.

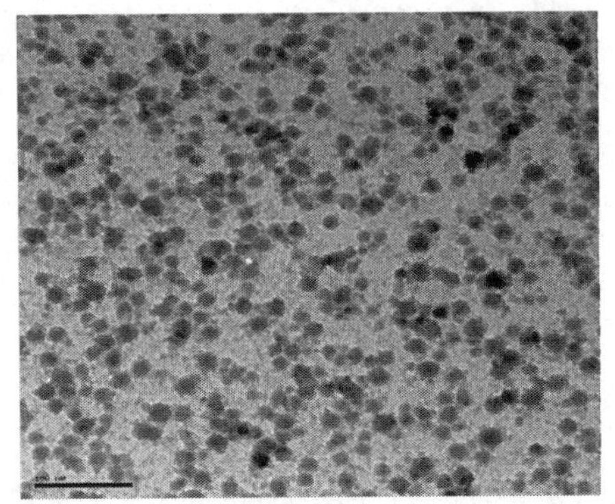

FIGURE 3. TEM image of 150 MeV Ag irradiated sample at a fluence of 3 x 10^{13} ions/cm^2.

The nanocrystal size is found to be around 20 nm which is different from XRD. This may be due to experimental error or strain in the system. It is possible that Ge nanocrystals with different average nanocrystal sizes and size distributions can be synthesized by varying ion fluence and energy.

The surface topography of Ge+SiO_2 films was investigated with AFM. It shows that the as-deposited films have less surface roughness than the irradiated ones, the roughness of films increased and a grain-like morphology was exhibited. The 3D AFM images of the as deposited and irradiated samples are shown in Fig. 4a and 4b respectively.

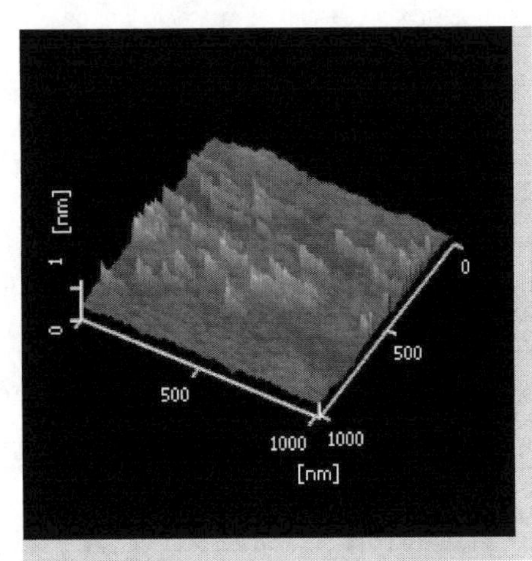

FIGURE 4a: Atomic force microscope image of as deposited sample.

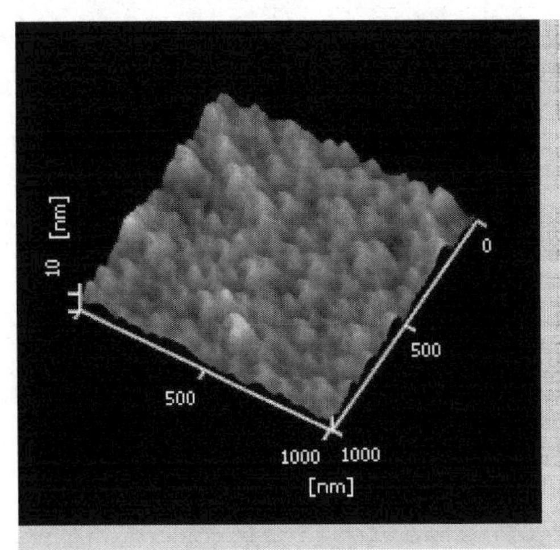

FIGURE 4b: Atomic force microscope image of irradiated sample.

The rms roughness values of as deposited and annealed films are 0.1 to 2.2 nm. Surface Morphological changes in the irradiated films can be observed clearly from the figures. While the as deposited film is more or less flat and featureless, the irradiated film shows an evolution of nano-sized grains.

When an energetic ion interacts with a solid, it undergoes a series of collisions with the atoms and electrons in the target. In these collision processes incident ion losses energy ranging from few to 100 keV/ nm depending upon the energy, mass, and density of the target. The energy loss rate dE/dx of an energetic ion moving through a solid is determined by screened Coulomb interactions with the substrate atoms and electrons. Ion losses its energy through two major processes i.e. nuclear or elastic collisions, in which energy is transferred to a target atom as a whole and electronic collisions, in which the moving particle energy is used in exciting or ejecting the atomic electrons of the target. Electronic collisions involve negligible deflection of the ion trajectory, and lattice disorder. When the projectile velocity v is much greater than that of an orbital electron, the influence of the incident particle on an atom (as a whole) may be considered as a sudden, small external perturbation. But the collision produces a sudden transfer of energy from the projectile to the target electrons. The stopping cross section increases with decreasing velocity because the particle spends more time in the vicinity of the atom, with slowing down. The deposition of energy into the film is predominantly through electronic energy loss. This energy deposition leads to crystallization similar to that of thermal crystallization and can be considered as the crystallization induced by thermal spikes. One can explain the basic mechanism of recrystallization and formation of nanocrystals in these samples, under ion-irradiation with the help of thermal spike theory [26]. When the swift heavy ion passes through the material the energy is transferred to the target lattice via electron–phonon coupling. It is known from the work of Meftah *et al.* [27] and Toulemonde *et al.* [28] that a track diameter of 10 nm is created for the Se value of ~11 keV/nm in silica. This large amount of energy is transferred to the target electrons in very short time, of the order of $\sim 10^{-12}$ seconds which leads to an increase of the lattice temperature above its melting point along the ion track. The temperature along the ion path goes to very high value which is sufficient to crystallize the Ge, which is embedded in Silica. This results in a modification of structure of films around the cylindrical zone and crystallization of the films.

CONCLUSIONS

Synthesis of Ge nanocrystals by using RF sputtering followed by ion irradiation has been investigated. The composite films were irradiated with 150 MeV Ag ions at a fixed fluence of 3×10^{13} ions/cm^2. As deposited films show amorphous nature but irradiation leads to crystallization of the films, which is clear from XRD and Raman spectroscopy. Surface morphology of the samples is studied by AFM. TEM images show clearly formation of nanocrystals and the average crystallite size of these Ge nanocrystals is found to be around 20 nm. The mechanism of Ion beam induced Ge recrystallization has been discussed.

ACKNOWLEDGMENTS

N.S.R. and GD thank CSIR, New Delhi for award of SRF. A.P.P. thanks Centre for Nanotechnoloy for project support. V S K thanks CSIR, New Delhi for award of JRF. We thank Dr D K Avasthi and Mr S A Khan for their kind help in during ion irradiation and fruitful discussions. We also acknowledge the help of Mr. Manohar Reddy and Mr. Durga Prasad for their kind support during TEM measurements, performed at the Centre for Nanotechnology established by DST New Delhi at University of Hyderabad.

*Corresponding author e-mail: appsp@uohyd.ernet.in
Tel: +91-40-23010181/23134316,
Fax: +91-40-23010181 / 23010227

REFERENCES

1. Y Maeda , *Phys Rev B* 51(3) (1995) 1658
2. M Zacharias, P M Fauchet . *J Non-Cryst Solids*; 227–230 (1998) 1058
3. C Bonafos , et al., *Nucl Instrum Methods Phys Res B*; 178 (2001) 17.
4. K Masuda, M Yamamoto, M Kanaya , Y Kanemitsu . *J Non-Cryst Solids* 299–302 (2002) 1079
5. A Stella, P Tognini, C E Bottani, P Milani, P Cheyssac, R Kofman R. *Thin Solid Films* 318 (1998) 100.
6. Q Wan, C L Lin, W L Liu, T H Wang. *Appl Phys Lett* 82(26) (2003) 4708
7. Z He , J Xu , W Li , K Chen, D Feng D. *J Non-Cryst Solids* 266–269 (2000) 1025.
8. Y Zhu, P P Ong *J Phys Condens Matter* 13 (2001) 4075.
9. J. Nakata, *Phys. Rev. B* 43 (1991) 14643
10. J. Nakata, *J. Appl. Phys.* 79 (1996) 682
11. T.Som, B. Satpati, O. P. Sinha, and D. Kanjilal, *J. Appl. Phys.* 98 (2005) 13532.
12. J. Linnros, G. Holmen and B. Sevensson, *Phys. Rev. B* 32 (1985) 2770
13. P. Songsiriritthigul and G. Holmén, *Nucl. Instr. and Meth. B* 120 (1996) 207
14. T. Mohanty, Mishra N C, Pradhan Asima and Kanjilal D, *Surf. and Coat. Technol.* 196 (2005) 34
15. Prajakta S. Chaudhari, Tejashree M. Bhave, D. Kanjilal and S.V. Bhoraskar, *J. Appl. Phys.* 93 (2003) 3486
16. G.A. Kachurin • S.G. Cherkova • D.V. Marin • A.G. Cherkov • V.A. Skuratov, Appl Phys A 98 (2010) 873
17. Shyama Rath, D. Kabiraj, D.K. Avasthi, A. Tripathi, K.P. Jain, Manoj Kumar, H.S. Mavi, A.K. Shukla, *Nuclear Instruments and Methods in Physics Research B* 263 (2007) 419
18. N. Srinivasa Rao, et al, *Nucl. Inst. Meth B* 264 (2007) 249–253.
19. N Srinivasa Rao, et al, *Radiation Effects and Defects in Solids.* Vol. 164, Nos. 7–8, July 2009, 452
20. D. K. Avasthi, Y. K. Mishra, D. Kabiraj, N. P. Lalla and J. C. Pivin, 2007 *Nanotechnology* 18 125604.
21. Y. K. Mishra, S. Mohapatra, Rahul Singhal, D. K. Avasthi, D. C. Agarwal, S. B. Ogale, 2008 *Appl. Phys. Lett.* 92, 043107
22. N. Srinivasa Rao, et al, *Nuclear Instruments and Methods in Physics Research B* 268 (2010) 1741.
23. M. C. Ridgway, G. de M. Azevedo, R. G. Elliman, C. J. Glover, D. J. Llewellyn, R. Miller, and W. Wesch, G. J. Foran, J. Hansen and A. Nylandsted-Larsen, *Phys. Rev. B* 71 (2005) 094107
24. Y. Sasaki and C. Horie, 1993 *Phys. Rev. B* 47, 3811
25. D. A. Orekhov, V. A. Volodin, M. D. Efremov, A. I. Nikiforov, V. V. Ul'yanov, and O. P. Pchelyakov, *Journal of Experimental and Theoretical Physics Letters*, Vol. 81, No. 7, (2005) 331
26. M. Toulemonde, C. Dufour, E. Paumier, *Phys. Rev. B* 46 (1992) 14362
27. A. Meftah, F. Brisard, M. Costantini et al., *Phys. Rev. B* 49 (1994) 12457
28. M. Toulemonde, J.M. Costantini, Ch. Dufour et al., *Nucl. Instrum. Methods B* (1996) 116 37

Ion Solid Interaction And Surface Modification At RF Breakdown In High-Gradient Linacs

Zeke Insepov,[a] Jim Norem,[a] Seth Veitzer [b]

[a] *Argonne National Laboratory, 9700 South Cass Avenue, Argonne, IL 60439, United States*
[b] *Tech-X Corp., 5621 Arapahoe Ave., Suite A, Boulder CO 80303, United States*

Abstract. Ion solid interactions have been shown to be an important new mechanism of unipolar arc formation in high-gradient rf linear accelerators through surface self-sputtering by plasma ions, in addition to an intense surface field evaporation. We believe a non-Debye plasma is formed in close vicinity to the surface and strongly affects surface atomic migration via intense bombardment by ions, strong electric field, and high surface temperature. Scanning electron microscope studies of copper surface of an rf cavity were conducted that show craters, arc pits, and both irregular and regular ripple structures with a characteristic length of 2 microns on the surface. Strong field enhancements are characteristic of the edges, corners, and crack systems at surfaces subjected to rf breakdown

Keywords: Ion solid interaction; unipolar arc formation; high-gradient rf linear accelerators; non-Debye plasma
PACS: 29.20.-c, 52.77.Dq; 34.35.+a, 79.20.Rf; 61.80.-x; 79.20.Rf; 34.80.Dp

INTRODUCTION

Energetic ion collisions with solid targets are an important area of research in basic science [1-8], as well as in numerous industrial applications [9-19]. Self-sputtering processes are also of fundamental interest in the development of high-gradient rf accelerators [1]. The Neutrino Factory and Muon Collider Collaboration is looking at developing low-frequency structures for muon cooling [2–6], the International Linear Collider is optimizing the performance of 1.3 GHz superconducting rf structures aimed at the design of a 1 TeV superconducting electron/positron collider [7], and the High Gradient RF Collaboration is studying high-frequency (f >10 GHz) structures aimed at an electron-positron collider operating at energies higher than 1 TeV [8].

A self-sustained self-sputtering occurring during high-current pseudospark operation ($\approx 10^4$ A/cm^2, $I > 10^3$ A) is shown in [9,10] to be a possible mechanism for superdense glow. The self-sputtered cathode atoms become ionized in the beam of electrons accelerated in the cathode sheath.

A review of the method of MD computer simulation and results obtained for the physics of sputtering is given in [20]. MD simulations of self-sputtering and sticking coefficients of low-energy ion were reported in [21].

The goal of this paper is to develop further a new unipolar arc plasma model in high-gradient linear accelerators.

SELF-SPUTTERING OF SOLID AND LIQUID COPPER SURFACES

Molecular dynamics (MD) simulations of the copper self-sputtering yields were performed at various temperatures (300, 800, and 1300 K) and electric field strengths and for a wide range of Cu ion energies from 50 eV to 50 keV. The details of the calculations can be found in our previous papers [11-13] where the obtained calculated yields were compared with the existing experimental and simulation results [20-22].

Our previous OOPIC simulations[1] showed that the copper surface can become very hot as a result of ion bombardment within a few nanoseconds in close vicinity to the plasma and can be melted. Therefore, a liquid Cu surface for MD simulation was prepared by heating a copper substrate to the melting temperature of 1310 K, equilibrating this configuration, and storing it as an input file for the sputtering yield calculations. The ion energies were selected between 50 and 150 eV, corresponding to the sheath potential value obtained by our OOPIC simulation. Figure 1 shows

[1] To be published elsewhere

Application of Accelerators in Research and Industry
AIP Conf. Proc. 1336, 345-348 (2011); doi: 10.1063/1.3586117

the results of our MD simulations of self-sputtering of the copper rf-cavity surface by accelerated Cu+ ions, at ion energies varied between 50 and 150 eV and surface temperatures of 300–1310 K. The most important result is that the self-sputtering yield at 800 K for an ion energy of 100 eV was found to be approximately 2.5 and increases as the melting point of 1310 K is reached.

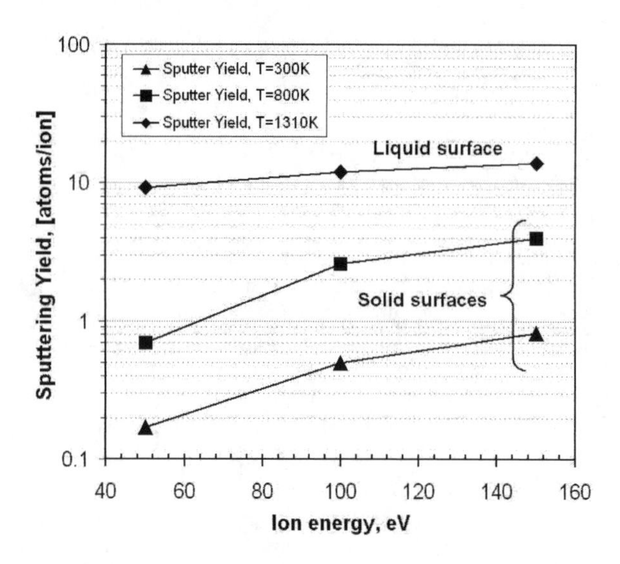

FIGURE 1. Dependence of the sputtering yield of a Cu surface bombarded with Cu+ ions on ion energy, for various copper surface temperatures, T = 300–1310 K.

The electric field affects the surface self-sputtering in two ways. One way is that the field accelerates the bombarding ion up to 100–150 eV, according to our OOPIC simulations. The second way is that a high electric field brings electric charges to the cavity surface, subjecting the surface to electrostatic tensile stress, thus making surface atoms less cohesive.

Figure 2 compares different currents leading to different mechanisms of surface erosion and heating that indicate an importance of the direct erosion of the surface by ion sputtering mechanism at surface temperatures above 500K.

Our simulation results show that both high temperature and high electric field can significantly increase the surface erosion. Both mechanisms will be important in a real rf cavity. We believe that the unipolar arc model explains the mechanism of surface damage in rf breakdown by increasing self-sputtering of hot copper surface residing at a very high local electric field (depending on the local curvature). This mechanism has never been addressed before in rf breakdown studies. Our simulation has showed—for the first time—that surface temperature and surface local field can significantly increase the erosion yield by a factor > 10 consistent with unipolar arcs. We note that the earliest unipolar arc model for Tokamak studies was developed based on an arbitrary assumption that the surface erosion rate was ~10 [11,12]. Here we confirm such high erosion rates at the rf cavity surfaces.

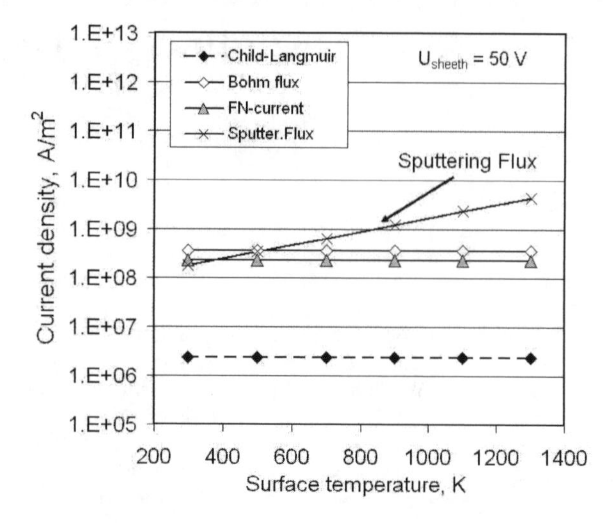

FIGURE 2. Comparison of different currents (electron and ion) calculated for the surface electric field generated by a unipolar arc with the sheeth potential of 50 V. The ion self-sputtering mechanism of surface erosion becomes dominant at surface temperatures above 500K.

NON-DEBYE PLASMA FORMATION

We assume that the trigger for breakdown events is the injection of high-density material above a field emitter, where the intense, field emission currents would break up and ionize the material to produce a plasma. A number of mechanisms could be involved. Ionization of neutral metallic gas has been modeled by using OOPIC Pro, assuming field-emitted electrons are produced below an inertially confined atomic gas [23,24]. Initial results show that the ionized electrons, as well as the majority of the field-emitted electrons, are accelerated through the plasma, producing a net positively charged plasma, which is slowly expanding because of its own charge.

OOOIC Pro is a particle-in-cell (PIC) physics simulation for 2D (x, y) and (r, z) geometries with 3D electrostatic and electromagnetic field solvers and Monte Carlo collision and ionization models. We model field emission at high current densities in a self-consistent way by calculating space charge fields in the presence of plasma ions and electrons. The ionization of copper and various secondary emission coefficients are contained in the code [24].

As the density of this ion cloud increases, the potential increases until it is able to trap both ionization and field-emitted electrons; but throughout the rf cycle, electrons are primarily lost to the far wall of the cavity, maintaining quasi-neutrality. As the arc evolves, simulations show that while the plasma density increases approximately exponentially, the electron and ion temperatures do not significantly increase. As the density passes 10^{24} charges/m^3, the Debye length approaches a few nanometers for electron temperatures on the order of 10 eV, and the PIC model becomes inappropriate.

For cold plasmas with densities on the order of 10^{25} m^{-3} and electric fields less than or equal to 1 GV/m, d is comparable to or smaller than λ_D, and the plasma is too dense to be subject to Debye screening. Under these conditions, a large area (square microns) would function like an active field emitter, subject to Fowler-Nordheim and thermal emission while being actively eroded by ion self sputtering. Modeling shows that electrons introduced into the arc would continue to ionize neutral atoms, increasing the flux of ions hitting the surface

FIELD ENHANCEMENT AND SURFACE FIELDS

Defining a field enhancement, β = (local field/average surface field), considerable data shows field enhancements in the range of 100–1000 [25,26] and areas down to a few square nanometers are responsible for triggering arcs.

In addition to the structure described above, we experimentally see a variety of sharp edges, corners, and cracks in the surface. We considered field enhancements that would be present in these edges, corners, and crack systems and how they might function as field emitters. Specifically, the electrostatic Laplace equations for surface structures were numerically solved by using a finite-element multiphysics simulation package COMSOL [29].

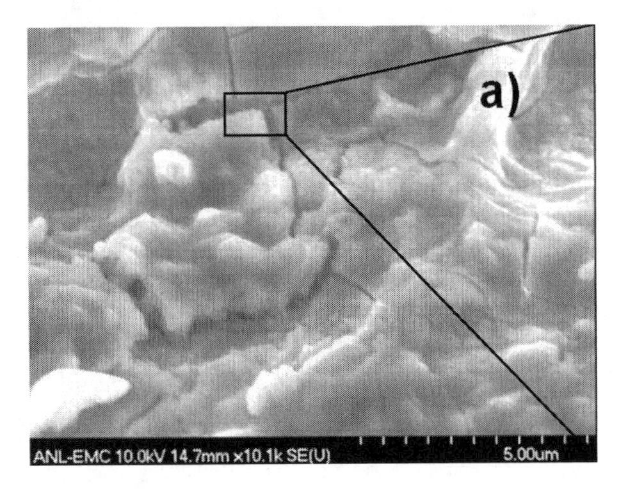

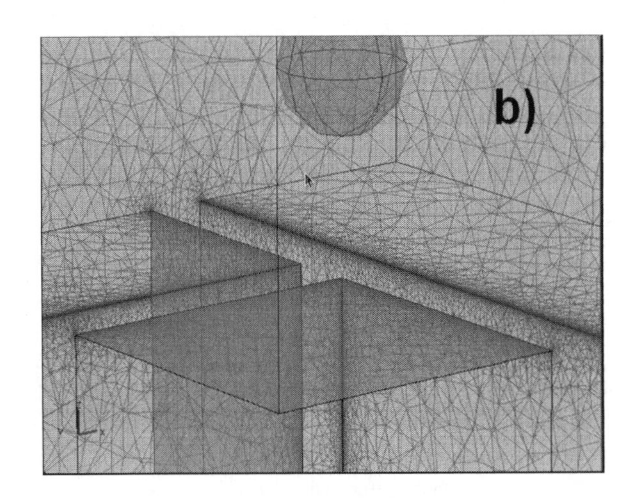

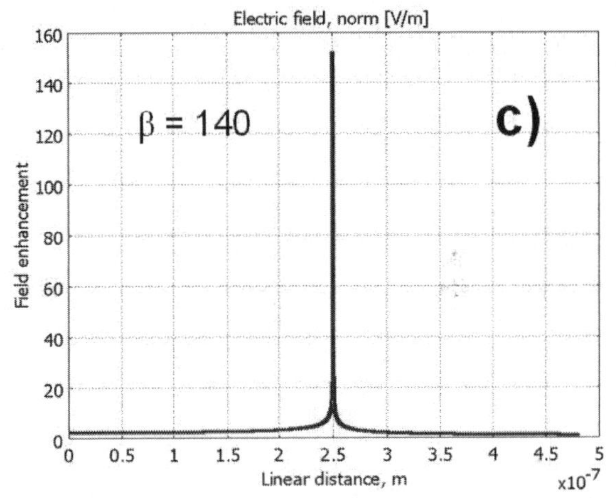

FIGURE 3. (a) SEM image of arc pit cracks on the breakdown surface from an rf cavity after a breakdown event showing considerable microstructure. The magnification in plot (b) shows our COMSOL simulation setting for a triple crack junction, and (c) shows the calculated field enhancement at the triple crack junction, with β = 140, consistent with cavity measurements.

These finite-element simulations showed that the calculated field enhancement can be comparable to the experimentally observed values $\beta \sim 140$; see Fig. 3c. From the electrostatic potential of 50–100 V produced by the OOPIC simulations, which is consistent with the burn voltage of copper arcs, this results in a surface field that increases rapidly with time to values on the order of 3–6 × 10^9, a range of surface fields that is capable of inducing enhanced field emission, tensile failure of materials, ion bombardment, and other mechanisms of damage, even in cold surfaces (however, these surfaces would be hot because of ion bombardment).

Experimentally, high-magnification scanning electron microscope (SEM) pictures of arc pits show hemispheres and ridges consistent with these high surface fields, in addition to cracks and pits with a variety of sharp edges at the limit of the resolution of the microscope. Mueller [28] and Lysenkov and Mueller [29] have also shown SEM pictures of a variety of physical shapes that produce high enhancement factors. We assume that these sharp edges contribute to the field enhancements and thus the field emission, and we calculate their properties.

SUMMARY

A new unipolar arc mechanism of vacuum surface breakdown in high-gradient rf linear accelerators was proposed. The unipolar arc mechanism can be realized through an intense surface self-sputtering by ions accelerated in plasma sheeth potential. Additionally, an intense surface field evaporation occurs due to a high electric field gradient. A non-Debye (dense) metal plasma can be formed in close vicinity to the surface that strongly affects surface atomic migration via ion bombardment, strong electric and magnetic fields, and high surface temperature. Micron-scale craters, pits, ripple structures with a characteristic length of 2 microns on the surface were observed with scanning electron microscope. Strong field enhancements was calculated for the edges, corners, and cracks at surfaces subjected to rf breakdown.

ACKNOWLEDGMENTS

This work was supported by the U.S. Dept. of Energy under Contract DE-AC02-06CH11357. The work of Tech-X personnel is funded by the U.S. Dept. of Energy under Small Business Innovation Research Contract No. DE- FG02-07ER84833.

REFERENCES

1. A. Hassanein, et al., Phys. Rev. STAB 9 (2006) 062001.
2. J. Norem, V. Wu, A. Moretti, M. Popovic, Z. Qian, L. Ducas, Y. Torun, N. Solomey, Phys. Rev. ST Accel. Beams 6 (2003) 072001.
3. J. Norem, Z. Insepov, I. Konkashbaev, Nucl. Instrum. Methods Phys. Res., Sect. A 537 (2005) 510.
4. Z. Insepov, J.H. Norem, A. Hassanein, Phys. Rev. ST Accel. Beams 7, 122001 (2004).
5. Feasibility Study-II of a Muon Based Neutrino Source, edited by S. Ozaki, R. Palmer, M. Zisman, and J. Gallardo, BNL-52623 (2001), http://www.cap.bnl.gov/mumu/studyii/
6. M.M. Alsharo'a et al., Phys. Rev. ST Accel. Beams 6 (2003) 081001.
7. http://www.linearcollider.org/cms/
8. http://www.slac.stanford.edu/grp/ara/HGCollab/HomePage/HGhome.htm.
9. A. Anders, S. Anders, M.A. Gundersen, A.M. Martsinovskii, IEEE Trans. Plasma Sci. 23 (1995) 275–282.
10. A. Anders, Cathodic Arcs: From Fractal Spots to Energetic Condensation, Springer Series on Atomic, Optical, and Plasma Physics, Springer (2008).
11. Z. Insepov, J. Norem, D. Huang, S. Mahalingam and S. Veitzer, Proceedings of PAC09, May 4–8, 2009, Vancouver Canada (2009)
12. Z. Insepov, J. Norem, D. Huang, S. Mahalingam and S. Veitzer, Proceedings of NuFact09 July 20–25, 2009, AIP Conference Proceedings 1222, (2009)
13. Z. Insepov, J. Norem, S. Veitzer, Nucl. Instr. Meth. B 268 (2010) 642–650.
14. F.R. Schwirzke, IEEE Trans. on Plasma Sci., 19 (1991) 690-696.
15. F.R. Schwirzke, X.K. Maruyama, IEEE Trans. on Plasma Sci., 21 (1993) 410-415.
16. Z. Insepov, M. Terasawa, K. Takayama, Phys. Rev. A 77, 062901 (2008).
17. R.L. Fleischer, P.B. Price, R.M. Walker, E.L. Hubbard, Phys. Rev. 156, (1967) 353–355.
18. A. Ootuka, K. Kawatsura, F. Fujimoto, K. Komaki, K. Ozawa, M. Terasawa, J. Phys. Soc. Jpn. 53, 1001 (1984).
19. W.H. Hayward, A.R. Wolter, J. Appl. Phys. 40 (1969) 2911.
20. H.M. Urbassek, Nucl. Instr. Meth. in Phys. Res. B 122 (1997) 427–441.
21. C.F. Abrams, D.B. Graves, J. Appl. Phys. 86 (1999) 2263–2267.
22 C. Steinbrüchel, Appl. Phys. Lett. 55,\ (1989) 1960.
23. G. Werner, Probing and Modeling Voltage Breakdown in Vacuum, Ph.D. thesis, Cornell University (2004).
24. Tech-X Software Inc.
25. J. Norem, Z. Insepov, D. Huang, S. Mahalingam, S. Veitzer, AIP Conf. Proc., Volume 1222, pp. 348-352...
26. A. Anders, S. Anders, and M. A. Gundersen, Phys. Rev. Lett. 71, 364 (1993).
27. COMSOL software
28. G. Mueller, University of Wuppertal, Private Communication (2007).
29. D. Lysenkov and G. Mueller, Int. J. Nanotech. 2, 239 (2005).

SECTION VIII – MEDICAL APPLICATIONS AND RADIOISOTOPES

Ion, X-Ray, UV And Neutron Microbeam Systems For Cell Irradiation

A.W. Bigelow, G. Randers-Pehrson, G. Garty, C.R. Geard, Y. Xu, A.D. Harken, G. W. Johnson, and D.J. Brenner

Center for Radiological Research, Columbia University
630 West 168th Street, 11th floor, New York, NY 10032

Abstract. The array of microbeam cell-irradiation systems, available to users at the Radiological Research Accelerator Facility (RARAF), Center for Radiological Research, Columbia University, is expanding. The HVE 5MV Singletron particle accelerator at the facility provides particles to two focused ion microbeam lines: the sub-micron microbeam II and the permanent magnetic microbeam (PMM). Both the electrostatic quadrupole lenses on the microbeam II system and the magnetic quadrupole lenses on the PMM system are arranged as compound lenses consisting of two quadrupole triplets with "Russian" symmetry. Also, the RARAF accelerator is a source for a proton-induced x-ray microbeam (undergoing testing) and is projected to supply protons to a neutron microbeam based on the $^{7}Li(p,n)^{7}Be$ nuclear reaction (under development). Leveraging from the multiphoton microscope technology integrated within the microbeam II endstation, a UV microspot irradiator – based on multiphoton excitation – is available for facility users. Highlights from radiation-biology demonstrations on single living mammalian cells are included in this review of microbeam systems for cell irradiation at RARAF.

Keywords: Radiation biology, microbeam, x-ray, UV, neutron
PACS: 87.53.-j

INTRODUCTION

Single-cell / single-particle microbeam irradiation technology at Columbia University's Radiological Research Accelerator Facility (RARAF) has evolved continuously starting with the user facility's initial aperture-based microbeam system[1] installed in February 1994. RARAF's first microbeam system pioneered numerous radiobiology studies in particle-induced DNA damage and repair mechanisms. The dominant interest among RARAF's users is basic research involving the bystander effect[2], where non-targeted cell nuclei manifest damage similar to targeted cell nuclei. In investigating cell-signaling pathways such as those potentially associated with the bystander effect, cell biologists at RARAF pursue fundamental understanding of low-dose effects. Highlights from radiation-biology demonstrations on single living mammalian cells are included in this review of microbeam systems for cell irradiation at RARAF.

MICROBEAM SYSTEMS AT THE RADIOLOGICAL RESEARCH ACCELERATOR FACILITY

Microbeam platforms at Columbia University's Radiological Research Accelerator Facility (RARAF) have evolved from an initial aperture-based microbeam (decommissioned) to an array of micron-sized irradiation probes. Our microbeam platforms are: microbeam II (electrostatic focusing system), the Permanent Magnetic Microbeam (PMM), an x-ray microbeam, the neutron microbeam (under development), and the UV microspot. All these systems have a common irradiation protocol that involves imaging, targeting and irradiating. In short: 1) cells are imaged and their center-of-gravity coordinates are registered within the control program, 2) cells are accurately positioned at the irradiation location, 3) cells are irradiated with a prescribed radiation dose, and 4) the next cell is brought to the irradiation location.

Application of Accelerators in Research and Industry
AIP Conf. Proc. 1336, 351-355 (2011); doi: 10.1063/1.3586118

Microbeam II

RARAF's principal microbeam system, microbeam II, uses an electrostatic quadrupole lens system capable of producing a submicron-diameter focused beam in air[3]. The electrostatic lens system is a compound quadrupole triplet lens system designed with Russian symmetry by A. Dymnikov[4]. Incentives for developing a microbeam system with a focused ion beam were: a) to overcome the penumbra or halo artifact associated with aperture-based microbeam systems and b) to improve beam rate at target. While a sharp submicron beam spot is compatible with irradiating sub-cellular structures, we demonstrate the horizontal-plane positioning stability of the vertical microbeam II ion beam in the section below.

Microbeam II Demonstration

To demonstrate the resolution of the microbeam II irradiator at RARAF, the letters "NIH" were irradiated onto a single live cell nucleus. The cells for this demonstration were HT1080 human Fibro Sarcoma cell nuclei containing GFP-tagged XRCC1 (single-strand DNA repair protein). These cells were provided by David Chen, UT Southwestern, Dallas, Texas; they were plated on microbeam dishes and kept under physiological conditions during the irradiation and imaging phases. A single cell nucleus was irradiated in an "NIH" dot-matrix stage-motion pattern with ~200 6-MeV alpha particles per spot. The pattern was reproduced via precision stage motions to irradiation locations. Typical spacing between points along each letter was 1 μm.

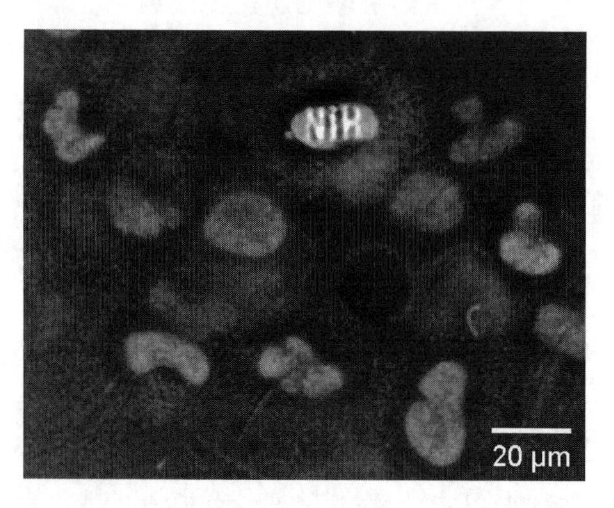

FIGURE 1. "NIH" written on a single cell nucleus with the RARAF microbeam. A tribute to our primary funding agency and a demonstration of our microbeam resolution.

As DNA damage occurred in the cell nucleus, XRCC1 repair protein formed foci at the damage sites. Immediately after the irradiation, a Z-stack of 21 wide-field fluorescent images was acquired using a water-dipping objective (60X 1.0NA), and a step size of $\Delta Z = 0.5$ μm, to reveal GFP concentrations in the nucleus exposed with the "NIH" pattern and in the neighboring (control) cell nuclei. Autoquant software was used to deblur the images in a post-processing step – result shown in Figure 1.

Permanent Magnetic Microbeam

The permanent magnetic microbeam (PMM)[5] shown in figure 2 is a second focused charged-particle microbeam developed at RARAF. It uses permanent magnetic quadrupole lenses (STI Optronics, Inc., Bellevue, WA) whose strengths are tunable by varying the insertion of permanent magnets into the shaped yokes[6]. Originally designed as a stand-alone microbeam with a ^{210}Po source, the system remains tethered to its accelerator-based test source of ions following media attention of the ^{210}Po-linked assassination of Alexander Litvinenko in November 2006[7] that led to heightened security and reduced availability of polonium. The use of an accelerator-based ion source allows for higher beam rate and, due to reduced chromatic aberrations, a smaller spot size than would be available with a ^{210}Po source. With the PMM fully functional as a second charged-particle microbeam line at RARAF, its design concept as a stand-alone microbeam is still valid and potentially more economical for radiation-research laboratories because it excludes purchase of a particle accelerator and support electronics. Contrary to electrostatic or electromagnetic lenses, the PMM requires neither high precision voltage/current supplies nor cooling. The use of permanent magnets also allows a much tighter pole-face gap, providing better optical properties compared to electromagnetic lenses.

Presently, the PMM is used for cell-irradiation experiments when development projects occupy the microbeam II endstation or beamline. Also, the PMM is used for development of point & shoot technology as well as microfluidics-based flow and shoot technology (FAST). Impetus for a point & shoot device is increased throughput. Pointing the beam with electromagnetic steering coils to cellular targets within a frame of view relies on electronic switching, which is inherently faster than moving a mechanical stage. Toward that end, a wide-field magnetic split-coil deflector system (Technisches Büro Fischer, Ober Ramstadt, Germany) was integrated into the PMM beamline. FAST developments are in collaboration with Dr. Daniel Attinger's Laboratory for Microscale

Transport Phenomena, Department of Mechanical Engineering, Columbia University. In-house software is in development for grabbing images with a high-speed camera of cell targets flowing through a microfluidic channel in order to track and predict the time and position of individual cells as they enter an irradiation zone to be irradiated by the point & shoot system.

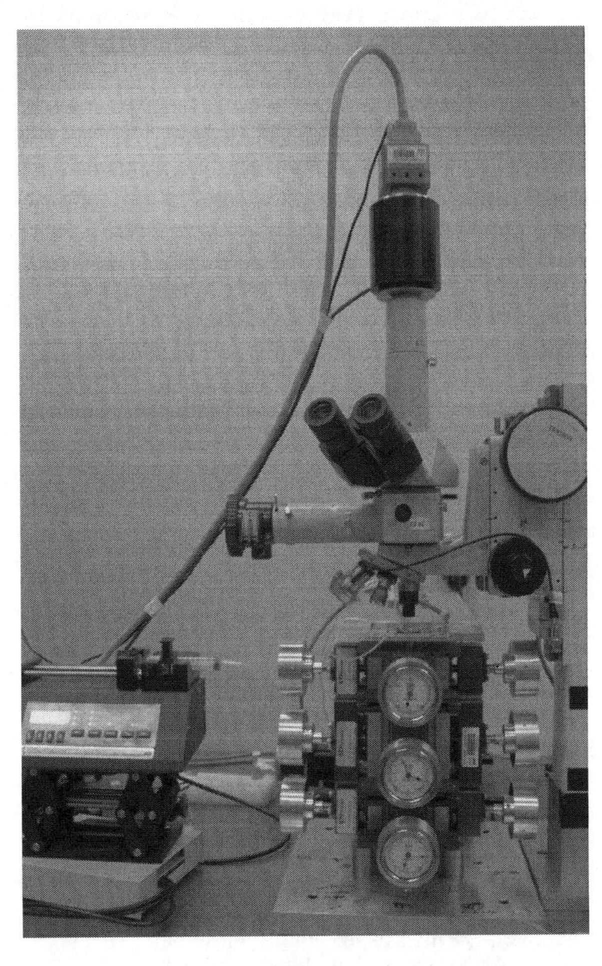

FIGURE 2. Photograph of the PMM endstation. Indexing knobs were added to the permanent magnet manipulators for improved tuning precision. The syringe pump in the lower left corner controls flow for FAST.

Room stability is a challenge for the PMM. The endstation was built in a 3rd floor laboratory, where beam position can shift by several microns as personnel move about the room. At present, the solution for this instability is to have both the microbeam operator and the biologist remain in a fixed location during the 20-minute long irradiation, while additional personnel are requested not to approach known locations where their presence would affect the ion beam position. Meanwhile, designs for a reinforced floor are being developed.

X-Ray Microbeam

In adding a soft x-ray microbeam for low linear energy transfer (LET) irradiation of cells at RARAF[8], the category of microbeam radiation has expanded beyond charged particles to include electromagnetic radiation. Our approach to x-ray production is through particle-induced x-ray emission (PIXE). In a two-stage process, an accelerator-produced proton beam impinges on a titanium target, embedded in a cooled copper rod, to produce x-ray emission, some of which is focused to a micron-sized diameter using a Fresnel zone plate diffractive optic (Zoneplates, Inc., London, UK). Cells are positioned at the x-ray focal point with an irradiation endstation modeled from the RARAF standard design developed on the microbeam II system.

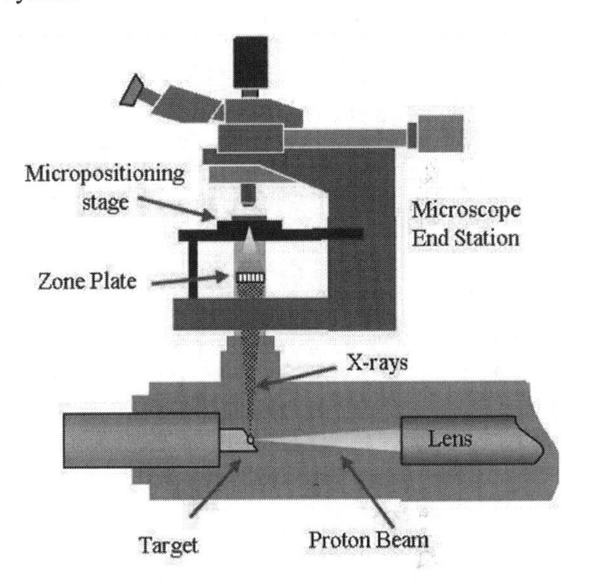

FIGURE 3. Schematic layout diagram of the x-ray microbeam endstation.

In our two-stage process to produce an x-ray microbeam, the first stage involves focusing a proton beam and the second stage involves focusing x-rays, as depicted in figure 3. In the first stage, an electrostatic quadrupole quadruplet lens focuses protons from an object distance of 5.77 m to an image distance of 14 cm, where a 100 micron diameter proton spot impinges on a titanium target face cut at a 70-degree angle to the proton incidence. With this geometry, part of the 4.5 keV Kα titanium characteristic x-rays are emitted in the vertical direction toward the zone plate and appear to originate from a 100 x 35 micron elliptical spot. In the second stage of x-ray microbeam production, x-rays exit the vacuum system through a beryllium window into a helium-filled chamber housing a Fresnel zone plate optic. The zone plate and order-selecting apertures are aligned through horizontal positioning adjustments on their mounts while optimizing x-ray fluence measurements with an ionization detector. The focused x-ray beam profile is measured using a knife-edge scan similar to the one used to determine beam profiles in the charged particle microbeams[3], the main difference being that here we monitor the beam fluence rather than the pulse height as the knife edge is scanned across the beam.

Currently, the available x-ray spot size is 8 x 3 microns, which is sufficient to irradiate a single cell nucleus. With an available dose rate of 1 mGy/s, initial biology experiments, expected to commence in Fall 2010, will focus is on low-dose and low-dose-rate applications, for example, bystander effect and adaptive response studies.

Neutron Microbeam

Complementing RARAF's microbeam irradiation platforms for charged particles and electromagnetic radiation, we are developing a revolutionary neutron microbeam that will have the capacity to deliver neutrons to cellular targets. The rationale for developing a neutron microbeam is to study exposure effects by neutrons with energy comparable to those within commercial nuclear reactors (< 50 keV), where personnel might be exposed.

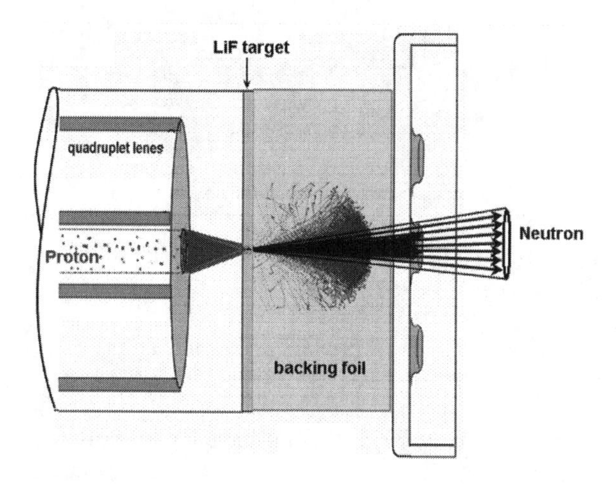

FIGURE 4. Schematic of Columbia University's neutron microbeam concept.

In our design, neutrons are formed through a $^{7}Li(p,n)^{7}Be$ nuclear reaction, which has a threshold at 1.881 MeV. An electrostatic quadupole quadruplet lens (positioned with object and image distances 2.5 m and 0.55 m, respectively) focuses 10 nA of 1.886 MeV protons to a 10 micron diameter spot at the lithium target where the nuclear reaction occurs – schematic shown in figure 4. The near-threshold reaction provides a relatively high neutron yield, as well as a very narrow forwardly-peaked angular distribution[9] with fluency parameters compatible with cellular irradiation: 21 degree maximum neutron angle, 16 micron neutron beam diameter at cell dish, and 31 keV mean neutron energy. The projected dose rate is 4.5 mGy/s. A thin Au/Al backing foil on the lithium target design stops forward-moving protons and removes them from the neutron beam.

Size of the neutron microbeam is determined by imaging alpha particle tracks on a customized thin $^{6}Li_2CO_3$-coated fluorescent nuclear track detector (FNTD) placed at the location for cellular irradiation. Alpha particles are produced through the $^{6}Li(n,\alpha)$ reaction in the FNTD coated layer. Alpha-particle tracks in FNTDs can be imaged with laser scanning microscopy techniques; the excitation wavelength is either 335 nm or 620 nm and the emission wavelength is 750 nm. In order to avoid the long turnaround time required for commercial FNTD imaging, we have obtained optics and a photomultiplier tube to allow in-house FNTD imaging with RARAF's multiphoton microscope[10]. The Titanium Sapphire laser excitation source of the multiphoton microscope is tunable over a wavelength range 680 - 1080 nm. Though this wavelength range excludes excitation at 620 nm, it offers three-photon excitation when the laser is tuned to 1000 nm – about three times the 335 nm excitation wavelength of an FNTD.

UV Microspot

Fulfilling requests from our facility users, we are integrating a UV microspot irradiator into our microbeam system. With our combination of a UV microspot and a charged-particle microbeam, RARAF is emerging as a unique facility for work that requires both photon and particle irradiations on the same platform. Our users have expressed needs to use an integrated UV microspot facility in two modes: 1) as a stand-alone UV microspot irradiator, and 2) as a probe in concert with ion-beam irradiation experiments.

Principles Of Operation

Traditional UV laser microbeam design involves an elongated laser path of exposure. Our system utilizes multiphoton excitation to produce a micro-volume of effective UV radiation. What makes our UV microspot unique is that it is integrated within the Microbeam II charged-particle cell-irradiation platform to provide a cocktail of photon and particle irradiations within one system.

Leveraging off of existing equipment, the tunable titanium-sapphire (Ti:S) laser from our multiphoton microscope[10] has the ability (through the multiphoton process) to produce a microspot of UV radiation at the focal point of the laser beam. In the two-photon process, two photons with a corresponding wavelength of 720 nm, for instance, can have a superposition of energy and act as one photon with a corresponding wavelength of 360 nm. This process is optimized at the focal point of a laser where the incident photons have a high probability of temporal and spatial coincidence.

The Ti:S laser for our multiphoton microscope covers a wavelength range of 680-1080 nm. These long wavelengths are minimally damaging to a tissue sample and, at the laser's focal point, the two-photon and the three-photon processes effectively provide (340-540 nm) and (227-360 nm), respectively. With this tool, we can implant a microspot of UV radiation into a tissue sample with 0.65 x 2.8 μm full width half maximum (FWHM) resolution and to depths of up to several hundred microns.

UV Microspot Demonstration

To demonstrate the resolution of RARAF's new UV microspot irradiator, the Columbia University crown logo was irradiated onto a live single cell nucleus. The cells for this demonstration were HT1080 human Fibro Sarcoma cell nuclei containing GFP-tagged XRCC1, described in the Microbeam II Demonstration section above. They were plated on Petri dishes and kept under physiological conditions during the irradiation and imaging phases.

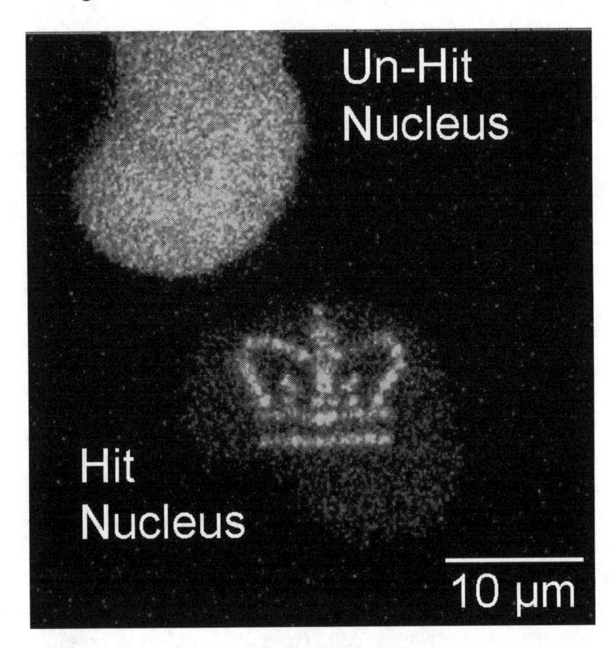

FIGURE 5. The Columbia University Crown logo irradiated by UV microspot onto a single cell nucleus.

Of the two cell nuclei visible in the multiphoton microscopy image shown in Figure 5, the lower cell nucleus was irradiated using a Titanium Sapphire laser tuned to 976 nm, which acts as 488 nm and 325 nm in the two- and three-photon modes, respectively. The crown pattern was produced via precision stage motions to 59 irradiation locations. Each location received 16 mW of laser power for 1 second to an elliptical volume 0.65 μm radial by 2.8 μm axial FWHM. Typical spacing between points was 1 μm.

As DNA damage occurred in the cell nucleus, XRCC1 repair protein formed foci at the damage sites. Following irradiation, multiphoton microscope z-stack imaging (ΔZ = 1 μm, Z range = 15 μm, λ = 976 nm, ~1sec/frame, 10X averaging) revealed GFP concentration in the nucleus exposed to the crown pattern and in the neighboring cell nucleus. With a 60X water-dipping objective, the image size is 58.1 μm wide by 55.8 μm high. To improve image quality, AutoQuant, an image debluring program, was used to surpass point-spread-function limitations.

ACKNOWLEDGMENTS

Supported by the National Institute of Biomedical Imaging and Bioengineering under Grant: NIBIB 2 P41 EB002033-14.

REFERENCES

1. G. Randers-Pehrson, C. R. Geard, G. Johnson, C. D. Elliston and D. J. Brenner, *Radiation Research* **156** (2), 210-214 (2001).
2. B. Ponnaiya, G. Jenkins-Baker, D. J. Brenner, E. J. Hall, G. Randers-Pehrson and C. R. Geard, *Radiation Research* **162** (4), 426-432 (2004).
3. A. W. Bigelow, D. J. Brenner, G. Garty and G. Randers-Pehrson, *IEEE Transactions on Plasma Science* **36** (4), 1424-1431 (2008).
4. A. D. Dymnikov, D. J. Brenner, G. Johnson and G. Randers-Pehrson, *Review of Scientific Instruments* **71** (4), 1646-1650 (2000).
5. G. Garty, G. J. Ross, A. W. Bigelow, G. Randers-Pehrson and D. J. Brenner, *Radiation Protection Dosimetry* **122** (1-4), 292-296 (2006).
6. S. C. Gottschalk, Dowell, D.H., and Quimby, D.C., *Nucl. Instr. & Meth. A* **507**, 181-185 (2003).
7. J. Harrison, R. Leggett, D. Lloyd, A. Phipps and B. Scott, *Journal of Radiological Protection* **27** (1), 17-40 (2007).
8. A. Harken, G. Randers-Pehrson and D. Brenner, *Journal of Radiation Research* **50** (Suppl.A), A119 (2009).
9. C. L. Lee and X. L. Zhou, *Nuclear Instruments and Methods in Physics Research Section B: Beam Interactions with Materials and Atoms* **152** (1), 1-11 (1999).
10. A. W. Bigelow, C. R. Geard, G. Randers-Pehrson and D. J. Brenner, *Review of Scientific Instruments* **79** (12), 123707 (2008).

X-Rays Compton Detectors For Biomedical Application

Paolo Rossi[a], Giuseppe Baldazzi[b,c], Andrea Battistella[d], Michele Bello[d], Dante Bollini[c], Valter Bonvicini[e], Cristiano Lino Fontana[a], Gisella Gennaro[f], Giuliano Moschini[a], Francesco Navarria[b,c], Alexander Rashevsky[e], Nikolay Uzunov[d,g], Gianluigi Zampa[e], Nicola Zampa[e], Andrea Vacchi[e]

[a]*Department of Physics of the Universityof Padua and INFN, via Marzolo 8, Padua 35131, Italy*
[b]*Department of Physics, University , Bologna, Italy*
[c]*INFN, Bologna, Italy*
[d]*National Laboratories of Legnaro, INFN, Legnaro (Padua), Italy*
[e]*INFN, Trieste, Italy*
[f]*Veneto Institute of Oncology IOV – IRCCS, Padua, Italy*
[g]*Faculty of Natural Sciences, Shumen University, Shumen, Bulgaria*

Abstract. Collimators are usually needed to image sources emitting X-rays that cannot be focused. Alternately, one may employ a Compton Camera (CC) and measure the direction of the incident X-ray by letting it interact with a thin solid, liquid or gaseous material (Tracker) and determine the scattering angle. With respect to collimated cameras, CCs allow higher gamma-ray efficiency in spite of lighter geometry, and may feature comparable spatial resolution. CCs are better when the X-ray energy is high and small setups are required. We review current applications of CCs to Gamma Ray Astronomy and Biomedical systems stressing advantages and drawbacks. As an example, we focus on a particular CC we are developing, which is designed to image small animals administered with marked pharmaceuticals, and assess the bio-distribution and targeting capability of these latter. This camera has to address some requirements: relatively high activity of the imaged objects; detection of gamma-rays of different energies that may range from 140keV (Tc99m) to 511 keV; presence of gamma and beta radiation with energies up to 2MeV in case of 188Re. The camera consists of a thin position-sensitive Silicon Drift Detector as Tracker, and a further downstream position-sensitive system employing scintillating crystals and a multi-anode photo-multiplier (Calorimeter). The choice of crystal, pixel size, and detector geometry has been driven by measurements and simulations with the tracking code GEANT4. Spatial resolution, efficiency and scope are discussed.

Keywords: Compton camera, Gamma detectors, Molecular imaging, Small animals SPET
PACS: 87.58.-b, 87.58.Ce, 87.58.Pm

INTRODUCTION

The Compton Concept has been rather extensively applied to satellite X-ray astronomy above all when energetic X-ray sources, above 100 keV, have to be imaged [1, 2, 3]. Focusing energetic X-rays is not possible and collimators or masks have to be employed. In satellite astronomy "coded mask telescopes" are largely used for low energy X-rays of less than 100 keV [3, 4], where a mask with a known, although arbitrary, pattern of opaque and transparent areas is projected on a downstream detector plane, singling out the radiation coming from an astronomical source of given angular direction. The tight requirement of having a light apparatus in a satellite together with a reasonable angular resolution discourages application of this method to high energy penetrating X-rays needing large screen thickness that moreover entails reduced detection efficiency. In this domain the Compton Telescope application is gaining momentum, being the preferred tool today. However, the Compton Concept has not yet proved to be the best solution in terrestrial detectors. We may quote for example only a quite few research projects aimed at demonstrating the adequacy of this concept to what is

Application of Accelerators in Research and Industry
AIP Conf. Proc. 1336, 356-360 (2011); doi: 10.1063/1.3586119

by large the most widespread application of gamma detection: i.e. the bio-medical imaging related to either human diagnostics or research in the field of new pharmaceuticals [5, 6, 7]. In this domain one may employ gammas with an energy in the range 100-1000 keV coming from a family of radioisotopes which are named "medical". This paper is concerned with the Compton Concept in general, the bio-medical application of it, and the development of an innovative Compton Camera for small animal imaging.

THE COMPTON CONCEPT

The bio-distribution and targeting capability of pharmaceuticals may be assessed in small animals by imaging gamma-rays emitted from radio-isotope markers.

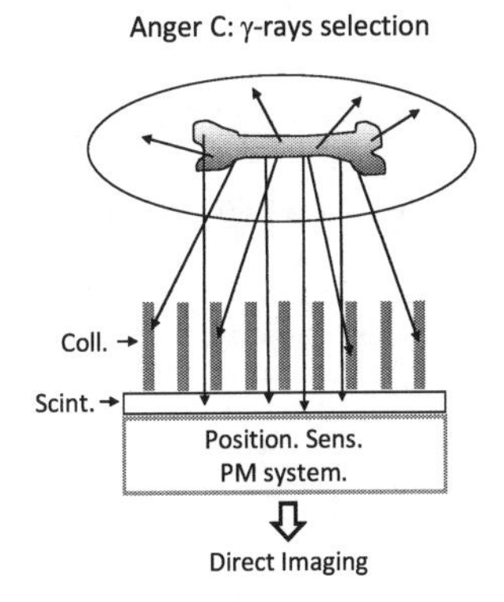

FIGURE 1. Compton concept (down) compared to the traditional Anger method employing collimators (up).

Orthogonal projections of this bio-distribution are traditionally obtained by employing lead collimators featuring a dense matrix of millimeter holes separated by septa, which involve the rejection of an extremely large fraction of gammas (CAC, Collimated Anger Cameras). Alternately, one may have a Compton Camera (CC) and establish the direction of the gamma by letting it interact with a thin solid, liquid or gaseous material (Tracker) and measure the scattering angle (Figure 1).

Finally the total energy of the scattered gamma is determined by a downstream Calorimeter. CCs allow higher gamma-ray efficiency and feature millimeter resolution also at higher energies, like those coming from I-131 (364 keV), beta+ emitters (511 keV) or even Co-58 (810 keV), where CAC would require a very large septa thickness. Higher energy gamma-rays are appealing because they feature reduced scattering in the animal body and hence a better imaging. Compton tomography luckily requires only a few views (for example rotating the apparatus in just four positions around the subject), as the "electronic collimation" that takes place in each position already extracts X-rays coming from many directions. Finally the compact geometry and lighter setup are suited for "small" subjects, like prostate, thyroid and of course mice.

Differences between solid (or even liquid) and gaseous tracker in a CC can be summarized as follows: recoiling electrons travel very little in solid, a few tens of microns, and their direction cannot be measured. On the other hand, one needs a large gaseous volume (about 1000 times more) to have comparable detection efficiency, and hence a larger encumbrance of the camera. Moreover, the measurement of electron tracks in space requires a 3D sophisticated detector, which is demanding in cost and operation [8]. On the other hand as solid or liquid Cameras are concerned, the image reconstruction from "incomplete" Compton scattered data, with no information on the recoiling electron, requires special algorithms. In fact we merely obtain the gamma-ray's outgoing direction and scattering angle and with them a "cone" of possible solutions of the incoming direction. At least three different cones are needed to have a unique solution. A special Filtered Back-Projection algorithm has been developed for this kind of image reconstruction in case of small animal tomography [9].

We shall consider henceforth a CC prototype [10] consisting of a Tracker based on the SDD (Silicon Drift Detector) built for the LHC-Alice experiment [11, 12], and a Calorimeter based on a scintillating crystal and a position sensitive multi-anode photomultiplier, model Hamamatsu H8500. Application to higher energy gamma-rays (usually up to 511 keV) has to be taken into account, and calorimeter thickness has to be sized accordingly.

EFFICIENCY AND SPATIAL RESOLUTION

The efficiency of a CC is: $E_{ff} = P_{tr} \cdot \frac{\Omega_{tr}}{4\pi} \cdot \frac{\Omega_{Ca}}{4\pi}$ where P_{Tr} is the probability that a gamma-ray interacts with the tracker, and $\Omega_{Tr}/4\pi$ and $\Omega_{Ca}/4\pi$ are the "solid angle fraction" covered by the Tracker and the Calorimeter. In Table 1 we have collected gamma detection efficiencies for some simplified configurations of interest relative to a silicon Tracker. As far as the detectors coverage is concerned, we have considered two cases: 1)a full "square" lateral cylindrical coverage:

$$\frac{\Omega}{4\pi} = 1 - 2 \cdot \frac{(1-\cos\theta)}{2} = 0.71, \theta = 45^\circ$$

2) a module 5x5 cm^2 again placed at 2.5 cm from the mouse center:

$$\frac{\Omega}{4\pi} = \frac{(1-\cos\theta)}{2} = 0.15, \theta = 45^\circ$$

This latter case roughly corresponds to a module consisting of a SDD Tracker detector and a scintillating crystal calorimeter read by a multi-anode PMT Hamamatsu 8500. Furthermore we have considered the solid angle fractions of Tracker and Calorimeter as equal. P_{Tr} in Table 1 is the probability that a gamma-ray does interact with a Silicon wafer of thickness L and has been roughly set equal to 1-exp(-L/λ) where λ is the interaction length of gamma-rays of given energy.

TABLE 1. Gamma detection efficiencies for some simplified configurations of interest relative to a silicon Tracker

Tr-Th L mm	γ En keV	Ω/4π	λ mm	P_{Tr}	Eff
1.0	140	0.71	3	0.28	0.142
1.0	511	0.71	5	0.18	0.091
1.0	140	0.15	3	0.28	0.006
1.0	511	0.15	5	0.18	0.004
0.3	140	0.71	3	0.10	0.048
0.3	511	0.71	5	0.06	0.029
0.3	140	0.15	3	0.10	0.002
0.3	511	0.15	5	0.06	0.001

One has to compare these values with those of a Collimated AC, where efficiency actually depends on the hole-septum geometry and the spatial resolution. From Figure 2 one can easily obtain the resolution and efficiency curves for a full coverage CAC camera consisting of two planes "sandwiching" the subject:

$$R_C{}^2 = \frac{d^2 \cdot (L+Z)}{L} \qquad R_T^2 = R_i^2 + R_C^2$$

$$Efficiency = 2 \cdot \frac{\Omega}{4\pi} \cdot F_{Area} = \frac{\pi}{2} \cdot \left[\frac{d^2}{L \cdot (t+d)}\right]^2$$

Being R_C, R_i and R_T respectively the Collimator, the Intrinsic (of the detector) and Total resolution.

Figure 3 shows the comparison of a CAC with a CC for a CC's typical efficiency of 10%. In this case, to

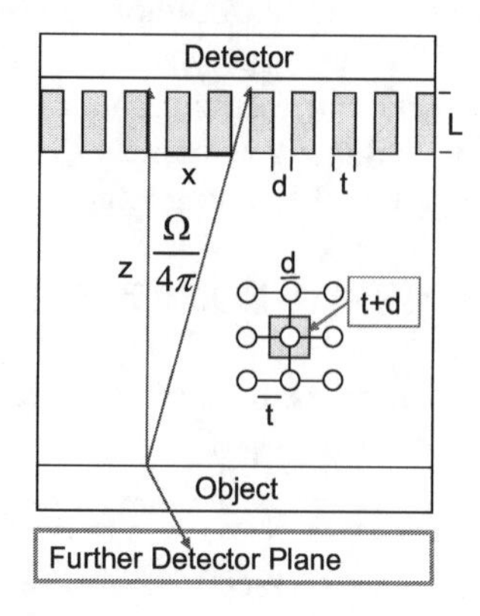

FIGURE 2. Imaging with a Collimated Anger Camera.

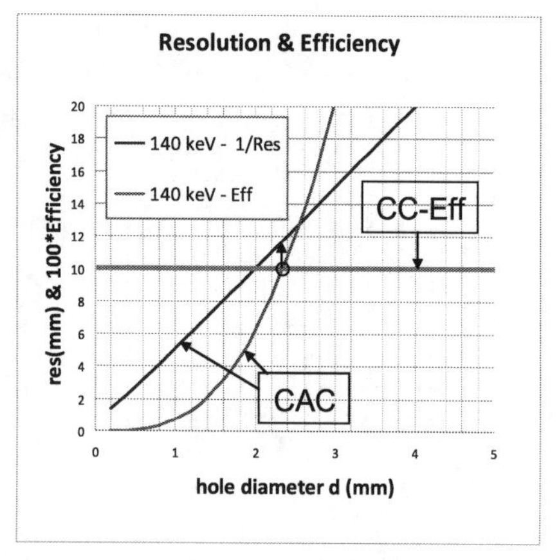

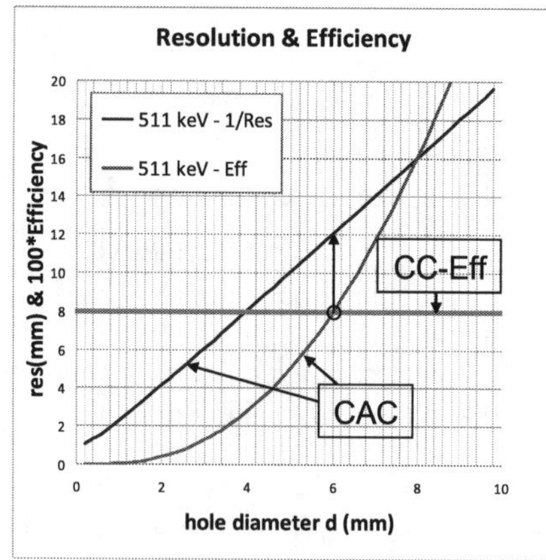

FIGURE 3. Resolution and efficiency curves as a function of the collimator hole diameter for a full coverage Collimated Anger Camera, "sandwiching" the subject and comparison with a Compton Camera for gamma-rays of 140 keV (left) and 511 keV (right). Distance mouse-coll=20 mm, thickness = 5 (20) mm, septa = 2 mm, , intrinsic res. = 1 mm. The green straight line refers to a typical CC efficiency.

reach a similar efficiency with a CAC, one should have a resolution of 12 mm FWHM for the favorable case of 140 keV, and even a worse one for higher energies, and substantially worse of what one can get with a CC. In fact the resolution of this latter can be kept to less than 6-8 mm for most of the cases (at energy of 511 keV) .

SIMULATION AND DETECTOR PERFORMANCE

The Compton Camera design has been developed with the help of a numerical simulation [10]. To this aim we have used the GEANT4 library (http://geant4.web.cern.ch/geant4/) and the Livermore low-energy package. The latter is based on a model of electromagnetic processes down to 250 eV and up to 100 GeV, which also describes scintillation and recoiling electrons coming from Compton scattering.

Gamma-rays of 140 keV and 511 keV have been tracked from a point in the center of a tissue cylinder, 30 mm in diameter, representing a mouse, through air, Tracker (1 mm Si), air again and the calorimeter, as described in Figure 4. The calorimeter is a total absorber. Gaussian errors on the Tracker and Calorimeter energy and position values have been added externally. Compton scattering in mouse and air before the Tracker has been included. The trajectory of the incident gamma was estimated from the energies and angles provided by GEANT4 and the added errors. Finally the distance (DIST) of this trajectory from the source point was calculated. Reproducing what a reconstruction program would do, we have employed the Compton formula for free electrons (with E_1 and E_2 energies of incoming and outgoing gamma-ray into and from the tracker):

$$\cos\theta = 1 - mc^2\left(\frac{1}{E_2} - \frac{1}{E_1}\right)$$

which unavoidably lacks high accuracy when applied to electrons that are bound to atoms (due to the Doppler Broadening).

DIST has been plotted for different values of the initial energy (140 keV, 511 keV), the Tracker spatial resolution (0, 30, 100, 200 microns) and the Calorimeter spatial resolution (0,1, 2, 5, 10, and 20 mm). Figure 5 shows the histograms of DIST in logarithmic scale for the 511 keV case. They have been obtained by varying the Calorimeter position error and normalizing the initial value to 10^5. Distributions are far from being Gaussian, featuring a very large tail. Different "tail" contributions have been parameterized by two numbers (in mm) that give the DIST values for which the histogram height is respectively 1/10 and 1/100 of the initial, highest value (intersections of the horizontal lines with the curves in Figure 5). These values are there referred to as Short-Tail and Long-Tail errors. Here we report some conclusions drawn from the simulations:

1) DIST may have substantial tails even without any added error, as illustrated by the curves in Figure 5. This means that the unavoidable smearing-out of the Compton angle due both to Doppler Broadening and the scattering in the materials upstream the tracker, and which GEANT4 accounts for, puts a lower limit on the overall accuracy of the method, and that a DIST smaller than 0.9 mm for 140 keV and 0.5 mm for 511 keV is not obtainable.

2) The Short-Tail does not depend much on both Tracker and Calorimeter spatial accuracy.

3) Cutting off gamma-rays with energy lower than the initial energy reduces the Long-Tail errors of both 140 keV (considerably) and 511 keV (slightly) gamma-rays, while leaving the Short-Tails values unaffected.

4) The Long-Tail is somehow sensible only to changes in the Calorimeter, but this is because the Calorimeter positioning is intrinsically much less accurate than that of the Tracker (a resolution in the worst case of 0.2 mm for the Tracker compared to 10 mm of the Calorimeter).

5) Both Short-Tail (slightly) and Long-Tail (significantly) values depend on the energy resolution of both Tracker and Calorimeter, because these resolutions largely affect the scattering angle).

6) As expected, 511 keV gammas allow for higher accuracy.

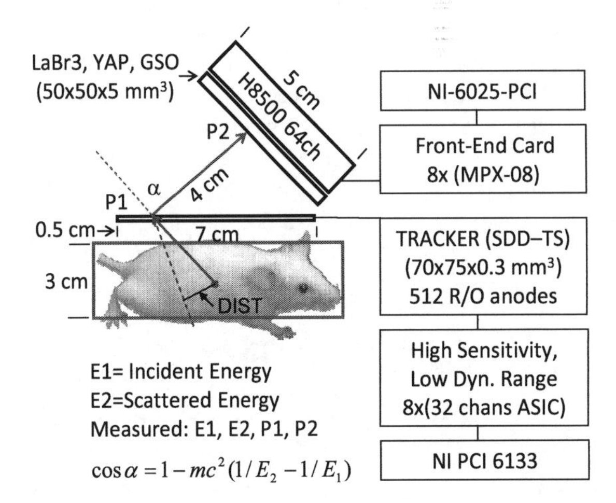

FIGURE 4. Imaging with a Collimated Anger Camer. Small Animal Compton Camera Prototype. Dimensions used in the GEANT4 simulation.

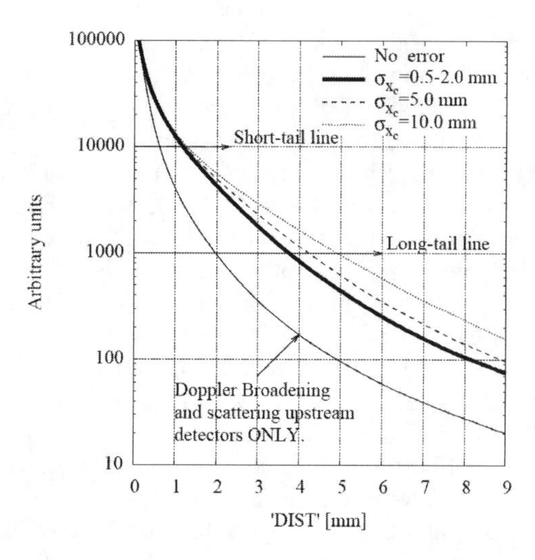

FIGURE 5. Distribution of the distance (DIST) of the reconstructed gamma trajectory from a point like source for the 511 keV case. The lowest Curv refers to a "no error" situation, including only Doppler Broadening and Compton Scattering in mouse and air. The others have: ΔE/E (Tracker) = 4%, ΔE/E (Cal.) = 7% , ΔP (Tracker) = 0.030 mm, and different spatial accuracies for the Calorimeter. Lines at a DIST probability of 1/10 and 1/100 are drawn, permitting the definition of "Short-Tail" and "Long-Tail" errors. The Short-Tail error for the lowest curve is 0.9 mm for 140 keV, and 0.5 for 511 keV gammas.

Crystals like NaI, LaBr3, GSO and CsI, in "continuous" or segmented geometry have been considered. They have been optically coupled to a multi-anode detector Hamamatsu H8500, allowing measurements of spatial resolution and efficiency. Measurements were performed using a Tc-99*, 0.5 mCi "Flat Field", obtained by elution from a standard Mo-99 generator. Although "Continuous" geometry in "integral configuration" that optimizes the optical coupling has been proven to reach millimeter resolution for 5 mm thick crystals [13], nevertheless segmented geometry looks the best when thicker crystals are employed to absorb higher energy gamma-rays. This is our choice for a detector that seeks to be a general purpose tool suitable for a wide range of energies, up to 511 keV.

CONCLUSION

The "Compton" concept has been already fruitfully demonstrated in satellite astronomy for measuring cosmic X-rays, above all when their energy is higher than 1 MeV, while conclusive assessment is lacking when this detection method is applied to small subject bio-medical imaging and actual studies are still under way. Positive facets are: 1) low encumbrance and weight of equipment, and 2) high detection efficiency. However, there are weak points such as: 1) physical limit in the obtainable spatial resolution due to the Doppler Broadening; 2) the measurement of the recoiling electron, if decided, would require demanding 3D tracking gaseous detector; 3) Tomography reconstruction applied to small subjects is more demanding and less effective when the recoiling electron is not known; 4) spatial resolution and efficiency favorably compete with those of collimated cameras only for energies higher than some hundreds of keV.

To support the measurements and help the design of a CC, 140 and 511 keV gammas have been simulated with GEANT4 and tracked from a point-like source inside a mouse through the Compton Camera. The performance has been evaluated by calculating the distance of the reconstructed trajectory from the point-like source. Detailed conclusions have been drawn before. Here we stress that no matter how small the detector's error is, Doppler Broadening and scattering in materials upstream the detector entail an unavoidable "blurring" that brings the overall best spatial accuracy to 0.9 mm for the 140 keV gammas and 0.5 mm for the 511 keV gamma-rays.

Segmented crystals with a stride comparable to the profitable accuracy, i.e. ~2 mm, coupled to a position sensitive detector with millimeter resolution provide the best solution when gamma-rays of a wide energy range have to be imaged.

REFERENCES

1. Imaging Compton Telescope (COMPTEL); http://www.gro.unh.edu/comptel/
2. A. Takada, et al. (SMILE), J. Phys. Soc. Jpn. 78 (2009) Suppl. A, pp. 161.
3. M. Forot et al. (INTEGRAL), The Astrophysical Journal, 668:1259Y1265, 2007 October 20
4. M. Feroci et al. (AGILE), Nucl. Instrum. And Meth. A 581(2007) 728-754
5. R. Pani et al, Nucl. Instrum. And Meth. A 571(2007)187
6. G. Moschini, et al., The Scintirad Project, Nat. Lab. of Legnaro (Italy), INFN-LNL-221 (2007)
7. R. Pani et al., *AIP,* CP1099 (2009) 488-491
8. A. Takada et al., NIM A 546 (2005) 258–262
9. C. Fontana et al., Communication to CAARI 2010.
10. P. Rossi, et al., Nucl Instr. And Meth. A (2010), doi:10.1016/j.nima.1010.07.018
11. A. Vacchi, et al., Nucl. Instr. And Meth. A 306 (1991) 187.
12. A. Rashevsky, et al., Nucl. Instr. And Meth. A 485 (2002) 54.
13. R. Pani et al., Nucl. Instrum. And Meth. A 572 (2007) 268

$[^{13}N]$ Ammonia Cardiac Program At West Virginia University Health Sciences

John M. Armbruster

IBA Molecular, NA, 3601 Morgantown Industrial Park, Morgantown, WV 26501

Abstract. Due to the shortage of the more traditional cardiac imagining isotopes, specifically, Technicium-99, the Cardiologists at WVU have had to look to alternative imagining techniques such as PET. This has led to a dramatic increase in the use of $[^{13}N]$ Ammonia PET scans at the Health Sciences Center. The patient load has gone from one to two patients one day a week to typically two to three patients, two days a week, with occasional add-on in-house patients; each patient typically requiring two target irradiations. In this paper, we will discuss the process that is being used to meet this increased demand from the production of the isotope through the final result.

Keywords: $[^{13}N]$ Ammonia, PET, Cardiology, Cyclotron
PACS: Replace 87.57.uk

INTRODUCTION

When the AECL shut down the nuclear reactor at Chalk River on May 14, 2009, it created a world-wide critical shortage of the isotope Technicium-99m (Tc-99m). Shortly thereafter, the FDA published some recommendations to suggest how Medical Centers could make up the shortfall. For coronary artery disease, the recommendations included Thallium-201 gamma scintigraphy, Rubidium-82 PET/CT, Ammonia N-13 PET/CT, echocardiographic stress, or electrocardiographic stress tests. At West Virginia University Health Sciences, this resulted in a dramatic increase in the number of $[^{13}N]$ Ammonia runs to satisfy the requirements of the Cardiology Physicians with the number of runs almost tripling and the time spent on $[^{13}N]$ Ammonia more than doubling. Prior to the end of 2009, the average number of runs per month was 13 runs, while since January 2010, the average number of runs increased to 35 runs. The time spent on $[^{13}N]$ Ammonia has correspondently increased from an average of ~20 hr per month to 46 hr per month.

PRODUCTION

The $[^{13}N]$ Ammonia is produced on a GE PETtrace Cyclotron using the ^{16}O (p,α) ^{13}N reaction on natural water with the target being loaded with ~1.5 ml of HPLC water. The target is then irradiated by 18 MeV protons at 30 μA for ~10 min to produce ^{13}N-NO_x. Upon completion of the target irradiation, the radioactive liquid is transported from the target to a cassette in the Ammonia Process Unit by means of a low pressure push of Helium Gas to a Hot Cell. The nitrates and nitrites are then reduced to ammonium ions by means of Devardas Alloy mixed with a small Sodium Hydroxide pellet. When the liquid reaches the process unit, it reacts with the Devardas Alloy and Sodium Hydroxide to generate a violent boiling reaction where a mixture of $[^{13}N]$ NH_3 gas and caustic Sodium Hydroxide is formed. The gas is then pushed into a trapping vessel by means of the Helium push gas where it is trapped in a saline solution. This solution is then pushed through a Sterility Filter to the final product vial. There is a "drop trap" in the cassette to prevent any drops of the reaction solution from reaching the trapping vessel. At the conclusion of the process, the final product vial is transferred from the process pig to a Capintec Dose Calibrator where the amount of activity is measured. After the activity is measured, it is then placed into the lead pig of the Biodex PET Dose Drawing station where the QC samples and final patient dose is drawn.

One of the major concerns during the production of the product is the presence of radioactive gases going out the exhaust of the waste vial. Initially, the stack monitoring detector was measuring on the order of 12 mCi of effluent coming out of the process unit per run. In an effort to reduce this effluent, a simple trap using Soda Lime has been installed in the outlet line from the waste vial. This resulted in a reduction of ~75% down to 3 mCi per run. More experiments

Application of Accelerators in Research and Industry
AIP Conf. Proc. 1336, 361-363 (2011); doi: 10.1063/1.3586120

are ongoing to further decrease this amount and eliminate any effluent from going up the exhaust stack.

QUALITY CONTROL

At the conclusion of the production process, the final product vial is transported from the process pig to the Capintec Dose Calibrator for activity measurement and then to the Biodex PET Dose Drawing Station. During this transfer, the product is visually inspected for any signs of discoloration, particulates, or non-clarity. Once the vial is in the Drawing Station, a small sample is drawn and tested for pH and any metallic impurities by means of an Aluminum Breakthrough Kit. After the patient dose of ~20 mCi is drawn, the remaining product is placed back into the Capintec where the radionuclidic purity is verified by measuring the activity at two times; typically 10 min. and the deviation from the expected activity is then calculated and recorded. Since the tests for Sterility and Pyrogens require long testing times, these are delayed until the completion of the day's production when all runs are tested at the same time. The Sterility is tested by injecting a small sample (0.2 ml) of the product into growth mediums of Tryptic Soy Broth and Fluid Thioglycollate which is then placed in an oven at 30 - 35°C for 2 weeks. At the end of this time, they are inspected for any signs of growth and the results noted. The test for Pyrogens is the gel clot using Limulus Amebocyte Lysafe (LAL). A small amount (0.2ml) of the product sample is injected into a tube containing the LAL while another sample (0.18 ml) is injected into another tube along with 180 μl of pre-mixed Control Standard Endotoxin E.Coli. The two samples are then placed in a water bath at 37.5°C for 1 hr. At the end of that time, they are inspected and the sample containing the CSE should be clotted while the other sample should still be liquid. The results are then noted on the batch record.

DOSE DRAWING

Once the production process is completed, the final product vial is transferred into the drawing station. After the initial QC testing (pH and Aluminum Breakthrough for metallic impurities) is completed, the actual patient dose is drawn by a qualified member of the PET scanner staff and the syringe is placed back into the Capintec to be measured. A typical dose of 15 mCi to 25 mCi, dependent upon the size of the patient, is drawn so that the required amount of activity after transport from the Laboratory to the Scanner Facility is sufficient for the study. By use of the Biodex PET Drawing Station located in a Hot Cell behind an L-block, the exposure to the Technician who is drawing the dose is minimal with the majority of the extremity exposure occurring when transferring the syringe with the activity into the Dose Calibrator and into the Syringe Shield. The shielded syringe is then placed on a shielded cart for transfer from the Laboratory to the Scanner Facility.

PATIENT TREATMENT

When the patient arrives at the facility, he is interviewed to verify that he has followed the pre-test instructions regarding medications, medical conditions, or habits that may impact the test. After taking the patient history, the Cardiology nurse will explain the stress portion of the exam and answer any questions the patient may have. The nurse will also evaluate the patient to determine if the pharmalogical stress agent can be used. If there are no contravening issues, the patient is positioned in the gantry with their arms lying above their head. A cardiac gating monitor is attached to the gantry and the EKG leads are attached to the patient. At this point, the production of the [^{13}N] Ammonia is begun. While the isotope is being produced, a CT scout scan of the upper chest region is done to assess the proper axial positioning of the patient. The CT portion of the study is done for the attenuation correction with no IV Contrast being used for CT Cardiac imaging. The patient is then positioned for the PET scan. A dose of 15-20 mCi of [^{13}N] Ammonia is injected as a bolus injection for the resting perfusion portion of the test with the PET scan is started three minutes post injection to allow for the circulation of the tracer to the heart. Production of another batch of [^{13}N] Ammonia is begun making sure that there will be at least 20 minutes between the rest injection and the stress injection. Another CT scout scan of the upper chest region is performed to assess proper axial positioning with yet another CT scan done for attenuation correction. Before starting the Lexiscan injection, the availability of the [^{13}N] Ammonia is verified. A Physician arrives to briefly examine the patient, the patient's history, and to obtain his consent. Cardiology will then start the Lexiscan injection, typically 0.4 mg in 5 ml over 10 seconds followed by a saline flush. The Cardiologist will then monitor vitals for the patient during and until the stress procedure is completed with a continuous 12 lead EKG being obtained throughout the stress portion of the exam. Another 15-20 mCi dose of [^{13}N] Ammonia is then injected as a bolus injection immediately followed by the flush of saline by the RN with the PET scan once more starting three minutes post injection. After the scan is completed, the patient is monitored and once stable, released from the PET facility.

RADIATION SAFETY

One of the main concerns with the increase in patient throughput is the increase in radiation exposure to the staff from handling the radioactive material during its production, testing, and dose drawing. Even prior to the increased usage of the [^{13}N] Ammonia, the exposure of the staff, especially the hand dose, was substantial. When these productions were done initially, it was not unheard of for the weekly hand dose for the individual preparing and drawing the dose to be approaching 1000 mRem with the main contribution to this exposure was the actual handling of the activity. Upon completion of the production process, the needles were removed from the final product vial using a set of 9" forceps and the vial moved from the collection pig to the Capintec Dose Calibrator, where the activity was measured. It was then returned to the Collection pig in preparation for the actual drawing of the patient dose. All transfers of the vial were done with the forceps. Initially, the final product vial was removed from the product pig by means of the forceps and the patient dose was hand drawn behind an L-Block to provide shielding for the Technician. Unfortunately, much of this had to be done without any shielding since the weight of any hand-held shielding around the Vial or the Syringe would cause difficulties performing the process. The first step in reducing the exposure for this process was to obtain a shielded dose drawing station from Biodex Medical Systems, Inc. This made it possible to perform the dose drawing process with the activity shielded at all times. To further reduce the exposure, much longer forceps (18" forceps) were obtained to be used when moving the final product vial. Then to further reduce the exposure, the actual dose drawing was spread out among various members of the PET Scanner Technicians. The final result is that the extremity radiation exposure for even the busiest weeks is now virtually indistinguishable from any normal week's extremity exposure.

CONCLUSION

The Cardiologist at WVU had been using the [^{13}N] Ammonia PET primarily for their overly large patients, patients with attenuation artifacts from routine myocardial perfusion studies (MPS) using Technetium, and Viability studies for some time. For the overly large patients, it was the preferred modality and for patients with some sort of attenuation artifact, it produced better results. It was initially used for Viability Studies, for patients with prior MPS's, and for patients where prior studies had some sort of issue. The Cardiologists have been happy with the results and found that they had few false positives using the [^{13}N] Ammonia. With the loss of availability of the more traditional imaging isotopes, they began using the [^{13}N] Ammonia PET studies on less problematic patients when they could get approval from the various insurance providers. All in all, they were happy with the results and will continue to use this modality although when the isotope shortage eases, they most likely will return to the more traditional modalities for the more routine studies. Even then, [^{13}N] Ammonia will still play an important part in the Cardiology Program at WVU.

ACKNOWLEDGMENTS

I would like to acknowledge Karyn Wallace and Dr. Anthony Morise from West Virginia University Health Center for their assistance.

Measuring Prompt Gamma Ray Emission During Proton Radiotherapy For Assessment Of Treatment Delivery And Patient Response

Jerimy C. Polf, Stephen Peterson, Daniel Robertson and Sam Beddar

Department of Radiation Physics
Univeristy of Texas M. D. Anderson Cancer Center
Houston Tx 77030, USA

Abstract. In this paper, we present results from a study of characteristic prompt gamma rays produced in patients during proton beam radiotherapy. A Monte Carlo study was performed calculating both proton dose deposition and prompt gamma ray production within a patient during proton beam irradiation. Form these calculations, comparisons were made between dose deposition and prompt gamma ray production within the patient. Also, calculations were performed of the production of characteristic gamma rays emitted from excited oxygen nuclei as a function of oxygen concentration within irradiated tumor tissue. Results show that there is a strong correlation between dose deposition and prompt gamma ray production within the patient, and that the production of prompt characteristic oxygen gamma rays is directly proportional to the oxygen concentration within the irradiated tissue.

Keywords: Proton radiotherapy, in-vivo dose verification, prompt gamma ray emission
PACS: 87.53.Kn, 87.55dh, 87.55.Kh, 87.55ne

INTRODUCTION

The goal of radiation therapy is to ensure local tumor control by delivering a prescribed dose of radiation to cancerous tissues while minimizing radiation-induced side effects to surrounding healthy tissue. The overall clinical outcome of radiation therapy depends strongly on both the accuracy of treatment delivery and the response of the tumor to the radiation. Proton beam radiation therapy is a highly conformal treatment method that provides the ability to precisely target dose delivery in the desired treatment volume, thus sparing dose to healthy tissues. However, the effectiveness of proton therapy, like the effectiveness of other forms of external beam radiation therapy, remains limited by uncertainties due to patient set-up errors and changes in patient anatomy, as well as a lack in our understanding of the response of tissue to proton beam irradiation. One possibility for overcoming these limitations in proton therapy delivery would be to provide feedback about the in-vivo distribution of dose and composition of tissues irradiated during treatment delivery. This type of personalized feedback could be used to track changes in internal anatomy or changes to elemental composition within tissues as a function of delivered dose over the course of treatment allowing for adjustment of the treatment as needed to ensure proper treatment delivery.

One suggested method for measuring and imaging in-vivo dose delivery is through the measurement of secondary gamma rays emitted from the patient during irradiation [1,2]. The two main sources of these secondary gamma rays are: 1) through emission of coincident gamma rays arising from the production of positron emitting isotopes created during irradiation known as "delayed emission," or 2) through the direct emission of characteristic gamma rays from an intact, excited nucleus, known as "prompt emission." Over the past decade, there has been a lot of work done on dose verification using PET imaging techniques to measure the delayed gamma signal from positron emitters created during irradiation [2]. More recently, several researchers have also begun to study the possibility of measuring in-vivo proton beam range and dose distribution through the measurement of prompt gamma rays [1, 3].

The purpose of this study was to determine the correlation (if any) between in-vivo dose deposition and prompt gamma ray production, as well as, studying the feasibility of determining the concentration of different elements with irradiated

Application of Accelerators in Research and Industry
AIP Conf. Proc. 1336, 364-368 (2011); doi: 10.1063/1.3586121

tissues by measuring characteristic prompt gamma ray emitted from the patient during proton beam irradiation. Since different atomic nuclei will produce characteristic prompt gamma rays with different energies, tissues with differing atomic compositions and concentrations will produce different prompt emission spectra when irradiated [3, 4]. We believe that by measuring the characteristic prompt emission from specific elements contained in tissue during a daily treatment fraction, it may be possible to track changes in the concentration of these elements over the course of treatment. This would in turn provide a method of measuring the in-vivo response of irradiated tissue to proton radiotherapy over the course of treatment. For this study we used a validated Monte Carlo (MC) model [4] of the prompt gamma ray emission occurring during proton radiotherapy to calculate the production of prompt gamma rays within a patient during proton beam irradiation. The calculations included comparisons of the spatial distribution of prompt gamma rays created within the patient due to the distribution of delivered dose. Additionally, calculations were performed to study the correlation between the production of characteristic prompt gamma rays from specific elements and the concentration of those elements within irradiated tissues. To test this concept, we studied the feasibility of tracking tumor "hypoxia" (reduced oxygen concentration) through the measurement of characteristic prompt gamma rays from oxygen emitted during irradiation. To accomplish this, we performed MC calculations of characteristic oxygen prompt gamma ray production in specially defined "hypoxic" tumor tissue within our model.

METHODS AND MATERIALS

An MC model of a clinical passive scattering proton therapy treatment nozzle built with the MCNPX Monte Carlo code [5] was used for our calculations of prompt gamma ray production in a patient. This model has been extensively validated for use in calculating proton dose deposition [6, 7] and has been used for preliminary studies of prompt gamma ray emission from tissue phantoms during proton therapy [3]. In the model, protons, neutrons and photons were tracked. Other secondary particles created during the simulations were not tracked, and instead were set to deposit their energy at the point of creation. Transport physics contained in the model included energy straggling, multiple Coulomb scattering, and elastic and inelastic nuclear interactions. For the dose deposition calculations, 1×10^7 incident proton histories were run and for gamma-ray production calculations, up to 1×10^9 incident proton histories were run to ensure a 1σ statistical uncertainty of less than 1% and 8% in the calculated data for dose and gamma-ray production, respectively. For full details of the transport physics used in the MC model, the reader is referred to the paper by Polf et al. [3] .

For this study we imported patient CT datasets into the Monte Carlo model. The CT data was imported as a fully voxelized geometry with each voxel of the CT defined as a separate cell within the geometry and assigned a tissue type and density based upon CT hounsfeld number-to-tissue conversion tables used for patient treatment planning. Tissues defined in the MC model were bone, fat, muscle, lung, and average soft tissue with the composition and density of all tissues except lung set according to definitions given in ICRU Reports 49 and 63 [8, 9]. All lung tissues in the CT scans were defined simply as air (80% N, 15% O, 3% H, 2% C, $\rho = 0.0012$ g/cm^3) for this preliminary study.

Two specific studies were preformed, first a comparison of the distribution of deposited dose and the distribution of prompt gamma ray production within the patient, and second the production of characteristic gamma rays from oxygen within the irradiated tissues. MC Calculations were performed by modeling the delivery of a broad 250 MeV proton beam onto the patient. Dose deposition and prompt gamma ray production were tallied using a 3-dimensional grid of 5 mm^3 voxels defined within the patient geometry in the MC model. Dose was recorded in units of MeV/gram in each voxel and prompt gamma ray production was recorded as total number of gamma rays per voxel. For studies of characteristic gamma rays from specific elements, only gamma rays corresponding to the energies of characteristic gammas from oxygen (7.12, 6.92, and 6.13 MeV), were recorded [3, 10]. Then, the material definition for a selected hypoxic region within a tumor identified in the patient CT was modified by reducing the percentage of oxygen contained in this tissue. A series of MC calculations of oxygen gamma ray production were performed with material definitions containing successively lower percentages of oxygen for this selected region. In this manner we were able to characterize the production of prompt gammas from oxygen as a function of oxygen concentration in tissue.

RESULTS

Comparison Of Dose And Prompt Gamma Production

Calculations of dose deposition and total prompt gamma ray production within a patient irradiated with a lateral 250 MeV proton beam are shown in figure 1.

The range of the proton beam in the patient is highly dependent on the density of tissues that it traverses.

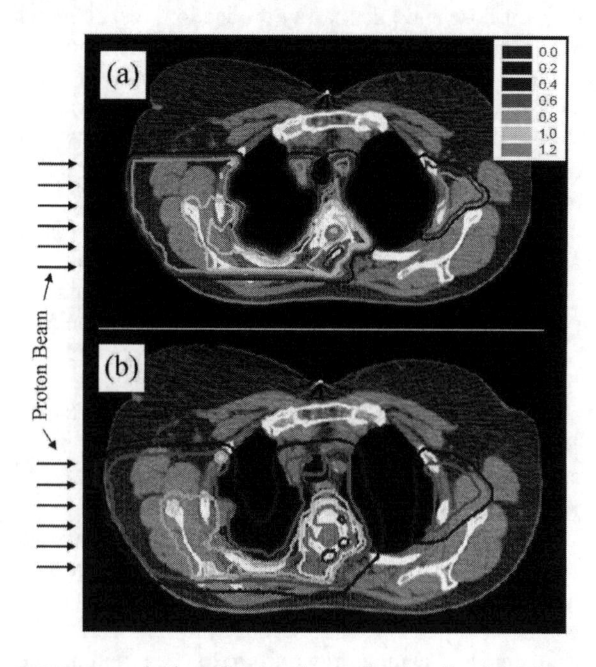

FIGURE 1. (color) Comparison of normalized (a) dose delivered to a patient and (b) prompt gamma ray production in the patient during irradiation with a laterally incident, 250 MeV proton beam.

As can be seen in Figure 1a, the proton beam range is much shorter in the more posterior portions of the patient, due to the presence of high density bone in the path of the beam. In fact, the most posterior portions of the beam are stopped in the vertebral body of the patient, while the beam traverses a greater distance more anteriorly due to less bone and a greater amount of low density lung tissue in the beam path. This same pattern is seen in the distribution of prompt gamma ray production shown in Figure 1b, indicating that the distribution of prompt gamma rays produced in the patient is strongly correlated to the distribution of deposited dose. While both distributions are similar, there are notable differences. We see that there is no dose tallied within the lungs, however, the prompt gamma production is not zero. Due to the low density of the lungs (p = 0.0012 g/cm3) in our model, the value of dose deposited was often less than 0.01 MeV/gram. However, the dose deposited, was calculated using only two decimal places in the MC model. This meant that voxels in the calculation grid with doses of less than 0.005 MeV/gram were rounded down to a value of zero. In contrast, for prompt gamma production, the number of gamma rays produced in each voxel were counted and reported in integral numbers, avoiding any rounding errors. Additionally, the prompt gamma production is seen to rise more sharply in boney regions than the proton dose. This is due to the fact that the prompt gamma ray production cross-section increases both near the end of proton range and with tissue density [3].

From these calculations, the average prompt gamma production rate in soft tissue was shown to be $\sim 10^{-5}$ per incident proton per 2.5 mm^3 CT voxel in our calculation grid. For a typical daily treatment dose (2 Gy), requiring the delivery of $\sim 10^{11}$ protons [11], we would estimate that $\sim 10^{6}$ prompt gammas would be emitted per CT voxel. Using a Compton camera specially designed for prompt gamma ray measurements [12] with a detection efficiency of 10^{-3}, we estimate that it may be feasible to measure approximately 1000 prompt gammas per voxel during a typical proton treatment delivery. If measurement of this number of gammas could be achieved, we believe it could be feasible to produce a statistically and clinically viable image of the prompt gamma emission.

Characteristic Prompt Gamma Ray Production From Oxygen

Figure 2 shows a 2D histogram image of the production of characteristic prompt gamma rays from oxygen in a patient irradiated with an anterior 250 MeV proton beam. From this histogram it is immediately apparent that characteristic oxygen gamma rays are produced in all tissues that were irradiated since all tissues contain oxygen. Production of gamma rays in the lung were extremely low however, due to the fact that the density of the lungs was 3 orders of magnitude lower than that of average soft tissue and that only 16% of the lung was composed of oxygen in the MC model. In figure 2a, the tumor appears as a large mass in the patients right lung (left hand side of histogram image). Also, the vertebral body, located posteriorly, appears prominently (darkest structure) in the histogram image. The appearance of the vertebral body is again due to the high density of bone and the fact that it is near the end of the proton beam range where the cross-section for prompt gamma ray production reaches its maximum [3].

Next, production of characteristic gamma rays as a function of percent oxygen in tissue was studied by purposely defining a rectangular section of the tumor as hypoxic. This was accomplished by assigning the CT voxels in this region to be filled with a modified definition of soft tissue which contains a lower percentage of oxygen than normal soft tissue. As can be seen in Figure 2b, the production of prompt gamma rays is markedly reduced when the percentage of oxygen is reduced from a normal 70% value to a value

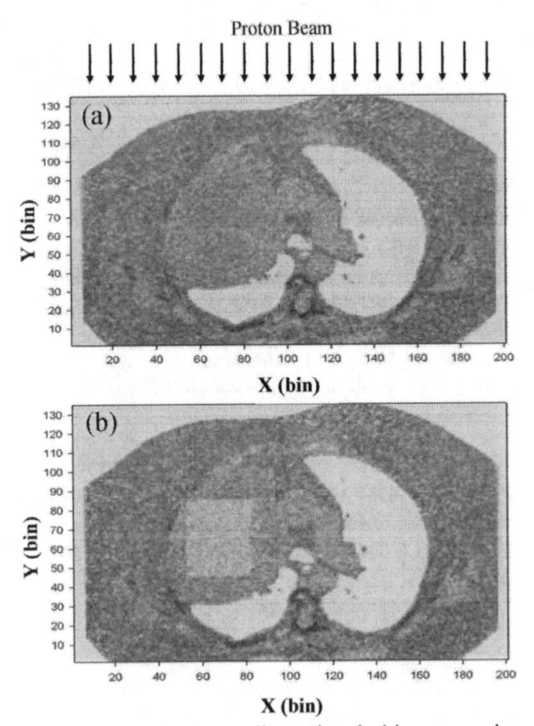

FIGURE 2. ((a) Two-dimensional histogram image representing characteristic prompt gamma ray production inside a patient irradiated with an anteriorly located 250 MeV proton beam. In the images, darker colored histogram bins represent higher characteristic oxygen gamma ray production. (b) A rectangular hypoxic region defined within the tumor is readily visible. In this image the oxygen concentration in the hypoxic region is reduce to 50% from a normal value of 70%.

of 50%. Figure 3 shows the characteristic oxygen gamma ray production for a series of calculations in which the percentage of oxygen in the hypoxic region of the tumor was steadily reduced. The results show a logarithmic dependence of the oxygen gamma ray production on the oxygen concentration. This provides us with preliminary evidence that it may be possible to track changes in the concentration of specific elements by measuring the characteristic prompt gamma rays emitted from irradiated tissues over the course of proton radiotherapy treatments.

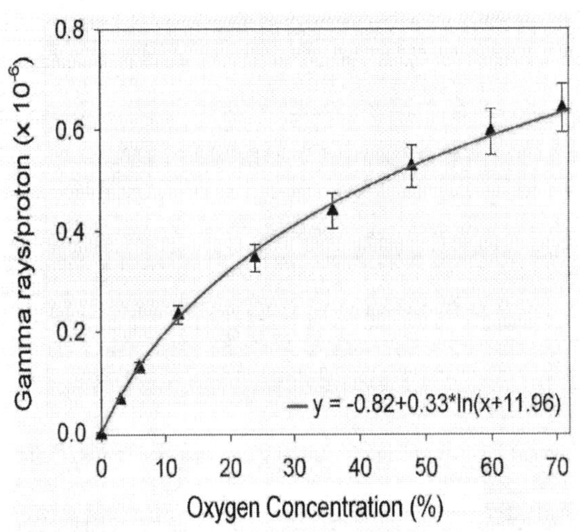

FIGURE 3. (color) Production of characteristic prompt gamma rays from oxygen as a function of oxygen concentration within a hypoxic tumor region irradiated with a 250 MeV proton beam. The line represents the logarithmic fit to the data using to the equation shown in the lower right of the graph. Error bars represent 5-8% statistical uncertainty in the MC calculated data.

DISCUSSION AND CONCLUSIONS

To our knowledge, this paper is the first to report on studies of prompt gamma ray production within a patient during proton beam irradiation. From results of this preliminary Monte Carlo study, we conclude that a strong correlation exists between dose deposition and the production of prompt gammas within a patient during proton irradiation. These results are in agreement with initial studies in tissue phantoms which also showed a correlation between prompt gamma ray emission and delivered dose [1, 13] as well as prompt gamma ray emission and tissue density. This suggests that if a proper device could be developed for measuring and/or imaging prompt gamma rays, this could be used as method for providing in-vivo verification of delivered dose.

Additionally, our results indicate that the production of characteristics prompt gamma rays from individual elements can be proportional to the percent concentration of that element within the irradiated tissue. We believe this shows that, through the development of proper measurement techniques, it may be possible to measure the in-vivo concentration of the major elemental constituents within the irradiated tissues on a daily basis over the course of an entire proton radiotherapy treatment regiment. This would then provide a method to track changes to the elemental composition of irradiated tissues, such as changes in hypoxia in tumors, which is known to increase radio-resistance in tumors [14, 15], and also to study the uptake and clearance of new drugs and nanoparticle-based radiosensitizing agents (i.e. gold nanoparticles). These studies could be used to characterize tumor responses to therapeutic radiation or to correlate changes in irradiated healthy tissues to the onset of treatment-related side effects. We believe that results of this preliminary study offer adequate justification for future work into the development of a clinically viable system for in-vivo dose verification through imaging of prompt gamma ray emission, as well as, further work into quantifying and characterizing the relationship between characteristic emission and elemental concentration within proton irradiated tissues.

ACKNOWLEDGMENTS

This work was supported in part by grant # IRG-08-061-01 from the American Cancer Society.

REFERENCES

1. Min, C.H., et al., App. Phys. Lett., 2006. 89: p. 183517:1-3.
2. Parodi, K., et al., Int. J. Radiat. Oncol. Boil. Phys., 2007. 68(3): p. 920-934.
3. Polf, J., et al., Phys Med Biol, 2009. 54(3): p. 731-743.
4. Polf, J.C., et al., Phys Med Biol, 2009. 54: p. N519-N527.
5. Waters, L.S., J.S. Hendricks, and G.W. Mckinney, *MCNPX User's Manual, Version 2.5.0.* 2005, Los Alamos National Laboratory: Los Alamos NM.
6. Harvey, M.C., et al., Med. Phys., 2008. 35(6): p. 2243-2252.
7. Polf, J.C., et al., Med. Phys., 2007. 34(11): p. 4213-4218.
8. ICRU, *Clinical Proton Dosimetry Part I: Beam Production, Delivery, and Measurement of Absorbed Dose*. 1993, International Commission on Radiation Units and Measurements: Bethesda, MD.
9. ICRU, *Nuclear Data for Neutron and Proton Radiotherapy and for Radiation Protection*. 2000, International Commission on Radiation Units and Measurements: Bethesda, MD.
10. NNDC, *National Nuclear Data Center*. 2007, Brookhaven National Laboratory.
11. Smith, A.R., et al., *17th International Conference on the Application of Accelerators in Research and Industry, Nov. 12-16*. 2003. Denton, TX: American Institute of Physics.
12. Peterson, S., D. Robertson, and J.C. Polf, Phys Med Biol, 2010. 55: (in press).
13. Styczynski, J., et al., Med. Phys., 2009. 36: p. 2425-2426.
14. Cox, J.D. and K.K. Ang, *Radiation Oncology: Rationale, Technique, Results*. 8th Ed. ed. 2003, St. Louis: Mosby-Elserver Scientific.
15. McAneney, H. and S.F. O'Rouke, Phys. Med. Biol., 2007. 52: p. 1039-1054.

Status Of The Dielectric Wall Accelerator For Proton Therapy

George J. Caporaso, Yu-Jiuan Chen, James A. Watson, Don T. Blackfield, Scott D. Nelson, Brian R. Poole, Joel R. Stanley and James S. Sullivan

P.O. Box 808, L-410
Lawrence Livermore National Laboratory
Livermore, California 94551 USA

Abstract. The Dielectric Wall Accelerator (DWA) offers the potential to produce a high gradient linear accelerator for proton therapy and other applications. The current status of the DWA for proton therapy will be reviewed. Recent progress in SiC photoconductive switch development will be presented. There are serious beam transport challenges in the DWA arising from short pulse excitation of the wall. Solutions to these transport difficulties will be discussed.

Keywords: Dielectric wall accelerators; hadron therapy; proton therapy; high gradient insulators; photoconductive switches; alternating phase focusing; alternating gradient quadrupole focusing
PACS: 41.75.Lx, 41.85.Ja

INTRODUCTION

The Dielectric Wall Accelerator (DWA), originally conceived of as a compact replacement for high current electron accelerators, is being actively developed for proton therapy for cancer[1]. This paper will review recent progress of the DWA development in three areas: testing of "large scale" high gradient insulators (HGI), improvements in SiC material for photoconductive switches, and beam transport in the DWA.

LARGE SCALE HGI TESTING

The original motivation for the short pulse version of the DWA was the observation that the surface flashover strength of vacuum insulators appears to have an inverse power law dependence on the duration of the applied voltage pulse. These observations also seem to hold for HGIs with increased flashover strengths as compared to conventional insulators[2]. However, due to the high gradients involved (10's of MV/m) this data set is largely derived from very small disk samples (~ 1 – 2 cm diameter and 0.2 – 2 cm high). In particular, the 100 MV/m result was obtained on a disk sample that was 1 cm in diameter with a total height of about 2.3 mm that was subjected to a several ns pulse[3].

The short pulse DWA concept requires that the *inner* surface of the HGI withstand high gradients for times on the order of 1 ns over an axial length on the order of 3 times the beam tube radius or a distance of 3 – 7 cm. Recently a capability to test larger, annular samples has been obtained. Figure 1. shows an annular sample similar to the one currently undergoing tests and one that is the same size used in the original test that yielded the 100 MV/m result.

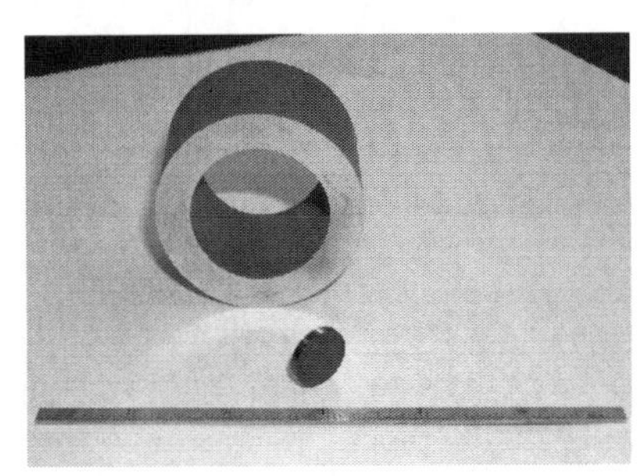

FIGURE 1. Comparison between the size disk sample used to obtain 100 MV/m and the annular sample currently undergoing tests. The disk sample is 2.3 mm long with a diameter of 10 mm. The annular sample is 35 mm long with a 40 mm inner diameter.

The test stand uses an Optoel pulser (see Figure 2.) that can deliver up to a 5 MV, 1 ns pulse into a matched (160 ohm) load[4]. A test chamber containing the annular sample has been mounted on the output

Application of Accelerators in Research and Industry
AIP Conf. Proc. 1336, 369-373 (2011); doi: 10.1063/1.3586122

stalk of the pulser. The exterior of the sample is surrounded by oil while the inner diameter is in a vacuum. Because of the shunt capacitance between the sample and the test chamber the peak voltage that can be delivered to the sample is about 3 MV.

FIGURE 2. Optoel pulser installed inside a screen room. Moveable x-ray shielding is visible on the left side of the picture near the output end of the tapered transmission line. The test chamber is mounted on the output end of this line. Peak voltages of approximately 3 MV can be applied across the HGI sample.

Current HGI Testing

The first annular sample is currently undergoing tests and has survived at voltage levels over 50% of the pulser maximum. We are currently verifying the calibrations of the voltage probes by several different methods before we can determine the breakdown strength.

SIC PHOTOCONDUCTIVE SWITCHES

SiC photoconductive switches operated below bandgap have distinct advantages for the DWA. Firstly, they can take full advantage of the high bulk breakdown strength of the material by placing electrodes on opposite sides of a planar wafer, and ideal configuration for a Blumlein switch. In this mode of operation the carriers are generated from impurities that are deliberately introduced into the SiC during crystal growth.

Two problems have been encountered during the development of these devices: structural defects in the crystal that have compromised the bulk breakdown strength and non-optimum concentration of the dopants. The structural defects encountered have been micropipes, a screw dislocation that propagates through the entire length of the crystal, and inclusions probably arising from the precipitation of the dopants at too high a concentration.

The concentration of the dopants affects the carrier concentration produced under laser illumination. A non-optimal concentration results in low quantum efficiency and increases the required laser power for good switching.

Current SiC Defect Status

The currently supplied material is micropipe-free down to magnifications of 25X. In the current switch configuration there is a large, calculable field enhancement at the electrode edges. Testing with the present material has pushed these enhanced fields to over 100 MV/m without failure in the substrates. We are currently working on improved switch package designs to minimize the field enhancements.

Current SiC Dopant Status

After several iterations with the vendor we now have a material that has a quantum efficiency 25X higher than that of the first generation material. We have demonstrated sub-ohm switch impedances for switches measuring 1 cm x 1 cm x 1mm thick. Figure 3 shows a comparison between the switch characteristics of the various generations of material.

Since the transmission line impedances in the Blumleins are of order 20 ohms, the demonstrated switch impedance is more than adequate for this application.

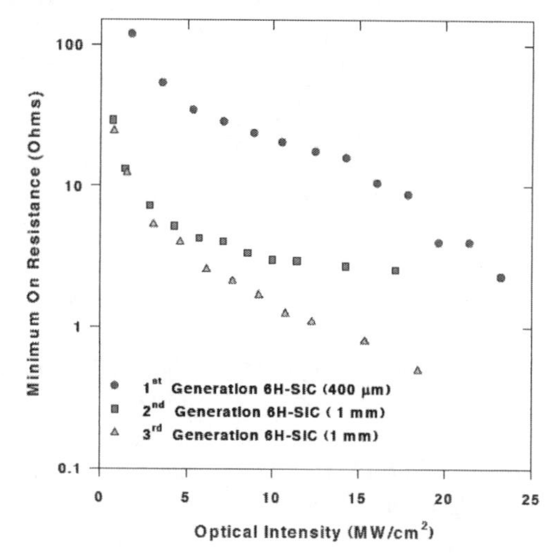

FIGURE 3. Measured minimum on resistance of SiC photoconductive switch is plotted vs. laser intensity for three generations of SiC material. The 3rd generation has a quantum efficiency 25X higher than the 1st generation material.

BEAM TRANSPORT IN THE DWA

The stacked Blumlein structure of the short pulse DWA leads to parasitic coupling of adjacent layers. This coupling distorts the otherwise flat temporal waveshape of a single Blumlein so that there is little or no flat top on the accelerating pulse[5]. This significantly complicates beam transport in the DWA. A charge bunch riding on the rising edge of the accelerating waveform will experience strong transverse defocusing forces. Along this rising edge the bunch is longitudinally stable[6].

Because the flat top is very short it is desirable to have a very short proton bunch riding on the longitudinally stable phase of the accelerating waveform to facilitate longitudinal bunch compression and small energy spread of the output beam. Because the gradient is potentially so high, the transverse defocusing force is also potentially very high. The DWA structure could be interrupted to insert focusing elements but this would significantly reduce the average accelerating gradient of the machine. There are at least two approaches to this problem that do not involve the use of external focusing elements.

Baseline Focusing Scheme

The baseline strategy is to provide one or more phase reversals of the proton bunch in the virtual accelerating wave. That is, the position of the bunch within the wave is varied so that it occupies the longitudinally stable, transversely unstable phase for some distance and then occupies the transversely stable, longitudinally unstable phase over another period. By alternating the phase, net focusing may be achieved both transversely and longitudinally[7].

A straw man design employing a commercially supplied RFQ injector was simulated. A synchronization and pulse selection system have been designed and will be implemented to phase lock an injected bunch from the RFQ to the DWA. The sensitivity of the output beam parameters as a function of the injector/DWA timing jitter was examined with simulations. Figure 4 shows the output phase space of the beam. The RFQ is assumed to provide a 2 MeV pulse of protons that is 200 ps wide with a total charge of 32 pC. The DWA has an average gradient of 60 MV/m in this design study. Only one phase reversal is needed to transport the beam without loss.

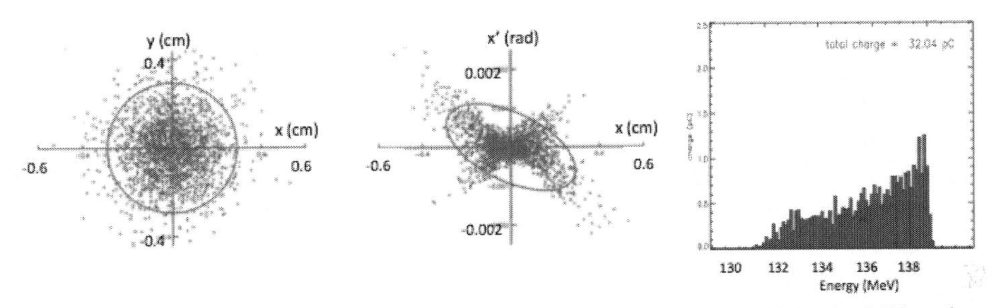

FIGURE 4. Calculated output phase space and energy spectrum at the end of a 2-meter DWA. An RFQ and matching section supply a 2 MeV, 200 ps proton bunch of 32 pC into the DWA.

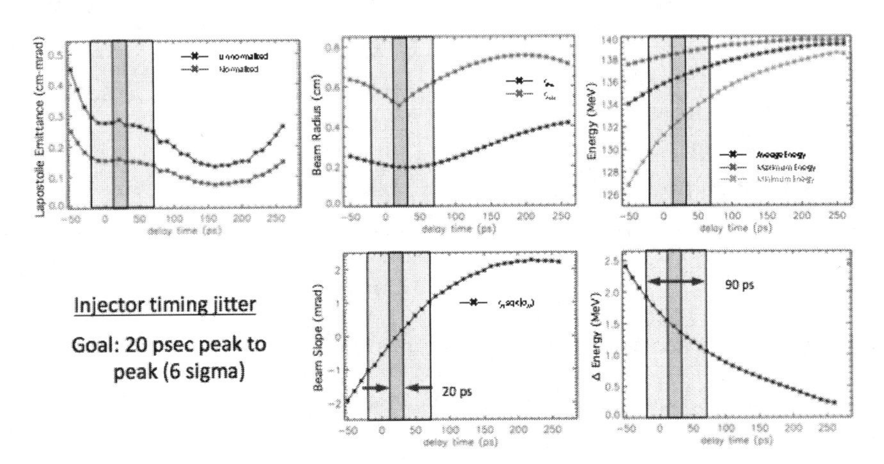

FIGURE 5. Output variation of the beam parameters as a function of injector/DWA timing shifts. Up to 90 ps (6- sigma) jitter is tolerable. The goal for the synchronization system is a 6-sigma jitter of 20 ps.

After a satisfactory tune was achieved all parameters in the DWA were frozen and the relative timing between the injected pulse and the DWA was varied. Figure 5 shows the output variation with respect to the timing variations. The initial goal for the 6-sigma jitter of the synchronization system was set at 20 ps prior to the sensitivity study. The transport study reveals that up to approximately 90 ps peak to peak jitter is tolerable.

Novel Focusing Scheme

If one considers deforming the conductors of the HGI according to the following prescription powerful focusing and/or deflecting fields can be obtained with minimal reduction to the accelerating gradient (the axis of the HGI is along the z-direction):

$$x = r\cos(\phi), \tag{1}$$

$$y = r\cos(\phi), \tag{2}$$

$$z - z_i = \Delta z\left(\frac{r}{b}\right)^n \cos(n\phi). \tag{3}$$

Here, z_i is the average axial coordinate of the ith conductor and b is it's inner radius, r, z and ø are the usual cylindrical coordinates and Δz is the maximum axial deviation of the conductor along it's inner radius. Figure 6 shows a design calculation for the case n=2 (quadrupole).

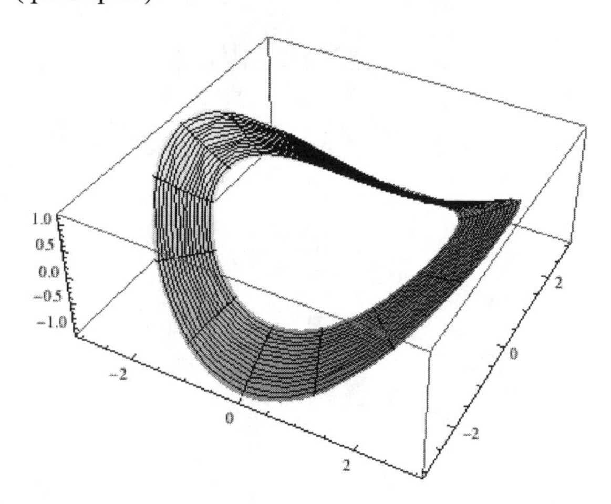

FIGURE 6. Conductor shape for the quadrupole case (n=2). The inner radius is 2 cm while the Δz for this case is 0.5 cm.

The deformed conductors basically lie along the equipotentials of a combined axial and multipole electric field distribution.

The order of the multipole generated by this conductor deformation is n. Conventional HGI's have n=0 while for n=1 the HGI will have a dipole field in addition to the accelerating field. If the electric field normal to the conductors is E_o, then the axial (accelerating) field is given by

$$E_z = \frac{E_o}{\sqrt{1+\left(\frac{n\,\Delta z}{b}\right)^2}}. \tag{4}$$

In particular, for n=2 we have quadrupole fields which are given (for a particular orientation of the HGI) as

$$E_x = \frac{E_o}{\sqrt{1+\frac{4\,\Delta z^2}{b^2}}}\frac{2\,\Delta z\,x}{b^2}, E_y = -\frac{E_o}{\sqrt{1+\frac{4\,\Delta z^2}{b^2}}}\frac{2\,\Delta z\,y}{b^2}. \tag{5}$$

These fields will be produced both in the static case where DC voltages are applied, and also in the traveling wave case if the gradient criterion is satisfied (the axial length of the excited wall must be > 3X the tube radius). Stacking a large number of these conductors with the same orientation can create a quadrupole lens. The orientation of these lenses can be alternated in the usual way to achieve alternating gradient quadrupole focusing.

An example of this arrangement is shown in Figure 7. K-V envelope equations are used to calculate the horizontal and vertical beam size including space charge (1 Ampere), emittance (10π mm-mrad), and acceleration produced by a peak gradient of 100 MV/m in a 2 cm radius DWA with a Δz of 0.42 cm. It is assumed that the bunch rides on the longitudinally stable phase so that the bunch is subject to a strong radial defocusing force from the accelerating waveform (assumed to be Sinusoidal). The quadrupole alternation period is varied throughout the accelerator.

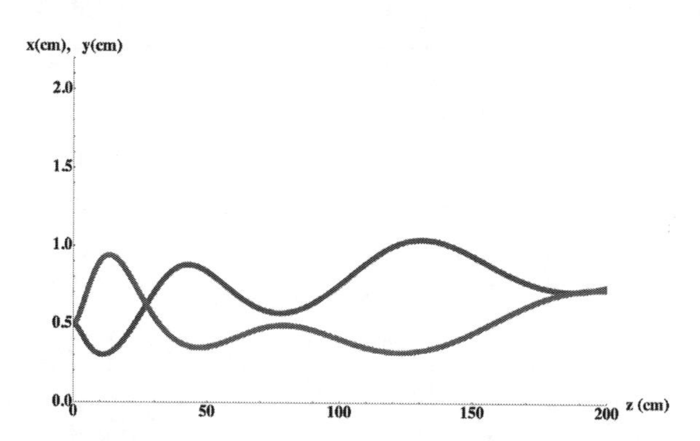

FIGURE 7 Horizontal and vertical beam envelopes for an accelerating 1 A proton beam with emittance and radial defocusing provided by the 100 MV/m accelerating gradient. Quadrupole HGI focusing is used to confine the beam.

SUMMARY

Progress on the DWA has been made in several areas. We now have device quality SiC material for switches. Our baseline beam transport strategy produces an output beam of sufficient quality provided that the injector/DWA timing can be suitably controlled, and we have developed an alternate focusing method based on deformed HGI's. In addition we have begun large sample testing of HGI's. While initial results are promising the system must be completely calibrated before we can determine a field stress limit.

ACKNOWLEDGMENTS

It is a pleasure to acknowledge the help provided by the staffs of the Beam Research Program at LLNL and of the Compact Particle Acceleration Corporation.

This work was performed under the auspices of the US Department of Energy by Lawrence Livermore National Laboratory under contract DE-AC52-07NA273444. Patents pending.

REFERENCES

1. G. J. Caporaso et al., *Phys. Medica **24**, 98 (2008)*
2. S. E. Sampayan et al., *IEEE Trans. Diel. Insul.* **7**(3) 334 (2000).
3. W. C. Nunnally et al., "Investigation of Vacuum Insulator Surface Dielectric Strength With Nanosecond Pulses," in *Pulsed Power Conference 2003 Digest of Technical Papers, 14th IEEE Int. PPC*, **1** (2003) pp. 301-304.
4. K. Batchelor et al., "A Laser Triggered Synchronizable, Sub-Nanosecond Pulsed Electron Source", in *Proc. Int. Conf. Future Accelerators*, Stony Brook, NY, 2001, pp. 214-229.
5. G. J. Caporaso, Y.-J. Chen and S. E. Sampayan, *Revs. Of Accel. Sci, and Tech.,* **2** (2000) 253-263.
6. T. P. Wangler, *Principles of RF Linear Accelerators*, New York, John Wiley & Sons, Inc. 1998, pp. 172-198.
7. Ibid, pp. 357-359.

Current State Of Proton And Carbon-Ion Radiotherapy At The Hyogo Ion Beam Medical Center (HIBMC)

Masao Murakami M.D.[ab], Yusuke Demizu M.D.[a], Yasue Niwa M.D.[a], Osamu Fujii, M.D.[a], Kazuki Terashima M.D.[a], Masayuki Mima M.D.[a], Daisuke Miyawaki M.D.[a], Ryohei Sasaki M.D.[a], Yoshio Hishikawa M.D.[ab] and Mitsuyuki Abe M.D.[a]

[a] *Department of Radiology, Hyogo Ion Beam Medical Center (HIBMC)*
[b] *Division of Ion Beam Therapy, Kobe University Graduate School of Medicine*

Abstract. HIBMC is the world's first facility to be able to use both proton (PRT) and carbon-ion radiotherapy (CiRT). The medically dedicated synchrotron can accelerate protons up to 230 MeV and carbon ions up to 320 MeV. From April 2001 to March 2010, the facility treated 3275 patients, with 2487 patients treated using PRT and 788 using CiRT. Particle radiotherapy was delivered to patients suffering from malignant tumors originating in the head and neck (502 patients), lungs (330), liver (539), prostate (1283), and the bone & soft tissue (130). The clinical results are as follows: (1) H & N tumors: The 2-year overall survival (OS) rates of patients with olfactory neuroblastoma, mucoepidermoid cancer, adenoid cystic cancer, adenocarcinoma, squamous cell carcinoma, and malignant melanoma was 100%, 86%, 78%, 78%, 66%, and 62%, respectively. (2) Lung cancer: For all 80 patients, the 3-year OS rate was 75% (Stage IA: 74%; Stage IB: 76%) and local control (LC) rate was 82% (IA: 87%; IB: 77%). Grade 3 pulmonary toxicity was observed in only 1 patient. These results are comparable to those obtained by surgery, and indicate proton therapy and carbon-ion therapy are safe and effective for stage I lung cancer. (3) Liver cancer: The 5-year LC rate for 429 tumor patient was 90%, and the 5-year OS rate for 364 patients was 38%. These results seem equivalent to those obtained by surgery or radio-frequency ablation. (4) Prostate cancer: In 290 patients treated by proton radiotherapy, five patients died from other disease in the median follow-up period of 62 months. Biochemical disease-free survival and OS rate at 5 years was 88.2% and 96.5%, respectively. Our proton radiotherapy showed excellent OS and biochemical disease-free survival rates with minimum late morbidities. PRT VS CiRT: From our retrospective analysis, it seems that there is no significant difference in the LC and OS rate in H&N, lung and liver cancer between PRT and CiRT.

Keywords: proton radiotherapy, carbon-ion radiotherapy, particle radiotherapy
PACS: ionizing radiations in radiation therapy, 87.53.-j

OUTLINE OF THE HYOGO ION BEAM MEDICAL CENTER (HIBMC)

Construction of a prefectural particle beam radiation medical center was planned as a leading project of the 'Hyogo Cancer Strategy' of Hyogo Prefecture, and the center was opened in May 2001, 9 years after the plan was proposed[1]. Approval for manufacture for the proton beam irradiation equipment was issued at the end of October 2002, and general medical practice for proton-beam radiotherapy started in April 2003. Proton beam and carbon-ion beam radiation were approved as advanced medical technology in July 2004 and May 2005, respectively, and hospital admission and examination fees were included in insurance coverage.

Distinguishing characteristics of HIMBC are (1) Japan's first municipal medical facility for ion-beam radiotherapy and (2) the first facility in the world to employ both proton and carbon-ion beam radiotherapies. Mission statements are: (1) to improve cancer cure rate and speedy return of cancer patients to normal life, (2) to target localized primary tumors with curative intent, (3) to create a hospital with a relaxed atmosphere to relieve patients' fear of cancer, (4) to open this hospital to the world, and (5) to dispatch information on ion beam radiotherapy worldwide.

Application of Accelerators in Research and Industry
AIP Conf. Proc. 1336, 374-376 (2011); doi: 10.1063/1.3586123

CRITERIA FOR TREATMENT, TREATMENT PLANNING AND CONFIRMATION

Treatment protocols are determined according to the criteria for treatment. The treatment criteria have been developed by committee, mainly consisting of specialists from each field. Preparation for ion beam radiotherapy takes a week to make a fixation device, a treatment plan using CT /MRI images and a treatment planning device, a bolus collimator, measurement of the dose, and conduct rehearsal. The dose distribution is verified immediately after the first irradiation utilizing autoactivation[2,3]

Particle Radiotherapy

The number of patients after 2003 has been increasing, expect for 2008, as follows: 250 (2003), 294 (2004), 360 (2005), 514 (2006), 594 (2007), 567 (2008), 636 (2009). The statistics as of March 2010 show a total 3275 patients with 2487 for proton beam (76%) and 788 for carbon-ion beam. After 2007, we started making treatment plans for all patients, except for prostate cancer patients, using both proton and carbon-ion beams, as part of the process of selecting the most appropriate beam type. The carbon-ion exhibits little lateral scattering, although the nucleus fragmentation reaction is observed, and often cuts better than proton beam providing better dose concentration. The most frequently treated disease is prostate cancer, 1283 cases (39,2%), followed by liver 539 (16.5%), head and neck 502 (15.3%), lung 330 (10.1%) and bone and soft tissue tumors 130 (4.0%). Recently, we treat locally advanced pancreatic cancer using proton beams in combination with chemotherapy. We are now expanding the indications for abdominal malignancies through a collaboration with Kobe University's Division of Hepato-Biliary-Pancreatic Surgery, conducting space placement surgeries in order to prevent damage in the digestive tract[4]

Treatment Results[5] for Head and Neck Tumors

Of the head and neck tumors, those considered to be radiation-resistant, such as malignant melanoma, adenoid cystic carcinoma, and adenocarcinoma, are treated more often than squamous cell carcinoma. Acute reaction during treatment is minimal, and oral intake is possible in almost all cases. The two-year survival rates for olfactory neuroblastoma, mucoepidermoid cancer, adenoid cystic carcinoma, adenocarcinoma, squamous cell carcinoma and malignant melanoma are 100%, 86%, 78%, 78%, 66%, 62%, respectively

Treatment Results For Lung Cancer

Treatment related to lung cancer has primarily been Stage I lung cancer, but we also treat T3/T4 cancers such as Pancoast lung cancer. Of 80 patients with Stage 1 lung cancer, the overall survival rate and the three-year local control rate were 75% and 82%, respectively, and three-year overall survival rate of T1N0M0 (IA) and T2N0M0 (IB) were IA 74%, IB 76%, and three-year local control rates were IA 87% and IB 77%[6]

Treatment Results For Liver Cancer

Of 429 patients with liver cancer, the five-year survival rate was 38%, which is equivalent to the results of resection and radiofrequency ablation

Treatment Results For Prostate Cancer

Acute reaction during or after proton radiotherapy offers little or no difficulty[7]. Of 291 patients with prostate cancer, five-year PSA control rate was 88.2%, five-year overall survival rate is 96.5%. The effect was confirmed to be equivalent to other treatments.

HIBMC'S MAJOR OBJECTIVES AND FUTURE PROSPECTS

HIBMC can use both proton and carbon-ion beams, but we don't yet have definitive criteria on whether to select proton or carbon-ion radiotherapy. At present, we usually decide upon proton or carbon-ion beam therapy by comparing their dose distributions to the target. Randomized clinical trials have not yet been conducted between proton and carbon-ion therapy. However, from our clinical results, it seems that there is no apparent difference in effectiveness, prognosis and late adverse events in H & N cancer, lung cancer, and liver cancer between the two treatment modalities. So far we have not recognized statistically significant difference between carbon ion and proton beam treatments in their incidence of producing visual loss through radiation-induced optic neuropathy during treatment of tumors adjacent to optic nerves[8]. In the treatment of skull base tumors, we have found that ion beam radiotherapy produced minimal symptomatic brain toxicities, but sequential evaluation with magnetic resonance imaging detected a greater incidence of radiation-induced brain changes[9]. We have to conduct a well-planned prospective, controlled clinical trial to clarify the difference between proton and carbon-ion radiotherapy.

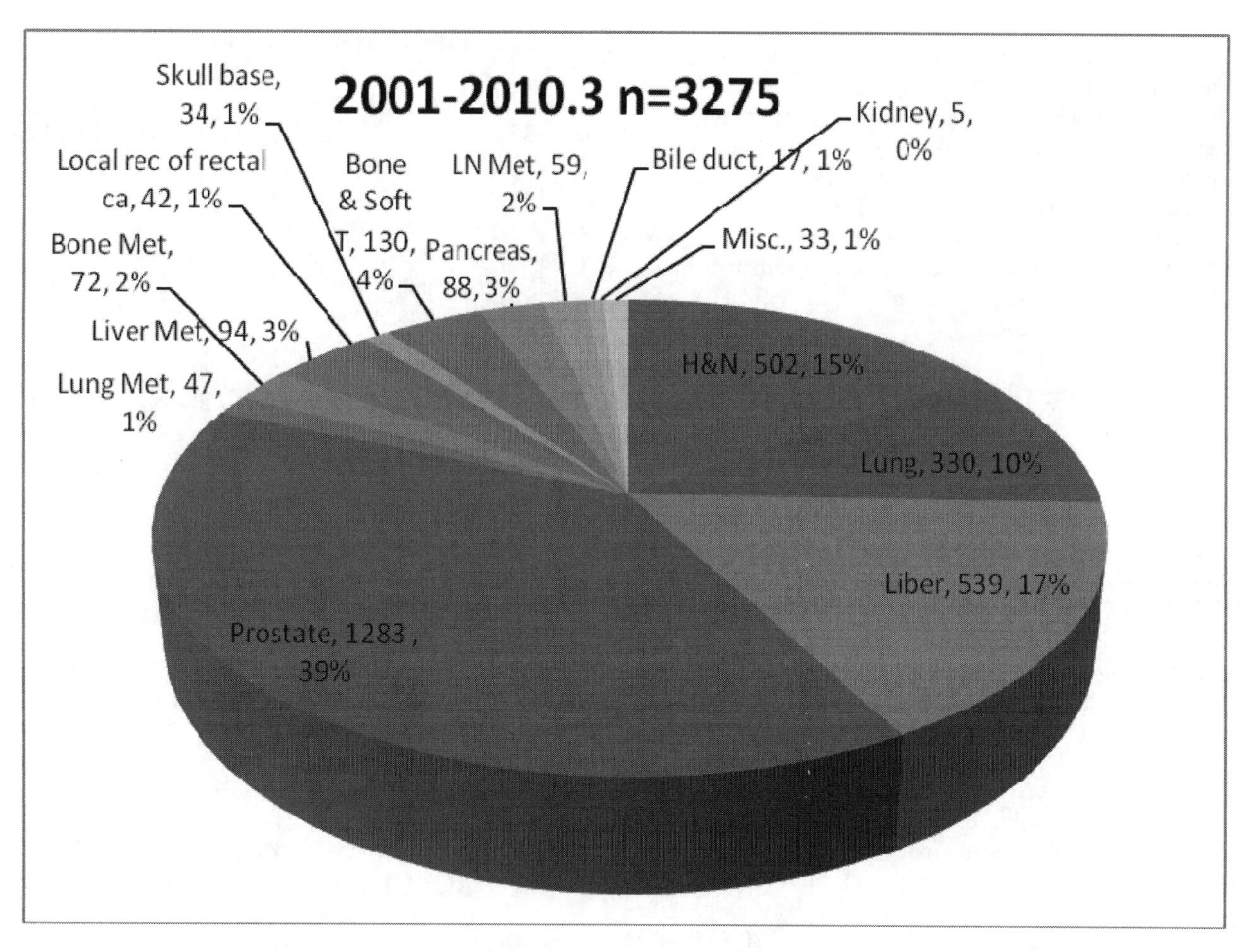

FIGURE 1. Diseases by Beam Type (2001-2010.3)

REFERENCES

1. Kagawa K, Murakami M, Hishikawa Y, et al. *Int J Radiat Oncol Biol Phys* 54:928-938 (2002)
2. Hishikawa Y, Kagawa K, Murakami M, et al. *Int J Radiat Oncol Biol Phys*. 54: 928-938 (2002)
3. Shimizu M, Sasaki R, Miyawaki D, et al, *Int J Radiat Oncol Biol Phys*. 75, 580-586 (2009)
4. Komatsu S, Hori Y, Fukumoto T, et al. *World J Gastroenterol*. 16: 1800-1803 (2010)
5. Murakami M, Demizu Y, Niwa Y, et al. "Current Status of the HIBMC and Results of Representative Diseases" in laser-driven-relativistic plasmas applied to science industry and medicine the 2nd international symposium edited by PR Bolton, et al. *AIP Conference Proceedings 1153*, American Institute of Physics, Melville, NY, 2009, pp. 400-415
6. Iwata H, Murakami, M, Demizu, Y, et al. Cancer 116:2476–85 (2010)
7. Mayahara H, Murakami M, Kagawa K, et al. *Int J Radiat Oncol Biol Phys*.69:434-443, (2007)
8. Demizu Y, Murakami M, Miyawaki D, et al. *Int J Radiat Oncol Biol Phys*.75:1487-1492 (2009)
9. Miyawaki D, Murakami M, Demizu Y, et al. *Int J Radiat Oncol Biol Phys*. 75: 378-384 (2009)

Fast Neutron Induced Autophagy Leads To Necrosis In Glioblastoma Multiforme Cells

Linda Yasui[1], Samantha Gladden[1], Christine Andorf[2] and Thomas Kroc[3]

[1]*Northern Illinois University, Department of Biological Sciences, DeKalb, IL 60115*
[2]*Northern Illinois Institute for Neutron Therapy, Fermi National Accelerator Laboratory, Batavia, IL 60510*
[3]*Fermi National Accelerator Laboratory, Batavia, IL 60510*

Abstract. Fast neutrons are highly effective at killing glioblastoma multiforme (GBM), U87 and U251 cells. The mode of cell death was investigated using transmission electron microscopy (TEM) to identify the fraction of irradiated U87 or U251 cells having morphological features of autophagy and/or necrosis. U87 or U251 cells were irradiated with 2 Gy fast neturons or 10 Gy γ rays. A majority of U87 and U251 cells exhibit features of cell death with autophagy after irradiation with either 10 Gy γ rays or 2 Gy fast neutrons. Very few γ irradiated cells had features of necrosis (U87 or U251 cell samples processed for TEM 1 day after 10 Gy γ irradiation). In contrast, a significant increase was observed in necrotic U87 and U251 cells irradiated with fast neutrons. These results show a greater percentage of cells exhibit morphological evidence of necrosis induced by a lower dose of fast neutron irradiation compared to γ irradiation. Also, the evidence of necrosis in fast neutron irradiated U87 and U251 cells occurs in a background of autophagy. Since autophagy is observed before necrosis, autophagy may play a role in signaling programmed necrosis in fast neutron irradiated U87 and U251 cells.

Keywords: fast neutrons, glioblastoma multiforme, necrosis, autophagy, radiation therapy
PACS: 87.55.ne

INTRODUCTION

Glioblastoma multiforme (GBM), the most commonly diagnosed brain cancer in adults, is the most malignant and aggressive type of primary human brain tumor. GBM is an anaplastic and highly infiltrative tumor which originates from glial tissues in the supportive structure of the brain (1). The average survival time for patients having GBM is 9-11 months. Standard treatment for GBM consists of maximal surgical resection, followed by post-operative radiotherapy, and concomitant or adjuvant chemotherapy with temozolomide (2). Complete surgical resection of GBM is often impossible due to diffuse, infiltrative growth, and therefore recurrence is high. In the end, all treatments are acknowledged as palliative, as they do not provide a cure for GBM. A pressing need for new and improved therapeutic approaches for the treatment of GBM, minimally to prolong life expectancy or perhaps even provide a cure, is imperative.

Radiation therapy is a standard therapy for GBM but these tumors are resistant to photon irradiation relative to fast neutron irradiation (3, 4). Therefore, fast neutron irradiation may offer a therapeutic advantage for GBM patients. Several decades ago, clinical trials for fast neutron radiation therapy of GBM were run (5, 6, 7, 8). Fast neutrons were found to sterilize GBM in a significant proportion of the patients as determined by histology data. However, in those early fast neutron trials, no survival benefit was found because patients suffered from fatal post-irradiation gliosis to normal brain tissue (9). More recently, improved methods of fast neutron dose delivery (modern treatment planning and image guided conformal fast neutron irradiation) were investigated at University of Washington (10). Although fast neutron treatment was clinically well tolerated, evidence of mild to moderate gliosis and microvascular sclerosis consistent with radiation injury in specimens collected at autopsy from the contralateral brain (received approximately 6-10 Gy) was still observed (10). Therefore, protection of normal brain tissue from an inflammation response remains a major issue associated with fast neutron treatment of GBM. Given that necrosis induces an inflammation response, the presence of necrosis, also associated with a poor tumor prognosis, is an obstacle for the use of fast neutron irradiation in the treatment

Application of Accelerators in Research and Industry
AIP Conf. Proc. 1336, 377-382 (2011); doi: 10.1063/1.3586124

of GBM. Reduction or elimination of radiation-induced necrosis may help improve patient outcome.

Radiation-induced cell death occurs via several different programmed cell death pathways (11). Apoptosis, autophagy and necrosis are 3 major programmed cell death pathways. These 3 modes of cell death are morphologically distinct and are induced by irradiation. It is important to distinguish between these modes of cell deaths because necrosis is often associated with some disastrous side effects of cancer therapy and can lead to local inflammation, presumably by spilling factors that alert the innate immune system (12). Several authors propose that there are functional interactions between apoptosis, autophagy and necrosis (13, 14, 15).

Autophagy is a normal, baseline cell survival activity that is involved in recycling internal cellular components, preventing accumulation of damaged proteins and organelles during stress or starvation (14, 16). The morphological features of autophagy or cell "self-eating" include membranous, cup-shaped pre-autophagosomes (phagophores) that encircle cytoplasmic components (organelles, etc) to form enclosed double-membraned autophagosomes (also known as autophagic vesicles). Autophagosomes fuse with lysosomes to form autolysosomes, the degradative form of autophagic vesicles. Lysosomal hydrolases degrade the vesicle contents (17) and in extreme cases, this can lead to the clearance of dead cells from the organism (18, 19). Autophagy is induced in breast, colon, prostate and brain cancer cells (16, 20) after treatment with chemotherapeutic agents and radiation (X-rays and fast neutrons) (21, 22).

Necrosis can be either passive and "accidental" or under genetic control of a programmed cell death process (23). Cells undergoing necrosis exhibit distinct morphological and biochemical modifications such as early loss of membrane integrity, cell lysis, swelling, dilation of cytoplasmic organelles, release of nuclear and cytoplasmic HMGB1, ATP depletion, failure of osmotic regulation and formation of cytoplasmic vesicles and random degradation of DNA (24). Necrosis is a common feature of human tumors and is a hallmark of GBM tumors (25). In irradiated GBM cells, the process of autophagy is known to be induced (21, 22), in the absence of classical features of apoptosis (GBM cells are intensely resistant to apoptosis) (25). In this study, we investigated the induction of necrosis by fast neutron irradiation, in the background of autophagy.

MATERIALS AND METHODS

Cell Culture

The GBM cell lines, U87 and U251, were maintained in DMEM-F12 (1:1) medium supplemented with 10% fetal bovine serum and penicillin/streptomycin as previously described (26, 27). Exponentially growing cells were used in all experiments.

γ Irradiation Of Cells

U87 or U251 cells were irradiated on monolayer at room temperature with γ rays using a routinely calibrated ^{137}Cs source.

Fast Neutron Irradiations

Adherent cells in T25 flasks were transported at room temperature by car to Fermi lab. Cells were irradiated with 2 Gy fast neutron irradiations at Fermilab as previously described (27). A fast neutron dose of 2 Gy was chosen to approximate the dose from one fraction of fast neutron radiotherapy. The experimental cell irradiation set-up at Fermi lab is similar to the set-up used for patient treatment. Fast neutrons are produced by bombarding a 22 mm thick beryllium target with a 66 MeV proton beam from a linear accelerator. The beryllium target is thick enough to absorb 49 MeV protons to produce neutrons. The rest of the energy is deposited in a 0.5 mm thick gold backing. The T25 flasks were placed vertically in a "tissue equivalent" polyethylene holder with the cell growth surface of the plates positioned at the plane of calculated dose. Polyethylene "build-up" material (2.6 cm) was placed between the beam source and the cells and additional buildup material (8 + 8 + 2.6 cm) was placed behind the cells to simulate tissue equivalent material. Radiation dose delivery from fast neutron irradiation is experimentally determined on a daily basis by calibration of the fast neutron beam following the AAPM neutron dosimetry protocol (AAPM report #7). Observations over 30 years show extremely small variations in manual and computer calculated dose. After irradiation, cells were transported back to NIU for 1 or 3 day incubation before processing samples for TEM.

Transmission Electron Microscopy (TEM)

TEM was performed in order to visualize autophagocytic vesicles (membrane enclosed vesicles containing cytosolic material or organelles) or swollen, vesiculated necrotic cells. Fast neutron (2 Gy) or sham irradiated U87 or U251 cells were collected 24 or 72 hours after irradiation and fixed for TEM. U87 or U251 cells, irradiated using 10 Gy γ rays, were collected for TEM 24 hours post irradiation. Preliminary experiments showed that a dose of at least 10 Gy γ rays was needed to induce necrosis in U87 and U251 cells. Fixation and embedding in Spurrs epoxy resin was performed as previously described (28). Randomly collected cell images were acquired using a Hitachi H-600 electron microscope, operated at 75 kV accelerating voltage. Approximately 200 images were acquired for this study. Negatives (Kodak EM film 4489, 3 ¼" X 4", Electron Microscopy Sciences, Hatfield, PA) were developed using Kodak D-19 developer for 4 minutes with agitation. Digital images of the negatives were obtained by scanning the negatives at a resolution of 2400 dpi and 8 bit data depth of approximately 50 MB. A 2400 dpi resolution optimally balances high resolution for image quality and the ability of Photoshop to handle images having a large file size. Digital images were saved as tagged image file format (tif) files. Figures were composed using Photoshop 6.0.

Ultrastructural criteria used to identify cell necrosis were cell swelling and increase in cell volume, perinuclear clustering of organelles, degeneration of mitochondria and organelle swelling, intracellular vesiculation, lack of formation of apoptotic bodies, translucent appearing cytoplasm, mottled appearing nucleus with chromatin condensation in ill defined masses and loss of membrane integrity (24). The morphological criterion used to identify the process of autophagy was the presence of darkly staining membranes surrounding cytoplasmic material since the multiple membranes surrounding autophagosomes have increased electron density probably due to the presence of having double- or multiple membranes. These structures were named autophagic vesicular organelles or AVOs.

Error Estimates And Statistical Analysis For Significance

Statistical inference using the Student t distribution was used to calculate the 90% confidence interval for the percent autophagic or necrotic cell values presented in figure 4.

RESULTS

TEM Of GBM Cells 1 Day Post 10 Gy γ Irradiation

U87 and U251 cells were fixed for TEM 1 day after 0 or 10 Gy γ irradiation and images of cells profiles were scrutinized for the presence of autophagosomes (indicative of autophagy) or cell swelling (indicative of necrosis) (Fig. 1 and 2).

Unirradiated U87 control cells only had 2 of the cell profiles show slight evidence of swollen organelles or necrosis. The remaining unirradiated control U87 cells appeared normal and did not exhibit any evidence of autophagy (Fig. 1a). All the unirradiated control U251 cells appeared normal without evidence of either autophagy or necrosis (Fig. 2a).

The majority of the U87 cell profiles, irradiated with 10 Gy γ rays showed pronounced evidence of autophagy by the appearance of autophagosomes and autophagophores (Fig. 1b). Nineteen percent of the U87 cells, irradiated with 10 Gy γ rays, did not contain evidence of autophagy. Fifteen percent of the autophagic, γ irradiated U87 cell profiles showed evidence of cell swelling or necrosis.

The majority of U251 cell profiles of cells irradiated with 10 Gy γ rays also showed evidence of autophagy. Only 7% of the autophagic U251 cells contained evidence of both autophagy and necrosis (Fig. 2).

TEM Of GBM Cells Irradiated With 2 Gy Fast Neutrons

In GBM cells that were irradiated with 2 Gy fast neutrons, the presence of autophagosomes was detected in irradiated U87 and U251 cells collected 24 hours after irradiation (Fig. 3). Many mitochondria in U87 and U251 cells appeared to be more elongated and swollen. Increased presence of membranous structures was found in the cytoplasm of fast neutron irradiated U87 and U251 cells, along with increased numbers of vacuoles and double membranes structures containing bits of cytoplasm (autophagosomes). U87 and U251 cells do not appear to be more translucent (as would be found in necrotic cells in the process of swelling) or significantly larger 1 day after 2 Gy fast neutron irradiation. The cell surface of both U87 and U251 cells did not exhibit increased membrane activity (more villi) or obvious disruptions after 1 day of fast neutron irradiation but small membrane enclosed bodies were observed.

By 3 days after 2 Gy fast neutron irradiation, U87 and U251 cells contained noticeably swollen mitochondria and highly vacuolated cytoplasms, as well as numerous autophagosomes (Fig 3). The cell surface of both cell lines exhibited increased number of projections and membrane enclosed bodies. Only U251 cells, 3 days after 2 Gy fast neutron irradiation, showed any perinuclear concentration of cytoplasmic material. In a U251 cell profile, the cell cytoplasm appeared quite translucent (indicator of necrosis) while also containing several autophagosomes (Fig. 3d).

Further in the same U251 cell profile, the nucleus contained with only minimal chromatin localized at the nuclear periphery (indication of necrosis). In Fig. 3f, an example of a lysed nucleus from a U251 cell, 3 days after 2 Gy fast neutron irradiation is shown.

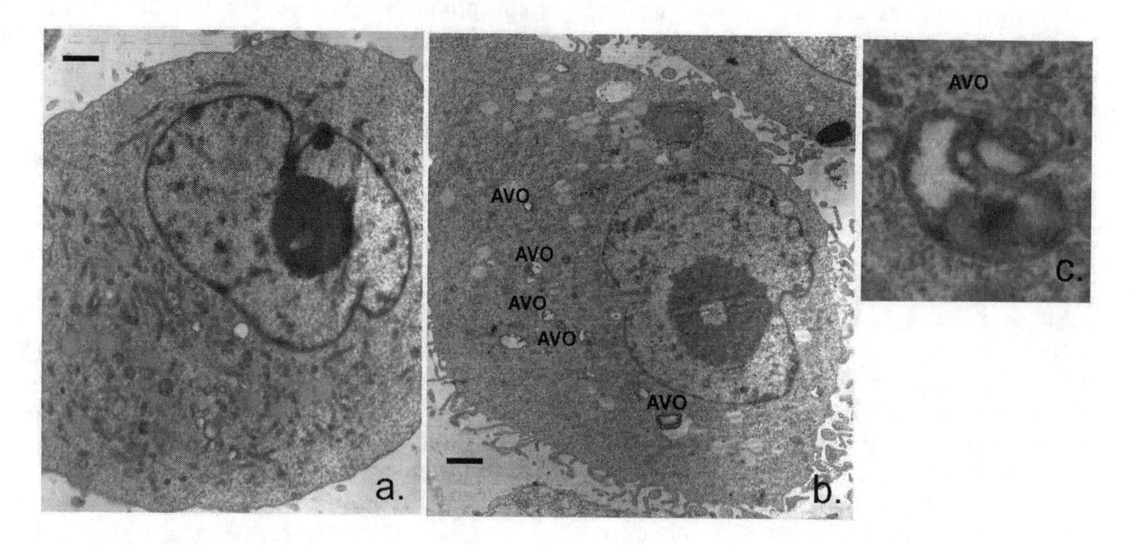

FIGURE 1. TEM of U87 cells collected 1 day after 10 Gy γ irradiation. A control, unirradiated U87 cell profile is shown in panel (a) and a cell profile of a U87 cell irradiated with 10 Gy γ rays is shown in panel (b). An enlarged image of an autophagosome is shown in panel (c). Bar equals 1 um.

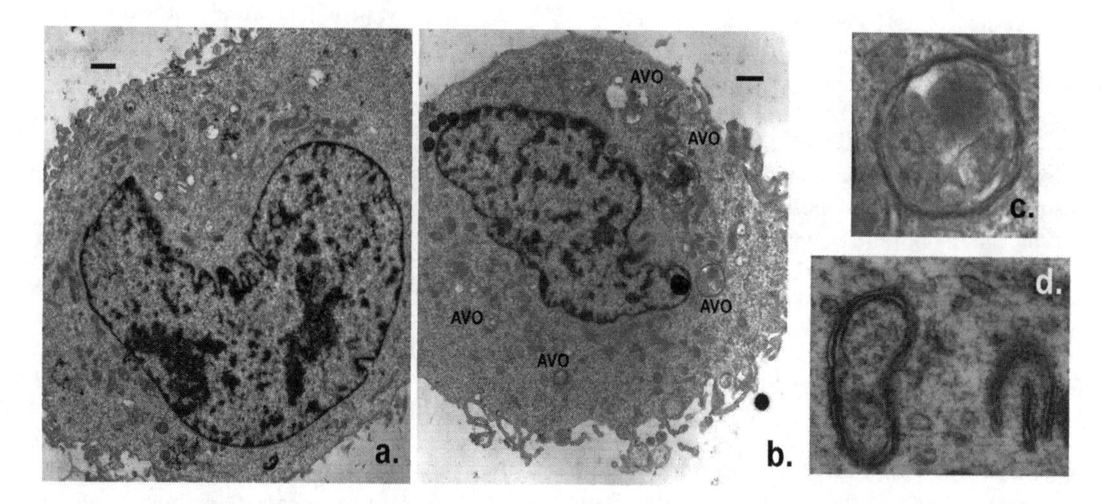

FIGURE 2. TEM images of U251 cells collected 1 day after 10 Gy γ irradiation. A control, unirradiated U251 cell profile is shown in panel (a) and a cell profile of a U251 cells irradiated with 10 Gy γ rays is shown in panel (b). Enlarged images of autophagosomes are shown in panels (c) and (d). Bar equals 1 um.

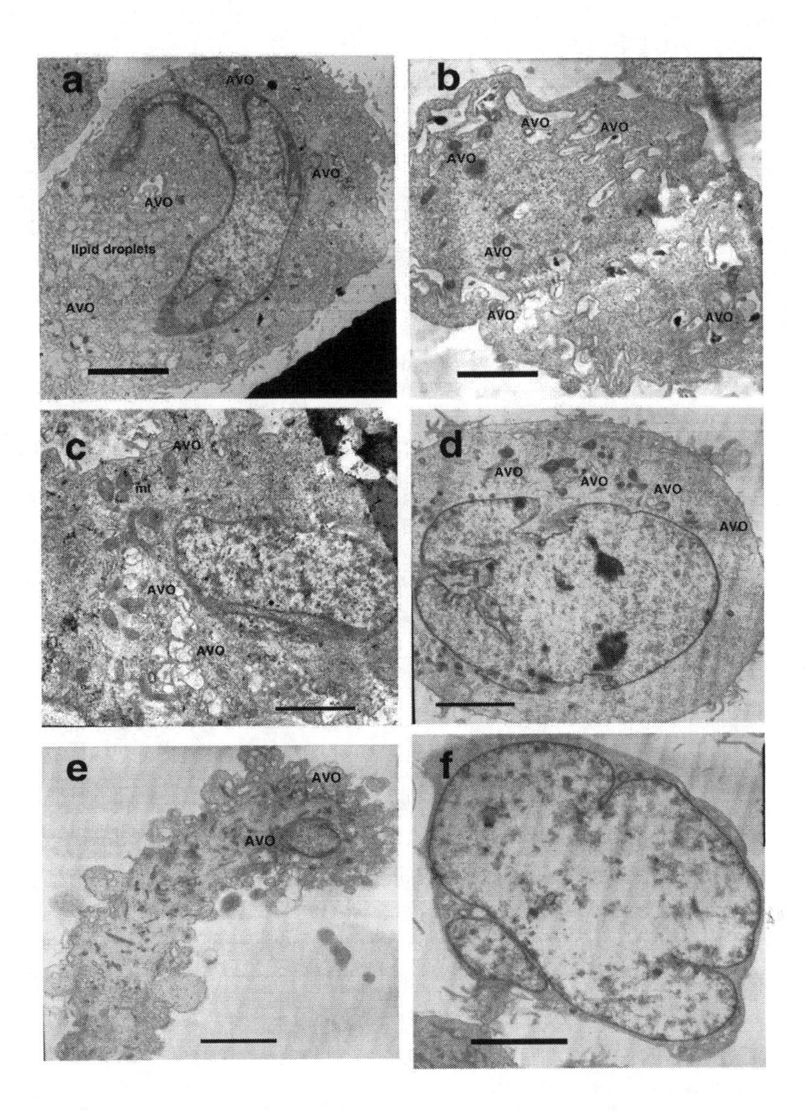

FIGURE 3. TEM of U87 (a, b) and U251 (c, d) cells irradiated with 2 Gy fast neutrons. U87 and U251 cells were collected and fixed for TEM 1 day after 2 Gy fast neutron irradiation (a, c) or 3 days after 2 Gy fast neutron irradiation (b, d). Lipid droplets are indicated in panel a. AVO indicates locations of autophagosomes. A U251 cell profile in panel e shows a highly vacuolated (necrotic) cell with membrane blebs. A translucent, swelling (or lysing) necrotic U251 cell with a swollen cell nucleus and loss of membrane integrity is shown in panel f. Bar = 3 um.

DISSCUSSION

Fast neutron irradiation is superior to γ irradiation in its ability to kill GBM cells. Using some previously determined clonogenic survival data (27), the relative biological effectiveness (RBE) of fast neutron irradiation was found to be 2.4 for both U87 and U251 cells.

Morphological evidence for autophagy was found in this TEM study. Both 10 Gy γ rays and 2 Gy fast neutrons induced autophagy (Fig. 1-4). Evidence for necrosis was also found in both U87 and U251 cells after fast neutron irradiation. Necrotic U87 and U251 cells were swollen and lost membrane integrity. Associated with cell swelling was an increase in cell size. Mitochondria were found to be swollen and cytoplasm translucent especially in U87 and U251 cells sampled at 3 days post 2 Gy fast neutron irradiation (Fig. 4). Interestingly, if fast neutron irradiated U87 or U251 cells were permitted to continue towards death by increasing the time cells were cultured at 37°C (3 days post 2 Gy fast neutron irradiation), the proportion of cells with evidence of necrosis was found to increase. The histogram data in figure 4 shows the increase (Fig 4c and 4d) and image data in figure 3 shows the morphological changes (Fig. 3b and 3d). These same cells also contained autophagosomes, indicating necrosis was taking place with autophagy.

Little is known about the role of programmed necrosis in radiation induced cell death. These data show fast neutron irradiation is effective at inducing necrosis in the background of autophagy in U87 and U251 cells, whereas γ irradiation is not. Since autophagy is observed to occur before fast neutron irradiation, we suggest that autophagy may signal programmed necrosis in fast neutron irradiated U87 and U251 cells, contrary to evidence in normal cells showing that autophagy inhibits programmed necrosis (30).

REFERENCES

1. M. Lino, A. Merlo. Curr. Opin. Cell Biol. 21 (2009) 1.
2. C. Nieder, M Adam, A.L. Grosu. Rev. Recent Clin.Trials 1 (2006) 43.
3. J.M. Akudugu, A. Binder, A. Serafine, J. Slabbert, A. Giese, L. Bohm, 2004. Life Sci. 75 (2004) 623.
4. Y. Taghian, H. Suit, F. Pardo, D. Gioioso, K. Tomkinson, W. Dubois, L. Gerwick, Int. J. Radiat. Oncol. Biol. Phys. 23 (1992) 55.
5. M. Catterall, H.J. Bloom, D.V. Ash, L. Walsh, A. Richardson, D. Uttley, N.F.C. Gowing, P. Lewis, B. Chaucer, Int. J. Radiat. Oncol., Biol., Phys. 6 (1980) 261.
6. P.D. Kurup, T.F. Pajak, F.R. Hendrickson, J.S. Nelson, J. Mansell, L. Cohen, M. Awschalom, I. Rosenberg, K. Ten Haken, Int. J. Radiat. Oncol., Biol., Phys. 11 (1985) 679.
7. G.E. Laramore, et al, Int. J. Radiat. Oncol., Biol., Phys.14 (1988) 1093.
8. G.E. Laramore, T.W. Griffin, A.J. Gerdes, R.G. Parker, Cancer 42 (1978) 96.
9. S.R. Saroja, J. Mansell, F.R. Hendrickson, L. Cohen, A. Lennox, Int. J. Radiat. Oncol., Biol. Phys. 27 (1989) 1296.
10. K.J. Stelzer, J.G. Douglas, D.A. Mankoff, D.L. Sibergeld, K.A. Krohn, G.E. Laramore, A.M. Spence, Neuro-oncol. 10 (2008) 88.
11. D. Eriksson, K. Eriklund, L. Johansson, T. Stigbrand, Radiation induced cell deaths. In: Stigbrand T, Carlsson J, Adams GP (Eds). Targeted radionuclide tumor therapy. Springer Science + Business Media, 2008.
12. P. Golstein, G. Kroemer, TRENDS Biochem. Sci. 32 (2007) 37.
13. K. Degenhardt, et al., Canc. Cell 10 (2006) 51
14. S. Jin, E. White, Autophagy 4 (2008) 563.
15. N. Zucchini-Pascal, G. de Sousa, G. Rahmani, Toxicol. 256 (2009) 32.
16. B. Levine, J. Yuan, 2005. J. Clin. Invest. 115 (2005) 2679.
17. E-L. Eskelinen, Autophagy 1 (2005) 1.
18. M.C. Maiuri, E. Zalckvar, A. Kimchi, G. Kroemer, Nat. Rev. Mol. Cell Biol. 8 (2007) 741.
19. A.L. Edinger, C.B. Thompson, Canc. Cell 4 (2003) 422.
20. Y. Kondo, T. Kanzawa, R. Sawaya, S. Kondo, Nat. Rev. Canc. 5 (2005) 726.
21. S. Daido, A. Yamamoto, K. Fujiwara, R. Sawaya, S. Kondo, Y. Kondo, Cancer Research 65 (2005) 4368.
22. S. Benzina, A. Altmeyer, F. Makek, P. Dufour, J-M. Denis, J. Gueulette, P. Bischoff, Canc. Letts. 264 (2008) 63.
23. J. Hitomi, D.E. Christofferson, A. Ng, J. Yao, A. Degeterev, R.J. Xavier, Cell 135 (2008) 1311.
24. P. Golstein, G. Kroemer, TRENDS Biochem. Sci. 32 (2007) 37.
25. F.B. Funari, et al, Genes & Dev. 21 (2007) 2683.
26. L.S. Yasui, A. Hughes, E.R. DeSombre, Int. J. Radiat. Biol. 77 (2001) 955.
27. L.S. Yasui, C. Andorf, L. Schneider, T. Kroc, A. Lennox, K.R. Saroja, Int. J. Radiat. Biol. 84 (2008) 1130.
27. C.C. Korte, L.S. Yasui, Scanning Micros. 7 (1993) 943.
28. Y. Iwadate, J-E. Mizoe, Y. Osaka, A. Yamamura, H. Tsuju, Int. J. Radiat. Oncol. Biol. Phys. 50 (2001) 803.
29. K. Degenhardt, et al, Canc. Cell 10 (2006) 51.

A New F-18 Labeled PET Agent For Imaging Alzheimer's Plaques

Padmakar V. Kulkarni[1], Neil Vasdev[2], Guiyang Hao[1], Veera Arora[1], Michael Long[1], Nikolai Slavine[1], Srinivas Chiguru[1], Bao Xi Qu[1], Xiankai Sun[1], Michael Bennett[1], Peter P. Antich[1], Frederick J. Bonte[1]

(1)UT Southwestern Medical Center, 5323 Harry Hines Blvd., Dallas, TX 75390, United States
(2)CAMH PET Centre & Department of Psychiatry, University of Toronto, Toronto ONT, Canada, M5T-1R8

Abstract. Amyloid plaques and neurofibrillary tangles are hallmarks of Alzheimer's disease (AD). Advances in development of imaging agents have focused on targeting amyloid plaques. Notable success has been the development of C-11 labeled PIB (Pittsburgh Compound) and a number of studies have demonstrated the utility of this agent. However, the short half life of C-11 (t1/2: 20 min), is a limitation, thus has prompted the development of F-18 labeled agents. Most of these agents are derivatives of amyloid binding dyes; Congo red and Thioflavin. Some of these agents are in clinical trials with encouraging results. We have been exploring new class of agents based on 8-hydroxy quinoline, a weak metal chelator, targeting elevated levels of metals in plaques. Iodine-123 labeled clioquinol showed affinity for amyloid plaques however, it had limited brain uptake and was not successful in imaging in intact animals and humans. We have been successful in synthesizing F-18 labeled 8-hydroxy quinoline. Small animal PET/CT imaging studies with this agent showed high (7-10% ID/g), rapid brain uptake and fast washout of the agent from normal mice brains and delayed washout from transgenic Alzheimer's mice. These promising results encouraged us in further evaluation of this class of compounds for imaging AD plaques.

Keywords: Alzheimer's disease, amyloid, PET imaging, animal model, quinoline
PACS: 87.19.xr, 87.14.em, 87.57.uk

INTRODUCTION

In 1906, Dr. Alois Alzheimer described the histopathological characteristics of the brain of a demented patient indicating the presence of abnormal deposits of proteins as the cause of the disease. Presence of extracellular amyloid plaques and intracellular neurofibrillary tangles are hallmarks of Alzheimer's disease (AD). According to amyloid cascade hypothesis, deposition of amyloid plaques is the early event triggering deposition of neurofibrillary tangles and neuronal degeneration leading to dementia. The process of neurodegeneration may be taking place several years before the clinical symptoms start appearing. Thus it would be critical to detect early changes in the brain for developing effective therapies. Conventional methods of imaging such as CT (Computed tomography) and MRI (magnetic resonance imaging) are not very helpful. Brain blood flow measurements by SPECT (single photon emission computed tomography) and glucose utilization by FDG (fluorodeoxy glucose) PET (positron emission tomography) are not very specific. The presence of plaques in AD patient brains is confirmed only at autopsy. It is desirable to develop noninvasive methods to directly image the presence and extent of amyloid plaques. Development of therapeutic and diagnostic agents is based on targeting the amyloid plaques. Derivatives of amyloid affinity dyes labeled with SPECT and PET radio nuclides are being evaluated as amyloid imaging probes (1-3). Amongst these, C-11 labeled Thioflavin derivative, popularly known as PIB (Pittsburgh compound B) has been extensively evaluated (2). PIB uptake has been shown to correlate to amyloid burden and differentiate between normal, MCI (mild cognitive impairment) and AD patients.

C-11 has a short half life (20 min) and thus its use is limited to sites with an onsite cyclotron. F-18 has many advantages compared to C-11; F-18 is a longer lived positron emitting isotope with a half-life of 109.7 min that facilitates radiosynthesis, and can be transported to an offsite facility. Fluorine-18 has lower positron energy than that of C-11 (0.635 vs. 0.96

Application of Accelerators in Research and Industry
AIP Conf. Proc. 1336, 383-385 (2011); doi: 10.1063/1.3586125

MeV). Thus, higher spatial resolution can be achieved with F-18 labeled agents. Because of these advantages and availability of PET/CT systems, F-18 labeled agents are being developed for imaging amyloid plaques (3).

Main features of amyloid imaging agents are: 1. they should cross the blood brain barrier (efficiently transported into the brain); 2. rapid washout from the normal brain; 3. have high affinity for the plaques. The UCLA PET group developed an F-18 labeled tracer ([F-18]-FDDNP) and first demonstrated the prospects of imaging amyloid plaques and neurofibrillary tangles in the brains of AD patients (1). Figure 1 shows structures of some of the agents that have been radiolabled for PET imaging of amyloid plaques.

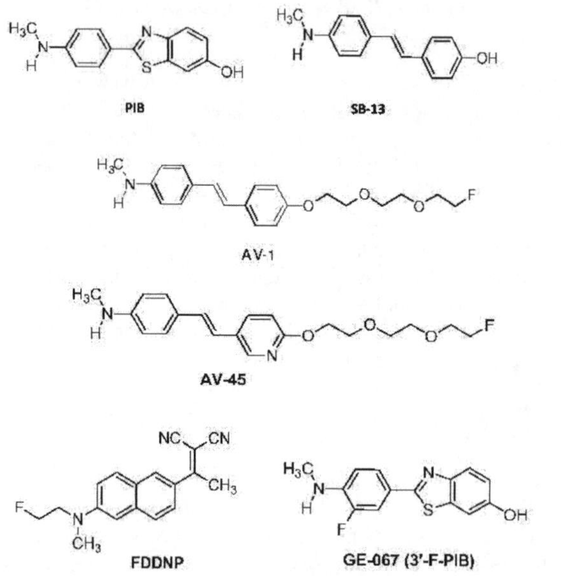

PIB

SB-13

AV-1

AV-45

FDDNP

GE-067 (3'-F-PIB)

FIGURE 1. Compounds developed as PET tracers for plaque imaging in human subjects. These are based on amyloid affinity dyes, Congo red and Thioflavin.

An F-18 labeled analog of PIB has been developed (GE-067) and is undergoing clinical trials with encouraging results. A University of Pennsylvania group has developed a number of radioiodine and fluorine labeled compounds that show great promise (3). One of these compounds, AV-45 (Flurbetapir®, Avid Pharmaceuticals Inc) is in phase II/III clinical trials and reported to have very high sensitivity and specificity in detection of amyloid plaques in humans.

Metal Chelators

Certain transition metals such as Zn, Cu, and Fe are elevated in the plaque deposits. Thus, these may be potential targets for developing diagnostic and/or therapeutic agents. Hydroxyquinoline (HQ) and its derivatives are neutral, lipophilic molecules and are weak metal chelators. They are under investigation for therapeutic applications based on extraction of metal ions and disaggregation of plaque (4-6). Opazo et. al. utilized this strategy in exploring iodine-123 labeled clioquinol as a biomarker for β-amyloid in human subjects (7). However, <1% of this activity was taken up by brain. We labeled clioquinol (5-chloro-7-iodo-8-hydroxy quinoline, CQ) with I-125 and showed [I-125] CQ to be specific for imaging β-amyloid plaque in histological samples (8). However, its uptake at two minutes post injection in mice brains (n=4, number of mice injected) was low (1-2% ID/g). Radioiodine labeled compounds are likely to undergo in vivo de-iodination, thus [I-123/124] CQ may not be optimum for in vivo imaging. On the other hand, F-18 labeled derivatives may be better suited due to their better in vivo stability, better transport across the blood brain barrier, as well as all of the aforementioned advantages of F-18. Thus, we synthesized F-18 labeled HQ and evaluated its pharmacokinetic properties in normal and experimental mice.

METHODS

Radiochemistry

8-Benzyloxy-2-chloroquinoline (3 mg) was heated at 135 0C for 15 minutes with dry fluorine-18 labeled potassium cryptand fluoride [K222)] [F-18] in DMSO. The labeled product was loaded onto a tC18 Sep-Pak® cartridge (Waters) and eluted with CH3CN. Catalytic de-protection was carried out with Pd(OH)2 and Pd/C under H2 balloon. After filtration, the labeled product was purified by reverse phase semi-preparative HPLC. Sep-Pak® cartridge (Waters) was used to further process the desired product (2-[F-18]-8-hydroxy quinoline, Figure 2). The purified product was formulated in PBS (phosphate buffered saline) for in vivo PET/CT imaging.

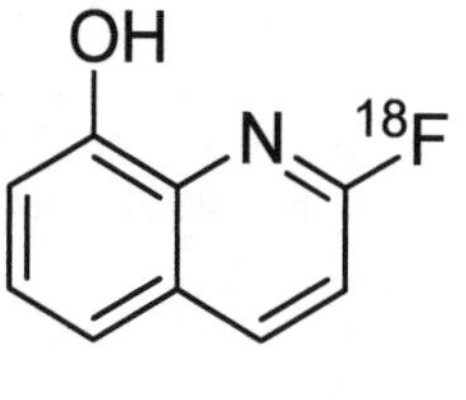

FIGURE 2. Structure of 2-Fluoro- [F-18]-8-hydroxy quinoline (2-F-HQ).

Animal Model

Aβ(1-42) was dissolved in PBS (phosphate buffered saline, pH 7.4) to a final concentration of 433 μg/mL (100 μM) with $ZnCl_2$ (100 μM) and magnetically stirred at 1200 RPM for 1 h at room temperature, after which time it became visibly cloudy. In order to complete the aggregation process, the mixture was incubated at 37°C for 48 h. Animals were anesthetized with a cocktail mixture containing (Ketamine/Xylazine/Acepromazine). Mice brains were injected with aggregated Aβ-42 peptide by Sterotaxic instrument. The Aβ injection location corresponded to the mouse brain hippocampus structure CA1. Animals were allowed to recover for 10-15 days, before imaging and histopathology.

Small Animal PET/CT Imaging

Small animal PET/CT imaging studies were performed using a Siemens Inveon® system (Siemens Medical Solutions Inc., Knoxville, TN, USA). The PET imaging was acquired directly following the acquisition of CT data. Immediately following the injection, a 30 minute dynamic scan was performed. The resulting quantitative data were expressed in Percent Injected Dose per Gram (%ID/g).

RESULTS

F-18 labeled 2-F-HQ was prepared in 30±4% (n=20) radiochemical yields (uncorrected for decay) in 70 minutes with radiochemical purity >99%. Specific activity of the final product was ≥ 1200 mCi/μmol and the log P was 2.07±0.03. Initial uptake in the mouse brain was very significant 8-10% ID/g within a minute; which is comparable to [C-11]-PIB. Washout from the normal mouse brain was very rapid; less than 20% of the compound was retained in the brain after 5 min (Figure 3). Washout of the tracer was delayed in experimental animals with intracranial injection of Aβ (1-42) peptide aggregates. Experimental animal brain activity was 177% of the control animals at 6.5 minutes and remained 140% up to 30 minutes (Figure 3). Congo red staining confirmed the presence of plaques. PET/CT imaging with [F-18] labeled radiotracer based on hydroxyl quinoline derivative showed rapid, significant brain uptake of the tracer and had fast washout from normal mice brains and had delayed washout from AD mice brains.

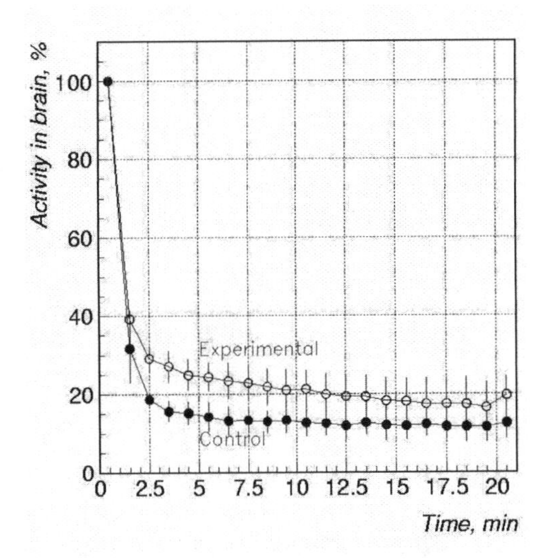

FIGURE 3. Activity retained in the brains of control and experimental mice (n=4) vs. time post injection in minutes. Initial brain uptake is considered 100% of the activity.

CONCLUSIONS

Experimental AD mouse model had delayed brain wash out of the tracer, and the presence of the Aβ aggregates was confirmed by histopathology. Further characterization and evaluation of [F-18]-labeled quinoline derivatives as amyloid imaging agents via PET/CT imaging are in progress.

ACKNOWLEDGMENTS

Supported by NIH, NIA Grant Number: 1RCAG036308-01 and Worsham Fund.

REFERENCES

1. J. Shoghi, G. W. Small, E. D. Agdeppa, et al. Amer. J. Geri. Psych. 46 (2002) 24
2. C. A. Mathis, Y. Wang, D. P. Holt, et al. J. Med. Chem. 46 (2003) 2740.
3. W. Zhang, M. P. Kang, S. Oya, et al. Nucl. Med. Biol. 34 (2007) 89.
4. A. I. Bush AI. Trends in Neurosci. 26 (2003) 207.
5. A. I. Bush J. Alzheimer's. Dis. 15 (2008) 223.
6. R. A. Cherney, C. A. Atwood, X. E. Xillinas, et al., Neuron. 30 (2001) 665.
7. C. Opazo, S. Luza, V. L. Villemagne, et al. Aging Cell 5 (2006) 69.
8. P.V. Kulkarni, C.R. Roney, V. Arora, et al. AIP Conf. Proc.1099 (2009) 492.

Lasers As Particle Accelerators In Medicine: From Laser-Driven Protons To Imaging With Thomson Sources

I. V. Pogorelsky[a], M. Babzien[a], M. N. Polyanskiy[a], V. Yakimenko[a], N. P. Dover[b], C. A. J. Palmer[b], Z. Najmudin[b], P. Shkolnikov[c], O. Williams[d], J. Rosenzweig[d], P. Oliva[e], M. Carpinelli[e], B. Golosio[e], P. Delogu[f], A. Stefanini[f], M. Endrizzi[g]

[a] *Brookhaven National Laboratory, Accelerator Test Facility, Upton NY 11973, USA*
[b] *The Blackett Laboratory, Imperial College London, SW7 2BW, United Kingdom*
[c] *State University of New York, Stony Brook, NY 11794, USA*
[d] *University of California, Los Angeles, Department of Physics and Astronomy, Los Angeles, CA 90095, USA*
[e] *Struttura Dipartimentale di Matematica e Fisica, Univ. di Sassari and INFN, Sezione di Cagliari 07100, Italy*
[f] *Dipartimento di Fisica "E. Fermi", Universit`a di Pisa, Largo B. Pontecorvo 3, 56127 Pisa and INFN, Sezione di Pisa 56127, Italy*

Abstract. We report our recent progress using a high-power, picosecond CO_2 laser for Thomson scattering and ion acceleration experiments. These experiments capitalize on certain advantages of long-wavelength CO_2 lasers, such as their high number of photons per energy unit and beneficial wavelength- scaling of the electrons' ponderomotive energy and critical plasma frequency. High X-ray fluxes produced in the interactions of the counter-propagating laser- and electron-beams for obtaining single-shot, high-contrast images of biological objects. The laser, focused on a hydrogen jet, generated a monoenergetic proton beam via the radiation-pressure mechanism. The energy of protons produced by this method scales linearly with the laser's intensity. We present a plan for scaling the process into the range of 100-MeV proton energy via upgrading the CO_2 laser. This development will enable an advance to the laser-driven proton cancer therapy.

Keywords: CO_2 laser, ion acceleration, Thomson scattering, electron beams, cancer therapy, x-ray imaging
PACS: 52.38.Kd, 42.55.Lt, 42.62.Be, 41.75.Lx, 07.85.Fv

INTRODUCTION

During the half century since invention of lasers, they have penetrated nearly every kind of human activity, including health care. Nowadays, laser technology continues to evolve, and new applications emerge. A shining example of this trend is the new generation of super-intense, ultra-short-pulse lasers that already revolutionized physical sciences, and enabled a novel approach to devising compact particle accelerators [1]. Medical applications for such advanced accelerators are being considered, among which the most promising is the laser method of hadron acceleration to therapeutic energies up to 250 Mev for treating most cancers [2]. This innovation opens the way to less expensive and high- productivity laser-based oncology clinics.

Another propitious example of downscaling the equipment whilst simultaneously adding new capabilities for medical- and biological-examinations is a linac-based, high-brightness Thomson X-ray source that will complement present-day synchrotron radiation sources.

Mainstream efforts in these directions rely on pico- and femto-second solid-state lasers based on chirped pulse amplification (CPA). In the present paper, we describe an alternative approach founded on the emerging technology of ultra-fast CO_2 gas lasers.

Although ultra-fast CO_2 lasers still are under development, they undoubtedly offer certain advantages due to their long wavelength, such as their high number of photons per unit energy, and wavelength scaling of the electrons' ponderomotive energy and critical plasma density. These features benefit correspondingly the two subjects addressed in this paper, viz. Thomson X-ray sources, and proton accelerators. Although their physical nature is quite different, the two techniques are united by the fact that the same type of picoseconds CO_2 laser can successfully implement them. We discuss below a

Application of Accelerators in Research and Industry
AIP Conf. Proc. 1336, 386-390 (2011); doi: 10.1063/1.3586126

report from the Brookhaven Accelerator Test Facility (ATF) on our progress in developing the picosecond CO_2 laser system and applying it to proton acceleration and X-ray generation.

PICOSECOND CO_2 LASER TECHNOLOGY

A high peak power, ultra-fast CO_2 laser system starts with a wavelength converter wherein a near-IR picosecond solid-state laser with $\lambda \approx 1$ μm produces a mid-IR 10-μm pulse. Various methods can be employed for this process including semiconductor optical switching, the Kerr effect in optically active dichroic material, or parametric generation in nonlinear crystals. Immediately after the wavelength conversion is made and a low-power $\lambda \approx 10$ μm seed pulse produced, it is amplified in a chain of high-pressure (~10 atm.) CO_2 laser amplifiers. The high pressure induces collisional broadening of the molecular gas spectrum to produce sufficient bandwidth (~THz) for amplifying the picoseconds pulse. Another way to develop a broader spectrum is to combine different CO_2 isotopes.

The ATF's laser system merges the approaches of picosecond pulse generation and amplification. First, we combined the outputs from a multi-nanosecond CO_2 laser oscillator with a picosecond Nd:YAG laser on a Germanium Brewster plate to produce an ~200 ps, 10-μm pulse by semiconductor optical switching. Combining this pulse with a Nd:YAG's 2nd harmonic in a Kerr cell filled with an optically active CS_2 fluid, we sliced out a 5-ps, 10-μm pulse at the ~0.1 MW peak power-level. This pulse is seeded into a 10-atm isotope-filled CO_2 regenerative amplifier where it is trapped for 10-12 round trips and then is released on reaching the ~1 GW level. A final high-pressure, large-aperture amplifier boosts the laser pulse to 1 TW that is used for experiments discussed below.

For future upgrades to the higher peak power and for a faster repetition rate, both needed for the applications described in the next two Sections, particularly for hadron laser therapy, we also can implement currently available CO_2 laser technology. Our proposed design of a prospective 100-TW, 10-Hz mid-IR laser system combines a parametric short-pulse generator, with commercial electric-discharge lasers (e.g., those manufactured by SDI Ltd., RSA and SOPRA, France), a high-pressure multi-isotope amplifier optically pumped with a chemical HF laser, and R&D on chirped pulse-compression to 200 fs

PROTON ACCELERATION

Previous research on laser-driven ion acceleration was inspired largely by the prospects of applying this technology to cancer therapy. The high medical value of hadron therapy in contrast to the prohibitive cost and complexity of conventional hadron accelerators and beam-transport systems creates a unique opportunity: Replace the large accelerator and beam transport with an ultra-compact, laser-driven accelerator. This small accelerator (of several ft^3) easily can be maneuvered and moved around the patient like an x-ray headpiece. The laser beam that drives the accelerator is transmitted through air and does not generate ionizing radiation until it interacts with a target where a therapeutic dose of hadrons is delivered. Mainstream research on laser-driven proton acceleration is based on a combination of CPA lasers and solid targets. The energy deposited by the focused laser beam converts a target (usually a metal or plastic thin-foil) into plasma. Hot electrons escape the plasma, pulling positive ions behind by the charge-separation field. This method allows only limited control over the spectral purity of the generated ion beams [3]. Further, the maximum cut-off energy scales as the square root of the laser's intensity ($I^{1/2}$) [4]. Accordingly, progressing from the presently achievable tens of MeV to the hundreds of MeV proton-energies required for treating most cancers requires radical improvements in laser technology and methods.

We pioneered a new approach in work carried out at BNL by a collaboration led by Imperial College, UK. It resolves the problem of power scaling by employing a CO_2 laser and a supersonic gas-jet, as a target. The wavelength of the CO_2 laser, ten times longer than that of solid-state lasers used elsewhere, offers a two-fold benefit. First, the hadron-energy scales with the square of the wavelength λ^2 [4, 5]. Second, the $1/\lambda^2$ scaling of the critical plasma-density means that the CO_2 laser can generate the plasma-acceleration medium in low-density hydrogen-gas jet targets.

Our CO_2 laser gas-jet plasma accelerator employs a new method of accelerating ions Radiation Pressure Acceleration or RPA [5]. It works as follows: A circularly polarized laser beam pushes a layer of critical density plasma, first within and then out of the gas target. The large radiation pressure speeds up the hadrons in the plasma layer at an enormous rate, ensuring a very compact accelerator. The advantage of the RPA mechanism lies in the much more favorable scaling of the ion energy with the laser's intensity (~I) [5, 6]. This excellent power scaling should engender a very compact accelerator generating high-energy hadrons, whilst the gas jet should assure good control over the repetition rate and purity of the produced beams. In addition, the RPA mechanism can produce monoenergetic hadron beams.

A recent proof-of-principle experiment conducted at the ATF resulted in a 1.2 MeV proton beam from a supersonic hydrogen jet using a 1-TW CO_2 laser. The experiment's impressive aspects are the monoenergetic beam we generated with an energy spread of only ±4% (see Fig.1) and the low power of the laser used. The intensity of this CO_2 laser was a hundredfold lower than the minimum required for producing ~1 MeV protons with a solid-state laser. The circular polarization of the laser beam, the occurrence of the effect close to the critical gas density, and especially, the monochromaticity of the spectrum indicate that the ATF experiment met all conditions for an RPA. The outcome of this experiment allows us to draft a roadmap towards higher proton energies (>200 MeV) via the laser upgrade, and to plan a full-energy, high-dose hadron therapy facility.

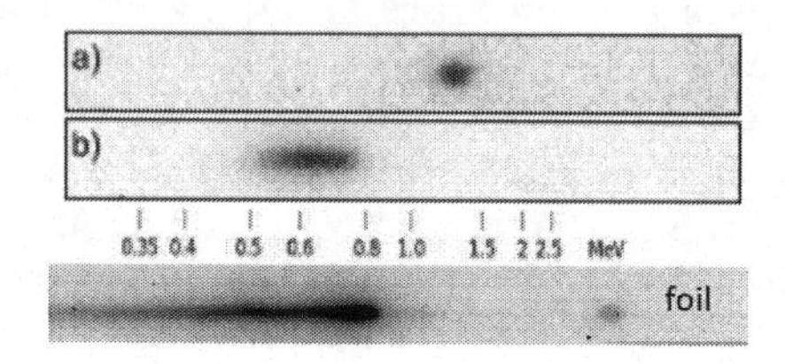

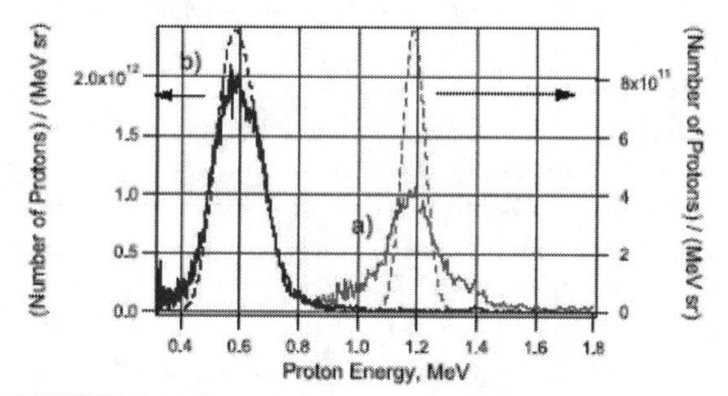

FIGURE 1. Ion spectra: A selection of images on a spectrometer screen (above) and corresponding plotted spectra after background subtraction (solid lines) and additional correction through deconvolution of the spectrometer instrument function (dashed lines).

THOMSON SCATTERING FOR BIOLOGICAL X-RAY IMAGING

In other series of experiments, we employed the same CO_2 laser, combined with a relativistic electron beam from the 70-MeV linac, for generating X-rays via Thomson scattering and investigating the applicability of such a source for biological imaging. To capture, in a single shot, a high-resolution X-ray image of a live object, four parameters are especially important: Short duration pulses; small source; high proton-flux adequate for detecting with realistic imaging techniques; and high monochromaticity. In counter-propagating laser- and electron-beam geometry, the X-ray pulse length is defined by the electron bunches' length, which is of the order of picoseconds. As long as the X-rays originating from very narrow laser- and electron beams that are focused and overlap (typically ~50 μm wide), and the test sample is placed at a sufficiently long distance away (~1 m), the point-source condition is satisfied. This is the prerequisite for obtaining high-resolution images. Earlier BNL experiments [7] demonstrated the straightforward advantage of using a CO_2 laser that delivers ×10 more protons per energy unit than does a solid-state 1-μm laser. This results in a proportionally higher number of scattered X-rays, thereby simplifying their detection. However, the monochromaticity of a Thomson source, which potentially is affected by multiple parameters of the electron- and laser beams, was not experimentally characterized until we undertook a recent series of collaborative experiments with UCLA, USA, and INFN, Italy.

We can simply but precisely calibrate the X-ray energy and monochromaticity by imaging the X-ray beam transmitted through a thin foil of a material with a K-edge energy cut-off close to the characteristic spectral maximum that coincides with the beam's axis. It is important to keep in mind that Thomson-scattered radiation propagates within a cone with opening angle of $1/\gamma$ around the direction of the electron beam, and exhibits the angular energy dependence

$$E_X \approx \frac{4\gamma^2}{\left(1+\gamma^2\theta^2\right)} E_L \qquad (1)$$

where γ is the electron Lorentz factor equal to $\gamma \approx 2E_e$, E_e is the electron energy measured in MeV, θ is the observation angle, and E_L is the laser's photon energy.

Therefore, a pinhole centered on the axis should define the monochromaticity of the selected X-ray beam unless other factors smear the perfect angular distribution in Eq.1, such as the finite electron- beam's energy spread and emittance. When the element K-edge spectral position is chosen so to attenuate the on-axis photons but transmit the off-axis red-shifted photons that fall below K-edge, the resulting image should be donut shaped. The sharpness or visibility of the donut's hole will serve as a characteristic of the monochromaticity of the source. Fig.2 shows a series of such images obtained through Fe-foil at variable electron energies. Simulations indicate that the X-ray energy spread along the beam's axis is ≤5% that we estimated was sufficient for attempting high-resolution X-ray imaging. Details about this experiment appear elsewhere [8].

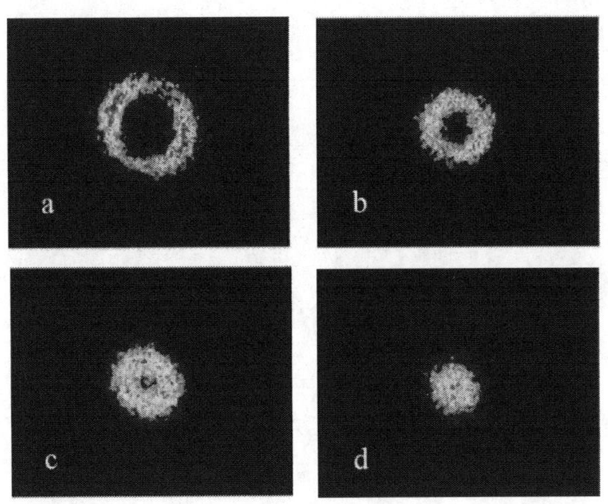

FIGURE 2. Images of Thomson scattered X-ray beam after passing through Fe foil obtained with (a) 68 MeV, (b) 66 MeV, (c) 65 MeV, and (d) 64 MeV electron beam; the hole's diameter is getting smaller for lower electron energies that are converted by Eq.1 to lower X-ray energies.

Having answered the question of monochromaticity, our next step was to demonstrate phase-contrast (PhC) imaging. This method enhances the resolution of low-contrast details by detecting the phase shift, together with absorption of X-rays [9]. An example of such an application is angiography. We first tested the value of the BNL Thomson source for PhC imaging with plastic wires placed at 2 m from the source, and 1 m from the detector (CMOS flat panel Hamamatsu C9728DK-10). When a partially coherent X-ray beam crosses the object, the wave-function undergoes attenuation and a shift in phase that both depend upon the complex refractive index of the material and its thickness. The amplitude of interference fringes produced on the image (shadow) of the object in the detector plane is particularly high around the borders of structures inside the object having a different refractive index than that of the bulk material. This phenomenon, illustrated in Fig.3, that increases the contrast above that of conventional radiographic images, is called the edge-enhancement effect.

To illustrate potential biological application of this new technique, we present in Fig.4 a single-shot image of a wasp taken at a 1-ps exposure time. In particular, the borders of low absorption details are visible due to the PhC effect.

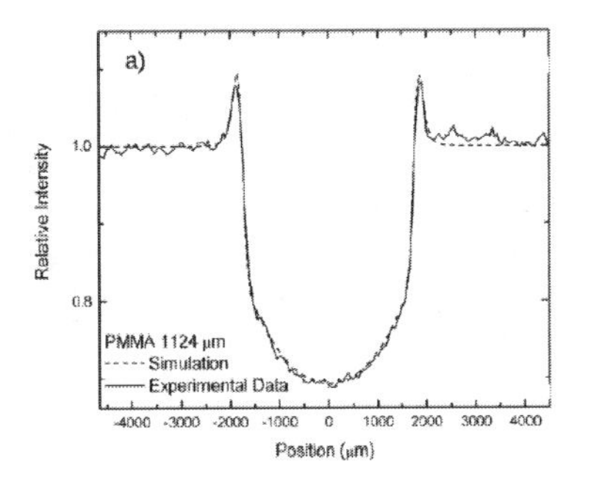

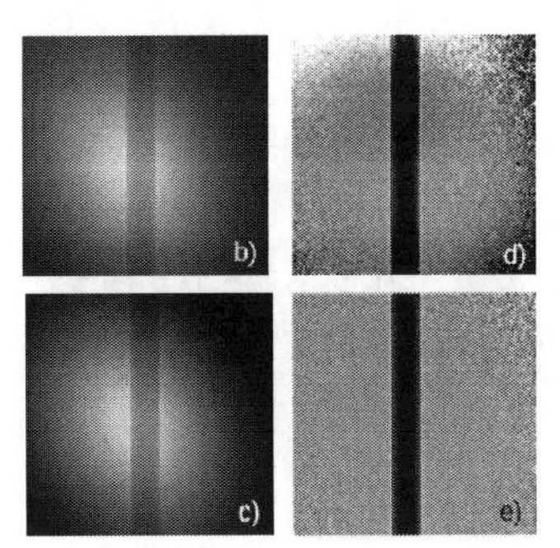

FIGURE 3. Experimental and simulated images for a 1-mm PMMA wire obtained with 10.8 keV X-rays from the ATF Thomson source; (a) experimental (solid line) and simulated (dashed line) cross-sectional scans of a raw wire image in a central position of the beam, experimental (b) and simulated (c) raw images, normalized experimental (d) and simulated (e) images.

CONCLUSIONS

We describe a novel long-wavelength ultra-fast laser technology that supports new applications in medicine and biology. Although the principles of this technology were established more than 20 years ago in seminal works of P.B. Corkum and A.J. Alcock [10, 11] only recently did this technique become viable through the work of two American research groups at BNL and UCLA [12]. It is based on converting picosecond solid-state laser pulses in the mid-IR and amplifying them in high-pressure CO_2 amplifiers to reach relativistic intensities (when a normalized laser vector-potential $a \geq 1$).

We illustrated capabilities of these new laser sources for medical and biological applications by two examples. The first was our demonstration of monoenergetic proton beams radiated from a hydrogen gas jet by laser-induced radiation pressure. The second example was single-shot phase-contrast imaging of fine biological details with a 1 ps exposure time using Thomson X-rays produced in counter-propagating a CO_2 laser beam and a relativistic electron beam. Both methods benefit from wavelength scaling from near- to mid-IR. While our method of biological X-ray imaging is ready for practical implementation, the availability of laser-accelerated ion beams for cancer therapy rests on the anticipated progression of ultra-fast CO_2 lasers to multi-terawatt peak power.

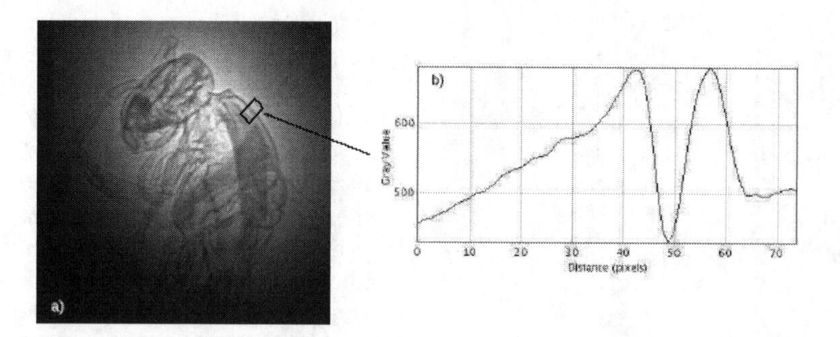

FIGURE 4. Single-shot image of a wasp (a); the inline phase contrast is clearly visible on the back of the insect (b).

ACKNOWLEDGMENTS

The authors thank D. Neely, J. Schreiber, G. Dudnikova, M. Ispirian, G. Andonian, E. Hemsinga, G. Priebe, K. Kusche, J. Park, and D. Davis for assisting in the reported research, supported by US DOE Grant DE-FG02-07ER41488, US DOE Grant DE-FG03-92ER40693, Libra Basic Technology Consortium and ONR Grant 140810463.

REFERENCES

1. V. Malka, J. Faure, Y.A. Gauduel, E. Lefebvre, A. Rousse, K.T. Phuoc, *Nature Phys.* **4**, 447-453 92008).
2. U. Linz, J. Alonso, *Phys. Rev. STAB* **10**, 094801 (2007).
3. S. Wilks, et al, *Phys. Plasmas* **8**, 542 (2001).
4. J. Fuchs, et al. *Nature Phys.* **2**, 48–54 (2006).
5. Esirkepov, M. Borghesi, S. V. Bulanov, G. Mourou, T. Tajima, *Phys. Rev. Lett.* **92**, 175003 (2004).
6. A. P. L. Robinson, et al, *New J. Phys.* **10**, 0130231 (2008).
7. I. V. Pogorelsky, et al., *Phys. Rev. STAB* **3**, 090702 (2000).
8. O. Williams, et al., *Nucl. Instum. Methods in Phys. Res. A* **608**, S18-S22 (2009).
9. R.A. Lewis, *Physics in Medicine and Biology* **49**, 3573 (2004).
10. A.J. Alcock, P.B. Corkum, *Can. J. Phys.* **57**, 1280-1290 (1979).
11. P.B. Corkum, *IEEE J. Quant. Electron.* **QE-21**, 216-232 (1985).
12. S.Y. Tochitsky, C. Filip, R. Narang, C.E. Clayton, K.A. Marsh, C. Joshi, *Opt. Lett.* **26**, 813-815 (2001).

Carbon Ion Radiotherapy At Gunma University: Currently Indicated Cancer And Estimation Of Need

Tatsuya Ohno[1], Takashi Nakano[1], Tatsuaki Kanai[1], Satoru Yamada[1]

[1]*Gunma University Heavy Ion Medical Center, Gunma University (3-39-22 Showa, Maebashi 371-8511, Japan)*

Abstract. Carbon ion radiotherapy for the first patient at Gunma University Heavy Ion Medical Center (GHMC) was initiated in March of 2010. The major specifications of the facility were determined based on the experience of clinical treatments at National Institute of Radiological Sciences (NIRS). The currently indicated sites of cancer treatment at GHMC are lung, prostate, head and neck, liver, rectum, bone and soft tissue. In order to evaluate the potential need for treatment in the region including Gunma prefecture and the adjacent 4 prefectures, an estimation model was constructed based on the Japanese cancer registration system, regular structure surveys by the Cancer Societies, and published articles on each cancer type. Carbon ion RT was potentially indicated for 8,085 patients and realistically for 1,527 patients, corresponding to 10% and 2% of the newly diagnosed cancer patients in the region. Prostate cancer (541 patients) followed by lung cancer (436 patients), and liver cancer (313 patients) were the most commonly diagnosed cancers.

Keywords: Carbon ion radiotherapy, Cancer patients, Indication, Estimation, Gunma University

INTRODUCTION

Charged particle therapy with protons and carbon ions allows uniquely precise delivery of a high dose to the target volume while sparing the surrounding normal tissue. Additionally, carbon ion beams deliver a larger mean energy per unit length of trajectory in the body than protons and X-ray beams. This unique property includes a high biological effect when used in radiotherapy. At present, there are a total of 8 facilities in operation for particle beam radiotherapy in Japan. These include the National Institute of Radiological Sciences (NIRS, carbon ions), Tsukuba University (PMRC, protons), National Cancer Center East (NCCE, protons), Hyogo Ion Beam Medical Center (HIBMC, carbon ions and protons), Wakasa-Wan Energy Research Center (protons), Shizuoka Cancer Center (protons), Minami-Tohoku Proton Center (protons), and Gunma University Heavy Ion Medical Center (GHMC, carbon ions).

The carbon ion radiotherapy project was launched at Gunma University in 2001, and collaboration with NIRS was initiated in 2004. Based on the design and R&D studies at NIRS, the construction (size: 45m x 65m) and operation costs of the accelerator system were reduced to one-third of those at NIRS while maintaining its high performance. It took about 3 years from the start of construction (February of 2007) to treatment of the first patient (March of 2010).

The present article describes the current indications of cancer for carbon ion radiotherapy at Gunma University as well as the estimation of the need in Gunma prefecture and adjacent prefectures.

Current Clinical Indications For Carbon Ion Radiotherapy At Gunma University

Carbon ion radiotherapy for the first patient at Gunma University was initiated in March of 2010. Our facility is the first such university hospital-based facility in Japan, is supported by the Japanese and local governments, and is a compact prototype of a commercial design ready for distribution. The major specifications of the facility were determined based on the experience of clinical treatments at NIRS. The main accelerator is a slow-cycling 20m-diameter synchrotron, and it accelerates carbon ions up to an energy range from 140 to 400 MeV per nucleon. A spiral wobbler system was adopted for the beam delivery system in order to improve the beam efficiency in a large irradiation field size. Of the 4 treatment rooms, one has a fixed horizontal beam line, one has a fixed vertical beam line, one has both fixed horizontal and vertical beam lines, and the remaining

Application of Accelerators in Research and Industry
AIP Conf. Proc. 1336, 391-396 (2011); doi: 10.1063/1.3586127

one is prepared for research and development of advanced scanning techniques for smaller targets.

Many phase I/II dose escalation and phase II studies for various tumor sites have been carried out at NIRS since 1994 (1). Although promising clinical outcomes have been reported from NIRS, it is of interest whether the efficacy of carbon ion radiotherapy from a single institution can be reproduced in other facilities when optimal doses and fractionations are used for a similar patient population. At GHMC, the efficacy and safety of carbon ion radiotherapy were reviewed for each tumor type, and then the best available dose and fractionation schedules determined at NIRS were adopted for our clinical protocols (Table 1).

Table 1. Ongoing clinical protocol at Gunma University Heavy Ion Medical Center

Site (protocol #)	Eligible patients	Treatment
Lung cancer (GUNMA0701)	· Histologically proven non-small-cell lung cancer · T1a-T2aN0M0 (peripheral type,TNM classification, 2009) · Inoperable or decline surgery	*T1a-b:* 52.8 GyE/4 fr/1 week *T2a:* 60.0 GyE/4 fr/1 week
Prostate cancer (GUNMA0702)	· Histologically proven prostate cancer · T1c-T3N0M0 (TNM classification, 2002) *Low risk group:* PSA <10ng/mL and Gleason score ≤6 and T1c-T2b N0 M0 *Intermediate risk group:* Other than low risk and high risk groups *High risk group:* PSA ≥20ng/mL or Gleason score ≥8 or T3 N0 M0	*Low risk group:* Carbon ion RT alone (57.6 GyE/16 fr/4 weeks) *Intermediated risk group* Hormone therapy (6-8 months) and Carbon ion RT *High risk group* Hormone therapy (2 years) and Carbon ion RT
Liver cancer (GUNMA0703)	· Histologically proven or compatible feature on CT/MRI with hepatocellular carcinoma · Single lesion · T1-3N0M0 (TNM classification, 2002) · Child-Pugh A or B	52.8 GyE/4 fr/1 week
Rectal cancer (GUNMA0801)	· Recurrent pelvic tumor after surgery for rectal cancer · No metastases other than pelvic tumor · Curative intent for primary surgery	73.6 GyE/16 fr/4 weeks
Head and neck cancer (GUNMA0901)	· Histologically proven non-squamous cell carcinoma · TanyN0M0 (TNM classification, 2002)	64.0 GyE/16 fr/4 weeks
Bone and soft tissue sarcoma (GUNMA0904)	· Histologically proven bone and soft tissue sarcomas · Stage IA-III (TNM classification, 2002)	*Standard:* 70.4 GyE/16 fr/4 weeks *Spine tumor:* 64.0 GyE/16 fr/4 weeks

Estimating The Need Of Carbon Ion Radiotherapy In Gunma And Adjacent Prefectures

Backgrounds and Methods

In Japan, 641,594 new cancer cases were diagnosed in 2003 (males 372,273 and females 269,220). By the Research Group for Population-based Cancer Registration in Japan, the number of cancer cases in 2020 was projected to be 838,000 (501,000 males and 337,000 females) (2). On the other hand, a structural survey by the Japanese Society for Therapeutic Radiation Oncology (JASTRO) has reported that the total numbers of new cancer patients and total cancer patients (new and repeat) treated with radiotherapy (RT) in 2005 were estimated at approximately 162,000 and 198,000, respectively, demonstrating an approximately 2-fold increase during the last decade (3). The increasing number of cancer patients undergoing RT has resulted in a greater focus on estimating the requirements, including the equipment, personnel, patient load, and geographic distribution, in order to identify and improve any deficiencies.

An estimation model of the need for particle radiotherapy was constructed as follows. First, the region including Gunma prefecture and the adjacent 4 prefectures (Tochigi, Niigata, Nagano, and Saitama) was specified. In 2003, the total population of the region was 15,748,026 (12.3% of the population of Japan). Based on the Japanese cancer registration, the estimated number of cancer patients of the region was 79,172 in 2003 (2). Other than GHMC, there is no facility offering particle beam RT in the region.

Second, the total number of patients was calculated for head and neck, rectum, liver, lung, prostate, and bone and soft tissue by using data on cancer incidence statistics and the size of the regional population (2). Third, certain proportions of the clinical attributes (stage, tumor size, histology, etc.) of each cancer site were determined on the basis of regularly structured surveys by the cancer society, cancer registration, published manuscripts, and textbooks (4). Regarding this process, Japanese data from around 1999 to 2003 were quoted as much as possible. The potential number of patients in the region who might benefit from carbon ion RT at GHMC was calculated in each site as described below.

Results

Head and neck

Salivary gland carcinoma

$82^{a)} \times 0.123^{b)} \times 0.752^{c)} = 8$

a): number of patients with salivary gland carcinoma, b): ratio of regional population among Japanese population, c): incidence of T1-4 and N0

Nasal and paranasal cavity carcinomas

$1447^{a)} \times 0.123^{b)} \times 0.904^{c)} = 161$

a): number of patients with nasal and paranasal cavity cancer, b): ratio of regional population among Japanese population, c): incidence of carcinoma

Nasal and paranasal cavity sarcoma and melanoma

$1447^{a)} \times 0.123^{b)} \times 0.096^{c)} = 17$

a): number of patients with nasal and paranasal cavity cancer, b): ratio of regional population among Japanese population, c): incidence of sarcoma and malignant melanoma

Rectum

Pelvic recurrence after surgery

$34649^{a)} \times 0.123^{b)} \times 0.798^{c)} \times 0.10^{d)} = 340$

a): number of patients with rectal cancer, b): ratio of regional population among Japanese population, c): indication for surgery, d): incidence of pelvic recurrence

Liver

Hepatocellular carcinoma of 3-5cm in size

$40485^{a)} \times 0.123^{b)} \times 0.878^{c)} \times 0.722^{d)} \times 0.232^{e)} = 735$

a): number of patients with hepatocellular carcinoma, b): ratio of regional population among Japanese population, c): incidence of stage I+II, d): incidence of 1-2 lesions, e): incidence of 3-5cm in size

Hepatocellular carcinoma of 5-10cm in size

$40485^{a)} \times 0.123^{b)} \times 0.878^{c)} \times 0.722^{d)} \times 0.163^{e)} = 515$

a): number of patients with hepatocellular carcinoma, b): ratio of regional population among Japanese population, c): incidence of stage I+II, d): incidence of 1-2 lesions, e): incidence of 5-10cm in size

Lung

Non-small cell lung cancer, T1N0M0

$78745^{a)} \times 0.123^{b)} \times 0.284^{c)} \times 0.610^{d)} = 1678$

a): number of patients with lung cancer, b): ratio of regional population among Japanese population, c): incidence of non-small-cell lung cancer stage I, d): incidence of T1

Non-small cell lung cancer, T2N0M0

$78745^{a)} \times 0.123^{b)} \times 0.284^{c)} \times 0.390^{d)} = 1073$

a): number of patients with lung cancer, b): ratio of regional population among Japanese population, c): incidence of non-small-cell lung cancer stage I, d): incidence of T2

Prostate

Prostate cancer, stage A-B

$40062^{a)} \times 0.123^{b)} \times 0.431^{c)} = 2124$

a): number of patients with prostate cancer, b): ratio of regional population among Japanese population, c): incidence of stage A-B in the era of PSA measurement

Prostate cancer, stage C

$40062^{a)} \times 0.123^{b)} \times 0.267^{c)} = 1316$

a): number of patients with prostate cancer, b): ratio of regional population among Japanese population, c): incidence of stage C in the era of PSA measurement

Bone and soft tissue

Bone sarcoma

$765^{a)} \times 0.123^{b)} \times 0.260^{c)} = 24$

a): number of patients with bone sarcoma, b): ratio of regional population among Japanese population, c): incidence of spine and inoperable cases of extremities

Soft tissue sarcoma

$2552^{a)} \times 0.123^{b)} \times 0.300^{c)} = 94$

a): number of patients with soft tissue sarcoma, b): ratio of regional population among Japanese population, c): incidence of retroperitoneum and pelvis

A total of 8,085 patients were potential candidates for carbon ion RT in the region, including Gunma prefecture and the adjacent 4 prefectures.

Table 2. Estimated number of patients for carbon ion RT in the region

Tumor site	Total no. of Japan (2003)	Cancer and clinical attributes	Potential no. of the region (2003)	Indication (%)	Realistic no. of the region (2003)
Head and neck	82	Salivary gland carcinoma, T1-4N0	8	50	4
	1,447	Nasal/paranasal cavity			
		Carcinoma	161	25	40
		Sarcoma or malignant melanoma	17	80	14
Rectum	34,649	Pelvic recurrence after radical surgery	340	25	85
Liver	40,485	Hepatocellular carcinoma			
		Stage I-II and 1-2 lesions and size: 3-5cm	735	25	184
		Stage I-II and 1-2 lesions and size: 5-10cm	515	25	129
Lung	78,745	Lung cancer			
		Non-small-cell lung cancer, T1N0M0	1,678	10	168
		Non-small-cell lung cancer, T2N0M0	1,073	25	268
Prostate	40,062	Prostate cancer			
		Stage A-B	2,124	10	212
		Stage C	1,316	25	329
Bone and soft tissue	765	Bone sarcoma			
		Spine or inoperable cases of extremities	24	80	20
	2,552	Soft tissue sarcoma			
		Retroperitoneum or pelvis	94	80	75
		Total	8,085		1,527

In order to realistically estimate the number of patients of the region, additionally indicated proportions (%) were assumed at GHMC based on the amplitude of advantage compared with other treatment modalities. For sarcomas of the head and neck, and bone and soft tissue, our estimation showed that 80% of the patients could be candidates for carbon ion RT, and especially in inoperable cases. For salivary gland carcinomas, high LET irradiation is effective and the proportion was set at 50%. For lung, liver, and prostate, due to their commonality, there are several other treatments that are recognized as first choice-treatment according to the guidelines. Thus, the proportion was estimated as 25% in most cases. Especially, for T1N0M0 non-small-cell lung cancer and stage A and B prostate cancer, nearly 10% of the patients would be candidates at present. The total realistic number of patients meeting the indications of GHMC was calculated to be 1,527 per year (Table 2).

DISCUSSION

According to our estimation model, carbon ion RT was potentially indicated for 8,085 patients and realistically for 1,527 patients, corresponding to 10% and 2% of the newly diagnosed cancer patients in Gunma and adjacent prefectures. It must be cautioned that the indicated proportion assumed by us was tentative and can be modified by existing standard treatments and future clinical evidence usually based on prospective clinical trials. In addition, we did not intent to estimate the number of patients only for carbon ion RT. The major reason was that only one carbon ion RT facility exists in the specified region.

There are many factors to influence the actual number of patients for treatment of particle RT. First, patients were supposed to receive carbon ion RT once during the course of their illness. In some proportion of the patients, tumor recurrence out of the target will occur even after successful local control, and a part of such tumors can be indicated for additional treatment with carbon ion RT. Second, there is no consideration for the general condition such as performance status, age, and preference of the patients. Third, geographical or economical reasons exist for not offering the treatment even though prospective patients might have great interest in carbon ion RT. Finally, combined therapies such as photon-carbon and proton-carbon were not considered in the present estimation. Along with the development of high-precision RT using photon, updated data will be necessary in the future.

Several reports are available regarding the estimation of the patient number suitable for particle beam RT using proton or carbon ions, such as that by the Italian National Center for Oncological Hadron Therapy (CNAO) (5,6). In their model, patients were separated into two categories according to proton or carbon ion RT indication. Category A, in which proton or carbon ion therapy can be regarded as the treatment of choice, included uveal melanoma, chondrosarcoma, chordoma, meningioma of the skull base, paraspinal tumors and retinoblastoma. Category B included tumors in which local control would be affected favorably in obtaining higher percentages of cure, such as brain, lung, rectum, prostate, bladder, liver, pancreas, and gynecological tumors. The estimation revealed that 830 patients, constituting 44% of the number of patients with these diagnoses in Italy per year, were candidates for elective proton therapy (category A), and that 15,853 patients were candidates for proton therapy in clinical trials (category B). It was estimated, in total, that about 16% of the irradiated patients were candidates for proton therapy. An estimate was also made for the number of Italian patients eligible for carbon ion therapy of those eligible for proton therapy. About 3,700 patients were considered as such candidates, constituting 23% of those considered candidates for proton therapy and 5% of all irradiated patients.

In a prospective nationwide study conducted by the MedAustron project in Austria, data of all new patients starting RT at all 12 RT facilities were analyzed (7). A total of 3,783 patients received RT between September and December of 2002, corresponding to 15,132 patients per year. Based on the published reports on particle beam RT, the proportion of tumor subgroups suitable for particle beam RT was estimated at Heidelberg, Milano, Lyon, Vienna, and Innsbruck Universities. The mean values of the university-specific percentages per tumor site were further calculated (chordoma and condrosarcoma of the skull base, 100%; uveal melanoma, 100%; liver cancer, 47%; salivary gland tumors, 45%; soft tissue sarcoma, 38%; prostate cancer, 24%; etc.). Based on the epidemiological cancer incidence, the number of potential patients was estimated as 2,044 per year, representing 5.6% of all newly diagnosed cancer patients and 13.5% of all irradiated cancer patients.

The French ETOILE project performed a ''one-day survey'' at five university hospitals, identifying 77 patients, mainly head and neck cancers (n=31), gliomas (n=/8), lung cancer (n=6), uterus (n=5), gastric (n=/5) and prostate (n=/3), being potential candidates for proton or carbon ion RT (8). This figure corresponded 14.5% of 532 patients receiving

RT and was extrapolated to 160,000 irradiated patients per year in France, of which 23,000 were potential candidates for proton or carbon ion therapy each year.

A group of Swedish radiation oncologists and hospital physicists have estimated the number of patients in Sweden suitable for proton beam therapy in a facility where one of the principal aims is to facilitate randomized and other studies in which the advantage of protons can be shown and the magnitude of the differences, compared with optimally administered conventional radiation treatment, also including intensity-modulated radiation therapy (IMRT) and brachytherapy, can be demonstrated (9). The most common cancer was lung cancer (350 patients), followed by prostate cancer (300 patients), breast cancer (300 patients), head and neck cancer (240 patients), and re-irradiation was estimated to be 150-400 patients per year. As a result, between 2,200 and 2,500 patients annually are eligible for proton therapy, and for these patients the potential therapeutic benefit is high, justifying the additional expense of this treatment. This number constitutes 14-15% of all irradiated patients annually.

Interestingly, all the above European investigations reached a similar conclusion, that 13-16% of all patients receiving RT were suitable for particle beam RT. In our study, the estimated number of patients undergoing RT in the region was 24,354 (12.3%) of all Japanese 198,000 cancer patients. The estimated number of 8,085 patients (potentially indicated) and 1,527 patients (realistically indicated) corresponded to 33% and 6% of the 2,4354 patients, respectively. The difference between our estimation and European investigations may originate from the fact that various methods for calculation and the relatively lower proportion of patients indicating for RT (31% for Japan, 40-60% for European countries). Nevertheless, we think that the current estimation provides an appropriate rate of particle RT application as a model region, and will be useful for future planning of allocation of particle RT facilities.

REFERENCES

1. Okada T, Kamada T, Tsuji H, et al. Carbon Ion Radiotherapy: Clinical Experiences at National Institute of Radiological Science (NIRS) J. Radiat. Res., 51: 355-364, 2010.
2. Cancer statistics in Japan: http://www.fpcr.or.jp/publication/statistics.html
3. Teshima T, Numasaki H, Shibuya H, et al. Japanese Society of Therapeutic Radiology and Oncology Database Committee. Japanese structure survey of radiation oncology in 2005 based on institutional stratification of patterns of care study. Int J Radiat Oncol Biol Phys. 72:144-152, 2008.
4. Toward the spread of particle beam radiotherapy. Study group of particle beam radiotherapy in Japan. 31-32, 2004 (in Japanese).
5. Orrecchia R, Krengli M. Number of potential patients to be treated with proton therapy in Italy. Tumori 84:205-208, 1998.
6. Krengli M, Orecchia R. Medical aspects of the National Centre for Oncological Hadrontherapy (CNAO-Centro Nazionale Adroterapia Oncologica) in Italy. Radiother Oncol. 73 (Suppl 2):/21-23, 2004.
7. Mayer R, Mock U, Jäger R, et al. Epidemiological aspects of hadron therapy: a prospective nationwide study of the Austrian project MedAustron and the Austrian Society of Radiooncology (OEGRO). Radiother Oncol. 73 (Suppl 2):24-28, 2004.
8. Baron MH, Pommier P, Favrel V, et al. A ''one-day survey'': As a reliable estimation of the potential recruitment for proton- and carbon-ion therapy in France. Radiother Oncol. 73 (Suppl 2):15-17,2004.
9. Glimelius B, Ask A, Bjelkengren GR, et al. Number of patients potentially eligible for proton therapy. Acta Oncologica. 44:836-849, 2005.

Proton Therapy At Siteman Cancer Center: The State Of The Art

Charles Bloch, Ph.D.

Washington University School of Medicine, 4921 Parkview Place, St. Louis MO 63110

Abstract. Barnes-Jewish Hospital is on the verge of offering proton radiation therapy to its patients. Those treatments will be delivered from the first Monarch 250, a state-of-the-art cyclotron produced by Still River Systems, Inc., Littleton, MA. The accelerator is the world's first superconducting synchrocyclotron, with a field-strength of 10 tesla, providing the smallest accelerator for high-energy protons currently available. On May 14, 2010 it was announced that the first production unit had successfully extracted 250 MeV protons. That unit is scheduled for delivery to the Siteman Cancer Center, an NCI-designated Comprehensive Cancer Center at Washington University School of Medicine. At a weight of 20 tons and with a diameter of less than 2 meters the compact cyclotron will be mounted on a gantry, another first for proton therapy systems. The single-energy system includes 3 contoured scatterers and 14 different range modulators to provide 24 distinct beam delivery configurations. This allows proton fields up to 25cm in diameter, with a maximum range from 5.5 to 32 cm and spread-out-Bragg-peak extent up to 20cm. Monte Carlo simulations have been run using MCNPX to simulate the clinical beam properties. Those calculations have been used to commission a commercial treatment planning system prior to final clinical measurements. MCNPX was also used to calculate the neutron background generated by protons in the scattering system and patient. Additional details of the facility and current status will be presented.

Keywords: Proton radiation therapy, superconducting synchrocyclotron
PACS: 87.56.bd; 87.56.-v

INTRODUCTION

In the United States, proton radiation therapy began at the Lawrence Berkeley Laboratory in 1954. The 184 inch cyclotron was a massive accelerator weighing approximately 4500 tons. Initial beam energy from that machine was 200 MeV deuterons, equivalent to 100 MeV protons. The radiation therapy program continued there using alpha particles from 1957 to 1992. During that time, proton radiation therapy was also initiated at the Harvard Cyclotron Laboratory (HCL).[1] The HCL synchrocyclotron was more compact, weighing approximately 700 tons, and capable of delivering 160 MeV protons. The program continued from 1961 through 2002 treating over 9000 patients during that period. While both of these programs were highly regarded and extremely successful, for many years the availability of proton therapy in the U.S. did not increase. The main reasons were that a) high-energy particle accelerators existed only in large research labs (not in hospitals), and b) treatments with these machines were limited to fixed horizontal beams (isocentric gantries were not available).

A major breakthrough came about in 1990 when the Loma Linda University Medical Center (LLUMC) introduced the world's first dedicated hospital based proton therapy facility including three isocentric gantries.[2] In this case, the accelerator was a 250 MeV synchrotron built in their hospital specifically for the purpose of treating patients. Beam was transported down a high-energy beam line to any one of 4 treatment rooms. Three of the rooms had 360 degree isocentric gantries that allowed the beam to be directed at a patient from any angle. This new facility eliminated what had been the two major limitations in the adoption of proton beams by the radiation therapy community. The LLUMC proton therapy facility was an enormous success, and continues to treat patients to this day.

Since the inception of the LLUMC facility, several new proton therapy facilities have been constructed in the U.S. Ion Beam Applications (IBA) designed a compact cyclotron for Massachusetts General Hospital (MGH) to replace the aging HCL facility.[3] This isochronous cyclotron weighed only 200 tons, less than a third of the machine it was replacing, and produced protons up to 235 MeV. That cyclotron was

Application of Accelerators in Research and Industry
AIP Conf. Proc. 1336, 397-400 (2011); doi: 10.1063/1.3586128

eventually installed at new facilities in Florida (2006), Oklahoma (2009), and Pennsylvania (2009), as well as facilities outside of the U.S. and more sites yet to open. While this progress is significant, it should be noted that the MGH facility did not open until 2001, more than a decade after LLUMC. Besides these IBA machines, the M. D. Anderson Cancer Center opened a proton therapy facility in 2006 based on a Hitachi designed synchrotron.[4] The only other U.S. facility to start treating patients during this 20 year period is in Indiana. In that case a radiation therapy clinic was built around the existing accelerator at the Indiana University Cyclotron Facility. While the accelerator at IUCF was originally a physics research machine, the retrofit included isocentric gantries from IBA, and the conversion to a radiation therapy clinic is 100%.

While these are significant accomplishments and progress is continuing, two new (and somewhat related) impediments are now identified as the major factors limiting the adoption of proton therapy: the size and the cost of the facilities. Approximate numbers are given in Table 1 for several new/planned facilities.[5]

TABLE 1. Approximate cost and size of recently completed and under construction proton therapy facilities.

Facility	Cost	Size (sq. ft.)
MDACC	$125M	94000
FPTI	$125M	98000
U. Penn	$140M	75000
Hampton U.	$207M	--
Scripps	$185M	102000

These facility costs, in excess of one hundred million U.S. dollars, can be prohibitive for many medical facilities in the U.S. While not entirely unrelated, the area required for installing such a facility, close to 100,000 sq. ft., can also be a significant obstacle for urban medical centers.

Still River Systems is poised to provide the next major breakthrough in the adoption of proton therapy. Their new superconducting synchrocyclotron combined with a unique beam delivery system greatly reduces the cost and size for new proton therapy facilities. Their first system is currently being installed at Barnes-Jewish Hospital (BJH) in St. Louis, MO and should be completed in the next year.

MONARCH 250

Still River System is producing the Monarch 250, a superconducting synchrocyclotron. This is an ultra-compact accelerator, weighing just 20 tons, one tenth the weight of the IBA machine. The beam energy is 250 MeV. At the heart of this accelerator is a 10 tesla superconducting magnet. This high field strength is what allows the accelerator to be so small. While synchrocyclotrons have some limitations when compared to isochronous cyclotrons, none pose a problem for proton therapy as clearly demonstrated by the longevity and success of the HCL synchrocyclotron. The main advantage of an isochronous cyclotron is higher beam current. However, proton radiation therapy does not require beam currents in excess of those easily produced by a synchrocyclotron.

Because of the enormous reduction in size and weight, the Still River design has the cyclotron mounted on the gantry itself. The significance is two-fold: 1) a separate room is not need to house the accelerator, reducing the square footage requirements and 2) this can be configured as a single-room facility, also reducing the square footage requirements as well as the initial cost. To date, all of the other treatment facilities planned in the last 20 years have been multi-room facilities with 2 to 4 treatment rooms plus an additional shielded room to house the accelerator. Such designs are a large part of the enormous facility costs and space requirements. By comparison, the proton therapy facility at BJH is just 8000 square feet, approximately one tenth the space of other facilities. The estimated cost for such a facility is $25M, also significantly below any other new proton facility.

While the treatment capacity of such a facility is also reduced, it allows hospitals to scale the facility to their patient load, instead of demanding that the patient load be adjusted to match the facility design. Relatively few radiation therapy departments treat 2000 patients per year with all modalities. Expecting to treat that many additional patients with the addition of a large proton facility is an unrealistic business plan for most radiation therapy departments. By contrast, the proton therapy facility at BJH will be successful treating just 200 patients per year, a small fraction of the current total patient treatments.

In addition to the unique accelerator, the Still River System has other unique features. The cyclotron is supported in the center of the gantry which has symmetric bearing mounts on both sides of the treatment room. While the gantry only rotates 180 degrees (as opposed to the conventional 360 degrees), all treatment angles are supported through the use of a computer controlled robotic couch. In addition to the cyclotron gantry, a separate inner treatment gantry with synchronized motion will support the patient specific treatment devices (apertures and compensators).

Another unique feature of the Still River system is the elimination of bending magnets and momentum analysis slits from the energy degrading system found with other cyclotron based facilities. This has significant benefits beyond the obvious space and cost savings. Foremost are the reduced beam current requirements. Other systems have significant beam loss in the momentum analysis system which increases beam current demands (to maintain adequate dose rates in the patient) and produces significant radiation background, especially from high-energy neutrons. For example, as much as 500nA of extracted beam may be required to provide 10 to 20 nA of the lowest energy beam through a momentum analysis system. By contrast, the Still River system will deliver most of its extracted beam to the patient thereby reducing the beam current requirements by an order of magnitude for the same dose rate. Eliminating this momentum analysis system entirely reduces the demand for high beam current which in turn makes the synchrocyclotron feasible. Eliminating the high beam current and the losses associated with it greatly reduces the neutron background and required shielding. The advantage this provides cannot be overstated.

Simplifying the energy selection system in this manner has some drawbacks, although they are small given the gains. Primary is the inferior distal dose fall-off for the low energy beams. A well designed energy selection system can produce a sharp distal dose fall-off of 2 to 3 mm (in water) for low energy proton beams. By comparison the highest energy beams will have a distal dose fall-off (from 80 to 20% of maximum dose) in 6 to 7 mm of water. The simplified design of the Still River system will have the same distal dose fall-off (~6mm based on Monte Carlo simulations) for all beam energies. While this may not be as good as the best proton facilities do, it still provides a depth-dose distribution far superior to conventional (photon or electron) treatments. The actual clinical advantage of sharper distal dose fall-offs for the lowest energy protons is not clear.

A final concern over this simplification of the energy selection system is the neutron exposure to the patient. To a large extent, this concern is unfounded. For the deepest tumors (e.g. prostate), all facilities bring the highest energy protons into the treatment room. The highest energy beams produce the most neutrons, and most of the energy loss (and therefore a significant fraction of the neutron production) is in the patient. In the case of the Still River system, shallower tumors will be treated by degrading the beam in a low-Z material upstream from the patient, rather than in the patient themselves. In this way, one can see that the neutron exposure to these patients will not be significantly different than what other facilities allow their prostate patients to receive. This has also been confirmed through Monte Carlo simulations.[6] While it may be desirable to lower the neutron exposure to patients who have shallow tumors (relative to the neutron exposure to patients who have deep tumors), it is not necessary, and it is part of what has made proton therapy cost-prohibitive to many hospitals.

STATUS

The first Still River system is currently being installed at Barnes-Jewish Hospital in St. Louis, MO. Significant progress has been made both at the Still River factory and at the hospital. At the factory, the entire gantry has been assembled and tested for mechanical rotational accuracy. (See Figure 1). After final adjustments measurements confirmed mechanical pointing error was less than 1 mm for all angles. The first cyclotron has been successfully tested. In Spring of 2010 over 1nA of 250 MeV beam current was extracted continuously. This was the culmination of many tests including magnetic field and cooling for the cyclotron as it is rotated from 0 to 180 degrees. Since then work has continued to maximize the performance of the accelerator. By Fall, beam current should be up over 10 nA, the beam modifiers (energy degraders, scattering system and range modulators) should be installed and measurements of clinical beams in a water phantom will begin.

FIGURE 1. Still River Systems gantry undergoing rotational tests at the factory.

Meanwhile at the hospital, civil construction has been completed with the exception of the work that needs to be completed after the cyclotron has been installed. This includes the treatment room and maze as well as waiting area, nursing station, recovery area for anesthesia patients, exam rooms and offices for staff. Significant hardware components have been delivered and installed. The first were the gantry bearing mounts which are embedded in the shielding walls on either side of the treatment room. Since then the gantry arms have been installed as well as the counterweights. (See Figure 2). Also scheduled for delivery are the robotic patient couch, imaging systems, and computer control systems. The cyclotron itself will not be delivered until final factory testing has been completed. After that, the final shielding wall will be put in place (left open for delivery and installation of the cyclotron) followed by the finishing of the treatment room itself. Beam commissioning will begin after all components are in place.

FIGURE 2. Gantry arms mounted in treatment vault at Barnes-Jewish Hospital in St. Louis. Counterweights (not visible) are attached to lower end of the arms which is below the treatment room floor level. Cyclotron and support structure are not yet mounted.

CONCLUSIONS

The Still River proton therapy system represents a major breakthrough in the field. Initial costs have been reduced by a factor of 5. Facility space requirements are down by as much as a factor of 10. Availability of single-room systems makes this the first fully scalable proton therapy facility. Minor compromises in the design and equipment performance are not expected to have major clinical significance for most cases but are directly responsible for the huge advantages described. Elimination of the last major impediments to widespread adoption of proton therapy is expected to begin the next era of proton therapy in St. Louis, MO.

REFERENCES

1. Wilson, Richard. *A brief history of the Harvard University cyclotrons*. Cambridge: Harvard University Press, 2004.
2. J.M. Slater, J.O. Archambeau, D.W. Miller, M.I. Notarus, W. Preston and J. D. Slater, *Int. J. Radiat. Oncol. Bio.l Phys.* **22**, 383–389 (1992).
3. Flanz J, Durlacher S, Goitein M, Levine A, Reardon P and Smith A. *Nucl. Instrum. Methods* B **99** 830–4 (1995).
4. Alfred Smith, Michael Gillin, Martin Bues, X. Ronald Zhu, Kazumichi Suzuki, Radhe Mohan, Shiao Woo, Andrew Lee, Ritsko Komaki, and James Cox, *Med. Phys*. **36**, 4068 (2009).
5. Huff, C. Catching the Proton Wave. *Hospitals & Health Nework*, p.62, 2007.
6. Zheng Y, Klein E, Chen K, Liu Y. *Med. Phys*. **37**, 3273 (2010).

Neutron Exposure Accelerator System For Biological Effect Experiments (NASBEE)

Masashi Takada, Mitsuo Suda, So Kamada, Takuya Hagiwara, Hitoshi Imaseki and Tsuyoshi Hamano

National Institute of Radiological Sciences, 4-9-1, Anagawa, Inage-ku Chiba, Japan 263-8555

Abstract. The neutron exposure accelerator system for biological effect experiments (NASBEE) was developed to study biological effects of fast neutrons. We have characterized the NASBEE neutron beams with neutron energy spectrum, absorbed dose energy distributions, and space distributions. The neutron energy spectrum shows 2.3 MeV as mean energy and 3.0 MeV as kerma of tissue-equivalent-plastic (A150) weighted mean energy, and the maximum neutron energy was determined to be 9 MeV. Neutron absorbed doses occupy 82% of the NASBEE neutron beam. NASBEE has been used to learn some of the outcomes of the biological effects of fast neutrons.

Keywords: Fast neutron, Neutron energy spectrum, Absorbed dose energy distribution, Tandetron accelerator, d-Be neutron production, Biological study
PACS: 29.20.Ej, 29.25.Dz, 29.30.Hs, 29.40.Mc, 29.40.Cs

INTRODUCTION

It is important to evaluate the biological effects of neutrons because the public is exposed to neutrons in a variety of settings, at several radiation fields, for example, around nuclear reactors and accelerators, at nuclear accidents such as the Tokaimura critical accident in Japan, at radiation therapy (Boron Neutron Capture Therapy: BNCT, charged particle therapy, and Intensity Modulated Radiation Therapy: IMRT), at aviation altitudes, and in the international space station. The radiation weighting factor, w_R for the protection quantity is based on relative biological effect (RBE) data of life shortening due to cancer induction from in vivo investigations. It can be argued that determination of the RBE of lethal tumors is more important for extrapolation of values in the estimation of risk for humans. But the International Commission on Radiological Protection (ICRP) and the International Commission on Radiation Units and Measurements (ICRU) give diversely scattered RBE values [1, 2]. The higher RBE of neutrons compared to that of protons and other charged particles, and the inverse dose-rate effect have never been understood clearly. Although neutron protection is important, the available data about the effects of neutrons are insufficient. We need to investigate a variety of biological objectives with regards to neutrons.

The National Institute of Radiological Sciences (NIRS) in Japan has started to acquire data about the biological effects of neutrons. NIRS completed a neutron study using a high-energy neutron source with a peak energy around 10 MeV produced by bombarding 25-MeV deuterons into a beryllium target five years ago [3-6]. Now, we want to examine fast neutrons with a peak energy around 2 MeV. To investigate the biological sensitivity to fast neutrons in vivo, we constructed the neutron exposure accelerator system for biological effect experiments (NASBEE) at NIRS [7]. NIRS has studied several biological effects: carcinogenesis (myeloid leukemia, cancers of the mammary gland, lung, liver, brain, and intestine), age-dependent cancer risk, and lifespan shortening using NASBEE neutron beams. To evaluate the biological effects, we need to know the neutron energy spectrum and absorbed dose distribution. We characterized the neutron energy spectrum, the absorbed dose energy distribution, and the space distribution in this study.

NEUTRON SOURCE

The d-Be reaction is known as an intense neutron source with strong forward angle dependence. We chose the d-Be reaction to produce a large exposure dose-rate fast-neutron beam with which we could perform short-term irradiation of neutrons to mice and rats, at a dose of 1 Gy for less than 30 minutes. The

Application of Accelerators in Research and Industry
AIP Conf. Proc. 1336, 401-405 (2011); doi: 10.1063/1.3586129

fast neutrons were produced by 4-MeV deuterons bombarding into the 3-mm-thick beryllium target, where 4-MeV deuterons deposit their full energies. We completed the fast neutron beam with high fluxes by installing a 2-MV Tandetron™ accelerator (High Voltage Engineering Europe, HVEE, B.V., The Netherlands), as shown in Fig. 1. The accelerator has a capability to produce its highest beam current up to several mA for protons and deuterons (4-MeV deuterons up to 500 μA and 4-MeV protons up to 800 μA). The deuteron beam was expanded using wobbler magnets to be 100 mm in diameter on the target. The 1.6-kW heat produced by the high current beam is cooled by coolant water (10-20 l/min.) attached downstream of the beryllium target. The target was cooled well but was broken several times due to blisters, incomplete elimination of extra heat, or heat expansion. The neutron beam was collimated using polyethylene blocks. The neutron dose rates are stable at the range of 2.5%, over 12, 24, and 30 cm in diameter at locations listed in Table 1. The dose rates were measured using an air-filled ion chamber (1 cm^3 volume, A150 tissue-equivalent plastic wall, IC17A, manufactured by Far West Technology). The ion chamber was exposure-dose calibrated with a ^{60}Co photon source. Intense neutron beams were measured using the air-filled ion chamber and found to be 7.62, 2.18, and 0.22 Gy/h at distance 71, 117, and 235 cm downstream from the target, respectively. Neutron exposure doses were calibrated using the ion chamber in every experiment normalized with deuteron beam charge.

NASBEE has two neutron irradiation rooms, as sketched in Fig. 1: one is a normal experimental room (conventional room) for in vitro irradiation and physics experiments, and the another is a specific pathogen free (SPF) room to protect mice and rats against bacteria and viruses. NASBEE in vivo experiments can clarify the biological effects of only fast neutrons under the SPF condition. Both rooms were constructed using the same configuration.

TABLE 1. Absorbed dose rates and beam size of NASBEE neutron beam. * indicates dose rates calculated with square law of distance from the target and --- indicates no measurement was made. IC, LLPC and LSO mean air-filled ion chamber, low-pressure proportional counter and organic liquid scintillator, respectively.

Distance from Be Target	Beam Diameter with +/- 2.5% of Beam Center	Absorbed Dose Rates (Gy/C)		
		IC	LLPC	OLS
71 cm	12 cm	5.27	--	--
174 cm	24 cm	1.56	N 1.51 G 0.33	1.61 *
248 cm	30 cm	0.604	--	--
248 cm	--	--	--	0.36

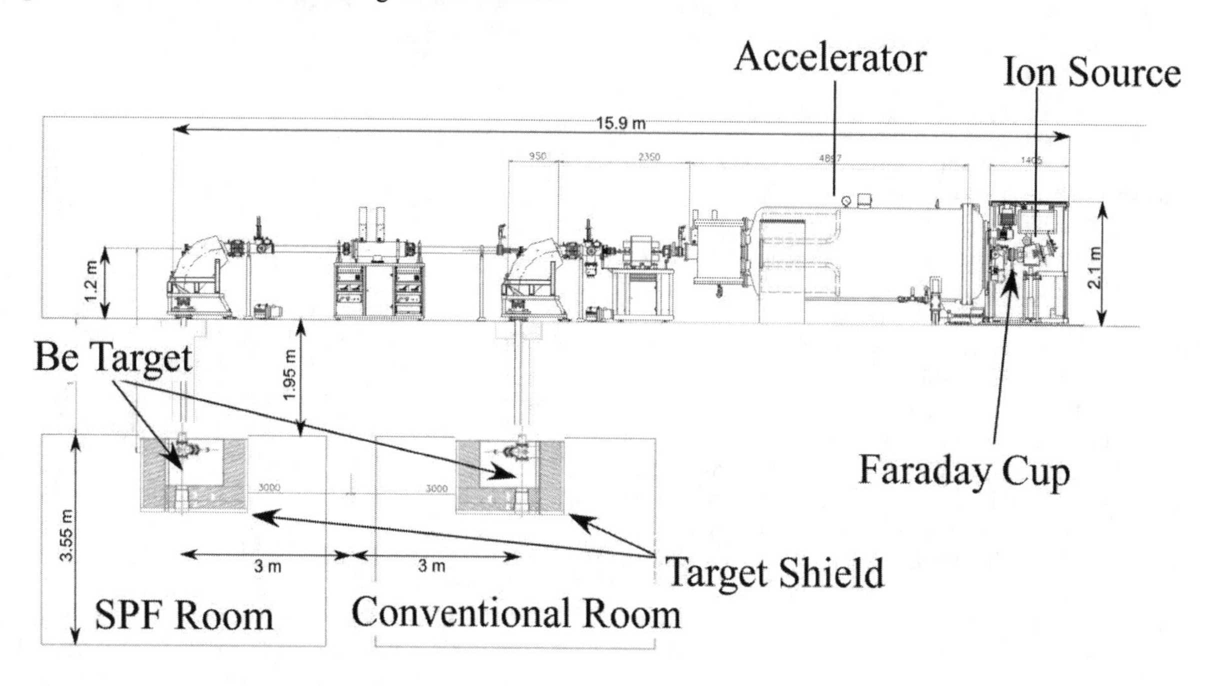

FIGURE 1. NASBEE configuration: Tandetron accelerator, and conventional and SPF irradiation rooms.

MEASUREMENT OF NEUTRON SPECIFICATIONS

Neutron Energy Spectrum

Characterization of the neutron energy spectrum is necessary for biological study with a wide energy range, down to a few hundred keV. The low energy neutrons have large RBE values [8]. The neutron energy spectra from 200 keV to 10 MeV were measured using an organic liquid scintillator, model EJ-399-06, of 1.27-cm length and 2.54-cm diameter with neutron-photon discrimination property. The liquid scintillator measures fast neutron energy spectra with higher energy resolutions than neutron measurements using metal activation methods and multi-moderator spectrometers. The energy spectrum is obtained using an unfolding method from measured neutron pulse heights. Neutron response functions of the liquid scintillator for the unfolding method were obtained based on the Monte Carlo calculation using the MCNPX code [9].

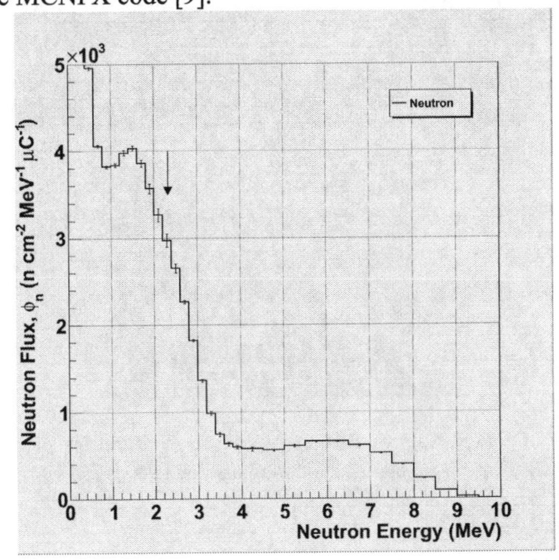

FIGURE 2. Measured neutron energy spectrum at NASBEE. An arrow indicates mean neutron energy in the energy spectrum.

Fig. 2 shows the neutron energy spectrum measured at NASBEE. The neutron energy spectrum consists of scattered neutrons from 0 to 1 MeV, neutrons from the stripping reactions of deuteron beams from 1 to 4 MeV, and neutrons produced by the $^{9}Be(d,n)^{10}B$ reaction (Q=+4.36 MeV). The neutron spectrum shows two peaks, at 1.7 and 6.2 MeV. Mean neutron energy represents the neutron field to be 2.3 MeV, while, the kerma of tissue-equivalent-plastic (A150) weighted mean energy is 3.0 MeV. The neutrons produced at the Be target are attenuated and scattered in the water and copper layers downstream of the target. The energy spectrum and average energies are useful for analyzing biological data.

Photons were also produced in the neutron beams. Several high-energy photons of 2.2, 6.0, 7.6 and 9.2 MeV were observed using a high-purity germanium detector. 2.2 MeV photons were produced from the thermal-neutron capture reaction of a hydrogen nucleus but no nuclei producing the other photon energies has been identified, yet.

Absorbed Dose Energy Distributions

Absorbed dose energy distributions of the NASBEE neutron beams were measured using low-pressure proportional counters (LLPCs) based on a microdosimetric technique. LLPCs measure particle energies deposited in 1-μm-diameter sphere tissues. We used two types of LPPC to obtain the dose distribution of neutrons and photons [10]. One LPPC is made from a wall of tissue equivalent plastic, Shonka A150, and the other is from a graphite wall. The A150-plastic wall counter has sensitivity to both photons and neutrons; however, the graphite-wall counter has to photons only, with negligible sensitivity to fast neutrons. From different responses of the counters, we were able to measure neutron and photon dose distributions separately.

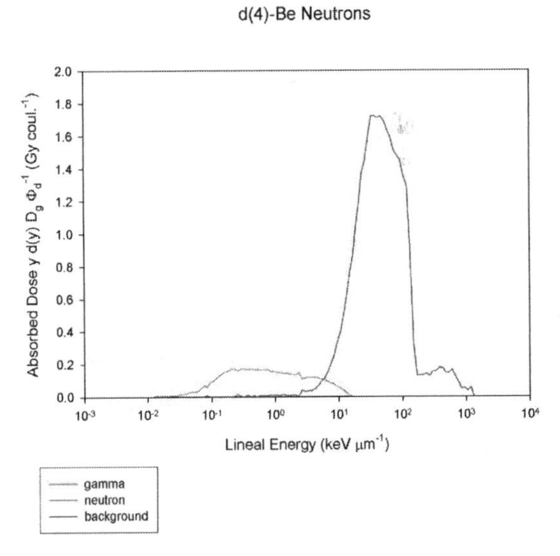

FIGURE 3. Neutron and photon absorbed dose distributions at NASBEE.

Fig. 3 shows the absorbed dose energy distributions of neutrons and photons. The photon distribution covers below 15 keV/μm lineal energy, while the neutron distribution is up to 2000 keV/μm. The neutron distribution consists of three components: recoiled protons up to 120 keV/μm, alpha particles produced by the C(n,α) reactions up to 400 keV/μm, and up to 2000 keV/μm heavy charged particles (carbon, nitrogen, and oxygen) knocked on by fast neutrons. Neutron and photon dose rates were obtained by integrating the dose energy distributions to be 1.51 and 0.33 Gy/C, respectively, normalized with incident deuteron beams. In the absorbed dose rates, neutrons occupy 82% and the rest is from photons. The dose rate agrees well with the absorbed dose measured using the ion chamber (1.56 Gy/C) in Table 1. Relative biological effectiveness (RBE), defined by biological response for radiations, is an important parameter to evaluate the radiation quality. RBE was estimated from microdosimetric spectrum using a 2 Gy biological response of fractional cell survivals; *r(y)*, which was given by Tilikidis et al. [11]. We estimated the value of RBE to be 3.5 for the NASBEE neutron beams. And the other hand, as a preliminary experiment, cell inactivation of V70 (fibroblast lung, Chinese hamster) cells was measured to obtain RBE value in vitro experiment. From 10% survival dose, D_{10} to be 7.45 and 2.10 Gy for X rays of 200 kV_p and NASBEE neutron beam, respectively, the value of RBE was obtained to be 3.54 [7]. The value of RBE obtained from the microdosimetric spectrum agrees well with the value from the in vitro experiment.

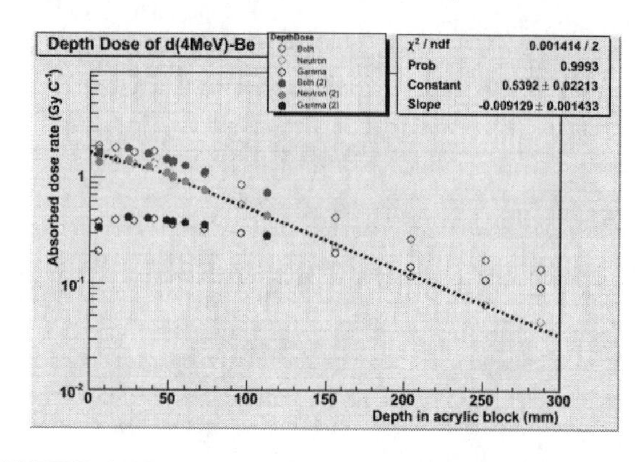

FIGURE 4. Neutron and photon absorbed doses in acrylic blocks.

Neutron and photon dose rates were measured according to depth in acrylic blocks, 30-cm cubes, to observe dose attenuation in large samples such as rats. Measured neutron and photon absorbed doses at various depths are plotted in Fig. 4. The neutron dose rates were constant below 3 cm depth from the front face and then were attenuated in an exponential manner, while the photon dose rates reached to a maximum at 30-mm depth and then were attenuated in an exponential manner. Over 20-cm depth, the photon doses are larger than the neutron doses. The RBE values were evaluated to be a constant 3.5 below 35 mm and then were decreased to 2.2 at 300 mm in depth.

Neutron Space Distributions

Neutron space distributions were observed using the ion chamber and the LPPCs at several distances from the target center (Table 1). The neutron irradiation areas are wider as the distance from the target is greater. The absorbed energy distributions vary only slightly within the neutron irradiation area.

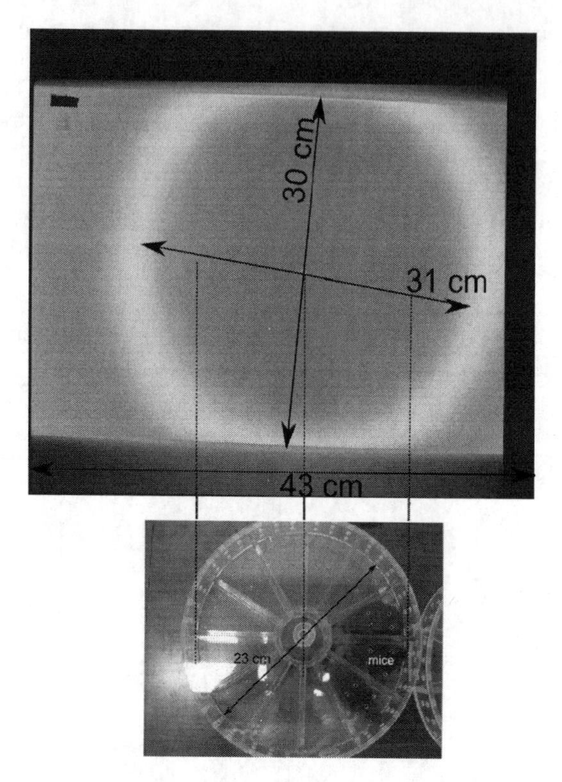

FIGURE 5. Neutron beam distribution measured by neutron activation and exposure to imaging plate (top), and acrylic holder of 12 mice (bottom).

Neutron space distributions were also measured simply using activated aluminum plates, as shown in Fig. 5. Reds indicate higher neutron fluxes than do blues. Neutrons were detected by activating aluminum plates with neutrons and then exposing the activated aluminum plates to imaging plates. The imaging plates detected radiation, mainly betas from the aluminum. This method has the advantage of a lack of sensitivity to photons.

New Neutron Irradiation Method

We have produced localized neutron beams to irradiate neutrons to only organs of interest in rats. We wanted to observe carcinogenesis without any interruption of carcinogenesis in other organs. Neutrons were localized with a 4-cm-wide slit using iron and polyethylene shields. Neutron absorbed doses behind the shields were decreased to 8% of those at the target center. The neutron effect outside of the interested organs was negligible. Using the localized neutron beams, we irradiated the lungs of rats to observe the biological effects of the beams without affecting any carcinogenesis of other organs.

CONCLUSION

We have characterized the NASBEE neutron beams with the neutron energy spectrum, absorbed dose energy distributions, and space distributions. These distributions are useful for biological study. We hope the results of this study will provide researchers outside of our institute an opportunity to determine some of the biological effects of fast neutrons.

REFERENCES

1. International Commission on Radiation Protection, ICRP Publ. 92, Pergamon Press, Oxford (2003).
2. International Commission on Radiation Protection, ICRP Publ. 60, Pergamon Press, Oxford (1990).
3. M.Takada et al., Nucl. Instrum. Meths., A545[3], 765-775 (2005)
4. M.Takada et al., Radiat. Prot. Dosim., 110[1/4], 601-606 (2004)
5. Y.Ishida et al, Journal of Radiation Research, 47[1], 41-47 (2006).
6. R.Kanda et al., International Journal of Radiation Biology, 82[7], 483-491 (2006).
7. M.Suda et al, Radiation Physics and Chemistry, 78[12], 1216-1219 (2009)
8. R.C.Miller et al, Radiat. Res. 117, 114-127 (1989).
9. M.Takada et al., J. Nucl. Sci. Technol. 47[10], 917-931 (2010)
10. M.Takada et al., Radiat. Prot. Dosim. 114[4], 481-490 (2005).
11. A.Tilikidis et al., Phy. Med. Biol. 41, 55-69 (1996)

Monte Carlo Simulations Of The Dose Distributions From Carbon Microbeams Used In An Experimental Radiation Therapy Method

I. Dioszegi[a], A. Rusek[b], B. R. Dane[c], I. H. Chiang[b], A. G. Meek[d] and F. A. Dilmanian[d,e]

[a]*Nonproliferation and National Security Department, Brookhaven National Laboratory, Upton, New York 11973, USA.*
[b]*NASA Space Radiation Laboratory, Brookhaven National Laboratory, Upton, NY 11973, USA*
[c]*Medical School, State University of New York at Stony Brook, Stony Brook, NY 11794, USA.*
[d]*Department of Radiation Oncology, State University of New York at Stony Brook, Stony Brook, NY 11794, USA.*
[e]*Medical Department, Brookhaven National Laboratory, Upton, NY 11973, USA.*

Abstract. Recent upgrades of the MCNPX Monte Carlo code include transport of heavy ions. We employed the new code to simulate the energy and dose distributions produced by carbon beams in rabbit's head in and around a brain tumor. The work was within our experimental technique of interlaced carbon microbeams, which uses two 90° arrays of parallel, thin planes of carbon beams (microbeams) interlacing to produce a solid beam at the target. A similar version of the method was earlier developed with synchrotron-generated x-ray microbeams. We first simulated the Bragg peak in high density polyethylene and other materials, where we could compare the calculated carbon energy deposition to the measured data produced at the NASA Space Radiation Laboratory (NSRL) at Brookhaven National Laboratory (BNL). The results showed that new MCNPX code gives a reasonable account of the carbon beam's dose up to ~200 MeV/nucleon beam energy. At higher energies, which were not relevant to our project, the model failed to reproduce the Bragg-peak's extent of increasing nuclear breakup tail. In our model calculations we determined the dose distribution along the beam path, including the angular straggling of the microbeams, and used the data for determining the optimal values of beam spacing in the array for producing adequate beam interlacing at the target. We also determined, for the purpose of Bragg-peak spreading at the target, the relative beam intensities of the consecutive exposures with stepwise lower beam energies, and simulated the resulting dose distribution in the spread out Bragg-peak. The details of the simulation methods used and the results obtained are presented.

Keywords: Monte Carlo simulation, MCNPX, Carbon radiotherapy, microbeam
PACS: 87.55.Gh, Simulation 87.55.K, Monte Carlo methods

INTRODUCTION

Radiotherapy of cancer using proton beams has been in use in the US and other countries for decades[1]. The more advanced method of carbon therapy has also been in use in Japan[2] and Germany[3] for over 20 years. Heavy ions have certain properties that make them especially suitable for radiosurgery. First, their depth-dose distribution exhibits a narrow but very high peak at the end of the trajectory (Bragg peak). Second, the radiation has a high relative biological effectiveness (RBE) mostly in its Bragg peak. The existing clinical carbon therapy facilities mostly use either passive beam spreading or use raster scanning with pencil beams of variable energies. Depending on the application the energy range of carbon ions used in radiotherapy can reach 400 MeV/A

"Interlaced carbon microbeams", a recently introduced experimental radiosurgery method[4,5] employs arrays of parallel, thin planes of carbon beams. Healthy tissue shows an extremely high tolerance for the microbeam, an effect first observed with synchrotron x-ray microbeams at the National Synchrotron Light Source (NSLS), BNL[6]. Figure 1 illustrates the concept of interlaced X-ray microbeams.

Application of Accelerators in Research and Industry
AIP Conf. Proc. 1336, 406-409 (2011); doi: 10.1063/1.3586130

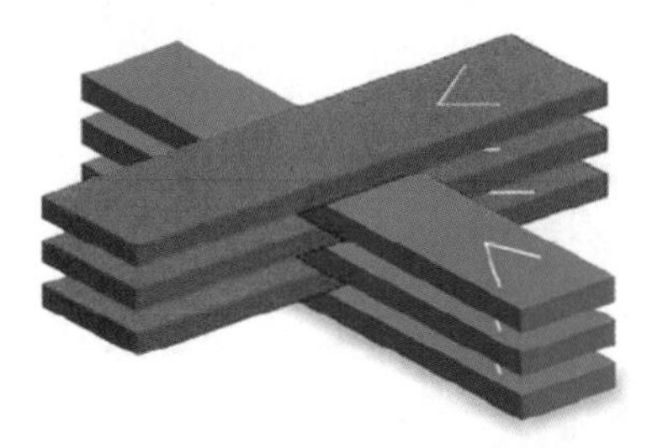

FIGURE 1. Schematic representation of interlaced x-ray microbeams. The individual arrays of thin microbeams spare the non-targeted tissues, while the target receives full dose.

Monte Carlo codes are generally essential for radiation therapy treatment planning. The widely used Monte Carlo radiation transport code MCNPX, which was not equipped with heavy ion transport routines, was recently upgraded to include that capability[7]. We are reporting the use of this new code for carbon microbeam therapy treatment planning.

MONTE-CARLO CALCULATIONS

Benchmarking Of Calculated Dose Distributions

The first General Release MCNPX version to include heavy ion transport was[8] version 2.6.0. In the present calculations we have used the higher test version, 2.6.f beta release version, which is identical to 2.6.0 in describing heavy ions except for minor corrections. The code's physics model allows the user to choose from a large selection of options to define the detailed models, including the Los Alamos Quark-Gluon String Model (LAQGSM) model. Very few bench-marks of the code have been published so far. In fact the NSRL is one of the only facilities in the U.S. capable or producing data necessary for bench-marking this very important capability of the code.

We benchmarked the code for carbon ions at 137, 200, and 293 MeV/A energies using a library of data[9] produced by the NSRL (Fig. 2). The calculations used the MCNPX code, while the experimental data was produced by measuring the transmitted energy of the incident heavy ion beam passing through a binary high-density polyethylene filter and measured by an ion chamber. These comparisons show a remarkable agreement between the calculations and measurements for all three incident beam energies and for the entire Bragg curve for each such energy except for the nuclear break-up tail data at the high energy ends of the curves. The code failure to produce the break-up tails at the two energies of 200 and 293 MeV/A is not surprising as it was indicated by the code's authors in the code's published report[7]. The subject, however, did not affect our work because very little nuclear break-up events are expected in the beam energy range of 130-140 MeV/A we used.

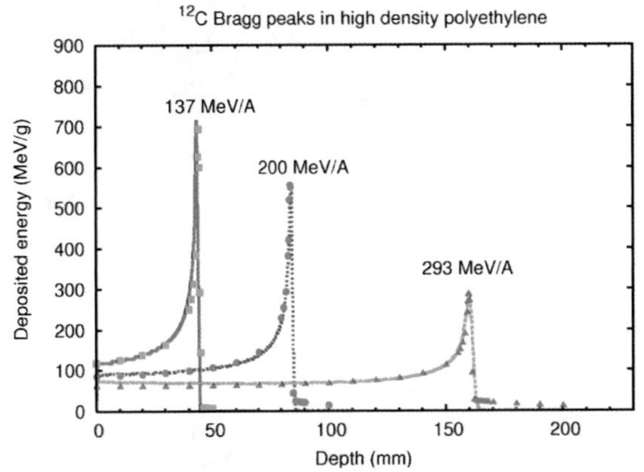

FIGURE 2. Comparison of the MCNPX simulations of deposited energy in high density polyethylene to the measured data. The low-energy parts of the experimental curves were normalized to the calculations. The resulting coincidences of the peaks of the experimental and measured Bragg curves are remarkably good.

Dose Calculations For Rabbit Brain Tumor

MCNPX calculations were first performed to determine the minimum and maximum beam energies needed to reach the appropriate tissue depths for covering the 14-mm-diameter target in rabbit's brain. Figure 3 shows the geometrical model used as the MCNPX input, where the different tissues, i.e., skin, fat, muscle, bone, brain and tumor, were modeled as cubes and a sphere, respectively. Although the tumor was expected to be only 5 mm in diameter, we chose a much larger target volume of 14 mm in diameter to cover for the uncertainties we had in the localization and targeting of the tumor.

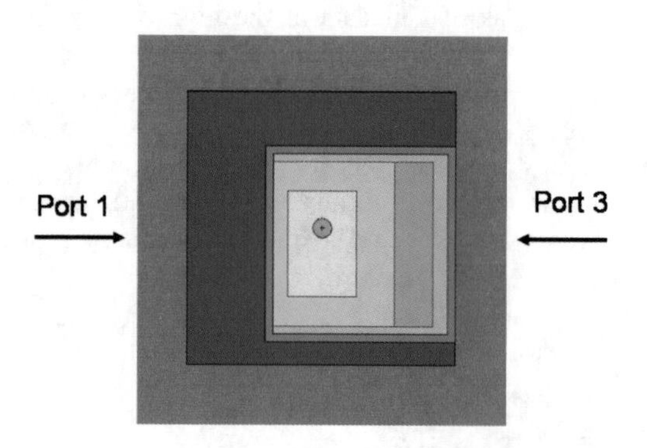

FIGURE 3. Simplified input geometry of the rabbit head and the Lucite boluses surrounded with air (red). The material around the brain tumor included brain, bone, muscle, fat, skin, and Lucite boluses (dark blue); the latter was used for adjusting the effective beam energy.

The following steps were involved in the simulation of the spread-out Bragg curve for the therapy. We first ran MCNPX calculations in small energy steps to determine beam energies for reaching the proximal and the distal sides of the 14-mm target in the subject model of Fig. 3. Next, we simulated the two Bragg curves for the irradiations through Ports 1 and 3 in Fig. 3 that extend to the distal ends of the subject from these two directions (Fig. 4). Because we intended to produce spread-out Bragg peaks of dose while the code provided us the values of energy per volume, at each step we converted the MeV/cm^3 values to those of MeV/g by dividing the numbers with the local tissue density. Similar calculations were made for the two side directions. We note that the boluses were used to allow us a single incident beam energy for all four directions, and would not have been necessary otherwise.

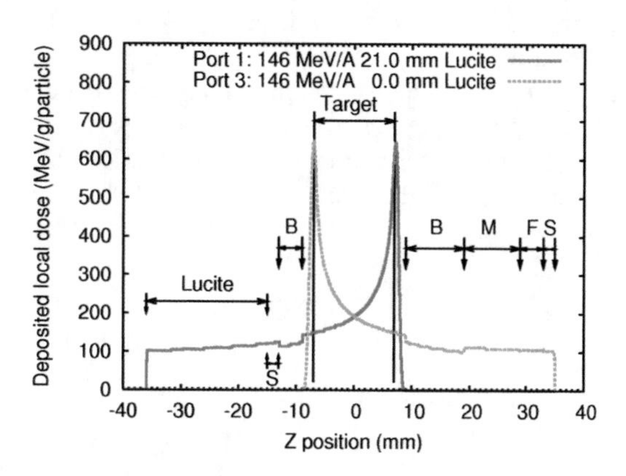

FIGURE 4. Bragg curve calculations for irradiation from Ports 1 and 3 of Fig. 3. The tissue-type labels are S: skin, F: fat, M: muscle, and B: bone.

Next, we developed a routine to calculate the weighting ratios necessary for spreading out the Bragg peak stepwise using 2-MeV steps of beam energy reductions. The results are shown in Fig. 5.

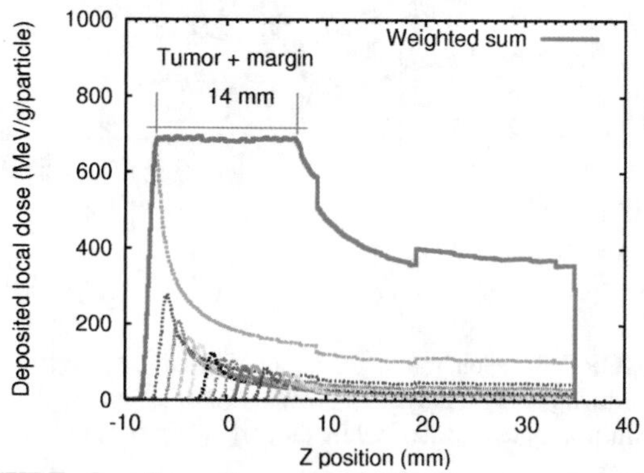

FIGURE 5. Spreading out of the Bragg curve using weighted sum of irradiations with decreasing values of the incident beam energy. The curve relates to Port 3 of Fig. 3.

Next, the calculated beam irradiation ratios for the beam steps of Fig. 5 were tested experimentally at NSRL using the beam measurement procedure with a binary filter described above. For this purpose, the circuit that stops the beam once a given accumulated dose is reached in the ion chamber was used to control the irradiation time for each step. The measured sum of the deposited energy is shown on Fig. 6. The small fluctuation (~5%) of the measured Bragg plateau (marked by the red arrow) can be understood, since certain experimental parameters (beam energy spread and distribution) were not exactly known and were modeled by a simple energy distribution in the model.

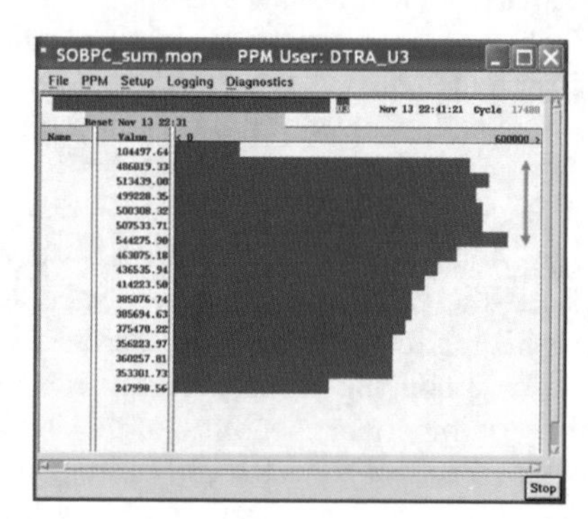

FIGURE 6. Experimental measurement of Bragg curve spreading in high density polyethylene. The screen capture displays the varying depth polyethylene absorber on the vertical axis, and the relative dose values are shown horizontally. The location of the spread Bragg peak is marked by the red arrow.

Angular Straggling

The next task of the simulation was to determine the broadening of the microbeams in the tissue due to angular straggling. The information was necessary to choose the spacing between microbeams for producing an overlap between the interlacing beam arrays at the proximal side of the target. The initial microbeam thickness (defined by the collimator opening) is 0.3 mm. We tallied the deposited local dose at the proximal side of the target in a direction perpendicular to the beam plane. Fig. 7 shows the results. The individual irradiation beam distributions are shown on the curves together with their weighted sum according to the required weights to achieve the Bragg peak spreading. The weighted sum (red line) has a full width half maximum (FWHM) of 0.4 mm.

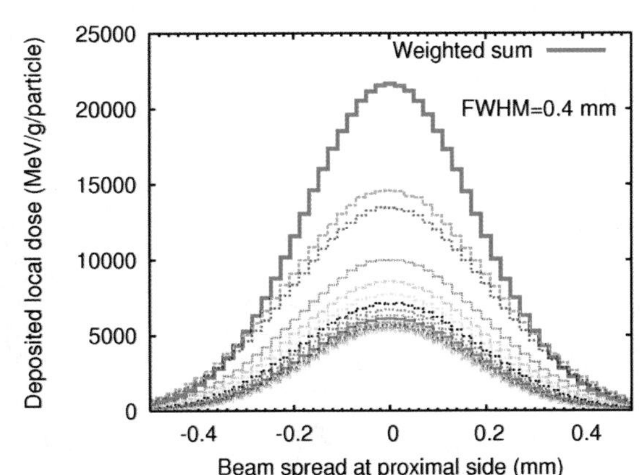

FIGURE 7. Angular straggling calculations for the a single carbon microbeam.

SUMMARY

We performed MCNPX calculations in the framework of the experimental radiosurgery method with interlaced carbon microbeams applied to the treatment of s rabbit brain tumor.. The benchmark calculations for the Bragg peak position in the tissue at a given incident beam energy showed a good agreement with experimental data. The nuclear breakup tail of the carbon beam's beyond the Bragg peek, however, was only sub-optimally reproduced, a discrepancy that was significant only at beam energies above 200 MeV/A.

For treatment planning we calculated the weighting factors needed to spread the Bragg peak over the tumor area. We verified our calculations by the experimental results obtained with a high density polyethylene stack chamber measurement. The results showed a flat plateau with small fluctuations.

We performed angular straggling calculations to show the effect of the spreading the beam in the tissue.

ACKNOWLEDGEMENTS

We thank Joseph Gatz III, Charles Pearson, and Michael Sivertz for assistance. Funding for this work was provided by Musella Brain Tumor Foundation, SB's Targeted Research Opportunities program, SB's Office of the Dean of School of Medicine, SB's Research Foundation funds (Allen G. Meek, MD, PI), and Voices against Brain Cancer.

REFERENCES

1. G. Coutrakon, et al., Med. Phys. 18 (1991) 1093.
2. Matsufuji, N., Fukumura, A., Komori, M., Kanai, T. and Kohno, T. *Influence of fragment reaction of relativistic heavy charged particles on heavy-ion radiotherapy.* Phys. Med. Biol. 48, 1605 (2003).
3. Gunzert-Marx, K., Schardt, D. and Simon, R. S. *Fast neutrons produced by nuclear fragmentation in treatment irradiations with 12C beam.* Rad. Prot. Dosim. 110, 595.
4. Schulz-Ertner D et al. *Results of carbon ion radiotherapy in 152 patients.* Int. J. Radiat. Oncol. Biol. Phys. 58 631–40 (2004).
5. Dilmanian et al., manuscript in preparation.
6. Dilmanian et al., PNAS 2006.
7. Michael R. James, G.W. McKinney, John S. Hendricks, Michael Moyers, *Recent enhancements in MCNPX: Heavy-ion transport and the LAQGSM physics model.* Nucl Instr. Meth. in Phys. Res. A 562 819-822 (2006).
8. D. Pelowitz (Ed.), MCNPX User's Manual Version 2.6.0, *Los Alamos National Laboratory report LA-CP-07-1473* (April 2008).
9. http://www.bnl.gov/medical/NASA/CAD/Bragg/Bragg.asp

Pencil Beam Scanning System Based On A Cyclotron

Toshiki Tachikawa[a], Hideki Nonaka[a], Yukio Kumata[a],
Teiji Nishio[b] and Takashi Ogino[b]

[a]*Sumitomo Heavy Industries, Ltd., 5-2 Soubiraki-cho, Niihama, Ehime 792-8588, Japan*
[b]*National Cancer Center Hospital East, 6-5-1 Kashiwanoha, Kashiwa, Chiba 277-8577, Japan*

Abstract. Sumitomo Heavy Industries, Ltd. (SHI) has developed a new pencil beam scanning system (PBS) for proton therapy in collaboration with National Cancer Center Hospital East (NCCHE). Taking advantage of the continuous beam from the cyclotron P235, the line scanning method is employed in order to realize continuous irradiation with high dose rate. 3D uniform and sphere field was irradiated and compared with the simulation.

Keywords: Cyclotron, Proton therapy, Scanning, Wobbling
PACS: 87.53.jw

INTRODUCTION

Several types of pencil beam scanning systems for proton therapy have been developed (ref.1 and 2) and applied to clinical treatment. SHI has developed a new pencil beam scanning system taking advantage of the continuous and high intensity beam from the cyclotron P235. The line scanning method is employed in order to realize continuous irradiation with high dose rate. A 3D irradiation volume is formed by superposing energy layers in depth direction, where the energy of each layer is adjusted with the energy selection system (ESS) at the exit of the cyclotron.

The SHI pencil beam scanning system (PBS) can be incorporated into either a dedicated scanning nozzle or a multi-purpose nozzle which enables also the wobbling irradiation technique.

NOZZLE SYSTEM

The multi-purpose nozzle has the function of both wobbling and scanning methods. Each of them can be selected without any exchange of devices in the nozzle. The irradiation method can be switched from wobbling to PBS by automatically retracting the devices not necessary to PBS (FIGURE 1). Only a patient collimator and bolus should be removed manually.

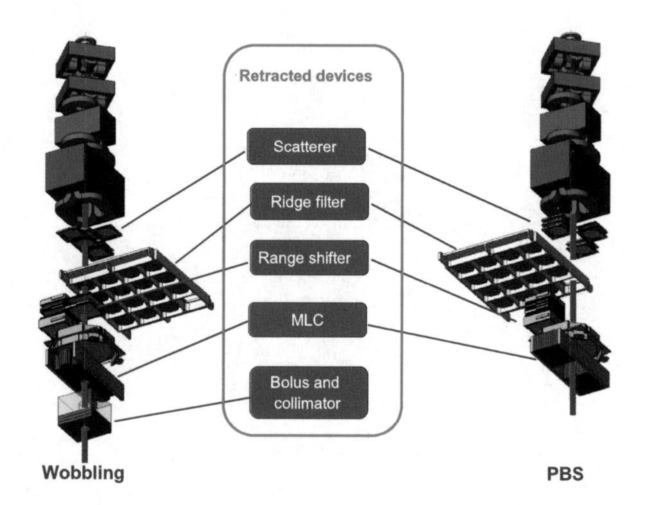

FIGURE 1. Function of the multi-purpose nozzle. Scanning mode and wobbling mode can be switched automatically.

The dedicated scanning nozzle does not have the devices necessary for wobbling and can be used for larger field size. Main specifications of both nozzles are summarized in TABLE 1.

Application of Accelerators in Research and Industry
AIP Conf. Proc. 1336, 410-412 (2011); doi: 10.1063/1.3586131

TABLE 1. Specifications of SHI nozzles

Specification	Dedicated scanning nozzle	Multi-purpose nozzle	
		Scanning method	Wobbling method
Max. field size	30cm x 40cm	24cm x 24cm	25c x 25cm
Max. range	32cm	32cm	26cm
Max. SOBP	28cm	28cm	15cm
Field uniformity	+/- 2.5%	+/- 2.5%	+/- 2.5%

LINE SCANNING TECHNIQUE

Line scanning is a method to scan the beam continuously with, typically, a square trajectory (FIGURE 2) taking advantage of the continuous beam from the cyclotron.

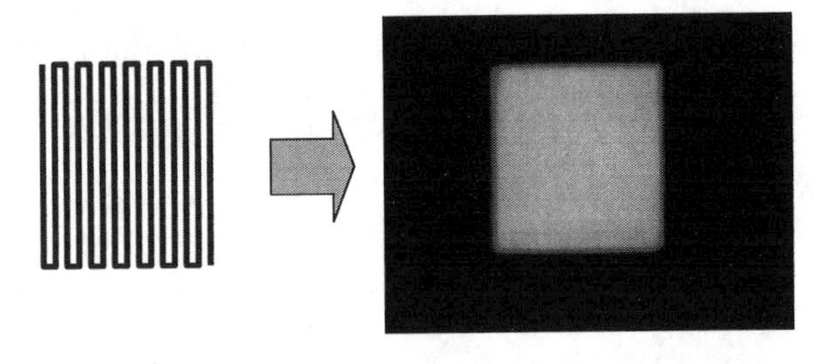

FIGURE 2. Typical scanning pattern of line scanning. Left figure: Scanning pattern for a square shape. Right figure:Dose distribution measured by a fluorescent plate and CCD camera

Due to the depth dose distribution of a proton pencil beam, formation of a 3D irradiation field requires a modulation of the dose distribution in each energy layer. This is done by varying the scanning speed and keeping the beam current constant. The dose at different layers is adjusted by either changing the beam intensity or changing the number of times the layer is rescanned.

Beam intensity is monitored by a non-interceptive ionization chamber at the exit of cyclotron and is stabilized within +/- 1% between 1nA and 300nA by adjusting the arc current of the ion source with a feedback loop. Scanning speed can be modulated with the response time of less than 1ms. Fast beam ON/OFF is done by switching the RF voltage of the cyclotron within 0.2ms.

The position of the scanned pencil beam is monitored in real time by a beam position monitoring system with a multi-strip ionization chamber (64 x 64 strips in case of the multi-purpose nozzle). The spatical resolution is better than 0.1mm and the sampling period is 12 micro seconds.

PERFORMANCE TEST OF LINE SCANNING

Test nozzle was installed in the gantry of NCCHE and a performance test has been conducted. A 2D intensity modulated field was irradiated and evaluated by the gamma index method (FIGURE 3).

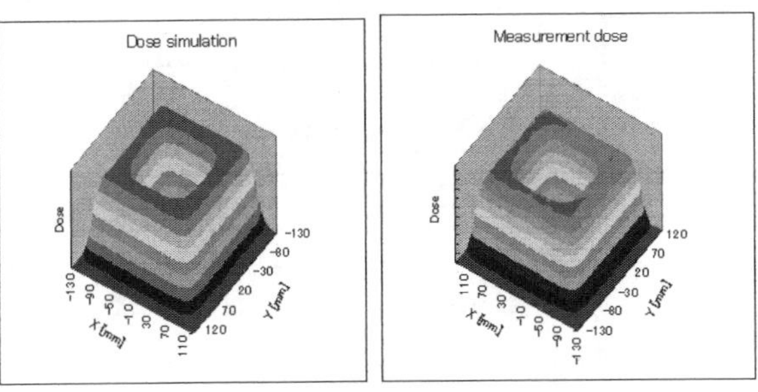

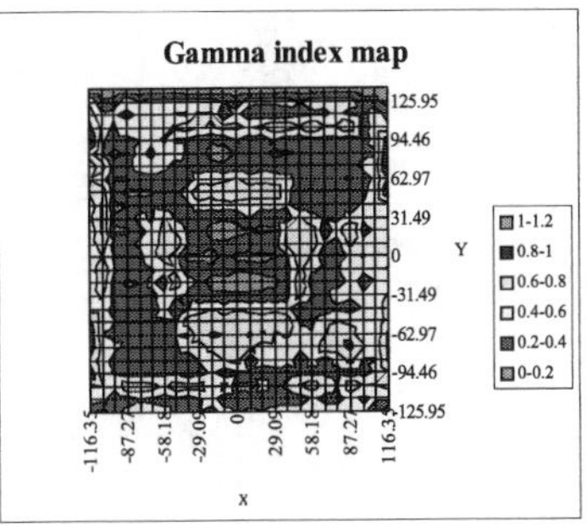

FIGURE 3. 2D intensity modulated field. A gamma index value of 99.7% (3mm/3%) was obtained.

An excellent result for the gamma index value of 99.7% (3mm/3% criterion) was obtained. Based on this result, a uniform 3D spherical field of 8cm in diameter was generated by superposition of 14 irradiated 2-dimensional intensity-modulated energy layers.. The maximum scanning speed was 10mm/ms. The dose distribution was measured with a water phantom and an array of 11 ionization chambers. The measured data were in good agreement with the similation (FIGURE 4).

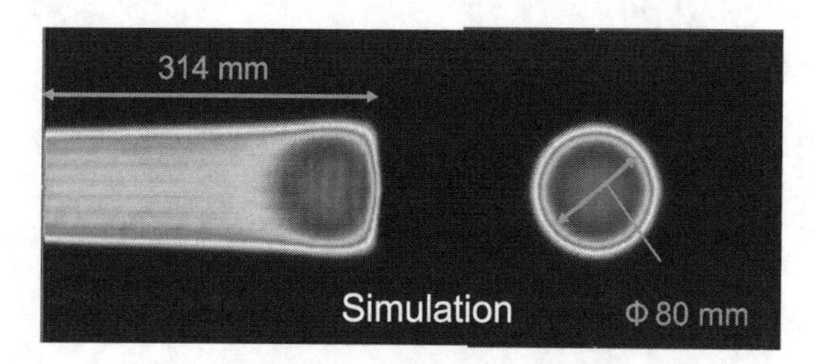

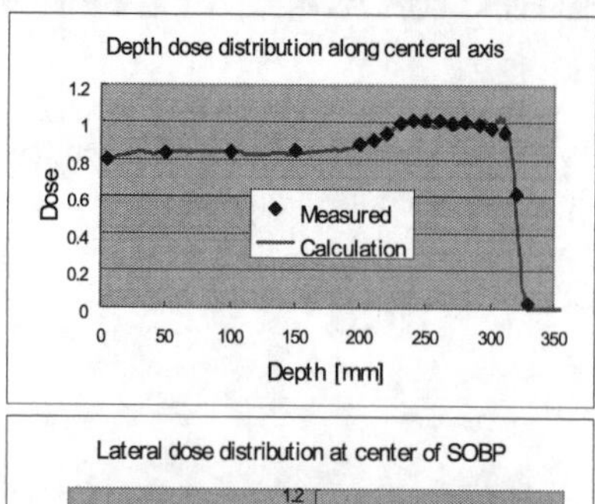

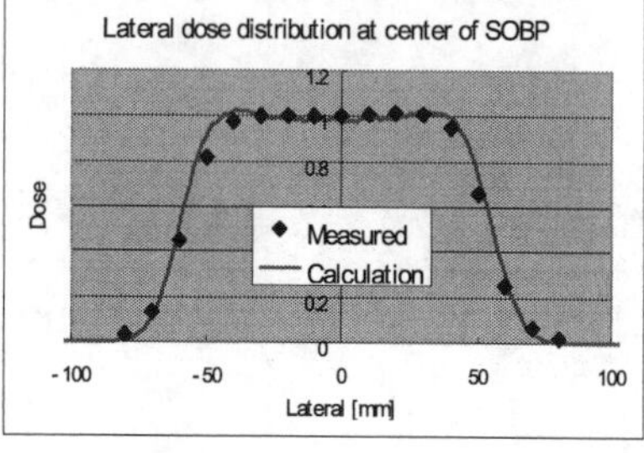

FIGURE 4. 3D spherical field irradiation of 8cm in diameter. The maximum range is 314mm and the number of energy layers is 14. Measured dose distributions along the irradiation axis and along lateral direction at the spread out Bragg peak (SOBP) center are compared with the simulation.

CONCLUSION

A continuous line scanning system for intensity modulation has been developed in collaboration with NCCHE. The characteristics of intensity modulation is the constant beam current and varying scanning speed. Excellent result of gamma index evaluation, 99.7% (3mm/3%), for 2D intensity modulation was obtained and 3D intensity modulation has been successfully demonstrated. Irradiation tests for more complex shapes or inhomogeneous dose distributions are planned in order to apply this technique for actual treatments.

REFERENCES

1. B. Marchand et.al., "IBA proton pencil beam scanning: An innovative solution for cancer treatment", Proceedings of EPAC2000, Vienna Austria, p.2539-2541
2. Alfred Smith et. al., "The M.D. Anderson proto therapy system", Med. Phys. 36(9), September 2009, p.4068-4083

Compact Electronic Gamma Source For Radiotherapy

A. X. Chen[a,b], A. J. Antolak[a], K. -N. Leung[c], T. N. Raber[a], D. H. Morse[a]

[a]*Sandia National Laboratories, PO Box 969, Livermore, CA, 94550*
[b]*Department of Mechanical Engineering, University of California, Berkeley, CA 94720*
[c]*Department of Nuclear Engineering, University of California, Berkeley, CA 94720*

Abstract. A compact mono-energetic gamma source is being developed to replace the radiological sources used in radiotherapy and other medical instruments. The electronic gamma source utilizes low-energy nuclear reactions to generate gammas in the 0.5 to 1.0 MeV energy range. Independent control of the ion current and energy is achieved by decoupling the RF-driven ion source and pyroelectric crystal-based acceleration systems The ions are accelerated to voltages above 100 keV and bombard a reaction target to produce gammas. Thermal management of the pyroelectric crystal-based accelerator is achieved by convective dielectric fluid flow around the crystal. This approach provides better temperature uniformity in the crystal and higher dielectric strength for suppressing voltage breakdown and enabling faster thermal cycling rates.

Keywords: LiTaO3 pyroelectric accelerator, gamma generator, RF ion source, 9Be(D,ng)10B reaction, radiotherapy
PACS: 29.25.Dx; 29.25.Dz; 77.70.+a

INTRODUCTION

According to the National Cancer Institute, half of all men and one third of all women will develop some form of cancer during their lifetime [1]. Radiotherapy has proved to be an effective non-surgical treatment against many kinds of tumors. In particular, gamma-based radiotherapy gives physicians unparalleled control of the dose delivery depth to allow precise attacks on tumor cells while preserving healthy cells. Current radiotherapy instruments such as gamma knives use an array of radioactive sources which are collimated onto a single point (tumor) for irradiation treatments [2]. Recently, however, national security issues regarding the use of high activity radiological sources in medical and commercial applications have been raised [3]. As a result, there is increased interest in developing electronic sources to replace the existing radiological sources since they can be turned off when not in use and can be designed to produce gammas with energies similar to existing instruments. In particular, the development of an electronic source producing mono-energetic gammas in the megavolt energy range would be a good replacement for the 60Co source currently used in radiotherapy or, potentially, to replace radiological sources used in other commercial applications such as oil well logging and industrial radiography.

GAMMA SOURCE CONCEPT

The gamma source concept is based on using low-energy nuclear reactions to produce the desired radiation by extracting ions from a suitable ion source and accelerating them across a single gap onto a production target. In the present system, a deuterium (D^+) ion beam bombards a ^{9}Be target to generate gammas with energies 410 keV, 718 keV, 1.03 MeV, 1.44 MeV, and 2.87 MeV; the 718 keV and 1.03 MeV gammas have the highest emission probabilities [4]. The D-^{9}Be reaction cross section (FIGURE 1) shows a general increasing logarithmic behavior from approximately 100 keV up to 700 keV. The magnitude of the cross section continues to increase with increasing deuteron energy, but the rate of increase is slower above ~300 keV, so the gamma source is being designed to operate around 300 - 400 keV.

The present source employs a pyroelectric crystal powering system to achieve the required voltage for optimum gamma production in a compact device. Pyroelectric crystal accelerators have previously been used to produce the high voltages necessary for D-D neutron production [5][6]. In the pyroelectric effect, a spontaneous polarization develops across the crystal due to a change in the crystal temperature which results in a charge on its polar face given by

Application of Accelerators in Research and Industry
AIP Conf. Proc. 1336, 413-417 (2011); doi: 10.1063/1.3586132

$$Q = \gamma A \Delta T \qquad (1)$$

where Q is the net charge, γ is the pyroelectric coefficient, A is the area of the crystal polar face, and ΔT is the temperature change [7]. The voltage produced is

$$V = \frac{\gamma A \Delta T}{C} \qquad (2)$$

where C is the capacitance of the system. It is important to note that the voltage may not be constant over time due to the functional dependence between the ion current load on the crystal and the rate of temperature change. Differentiating [1] with respect to time gives the relationship between the current and rate of temperature change,

$$I_a = \frac{dQ}{dt} = \gamma A \frac{dT}{dt} \qquad (3)$$

where I_a is the available current. Assuming negligible losses, the voltage can be kept constant if $I_a = I_{load}$ by controlling I_{load} through a separate ion source.

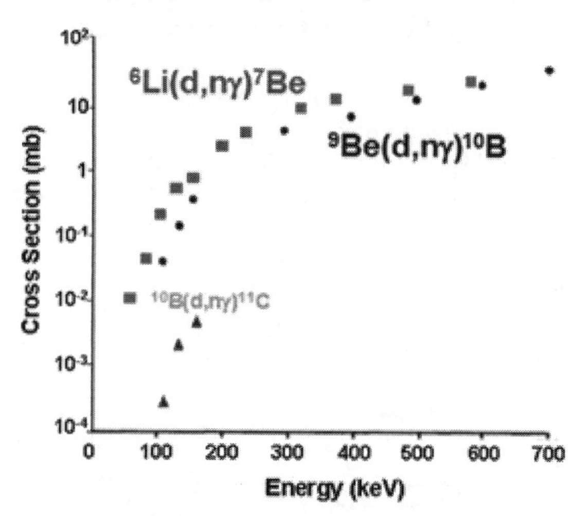

FIGURE 1. Reaction cross-sections of several nuclear reactions for gamma production [4].

GAMMA SOURCE DESIGN

The prototype gamma source in (FIGURE 2) consists of three major components: 1) RF ion source, 2) pyroelectric crystal acceleration system, and 3) gamma production target. In the prototype, a 10" diameter stainless steel chamber is used to house the crystal accelerator and beryllium target. The deuterium ion source is mounted on a 6" Conflat flange attached to the chamber. A turbo-molecular pump coupled with a roughing vacuum pump is used to provide vacuum for both the chamber and ion source.

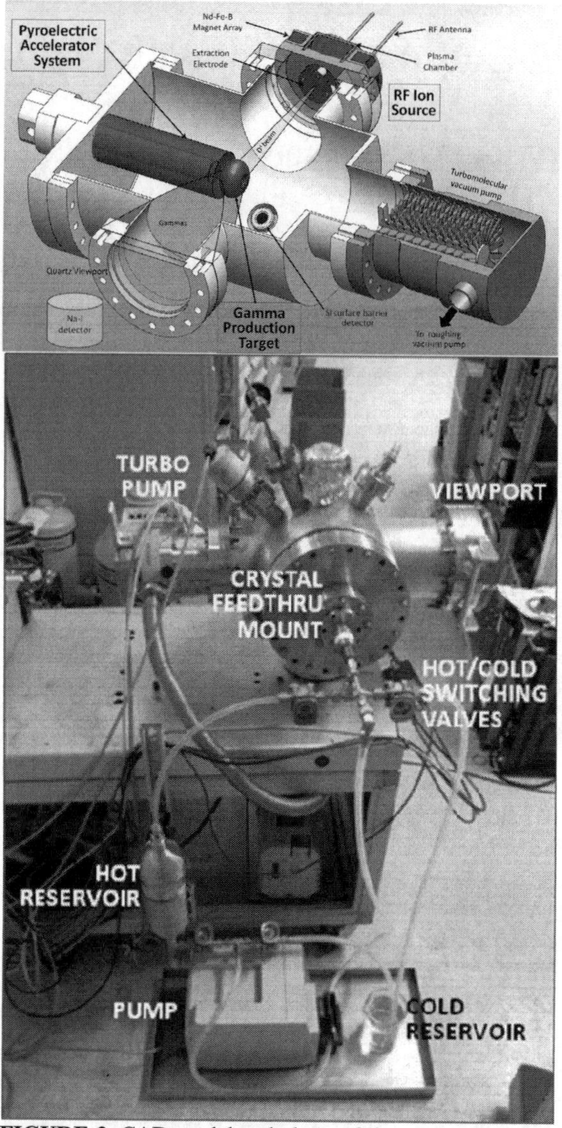

FIGURE 2. CAD model and photo of the prototype gamma source with dielectric fluid heating system.

A needle valve is used to adjust the pressure for the ion source. The ion source and acceleration regions are separated to allow differential pumping to the desired pressure for optimal ion source plasma operation while the pyroelectric accelerator chamber can be pumped to a pressure several orders of magnitude lower to generate higher voltages and minimize arc discharges. Tests made with a 1 mm diameter aperture showed that the pyroelectric accelerator chamber pressure can be maintained in the 10^{-5} Torr range while the plasma chamber is operating in the 10^{-2} Torr range. Additionally, the current load on the pyroelectric crystal accelerator can be varied by changing the ion source parameters, allowing independent control of the accelerator current and voltage.

The prototype gamma source implements an RF discharge plasma ion source to produce the D^+ ion beam [8]. Mass spectrometer measurements of the hydrogen species fractions in the RF discharge ion source show a high ratio of atomic compared to molecular ions (FIGURE 3). This is ideal for the gamma source because D^+ ions will achieve the full potential energy from the pyroelectric crystal accelerator, while D_2^+ and D_3^+ will produce deuterons with 1/2 and 1/3 of the full energy per nucleon, respectively, resulting in lower gamma yield.

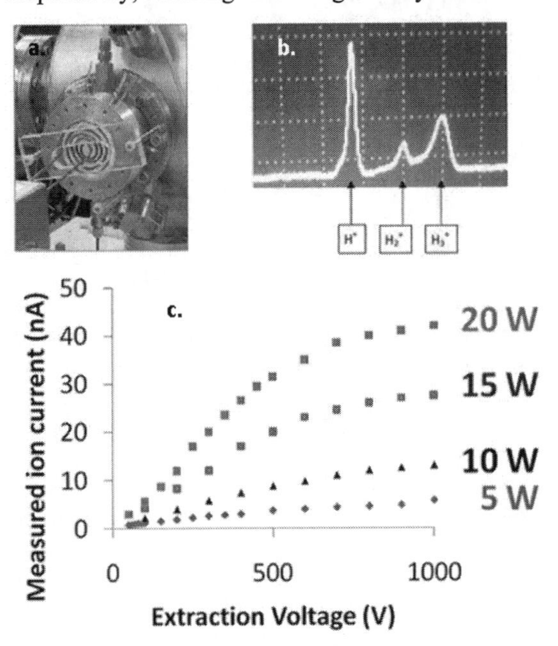

FIGURE 3. a) RF discharge ion source, b) mass spectrum of the hydrogen plasma, and c) ion source current as a function of extraction voltage and RF power.

During operation of the ion source, a 13.56 MHz RF field is coupled to a quartz chamber that is continuously filled with deuterium gas regulated to a pressure of 10-50 mTorr. A circular array of Nd-Fe-B magnets is placed around the chamber for better plasma confinement. Ceramic standoffs are used to separate the extraction electrode from the ion source. Deuterium ions are extracted with low (0.1-1.0 kV) extraction voltage and low (5-20 W) RF power. The available ion current, I_a, depends on several parameters including extraction voltage, RF power, and plasma density [9]. The ion source current was measured with a Faraday cup placed at the target location (FIGURE 3). Measurements indicated that the current is in the range of 0.1 to 10's of nano-amperes for the above described conditions.

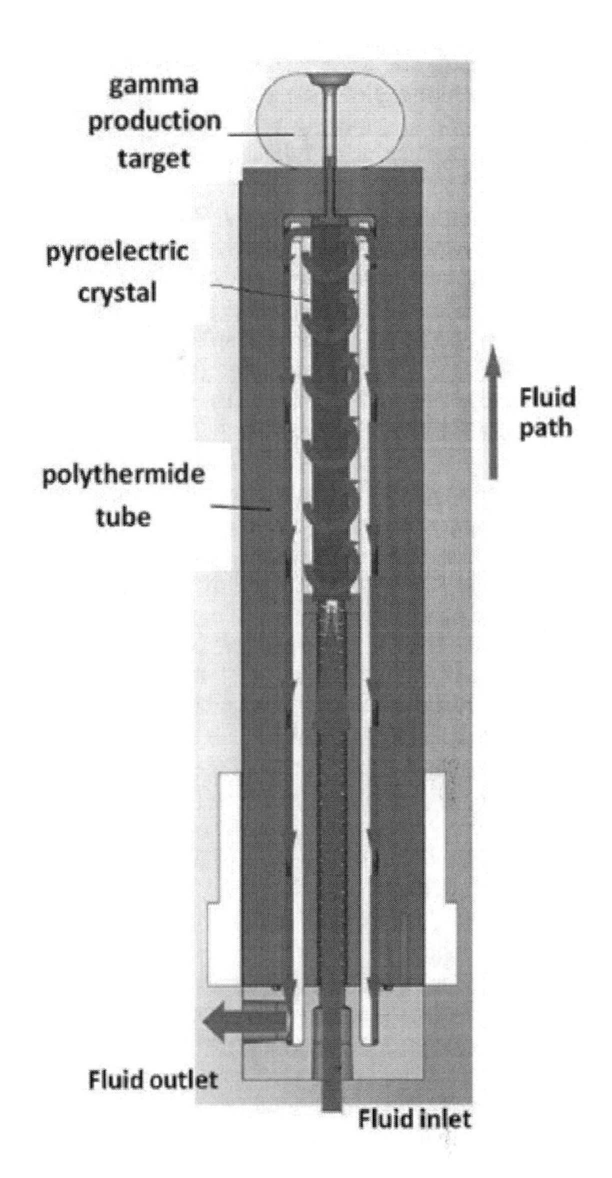

FIGURE 4. Cross-sectional schematic of the pyroelectric crystal acceleration system.

The pyroelectric crystal acceleration system is comprised of a 10 cm long x 1 cm diameter $LiTaO_3$ crystal housed in a polythermide tube as shown in FIGURE 4. The use of a long crystal reduces the electric field gradient of the accelerator system, providing better voltage holding and stability of operation. Using Equations 1 and 2 above, the theoretical surface charge and voltage for a $LiTaO_3$ crystal of these dimensions are ~15 nC and 48.8 kV, respectively, for a 1°C temperature change in an isolated environment. The theoretical gamma source strength with the D-Be nuclear reaction is on the order of 0.1 μCi, assuming operation at -300kV and a deuterium ion current of 20nA. Heating and cooling of the crystal is provided by continuous flow of a dielectric fluid (Fluorinert^{TM} FC-70) around the entire crystal. One end of the crystal polar face is grounded via a metal spring structure connected to the vacuum chamber, while the other end is electrically connected to a sealing screw which protrudes out of the assembly for attachment of the beryllium target. The entire accelerator system is flanged inside the 10" vacuum chamber through an Ultra-Torr® adapter.

Electrostatics modeling of the pyroelectric accelerator system and target based on Equation 1 and the specific geometry predicts that -300 kV can be achieved with a temperature change of approximately 150°C. During operation, a heated reservoir of dielectric fluid is pumped into the polythermide tube, with a volumetric flow rate of approximately 1 L/min, heating the crystal and generating sufficient voltage for the D-Be nuclear reaction. The system temperature is monitored with thermocouples placed in the fluid reservoir and at the grounded end of the pyroelectric crystal. A LabVIEW™ program is used to control the flow rate and reservoir temperature giving the desired transient fluid temperature profile which depends on the bath temperature, thermal capacitance of the fluid hardware and crystal assembly, and flow rate of the fluid. The temperature profile of the crystal can be modeled as a one-dimensional transient conduction problem with convective and insulated boundary conditions subject to a step increase in fluid temperature [10]. The analytical solution for the temperature at the center of the crystal (FIGURE 5) shows that it takes ~7 seconds to approach within 5% of the fluid temperature. A separate fluid loop is connected to the pyroelectric accelerator to provide cold dielectric fluid into the system and return the crystal to ambient temperature. The dielectric fluid heating scheme provides better temperature uniformity throughout the crystal compared to only heating it from one end.

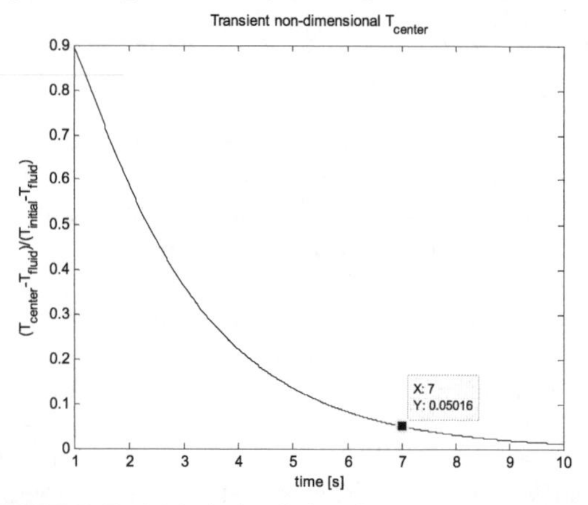

FIGURE 5. Analytical solution for the temperature at the center of the crystal as a function of time. Data prior to one second is discarded due to large errors associated with very low Fourier numbers.

A 0.010" thick x 1" diameter beryllium foil is used for the gamma production target. The foil is attached to a 1.18" diameter x 0.5" long hollow aluminum cylinder fastened to the sealing screw protruding from the pyroelectric crystal. A pair of magnets are attached to the inside of the cylinder to suppress back streaming electrons that are produced when incoming ions strike the target.

EXPERIMENTAL RESULTS

The ion source was initially operated with argon plasma to measure the voltage produced by the pyroelectric crystal and to calibrate the detectors. In this case, Ar^+ ions were accelerated onto a stainless steel target and produced back streaming electrons at energy equal to the pyroelectric crystal voltage. The electron energy was measured by a Si surface barrier detector mounted inside the chamber with line of sight on the incident face of the target. The surface barrier detector was initially calibrated using a metal target biased to -70 kV in place of the pyroelectric accelerator and beryllium target assembly. The resulting electron energy spectra from the Ar^+ ion bombardment of the metal target showed the main peak corresponding to the bias voltage and subsequent peaks caused by pileup in the surface barrier detector. The pileup can be reduced by using a very small extraction aperture to reduce the electron intensity. A 5" x 5" NaI detector was also externally mounted near the chamber viewport to measure the X-ray endpoint energy of the back streaming electrons as they strike the chamber wall. Unfortunately, no X-rays were measured by the NaI detector due to the very low signal-to-noise ratio.

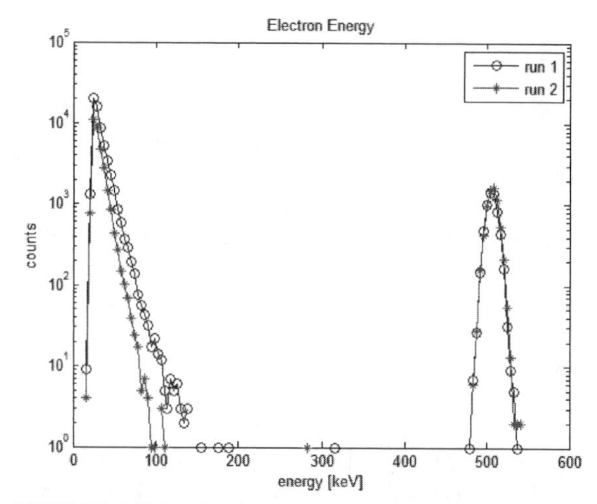

FIGURE 6. Si barrier detector measurement of the electron energy achieved for the two runs. A pulser representing the 508 keV energy peak was used as the reference energy. Run 1: crystal was heated from 25°C to 80°C. Run 2: crystal was heated from 25°C to 60°C.

In the current configuration, the D-Be gammas are only produced during the heating phase in which the beryllium target becomes negatively charged. The crystal was heated rapidly from approximately 25° C to 80°C and 60°C by pumping the hot dielectric fluid into the pyroelectric system. The fluid temperature around the crystal reached the high steady state temperature within 2-3 seconds, which is well within the time frame of the analytical transient temperature solution. Backstreaming electron energy measurements were obtained with the Si surface barrier detector during the heating phase. The spectrum (FIGURE 6) shows a maximum electron energy of ~95 keV and ~130 keV was achieved for the two runs. Additional experiments to quantify the acceleration voltage, characterize the gamma spectrum, and improve device performance are in progress.

CONCLUSION

A prototype electronic gamma source has been built and preliminary operational tests performed. The source utilizes a pyroelectric crystal to accelerate deuterium ions onto a beryllium reaction target to produce gamma-rays in the megavolt energy range. Thermal management of the pyroelectric accelerator is achieved by convective flow of a dielectric fluid around the entire crystal providing better temperature uniformity in the crystal, suppression of voltage breakdowns between the crystal polar faces, and faster thermal cycling rates especially for long crystal geometries. Operational tests indicate that improved performance could be attained by synchronizing the RF power with the ion beam extraction pulse and by improving the target geometry. For example, a torus-like geometry could give lower potential gradients in the vicinity of the target and reduce the secondary electron emission from the target. While >100 keV voltages have already been demonstrated with the current system, achieving an operational energy range of 300 – 400 keV is more challenging. Options being explored are operating at higher temperatures (near the Curie temperature of the crystal) to increase the magnitude of the pyroeletric coefficient, using different crystal geometries which affects the generated spontaneous surface charge, and testing other types of pyroelectric crystal materials.

ACKNOWLEDGMENTS

The author would like to thank Professor Ralph Greif and Samuel S. Mao for useful discussions. This work was funded by DOE/NA-22 Office of Nonproliferation Research and Development. Sandia National Laboratories is a multi-program laboratory operated by Sandia Corporation, a wholly owned subsidiary of Lockheed Martin Company, for the U.S. Department of Energy's National Nuclear Security Administration under contract DE-AC04-94AL85000.

REFERENCES

1. American Cancer Society (http://www.cancer.org/Cancer/CancerBasics/lifetime-probability-of-developing-or-dying-from-cancer).
2. http://www.gammaknife.com.
3. Findings of the President of the Conference, Proceedings of the International Conference on Security of Radioactive Sources, Vienna, Austria, IAEA, (2003).
4. F.E. Cecil, R. F. Fahlsing and R. A. Nelson, Nucl. Phys. 376, 379-388 (1982).
5. D. Gillich, A. Kovanen, B. Herman, T. Fullem, Y. Danon, Nuc. Instr. Methods Phys. Res. A 602 pp. 306-310 (2009)
6. V. Tang, G. Meyer, S. Falabella, G. Guethlein, S. Sampayan, P. Kerr, B. Rusnak, J.D. Morse, J. Appl. Phys. 105, 026103 (2009)
7. A.M. Glass Phys. Rev. Vol. 172, Issue 2 pp. 564-571 (1968).
8. Y. Wu, J.P. Hurley, Q. Ji, J. Kwan, K.N. Leung, AIP Conf. Proc. Vol. 1099, Issue 1 pp. 614-618 (2009).
9. X. Jiang, Q. Ji, A. Chang, K.N. Leung, Rev. of Sci. Instr. Vol. 74, Issue 4 pp. 2288-2293(2003).
10. F.P. Incropera, D.P. DeWitt, "Transient Conduction," in *Fundamentals of Heat and Mass Transfer*, New York, John Wiley and Sons, 2002, pp. 254-260 .

A Novel Linear Accelerator For Image Guided Radiation Therapy

Xiaodong Ding and Salime Boucher

RadiaBeam Technologies, 1717 Stewart St., Santa Monica, CA 90404, United States

Abstract. RadiaBeam is developing a novel linear accelerator which produces both kilovoltage (~100 keV) X-rays for imaging, and megavoltage (6 to 20MeV) X-rays for therapy. We call this system the DEXITron: Dual Energy X-ray source for Imaging and Therapy. The Dexitron is enabled by an innovation in the electromagnetic design of the linac, which allows the output energy to be rapidly switched from high energy to low energy. In brief, the method involves switching the phase of the radiofrequency (RF) power by 180 degrees at some point in the linac such that, after that point, the linac decelerates the beam, rather than accelerating it. The Dexitron will have comparable cost to other linacs, and avoids the problems associated with current IGRT equipment.

Keywords: medical linac, IGRT, accelerator, energy switching, imaging, radiation therapy
PACS: 41.75.Ht ; 29.20.Ej; 87.56.bd; 29.27.-a

INTRODUCTION

Conventional linear accelerators used for radiation therapy only provide MV treatment beams. Tumor images are taken prior to treatment and marks are placed on the skin to align the treatment field. Then, once the patient is in place in the radiation treatment room, a “portal image” is taken with the MV treatment beam. The portal image is used for alignment of the patient with respect to the treatment beam, however it suffers from low contrast and poor signal-to-noise ratio. In addition, the tumor position may move during patient transportation from the imaging room to the treatment room, because of breathing, organ movement, or amount of water in the bladder. To account for this, larger planning target volume (PTV) margins are used to compensate for potential errors in localization, which results in unnecessary doses to healthy tissue. Accuracy of the localization of the PTV becomes even more important with advanced radiotherapy techniques, such as conformal radiotherapy and Intensity Modulated Radiation Therapy (IMRT), where high-dose radiation volumes are designed to drop steeply outside the tumor.

Image Guided Radiation Therapy (IGRT) has recently become one of the most advanced tools for cancer treatment, as it allows precise dose conformation to the tumor volume, and improved sparing of healthy tissue. IGRT combines imaging and radiation therapy equipment, to provide images of the patient’s organs in the treatment position, at the time of treatment. This technique also allows radiation delivery to be gated to the patient’s respiration, to further improve the accuracy of PTV localization [1]. New approaches to IGRT are exploring the possibility of real-time tracking of the tumor position and shape by adjusting the multileaf collimator or linac position based on pre-treatment and intra-treatment images.

There are currently two primary ways to take images during radiation treatment: 1) add a kilovoltage x-ray source and detector on the gantry along with the megavolt treatment beam system; or 2) use the megavoltage treatment beam to take images. However both options have disadvantages. Usually the kilovoltage x-ray source and detectors must be put on retractable arms, thus the x-ray image is not taken from the same angle as the treatment beam. In order to properly determine tumor position, the gantry must be rotated around the patient, which adds time to the treatment. The tumor position could also move slightly during gantry rotation, which would cause inaccuracy during treatment. Furthermore, adding an additional kV X-ray source on the gantry makes the system heavier and complicates gantry rotation and patient loading/unloading.

On the other hand, megavoltage treatment beams have very low contrast for the tissues of interest in cancer imaging. High quantum efficiency detectors and sophisticated image processing software must be used to make the image possible, but still the image

Application of Accelerators in Research and Industry
AIP Conf. Proc. 1336, 418-422 (2011); doi: 10.1063/1.3586133

quality is inferior. Also, the imaging process is more complicated and the detectors are more expensive.[2]

In this proposed project, Radiabeam will develop the Dexitron, a new x-ray source which will provide kV and MV X-ray beams from a single source. The Dexitron is able to switch between high-energy treatment mode and low-energy imaging mode in less than 100 ms, using a novel phase switching system incorporated into the electromagnetic design of the linac.

TECHNICAL APPROACH

The goal of the first step of the project will be to demonstrate the feasibility of the Dexitron. Specifically, the design of the linac will be generated and verified with computer simulations, and a proof-of-principle test unit will be fabricated and tested. RadiaBeam follows an applications-oriented design approach, in which we work backwards from the end-use of the accelerator to develop the design of the various sub-systems and components. First, we must choose the parameters required by the end user; initial parameters have been used in this proposal, however they will be optimized in consultation with our consultants. Next, the X-ray conversion target (and necessary hardening filters) must be designed with Monte-Carlo simulations to determine the required output beam current of the linac. Then the linac is designed to deliver the desired output energy, current and spot size to the target. The output of the linac design will be dimensions of the linac itself, and a parts list of required components. Finally, since the key innovation of our approach – the phase switching cavity – has never been built, we will fabricate a test unit and verify its performance.

Task 1: Optimize Parameters

The first task in the project will be to optimize the parameters of the low-energy imaging mode and the high-energy treatment mode. This will start with the selection of the energy. The treatment beam energy has been referred to as 6 MeV and 18 MeV, as these are commonly used energies for radiation therapy. Other energies between 6 MeV and 18 MeV (e.g. 15 MeV) would be easily achieved. The selection of energies for treatment beam depends on the intended application of the Dexitron. Once the treatment energies have been optimized, the dose rate must be chosen to match the output of other IGRT systems (both treatment and imaging beams). Proposed beam parameters of the first prototype Dexitron are listed in table 1.

TABLE 1. Proposed beam parameters of the first prototype Dexitron

	Treatment mode		Imaging mode
Electron Beam energy	6MeV	18MeV	150 keV
Average beam current	100 μA	10 μA	5 mA
Average beam Power	600 W	180 W	750 W
Dose rate at 1 meter	~ 10 Gy/min	~ 10 Gy/min	~ 20 mGy/min

The imaging energy must be chosen to provide good image quality while at the same time producing sufficient dose rate and being feasible from the accelerator engineering standpoint. Low energy (i.e. 150 keV) will provide optimal image quality, but lower X-ray dose rate for the same beam current, than intermediate energy X-rays (e.g. 500 keV to 1 MeV). In addition, the stability of the switching mechanism should be evaluated for different energies, which will be done in the following tasks. Finally we will also investigate the possibility of a varying the imaging energy (e.g. between 100 – 150 keV), which clearly would have clinical advantages. This may be possible by using the beam loading effect. [3]

Switching between 6 MeV and 18MeV treatment beams is achieved by utilizing the beam loading effect. The energy gain V in a linac is determined by the guide design (shunt impedance), the input RF power level, and the electron gun current: $V = F\sqrt{P_0 r_s l} - G r_s l i$, where P0 is the input RF power, $r_s \equiv E^2/2\tau P_0$ is the guide shunt impedance, i is the beam current, l the effective length of the accelerating cavity, E is the peak field gradient on-axis, and F and G are constants dependent on the design (geometry) of the accelerator. With a specially designed accelerator, one can optimize the regime where multiple energy outputs are possible.

It should be noted that the final RF power and current parameters must be refined during linac design. Detailed study of all beam parameters, including spot size and energy spectrum will be performed in task 3. Switching between MV treatment beam and kV imaging beam for the same accelerator will be achieved by adding a special switching cavity, which will also be designed in task 3

Task 2: Determine X-Ray Conversion Target Geometry

Generally, the X-ray converter for low-energy X-rays (i.e. the anode in an X-ray tube) is different than the converter for high-energy X-rays. In both cases, the beam is directed at a high-Z (e.g. Tungsten or

Tantalum) target. For electron beams below ~500 keV, more X-ray photons are backscattered from the target than transmitted; thus the target is placed at an angle and the X-ray beam emerges approximately perpendicular to the electron beam. For higher energy electrons, the converted X-rays are more forward directed, and so a transmission target is used.

For the Dexitron, we must develop a solution which allows both low-energy and high-energy X-rays to be produced from the same location. The high-energy electron beam prefers a transmission target, while low-energy kV beam prefers a backscatter target. However initial calculations shows that it is possible that a transmission target can produce sufficient X-ray flux with the low energy electron beam for imaging with an optimized target thickness. This will be verified through Monte-Carlo simulations during this portion of the task. The higher energy beams will require different target thicknesses and backing materials than the lower energy target. Each target will be optimized with Monte Carlo simulations.

In addition to the individual target optimization, a method must be designed to switch targets as the linac energy is switched. This can be achieved by placing the three targets (for 18 MeV, 6 MeV and 150 keV) on a carousel that rotates to switch targets. The target will switch when the energy switches. The conceptual design of the carousel will also be generated in this task.

Task 3: Design Accelerator Guide And Switching Cavity

The Dexitron will be a specially designed side-coupled standing wave RF linear accelerator guide (linac guide). The $\pi/2$ mode, side-coupled standing wave RF accelerating structure is the most popular accelerator structure in radiation therapy industry, due to its superior stability and efficiency [4]. In the side-coupled structure there are separate accelerating cavities and coupling cavities. The electron beam is injected from the electron gun to the beam channel at one end of the structure and is accelerated by the RF electromagnetic waves in the accelerating cavities.

Figure 1 shows the field distribution of a typical linac operated in $\pi/2$ mode for generating a MV treatment beam. One can see that the phase of the RF waves changes $\pi/2$ radians between the coupling cavity and its adjacent accelerating cavities. For example, the arrow in cavity A1 points to the right, meaning it has a phase of 0; The arrow in cavity C1, right next to A1, points up, meaning it has a phase of $\pi/2$ radians (or 90°); A2, the other cavity adjacent to C1, has a phase of π radians (or 180°); C2 has a phase of $3\pi/2$ radians (270°), A3 has a phase of 2π radians (360°), and so on. From this arrangement, one can see that the phase difference between adjacent accelerating cavities (such as A1 to A2, A2 to A3, etc...) is π radians.

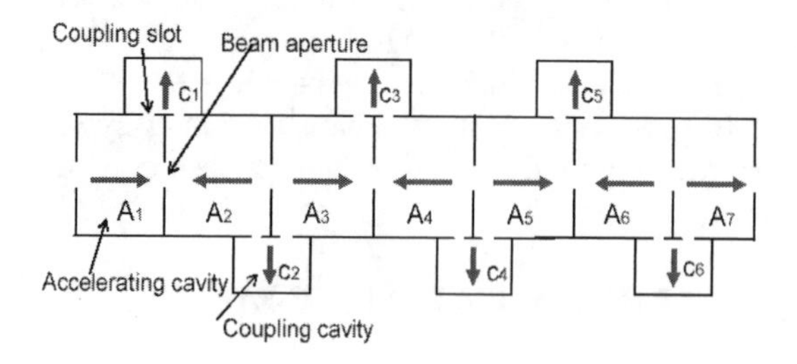

FIGURE 1. RF Electromagnetic field phase distribution in a regular MeV treatment linac.

Compared to Figure 1, Figure 2 shows the Dexitron in treatment mode, with an additional special side coupling cavity called the "π-phase side cavity" or Sπ. It is added between the 4th and 5th main cavity. This is the key innovation of the Dexitron that makes the phase switching possible. In treatment mode, Sπ is shut off, and the accelerator acts like a normal treatment beam accelerator. The output beam can be 6MeV to 18MeV.

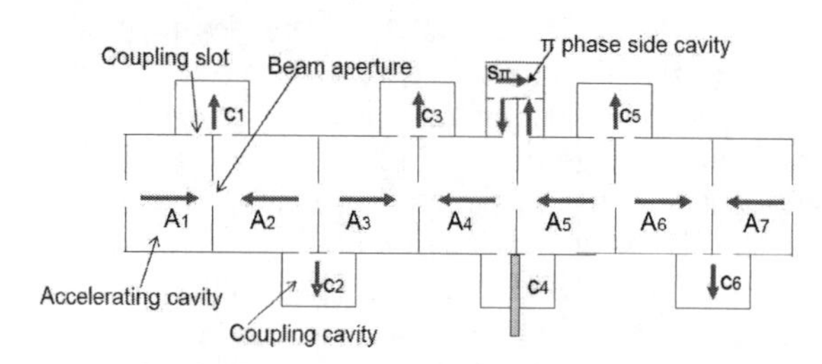

FIGURE 2. RF field phase distribution for MV treatment mode of the Dexitron. (Sπ is off)

In imaging mode, Sπ is turned on and the normal coupling cavity C4 is turned off, as shown in Figure 3. The RF electromagnetic waves flow through Sπ instead of C4 from A4 to A5. Sπ has a special internal structure that changes the RF wave's phase by π radians from its adjacent accelerating cavities, rather than $\pi/2$ radians for a regular side cavity. As a result, A5 will have a phase difference of 2π radians from A4 in imaging mode. This phase change makes the electron beam 180° "out of phase" with the RF electromagnetic wave in cavity A5, A6 and A7. Thus the beam is decelerated in the second section of the linac, down to an energy of a few hundred keV.

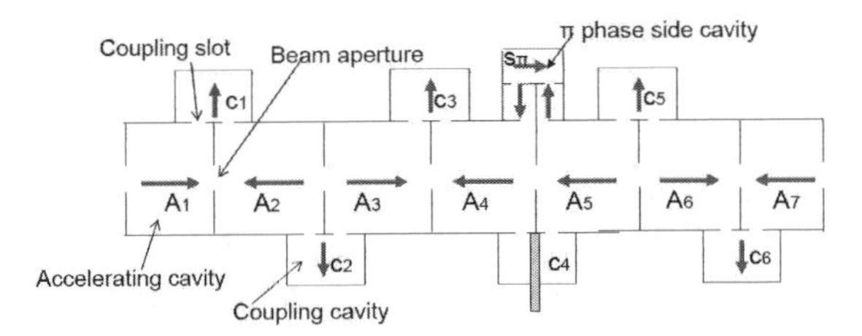

FIGURE 3. RF Electromagnetic field phase distribution for kV imaging mode of the Dexitron. (Sπ is on)

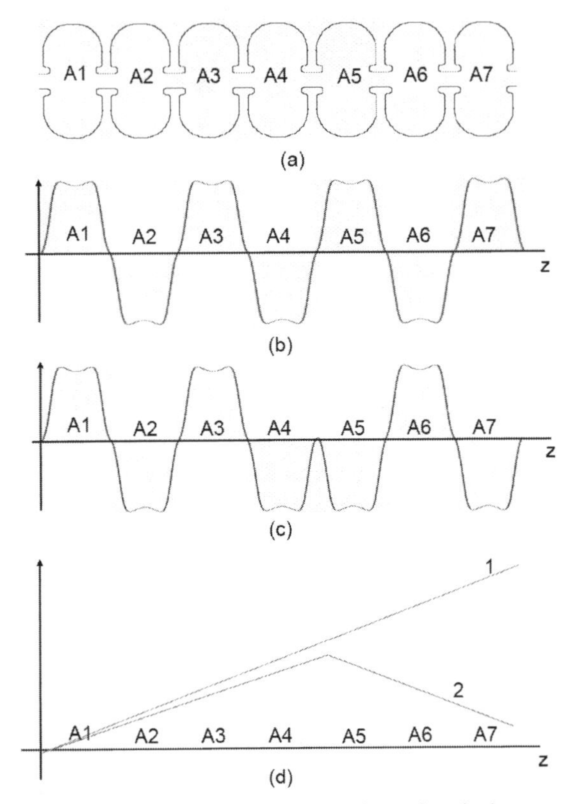

FIGURE 4. Phase switching technique for dual energy linac:

This phase switching technique is also shown in Figure 4. The key innovation of this phase switching technology is the design of the "π-phase side cavity" or Sπ cavity. The basic structure of the Sπ cavity is made up of three coupled side cavities. Each cavity has the same resonant frequency as the operating frequency. The phase shift between each cavity is still π/2. The innovation of this scheme is that the RF phase change remains π/2 between adjacent cavities for both imaging and treatment operation, which means the accelerator can have very stable operation in both modes.

a) Cut-off section of a seven-cavity linac guide;

b) On-axis electric field distribution for treatment mode;

c) On-axis electric field distribution for imaging mode, showing the 180 degree phase switch on the 5th cavity;

d) Electron energy gain along the beam line axis. 1) Red line 1 shows in treatment mode, energy gains from A1 to A7 to MeV. 2) Blue line 2 shows imaging mode, energy gains from A1 to A4 to MeV, then down from A5 to A7 to keV.

The above description uses the seven-cavity system as an example. In order to reach 18 MeV, the guide will be longer and the number of cells will be about 20 to 25 for an S-band RF structure. However the Sπ cavity will be still located close to the middle of the structure.

The detailed design of the accelerating cavities, regular coupling cavities and the special coupling Sπ cavity will be done by using electromagnetic field simulation codes. The design of accelerating cavities and regular coupling cavities with these codes is mature and well documented. RadiaBeam has extensive experience in this area. We will focus on the design of the special coupling cavity Sπ and study its influence on the two adjacent main line cavities. The starting point will be the 2D code SUPERFISH [5], which is very helpful in finding the dimension of a cavity for a given operation frequency. The 3-D code HFSS [6] will be used to obtain a more complete picture of the electromagnetic field distributions in the accelerating cavities and coupling cavities.

The dimensions of the three cavities inside Sπ are determined by the resonant frequency at the π/2 mode. Renderings of the Dexitron showing the special Sπ are shown in Figure 5. We will study the internal coupling between the three cavities inside Sπ, as well as the coupling between Sπ and the two adjacent main cavities. All these parameters can be studied with computer simulation codes.

We should point out that several other companies have developed different schemes to produce dual energy X-rays [7], however the energy switching is not capable of large variations in energy. The output energy in the low energy mode is usually 4 to 6 MeV, which is not ideal for medical imaging. The proposed design of Sπ cavity is innovative.

NEXT STEP

Following the detailed design efforts described above, we will build a test piece with two main accelerating cells, one normal side cell and the special π phase switching cell to test the phase switching technique. The geometry of the test piece will be based on the HFSS 3D simulation described above.

We envision that the next goal of the project will be to build and test a prototype Dexitron, and verify its performance at both low and high energy. As such, we must develop a list of parts and equipment required for the prototype so that an accurate budget can be proposed. Preliminary drawings of the Dexitron linac cells will be made to use for estimation by machinists and braze shops. Quotes will be requested from vendors for major components, such as the klystron and modulator.

SUMMARY

In conclusion, there are already a number of exciting applications of the dual-energy, kV/MV linac that we propose here. While we have mentioned several examples of potential improvements to current IGRT systems, it is clear that the ability to quickly switch between a large range of energies, from ~100 kV to ~20 MV, will have a wide-ranging impact on the field of radiation therapy.

RadiaBeam's design, by adding a special 180° phase switching cavity in the linac structure, is capable of a producing a large energy difference (~100 kV to ~20 MV) with stable operation. It can achieve "beams-eye-view" for IGRT treatment. The device may reduces many problems in patient positioning and tumor delineation during radiotherapy treatment.

REFERENCES

1. C.C. Ling et al. From IMRT to IGRT: frontierland or neverland? Radiotherapy and Oncology (2006).
2. E J Orton et al. Megavoltage image contrast with low-atomic number target materials and amorphous silicon electronic portal imagers. 2009 Phys. Med. Biol. 54 1275-1289.
3. R.B. Miller, Electronic Irradiation of Foods, pp. 164-165
4. R.H. Miller, Comparison of standing-wave and traveling-wave structures. SLAC PUB-3935 (1986) .
5. J.H. Billen and L.M. Young, Poisson Superfish LA-UR-96-1834
6. Ansoft HFSSTM (http://www.ansoft.com/hfss/).
7. US patents 4286192, 4382208, 4629938, 4746839, 6366021

Accelerator Production of ^{225}Ac For Alpha-Immunotherapy

J.W. Weidner, F.M. Nortier, H.T. Bach, K.D. John, A. Couture, J.L. Ullmann, M.E. Fassbender, G.S. Goff, W. Taylor, F. Valdez, L.E. Wolfsberg, M. Cisneros, D. Dry, M. Gallegos, R. Gritzo, L.J. Bitteker, S. Wender, R.S. Baty

Los Alamos National Laboratory, P.O. Box 1663, Los Alamos, NM 87545

Abstract. ^{225}Ac has tremendous potential for the treatment of metastatic cancer due to the four alpha-particles emitted during its decay to stable ^{209}Bi. Additionally, it is one of the few alpha-emitters being considered for clinical trials. The anticipated ^{225}Ac demand for these trials is expected to far exceed the annual worldwide supply of approximately 1,000 mCi/yr. Consequently, the DOE Office of Science has funded investigations into accelerator-based production of ^{225}Ac. Existing ^{232}Th(p,x)^{225}Ac cross section data indicate that up to 480 mCi/day of ^{225}Ac could be created by bombarding a thick target of natural thorium with 100 MeV protons at the Los Alamos Isotope Production Facility. To verify these predictions, experiments are underway at the Los Alamos Neutron Science Center to measure the ^{232}Th(p,x)^{225}Ac production cross sections for protons in the energy range 40-200 MeV, and at 800 MeV. For 800 MeV protons, preliminary results indicate that the ^{225}Ac production cross section is 12.4 ± 0.6 mb and the ^{225}Ra production cross section is 3.2 ± 0.2 mb. Moreover, preliminary results suggest that the ^{227}Ac production cross section is 16 ± 1 mb. Experiments to measure these same cross sections at proton energies below 200 MeV are planned for the last half of calendar year 2010.

Keywords: ^{225}Ac, cross section, alpha-emitter, cancer, ^{232}Th
PACS: 87.53.Jw

INTRODUCTION

In 2009, the Nuclear Science Advisory Committee (NSAC) published a report on the National Isotopes Production and Application Program. In it, the NSAC Isotopes subcommittee (NSACI) made six prioritized recommendations for the program. The most important recommendation was to invest in new production approaches of alpha emmiters with highest priority going towards ^{225}Ac[1]. The NSACI made research into new production methods of ^{225}Ac their highest-priority recommendation for two reasons. First, during its decay to the stable isotope ^{209}Bi, ^{225}Ac emits four alpha particles (see Figure 1), which are far more effective in treating cancer than other forms of radiation. Additionally, the alpha-emitting daughters of ^{225}Ac are all relatively short-lived – an important characteristic of any therapeutic radioisotope.

Second, the world's current annual supply of ^{225}Ac is approximately 1000 mCi/yr and comes from milking ^{229}Th cows, which in turn come from ^{233}U sources. Oak Ridge National Laboratory currently supplies approximately 600 mCi/yr[2] of ^{225}Ac.

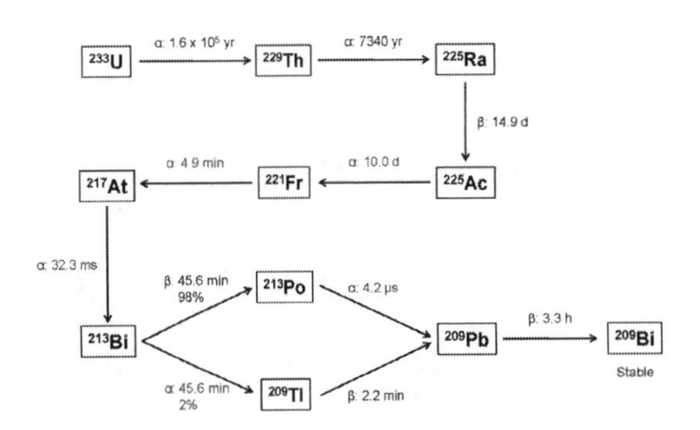

FIGURE 1. The ^{233}U decay chain, which produces ^{225}Ac and its short-lived, alpha-emitting daughters.

Clinical trials are currently underway for ^{225}Ac and the therapeutic demand for this isotope is expected to exceed the world's supply only a few years into these trials. Moreover, the DOE has a 2012 deadline for down-blending all of its ^{233}U stocks, which would eliminate the world's largest source of ^{225}Ac. Therefore, the NSACI thought it imperative that alternate methods of ^{225}Ac production be developed immediately.

Application of Accelerators in Research and Industry
AIP Conf. Proc. 1336, 423-425 (2011); doi: 10.1063/1.3586134

EXPERIMENTAL PROCEDURE

An experiment to measure the ^{225}Ac production cross section via proton interactions with a natural thorium target was conducted in December 2009 at the Los Alamos Neutron Science Center (LANSCE). The LANSCE facility includes an 800 MeV proton accelerator, which was utilized for this experiment. Three natural thorium targets approximately 2.5 cm square with a thickness of 50 mg/cm^2 were mounted on separate acrylic sheets and placed on an open-air table in the LANSCE Weapons Neutron Research Facility Blue Room as shown in Figure 2. A similar-sized aluminum foil was placed downstream of the thorium targets for later calculation of the proton current impinging upon the thorium targets. All foils were then irradiated with approximately 80 nA of 800 MeV proton current for one hour.

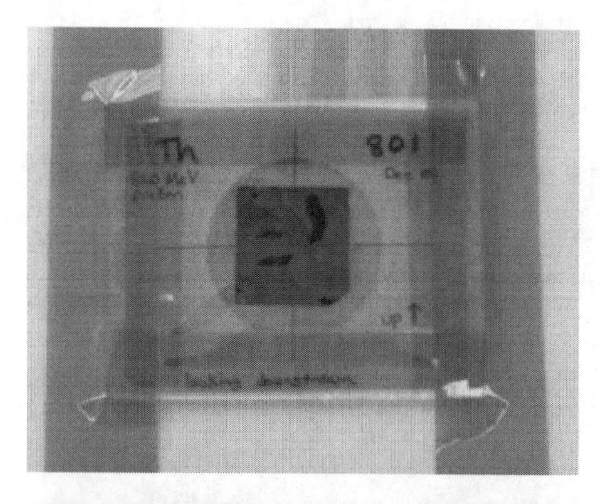

FIGURE 2. Mounting and set up of natural thorium foil targets for 800 MeV proton irradiation.

Following proton irradiation, the targets were transported to the Chemistry Division Actinide Research Facility (ARF). There, two thorium foils were prepared for gamma-counting while the remaining thorium foil was dissolved in a solution of 8M nitric acid with a few drops of HF added to speed the dissolution process. This solution was then passed through a resin column to isolate the isotopes of radium and actinium in one 63.97 mL liquid sample. From this, 50 μL were deposited on a platinum disc in small stipples and evaporated in preparation for alpha-counting.

The two remaining thorium foils were placed on shielded HPGe detectors for gamma counting. One foil was counted on a beryllium-windowed LO-AX detector while the other was counted on a standard coaxial detector. Initial count times were on the order of 10 minutes and the initial count frequency was several times per day. Over the course of the following weeks and months, the count time was gradually increased to a maximum of ten hours and the count frequency was gradually decreased to bi-weekly.

RESULTS

Not unexpectedly, the 800 MeV protons created an extraordinary variety of fission products within the two thorium foils. Some of these intense gamma-emitters interfered with the photon energies used to identify and quantify both ^{225}Ac and ^{225}Ra. This made the determination of their initial activity extremely challenging with the equipment available. Consequently, the gamma signature of ^{221}Fr, the daughter product of ^{225}Ac, was analyzed in order to determine the activity of both ^{225}Ac and ^{225}Ra at end of bombardment (EOB). Fortunately, ^{221}Fr emits a 218 keV gamma that was not interfered with by other radioisotopes. Moreover, its 4.9 minute half-life causes it to be in radioactive equilibrium with its parent within one hour after EOB. Thus, the ^{225}Ac activity at EOB was found to be 2116 ± 90 Bq. This was in good agreement with the alpha-counting results, which found the ^{225}Ac activity at EOB to be 2150 Bq. Additionally, the gamma activity of ^{225}Ac after EOB was being fed by the decay of ^{225}Ra. Using this differential relationship, the activity of ^{225}Ra at EOB was measured to be 370 ± 30 Bq. Lastly, preliminary alpha-counting measurements taken 210 days after EOB for the Ra/Ac sample suggest that the ^{227}Ac initial activity is on the order of 0.002 Bq.

Next, the proton fluence impinging upon the thorium targets was calculated based upon the ^{22}Na and ^{24}Na activity in the aluminum foil. Using cross sections of 15.04 mb and 10.75 mb, respectively, and the thin-target yield equation below,

$$A = \sigma n \varphi \Delta x (1 - e^{-\lambda t}) \quad (1)$$

the proton current was found to be $(4.6 \pm 0.2) \times 10^{11}$ protons/s. Reapplying equation (1), the production cross section for ^{225}Ac is determined to be 12.4 ± 0.6 mb and the ^{225}Ra production cross section is calculated to be 3.2 ± 0.2 mb. Finally, preliminary measurements suggest that the ^{227}Ac production cross section is 16 ± 1 mb.

ACKNOWLEDGMENTS

We wish to thank the Office of Nuclear Physics in the DOE's Office of Science for funding this work.

REFERENCES

1. NSAC Isotopes Subcommittee, "Compelling Research Opportunities Using Isotopes", One of Two NSAC Charges on the National Isotopes Production and Application Program, 2009, pp. 42.
2. J. Norenberg, P. Stapples, R. Atcher, R. Tribble, J. Faught and L. Riedinger, "Report of the Workshop on the Nation's Need for Isotopes: Past and Present", *DOE/SC-0107*, August 5-7, 2008, pp. 22.

Design Features Of Microfluidic Reactor For [18F]FDG Radiopharmaceutical Synthesis

J.H Oh[a], B.N Lee[b], K.R Nam[a], G.A Attla[a], K.C Lee[c], J.S Cjai[a,b]

[a] *Department of Energy Science, Sungkyunkwan University, 300 Cheoncheon dong Suwon Gyeonggi Province 440-746 Korea*
[b] *School of Electrical and Computer Engineering, University, 300 Cheoncheon dong Suwon Gyeonggi Province 440-746 Korea*
[c] *KIRAMS, Seoul, Korea*

Abstract. Microfluidic reactor exhibits advantages for radiopharmaceutical synthesis. Microfluidic chips can reduce the time for radiosynthesis using tiny quantities of chemical compounds. It also has a good heat transfer, performance and provides an integrated system including synthesis, separation, and purification. These advantages make FDG production. So we have designed a microreactor chip which included the whole chemical processing; water evaporation, solvent exchange, radiofluorination and so on. It was designed by using a commercial 3D CAD modeling program CATIA V5, heat transfer performance was analyzed by ANSYS, and CFX was used for analyzing fluid performance. This paper described the design of FDG synthesis system on a microchip, the relevant locations of its parts, both heat and fluid performance efficiency analysis.

Keywords: Microfluidic, PET, FDG synthesis, Radiopharmaceutical
PACS:

INTRODUCTION

PET/CT using the glucose analogue FDG is emerging as the most important diagnostic imaging tool in oncology. More than 2000 PET/CT scanners are operational worldwide and its unique role for diagnosing, staging, restaging and therapeutic monitoring in cancer is undisputed [1]. FDG (2-[^{18}F]-Fluoro-2-deoxy-d-glucose) is the most widely used radiopharmaceutical compound as a functional imaging agent that permits characterization of tumor metabolism and played an important role in the staging of various malignant tumors, [2]. Various techniques have been developed and used to synthesize FDG. Microfluidic chip-based analysis has made great progress recently , due to its outstanding advantages including high analysis speed and low sample and reagent consumption, as well as its great potential to allow automation and integration of the complete analytical system on a single chip[3-7].In comparison to the conventional production method, the microreactor can produce a new batch of FDG as the patient need basis. In this study, we consider the mechanical structure of microfluidic reactor. CATIA V5 which is a strong CAD program was used in the design of this microreactor. During FDG synthesis, there is an elevated heat and pressure, so we focused on analyzing its mechanical structure. Heat transfer was solved by ANSYS steady-state thermal solver. Fluid dynamics was solved by ANSYS CFX. In this study, we proposed a new analytical method for microfluidic chip in respect to the mechanical structure.

CHEMICAL PROCESS

Fluorine-18 is produced by an ^{18}O target in a cyclotron by the ^{18}O (p,n)^{18}F reaction. Fluorine has high hydration energy, so water is not acceptable solvent in this synthesis. Water should be evaporated to make pure F-18. The most useful way to be isolated is to use an anion exchange resin [9] often with a Sep-Pak Light Accell QMA cartridge which is a light QMA (Quaternary ammonium anion exchanger) [9-11].

The F-18 is retained via an ion-exchange reaction and [O-18] water allowed to flow through. The retained F-18 is then eluted with acetonitrile solution of Kryptofix and potassium carbonate. Kryptofix 2.2.2 is cyclic crown ether, which binds to the potassium ion

Application of Accelerators in Research and Industry
AIP Conf. Proc. 1336, 426-429 (2011); doi: 10.1063/1.3586135

preventing the formation of [F-18]KF and thus, potassium acts as the counter ion of F-18 to enhance its reactivity but does not interfere with the synthesis. Since Kryptofix 222® causes apnoea and convulsion, there should be minimum amount of Kryptofix in the final [F-18]FDG products.

After that, the precursor is added to the F-18. The choice of precursor depends on the ease of preparation, ease of producing the final product, consistency, yield, and so on. Precursor molecule in synthesis of [F-18]FDG is 1,3,4,6-O-acetyl-2-O-trifluoromethanesulfonyl-beta-D-mannopyranose (mannose triflate).

Mannose triflate structure is similar to that of FDG, except with a triflate group at the 2-position of carbon and acetyl groups at 1,3,4,6-position of carbons via ester bonds, which can be readily broken at a higher or lower pH. The use of acetyl groups is to protect the hydroxy groups so that fluorination would not occur at these positions.

F-18 ion approaches the mannose triflate at back side of 2-position of carbon, while the triflate group leaves the protected mannose. A small stream of nitrogen is introduced above the solution. After the nucleophilic replacement of the triflate group by [F-18]F^- (Figure 2), the acetyl groups can be easily removed by hydrolysis to form [F-18]FDG. Basic hydrolysis step is more commonly used. Alkaline hydrolysis does not produce any ClDG(2-deoxy-2-chloro-d-glucose) but it is produced during acidic hydrolysis with HCl.[11-12]

One of the improved base hydrolysis is to absorb the 1,3,4,6-acetyl protected F-18 labeled 2-deoxyglucose on

A TC-18 reverse phase cartridge. All other impurities can be removed by rinsing heavily with water. Sodium hydroxide is added to the cartridge so that the base hydrolysis occurs on the cartridge surface. The final [F-18]FDG product can be eluted with water while the unhydrolysed or partially hydrolysed 1,3,4,6-acetyl protected F-18 labeled 2-deoxyglucose remains on the cartridge .[13]

DESIGN OF MICRO REACTOR

Microfludic Reactor is a square shaped chip has a size of 3cm width, 2cm length and 2mm thickness. The chip is made up of glass but it can be of PDMS. Size of Fluid path is 100um x 100um. As shown figure 1, 1 is the entry for F-18 produced in the cyclotron. 2

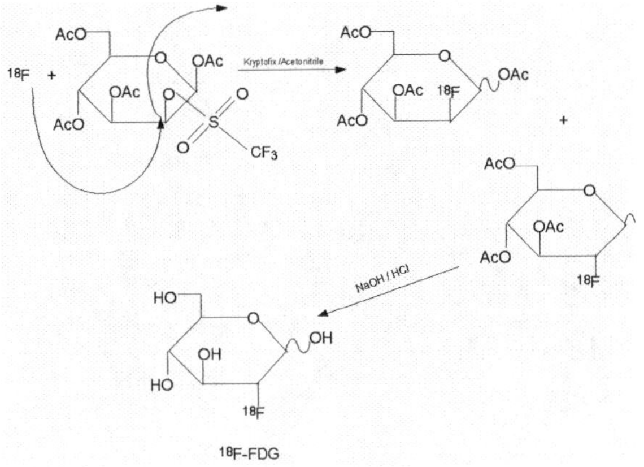

FIGURE 1. Synthesis of [F-18]FDG by nucleophilic substitution.[14]

is outlet of the evaporated water including O-18. To accomplish this step, a light QMA cartridge will be used (part A). This cartridge dimension would be 2mm x 2mm surface, the role of the cartridge is to remove the water. After Kryptofix 2.2.2 in K_2CO_3 enter through inlet 3, pure F-18 and Kryptofix 2.2.2 will be mixed in the cartridge. These processes will be controlled by electronic sensor. A number of electronic valves are installed beside path and cartridges. They are connected with a microwire to an automatic controller for FDG systhesis.

When the mixture is produced at the first step in part A, valves will be opend. At the same time there is another water evaporation process for any tiny quantity of water in part B using CH_3CN which enters through inlet 4. After that, mixture of FDG precursor, mannose triflate, and CH_3CN will enter from inlet 5. This mixture is mixed with [F-18] in Krytofix 2.2.2 and is followed by the most important step; the radiolabeling process. To get a good performance, fluid path was designed to be zigzag like shape composed of 6 channels, allowing more slow fluid velocity and increasing of synthesis time.

Subsequent to the labeling step, NaOH will pass through entry 6 for basic hydrolysis for removing acetyl group in mannose triflate. This reaction takes place on the Tc-18 cartridge. On the Tc-18 cartridge (part D), two processes occur: one is the hydrolysis and the other step is the purification. The reactants used for FDG synthesis will exit through outlet 7 after a reaction accomplished. Finally the pure FDG will pass through outlet 8.

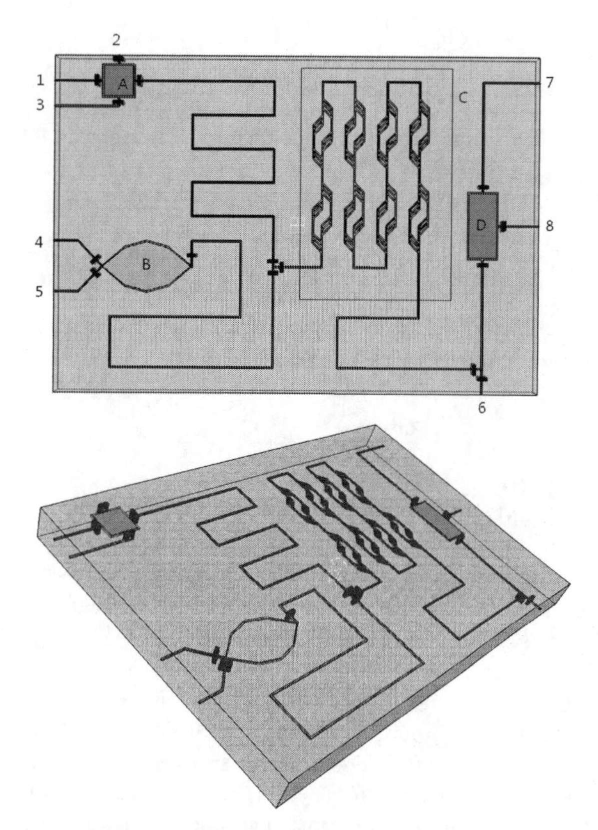

FIGURE 2. Design view of whole Microfludic reactor

MECHANICAL ANALYSIS

Thermal Analysis

The most important part in the FDG synthesis is heat transfer. So we analyzed using ANSYS steady-state thermal solver.

Microreactor is divided into 4 sectors to give temperature respectively at first, Part B and D is analyzed at a room temperature, part A at 120°C for removing tiny amount of water after the evaporation step in A. while Part C is analyzed at 105°C to synthesize FDG. As shown on figure 3, the microchip has good heat transfer in order to maintain the temperature facilitating FDG synthesis.

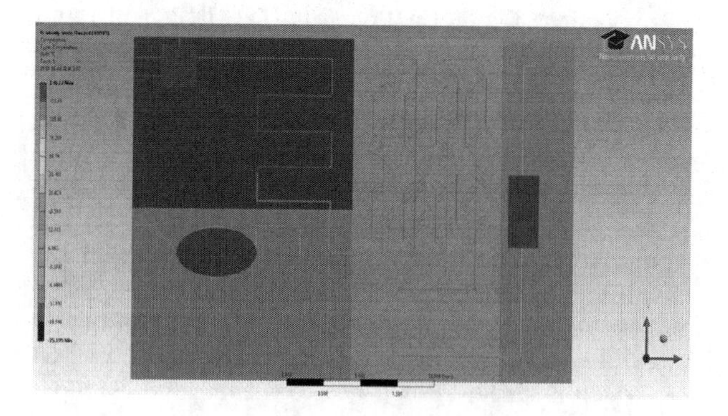

FIGURE 3. Thermal analysis for Microreactor with fluid path. Temperature is given 25-120 with heat plate.

Fluid Analysis

We analyzed fluid dynamics using CFX module in Ansys. So we can make a result streamline, vector value of microreactor. In this analysis, reference flow rate is 250 μL per min. So we calculated input velocity which is 4.17m/s in Inlet 1. Total volume of fluid path is near 2mL. Entire process timing is estimated to be about 10 minutes. Water flow is shown in figure 4. As a result, we can see streamline of water in the fluid path. In the cartridges and the synthesis space, the velocity is minimum which increase synthesis efficiency inside the capillary. The numerical value is also proper to glass structure.

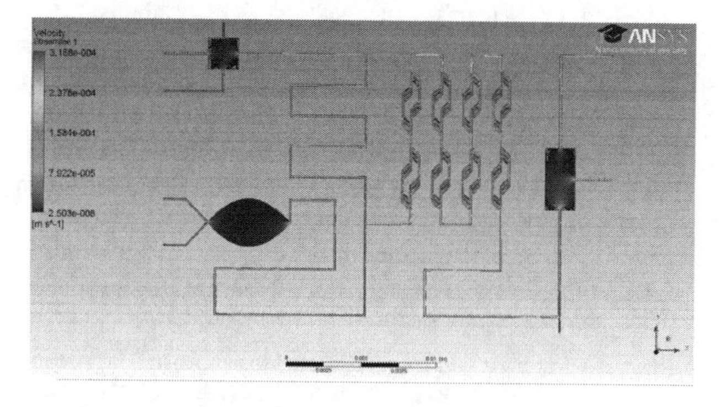

FIGURE 4. Streamline view of Fluid dynamics in Microreactor. Input velocity is 4.17mm/s

CONCLUSION

We have developed a multistep microreactor that was designed for the sequential chemical synthesis of [18F]FDG in one chip of 3 * 2 * 0.2 cm including the whole FDG synthesis process from water evaporation to purification along a fluid path of 100μm * 100μm. There are 5 inlets, 3 outlets and 2 cartridges in this reactor. It is a reasonable structure for FDG synthesis in one chip. From a point of view fluid dynamics, velocity is from 0.02503 μm/sec to 0.3158 mm/sec using 250 μL per min flow rate using a syringe pump. Total reaction time is estimated to be about 10 minutes. This reaction time is very important ensuring high efficiency in producing FDG. The next step of this development is fabricating the microchip and carrying out the chemical experimentation with the final optimization to obtain the highest possible FDG yield.

ACKNOWLEDGMENTS

This research was supported by the Converging Research Center Program funded by the Ministry of Education, Science and Technology (2010K001052). Also Department of Energy Science of Sung Kyun Kwan University supported this work.

REFERENCES

1. Giancarlo Pascalia, Grazia Mazzone, Giuseppe Saccomannic, Clementina Manerac and Piero A. Salvadoria, Nuclear Medicine and Biology 37 (2010) 547–555.
2. Koichiro Abea,Shingo Babaa, Koichiro Kanekoa, Takuro Isodaa,Hidetake Yabuuchia, Masayuki Sasakib, Shuji Sakaib, Ichiro Yoshinoc and Hiroshi Hondaa, Clinical Imaging 33 (2009) 90–95.
3. T. Vilkner, D. Janasek, A. Manz, Anal. Chem. 76 (2004) 3373.
4. P.S. Dittrich, K. Tachikawa, A. Manz, Anal. Chem. 78 (2006) 3887.
5. Qun Fang, Xiao-Tong Shia, Wen-Bin Dua, Qiao-Hong Hea, Hong Shena, Zhao-Lun Fanga, TrAC Trends in Analytical Chemistry, Volume 27, Issue 6, June 2008, Pages 521-532.
6. D J Schlyer, M A Bastos, D Alexoff, A P Wolf, Int J Rad Appl Instrum A. 1990;41(6):531-3.
7. Gatley S.J., Shoughnessy W.J. (1982). Int J Appl Radiat Isot 33:1325–1330.
8. Schlyer D.J., Bastos M.A., Alexoff D.,Wolf A.P. (1990) Int J Appl Radiat Isto [A]41:531–533.
9. Hamacher K., Coenen H.H., Stocklin G., J Nucl Med 1986;27(2):235-8.
10. Schlyer D.J., Ann Acad Med Singapore 2004;33(2):146-54.
11. F. Füchtner, J. Steinbach, P. Mäding and B. Johannsen, *Appl. Radiat. Isot.* **47** (1996), pp. 61–66.
12. J. Koziorowski, Applied Radiation and Isotopes 68 (2010) 1740–1742.
13 Saha GB. Synthesis of PET Radiopharmaceuticals. Basics of PET Imaging, Physics, Chemistry and Regulations. NY Springer Publishing , 2005:113.
14 S Yu. Biomed Imaging Intery J 2006: 2(4):e57.

SECTION IX – NUCLEAR-BASED ANALYSIS

Nuclear Tools For Oilfield Logging-While-Drilling Applications

Jani Reijonen

Schlumberger PTC, 20 Wallace Rd., Princeton Junction, NJ 08550

Abstract. Schlumberger is an international oilfield service company with nearly 80,000 employees of 140 nationalities, operating globally in 80 countries. As a market leader in oilfield services, Schlumberger has developed a suite of technologies to assess the downhole environment, including, among others, electromagnetic, seismic, chemical, and nuclear measurements. In the past 10 years there has been a radical shift in the oilfield service industry from traditional wireline measurements to logging-while-drilling (LWD) analysis. For LWD measurements, the analysis is performed and the instruments are operated while the borehole is being drilled. The high temperature, high shock, and extreme vibration environment of LWD imposes stringent requirements for the devices used in these applications. This has a significant impact on the design of the components and subcomponents of a downhole tool. Another significant change in the past few years for nuclear-based oilwell logging tools is the desire to replace the sealed radioisotope sources with active, electronic ones. These active radiation sources provide great benefits compared to the isotopic sources, ranging from handling and safety to nonproliferation and well contamination issues. The challenge is to develop electronic generators that have a high degree of reliability for the entire lifetime of a downhole tool. LWD tool testing and operations are highlighted with particular emphasis on electronic radiation sources and nuclear detectors for the downhole environment.

Keywords: oilfield logging tools, nuclear measurement, radiation generator, logging while drilling, radiation detection
PACS: 29.20.Ba, 29.25.Dz, 29.30.Kv, 29.40.-n, 93.85.Fg, 93.85.Tf

INTRODUCTION

The well-logging application is one of the harshest applications of complex electromagnetic, chemical, and nuclear systems in use in the industry today. Two main methods of well-logging that are used today are wireline logging (WL) and logging while drilling (LWD). Of these two, the most widely utilized logging method is wireline logging, in which the tool is attached to a cable on the surface. Telemetry and tool power in WL is via the suspension cable, facilitating high data transfer speeds and relatively large tool power consumption. High vibration and shock are rarely an issue with WL operation and result primarily from mishandling of the tool or some other non-typical form of tool handling and operation.

LWD operation is based on formation analysis while the borehole is being drilled. This technology has been developed over the past 15 years, and is currently deployed routinely in offshore and other highly deviated well operations. In this method, the tool string is powered and the measurement data is transferred uphole by the flowing drilling mud. This means that the downhole power and the data transfer for the various measurement instruments is very limited. A typical LWD logging operation varies from a few tens of hours to a few hundred hours, thus an extremely high degree of reliability is required.

Schlumberger has multiple nuclear well logging tools for both logging applications. Some tools use gamma detectors to detect the natural gamma background in the formation, while others use radioisotope sources, such as ^{137}Cs and AmBe for gamma ray and neutron generation, respectively, to evaluate the porosity, density, and fluid saturation of the formation. The most advanced nuclear tools utilize electronic pulsed neutron generators (PNGs) to generate neutrons. In the Schlumberger portfolio these include Wireline's reservoir saturation tool (RST) and accelerator porosity sonde (APS) and the EcoScope multifunction tool for LWD operations.

The use of electronic neutron generators in LWD presents a unique challenge for mechanical and electronic engineering. The longer development time of these PNG-based tools is offset by the superior measurements they can provide in comparison to radioisotope-based neutron logging tools. Another compelling reason for the PNG-based systems is the

Application of Accelerators in Research and Industry
AIP Conf. Proc. 1336, 433-436 (2011); doi: 10.1063/1.3586136

realization from various national security and regulatory agencies of the potential dangers of the radioisotopes used in well logging. Specifically, the National Academies of Sciences[1] in the US has identified well logging AmBe sources[1] as a potential proliferation threat. The need to develop well logging tools with sourceless technology has been clearly recognized by some of the oil field service companies, including Schlumberger[2]. Currently there is only one LWD tool, industry-wide, in active operation that utilizes PNGs as a neutron source, namely, Schlumberger's EcoScope multi function LWD tool[3-5]. This presentation highlights the basic nuclear measurements of a oilwell logging tool, and furthermore, introduces the basic nuclear components of the EcoScope LWD tool.

NUCLEAR LOGGING MEASUREMENTS

Nuclear well logging is an integral part of modern well logging operations[6]. For cased hole logging, nuclear methods are particularly well adapted and thus widely used. Typical neutron-based measurements include neutron porosity, thermal neutron capture cross-section, i.e., sigma measurement and spectroscopy for elemental analysis. The purpose of these measurements is to 1) find out the porosity of the formation, 2) identify the existence of water and hydrocarbons (oil or gas), and 3) provide elemental information of the formation. Even though the AmBe-based nuclear tools can measure the neutron porosity, the accurate formation sigma and clean thermal capture measurement can be only done with PNGs.

The neutron porosity or hydrogen index measures directly the pore space. In nonshale formations, the pore space is filled with water, gas, or oil. In this case, the hydrogen is contained only in the pore space. The faster the neutron slows down, the more hydrogen (or larger pore space) the formation has. Thus the hydrogen index measurement is essentially a measurement of the amount of slowed-down neutrons. In order to cancel out the borehole effects, two or more neutron detectors with different spacings are used. The relative count rate of these detectors allows the formation events to be separated from the borehole events.

Sigma, or thermal neutron capture cross-section, measures the gamma rays induced by the thermal neutron capture. The elements of interest are chlorine, iron, and various rare earth elements. The sigma primarily measures the existence or the lack of salt water, so low sigma indicates high hydrocarbon concentration. The measurement is typically done with gamma detectors operating in time domain, i.e. the counts are recorded as a function of time.

On the other hand, the thermal neutron capture gamma ray energy spectra can be used in determining the elemental composition of the formation. The elements of interest are hydrogen, silicon, calcium, chlorine, iron, sulfur, gadolinium, potassium, titanium, and magnesium. Although some of the elements are found only in trace quantities, the high neutron capture cross-section (such as for gadolinium), makes them detectable in the gamma spectrum. Deconvolution of the spectra into its constituent single element spectra is a starting point in the analysis of formation lithology.

Besides the sigma and thermal neutron capture measurements, a unique feature of the PNG-based neutron measurement (and the 14 MeV neutron energy) is its capability to directly measure the inelastic gamma from carbon and oxygen. Although the measurement is relatively slow, the direct measurement can be used in determining the relative amounts of hydrocarbons and water in the formation.

Typically, the nuclear tool is used in combination with other tools, such as formation resistivity, gamma-gamma density, and natural gamma ray. A combination of the measured data from multiple tools combined with depth data provides detailed information of the formation as well as the presence of liquids and/or gases in the formation. An example of data derived from a typical logging tool combo consisting of neutron porosity, gamma density, natural gamma background, and resistivity measurement is shown in Figure 1.

The three zones of interest are shown in Fig. 1, from the left, the natural gamma background log, neutron porosity and gamma density log and resistivity log. Low count rate of natural gamma rays indicates low concentration of natural gamma ray sources, ^{40}K, ^{232}Th and ^{238}U. These elements are typically present in shale, which is regarded as a cap-formation. Low natural gamma ray count rate indicates potential production formation. In the case of neutron porosity, the shale is identified as a high neutron scattering (or neutron slowing-down) region due to its bound water content. In a porous region filled with gas, the scattering is low due to low hydrogen density in gas. This scattering will increase in liquid oil and water registering higher neutron porosity than in gas. Gamma density measures the matrix density of formation, thus in shale the density is high while in porous gas regions the density drops. Finally, the resistivity indicates a high-resistivity region associated with hydrocarbons, differentiating with low-resistivity of salty water in shales or in porous media.

[1] The other widely used source at the well logging applications is ^{137}Cs source. This is qualified as a category 3 source by IAEA, thus regarded as being a smaller risk than category 2 AmBe source.

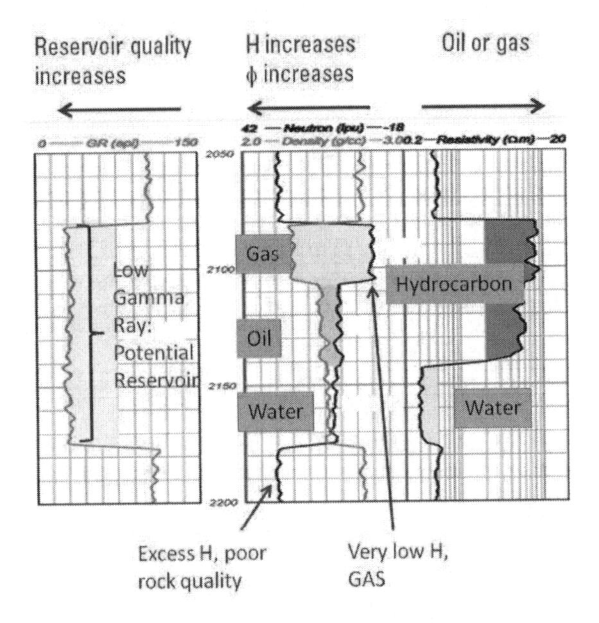

FIGURE 1. Example data of typical neutron porosity, gamma density, and resistivity measurement. The relative changes of the response with these tools indicate the changes in formation and the existence of oil, gas, or water[2].

In the idealized case shown above, the formation interpretation is relatively straightforward. On the other hand, in a real well logging environment the interpretation can be extremely complicated, enabled only by utilizing data from multiple measurements.

LOGGING WHILE DRILLING TOOL AND ITS NUCLEAR COMPONENTS

The operation requirements of an LWD tool are very different from traditional WL tools. The operation environment with high shock (10's of g's in WL, 100's of g's in LWD), high vibration and high temperature (up to 200 °C without dewars) requiresrigorous testing of the tools and the various components in a laboratory environment.

A schematic drawing of an LWD tool partial nuclear section is shown in Fig. 2, where the PNG, gamma, and neutron detectors are shown. Schlumberger's Princeton Technology Center (PTC) has a long history of manufacturing and developing neutron generator tubes (Minitrons)**[7,8]**, PNGs (neutron generator tubes integrated with HV power supplies), and scintillator-based gamma detectors with integrated photo multiplier tubes**[9,10]**. Some of the historical developments of neutron generator tubes are highlighted in Table 1. The development from large glass insulator tubes to small, ruggedized ceramic envelopes with reliable operation in high shock and temperature environment is listed in the table.

[2] Chart courtesy of Brad Roscoe, Schlumberger Doll Research

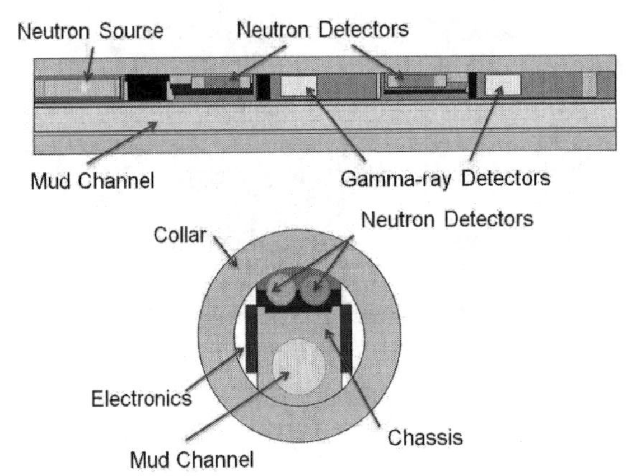

FIGURE 2. Schematic representation of a logging while drilling tool. The axial placement of the radiation source and the far and near radiation detectors are shown above, and the radial placement of the detectors are shown below[3].

TABLE 1. History of Schlumberger Neutron Generators[4]

Year	Type	Size [in]	Output[n/s]
1954	Glass	2.5 x 14	10^5
1955	Glass	1.25 x 12	$7x10^7$
1962	Glass	1.25 x 9	$9x10^7$
1967	Glass	1 x 7	$1x10^8$
1976	Glass	1 x 6	$1.3x10^8$
1980	Glass	1 x 7	$2x10^8$
1983	Ceramic	1 x 7	$2x10^8$
1992	Ceramic	1 x 7	$3x10^8$

Neutron generators currently use a hot-cathode ionizer with an all-ceramic HV insulator structure in shock dampened housing. The HV power supply section consists of a slim HV generator enabling placement in a 1 11/16-in housing.

Detector packaging includes the scintillation crystal, ruggedized photo multiplier tube with multiplier stage's mechanical design optimized for high shock environment and the associated electronics in housing with high g-load, mechanical shock dampening, see Figure 3.

[3] Schematic drawing courtesy of Brad Roscoe, Schlumberger Doll Research. Not dimensionally accurate.
[4] Table courtesy of Joel Groves, Schlumberger Princeton Technology Center

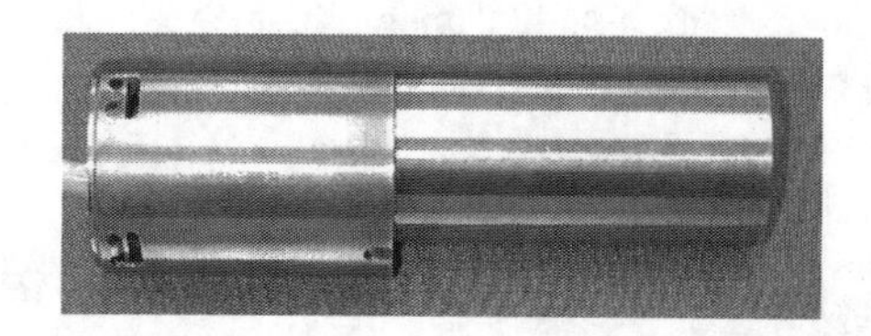

FIGURE 3. Gamma ray detector package manufactured at Schlumberger Princeton Technology Center.

These units are tested in high temperature, high shock and vibration environments at PTC. They are custom-built for the oil-well logging applications and are not available to external customers.

SUMMARY

Overcoming the challenge of developing a ruggedized PNG has enabled Schlumberger's EcoScope tool to enhance the nuclear measurement capabilities of an oilfield LWD tool. The use of pulsed neutron beam, instead of steady-state AmBe, allows the tool to measure not only the neutron porosity, but also sigma (formation capture cross-section) and formation elemental composition to calculate the formation lithology. The use of PNG in LWD tool provides tangible benefits for personnel safety during the source-loading phase of the tool preparation, minimizing the environmental impact of lost and potentially broken radioactive sources in the borehole and eliminating radioactive source proliferation issues on surface. All this is in addition to benefits for the client in the form of faster logging speeds and more accurate data due to higher neutron output of a PNG based neutron generator in comparison to a more traditional radioactive AmBe neutron source.

ACKNOWLEDGMENTS

Designed around a PNG, the EcoScope service uses technology resulting from a collaboration between Schlumberger and Japan Oil, Gas and Metals National Company. The author would like to thank Schlumberger Advisors Brad Roscoe, Joel Groves, and Peter Wraight for their help in writing this paper.

REFERENCES

1. "Radiation Source Use and Replacement," National Research Counsel of the National Academies, ISBN: 0-309–11015–7 (2008).
2. J.D. Atkin, R. Adolph, M. Evans, N. Wijeyesekera, R. McGowan and D. Mackay, "Radiation Sources in Drilling Tools: Comprehensive Risk Analysis in the Design, Development and Operation of LWD Tools", SPE 73896 (2002).
3. G. Weller, T. El-Halawani, I. Tribe, K. Webb, C. Stoller, S. Galvin and G. Scott, "A New Integrated LWD Platform Delivers Improved Drilling Efficiency, Well Placement, and Formation Evaluation Services", SPE 96652 (2005).
4. E. Mirto, G. Weller, T. El-Halawani, J. Grau, M. Berheide, F. Allioli, M. Evans, R. Berto, M. Borghi, M. Firinu and M. Giorgioni, "New Developments in Sourceless Logging-While-Drilling Formation Evaluation: A Case Study From Southern Italy", SPE 100351 (2006).
5. K. H. Al-Azmi, H. Al-Enezi, R. Kotecha, S. Al-Sabea, E. Archibong, A. Al-Khaledi and O. Oyeyemi, "From Issues to Solutions – Introducing the Multi Function Logging While Drilling Tool for Reservoir Characterization in the Greater Burgan Field of Kuwait Oil Company", SPE 126044 (2009).
6. D.V. Ellis: "Well Logging for Earth Scientist," Elsevier, New York 1987.
7. H.G. Pfutzner, J.L. Groves, and M. Mahdavi, "Performance characteristics of a compact D-T neutron generator system," Nucl. Instr. and Meth. B 99, (1995) 516
8. F.K. Chen and A.D. Liberman, U.S. Patent No. 5,293,410 (8 March 1994).
9. J.S. Pietras and S.R. Smith, "Photomultiplier Tubes and Detector Packaging for Hostile Environments," IEEE Trans. Nucl. Science, Vol. 35, No. 1 (1988).
10. B.A. Roscoe, J.A. Grau, R.A. Manente, C.L. Melcher, C.A. Peterson, J.S. Schweitzer, and C. Stoller, "Use Of GSO For Inelastic Gamma-Ray Spectroscopy Measurements in the Borehole," IEEE Trans. Nucl. Science, Vol. 39, No. 5 (1992).

Calibration Of A 14 MeV Neutron Generator With Reference To NBS-1

Craig R. Heimbach

National Institute of Standards and Technology
100 Bureau Drive MS 8461, Gaithersburg MD 20899
craig.heimbach@nist.gov
Currently a Guest Researcher at NIST, with same address and email

Abstract. NBS-1 is the US national neutron reference source. It has a neutron emission rate (June 1961) of 1.257×10^6 n/s[1,2,3] with an uncertainty of 0.85% (k=1). Neutron emission-rate calibrations performed at the National Institute of Standards and Technology (NIST) are made in comparison to this source, either directly or indirectly. To calibrate a commercial 14 MeV neutron generator, NIST performed a set of comparison measurements to evaluate the neutron output relative to NBS-1. The neutron output of the generator was determined with an uncertainty of about 7% (k=1). The 15-hour half-life of one of the reactions used also makes possible off-site measurements. Consideration is given to similar calibrations for a 2.5 MeV neutron generator.

Keywords: neutron, calibration, NBS-1, DT, DD, 14 MeV
PACS: 06.20.fb 06.60.Mr

INTRODUCTION

Neutron emission-rate calibrations performed at the National Institute of Standards and Technology (NIST) are made in comparison to NBS-1, either directly or indirectly. NBS-1 is a Radium-Beryllium (γ,n) source selected for its stability and long (1600 year) half-life. It was fabricated in the 1950's, calibrated for its absolute emission rate[1,2,3,4], and used by NIST ever since. When presented with the challenge of measuring the neutron output of a DT (deuterium-tritium) neutron generator (Physics International P325[5]), a comparison against NBS-1 was used. Direct comparisons are difficult because of the low-energy and uncertain spectrum of NBS-1[6]. An intermediary ^{252}Cf source was used in a chain of comparisons: NBS-1→ ^{252}Cf → DT.

Three comparisons were attempted for the ^{252}Cf → DT link: a ^{235}U fission chamber, and Al and Cu activation. The NBS-1→^{252}Cf link was done with a manganese bath comparison following standard NIST procedures. Both the neutron generator and the ^{252}Cf source are housed in the NIST Californium Neutron Irradiation Facility (CNIF) where the measurements were made. See Figure 1. The neutron generator was mounted vertically, with the electronics end pointing up, and with the targets below, at a height of 3.2 m above the floor. The ^{252}Cf source was 2.6 m above the floor.

Issues in the calibration included the low output of the generator, lack of a real-time fluence monitor, room scatter, uncertainties in the cross sections and the ^{252}Cf neutron spectrum, and the unknown interior design of the neutron generator.

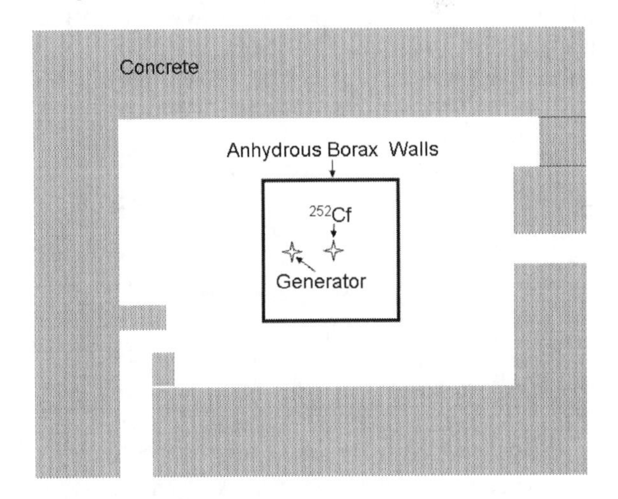

FIGURE 1. Top view of Californium Neutron Irradiation Facility (CNIF).

Application of Accelerators in Research and Industry
AIP Conf. Proc. 1336, 437-443 (2011); doi: 10.1063/1.3586137
2011 American Institute of Physics 978-0-7354-0891-3/$30.00

FISSION CHAMBER CALIBRATION

NIST had previously participated in an international calibration exercise for 14 MeV neutrons with its ^{235}U fission chamber[7]. The active material of the chamber is uranium, enriched to 93.18% ^{235}U. The chamber may be rotated, and was run in two orientations, rotated by 180 degrees, to reduce positioning uncertainty. The average of the two orientations is used. In practice, no differences were found between the two orientations, implying a well-placed center.

^{235}U has a large thermal neutron cross section, so the entire outer shell of the chamber is wrapped in cadmium. The CNIF exposure room is lined with anhydrous borax to suppress thermal neutrons.

^{252}Cf Measurements

Measurements were made at 58 to 153 cm from the ^{252}Cf source and at 46 to 180 cm from the DT neutron generator. Standard corrections were made for room return[8]. A calibration factor for the detector was determined from the ^{252}Cf measurements, and a cross section ratio was used to determine the fluence for the 14 MeV measurements.

Figure 2 gives the counts per neutron for each fission chamber measurement. The data is multiplied by the source-detector distance squared and then plotted against the source-detector distance squared to approximate a straight line. All separations were horizontal (to the right in Figure 1). Fission spectra were extrapolated to zero energy to account for counts below the discriminator setting. Dead time was monitored both with the multichannel analyzer used to take the data and with a pulser, and was held to less than 1% in all cases except one, where it was 1.4%. Pulser times were used to compute counts per second. Differences between analyzer times and pulser times were negligible.

The divergence between the measured and calculated count rates is due to the uncertainty in the composition of the concrete of the CNIF.

14 MeV Measurements

After the calibration with ^{252}Cf, similar measurements were made with the 14 MeV generator, both in the horizontal and vertical directions. The generator was run at the same power settings for all data reported here (70 kV, 50 μA); consistency of output was monitored by a ZnS neutron detector. The number of counts attainable with the generator was substantially less than that from the ^{252}Cf source, because the source has a higher intensity and because the generator runs were limited to a few hours each.

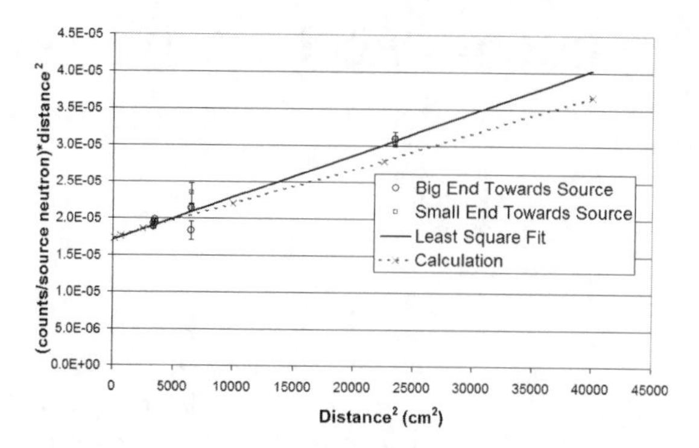

FIGURE 2. Scaled fission chamber measurements with ^{252}Cf.

As with the ^{252}Cf data, a least squares fit of the form $A+Bx^2$ was done to extrapolate to zero distance. The horizontal and vertical A values should not be equal. Based on DT reaction kinematics, the forward (vertically down) yield should be 4% higher than the horizontal yield. The fission chamber measurements indicated a 2% difference with a 2.5% (k=1) uncertainty.

Calibration

The neutron fluence from the 14 MeV generator may be determined from the relative counts for the fission chamber in the ^{252}Cf and 14 MeV fields and the neutron cross section ratio (93.18% enriched uranium).

The generator emission rate is expressed in terms of the ^{252}Cf emission rate.

$$\Phi_{14MeV} = \frac{fc(14MeV)}{fc(Cf)/\Phi_{Cf}} \cdot \frac{\sigma_{Cf}}{\sigma_{14MeV}} \quad (1)$$

where Φ_{14MeV} and Φ_{Cf} are the 14 MeV and ^{252}Cf neutron emission rates, respectively. σ_{Cf} and $\sigma_{14\ MeV}$ are the fission cross sections adjusted for uranium enrichment. fc is the fission chamber count rate in each field. Numerical results for the generator output are listed in Table 3, along with the results for the aluminum calibration. In the Table, A and B refer to Type A (statistical) and Type B (other) uncertainty evaluations.

TABLE 1. Uncertainty budget for fission chamber measurements.

Source of Uncertainty	Uncertainty (k=1, %)
^{252}Cf Emission Rate (B)	1.3
Fit Uncertainty (Room Return) ^{252}Cf (A)	0.6
Fit Uncertainty (Room Return) 14 MeV (A)	1.8
Scattering in Fission Chamber (25% of correction) (B)	3.2
Air Scatter (B)	0.02
Cross Section Ratio (^{252}Cf / 14 MeV) (B)	2.0
Combined Standard Uncertainty	4.4

ALUMINUM CALIBRATION

Both the aluminum (n,p) and (n,α) reactions were used for calibration. The (n,p) reaction is suitable for calibration, but too short (9.458 min half-life) to act as a monitor foil for irradiations longer than about 20 minutes. The 15 hour (n,α) reaction is useful for both purposes. The half-life also makes it useable for NIST to verify the neutron output from 14 MeV generators at off-site locations. Due to the low output of the generator, a hollow 6063 aluminum cylinder was used to surround the sources for both the ^{252}Cf and the DT irradiations. All counting was done with a 7.6 cm x 7.6 cm (3"x3") NaI detector. To some extent, the uncertainties in cross section cancel because the same reaction is used for each source, but the cancellation is not exact because the spectra are not the same. Recent[9] reductions in the uncertainty of the Al(n,α) cross section made this reaction useful.

The (n,p) reaction has a higher activity immediately after irradiation. Counting continues for at least two half-lives of the (n,α) reaction.

The method of counting chosen was to monitor the count rate above a threshold on a NaI detector. A discriminator is set to count only gamma rays above 661 keV. This is easily verified with a ^{137}Cs check source and includes all the peaks listed in Table 2. The techniques used are modeled after Reference 12. The advantage of this procedure is a reduced sensitivity to gain drift in the gamma-ray counting system. A disadvantage is that peaks from different reactions are not distinguished.

An aluminum cylinder (17.8 cm long by 11.4 cm outer diameter by 0.30 cm thick (7 in. by 4.5 in. by 0.12 in.)) was positioned around the generator, centered on the source. This ensured a substantial count rate and avoided much of the room wall scattering which added uncertainty to the fission chamber calibration. It did make the aluminum calibration more sensitive to positioning. As a check of the positioning and the size of the neutron source, a set of stacked aluminum rings was constructed which matched the cylinder, but which could be counted individually.

TABLE 2. Aluminum reactions used for calibration.

Reaction	Half-life[10]	Gamma Energy (keV)	Yield (%)	Cross Section for 14 MeV[10] (mb)	Cross Section for ^{252}Cf[11] (mb)
^{27}Al(n,p)^{27}Mg	9.458 m (12)	843.76	71.8	74.9 ± 1.5%	4.880 ± 2.14%
		1014.46	28.0		
^{27}Al(n,α)^{24}Na	14.9512 h (3)	1368.63	100.	121.7 ± 0.47%	1.016 ± 1.28%
		2754.03	99.94		

Uncertainties represent one standard deviation.

Source Positioning

MCNP calculations were performed for the induced radioactivity in an aluminum ring from a source in the center. Comparison calculations were performed with the source offset horizontally and/or vertically.

For horizontal (radial) displacements of up to 1 cm, the activation of rings was not sensitive to displacement. Vertical displacements displace radioactivity from the center ring to side rings.

The ^{252}Cf measured data matches the calculation quite well. This indicates that the ^{252}Cf source is small with respect to cylinder dimensions and was well-centered for the activation.

Measurements made with the DT generator showed a slight reduction in activity in the region of the cylinder towards the floor, possibly due to absorption within the materials of construction. No correction was made, but the uncertainty in total and 0 degree fluence was increased by 2.4% due to this asymmetry.

Cylinder Sampling Of Total Fluence

There is a slight offset in the angular distribution of neutrons due to the reaction kinematics of the DT reaction, which produces a slightly higher fluence and a slightly higher neutron energy in the forward direction. If a detector is horizontal to the tube, the angle of interest is 90 degrees, and the source emission rate should be used as measured by the aluminum. If the detector is directly below the generator, the angle of interest is 0 degrees and the emitted fluence should be 4.18 percent higher than the isotropic average.

Energy

The neutron energy from the generator depends on the angle of emission. For analysis of the cylinder, and for neutrons exiting at 90 degrees, 14.1 MeV is used. For neutrons emitted at 0 degrees, 14.65 MeV is used.

Sensitivity To Placement On NaI

The cylinder is counted on a NaI detector. The center of the cylinder is aligned with the center of the NaI. A set of MCNP calculations were performed to investigate the sensitivity of the NaI counts to cylinder vertical displacement. The calculations were made for a point source placed to the side of the vertical midplane, and for vertical offsets.

Two integrations were performed for this data. The first integrated the results over a 17.8 cm (7 inch) vertical line. This gives sensitivity as a function of displacement for a cylinder of uniform activation. The second integration was similar to the first, with the exception that the line was weighted to simulate actual cylinder activation. In each case, the effect of a 2 mm offset, the largest expected, is negligible. Measurements of deliberately offset rings confirmed the calculations.

The activity is fit to a double exponential

$$Activity = A_1 e^{-\lambda_1 t} + A_2 e^{-\lambda_2 t} + Background \quad (2)$$

where λ_1 and λ_2 are the known decay constants for ^{27}Mg and ^{24}Na, respectively. The A_1 and A_2 values may be determined to give neutron fluence. Special care must be taken to separate the background count rate from A_2. If the post-irradiation count is a few hours or less, the 15-hour Na does not decay sufficiently to be able to separate background. Longer counts are simple to do and appropriate. The cylinder may then be removed and background counted.

The determination of a calibration factor is shown in Table 3. Calibration factors were computed for both the (n,p) and the (n,α) reactions. The impulse fluence is the fluence that would be derived if all the neutron were delivered at the end of irradiation (EOI). The averaged fluence is the fluence corrected for decay during the exposure.

The activity at the end of the irradiation is decay-corrected back to the time when the irradiation was stopped and corresponds to the time of the delivery of the effective fluence. The calibration factors are appropriate for count rates decayed back to the end of a constant exposure and do not account for decay during exposure.

The aluminum decay during the run gives an effective fluence of

$$\Phi_{eff} = \Phi_{true} \cdot \frac{\left(1 - e^{-\lambda t}\right)}{\lambda t} \quad (3)$$

where λ is the decay constant for the reaction and t is the duration of the run. This effect is substantial (a factor of 1.7) for the (n,p) reaction on a 15 minute run.

The two aluminum calibration results differ by 6%, which is marginally consistent with the uncertainty budget in Table 4. A and B refer to type A and Type B uncertainty evaluations. Agreement is consistent with the fission chamber results.

TABLE 3. Calibration results

	Fission Chamber	$^{27}Al(n,p)^{27}Mg$	$^{27}Al(n,\alpha)^{24}Na$
Duration of exposure(s)		900	900
Count Rate at EOI (c/s)		3156.59 ± 0.47	207.86 ± 0.02
Impulse fluence (neutrons)		4.857E10	7.209E10
Fluence (neutrons) averaged during run		8.005E10	7.525E10
Fluence rate (n/s) (90°)	8.50E7	8.894E7	8.361E7
Fluence Rate (n/s) (0°)	8.58E7	9.262E7	8.710E7

TABLE 4. Uncertainty Budget for aluminum activation.

Source of Uncertainty	$^{27}Al(n,p)^{27}Mg$ (%, k=1)	$^{27}Al(n,\alpha)^{24}Na$ (%, k=1)
^{252}Cf source fluence rate (B)	1.3	1.3
Positioning on Cf (B)	0.01	0.01
A fit (A)	0.6	2.3
Positioning on NaI (B)	0.1	0.1
Half life[a] (B)	0.1	0.0
Cross Section (B)	2.14	1.28
Positioning on generator (B)	2.4	2.4
A fit (A)	0.03	0.6
Positioning on NaI (B)	0.1	0.1
Half life[a] (B)	0.1	0.0
Cross Section (B)	1.5	0.47
Combined Standard Uncertainty	3.8	3.8

a. Half life comes into play twice: in decay during run, and in decay after run. In neither case is it significant.

COPPER CALIBRATION

As with aluminum, copper has two reactions with potential for the calibration: $^{65}Cu(n,2n)^{64}Cu$ and $^{63}Cu(n,2n)^{62}Cu$. The copper used was CDA Alloy 101 Oxygen Free High Conductivity (OFHC), with the same dimensions as the aluminum.

$^{65}Cu(n,2n)^{64}Cu$ results in ^{64}Cu, with a 12.7 hour half-life. Unfortunately, there are two pathways to this end product $^{65}Cu(n,2n)^{64}Cu$ and $^{63}Cu(n,\gamma)^{64}Cu$. The first of these dominates at 14 MeV, but the second dominates in ^{252}Cf and also has a substantial low-energy cross section, making it sensitive to scattered neutrons. Fluence as determined by these reactions differed by 8 % from the fission chamber measurements, but required about a 30% correction for scattering in the copper. A thinner copper cylinder would have required smaller scattering corrections.

The $^{63}Cu(n,2n)^{62}Cu$ reaction is frequently used for measuring 14 MeV fluence. It has the advantage of a large cross section with a high count rate (mostly 0.511 MeV). Its cross section threshold is also just under 14 MeV, allowing it to reject scattered neutrons. Its 9.7 minute half-life renders it unsuitable for monitoring lengthy irradiations. It also has several competing reactions when activated by ^{252}Cf, which interfere with the transfer calibration as practiced here. Half-life and emitted gamma-ray energy may be used to separate these competing reactions, although the use of gamma-ray energy reduces the stability of the system as used here, which counts all gamma rays above a threshold. In particular, the $^{63}Cu(n,\alpha)^{60m}Co$ reaction has a half-life close to that of the (n,2n) reaction, making it difficult to separate the two. The results as determined from $^{63}Cu(n,2n)^{62}Cu$ are about 17% below that determined from the fission chamber and the aluminum techniques discussed above.

DD NEUTRON CALIBRATION

NIST also possesses a DD (deuterium-deuterium) neutron generator, producing neutrons of about 2.5 MeV. Its intensity is substantially lower than that of the DT generator. No measurements were made, but a preliminary investigation into activation techniques discovered some potential reactions. The reactions considered are listed in Table 6.

Fission reactions were eliminated on the basis of administrative difficulties in handling the materials. Niobium, iron, and nickel were eliminated on the basis of long half-life. Rhodium has a low gamma-ray energy which makes issues of counting complex (though a $^{252}Cf \leftrightarrow$ 2.5 MeV comparison would overcome much of this, and counting of x-rays would improve efficiency). Sulfur emits a beta, which would make counting difficult. This leaves indium, titanium, and zinc as viable candidates.

Indium has some competing reactions which would appear in a ^{252}Cf irradiation, particularly $^{115}In(n,\gamma)^{116}In$, but the 54 minute half-life should make it distinguishable from the 4.5 hour half-life of the main reaction. Zinc has a competing $^{68}Zn(n,\gamma)^{69}Zn$ reaction with a 13.8 hour half life, though this should be down by about a factor of 20, mostly because of isotopic composition. Titanium has a few competing reactions: $^{48}Ti(n,p)^{48}Sc$ (1.8 day half-life), $^{48}Ti(n,np)^{47}Sc$ (3.35 d), and $^{50}Ti(n,\alpha)^{47}Ca$ (4.5 day). Interferences seem to be at the percent level. The $^{48}Ti(n,np)^{47}Sc$ reaction would be especially insidious because it has the same product as the desired reaction, but calculated activity would be only about 0.2% of the main activity.

Indium is the most sensitive reaction. Calibrations involving off-site DD generators may be better served with a reaction with a longer half life.

TABLE 5. Copper reactions involved in calibration.

Reaction	Half-life	Gamma Energy (keV yield)	Cross Section (14 MeV)[10] (mb)	Cross Section (Cf)[11] (mb)
$^{63}Cu(n,2n)^{62}Cu$	9.74 m (2)	511. 195%	506 ± 1.5%	0.184 ± 3.98%
		875.68 0.15%		
		1173.02 0.34%		
$^{63}Cu(n,\alpha)^{60m}Co$	10.467 m (6)	58.6 2.0%	8.2[a] ± 36%	0.527[a] ± 40%
		826.1 0.008%		
		1332.5 0.24%		
$^{65}Cu(n,\gamma)^{66}Cu$	5.120 m (14)	1039.2 9%	0.52[b]	4.62[b]
$^{65}Cu(n,\alpha)^{62}Co$	1.5 m (4)	1172.9 11.1%	5.4[a] ± 16%	0.027[a] ± 30%
		2301.8 14.7%		
		2345.7 1.34%		
		3158 0.84%		
$^{65}Cu(n,\alpha)^{62m}Co$	13.91 m (5)	1163.4 68%	6.4[a] 9.8%	0.014[a] ±30%
		1172.9 98%		
$^{65}Cu(n,p)^{65}Ni$	2.5172 h (3)	368 4.5%	25.0[b]	0.57[b]
		1115 16%		
		1481 25%		
$^{65}Cu(n,2n)^{64}Cu$	12.700 h (2)	1345.84 0.473%	952 2.8%	0.658 ±2.2%
$^{63}Cu(n,\gamma)^{64}Cu$	12.700 h (2)	1345.84 0.473%	0.727 1.5%	10.44 ±3.24%

a. My estimates from data in Ref 12.
b. Using ENDF/B-VII[13].

TABLE 6. Some reactions of potential use for measurement of 2.5 MeV neutrons.

Reaction	Half-life	Thermal sigma (b)	Natural Abundance (percent)	Gamma Energy (keV)	Gamma yield (%)	Cross Section at 2.5 MeV (b)	Activity Relative To ^{115}In
$^{235}U(n,f)$		507	1			1.29	
$^{239}Pu(n,f)$		698				1.5	
$^{93}Nb(n,n')^{93m}Nb$	16.13 y	1.02	100	16.6	9.2	0.0075	0.000
$^{103}Rh(n,n')^{103m}Rh$	56.1 m		100	39.8	0.068	0.032	0.018
$^{237}Np(n,f)$		0.019				1.67	
$^{115}In(n,n')^{115m}In$	4.49 h		95	336.2	45.9	0.013	1.000
$^{238}U(n,f)$	40.27 h	1.00E-05	99			0.54	
$^{232}Th(n,f)$	12.75 d		100				
$^{54}Fe(n,p)^{54}Mn$	312.1 d		6	834.8	100.	0.064	0.000
$^{58}Ni(n,p)^{58}Co$	70.8 d	1.00E-13	68	810.8	99.5	0.108	0.034
$^{47}Ti(n,p)^{47}Sc$	3.35 d	4.00E-16	7	159.4	68.3	0.024	0.011
$^{32}S(n,p)^{32}P$	14.28 d		95	beta$^-$	100.	0.089	0.195
$^{64}Zn(n,p)^{64}Cu$	12.7 h		49	1345.9	0.47	0.011	0.002

REFERENCES

1. J. A. DeJuren, D. W. Padgett, L. F. Curtiss, *Absolute Calibration of the National Bureau of Standards Photoneutron Standard: I*, Journal of Research of the National Bureau of Standards, Vol. 55, No. 2, August 1955.
2. J. A. DeJuren and J. Chin, *Absolute Calibration of the National Bureau of Standards Photoneutron Standard: II, Absorption in Manganese Sulfate*, Journal of Research of the National Bureau of Standards, Vol. 55, No. 6, December 1955.
3. R. H. Noyce, E. R. Mosburg Jr., S. B. Garfinkle, and R. S. Caswell, *Absolute Calibration of the National Bureau of Standards Photoneutron Source: III, Absorption in a Heavy Water Solution of Manganous Sulfate*, Reactor Sci. Technol. (J. Nucl. Energy, Part A/B) Vol. 17, (1963) pp 313-319.
4. J.M. Adams, *Present and Future Trends for Neutron Source Calibrations at The National Institute of Standards and Technology*, Nucl. Inst. Meth. B 213 (2004) 218-222.
5. Any mention of commercial products or reference to commercial organizations is for information only; it does not imply recommendation or endorsement by NIST nor does it imply that the products mentioned are necessarily the best available for the purpose.
6. C.R. Heimbach, *The Neutron Spectrum of NBS-1*, Journal of ASTM International (JAI), Vol. 3, Issue 4, Apr 2006
7. D.M. Gilliam, C.M. Eisenhauer, J.S. Nico, and M.S. Dewey, *Comparison of Measurements in Monoenergetic Fast Neutron Fields, NIST Results and Estimated Uncertainties*, Private Communication, Oct 2002
8. C.M. Eisenhauer, J.B. Hunt, and R.B. Schwartz, *Calibration Techniques for Neutron Personnel Dosimetry*, Rad. Prot. Dosim., Vol 10, No 1-4, pp 43-57 (1985)
9. The International Reactor Dosimetry File (IRDF) uncertainty in the n,α cross section at 14 MeV was reduced from 15.6% to 0.5 % in going from IRDF-90 to IRDF-2002.
10. International Reactor Dosimetry File (IRDF) -2002. http://www-nds.iaea.org/irdf2002/index.htmlx.
11. W. Mannhart in International Atomic Energy Agency Technical Report Series No. 452 (2006).
12. A.D. Lieberman et. al., *A Method to Determine the Absolute Neutron Output of Small D-T Neutron Generators*, Nucl. Instr. Meth. B79 (1993) 574-578.
13. Evaluated Nuclear Data File. ENDF/B-VII http://www.nndc.bnl.gov/exfor/endf00.jsp

Positron Annihilation Studies In Polymer Nano-Composites

H.M. Chen[a], Somia Awad[a], Y.C. Jean[a], J. Yang[b], L. James Lee[b]

[a]*Department of Chemistry, University of Missouri-Kansas City, Kansas City, MO 64112, USA*
[b]*Department of Chemical and Bimolecular Engineering, The Ohio State University, Columbus, OH 43210, USA*

Abstract. Positron annihilation spectroscopy coupled with a variable mono-energy positron beam has been applied to study nanoscale polymeric nanocomposites. New information about multilayer depth profiles and structures, interfacial free-volume and open space properties have been obtained in polystyrene/carbon nano fiber composites. The S parameter in Doppler Broadening Energy Spectra combined slow positron beam is used to quantitatively represent the free volume, open spaces, and interactions in the interface between polystyrene matrix and carbon nanofibers.

Keywords: Positron, Carbon nanofiber, nanocomposite
PACS: 81.07.-b, 82.30.Gg, 82.80.-d, 78.66.Qn

INTRODUCTION

Inorganic-polymer nanocomposites have been investigated extensively due to their advanced mechanical, thermal, and physical-chemical properties [1-3]. Those advanced properties are from the interaction between polymers and nano fillers in the interfacial area, which has a much larger percentage of the composite than that in conventional filler composites. However, interactions between inorganic and its surrounding matrix polymer have not been fully understood due to the limited probe available to detect only at the interface area without interference from other layers in the materials. The traditional indirect way to study the interface is to change the percentage of fillings to observe the property changes which assume those changes are from the interfacial effect. Recently, we reported a study of positron annihilation lifetime in a series of polystyrene/carbon nanofiber (PS/CNF) nanocomposite which have a strong interfacial interaction and found that T_g increases as a function of percentage of CNF [4-5]. In this paper, we report a study of the interfacial properties using a variable mono-energy positron beam which can directly measure the depth profile in the nanocomposite from the free volume results.

EXPERIMENTS

Polystyrene (PS) (Mn=86,000, Total Petrochemicals, formerly Atofina, Houston, Texas) was used as received. Vapor grown carbon nanofibers (CNF) (PR-24-PS, Applied Science Inc. Cedarville, Ohio) were pyrolytically stripped to remove the surface organic contamination. The average diameter of these CNF was about 100 nm and the originally length range was from 30 to 100 μm. The PS/CNF nanocomposites were prepared at 0, 1, 3, 5, 7, 9 wt% via melt blending which was performed using a DACA micro-compounder. The compounding temperature, rotor speed and compounding time were set at 200 °C, 250 rpm and 5 min, respectively. All of the samples were annealed at 160 °C in vacuum oven for 24 hrs before further experiments.

The Doppler Broadening Energy Spectra (DBES) of positron annihilation were measured as a function of positron incident energy (0-30 keV) using the slow positron beam at the University of Missouri – Kansas City. The energy resolution of solid-state detector was 1.5 keV at 511 keV (under vacuum 10-7 torr). Each spectrum contained 1 million counts at 511 keV peak with a counting rate of 2500 cps. The S parameter, a qualitative measurement of free volume, was calculated from the DBES spectrum. It is defined as the ratio of the central area (509.5 – 512.5 keV) to the total peak counts after the background is properly subtracted. The detailed description of S parameter can be found in our previous work [4,5]. The VEPFIT program was used to fit the layered structure from S parameter [6].

RESULTS AND DISCUSSION

In DBES, S parameter represents a quantity related to the amount and size of free volume contributed from positron annihilation in polymers [7]. Figure 1

Application of Accelerators in Research and Industry
AIP Conf. Proc. 1336, 444-447 (2011); doi: 10.1063/1.3586138

shows the S results of DBES experiments for PS and CNF nanocomposite as a function of positron incident energy in different % CNF. The mean depth of the polymer is labeled at the top x axis of the figures based on a conversion equation, which can be found elsewhere [8].

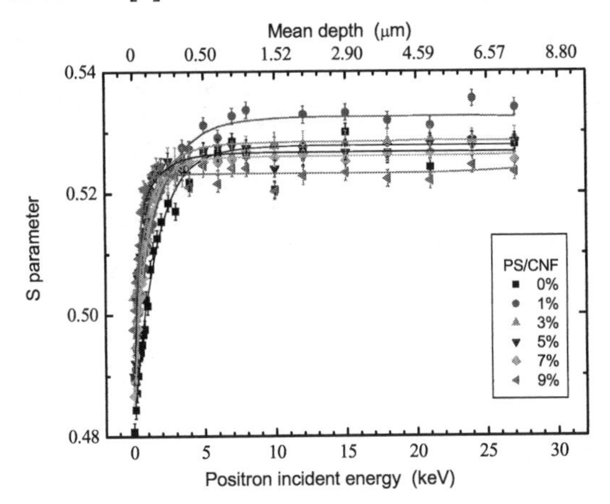

FIGURE 1. S parameter vs positron incident energy (depth) in PS/CNF nanocomposite with the different CNF percentage. Lines are fitted results from VEPFIT program using a two-layer model.

The observed sharp increase from the surface to the bulk is a general feature of polymeric materials due to the back diffusion or scattering of positron and positronium [7]. Within 100 nm from the surface, some positrons and positronium may diffuse back or scatter to the vacuum. It provides the information of the diffusion lengths of the positron and positronium.

The model of a two-layer structure, surface layer and bulk, was fitted using the VEPFIT [6] and the fitted results are shown as the lines in Figure 1, and the fitted results of layer thickness, S parameter and positron diffusion length in each layer are listed in Table 1.

Positronium diffusion length in polymers is typically very small, on the order of 1-10 nm, while positron diffusion length in polymers is on the order of 100 nm [9]. When the positron energy increases to a certain value, the S parameter will reach a plateau. It represents the bulk property of PS or PS/CNF. From Table 1, we found that the S parameter in the bulk, S2, (representing free volume) increases with the small amount of CNF (1%). Because there is no free volume in CNF, the source of this increase of S parameter is from the interface between PS and CNF. It is possible that there exists open space in the interface between CNF and polymer matrix even PS and CNF may form a strong bonding [1]. The results are consistent with our previous positron lifetime results [4], which show larger free volumes in the interface. It can be explained that when the interface bonding forms, there will be space limitation from the CNF particle to form a close-packed polymer-polymer structure as free volume in the neat polymer. Those open spaces are the relatively rigid structure in the interface and cannot be completely filled by the moving polymer chains. Therefore they are observed as larger S value than that in the neat PS matrix. However, with further increase of CNF contents, S parameter starts to decrease. It can be explained that the presence of additional CNF particles, which have no free volume, fills the interfacial spaces between CNF and PS.

TABLE 1. VEPFIT results of S parameters, surface thickness, and positron diffusion lengths (D_1, D_2) in different percentage of CNF in polystyrene nanocomposite

CNF percentage	S_1 Surface	S_2 Bulk	Layer Boundary (nm)	D_1(nm) Surface	D_2(nm) Bulk
0 %	0.5047 ± 0.0004	0.5279 ± 0.0004	49.8 ± 15.6	21.8 ± 5.1	100.7 ± 23.9
1%	0.5156 ± 0.0004	0.5327± 0.0004	57.7 ± 13.4	16.2 ± 4.8	141.0 ± 45.1
3%	0.5159 ± 0.0004	0.5285± 0.0004	31.2 ± 6.5	14.8 ± 5.3	100.8 ± 18.6
5%	0.5157 ± 0.0004	0.5267± 0.0004	8.7± 2.7	15.9 ± 3.2	98.2 ± 21.2
7%	0.5052 ± 0.0004	0.5261± 0.0004	7.8 ± 2.6	5.4 ± 1.5	48.9 ± 19.7
9%	0.5101 ± 0.0004	0.5232± 0.0004	7.6 ± 2.1	3.3 ± 1.1	42.3 ± 17.5

Another interesting result is that the surface layer boundary is decreased with the increase of CNF. It can be seen in Figure 1 there is a sharp increase of S parameter near the surface. The VEPFIT results quantitatively show the layer thickness in Table 1. It can be explained that in the neat PS, the surface contains a flexible chain layer and when adding CNF, the chains motions are restricted by the carbon nanofibers and the flexibility of the polymer chains become reduced. This results in a reduction of surface layer thickness as a function of CNF % in PS.

In order to quantitatively show the difference of S parameter between pure PS and nano composites, we calculated the difference of S using the follow equation,

$$\Delta S = S_{PS/CNF} - S_{PS} \quad (1)$$

Where $S_{PS/CNF}$ is the S parameter of nano composite and S_{PS} is the S parameter of neat polystyrene. We plot the $-\Delta S$ vs positron incident energy in Figure 2. We observe that $-\Delta S$ increases as a function of content of CNF in PS. We attempt to quantify the magnitude of $-\Delta S$ as an indicator of free volume and open space. We fit $-\Delta S$ vs depth d from the surface to the bulk in an exponential function as

$$-\Delta S = -\Delta S_{\infty} - \Delta S_0 e^{-\varepsilon d} \qquad (2)$$

Where $-\Delta S_{\infty}$ is the $-\Delta S$ at $d \rightarrow \infty$, which is the bulk and it reflects the level of change of S parameter for a particular percentage of CNF. ε is a coefficient for the S parameter change with respect to the depth. It reflects the $-\Delta S$ increasing rate along the depth. The fitted results are listed in Table 2. It is seen that ε decreases with the increase of % CNF. If we define the half depth as $d_{1/2} = ln2 / \varepsilon$, where $-\Delta S$ changed to half from surface to bulk, the $d_{1/2}$ will increase with the increase of wt% CNF. Those results indicate that depths of CNF vary from the surface in terms of free volumes and interfacial spaces with the addition of CNF particles to PS. The addition of CNF affects the surface layer thickness varying between 1 μm and 3 μm.

Because the CNF is a fiber, the composite can be regarded as a two-dimensional structure. A schematic drawing is shown in Figure 3 with the CNF diameter as 100 nm. Because the density of CNF is 2.0 g/cm^3 and the density of PS is 1.1 g/cm^3, the volume percentage should be about half of the weight percentage for the PS/CNF composite. Assuming it is a hexagonal close packing, which is the most efficient packing, the average distance (labeled d in Figure 3) between two carbon fibers can be estimated from a simple calculation of area ratio, which should be equal to the CNF volume percentage. For 1 wt% CNF, the distance is around 1.25 μm, while for 9 wt%, the distance is around 0.35 μm. If we applied the obtained surface layer $d_{1/2}$ here, even 1 wt% of CNF, the CNF affected the surface distance, $d_{1/2}$ = 1.16 μm, has almost spread to whole polymer matrix between two CNF (1.25 μm). This implies that the entire polymer matrix has been affected even by the addition of 1 wt% CNF. Hexagonal close packing has the largest average d for a certain percentage of CNF. For example, if it is a square packing, the d between two carbon nanofibers will be 1.15 μm for 1 wt% CNF. This also can be used to explain why in the bulk the 1% nanocomposite has the highest S parameter. When CNF content increases further, it starts to fill the interfacial spaces but not to increase the interface percentage. This results in the decrease of S parameter at the high CNF concentrations.

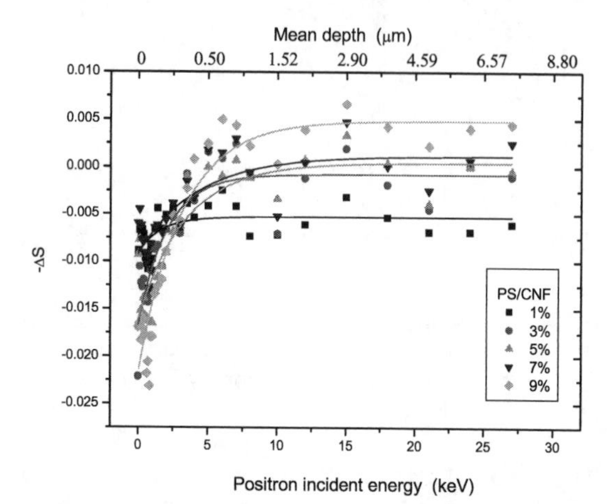

FIGURE 2. -ΔS parameter versus positron incident energy and depth in polystyrene/CNF nanocomposite with the different CNF percentage. Lines are fitted results from Equation 2.

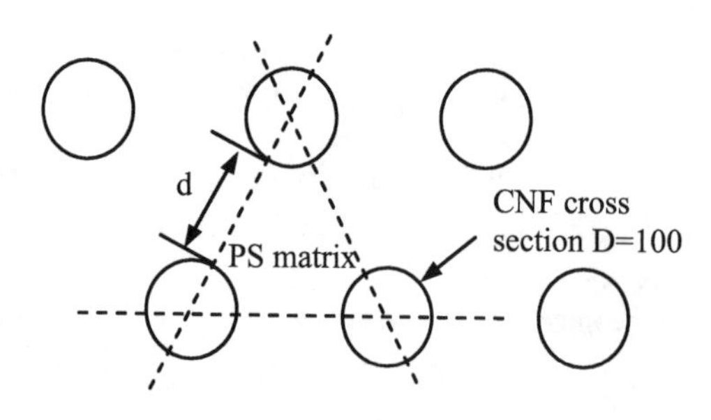

FIGURE 3. PS/CNF nanocomposite schematic presentation.

TABLE 2. The fitted results of -ΔS curve by Equation 2

CNF percentage	$-\Delta S_{\infty}$	$-\Delta S_0$	ε (μm^{-1})	$d_{1/2}$ (μm)
1%	-0.00534 ± 0.00095	0.0037 ± 0.0010	0.596 ± 0.045	1.16
3%	-0.00080 ± 0.00014	0.0159 ± 0.0015	0.480 ± 0.092	1.44
5%	0.00046 ± 0.00029	0.0165 ± 0.0016	0.295 ± 0.081	2.34
7%	0.00116 ± 0.00036	0.0104 ± 0.0014	0.323 ± 0.095	2.14
9%	0.00489 ± 0.00130	0.0267 ± 0.0016	0.238 ± 0.053	2.91

CONCLUSION

We have investigated PS/CNF nanocomposites using the Doppler Broadening Energy Spectroscopy coupled with a variable mono-energy positron beam. S parameters are used to study the surface and interfacial free volumes and open spaces and the surface thickness. The positron annihilation spectroscopy coupled with a variable mono-energy positron beam is a useful tool to study the surface and interface of polymer nanocomposite systems.

ACKNOWLEDGMENTS

This research is supported by the NSF sponsored Nanoscale Science and Engineering Center for Affordable Nanoengineering of Polymeric Biomedical Devices (NSEC-CANPBD) and by the National Institute of Standards and Technology.

REFERENCES

1. M.T.Octavio, S.V.Saul, and F.R.V. Luis, *Macromol. Mater. Eng.* **291**, 1547-1555 (2006)
2. M.Kato, A.Usuki, and A.Okada, *J. Appl. Polym. Sci.* **66**, 1781-1785 (1997)
3. R.D. Priestley, C.J. Ellison, L.J. Broadbelt, and J.M.Torkelson, *Science* **309**, 456-458 (2005)
4. H.M.Chen, L.J. Lee, J.Yang, X.Gu, and Y.C.Jean, *Materials Science Forum*, **607**, 177-179 (2009)
5. H.M.Chen, Y.C.Jean, L.J. Lee, J.Yang, and J. Huang, X.Gu, *Phys. Status Solidi C* **6**, 2397-2400 (2009)
6. A. Van Veen, H. Schut, J. de Vries, R.A. Hakvoort, and M.R. IJpma, in *Positron Beams for Solids and Surfaces*, edited by P.Z. Schutlz et al., AIP Conference Proceedings 218, AIP, New York, 1990, pp. 171-196
7. Y.C. Jean, R. Zhang, H. Cao, J.P. Yuan, and C.M. Huang, *Phys. Rev. B*, **67**, 125404-125010 (1997).
8. P.J. Schultz, K.L. Lynn, *Rev. Mod. Phys.* **60**, 701-779 (1988).
9. Y.C. Jean, *Macromolecules*, **29**, 5756-5757 (1996).

Photon Activation Analysis Of Light Elements Using "Non-Gamma" Radiation Spectroscopy - The Instrumental Determination Of Phosphorus

Christian Segebade[a] and Wolf Goerner[b]

[a]*Idaho Accelerator Centre (IAC), Idaho State University, 1500 Alvin Ricken Drive, Pocatello, ID 83201, U.S.A.*
[b]*Bundesanstalt für Materialforschung und –prüfung, unter den Eichen 86, 12205 Berlin, Germany*

Abstract. Unlike metal determinations the analysis of light elements (e.g., carbon, oxygen, phosphorus) is frequently problematic, in particular if analysed instrumentally. In photon activation analysis (PAA) the respective activation products do not emit gamma radiation in the most cases. Usually, annihilation quanta counting and subsequent decay curve analysis have been used for determinations of C, N, O, and F. However, radiochemical separation of the respective radioisotopes mostly is indispensable. For several reasons, some of the light elements cannot be analysed following this procedure, e.g. phosphorus. In this contribution the instrumental PAA of phosphorus in organic matrix by activation with bremsstrahlung of an electron linear accelerator and subsequent beta spectroscopy is described. The accuracy of the results was excellent as obtained by analysis of a BCR Reference Material.

Keywords: Photon activation analysis, linear accelerator, light elements, coincidence counting, beta spectroscopy
PACS: 23.20.Lv; 24.30.Cz; 25.20.-x

INTRODUCTION

Using standard classical analytical methods trace determinations of light non-metal elements cannot be performed instrumentally in the most cases. Using radioanalytical methods for the determination of e.g. C, N, O, F, P, S normally radiochemical separation of the analytes from the sample matrix is required, too. Using instrumental X-ray fluorescence spectroscopy instrumental determinations of some of the light elements are possible, but normally trace determinations cannot be carried out. However, in several cases instrumental photon activation analysis (IPAA) can be used to determine C, N, O and P instrumentally at trace level. Since the photoreaction products of these elements decay by β^+ emission only classical γ-spectrometry cannot be applied to separate the analytes spectroscopically. There are two options to overcome this barrier. First, sequential multiple integration of the 511 keV annihilation line and subsequent analysis of the decay curve; second, measurement of the β^+ continuum exploiting the difference of the energy maxima of the respective activation products. Combining both options one can establish a method for instrumental analysis of the light elements mentioned above. Whilst element determination by decay curve analysis is a well-established standard method [1] instrumental analysis using β-ray spectroscopy has been reported rarely up to present [2]. Thus, the determination of phosphorus is emphasised in this paper. The small difference between the half lives of ^{15}O (122.2s) and ^{30}P (149.9s) does not allow a successful deconvolution of the respective decay curves.

DETERMINATION OF PHOSPHORUS IN BIOLOGICAL MATERIAL

Exposure Energy

Preliminary tests indicated 17 MeV bremsstrahlung as optimum between appreciable product activity yield and minimum influence of interfering reactions, ^{12}C(γ,n)^{11}C in particular; being one of the most abundant components in organic/biological material excessive integral background activity would be produced by carbon if irradiated at incident energies exceeding the reaction threshold of ca. 19 MeV. The oxygen reaction threshold is lower (see Tab. 1), but due to the irregular "Quasi-Non-GDR" type of the cross section function of ^{16}O(γ,n)^{15}O [3] the activity yield beneath 20 MeV is small, thus can be neglected in many cases; the difference in production rates O vs. P at 20 MeV

Application of Accelerators in Research and Industry
AIP Conf. Proc. 1336, 448-451 (2011); doi: 10.1063/1.3586139

irradiation energy is less than 1:600 (obtained experimentally).

β^+ Energy Distribution And Half-life Of The Product Nuclides

Spectrometry of electrons/positrons is performed most efficiently using large plastic scintillation detectors. These have a poor energy resolution capacity only. However, since the energy distribution β^+/β^- radiation describes a broad continuum high resolution spectroscopy is not required. The contribution of P to the integral β^+ continuum can be determined by evaluating the high energy part of the continuum. At excessively high oxygen contents the integrated region of interest (ROI) should range from ca. 1.8 MeV to ca. 3.5 MeV; at lower oxygen level the ROI might start at 1.2 MeV (the maximum of the ^{15}N β^+ continuum). If the activation products cannot be separated by β spectroscopy (e.g. N and O) the difference between their half-lives must be exploited for their instrumental separation.

Sample Preparation

To minimise the β absorption in the sample matrix and, yet more important, the absorptivity difference between the sample and the calibration material (also referred to as reference, standard or comparator) it is essential to provide small and quasi-equal thickness and average atomic number for both specimens. By experiment a thickness of 0.5 mm (about 150 $mg \cdot cm^{-2}$) was found optimal regarding appreciably high count rate at minimum "smearing" of the electron energy distribution. Thus, the calibration specimens were prepared mixing pure Red Phosphorus or High Purity $(NH_4)_2HPO_4$ (the latter to provide a combined P and N standard) with chromatography cellulose powder of high purity, too. Thus, the matrices of both the samples under study (four different biological BCR Certified Reference Materials) and the synthetic calibration samples had very similar matrix compositions. Pure cellulose pellets were prepared as blank specimens. The spectra after activation of these served for monitoring the actual activation energy. However, the contribution of the matrix absorption to the total uncertainty budget is significant in any case. Thus, using a set of samples with stepwise increasing area densities at constant amount of P was analysed to obtain a respective correction factor F (ideally equals unity).

Activation

A 35 MeV S-Band electron linear accelerator (LINAC) was used for bremsstrahlung exposure. This machine was equipped with a beam-scanning device that provided a well-defined, homogeneous photon field so that sample and comparator material are activated uniformly [4]. The electron energy of the LINAC was set to 17 MeV in order to avoid activation of C and O. Sample and reference sample were stacked in a rabbit and transported to the LINAC by a quick pneumatic tube system. The standard exposure time was 300 s. The typical cooling period was about 60 s.

Spectroscopy

Two plastic scintillation detectors were applied for positron measurement. A pulse routing device was used to enable simultaneous operation of both measuring channels at common dead-time. Both detectors had equal efficiencies which was demonstrated by measurement of a ^{90}Sr source with a statistics better than 0.2%.The typical counting period was 200s.The resulting spectra were stored in an MCA. In this way the combined decay-and dead-time correction is automatically carried out by the counting electronics. There is no need to register operational periods (decay period, duration of counting). This facilitates the procedure tremendously.

Data Processing

The formula for calculating the P content C_{sm} of the sample is

$$C_{sm} = \frac{I_{sm} \cdot m_{ref} \cdot \eta_{ref}}{I_{ref} \cdot m_{sm} \cdot \eta_{sm}} \cdot F \cdot C_{ref} \quad (1)$$

$I_{sm(ref)}$ = number of counts in ROI of sample (reference)
$m_{sm(ref)}$ = mass of sample (reference)
$\eta_{sm(ref)}$ = efficiency of detector counting sample (reference)
F = correction factor
C_{ref} = P content of reference sample

Uncertainty Budget

As Eq. (1) is of pure multiplication/division type, one may express the budget in the form of relative uncertainties:

$$\frac{u(C_m)}{C_{sm}} = \frac{u(I_{sm})}{I_{sm}} + \frac{u(I_{ref})}{I_{ref}} + \frac{u(C_{ref})}{C_{ref}} + \frac{u(F)}{F} + \frac{u(\eta_{ref}/\eta_{sm})}{\eta_{ref}/\eta_{sm}} \quad (2)$$

By experiment F and η_{ref}/η_{sm} were found to nearly equal unity. Moreover, by a long-term experiment no significant drift of the detector photomultipliers was stated. Thus, Eq. 2 can be simplified to:

$$\frac{u(C_{sm})}{C_{sm}} = \frac{1}{\sqrt{I_{sm}}} + \frac{1}{\sqrt{I_{ref}}} + \frac{u(C_{ref})}{C_{ref}} + u(F) + u(\eta_{ref}/\eta_{sm}) \quad (2a)$$

In a series of analyses of a synthetic material batch the total uncertainty budget was found

$$\frac{u(C_{sm})}{C_{sm}} = 0.037$$

RESULTS AND DISCUSSION

BCR-CRM's 060 to 063 (different biological matrices) were analysed in a first run. Acceptable results were obtained at P levels between 0.1% and 1%. Determination of P at levels of 0.1% and lower needs repeated runs of the procedure. Because the P value of BCR-CRM-063 "Milk Powder" is certified it was analysed further (see Fig. 1).

An excellent agreement with the certified P value was found. The obtained uncertainty (k=2) of the average is ±1.5% (relative). It is smaller than the value ±3.7% in the uncertainty budget. The latter value may be justified in order to cover Type B uncertainties (preparation of the samples/reference specimens). Improvement could be achieved by direct gravimetric synthesis, mixing and pressing comparator by comparator combined with consecutive monitoring comparator against comparator. The realisation of larger P contents (about 10%) in the reference specimens would reduce their contribution to the Type A uncertainty (counting statistics). Moreover, the number of repetitious determinations could be enhanced up to 25 which would need only 4 hours of experimental time.

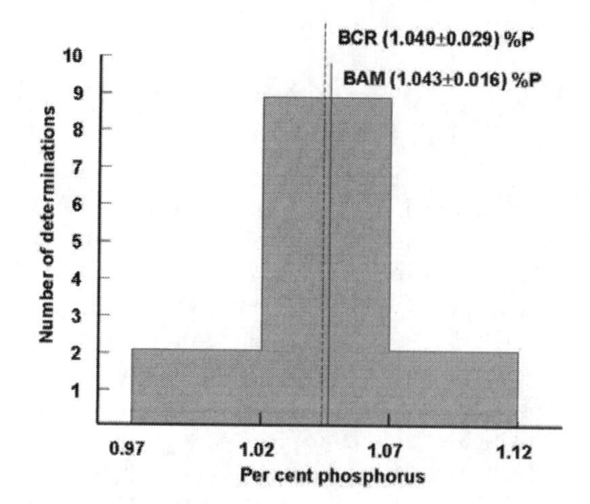

FIGURE 1. Histogram presentation of the phosphorus values of BCR CRM 063 (Milk Powder)

The future usage of two detector pairs will enlarge the counting budget by a factor of 2 and further reduce Type B uncertainty due to varying sample-detector-geometry. All these measures are supposed to allow extended uncertainties of the mean < 1% (relative).

CONCLUSION

The instrumental determination of P in biological and organic material by β^+- spectrometry after photo-activation has proven to yield reliable results at contents of 0.5% (abs.) and above. Further experimental measures can improve the precision and make limits of detection well below 100µg/g available. This non-destructive procedure has a productivity of 50 single determinations per day. It can be developed to a valuable reference method for validation and certification.

FUTURE PERSPECTIVE

TABLE 1. Photonuclear reactions of C, N, O and P

Element	Reaction	Threshold Energy, MeV	Half-life, s	$E\beta^+_{max}$, MeV
Carbon	$^{12}C(\gamma,n)^{11}C$	18.7	1223.1	0.961
Oxygen	$^{16}O(\gamma,n)^{15}O$	15.7(small σ) 20.0	122.2	1.73
Nitrogen	$^{14}N(\gamma,n)^{13}N$	10.6	597.9	1.19
Phosphorus	$^{31}P(\gamma,n)^{30}P$	12.3	149.9	3.24

A procedure for the simultaneous instrumental analysis of C, N, O and P can be developed. Regarding the physical data of the product nuclides involved (see Tab. 1) the determination of phosphorus using photo-activation is straightforward, e.g. in organic matter, making use of the extremely high maximum energy of the β^+ continuum of ^{30}P.

The known phosphorus value will facilitate the determination of N and O improving the quality of the respective results significantly. Analogously, a predetermined N content will reduce the uncertainty of the C value. The most fragile evaluation paths occur during O determination in the presence of P, and for C besides P and N. However, following the flow scheme shown in Fig. 2 a successful determination of the four elements will be possible even in these cases.

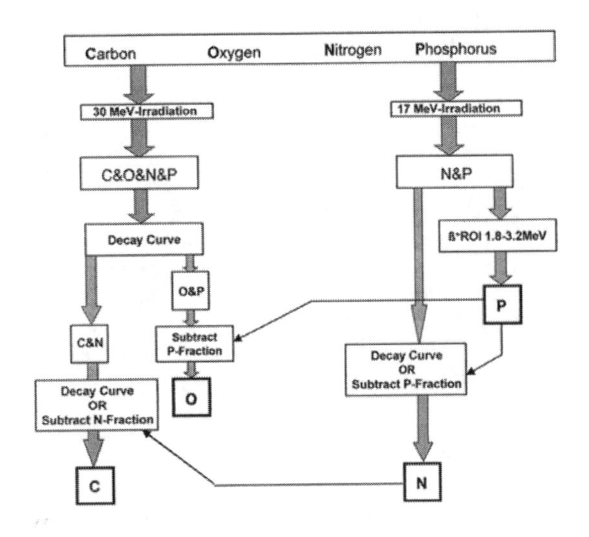

FIGURE 2. Scheme of instrumental separation of C, N, O and P

REFERENCES

1. Ch. Engelmann, J. Radioanal. Chem. 55 (1980) 379-390.
2. Ch. Engelmann, A.C. Scherle, Intern. J. Appl. Radiation Isotopes 22 (1971) 65-68.
3. B. Bülow, B. Forkman, Photonuclear Cross Sections, in: Handbook of Nuclear Activation Cross Sections, IAEA Technical Reports Series, No. 156 (1974).
4. W. Goerner, A. Berger, K.H. Ecker, O. Haase, M. Hedrich, C. Segebade, G. Weidemann, G. Wermann, J. Radioanal. Nucl. Chem. 248 (2001), 45-51.

An Overview Of Activation Analysis Techniques And Applications

Douglas P. Wells and Christian R. Segebade

Idaho State University , Idaho Accelerator Center, Pocatello, ID 83201 USA

Abstract. Activation analysis can be performed using three different techniques, namely: activation with neutrons, with high-energy photons, and with charged particles. The different features as well as the advantages and drawbacks of these separate techniques will be discussed in this paper. During exposure radionuclides are produced in the activated sample and, the turn, the radiation emitted during radioactive decay (photons in the most cases) is measured with appropriate spectrometers. By evaluating the produced spectra, the elemental components of the activated material can be quantitatively ascertained. Through means of activation analysis, basically any kind of material can be studied. And frequently the sample can be analysed *as is* (instrumentally without complicated radiochemical separation steps), and in favourable cases, the analysis may be conducted non-destructively, i.e. without removing pieces of the interrogated sample. Several common applications will be discussed in this paper.

Keywords: Activation Analysis, Instrumental Analysis, Neutrons, Charged Particles, Photons
PACS: 23.20.Lv; 24.30.Cz; 25.20.-x

INTRODUCTION

Among the many analytical methods, activation analysis (AA) techniques are the only ones which are based upon the nuclear reaction. The material under study is exposed to high-energy radiation some of which is absorbed by a nucleus in the sample. For the most part, a radioactive nuclide is produced, and upon decaying, the delayed radiation can be measured using appropriate radiation detectors. By evaluating the spectrum qualitative/quantitative analyses of the sample under study can be performed. Uncharged (neutrons, photons) or charged particles, e.g. protons, deuterons can be used for activation. Usually thermal neutrons from nuclear research reactors have been used since this technique offers the highest average sensitivity due to the higher cross sections.

It is interesting to note that these three techniques (activation analysis using neutrons, NAA, charged particles, CPAA and high energy photons, PAA) were established contemporaneously in late 1930s [1,2,3]. In practice, these analytical techniques often complement one other.

ACTIVATION ANALYSIS TECHNIQUES

Common Features

The three kinds of AA share several features. The most relevant of these are:

1. Elements only can be detected, chemical speciation is normally not possible.
2. Analyses can be carried out either instrumentally or through radiochemical separation.
3. In many cases the analytical procedure is free from blank problems.
4. Simultaneous multi-element analyses are possible.
5. The quantitative evaluation is based upon the general radioactivation equation (Eq. 1).
6. For quantitative evaluation calibration materials normally have to be used.

Since open radioactivity is produced a fully licensed Type C radiochemical laboratory within a Controlled Area must be available, and the laboratory personnel must have successfully completed a special training for handling radioactive matter and radiation protection. The respective local legislation/regulations apply.

Application of Accelerators in Research and Industry
AIP Conf. Proc. 1336, 452-457 (2011); doi: 10.1063/1.3586140

$$A(t_i,t_d) = \frac{m \cdot L \cdot \Theta}{A_r} \int_{E_{th}}^{E_{max}} \varphi(E) \cdot \sigma(E) dE \cdot (1 - e^{-\lambda \cdot t_i}) \cdot e^{-\lambda \cdot t_d} \quad (1)$$

where:

A	Radioactivity induced by nuclear activation (s^{-1})
t_i	Exposure period (s)
t_d	Decay period (s)
m	Mass of the irradiated element (g)
L	Avogadro's constant (mol^{-1})
Θ	Abundance of the isotope activated
A_r	Relative atomic mass of the activated element ($g \cdot mol^{-1}$)
E_{th}	Threshold energy of the nuclear reaction (MeV)
E_{max}	Maximum energy of the activating radiation (MeV)
$\varphi(E)$	Energy-differential activating particle flux density ($cm^{-2} \cdot s^{-1}$)
$\sigma(E)$	Energy-differential activating cross section (cm^2)
λ	Decay constant of the radionuclide produced ($=\ln(2)/T$, s^{-1})
T	Half life of the product nuclide (s)

Fields Of Application

Activation analysis has been applied mainly in the following areas: environmental science, geo- and cosmochemistry, oceanography, analysis of industrial raw and end products, high purity material analysis, analysis of organic matrices (biological and medical material), forensic science, analysis of objects of art and archaeology, certification of candidate reference materials.

In the following, the three kinds of AA will be described only briefly. Detailed evaluations can be found in the scientific literature, e.g. Ref. [4,5] (PAA), Ref.[6,7,8] (NAA), Ref. [9,10] (CPAA), respectively.

Photon Activation Analysis

Upon the energetic photon interrogating the sample, the target nucleus can be activated through a photonuclear process, usually through means of neutron knockout. Activation is induced by higher energy photons, having energy of at least 10 MeV. The photonuclear reaction data of the elements advocate for an activation energy around 30 MeV. Bremsstrahlung sources such as linear accelerators or microtrons are mostly employed. Favourable irradiation parameters are: 30 MeV electron energy at 100–150 μA mean electron beam current. With the help of suitable radiation spectrometers, e.g. high-resolution germanium detectors photon spectra (γ–rays, annihilation radiation and characteristic X-rays) can be measured. In the determination of some Light Elements (C, N, O, F) radiochemical separation of the analyte nuclides from the sample matrix is mostly required. In rare cases, β counting can be applied [11].

Neutron Activation Analysis

Neutrons of different energies are used for radioactivation. Mostly thermal neutrons (Energy: 0.025 eV corresponding to $v=2200\ ms^{-1}$) produced in nuclear research reactors are used. And through which, high neutron flux densities can be obtained. Moreover, at this energy activation cross sections are extremely high for many elements, thus very low detection limits can be achieved. Neutron emitting radionuclide sources have also been used for activation [12,13]. Finally, accelerator neutron sources have been used for special application cases, e.g., determination of oxygen with 14 MeV neutrons from a D,T neutron generator [14]. Mostly instrumental analysis has been performed using NAA, but in several cases radiochemical separation of the analytes before or after activation is either required or more favourable than the instrumental approach.

Activation Analysis With Charged Particles

Mostly protons, deuterons or ^{3}He nuclei from cyclotrons have been used for activation at various energies; light elements, e.g. B, C, N, O have been determined predominantly; the analysis of heavier elements is possible but frequently problematic due to the multitude of nuclear reactions and, consequently, the complexity of the spectra obtained; depending on the irradiation parameters (e.g. kind of the activating particle) irradiation can be carried out in air, in vacuum or under light inert gases, e.g. helium; Analyses of light elements normally have been performed including chemical separation of the activated analyte; multi-element determinations can be performed in rare cases only. Complication might arise due to the rapid energy loss (and the accordingly rapid change of reaction cross sections) of the activating particles on their path through the sample matrix.

Comparison of Features

In the following, the different analytically relevant characteristics of the techniques are compared.

Matrices Analysed

First, in NAA and PAA any material – solid, liquid, or gaseous – can be analysed, while for the CPAA case only solids can be studied. Secondly, for the case of

PAA, the heat transfer to the sample is usually small and therefore is inconsequential. With the reactor neutrons in NAA, however, the transfer of radiation and thermal energy is of major concern. The samples therefore need to be encapsulated accordingly, e.g. in sealed high purity quartz vials. Using other neutron sources heat transfer usually is negligible. In CPAA, moreover, measures against heat damage are imperative. Thirdly, total *activity of the matrix* after activation; this is dependent on the behaviour of the major/minor sample components under activating particle bombardment. In PAA elevated matrix concentrations of e.g. C, Ni, Sn, Pb might lead to excessive matrix activities while for the case of NAA high concentrations of Na, Br, and noble metals may cause matrix activity problems. CPAA mostly is carried out radiochemically, hence matrix activities usually are of minor concern. Fourthly, *the homogeneity of the material*; this is of major relevance for each analytical method, instrumental ones in particular. In routine PAA and NAA minor inhomogeneity can be handled satisfactorily. In CPAA, however, only a thin sample layer is activated. Thus, the overall homogeneity is of significance.

Elements Detected

Which elements can be detected/determined (see Eq. 1) depends on the irradiation parameters and the behaviour of the respective target nucleus and on which activating particle bombards the sample. One advantage of thermal NAA is the fact that (n,γ) is the dominant reaction type; others occur in rare cases only. At higher irradiation energies, regardless of the kind of the activating particle, more reaction types are likely to occur. Both in PAA and CPAA the multitude of different reaction types lead to frequent interference among neighbouring elements in the Chart of Nuclides, at high Z numbers in particular. Thus, CPAA is mostly used for the determination of light elements. Nonetheless, through the complementary activation analysis techniques nearly all elements of the Periodic Table can be analysed.

Calibration Material

(CM; also referred to as "Reference material" or "Standard") Usually CMs contain the elements to be determined in known concentrations and are activated together with the samples under study (in PAA and NAA) or sequentially (CPAA). The use of CMs is indispensable in AA (see the section Common Features above). CMs have to undergo the same irradiation conditions as the samples. Since for neutron activation the physical parameters involved mostly are known with good precision a mono-standard method (k_0 technique [15]) can be applied. Frequently Certified Reference Materials (CRM) are used for calibration. For different reasons, in instrumental multi-element analysis synthetic CMs often are preferable [16,17]. CRMs should be used mainly for quality control.

Generally, the CM should have a matrix composition similar to that of the material under study. This is particularly true for CPAA.

Sample Handling

In NAA using reactor neutron exposure in the core material samples must be prepared so that no contamination of the core can occur; samples must be packed accordingly before being transferred to the exposure position, usually done with a pneumatic transfer tube system. The use of a pneumatic sample transfer is also recommendable in PAA. In CPAA a sample transfer as used in NAA and PAA normally cannot be used if the exposure takes place in a vacuum or inert gas chamber. Moreover, in CPAA the overall sample geometry is crucial; any geometrical inconsistency between the sample and the CM might lead to fatal inaccuracy of the results.

In some cases eventual superficial radioactive contamination can be removed before counting. In radiochemical AA the sample preparation for counting depends upon the chemical treatment procedure in each individual case. In the instrumental approach samples usually are repacked before counting unless they were irradiated in vessels of materials that are not activated during particle bombardment, e.g. high purity quartz. Equal sample geometry has to be provided for the material analysed and the CM. In PAA it is useful to make samples into thin pellets if low energy photons (e.g. X-ray fluorescence) are measured.

Irradiation

The irradiation parameters (e.g. energy and beam current in PAA and CPAA, sample location in the activating particle field, exposure period) have to be selected depending upon the individual analytical requirements. According to Eq. 1 equal irradiation doses received by samples and CMs and thus homogeneous activating radiation fields are required. However, strictly speaking, these are not achievable in any case, and especially for PAA and CPAA. In reactor NAA if irradiation is carried out inside the core the neutron field homogeneity is acceptable for routine analysis. In PAA and CPAA, however, one has to account for pronounced radial and axial beam intensity gradients, thus internal or external particle flux monitors must be used during exposure to account for particle flux gradients [18, 19]. In PAA it is possible to improve the homogeneity of the bremsstrahlung field with the help of an oscillating beam scanner [20].

Radiation Spectrometry

Photon spectroscopy, mostly by using high resolution semiconductor detectors, has been applied in more than 99% of all AA application cases. The respective calibration function is quasi-linear in the energy region of interest in activation analysis (5-

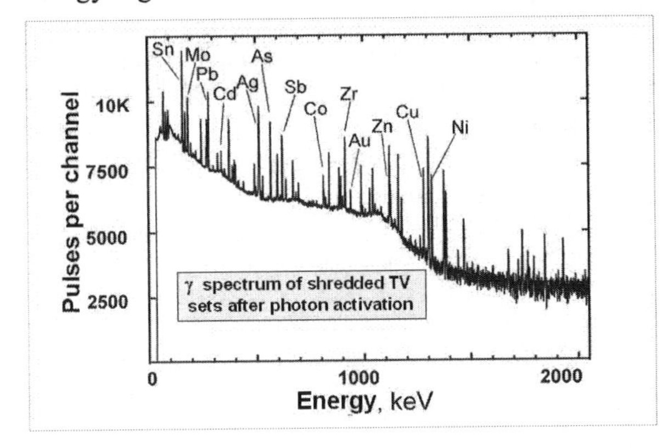

FIGURE 1. γ spectrum of electronic waste after photoactvation

3000 keV). Thus, simultaneous multi-element analyses are possible. The resulting spectra (see Fig. 1) are processed by computer [21].

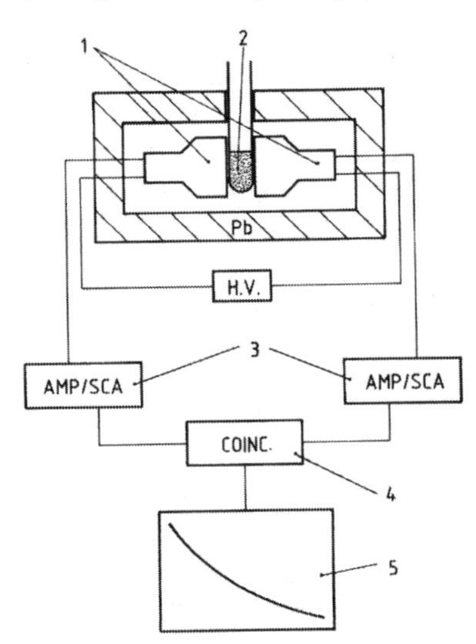

FIGURE 2. Setup for coincidence spectrometry; 1: Scintillation crystals, 2: Sample, 3: Amplifier/single channel analyser, 4: Digital coincidence unit, 5: Pulse counter/multichannel analyser

In PAA low energy photons (e.g. intense X-ray fluorescence) can be measured favourably using low energy photon spectrometry (LEPS, see Ref. [22]). A special spectrometer design is used for annihilation quanta counting (Fig. 2).

This is applied to the analysis of light elements using radiochemical PAA or CPAA. Most of the nuclides produced through photon or charged particle activation do not emit γ-rays but decay only by emission of a positron, which subsequently annihilates upon encountering an atomic electron, and in the annihilation process, two 511 keV photons are emitted back to back. The activation product is separated from the sample matrix and counted with the double crystal spectrometer in coincident mode. A decay curve is recorded and analysed by computer [23].

Data Processing

Many programs for spectrum processing (peak search, calculation of energy and net peak, assignment to the emitting radionuclide etc. [24]) are commercially available; frequently these are provided by the spectrometry hardware manufacturer. However, there are comparatively few programs on the market that calculate the final analytical results. This is due to the multitude of parameters to be included that are different in each individual analysis procedure. Thus, "home made" evaluation programs have been used frequently. Nowadays, however, efficient programming software being available evaluation programs can be developed that are usable quasi universally for instrumental AA [25].

Application Examples

In the following several typical examples of application of the different activation analysis techniques are discussed.

Photon Activation Analysis

A paper dealing with the *instrumental PAA of large material samples* was published in 2007 [26]. The average elemental composition of shredded electronic waste material (TV sets) was studied. Since the average grain size of the material was up to 1.5 cm only a large (1 to 1.5 litre) sample could be regarded representative for the entire material lot. Exposures were carried out with a 30 MeV electron beam from a linear accelerator equipped with an electron beam scanner providing a large and comparatively homogeneous bremsstrahlung field [20]. Measurements were performed with a standard HPGe photon spectrometer

(Fig.1). The results were mostly in satisfactory agreement with those obtained by INAA.

IPAA of elements whose reaction products do not emit γ spectra are rarely achievable and only in isolated and very favourable cases only. The *determination of nitrogen in boron carbide by instrumental PAA* was reported by Merchel and Berger in 2007 [27]. Irradiations had to be performed at a bremsstrahlung energy well below 20 MeV which is the activation threshold of the matrix carbon. The matrix boron does not undergo any photonuclear reaction at this energy. Measuring the annihilation quanta due to ^{13}N a decay curve was recorded and analysed subsequently. The result obtained (2260±100 µg/g) was in satisfactory agreement with that obtained by classical gas fusion analysis (GFA; 2303±64 µg/g).

Neutron Activation Analysis

In a paper on the INAA of tooth samples [28] the authors demonstrated a proper strategy of analytical quality assurance. Major (Na, Mg, Ca) and trace (Cl, Mn, Zn, Sr) constituents of different kinds of human tooth material (deciduous and permanent, enamel and dentine fraction, respectively) were determined so as to find significant differences between the mentioned material groups and eventual correlation to epidemiological facts. Several elements of interest (e.g. Fe, Se, Rb) could not be detected due to the huge bremsstrahlung background produced by β^- emission of ^{32}P being generated by activation of the matrix phosphorus which is a major component of bone material (Calcium hydroxy apatite). As to the quality assurance, a synthetic multi-element CM was prepared from high purity element solutions. A certified reference material (NIST SRM 1486 "Bone Meal") was also analysed to check the accuracy of the results obtained. The agreement between the data found and the certified values was excellent.

The suitability of accelerator-based NAA for instrumental determination of major and minor components using a 14 MeV D-T neutron generator was demonstrated in a paper about an analytical study of industrial raw material [29]. Several components of calcium carbide used for acetylene production had been suspected to be responsible for frequent spontaneous explosions during the process. Mg, Al, Si, P, Ca, Ti, V, Fe, Sr, and Ba were determined simultaneously. In particular, Si and P were moved into focus since they might give rise to the generation of highly explosive silanes and phosphines, respectively. 20 g batches were irradiated for 600s at a neutron flux density of ca. 10^9 $cm^{-2}s^{-1}$. A synthetic mixture of compounds of the respective elements was used as a CM. Concentration values in the trace level could be obtained for V, Sr and Ba only.

Activation Analysis With Charged Particles

The complexity of CPAA was lined out by Fukushima *et al.* [30] regarding the multitude of nuclear reactions at elevated activating particle energies. C, N and O were determined by radiochemical analysis using protons, deuterons and ^3He ions at various energies. Particular attention was directed to the respective chemical separation procedures. Three different techniques for ^{11}C isolation were described in detail, one of them basing upon the classical oxidising fusion using Pb_2O_3 plus B_2O_3. Oxidising dissolution using HF – HNO_3 – KJO_4 and dissolution in NaOH followed by subsequent oxidation with $KMnO_4$ were also used. Separation of ^{13}N from Al was performed by dissolution of the matrix and precipitation of the analyte as ammonium tetraphenyl borate. Heat extraction and conversion to $^{13}N_2$ subsequently scavenged in 5Å molecular sieve was also used. A distillation/precipitation (as KBF_4) technique for separation of ^{18}F was described in detail, too. The detection limits for the Light Elements studied were found in the low ng/g range.

The determination of Cd, Tl, and Pb in certified reference materials using both instrumental and radiochemical proton activation at different energies was described in a paper by Wauters *et al.* [31]. The unavoidable multiple interference by competing reactions were evaluated. A few reactions were found quasi interference-free thus were found preferable for analytical use. Ni foil was used to monitor the proton flux of the cyclotron used. ^{57}Ni ($E\gamma$ = 1378 keV) was used as monitoring nuclide. In the radiochemical approach samples were fused in $NaOH$-Na_2CO_3 and dissolved subsequently in hydrochloric acid. Components were separated using column chromatography and measured using high resolution Ge spectrometry. The results obtained were in remarkably good agreement with the respective certified data. Detection limits were found in the low ng/g range.

CONCLUSION

Activation analysis, NAA in particular, regarding detection limits was superior to all other methods of elemental analysis for a long period of time. However, the situation has changed in the last 2 – 3 decades; other methods, such as ICP-MS, offer comparable or even lower detection limits, frequently with superior overall quality of results and precision. Further, no undesirable radioactive waste is produced. Nevertheless, some features of AA still are unique, e.g. non-destructive trace analyses of bulky samples. This is of particular advantage, such is for studies of precious objects and matrices difficult to process chemically, as well as some refractory metals and their chemical compounds. Instrumental analyses without chemical preparation/separation steps can be performed as well as fast analyses of material streams. Thus, AA will still be indispensable in the foreseeable future.

REFERENCES

1. G. v. Hevesy and H. Levy, *Math.-Fys. Meddelelser* 14 (1936), 3-34.
2. G.T. Seaborg and J.J. Livingood, *J. Amer. Chem. Soc.* 60 (1938), 1784-786.
3. K.B. Meshiborskaya, *Report State Press for Literature of Atomic Science and Technology* 1960 (1961), 1-10.
4. C. Engelmann, Photon Activation Analysis, in *Advances in Activation Analysis,* eds. J.M.A. Lenihan and V.P. Guinn, Academic Press, New York (1972), 1-55.
5. C. Segebade, H.-P. Weise, and G.J. Lutz, Photon Activation Analysis, W. de Gruyter, Berlin New York (1988).
6. R.C. Koch, Activation Analysis Handbook, Academic Press, New York London (1960).
7. J. Hoste, J. Op De Beeck, R. Gijbels, F. Adams, P. van den Winkel, and D. de Soete, Instrumental and Radiochemical Activation Analysis, CRC – Butterworths & Co (1971), 9-85.
8. G. Pfrepper, W. Görner, and S. Niese: *Spurenelementbestimmung durch Neutronenaktivierungsanalyse*, Akademische Verlagsgesellschaft Geest & Portig K.G. Leipzig (1981).
9. C. Engelmann and P. Albert, Activation analysis with charged particles, in: *Techniques of Metal Research*, Vol. III, Pt. 2, Interscience Publishers-Wiley & Sons, New York (1970), 803ff.
10. J. Hoste, J. Op De Beeck, R. Gijbels, F. Adams, P. van den Winkel, and D. de Soete, Instrumental and Radiochemical Activation Analysis, CRC – Butterworths & Co (1971), 104-120.
11. C. Segebade and W. Görner, *Photon activation analysis of light elements using "Non-Gamma" radiation spectroscopy*, this conference
12. E. Ricci and T.H. Handley, Anal. Chem. 42 (1970), 378ff.
13. J.T. Gerard and J.L. Pietruszewski, Anal. Chem. 50 (1978), 906ff.
14. S.S. Nargolwalla and E.P.Przybylowicz, Activation Analysis with Neutron Generators, Interscience Publishers-Wiley & Sons, New York (1971).
15. F. De Corte and A. Simonits, *Vade Mecum for k_0 – Users*, Geleen DSM Research Report (1994).
16. K. Masumoto, KEK Ibaraki, Japan, private communication
17. C. Segebade, K. Masumoto, O. Haase, P. Springorum, S. Hunt, Win Tin, A. Berger, and M. Radtke, *Polymer based synthetic multi-element calibration materials*, 21st Seminar on Activation Analysis, Mainz, Germany, March 21st – 23rd 2007; published in summary form only.
18. C. Segebade, H.-P. Weise, and G.J. Lutz, Photon Activation Analysis, W. de Gruyter, Berlin New York (1988), 416-419.
19. C. Engelmann and P. Albert, Activation analysis with charged particles, in: *Techniques of Metal Research*, Vol. III, Pt. 2, Interscience Publishers-Wiley & Sons, New York (1970), 806-811.
20. W. Görner, O. Haase, and C. Segebade, J. Radioanal. Nucl. Chem. 271 (2007), 199-201.
21. C. Segebade, H.-P. Weise, and G.J. Lutz, Photon Activation Analysis, W. de Gruyter, Berlin New York (1988), 93-158.
22. Ibid., p131-132.
23. G. Friedlander, *Nuclear and Radiochemistry, 2nd Ed.*, Wiley & Sons, New York (1964).
24. W. Westmeier, Nucl. Instr. Meth. A242 (1986), 437-466.
25. Sun Zaijing, Standardizing activation analysis: New software for photon activation analysis, this conference.
26. C. Segebade, P. Bode, and W. Görner, J. Radioanal. Nucl. Chem. 271 (2007), 261-268.
27. S. Merchel and A. Berger, Anal. Bioanal. Chem. 388 (2007), 385-389.
28. M.A.B. Soares, E.M. Adachi, and M. Saiki, J. Radioanal. Nucl. Chem. 276 (2008), 49-52.
29. C. Segebade, Journ. Radioanal. Chem. 43 (1978), 243-251
30. H. Fukushima, T. Kimura, H. Hamaguchi, T. Nozaki, Y. Itoh, and Y. Ohkubo, J. Radioanal. Nucl. Chem. 112 (1987), 415-423.
31. G. Wauters, C. Vandecasteele, K. Strijckmans, and J. Hoste, J. Radioanal. Nucl. Chem. 112 (1987), 23-31

Oxygen-Atom Defects In 6H Silicon Carbide Implanted Using 24- MeV O^{3+} Ions Measured Using Three-Dimensional Positron Annihilation Spectroscopy System (3DPASS)

Christopher S. Williams[1], Xiaofeng F. Duan[2], James C. Petrosky[1] and Larry W. Burggraf[1]

[1]Air Force Institute of Technology, AFIT/ENP, 2950 Hobson Way, Wright-Patterson Air Force Base , OH 45433
[2]Air Force Research Laboratory's (AFRL) Department of Defense Supercomputing Resource Center (DSRC), Wright-Patterson Air Force Base, OH 45433

Abstract. Three dimensional electron-positron (e^--e^+) momentum distributions were measured for single crystal 6H silicon carbide (SiC); both virgin and having implanted oxygen-atom defects. 6H SiC samples were irradiated by 24-MeV O^{3+} ions at 20 particle-nanoamps at the Sandia National Laboratory's Ion Beam Facility. O^{3+} ions were implanted 10.8 μm deep normal to the (0001) face of one side of the SiC samples. During positron annihilation measurements, the opposite face of the 254.0-μm thick SiC samples was exposed to positrons from a ^{22}Na source. This technique reduced the influence on the momentum measurements of vacancy-type defects resulting from knock-on damage by the O^{3+} ions. A three-dimensional positron annihilation spectroscopy system (3DPASS) was used to measure e^--e^+ momentum distributions for virgin and irradiated 6H SiC crystal both before and following annealing. 3DPASS simultaneously measures coincident Doppler-broadening (DBAR) and angular correlation of annihilation radiation (ACAR) spectra. DBAR ratio plots and 2D ACAR spectra are presented. Changes in the momentum anisotropies relative to crystal orientation observed in 2D ACAR spectra for annealed O-implanted SiC agree with the local structure of defect distortion predicted using Surface Integrated Molecular Orbital/Molecular Mechanics (SIMOMM). Oxygen atoms insert between Si and C atoms increasing their separation by 0.9 Å forming a Si-O-C bond angle of ~150°.

Keywords: ion implantation, positron annihilation, ACAR, DBAR, oxygen defects
PACS: 29.40.Gx, 78.70.Bj

INTRODUCTION

Wide band-gap semiconductors such as silicon carbide (SiC) have been extensively studied in the past few decades for use in specialty electronic devices. As good quality SiC materials became available, they gained utility in several applications: micro-structures, opto-electric devices, high temperature electronics, radiation hard electronics and high power/frequency devices [1]. Popularity of SiC is the result of several favorable properties important for devices used in harsh environments: low density, high strength, low thermal expansion, high thermal conductivity, high hardness and superior chemical inertness [1]. Reliably characterizing deep-level defects formed in bulk, as-grown material is critical to many of these applications of SiC devices.

We have developed a new positron spectroscopy method to characterize the electronic structure of defects in semiconductor materials. This 3DPASS method simultaneously measures angles and energies for the coincident photons produced by e^--e^+ annihilation events [2-3]. The 3DPASS measurement method combined two historically independent techniques: 2D ACAR, measuring momentum perpendicular to the direction of gamma propagation, and two-detector DBAR, measuring momentum in the direction parallel to gamma propagation. The 3DPASS method measures coincident 2D ACAR and two-detector DBAR spectra, such that it is possible to determine the 3D e^--e^+ momentum distribution in the sample. Comparison between 3DPASS measurements of defect-free material and samples with known defect types provide information on the defect characteristics.

Application of Accelerators in Research and Industry
AIP Conf. Proc. 1336, 458-462 (2011); doi: 10.1063/1.3586141
2011 American Institute of Physics 978-0-7354-0891-3/$30.00

Comparisons of 3DPASS with quantum calculations permit the geometry of the defect to be confirmed.

This research investigated the effect of oxygen-atom defects on the momentum distribution in 6H SiC. Oxygen defects were created by irradiation using 24-MeV O^{3+} ions. 3DPASS measurements were made on virgin 6H SiC and O-ion irradiated 6H SiC samples both before and after annealing. 2D ACAR spectra were reconstructed to display the e^{-}-e^{+} momentum distribution in the plane perpendicular to the propagation direction of the annihilation photons and DBAR results are presented to display changes in the distribution of the parallel momentum component. Momentum anisotropies present in the 2D ACAR spectra are compared to the O-atom defect structure predicted using hybrid quantum mechanics/molecular mechanics (QM/MM) calculations.

ION IRRADIATION AND POSITRON INTERACTION DEPTH

SiC samples were irradiated at the Ion Beam Laboratory (IBL) at Sandi National Laboratory (SNL). The ion beam produced 24-MeV O^{3+} ions at 20 particle-nanoamps, resulting in 1.25×10^{11} O^{3+} ions/sec implanted normal to the (0001) surface of 6H SiC samples. The beam spot-size measured 0.5 cm x 0.5 cm. Four, 1 cm^2 6H SiC samples described in [3] were irradiated with different irradiation times, 1.0, 10.0, 100.0 and 1000.0 sec, resulting in the following fluences: 3.12×10^{11}, 3.36×10^{12}, 3.13×10^{13}, and 3.12×10^{14} ions/cm^2. The damage produced in the 6H SiC samples was clearly visible for the three most highly irradiated samples as a dark brown region, which decreased in opacity as the O-ion fluence decreased. The darkening significantly faded when the samples were annealed at 1000°C. These ion fluences were similar to the conditions Uedono et al [4] used when they studied near-surface vacancy-type defects in 6H SiC. They implanted 10^{13} and 10^{15} 150-keV O^{+} ions and observed significant changes in valence electron momentum measured by the S parameter in their DBAR analysis.

The depth of the 24-MeV O^{3+} ion implantation was calculated to be 10.8 ± 0.5 μm using the algorithm Stopping and Range of Ions in Matter (SRIM) 2008 [5], which is a penetration of just a small fraction of the 254-μm thickness of the SiC sample. According to SRIM calculations, each ion produced 1,796 vacancy-type defects in the ion tracks through the SiC lattice. Lateral spreading of the ions from the beam's focal direction was calculated to be approximately 3 μm, so the ion tracks were strongly oriented with respect to the SiC lattice. Since the effect of the O atoms on the crystal lattice of the SiC was of primary interest rather than the damage the oxygen ions created, the irradiating O ions were directed toward one face of the crystal sample, while subsequently the analyzing positrons entered from the opposite crystal face. This allowed the positrons to preferentially sample the O–atom perturbation of the lattice structure, instead of the plethora of defects produced by high energy oxygen ions.

In order to determine the range of the positrons in SiC, a simulation was executed in GEANT4 [6-7]. GEANT4 is a software program written in C++ which utilizes a high-fidelity Monte Carlo simulation to model the transport and interaction of particles in matter. Based on the GEANT4 simulations, positrons annihilate at a median depth of approximately 105 μm in 6H SiC. In order to optimize the overlap of thermalized positrons with oxygen ion defects prior to analysis, the thickness of the sample was decreased from 254 μm to 105 μm by ion milling. SiC samples were dry etched by personnel of the AFRL Sensors Directorate to remove 140 ± 5 μm of the 6H SiC from the side of the sample opposite to the face in which the ions were implanted so that the layer of implanted O atoms was located 105 μm from the dry etched surface. This allowed us to make coincident, simultaneous 2D ACAR and CDBAR measurements on samples having O atoms deposited at the median positron penetration depth.

THEORETICAL STRUCTURE PREDICTION

Defect structure is largely a local electronic property as compared to the electron momentum associated with defects, which sums significant contributions from a larger portion of the crystal lattice. The electronic/positronic wavefunction that describes both the defect electronic structure and its electronic momentum prior to e^{-}-e^{+} annihilation has symmetry consistent with both the local defect structure and the lattice structure in which the defect is embedded. In virgin SiC, positrons preferentially reside on the more electronegative carbon atoms, which locate the center of ACAR measurement symmetry. For O-defected SiC, positrons preferentially localize on the electronegative oxygen atoms, shifting the center of symmetry for ACAR spectra.

The result of O-atom interactions with the Si and C atoms in the SiC lattice were modeled using the SIMOMM QM/MM method [8] to calculate the optimized geometry for an O atom interstitially residing inside of bulk SiC. The SIMOMM method is a two-step process which first computes electrostatic forces of large molecular clusters of the bulk crystal onto a small molecular cluster using molecular mechanics and then uses those forces as the boundary conditions to solve for the wavefunctions of the smaller embedded cluster using quantum mechanical (QM) calculations carried out at an appropriate QM theory level. Based on these calculated results, the O atom inserts into the Si-C bond, locally perturbing the SiC lattice. Oxygen atoms insert between Si and C atoms breaking the Si-C bond as indicated by increasing the Si-C separation by 0.9 Å to form a Si-O-C bond having bond angle of about 150°. The oxygen atom protrudes toward the center of an adjacent lattice interstice. The minimum-energy optimized geometry of the defect structure predicted by the QM/MM calculations is shown in figure 1. SIMOMM quantum calculations performed using DFT B3LYP/6-31G* and MP2/6-31G* gave similar results, the largest difference between these calculations being that the bond angle for Si-O-C was predicted to be 153° by DFT and 147° by MP2. These results were compared to measured spectra from the O^{3+} implanted 6H SiC using 3DPASS in which the center of symmetry became the O atom for the 3DPASS measurements of the distorted defect structure.

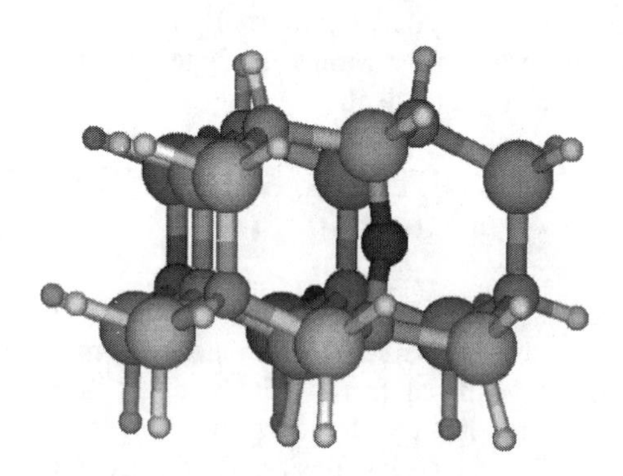

FIGURE 1. Optimized geometry of distorted 6H SiC embedded cluster from SIMOMM calculation. Small light gray terminating spheres represent cluster boundary, large light-colored spheres represent Si atoms, darker small spheres represent C atoms and the black sphere with represents the O atom.

EXPERIMENTAL RESULTS

3DPASS measurements were performed on the virgin 6H SiC, on 6H SiC samples that were O-ion irradiated normal to the (0001) surface and then on O^{3+} ion irradiated 6H SiC samples following annealing at 1000°C. The 3DPASS datasets were measured for 6H SiC samples with the (0001) surface oriented at 45° relative to the centerline direction of photon detection. Efficiency-corrected 2D ACAR spectra, reconstructed from the 3DPASS datasets are presented in figure 2(a)-(c).

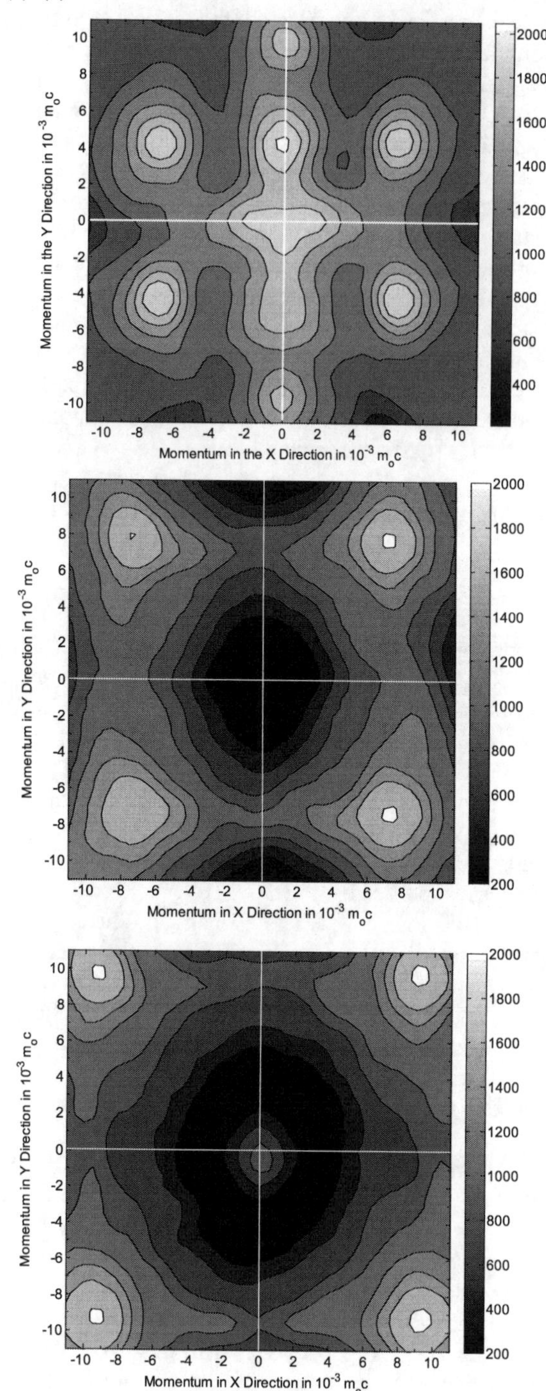

FIGURE 2. Top: (a) Efficiency-corrected 2D ACAR spectrum for virgin 6H SiC (oriented 45° with respect to normal to the (0001) surface). Middle: (b) Efficiency-corrected 2D ACAR spectrum for irradiated, un-annealed 6H SiC. Bottom: (c) Efficiency-corrected 2D ACAR spectrum for irradiated 6H SiC, annealed at 1000°C for 30 minutes.

The 2D ACAR spectrum for virgin 6H SiC (figure 2(a)) compares well with the measured and computed results by Kawasuso et al. [9] when the different measurement orientation is considered. The major anisotropy features in figure 2(a) centered at (7.0,4.5), (7.0,-4.5), (-7.0,-4.0), and (-7.5,4.0) are nearly identical to anisotropy features in Kawasuso's calculated and measured spectra. The ACAR features representing momentum maxima are consistent with summed momenta contributions along C-Si bonding directions in the measurement plane perpendicular to the gamma propagation direction. These features are very different from the ACAR spectra that Kawasuso et al. reported for Si vacancies that persist above 1000°C annealing temperature.

2D ACAR spectra for the O-ion irradiated samples exhibit differences and similarities as compared to the ACAR spectrum for virgin 6H SiC. Projections of momentum along the y-axis in the virgin 6H SiC spectrum disappear, the direction of the spots in the four quadrants change and the magnitude of the momentum increases for the quadrant spots in irradiated SiC. The spectra for the O-ion irradiated samples are consistent with the geometry of the O atom defect predicted by our modeling results.

The 2D ACAR spectra of O-ion irradiated 6H SiC is dominated by the four major anisotropy features. The direction of the four observed momentum anisotropies in the spectra for the O-ion irradiated samples (figures 2(b) and 2(c)) changed from ~33° relative to the x-axis in the virgin SiC 2D ACAR spectrum to ~45°, consistent with a shift in the positron localization. Projections of momentum along the y-axis vanish as these momentum contributions are spoiled by the local SiC distortion and the shift in positron localization from carbon-atom centers to oxygen in interstitial locations. The four momentum anisotropies in the ion irradiated, un-annealed spectrum (figure 2(b)) approximately corresponded to the four features from the virgin spectrum centered at (7.0,4.5), (7.0,-4.5),(-7.0,-4.0), and (-7.5,4.0) except that the momentum is increased and the angle with respect to the x-axis changed from ~33°, the projection of the Si-C-Si bond angle in measurement plane, to ~45° as the center of symmetry is shifted to the oxygen atom. Additionally, the magnitude of e^--e^+ momentum in the plane perpendicular to the propagation direction of the annihilation photons increased for the quadrant features from $8.1 \times 10^{-3}\ m_oc$ to $10.6 \times 10^{-3}\ m_oc$ consistent with a positron localized on more electronegative oxygen atom.

We attribute the ACAR spectra for O-ion irradiated samples to O atoms in interstitial sites rather than O atoms in atom vacancies produced by ion irradiation, or to vacancy defect sites produced by O-ion bombardment. Substitution defects are expected to produce quantitative differences but not the qualitative spectra differences such those as we observed. The 2D ACAR spectrum for the irradiated and annealed sample, shown in figure 2(c), is qualitatively similar in shape to that of the un-annealed sample in Figure 2(b), suggesting that vacancy-type defects contributed little to these ACAR spectra. Only quantitative changes are observed in the four major anisotropy features when a majority of the vacancy-type defects were annealed. Annealing increased the magnitude of the momentum for the quadrant features and produced a small feature near zero momentum.

In the spectrum for the annealed sample, the magnitude of the momentum of the four observed features increased $2.8 \times 10^{-3}\ m_oc$, from 10.6 to $13.4 \times 10^{-3}\ m_oc$ along the same 45° direction in the measurement plane. Since the direction of these momentum features correlated with the same orientations in the ion irradiated, annealed 6H SiC, the location of the O atom relative to the Si-C structure did not significantly change due to the annealing process. Relaxation of the Si-C structure communicating with the O atom may account for the differences. The small feature near zero momentum and the increase in feature momentum for the annealed sample represents a lattice relaxation that allows the O-atom valence electrons to communicate more efficiently with electrons of the SiC.

The ratio curve for the O^{3+} ion irradiated 6H SiC, both un-annealed and annealed, relative to the virgin sample was calculated by normalizing the DB lineshapes for the irradiated samples to the virgin sample, discussed in [3]. As shown in the ratio curve displayed in figure 3, more positrons interacted with low momentum valence electrons than higher momentum electrons in the ion irradiated, annealed sample compared to the virgin 6H SiC but to a lesser degree than the ion irradiated, un-annealed sample. More e^--e^+ annihilations were also observed in the ion irradiated, annealed 6H SiC in the valence and core electron momentum region compared to the virgin 6H SiC, once again to a lesser degree than the ion irradiated, un-annealed sample. Additionally, there were more positron interactions with core electrons after ion irradiation and annealing relative to the un-annealed, but still less than the virgin 6H SiC sample. This implied more core electrons were readily available to interact with after the annealing, which indicated the O atoms that lodged into the vacancy-type defects due to the ion irradiation dislodged and most of the vacancies were filled with the Si and C interstitials. Not all vacancies were filled though because the ratio in the core electron region was still below the virgin sample.

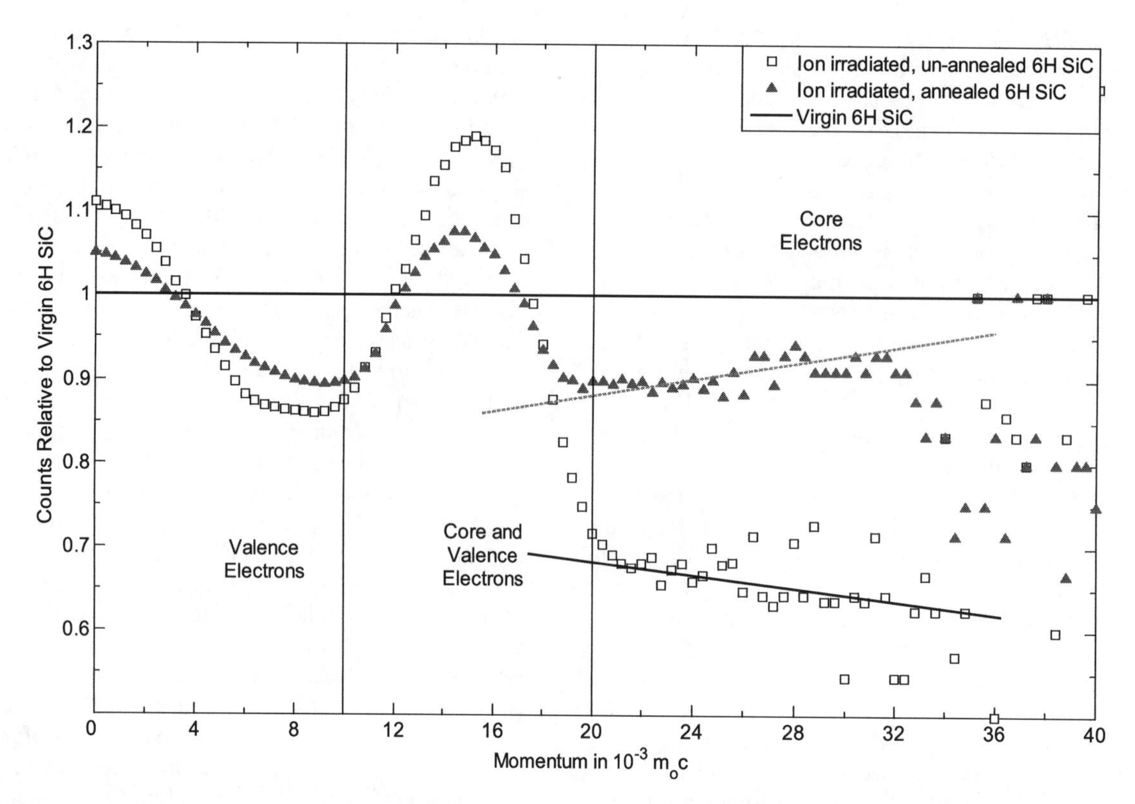

FIGURE 3. Ratio curve illustrating ion irradiated, un-annealed 6H SiC (squares) and ion irradiated, annealed 6H SiC (triangles) frequency of counts, relative to the virgin 6H SiC sample (straight line at Y = 1). Dashed lines indicate least squares linear fit to count distribution in core electron region.

CONCLUSIONS

Both the ACAR and DBAR experimental results and theoretical predictions show that O-atom defects are dominated by an interstitial O atom which couples to the SiC electron momentum distribution. This prediction was qualitatively consistent with 2D ACAR spectra for O^{3+}-ion irradiated samples, both with and without annealing. O-ion irradiation of 6H SiC resulted in interstitial O atoms which significantly distorted the local SiC structure, both with and without annealing. Based on these QM/MM calculated results, the O atom inserts into the Si-C bond, locally perturbing the SiC lattice. Oxygen atoms insert between Si and C atoms to form a Si-O-C bond having bond angle of about 150° breaking the Si-C bond as indicated by the increase of the Si-C separation by 0.9 angstroms. This process may require thermal activation of annealing to complete.

ACKNOWLEDGMENTS

Views expressed in this article are those of the authors and do not reflect the official policy or position of the Air Force, Department of Defense, or the U. S. Government.

REFERENCES

1. G. Harris, *Properties of Silicon Carbide*, United Kingdom, Short Run Press Ltd, 1995.
2. C. Williams, L. Burggraf, P. Adamson, J. Petrosky, Oxley M, *J. Phys Conf. Ser.* 225 012058 (2010)
3. C. Williams, L. Burggraf, P. Adamson, J. Petrosky, submitted to *Nucl. Instr. Meth. Phys. Res.* A, Jun 2010
4. A. Uedono et al.. *Appl. Phys.* 86, 5392-8 (1999)
5. J. Ziegler, J. Biersack, M. Ziegler. *The Stopping and Range of Ions in Solids*. New York, Pergamon Press, 2009
6. S. Agostinelli et al., *Nucl. Instr. and Meth.* Phys. Res. 506, 250-303 (2003)
7. J. Allison et al., *IEEE Trans. Nucl. Sci.* 53, Num. 1, 270-278 (2006)
8. J.R. Shoemaker, L.W. Burggraf, M.S. Gordon, *J. Phys. Chem.* 103, 3245 (1999)
9. A. Kawasuso, T. Chiba, T. Higuchi, *Phys. Rev.* B 71, 193204:1-4, 2005

Directed Neutron Beams From Inverse Kinematic Reactions

J.R. Vanhoy[a], N.A. Guardala[b], G.A. Glass[c]

[a]*Department of Physics, 572C Holloway Road, US Naval Academy, Annapolis, MD 21402 USA*
[b]*Positive Ion Accelerator Facility, NSWC-Carderock Division, West Bethesda, MD 20817 USA*
[c]*Louisiana Accelerator Center, Physics Department, University of Louisiana at Lafayette, Louisiana 70504 USA*

Abstract. Kinematic focusing of an emitted fairly mono-energetic neutron beam by the use of inverse-kinematic reactions, i.e. where the projectile mass is greater than the target atom's mass, can provide for the utilization of a significant fraction of the fast neutron yield and also provide for a safer radiation environment. We examine the merit of various neutron production reactions and consider the practicalities of producing the primary beam using the suitable accelerator technologies. Preliminary progress at the NSWC-Carderock Positive Ion Accelerator Facility is described. Possible important applications for this type of neutron-based system can be both advanced medical imaging techniques and active "stand-off" interrogation of contraband items.

Keywords: Neutron beam, Heavy Ion Reactions, Directed Beams, Inverse Kinematics
PACS: 29.25.Dz, 29.20.Ba, 29.27.Fh.

INTRODUCTION

Advanced techniques of neutron-based medical imaging [1,2,3] and contraband interrogation /inspection require uni-directional neutron beams in order to be highly effective and productive in locating and identifying objects buried beneath layers of material. Commonly used neutron production reactions produce beams which are typically isotropic in their intensities. Many neutrons are wasted because forming a beam requires absorption of all neutrons not traveling in the desired direction. Large radiation fields (in some cases consisting of both penetrating gamma rays as well as fast neutrons) that occur in the vicinity of the neutron production target are hazardous to both the individuals involved with beam production and others who are in close proximity of the areas where the objects are located. It is very desirable to limit the neutron radiation exposure to those individuals and also achieve a good deal of compactness of the neutron production system. Kinematic focusing by the use of inverse-kinematic reactions can provide a more efficient utilization of the neutron yield and provide a safer environment.

CANDIDATE REACTIONS

Kinematics

Formulae and tools for calculating reaction kinematics can be found in many references, among which are Marion and Young [4] and online at the Nuclear Reactions Video Low-Energy Knowledge Base [5].

In forward reactions, the light reactant serves as the projectile. The center of mass (cm) velocity of the reaction tends to be smaller or comparable to the speed of the created neutron in the cm system, and as a result, lab-frame neutrons emerge at all angles. One can overcome this problem by greatly increasing the kinetic energy of the incident beam, but the size of the accelerating structure grows rapidly and the neutron energy increases beyond the useful range.

In inverse kinematics reactions, where the heavy ion is used as the projectile and strikes the low-mass production target, the cm velocity dominates the cm neutron speed just above threshold and the reaction products tend to emerge in the forward angles.

Similar techniques were under development at the Kharkov Institute of Physics and Technology in Ukraine [6,7]. We examine the potential for using a tandem electrostatic accelerator as these tend to be more compact, lighter weight, and easier to maintain than linacs or small heavy-ion cyclotrons [8,9].

Application of Accelerators in Research and Industry
AIP Conf. Proc. 1336, 463-468 (2011); doi: 10.1063/1.3586142

A search was made for neutron-producing reactions of hydrogen, helium, lithium, and beryllium reactions on stable isotopes with atomic number less than 32 using the 2003 Atomic Mass Evaluation [10, 11]. We retained those reactions with negative Q-values no less than -6 MeV. This Q-value cutoff provides a list of candidate reactions that are accessible with a medium-sized tandem accelerator. Results are presented in Table 1.

TABLE 1. Candidate Neutron Production Reaction Summary. Information presented includes Q-value, forward and inverse thresholds, the coulomb barrier height as a guide to the anticipated cross section, neutron energies obtained at threshold for the forward and inverse situation, and the tandem accelerator terminal potential required for various charge states at the inverse threshold. See text for detailed discussion of columns.

Reaction	Ion	Q (MeV)	Forward Threshold (MeV)	Inverse Threshold (MeV)	Coulomb Barrier Height (MeV)	Fraction Of Barrier Height	Neutron Energy at FWD Threshold (MeV)	Neutron Energy at INV Threshold (MeV)	Terminal Potential for Charge State At INV Threshold (MV)		
									1+	2+	3+
(d,n)	^{3}He	-5.328	8.89	13.31	0.74	7.2	0.713	1.600	6.7	4.4	
(d,n)	^{4}He	-4.190	6.30	12.52	0.70	6.0	0.353	1.396	6.3	4.2	
(p,n)	^{3}H	-0.764	1.02	3.05	0.41	1.9	0.064	0.573	1.5	1.0	0.8
(p,n)	^{7}Li	-1.644	1.88	13.09	1.03	1.6	0.030	1.439	6.5	4.4	3.3
(p,n)	^{9}Be	-1.850	2.06	18.40	1.30	1.4	0.021	1.666	9.2	6.1	4.6
(^{3}He,n)	^{6}Li	-1.975	2.97	5.91	1.84	1.1	0.111	0.440	3.0	2.0	1.5
(d,n)	^{16}O	-1.624	1.83	14.52	2.12	0.8	0.011	0.723	7.3	4.8	3.6
(d,n)	^{16}O	-3.194	3.80	20.13	4.04	0.8	0.032	0.899	10.1	6.7	5.0
(p,n)	^{14}C	-0.626	0.67	9.32	1.76	0.4	0.003	0.584	4.7	3.1	2.3
(^{3}He,n)	^{12}C	-1.147	1.44	5.71	3.22	0.4	0.019	0.307	2.9	1.9	1.4
(^{3}He,n)	^{14}N	-0.957	1.16	5.40	3.63	0.3	0.012	0.263	2.7	1.8	1.4
(^{3}He,n)	^{36}Ar	-1.312	1.42	16.96	7.59	0.2	0.003	0.405	8.5	5.7	4.2
(d,n)	^{12}C	-0.281	0.33	1.96	1.69	0.2	0.003	0.121	1.0	0.7	0.5
(^{3}He,n)	^{32}S	-0.779	0.85	9.03	6.93	0.1	0.002	0.238	4.5	3.0	2.3
(d,n)	^{36}Ar	-0.367	0.39	6.92	3.95	0.1	0.001	0.174	3.5	2.3	1.7
(^{3}He,n)	^{28}Si	-0.570	0.63	5.86	6.25	0.1	0.002	0.172	2.9	2.0	1.5

DISCUSSION

The candidate reaction summary in Table 1 provides 16 reactions. Information presented includes Q-value, forward and inverse thresholds, the Coulomb barrier height as a guide to the anticipated cross section, neutron energies obtained at threshold for the forward and inverse situation, and the tandem accelerator terminal potential required for various charge states at the inverse threshold. The table contains the commonly used forward reactions: ^{7}Li(p,n), ^{9}Be(p,n), and ^{3}H(p,n).

Cross sections for these reactions would depend on Coulomb barrier heights, Q-values, and the location of resonances. The Coulomb barrier height of the entrance channel is estimated as $Z_1 \bullet Z_2/(A_1^{1/3} + A_2^{1/3})$ MeV in the table. Cross sections are anticipated to be larger if the entrance channel kinetic energy in the cm system is greater than the Coulomb barrier height. Reactions at the top of the table should be prolific neutron producers while those at the bottom of the table are seriously hindered by the height of the barrier. Reactions with larger ratios of heavy/light masses will have stronger forward focusing.

At threshold, emerging neutrons have kinetic energy because of the cm motion. Values for the neutron energies at threshold are given for the forward and inverse reaction. The inverse neutron energies are much larger than those in the forward reaction situation, as would be expected due to the much larger cm momentum. The neutron energies tend to range from several hundred keV to a couple of MeV, and imply the interrogation technique will rely upon a mixture of neutron capture and inelastic scattering in the sample under study.

EXPERIMENTAL CONSIDERATIONS

Accelerator Facility

The Positive Ion Accelerator Facility at NSWC-Carderock is a National Electrostatics Corp 9SDH-2 Pelletron with 3.4 MV terminal potential. Ions may be prepared from an Alphatross™ rf source, a SNICS sputter source, and in the near future a bright TORVIS™ source. The two charging chains allow it to inject ~120 μA of beam. Carbon stripper foils are generally used and accelerated ions of charge states 1^+ to 3^+ are possible, although better overall performance is obtained with gas stripping at lower ion energies.

The last three columns of Table 1 provide the required terminal potentials to achieve the threshold for the inverse kinematic reactions. While most of the candidate reactions are possible on the Carderock machine, the $^{12}C + d$ reaction was chosen in the initial study because of the prolific nature of carbon beams from the SNICS source and the flexibility of accelerating carbon beams. Many ^3He candidates appear in the Table. These reactions would be difficult to implement because they would require a gas jet target with a very valuable gas. Tritium beams present regulatory problems. The remaining candidate reactions require a production target of hydrogen or deuterium. Possible options are a gas target, a metal hydride target, a thin foil, or a thick solid compound.

Neutron Production Targets

High-pressure gas-cell targets are not feasible because these require an entrance foil vacuum window, typically tantalum or molybdenum, and the range of the incident ions (~2 um [12]) is insufficient to penetrate practical foil thicknesses. A windowless gas-jet target would avoid this problem. The range of ions in a gas is rather large, tens of of millimeters, but this is insufficient to overcome the low particle density, ~10^{17}-10^{18} atoms/cm^3. [13,14,15].

Hydrogen and deuterium can be efficiently stored in materials such as titanium, uranium, and erbium [16,17]. These materials have high decomposition temperatures and are easy to cool. Hydrogen number densities of 6-9 • 10^{22} atoms/cm^3 can be obtained. This class of targets does have limitations. Heavy ion projectiles slow down rather quickly in these materials. Ions travel only 2-4 μm in these materials while exceeding the threshold energy for producing neutrons. This severely limits the reaction rate.

In order to increase the ion range but maintain a high number density, a low atomic mass target must be used. Some groups have had success with ices coated on metal substrates, but a hydrocarbon is the easiest to manage [18,19,20,21]. Hydrogen number densities are ~7 • 10^{22} atoms/cm^3 and the range of ions while above threshold energy is ~10 μm. Melting point temperatures for hydrocarbons are rather low, ~100 C, but the use of cross-linked polyethylene chain material will reduce the tendency of the target material to flow. Polyethylene may be used in either a thick or thin target design. For reactions with ~MeV inverse reaction thresholds, the thick target is appropriate. For reactions with ~10 MeV thresholds, a thin target design may be more appropriate in order to dump less energy in the polyethylene -- the lower energy ions serve no purpose in neutron production and only heat the material.

Production Rates

Several issues determine if the reaction will form a practical beam; 1) neutron production rate, 2) angular extent of the neutron beam, and 3) energy spectrum of the neutrons.

A standard expression for evaluating the infinitesimal event rate dI (counts/sec) in a 'detector' subtending a solid angle $d\Omega$ is:

$$dI = I_o \; n \; \frac{d\sigma}{d\Omega} \; x \; d\Omega, \qquad (1)$$

where I_0 is the incident projectile rate (particles/sec), n and x are the number density (atoms/cm^3) and thickness of the target (cm). This formula can be directly applied to a thin target. When a very thick target is used, the energy of the incident beam is reduced as it penetrates deeper into the material and one must take into account the energy variation in the cross section. Making a slight change in notation $d\Omega \rightarrow \Omega$, and $x \rightarrow dx$ for a finite solid angle and an infinitesimal thickness, the integral over target thickness can be changed into an integral over beam energy by making use of the stopping power -- energy loss by the projectile per distance dE/dx = S. After changing the variables one obtains:

$$I(\theta) = I_o \; n \int_{E_{beam}}^{E_{thres}} \frac{d\sigma}{d\Omega} \frac{dE}{S} \; \Omega, \qquad (2)$$

where the limits of integration are the energy E_{beam} at which the projectile enters the target surface down to the threshold energy E_{thres}. The stopping power depends on the energy of the projectile so that the integral must be done numerically. Parameterizations for S(E) may be found in a reference such as Anderson & Ziegler "Hydrogen: Stopping Powers and Ranges in All Elements" [22] or on the "Stopping and Ranges of Ions in Matter" website [12]. Cross sections are assumed to be isotropic in the center-of-mass. Intensities must then be transformed into the laboratory system using the angle and solid angle transformations given in Ref [4]. The integrand of expression (2), properly transformed, provides the energy spectrum at a given angle.

Several interesting features were discovered. It is easily appreciated that the strongest 'focusing' occurs near the threshold energy, but very few neutrons are produced due to the small cross section. The incident ion beam energy must be chosen significantly above threshold to obtain an appreciable yield. One must balance increased yield against opening angle of the neutron beam.

SPECIFIC REACTIONS

The $^2H(^{12}C,n)$ reaction

The carbon-12 on deuterium reaction is examined first because large beam currents are easy to obtain, the Carderock 9SDH can accelerate several carbon charge states to a variety of energies, a carbon beam provides a severe test of production target performance, and allows one to diagnose detector response and efficiency issues. For the purposes of predictions, cross section values from Reference [23] were used. The kinematics of the forward and reverse reactions are compared in Table 2.

TABLE 2. Neutron beam opening angles and energies for the ^{12}C – deuterium reaction. The pairs of energies chosen illustrate the extent of focusing above the reaction threshold

$^{12}C(d,n)$			$^2H(^{12}C,n)$		
d Energy (MeV)	n Angle Range (deg)	n Energy Range (MeV)	^{12}C Energy (MeV)	n Angle Range (deg)	n Energy Range (MeV)
0.40	0-180	0.03-0.09	2.0	0-11	0.1-0.2
1.0	0-180	0.4-0.7	2.5	0-43	0.0-0.4
2.0	0-180	1.0-1.7	3.0	0-60	0.0-0.6

Initial development uses an erbium deuteride target because it is thermally rugged and easy to fabricate while we await development of a cross-linked deuterated polyethylene target. A comparison of merits of various deuterium targets is provided in Table 3.

TABLE 3. Comparison of deuterium targets for an 8 MeV carbon beam. An energy of 8 MeV is chosen because there is a substantial neutron production rate.

Target Material	Range (μm)	Distance 8 – 1.9 MeV (μm)	Number density n (atoms/cm^3)
D_2 gas at STP	41	23	$5 \cdot 10^{19}$
Deuterated polyethylene	12	7.3	$8 \cdot 10^{22}$
Erbium deuteride	4.7	3.1	$6 \cdot 10^{22}$

Calculations of the angular emission of neutrons performed taking into account the stopping power and energy dependence of the cross section are shown in Fig 1. Experimental measurements have confirmed the near-constant yield as a function of angle in the

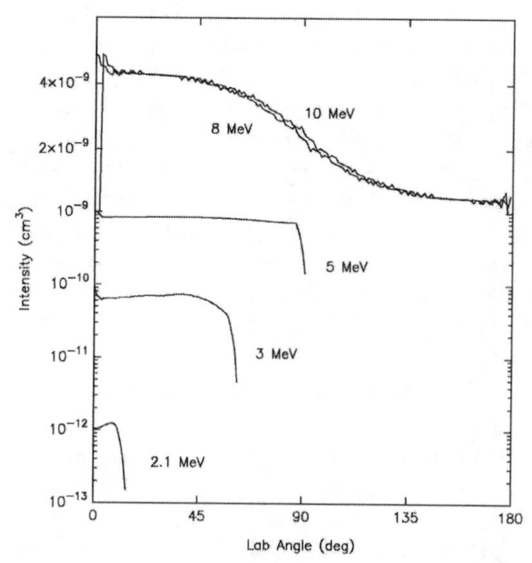

FIGURE 1. Predicted neutron intensity as a function of observation angle from a thick ErD_2 target for a selection of carbon beam energies. The vertical axis has a split log/lin scale and intensity is the integrand of Eq (2).

forward direction and "room background" in the rear direction. Carbon beams of 8 and 10 MeV produce neutrons at all angles although the yield at backward angles is greatly reduced. The production rate for 8 and 10 MeV curves are very similar because of the decreasing cross section in the Jaszczak data set [23] at those energies. For lower carbon beam energies, there exists a maximum lab angle for neutron emission.

Sample calculated energy spectra are shown in Fig 2. Despite the thick production target, neutrons are reasonably monoenergetic because the greatest numbers of neutrons are generated at the highest beam energies due to the rapidly increasing cross section above threshold.

The $^1H(^7Li,n)$ And $^2H(^{36}Ar,n)$ Reaction

The $^1H(^7Li,n)$ reaction has a higher projectile/target mass ratio of 7 and will produce stronger forward-focusing. The extent of compression is demonstrated in Table 4. Calculations of the angular emission of neutrons are shown in Fig 3. These kinematic considerations and the factor of 3 larger cross sections available with p + ^7Li may be offset with the difficulties encountered producing the lithium beam. While hundreds of microamps of carbon beam are easily produced in a sputter source, lithium beams are limited to ~0.8 μA [24]. High-current lithium sources are under development for

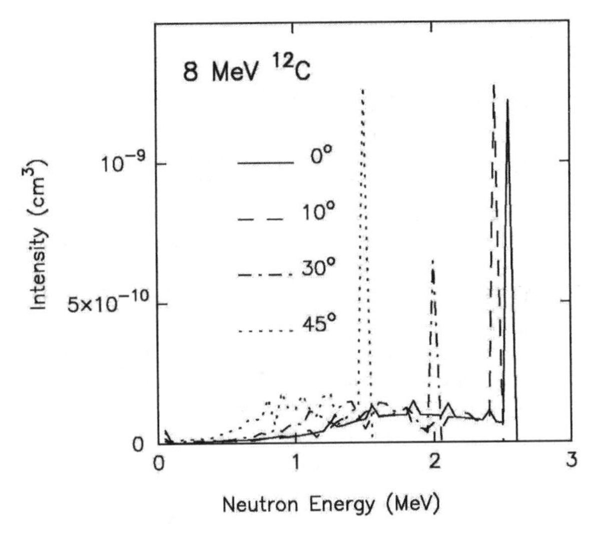

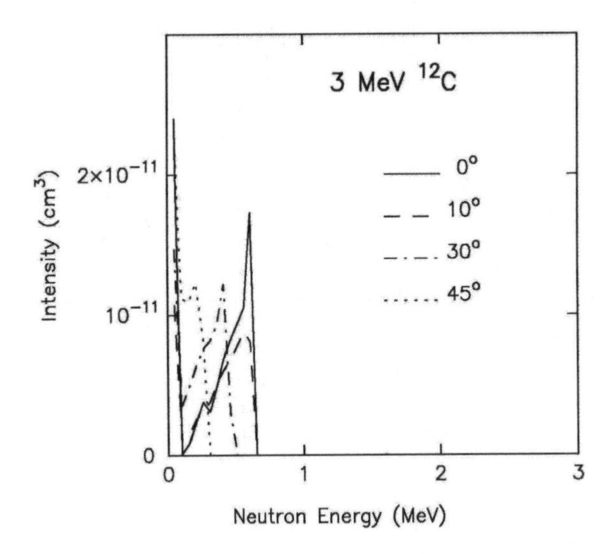

FIGURE 2. Calculated neutron spectra for two carbon beam energies. Vertical axis scale taken from the integrand of Eq (2).

TABLE 4. Neutron beam opening angles and energies for the ^{7}Li – hydrogen reaction. Equivalent pairs of energies are chosen.

^{7}Li(p,n)			^{1}H(^{7}Li,n)		
d Energy (MeV)	n Angle Range (deg)	n Energy Range (MeV)	^{12}C Energy (MeV)	n Angle Range (deg)	n Energy Range (MeV)
1.90	0-47	0.0-0.09	13.2	0-3	1.3-1.6
1.94	0-180	0.0-0.15	13.5	0-9	1.0-2.0
2.02	0-180	0.0-0.26	14.0	0-14	0.9-2.4

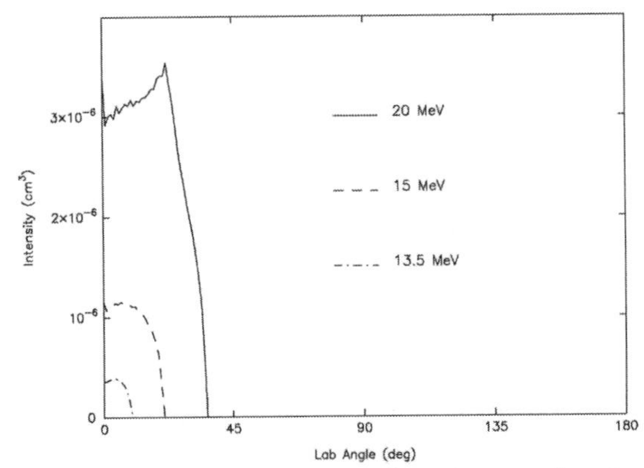

FIGURE 3. Predicted neutron intensity as a function of observation angle for ^{7}Li on a thick polyethylene target. The confinement of neutrons to forward angles over a wide range of bombarding energies is quite dramatic.

diagnostic measurements in plasmas [25], but are not available as particle accelerators ion sources at the present time.

The ^{2}H(^{36}Ar,n) reaction is another curious entry in the candidate reaction summary table, with a much larger mass ratio of 18. Measured cross sections are not known for this reaction, but are likely much smaller than all other entries in the table. Argon-36 has 0.34% isotopic abundance and this may prove to be a rather expensive reaction to investigate.

SUMMARY

Medical imaging and contraband interrogation require uni-directional neutron beams. Commonly used neutron source reactions require large amounts of shielding to form a beam and create a safety work area. Neutron focusing by use of inverse-kinematic reactions may provide a more efficient utilization of the neutron yield. A search identified 16 reactions of interest. The practicalities of implementing these reactions were discussed. The most fruitful reactions for a modest-size tandem electrostatic accelerator may be the carbon and lithium reactions on hydrogen and deuterium solid targets.

ACKNOWLEDGEMENTS

This work was funded in part by the Internal Research Investment of the Naval Surface Warfare Center, and the Office of Naval Research.

REFERENCES

1. C.E. Floyd Jr, J.E Bender, A.C. Sharma, A.J. Kapadia, J. Xia, B. Harrawood, G.D. Tourassi, J.Y. Lo, A. Crowell and C. Howell, *Phys. Med. Biol.* **51**, 3375 (2006).
2. C.E. Floyd Jr, et al., *Phys. Med. Biol.* **53**, 2313 (2008).
3. J.E. Bender, A.J. Kapadia, A.C. Sharma, G.D. Tourassi, B.P. Harrawood, C.E. Floyd, *Med. Phys.* **34**, 3866 (2007).
4. J.B. Marion and F.C.Young, *Nuclear Reaction Analysis: Graphs and Tables*, Amsterdam: North-Holland Publishing Co, 1968.
5. V. Zagrebaev and A. Kozhin, *Nuclear Reactions Video*, JINR Report No. E10-99-151, Dubna, 1999. and http://nrv.jinr.ru/nrv
6. V.M. Sanin, V.A. Bomko, B.V. Zaitsev, A.P. Kobets, Yu.P. Mazalov, Yu.V. Meleschkova, B.I. Rudjak, *Problems of Atomic Science and Technology*, **3**, 99 (1999).
7. V.A.Bomko, N.P.Diky, A.P.Kobets, Yu.P. Mazalov, Z.E.Ptukhina, B.I.Rudjak, V.M.Sanin, *Problems of Atomic Science and Technology*, **4**, 20 (1999).
8. B. Gikal, et al., *Phys. Part. and Nucl. Lett.*, **5**, 642 (2008).
9. B. N. Gikal, A. V. Tikhomirov, M. V. Khabarov, and O. A. Chernyshev, *Phys. Part. and Nucl. Lett.*, **5**, 392 (2008).
10. G. Audi, A.H. Wapstra, and C. Thibault, *Nucl. Phys.* **A 729**, 337 (2003).
11. A.H. Wapstra, G. Audi, and C. Thibault, *Nucl. Phys.* **A 729**, 129 (2003).
12. J.F. Ziegler, *The Stopping and Range of Ions in Matter*, http://www.srim.org. (2008).
13. H.W. Becker, L. Buchmann, J. Gorres, K.U. Kettner, H. Krawinkel, C. Rolfs, P. Schmalbrock, H.P. Trautvetter, and A. Vlieks, *Nucl. Instrum. Meth.* **198**, 277 (1982).
14. A. Redder, H.W. Becker, H. Lorenz-Wirzba, C. Rolfs, P. Schmalbrock, H.P. Trautvetter, *Z. Phys.*, A **305**, 325 (1982).
15. C.E.Rolfs and W.S.Rodney, *Cauldrons in the Cosmos*, Chicago: Univ of Chicago Press, 1988.
16. W. Grochala and P.P. Edwards, *Chem. Rev.*, **104**, 1283-1315, (2004)
17. I. Gabisa, E. Evarda, A. Voyta, I. Chernovb, Yu. Zaikab, *J. of Alloys and Compounds,* **356**, 353 (2003) .
18. G.T.J. Arnison, *Nucl. Instrum. Meth.*, **40**, 359 (1966)
19. M. Kusuhara, *Nucl. Instrum. Meth.*, **83**, 328 (1970)
20. L.M. Makosky and C. Hojvat, *Nucl. Instrum. Meth.* **74**, 342, (1969)
21. M.A. Olivo and G.M. Bailey, *Nucl. Instrum. Meth.*, **57**, 353 (1967).
22. H.H. Anderson and J.F. Ziegler, *Hydrogen Stopping and Ranges in All Elements, Vol. 3*, Oxford: Pergamon Press, 1977.
23. R.J. Jaszczak, R.L. Macklin, J.H. Gibbons, *Phys. Rev.*, **181**, 1428 (1969).
24. R. Middleton, "A Negative Ion Cookbook", University of Pennsylvania 1990 (unpublished) available online at http://www.pelletron.com/cookbook.pdf
25. R. Sah, M. Neubauser, A. Dudas, J.W. Kwan, "High Current Density Lithium Ion Source", in *Proceedings of the 1st International Particle Accelerator Conference IPAC10*, edited by A. Noda, Ch. Petit-Jean-Genaz, V. Schaa, T. Shirai, A. Shirakawa, Asian Committee for Future Accelerators, 4229 (2010) available online at http://epaper.kek.jp/IPAC10 .

Photon Activation Analysis Of Light Elements Using "Non-Gamma" Radiation Spectroscopy - The Instrumental Determination Of Phosphorus

Christian Segebade[a] and Wolf Goerner[b]

[a] *Idaho Accelerator Centre (IAC), Idaho State University, 1500 Alvin Ricken Drive, Pocatello, ID 83201, U.S.A.*
[b] *Bundesanstalt für Materialforschung und –prüfung, unter den Eichen 86, 12205 Berlin, Germany*

Abstract. Unlike metal determinations the analysis of light elements (e.g., carbon, oxygen, phosphorus) is frequently problematic, in particular if analysed instrumentally. In photon activation analysis (PAA) the respective activation products do not emit gamma radiation in the most cases. Usually, annihilation quanta counting and subsequent decay curve analysis have been used for determinations of C, N, O, and F. However, radiochemical separation of the respective radioisotopes mostly is indispensable. For several reasons, some of the light elements cannot be analysed following this procedure, e.g. phosphorus. In this contribution the instrumental PAA of phosphorus in organic matrix by activation with bremsstrahlung of an electron linear accelerator and subsequent beta spectroscopy is described. The accuracy of the results was excellent as obtained by analysis of a BCR Reference Material.

Keywords: Photon activation analysis, linear accelerator, light elements, coincidence counting, beta spectroscopy
PACS: 23.20.Lv; 24.30.Cz; 25.20.-x

INTRODUCTION

Using standard classical analytical methods trace determinations of light non-metal elements cannot be performed instrumentally in the most cases. Using radioanalytical methods for the determination of e.g. C, N, O, F, P, S normally radiochemical separation of the analytes from the sample matrix is required, too. Using instrumental X-ray fluorescence spectroscopy instrumental determinations of some of the light elements are possible, but normally trace determinations cannot be carried out. However, in several cases instrumental photon activation analysis (IPAA) can be used to determine C, N, O and P instrumentally at trace level. Since the photoreaction products of these elements decay by β^+ emission only classical γ-spectrometry cannot be applied to separate the analytes spectroscopically. There are two options to overcome this barrier. First, sequential multiple integration of the 511 keV annihilation line and subsequent analysis of the decay curve; second, measurement of the β^+ continuum exploiting the difference of the energy maxima of the respective activation products. Combining both options one can establish a method for instrumental analysis of the light elements mentioned above. Whilst element determination by decay curve analysis is a well-established standard method [1] instrumental analysis using β-ray spectroscopy has been reported rarely up to present [2]. Thus, the determination of phosphorus is emphasised in this paper. The small difference between the half lives of ^{15}O (122.2s) and ^{30}P (149.9s) does not allow a successful deconvolution of the respective decay curves.

DETERMINATION OF PHOSPHORUS IN BIOLOGICAL MATERIAL

Exposure Energy

Preliminary tests indicated 17 MeV bremsstrahlung as optimum between appreciable product activity yield and minimum influence of interfering reactions, $^{12}C(\gamma,n)^{11}C$ in particular; being one of the most abundant components in organic/biological material excessive integral background activity would be produced by carbon if irradiated at incident energies exceeding the reaction threshold of ca. 19 MeV. The oxygen reaction threshold is lower (see Tab. 1), but due to the irregular "Quasi-Non-GDR" type of the cross section function of $^{16}O(\gamma,n)^{15}O$ [3] the activity yield beneath 20 MeV is small, thus can be neglected in many cases; the difference in production rates O vs. P at 20 MeV

Application of Accelerators in Research and Industry
AIP Conf. Proc. 1336, 469-472 (2011); doi: 10.1063/1.3586143

irradiation energy is less than 1:600 (obtained experimentally).

β^+ Energy Distribution And Half-life Of The Product Nuclides

Spectrometry of electrons/positrons is performed most efficiently using large plastic scintillation detectors. These have a poor energy resolution capacity only. However, since the energy distribution β^+/β^- radiation describes a broad continuum high resolution spectroscopy is not required. The contribution of P to the integral β^+ continuum can be determined by evaluating the high energy part of the continuum. At excessively high oxygen contents the integrated region of interest (ROI) should range from ca. 1.8 MeV to ca. 3.5 MeV; at lower oxygen level the ROI might start at 1.2 MeV (the maximum of the ^{15}N β^+ continuum). If the activation products cannot be separated by β spectroscopy (e.g. N and O) the difference between their half-lives must be exploited for their instrumental separation.

Sample Preparation

To minimise the β absorption in the sample matrix and, yet more important, the absorptivity difference between the sample and the calibration material (also referred to as reference, standard or comparator) it is essential to provide small and quasi-equal thickness and average atomic number for both specimens. By experiment a thickness of 0.5 mm (about 150 mg·cm^{-2}) was found optimal regarding appreciably high count rate at minimum "smearing" of the electron energy distribution. Thus, the calibration specimens were prepared mixing pure Red Phosphorus or High Purity $(NH_4)_2HPO_4$ (the latter to provide a combined P and N standard) with chromatography cellulose powder of high purity, too. Thus, the matrices of both the samples under study (four different biological BCR Certified Reference Materials) and the synthetic calibration samples had very similar matrix compositions. Pure cellulose pellets were prepared as blank specimens. The spectra after activation of these served for monitoring the actual activation energy. However, the contribution of the matrix absorption to the total uncertainty budget is significant in any case. Thus, using a set of samples with stepwise increasing area densities at constant amount of P was analysed to obtain a respective correction factor F (ideally equals unity).

Activation

A 35 MeV S-Band electron linear accelerator (LINAC) was used for bremsstrahlung exposure. This machine was equipped with a beam-scanning device that provided a well-defined, homogeneous photon field so that sample and comparator material are activated uniformly [4]. The electron energy of the LINAC was set to 17 MeV in order to avoid activation of C and O. Sample and reference sample were stacked in a rabbit and transported to the LINAC by a quick pneumatic tube system. The standard exposure time was 300 s. The typical cooling period was about 60 s.

Spectroscopy

Two plastic scintillation detectors were applied for positron measurement. A pulse routing device was used to enable simultaneous operation of both measuring channels at common dead-time. Both detectors had equal efficiencies which was demonstrated by measurement of a ^{90}Sr source with a statistics better than 0.2%.The typical counting period was 200s.The resulting spectra were stored in an MCA. In this way the combined decay-and dead-time correction is automatically carried out by the counting electronics. There is no need to register operational periods (decay period, duration of counting). This facilitates the procedure tremendously.

Data Processing

The formula for calculating the P content C_{sm} of the sample is

$$C_{sm} = \frac{I_{sm} \cdot m_{ref} \cdot \eta_{ref}}{I_{ref} \cdot m_{sm} \cdot \eta_{sm}} \cdot F \cdot C_{ref} \quad (1)$$

$I_{sm(ref)}$ = number of counts in ROI of sample (reference)
$m_{sm(ref)}$ = mass of sample (reference)
$\eta_{sm(ref)}$ = efficiency of detector counting sample (reference)
F = correction factor
C_{ref} = P content of reference sample

Uncertainty Budget

As Eq. (1) is of pure multiplication/division type, one may express the budget in the form of relative uncertainties:

$$\frac{u(C_m)}{C_{sm}} = \frac{u(I_{sm})}{I_{sm}} + \frac{u(I_{ref})}{I_{ref}} + \frac{u(C_{ref})}{C_{ref}} + \frac{u(F)}{F} + \frac{u(\eta_{ref}/\eta_{sm})}{\eta_{ref}/\eta_{sm}} \quad (2)$$

By experiment F and η_{ref}/η_{sm} were found to nearly equal unity. Moreover, by a long-term experiment no significant drift of the detector photomultipliers was stated. Thus, Eq. 2 can be simplified to:

$$\frac{u(C_{sm})}{C_{sm}} = \frac{1}{\sqrt{I_{sm}}} + \frac{1}{\sqrt{I_{ref}}} + \frac{u(C_{ref})}{C_{ref}} + u(F) + u(\eta_{ref}/\eta_{sm}) \quad (2a)$$

In a series of analyses of a synthetic material batch the total uncertainty budget was found

$$\frac{u(C_{sm})}{C_{sm}} = 0.037$$

RESULTS AND DISCUSSION

BCR-CRM's 060 to 063 (different biological matrices) were analysed in a first run. Acceptable results were obtained at P levels between 0.1% and 1%. Determination of P at levels of 0.1% and lower needs repeated runs of the procedure. Because the P value of BCR-CRM-063 "Milk Powder" is certified it was analysed further (see Fig. 1).

An excellent agreement with the certified P value was found. The obtained uncertainty (k=2) of the average is ±1.5% (relative). It is smaller than the value ±3.7% in the uncertainty budget. The latter value may be justified in order to cover Type B uncertainties (preparation of the samples/reference specimens). Improvement could be achieved by direct gravimetric synthesis, mixing and pressing comparator by comparator combined with consecutive monitoring comparator against comparator. The realisation of larger P contents (about 10%) in the reference specimens would reduce their contribution to the Type A uncertainty (counting statistics). Moreover, the number of repetitious determinations could be enhanced up to 25 which would need only 4 hours of experimental time.

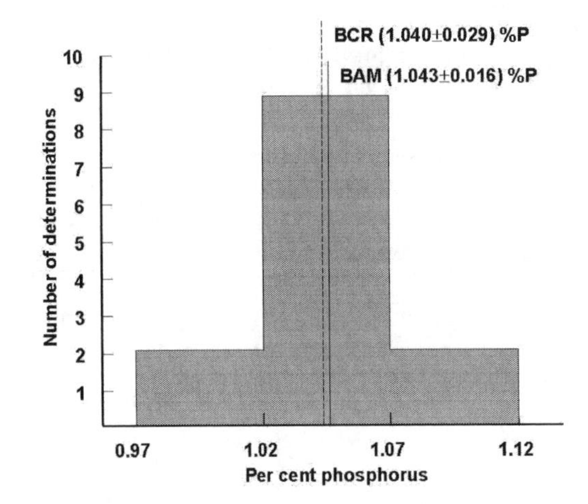

FIGURE 1. Histogram presentation of the phosphorus values of BCR CRM 063 (Milk Powder)

The future usage of two detector pairs will enlarge the counting budget by a factor of 2 and further reduce Type B uncertainty due to varying sample-detector-geometry. All these measures are supposed to allow extended uncertainties of the mean < 1% (relative).

CONCLUSION

The instrumental determination of P in biological and organic material by β^+- spectrometry after photo-activation has proven to yield reliable results at contents of 0.5% (abs.) and above. Further experimental measures can improve the precision and make limits of detection well below 100µg/g available. This non-destructive procedure has a productivity of 50 single determinations per day. It can be developed to a valuable reference method for validation and certification.

FUTURE PERSPECTIVE

TABLE 1. Photonuclear reactions of C, N, O and P

Element	Reaction	Threshold Energy, MeV	Half-life, s	$E\beta^+_{max}$, MeV
Carbon	$^{12}C(\gamma,n)^{11}C$	18.7	1223.1	0.961
Oxygen	$^{16}O(\gamma,n)^{15}O$	15.7(small σ) 20.0	122.2	1.73
Nitrogen	$^{14}N(\gamma,n)^{13}N$	10.6	597.9	1.19
Phosphorus	$^{31}P(\gamma,n)^{30}P$	12.3	149.9	3.24

A procedure for the simultaneous instrumental analysis of C, N, O and P can be developed. Regarding the physical data of the product nuclides involved (see Tab. 1) the determination of phosphorus using photo-activation is straightforward, e.g. in organic matter, making use of the extremely high maximum energy of the β^+ continuum of ^{30}P.

The known phosphorus value will facilitate the determination of N and O improving the quality of the respective results significantly. Analogously, a predetermined N content will reduce the uncertainty of the C value. The most fragile evaluation paths occur during O determination in the presence of P, and for C besides P and N. However, following the flow scheme shown in Fig. 2 a successful determination of the four elements will be possible even in these cases.

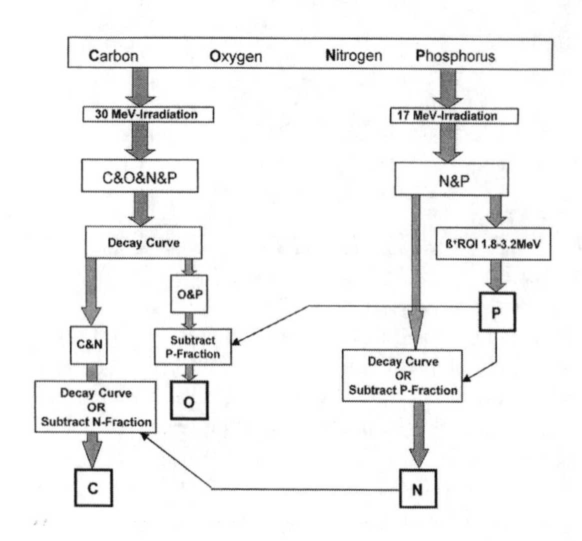

FIGURE 2. Scheme of instrumental separation of C, N, O and P

REFERENCES

1. Ch. Engelmann, J. Radioanal. Chem. 55 (1980) 379-390.
2. Ch. Engelmann, A.C. Scherle, Intern. J. Appl. Radiation Isotopes 22 (1971) 65-68.
3. B. Bülow, B. Forkman, Photonuclear Cross Sections, in: Handbook of Nuclear Activation Cross Sections, IAEA Technical Reports Series, No. 156 (1974).
4. W. Goerner, A. Berger, K.H. Ecker, O. Haase, M. Hedrich, C. Segebade, G. Weidemann, G. Wermann, J. Radioanal. Nucl. Chem. 248 (2001), 45-51.
5.

Standardizing Activation Analysis: New Software for Photon Activation Analysis

Z. J. Sun[1,2], D. Wells[1,2], C. Segebade[3], J. Green[1,2]

(1)Idaho Accelerator Center, Idaho State University, 921 S. 8th Ave. Pocatello, ID 83209-8263, U.S.A
(2)Department of Physics, Idaho State University, 921 S. 8th Ave. Pocatello, ID 83209-8106, U.S.A.
(3)Federal Institute for Materials Research and Testing, Unter den Eichen 87, Berlin 12205, Germany

Abstract. Photon Activation Analysis (PAA) of environmental, archaeological and industrial samples requires extensive data analysis that is susceptible to error. For the purpose of saving time, manpower and minimizing error, a computer program was designed, built and implemented using SQL, Access 2007 and asp.net technology to automate this process. Based on the peak information of the spectrum and assisted by its PAA library, the program automatically identifies elements in the samples and calculates their concentrations and respective uncertainties. The software also could be operated in browser/server mode, which gives the possibility to use it anywhere the internet is accessible. By switching the nuclide library and the related formula behind, the new software can be easily expanded to neutron activation analysis (NAA), charged particle activation analysis (CPAA) or proton-induced X-ray emission (PIXE). Implementation of this would standardize the analysis of nuclear activation data. Results from this software were compared to standard PAA analysis with excellent agreement. With minimum input from the user, the software has proven to be fast, user-friendly and reliable.

Keywords: Photon Activation Analysis (PAA), Photon Activation Analysis System (PAAS), Nuclear Analysis, Trace Elements Analysis
PACS: 82.80.Jp; 29.85.Fj; 82.50.Kx; 82.80.Ej

INTRODUCTION

Photon activation analysis (PAA) has been used for various multi-elemental analyses on biological, archaeological and environmental samples for decades. The PAA procedure usually includes: irradiation, measurement, peak reading and PAA analysis. Irradiation is normally performed by the high energy bremsstrahlung photon beam created by the electron LINAC through a high Z photon converter. After the irradiation, the sample is cooled down for a while before putting it into the high purity Germanium detector (HpGe) for measurement. The gamma spectrum collection is completed automatically by a computer and a series of electronics. Computer programs such as Gamma-W, Gennie 2K, etc., read the peak area of the spectrum by fitting the peaks with the Gaussian function modified by suitable auxiliary functions. Thus all the peaks including the multiplet can be automatically analyzed. After that all the peak and irradiation information is sent to the PAA analysis software for further concentration calculation.

The PAA analysis program was written via BASIC language in the 1970s. It is not a structured program and needs to be operated on the disk operation system (DOS). Some descriptions and commands in the program were written in German. To make the situation worse, the output of the program does not contain the uncertainty values. So, after forty years of development on both PAA and computer technology, the programs are not adequate anymore. A new program is needed to standardize the PAA analysis.

The drawbacks of the old program leads to the expectations of the new software: the program must have a friendly human-machine interface and should be able to run on the popular windows systems; the algorithm of the program should be well structured for the easy modification in the future; the output of the program must include the uncertainty propagation of the analysis; and finally, the program should be able to be accessed by the web browser if possible, which gives a possibility to use it anywhere the internet is accessible.

Application of Accelerators in Research and Industry
AIP Conf. Proc. 1336, 473-478 (2011); doi: 10.1063/1.3586144

DESIGN OF THE PHOTON ACTIVATION ANALYSIS SYSTEM

Based on the idea to manage the control file and the calculation process separately, a new computer program for photon activation analysis was designed, built and implemented at the Idaho Accelerator Center in 2010. It was named the Photon Activation Analysis System (PAAS) which accomplished the initial expectations of the PAA software.

1. Operating mode and language tools

PAAS is a database processing system. All the databases are stored in a data server. The operating modes of the PAAS are illustrated in figure 1. It can be operated in two modes: Client /Server (C/S) mode and Browser/ Server (B/S) mode.

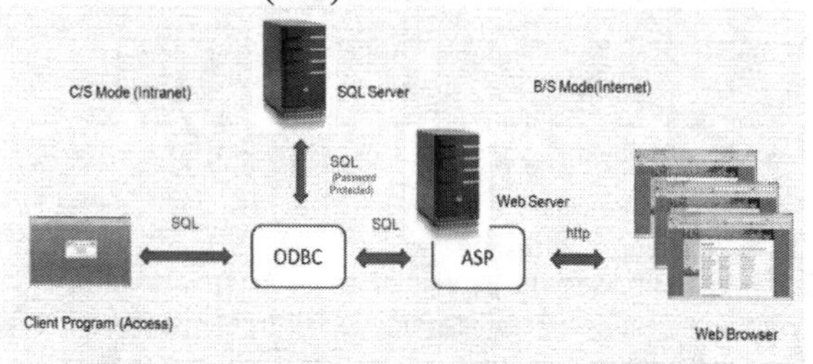

FIGURE 1. Operating Mode of PAAS

In C/S mode, the client program is run on one's personal computer. The client program gives orders to the open database connection (ODBC) and runs queries on the data server via the structured query language (SQL). After searching the databases and calculating the corresponding variables, the data server feeds the results back to the client program.

In B/S mode, the client program is a web browser, such as Firefox, Chrome, etc. When the browser visits the active server pages (ASP) on the web server, the ASPs are executed and ask queries to the data server through ODBC. The queries and calculations are performed by the data server and the information is sent back to result web pages. Simply by manipulating the web pages, the remote computer can input the data and get the output result. The link between the ODBC and the data server can be password protected in order to avoid random attacks from the internet.

As the operating modes show, it is necessary to use SQL, MS Access and Asp.net in designing PAAS. The program has been compiled in the MS Visual Studio 2010. Using Microsoft products solves the compatibility issues with windows and applying SQL secures the structured algorithm of the program. Some control and switch functions are realized through Visual Basic and C# for convenience. However, those VB and C# statements do not influence the main structure of the program.

2. Calculations

From the position of the energy lines, one can recognize the photonuclear reaction and deduce the target and product nuclides. From the intensity of the energy lines, one can calculate the quantitative information of the target nuclides. This is the fundamental principle of PAA. Although absolute evaluation might be possible in activation analysis[1], relative evaluation has advantages and is typically used in real experiments[2]. By matching the spectra of the sample (S) with that of the reference, one can figure out the relative concentration ratio of a certain element. At present, relative PAA analysis can be realized by three approaches: the analysis without monitor, the internal monitor method and the external monitor method. According to these methods, the PAAS divides into three parts with respective calculations and algorithms. However, the input and output interfaces of these parts are combined into one in the program.

In the method of analysis without monitor, the sample and the reference with the same geometry are put in the photon beam without any photon flux monitor. One assumes that the sample and the reference get the same amount of photon flux in exposure. According to Dr. Segebade's *Photon Activation Analysis*, the concentration of a certain isotope in the sample c_S is calculated by equation (1):

$$c_S = c_R \cdot \frac{M_R}{M_S} \frac{I_S}{I_R} \frac{e^{-\lambda T_{DR}}}{e^{-\lambda T_{DS}}} \frac{1-e^{-\lambda T_{CR}}}{1-e^{-\lambda T_{CS}}} \qquad (1)$$

In the equation above, c_S is the concentration of a certain target isotope in the sample, c_R is the concentration of the same target isotope in the reference; M_R is the total mass of the reference, M_S is the total mass of the sample; I_S is the net peak intensity of the product isotope in the sample, I_R is the net peak intensity of the corresponding product isotope in the reference; T_{DR} and T_{DS} are the respective decay times of the reference and the sample; T_{CR} and T_{CS} are the respective measurement times of the reference and the sample; λ is the decay constant of the certain product isotope.

[1] For NAA absolute evaluation, See Krane, Kenneth, *Introductory Nuclear Physics*, John Wiley & Sons, Inc, New York, 1988, p 462-465, p789-794.

[2] Miloslav, Rakovic, *Activation Analysis*, Iliffe Book LTD, London, 1970, p 2-p 7

Generally, the uncertainty of the decay constant is quite small. The uncertainty from the natural abundance is not precisely quantifiable. So we exclude them from the error propagation. Assuming time is recorded without any uncertainty. The error will propagate as equation (2),

$$\Delta C_S = C_S \cdot \sqrt{(\frac{\Delta C_R}{C_R})^2 + (\frac{\Delta M_R}{M_R})^2 + (\frac{\Delta M_S}{M_S})^2 + (\frac{\Delta I_R}{I_R})^2 + (\frac{\Delta I_S}{I_S})^2} \quad (2)$$

In the real experiments, reference concentration usually has 2- 5% error, mass might have 0.1% error or less, and the net peak intensity might have 5% error or more. Therefore, the result uncertainty primarily originates from the measurement fluctuation and the concentration error of the reference.

Analysis without monitor is an idealization. In practice, photon flux monitor is frequently used to track the flux ratio between the sample and the reference, either internally or externally. An internal monitor can be any isotope which exists both in the reference and in the sample. The concentrations of the internal monitor are known. In internal monitor method, one does not need to know the mass of the sample and the reference since this information will be cancelled out in calculation. Equation (1) changes its form as

$$c_S = f_i \cdot c_R \cdot \frac{M_R}{M_S} \frac{I_S}{I_R} \frac{e^{-\lambda T_{DR}}}{e^{-\lambda T_{DS}}} \frac{1-e^{-\lambda T_{CR}}}{1-e^{-\lambda T_{CS}}} \quad (3)$$

Where

$$f_i = \frac{\phi_R}{\phi_S} = \frac{M_S}{M_R} \frac{c_{Si}}{c_{Ri}} \frac{I_{Ri}}{I_{Si}} \frac{e^{-\lambda_i T_{DS}}}{e^{-\lambda_i T_{DR}}} \frac{1-e^{-\lambda_i T_{CS}}}{1-e^{-\lambda_i T_{CR}}} \quad (4)$$

f_i is the flux ratio between the reference and the sample for the internal monitor method.

The corresponding error propagation equation is

$$\Delta C_S = C_S \cdot \sqrt{\begin{array}{l}(\frac{\Delta C_R}{C_R})^2 + (\frac{\Delta C_{Ri}}{C_{Ri}})^2 + (\frac{\Delta C_{Si}}{C_{Si}})^2 + (\frac{\Delta I_R}{I_R})^2 + \\ (\frac{\Delta I_S}{I_S})^2 + (\frac{\Delta I_{Ri}}{I_{Ri}})^2 + (\frac{\Delta I_{Si}}{I_{Si}})^2\end{array}} \quad (5)$$

In external monitor method, flux monitors (usually nickel foils) are placed in both the front and back of the sample or the reference, equation (1) changes into

$$c_S = f_e \cdot c_R \cdot \frac{M_R}{M_S} \frac{I_S}{I_R} \frac{e^{-\lambda T_{DR}}}{e^{-\lambda T_{DS}}} \frac{1-e^{-\lambda T_{CR}}}{1-e^{-\lambda T_{CS}}} \quad (6)$$

Where

$$f_e = \frac{\phi_R}{\phi_S} = \frac{\phi_{R1} + \phi_{R2}}{\phi_{S1} + \phi_{S2}} = \frac{\frac{I_{R1}}{M_{R1}(1-e^{-\lambda_m T_{CR1}})e^{-\lambda_m T_{DR1}}} + \frac{I_{R2}}{M_{R1}(1-e^{-\lambda_m T_{C,R2}})e^{-\lambda_m T_{DR2}}}}{\frac{I_{S1}}{M_{S1}(1-e^{-\lambda_m T_{CS1}})e^{-\lambda_m T_{DS1}}} + \frac{I_{S2}}{M_{S1}(1-e^{-\lambda_m T_{CS2}})e^{-\lambda_m T_{DS2}}}} \quad (7)$$

The error propagation is more complicated than before,

$$\Delta C_S = C_S \cdot \sqrt{\begin{array}{l}(\frac{\Delta C_R}{C_R})^2 + (\frac{\Delta M_R}{M_R})^2 + (\frac{\Delta M_S}{M_S})^2 + \\ (\frac{\Delta I_R}{I_R})^2 + (\frac{\Delta I_S}{I_S})^2 + (\frac{\Delta f_e}{f_e})^2\end{array}} \quad (8)$$

Where

$$\Delta f_e = f_e \sqrt{\begin{array}{l}(\frac{\Delta R1 + \Delta R2}{\frac{I_{R1}}{M_{R1}(1-e^{-\lambda_m T_{CR1}})e^{-\lambda_m T_{CR1}}} + \frac{I_{R2}}{M_{R2}(1-e^{-\lambda_m T_{CR2}})e^{-\lambda_m T_{CR2}}}})^2 + \\ (\frac{\Delta S1 + \Delta S2}{\frac{I_{S1}}{M_{S1}(1-e^{-\lambda_m T_{CS1}})e^{-\lambda_m T_{CS1}}} + \frac{I_{S2}}{M_{R2} \cdot -e^{-\lambda_m T_{CS2}} e^{-\lambda_m T_{CS2}}}})^2\end{array}} \quad (9)$$

And

$$\Delta R1 = \frac{I_{R1}}{M_{R1}(1-e^{-\lambda_m T_{CR1}})e^{-\lambda_m T_{CR1}}} \cdot \sqrt{(\frac{\Delta I_{R1}}{I_{R1}})^2 + (\frac{\Delta M_{R1}}{M_{R1}})^2}$$

$$\Delta R2 = \frac{I_{R2}}{M_{R2}(1-e^{-\lambda_m T_{CR2}})e^{-\lambda_m T_{CR2}}} \cdot \sqrt{(\frac{\Delta I_{R2}}{I_{R2}})^2 + (\frac{\Delta M_{R2}}{M_{R2}})^2}$$

$$\Delta S1 = \frac{I_{S1}}{M_{S1}(1-e^{-\lambda_m T_{CS1}})e^{-\lambda_m T_{CS1}}} \cdot \sqrt{(\frac{\Delta I_{S1}}{I_{S1}})^2 + (\frac{\Delta M_{S1}}{M_{S1}})^2}$$

$$\Delta S1 = \frac{I_{S2}}{M_{S2}(1-e^{-\lambda_m T_{CS2}})e^{-\lambda_m T_{CS2}}} \cdot \sqrt{(\frac{\Delta I_{S2}}{I_{S2}})^2 + (\frac{\Delta M_{S2}}{M_{S2}})^2} \quad (10)$$

3. Algorithms

Figure 2 shows the algorithm structure of the program for the analysis without monitor. A gray rectangle represents an input table. The yellow rectangle is the library table. All tables are actually stored in the remote data server. We then manipulate the input data and library tables by the local client program through SQL. After executing an SQL query, a query table is generated, which is represented by a white rectangle. After several intermediate steps, the result query creates a printable report, which is represented by the green rectangle in the figure.

The library file is the most decisive table in the program. It includes the photonuclear reactions and the energy lines information of the corresponding product nuclides. If the information in the library is not accurate, all the following calculations are built on sand.

Peak data from peak reading software (Gamma W etc.) are placed into two input data tables: R_peak and S_peak, which are of the reference and of the sample respectively. The R_ppm table has concentration information of the reference. The Mass table has mass data. The Time table has irradiation, cooling and counting time information. The unit of energy is *keV*, peak area is in *counts*, concentration is in *ppm* and mass is in *mg*. Time parameters always use *second* as unit, even including the half life of nuclear decay.

The first step is peak recognition. It is conducted by the cross queries among the input peaks (R_Peak and S_Peak) and the library file (Lib). The program checks the energy of each record in the peaks table. If the energy is "close" enough to the character energy line of certain decay in the library, the program automatically keeps this record and assigns that certain decay and the corresponding half life to the record. For instance, the SQL statement to recognize the peak of the sample is:

```
SELECT lib.Energy, S_Peak.energy,
lib.Nuclide_before, lib.Nuclide_after,
S_Peak.Area AS P_S, S_Peak.Dev_Area AS Dev_P_S,
log(2)/(lib.half_life) AS Lamda,
lib.atomic_number
FROM lib, S_Peak
WHERE lib.energy Between [S_Peak.energy]-
[S_peak.Dev_Energy]-1.5 And
[S_Peak.energy]+[S_peak.Dev_Energy]+1.5
ORDER BY lib.atomic_number;
```

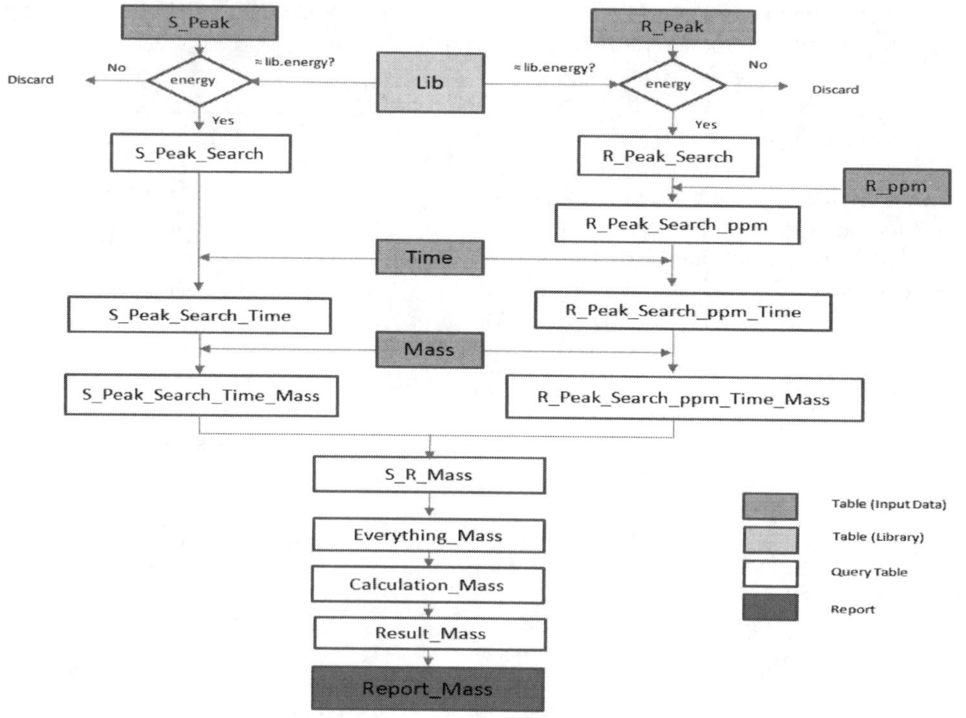

FIGURE 2. Algorithm of PAAS for the Analysis Without Monitor.

The above statement searches across the lib table and the S_peak table. If lib.energy is in the region `[S_Peak.energy]-[S_peak.Dev_Energy]-1.5` to `[S_Peak.energy]+[S_peak.Dev_Energy]+1.5`, the record is kept and the library.energy is the new energy line of the record. Besides the standard deviation, we add 1.5 keV on each direction of the deviation given by the fluctuation of the calibration. For gamma rays, add 1.5-2 keV will not cause excessive cross recognitions. But for X rays, 0.5-1 keV is recommend to add on each deviation direction since many X ray lines are close together.

After filtering the S_Peak and R_Peak tables, we have two new query tables: S_Peak_Search and R_Peak_Search. All the records in those two tables are valuable for PAA and several parameters are then assigned to the records by the cross queries among different tables. After cross queries the R_ppm table, the record gets the concentration information of the reference. After cross queries with the Time table and the Mass table, all the records will have time and mass values. Then the two tables from the reference and the sample will combine into a big joint table (SR_mass) according to the same energy lines of the records.

The SR_mass table has net peak area (P_R, P_S), but not net peak intensity (I_R, I_S). Further calculation leads to the query table everything_mass, which divides the net peak area with the counting time to get the net peak intensity. The final calculation is finished by the SQL statement below, following equation (1) and equation (2):

```
SELECT EVERYTHING_Mass.energy,
EVERYTHING_Mass.Nuclide_before,
EVERYTHING_Mass.Nuclide_after,
EVERYTHING_Mass.P_R, EVERYTHING_Mass.I_R,
EVERYTHING_Mass.Dev_I_R, EVERYTHING_Mass.Lamda,
EVERYTHING_Mass.atomic_number,
EVERYTHING_Mass.c_R, EVERYTHING_Mass.Dev_c_R,
EVERYTHING_Mass.T_Rz, EVERYTHING_Mass.T_RD,
EVERYTHING_Mass.T_RC,
EVERYTHING_Mass.R_Peak_Search_PPM_Time_Mass.mass,
EVERYTHING_Mass.R_Peak_Search_PPM_Time_Mass.Dev_ma
ss, EVERYTHING_Mass.P_S, EVERYTHING_Mass.I_S,
EVERYTHING_Mass.Dev_I_S,
EVERYTHING_Mass.S_Peak_Search_Time_Mass.mass,
EVERYTHING_Mass.S_Peak_Search_Time_Mass.Dev_mass,
EVERYTHING_Mass.T_Sz, EVERYTHING_Mass.T_SD,
EVERYTHING_Mass.T_SC,
EVERYTHING_Mass.c_R*(EVERYTHING_Mass.R_Peak_Search
_PPM_Time_Mass.mass/EVERYTHING_Mass.S_Peak_Search_
Time_Mass.mass)*(EVERYTHING_Mass.I_S/EVERYTHING_Ma
ss.I_R)*(Exp(-
EVERYTHING_Mass.Lamda*EVERYTHING_Mass.T_RD)/Exp(-
EVERYTHING_Mass.Lamda*EVERYTHING_Mass.T_SD))*((1-
Exp(-
EVERYTHING_Mass.Lamda*EVERYTHING_Mass.T_RC))/(1-
Exp(-EVERYTHING_Mass.Lamda*EVERYTHING_Mass.T_SC)))
AS c_S,
c_S*Sqr((EVERYTHING_Mass.Dev_c_R/EVERYTHING_Mass.c
_R)^2+(EVERYTHING_Mass.R_Peak_Search_PPM_Time_Mass
.Dev_mass/EVERYTHING_Mass.R_Peak_Search_PPM_Time_M
ass.mass)^2+(EVERYTHING_Mass.S_Peak_Search_Time_Ma
ss.Dev_mass/EVERYTHING_Mass.S_Peak_Search_Time_Mas
s.mass)^2+(EVERYTHING_Mass.Dev_I_S/EVERYTHING_Mass
.I_S)^2+(EVERYTHING_Mass.Dev_I_R/EVERYTHING_Mass.I
_R)^2) AS Dev_c_S
FROM EVERYTHING_Mass
ORDER BY EVERYTHING_Mass.atomic_number;
```

As one may notice in the above SQL statement, not only is the concentration of the sample calculated, but also the standard deviation of the concentration is given, which was not considered by the previous analysis software. All the records are ordered by the atomic numbers, so the order of nuclides in the result is the same as that of the period table.

To report the result, one merely needs to pick up several columns that one is interested in from the calculation table and send the data to a printable report file. The decimal place of the outputting numbers can be controlled in the report. Generally, we select three numbers behind the decimal point.

The algorithms of the internal monitor method and the external method are similar to that of the analysis without monitor, except that one needs to evaluate the flux monitor factors during the calculation.

IMPLEMENTATION OF THE PHOTON ACTIVATION ANALYSIS SYSTEM

The Brower/Server (B/S) mode of the program is particularly developed for internet access. The whole web site is integrated with a login security system. One needs a username and respective password to enter the system. Figure 3 is the process flow chart of the program in B/S mode.

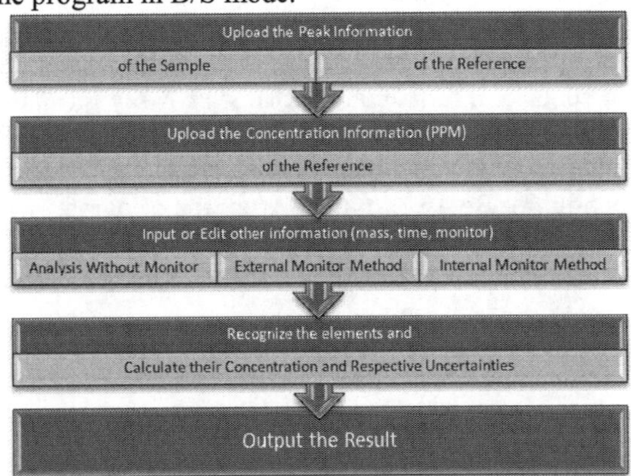

FIGURE 3. Process flow of the PAAS in B/S mode.

The first step is to upload the peak files and ppm file. Peaks and ppm files are in the excel format, which is common in the science community. After uploading, one needs to click the publish bottom and then the program will delete the old data (if any) and publish one's new data instead. On the publishing pages, the system will show the upload data with "edit" and "delete" options. If needed, one still can edit or delete any record at this time.

Now, the system has all the information it needs to identify the peaks and calculate the respective concentration values. Clicking next will lead to the final result page. On this page, the system gives all the isotope concentrations in the sample, their respective uncertainties, and the time parameters of the experiment.

The Client/Server (C/S) mode of the program is designed for intranet use. Compare with the B/S mode program, it is fast, safe and easy to modify for future expansion.

CONCLUSION AND FUTURE WORK

PAAS fulfilled our initial expectations of photon activation analysis software. It is user-friendly, reliable and has a clearly structured algorithm. It offers the function of multi-element recognition and performs concentration calculations through the SQL language based solely on comparison, whether the comparison is between the library and the input peak data or between input peak data of the reference and that of the sample. Furthermore, it also has error computation. However, compared with the DOS-based analysis software written in the 1970s, the program requires more support from the operating environment, either from software or hardware. Fortunately, most computers today can satisfy those requirements without any difficulty.

The library file decides the function of the program. If the library could be expanded to include neutron reactions, PAAS could be easily transferred to neutron activation analysis software. If the library file also contains information on character X ray lines, the program could even be used on PIXE. Furthermore, it is also possible to directly transfer the peak-reading function to PAAS. In that case, the program will combine the peak reading and analysis functions and become exclusive software in activation analysis.

ACKNOWLEDGMENTS

Photon activation analysis experiments were conducted in the Idaho Accelerator Center at the Idaho State University (ISU). The PAAS program was complied on the high performance computer in the Department of Physics at the ISU. This work was funded by the U.S. Department of Defense under grant No. DF-FC07-06ID14780.

REFERENCES

1. C. Segebade, H.-P. Weise, G.J. Lutz, *Photon Activation Analysis*, Walter de Gruyter, Berlin, 1988.
2. J. Ni, R.C. Block, X.G. Xu, *Applied Radiation and Isotopes*, 53, 535–540 (2000).
3. M. Yagi, K. Masumoto, *Journal of Radioanalytical and Nuclear Chemistry*, 84, 369–380 (1984).
4. N. Jerke, *Microsoft Office Access 2003 : professional results*, McGraw-Hill/Osborne, , New York, 2003.
5. M. Egea, et al., *Microsoft SQL server 2005: database essentials: step by step*, Microsoft Press, Redmond, WA, 2007
6. The Lund/LBNL Nuclear Data Search: http://nucleardata.nuclear.lu.se/NuclearData/toi/

A Permanent-Magnet Microwave Ion Source For A Compact High-Yield Neutron Generator

O. Waldmann and B. Ludewigt

Lawrence Berkeley National Laboratory, 1 Cyclotron Road, Berkeley, CA 94720

Abstract. We present recent work on the development of a microwave ion source that will be used in a high-yield compact neutron generator for active interrogation applications. The sealed tube generator will be capable of producing high neutron yields, $5\text{x}10^{11}$ n/s for D-T and $\sim 1\text{x}10^{10}$ n/s for D-D reactions, while remaining transportable. We constructed a microwave ion source (2.45 GHz) with permanent magnets to provide the magnetic field strength of 87.5 mT necessary for satisfying the electron cyclotron resonance (ECR) condition. Microwave ion sources can produce high extracted beam currents at the low gas pressures required for sealed tube operation and at lower power levels than previously used RF-driven ion sources. A 100 mA deuterium/tritium beam will be extracted through a large slit (60x6 mm^2) to spread the beam power over a larger target area. This paper describes the design of the permanent-magnet microwave ion source and discusses the impact of the magnetic field design on the source performance. The required equivalent proton beam current density of 40 mA/cm^2 was extracted at a moderate microwave power of 400 W with an optimized magnetic field.

Keywords: Neutron generator, Microwave ion source, Active interrogation
PACS: 29.25.Dz, 52.50.Sw, 52.80.Pi

INTRODUCTION

A high-yield, sealed-tube neutron generator (Fig. 1) was developed at the Lawrence Berkeley National Laboratory to address the needs of homeland security, non-proliferation, and nuclear safeguards applications.

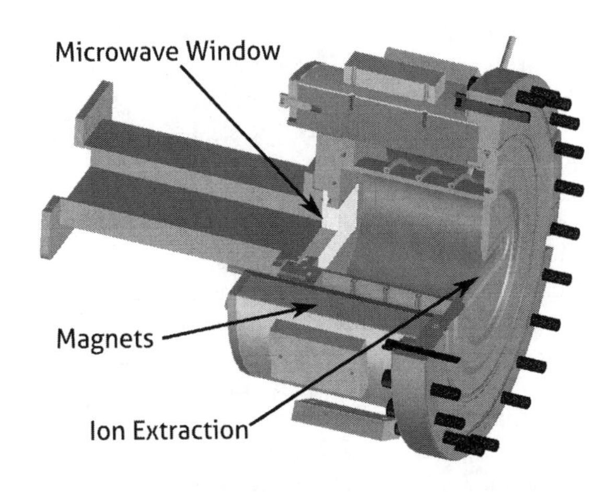

FIGURE 1. Microwave ion source. The plasma chamber is 100 mm long and 90 mm in diameter with a 60x6 mm^2 ion extraction slit. It is surrounded by seven rectangular permanent magnet bars. A protective sheet can be inserted in front of the microwave window.

This generator was designed to produce high neutron yields, exceeding $5\text{x}10^{11}$ n/s in D-T operation and reaching about $1\text{x}10^{10}$ n/s in D-D operation [1] when operating at 100 kV acceleration voltage and a 100 mA beam current impinging on a beam-loaded neutron production target. A V-shaped target geometry in combination with beam extraction from the ion source through a large 60x6 mm^2 slit serves to distribute the beam current and heat load over a larger area for better cooling of the target surface. Based on the assumption of a fully D-T loaded titanium target layer the amount of tritium in the generator is estimated to be roughly 200 Ci. However, this number depends on the thickness of the loaded target layer and could possibly be significantly lower for other targets.

Originally, this generator was equipped with an RF-driven plasma ion source that was specifically designed to operate at the low gas pressures required for sealed-tube generators. Initial testing of the generator was performed in D-D operation. While the RF-driven ion source was capable of providing the high beam current [1] the lifetime of the source was limited due to frequent sputtering. In recent years permanent-magnet microwave ion sources were developed [2-5] that offer significant advantages. They operate at low gas pressures of around 0.3 Pa,

Application of Accelerators in Research and Industry
AIP Conf. Proc. 1336, 479-482 (2011); doi: 10.1063/1.3586145

produce high plasma densities and a high atomic species fraction at significantly lower power levels than RF-driven ion sources, and offer long lifetimes.

This paper describes the development and design of a permanent-magnet microwave ion source for the high-yield neutron generator. The magnetic field design and its impact on the source performance is discussed, and initial results are presented.

MICROWAVE ION SOURCES

Microwave-driven ion sources offer high extracted beam current densities, high atomic fractions, and high power efficiencies at low gas pressure [6]. Efficient ionization and high plasma densities are achieved when the electron-cyclotron resonance condition is met, i.e., at a magnetic field of B = 87.5 mT for 2.45 GHz microwave frequency operation. Permanent-magnet ion sources, without field coils and their power supplies, are compact and easy to operate, which makes them attractive for use in a neutron generator.

The ion source described here, and schematically shown in Fig. 2, is based on an early proton microwave ion source that employed magnetic field coils [5]. In our design, the magnetic field is generated and shaped by permanent magnets and a ferromagnetic extraction electrode such that the value of 87.5 mT is reached very close to the extraction slit and at the vacuum window . Seven bars of permanent, rectangular NdFe-magnets surround the cylindrical plasma chamber (100 mm long and 90 mm in diameter, made of aluminum) to generate the magnetic field lines parallel to of the ion source axis. The plasma electrode extends radially beyond the magnets and matches the geometry of the existing neutron generator.

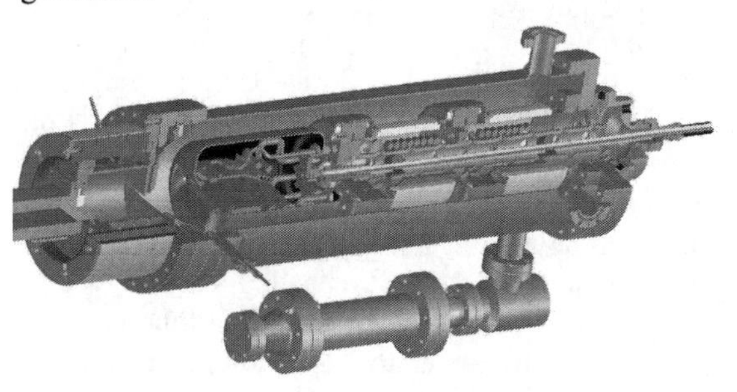

FIGURE 2. Neutron generator with microwave ion source. Dimensions 1 m in length and 0.3 m in diameter, weighting approx. 30 kg.

This design avoids the need for extending the permanent magnets beyond the plasma electrode and allows us to simply attach the source with a conflat flange to the existing generator tube. The ferromagnetic plate also greatly lowers the magnetic field in the extraction and acceleration region thus minimizing its impact on secondary, backstreaming electrons and avoiding possible discharge problems. The magnetic design was performed based on three-dimensional simulations with the code RADIA [7] that included the permanent magnets and all ferromagnetic materials. As can be seen in Fig. 3, the agreement between the simulated and the measured magnetic field along the axis of the source is very good.

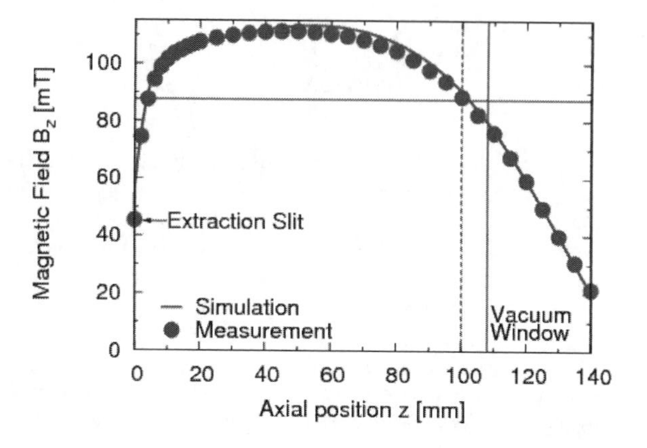

FIGURE 3. Magnetic Field. Measurement is shown in dots while the RADIA simulation is shown as a line. The dotted line marks the length of the chamber.

EXPERIMENTAL SET-UP

For testing and experimental characterization the microwave-driven ion source was mounted on a vacuum chamber equipped with an extraction system consisting of three electrodes and a Faraday cup for measuring the extracted beam current. Currents collected by the extraction electrodes were monitored to detect possible beam optics problems. Hydrogen gas was fed directly into the source chamber and pumped away through the vacuum chamber. Typical gas pressures inside the source were in the 0.1 to 0.4 Pa range and in the extraction section about $5x10^{-4}$ Pa. The pressures were measured with a baratron cell and an ionization gauge, respectively.

The microwave generator provided up to 1.5 kW of continuous wave or pulsed microwave power that was guided through a circulator, a two-stub tuner, and a ridged waveguide [8] for optimum coupling of the microwave power into the source. Forward and reflected powers were monitored.

In the neutron generator the 100 mA D-T beam current will be extracted through a 60x6 mm^2 slit corresponding to an average beam current density of about 25 mA/cm^2. However, in these experiments a proton beam was extracted through a 3 mm diameter aperture instead of the slit.

RESULTS

Magnetic Field Variation

In order to study the impact of magnetic field variations on source performance and to find the optimum magnetic configuration, we added a coil and ferromagnetic steel plates to the ion source. The coil was installed at the back of the source for changing the field predominantly at the microwave window and the steel plates were mounted outside of and parallel to the magnet bars as seen in Fig. 1 (pink blocks). By moving the plates radially in and out the magnetic field throughout the source volume is decreased or increased. Using the plates and the coil together, the shape of the magnetic field as produced by the permanent magnets and the ferromagnetic steel extraction plate can be varied. Fig. 4 shows the change in extracted beam current with coil current for a set of different radial positions of the field weakening plates.

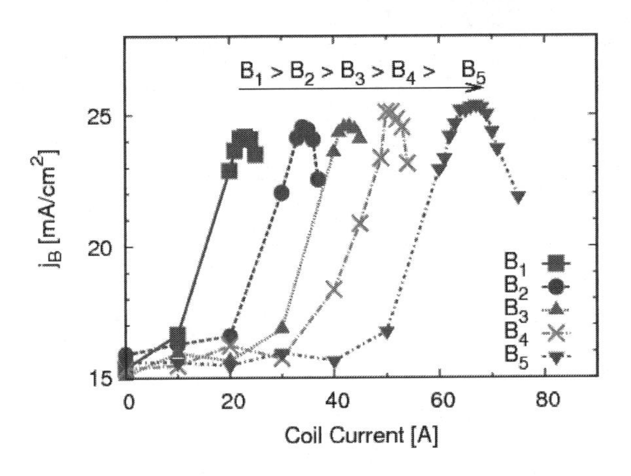

FIGURE 4. Extracted beam current densities for different magnetic fields. The permanent magnetic field is varied by different radial position of the steel plates while the field at the window is increased with the coil current.

In all cases the beam current was nearly doubled by increasing the coil current, i.e., the magnetic field at the window. The shift in optimal coil current for the different cases reflects the varying decreases in field strength due to the steel plates being at different radial positions. Fig. 5 shows simulations and measurements of magnetic fields along the axis of the source for the coil currents that give the maximum beam current for each radial position of the field weakening plates. As seen in Fig. 5 there is a disagreement between simulation and measurement at distances more than 20 mm away from the microwave window. This was not further investigated, since this study focuses on the field in the vicinity of that window. The magnetic fields differ quite strongly throughout most of the ion source but for all cases the ECR condition is satisfied (B_z = 87.5 mT) at the microwave window indicating that this is a crucial requirement for good source performance. We are now adjusting the permanent magnet configuration of our ion source to fine-tune the magnetic field at the window.

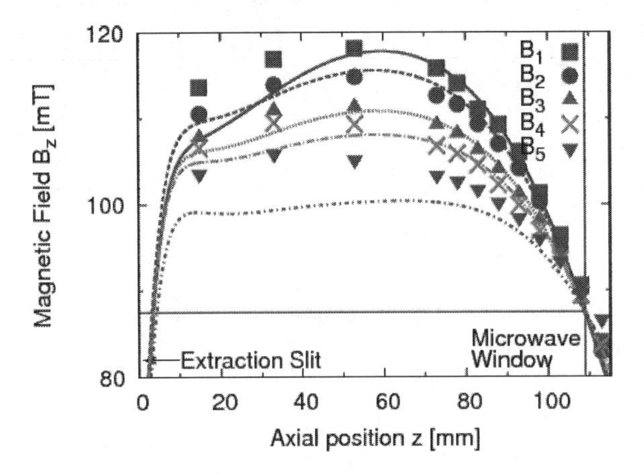

FIGURE 5. Magnetic fields for the peak current densities shown in Fig. 4. The measurements are shown as dots while the RADIA simulations are shown as lines. The ECR condition (87.5 mT) and the position of the microwave window are marked in black.

Variation Of Microwave Power And Window Material

The extracted beam current density was measured as a function of microwave power for two different window materials at a source pressure of 0.3 Pa, shown in Fig. 6. For both materials, aluminum nitride (AlN) and quartz, the current density increases linearly with power. The proton current density corresponding to the neutron generator design value of a D-T current density of 25 mA/cm^2 is 40 mA/cm^2. This value was reached at 400 W with an AlN window. When using a quartz window the required microwave power is about 30% higher.

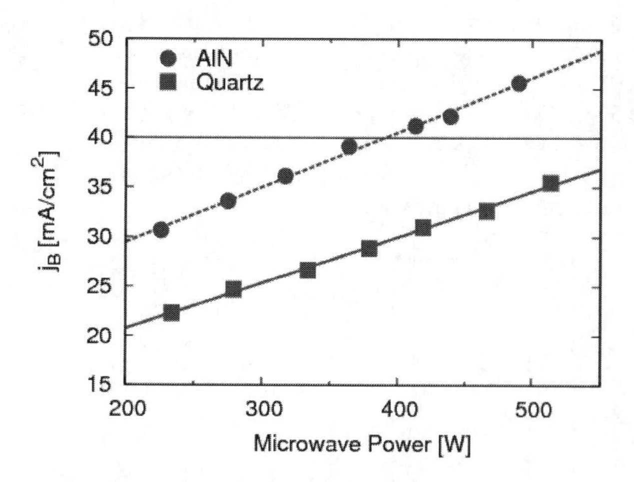

FIGURE 6. Change of extracted beam current with microwave power for an aluminum nitride and a quartz window. The target value of the beam current density is indicated by the black line.

CONCLUSION

A permanent-magnet microwave ion source was developed for a compact high-yield neutron generator. The required beam current density was extracted at a moderate microwave power of 400 W at the desired low gas pressures. This compares favorably with the power requirements (~1500 W) of a previously developed RF-driven source [9,1]. It was further found that the use of an aluminum-nitride vacuum window led to an approximately 30% higher extracted beam current than measured for a quartz window.

By varying the magnetic field, we demonstrated that meeting the ECR condition directly at the microwave window is crucial for good source performance. This finding helped to optimize the permanent magnet configuration.

ACKNOWLEDGMENTS

The authors would like to thank Qing Ji for many helpful discussions. This work was supported by the Office of Proliferation Detection (US DOE, NA-221) and the Director, Office of Science of the US Department of Energy at the Lawrence Berkeley National Laboratory under contract number DE-AC02-05CHI1231.

REFERENCES

1. B. Ludewigt et al., Nucl. Inst. and Methods in Phys. Res. B, **261**, 830-834 *(2007)*
2. W. Cornelius, Proc. Of the IEEE Particle Accelerator Conf., 1884-188 *(1999)*
3. Z. Song et al., Rev. Sci. Inst. **77**, 03A305 *(2006)*
4. Z. Yao et al.,., Rev. Sci. Inst. **79**, 073304 *(2008)*
5. O. Tuske et al., Rev. Sci. Inst. **79**, 02B710 *(2008)*
6. S. Jing et al., Rev. Sci. Inst. **72**, 92-95 *(2001)*
7. O. Chubar et al., J. Synchrotron Rad. **5,** 481-484 *(1998)*.
8. T. Taylor and J. F. Mouris, Nucl. Inst. and Methods in Phys. Res. A, 336, 1-5 (1999)
9. J. Reijonen et al., Appl. Rad. And Isotopes **63**, 757-763 *(2005)*

Testing The High-Energy Prompt Neutron Signature At Low Beam Energies

Scott J. Thompson[a], Mathew T. Kinlaw[a] and Alan W. Hunt[b]

[a]*Idaho National Laboratory, 2525 N Fremont Ave., Idaho Falls, ID, 83415*
[b]*Idaho Accelerator Center, 1500 Alvin Ricken Dr., Pocatello, ID, 83209*

Abstract. Prompt fission neutrons continue to be examined as a signature for detecting the presence of fissionable material. This technique exploits the neutron energy limitations inherent with photonuclear emissions from non-fissionable material, allowing prompt fission neutrons to be identified and engaged for detecting nuclear material. Prompt neutron signal measurements were acquired with bremsstrahlung endpoint energies of 6 MeV for 18 targets comprised of both fissionable and non-fissionable material; delayed neutron measurements were also collected as a reference. The ^{238}U target was also shielded with increasing thicknesses of lead or borated polyethylene to compare the resulting detection rates of the prompt and delayed fission neutron signals.

Keywords: Nuclear Material Detection, Active Inspection, Photofission, Prompt Neutrons
PACS: 25.85.Jg, 29.40.-n, 29.30.Hs

INTRODUCTION

Collaborative efforts between Idaho National Laboratory and Idaho State University's Idaho Accelerator Center continue to explore active inspection techniques for nonproliferation and national security applications. These techniques typically employ bremsstrahlung photon beams to stimulate materials of interest and monitor the resulting emissions for fissionable material signatures, which include prompt neutrons, delayed neutrons, and delayed gamma-rays[1, 2, 3]. In contrast to delayed neutrons, which are essentially exclusive to the fission process, prompt neutron emission is a readily available decay channel for many commonly-encountered isotopes. Prompt neutrons emitted during fission are produced with energy distributions described by a Watt spectrum, with allowable energies exceeding 12 MeV; however, neutron emitted via (γ,n) reactions are limited to maximum allowed energies that can be approximated by[4]:

$$E_{\max} = \frac{A-1}{A}\left(k_{\max} - Q\right),$$

where A is the atomic number of the target nucleus, $k_{\max}$ is the bremsstrahlung endpoint energy, and Q is the reaction threshold energy. An energy limit can then be defined above which a statistically significant number of detected prompt neutrons indicate the presence of fissionable material. As an example, for many potential applications, ^{207}Pb, with a (γ,n) reaction threshold of 6.7 MeV, will generally emit neutrons with energies higher than most standard nonfissionable materials[1]. Hence, the high-energy prompt neutron signature would be defined as all photoneutron emissions with energies

$$E_n \geq \frac{206}{207}\left(k_{\max} - 6.7\text{MeV}\right).$$

The Idaho Accelerator Center has developed a neutron time-of-flight (nToF) spectrometer specifically for measuring and exploiting the aforementioned prompt fission neutron signatures[5]. The spectrometer has proven to be an effective detection system, although an a priori understanding of a well-defined flight path and timing calibration are required for precise measurements[5]. A similar detection technology has also been examined which employs liquid scintillators and a combination of pulse shape discrimination and pulse height analysis to determine neutron energies as a way to exploit the prompt signature[6]. This system is considerably less efficient than the nToF spectrometer-based system and requires a much more complex supporting hardware/software suite. Further, simple neutron counting techniques can be used, eliminating the

Application of Accelerators in Research and Industry
AIP Conf. Proc. 1336, 483-486 (2011); doi: 10.1063/1.3586146

necessity for strict timing resolution and subsequent energy discrimination, by maintaining incident photon energies below the 6.7 MeV threshold. While photofission reaction cross-sections drop rapidly with decreasing incident photon energies, eliminating the requirement for nanosecond-scale timing resolution allows for much wider bremsstrahlung pulses and increased beam currents to compensate for the decrease in reaction rates.

This article describes a series of measurements assessing the high-energy prompt fission neutron signature at bremsstrahlung endpoint energies of 6 MeV. Several targets, comprised of both fissionable and non-fissionable materials, were investigated using a linear electron accelerator and simple ^{3}He-based neutron counters. Simultaneous measurements of delayed neutron emissions were obtained with the same neutron detection system. Delayed neutron emissions from photofission are well known as a reliable signature of fissionable materials; hence, these parallel measurements provide a reference for signature sensitivity assessment[2].

MEASUREMENTS

The bremsstrahlung inspection beam was generated by impinging 6 MeV electrons from the linac upon a 4.2 g/cm^2 tungsten radiator. The accelerator was operated at 100 Hz, producing 4μs wide pulses, each of which contained approximately 150 nC of charge. Remaining electrons were swept from the beam using a 600 G magnet. The resulting bremsstrahlung beam passed through a 1.83 m thick wall separating the accelerator hall from a shielded cell containing the target and neutron detection system. This wall consisted of earth sandwiched between 61 cm of concrete on either side. The beam was collimated on either side of the wall by 15.24 cm thick lead collimators, the primary of which had a diameter of 1.2 cm while the secondary's diameter was 3.7 cm. The collimated beam produced a circular beam spot at the target position with an approximate diameter of 6.5 cm.

A total of 6 neutron detectors were positioned approximately 1 m from the target at an angle of 90 degrees with respect to the beam direction. A schematic of the experimental setup is shown below in Figure 1. Each detector consisted of a 24.05 cm long (active region) ^{3}He gas filled proportional counter with an external diameter of 2.54 cm, and filled to a pressure of 1013.25 kPa (10 ATM). The tubes were surrounded with 2.54 cm of high-density polyethylene to moderate incoming neutrons to energies at which the signal-producing $^3He(n,p)^3H$ reaction becomes more probable[7]. Each detector was also surrounded by a 110 μm thick layer of natural cadmium and a 9.51 mm thick layer of 25.3 percent by mass borated silicone elastomer (Boroflex) to suppress thermal neutron backgrounds. These detectors were designed to operate in high-flux, active inspection environments and have been used extensively for measuring fundamental nuclear data constants pertaining to delayed neutron emissions[8, 9].

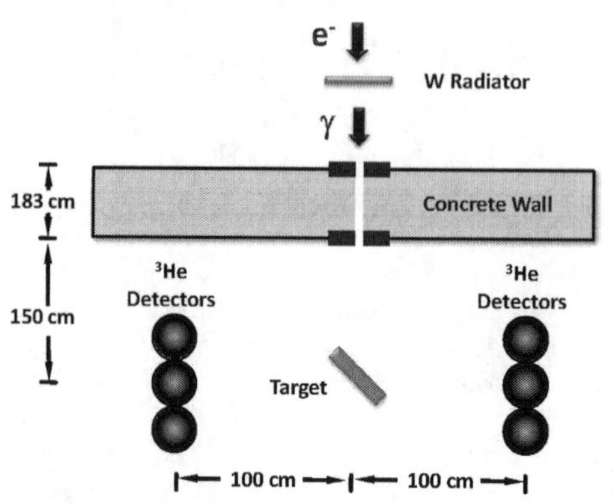

Figure 1. A schematic of the experimental setup for the measurements.

The detector signals were split and passed to 2 Canberra model 578 multichannel scalers (MCS) functioning as latched scalers with 2048 channels. The first MCS monitored the high-energy prompt signal, measuring counts during the first 2 ms following a pulse. The high-energy prompt signal was then obtained by summing the counts between approximately 10 and 500 μs following the pulse. The second MCS, configured to measure the delayed neutron signal, was set to span the entire duration (~10 ms) between pulses; the delayed neutron fission signature was obtained by summing all counts after approximately 1 ms. Figure 2 contains plots demonstrating the dual MCS setup and the summing regions for the two signatures.

The first measurements compared the two signatures for 18 targets consisting of fissionable and non-fissionable materials, listed below in Table 1. Figure 3 compares the high-energy prompt and delayed neutron fission signal strengths for each target, demonstrating the advantage of the high-energy prompt neutron fission signature of ^{238}U and ^{232}Th over the corresponding delayed signature for the two isotopes. The delayed neutron signal strength is greater than the prompt signal from ^{239}Pu; however, this elevated delayed signal is most probably due to a ~5% isotopic abundance of ^{240}Pu contamination in the target and because the delayed region is ~10 times larger than the prompt region[10]. Three non-fissionable targets (Be, Li, and ^{2}H) produced elevated prompt neutron count rates and could possibly produce false positives in detection systems utilizing this technique.

Beryllium, with a (γ,n) threshold of 1.9 MeV, emits neutrons similar to fissionable materials at all bremsstrahlung endpoint energies above the photofission thresholds; however, lithium and deuterium signal interference is inherent only to these types of simple counting, low beam energy detection methods. At higher beam energies, photoneutron emission energies from these materials fall below the high-energy signal threshold due to conservation of momentum.

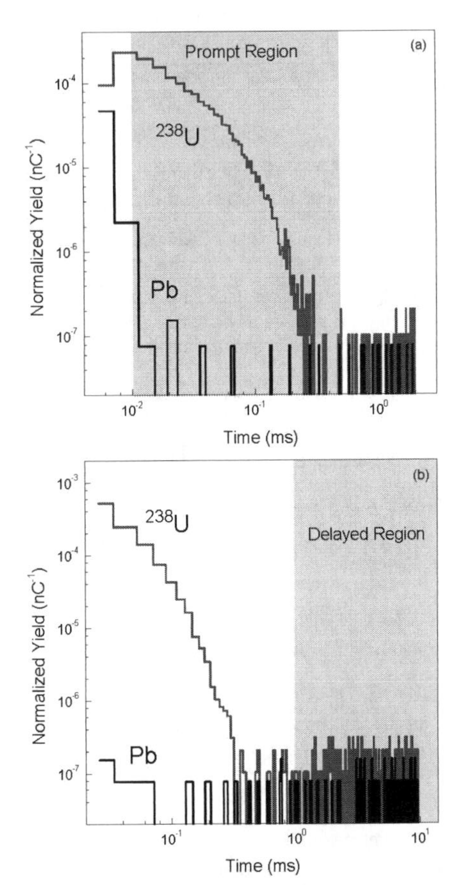

FIGURE 2. Examples of the dual multichannel scaler setup implemented in the presented measurements. Plot (a) demonstrates the MCS configuration and summing region used to measure the high-energy prompt signature, while Plot (b) shows the delayed signature collection setup.

Additional measurements tested the signature sensitivities with targets shielded by both high and low-Z materials. The depleted uranium (fission signal) and lead (background signal) targets were surrounded by increasing thicknesses of both lead and 5% borated polyethylene. Lead shielding attenuates the investigating bremsstrahlung beam, reducing induced reaction rates, and has small effect on neutron emissions. Conversely, borated polyethylene has little effect on the investigating photons, but moderates and captures neutron emission before they reach the detectors. The irradiation and recording times for these measurements were dependent upon shielding thickness and ranged from 10 to 60 minutes to ensure acceptable counting statistics were achieved.

TABLE 1. A list of the targets for which the high-energy prompt and delayed neutron fission signatures were measured. Reaction threshold values were taken from the Centre for Photonuclear Experiments Data's (CDFE) Ame2003 atomic mass evaluation (II) data library[11].

Target	Thickness /Amount	Lowest (γ,n) Threshold
DU Plates	36.4 g/cm^2	6.2 MeV
Pu Discs	12 g	6.2 MeV
Th Cylinders	~ 4 cm	5.7 MeV
Natural Pb Brick	28.8 g/cm^2	6.7 MeV (^{207}Pb)
Be Cylinders	14.1 g/cm^2	1.7 MeV
Steel Brick	19.9 g/cm^2	7.6 MeV (^{57}Fe)
Al Brick	13.7 g/cm^2	13.1 MeV
Cu Brick	45.42 g/cm^2	9.9 MeV (^{65}Cu)
W Brick	146.69 g/cm^2	6.3 MeV (^{183}W)
Ethanol	Ø8.9 cm bottle	4.1 MeV (^{17}O)
Borated Poly	5.08 g/cm^2	4.9 MeV (^{13}C)
Li_6CO_2 Poly	2.54 g/cm^2	5.7 MeV (^{6}Li)
Ream of Paper	4 g/cm^2	4.9 MeV (^{13}C)
Osmium Oxide	~ 2 g	5.9 MeV (^{189}Os)
Water	Ø8.9 cm bottle	4.1 MeV (^{17}O)
Heavy Water	Ø12 cm bottle	2.2 MeV (^{2}H)
Graphite Block	12.47 g/cm^2	4.9 MeV (^{13}C)
Concrete Brick	12.1 g/cm^2	7.9 MeV (^{43}Ca)

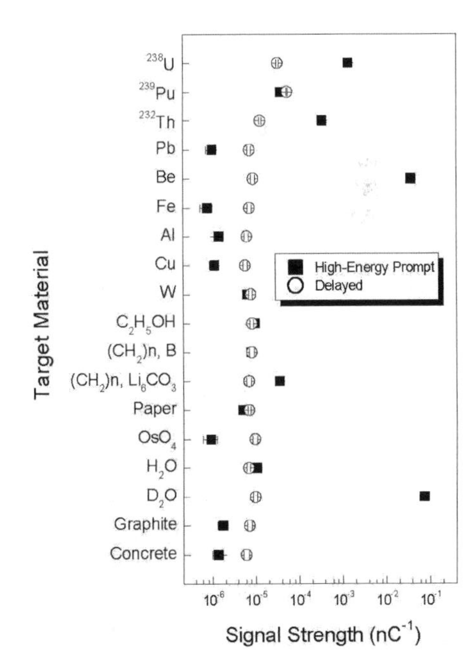

FIGURE 3. Measured high-energy prompt and delayed neutron fission signatures for the targets described in Table 1. Values have been normalized to total electron charge incident upon the tungsten radiator. Error bars represent 1σ statistical uncertainty.

Figure 4 presents the calculated statistical confidences for the measured signatures versus increasing shielding thickness. The high-energy prompt signature outperformed the delayed signature for both types of shielding. In the lead case, the high-energy prompt signature remained nearly an order of magnitude above the delayed at shielding thicknesses up to 7.62 cm. At greater thicknesses, the prompt signal continues to decline in a near exponential fashion, while the delayed fission signal plateaus at approximately 2σ above background. In the borated polyethylene case, the delayed neutron signal initially declines at the same rate as with the lead shielding; however, the decline in statistical confidence continues, falling below 1σ between 5.08 and 7.62 cm. Conversely, the high-energy prompt fission signal sensitivity declines at a much slower rate than with lead shielding case. At the thickest level of shielding measured (15.24 cm), the fission signal remains greater than 20σ above background.

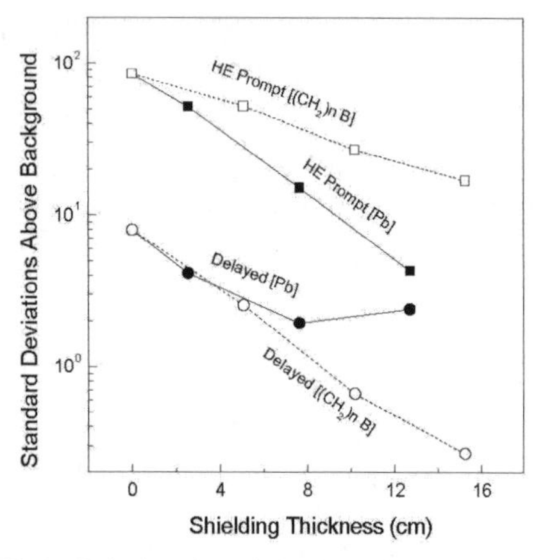

FIGURE 4. Calculated statistical confidence for the high-energy prompt and delayed neutron fission signature measurements with varying lead and borated polyethylene shielding thickness. The depleted uranium and natural lead targets described in Table 1 were implemented for the fission and background signals respectively.

CONCLUSIONS

The results of the measurements presented here demonstrate the simplicity and efficacy of implementing the high-energy prompt fission neutron signature in detection applications that require a lower investigating beam energy. At a bremsstrahlung endpoint of 6 MeV, the prompt signal is considerably stronger than the corresponding delayed signal for ^{238}U and ^{232}Th, while the delayed signal is slightly stronger for ^{239}Pu. The latter is somewhat misleading, as there existed a strong passive neutron background, attributed to ^{240}Pu in the target, and a factor of ~10 wider counting window for the delayed signal. Further, results of a basic shielding study support the justification of utilizing prompt neutrons to detect fissionable materials. When the ^{238}U target was shielded with increasing thicknesses of lead, the prompt signal remained nearly an order of magnitude above the delayed signal with shielding thicknesses up to 7.62 cm. Shielding the same target with borated polyethylene resulted in a delayed signal less than 1σ above background for thickness greater than 5.08 cm, while the prompt signal remained above 20σ for borated polyethylene thicknesses up to 15.24 cm.

ACKNOWLEDGMENTS

This work was supported in part by the U.S. Department of Energy National Nuclear Security Administration Office of Nonproliferation and International Security, Next Generation Safeguards Initiative. Idaho National Laboratory is operated for the U.S. Department of Energy by Battelle Energy Alliance under DOE contract DE-AC07-05-ID14517.

REFERENCES

1. S.J. Thompson, M.T. Kinlaw, J.F. Harmon, D.P. Wells, E.B. Farfan, and A.W. Hunt, *Appl. Phys. Letters* **90**, 074106 (2007).
2. M. T. Kinlaw and A. W. Hunt, *Appl. Phys. Letters* **86**, 254104 (2005).
3. E.T.E. Reedy, S.J. Thompson, and A.W. Hunt, *Nucl. Instr. Meth. A* **606**, 811(2009).
4. T. Gozani, "Active Nondestructive Assay of Nuclear Materials". Technical Report NUREG/CR-0602, United States Regulatory Commission, 1981.
5. S. J. Thompson, "A High-Energy Prompt Neutron Signature for the Detection of Fissionable Materials", Ph.D. Thesis, Idaho State University, 2010.
6. M.T. Kinlaw, J.T. Johnson, S.M. Watson, A.W. Hunt, and S.J. Thompson, "Laboratory Directed Research and Development FY-2009 Annual Report". INL External Report INL/EXT-09-17391, 2010.
7. G. F. Knoll, *Radiation Detection and Measurement,* New York: John Wiley and Sons, Inc., 2000.
8. J.L. Jones, W.Y. Yoon, K.J. Haskell, D.R. Norman, J.M. Hoggan, C.E. Moss, C.A. Goulding, C.L. Hollas, W.L. Meyers, and E. Franco, "Photofission-Based, Nuclear Material Detection: Technology Demonstration", Technical Report INEEL/EXT-02-01406, Idaho National Engineering and Environmental Laboratory, 2002.
9. M. T. Kinlaw, "Delayed Neutron Group Parameters from Photofission", Ph.D. Thesis, Idaho State University, 2007.
10. U.S. Department of Energy, "Uncertainty Analysis of the SWEPP Plutonium Sources (NADS and ZPPRS)", Technical Report EDF-3153, Idaho National Engineering and Environmental Laboratory, 2002.
11. A.H. Wapstra, G. Audi, and C. Thibault, *Nucl. Phys. A*, **729**, 337-676 (2003)

Studies Of Oxidation And Thermal Reduction Of The Cu(100) Surface Using Positron Annihilation Induced Auger Electron Spectroscopy

N.G. Fazleev,[a,b] M.P. Nadesalingam,[a] W. Maddox,[a] A. H. Weiss[a]

[a]*Department of Physics, University of Texas at Arlington, Arlington, Texas 76019, USA*
[b]*Department of Physics, Kazan State University, Kazan 420008, Russian Federation*

Abstract. Positron annihilation induced Auger electron spectroscopy (PAES) measurements from the surface of an oxidized Cu(100) single crystal show a large increase in the intensity of the annihilation induced Cu M2,3VV Auger peak as the sample is subjected to a series of isochronal anneals in vacuum up to annealing temperature 300°C. The PAES intensity then decreases monotonically as the annealing temperature is increased to ~550°C. Experimental positron annihilation probabilities with Cu *3p* and O *1s* core electrons are estimated from the measured intensities of the positron annihilation induced Cu $M_{2,3}VV$ and O KLL Auger transitions. PAES results are analyzed by performing calculations of positron surface states and annihilation probabilities of the surface-trapped positrons with relevant core electrons taking into account the charge redistribution at the surface and various surface structures associated with low and high oxygen coverages. The variations in atomic structure and chemical composition of the topmost layers of the oxidized Cu(100) surface are found to affect localization and spatial extent of the positron surface state wave function. The computed positron binding energy and annihilation characteristics reveal their sensitivity to charge transfer effects, atomic structure and chemical composition of the topmost layers of the oxidized Cu(100) surface. Theoretical positron annihilation probabilities with Cu *3p* and O *1s* core electrons computed for the oxidized Cu(100) surface are compared with experimental ones. The obtained results provide a demonstration of thermal reduction of the copper oxide surface after annealing at 300°C followed by re-oxidation of the Cu(100) surface at higher annealing temperatures presumably due to diffusion of subsurface oxygen to the surface.

Keywords: Positron, Annihilation, Spectroscopy, Oxidation, Copper, Beam, Auger
PACS: 78.70.Bj; 71.60.+z; 68.47.Gh; 68.47.De

INTRODUCTION

The study of oxidation of single crystal metal surfaces is important in understanding the corrosive and catalytic processes associated with thin film metal oxides. Copper has received much attention due to the use of copper and its oxides in a wide variety of applications such as gas sensors, metal to metal adhesion, various biomedical and catalytic applications [1-6]. A better understanding regarding various stages of copper oxidation would be of considerable interest in order to not only improve upon these technological processes but to possibly discover new technologies. The process by which oxide layers form on metal surfaces, however, is still not well understood. Recently, Kangas et al. explored reconstructed phases involving high coverage oxygen adsorption based on missing row reconstruction [7, 8]. From their results, they concluded that sub-surface oxygen adsorption is likely to take place at high coverages due to low transition barriers. Unlike conventional Auger techniques that average over several atomic layers, positron annihilation induced Auger electron spectroscopy (PAES) has a very high degree of top layer selectivity due to the fact that the positrons are trapped just outside the surface prior to annihilation, hence is ideal to probe the near surface region and study the initial stages of oxidation [9, 10]. In this paper we present the results of PAES studies of the oxidation and reduction behavior of copper oxide. PAES results are analyzed by performing calculations of positron surface states and annihilation probabilities of positrons trapped at the oxidized Cu(100) surface with relevant core electrons taking into account the charge redistribution at the surface and various surface structures associated with the low and high oxygen coverages. Such studies are indispensable for clarifying localization and spatial extent of the positron surface state wave function at the oxidized surface and the surface atomic structure and chemical

Application of Accelerators in Research and Industry
AIP Conf. Proc. 1336, 487-492 (2011); doi: 10.1063/1.3586147

composition dependences of positron annihilation characteristics.

EXPERIMENT

A Cu(100) single crystal (purity 99.999%) was used as a substrate. The sample 10 mm in diameter and 1mm in thickness was polished on one side with roughness less than 0.3 micron. Before mounting to the sample holder, the Cu single crystal was degreased ultrasonically in acetone and rinsed with ethyl alcohol and deionized water. The sample was then mounted on the sample holder, rinsed again with deionized water and dried using nitrogen gas (N_2). The base pressure of the PAES system was maintained at 1.0×10^{-9} Torr. The Cu(100) surface was cleaned in the vacuum chamber using a sequence of sputtering and low temperature (at 300°C) annealing cycles to remove impurities from the sample surface prior to the oxidation process. Oxidation of the Cu(100) surface to form a Cu_2O layer was performed as discussed in Refs. 11 and 12. In our experiment, a pressure of 5.0×10^{-4} Torr was selected as the oxygen (high purity 99.999%) partial pressure and the substrate temperature was maintained at ~ 590-595°C to obtain only a Cu_2O layer on the Cu(100) surface. After oxidation the sample substrate was annealed by a sequence of isochronal annealing cycles in vacuum at increasing temperature, holding the sample at each different temperature for 2 minutes. PAES spectra were acquired at room temperature after each thermal anneal cycle. The sequence of oxidation and annealing cycles was repeated three times in order to check the reproducibility. Before each PAES spectra NaI gamma ray spectrum was obtained in order to calculate the positronium (Ps) fraction.

EXPERIMENTAL RESULTS

The PAES spectra from the clean and oxidized Cu(100) surfaces are shown in figures 1(a) and 1(b), respectively. Two peaks are observed at ~60 eV and ~103 eV corresponding to the Cu $M_{2,3}VV$ and Cu M_1VV Auger transitions, respectively. It follows from figures 1(a) and 1(b) that the Cu $M_{2,3}VV$ Auger peak intensity for the oxidized Cu(100) surface is reduced by ~ 93% compared to the clean surface. The higher energy range of the PAES spectra corresponding to the energy range of the O KLL Auger transition at ~503 eV is shown in the inset of FIGURE 1 that shows clearly that the O KLL Auger peak intensity for the oxidized Cu(100) surface is increased by ~200% compared to the clean Cu(100) surface.

The experimental annihilation probabilities of positrons trapped at the clean and oxidized Cu(100) surfaces with Cu *3p* and O *1s* core-level electrons calculated from the PAES intensities are plotted as a function of thermal annealing temperature in FIGURE 2 and 3, respectively.

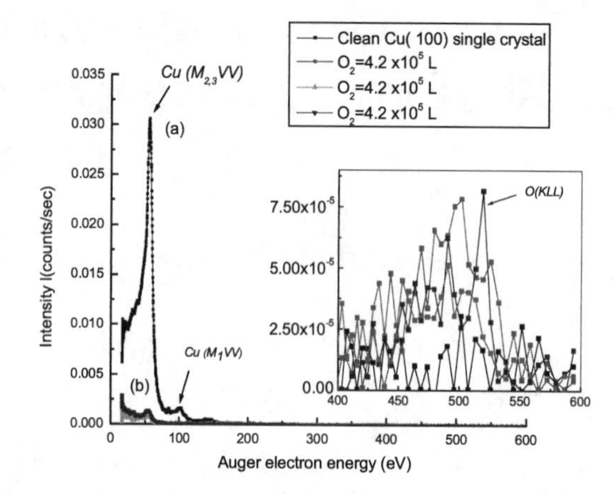

FIGURE 1. PAES spectra of Cu (100) single crystal (a) before and (b) after oxidation.

It follows from FIGURE 2 that the experimental annihilation probability of positrons trapped at the oxidized Cu(100) surface with the Cu *3p* core-level electrons increases with annealing temperature in comparison to its value for the bulk oxide surfaces, reaching a peak value of 4.6% at ~300°C.

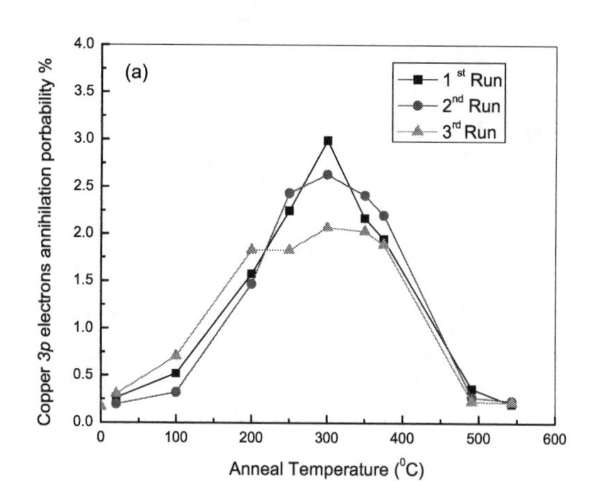

FIGURE 2. Experimental annihilation probability of positrons trapped at the previously oxidized Cu (100) single crystal with Cu *3p* core-level electrons as a function of thermal annealing temperature.

This value is close to the value of Cu *3p* core annihilation probability obtained previously for the clean Cu(100) surface [13, 14]. It also follows from FIGURE 2 that further annealing of the oxidized Cu(100) at higher temperatures leads to a decrease of the experimental probability of annihilation of surface-trapped positrons with Cu *3p* core-level electrons to the value of ~0.25%. It follows from the plot presented in FIGURE 3 that the experimental annihilation probability of positrons trapped at the oxidized Cu(100) surface with O *1s* core electrons is the smallest at 300°C and it increases as the annealing temperature is increased further to 500°C.

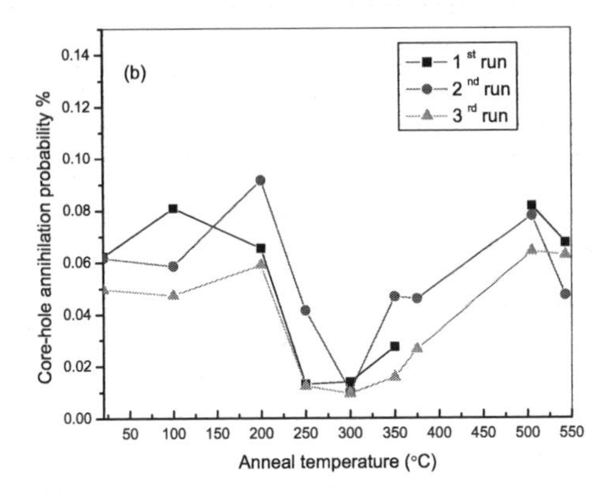

FIGURE 3. Experimental annihilation probability of positrons trapped at the previously oxidized Cu (100) single crystal with O *1s* core-level electrons as a function of thermal annealing temperature.

During the experiment the intensity of C KLL Auger peak at ~272 eV remained virtually constant.

THEORY

Positron Potential

The positron potential $V^+(\mathbf{r})$ at a metal (or semiconductor) surface contains an electrostatic Hartree potential, $V_H(\mathbf{r})$, and a correlation part, $V_{corr}(\mathbf{r})$. $V_H(\mathbf{r})$ is constructed as a superposition of atomic Coulomb potentials. Atomic calculations were performed self-consistently within the local-spin-density approximation [15] using the exchange-correlation functional from Ref. [16]. Following [17] and [18], atoms were placed in a ''stabilizing'' potential well to account for charge rearrangement effects on electron distribution functions and positron potential at the surface. The height of the potential well was adjusted for the surface to provide the positron potential at the surface that gives the proper positron work function. Atomic wave functions then provided the overlapping electron densities and corresponding atomic Coulomb potentials via Poisson's equation. $V_{corr}(\mathbf{r})$ is described using the local density approximation (LDA) in the bulk region [19]. The positron-electron correlation potential outside the surface $V_{corr}(\mathbf{r})$ is described by the image-type potential constructed to have the same corrugations as the total electron density. In the LDA, $V_{corr}(\mathbf{r}; n) = V^{EG}_{corr}(n)[f(n, \varepsilon_g)]^{1/3}$, where $V^{EG}_{corr}(n)$ is the correlation energy of a positron in a homogeneous electron gas of density n [20]. The function $f(n, \varepsilon_g)$ is a reduction factor which accounts for the diminished screening response of semiconductors to charged particles due to the existence of a band gap [21]. For metals $f = 1$. The "gap parameter" ε_g describes the effect of the band gap on the electron-positron correlation. The joining of the image potential to the LDA correlation potential is done by taking $V_{corr}(\mathbf{r})$ to be the larger of the two at each point outside the surface. The positron is assumed to be in the ground state and delocalized in the XY plane of the surface, and have a crystal momentum in this plane $\mathbf{k} = 0$.

Surface States

Positron surface states were obtained by solving Schrödinger's equation numerically taking into account discrete-lattice effects [18, 22]. The boundary conditions were chosen such that the positron wave function vanishes sufficiently far into the bulk and the vacuum. Experimental studies of surface structure of the oxidized Cu suggest that other oxide phases, such as CuO could possibly coexist on the surface [23].

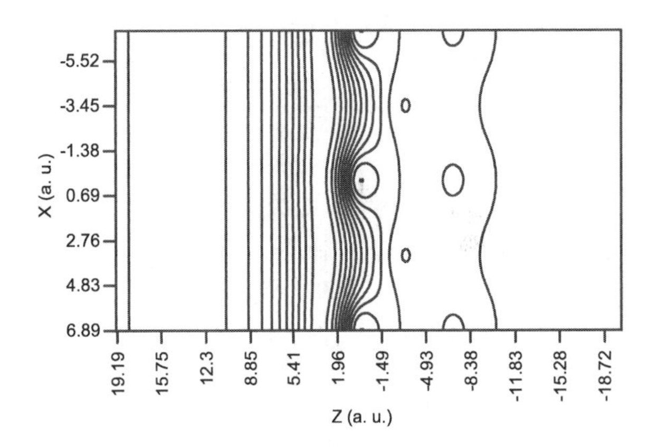

FIGURE 4. Contour plot in the X-Z plane for Y = 0 of the positron surface state wave function at the clean Cu(100) surface. Contours are separated by 0.0033 atomic units.

Calculations of positron surface states were performed for the (100) surfaces of Cu, Cu_2O, and CuO, both clean and covered with oxygen atoms adsorbed on-surface sites in the ordered p(4×4), p(2×2), c(2×2), and p(1×1) structures corresponding to 1/8, 1/4, 1/2, and one monolayer (ML) oxygen coverages. Positron surface states were also calculated for the oxidized Cu(100) surface with on-surface and sub-surface oxygen coverages (with total oxygen coverages up to 1 ML). In the latter case the adsorbed oxygen atoms occupy on-surface hollow and sub-surface octahedral and tetrahedral sites.

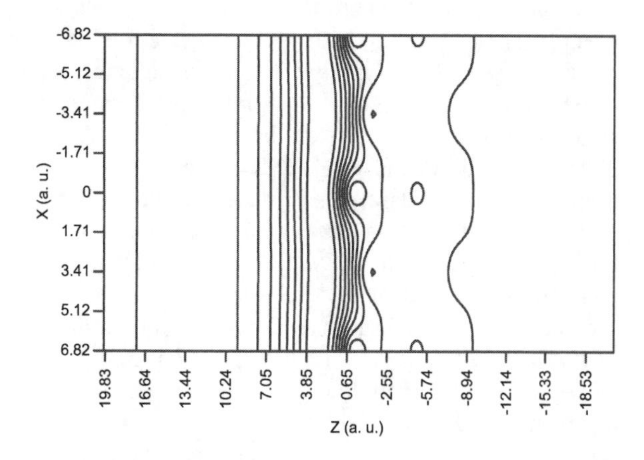

FIGURE 5. Contour plot in the X-Z plane for Y = 0 of the positron surface state wave function at the Cu(100) surface with the ordered c(2×2) structure of the adsorbed oxygen atoms. Contours are separated by 0.004 atomic units.

The adsorbate positions were obtained from studies of the interaction of atomic oxygen with the clean and oxidized Cu(100) surfaces by means of the periodic slab model density functional theory. All surfaces were fully relaxed in calculations. The position of the image plane for each surface structure was determined from the jellium calculations.

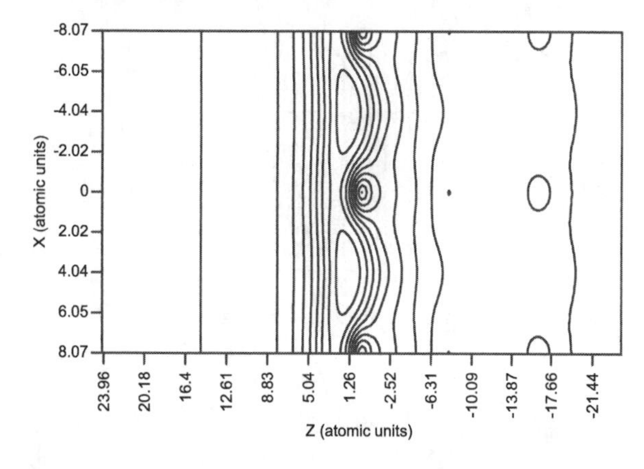

FIGURE 6. Contour plot in the X-Z plane for Y = 0 of positron surface state wave function at the (100) surface of Cu_2O when oxygen atoms form the topmost layer. Contours are separated by 0.0033 atomic units.

It follows from the calculations performed for the oxidized Cu(100) surface that the positron potential contains corrugations extending through the adsorbate overlayer into the regions between O and Cu atoms. These corrugations form the potential well on the vacuum side of the topmost layer of atoms of the surface responsible for trapping of the positron. The computed binding energies, E_b , of the surface trapped positrons are shown in Table 1. It follows from these results that adsorption of oxygen on the Cu(100) surface causes changes in E_b that reflect changes in the depth of the correlation well at the surface due to the O adsorption.

TABLE 1. Theoretical positron binding energy, E_b, and positron annihilation probabilities with relevant core electrons, $\rho_{n,l}$, at the (100) surfaces of Cu, Cu_2O, and CuO, both clean and covered with 0.125, 0.25, 0.5, and one monolayer (ML) of adsorbed oxygen.

System	E_b (eV)	$\rho_{n,l}$ (%)	
		Cu *3p*	O *1s*
Cu(100)	2.77	6.99	
Cu(100) + O (1/8 ML)	3.14	4.81	0.03
Cu(100) + O (1/4 ML)	3.12	4.17	0.07
Cu(100) + O (1/2 ML)	3.08	2.99	0.14
Cu(100) + O (1 ML)	2.94	1.60	0.23
CuO(100)	2.76	2.23	0.32
Cu_2O(100)	3.02	3.98	0.17
Cu_2O(100) + O (1/2 ML)	3.11	3.14	0.21

The positron surface state wave function is localized mostly on the vacuum side of the surfaces and delocalized in the surface plane (see Figures 4-6).

TABLE 2. Theoretical positron annihilation probabilities with relevant core electrons, $\rho_{n,l}$, at the oxidized Cu(100) surface with combined (on-surface and sub-surface) oxygen coverage.

Oxygen coverage (ML)			Type of adsorbed oxygen	$\rho_{n,l}$ (%)	
On-surface	Sub-surface	Total		Cu *3p*	O *1s*
0.5	0.25	0.75	Octahedral p2×2	2.412	0.132
0.5	0.5	1	Octahedral c2×2	2.282	0.143
0.5	0.5	1	Tetrahedral c2×2	2.580	0.144
0.5	0.5	1	Cu_2O	2.144	0.234

Calculations confirm that the oxygen overlayer pushes the positron surface state wave function away into vacuum from the layer of Cu atoms reducing its overlap with Cu atoms (see Figures 4-6).

Positron Annihilation

Positron annihilation rates $\lambda_{n,l}$ with specific core electrons, described by quantum numbers n and l, and the total annihilation rate λ (and corresponding positron surface state lifetime τ_s) are calculated taking the electron-positron correlation effects explicitly into account using the local density approximation (LDA). The measured PAES intensities are proportional to the probability $\rho_{n,l}$ that a positron in a surface state annihilates with an electron in the core level *n,l*. Theoretical positron annihilation probabilities $\rho_{n,l}$ with specific core-level electrons are obtained dividing $\lambda_{n,l}$ by λ. The results of calculations of $\rho_{n,l}$ are shown in Tables I and II. It follows from these results that the Cu *3s* and *3p* core annihilation probabilities that determine the intensity of the Cu $M_{2,3}VV$ Auger signal decrease with the increase of the adsorbed oxygen coverage in agreement with experiment. Note that in agreement with PAES data the O *1s* core annihilation probability that determines the intensity of the O KLL Auger signal increases with the increase of the adsorbed oxygen coverage. The measured PAES intensities are proportional to the probability $\rho_{n,l}$ that a positron in a surface state annihilates with an electron in the core level *n,l*. Theoretical positron annihilation probabilities $\rho_{n,l}$ with specific core-level electrons are obtained dividing $\lambda_{n,l}$ by λ. The results of calculations of τ_s and $\rho_{n,l}$ are shown in Tables I and II. It follows from these results that the Cu *3s* and *3p* core annihilation probabilities that determine the intensity of the Cu $M_{2,3}VV$ Auger signal decrease with the increase of the adsorbed oxygen coverage in agreement with experiment. Note that in agreement with PAES data the O *1s* core annihilation probability that determines the intensity of the O KLL Auger signal increases with the increase of the adsorbed oxygen coverage.

DISCUSSION

Based on a comparison of the theoretical and experimental results we posit that after oxidation, a layer consisting of oxide phases Cu_2O and CuO are formed on the surface of Cu (100) single crystal in a way such that a layer of oxygen covers the surface of Cu atoms. Both the experimental and theoretical results indicate that the intensity of the positron annihilation induced Cu $M_{2,3}VV$ Auger signal of the Cu(100) surface decreases significantly after oxidation. As a result of annealing up to 300°C, oxygen atoms desorb or diffuse into the bulk and Cu atoms appear on the surface. It has been shown that the Cu *3p* core annihilation probability increases with the decrease of the oxygen coverage which is consistent with the observed increase of the intensity of the positron annihilation induced Cu $M_{2,3}VV$ Auger signal. The obtained results are also consistent with the results of studies of oxidation and reduction of copper oxides performed using XPX and x-ray-excited Auger electron spectroscopy [24] as well as using O^{16} oxygen resonance measurement and transmission electron microscopy [25]. Slightly lower value of the experimental annihilation probability of positrons trapped at the clean Cu(100) surface with Cu *3p* core electrons when compared to the obtained theoretical one could be attributed to the presence of carbon and hydroxide species [26] on the surface and the use of LDA in calculations that is known to slightly overestimate the positron annihilation characteristics [27].

We interpret the decrease in the PAES intensity observed after the annealing temperature exceeds 300°C as being due to the diffusion of oxygen from the bulk back to the surface where it would produce a reoxidation of the surface. Calculations indicate that this reoxidation would lead to a large decrease in the Cu *3s* and *3p* core annihilation probabilities and hence to a large decrease in the Cu $M_{2,3}VV$ PAES intensity in agreement with the experimental results obtained for the higher annealing temperatures. In addition, the calculations also show that reoxidation would lead to an increase in the O *1s* core annihilation probability and hence to an increase in the O KLL PAES intensity also in agreement with the trends observed in the experimental results.

CONCLUSIONS

PAES results have provided direct measurements of initial stages of oxidation and thermal reduction of the Cu(100) surface. The PAES Cu signal from the oxidized Cu(100) surface has been found to be only ~1/10 that of a clean surface.

It has been shown that annealing of the oxidized Cu(100) surface in vacuum results in an increase in the PAES Cu signal back to the clean surface level at 300°C. These results confirm that vacuum annealing at ~300°C regenerates the oxidized surface leading to the formation of Cu metal at the surface.

It has been shown that further vacuum annealing at higher temperatures leads to a decrease of the PAES Cu signal down to the level seen for the oxidized surface suggesting reoxidation of the Cu(100) surface due to oxygen diffusing to the surface from the bulk.

PAES results have been analyzed by performing calculations of positron surface states and annihilation probabilities of surface-trapped positrons with relevant core electrons. The performed calculations indicate that the largest decrease in the Cu $M_{2,3}VV$ Auger signal is seen for the one monolayer oxygen coverage of the Cu(100) surface.

The obtained theoretical results have been found to be consistent with the experimental PAES data.

ACKNOWLEDGMENTS

This work was supported in part by the National Science Foundation Grant #DMR-0907679, the Robert A. Welch Foundation Grant Y-1100, and the Ministry of Education and Science of Russian Federation (Grant 2.1.1/2985, Program "Development of Scientific Potential of Higher Education").

REFERENCES

1. S. T. Shishiyanu, T. S. Shishiyanu, and O. I. Lupan, Sensors and Actuators B: Chemical 113 (2006) 468.
2. Y.-Y. Su and R. M. Shemenski, Applied Surface Science 161 (2000) 355.
3. K. Cho and E. C. Cho, Journal of Adhesion Science and Technology 14 (2000) 1333
4. M. Lebbai, J.-K. Kim, W. K. Szeto, M. M. F. Yuen, and P. Tong, Journal of Electronic Materials 32 (2003) 558.
5. C.C. Shih, C.-M. Shih, Y.Y. Su, L. H. J. Su, M.S. Chang, and S.J. Lin, Thrombosis Research 111 (2003) 103.
6. C. C. Shih, C. M. Shih, Y. Y. Su, R. A. Gerhardt, and S. J. Lin, Journal of Biomedical Materials Research Part A 74A (2005) 325.
7. T. Kangas and K. Laasonen, Surface Science 602 (2008) 3239.
8. T. Kangas, K. Laasonen, A. Puisto, H. Pitkanen, M. Alatalo, Surf. Sci. 584 (2005) 62.
9. A.R. Koymen, K.H. Lee, G. Yang, K.O. Jensen, Phys. Rev. B 48 (1993) 2020.
10. A.H. Weiss, D. Mehl, A.R. Koymen, K.H. Lee, C. Lei, J. Vac. Sci. Technol. 8 (1990) 2517.
11. G. Honjo, Soc. of Japan Jour. 4 (1949) 330.
12. G. Zhou, J.C. Yang, Appl. Surf. Sci. 222 (2004) 357.
13. A.H. Weiss, Positron Annihilation Induced Auger Electron Spectroscopy, in Positrons at Metallic Surfaces, Solid State Phenomena Volumes 28 & 29, edited by A. Ishii, Trans Tech Publications, Aedermannsdorf Switzerland, (1993) pp.317-340
14. A. Weiss, Positron-annihilation-induced Auger electron spectroscopy, in Positron Spectroscopy of Solids, Proceedings of the International School of Physics "Enrico Fermi," IOS Press, Amsterdam, (1995), pp. 259-284.
15. O. Gunnarsson, B.I. Lundqvist, Phys. Rev. B 13 (1976) 4274.
16. D.M. Ceperley, B.J. Adler, Phys. Rev. Lett. 45 (1980) 566.
17. M. Weinert, R.E. Watson, Phys. Rev. B 29 (1984) 3001.
18. N.G. Fazleev, J.L. Fry, K. Kuttler, A.R. Koymen, A.H. Weiss, Phys. Rev. B 52 (1995) 5351.
19. E. Boronski, R.M. Nieminen, Phys. Rev. B 34 (1986) 3820.
20. J. Arponen, E. Pajanne, Ann. Phys. 121 (1979) 343.
21. W. Brandt, J. Reinheimer, Phys. Rev. B 2 (1970) 3104.
22. R.M. Nieminen, M.J. Puska, Phys. Rev. Lett. 50 (1983) 281; 53 (1984) 1298.
23. J. Li, G. Vizkelethy, P. Revesz, J.W. Mayer, K.N. Tu, J. Appl. Phys. 69 (1991) 15.
24. S. Poulston, P.M. Parlett, P. Stone, M. Bowker, Surf Interface Anal, 24 (1996) 811.
25. J. Li, G. Vizkelethy, P. Revesz, J.W. Mayer, K.N. Tu, J. Appl. Phys. 69 (1991) 15.
26. S. Poulston, P.M. Parlett, P. Stone, M. Bowker, Surf Interface Anal, 24 (1996) 811
27. M.J. Puska, Phys. Status Solidi A 102 (1987) 11.

Positron Annihilation Induced Auger Electron Spectroscopic Studies Of Reconstructed Semiconductor Surfaces

N.G. Fazleev,[a,b] J.A. Reed,[a] S.G. Starnes,[a] A.H. Weiss[a]

[a]*Department of Physics, University of Texas at Arlington, Arlington, Texas 76019, USA*
[b]*Department of Physics, Kazan Federal University, Kazan 420008, Russian Federation*

Abstract. The positron annihilation induced Auger spectrum from GaAs(100) displays six As and three Ga Auger peaks below 110 eV corresponding to $M_{4,5}VV$, M_2M_4V, $M_{2,3}M_{4,5}M_{4,5}$ Auger transitions for As and $M_{2,3}M_{4,5}M_{4,5}$ Auger transitions for Ga. The integrated Auger peak intensities have been used to obtain experimental annihilation probabilities of surface trapped positrons with As *3p* and *3d* and Ga *3p* core level electrons. PAES data is analyzed by performing calculations of positron surface and bulk states and annihilation characteristics of surface trapped positrons with relevant Ga and As core level electrons for both Ga- and As-rich (100) surfaces of GaAs, ideally terminated, non-reconstructed and with (2×8), (2×4), and (4×4) reconstructions. The orientation-dependent variations of the atomic and electron densities associated with reconstructions are found to affect localization of the positron wave function at the surface. Computed positron binding energy, work function, and annihilation characteristics demonstrate their sensitivity both to chemical composition and atomic structure of the topmost layers of the surface. Theoretical annihilation probabilities of surface trapped positrons with As *3d*, *3p*, and Ga *3p* core level electrons are compared with the ones estimated from the measured Auger peak intensities.

Keywords: GaAs, Surface, Reconstruction, Positron, Annihilation, Spectroscopy, Beam, Auger.
PACS: 78.70.Bj, 71.60.+z, 68.47.Fg, 68.35.Bs, 68.35.bg

INTRODUCTION

The (100) surface of GaAs exhibits a variety of scientifically and technologically important properties. Unlike the stable, unreconstructed (110) surface of GaAs, the (100) termination of this crystal exhibits a variety of surface reconstructions roughly classified as As-rich and Ga-rich phases [1]. Because of the variety of reconstructions of this surface, its physical properties are significantly affected by its atomic composition [2]. Most of the effort has been devoted to the determination of a complex atomic arrangement of the reconstructed GaAs(100) surface. However, a full understanding of all the issues related to its surface electronic structure such as surface chemical reactions, Schottky barrier formation, and morphology of the metal-semiconductor interfaces has not yet been reached [2]. Recent studies of the Si(100) and Si(111) surfaces using positron annihilation induced Auger electron spectroscopy (PAES) have provided direct evidence for the effect of the surface reconstruction on PAES intensities [3]. In this paper, we present the results of studies of the reconstructed (100) surface of GaAs using PAES. The PAES spectrum measured from GaAs(100) is used to obtain experimental annihilation probabilities of surface trapped positrons with As *3d* and *3p* and Ga *3p* core level electrons. We analyze PAES data by performing quantum mechanical calculations of positron surface and bulk states and positron annihilation characteristics for both As- and Ga-rich GaAs(100) surfaces with different reconstructions. Theoretical positron annihilation probabilities are compared with the ones estimated from the measured Auger peak intensities.

EXPERIMENT

The PAES measurements were made on the high resolution PAES spectrometer at the University of Texas at Arlington. Details of the high resolution PAES system have been reported elsewhere [4]. The GaAs(100) sample (n-type, Si doping level ~ 1×10^{18} cm^{-3}) was etched in a 50% solution of hydrofluoric acid and de-ionized water before mounting in the sample chamber. The sample was maintained at a pressure of 2×10^{-10} – 4×10^{-10} Torr during data

Application of Accelerators in Research and Industry
AIP Conf. Proc. 1336, 493-496 (2011); doi: 10.1063/1.3586148

collection. The sample cleanliness was monitored using EAES which showed no measurable carbon or oxygen contamination of the surface over the period of time required for data acquisition in each cycle. By controlling the sputtering time and annealing temperature, the surface structure and composition were kept consistent for each data collection cycle. EAES was used to monitor the As $M_{45}VV$/Ga $M_{23}M_{45}M_{45}$ peak to peak ratio to determine the stoichiometry of the first few layers of atoms. A 20-minute sputtering and 10-minute annealing cycle was found to result in an As/Ga EAES peak to peak ratio of 1.6 ± 0.2, which remained consistent over the total period of PAES data collection. The high resolution Auger spectrum from GaAs(100) is shown in Figure 1. It displays six As peaks and three Ga peaks below 110 eV corresponding to $M_{4,5}VV$, M_2M_4V, $M_{2,3}M_{45}M_{45}$ Auger transitions for As and $M_{2,3}M_{4,5}M_{4,5}$ Auger transitions for Ga. Several Auger peaks from each element of GaAs result from annihilation of the same core level electron. The background due to detector dark counts, gamma ray induced counts, and Auger electrons that have lost energy as they exit the surface has been measured and subtracted from the data.

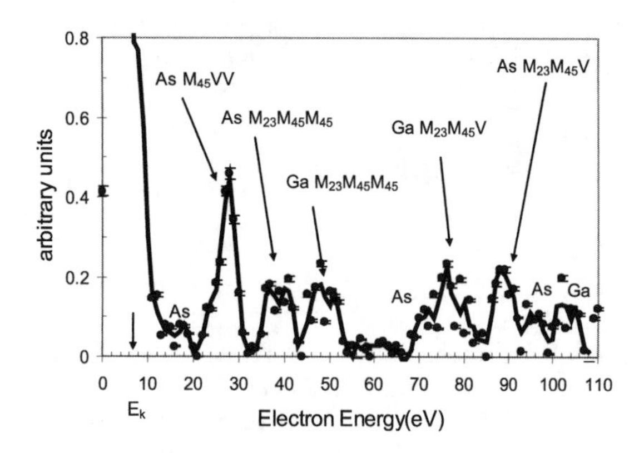

FIGURE 1. PAES spectrum from the GaAs(100) surface. The secondary electron cutoff is at $E_k = 10$ eV, the positron beam energy. The sample bias was +1.5 eV.

Experimental positron annihilation probabilities, $p_{n,l}$, with relevant core electrons are computed using the equation [3]:

$$p_{n,l} = N_{am}\, \delta_\Omega(E) / N_{sec} [(1 - f_{Ps}) W_{n,l} T]\, \Omega/4\pi. \quad (1)$$

Here N_{am} is the Auger peak intensity, N_{sec} is the secondary electron intensity used to normalize to the incident beam flux. The quantity, $\delta_\Omega(E)$, is the total integrated secondary electron yield produced by positrons of incident energy E. $W_{n,l}$ is the probability that ionization of the core level n,l results in an Auger transition and is unity for the core levels involved in this experiment. The transmission probability, T, accounts for the attenuation of Auger electrons traversing the surface layer and is calculated for the solid angle $(\Omega/4\pi)$ subtended by the acceptance apertures of the cylindrical mirror analyzer (CMA) [5]. The number of incident positrons that form positronium and are unavailable to annihilate with core electrons is accounted for by the factor $(1 - f_{Ps})$. Finally, $(\Omega/4\pi)$ is the fraction of the solid angle subtended by the CMA; this factor accounts for the fraction of emitted Auger electrons measured by the CMA. The parameters used in calculations of the experimental positron annihilation probabilities are the following: $\delta_\Omega(E) = 0.0092$ [6], $\Omega/4\pi = 0.035$ [5]. The intensities of Auger peaks were obtained by summing the counts in each peak region. The secondary electron yield proportional to the incident beam flux is obtained by measuring the secondary electron distribution with the 10 eV positron beam and a sample bias of $V_S = 0$ V to avoid changes in the PAES intensity and the secondary electron distribution. The integration ranges used to obtain the Auger intensity and the secondary electron peak intensity and the calculated values for T are given in Table 1. Experimental annihilation probabilities for the As *3d*, *3p*, and Ga *3p* core levels estimated from the labeled peak intensities in Figure 1 are equal to 0.51%, 0.92% and 0.33%, respectively.

TABLE 1. Integration ranges, transmission probabilities, and intensities.

Peak	Integration Range (eV)	Transmission Probability	Intensity
A As 3d	21 – 31	0.64	0.00830
B As 3p	33 – 43	0.56	0.00324
C Ga 3p	44 – 54	0.56	0.003813
D Ga 3p	73 – 83	0.54	0.000495
E As 3p	85 – 95	0.54	0.000661

The experimental positron work function, Φ_p, is equal to -0.6 ± 0.1 eV.

THEORY

The positron potential $V^+(\mathbf{r})$ at a semiconductor surface contains an electrostatic Hartree potential $V_H(\mathbf{r})$ and a correlation part $V_{corr}(\mathbf{r})$. $V_H(\mathbf{r})$ is constructed as a superposition of atomic Coulomb potentials. Atomic calculations were performed self-consistently within the local-spin-density approximation [7] using the exchange-correlation functional from Ref. [8]. To account for the effects of charge redistribution at the surface, we followed the method of Weinert and Watson [9] and placed As and Ga atoms in a "compensating" potential well extending from the atom center out to one Wigner-Seitz radius and then linearly ramping to a value of 0.00 Ry at twice the Wigner-Seitz radius and beyond. The modified atomic wave functions then provided the overlapping electron densities and the corresponding modified atomic Coulomb potentials via Poisson's equation. $V_{corr}(\mathbf{r})$ is described in the bulk region using the local density approximation (LDA): $V_{corr}(\mathbf{r};\ n) = V^{EG}_{corr}(n)\times[f(n,\ \varepsilon_g)]^{1/3}$, where $V^{EG}_{corr}(n)$ is the correlation energy of a positron in a homogeneous electron gas of density n [10-12].

TABLE 2. Theoretical positron surface state binding energy, E_b, positron work function, Φ_p, and annihilation probability of the surface trapped positron with core-level electrons for the As-rich GaAs(100) surface.

System	Eb (eV) (theory)	Φp (eV) (theory)	Annihilation probability (%) Core level			
			As 3p	As 3d	Ga 3p	Ga 3d
GaAs(100) nonreconstructed	2.63	–0.07	0.50	1.63	0.08	0.37
GaAs(100)-c(2×8) reconstructed	2.54	–1.17	0.43	1.30	0.18	0.76
GaAs(100)-(2×4) reconstructed	2.58	–0.95	0.36	1.15	0.16	0.71
GaAs(100)-(4×4) reconstructed	2.59	–0.68	0.66	2.06	0.01	0.04
GaAs(100)-Experiment		–0.6±0.1	0.51	0.92	0.33	

The function $f(n,\ \varepsilon_g)$ accounts for the diminished screening response of semiconductors to charged particles due to the existence of a band gap [11]. Outside the surface $V_{corr}(\mathbf{r})$ is described by the image-type potential that is constructed to have the same corrugations as the total electron density [3, 13].

TABLE 3. Theoretical positron surface state binding energy, E_b, positron work function, Φ_p, and annihilation probability of the surface trapped positron with core-level electrons for the Ga-rich GaAs(100) surface.

System	Eb (eV) (theory)	Φp (eV) (theory)	Annihilation probability (%) Core level			
			As 3p	As 3d	Ga 3p	Ga 3d
GaAs(100) nonreconstructed	1.76	–0.72	0.50	1.63	0.53	2.27
GaAs(100)-c(2×8) reconstructed	1.77	–0.66	0.43	1.30	0.15	0.62
GaAs(100)-(2×4) reconstructed	1.80	–0.17	0.36	1.15	0.14	0.61
GaAs(100)-(4×4) reconstructed	1.39	–0.74	0.66	2.06	0.50	2.22
GaAs(100)-Experiment		–0.6±0.1	0.51	0.92	0.33	

The joining of the image potential to the LDA correlation potential is done by taking $V_{corr}(\mathbf{r})$ to be the larger of the two at each point outside the surface.

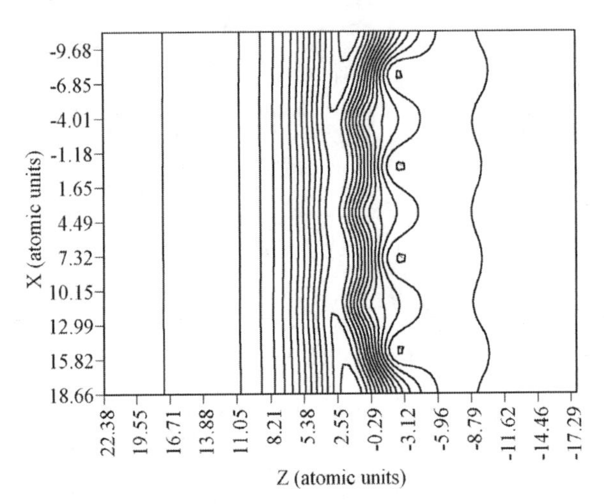

FIGURE 2. Contour plot in the X-Z plane for Y = 0 of the positron surface state wave function at the As-rich reconstructed GaAs(100)–(2×4) surface. Contours are separated by 0.0012 atomic units.

Positron surface states were obtained by solving Schrödinger's equation numerically [3, 13] with the boundary condition that the positron wave function vanishes sufficiently far into the bulk or the vacuum. Calculations were performed for both Ga- and As-rich, ideally terminated, non-reconstructed GaAs(100) surfaces, and for the GaAs(100) surfaces with (2×8), (2×4), and (4×4) reconstructions. The computed binding energies, E_b, for a positron trapped at the reconstructed As- and Ga-rich GaAs(100) surfaces are given in Tables II and III, respectively.

These values are comparable to E_b computed for the (100) surfaces of Si and Ge [3, 14]. The positron work function, Φ_p, was calculated for the GaAs(100) surfaces with the positron potential used in surface state calculations by imposing periodic boundary conditions deep inside the surface and assuming $\mathbf{k}$=0 to be the lowest Bloch state. The computed Φ_p^{theor} for both As- and Ga-rich GaAs(100) surfaces are also given in Tables II and III, respectively. It follows from the results presented in these Tables that the computed values of E_b and Φ_p are sensitive to the surface atomic structure and chemical composition of the top layers of the reconstructed GaAs(100) surface. Calculations confirm (see Figures 2 and 3) that localization and spatial extent of the positron surface state wave function are strongly affected by surface reconstructions and elemental content of the topmost layers of GaAs.

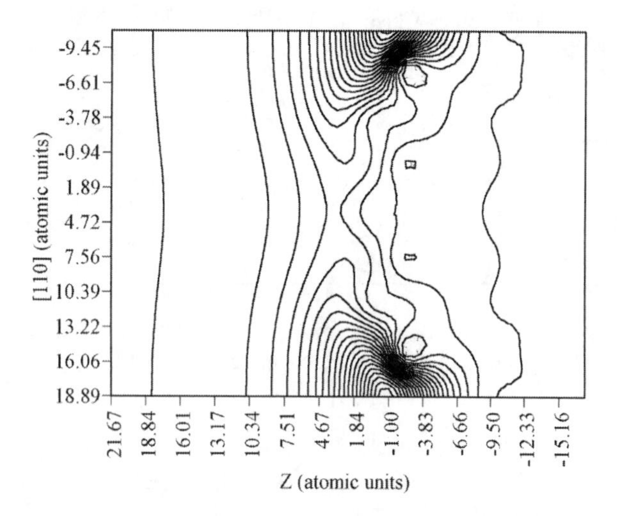

FIGURE 3. Contour plot in the X-Z plane for Y = 0 of the positron surface state wave function at the Ga-rich reconstructed GaAs(100)–(2×4) surface. Contours are separated by 0.0014 atomic units.

The probability of annihilation of surface trapped positrons, $p_{n,l}$, with core electrons characterized by quantum numbers n and l, is obtained by dividing the positron annihilation rate $\lambda_{n,l}$, by the total positron annihilation rate λ, calculated using LDA: $p_{n,l} = \lambda_{n,l} / \lambda$ [3, 13]. Theoretical $p_{n,l}$ given in Tables II and III are also found to be sensitive to reconstruction and chemical composition of the topmost layers of the GaAs(100) surface. It follows from Tables II and III that the experimental positron work function and annihilation probabilities of surface trapped positrons with As *3d*, *3p*, and Ga *3p* core electrons agree best with the theoretical Φ_p and $p_{n,l}$ computed for the Ga-rich reconstructed GaAs(100)-(2×8) surface: -0.66 eV, 0.39%, 1.18%, and 0.15%. These results indicate that PAES could serve as an important surface diagnostic tool capable of distinguishing different surfaces and defining their state of reconstruction.

CONCLUSIONS

The high-resolution PAES spectrum from GaAs(100) has been shown to display six As and three Ga Auger peaks below 110 eV. Experimental annihilation probabilities of surface trapped positrons with As *3d* and *3p* and Ga *3p* core electrons have been obtained from the measured Auger peak intensities. PAES data has been analyzed by performing calculations of positron surface and bulk states and annihilation characteristics of surface trapped positrons with relevant core level electrons for both Ga- and As-rich GaAs(100) surfaces with different reconstructions. The computed positron binding energy, work function, and annihilation characteristics have been found to be sensitive to charge transfer effects, reconstruction and chemical composition of the topmost layers of GaAs(100). A comparison of theoretical and experimental results for the positron binding energy, work function, and annihilation probabilities of surface-trapped positrons with As *3d*, *3p*, and Ga *3p* core electrons obtained for different reconstructed As- and Ga-rich GaAs(100) surfaces has been found to help to identify the surface atomic structure and chemical composition of the topmost layers of the reconstructed GaAs(100) surface.

ACKNOWLEDGMENTS

This work was supported in part by the National Science Foundation Grant #DMR-0907679, the Robert A. Welch Foundation Grant Y-1100, and Ministry of Education and Science of Russian Federation within the Program "Development of scientific potential of higher education" (Grants 2.1.1/2985 and 2.1.1/3199).

REFERENCES

1. W. Mönch, Semiconductor Surfaces and Interfaces, 2nd edition, Berlin: Springer Verlag, 1995.
2. I. M. Vitomirov, A. Raisanen, A. C. Finnefrock, R. E. Viturro, L. J. Brillson, P. D. Kirchner, G. D. Pettit, and J. M. Woodall, Phys. Rev. B 46, 13293 (1992).
3. N.G. Fazleev, J.L. Fry, and A.H. Weiss, Physical Review B 70, 165309 (2004); N. G. Fazleev, J. L. Fry, J. Kim, and A. H. Weiss, Physical Review B 68, 245300 (2003).
4. S. Yang, H. Q. Zhou, E. Jung, P. H. Citrin, A. H. Weiss, Rev. Sci. Instrum. 68, 10 (1997).
5. H. Q Zhou, Ph. D. dissertation, The University of Texas at Arlington, 1996.
6. Rulon Mayer and Alex Weiss, Phys. Rev. B 38, 11927 (1988).
7. O. Gunnarsson, B.I. Lundqvist, Phys. Rev. B 13, 4274 (1976).
8. D.M. Ceperley, B.J. Adler, Phys. Rev. Lett. 45, 566 (1980).
9. M. Weinert, R.E. Watson, Phys. Rev. B 29, 3001 (1984).
10. J. Arponen, E. Pajanne, Ann. Phys. 121, 343 (1979).
11. W. Brandt, J. Reinheimer, Phys. Rev. B 2, 3104 (1970).
12. E. Boronski, R.M. Nieminen, Phys. Rev. B 34, 3820 (1986).
13. N.G. Fazleev, J.L. Fry, K. Kuttler, A.R. Koymen, A.H. Weiss, Phys. Rev. B 52, 5351 (1995).
14. N.G. Fazleev, H. Jung, A.H. Weiss, Nuclear Ins. and Methods in Physics Research B 26, 1879 (2007).

Using Photon Activation Analysis To Determine Concentrations Of Unknown Components In Reference Materials

Jaromy Green[1], Sun Zaijing[1], Doug Wells[1,2], Herb Maschner[3]

[1] *Idaho State University—Physics Department, 921 S. 8th Avenue, Stop 8106, Pocatello, ID 83209, USA*
[2] *Idaho Accelerator Center, 1500 Ricken Drive, Pocatello, ID 83201*
[3] *Idaho State University—Anthropology Department, 921 S. 8th Avenue, Stop 8005, Pocatello, ID 83209*

Abstract. Using certified multi-element reference materials for instrumental analyses one frequently is confronted with the embarrassing fact that the concentration of some desired elements are not given in the respective certificate, nonetheless are detectable, e.g. by photon activation analysis (PAA). However, these elements might be determinable with sufficient quality of the results using scaling parameters and the well-known quantities of a reference element within the reference material itself. Scaling parameters include: activation threshold energy, Giant Dipole Resonance (GDR) peak and endpoint energy of the bremsstrahlung continuum; integrated photo-nuclear cross sections for the isotopes of the reference element; bremsstrahlung continuum integral; target thickness; photon flux density. Photo-nuclear cross sections from the unreferenced elements must be known, too. With these quantities, the integral was obtained for both the known and unknown elements resulting in an inference of the concentration of the unreported element based upon the reported value, thus also the concentration of the unreferenced element in the reference material. A similar method to determine elements using the basic nuclear and experimental data has been developed for thermal neutron activation analysis some time ago (k_0 Method).

Keywords: Photon activation analysis, PAA, non-destructive testing, photons, accelerator
PACS: 25.20.Lj; 29.20.Ej; 29.30.Kv;

INTRODUCTION

Historical and cultural artifacts can often help scientists understand the past. Thus, there is a significant interest in determining the origins, as well as the composition, of these items. Some of the methods that are used in accomplishing this are x-ray fluorescence, laser ablation, and neutron activation analysis (NAA). While these methods provide accurate results, the analyzed samples are often damaged or destroyed in the chemical processes needed for pre-analysis. Additionally, the results may only represent the surface concentrations and not the object as a whole.

An alternative method called photon activation analysis (PAA) has been designed to determine concentrations by way of activation. In this method, multiple elements can be determined simultaneously, the process is considered non-destructive, and it requires no time-consuming chemical separation procedures. Additional benefits include photons penetrating deeper into objects, the determination of elemental concentrations of lighter elements, portability of electron accelerators[1], and a greater number of heavy elements that can be analyzed.

Christian Segebade[2] pioneered the development of PAA in the early 1970s and has used it extensively to analyze everything from fake Roman coins to the sword of King Edward III. Many other groups around the world also use PAA in such roles as determining concentrations in: river sediment samples[3], blood samples[4], a meteorite[5], and biological samples[6,7,8].

In the experiment in this paper, isotopes of As, I, Nb, Y, and Zr were detected in two reference samples even though ppm concentrations were not reported. The effort of this paper is to determine whether these concentrations of elements can be determined by comparing them to known ppm values of other elements within the standards themselves.

Application of Accelerators in Research and Industry
AIP Conf. Proc. 1336, 497-501 (2011); doi: 10.1063/1.3586149

PAA THEORY

The PAA procedure is as follows. Electrons from a linear accelerator strike a converter and produce high-energy photons. Samples placed in the beamline absorb some of the photons by way of nuclear reactions. The most common nuclear reaction that occurs is A(γ,n)B, where a neutron is knocked out of the nucleus by the photon. Another common reaction is A(γ,p)C in which a proton is knocked out of the nucleus. If heavy elements (such as uranium) are present, the reaction A(γ,f) can occur in which the photon fissions the nucleus into two or more parts. Additional nuclear reactions can occur, although the probability is much less.

After a sample is irradiated, the activity of an element within that sample is given by[2]

$$A = \left(\frac{m \cdot L \cdot h}{A_R} \right) \cdot \varphi \cdot \sigma_{eff} \cdot \left(1 - e^{-\lambda \cdot T_i}\right) \qquad (1)$$

where A is the activity of the product, m is the mass of the target element, L is Avogadro's number, h is the abundance of the target isotope, A_r is the atomic mass of the target isotope, φ is the integral photon flux density, σ_{eff} is the integral effective cross section of the regarded nuclear reaction, λ is the decay constant of the product, and T_i is the duration of the irradiation.

The net peak area of a specific gamma energy is given by the equation[2]

$$I = A \cdot \theta \cdot \eta \cdot \frac{e^{-\lambda \cdot T_D}}{\lambda} \cdot \left(1 - e^{-\lambda \cdot T_C}\right) \qquad (2)$$

where I is the net peak area of the specific gamma energy, θ is the branching ratio of the specific gamma energy, λ is the decay constant of the product, A is the activity of the product, T_D is the decay period between the end of the irradiation and the beginning of the measurement, η is the detection efficiency for the regarded gamma energy, and T_C is the counting time.

A reference standard is always irradiated with the samples of interest, giving rise to an activity (A_R) and a net peak area (I_R) associated with the reference sample in addition to the activity (A_S) and net peak area (I_S) associated with the unknown sample.

Dividing I_S by I_R , and knowing that the two flux values φ_R and φ_S are equal, results in a ratio of the sample mass to the reference mass:

$$\frac{m_s}{m_R} = \frac{I_s}{I_R} \frac{A_{r,S}}{A_{r,R}} \frac{\lambda_S}{\lambda_R} \frac{\theta_R}{\theta_S} \frac{\eta_R}{\eta_S} \frac{h_R}{h_S} \frac{\sigma_R}{\sigma_S} \cdot \left[\frac{e^{-\lambda_R \cdot T_{D,R}}}{e^{-\lambda_S \cdot T_{D,S}}} \right] \times \left[\frac{1 - e^{-\lambda_R \cdot T_{C,R}}}{1 - e^{-\lambda_S \cdot T_{C,S}}} \right] \cdot \left[\frac{1 - e^{-\lambda_R \cdot T_{i,R}}}{1 - e^{-\lambda_S \cdot T_{i,S}}} \right] \qquad (3)$$

where m_S is the mass of the element of interest in the sample and m_R is the mass of the element of interest in the standard.

The following equation was used to determine values for σ_{eff}

$$\sigma_{eff} = \frac{\int_{E_{thresh}}^{E_{max}} \sigma(E_\gamma) \frac{1}{E_\lambda} d(E_\gamma)}{\int_{E_{thresh}}^{E_{max}} \frac{1}{E_\gamma} d(E_\gamma)} \qquad (4)$$

where $\sigma(E_\gamma)$ is the fundamental resonance absorption cross section, E_{max} is the maximum energy of the incoming photon (23 MeV), and E_{thresh} is the minimum energy at which incoming photons cause nuclear reactions to occur. The factor of $1/E_\gamma$ represents an approximate relative bremsstrahlung flux.

For cases in which experimental $\sigma(E_\gamma)$ was not known, a calculation of $\sigma(E_\gamma)$ was performed using the equation

$$\sigma(E_\gamma) = \frac{\sigma_m}{1 + \frac{\left(E_m^{\ 2} - E^2\right)^2}{E^2 \Gamma^2}} \qquad (5)$$

where E_m is the resonance energy, σ_m is the peak cross section, and Γ is the full width at half-maximum. These three Lorentz parameters vary depending on the isotope of interest and were taken directly from Berman and Fultz[9].

MEASUREMENTS

The accelerator used in this experiment was a LINAC (operating at 80 μA) that accelerated electrons to 23 MeV. The electrons crashed into a tungsten target (thickness of 2.3 mm) and produced irradiating bremsstrahlung x-rays.

In order to ensure equal photon flux, all samples (twelve in total) were placed evenly around the outer edge of a Teflon wheel. The wheel was attached to a machine that rotated at 36 revolutions per minute (see Figure 1). The wheel was positioned so that the samples were 1 m away from the converter but still along the axis of the photon beam. Irradiation time lasted twelve hours.

Due to the lengthy irradiation time, the samples remained in place after the LINAC was turned off. After twelve hours the samples had decayed enough to be re-located to a shielded area. Ten days to two weeks from the end of irradiation the samples were counted one at a time using a high purity germanium detector that was surrounded by lead. The samples were placed at varying distances from the detector, depending on the activity. Counting times were typically 45 minutes to one hour. Spectra were produced that were later analyzed to determine count rate and, ultimately, concentrations.

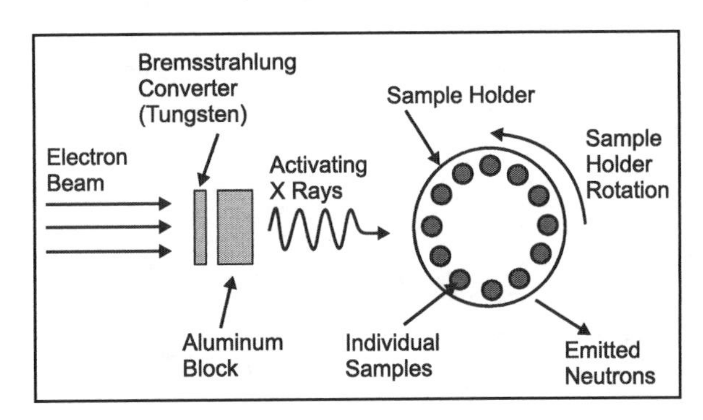

FIGURE 1. A typical setup for photon activation analysis.

Of note is the fact that some peaks show up in the spectra from one sample but not from the other. This is due to varying elemental compositions as well as differences in count decay time. In addition, peaks often show up but they are small. The errors associated with these small peaks are significant and carry through the calculations, with the end result being sizeable errors associated with the inferred concentration values.

The first standard used in this experiment was National Bureau of Standards (NBS) reference material 688—basalt rock. The second was a National Institute of Standards and Technology (NIST) reference material 278—obsidian rock. Each reference standard (pre-ordered as a powder) was put into a marked and sealed plastic bag. (Sample size was between 3g and 6 g and the bag size was approximately 2 in × 4 in). The bags were folded to resemble the general shape of the other samples (arrowheads, in this case, that were irradiated at the same time) and were placed upright in Teflon containers (approximately 2 in diameter and 2.5 in tall) before being positioned on the Teflon wheel.

RESULTS

The first step in identifying concentrations of the unreported elements was to calculate the energy-weighted effective photo-nuclear absorption cross sections. Yields from the activation of Rb and Mn were chosen as the normalization standards. The major criteria in deciding which element(s) to choose as the standards was mass, since it was assumed that isotopes similar in mass would have similar values for cross section, threshold energy, and peak energy. Because of these similarities, it was assumed that the inferred concentration values would consequently be more accurate. Then, the inferred ppm values were calculated for the unknown elements as well as for the known.

The spectra from the NIST sample had four measured peaks that arose from elements that were not reported in the NIST sheet. Table 1 lists the calculated photo-neutron cross sections and inferred values for the unreported elements. Mn and Rb were both used as standards of inference for this sample, for comparative purposes.

Also listed in Table 1 are elements that had reported and inferred values for comparative purposes. Ca was one of the reported elements that was measured in the NIST sample—its inferred concentration was statistically close to the reported concentration. When Rb was used as the standard of inference, Mn also had an inferred concentration that was statistically close to the reported.

The spectra from the NBS reference sample also had four energy peaks that arose from elements that were not reported. Table 2 lists the calculated photo-neutron cross sections and inferred values for the unreported elements. Mn was used as the standard of inference for this sample.

Also listed in Table 2 are elements that had reported and inferred values for comparative purposes. In this case, due to the high concentration in the NBS sample, three distinct Ca lines showed up in the spectrum. While concentrations were inferred from all three of these peaks, some of the associated errors were significant due to the measured peaks being so small in area.

The fact that some of the inferred concentrations were close to the reported values despite significant differences in mass could indicate that the reference element does not have to be of similar mass to the element being inferred. This could mean that one standard element could be chosen to infer values for all the unknowns within a sample. If this turns out to be the case, analysis cost could decrease significantly due to the lack of need for expensive reference samples and due to the decrease in the amount of time needed to identify peaks and to interpret data.

In contrast, the reported and inferred concentrations of some of the elements measured in both the NIST and the NBS samples, such as Na, Ce, Co, and Cr, were not statistically close to the reported values. Na in particular resulted in inferred concentrations that were 2-3 times lower than what they should have been for both reference materials. This could indicate some error in the σ_{eff} calculation or with the original data that was used in the calculation. In addition, the large error could indicate a possible error in the gamma ray efficiency correction.

The fact that several of the calculations returned inferred elemental concentrations that were not statistically close to the reported values could represent a limitation with the inferential method outlined here. More work, time, and research are needed to identify and solve the problems associated with this method.

TABLE 1. Reported and unreported elements detected in NIST sample 278—obsidian rock. Rb and Mn were used as the standard of inference. σ_{eff} = calculated photonuclear absorption cross section.

	Element	Reaction pathway	Observed energy (keV)	σ_{eff} (mb)	Inferred conc. from Rb (ppm)	Inferred conc. from Mn (ppm)	Reported conc. (ppm)
Unreported	As	$^{75}As(\gamma,n)^{74}As$	596	57.9±6.56	2.22±0.644	2.64±0.83	-
	Nb	$^{93}Nb(\gamma,n)^{92}Nb$	934	78.3±7.83	11.2±0.943	13.3±2.10	-
	Y	$^{89}Y(\gamma,n)^{88}Y$	898	88.1±3.27	37.2±4.02	40.3±5.8	-
			1836		27.6±3.74	32.8±5.4	-
	Zr	$^{90}Zr(\gamma,n)^{89}Zr$	909	96.6±3.50	222±9.57	264±28	-
Reported	Ca	$^{48}Ca(\gamma,n)^{47}Ca$	1297	36.7±4.64	6780±787	8060±1560	7030±10
	Ce	$^{142}Ce(\gamma,n)^{141}Ce$	146	74.7±7.25	125±23.2	148±34	62.2
		$^{140}Ce(\gamma,n)^{139}Ce$	166	141±4.02	89.3±5.39	105±12	
	Cs	$^{133}Cs(\gamma,n)^{47}Cs$	667	124.5±13.0	7.66±0.732	9.11±1.52	5.5
	Mn	$^{55}Mn(\gamma,n)^{54}Mn$	835	32.9±2.19	369± 23	-	403±15
	Na	$^{23}Na(\gamma,n)^{22}Na$	1274	5.97±0.602	15600±673	18600±2400	35900±371
	Rb	$^{85}Rb(\gamma,n)^{84}Rb$	882	79.0±7.90	-	152±22	128±10

TABLE 2. Unreported and reported elements detected in NBS sample 688—basalt rock. Mn was used as the standard of inference. σ_{eff} = calculated photonuclear absorption cross section.

	Element	Reaction pathway	Observed energy (keV)	σ_{eff} (mb)	Inferred concentration (ppm)	Reported concentration (ppm)
Unreported	I	$^{127}I(\gamma,n)^{126}I$	388	103.5±12.8	8.74±3.33	-
	Nb	$^{93}Nb(\gamma,n)^{92}Nb$	934	78.3±7.83	1.96±2.00	-
	Y	$^{89}Y(\gamma,n)^{88}Y$	898	88.1±3.27	21.5±8.38	-
			1836		35.1±10.8	
	Zr	$^{90}Zr(\gamma,n)^{89}Zr$	909	96.6±3.50	99.0±9.74	-
Reported	Ca	$^{48}Ca(\gamma,n)^{47}Ca$	489	36.7±4.64	99900±24200	87000
			808		79900±35700	
			1297		92100±6000	
	Co	$^{59}Co(\gamma,n)^{58}Co$	811	36.6±3.51	25.9±8.86	49.7
	Cr	$^{52}Cr(\gamma,n)^{51}Cr$	320	37.6±4.00	191±33.0	332
	Mn	$^{55}Mn(\gamma,n)^{54}Mn$	835	32.9±2.19	-	1290±20
	Na	$^{23}Na(\gamma,n)^{22}Na$	1274	5.97±0.602	6950±478	15900±223

RESULTS

PAA has been used to infer unknown concentrations in reference samples by using known quantities within those samples. Some of the inferred results were statistically close to reported values. Ca and Mn, for example, had errors of 4% and 8% for the NIST sample. This could indicate a potential method of determining concentrations of elements within a sample without needing to analyze a sample multiple times for each element. However, some of the inferred results were not statistically close to reported values, and could represent a severe limitation to using this method. Na, as an example, resulted in an inferred concentration that was 2-3 times lower (>50% error) than the reported values. Further work is needed to identify whether this method is accurate and useful.

ACKNOWLEDGMENTS

This work funded by Department of Defense under grant # DF-FC07-06ID14780.

REFERENCES

1. J. Ni, G. Xu, and R.C. Block, J. Radioanal. Nuc. Chem., 245 (2000) 501.
2. C. Segebade, H.P. Weise, and G.J. Lutz, Photon Activation Analysis, New York: Walter De Gruyter, 1988, 705 p.
3. J. Ni, G. Xu, and R.C. Block, J. Radioanal. Nuc. Chem., 245 (2000) 501.
4. M.C. Cantone et al., J. Radioanal. Nuc. Chem., 271 (2007) 559.
5. Z. Randa, J. Kucera, and L. Soukal, J. Radanal. Nuc. Chem., 257 #2 (2003) 275.
6. Z. Randa, J. Kucera, J. Mizera, and J. Frana, J. Radioanal. Nuc. Chem., 271 #3 (2007) 589.
7. M. Fukushima, H. Tamate, and Y. Nakano, J. Radanal. Nuc. Chem., 244 #1 (2000) 55.
8. Z. Randa and J. Kucera, J. Radanal. Nuc. Chem., 259 #1 (2004) 99.
9. B.L. Berman and S.C. Fultz, Rev. Mod. Phys, (47) #3 (1975) 713.

Cu-67 Photonuclear Production

Valeriia Starovoitova[1], Davy Foote[1,2], Jason Harris[1], Vakhtang Makarashvili[1,3], Christian R. Segebade[1], Vaibhav Sinha[1], Douglas P. Wells[1]

[1]-IdahoAccelerator Center, Idaho State University, 1500 Alvin Ricken Drive, Pocatello, ID 83201, USA
[2]-University of Nebraska-Lincoln, 208 Jorgensen Hall, 855 N. 16th Street, Lincoln, NE 68588-0299, USA
[3]-Argonne National Laboratory, 9700 S. Cass Avenue, Argonne, IL 60439, USA

Abstract. Cu-67 is considered as one of the most promising radioisotopes for cancer therapy with monoclonal antibodies. Current production schemes using high-flux reactors and cyclotrons do not meet potential market need. In this paper we discuss Cu-67 photonuclear production through the reaction Zn-68(γ,p)Cu-67. Computer simulations were done together with experiments to study and optimize Cu-67 yield in natural Zn target. The data confirms that the photonuclear method has potential to produce large quantities of the isotope with sufficient purity to be used in medical field.

Keywords: Medical isotope production, photon activation analysis, linear accelerator, Cu-67
PACS: Replace 82.80.Ej

INTRODUCTION

Copper-67 is a β^- emitter with an average energy 400 keV, which is a medium energy for cancer therapy, suitable to treat mm-size tumors. With the half-life of 61.8 hours and low energy gamma emission lines (93 keV and 185 keV) Cu-67 has low long-term effects on patients. As a simultaneous β^- and γ emitter, it can be used for both therapy and diagnostics. Compared to similar radiotherapy sources, Cu-67 has the potential to be quite inexpensive. Cu-67, for the above reasons, is one of the most promising isotopes to expand the scope of radio immunotherapy. Cu-67 can be produced through many different reactions, the most common is the Zn-68(p,2p)Cu-67 cyclotron reaction (see Table 1).

Generally, a highly refined Zn-68 target is bombarded with protons from the cyclotron and decays to Cu-67. This method has the advantage that cyclotrons are fairly easy to produce and operate (at Gallium and copper can be separated chemically, but Stoll et. al.[1] reported that at cyclotron energies below 50 MeV, the resulting yield of Cu-67 is too low to be practical. In this report we discuss another Cu-67 production method, namely, the photonuclear reaction Zn-68(γ,p)Cu-67. least in comparison to linear accelerators and especially reactors). However, a problem arises in the production of Ga-66 and Ga-67 from (p,2n) and (p,n) reactions on Zinc. These two reactions are both energetically more favorable than the one that produces Cu-67. Furthermore, Ga-67 has a similar half-life and produces nearly identical gamma lines to those of Cu-67.

TABLE 1. Photonuclear reactions in natural zinc target

Reaction	Threshold Energy, MeV	Abundance of the target isotope	Half-life
$^{64}Zn(\gamma,2n)^{62}Zn$	21.0	48.6%	9.13 hours
$^{66}Zn(\gamma,n)^{65}Zn$	11.1	27.9%	244 days
$^{66}Zn(\gamma,np)^{64}Cu$	20.3	27.9%	12.7 hours
$^{68}Zn(\gamma,p)^{67}Cu$	9.99	18.8%	61.8 hours
$^{70}Zn(\gamma,n)^{69m}Zn$	9.22	0.62%	13.9 hours

Application of Accelerators in Research and Industry
AIP Conf. Proc. 1336, 502-504 (2011); doi: 10.1063/1.3586150

YIELD CALCULATIONS AND MEASUREMENTS ON THIN FOILS

The experiment was done at the Idaho Accelerator Center, using a 5kW RF linear electron accelerator with 30 MeV electrons incident on a series of tantalum converters to produce bremsstrahlung photons. A natural zinc foil target (1 cm x 1cm x 1 mm) was placed behind the Al hardener and irradiated for 10 hours (see Figure 1, top). For natural zinc, there are five reactions with a combination of high initial count rates and relatively long half-life (see Table 1). There will of course be all the stable isotopes of zinc present as well, along with Cu-65.

The activity of all these isotopes was calculated as

$$A(t) = \frac{m \cdot N_A \cdot h}{A_r} \int \sigma(E) \cdot \varphi(E) dE \cdot (1 - e^{-\lambda t})$$

where m is the sample mass, N_A – Avogadro's number, h – target isotope abundance in the sample, A_r – molar mass of the target, φ(E) is the photon energy spectrum, σ(E) is reaction cross-section, λ –decay constant of the isotope of interest[2]. Differential photon flux density, φ(E), was simulated using MCNPX software. This was done for the electron beam profile given since the electrons do not constitute a monochromatic beam. The electron beam profile and the resulting bremsstrahlung photon spectrum (per 50 keV) are shown below in Figure 2.

Activities of the four longest-lived isotopes of Zn and Cu were calculated and measured experimentally for two irradiation times: 5 hours and 10 hours. There is some consistent disagreement between the data and the experiment, most likely caused by an error in flux estimation (see Table 2). However, a possibility to achieve about 3 μCi Cu-67 activity per gram per hour per 5 kW was clearly demonstrated.

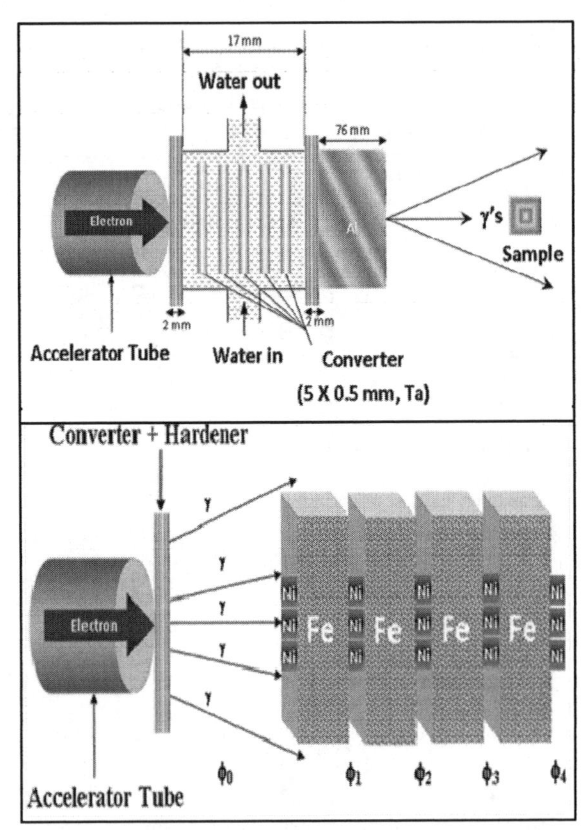

Figure 1. Top: Experimental setup for the thin foil experiment; Bottom: Experimental setup for the iron brick experiment

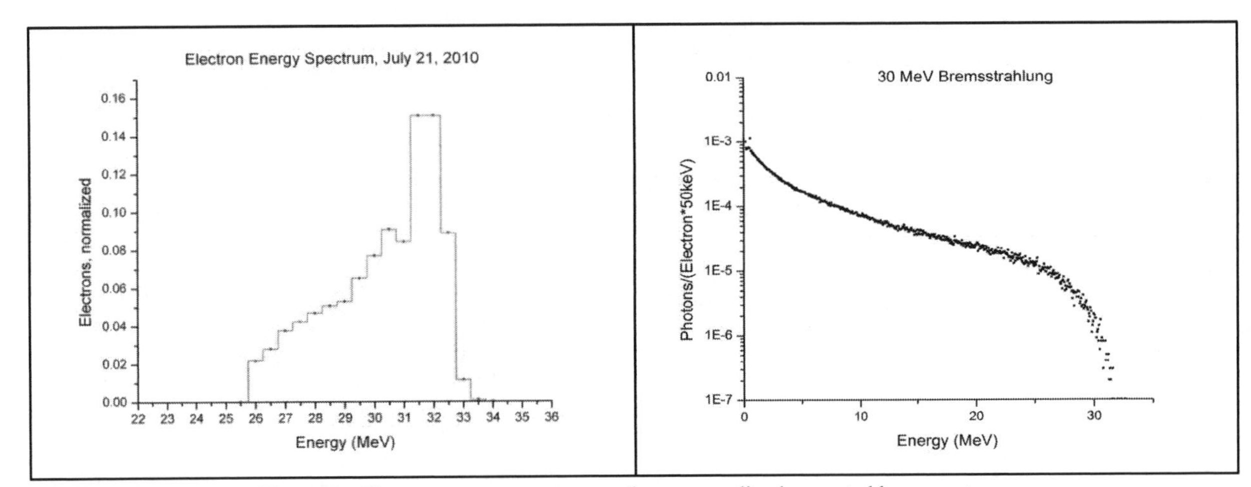

Figure 2. Electron energy spectrum and corresponding bremsstrahlung spectrum

TABLE 2. Calculated and measured activities for Zn and Cu isotopes.

Product	5.25hr Calc (μCi)	5.25hr Meas (μCi)	10hr Calc (μCi)	10hr Meas (μCi)
^{65}Zn	1.18	3.43±0.03	2.24	6.404±0.05
^{69m}Zn	2.93	6.66±0.04	5.01	11.51±0.08
^{64}Cu	2.37	6.87±1.06	2.41	N/A
^{67}Cu	4.44	14.30±1.44	8.64	31.02±1.49

TABLE 3. Activity of Ni monitors (μCi per g per 5kW per 2 hours)

1 cm		3.5 cm (after first Fe block)		6 cm (after second Fe block)		8.5 cm (after third Fe block)	
Central voxel	Adjacent voxel	Central voxel	Adjacent voxel	Central voxel	Adjacent voxel	Central voxel	Adjacent voxel
115.9±2.6	42.9±1.5	41.6±0.7	17.7±0.9	22.2±0.3	10.0±0.4	17.7±0.3	8.5±0.4

TARGET OPTIMIZATION

To make more precise estimation of production rate with varied thickness target geometry for Cu-67 production, an iron block experiment has been performed (see Figure 1, bottom). The transmitted flux is comparable in the case of material made from Zn and Fe because both of the material have comparable atomic number (Z=26 vs. Z=30) and density (ρ=7.15 g/cm^3 vs. ρ=7.87 g/cm^3). MCNPX simulations have been performed to calculate photon flux 1 cm, 3.5 cm, 6 cm, and 10 cm away for central and adjacent voxels (1cm x 1 cm).

Iron blocks of known thicknesses (2.54 cm) were placed after the hardener with nickel flux monitors on them according to the designed geometry for central and adjacent voxels (1cm x x1cm). The Ni monitors' activities were measured using HPGE gamma ray spectroscopy. Table 3 summarizes the calculated and measured activity values for Ni monitor foils, which are in a very good agreement. Assuming a rectangular prism target with a 3cm x 3cm cross-section and 15.5 cm in length, the resulting activity is 0.05 mCi/g/ 5kW/ hour (Figure 2). This is about a factor of 50 less than the yield reported by Danon et al.[3] in a similar experiment.

CONCLUSIONS

We have shown that with a 5 kW 30 MeV linear accelerator it is possible to achieve 0.05 mCi/g/ 5kW/hour Cu-67 activity. This number can be significantly scaled up by removing aluminum beam hardener and moving the target closer to the converter. It can also be placed directly into the electron beam. Using more powerful accelerators would also help, however one needs to be careful about the low melting point of natural Zn. Zn might need to be replaced by ZnO or other Zn compounds. Using an enriched Zn-68 source would not only boost the production yield by a factor of five, but also eliminate the Cu-65 impurities. The economic incentives for using a substantially more expensive enriched Zn-68 source to achieve higher specific activity should be investigated. Different target geometries need to be explored as well.

REFERENCES

1. T.Stoll, S. Kastleiner, Yu. N. Shubin, H. H. Coenen, and S. M. Qaim. Excitation Functions of Proton Induced Reactions on ^{68}*Zn* from Threshold up to 71 MeV, with Specific Reference to the Production of ^{67}*Cu*. *RadiochimicaActa*, 90:309–313, 2002.
2. Christian Segebade, Hans-Peter Weise, and George J. Lutz. *Photon Activation Analysis*. Walter de Gruyter, 1988
3. Y. Danon, R.C. Block, R. Testa, H. Moore, Medical Isotope Production Using A 60 MeV Linear Electron Accelerator, NS Transactions vol. 98, pp 894-895, 2008 ANS national meeting, Anaheim, CA, 2008

Positron Annihilation And Electrical Conductivity Studies Of Cu Substitution Effect In $Li_{0.5-0.5x}Cu_xFe_{2.5-0.5x}O_4$ Spinel Ferrite

M.O. Abdel-Hamed and E.E. Abdel-Hady

Physics Department, Faculty of Science, Minia University, Minia 61519 *Egypt*
mazosman2005@yahoo.com

Abstract. Positron annihilation lifetime (PAL), and electrical conductivity studies were carried out on a series of polycrystalline spinel ferrite with composition $Li_{0.5-0.5x}Cu_xFe_{2.5-0.5x}O_4$ where (x = 0.2, 0.4, 0.6, 0.8, 1). The radii of the tetrahedral A-sites (r_A) and the octahedral B-sites (r_B) were studied as a function of composition. The infrared spectra have been analyzed in the frequency range 200-1000 cm^{-1} and two bands were observed. The bands around 589 cm^{-1} (υ_A) and 400 cm^{-1} (υ_B) were assigned to the tetrahedral and octahedral complexes, respectively. The compositional dependence of the dc electrical conductivity σ_{dc} showed a maximum at x= 0.4 then decreases with increasing Cu^{2+} ions content. For lifetime measurements, a conventional fast –fast coincidence system was used. The variations in positron lifetime parameters with Cu^{2+} ions substitution are attributed to positron trapping in A- and B- sites.

Keywords: Positron Annihilation, Electrical Conductivity, IR-Spectra, Lifetime, Spinel Ferrite
PACS: 78.70.Bj, 75.50.Gg

INTRODUCTION

Spinel ferrites are important class of magnetic materials used in many technological applications. Their properties are very sensitive to the presence of impurities. Lithium ferrites are known for its highest Curie temperature and are important for microwave applications as a low-cost substitution for other spinel ferrites [1]. The crystal structure, microstructure and cation distribution have a notable effect on the ferrite properties and its applied aspects. Several studies [2-4] have been reported with addition of divalent, trivalent and tetravalent ions to monitor various parameters for obtaining a high–quality ferrite for any particular applications. The increase in the concentration of a divalent metal ion gives rise to structural changes within the lattice without affecting the spinel structure of the ferrite as a whole [5]. Defects distrib-ution plays a dominant role in tuning the properties of ferrites. The atomic rearrangements in solids after doping will create vacancies in the structure and should be tested. Positron annihilation study is especially useful in such situations because it is highly sensitive to the change in properties of structural defects in solids [6]. In the present work the effect of Cu^{+2} ions substitutions on the cation distribution, the lifetime spectra and electrical conductivity of spinel system $Li_{0.5-0.5x}Cu_xFe_{2.5-0.5x}O_4$ have been studied.

EXPERMENTAL

High purity oxides of Fe_2O_3, CuO and Li_2CO_3 were mixed together in molar ratio to prepare a series of spinel ferrite with the composition $Li_{0.5-0.5x}Cu_xFe_{2.5-0.5x}O_4$ with x= (0.2, 0.4, 0.8 and 1). The powders were grounded for 4 hours and presintered at 1270K for 6 hours and then left to be slowly cooled to room temperature within 48 hour. The presintered mixtures were grinding again for 4 hours and then pressed into pellets under pressure 5 ton/cm^2. Finally the samples were sintered at 1470K far 6h and again slowly cooled to room temperature. The x-ray diffraction analysis confirmed that, all the prepared samples have single phase spinel cubic structure. The pellets were polished and their surfaces were coated by a thin layer of silver paste as good contacting material for the dc electrical conductivity. The dc conductivity was measured under vacuum by using the two–probe method. The infrared spectra were recorded on a Perkin–Elmer infrared spectrometer with range from 200 to 1000 cm^{-1}. The positron annihilation lifetime measurements were made using a standard fast–fast coincidence timing technique. The sample/positron source/sample

Application of Accelerators in Research and Industry
AIP Conf. Proc. 1336, 505-508 (2011); doi: 10.1063/1.3586151

sandwich was put in a glass tube in order to perform the PAL measurements in vacuum at room temperature. The accumulation time of the PAL spectra (4 h) provided excellent counting statistics namely 10^6 counts in the peak channel. The lifetime spectra were analyzed to finite term lifetimes using the PALSFit program [7] with source correction.

RESULTS AND DISCUSSION

Analysis of the X-ray diffraction patterns gives only one cubic spinel phase for all samples. The radii of the tetrahedral A- sites (r_A) and of the octahedral B-sites (r_B) can be calculated using the relations [8]:

$$r_{tetra} = a\sqrt{3}(u-0.25) - R_o, \quad (1)$$

$$r_{octa} = a(\frac{5}{8} - u) - R_o \quad (2)$$

Where a, is the lattice parameter, u is the oxygen positional parameter (u = 0.382) [8] and R_o is the oxygen ionic radius. Figure 1 shows that, both r_A and r_B increase with increasing of x, this behavior may be ascribed to the largest radius of substituted Cu^{2+} ions (r_{Cu} =0.72 Å) [9] compared to Li^{+1} (r_{Li} =0.68 Å) and Fe^{3+} (r_{Fe} =0.64 Å).

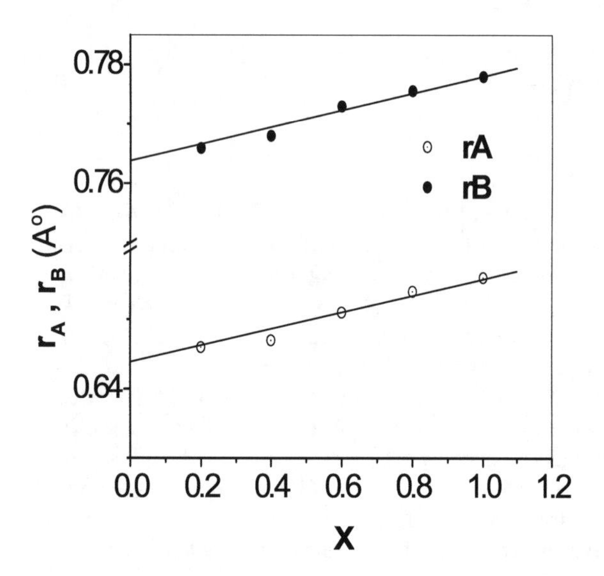

FIGURE 1. The composition dependence of the tetrahedral and octahedral ionic radii (r_A and r_B) for $Li_{0.5-0.5x}Cu_xFe_{2.5-0.5x}O_4$.

The infrared studies can give information about the ion valence and its position in the crystal lattice. For spinel ferrite two absorption bands υ_A and υ_B are usually observed at high and low frequency rang. These bands are related to the Fe^{3+}-O^{2-} bonds on A- and B- sites, respectively [11-13]. Waldron [11] and Hafner [12] attributed the band at 600 cm^{-1} to stretching group. These two bands were observed for all the studied samples in the range 578-583 cm^{-1} and 391 – 400 cm^{-1}, respectively as shown in Table 1.

TABLE 1: IR – absorption bands in cm^{-1} for $Li_{0.5-0.5x}Cu_xFe_{2.5-0.5x}O_4$ spinel ferrite.

	Tetrahedral site A-site	Octahedral site B- site		
	Fe^{3+}-O^{2-}	Li^{1+}- O^{-2}	Fe^{3+}- O^{-2}	vibration lattice
X	υ_A	υ_L	υ_B	υ_V
0.2	589	449	399	-
0.4	584	448	400	-
0.6	575	-	390	-
0.8	579	-	400	280
1.0	573	-	392	296

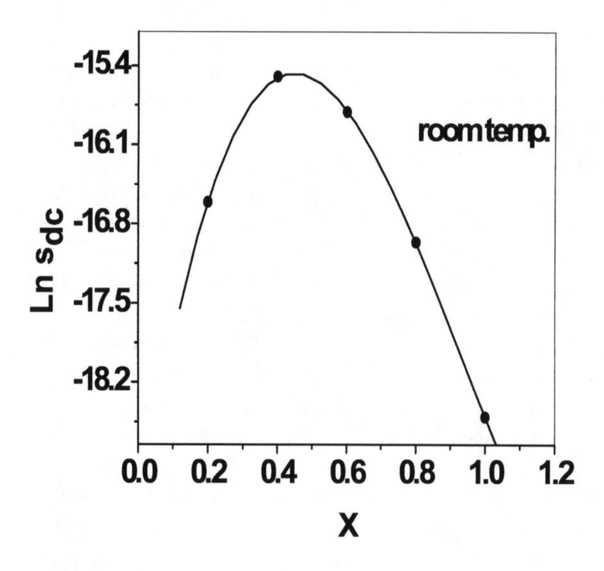

FIGURE 2. The composition dependence of the dc-conductivity at room temperature with Cu^{2+} ions concentration (x).

Figure 2 illustrates the compositional dependence of the dc-conductivity for the studied samples at room temperature. It is shown that, the dc conductivity increases with increasing Cu^{2+} ions substitution and reaching a maximum value at $x \leq 0.4$ after which it decreases. This behavior was reported for $Co_{0.6}$ $Zn_{0.4}$ Cu_xFe_{2-x} O_4 and $BaNi_{2-x}Zn_xFe_{12}O_{22}$ ferrites [14,15]. It is known that, [9] the Li^+ ions prefer to occupy the octahedral B- sites while the Fe^{3+} and Cu^{2+} ions are distributed between both A and B- sites. It was reported also that, some of Fe^{3+} and Cu^{2+} ions were oxidized into Fe^{2+} and Cu^{1+} ions respectively during sintering process at elevated firing temperature. In this case the conduction mechanism process perhaps occurs as a result of the electron hopping between Fe^{2+} and Fe^{3+} and Cu^{2+} and Cu^{1+} ions at octahedral B-sites [16]. For lithium ferrite, it was reported that, [3], the presence of Li^{1+} at octahedral sites act as electrostatic traps for the hopping electron between Fe^{3+} and Fe^{2+} ions. This reveals the lower value of the electrical conductivity for the samples with $x< 0.4$. Therefore it is expected that, the replacement of Li^{1+} by Cu^{2+} ions will lead to the increase of the conductivity. The increase of the conductivity may be also due to the contribution from the electron exchange between Cu^{2+} and Cu^{1+} ions. For the samples with $x> 0.4$ it was observed that, the conductivity decreases with further increase of x which may be due to the reduction in the amount of iron ions available for hopping process at octahedral.

The lifetime spectra collected for the ferrite samples were analyzed in terms of three components τ_1, τ_2 and τ_3 with relative intensities I_1, I_2 and I_3. The longest component τ_3 (1.6 ns) with small intensity ($I_3 \sim$ 2%) is attributed to a small fraction of orthopositronium atoms formed and trapped at the crystalline grain boundaries [8]. The relationship between the positron annihil-ation parameters (τ_2 & I_2) and Cu^{2+} ions concentration is shown in Figure 3. The shortest lifetime τ_1 (data not shown) varies with the change of Cu^{2+} concentration is in fact related to that of the other lifetime parameters τ_2 and I_2. From Figure 3 it can be seen that, τ_2 decreases with the Cu concentration up to x=0.4 after that it starts to increase. The Li-Cu ferrite is an inverse spinel ferrite and the copper ions are distributed over the A and B sites. It was reported [7] that, for inverse spinel ferrites the ionic distribution results in a charge imbalance in the two octants (A and B) with excess negative charge in the octahedral sites.

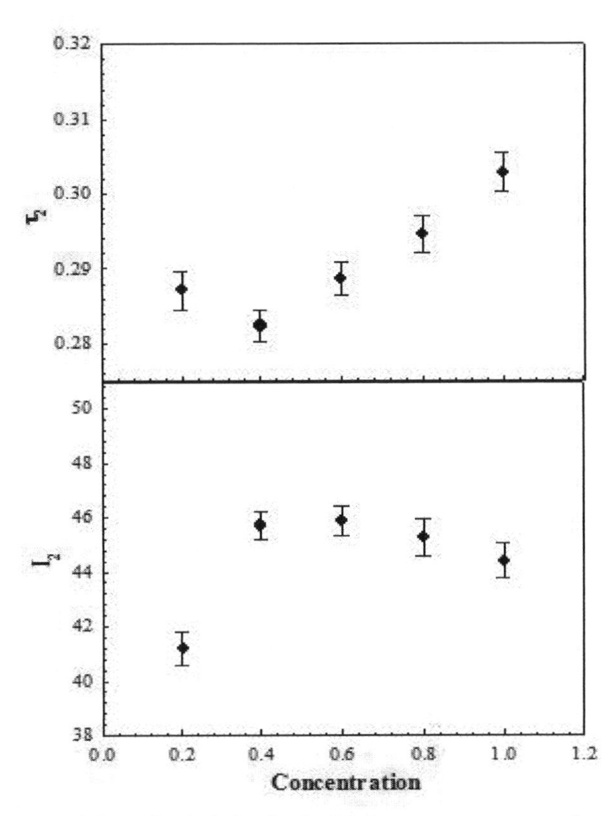

FIGURE 3. The variation in the lifetime parameters τ_2 and I_2 with Cu^{2+} ions concentration (x).

Positrons will, therefore, show preferential annihilation in these octahedral octants, even in the absence of any induced vacancies. From Figure 1 it can be seen that, the radius of the tetrahedral site (0.65 Å) is small compared to the Cu^{+2} radius (0.72Å) so that, several of the tetrahedral sites be vacancy because many of them will not be occupied by Cu^{+2} ions. Although the radius of the octahedral sites (0.77 Å) is larger in size than the Cu^{+2} ions radius, some of the octahedral sites remain unoccupied where the positrons get trapped. The variation of the relative intensity I_2 with the Cu^{2+} contents shows a Sharp increase in I_2 up to x=0.4 after that a small decrease can be noticed. At a small concentration of Cu^{2+} $x\leq$ 0.4, the Cu ions cannot replace the Fe^{3+} at the A-sites due to the difference in radius leave the latter as vacancies where positrons can be annihilated. At the B-sites the Cu^{2+}ions were may be capable of replacing Li^{1+} ions, which act as electrostatic traps for the hopping electron between Fe^{3+} and Fe^{2+}ions, and a small amount of Fe^{3+} ions; this can explain why the conductivity is high at $x\approx0.4$. Further substitution of Fe^{3+} ions by Cu^{2+} ions ($x\geq 0.4$) leads to an increase in τ_2 and a decrease in I_2 because the defects grow in size by a way of agglomeration. At those concentrations ($x\geq 0.4$) the Cu^{2+} ions were able to replace Fe^{3+} ions especially at B-sites which lead to a decrease in the electrical conductivity. The S-W plot in Figure 4 shows a linear dependence. The linear relation of (S-W) points represents the same defect structure.

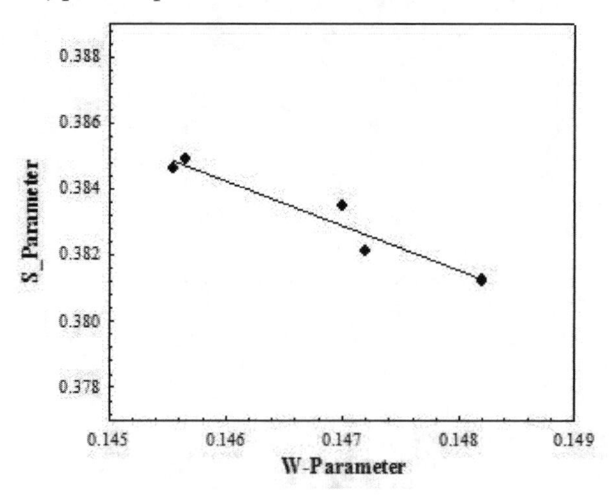

FIGURE 4: S-W Plot for with Cu^{2+} ions concentration (x).

CONCLUSION

The composition dependence of dc- condu-ctivity is divided into two regions. In the region for $x \leq 0.4$, the electron exchange $Fe^{3+} \leftrightarrow Fe^{2+}$ is favorable, while for $x > 0.4$, the electron exchange $Cu^{2+} \leftrightarrow Cu^{+}$ is responsible. The cu substitution has a significant effect on the electrical conductivity of the Li-ferrite which leads to the capability to produce a special material for a certain industrial applications. The dc conductivity results can be explained in terms of the hopping model. The IR spectra confirmed the formation of spinel structure and gave information about the distribution of ions between the two sites, tetrahedral (A) and octahedral (B). The variations in positron lifetime parameters with Cu^{2+} ions substitution are attributed to positron trapping in A- and B- sites. The linear relation of (S-W) points represents the same defect structure.

REFERENCES

1 B.K. Kuaner, G.P. Srivastava, J. Appl. Phys. 75, 10 (1994).

2. M. Amer, S. Ata-Allah, T. Meaz, S. Aboul- Enein, M.O. Abd-El- Hamid, Turk. J. Phys. 29, 163 (2005).
3. S.A. Mazen, H.A. Dawoud, Materials Chemistry and Physics 82, 557 (2003).
4. P.V. Reddy, M.B. Reddy, V.N. Muley, Y.V. Ramana, J. Mater Sci. Lett. 7, 1243 (1988).
5. P.V. Reddy, M. Salagram, Phys. Status Solidi (a) 100, 639 (1987).
6. T. Hyodo, Y. Kobayashi, Y. Nagashima, H. Saito (Eds.), Proceedings of the 13th International Conference on Positron Annihilation, Mater. Sci. Forum 445 (2004).
7. J. V. Olsen , Kirkegaard P, Pedersen N J, and Eldrup M Phys. Status Solid *C* **4**, 4004 (2007).
8. S. Ghosh, P.M.G. Nambissan, R. Bhattacharya, Phys.Lett.A 325, 301 (2004).
9. A.M. Abo El Ata, S.M. Attia, D. El Kony, A.H. Al-Hammadi, J. Magn. Magn. Mater. 295, 28 (2005).
10. F. S. Garasso, Structure Properties and Preparation of Proveskite-Type Compounds, Pegamon Press, Oxford, (1969).
11. R.D. Waldron, Phys. Rev. 1727 (1955).
12. S.T. Hafner, Z. Krist 115, 331 (1961).
13. M.A. El-Hiti, A.L. El Shora, S.M. Hammad, Mater. Sci. Technol. 13, 625 (1997).
14. S.A. Alofa, O.M. Hemeda, M.A. Amer, Asian J. Phys. 2, 113 (1993).
15. A.M. Abo El Atta, M.A. Ahmed, J. Magn. Magn. Mater. 208, 27 (2000).
16. G.H. Jonker, J. Phys. Chem. Solids 9, 105 (1959).

Correlation Of Ortho-Positronium Intensity With Doppler Broadening For Rubber Above And Below The Glass Transition Temperature

Amanda Towry[1*] and C. A. Quarles[2]

[1] *New Mexico State University, Alamogordo, NM 88310*
[2] *Texas Christian University, Fort Worth TX 76109*

Abstract. Previous research[1,2] has demonstrated a correlation between the Doppler broadening S parameter and the intensity of the ortho-positronium lifetime component in polymers that depends on the composition of the polymer. On the other hand, rubber polymers do not show this correlation and behave more like liquids for which the S parameter is essentially independent of the ortho-positronium intensity. The bubble model has been suggested as an explanation of the lack of correlation in analogy with liquids, but the bubble model applied to rubber is controversial. There are two important differences between the rubber and the polymers samples: first, the rubber samples at room temperature were all above the glass transition temperature (T_G). Second, the rubber samples all contained sulfur and were vulcanized. These differences were investigated by first measuring the S parameter for six rubber samples below T_G where the bubble model would not be expected to work. Second, raw rubber samples that did not contain any sulfur and were unvulcanized were studied at room temperature. The results show that the lack of correlation between the S parameter and the ortho-positronium intensity previously observed for vulcanized rubber is due to the inhibition of positronium formation by the sulfur in the vulcanized rubber samples rather than to the rubber being above T_G.

Keywords: positron spectroscopy, Doppler broadening, rubber, polymer, S parameter, ortho-positronium
PACS: 81.05 Lg; 78.70 Bj;61 41 +e;83.80 Va

INTRODUCTION

Recent studies have shown an interesting linear correlation between the Doppler broadening S parameter and the ortho positronium (oPs) pickoff intensity for a variety of polymers.[1,2] The S parameter is found to be proportional to the oPs intensity for chemically similar polymers such as hydrocarbons, polymers which contain oxygen or nitrogen or fluorine. Vulcanized rubbers and liquids of similar chemical composition, on the other hand, show no correlation. Typically for these materials the S parameter is relatively constant as the oPs intensity varies from sample to sample. Initially, it was suggested[2] that the bubble model of positronium annihilation might be applicable to the vulcanized rubber samples since they behaved like liquids for which the bubble model has been successfully applied.[3] However, there is other evidence that the bubble model is not applicable to rubbers.[4]

There are two significant differences between the rubber samples and the polymers previously studied. First, all the rubber samples are above the rubber glass transition temperature (T_G) at room temperature where the measurements were made. The polymers, on the other hand, were rigid and typically below their T_G. Second, the rubbers all contained sulfur and other vulcanization additives and were vulcanized. In the vulcanization process, the sulfur atoms form cross links between carbon atoms in the rubber polymer chains. So, the purpose of this work was first to investigate whether the proportionality between S parameter and oPs intensity would be seen in the rubber if the rubber samples were studied below the rubber T_G. Below T_G the rubber samples are rigid and the bubble model would not be applicable. So, it might be expected that the proportionality between S parameter and oPs intensity would show up in this situation. Second, raw unvulcanized rubber samples, with no sulfur added, were studied at room temperature to see if the presence of sulfur in the samples could be the reason for the observed independence of S parameter and oPs intensity.

Application of Accelerators in Research and Industry
AIP Conf. Proc. 1336, 509-512 (2011); doi: 10.1063/1.3586152

EXPERIMENTAL DETAILS

Six vulcanized rubber samples were selected from samples on hand in the lab from previous studies of rubber and carbon black. The samples were vulcanized for us by Sid Richardson Carbon Black Co in the form of one inch cylinders about ½ inch thick. The raw unvulcanized samples were rough pieces of raw rubber from which we could cut a smooth sample for contact with the source. The rubber samples were separated from the source by an 8 micron Kapton film so that the samples would not stick to the source itself. The polymer samples were all obtained from Goodfellow Inc. and were typically one inch squares about 1 mm thick. Several pieces could be stacked on each side of the source for the experiments.

The positron source was Na-22 in the form of NaCl deposited on Kapton film with a strength of ~ 30 micro Curies. This source was used because it was flat and the sample-source sandwich could be pressed together so that there was no air space between the source and sample. This was important when the sample and source were placed in liquid nitrogen. Initially, we had used a source deposited on Titanium film, but the construction of that source left a few tenths of mm space between the sample and source which filled with liquid when placed in liquid nitrogen. In that case the lifetime results were dominated by the positron annihilation in the liquid nitrogen layer rather than the sample.

The positron annihilation lifetime is measured with two Photonis XP2020/URQ photomultiplier tubes with barium fluoride scintillators. The output signals from the phototubes are connected to Ortec 583 constant fraction differential discriminators one of which is set to select the 1.27 MeV gamma ray from the Na-22 source for the start signal and the other is set for the 511 keV positron annihilation gamma ray for the stop signal. The start and stop signals go to an Ortec time-to-amplitude converter whose output is fed to an Ortec Trump Multi Channel Analyzer (MCA) installed in a PC computer and operated with Maestro software.

The lifetime data has been analyzed using the program LT (version 9).[5] The data was analyzed with three lifetime components. The shortest component was fixed to 125 ps to stabilize the fitting process and reduce dependence on the details of the resolution model. The longest lifetime is attributed to ortho-positronium pickoff and is of main interest in this experiment. The intermediate lifetime is due to positron trapping and direct positron annihilation in the bulk while the shortest lifetime is due in part to para-positronium. No source correction was considered to be necessary in the analysis since only the oPs lifetime and relative intensity from sample to sample were of interest and any source correction would be the same for all the samples studied. One Gaussian function was sufficient to describe the system resolution and the range of data analyzed extended from a counting level about half the peak to well into the region of the constant random background. About 1 million events were obtained in the lifetime spectrum for each run.

A liquid nitrogen cooled high purity germanium detector, a 5.08 cm coaxial crystal with an aluminum cover, was used for the Doppler broadening experiment. The output signal from the detector is sent to an Ortec 571 amplifier and then fed to a second Trump MCA. Data is typically collected for both the lifetime and Doppler broadening experiments at the same time. Doppler broadening data consists of counts vs gamma ray energy. The shape of the annihilation gamma ray peak at 511 keV is described by the S parameter: the ratio of the number of counts within about 1 keV of the peak center to the total number of counts in the peak. The S parameter measures the relative number of positron annihilations with very small relative momentum between the positron and electron. S is higher when more positronium is formed. Also of interest is the so-called W parameter: the ratio of the counts in a selected region of the wings of the peak to the total number of counts in the peak. A comparison of the S and W parameters for a sample can suggest the relative number of positron annihilations with electrons in the sample that have low momentum (such as positronium pickoff or direct positron annihilation with valence electrons) compared with positron annihilations with higher momentum core electrons in the sample. Thus an S versus W plot can be used to determine whether the nature of trapping sites in various samples is similar.

RESULTS AND DISCUSSION

In Figure 1 the main results from the first part of the experiment are shown. The S parameter is plotted versus the oPs pickoff intensity for six vulcanized rubber samples. The measurements at room temperature, above T_G, are shown as solid points. The open points are measurements made below T_G with the sample immersed in liquid nitrogen. The experimental errors in S are of the order of the size of the data points. While the T_G of each sample is different, liquid nitrogen temperature is well below the T_G for each sample. Previous studies of positron annihilation in several rubbers as a function of temperature[6] has shown that the oPs intensity decreases from room temperature to T_G and remains relatively constant below T_G. However, not all rubbers behave the same

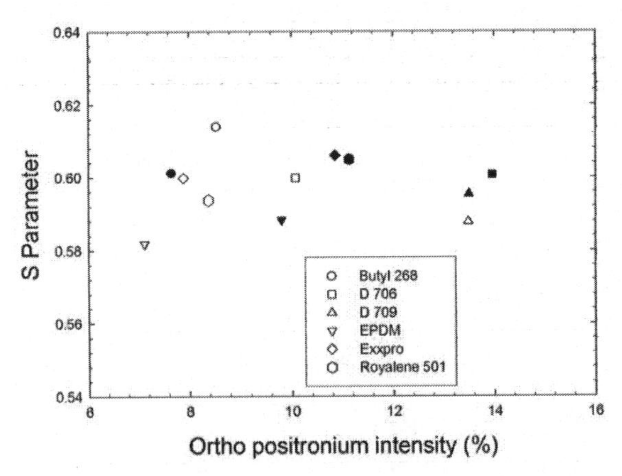

FIGURE 1. S parameter versus ortho positronium intensity for vulcanized rubber samples. The open points are at liquid nitrogen temperature and below the glass transition temperature of each sample. The solid points are at room temperature. The error in the S parameter is of the order of the size of the point.

way and the change in oPs intensity as temperature decreases is less significant for some rubbers. As can be seen, the data at room temperature show no correlation of S parameter with oPs intensity, as had been previously reported.[2] While there is some scatter in the data from sample to sample, the S parameter is seen to remain independent of oPs intensity whether the sample is at room temperature or at liquid nitrogen temperature, that is, whether the sample is above or below T_G. If the bubble model were the correct explanation for the lack of correlation above T_G, then one would have expected to see a linear correlation below T_G similar to that seen in rigid polymers where the bubble model could not be acting. Since there is no significant change in the S parameter versus oPs intensity above or below T_G, we conclude that the bubble model in not the explanation of the behavior.

In Figure 2 the results for the second part of the experiment are shown. Again the S parameter is plotted versus oPs intensity, but this time the polymer samples and the raw unvulcanized rubber samples are compared. The data are all at room temperature, above T_G for the rubber which is soft while the polymers are all rigid. The polymers are all pure hydrocarbons: low density polyethylene (LDPE), high density polyethylene (HDPE), ultra high molecular weight polyethylene (UHMWPE), polystyrene (PS), polypropylene (PP) and tri methyl pentene (TPX). Polymer results lie along the expected linear correlation line. The raw rubber samples are polyisoprene or natural rubber (NR), solution styrene butadiene rubber (D706) and tin coupled solution styrene butadiene rubber (SnSSBR).

In Figure 2 the correlation between the S parameter and oPs intensity for the hydrocarbon polymers is clearly seen. The regression lines in the figure are drawn through the polymer samples and through the rubber samples respectively. The raw rubber samples also clearly show the expected correlation, but there is more scatter than seen with the polymers.

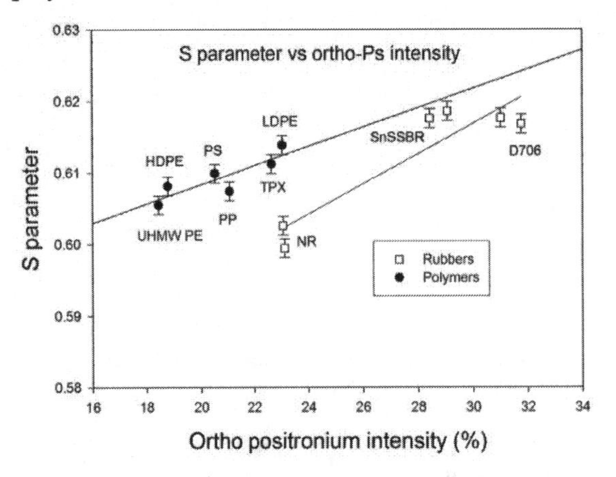

FIGURE 2. S parameter versus ortho positronium intensity for both polymer (solid circles) and unvulcanized rubber (open squares) samples. The two lines are linear regressions through the polymers and the rubbers respectively.

It is likely that natural rubber has significant impurities which may cause the S parameter to be somewhat lower than might be expected. The SnSSBR rubber seems to lie close to the polymer hydrocarbon line. The D706 rubber seems to be in better line with the natural rubber than the polymer line. The correlation seen in Figure 2 indicates that it is the sulfur (or perhaps also the other vulcanization additives) added to raw rubber for vulcanization that is responsible for the lack of correlation observed in refs. 1 and 2 for vulcanized rubber. Previous work by Wang, et al[7] has studied the effect of sulfur concentration in rubber on cross link density. The positron spectroscopy studies as a function of sulfur concentrations showed that sulfur acts as a positronium inhibitor. Sulfur reduces both the oPs lifetime and the oPs intensity. Thus the presence of sulfur in the vulcanized samples inhibits the formation of positronium and effectively decouples the S parameter from the oPs intensity leading to the observed lack of any correlation.

In Figure 3 the S parameter is plotted versus the W parameter. The rubber samples lie along the regression line indicating that the trapping sites for positrons and positronium are essentially the same type in all the raw rubber samples. There is more scatter in the polymer data, although the three polyethylene samples seem to lie along their own line (not shown) suggesting common trapping sites.

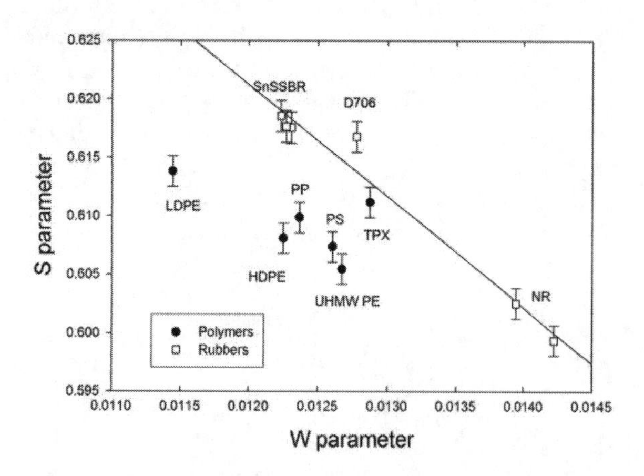

FIGURE 3. S parameter versus W parameter for polymers (solid circles) and unvulcanized rubbers (open squares). The line is a linear regression through the rubber points.

CONCLUSIONS

The results of the two experiments show that the lack of correlation between S parameter and oPs intensity observed for vulcanized rubber is not due to the bubble formation in the samples above T_G. The lack of correlation is also seen for samples well below T_G which would not be expected if the bubble model were applicable. On the other hand, the correlation between S parameter and oPs intensity is observed for raw rubber samples. This indicates that the correlation is suppressed by the addition of sulfur (and perhaps other additives) which is mixed into the rubber to enable the vulcanization process. In other work[7] sulfur has been shown to inhibit positronium formation so the vulcanization process effectively decouples the oPs intensity from the S parameter.

ACKNOWLEDGMENTS

The research was supported by the National Science Foundation Research Experience for Undergraduates Program under Grant #PHY-0851558.

* REU participant at TCU, summer 2009, supported under NSF Grant # PHY-0851558.

REFERENCES

1. K. Sato, et al., Phys. Rev. B71 (2005) 012201.
2. C. Quarles, et al., Nucl. Inst. Meth. Phys. Res. B 261(2007) 875.
3. Y.C. Jean, P.E. Mallon, DM. Schrader (Eds.), Principles and Applications of Positron and Positronium Chemistry, World Scientific Publishing Co., 2003.
4. V. P. Shantarovich, J. Nuc. Radiochemical Sciences 7 (2006) R37.
5. J. Kansy, Nucl. Instrum. Methods Phys. Res. A 374 (1996) 235.
6. J. Wang and C. A. Quarles, Radiation Physics and Chemistry 68 (2003) 527.
7. J. Wang, V. Jobando and C. A. Quarles, Nucl. Inst. Meth. Phys. Res. B241 (2005) 271.

Positron Annihilation Spectroscopy Of High Performance Polymer Films Under CO_2 Pressure

C.A. Quarles,[1] John R. Klaehn,[2] Eric S. Peterson[2] and Jagoda M. Urban-Klaehn[3]

[1]*Texas Christian University, Fort Worth TX 76109*
[2]*Idaho National Laboratory, Idaho Falls ID 83415-2208*
[3]*Pajarito Scientific Corporation, Idaho Falls ID 83404*

Abstract. Positron annihilation Lifetime and Doppler broadening measurements are reported for six polymer films as a function of carbon dioxide (CO_2) absolute pressure ranging from 0 to 45 psi. Since the polymer films were thin and did not absorb all positrons, corrections were made in the lifetime analysis for the absorption of positrons in the positron source and sample holder using the Monte Carlo transport code MCNP. The studied polymers are found to behave differently from each other. Some polymers form positronium and others, such as the polyimide structures, do not. For those polymers that form positronium an interpretation in terms of free volume is possible; for those that don't form positronium, further work is needed to determine how best to describe the behavior in terms of the bulk positron annihilation parameters. A few of the studied polymers exhibit changes in positron lifetime and intensity under CO_2 pressure which may be described by the Henry or Langmuir sorption models[3], while the positron response of other polymers is rather insensitive to the CO_2 pressure. The results demonstrate the usefulness of positron annihilation spectroscopy in investigating the sorption of CO_2 into various polymers at pressures up to about 3 atm (45psi).

Keywords: positron lifetime spectroscopy, Doppler Broadening, polymer film, CO_2 pressure
PACS: 81.05 Lg; 78.70 Bj;61 41 +e;83.80 Va

INTRODUCTION

There is an interest in tailoring the properties of polymer membranes to vary the absorption or the permeability of the polymer film to various gases such as methane and CO_2 under a wide variety of temperatures and pressures. It is generally recognized that many of the important properties of polymers are governed by the polymer free volume.[1,2,3,4,5,6] The free volume can be defined in various ways, but a simple definition is $FV = (V - V_O)$ where V is the actual volume and V_O is the occupied volume, often related to the van der Waals volume of the molecules in the polymer. Physically, one can think of the free volume in a polymer as due to the voids between the polymer chains. For gas molecules to penetrate into the polymer film, there must be sufficient void-space to accommodate the molecule. There can be connected free volume voids or the voids can be effectively isolated from each other, that is there can be films with a significant free volume, but low permeability.

Direct measurement of free volume is difficult. But for void sizes a bit larger than a few atomic diameters, the method of positronium annihilation lifetime spectroscopy can provide a measure of the free volume.[1] When positrons from a radioactive source enter a sample, they thermalize in a few picoseconds by scattering from atomic electrons and can then form positronium by combining with one of the free electrons produced in the so-called spur formed in the thermalization process. Positronium is a neutral particle about the size of a hydrogen atom and is formed in two spin states: para and ortho. In the para state the positron and electron annihilate with a lifetime of 125 ps. The ortho state lives much longer (142 ns in vacuum) and can trap in a free volume void. The neutral ortho-positronium atom then bounces off the walls of the void and eventually the positron is picked off by an electron in the wall. This pickoff lifetime is typically of the order of 2 ns and depends in detail on the void size and the chemistry of the sample. So for polymer films that form positronium, it is possible to study the free volume under a wide variety of experimental conditions. For polymers where the formation of positronium is inhibited or the free volume void size is too small to trap positronium, the application of positron lifetime spectroscopy techniques needs further study. Nevertheless, some aspects of positron spectroscopy, such as Doppler

Application of Accelerators in Research and Industry
AIP Conf. Proc. 1336, 513-518 (2011); doi: 10.1063/1.3586153

broadening spectroscopy, may be useful even when positronium is not formed.

The sorption of a gas into a polymer has been previously investigated[3] and the process was discussed in terms of Henry and Langmuir sorption. In the Henry-type sorption, the gas molecules are absorbed in the polymer and the free volume size may become larger, increasing both the ortho Positronium intensity and lifetime. For the Langmuir-type sorption, the gas molecules fill the preexisting free volume and so the ortho Positronium intensity and lifetime are decreased. Cases of mixed sorption were observed where the ortho Positronium intensity and lifetime first decreases and then increases with pressure suggesting that the gas fills the pores at lower pressure and then is dissolved in the polymer at higher pressure.

EXPERIMENTAL DESCRIPTION

The thin films investigated in the work reported here were: TPX (polymethylpentene from Mitsui Chemicals), Ultem 1000 (polyetherimide from GE plastics), Kapton (polyimide from Dupont), Pyre RC ML 5083 (polyimide from Industrial Summit Technology Co.), PBI (polybenzimidazole from PBI Performance Products, Inc.) and PBI-TMS (bis (trimethylsilylmethyl)-polybenzimidazole, fabricated at INL[7]).

The sample-source is prepared as a stack beginning with a substrate of titanium foil (0.7 mm), then a stack of polymer film, the source (NaCl between titanium film of 0.5 mil), another polymer stack (if enough sample is available) , and a final titanium foil. The sample-source stack is held together with a small metal clip. The intention is that all positrons from the source annihilate in the source itself, the polymer sample, or the titanium substrate foil if the sample is too thin to stop all the positrons.

The positron annihilation lifetime is measured with two Photonis XP2020/URQ photomultiplier tubes with barium fluoride scintillators. The start and stop signals from the phototubes go to Ortec 583 constant fraction differential discriminators which are set to select the 1.27 MeV gamma ray from the Na-22 source for the start signal and the 0.511 MeV gamma ray for the stop signal. The signals go to an Ortec time to amplitude converter and then to an Ortec Trump Multi channel Analyzer (MCA) installed in a PC computer and operated with Maestro software. The Doppler broadening experiment is done with a liquid nitrogen cooled high purity germanium detector (HPGe) which is a 5.08 cm coaxial crystal with an aluminum cover. The output signal is processed by an Ortec 571 amplifier and then sent to a second Trump MCA. Data is typically collected for both the lifetime and Doppler broadening experiments at the same time. The detectors are placed with respect to the sample-source to give a suitable counting rate. Typically we would like to have the lifetime detectors as close to the source as possible, but the HPGe detector is set some distance from the source to reduce the dead time and adjust the counting rate so that it is commensurate with the much lower lifetime counting rate.

After measurement on the films as received, the films were placed in the vacuum oven and heat treated to 130 C for several hours. Then a CO_2 atmosphere was introduced to the films. Lifetime and Doppler runs were then made on the CO_2 treated sample in air at standard temperature and pressure. The data runs with CO_2 pressure varying from 0 to 30 psi were made at room temperature in a sample chamber that was a 2.75 Conflat 6" nipple with a CO_2 gas inlet on one end-flange and a vacuum to 30 psi gauge on the other end-flange. The two scintillation detectors were placed on either side of the nipple with the sample-source sandwich inside. The counting rate was about 30 cps so it required about 8-9 hours to make one lifetime run with 10^6 events.

ANALYSIS AND DISCUSSION

The lifetime data were analyzed using the program LT (version 9)[8] with three lifetime components, fixing the shortest component to 125 ps. This serves to stabilize the fitting process and make the analysis less dependent on the details of the resolution model. A resolution assumption of one Gaussian was used and the range of data was selected to extend from a counting level about half the peak to well into the region of the constant random background. Lifetime runs were made to obtain about 1 million events in the lifetime peak.

The lifetime data was analyzed with a source correction determined using the Monte Carlo code MCNP.[9] The lifetime was measured from titanium foil alone, placing the source that is deposited between two 0.0005 mil titanium thin films between two titanium foils of thickness about 1mm. The MCNP results were used to determine what fraction of the positron spectrum annihilates in the different parts of the sample: the NaCl, the titanium foil covering the NaCl, the sample film stack, and the titanium backing sandwiching the sample stack. This procedure establishes the source correction. Both the total correction fraction as well as the lifetime components determined for titanium are fixed in the lifetime fit program. It helps to have a stack of sample films on both sides of the source and the two stacks thick enough to minimize the source correction.

The Pyre ML5083 film (55 mil) and the PBI film (30 mil) were stacked only on one side of the source and both had a source correction of 61%. The PBI-TMS sample was two stacks of 20 and 25 mils, Kapton was two stacks of 51 mil each, TPX was two stacks of 104 mil each, and Ultem 1000 was two stacks of 82 mils. These four samples had a source correction of 28%.

In the Doppler broadening experiments, data were collected to obtain 1-4 million events in the 511 keV peak. The analysis of the Doppler broadening spectrum was done with Sigma Plot using an analysis program written in our lab. This program analyzes the Doppler broadened 511 keV annihilation gamma ray and determines the S parameter and W parameters for the peak. The S (or shape) parameter is determined as the ratio of the number of counts within about 1 keV of the 511 peak center to the total number of counts in the peak. The W parameter is the ratio of the number of counts in the wings of the peak, WL being to the left or low energy side and WR to the right or high energy side of the peak. Generally the S parameter is larger when there are more positrons annihilating with low momentum electrons such as valence electrons, and the W parameter is larger when there are more annihilations with higher momentum or core electrons. In the analysis, the user selects the peak channel and the range of channels for determination of the S and W parameters and a step-like background is calculated from the spectrum and subtracted before analysis.[10] In the analysis below we treat only the W parameter from the higher energy side of the 511 peak (the WR region). No correction has been made for the contribution from positron annihilation in the titanium foils of the source or the titanium substrate if the sample stack was too thin to completely stop all the positrons.

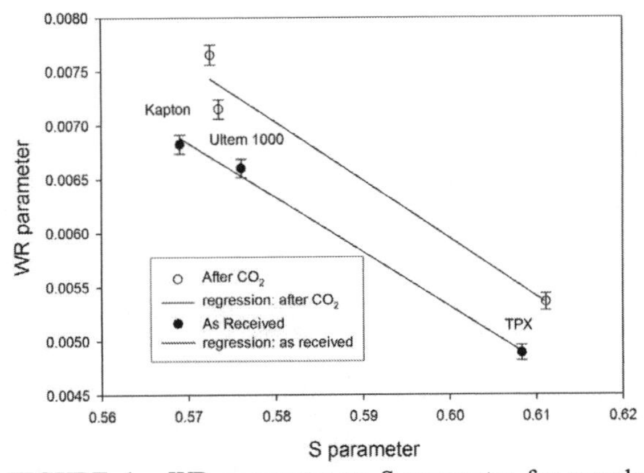

FIGURE 1. WR parameter vs S parameter for samples before and after CO_2 treatment, indicating the change in positron or positronium trapping sites after CO_2 loading.

The Doppler broadening results are shown in Figures 1 and 2. In Figure 1, the WR parameter is plotted versus the S parameter for several samples. This type of plot can be used to interpret the nature of trapping sites.

For the samples shown, Kapton does not form positronium; Ultem 1000 forms a small amount of positronium and TPX forms the most positronium of the samples studied. The two regression lines shown indicate a change in the nature of the trapping sites when CO_2 is in the polymer. The first regression line is drawn through the "as received" data points, presumably with any free volume voids in the sample containing air or mostly nitrogen. The second line drawn through the data points for the samples that have been heated, pumped on and exposed to CO_2 is displaced significantly from the "as received" line. This indicates that each of the samples, regardless of whether positronium is formed in the samples, have absorbed CO_2 into the free volume voids. The presence of CO_2 changes the nature of the positron or positronium trapping sites due to the substitution of air by CO_2 which may be due to the increase in oxygen at the trapping sites.

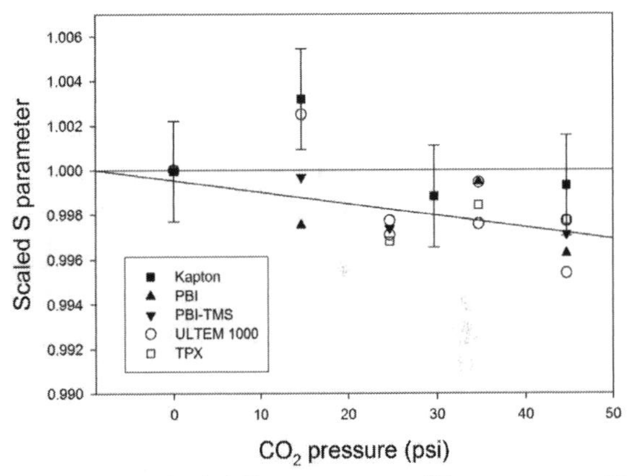

FIGURE 2. Scaled S parameter vs CO_2 pressure. The sloping line is the best fit to the data suggesting a small decrease in S parameter with increasing CO_2 pressure for all samples other than Kapton.

In Figure 2, the S parameter, normalized to the value at 0 psi, is shown. Representative error bars are shown on the Kapton data. The slopping line is the best fit to the data and shows that on the average the data decreases with CO_2 pressure in all samples. The most significant decrease is observed for the Ultem 1000 sample,while the Kapton sample is consistent with no change. Again, with the possible exception of the polyimide Kapton, variation of the CO_2 pressure is reflected in the S parameter. A decrease in the S parameter can be interpreted as a decrease in the number of annihilations with low momentum electrons or as a corresponding increase in the annihilations with higher momentum core electrons. As more oxygen atoms are forced into the free volume voids, the probability of increased annihilation with oxygen core electrons could lead to the observed decrease of the S parameter with increasing CO_2 pressure.

Results for the ortho positron (oPs) annihilation intensity and lifetime are shown in Figures 3 – 4 for the three samples which form positronium as exhibited by the presence of a long (~1.5 - 2 ns) lifetime: TPX, PBI-TMS and Ultem 1000. The polyimide films such as Kapton and Pyre ML5083 do not form positronium. PBI by itself shows no positronium formation, but does form positronium when combined with N-substituted trimethysilylmethylene (TMS) functional groups.

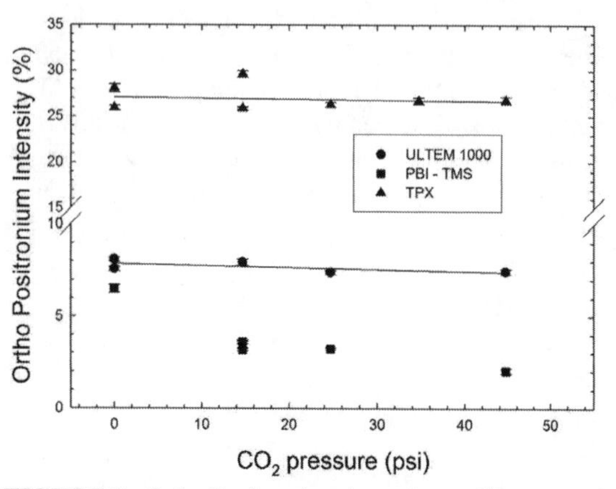

FIGURE 3. Ortho Positronium intensity vs CO_2 pressure.

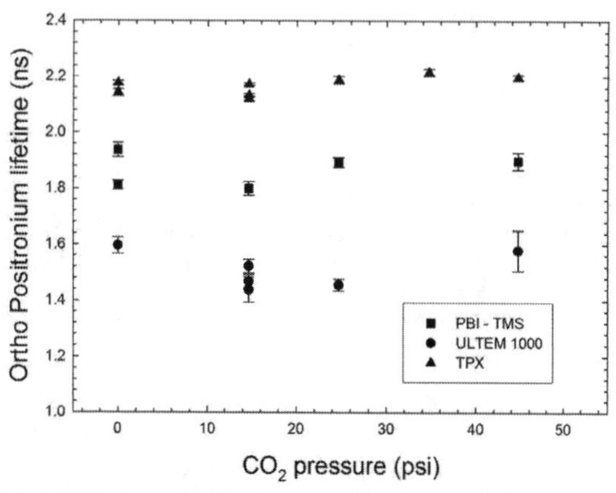

FIGURE 4. Ortho Positronium lifetime vs CO_2 pressure.

Generally the oPs intensities in Figure 3 for the TPX and PBI-TMS samples do not show a significant change with CO_2 pressure. In a few cases the differences between runs at the same CO_2 pressure are larger than the errors, suggesting that the history of the sample may be important. The oPs intensity for Ultem 1000 shows a distinct decrease with increasing CO_2 pressure. This change in oPs intensity for Ultem 1000 is the most significant change observed among all the samples.

The oPs lifetimes are shown in Figure 4. TPX shows a small increase as the pressure goes above 1 atm or perhaps a small dip at the 14.7 psi level. The oPs lifetime for PBI-TMS is consistent on average with no dependence on CO_2 pressure, although the data do tend to vary more than the errors would suggest again indicating some possible dependence on the history of the sample. For the Ultem1000 sample, the change in oPs lifetime is much less than the change observed in the oPs intensity. The lifetime generally appears to first decrease as the CO_2 pressure increases, but then the point at the highest pressure is somewhat higher than average.

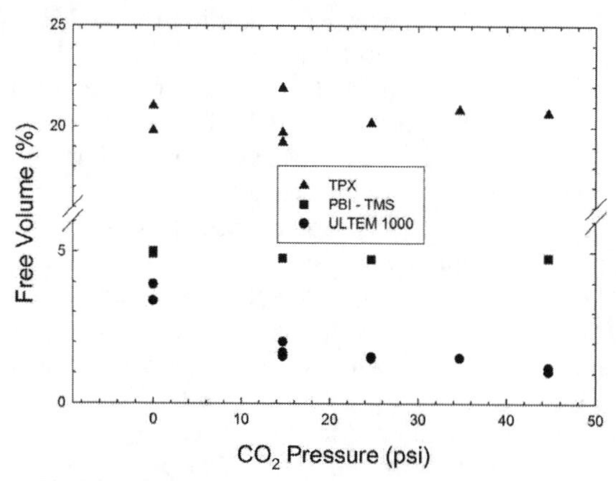

FIGURE 5. Percent free volume vs CO_2 pressure.

The percent free volume is calculated from: $\%FV = 0.0018*4/3*\pi(R_0)^3$.[1] R_0 is the radius of the free volume void and is calculated from the oPs lifetime using the Tao-Eldrup formula.[11,12] The results for the percent free volume in the three samples which form positronium are shown in Figure 5. The percent free volume for TPX and PBI-TMS change very little with CO_2 pressure. TPX tends to increase slightly on average with pressure and PBI-TMS tends to decrease, but both are certainly consistent with no change. The percent free volume in Ultem 1000, on the other hand, significantly decreases with CO_2 pressure.

Figure 6 shows the average positron annihilation lifetime versus CO_2 pressure for the six samples studied. The average lifetime is simply the weighted average of the intensity and the lifetime of the components in the fit. There were three lifetime components for the samples that form positronium (TPX, Ultem 1000 and PBI-TMS), and two lifetime components for the others (Kapton, Pyre RC ML 5083, PBI). The lines are regression lines for each set of data. The average lifetime for TPX is much larger than the other samples due to the large oPs component (~25% oPs) and does not change significantly with CO_2 pressure. When the oPs component is small, such as in Ultem 1000 and PBI-TMS, or when it is not present, the average lifetime is rather close to what is usually referred to as the bulk annihilation lifetime. This lifetime is due to direct annihilation of positrons with the bulk and to positron trapping in the distribution of free volume voids, which may be too small to allow for positronium formation and trapping. The average lifetime for both Ultem 1000 and PBI-TMS follows the same trend line and tends to decrease slightly with increasing CO_2 pressure. Finally the average lifetimes for the three samples which do not form positronium show essentially no change with CO_2 pressure.

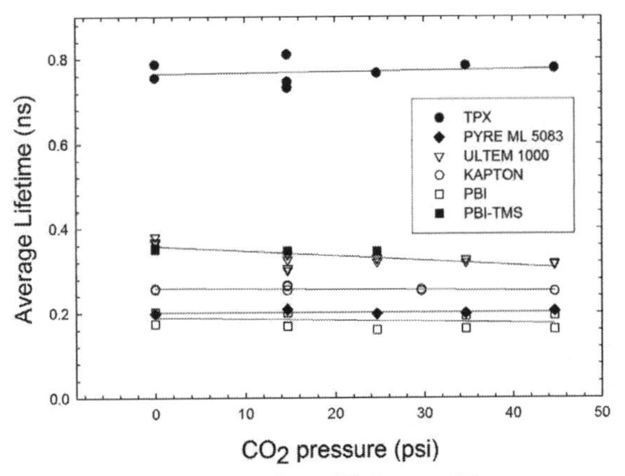

FIGURE 6. Average positron lifetime vs CO_2 pressure. The lines show the best fit to each curve (except PBI-TMS).

CONCLUSIONS

The Doppler broadening results demonstrate that the presence of CO_2 in the samples changes the nature of the positron or positronium trapping sites. This is seen in the polyimide samples such as Kapton as well as the samples that form positronium. The general trend of the S parameter is to decrease slightly with increasing CO_2 pressure. This decrease is less evident in the polyimides and more evident in the samples which form positronium. A reduction in S parameter may indicate that more positrons annihilate with oxygen in the free volume voids. In general, these results indicate that it is very useful to have both Doppler broadening and lifetime data to help understand how CO_2 is going into the polymer.

The lifetime results differ significantly for different polymers. The main difference is in the formation and trapping of positronium which manifests as a third, longer lifetime component in the lifetime spectrum. It is well known that positronium does not form in the polyimide structures such as Kapton. Presumably the polyimide polymer chains are aligned such that there is not a large enough free volume void between the polymer chains to trap positronium. By N-substituting TMS to the PBI, some positronium can be trapped, presumably due to the voids created by the within the TMS component. Among the three samples in which positronium formed, there is much variation. In both TPX and PBI-TMS, the data are consistent with no change in oPs lifetime or intensity as the CO_2 pressure is varied. For Ultem 1000, on the other hand, the intensity of oPs formation is much less than TPX, but there is a significant change in oPs intensity with CO_2 pressure, The oPs lifetime, however, is consistent with no change with pressure or with initially a small decrease as the pressure increases followed by a gradual increase.

Applying the sorption description to the three samples that form positronium, we find only small changes in free volume for the TPX and PBI-TMS samples. The TPX free volume tends to increase very slightly over the CO_2 pressure range. For PBI-TMS the free volume decreases very slightly over the CO_2 pressure range, but the data in this case is certainly consistent with no change at all. So, perhaps TPX is following the Henry-type process and tending to absorb the gas molecules while PBI-TMS is following the Langmuir-type sorption. Or perhaps both processes are going on and there is a competition between gas molecules simply filling the pores and being absorbed such that little effect is actually observed. The only sample to show any significant free volume dependence on CO_2 pressure is the Ultem 1000 sample where the free volume clearly decreases. This suggests that the Langmuir-type sorption dominates for this sample.

Finally, we have looked at the average lifetime for all the samples. The only samples which show any variation in average lifetime with CO_2 pressure are Ultem 1000 and PBI-TMS. The TPX sample and the other three samples which do not form positronium show no change in average lifetime as CO_2 pressure is varied. In the samples which form positronium, the average lifetime is, of course, weighted by the longer oPs lifetime. In the three samples which do not form positronium, the data can be typically interpreted as an average lifetime resulting from positron annihilation in the bulk and positron trapping in free volume. The data are not precise enough at this point to separate these two lifetime components, although in future studies it may be possible to fit the data to two lifetimes (not fixing the shortest lifetime to 125 ps) and investigate variation of the longer of the two lifetime components, which may depend on positron trapping in the available free volume.

REFERENCES

1. C. Jean, P.E. Mallon, DM. Schrader (Eds.), Principles and Applications of Positron and Positronium Chemistry, World Scientific Publishing Co., 2003.
2. A. Thornton, K. M. Nairn, A. J. Hill and J. M. Hill, Journal of Membrane Science doi:10-1016/j.memsci.2009. 03.053.
3. Ito, H. F. M. Mohamed, K. Tanaka, K. Okamoto, K. Lee, Journal of Radioanalytical and Nuclear Chemistry, Vol. 211, No. 1 (1996) 211-218.
4. Dlubek, J. Pionteck, M. Q. Shaikh, L. Haussler, S. Thanert, E. M. Hassan, R. Krause-Rehberg, e-Polymers 2007, No. 108, ISSN 1618-7229.
5. C. Jean, H. Chen, L. J. Lee and J. Yang, Acta Physica Polonica A 113 (2008) 1385.
6. H. Kim, W. J. Koros and D. B. Paul, Journal of Membrane Science 282 (2006) 32-43.
7. R. Klaehn, T. A. Luther, C.J. Orme, M. G. Jones, A. K. Wertsching, and E. S. Peterson, *Macromolecules*, Vol. 40 (2007) 7487.
8. Kansy, Nucl. Instrum. Methods Phys. Res. A 374 (1996) 235.
9. M. Urban-Klaehn, R. Spaulding, A. W. Hunt and E. S. Peterson, phys. stat. sol. (c) 4 (2007) 3732-3734.
10. Dryzek and C. A. Quarles, Nucl. Instrum. Methods Phys. Res. A 378 (1996) 337-342.
11. Eldrup, D. Lightbody, J. N. Sherwood, J. Chem. Phys. 63 (1981) 51.
12. J. Tao, J. Chem. Phys. 56 (1972) 5499.

Measurement Of The Neutron Spectrum Of A DD Electronic Neutron Generator

David L. Chichester, James T. Johnson and Edward H. Seabury

Idaho National Laboratory, 2525 N. Fremont Avenue, Idaho Falls, Idaho 83415

Abstract. A Cuttler-Shalev (C-S) ^{3}He proportional counter has been used to measure the energy spectrum of neutrons from a portable deuterium-deuterium electronic neutron generator. To improve the analysis of results from the C-S detector digital pulse shape analysis techniques have been used to eliminate neutron recoil artifacts in the recorded data. Data was collected using a 8-GHz, 10-bit waveform digitizer with its full scale corresponding to approximately 6-MeV neutrons. The measurements were made with the detector axis perpendicular to the direction of ions in the ENG in a plane 0.5-m to the side of the ENG, measuring neutrons emitted at an angle from 87.3° to 92.7° with respect to the path of ions in the ENG. The system demonstrated an energy resolution of approximately 0.040 MeV for the thermal peak and approximately 0.13 MeV at the DD neutron energy. In order to achieve the ultimate resolution capable with this type of detector it is clear that a higher-precision digitizer will be needed.

Keywords: Neutron spectrometry, digital pulse shape analysis, DD-fusion, electronic neutron generator
PACS: 25.45z, 29.20.Ej, 29.25.Dz, 29.30.Hs, 84.30.Sk

INTRODUCTION

Neutron spectrometry can provide a wealth of information about the structure of nuclei and the characteristics of nuclear reactions.[1] At a more applied level neutron spectrometry can provide primary and supplemental information as a nondestructive measurement technique to support industrial, medical, and security related activities. Unfortunately, it is difficult to accurately measure the energy of neutrons with both sufficient detection efficiency and sufficient energy-resolution to make meaningful and timely analyses in many cases. Time of flight measurements may be made using a nuclear reactor with a chopper wheel or with some neutron-producing particle accelerators; however, the infrastructure required for these measurements is often prohibitive for low turnaround or field applications. Detectors employing organic scintillators, both liquids and solids, with the ability to use pulse shape discrimination to separate neutron events from photon events have found some use for laboratory measurements. These detectors have reasonably sufficient detection efficiencies for many measurements but their energy resolution is typically limited to a few hundred keV.[2] The energy resolution achievable in these detectors is limited due to the need to employ sensitive mathematical unfolding algorithms to 'deconvolve' measured spectra, using simulated and/or measured response-function definitions to iteratively determine the spectrum of an incident neutron field.

An alternate detector approach for neutron spectrometry that has found some utility is the use of a ^{3}He-ionization chamber coupled with pulse-height analysis, recording the energy of incident neutrons in a technique not unlike that used in scintillator-based photon spectrometry.[1-5] Detectors based on this approach are frequently referred to as Cuttler-Shalev (C-S) spectrometers. Neutron detection in this approach is a result of neutron absorption in the ionization chamber's ^{3}He fill gas in the ^{3}He(n,p)^{3}H reaction, the resulting proton and tritium atom proceed to ionize the ionization chamber's fill gas (which also often contains a denser gas such as methane, to reduce the path length of the charged particles) which then produce an electronic signal as these electrons and ions are collected.[6] Neutron capture in ^{3}He is exothermic with a Q-value of 0.764 MeV; the amplitude of signal pulses resulting from neutron capture corresponds to the sum of this energy plus the incident neutron's energy. These detectors are very sensitive to microphonic noise and electromagnetic interferences but under ideal conditions neutron energy measurements have been made with a resolution of <15 keV, which is the limiting factor for the low-energy range of spectrometry for these detectors. The energy resolution of these detectors

Application of Accelerators in Research and Industry
AIP Conf. Proc. 1336, 519-523 (2011); doi: 10.1063/1.3586154

degrades at higher energies; a good study of the relationship between neutron energy and resolution for C-S spectrometers is provided in the work of Ohm et al. and Owen et al.[7,8] The C-S spectrometer energy resolution performance has been reported several times in the literature by several groups and includes observations of E_n = thermal (FWHM = 0.013 MeV) [9], E_n = thermal (FWHM = 0.013 MeV) [10], E_n = 1 MeV (FWHM = 0.025 MeV) [9], E_n = 1 MeV (FWHM = 0.027 MeV) [8], E_n = 1 MeV (FWHM = 0.029 MeV) [10], E_n = 1.27 MeV (FWHM = 0.065 MeV) [2], and E_n = 2.450 MeV (FWHM = 0.046 MeV) [11].

A key aspect of the use of a C-S spectrometer is the application of rise-time pulse shape discrimination to distinguish neutron capture events leading to full-energy deposition in the detector from elastic scattering events and from wall-effect events, where one or both of the neutron capture reaction daughter products hits a wall before depositing all of its energy into the chamber gas.[1,10,12,13] Until recently most measurements systems used with C-S spectrometers used analog pulse analysis techniques to execute the rise-time pulse shape analyses. Continued development of laboratory electronics has also lead to the use of high-speed waveform digitizers has presented new options for collecting and analyzing this data. An early example of the application of the use of waveform digitization with a C-S spectrometer was presented by Takahashi et al. who used a 10-bit, 100-MHz digitizer to record the thermal neutron capture peak in a C-S style neutron spectrometer and then analyzed their data using digital pulse shape analysis (dPSA) techniques.[14] This group did not perform rise-time analyses for artifact rejection and studied only the detection characteristics of their system for analyzing the thermal neutron peak.

In this paper we present research examining the use of dPSA techniques to analyze data from a C-S neutron spectrometer measuring neutrons from a deuterium-deuterium (DD) electronic neutron generator (ENG). A DD-ENG produces pseudo-monoenergetic neutrons, with the actual energy of the neutron spectrum a) varying with respect to angle of the ENG's internal ion beam and b) having a distribution as a result of ion straggling within the ENGs metal hydride target. The kinematics of the reaction are well known and have been described in the literature.[15-17] The angular dependence of neutron yield versus angle is shown in FIGURE 1 together with a plot of neutron energy versus angle, for cases with 50-keV and 100-keV incident deuteron energies. Also shown in this figure is the impact a thick target has on the neutron energy. At the 90° observation angle (perpendicular to the ENG) the neutron yield is relatively constant over a short range of ± 10°. Also, over this narrow range the relation between neutron energy (2.46 MeV at 90°) and angle is nearly the same for incident deuteron energies from 0.050-MeV to 0.100-MeV and for thin and thick targets.

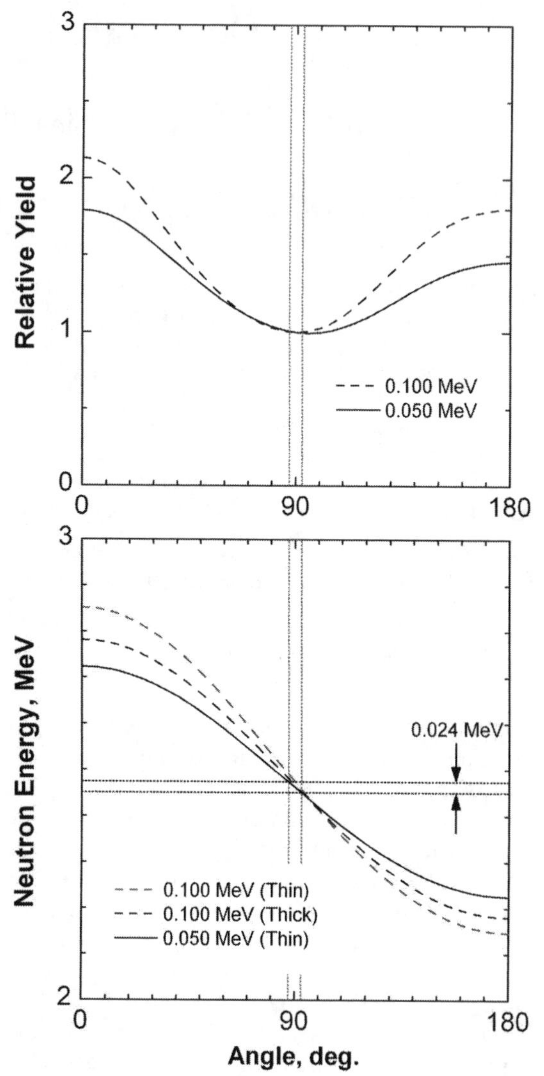

FIGURE 1. The upper figure shows the theoretical relative yield of neutrons as a function of angle for 0.050-MeV and 0.100-MeV deuterons striking a deuterated target inside an ideal ENG. The lower figure shows the theoretical energy of neutrons as a function of angle for 0.050-MeV deuterons striking a thin target and for 0.100-MeV deuterons striking both thin and thick targets. The vertical dotted lines represent the emitted neutron angle range detectable by the spectrometer used here. The horizontal dotted lines in the figure on the bottom represent the approximate spread in energy of neutrons incident on the C-S spectrometer as used in the measurement reported here.[15-17]

reasonable conversion efficiency, high positron work function and relatively low cost.

Depending on the type of moderator, different conditioning methods exist for improving the conversion efficiency and work function. If the moderator is formed by filaments, foils or meshes conditioning requires warming of the material to a high temperature, this induces annealing of the material allowing tungsten to re-crystallize thus removing defects, and cleaning the surface of contaminants that might be present.

EXPERIMENTAL

Sample Preparation

Tungsten Thin Films

As previous studies show [3], W thin films can be used as positron moderators. In this work we prepared the W films by sputtering of a W target. Technical details on the sputtering system are as follows; the thin film growth sputtering system is composed by a 20" diameter by 20" height cylindrical sample-preparation chamber, which has been designed and constructed at IFUNAM [11]. Vacuum system consists on a 150 lt/s turbo pump, allowing to achieve a base pressure of 6×10^{-6} torr. It is possible to install and use up to five magnetron sputtering sources in the chamber. In order to produce the tungsten thin films we employed a circular magnetron sputtering source of 1" diameter. Process gas is argon (HP graded), pressure is recorded and controlled by means of a system based on a capacitive sensor (MKS baratron) and a proportional gauge.

Substrate is placed at the geometric center of the chamber; but distance between sputtering target and substrate planes can be adjusted. For this work the distance was fixed at 11 cm, it has been observed in this sputtering system that uniformity of the films depend strongly on the distance to the sputtering target. At this working distance films are expected to be near uniform. Different thicknesses of the tungsten films can be achieved, once the experimental conditions are chosen, thickness will depend on the exposure time. It is possible to grow tungsten films with thicknesses ranging from 10 nm in about 4 minutes up to $0.5\,\mu$m in about 1 hour.

In order to use tungsten films as moderators, these films are needed to be self supported but films are also expected to be brittle. Films were grown in top of soluble substrates, so no mechanical efforts are exerted when separating tungsten films from substrates. We used NaCl crystals and deposited KI by filament evaporation on glass as soluble substrates.

Tungsten Nanopowder In Parlodion Films

As a proposal for positron moderator we are considering the use of Tungsten nanopowder embedded in Parlodian™ membranes. Parlodian™ is composed by Nitrocellulose in Amyl Acetate, typically sold as a 2% solution. Common applications of this substance include the casting of films for Transmission Electron Microscopy (TEM) grids, making replicas on metallurgical or ceramic surfaces. In Electron Microscopy it is common to find a variety of applications of Parlodian™ films, since it is used as substrate for direct deposition of biological materials. Tungsten used in this proposed moderator corresponds to Tungsten (IV) oxide, in the form of a nanopowder. Size of the W particles is < 100 nm, measured by TEM, as reported and certified by Aldrich Chemistry.

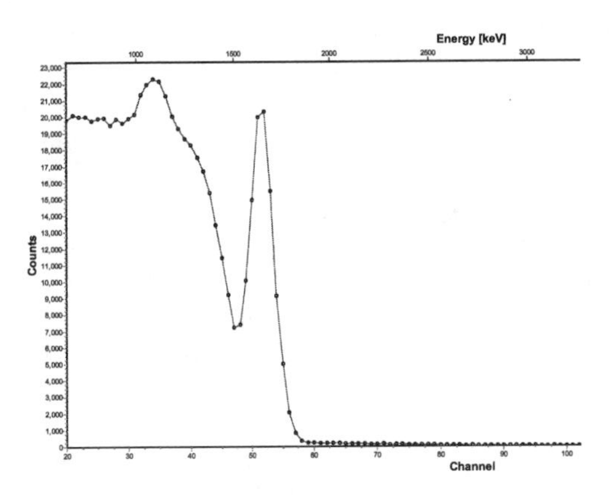

FIGURE 1. RBS spectrum of one of the tungsten films deposited on a NaCl substrate. The narrow peak on the right corresponds to the tungsten film. It s clearly separated from the substrate on the left, seen as a broad distribution. The structure in the form of a peak on the substrate region corresponds to C contamination. Bombardment energy employed while acquiring the spectrum was 1900 keV.

In order to prepare one positron moderator, we added 10 mg of W nanopowder in 100 μl of Parlodian™. By means of an ultrasound cleaner the W nanopowder was dissolved and homogenized in the Parlodian™ solution. And then, following a standard method for casting the membrane, the solution was poured into distilled water. Hydrophobic properties of the Parlodian™ allow it to form the membrane with the W nano particles embedded into it. Once the membrane is cast, it is taken out of the water and dried. The amount of material used to cast the polymer, allowed to have rectangular membranes of about 8 mm × 5 mm for each positron moderator.

For a first approach in building a positron moderator we decided to use Parlodian™ because it is commonly used in electron microscopy studies, and characterization of our positron moderators involve a series of microstructure studies. We are aware that it is possible to construct electroactive polymer films [12], which for holding the W particles and moderating positrons could be much better. In the same fashion, at this point we are employing WO_3 nanoparticles, but for a more advanced moderator it is most likely that W mono crystalline powder will be used.

RESULTS

Constructed moderators were characterized by different techniques. In the case of the tungsten thin films, parameters of interest are surface structure of the films and thickness. In order to characterize the surface structure measurements were performed by different microscopy techniques (not shown here). Only the thicker films (about $0.1\mu m$) start showing some crystallization structure, thinner films exhibit only the shape of the substrate in which they are deposited. The other parameter of interest, thickness of the films, has been characterized by IBA techniques, in particular RBS (Rutherford BackScattering). Different deposition times under same conditions, lead to different thicknesses of the films. It has been possible to establish parameters to grow films ranging from a few nanometers to a fraction of a micrometer. Uniformity of the films has also been tested in the same way, by performing a scan of the films. An example of the RBS analysis of a tungsten thin film is presented in Figure 1. Thickness is important due to the range and interaction of antiparticles in matter, films too thin will not thermalize sufficient positrons. While films too thick would not allow positrons to go through the material.

In the case of Parlodian™ membranes with embedded tungsten nanopowder, microscopy techniques were employed to characterize them. In this case, parameters of interest are: Uniformity and thicknes of the membranes, and tungsten nanopowder distribution. We started by using SEM in order to acquire some information on the candidate moderator structure, but resolution of the microscopy was on the limit, and therefore no clear information on the structure could be acquired. Despite this fact, evidence on the formation of clusters of tungsten nanopowder was found. Next study was performed by AFM. Uniformity of the membranes allowed us to get some useful surface information by employing this analysis

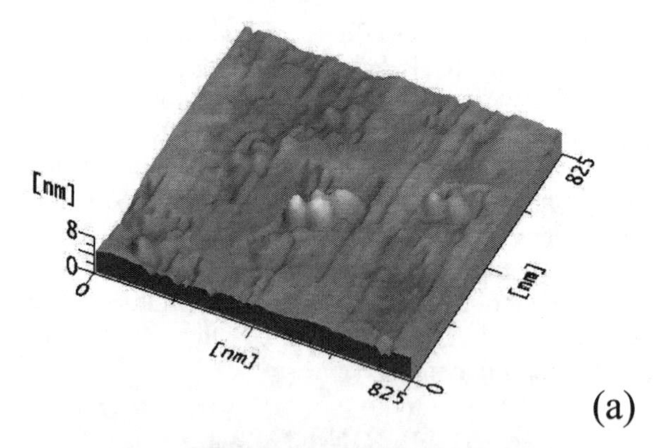

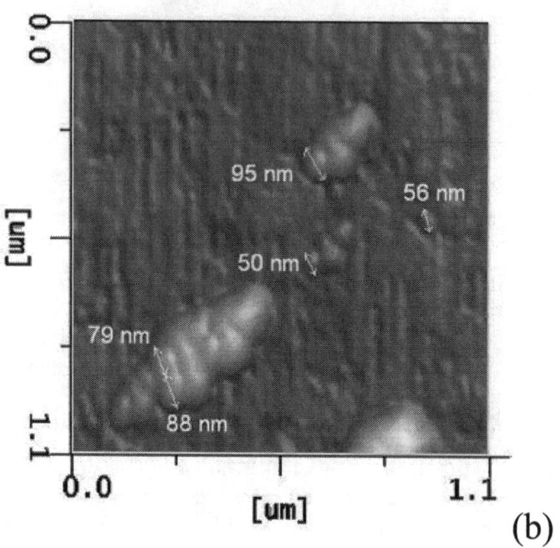

FIGURE 2. (a) AFM analysis of the Parlodian™ membranes with embedded tungsten nanopowder. From the images results clear that instead of completely dissolving the tungsten on the membrane it is forming clusters of several nano particles. Images also show that the nanopowder distributes both on the surface of the membrane and within the membrane. (b) Same as before, but with dimensions of the tungsten nanoparticles added on top of the image.

technique; results are shown in Figure 2. As predicted by the SEM measurements, the tungsten nanopowder is forming clusters of nanoparticles. It is still unclear if clusters were formed before or after membrane was casted, but seems to be strongly related with the dissolving process of the tungsten nanopowder. Refinement of the nanopowder-dissolving and membrane-casting technique will allow us to produce candidate moderators in which the nanoparticles will be uniformly integrated within the membrane, and we will explore the possibility of having the nanoparticles distributed only over one of the membranes surface. Currently, TEM studies are underway, which will also provide information on the distribution of the nanopowder in the membrane.

Due to a tight time schedule, measurement of the positron conversion efficiency has not yet been performed, but it is planned to be done in collaboration with Missouri University of Science & Technology (MS&T), where a positron production system based on a ^{22}Na radioactive source is working at this time. The system allows the moderator, which is currently in use to be replaced by our candidate moderators without requiring any other modification. Since activity, and solid angle of the radioactive source are known, and by using a channeltron detector, available in the positron transport system, it will be possible to determine the ratio of high energy positrons to moderated positrons. Before performing such measurements we still need to refine the candidate moderator production process and completely characterize the produced moderators with the analysis techniques employed for this work, so it will be possible to later reproduce moderators with the highest conversion efficiency.

REMARKS

Two methods for constructing tungsten based positron moderators have been proposed, one method in which moderators are in the form of thin films, growth by sputtering of a tungsten target and deposition of it on top of soluble substrates. And a second method in which moderators are built by embedding tungsten nanopowder in a polymer membrane. Parameters of interest of the proposed positron moderators were characterized by means of IBA and microscopy techniques. A range for the thickness of the tungsten films has been established, and has been related to the deposition time, for given deposition parameters. And surface structure has been characterized, showing that for thicker films crystalline structure is present. In the case of the membranes, microscopy studies show that the embedded nanopowder has not been uniformly distributed, possibly due to the dissolving process.

It is still necessary to perform measurements of the positron conversion efficiency of the proposed moderators, which will be done in a near future.

ACKNOWLEDGMENTS

This work was supported by UNAM-DGAPA-PAPIIT, under contract IN-105510. This work was partially supported by CONACyT. The authors acknowledge the technical support of: J. Cañetas, C. Valencia, D. A. Quiterio, and M. Galindo.

REFERENCES

1. E. M. Gullikson, A. P. Mills Jr, Phys. Rev. Lett. **57** (1986) 376.
2. J. Störmer, A. Goodyear, W. Anwand, G. Brauer, P. G. Coleman, W. Triftshäuser, J. Phys.: Condens. Matter **8** (1996) L89.
3. K. G. Lynn, B. Nielsen, J. H. Quateman, Appl. Phys. Lett. **47** (1985) 239.
4. A. Zecca, L. Chiari, A. Sarkar, S. Chattopadhyay, M. J. Brunger, Nucl. Instr. Meth. B **268** (2010) 533.
5. H. M. Weng, C. C. Ling, C. D. Beling, S. Fung, C. K. Cheung, P. Y. Kwan, I. P. Hui, Nucl. Instr. Meth. B **225** (2004) 397.
6. N. Zafar, J. Chevallier, G. LAricchia, M. Charlton, J. Phys. D: Appl. Phys. **22** (1989) 868.
7. V. V. Plokhoi, Ya. Z. Kandiev, S. I. Samarin, G. N. Malyshkin, G. V. Baidin, I. A. Litvinenko, V. P. Nikitin, Nucl. Instr. Meth. A **448** (2000) 94.
8. L. Liszkay, Zs. Kajcsos, M. F. Barthe, L. Henry, G. Duplâtre, A. Nagy, Appl. Surf. Sci. **194** (2002) 16.
9. P. J. Schultz, K. G. Lynn, Rev. Mod. Phys. **60** (1988) 701.
10. P. G. Coleman (Ed.), Positron Beams and Their Applications, World Scientific, 2000.
11. H. Cruz-Manjarrez (Private Communication).
12. M. N. Simon, B. Y Lin, H. S. Lee, T. A. Skothelm, J. S. Wall, Procs. of the XIIth International Congress for Electron Microscopy, 1990, 290.

Compact Permanent Magnet Microwave-Driven Neutron Generator

Qing Ji

Lawrence Berkeley National Laboratory, 1 Cyclotron Rd., MS5R0121, Berkeley, CA 94720

Abstract. Permanent magnet microwave-driven neutron generators have been developed at Lawrence Berkeley National Laboratory. The 2.45 GHz microwave signal is directly coupled into the plasma chamber via a microwave window. Plasma is confined in an axial magnetic field produced by the permanent magnets surrounding the plasma chamber. The source chamber is made of aluminum with a diameter of 4 cm and length of 5 cm. A stack of five alumina discs, which are 3 cm in diameter and total length of 3 cm, works as microwave window. Three permanent ring magnets are used to generate the axial magnetic field required for the microwave ion source. Both hydrogen and deuterium plasma have been successfully ignited. With 330W of microwave power, source chamber pressure of 5 mTorr, and an extraction aperture of 2 mm in diameter, the deuterium ion beam measured on the target was approximately 2.5 mA. Over 90% of the ions are atomic. With the ion source at ground potential and titanium target at –40 kV, the analysis of the activated gold foil and calibrated neutron dose monitor both indicated that roughly 10^7 n/s of D-D neutrons have been produced. The D-D neutron yield can be easily scaled up to 10^8 n/s when the titanium target is biased at –100 kV.

Keywords: microwave ion source, neutron generator.
PACS: 41.75.Ak, 29.25.Dz

INTRODUCTION

Active neutron interrogation has been demonstrated to be an effective method of detecting fissile material due to the ability of neutrons to penetrate substantial amounts of material, high-Z shielding material in particular, and to stimulate fission signatures. The basic principle of a neutron generator is to bombard an energetic ion beam of either deuterium or tritium onto a target. Neutrons are produced via the D+D, D+T, or T+T reactions if the target surface is loaded with the D or T molecules. There are some advantages of employing D-D instead of D-T generators in the field: a D-D neutron generator does not contain tritium and thus avoids associated safety requirements and operational restrictions, making it more suitable for applications in the field; the lower energy of the source neutrons, 2.8 MeV compared to 14.1 MeV, makes it possible to exploit an additional SNM signature. Fission neutrons with energies of greater than that of the source neutrons (2.8 MeV) provide an unambiguous signal indicating the presence of SNM that can be detected with suitable neutron detectors in future systems. However, the comparable D-D yield is about 1/100 of the D-T yield due to the lower cross sections. Therefore, compact, low-power consumption, higher intensity neutron sources that can produce comparable D-D neutrons are of great interest for applications where human portable equipment is needed, such as the interrogation of areas where larger vehicle mounted equipment cannot reach or portability is essential.

Current commercially available small neutron generators, such as borehole inspection neutron generators, have certain intrinsic limitations mainly related to the ion source. The D-D neutron yield of such generators is on the order of 10^6 n/s. The ion source used in commercial neutron generators is the Penning ion source[1]. It produces less than 10% of atomic ion fraction and its current density is approximately 2 mA/cm^2 at around 10 W, which is difficult to scale up. Rf-driven multicusp ion source based compact neutron generators have been developed in Lawrence Berkeley National Laboratory for various applications, such as material analysis, boron neutron capture therapy, and cargo screening for homeland security[2-4]. The RF-driven ion source is capable of producing higher intensity of proton ions; however, it is typically operated at around 10 mTorr and the gas pressure required to ignite the RF discharge can be in the tens of mTorr range, which is not desirable in sealed source operation.

Application of Accelerators in Research and Industry
AIP Conf. Proc. 1336, 528-532 (2011); doi: 10.1063/1.3586156

Microwave-driven ion sources are widely used for the production of high quality ion beams for accelerators, atomic physics research, and industrial applications. They are typically easy to operate and were found to have a long lifetime. Extensive development and refinement of microwave-driven ion sources has taken place over the last three decades. Either solenoid coils or permanent magnets are used to generate the required magnetic field for establishing the electron resonance condition at the driving frequency (typically 875 gauss for 2.45 GHz)[5-11]. Microwave-driven ion sources can be operated at low gas pressures (in the 1-2 mTorr range) and do not require high pressure for ignition, in contrast to the previously developed RF-driven ion source. Microwave-driven ion sources are power efficient and can produce a high fraction of atomic ions. Neutron generators based on permanent-magnet microwave ion sources are expected to provide reliable, long lifetime D-D and D-T neutron production.

In contrast to the Penning ion sources used in commercial generators that produce mostly molecular ions, microwave-driven ion sources have the advantages of high power efficiency and high fraction of atomic ions. For generating hydrogen ions, a beam current density of over 150 mA/cm^2 has been reported by several groups at a microwave power of around 400 W with an atomic ion fraction of 80-90% [5, 8, 11]. This current density greatly exceeds the requirement of most neutron generator applications. A compact, moderate yield D-D neutron generator based on a permanent-magnet microwave plasma source has been developed at Lawrence Berkeley National Laboratory. In this article, ion source characteristics and neutron yield measurement of this neutron generator are reported.

EXPERIMENTAL SETUP

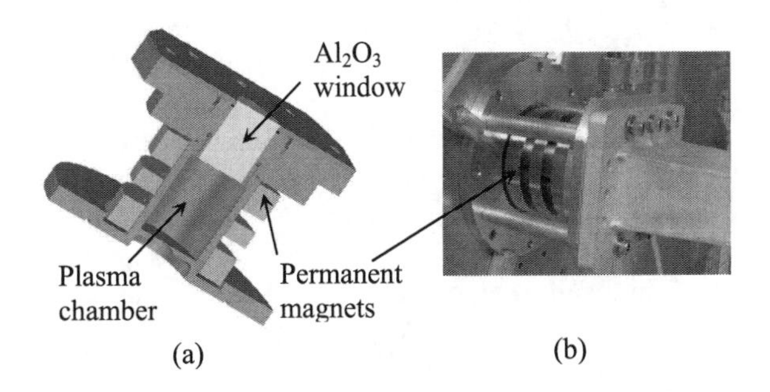

FIGURE 1. (a) Schematic drawing and (b) Photograph of the permanent magnet microwave ion source. The source chamber is 4 cm in diameter, and 5 cm in length.

The permanent magnet microwave source employed in the neutron generator is shown in Figure 1. The source chamber is made of aluminum, 4 cm in diameter and 5 cm in length. A stack of aluminum discs, which are 3 cm in diameter and total length of 3 cm^5, works as the microwave window. Permanent magnet stacks with three NdFeB ring magnets have been designed to generate an axial magnet field in the microwave plasma source. These ring magnets have the same outer diameter of 8 cm, and inner diameter of 5 cm. Their thicknesses are 1.7 cm, 1.0 cm, and 1.7 cm, respectively, and the spacing between magnets is 0.8 cm. Both the simulated and measured axial magnetic fields as a function of position are shown in Figure 2. At the microwave window and beam extraction aperture, the axial magnetic field is 875 Gauss.

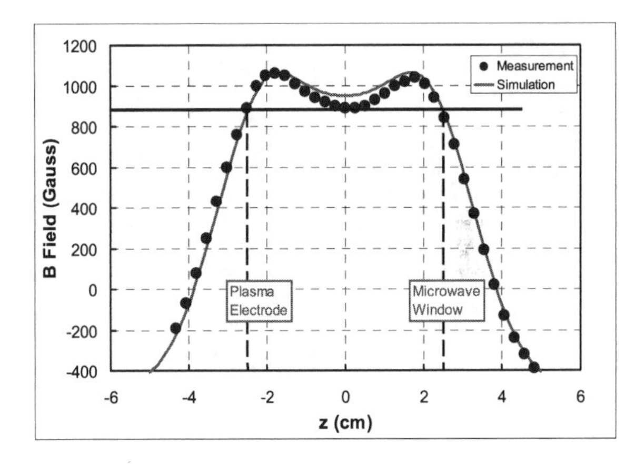

FIGURE 2. Axial magnetic field distribution in the microwave ion source. The axial magnetic field is 875 Gauss at both the microwave window and beam extraction aperture.

In this case, plasma is confined in an axial magnetic field produced by the permanent magnets surrounding the plasma source chamber. The electrons in the plasma are resonantly energized when the microwave passes the resonance zone where the magnetic field corresponding to an electron gyro-frequency equals the frequency of the microwave. Both hydrogen and deuterium plasmas have been successfully ignited.

The ions are extracted from an aperture of 2 mm in diameter and accelerated to the required kinetic energy before they bombard the target. The ion source is at ground potential, and the Mo target and a secondary electron suppressor inside the main vacuum chamber are biased at negative potentials. Polyethylene shielding was installed around the neutron generator to minimize the neutron dose outside of the experimental area. Figure 3(a) shows the permanent magnet microwave-driven neutron generator housed in the polyethylene shielding box. A neutron dose monitor is positioned outside the polyethylene shielding box. The deuterium beam energy is limited to below 40 keV to keep the neutron production lower than 10^7 n/s.

Measurement of the yield of the D-D neutron generator has been accomplished through use of the moderate-and-capture technique. The detector is a 6-inch diameter by 6-inch high cadmium-clad paraffin moderator, cast in two 3-inch high pieces such that an activation element can be placed in a centrally located cavity, as shown in Figure 3(b). The activation element is a thin 2-inch diameter Au foil, weighing 2.0 grams. The neutron capture reaction in Au produces Au-198, an isotope with a 64.6 hour half-life which decays with emission of a high-yield gamma-ray at 498 keV. The 498 keV gamma-ray is later measured with a low-BKG HPGe gamma-ray spectrometer.

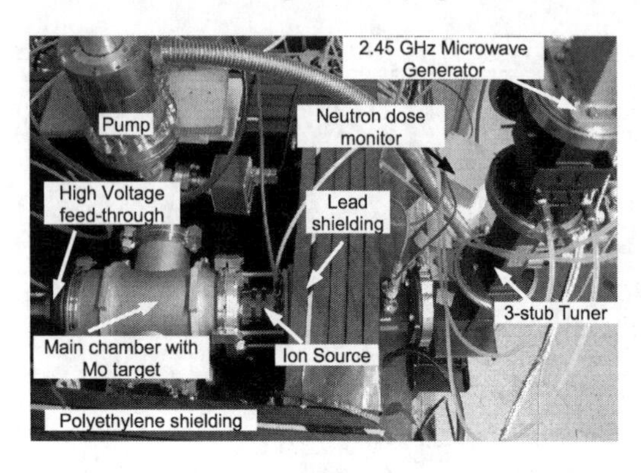

(a)

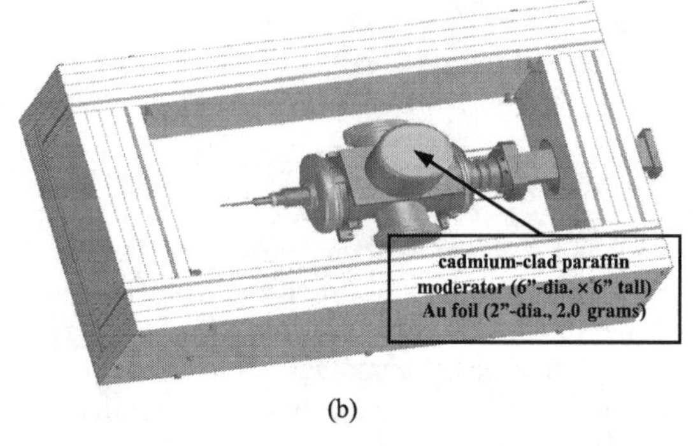

(b)

FIGURE 3. (a) The permanent magnet microwave-driven neutron generator in the polyethylene shielding box. (b) Gold foil activation is used to measure the neutron yield produced by the generator. The Au foil (2"-diameter, 2.0 grams) is placed in a cadmium-clad paraffin moderator 6" tall and 6" in diameter.

The response of this analysis system was previously calibrated through exposure to an NIST-calibrated (alpha,n) neutron source. Since exposures have been an hour or less, Au-foil activation is a faithful integrator, and produces a result that describes the time-averaged neutron yield. Response of the system to neutrons in the energy range from 0.1 to 10 MeV is relatively constant; hence, the detector is sensitive to scattered neutrons. A correction factor obtained from MCNP simulations has been applied to the reported yield values.

RESULTS AND DISCUSSIONS

Figure 4 shows the mass spectrum of the hydrogen plasma generated by the microwave plasma source. Over 90% of H^+ at microwave power of 350W and source pressure of 3 mTorr was observed. The atomic ion fraction varies from 88% to 92% with the source pressure ranging from 2 mTorr to 8 mTorr.

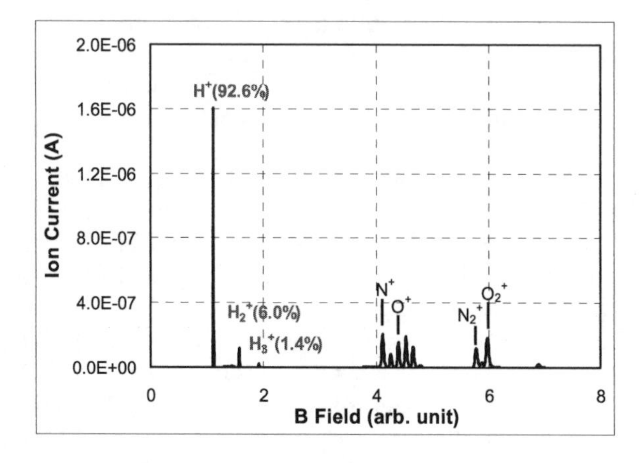

FIGURE 4. Mass spectrum of the hydrogen plasma generated by the microwave ion source. Over 90% of atomic hydrogen ions have been produced.

With 330 W of microwave power, source chamber pressure of 5 mTorr, and an extraction aperture of 2 mm in diameter, the deuterium ion beam measured on the target was approximately 2.5 mA. The current density increases almost linearly with the source pressure (Figure 5). At 2 mTorr, the current density at power of 330 W, with an extraction aperture of 2 mm in diameter, was approximately 26 mA/ cm^2, and became 94 mA/cm^2 at 5.7 mTorr.

Because the polyethylene shielding acts to thermalize neutrons emitted from the neutron generator, the thermal neutron flux at any given point may be higher relative to the case with no shielding. This can result in an overestimate of the neutron yield as obtained through neutron activation analysis. MCNP simulations were performed to calculate the shielding correction factor for application to neutron activation analysis of gold foil.

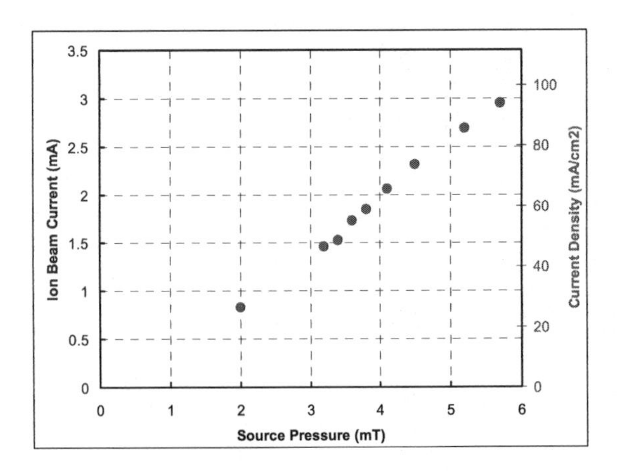

FIGURE 5. The ion beam current and current density of the extracted deuterium ion beam from the microwave ion source as a function of source pressure. The extraction aperture is 2 mm in diameter, and the microwave power is 330W.

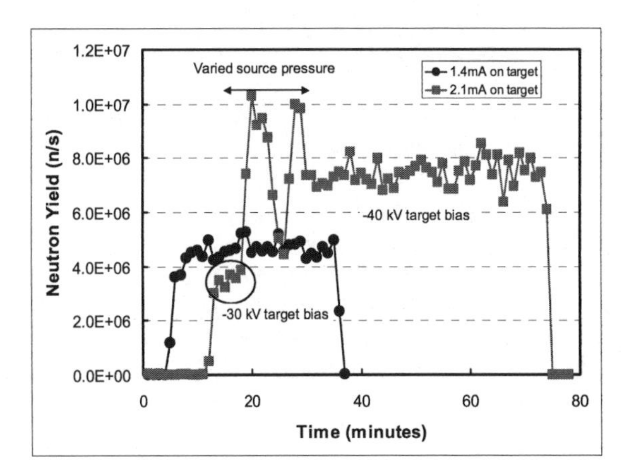

FIGURE 6. Typical runs of the permanent magnet microwave-driven neutron generator. The neutron yield varies from 5×10^6 n/s to 7.4×10^6 n/s at different source conditions.

The experimental area was previously modeled for simulation of radiation dose rates and to calculate required shielding thicknesses for adequate radiation protection. This model was altered through the addition of the detector setup, a gold foil sandwiched between cylindrical paraffin sample holders. The paraffin structure is encased in 1 mm of cadmium. The simulation results indicate that the presence of polyethylene shielding in the experimental area increases the neutron flux at the detector foil by a factor of 1.2. Neutron yield estimates obtained from the activity of the gold foil after irradiation are corrected by this factor for a more accurate result.
Figure 6 shows two typical runs of neutron production of the microwave-driven neutron generator. The neutron dose monitor positioned outside of the neutron generator has been calibrated using an AmBe source with a known intensity. The calibration factor is 1.23×10^7 n/s per mR/hr. The neutron yields listed in Figure 7 have been scaled from the real-time data of the neutron dose monitor. At a beam current of 1.4 mA, neutrons at a rate of approximately 5×10^6 n/s were generated. Compared with the gold foil activation result, which was 9.16×10^6 n/s, a factor of two discrepancy exists. More calibrations will be carried out in the future. At a microwave power of 350W, deuterium ion current of 2.1 mA, deuterium energy of 40 keV, neutron yield of approximately 7.4×10^6 n/s was produced on Mo target. The neutron production rate can be easily extrapolated to a higher beam energy, e.g. 100 keV, which turns out to be 1.18×10^8 n/s.

SUMMARY

A moderate yield D-D neutron generator based on a compact permanent-magnet microwave plasma source has been developed at Lawrence Berkeley National Laboratory. It aims at enabling portable interrogation technologies for field operation, offering inspection agencies or event responders the capabilities to examine suspicious containers for SNM in locations where equipment currently doesn't exist. Compared with RF-driven ion sources and Penning ion sources that are commonly used in neutron generators, microwave ion sources are power efficient, exhibit high atomic ion fraction, operate at low gas pressure and have been found to have long lifetime. With 330W of microwave power, source chamber pressure of 5 mTorr, and an extraction aperture of 2 mm in diameter, the deuterium ion beam measured on the target was approximately 2.5 mA. Over 90% of the ions are atomic. With the ion source at ground potential and titanium target at –40 kV, the analysis of the activated gold foil and calibrated neutron dose monitor both indicated that roughly 10^7 n/s of D-D neutrons have been produced. The D-D neutron yield can be easily scaled up to 10^8 n/s when the titanium target is biased at –100 kV.

ACKNOWLEDGMENTS

This work is supported by NA22 of NNSA under the Department of Energy contract No. DE-AC02-05CH11231. The author would like to thank Dr. Shixiang Peng for her helpful discussions on the microwave ion source; thank Amy Sy and Dr. Tak Pui Lou for the MCNP calculations; thank Dr. Al Smith for the neutron yield measurement; thank David Rogers and Jeff Bramble for neutron dose calibration; and thank Steve Wilde, Michael Dickinson, and Tom McVeigh for their technical support.

REFERENCES

1. D. Chichester, *Nucl. Instru. and Meth. in Phys. Res. B,* **241**, 753-758(2005).
2. J. Reijonen *et al*, *Appl. Rad. and Iso.* **63**, 757-763(2005).
3. J. Reijonen, *Nucl. Instru. and Meth. in Phys. Res. B,* **261**, 272-276(2007).
4. B. A. Ludewigt, R. P. Wells, and J. Reijonen, *Nucl. Instru. and Meth. in Phys. Res. B,* **261**, 830-834(2007).
5. Z. Song *et al*, *Rev. Sci. Instrum.,* **77**, 03A305(2006).
6. J. Sherman *et al*, *Rev. Sci. Instrum.* **69**, 1003-1008(1998).
7. F. Welton *et al*, *Rev. Sci. Instrum.,* **67**, 1634-1637(1996).
8. Taylor and J. F. Mouris, *Nucl. Instru. and Meth. in Phys. Res. A,* **336**, 1-5(1993).
9. R. Gobin *et al*, *Rev. Sci. Instrum.*, **77**, 03B502(2006).
10. V. P. Derenchuk, *Rev. Sci. Instrum.,* **75**, 1851-1853(2004).
11. J. W. Kwan *et al*, "A 2.45 GHz High Current Ion Source for Neutron Production", 17th International Workshop on ECR Ion Sources and Their Applications September 17-21, 2006, IMP, Lanzhou, China.

Ion Beam Collimation For Improved Resolution In Associated Particle Imaging

Amy Sy[a,b] and Qing Ji[a]

[a]*Lawrence Berkeley National Laboratory, 1 Cyclotron Road, Berkeley, CA 94720, USA*
[b]*Department of Nuclear Engineering, University of California, Berkeley, Berkeley, CA 94720, USA*

Abstract. Beam spot size on target for a Penning ion source has been measured under different source operating pressures as a function of the extraction channel length and beam energy. A beam halo/core structure was observed for ion extraction at low extraction voltages, and was greatly reduced at higher beam energy. Collimation through use of longer extraction channels results in reduced ion current on target; the resultant reduction in neutron yield for an API system driven by such an ion source can be compensated for by use of even higher beam energies.

Keywords: Penning ion source, Associated particle imaging, Neutron generator
PACS: 29.25.Dz, 52.50.Dg, 52.80.Sm, 29.25.Lg

INTRODUCTION

Associated particle imaging (API) is a method of active interrogation suitable for explosives detection that has been developed for homeland security applications. API utilizes the D-T fusion reaction for neutron production; these neutrons are used to interrogate containers and packages for illicit materials; the resultant gamma rays from neutron interactions are used to characterize the interrogated material. Imaging is achieved through determination of neutron interaction sites; because the neutrons are generated isotropically, determining the neutron trajectory through the associated alpha particle is critical to image resolution.

Neutron generators for API typically involve the acceleration of deuterium and tritium ions onto a stationary target. Image resolution is dependent on the beam spot on target; smaller beam spot sizes give reduced uncertainty in determining alpha particle and neutron trajectories. Time-of-flight analysis through coincidence detection of alpha particles and gamma rays imposes limitations on the neutron yield; the maximum neutron yield is 10^8 n/s to avoid accidental coincidence events [1].

Commercially-available neutron generators for API employ Penning ion sources [2] and are characterized by 2 mm diameter ion beams on target. Smaller beam spot sizes can be achieved with additional focusing lenses, but this is undesirable for field-portable systems. Penning discharges also exhibit high molecular ion fractions that limit the neutron yields obtainable for given ion energy and current. Radio-frequency driven ion sources with high atomic fraction have been developed for API [3] and have achieved beam spot sizes of less than 1 mm, but these sources require higher power and relatively high operating pressures that are also undesirable for field-operable units.

This work focuses on improving the beam spot on target without additional focusing lenses for a Penning source for use with API. The Penning source was obtained from Thermo Fisher Scientific. Beam collimation and smaller extraction apertures will result in reduced beam spot size at the expense of ion current and neutron yield. The extractable current from an ion source is governed by the Child-Langmuir law,

$$I = a^2 \chi \frac{V^{3/2}}{d^2}, \quad (1)$$

where a is the aperture radius, V is the extraction voltage, d is the extraction gap length, and χ is constant for a given ion species. Reducing the aperture radius by a factor of 2, while increasing the voltage and gap length each by a factor of 2, gives a total reduction in current by about a factor of 8. Because the D-T fusion cross-sections increase non-linearly with higher atomic ion energies, the increased beam energy can compensate for this reduction in current and the resultant decrease in neutron yield. The atomic ion fraction in a Penning discharge is 10-20%. Figure 1 shows a mass spectrum for source

Application of Accelerators in Research and Industry
AIP Conf. Proc. 1336, 533-537 (2011); doi: 10.1063/1.3586157

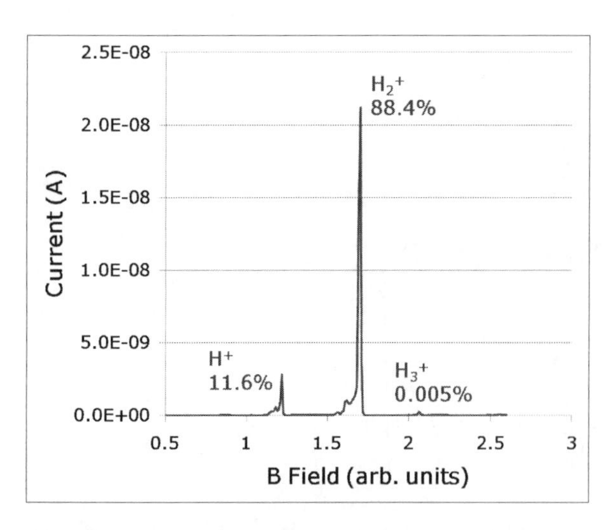

FIGURE 1. Mass spectrum obtained for Penning source operation with hydrogen gas. Over 85% of the ions generated in the Penning discharge are diatomic.

operation with hydrogen; over eighty five percent of ions generated in this discharge are diatomic. Molecular ions can dissociate upon target impact, and the resultant atomic species will carry half of the beam energy. Increasing the beam energy from 37.5 keV to 75 keV gives a nine-fold increase in the cross-section; further increasing the beam energy to 150 keV results in an additional four-fold increase [4]. The resultant neutron yields may then be sufficient for API.

EXPERIMENTAL SETUP

Photographs of the experimental setup are shown in Figure 2.

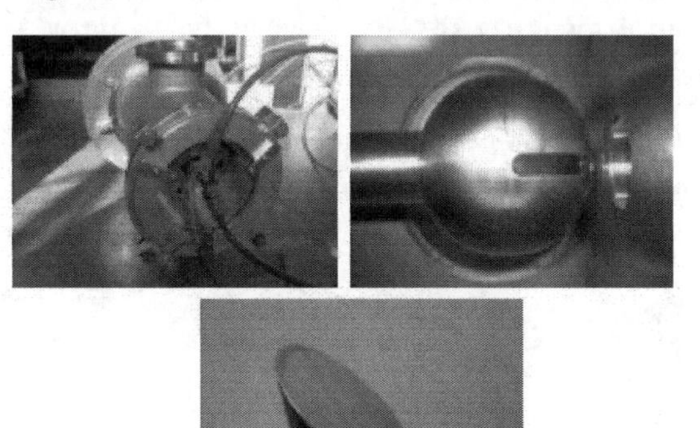

FIGURE 2. Experimental setup for ion extraction. (Clockwise from top left) The Penning source mounted onto the vacuum chamber; the electron suppression shroud; the angled target for beam spot observation during ion extraction.

The Penning source was mounted on one end of the vacuum chamber and held at ground potential; the target and electron suppression shroud were mounted on the opposite end. A 1.4 kV voltage drop between the shroud and target was implemented through the use of reverse-biased Zener diodes. The extraction gap length, the distance between the extraction electrode of the Penning source and the shroud, was adjusted with the applied extraction voltage. Ion current and beam spot size were measured for extraction channel lengths of 5.3, 8.1 and 11.2 mm. Beam profiles were analyzed for extracted ion energies below 50 keV through use of the image analysis program ImageJ.

Penning Source Operation

Penning sources for use as neutron generators have been described [5, 6] and are desirable for their relative simplicity and low power operation. A typical Penning source configuration includes two cathode plates surrounding a cylindrical anode. Plasma generation occurs within the cylindrical region between the anode and cathodes; a positive applied bias on the anode causes electrons to oscillate between the cathodes. External magnets create an axial magnetic field that forces the electrons to gyrate along their trajectories, effectively increasing the path length for ionization; the magnetic field prevents electrons from being lost to the anode. Ions are generated through collisions with the electrons and are extracted through an aperture in one of the cathodes.

The potential distribution within a Penning discharge varies with the source pressure [7] and influences the energy spread of ions within the beam. The beam spot on target can then vary with the source operating pressure. For "low-pressure" (below 1.0 mTorr) operation, the potential along the central axis and the anode potential can differ by as much as fifty percent; though the majority of ions are generated at the anode potential [8], the 1-2% of ions generated in a region of lower potential are responsible for the energy spread of the beam. For "high-pressure" operation, the potential difference between the central axis and anode is much less pronounced; for pressures above 1.0 mTorr, the maximum potential difference is ten percent. Beam spot on target was observed for low- and high-pressure operation of the Penning source during ion extraction with an angled target for beam energies below 50 keV. Beam spot size was measured through observation of burn marks from beams extracted onto flat targets for extracted ion energies above 50 keV.

RESULTS AND DISCUSSION

Images of the beam spot on target and beam profile plots for high- and low-pressure operation are shown in Figures 3, 4 and 5.

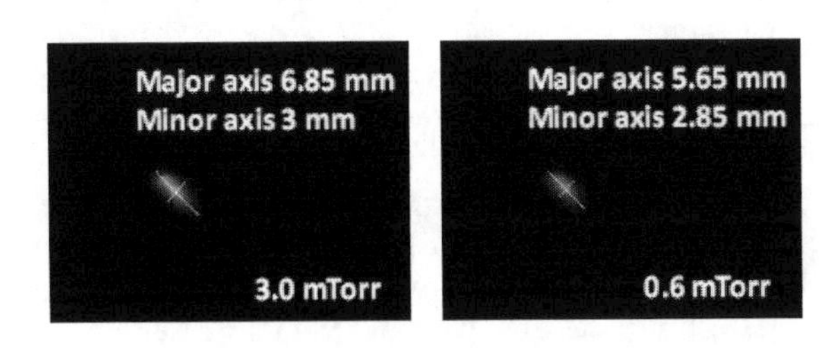

FIGURE 3. Beam spot images for high- and low-pressure operation. The extraction aperture is 2 mm in diameter and the extracted ion energy is 30 keV. Beam "core" is more intense for high-pressure source operation.

Hydrogen gas pressure was 3.0 mTorr for high-pressure operation and 0.6 mTorr for low-pressure operation. For anode voltage of 2.4 kV, typical discharge current obtained was 1.8 mA and 1.2 mA for high- and low-pressure operation, respectively. Ion beams for high-pressure operation are characterized by an intense beam core and faint beam halo. The beam halo is reduced with increased extraction voltage, resulting in smaller beam spot sizes as shown in Figure 4.

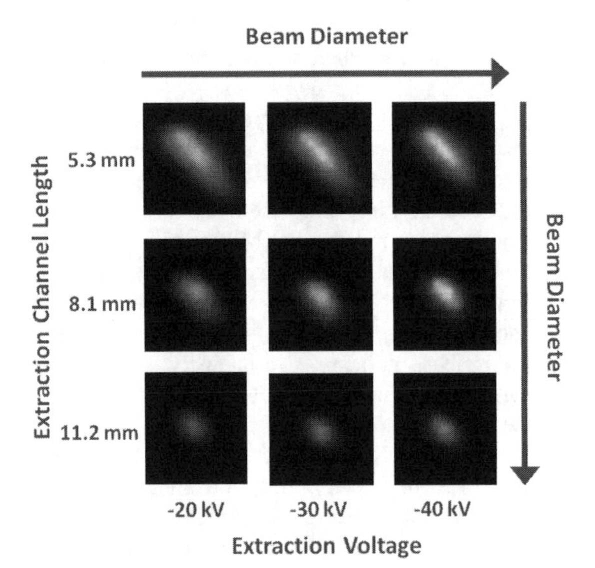

FIGURE 4. Beam spot images for 3.0 mTorr source operating pressure and 2 mm diameter extraction aperture. The extraction gap length was 3 cm. Beam diameter decreases with increasing beam energy and extraction channel length.

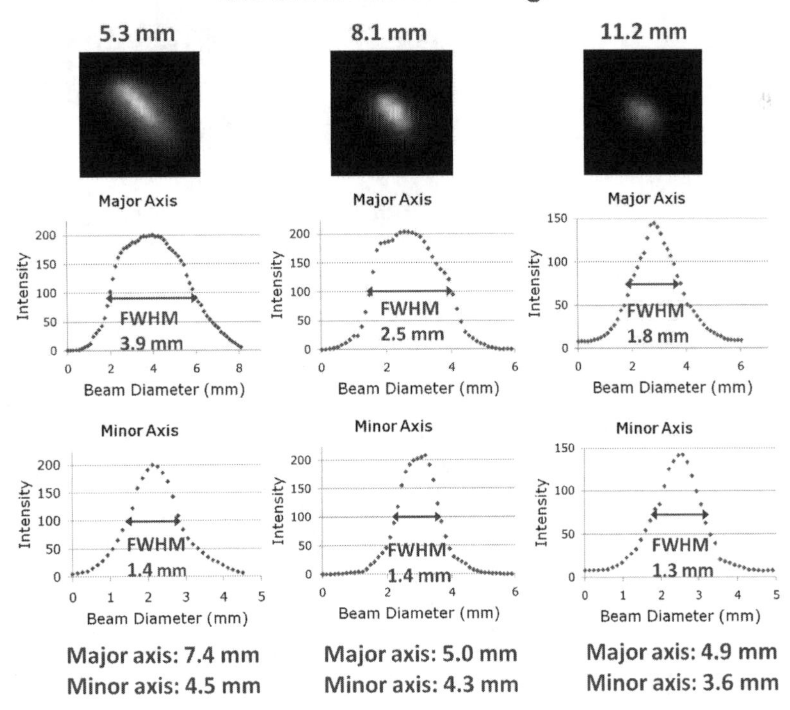

FIGURE 5. Beam spot images and beam profile plots for 3.0 mTorr source operating pressure, 2 mm diameter extraction aperture, and 40 keV extracted ion energy. The extraction gap length was 3 cm.

Because the energy spread of ions within the extracted beam is dependent on the discharge and not on the applied extraction voltage, increasing the beam energy for given discharge parameters results in extracted ion beams where the maximum difference in ion energy is a smaller percentage of the total beam energy. Beam halo and core are thus more prominent for lower beam energies. Similar behavior was observed in beams during low-pressure operation, with a distinct decrease in beam core intensity as compared to high-pressure operation. Major and minor axis beam diameter data is summarized in Table 1. High- and low-pressure operation were investigated to examine effects of pressure on the beam spot on target; while a decrease in beam core intensity was observed for low-pressure operation, no definitive trend in beam diameter was seen for the two pressure regimes as the channel length increased. For the longest extraction channel, 11.2 mm, very little difference in beam diameter was observed. Subsequent tests ran the Penning source in the high-pressure mode to maximize the discharge and extracted currents.

Extracted current for beam energies up to 81 keV is plotted in Figure 6.

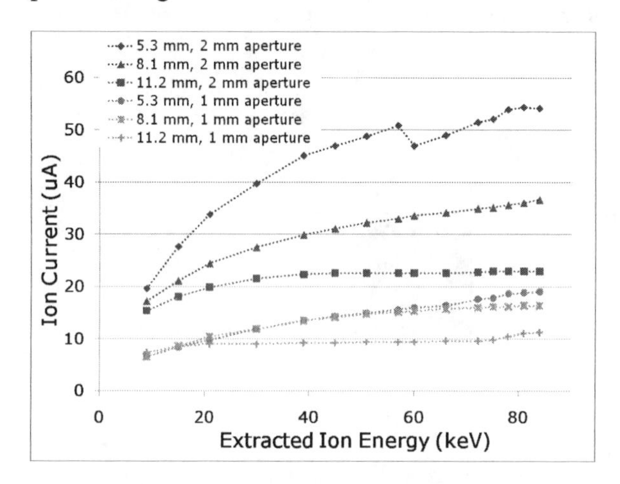

FIGURE 6. Ion current for 1 and 2 mm diameter extraction apertures.

The extracted ion current with 1 mm diameter aperture is greater than twenty-five percent of that for the 2 mm aperture for constant extraction voltage, indicating that the current density for the two cases is not constant. This will be further explored. Longer extraction channels result in substantial current attenuation.

The beam halo previously observed at low extraction voltages is greatly reduced, which is supported by observation of burn marks. A photograph of the burn mark obtained for beam extracted onto a flat target for the 5.3 mm extraction channel is shown in Figure 7. The diameter of the largest burn pattern is 2.7 mm. Similar results were obtained for the 8.1 mm channel; diameter of the largest burn pattern was 2.3 mm. The diameter of the burn mark for both channel lengths is slightly larger than the extraction aperture; we note that the largest burn pattern may be due to heat deposition from the energetic ions. Though beam collimation has not resulted in a substantial decrease in beam diameter, it may increase the lifetime for source operation; prolonged operation of the Penning source can result in aperture erosion due to energetic ion bombardment.

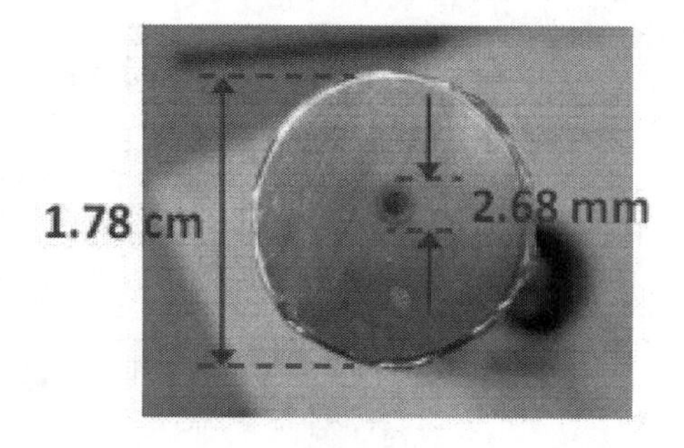

FIGURE 7. Burn mark for beam extracted with 2 mm aperture and 5.3 mm extraction channel. The extracted ion energy is 81 keV.

TABLE 1. Beam Diameter Data for High- and Low-Pressure Source Operation. Beam energy is 30 keV, and extraction aperture is 2 mm in diameter. All units in mm.

	Extraction Channel Length	Major Axis Diameter	Major Axis FWHM	Minor Axis Diameter	Minor Axis FWHM
High-Pressure	5.3	8.5	4.0	5.0	1.6
3.0 mTorr	8.1	4.9	2.1	4.0	1.4
	11.2	4.7	1.8	3.8	1.4
Low-Pressure	5.3	7.7	3.2	4.4	1.6
0.6 mTorr	8.1	5.5	2.3	4.3	1.5
	11.2	4.2	2.5	3.8	1.8

SUMMARY

Beam spot size on target for hydrogen ion beams extracted from a Penning ion source has been measured for various extracted ion energies and extraction channel lengths. Beam halo was observed for low extracted ion energies, but is greatly reduced at higher energies. Longer extraction channels result in smaller beam diameter and reduced ion current on target. For beam energy of 81 keV, the beam diameter on target was slightly larger than the 2 mm extraction aperture.

Future work will focus on beams extracted with a 1 mm extraction aperture. The beam energy will be further increased to 160 keV. Beam spot size for mixed-gas operation with hydrogen and deuterium will be explored to simulate beam spot sizes obtainable for deuterium-tritium gas mixtures.

ACKNOWLEDGMENTS

The authors would like to thank Jack Hoffman from Thermo Fisher Scientific for providing the Penning ion source, Paul Hurley for technical discussion, and Steve Wilde, Michael Dickinson, and Tom McVeigh for their technical support. This work was supported by DOE-DIA under contract number HHM40271611 and NA22 of NNSA under the Department of Energy contract No. DE-AC02-05CH11231.

REFERENCES

1. Y. Wu, J.P. Hurley, Q. Ji, J. Kwan, K.N. Leung, *IEEE Transactions on Nuclear Science* 56(3), art. no. 5076122, p. 1306-1311, 2009.
2. D.L. Chichester, M. Lemchak, J.D. Simpson, *Nucl. Instr. and Meth. in Phys. Res. B* **241** (2005) 753-758.
3. Y.Wu, J.P. Hurley, Q. Ji, J.W. Kwan, K.N. Leung, *Rev. Sci. Instrum.* **81**, 02B908 (2010).
4. H-S. Bosch, G.M. Hale, *Nucl. Fusion* **32** 611 (1992).
5. J.L. Rovey, B.P. Ruzic, T.J. Houlahan, *Rev. Sci. Instrum.* **78**, 106101 (2007).
6. B.K. Das, A. Shyam, *Rev. Sci. Instrum.* **79**, 123305 (2008).
7. P. Rohwer, H. Baumann, W. Schütze, K. Bethge, *Nucl. Instrum. Meth.* **211** (1983), p. 543.
8. P. Rohwer, H. Baumann, W. Schütze, K. Bethge, *Nucl. Instrum. Meth.* **204** (1982), p. 245.

X-Ray Measurements Of A Thermo Scientific P385 DD Neutron Generator

C.J. Wharton[a], E.H. Seabury[a], D.L. Chichester[a], A.J. Caffrey[a], J. Simpson[b] and M. Lemchak[b]

[a]*Idaho National Laboratory, 2525 N. Freemont Avenue, Idaho Falls, Idaho 83415*
[b]*Thermo-Fisher Scientific/MF Physics, 5074 List Drive, Colorado Springs, Colorado 80919*

Abstract. Idaho National Laboratory is experimenting with electrical neutron generators, as potential replacements for californium-252 radioisotopic neutron sources in its PINS prompt gamma-ray neutron activation analysis (PGNAA) system for the identification of military chemical warfare agents and explosives. In addition to neutron output, we have recently measured the x-ray output of the Thermo Scientific P385 deuterium-deuterium neutron generator. X rays are a normal byproduct from neutron generators, but depending on their intensity and energy, x rays can interfere with gamma rays from the object under test, increase gamma-spectrometer dead time, and reduce PGNAA system throughput. The P385 x-ray energy spectrum was measured with a high-purity germanium (HPGe) detector, and a broad peak is evident at about 70 keV. To identify the source of the x rays within the neutron generator assembly, it was scanned by collimated scintillation detectors along its long axis. At the strongest x-ray emission points, the generator also was rotated 60° between measurements. The scans show the primary source of x-ray emission from the P385 neutron generator is an area 60 mm from the neutron production target, in the vicinity of the ion source. Rotation of the neutron generator did not significantly alter the x-ray count rate, and its x-ray emission appears to be axially symmetric. A thin lead shield, 3.2 mm (1/8 inch) thick, reduced the 70-keV generator x rays to negligible levels.

Keywords: PGNAA, electrical neutron generator
PACS: 25.40.-h, 29.25.Dz

INTRODUCTION

Idaho National Laboratory (INL) is testing the Thermo Scientific P385 deuterium-deuterium (DD) neutron generator as a potential replacement for the californium-252 neutron source in the Portable Isostopic Neutron Spectroscopy (PINS) prompt gamma-ray neutron activation analysis (PGNAA) system[1]. PINS is used by military units for the nondestructive evaluation of suspect chemical warfare and explosive munitions. During normal neutron generator operation, x rays from the generator produced up to 90% dead time in the PINS HPGe gamma-ray spectrometer, greatly slowing data acquisition. We have carried out a set of measurements to characterize the x-ray energy and spatial distributions, as described below.

Figure 1 shows the low energy regions of two spectra taken using a high purity germanium (HPGe) detector, one with lead wrapped around the length of the neutron generator tube, and the other without lead. A broad x-ray peak centered at 70 keV is readily apparent in the bare-tube spectrum, and this peak was reduced greatly by the lead shielding. The neutron generator was operating at 100% duty cycle with a target current of 80 μA and an accelerating voltage of 130 kV during both measurements.

In order to more accurately identify the source of the x rays within the neutron generator tube, the lead shield was removed and the neutron generator was scanned by collimated scintillation detectors.

Application of Accelerators in Research and Industry
AIP Conf. Proc. 1336, 538-540 (2011); doi: 10.1063/1.3586158

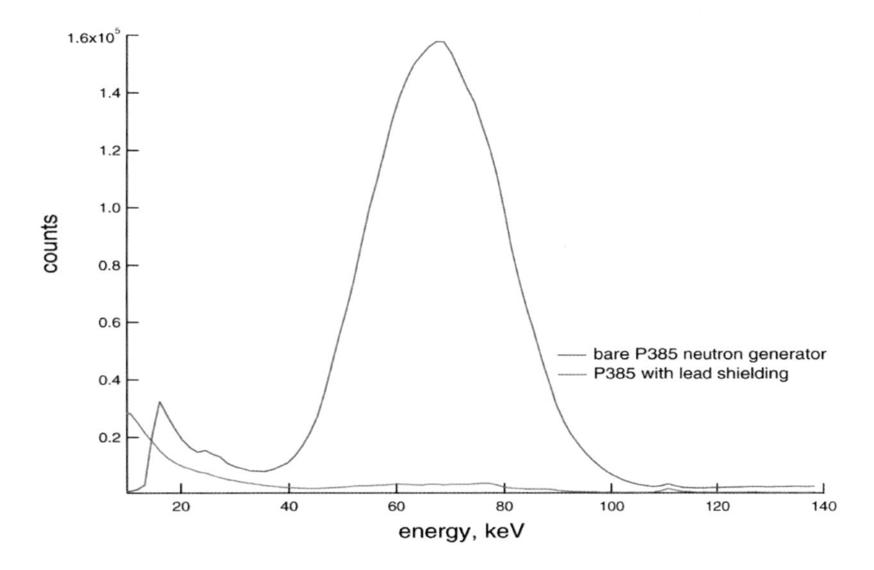

FIGURE 1. Comparison of two 100-second spectra measured with an HPGe detector using the P385 generator as the neutron source.

X-RAY SCANNING METHODS

During the scans, the neutron generator was mounted on a sliding track that was equipped with an optical linear encoder and digital readout, and the encoder precision is 0.5 microns. The generator was viewed through a 1.27-mm by 25.4-mm (0.05-inch by 1.0-inch) slit in a set of tungsten shielding blocks. The long axis of the slit was transverse to the long axis of the generator tube[2]. The generator was scanned with both a plastic scintillation detector and a sodium iodide detector, and both detector crystals were 25.4-mm diameter by 25.4-mm long (1.0-inch diameter by 1.0-inch long). Prior to scanning, the neutron generator target plane, marked on the outside of the generator tube with a painted line, was aligned with the viewing slit by replacing a detector with a laser pointer, and the linear encoder position was zeroed at this location.

During the scans, the neutron generator was operated at 100% duty cycle with a target current of 60 μA and an accelerating voltage of 90 kV.

SCANNING RESULTS

The scan results are presented in Figure 2, and they are discussed below.

Plastic Scintillator Measurements

An initial scan of the P385 neutron generator tube was performed using the plastic scintillation detector. The signal was passed through an amplifier, constant fraction discriminator, and scaler to determine the number of counts recorded in the detector in the energy range of the x rays. Eight ten-second counting measurements were taken at various distances along a 190 millimeter length of the generator, from the end of the generator housing without cable connectors to a point approximately 60 millimeters beyond the target plane. The number of counts at each distance was recorded and averaged. For much of the scan, the distance between measurement points was 2.5 millimeters, but in areas where a large increase in counts was seen the distance between measurement points was decreased to 0.5 millimeters. The results of the scan show the primary source of x-ray emission from the P385 neutron generator is an area 60 mm to the left of the neutron production target in the vicinity of the ion source.

In addition to the linear scan of the neutron generator, six measurements were performed with the detector stationary with its strongest x-ray emission point at -60 mm aligned with the tungsten collimator slit. The generator housing was rotated 60° about its long axis between these measurements. No significant change in the count rate was observed as a function of rotation angle.

NaI(Tl) Detector Scan

The plastic scintillator detector responds to both photons and fast neutrons, of course. To minimize the neutron signal, a second scan of the P385 neutron generator tube was performed using a 1-inch by 1-inch sodium iodide scintillation detector connected to an amplifier and multichannel analyzer. Data from the sodium iodide detector were collected for sixty live-seconds at each point along the neutron generator tube. Only counts with energies between 34 keV and 120 keV, the energy range of the x-rays produced by the generator, were included in the sodium iodide scan tallies.

Again, the measurements were taken at various distances along a 190 millimeter length of the generator tube, from the end of the generator housing without cables to a point approximately 60 millimeters beyond the target plane. For much of this scan, the distance between measurement points was set at 5 millimeters, but in areas where a large increase in counts was determined the distance between measurement points was decreased to 0.5 millimeters.

By comparison the results of the sodium iodide detector with those of the plastic scintillator, it appears that many of the counts observed in the plastic scintillator were the results of neutron interactions, rather than x rays. However, both scans show interesting structure in an area 60 mm to the left of the neutron production target near the generator tube ion source, and we believe that most of the x rays are produced there.

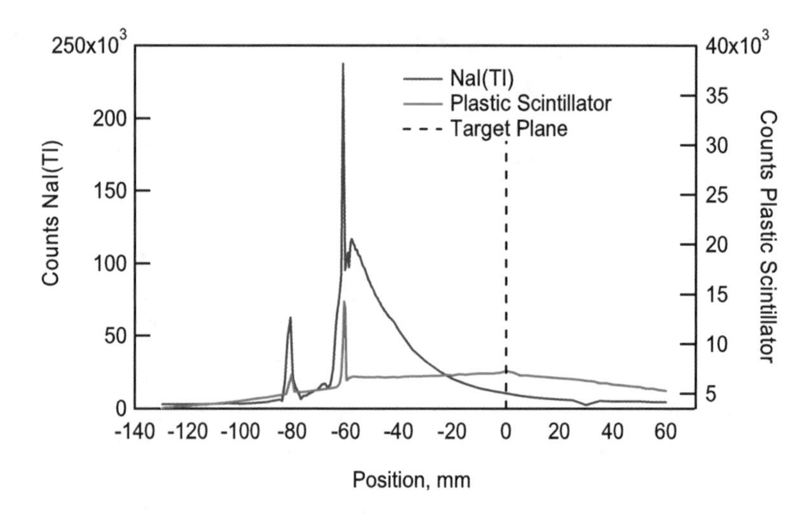

FIGURE 2. Results of scintillation detector scans along the generator's long axis.

CONCLUSIONS

The primary source of x-ray emission from the P385 neutron generator is an area 60 mm to the left of the neutron production target, as viewed in Figure 2, in the vicinity of the ion source. There is a second strong emission area 80 mm from the target corresponding to the back of the ion tube. Rotation of the neutron generator did not significantly alter the x-ray count rate, and the x-ray emission is appears to be axially symmetric within the neutron generator.

A 3.2 mm (1/8-inch) thick lead shield wrapped around the generator housing reduced x rays counted by the HPGe detector to negligible levels, as expected from attenuation calculations, and the HPGe spectrometer fractional dead time returned to normal. Currently, we are designing a bismuth x-ray shield that will fit inside the generator housing.

ACKNOWLEDGMENTS

This research at Idaho National Laboratory was supported by the U.S. Army Project Manager for Non-Stockpile Chemical Material, under U.S. Department of Energy Field Office Idaho contract DE-AC07-05ID14517.

I would also like to thank Rahmat Aryaeinejad and Karen Wendt for their assistance.

REFERENCES

1. A.J. Caffrey et al., *IEEE Trans. Nucl. Sci.* **39** (1992) 1422.
2. The P385 neutron generator's housing is a stainless steel cylinder, 26 inches long and 4 inches in diameter. It has been measured to generate ~5 X 10^6 neutrons per second.

Positronium Formation Of Glyeisdyl Methacrylic Acid (GMA)/Styrene Grafted On PVDF Membrane For Fuel Cells

E. E. Abdel-Hady[a], M. M. Eltonny[b], M. O. Abdel-Hamed[a]

[a] Physics Dpt, faculty of science, Minia university, Minia 61519 , Egypt
[b] Polymer Dpt, Atomic Energy Authority, Cairo, Egypt
mazosman2005@yahoo.com

Abstract. Simultaneous gamma irradiation was used effectively for grafting of glycidyl methacrylic acid and styrene onto Poly vinyldine fluoride (PVDF). Membranes were characterized by thermal gravimetric analysis (TGA) and scanning electron microscopy (SEM).The properties of the obtained membranes were evaluated in terms of proton conductivity, methanol permeability and positron annihilation lifetime (PALS) parameters. The high probability of Positronium formation enables the application of PALS to the study of free volume. Good property values approved the applicability of the membrane from the cost benefit point of view.

Keywords: Grafting, Positronium, Proton Conductivity, Irradiation dose
PACS: 82.47.Gh, 78.70.Bj

INTRODUCTION

Polymer electrolyte membrane fuel cells (PEMFC) are promising clean power sources for automotive and portable applications [1, 2]. For high efficiency polymer electrolyte fuel cells (PEFCs), polymer membranes with higher proton conductivity and lower gas permeability are desired. It was reported that [3], gas permeation is governed by the free volume. The positron annihilation lifetime (PAL) technique has been used as a powerful probe for the characterization of the free volume in polymers [4]. Alternative polymer electrolyte membranes (PEMs) and synthetic procedures are being sought. One of those methods is the radiation grafting of base polymers to produce co-polymers which may then functionalized. The irradiation of polymers creates active radical centers which can be used to initiate the polymerization of monomers leading to the formation of graft polymers [5]. Poly vinyldine fluoride (PVDF) is a kind of partial fluorinated polymer with excellent stability and has been widely used in the battery field [6-9]. It is also used for preparation of proton exchange membranes. Scrosati and co-workers [10] reported a kind of membrane based on PVDF by doping with a microporous ceramic. Lehtinen et al. [11] prepared poly(vinylidene fluoride) grafted polystyrene sulfonated acid (PVDF-*g*-PSSA) proton exchange membranes based on a radiation-grafting technique, and they found that PVDF-*g*-PSSA membranes have a higher proton conductivity, water-uptake and lower oxygen solution ability compared to Nafion membranes. The present study is focused on the evaluation of an inexpensive polymer electrolyte membranes based on radiation induced glycidyl methacrylic acid and styrene grafted onto Poly vinyldine fluoride (PVDF-g-GMA/St). This membrane could be improved to be an alternative to Nafion membrane in a fuel cell. PVDF was chosen because as a partially fluorinated polymer it is considerably cheaper than a fully fluorinated material, melt processable, mechanically strong, readily available, may undergo cross-linking when irradiated and it should still be resistant enough to withstand the conditions in a fuel cell [12].

EXPERIMENTAL

The simultaneous grafting was carried out to prepare the membranes by putting the PVDF films and styrene in a glass ampoule. A glycidyl methacrylic acid with different ratios relative to styrene was added to the glass ampoule content. The ampoule was subsequently irradiated for desired dose (5, 10, 20, 25 and 30 kGy). After the grafting reaction, the film was rinsed with benzene and soaked therein overnight to

Application of Accelerators in Research and Industry
AIP Conf. Proc. 1336, 541-544 (2011); doi: 10.1063/1.3586159

remove homopolymer from the film. After drying to a constant weight, the degree of grafting was calculated using the equation:

$$\text{Degree of grafting \%} = \frac{W_g - W_o}{W_o} X100 \qquad (1)$$

where W_o and W_g are the weight of initial and grafted film, respectively. The films were sulfonated with 0.2M chlorosulfonic acid and dichloroethane for an initial period of 5 h at 50°C, followed by overnight immersion at room temperature to convert them to proton conducting membranes. Thermo-gravimetric analysis (TGA) was recorded on a thermal analysis system at a heating rate of 10 °C ·min^{-1}. The proton conductivity measurements of the sulfonated grafted membranes were derived from AC impedance spectroscopy measurements over a frequency range of 50 to 10^6 Hz with an oscillating voltage of 50–500 mV, using a system 3532 Hioiki bridge LCR hitester. The sample was equilibrated at 100% RH at ambient atmospheric pressure and clamped between two electrodes. The proton conductivities of the samples were measured through the membrane and were calculated from the impedance data, using the relationship:

$$\sigma = \frac{L}{RS} \qquad (2)$$

where σ is the proton conductivity (in S/cm), L is the membrane thickness, S is the area for protons to transport through the membrane (in cm^2) and R is derived from the intersection of the semicircle on a complex impedance plane with the Re (Z) axis. Methanol permeability measurement was also carried out using a liquid diffusion cell composed of two compartments containing solution A and B. One compartment A (V_A=150ml) was filled with 10 mol·L^{-1} methanol solution, the other compartment (V_B=150ml) was filled with deionized water only. The tested membrane was immersed in deionized water for hydration before measurements and then vertically placed between the two compartments by a screw clamp. Amount of methanol diffused from compartment A to B across the membrane was measured over time. The methanol permeability P was calculated by the following equation:

$$P = \frac{k.VB.L}{A.CA} \qquad (3)$$

where k was the slope of the straight- line plot of methanol concentration in solution B versus permeation time, V_B, L and A were the volume of solution B, the thickness and the effective area of the tested membrane, respectively. The water-uptake of the membranes was defined as the ratio of the mass of the absorbed water to that of the dry membrane. It can be calculated from the equation:

$$water\,uptake = \frac{W_w - W_d}{W_d} \times 100 \qquad (4)$$

where W_w and W_d are the weight of swollen and dry grafted membranes, respectively. A conventional of fast-fast coincidence circuit of PAL spectrometer with a time resolution 240 ps was used to record all PAL spectra .The PAL spectra were analyzed into three components (τ_1, τ_2, and τ_3) and their intensities (I_1, I_2 and I_3) using PALSfit program [13].

RESULTS AND DISCUSSION

Figure 1 shows the thermogravimetry analysis of the grafted membranes. Clearly, the weight loss rate increases as the GMA/ST content increases. No weight change up to 180 °C after that a weight lose can be noticed which may be due to the absorbed humidity as well as possible condensation. Further weight loss above 180 °C is due to condensation which is followed with degradation above 200°C. The TG thermograms show that (PVDF-*g*-GMA)/ST) is thermally stable up to approximately180°C and then the material decomposes.

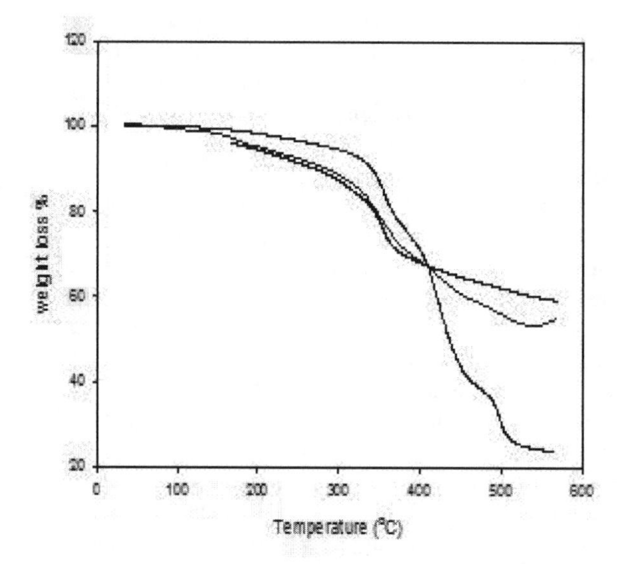

FIGURE 1. Thermogram of GMA/ST grafting onto PVDF with different grafting yields.

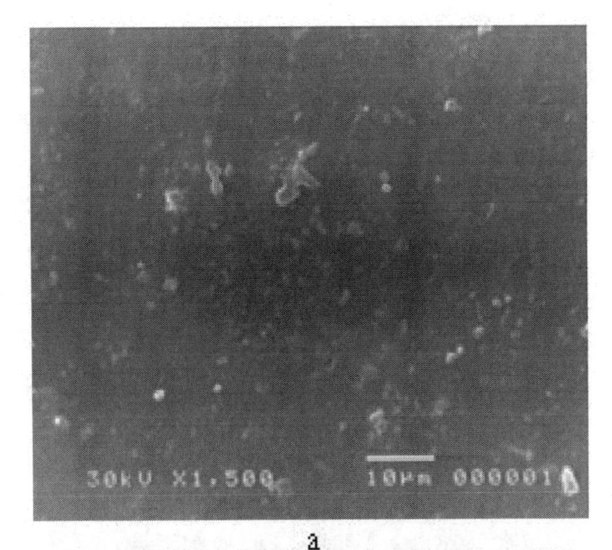

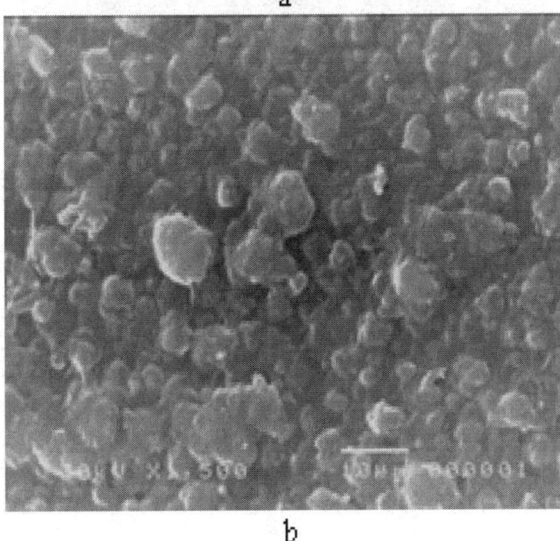

FIGURE 2. SEM of GMA/ST onto PVDF: a) 4% grafting (dose 1 Mrad) b) 58 % grafting (dose 3Mrad)

Figure 2 (a) shows the image of (PVDF -g-GMA)/ST) membrane in which a less numbers of white spots while black media is predominant. This is due to less grafting of GMA/ST onto PVDF film. Figure 2 (b) shows small size of white spots is distributed in regular manner through the PVDF matrix which describes high grafting yield of GMA/ST onto PVDF. The proton conductivity and Methanol permeability for PVDF-g-(GMA)/-ST) are illustrated in Table 1. From this table, one can noticed that the grafting membrane has relative good proton conductivity (23-29 $mS{\cdot}cm^{-1}$) and methanol permeability values ($1.29\text{-}2.11x10^{-6}$) comparing to Nafion 117 (54 $mS{\cdot}cm^{-1}$ & $1.65x10^{-6}$ $cm^{2}{\cdot}S^{-1}$), respectively. It is well known that the proton conductivity and mechanical stability of a membrane are related strongly to the presence of water. It is necessary to have an adequate level of water uptake to promote good proton conductivity. However, water uptake should be minimized to provide the membrane with mechanical stability and to avoid swelling. In the presence of water, the hydrophilic domains are hydrated and provide good proton conductivity, while the hydrophobic domains provide the membranes with good morphological and mechanical stability. Consequently, it is very import-ant to maintain a sulfonated membrane with an appropriate level of water uptake. As the dose rate increases, the water uptake of the studied membranes was ranged from 0.33 to 0.67.

For high efficiency polymer electrolyte fuel cells (PEFCs), polymer membranes with higher proton conductivity and lower gas permeability are desired. The free volume governs gas permeation in dry and wet polymer membranes [3]. The free volume was evaluated from the o-Ps lifetimes. Table 2 lists the positron annihilation parameters for the dry and wet

TABLE 1. The proton conductivity and Methanol permeability for PVDF-g-(GMA)/ST

Sample	Dose Mrad	Proton conductivity (Room temp) ($mS.cm^{-1}$)	Methanol permeability $x10^{-6}$ ($cm^{2}.S^{-1}$)
1	2	10.3	2.30
2	3	23.2	1.29
3	4	28.7	1.97
Nafion117	-	54 at 35°C	1.65

TABLE 2. The positron annihilation parameters for PVDF-g-(GMA)/ST

Sample	Dose (Mrad)	τ_3(ns) dry	τ_3 wet	I_3 dry	I_3 wet	V_F(Å) dry	V_F wet	F.F. vol. dry	F.F. vol. wet
1	2	2.25	2.21	7.61	7.98	120.39	116.81	9.17	9.32
2	3	2.18	1.92	6.99	8.69	113.64	89.68	7.95	7.79
3	4	2.10	2.03	7.34	6.60	106.10	99.25	6.87	6.55

PVDF-g-GMA/St membranes at different irradiation doses. The free volume sizes for the PVDF samples under investigation are ranged from 120 to 106 Å^3 which are smaller than that of Nafion (163 to 191 Å^3), which indicates that PVDF membranes have relatively low permeability. For the dry and wet membranes, the change in the free volume size, and fractional free volume as a function of irradiation dose is a result of new bonds formation during the grafting process, this leads to high grafting yield ratio and good proton conductivity values. As can be also seen from table 2, there is a small difference between the annihilation parameters for wet and dry membranes which may be due to the water sorption.

CONCLUSION

The TG thermograms show that (PVDF -g-GMA)/ST) is thermally stable up to approximately 180°C and then the material decomposes. The free volume sizes for the PVDF samples are smaller than that of Nafion, which indicate that PVDF membranes have relatively low permeability. For the dry and wet membranes, the change in the free volume size and fractional free volume as a function of irradiation dose is a result of new bonds formation during the grafting process. The grafting membrane has relative good proton condu-ctivity (23-29 mS.cm^{-1}) and methanol permeability values (1.29-2.11x10^{-6}) comparing to Nafion 117 (54 mS.cm^{-1} & 1.65x10^{-6} cm^2.S^{-1}), respectively. Good pro-perty values approved the applicability of the mem-brane from the cost benefit point of view.

ACKNOWLEDGMENTS

The authors wish to sincerely thank the STDF of Egypt, Project (ID220), for financially supporting this work.

REFERENCES

1. B. C. H. Steele, A. Heinzel, Nature **414**, 345 (2001).
2. M. Rikukawa, K. Sanui, Prog. Polym. Sci. **25**, 1463 (2000).
3. H. F. Mohamed, K. Ito, Y. Kobayashi, N.Takimoto, Y.Takeoka, A. Ohira Polymer 49,3091 (2008).
4. Y. C. Jean J.microchem.**42**,72(1990).
5. S. D. Flint, R. C. Slade, Solid State Ionics 97, 299 (1997).
6. N.P. Chen, L. Hong, Polymer 45, 2403 (2004).
7. M.A. Navarra, S. Materazzi, S. Panero, B. Scrosati, J. Electrochem. Soc.150, A1528 (2003).
8. G.K.S. Prakash, et al., J. Fluorine Chem. 1251217, (2004).
9. S. Holmberg, T. Lehtinen, J. N¨asman, J. Mater. Chem. 6, 1309 (1996).
10. M.A. Navarra, S. Materazzi, S. Panero, B. Scrosati, J. Electrochem. Soc. 150, A1528 (2003).
11. T. Lehtinen, G. Sundholm, S. Holmberg, F. Sundholm, P. Bjornbom, M. Burdell, Electrochim. Acta 43, 1881 (1998).
12. J.A. Brydson, Plastics Materials, 6th edn., Butterworth-Heinemann, Oxford, 1995.
13. J V Olsen. , Kirkegaard P, Pedersen N J, and Eldrup M *Phys. Status Solidi C* **4** 4004 (2007).

SECTION X – NUCLEAR PHYSICS

The Neutron Time-Of-Flight Facility n_TOF At CERN: Phase II

F. Gunsing[1], S. Andriamonje[2], J. Andrzejewski[3], L. Audouin[4], V. Bécares[5], F. Bečvář[6], F. Belloni[1], B. Berthier[4], E. Berthoumieux[1], M. Brugger[2], M. Calviani[2], F. Calviño[7], D. Cano-Ott[5], C. Carrapiço[8], P. Cennini[2], F. Cerutti[2], E. Chiaveri[2], M. Chin[2], N. Colonna[9], G. Cortés[7], M.A. Cortés-Giraldo[10], I. Dillmann[11], C. Domingo-Pardo[12], I. Duran[13], M. Fernández-Ordóñez[5], A. Ferrari[2], K. Fraval[1], S. Ganesan[14], I. Giomataris[1], G. Giubrone[15], M.B. Gómez-Hornillos[7], I.F. Gonçalves[8], E. González-Romero[5], F. Gramegna[16], C. Guerrero[5], S. Harrisopulos[17], M. Heil[12], K. Ioannides[18], E. Jericha[19], Y. Kadi[2], F. Käppeler[20], D. Karadimos[18], M. Krtička[6], E. Lebbos[2], C. Lederer[21], H. Leeb[19], R. Losito[2], M. Lozano[10], J. Marganiec[3], S. Marrone[9], T. Martinez[5], C. Massimi[22], P.F. Mastinu[16], M. Meaze[9], E. Mendoza[5], A. Mengoni[23], P.M. Milazzo[24], M. Mosconi[25], R. Nolte[25], T. Papaevangelou[1], C. Paradela[13], A. Pavlik[21], J. Perkowski[3], R. Plag[12], J. Praena[10], J.M. Quesada[10], T. Rauscher[26], R. Reifarth[12], F. Roman[2,27], C. Rubbia[2,28], R. Sarmento[8], G. Tagliente[9], J.L. Tain[15], D. Tarrìo[13], L. Tassan-Got[4], L. Tlustos[2], G. Vannini[22], V. Variale[9], P. Vaz[8], A. Ventura[23], V. Vlachoudis[2], R. Vlastou[29], Z. Vykydal[30], A. Wallner[21], C. Weiß[19], The n_TOF Collaboration (http://www.cern.ch/ntof)

[1] CEA, Irfu, F-91191 Gif-sur-Yvette, France
[2] European Organization for Nuclear Research (CERN), Geneva, Switzerland
[3] Uniwersytet Łódzki, Lodz, Poland
[4] Centre National de la Recherche Scientifique/IN2P3 - IPN, Orsay, France
[5] Centro de Investigaciones Energeticas Medioambientales y Technologicas (CIEMAT), Madrid, Spain
[6] Charles University, Prague, Czech Republic
[7] Universitat Politecnica de Catalunya, Barcelona, Spain
[8] Instituto Tecnológico e Nuclear (ITN), Lisbon, Portugal
[9] Istituto Nazionale di Fisica Nucleare, Bari, Italy
[10] Universidad de Sevilla, Spain
[11] Physik Department E12 and Excellence Cluster Universe, Technische Universität München, Garching, Germany
[12] GSI Helmholtzzentrum für Schwerionenforschung GmbH, Darmstadt, Germany
[13] Universidade de Santiago de Compostela, Spain
[14] Bhabha Atomic Research Centre (BARC), Mumbai, India
[15] Instituto de Fìsica Corpuscular, CSIC-Universidad de Valencia, Spain
[16] Istituto Nazionale di Fisica Nucleare, Laboratori Nazionali di Legnaro, Italy
[17] National Centre of Scientific Research (NCSR), Demokritos, Greece
[18] University of Ioannina, Greece
[19] Atominstitut, Technische Universität Wien, Austria
[20] Karlsruhe Institute of Technology, Campus Nord, Institut für Kernphysik, Karlsruhe, Germany
[21] Fakultät für Physik, Universität Wien, Austria

Application of Accelerators in Research and Industry
AIP Conf. Proc. 1336, 547-551 (2011); doi: 10.1063/1.3586160

[22] Dipartimento di Fisica, Università di Bologna, and Sezione INFN di Bologna, Italy
[23] Agenzia nazionale per le nuove tecnologie, l'energia e lo sviluppo economico sostenibile (ENEA), Bologna, Italy
[24] Istituto Nazionale di Fisica Nucleare, Trieste, Italy
[25] Physikalisch-Technische Bundesanstalt (PTB), Braunschweig, Germany
[26] Department of Physics and Astronomy - University of Basel, Basel, Switzerland
[27] Horia Hulubei National Institute of Physics and Nuclear Engineering - IFIN HH, Bucharest - Magurele, Romania
[28] Laboratori Nazionali del Gran Sasso dell'INFN, Assergi (AQ),Italy
[29] National Technical University of Athens (NTUA), Greece
[30] Institute of Experimental and Applied Physic (IEAP), Czech Technical University (CTU), Prague, Czech Republic

Abstract. Neutron-induced reactions are studied at the neutron time-of-flight facility n_TOF at CERN. The facility uses 6~ns wide pulses of 20 GeV/c protons impinging on a lead spallation target. The large neutron energy range and the high instantaneous neutron flux combined with high resolution are among the key characteristics of the facility. After a first phase of data taking during the period 2001-2004, the facility has been refurbished with an upgraded spallation target and cooling system for a second phase of data taking which started in 2009. Since 2010, the experimental area at 185 m where the neutron beam arrives, has been modified into a worksector of type A, allowing the extension of the physics program to include neutron-induced reactions on radioactive isotopes.

Keywords: n_TOF, CERN, neutrons, neutron induced reactions, neutron time-of-flight facility
PACS: 21.10.Ma, 25.40.Ny, 25.85.Ec, 28.20.Np, 28.20.Pr, 29.25.Dz

INTRODUCTION

The neutron time-of-flight facility n_TOF at CERN provides a pulsed neutron beam of high intensity, with neutron energies ranging from the meV to the GeV range. Neutron induced nuclear reactions are studied in this energy range, mainly neutron capture and neutron-induced fission reactions. Neutron-induced nuclear reactions play an important role in several research fields. In nuclear structure studies the neutron may serve as probe of excited nuclei revealing separate highly excited levels just above the neutron binding energy, which allows to derive an important calibration point in level densities [1,2]. In astrophysics the formation of the vast majority of all known nuclei with masses heavier than iron are formed in stellar environments mainly through slow and rapid neutron capture processes. Neutron-nucleus reaction data are an essential ingredient to calculate production rates in stellar nucleo-synthesis and refine the understanding of stellar evolution processes [3,4]. Finally neutron-induced reactions, or nuclear data in general, play a key role in the safety and criticality assessment of nuclear technology, not only for existing power reactors but also for radiation dosimetry, the transmutation of nuclear waste, alternative reactor fuel cycles, or future reactors like Generation IV [5,6].

Efforts concerning nuclear data have been ongoing since the early 50's and are growing since a decade or so, related to the renewed interest in new power reactors, and concern both reaction modeling and evaluation activities as well as nuclear data measurements and facilities. In this perspective the construction and commissioning of the neutron time-of-flight facility at CERN, Switzerland, after an initial proposal of Rubbia [7], was finished in 2001 when the facility became operational [8]. After a first phase of data taking from 2001-2004, the facility was obliged to stop for a refurbishment of the spallation target. Since the end of 2008 the facility has resumed operation and a new program of measurements was launched in 2009. Since 2010 the facility's experimental area, located at 185 m from the neutron production target, has been modified in order to comply with radioprotection regulations into what is called a worksector of type A, allowing the use of radioactive samples in the neutron beam.

THE NEUTRON TIME-OF-FLIGHT FACILITY N_TOF AT CERN

The facility uses a 6 ns wide, 20 GeV/*c* proton beam with up to 7 x 10^{12} protons per pulse, impinging on a lead spallation target, and yielding about 300 neutrons per incident proton. The time between proton pulses is an integer multiple of 1.2 seconds which allows to cover the neutron energy range down to subthermal energies without overlapping of slow neutrons in subsequent cycles. A water layer surrounding the upgraded spallation target serves as a coolant. This layer is 1 cm thick on the side of the neutron beam guide. An additional layer of 4 cm of water, separated from the cooling circuit, is placed in front of the cooling layer and acts as a moderator. The moderator can be optionally changed to a different material. In 2010 we have used for example water with a saturated, thus constant, ^{10}B-solution in order to reduce the number of 2.23 MeV gamma rays from hydrogen capture which otherwise constitutes an important contribution to the background due to in-beam gamma rays. This influences the energy distribution of the neutron flux only noticeably below 1 eV. The spallation target was a block of 80x80x60 cm^3 in the first phase, and was replaced by a cylindrical lead target of 40 cm diameter and 60 cm length for the second phase. A schematic view of the new spallation target is shown in figure 1. In addition a new ventilation system was installed and a new target cooling circuit was developed. This allows among others to monitor and control the temperature, oxygen content and conductivity of the water.

From the lead spallation target, an evacuated neutron beam tube leads to the experimental area located at 185 m from the spallation target. A 1.5 T sweeping magnet is placed at a distance of 145 m to remove residual charged particles from the neutron beam. A 3 m thick off-beam iron shielding was placed just after the magnet to remove negative muons. A first collimator with an inner diameter of 11 cm is placed at 135 m while a second collimator with a variable diameter of either 1.8 or 8 cm is situated at 175 m from the production target. This collimation results in a Gaussian beam profile in the experimental area at 185 m. The neutron beam line has been extended for an additional 12 m beyond the experimental area to minimize the background from back-scattered neutrons. For the 1.8 cm second collimator the Gaussian profile was simulated and measured and has a standard deviation of 0.77 cm. At present the facility is mainly used for capture and fission measurements. A full description of the characteristics and performances of the facility is described elsewhere [9]. The neutron beam characteristics in terms of energy distribution of the neutron flux and the energy resolution have been accurately determined by a combination of simulations and measurements.

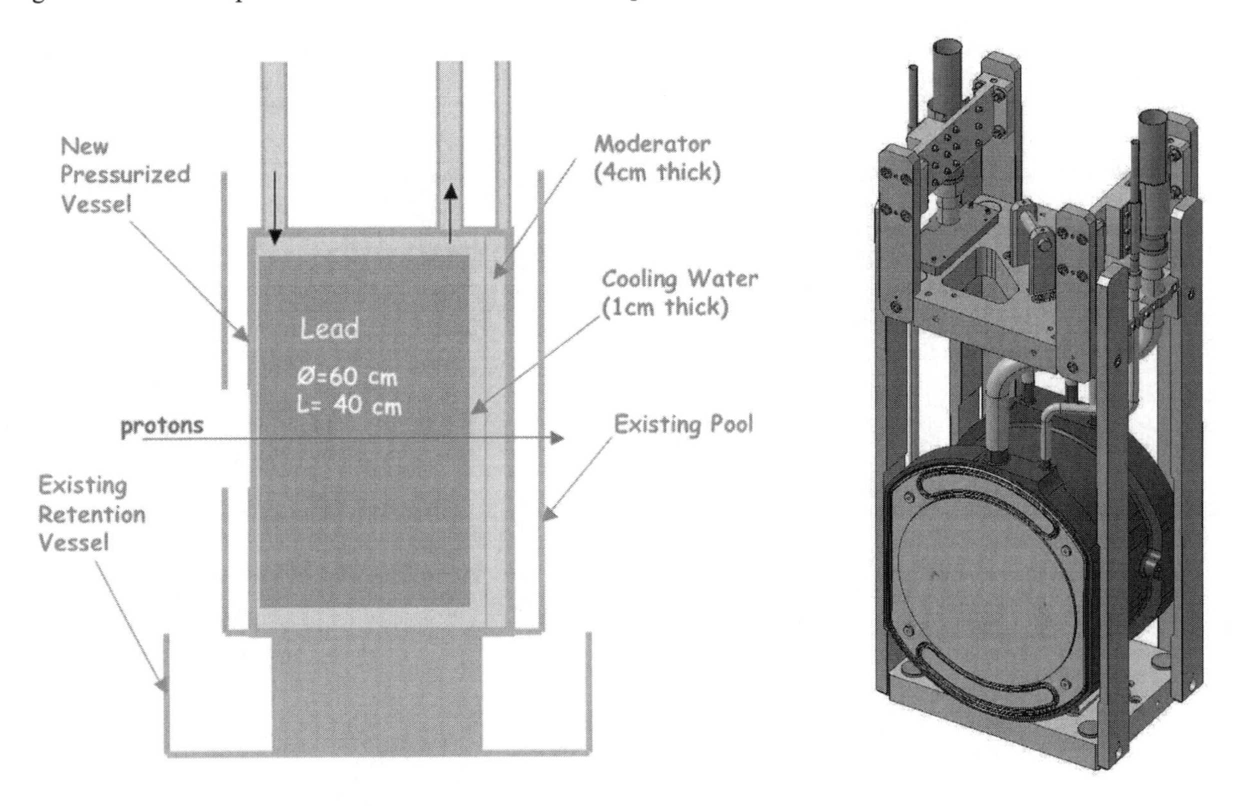

FIGURE 1. A schematic view of the upgraded lead spallation target as installed for the n_TOF facility.

DETECTORS AND EQUIPMENT

For the capture measurements two different detector systems are being used. In-house developed deuterated benzene C_6D_6–based gamma-ray detectors contained in a cylindrical low mass carbon fibre housing [10] are used as total energy detectors. In addition a total absorption calorimeter consisting of a

4π array with 40 BaF_2 crystals is also used for capture measurements [11].

Fission experiments have been performed with two different detector systems in the first phase. Two fission ionization chambers (FIC) used deposits of fissile isotopes on 100 μm thick aluminum foils [12]. The second type of fission detector is based on parallel plate avalanche counters (PPACs) [13], developed with target deposits on 1.5 μm thin mylar or 2 μm thick aluminium foils, allowing to detect the two fission fragments in coincidence. This detector will be used also in phase II. A third type of fission detector, based on the MicroMegas principle, is currently being developed [14].

The energy distribution of the relative incident neutron flux is continuously measured during the experiments with an neutron flux monitor SiMon [15], consisting of a ^{6}Li deposit on a mylar foil and 4 off-beam silicon detectors recording the particles from the standard ^{6}Li(n,^{3}H)α cross section. An additional transparent in-beam flux detector has been developed to overcome difficult angular distribution corrections needed above 1 keV for the SiMon detector. The new gaseous detector is based on standard cross sections like ^{235}U(n,f) and ^{10}B(n,α)^{6}Li and combines thin deposits with 25 μm thin MicroMegas-based detectors.

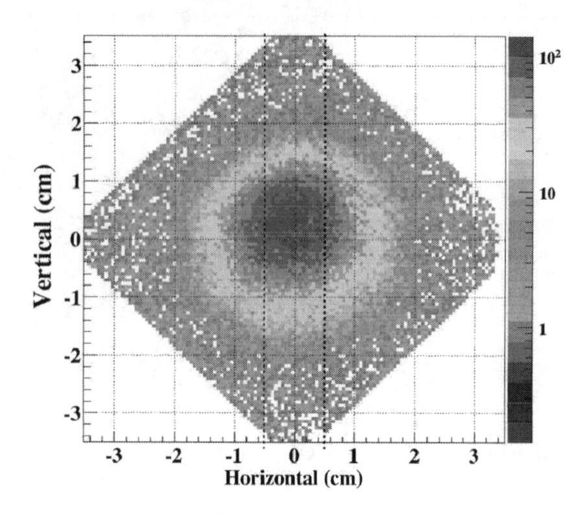

FIGURE 2. An example of a beam image obtained with the XY-MicroMegas detector, representing the spatial image of the neutron beam intensity. Images can be reconstructed as a function of the neutron time-of-flight.

In a dedicated experiment the spatial beam profile has been measured with two different devices. We have used a Medipix detector [16] with LiF and polyethylene converters to measure the distribution of the neutrons across the beam area. We also used a two dimensional MicroMegas-based detector with 106 x 106 perpendicular strips on a 6 x 6 cm^2 active area to determine the beam profile with a resolution of about 0.5 x 0.5 mm^2 [17]. In figure 2 the image of the beam, which can be obtained as a function of time-of-flight, is shown as an example and in figure 3 a typical reaction chamber containing MicroMegas detectors is shown.

The data acquisition system [18] is based on Acqiris flash ADCs with 8 or 16 Mb memory, 8 bit or 10 bit amplitude resolution, and a minimum sampling interval of 1 ns. For each detector its full signal is recorded, starting from time the incident protons create the neutrons up to 8 or 16 ms, allowing to go in energy down to the thermal region. After zero suppression the detector signals for each event are transferred for permanent storage to CERN's data storage facility CASTOR. With dedicated pulse shape analysis routines the time-of-flight and signal amplitude is extracted from the signals of each detector event and recorded for further analysis.

MEASUREMENTS

During the first phase from 2001 to 2004 capture and fission data for a large number of isotopes have been taken. Capture measurements with the C_6D_6 detectors concerned ^{24,25,26}Mg, ^{56}Fe, ^{90,91,92,93,94,96}Zr,

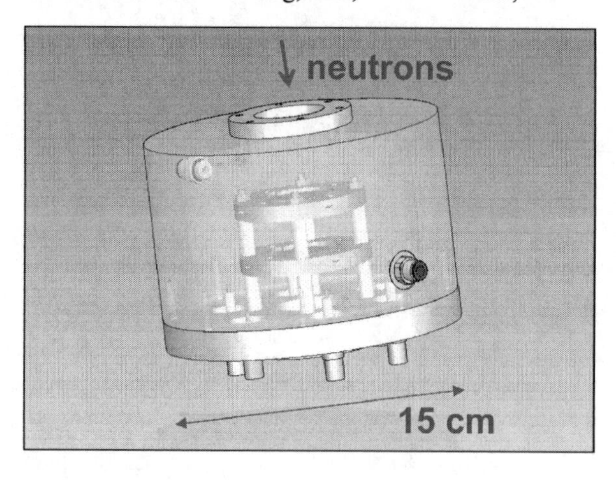

FIGURE 3. A schematic view of a reaction chamber, here equipped with two MicroMegas detectors.

^{139}La, ^{151}Sm, ^{186,187,188}Os, ^{197}Au, ^{204,206,207,208}Pb, ^{209}Bi, and ^{232}Th. The BaF_2 4π calorimeter has been used for measurements of ^{197}Au, ^{233}U, ^{234}U, and ^{237}Np, ^{240}Pu, and ^{243}Am. The fission cross sections measured with the FIC-0 detector were the isotopes ^{232}Th, ^{234}U, ^{235}U, ^{236}U, ^{238}U, and ^{237}Np, while the isotopes ^{233}U, ^{235}U, ^{238}Um ^{241}Am, ^{243}Am, and ^{245}Cm were measured in the ISO-2919 compliant FIC-1 detector. With the fission detectors based on Parallel Plate Avalanche Counters (PPACs) cross sections of ^{nat}Pb, ^{209}Bi, ^{232}Th, ^{237}Np, ^{233}U, ^{234}U, ^{235}U, ^{238}U have been measured. The analysis of most of these measurements is completed and the results have been published. The most up-to-date list of publications can be found on the n_TOF web site [19].

For the second phase a list of possible measurements has been established of which several have already been translated in accepted proposals and are ongoing. After a period of commissioning measurements in 2009, a program of capture measurements on the stable isotopes of Fe and Ni has been started and is at present ongoing. The primary interest for these nuclei is for stellar nucleo-synthesis but also as structure materials for nuclear technology. In addition, and as part of a contract within the 7th European Framework programme, capture measurements on ^{241}Am are currently ongoing and capture measurements on ^{238}U are foreseen in the nearby future. Fission cross sections measurements are foreseen for ^{240}Pu and ^{242}Pu and also the measurement of the angular distribution of fission fragments of ^{232}Th. A main challenge remains the measurement of capture in presence of fission.. In this respect, a first development of a fission tagging detector based on a MicroMegas detector in combination with the 4π BaF_2 total absorption calorimeter has been set up for a measurement. Finally, we intend to develop an alpha particle detector for a measurement of the $^{33}S(n,\alpha)$ cross section. This pilot experiment is planned as an extension of the fission and capture experiments that have been performed up to now.

CONCLUDING REMARKS

The n_TOF facility at CERN is operational again since the end of 2008 after an upgrade of the lead spallation target and the associated cooling and ventilation system. The experimental area has been modified in order to comply with radioprotection and safety regulations, which is called work sector of type A at CERN. This crucial modification opens the way for measurements on radioactive targets exploiting fully the strength of the facility.

ACKNOWLEDGMENTS

For the realization of the second phase of the n_TOF facility we would like to acknowledge the commitment and continuous technical support from CERN, and in particular from Y. Body, R. Mollay, P. Carbonez, J. Lendaro, M. Lazzaroni, T. Otto, D. Grenier, S. Girod, S. Sgobba, L. Marques-Ferreira and D. Duarte-Ramos.

REFERENCES

1. T. Von Egidy and D. Bucurescu, *Phys.Rev. C* **72**, 044311 (2005)
2. G. A. Mitchell and H. A. Weidenmüller, *Rev. Mod. Phys.* **81** (2009) 539
3. F. Käppeler, *Progr. Particle Nucl. Phys.* **43** (1999) 419
4. G. Wallerstein et al., *Rev. Mod. Phys.* **69** (1997) 995
5. G. Aliberti et al., *Nucl. Sci. Eng.* **146** (2004) 13
6. A. Nuttin et al., *Progr. Nucl. Energy* **46** (2005) 77
7. C. Rubbia et al., *A high Resolution Spallation driven Facility at the CERN-PS to measure Neutron Cross Sections in the Interval from 1 eV to 250 MeV*, report CERN/LHC/98-02 (EET) + Add.1
8. C. Borcea et al., *Nucl. Instr. and Meth. Phys. Res. Sect. A* **513** (2003) 524
9. U. Abbondanno et al., *CERN n TOF Facility: Performance Report*, Tech. Rep. CERN-SL-2002-053 ECT (2003)
10. R. Plag et al., *Nucl. Instr. and Meth. Phys. Res. Sect. A* **496** (2003) 425
11. C. Guerrero et al., *Nucl. Instr. and Meth. Phys. Res. Sect. A* **608** (2009) 424
12. M. Calviani et al., *Phys. Rev. C* **80** (2010) 044604
13. C. Paradela et al., *Phys. Rev. C* **82** (2010) 034601
14. S. Andriamonje et al., *J. Instrum.* **5** (2010) P02001
15. S. Marrone et al., *Nucl. Instrum. Methods Phys. Res. Sect. A* **517** (2004) 389
16. http://medipix.web.cern.ch/MEDIPIX/
17. S. Andriamonje, et al., in Proc. Int. Conf. Nucl. Data for Sci. and Tech. ND2010 (2010) #1408
18. U. Abbondanno et al., *Nucl. Instrum. Methods Phys. Res. Sect. A* **538** (2005) 692
19. http://www.cern.ch/ntof

Calorimetry At Very High Energy Colliders

Mickey Chiu

Brookhaven National Laboratory, Upton NY 11973

Abstract. The capability of hadron colliders has increased to where it will soon be possible to collide protons at center of mass energies of 14 TeV with the advent of the LHC. With increasing collision energy, calorimeters become ever more essential components of a detector, and collaborations often choose very different technologies to meet their goals. From the perspective of a high energy particle and nuclear physicist, a survey is presented of the differences in design considerations and actual performance of the wide variety of calorimeters used in modern hadron colliders such as the Tevatron, RHIC, and LHC. The lessons learned and some ideas for future development of calorimetry will also be discussed.

Keywords: Calorimeters, Tevatron, LHC, RHIC, Atlas, PHENIX
PACS: 07.20.Fw, 29.40.Vj

CALORIMETERS AT HADRON COLLIDERS FOR PARTICLE AND NUCLEAR PHYSICS

As the energy of colliders has increased, calorimeters have become the cornerstone of particle physics experiments, and in particular hadronic colliders such as proton-proton colliders. Calorimeters are detectors which make a destructive measurement of the particle in order to measure the energy, impact position, and sometimes the time of arrival of a particle. There are a wide variety of different types of calorimeters, and it is beyond the scope of this article to describe them all, and we refer the interested reader to the textbook by R. Wigmans [1] or to the review by Fabjan and Ludlam[2]. Calorimeters are ideally suited for the higher collision energies of a collider because it becomes more and more difficult to get good resolution from tracking, which typically has a momentum resolution that scales like $\Delta p/p \propto p$, while for calorimeters the energy resolution typically scales like $\Delta E/E \sim 1/\sqrt{E}$. Thus, the precision of calorimeters improves with energy while the momentum measurement of a charged hadron from tracking degrades. In addition, it has become increasingly less important in particle physics to measure the 4-vector of every particle, and instead rely on more global information, like the energy flow. This is because the connection to the theory is best done through the outgoing quarks and gluons from the collisions, which then fragment into collimated sprays of particles called jets. While the fragmentation process is interesting in and of itself, it is a distraction from the point of view of the fundamental connection to theory, which is expressed in terms of Feynman diagrams with outgoing elementary particles.

While for particle physics experiments calorimeters have become the central tool for the above reasons, in high energy nuclear physics colliders (i.e., heavy ion colliders), where fully-ionized nuclei are the projectiles, the calorimeters share equal weight with tracking detectors for two main reasons. One is that the large backgrounds produced in a heavy ion event, where several hundreds of nucleons may collide, make jet reconstruction algorithms somewhat difficult except at the very highest transverse momentum p_T. To quantify this, in the 5% most central Au+Au collisions at RHIC, PHENIX measures a transverse energy density $dE_T/d\eta$ = 600 GeV [3]. One can reduce this by an order of magnitude through a decrease in the cone size. In addition, it is the fluctuations in this background which determine how well the jet energy scale correction can be made. Still, this sets a minimum jet p_T of about 20 GeV from which reconstruction algorithms can be expected to work reliably.

Additionally, in a heavy ion collision, one is more concerned with the properties of the quark gluon plasma, which one probes with relatively low energy phenomena. For instance, the temperatures expected to be reached are about 300-600 MeV (note that 1

Application of Accelerators in Research and Industry
AIP Conf. Proc. 1336, 552-555 (2011); doi: 10.1063/1.3586161

MeV is 10^{10} K). One of the main signatures of a thermalized bath of such hot quark-gluon matter would be outgoing thermal photons which have an approximately exponential spectrum with a mean energy of only a few GeV. Coupled with the requirement that the occupancy of any channel must be kept below a few percent, this drives the design of a calorimeter for heavy ion physics to have a very high segmentation, and a good resolution down to the very lowest energies of a few hundred MeV.

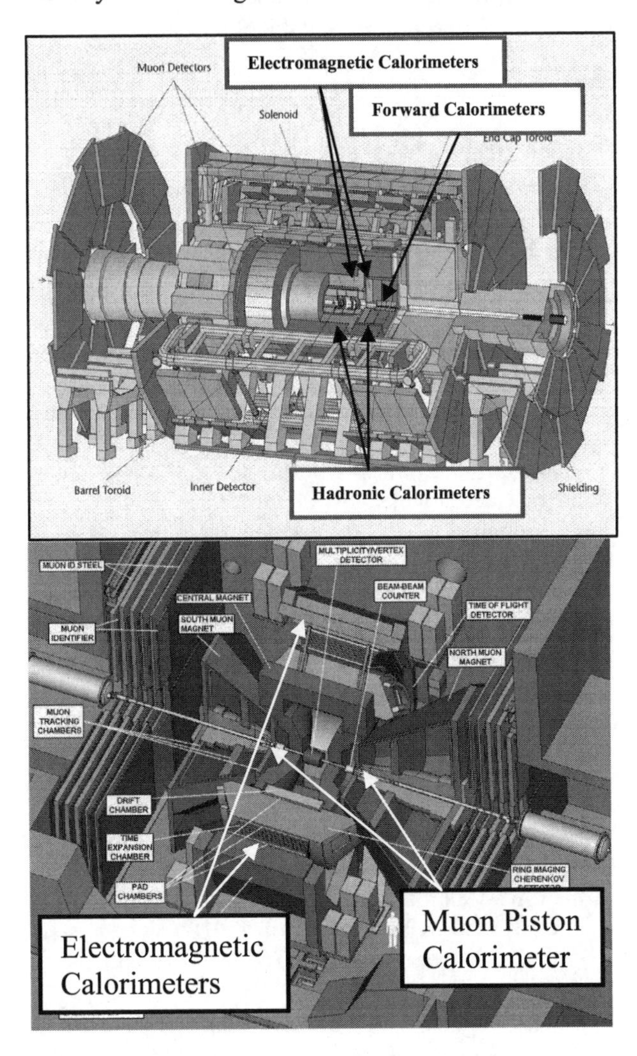

Figure 1. Examples of two quite different detectors, the Atlas detector (left) which has been optimized for Higgs searches and for new physics, and the PHENIX detector (right), which is a nuclear physics detector optimized for characterizing the properties of the quark-gluon plasma.

Atlas Calorimeters: Example For High Energy Physics

In a modern day particle physics experiment at a collider, a calorimeter has to provide a very fast event selection (trigger), lepton identification, and energy measurement of jets, leptons, and energy flow. The Atlas calorimeter [4,5] consists of many subsystems. The main electromagnetic calorimeter (EMCAL) is made of Liquid Argon (LAr) active elements with lead absorbers. It is broken up into two sections, a barrel EMCAL covering pseudorapidity $|\eta|<1.475$, and an endcap EMCAL covering $1.375<|\eta|<3.2$. The lead absorber plates are spaced about 2mm apart, but with a spacing that changes in order to generate a uniform response no matter what impact angle the particle has with the calorimeter. In addition, the absorber plates have an accordion shape which allows the response to be relatively uniform across the polar angle.

One of the disadvantages of liquid argon is the necessity of a cryostat, which introduces additional material far in front of the calorimeter and which often reduces the energy resolution due to early interactions of particles in the cryostat wall. To reduce this effect, a presampler covering $|\eta|<1.8$ was installed just behind the cryostat. The task of the presampler is to estimate, event by event, the energy lost in the material upstream of the calorimeter, in particular the cryostat wall and the magnet coil. These dead areas have a large effect on the energy resolution for electromagnetic particles, particularly electrons, since the material is far from the calorimeter and bremmstrahlung effects on electrons can be large. The energy resolutions of the different Atlas EMCALs have a range for $\Delta E/E = (10\text{-}15)\%/\sqrt{E} \oplus 0.7\%$.

The hadronic calorimeter system consists of three sections. The barrel consists of a sampling calorimeter composed of iron absorber and scintillation plastic material as the active material (Fe-Sc). There are also two LAr endcaps, one covering $1.5<|\eta|<3.2$ which backs up the endcap LAr EMCAL and two very forward calorimeters cover $3.1<|\eta|<4.9$. The central Fe-Sc hadronic calorimeter is the lone exception to LAr calorimetry in Atlas and was chosen due to cost, radiation hardness, and the requirements for the desired resolution. Hadronic calorimeters are designed for measuring the energies of hadrons, which create showers with large fluctuations. Hadronic calorimeters therefore need to be much deeper than electromagnetic calorimeters, and cannot be sampled as frequently in depth, so that they typically have worse resolutions. The Atlas hadronic calorimeters have a resolution of $\Delta E/E=(50\text{-}100)\%/\sqrt{E}$ for jets.

At field strengths of 10 kV/cm and for a 2 mm gap size in the liquid Argon regions, electrons will take 500 ns to cross the gap. This corresponds to 20 LHC bunch crossings, which is too long. In order to produce faster signals, ATLAS applies bipolar shaping with a shaping time of 45 ns. Initially this might seem to have a lower signal to noise ratio since only a fraction of the ionization charge produced by the shower particles are collected. However, at the LHC at design luminosities there are expected to be ~25 collisions per bunch crossing, with a crossing coming every 25 ns. For long integration times one gets a large noise contribution from pile-up noise, which results from particles from the different interactions piling up in the calorimeter. The shorter shaping time thus improves the signal to noise ratio and also makes the signals fast enough for triggering which is essential for a calorimeter.

PHENIX Calorimeters: Example For Heavy Ion Physics

As mentioned above, heavy ion experiments have requirements that can be quite different from those of a high energy particle collider experiment. These requirement differences are driven by the much higher multiplicities in a nucleus-nucleus collision, the requirement to measure lower pT particles, and the desire to measure each particle separately. The PHENIX experiment, one of the two large detectors at the Relativistic Heavy Ion Collider, has good examples of calorimeters optimized for the nuclear environment. PHENIX has two large EMCal systems consisting of a total of 24768 individual detector modules divided between the Lead-Scintillator (PbSc) sampling calorimeter of a shashlik design, which provides six sectors of azimuthal coverage, and the Lead-glass (PbGl) homogenous calorimeter, comprising two sectors. Each sector covers $|\eta|<0.35$ and $\Delta\phi=\pi/8$ [6].

The shashlik design was pioneered in Russia (shashlik means "sandwich" in Russian). It consists of alternating plates of lead and scintillator resembling a sandwich. The lead and scintillator are drilled with holes which are then threaded with wavelength shifting fibers running down the length of the stack. Photons from the scintillator are then guided down the fibers to be read out by a photomultiplier tube (PMT) at the end. The advantage of this design is that it allows one to easily tile the stacks to seamlessly cover a large contiguous area.

The calorimeters in PHENIX form two almost back-to-back arms each covering $\Delta\phi=\pi/8$. Both subdetectors are read out with PMT's and have both good energy and timing resolution, but their design is quite different. Their energy resolutions are $\sigma(E)/E \sim 6\text{-}8\%/\sqrt{E}$ with a constant term of a few percent, and both have an excellent granularity of $\Delta\eta\times\Delta\phi = 0.01\times0.01$ to minimize the double occupancy effects from the ~700 particles produced at $\eta=0$ in a central heavy ion collision. This is six times smaller than the Atlas EMCal granularity of $\Delta\eta\times\Delta\phi = 0.025\times0.025$. This fine granularity allows one to separate the two photons from a π° decay up to momenta of 25 GeV, so that direct photons and π° can be separately identified statistically up to that energy. This is important in heavy ion collisions since direct photons interact electromagnetically and therefore are not modified by the hot medium, while π°'s interact strongly and are highly modified.

In the even more forward region at $|\eta|>3$, the only free space for a calorimeter consisted of a 45 cm diameter, 43 cm deep hole in the front of the muon piston steel, which is about 200 cm from the interaction point. Instrumenting this region demanded a very compact calorimeter design. Lead-tungstate scintillating crystals ($PbWO_4$) were developed during the 1990's for the CMS EMCal and PHOS EMCal in the ALICE Detector. This type of detector results in very compact showers, i.e., very small Moliere radii of 2.2 cm and short L_{rad} of 0.89 cm.

Over the years 2006-2007, the Muon Piston Calorimeter (MPC) was installed into PHENIX. The MPC has become very important since particles in that forward region near the beam-pipe come from the lowest momentum fractions of the partons in the nuclei. The MPC consists of 416 $PbWO_4$ crystals with Avalanche Photodiode (APD) readout and it covers the pseudorapidity range of $3.1<|\eta|<3.7$ [7]. APDs were chosen for their compact size, their insensitivity to magnetic fields, and their low heat generation. The crystals cover $\Delta\eta\times\Delta\phi = 0.01\times0.01$ and are 20 L_{rad} and 0.8 nuclear interaction lengths (L_{int}) deep, which makes them good photon detectors and relatively hadron-blind, a necessity since there is no tracking upstream of this calorimeter. The energy resolution of the MPC is fairly good at $\Delta E/E \approx 14\%/\sqrt{E} \oplus 5\%$.

SUMMARY

As the energy of hadron colliders has increased, the calorimeter has become an ever more important component of the overall detector. This is especially true in high energy experiments where one may be less interested in the details of every particle created in the collision and more interested in the overall energy flow. In this case a good hermeticity is more important than a fine segmentation. This can be contrasted with a calorimeter optimized for nucleus-nucleus collisions, which need a much higher segmentation to handle the very high multiplicities in the events so that one can measure the 4-vectors of each particle in the event. However, as collider energies increase, perhaps with the advent of even higher energy next-generation machines, the events in p+p collisions look more like high multiplicity nucleus-nucleus collisions. High event rates lead to multiple events per bunch crossing, and the higher energy results in more particles being created per event. Combined with the need to make ever more precise measurements of jet energies, these new features of the environment at future colliders has spurred new research directions for future calorimeters and detector designs toward a merger between the current high energy and nuclear concepts. The next generation of calorimeters will have greater segmentation and the overall detectors will be optimized to measure the 4-vectors of every particle, moving away from the low granularity of the high energy calorimeter.

ACKNOWLEDGMENTS

The Author would like to acknowledge all his collaborators on PHENIX and particularly those at the University of Illinois, Kurchatov, Hiroshima, SUNY Stony Brook, and UC Riverside on the Muon Piston Calorimeter project.

REFERENCES

1. Richard Wigmans, *"Calorimetry: Energy Measurement in Particle Physics"*, International Series of Monographs in Physics, 2000.
2. C. W. Fabjan and T. Ludlam, "*Calorimetry in High Energy Physics*", Annual Review of Nuclear and Particle Science Vol. 32: 335-389
3. K. Adcox, et. al., "Formation of Dense Partonic Matter in Relativistic Nucleus-Nucleus Collisions at RHIC: Experimental Evaluation from the PHENIX Collaboration," *Nucl. Phys. A,* **757**, 184-283 (2005)
4. ATLAS Collaboration (A. Aireptian et al), ATLAS Calorimeter Performance, *Technical Design Report*, CERN-LHCC-96-40, Dec, 1996, 198pp.
5. ATLAS Collaboration, ATLAS Tile Calorimeter, *Technical Design Report*, CERN-LHCC-96-42, Dec, 1996, 347pp.
6. L. Aphecetche et al, *Nuclear Instr. And Methods A* **499**, 521-536 (2003).
7. M. Chiu (for the PHENIX Collaboration), "Calorimetery Upgrade at Forward Rapidities for the PHENIX detector", AIP Conf. Proc. **842**, 1091-1093 (2006)

A FOrward CALorimeter Upgrade For PHENIX

Richard S. Hollis (for the PHENIX Collaboration)

Dept. Physics and Astronomy, University of California, Riverside, CA 92521

Abstract. Over the past few years, the PHENIX detector has undergone several upgrades in the forward region ($1<|\eta|<4$), initially covered only by the muon arms. The focus of these upgrades is toward a better understanding of the Color-Glass Condensate and the interplay between the different components of the proton's spin valence/sea quark and gluon contributions. This paper highlights the newly proposed forward calorimeter detector, FOCAL. FOCAL is a tungsten-silicon sampling calorimeter with high position and energy resolution, covering a pseudorapidity of $1.6<\eta<2.5$. This future detector aims to constrain the current view of gluon saturation at small x in the Color-Glass Condensate framework, through isolation of direct photons at high-p_T over a broad range of pseudorapidity.

Keywords: Calorimeter, FOCAL, RHIC, PHENIX, d+Au, Color Glass Condensate
PACS: 25.75.-q, 25.75.Bh, 25.75.Cj

INTRODUCTION

In collisions of high energy nuclei, for example Au+Au, a hot, dense, and opaque medium is created with remarkably complex properties [1]. This state, theorized to be a state of matter known as the strongly interacting Quark-Gluon Plasma (sQGP), is characterized by the energy loss of highly energetic partons traversing it. The colliding system is a complex state of matter as it comprises both initial (cold-nuclear matter) and final (due to the sQGP) state effects. To fully understand the sQGP, we must be able to decouple these broad classifications (initial or final), thus it is imperative that we study the underlying initial state effects. Therefore we have to turn off the final state effect and examine the initial state effects alone; one way to accomplish this is to collide protons (or deuterons) with gold nuclei. Any initial state effects due to the size of the gold nucleus (such as the Cronin Effect – which has been observed) are manifest, although energy loss effects associated with gold-gold collisions are absent. The energy loss of high energy particles has already been found to be absent in the initial state which reinforces the sQGP picture in gold-gold collisions. Moreover, the initial-state results in the forward region can be interpreted as being consistent with another state of matter, known as the Color-Glass Condensate (CGC) (for a review see Ref. [2]). In this model, the wave functions from low-x gluons overlap and the number of gluons becomes saturated in the heavy nucleus; x is the fraction of momentum carried by the gluon (parton). However, the current measurements with charged particles do not provide evidence for detailed modification of the parton distribution functions (pdfs), particularly those of the gluon, within the nucleus in these extreme conditions. The FOCAL will shed some light on this important physics area by measuring the energy and angular dependence of direct photons from d+Au collisions. In the CGC framework, the pdfs are modified in the nucleus, leading to a modification of the production cross-section of direct photons. Once created, the direct photons are not further modified, unlike partons (as measured by hadrons) which have associated fragmentation functions, thus such measurements can be directly linked to the modification of the pdfs.

THE PHYSICS CHALLENGE

As explained, it is important to measure direct photons as these give us a handle on the gluon pdfs. As the region of interest lies firmly at low-x it is important that this measurement be performed at as low an x as possible. At RHIC low-x can be reached over a wide range of pseudorapidity, from the central region to the forward region. However, in the central region (η~0), the only measurements possible at low-x are at very low momentum, which ultimately excludes a direct photon measurement due to kinematic constraints. In the forward region (η~2 or higher), these constraints are less important as forward jets must have originated from the interaction of a low-x and a higher-x parton. In such interactions, which

Application of Accelerators in Research and Industry
AIP Conf. Proc. 1336, 556-559 (2011); doi: 10.1063/1.3586162

create direct photons, the low-x parton is readily assigned to the gluon which is associated to the photon. The gluon nuclear pdfs in the low-x region are relatively unknown and have large uncertainties such that new measurements in the region of $x \sim 10^{-2}$-10^{-3} will undoubtedly restrict models trying to describe the data [3].

These direct-γ events are extremely rare ($\sigma_{pp} \sim 0.0027$ mb) and their detection is additionally hindered by the fact that, in the kinematic region we are investigating, decayed π^0 and η mesons mostly appear as single photons. The primary objective of this detector is to isolate the direct photons from background sources such as π^0 and η.

THE FOCAL DETECTOR

The FOCAL detector (see Fig. 1) will sit in the region fiducially covering $1.6<\eta<2.5$, i.e. on the deuteron-going side. FOCAL is constructed from 52 modular bricks, each with an active transverse area of 6.2cm×6.2cm and segmented into 4×4 read-out pads. Each brick consists of 21 tungsten layers (each 4mm thick) which are interspaced with silicon used for read-out After layers 2, 3, 4, and 5 additional layers of silicon strips are added for high-resolution position readout (two orthogonal layers of 6.2cm×0.05cm each). The read-out consists of a summed signal from the pads in three segments of towers (7 layers each); all strips are individually read-out. This sampling calorimeter is a novel device as it couples an energy measurement (from a sum of energy in the pads) to a tracking device (through the spatial measurements of the silicon pads and strips). Although FOCAL is not projective by design, the fine tracking elements allow for projective-like tracking across the whole calorimeter. The strips included in the design were carefully positioned to allow for the measurement of the electromagnetic shower in its infancy. This early development, with a smaller average dispersion, helps with the identification of close photon showers, characteristic of $\pi^0 \rightarrow \gamma\gamma$ decays. Using the strips, we find that it is possible to reconstruct the invariant mass of two tracks, even when that is not possible with the silicon pads.

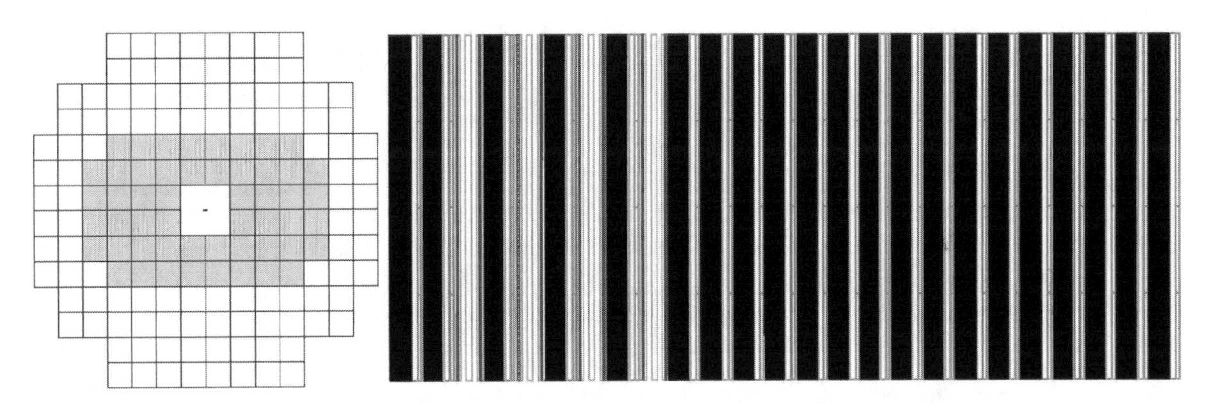

FIGURE 1. Schematic layout of FOCAL. The left shows the transverse view of the detector; the grey region highlights the region instrumented with silicon. The right shows the side view. Dark bands are the tungsten activator layers, behind which are silicon pads. Larger gaps behind the 2nd, 3rd, 4th, and 5th layers house the strip readout.

PERFORMANCE

Track reconstruction with FOCAL comprises of two independent procedures. First, a clustering algorithm finds regions of the detector through which an electromagnetic track passes. Here, neighboring towers are added to seed towers in each segment as long as the additional energy is smaller than the seed (otherwise this indicates the presence of two tracks). The typical cluster size in each segment is 3×3 or 5×5 towers. Once the clusters are found the energy of the track is estimated using the measured energy and energy profile. Note that no energy information can be obtained from the tungsten; the total energy deposited in the silicon represents just 2.5% of the total energy of the impinging track, and the rest is lost to the tungsten. The information from this procedure also includes the angular direction of the track and a measure of the likelihood that the shower is electromagnetic or hadronic in nature.

To determine whether the track is consistent with a π^0 decay or a single photon, the strips are used to measure two photon tracks. In order to simplify the tracking algorithm, a Hough transformation is made to convert the spatial positions into parameter space.

Here, the parameters (intercept and slope) are the same for the same track; in real space the *x*- (and/or *y*-) position gradually increase with *z*. All hits in Hough-space are placed into a histogram, where the most frequently observed slope is considered the primary track (see Fig. 2 left). Once this is found, a searching algorithm is applied to test whether a second peak is evident within a small region of interest in the region of the track. Fig. 2 right shows the results of the algorithm for many events. The single-γ simulation shows a single-peak distribution, where the peak is close to zero. Two tracks are required to form an invariant mass distribution; therefore both the first and second "track" is from the same shower. For π^0s, both a correct-mass peak and a low-mass (single-γ-like) distribution are seen. The latter case represents π^0 decays which have a very asymmetric energy (where either the energy of one photon is small or is outside of the acceptance of the detector), thus only one shower is detected.

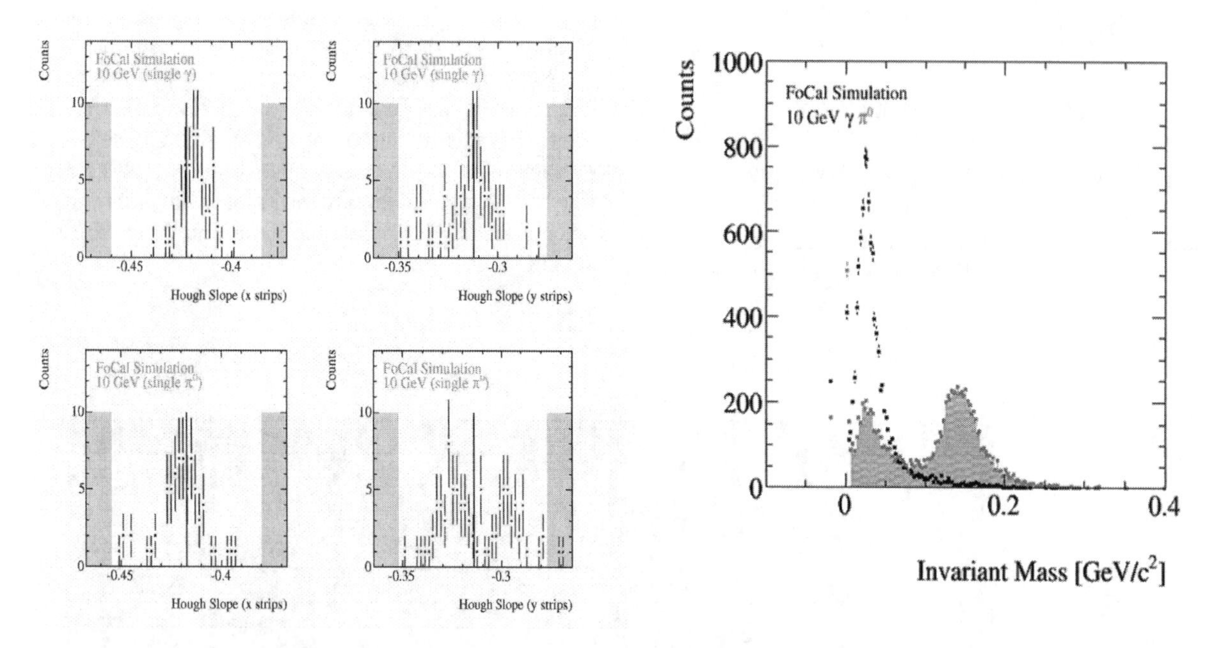

FIGURE 2. In the left multi-panels: reconstruction of single γ (top) and single π^0 (bottom) in FOCAL. The leftmost (rightmost) of the smaller figures shows the Hough parameter frequency in the *x*- (*y*-) direction. The line represents the center of gravity of the track as found by the pad layers. The grey bands show the start of the excluded region from the region of interest. For the π^0 distributions, in this case, two peaks are clearly observed. In the right panel, the resultant invariant mass distribution for 10 GeV single γ (black histogram) and π^0 (grey histogram) into the FOCAL acceptance is shown. Negative values in the invariant mass distribution represents the times when only a single track is found.

Dependent on energy and pseudorapidity, the tagging efficiency from this procedure is about 75%-95% for γ (for invariant mass less than 0.1) and 55%-65% for π^0 (for invariant mass between 0.1 and 0.2). Further quality control constraints applied lead to a ratio of $\gamma/\pi^0\sim5$, even if the mass is not precisely constructed. Within the environment of p+p or d+Au collisions, additional background from charged particles and merging tracks contaminates the track candidates. The electromagnetic shower development can be predicted based on the reconstructed energy, with the prediction compared to each single event. The likelihood that this is an e-m shower (rather than a hadronic shower) is quantified via a χ^2 distribution, which removes almost all of the charged background.

SUMMARY

The FOCAL detector (within PHENIX) has been proposed to examine the theorized Color-Glass Condensate in d+Au collisions at RHIC. Currently at the proposal stage, the FOCAL is a novel detector, which has the ability to separate single photon showers and double photon showers associated with π^0, up to energies of ~50GeV.

ACKNOWLEDGMENTS

We thank the staff of the Collider-Accelerator and Physics Departments at BNL for their vital contributions. We acknowledge support from the Office of Nuclear Physics in DOE Office of Science and NSF (U.S.A.), MEXT and JSPS (Japan), CNPq and FAPESP (Brazil), NSFC (China), MSMT (Czech Republic), IN2P3/CNRS and CEA (France), BMBF, DAAD, and AvH (Germany), OTKA (Hungary), DAE and DST (India), ISF (Israel), NRF (Korea), MES, RAS, and FAAE (Russia), VR and KAW (Sweden), U.S. CRDF for the FSU, US-Hungary Fulbright, and US-Israel.

REFERENCES

1. K. Adcox, *et al* [PHENIX Collaboration], *Nucl. Phys. A757*, 2005, pp. 184-283.
2. L. McLerran, *arXiv:hep-ph/0402137v2*, 2004.
3. K. J. Eskola, H. Paukkunen, and C. A. Salgado, *Nucl. Phys. A830*, 2009, pp. 599–602.

J-PARC Status, Nuclear and Particle Physics

Susumu Sato

Advanced Science Research Center / J-PARC Center, Japan Atomic Energy Agency, Shirakata-shirane 2-4, Tokai, Ibaraki, 319-1195 Japan

Abstract. J-PARC accelerator research complex, consisting of LINAC, RCS and MR synchrotron, has successfully produced neutron, muons, kaons, and neutrinos by steady commissioning since November 2006. There are three experimental facilities, and for nuclear and particle physics, nine experiments are approved in a hadron physics facility, and one experiment is approved in a neutrino physics facility. Those experiments and status of J-PARC are described in this paper.

Keywords: Spectrometers, Beam characteristics, Hyperons, Exotic Baryons, Standard Model
PACS: 29.30.Aj; 29.27.Fh; 14.20.Jn; 14.20.Pt

J-PARC ACCELERATOR COMPLEX

The J-PARC accelerator research complex [1] comprises the 181 MeV (currently being upgraded to 400 MeV) LINAC , 3GeV Rapid Cycle Synchrotron (RCS) , and 50 GeV (so far 30 GeV) Main Ring Synchrotron (MR). The complex is controlled through more than several hundreds of beam monitoring detectors [2], and its control system. Typical parameters and monitoring devices of the accelerator complex are shown in Figure 1. In J-PARC, three experimental facilities are in operation; materials and life science facility (MLF), neutrino facility (NU), and hadron physics facility (HD, see Figure 2). Since November 2006, beam commissioning of the accelerator, which is done in step by step from LINAC, RCS and to MR, has been performed, and then productions of neutrons, muons, kaons, and neutrinos were performed in May, September 2008, February, and April 2009, respectively. After summer shutdown in 2010, slow extraction experiment (beginning with the E19 experiment) in the HD facility is aiming for operation with more than 5 kW power and more than 15% duty factor, and the fast extraction experiment (namely T2K experiment) in the NU facility is aiming for 100kW operation toward the design power.

NUCLEAR AND PARTICLE PHYSICS

Currently (as of the 9th Program Advisory Committee (PAC) for nuclear and particle experiments, held in January 2010), nine experiments are approved in the hadron physics facility, and one experiment (T2K experiment, for neutrino oscillation) is approved in the neutrino facility. After the 1st stage approval (where scientific merit is high), the 2nd stage approval is given as a green light for each of these ten experiments to proceed, based on the technical achievability, the reliability of cost estimate, and the other various aspects of feasibility. Table 1 shows experiments which have the 2nd stage approvals as of the 9th PAC. In the following section, numerical evaluations for experiments are based on the each proposal [3] submitted to the PAC.

Hadron Physics Experiment With Hyperon And Searching For Exotic States

The E19 Experiment

The E19 experiments will measure exotic particle through $p\,(\pi^-, K^-)X$ reaction, where X is expected to be Θ^+ pentaquark. The first observation of the Θ^+ baryon with positive strangeness S = +1 was reported through photon induced reaction and has been supported by several collaboration [4]. As statistical limit in positive results and null results are also reported, therefore confirmation of the existence (or non-existence) of the Θ^+ is still controversial [5]. Especially the meson induced reaction using a proton target is still unique to KEK-PS and J-PARC. In J-PARC, experiment will be performed at the K1.8 beam line. The π^- beam with 1.87, 1.92 and 1.97

Application of Accelerators in Research and Industry
AIP Conf. Proc. 1336, 560-564 (2011); doi: 10.1063/1.3586163

GeV/c momentum will be used to a liquid hydrogen target of 12.5 cm thickness (re-use of the one constructed for the KEK-PS E559 experiment [6]). The momentum of scattered K^- is measured in the range of 0.7 to 0.95 GeV/c with SKS spectrometer [7]. The intensity of the π^- beam is determined by the rate capability of the detector system and is assumed to be 1×10^7 per 4 second cycle. Required intensity of primary proton beam to achieve this is less than 10% of the design intensity of J-PARC. The SKS spectrometer has a capability to measure the width of Θ^+ with a highest resolution, which is estimated to be 2.5 MeV. If the target is irradiated with 4.8×10^{11} pions for each momentum (which is 100 times more than E522) and assuming 1.9 $\mu b/sr$ for production cross section of Θ^+, we will obtain 1.2×10^4 events for the momentum setting. The E19 experiment has started beam commissioning run since October 2009, and during calibration run the SKS spectrometer showed Σ^- reconstruction with ΔM =1.66 ± 0.05 (FWHM) MeV in $\pi^- p \rightarrow K^+\Sigma^-$ reaction. E19 is scheduled to begin data taking in October 2010.

The E05 Experiment

The E05 experiment is proposed to get spectroscopic information of Ξ-hypernucleus, ${}^{12}_{\Xi}Be$, through ${}^{12}C(K^-,K^+)$ reaction. The Ξ single particle potential obtained from the observation of Ξ-hypernucleus states gives not only the information of ΞN interaction but also insight into the high-density hadronic matter with strangeness. Experimental method and apparatus are following. Missing mass, M, defined as $M^2=(E_B+m_T-E_S)^2-(\vec{p}_B-\vec{p}_S)^2$, where suffixes B, T, and S mean beam, target and scattered particles, respectively. Beam momentum is analyzed by QQDQQ magnets and four sets of tracking detector in the beam line, and expected momentum resolution, $\Delta p/p$, is 1.4×10^{-4} in rms. Scattered beam (K^+) momentum is measured by the SKS magnet with some modification.

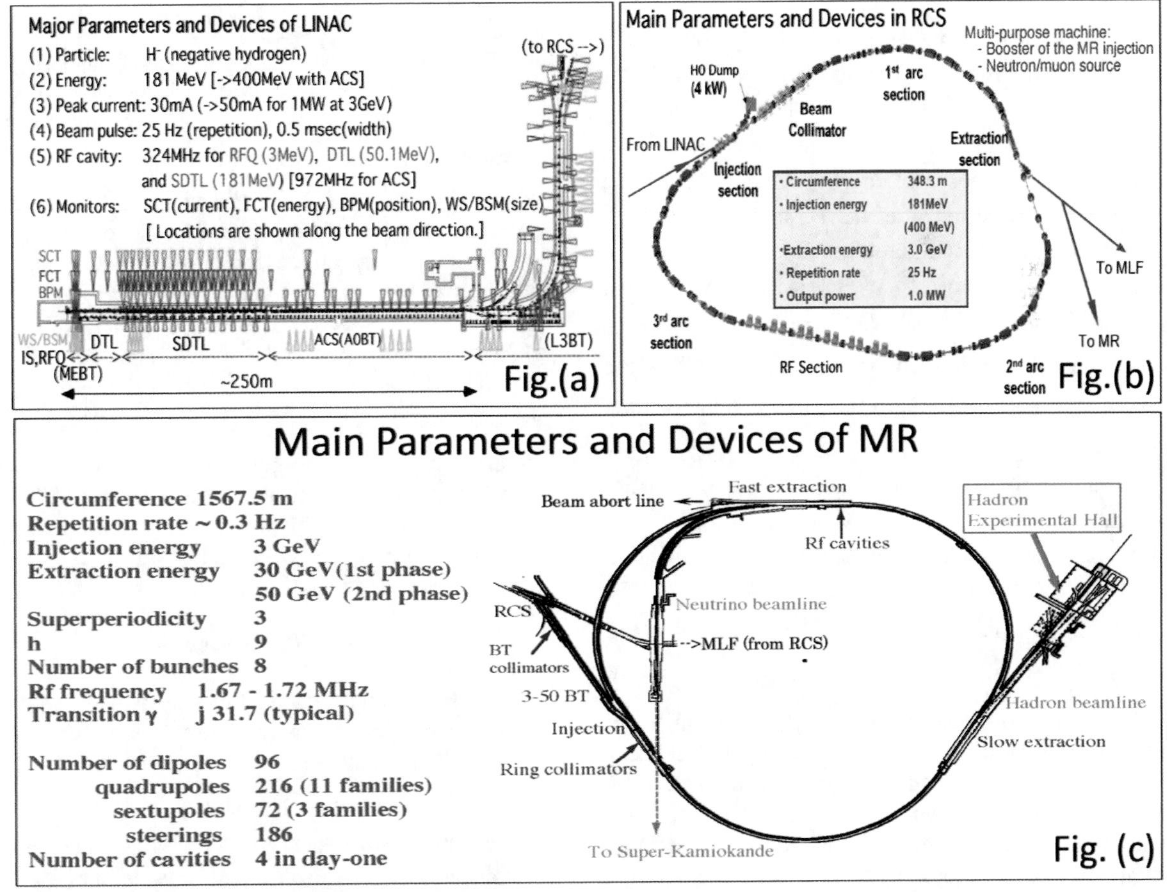

FIGURE 1. The J-PARC Accelerator Complex and Its Major Parameters and Devices for (a) LINAC, (b) RCS synchrotron, and (c) MR synchrotron.

Because K^+ momentum corresponding to the production of Ξ-hypernuclei is around 1.3 GeV/c and the SKS maximum magnetic field of ~2.7 T does not allow to put the central ray at 1.3 GeV/c, additional dipole magnet with ~1.5 T will be placed at the entrance of the SKS magnet. This is called as "SksPlus" spectrometer, which has a solid angle of ~30 msr with angular range upto 10 degrees, and momentum resolution, $\Delta p/p$, is 0.17% (FWHM). As missing mass energy resolution is mainly determined by the K^+ spectrometer and energy loss straggling in the target, the maximum target thickness is 5-6 g/cm^2 to keep the resolution within the acceptable level, 3MeV. For the number of K^-, we can expect 3.7×10^{10} [/day] for 1.4×10^6 [/spill] when we choose flat-top length of 0.8 second. And the yield of $^{12}_{\Xi}Be$ is estimated to be 6.3 [/day] ~ 190 [/month], which means we need ~ 1 month of data taking.

The E13 Experiment

The E13 experiment is to study structure of several light hyper nuclei ($^4_\Lambda H$, $^7_\Lambda Li$, $^{10}_\Lambda B$, $^{11}_\Lambda B$, *and* $^{19}_\Lambda F$) by high-precision γ spectroscopy. Newly-constructed large germanium detector array is dedicated for detection of γ rays from hypernuclei. Exited states of these hypernuclei are produced by (K^-, π^-) reaction. The purposes of the experiment are (1) the first precise measurement of the Λ-spin-flip B(M1) to investigate the magnetic moment of a Λ in a nucleus, and (2) further study of ΛN interaction to establish the spin-dependent ΛN interaction strengths and to clarify the $\Lambda N-\Sigma N$ coupling force as well as the charge symmetry breaking effect in ΛN interaction. For the experimental setup, K^- beam of 1.5 GeV/c is used at K1.8 beam line where two-stages mass separators provide lower π/K ratio (less the 0.5) to minimize the radiation damage to the germanium detectors. The kaon beam is irradiated to various targets; liquid ^4He (2.0 g/cm^2), $^{nat}Li_2O$ (17.2 g/cm^2 for ^7Li), ^{10}B metal (20 g/cm^2), ^{11}B metal (20 g/cm^2), and ^{19}F in teflon (15.2 g/cm^2 for ^{19}F). Just upstream and downstream of the target, aerogel Cerenkov detectors are used to idendify kaons in beam and pions in scattered particles. Scattered pions are momentum-analyzed by the SKS magnet and tracking chambers. Including energy loss effect in the thick target (typically 20 g/cm^2), expected mass resolution is 5.9 MeV (FWHM), which is enough to tag hypernuclear bound state region. The γ rays from hypernuclei are detected Ge detector, which is surrounded by PWO counter for background suppression, and estimated photo-peak efficiency is about 5 % for the realistic source point distribution in the Li_2O target. About 1000 hours of beam time is requested for the physics purpose described above, when assuming K^- beam intensity is 0.5×10^6 per spill for 1.5 GeV.

The E03 Experiment

The E03 experiment is to perform the first measurement of Ξ-atomic X rays from Fe target at K1.8 beam line. Physics interest is in Baryon-Baryon interaction at $S=-2$ sector, which is important to understand the properties of neutron stars of which density is so high that significant amount of hyperons is expected to appear in the core. Measurement of X-ray from Ξ-atom is promising approach to study optical potential in nuclei; this approach has been successfully used in cases of negative charged hadrons ($\pi^-, K^-, \bar{p}$, *and* Σ^-). The X-ray energy shift gives information on the real part of the optical potential, while X-ray width and yield are relevant to the imaginary part. K^- in the beam is identified by time of flight counters and an aerogel Cherenkov counter. The scattered K^+ particles are detected by KURAMA spectrometer, which has large acceptance of 0.2 sr, which allows to maximize the yield of Ξ^-. X-ray is detected by Hyperball-J, which consists of about forty Ge detectors surrounded by fast PWO counters for background suppression. Assuming 1.8 GeV/c K^- beam with an intensity of 1.4×10^6 per spill (4 sec/spill with 1.2 seconds flat-top) for 800 hours, expected X-ray yield is 2500 counts in 7.5×10^5 stopped Ξ^- on Fe target. The 2500 counts give statistical energy shift accuracy of better than 0.04 keV (~0.05 keV with systematic errors), which is sensitive enough to observe expected energy shift (~1keV) with reasonable accuracy.

The E07 Experiment

The E07 experiment is for a systematic study of double strangeness nuclei with 10 times higher statistics than previous experiment (KEK-E373 [8]). Expectation is to observe 10^4 stopping Ξ^- hyperons in the emulsion via quesi-free (K^-, K^+) reactions on a diamond target. And this will provide one thousand events showing formation of double strangeness nuclear systems, then among them we will detect one hundred nuclear fragments with double strangeness to make a chart of S = -2 nuclei. Requested beam is 600 hours with 3×10^5 K^-per spill. Experimental setup is following: (K^-, K^+) reaction, $K^- + "p" \rightarrow K^+ + \Sigma^-$, is detected by KURAMA spectrometer. Both (1) between the target and the emulsion and (2) just after the emulsion, silicon strip tracking detectors are placed to tag the Ξ^- hyperons produced in the target and stopped in the emulsion. Advanced detectors and automated scanning system enable us to obtain ten times more events than KEK-E373 within a few years of data analysis.

The Other Approved Experiments

Other approved experiments are: E10 experiment for Production of Neutron rich Λ-hypernuclei with the double charge-exchange reaction, E15 experiment as a search for deeply-bound kaonic nuclear states by in-flight $^3He\,(K^-, n)$ reaction, and E17 experiment for precision spectroscopy of Kaonic $^3He\ 3d \rightarrow 2p$ X-rays.

Particle Physics Experiment, and Neutrino Experiment

The E14 ("KOTO") Experiment

The E14 experiment (KOTO experiment) aims to discover a CP-violating decay mode, $K_L^0 \rightarrow \pi^0\ \nu\,\overline{\nu}$. The CP-violating parameter, η, in the CKM matrix can be determined from the branching ratio of the decay with a theoretical uncertainty of 1-2%. The branching ratio is expected to be $(2.8\pm0.4)\times10^{-11}$ in the Standard Model prediction. The signature of the decay distinct from other backgrounds are that only two gammas from a π^0 exist and the transverse momentum of the π^0 is large due to the missing neutrinos. A hermetic detector ensures only two gammas and nothing else, and the small-diameter beam enables the π^0 reconstruction to constrain the decay vertex of the π^0. The collimators need to be designed to reduce the ratio of the halo-neutron and K_L fluxes to be 7×10^{-4}, which is one of the key performances to determine the sensitivity of the experiment. The beam line was designed to realize a small-cross-section neutral beam with a solid angle of 7.8 μsr using two collimators and a sweeping magnet. The length of the beam line is 20m, where short-lived particles are decayed. Only gammas, neutrons, and K_L's are remained at the beam line exit. The gamma flux is reduced with a 7-cm-long lead absorber at the upstream of the beam line. The initial goal is to make the first observation of the decay with the expectation to observe 3.5 Standard Model events with 1.8×10^{21} protons (equivalent to 2×10^{14} protons/spill $\times$ 3×10^7 seconds of running, where 3.3 sec/spill) on target in total.

The E11 ("T2K") Experiment

The E11 experiment (T2K experiment) is a long baseline neutrino oscillation experiment, using fast extracted proton beam from MR synchrotron and existing Super-Kamiokande (SK) detector. One of the main goals of this experiment is the discovery of $\nu_\mu \rightarrow \nu_e$. A factor of 20 improvement in sensitivity over the present upper limit is possible, then the goal is to extend the search down to $\sin^2 2\theta_{13} \sim 2\sin^2 2\theta_{\mu e} > 0.008$. To achieve this goal, it is requested to have total integrated beam power larger than $0.75\ MW \times 15000h$ at any proton energy between 30 and 50 GeV. The SK detector is the world largest water Cherenkov detector as the far detector, and has excellent energy resolution and e/μ identification capability in low energy neutrino reaction, backed by about twenty years of experience of the water Cherenkov detector. The T2K accumulated neutrino beam data corresponding to 3.35×10^{19} PTO (proton on target) with beam operation until June 26 in 2010, and 22 neutrino events have been observed [9] by the SK in the analysis so far with data accumulated through the middle of May.

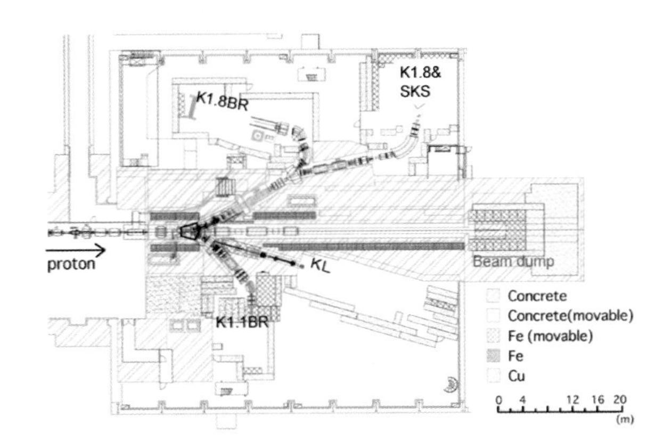

FIGURE 2. Beam Lines of the J-PARC Hadron Physics Facility (Hadron Experimental Hall).

TABLE 1. Experiments in J-PARC approved at the PAC (approved both as the 1st and 2nd stage (for the meaning of "stage", see the text), after the 9th PAC held in January 2010).

ID	Description of Experiment	Beam line
E03	Measurement of X rays from Ξ Atom	K1.8
E05	Spectroscopic Study of Ξ-Hypernucleus, ${}^{12}_{\Xi}Be$, via the ${}^{12}C(K^-,K^+)$ Reaction	K1.8
E07	Systematic Study of Double Strangeness System with an Emulsion-counter Hybrid Method	K1.8
E10	Production of Neutron-Rich Lambda-Hypernuclei with the Double Charge-Exchange Reaction	K1.8
E11	Tokai-to-Kamioka (T2K) Long Baseline Neutrino Oscillation Experiment	Neutrino
E13	Gamma-ray spectroscopy of light hypernuclei	K1.8
E14	$K_L \rightarrow \pi^0 \nu\bar{\nu}$ Experiment at J-PARC	K0
E15	A Search for deeply-bound kaonic nuclear states by in-flight ${}^3He(K^-, n)$ reaction	K1.8BR
E17	Precision spectroscopy of Kaonic ${}^3He$ $3d \rightarrow 2p$ X-rays	K1.8BR
E19	High-resolution Search for Θ^+ Pentaquark in $\pi^- p \rightarrow K^- X$ Reactions	K1.8

ACKNOWLEDGMENTS

The author expresses appreciation to the J-PARC center and its accelerator group both in JAEA and in KEK, for their steady efforts toward improvements of the accelerator research complex.

REFERENCES

1. J-PARC Design Report (JAERI-Tech 2003-044, KEK Report 2002-13).
2. For LINAC: S. Sato et al., Proceedings of the 10th European Accelerator Conference (EPAC), 1151-1153 (2006). S. Sato et al., Proceedings of the 22nd Particle Accelerator Conference (PAC), IEEE, 4072-4074 (2007). For RCS: N. Hayashi et al., Proc. of the 11 the EPAC, 1125-1127 (2008). For MR: K. Satou, Proc. of the 42th ICFA workshop (Hadron Beam 2008) 472-474.
3. Documentations of Program Advisory Committee (PAC) for Nuclear and Particle Physics Experiments at the J-PARC 50 GeV Proton Synchrotron, is found at http://j-parc.jp/NuclPart/PAC_for_NuclPart_e.html, which is linked also from the J-PARC home page http://www.j-parc.jp.
4. T. Nakano et al. Phys. Rev. Lett., 91, 012002 (2003).
5. K. H. Hicks. Prog. Part. Nucl. Phys., 55, 647 (2005).
6. 12-GeV PS 2008 Review Report (KEK), p17, PS_Reviews/PS_Review_2008.pdf, Reports/E559.pdf under http://www-ps.kek.jp/kekps/eppc/Review08/.
7. T. Fukuda et al. Nucl. Inst. Meth, A361, 485-496 (1995).
8. K.Nakazawa, Nulc. Phys. A585 (1995) 75c, Nulc. Phys.A636 (1998) 345c.
9. J-PARC Project Newsletter, No 41, July 2010, http://j-parc.jp/index-e.html.

Forward Drift Chamber for the GlueX Experiment at the 12 GeV CEBAF Machine

Lubomir Pentchev[a] and Benedikt Zihlmann[a] for the GlueX collaboration

[a]*Thomas Jefferson National Accelerator Facility (12000 Jefferson Ave., Newport News, VA 23606)*

Abstract. The GlueX experiment will search for exotic mesons produced by 9 GeV linearly polarized photons from the upgraded CEBAF machine. It is critical to detect and measure the four-momenta of all the charged particles and photons resulting from the decays of the mesons. The solenoid-based detector system includes tracking detectors and calorimeters. The Forward Drift Chamber, FDC, consists of 24 circular planar drift chambers of 1m diameter. Additional cathode readout is required to achieve efficient pattern recognition. The detection of photons by the electromagnetic calorimeters imposes constraints on the amount of material used in the FDC. The specific features of the detector and the readout electronics will be described. Results from the tests of the full scale prototype will be presented, as well.

Keywords: hybrid mesons, drift chambers

INTRODUCTION

The quark model of the mesons as bound states of a quark and an antiquark is based on the success in explaining the meson spectroscopy using the quantum numbers of the two quarks only. Quantum Chromodynamics (QCD) gives a richer picture of the mesons as strongly interacting quarks and gluons. The inclusion of gluonic degrees of freedom results in predictions for mesons in which the gluons are excited and contribute to the quantum numbers of these so called hybrid mesons. Some of the hybrid mesons have exotic quantum numbers not allowed by the quark-antiquark model. The existence of hybrid mesons is supported by Lattice QCD calculations [1] that made significant progress during the last years. Despite of the extensive experimental search, the results are controversial; still there is evidence [2] for a few isovector exotic states. The GlueX experiment [3] will search for such hybrid mesons with masses up to 2.8 GeV.

The identification of the hybrid mesons with exotic quantum numbers requires measurements of the four-momenta of all the charged particles and photons from the decay of the mesons. This necessitates a hermetic detector with a flat acceptance with respect to the meson decay products. The GlueX detector presently under construction will be installed in the newly built Hall D of the upgraded CEBAF machine at Jefferson Lab. The experiment will use 9 GeV linearly polarized photons produced from the 12 GeV electron beam. The polarization as an additional degree of freedom is critical for identifying the quantum numbers of the produced mesons.

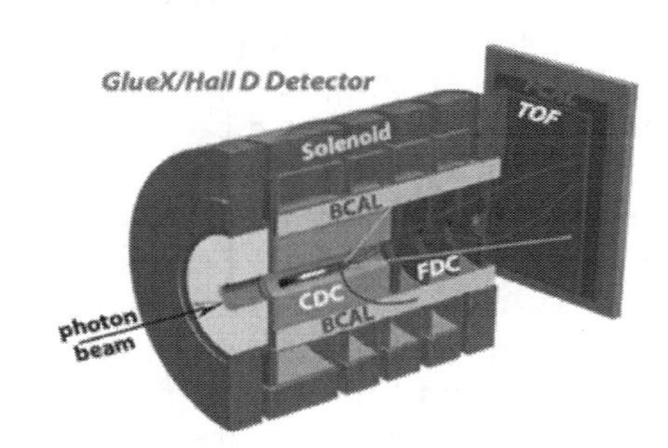

FIGURE 1. GlueX detector: Solenoid, Central Drift Chamber, Forward Drift Chamber, Barrel Calorimeter, Forward Calorimeter, Time of Flight.

The GlueX detector (Fig.1) consists of a tracking system that includes the Central Drift Chamber (CDC) and the Forward Drift Chamber (FDC), and two electromagnetic calorimeters, the Barrel Calorimeter (BCAL) and the Forward Calorimeter (FCAL). All the detectors, except FCAL, are placed inside the bore of a solenoid with a longitudinal magnetic field of 2T. The FDC tracks the charged particle in the forward direction. The CDC employs 1.6cm diameter straw tubes used for tracking at large angles and for low energy charged particle identification. The BCAL is a scintillation-fiber/lead calorimeter, the FCAL is a lead-glass calorimeter, both with high energy (~6%/√E) and

Application of Accelerators in Research and Industry
AIP Conf. Proc. 1336, 565-568 (2011); doi: 10.1063/1.3586164

position (~6mm/√E) resolution, aiming to register the photons in the full 4π acceptance. The Time of Flight scintillator wall will be used for particle identification and timing.

FORWARD DRIFT CHAMBER DESIGN

The FDC will consist of four cylindrical packages, each having 6 drift chamber layers. Each layer (Fig.2) consists of two cathode strip planes on both sides of the wire plane. Each wire plane has alternating sense and field wires 5mm apart. The main design parameters are summarized in Table 1.

TABLE 1. Main drift chamber parameters

Parameter	Value
Sense wire diameter	20 μ
Field wire diameter	80 μ
Cell size	10x10 mm^2
Cathode strip pitch	5mm
Active area diameter	1m
Strip to wire angle	75°
Gas mixture	Ar/CO_2
Gas gain	2 - 8 10^4
Total number of drift layers	24
Total thickness of active area	1.3% R.L.

The most critical requirement in the FDC design is to minimize the amount of the material not only in the active area of the chamber but also at the periphery: the frames, supporting systems and cables. At the same time the mechanical structure must be robust enough to minimize the deformations and to allow for good gas tightness. In an early design the chamber frame was made out of solid G10 material. Monte Carlo simulations showed a significant fraction of the photons from meson decays being converted by the frames and not properly reconstructed in the BCAL or FCAL due to the strong magnetic field. This required a design change in which the frames are composite: an outer thin G10 skin with an inner part made out of Rohacell material, reducing the total frame thickness by a factor of two. However, a full scale prototype built in this manner showed problems with gas tightness and robustness of the whole construction. This was mitigated in the final design, where a small fraction of the frame around the gas spacers is kept solid G10.

FIGURE 2. One drift chamber layer, from right to left: cathode strip plane, wire plane, spacer ring, second cathode strip plane, ground

The high particle densities - especially in the forward direction - require good pattern recognition. To achieve this, each drift layer has cathode strips on both sides of the wire plane. For a standard drift chamber, using the drift time from one wire plane only would give two lines parallel to the wire as possible hit positions. The cathode information allows to reconstruct the position of the gas avalanche along the wires. Thus, apart from the left/right ambiguity, each chamber layer provides a spatial point that contributes greatly to the identification of the charged tracks in the presence of a high magnetic field and to the reconstruction of the particle momenta. A resolution of 200μm is required for both wire and strip coordinates in order to achieve the physical goals of the experiment.

A gas mixture of Ar/CO_2 is our preference, as both gases are non-flammable and neutral. The exact percentage of the components will be determined experimentally based on the stability of operation and the optimum wire and strip resolutions. High argon percentage results in a weaker dependence of the drift velocity on the electric field and better resolution. However, in the presence of a magnetic field one would prefer lower drift velocities and therefore higher CO_2 concentration, to minimize the Lorentz force effects. Detailed simulations and experimental investigations are underway to quantify the magnetic field effects and study the stability of operation in the case of low CO_2 admixture.

The readout of the full FDC detector will require 2300 wire channels and 10,300 cathode strip channels. It is critical for both, wire and strip resolutions, to have a low noise-to-signal ratio even with a large detector capacitance (up to 100pF for the strips). For that purpose, an eight-channel preamplifier employing a novel ASIC was designed [4], providing good readout density at reasonable cost. The preamplifier has a peaking time of 11ns. Noise of about 2000e+20e/pF has been measured at a gain of 3mV/fC. The ASIC also includes a discriminator that will be used for the wire readout, connected to 64-channel TDC. In case of the strip readout, the differential outputs of the preamplifier are fed into a 72-channel flash ADC [5] with 8ns sampling time.

STUDIES WITH FULL-SCALE PROTOTYPE

Three full-scale drift chamber layers have been built to investigate the chamber performance. The results presented here were done with 40/60% Ar/CO_2 mixture using cosmic tracks, although other studies were done with radioactive sources and other gas mixtures.

The use of flash ADCs provides the opportunity to implement powerful methods to reduce the noise. The noise in adjacent channels with similar strip lengths has similar time structure and magnitude. This allows subtraction of the common noise from the signal using the channels not affected by the hit. By reading the flash ADC samples and applying such a procedure off-line, we were able to achieve a position resolution from the strips as good as 100μ m (Fig.3). Since the avalanches are created near the sense wires, the strips can be used to reconstruct the wire positions. That allows the estimation of the strip resolution. The strips are oriented at 75° with respect to the wires and a geometrical factor was taken into account to estimate the resolution in reconstructing the position along the wires. The position is reconstructed by calculating the center-of-gravity of the signals above a certain threshold induced on at least five adjacent strips. Thus the resolution is dominated by the signal-to-noise ratio. One can see from Fig.3 that the resolution improves by increasing the signal amplitudes (via high voltage), which means that the signal-to-noise ratio increases at the same time.

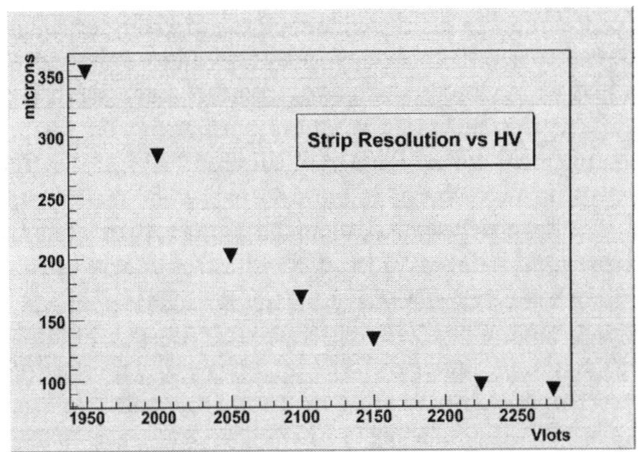

FIGURE 3. Preliminary result for the cathode strip resolution in reconstructing the position along the wires, for different sense wire voltages using cosmics (Ar/CO_2 40/60%, plateau starts at 2000V, field wire at -500V).

To estimate the wire resolution, two chamber layers were placed one on the top of the other, with wires oriented in the same direction. The difference between positions reconstructed from the two chambers was compared with the same difference calculated from the angle of the cosmic track reconstructed with an external tracking system. Because of the small distance between the chambers as compared to the base of the external system, the resolution of the latter does not contribute to the presented results. The estimated wire resolution is plotted at Fig.4 as a function of the distance to the wire in one of the chambers, averaged over the same distance in the other chamber. The measured resolution reaches the design value. However, an additional contribution is expected from the magnetic field effects. Possible improvements of the time-to-distance relation and of the gas mixture are studied using Garfield code [6] and tested experimentally.

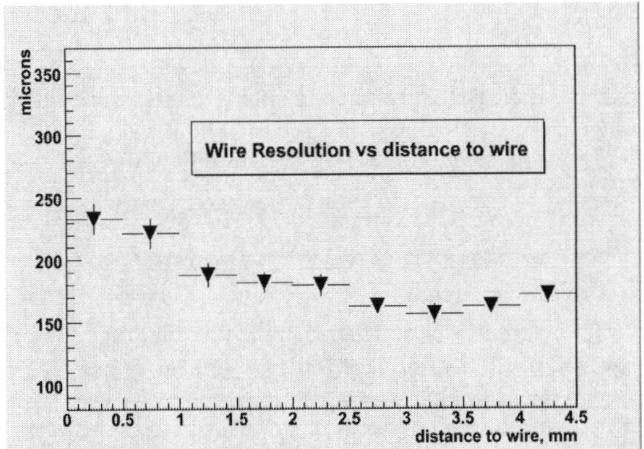

FIGURE 4. Preliminary result for the wire resolution using cosmics (Ar/CO_2 40/60%, 2225/-500V sense/field wire voltage).

We found that the chamber required higher than expected (from previous studies) high voltage to achieve maximum efficiency. Further investigations showed that the plateau starts earlier for hits with smaller drift times. This is illustrated at Fig.5 where the mean wire charge is plotted against the drift time. The wire charge was calculated from the maximum wire amplitude, as recorded with the flash ADC, taking into account the preamplifier gain. Thus, it represents the charge integrated over the peaking time of the circuit, rather than the total charge. The drop of the charge with increasing drift time indicates that the electrons recombine while drifting to the anode wire. Garfield simulations including only a 0.1% oxygen contamination (Fig.5) can explain the above behavior. Indeed, the above percentage was confirmed by measuring the oxygen contamination of the gas in the chamber. The contamination is a result of poor gas tightness due to the fragile G10/Rohacell frames. A mechanical prototype with a solid G10 ring at the position of the gas spacers was constructed. This increased the frame thickness by only a small fraction but at the same time guaranteed the gas tightness of the chamber.

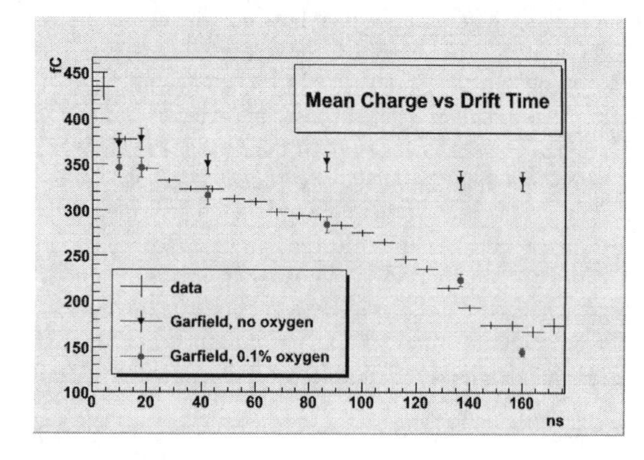

FIGURE 5. Preliminary results: Mean charge, as obtained from the flash ADC maximum amplitude, versus drift time, compared to Garfield [6] simulations without and with 0.1% oxygen contamination (2225/-500V sense/field wire voltage, Ar/CO_2 40/60%).

SUMMARY AND CONCLUSIONS

The GlueX experiment will search for hybrid mesons with exotic quantum numbers not allowed by the quark-antiquark meson model. The Forward Drift Chamber was designed to track charged particles with high resolution needed to identify the decay products of the hybrid mesons. The information from both, the wires and the two cathode strip planes, allows reconstruction of a spatial point within one chamber layer, improving the pattern recognition. The mechanical design reflects the delicate balance between robustness and low detector thickness minimizing the loss of detected photons in the e.m.-calorimeters due to conversions. Low noise to signal ratio was achieved by using novel ASIC preamplifiers and flash ADCs. Studies with a full-scale prototype demonstrated that the design strip and wire resolutions can be achieved. It was found that the oxygen contamination of the gas was responsible for the poorer signals at larger drift distances. The investigations with the prototype helped to improve significantly the FDC mechanical and electrical design. The FDC production will start at the end of 2010.

ACKNOWLEDGMENTS

Authored by Jefferson Science Associates, LLC under U.S. DOE Contract No. DE-AC05-06OR23177. The U.S. Government retains a non-exclusive, paid-up, irrevocable, world-wide license to publish or reproduce this manuscript for U.S. Government purposes. The authors are thankful to all the people from the FDC group of the GlueX collaboration for their significant contribution in the results presented here.

REFERENCES

1. J. J. Dudek, R. G. Edwards, M. J. Peardon, D. G. Richards, and C. E. Thomas, *Phys. Rev. Lett.* **103**, 262001 (2009).
2. V. Crede and C. A. Meyer, *Prog. Part. Nucl. Phys.* **63**, 74 (2004).
3. The GlueX Collaboration, "Mapping the Spectrum of Light Quark Mesons and Gluonic Excitations with Linearly Polarized Photons", GlueX-doc-1226, (January 2006); (http://argus.phys.uregina.ca/cgi-bin/public/DocDB/ShowDocument?docid=1226).
4. N.Dressnandt, N. Doshi, M. Newcomer "GAS II: A versatile wire chamber readout ASIC", IEEE Trans. Nucl. Sci. 2009, NSS, (2009).
5. G. Visser, D. Abbot, F. Barbosa, C. Cuevas, H. Dong, E. Jastrzembski, B. Moffit, B. Raydo, "A 72 Channel 125 MSPS Analog-to-Digital Converter Module for Drift Chamber Readout for the GlueX Detector", to be submitted to IEEE NSS 2010 conference.
6. R.Veenhof, GARFIELD, a drift chamber simulation program, Version 9, CERN (2007).

Large-Area Fast-Timing Systems In STAR

W.J. Llope for the STAR Collaboration

Rice University, Houston, TX, 77005

Abstract. The STAR experiment at RHIC concentrates on the tracking of charged hadrons via ionization, and the detection of electrons and photons via calorimetry, in a wide and azimuthally complete acceptance. STAR's ability to directly identify charged hadrons was initially limited to low momenta. This has been addressed via the construction of a large-area Time-of-Flight (TOF) system based on small Multigap Resistive Plate Chambers (MRPCs). The installation of the STAR TOF system was completed last fall. The full system is ran well in the recent RHIC Run 10. The operation of the system, and its performance for particle identification, during RHIC Runs 9 and 10 will be described. STAR's ability to identify muons is also extremely limited. Another large-area TOF system based on much larger MRPCs is envisioned. This system will be located outside the STAR magnet and is called the Muon Telescope Detector (MTD). Several different prototype MTD systems were operated in Runs 7 through 10, and a patch of near-final MTD detectors is under construction for use in the upcoming Run 11. The performance of the MTD prototype detectors, and the design of the Run 11 installation and the full system, will be described.

Keywords: Particle Identification; Time of flight; Multigap Resistive Plate Chamber, STAR
PACS: 06.60.Jn; 07.50.Ek; 29.40.Mc

INTRODUCTION

The STAR [1] experiment at RHIC [2] has installed a large area Time Of Flight (TOF) system based on Multigap Resistive Plate Chambers (MRPCs). We have also constructed and operated prototypes of another large-area system called the Muon Telescope Detector (MTD). Views of these detectors are shown in Figures 1 and 2. Additional details on both systems are available in Ref. 3.

FIGURE 1. A view of the interior of one of the TOF trays.

FIGURE 2. A view of the interior of the prototype MTD detector used in RHIC Runs 9 and 10.

ACKNOWLEDGMENTS

The author thanks the organizers for the invitation and the interesting program. This work is supported by the U.S. Department of Energy under the grant DE-FG02-10ER41666.

REFERENCES

1. http://www.star.bnl.gov/
2. http://www.bnl.gov/rhic/
3. W.J. Llope for the STAR Collaboration, Nucl. Inst. and Methods in Physics Research A, in press; http://dx.doi.org/10.1016/j.nima.2010.07.086

Application of Accelerators in Research and Industry
AIP Conf. Proc. 1336, 569-569 (2011); doi: 10.1063/1.3586165

Study Of The Scattering Of Halo Nuclei Around The Coulomb Barrier

L. Acosta[a], M. Cubero[b], D. Escrig[b], J.P. Fernández-García[c], J.A. Lay[c], A.M. Moro[c], M. Rodríguez-Gallardo[b,c], A.M. Sánchez-Benítez[a], M. Alcorta[b], M.A.G. Álvarez[c,d], M.V. Andrés[c], C. Angulo[e], M.J.G. Borge[b], L. Buchmann[f], J. Cabrera[e], S. Cherubini[g], M.A. Cortés[c], P. Demaret[e], C.G. Diget[h], A. Di Pietro[g], J.M. Espino[c], P. Figuera[g], L. M. Fraile[i], M. Freer[j], B. Fulton[h], H.O.U. Fynbo[k], D. Galaviz[b,l], J.E. García-Ramos[a], M.E. Gómez[a], J. Gómez-Camacho[c,d], M. Gulino[g], O.R. Kakuee[m], M. Madurga[b], A. Maira-Vidal[b], I. Martel[a], C. Metelko[j], A. Musumarra[g], I. Mukha[c], F. Pérez-Bernal[a], J. Rahighi[m], G.Randisi[g], E. Reillo[b], J. Rodríguez-Quintero[a], K. Rusek[n], V. Scuderi[g], A. Shotter[f], D. Smirnov[o], O. Tengblad[b], P. Van Duppen[o], P. Walden[f], V. Ziman[j]

[a]Dpto. de Física Aplicada. Universidad de Huelva E-21071 Huelva, Spain
[b]Instituto de Estructura de la Materia, CSIC, E-28006 Madrid, Spain
[c]Dpto. de Física Atómica, Molecular y Nuclear, Universidad de Sevilla, E-41080 Sevilla, Spain
[d]Centro Nacional de Aceleradores, E-41092 Sevilla, Spain
[e]Centre de Recherches du Cyclotron, Université Catholique de Louvain, B-1348 Louvain-la-Neuve, Belgium
[f]TRIUMF, V6T2A3 Vancouver B.C., Canada
[g]INFN Laboratori Nazionali del Sud and Sezione di Catania, I-95123 Catania, Italy
[h]Department of Physics, University of York, YO10 5DD Heslington, York, U.K.
[i]Dpto. de Física Atómica, Molecular y Nuclear, Universidad Complutense, E-28040, Madrid, Spain
[j]School of Physics and Astronomy, University of Birmingham, B15 2TT Birmingham, United Kingdom
[k]Institut for Fysik og Astronomi, Aarhus Universitet, Dk-8000 Aarhus C, Denmark
[l]Centro de Fisica Nuclear da Universidade de Lisboa, 1649-003 Lisbon, Portugal
[m]Van der Graaff Laboratory, Nuclear Research Centre, AEOI, PO Box 14155-1339, Tehran, Iran
[n]The Andrzej Soltan Institute for Nuclear Studies, 00-681 Warsaw, Poland
[o]Instituut voor Kern-en Stralingsfysica, University of Leuven, B-3001 Leuven, Belgium

Abstract. During the past ten years the present collaboration has carried out several experiments related with the study of radioactive nuclei. One of the topics in which we have centered our research, is the scattering of halo nuclei at energies around the Coulomb barrier. As part of this study, we present in this work a review of the results obtained from the scattering of ^{6}He, ^{11}Be and ^{11}Li. The presence of a "halo" in these exotic nuclei is found to have a striking effect on the dynamics of these reactions, making their study an interesting experimental problem and a challenge for existing reaction theories.

Keywords: Halo nuclei, elastic scattering, breakup products
PACS: 24.10.Eq, 24.50.+g, 25.45.De, 25.60.Gc.

INTRODUCTION

The increasing number of experiments performed at many facilities around the world has contributed to unveil the properties of exotic nuclei. Among these remarkable nuclei we find the so-called "halo nuclei" that are characterized by a very unusual structure, consisting of two distinct parts: a central core, usually

Application of Accelerators in Research and Industry
AIP Conf. Proc. 1336, 570-572 (2011); doi: 10.1063/1.3586166

composed by a stable nuclei, and a halo of neutrons or protons orbiting at a large distance of the central core [1]. Therefore, a common property of the halo nuclei is their large nuclear radius. Moreover, the halo nucleons are weakly bound to the core, giving rise to a large dissociation probability in reactions with other nuclei, even at energies below the Coulomb barrier. In this work we will show part of our results related with the study of the dynamics of the halo nuclei ^{6}He, ^{11}Be and ^{11}Li by heavy targets at Coulomb barrier energies. All of them are neutron halo nuclei and, in the case of ^{11}Li and ^{6}He, they are examples of Borromean nuclei [2] (*i.e.* three-body systems where the two-body sub-systems are unbound). Regarding ^{11}Be, it is a very weakly-bound one-neutron halo nucleus with a deformed ^{10}Be core.

EXPERIMENTAL SETUPS

The setups for the referred experiments, present some similarities and differences. A common feature is the kind of detectors used. In general we used strip detectors with different shapes, according with the requirements of each measurement. To obtain mass and charge separation, we normally built telescopes in order to employ the energy loss technique. Using small telescopes we can ensemble complex geometries to cover specific angular ranges, to finally obtain differential cross sections (*i.e.* elastic, breakup, etc) for each beam and energy [3].

SPECTROSCOPIC ANALYSIS

Once the data related with each experiment are stored, we start a spectroscopic analysis, normally generating spectra of one and two dimensions, where the important variable is the energy signal registered in the components of each telescope. If we plot the energy registered in the ΔE-detector as a function of the total energy (i.e. $E+\Delta E$) we can obtain spectra which allow the separation of the different isotopes produced from the projectile-target interaction. Once the reaction channels of interest are identified, we proceed to isolate and integrate them, in order to obtain the experimental associated cross sections.

RELEVANT RESULTS

We have performed two experiments at the CRC-Louvain-la-Neuve facility in which the elastic scattering of ^{6}He by ^{208}Pb was measured, for different energies around the barrier [4,5], as well as the ^{4}He products coming from this reaction [6]. In Fig. 1a, we show the measured elastic cross section (E_{lab} = 22 MeV) along with theoretical calculations, namely, an optical model fit of the data, and three-body and four-body CDCC calculations [7,8].

In the case of the ^{11}Be, we carried out one experiment using a beam at 32 MeV at the REX-ISOLDE-CERN facility, Geneva [9], obtaining experimental results about the quasi-elastic scattering on a ^{120}Sn target. At the same time, we registered ^{10}Be as reaction fragment produced from the interaction of the ^{11}Be projectiles with the target. The quasi-elastic measurements are compared in Fig. 1b, with coupled-channel and CDCC calculations. More results from this campaign of experiments can be consulted in Ref. [10].

In a more recent experiment performed at TRIUMF Vancouver facility, we measured the scattering of ^{9,11}Li on ^{208}Pb at two different energies, again at Coulomb barrier. The analysis of the obtained data is in progress. Some preliminary results can be found in Fig. 1c, where also are included theoretical interpretations based in a semi-classical model and CDCC calculations [5,7].

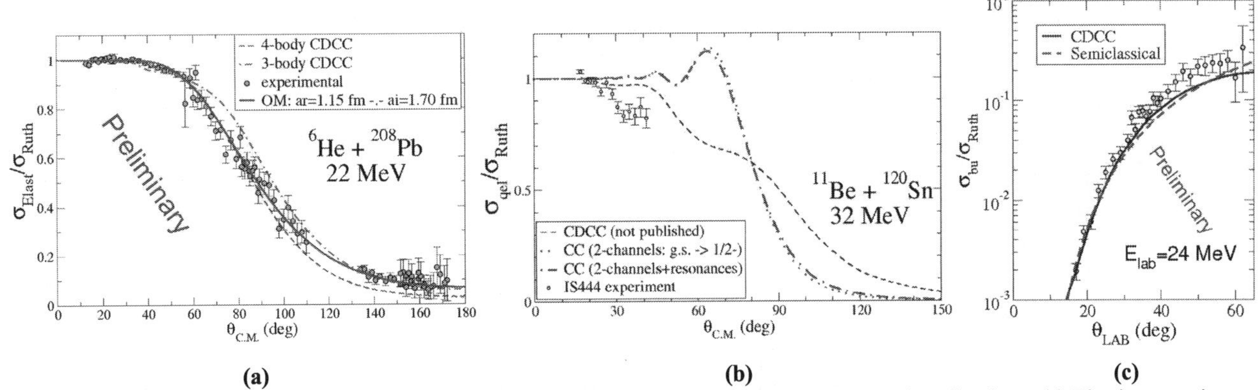

FIGURE 1. Some results from the reaction channels studied and theoretical calculations to describe them. (a) Elastic scattering of ^{6}He on ^{208}Pb at 22 MeV, results from [5] forward and backward angles and "preliminary" results from PH215 Experiment at angles around 90° [3]. (b) Quasi-elastic scattering from ^{11}Be on ^{120}Sn at 32 MeV taken from [9]. (c) Preliminary results from the breakup probability of ^{11}Li+^{208}Pb reaction at 24 MeV.

CONCLUSIONS

From this extended study about the scattering of halo nuclei around the Coulomb barrier it is relevant to mention that, the behavior of the cross section is remarkably different from that of stable nuclei. The elastic angular distribution departs significantly from the Fresnel-like behavior which characterizes the scattering of heavy ions at Coulomb-barrier energies.

Furthermore, comparing our results with the typical Rutherford pattern, we could find a lot of differences starting in the region of the nuclear rainbow [11]. In the case of ^{6}He at 22 MeV where we have a wide angular distribution, it is easy to observe the discrepancies between halo and stable nuclei.

For the ^{11}Be data, the absorption process [12] appears even at very forward angles, and presently the collaboration is investigating its origin.

Finally, the first results obtained from the analysis of ^{11}Li show a strong probability of breakup reaction, which appears even at very forward angles.

ACKNOWLEDGMENTS

This work has been supported by the Spanish MCyT projects FPA2005-04460, FPA2005-02379, FPA-2000-1592-C03-02, FPA2003-05958, FPA2002-04181-C04-02/03, FPA2006-13807-c02-0, FPA2009-07653, by the Spanish Consolider-Ingenio 2010 Programme CPAN (CSD2007-00042), by the contract number HPRI-CT-1999-00110 and the Belgian program P5/07 of the Belgian-state Federal Services for Scientific, Technical and Cultural Affairs. L. Acosta wants to thank to the CPAN-postdoctoral program for the support granted to realize this work.

REFERENCES

1. P. Hansen, B. Jonson, *Europhys. Lett* **4**, 409 (1987).
2. K. Katō, S. Aoyama, S. Mukai, K. Ikeda, *Nucl. Phys. A* **558**, 29c-4c (1995).
3. L. Acosta, "Estudio de las reacciones inducidas por núcleos halo a energías cercanas a la barrera de Coulomb", Ph.D. Thesis, University of Huelva, 2009. http://hdl.handle.net/10272/2794 ISBN: 978-84-92944-62-0 (2009).
4. O. Kakuee, et. al., *Nucl. Phys. A* **339** 728 (2003).
5. A. Sánchez-Benítez, et. al., *Nucl. Phys. A* **803** 30-45 (2008).
6. D. Escrig, et. al., *Nucl. Phys. A* **792** 02-017, (2007).
7. K. Rusek, I. Martel, J. Gómez-Camacho, A. Moro, R. Raabe, *Phys. Rev. C* **72**. 037603 1-4 (2005).
8. M. Rodríguez-Gallardo, et. al., *Phys. Rev. C* 77 (064609) 1-9 (2008).
9. L. Acosta, et. al., *Eur Phys. J. A* **42** 461-464 (2009).
10. A. Di Pietro, et al., *Phys. Rev. Lett.* (Accepted)
11. N. Keeley, et. al., *Nucl. Phys. A* **571** 326, (1994).
12. E.F. Aguilera, et. al., *Phys. Rev. C* **63** (061603(R)) 1-4 (2001).

Investigation Of Exotic Shapes, Correlations And Isomers In Nuclei With Large Compton Suppressed Clover Array

R. Palit

Tata Institute of Fundamental Research, Mumbai – 400 005, India

Abstract. An array is designed for 24 Compton suppressed clover detectors. Recent results from the array will be presented which highlights the structure of nuclei in A ~ 110 and 130 region and investigation of different reactions dynamics. The perspectives offered by the forthcoming operation of the clover array coupled to a new digital data acquisition system are also discussed briefly.

Keywords: Yrast and near yrast gamma ray spectroscopy, INGA, TIFR digital data acquisition
PACS: 21.10.Hw; 21.60.-n; 21.60.Jz

INTRODUCTION

The Discrete gamma ray spectroscopic studies of nuclei continue to provide new insight concerning the simple excitation modes of nuclei and various types of coupling between these modes. World-wide there are enormous efforts going on in studying these modes of nuclei at different regions of the spin-excitation diagram across the landscape using different types of reactions and new detector facilities. A large array of Compton suppressed clover detectors (INGA) is operational which moves between the three accelerator facilities within India and used for investigation of nuclear structure and reaction dynamics. INGA is a collaborative project between TIFR, IUAC, IUC-DAE, SINP, BARC, VECC and Indian Universities. The configuration of the clover array coupled with the new digital data acquisition system for the upcoming INGA campaign will be presented along with description of the future possibilities in nuclear structure and reaction studies. Results from our previous campaigns on medium and high spin states of nuclei in A ~ 110 and 130 region populated through different nuclear reactions will be presented.

Implementation Of DDAQ For The Clover Array

The array is designed for 24 Compton suppressed clover detectors arranged in a spherical geometry as shown in Fig. 1 with 6 detectors at 90° and 3 detectors each at 23°, 40°, 65°, 115°, 140° and 157° with respect to the beam direction. The distance from the target to

FIGURE 1. . (Colour online) Picture of the Indian National Gamma Array with the Compton suppressed clover detectors in Hall-2 of the TIFR-BARC Pelletron-LINAC facility at Mumbai.

crystal is 25 cm and the overall photo-peak efficiency is around 5% at Eγ ~ 1 MeV. A PCI-PXI based digital data acquisition system with 112 channels has been implemented for this array using Pixie-16 module by XIA LLC. the DDAQ is to provide good energy and timing resolution for all 96 channels of clovers vetoed with the BGO shields operating at 12 kHz count rate for each crystal. It can operate both in trigger-less as well as multi-fold coincidence mode. It has also the capabilities to couple other ancillary detectors like

Application of Accelerators in Research and Industry
AIP Conf. Proc. 1336, 573-575 (2011); doi: 10.1063/1.3586167

NaI(Tl) gamma ray multiplicity filters and CsI(Tl)/Si charged particle detectors with the clover array. One and two fold data has been acquired with different radioactive sources for testing of this system for coincidence measurements, energy and timing resolution studies for the clovers. A peak-to-total ratio of ~ 40% has been achieved with the Compton suppression for ^{60}Co source in the add-back mode. FWHM of 2.2keV was obtained for 1.33MeV gamma-ray energy for the individual crystals of the clovers at 20kHz count rate.

Dipole Bands In Odd-Odd Indium Isotopes

Regular sequences of M1 transitions are observed for 56<N<62 in A ~ 110 region, but gradually disappear with decreasing neutron number. Doublet dipole bands are known in odd-odd Rh and Ag isotopes, not much is known in In isotopes [1]. Tri-axial relativistic mean field calculations indicate multiple chiral rotations in A ~ 110 region. In order to understand the nature of the magnetic dipole bands and search for chiral rotations, a systematic investigation of odd-odd In isotopes are carried out. The excited states of ^{112}In were populated through ^{100}Mo(^{16}O,p3n) reaction at 85 MeV beam energy and gamma rays were measured with 18 CS-clover detector array. Polarization and lifetime measurements have been carried out for the excited states of ^{112}In. The B(M1) transition strengths in the strongly populated dipole band with gamma ray sequence of 128 – 178 – 272 – 393 – 554 – 708 – 738 keV are measured with Doppler Shift Attenuation Method (DSAM) technique and decrease with increasing spin [2]. The typical lineshape observed for 393 keV transition is shown in Fig 2. A quasi-particle configuration $\pi g_{9/2} \otimes \nu((h_{11/2})^2 d_{5/2}/g_{7/2})$ is used in the tilted axis cranking (TAC) calculation for the dipole band. A minimum is found at deformation of ε_2=0.12 and γ=6°. The extracted B(M1) strengths show a decreasing trend with increasing spin and thereby confirms the shears mechanism proposed for generation of angular momentum in this band [3]..

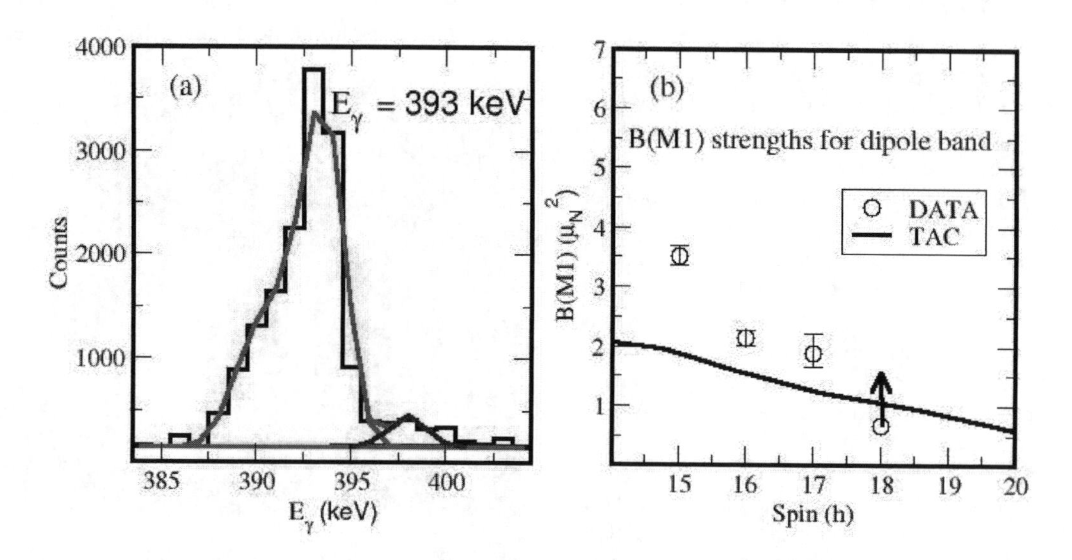

FIGURE 2. (Color online) (a) Lineshape of 393 keV transition observed at 148° detector and (b) Plot of B(M1) vs spin for B3 band of ^{112}In along with TAC calculation.

Medium And High Spin States In Nuclei With A ~ 130

Nuclei in the A ~ 130 region show a rich variety of phenomena at medium and high spin. A series of experiments have been carried out to study the Xe, Cs, Ba, Ce and Nd isotopes below the N = 82 shell closure. Low-Ω $h_{11/2}$ protons and high-Ω $h_{11/2}$ neutrons develop gamma instability in these nuclei. Low lying structure of N = 79 isotones of Ba and Nd indicate a triaxial deformation near their ground state based on the comparison of particle rotor model calculations [4]. Neutron deficient isotopes near Z = 54 are predicted to show softness towards octupole deformations at low spin. The population of the $\pi g_{7/2}$ and $\pi d_{5/2}$ bands relative to $\pi h_{11/2}$ increases for heavier odd-Cs isotopes. Due to occupation of protons in lowest Ω orbitals, the wave functions are highly localized and octupole overlap matrix elements got enhanced. This causes E1 transitions between $\pi d_{5/2}$ and $\pi h_{11/2}$ bands. From the ratio of intensities of E1,

E2 transitions and deformation of ^{130}Xe, the B(E1) values of ^{131}Cs has been found to be ~ 6.0 x 10^{-5} W.u. indicating an enhancement of octupole correlations [5] similar to that observed in neutron rich ^{141,143}Cs, ^{143,145}Ba isotopes, and neutron deficient ^{117}Xe and ^{125}Ba nuclei.

FUTURE PERSPECTIVE

Coupling of the digital DAQ with the clover array will give the opportunity to handle optimum count rate for gamma-gamma coincidence data. This will facilitate the investigation of high spin states to probe wide variety of phenomena like anti-magnetic rotations, chiral rotations, chopstick mode and tetrahedral shape in different nuclei. In addition, a moderate array of charged particle detectors are planned to be coupled with the Clover array. This ancillary detector array anlong with the new digital DAQ will improve the capability of the clover array for the channel selection in fusion reactions for high spin spectroscopy, search of isomers and reaction dynamics study in case of deuteron, tritium and alpha breakup of the weakly bound nuclei [6].

ACKNOWLEDGMENTS

I'd like to thank my colleagues A.Y. Deo, Z. Naik, T. Trivedi, V.V. Parkar, S. Sihotra, S. Kumar, S. Sharma, S. Saha, J. Sethi, B. S. Naidu, P.B. Chavan, S. Jadav, P. K. Joshi, P. Verma, S. Sinha, D. Mehta, A.K. Jain, R.K. Bhowmik, A.K. Sinha, S. Bhattacharya, H.C. Jain, and all the members of Indian National Gamma Array (INGA) collaboration.

REFERENCES

1. J . Meng, et al., Phys. Rev. C73, 037303 (2006).; T. Koike, *Nucl. Phys. A 834, 36c (2010)*
2. T.K. Alexander, J.S. Forster, Adv. Nucl. Phys. 10, 197 (1978).
3. A.Y. Deo, et al., Phys. Rev. C 79 067304 (2009); C.-Y., et al, Nucl. Phys. A834, 84c (2010).
4. S. Kumar, et al., Phys. Rev. C 78 014306 (2008).
5. S. Sihotra, et al., Phys. Rev. C 78, 034313 (2008).
6. V. Parkar, et al., *Submitted to Phys. Rev.C.*

Radioactive Ion Beam Production Capabilities At The Holifield Radioactive Ion Beam Facility

J. R. Beene[a], D. T. Dowling[a], C. J. Gross[a], R. C. Juras[a], Y. Liu[a], M. J. Meigs[a], A. J. Mendez II[a], W. Nazarewicz[a], J. W. Sinclair[a], D. W. Stracener[a], B. A. Tatum[a]

[a]*Oak Ridge National Laboratory, Oak Ridge, TN 37831*

Abstract. The Holifield Radioactive Ion Beam Facility (HRIBF) is a national user facility for research with radioactive ion beams (RIBs) that has been in routine operation since 1996. It is located at Oak Ridge National Laboratory (ORNL) and operated by the ORNL Physics Division. The principal mission of the HRIBF is the production of high quality beams of short-lived radioactive isotopes to support research in nuclear structure physics and nuclear astrophysics. HRIBF is currently unique worldwide in its ability to provide neutron-rich fission fragment beams post-accelerated to energies above the Coulomb barrier for nuclear reactions. HRIBF produces RIBs by the isotope separator on-line (ISOL) technique using a particle accelerator system that consists of the Oak Ridge Isochronous Cyclotron (ORIC) driver accelerator, one of the two Injectors for Radioactive Ion Species (IRIS1 or IRIS2) production systems, and the 25-MV tandem electrostatic accelerator that is used for RIB post-acceleration. ORIC provides a light ion beam (proton, deuteron, or alpha) which is directed onto a thick target mounted in a target-ion source (TIS) assembly located on IRIS1 or IRIS2. Radioactive atoms that diffuse from the target material are ionized, accelerated, mass selected, and transported to the tandem accelerator where they are further accelerated to energies suitable for nuclear physics research. RIBs are transported through a beam line system to various experimental end stations including the Recoil Mass Spectrometer (RMS) for nuclear structure research, and the Daresbury Recoil Separator (DRS) for nuclear astrophysics research. HRIBF also includes two off-line ion source test facilities, one low-power on-line ISOL test facility (OLTF), and one high-power on-line ISOL test facility (HPTL). This paper provides an overview and status update of HRIBF, describes the recently completed \$4.7M IRIS2 addition and incorporation of laser systems for beam production and purification, and discusses a proposed replacement of the ORIC driver accelerator.

Keywords: HRIBF, RIB, ISOL, ORNL, accelerator
PACS: Radioactive beams, 29.38.-c

INTRODUCTION

The Holifield Radioactive Ion Beam Facility (HRIBF) is a national user facility, funded by the DOE Office of Nuclear Physics, for research with radioactive ion beams (RIBs). It is located at Oak Ridge National Laboratory (ORNL) and operated by the ORNL Physics Division. The principal mission of HRIBF is the production of high quality beams of short-lived radioactive isotopes to support research in nuclear structure physics and nuclear astrophysics for an international RIB science community. HRIBF provides world-class research tools for the more than 200 researchers that utilize HRIBF beams each year, and the facility's users group now has 570 members. The facility presently operates on a 5 day 24 hour schedule with approximately 4000 to 4500 total research hours per year. HRIBF is currently unique worldwide in its ability to provide neutron-rich fission fragment beams post-accelerated to energies above the Coulomb barrier for nuclear reactions. Scientists and engineers at HRIBF are internationally recognized for developing new technologies and pioneering techniques in RIB production and purification. Other key objectives of HRIBF are to establish and maintain an international user base for RIB science, and we apply our core capabilities in nuclear science to problems of relevance to society.

HRIBF produces RIBs by the isotope separator on-line (ISOL) technique (Figure 1.1) using a particle accelerator system that consists of the following primary components: the Oak Ridge Isochronous Cyclotron (ORIC) driver accelerator [1], one of the two Injectors for Radioactive Ion Species (IRIS1 [2] and IRIS2) production systems, and the 25-MV

Application of Accelerators in Research and Industry
AIP Conf. Proc. 1336, 576-580 (2011); doi: 10.1063/1.3586168

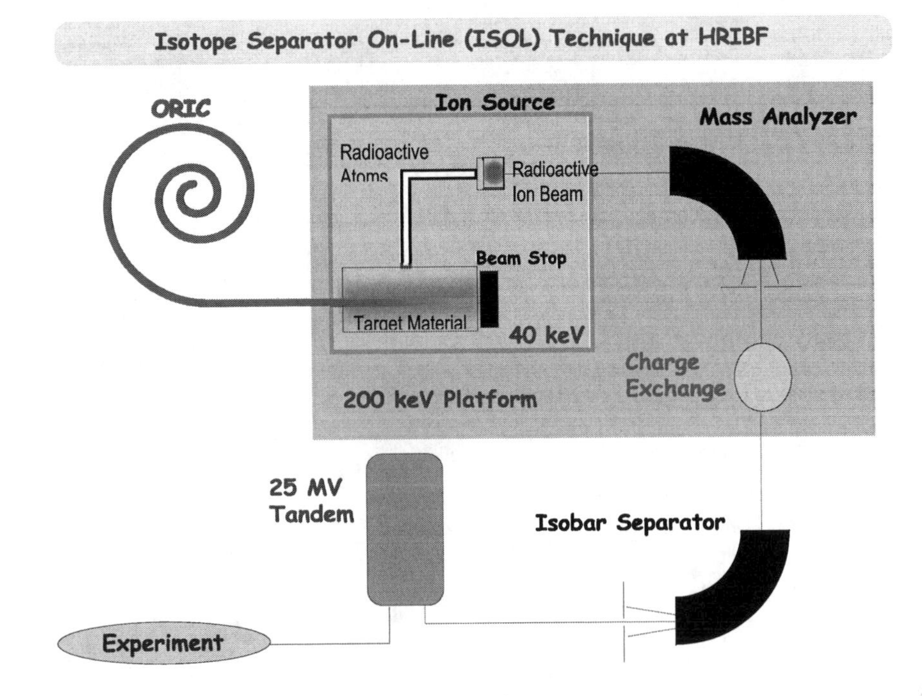

FIGURE 1.1 HRIBF RIB Production Process

tandem electrostatic accelerator [3] that is used for RIB post-acceleration.

The ORIC provides a light ion beam (proton, deuteron, or alpha), which is directed onto a thick target mounted in a target-ion source (TIS) assembly located on IRIS1 or IRIS2. Radioactive atoms that diffuse from the target material are ionized, accelerated, mass selected, and transported to the tandem accelerator where they are further accelerated to energies suitable for nuclear physics research. RIBs are transported through a beam line system to various experimental end stations including the Recoil Mass Spectrometer (RMS) [4] for nuclear structure research, and the Daresbury Recoil Separator (DRS) [5] for nuclear astrophysics research. Figure 1.2 provides a three-dimensional view of the facility showing major component locations. HRIBF can presently provide more than 175 post-accelerated RIB species (32 proton-rich and 143 neutron-rich species) plus 26 additional non-post-accelerated RIBs [6].

The HRIBF staff strives for continual improvement in capabilities, reliability, and operational efficiency of the facility. Some fifteen years ago when HRIBF came on line, the need for on-going facility improvements was identified and a strategic upgrade plan was developed [7]. The first two phases of an accelerator/RIB production system enhancement have been completed: the High Power Target Laboratory (HPTL) and the Injector for Radioactive Ion Species 2 (IRIS2). In parallel, tremendous accomplishments have been realized in the area of ISOL development [8]. The remainder of this paper highlights the two major upgrades that have been completed, describes plans for a third, and discusses some important research in radioactive beam purification.

High Power Target Laboratory (HPTL)

The most challenging aspect of operating a RIB facility typically involves the RIB target and ion source production system. It is crucial for an ISOL facility to have a location for on-line development of new targets, ion sources, and beam production and purification techniques, as well as a means to systematically improve the robustness of ion source components. The latter is vital because equipment used for RIB production necessarily operates in a harsh environment consisting of high voltage (up to

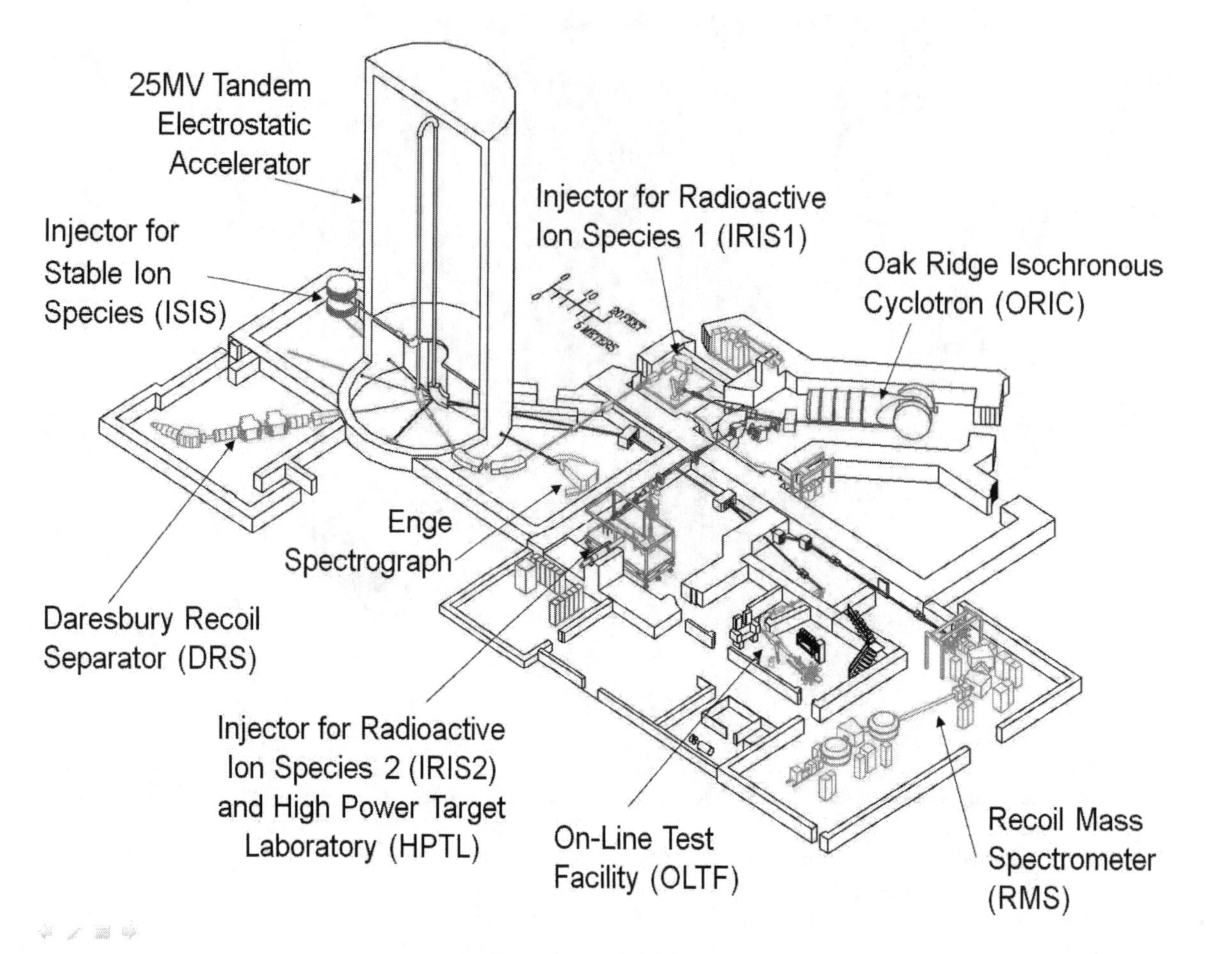

FIGURE 1.2 HRIBF Configuration

300 kV dc), high temperatures (2000 C), and severe radiological conditions, all of which can lead to component failure. During the first decade of HRIBF operation, the IRIS1 production system had to also be used for high power development activities, thus creating some risk to that system as well as causing delays between campaigns for the experimental physics program. To address this problem, a proposal was submitted to DOE-NP in January 2003 for the High Power Target Laboratory (HPTL), an upgrade project to enable us to carry out an aggressive program of R&D at full driver-beam power, without impacting our single on-line RIB production station. This $4.75M project was funded from 2003-2005. The project was completed early in 2006 and the HPTL has been in routine operation since that time. Optimum development time is 300-400 hours per year. Key features of the HPTL Project were the construction of a new, heavily shielded production vault within the existing building, a lightly shielded instrumentation room for housing controls, electronics, and utilities, a new ORIC beamline, and a high voltage platform structure with a target station and low energy analysis beamline. [9]

Injector For Radioactive Ion Species 2 (IRIS2)

The second phase of the HRIBF upgrade strategy was to construct a second RIB production system to provide redundancy, scheduling flexibility, and a more versatile configuration that will allow for the implementation of new production and purification techniques. The Injector for Radioactive Ion Species 2 (IRIS2) Project is a $4.735M upgrade funded from 2006 through 2009, with project completion in 2010. This system is co-located with the HPTL to make optimum use of the collective investment and minimize civil construction. An extension was added to the HPTL production platform system to accommodate a beam line that includes a nominal $\Delta m/m=1/1000$ first stage mass analysis system and a location for either a cesium-vapor charge exchange cell or negative-ion beam cooler. The mass-selected RIB is then transported through a new beamline that was constructed from the IRIS2 platform to the existing isobar separator beamline associated with IRIS1. Thus isobar-separated RIBs from either IRIS1 or IRIS2 can now be post-accelerated by the tandem, or delivered to the recently implemented Low Energy Radioactive Ion Beam (Decay) Spectroscopy Station (LeRIBSS) [10].

Driver Accelerator Upgrade

The third phase of the HRIBF upgrade strategy is to improve the performance and reliability of the driver accelerator. Replacement of the 50-year-old ORIC with a commercial accelerator offers the greatest return on investment when considering the enhancement to the RIB production capabilities, overall facility reliability, and operating cost. This combination would greatly increase the scientific reach of the facility, allowing the facility to remain competitive on an international basis and accommodate a larger user community. Several alternatives have been considered, including an electron driver to produce RIBs by photo-fission in an actinide target [11]. However, the preferred alternative, strongly supported by the HRIBF user community [12], is to replace ORIC with a high-power, dual-extraction-port 70 MeV commercial cyclotron which would substantially increase the production rate of fission fragments, extend existing proton-rich capabilities, and provide the capability to produce long-lived isotopes for medical research [13].

Although replacement of ORIC is desirable, the earliest that the new accelerator could be available is 5-7 years. Consequently ORIC refurbishments are in progress or planned to ensure its reliability for another 5-10 years. Planned upgrades to ORIC are funded through the HRIBF base capital and AIP funds, as well as a dedicated $2.5M ARRA project. The original ORIC trimming coil assemblies, containing ten concentric coils of which three have failed over the years due to un-repairable water leaks, are presently being replaced to restore the original capabilities. Replacement of the 1938-vintage main field motor-generator set with a dc power supply, upgrades to the rf system, and replacement of vacuum pumps are also high priority tasks.

Laser Systems

Laser-based beam purification techniques are being developed at HRIBF for the generation of isobarically pure RIBs: the production of positively-charged RIBs using a Resonant Ionization Laser Ion Source (RILIS), and purification of negative RIBs by photo-detachment using a gas filled RFQ ion beam cooler.

In the RILIS, atomic species can be selectively ionized by laser radiation via stepwise resonant excitation followed by ionization in the final transition. Because each element has unique atomic energy levels, RILIS is highly selective and can be used for many elements. RILIS may be required for a number of elements that are of particular interest for HRIBF research programs, including Ni, Sr, As, Se, Sb, and Te. The beams of short-lived isotopes of these elements have large isobaric contaminations, and no effective purification method is currently available at HRIBF. RILIS is of particular use in purifying the most neutron-rich beam of a given mass. This is usually both the isotope of greatest physics interest and the isotope with lowest reaction yield. A RILIS has been developed for the HRIBF research program [14] consisting of a hot-cavity ion source and three all-solid-state, pulsed Ti:Sapphire lasers to provide up to three-step, three-photon ionizations. In off-line studies in collaboration with the University of Mainz, resonant ionization schemes suitable for Ti:Sapphire lasers have been established for more than 10 elements, with ionization efficiencies ranging from about 1 % to 40% [15]. The Ti:Sapphire laser system has recently been tested at the HRIBF off-line ion source test facility to study resonant ionization of three new elements, Sr, Sb, and Se. Three-photon resonant ionization of Sr was obtained and new ionization schemes for Sr, using Rydberg states in the final transition, were identified. However, no photo-ionization of Sb and Se was observed. It was due to low fourth harmonic power (20 mW at 215 nm) for the first excitation step and no spectroscopic data for the second and final transitions. After the initial tests, a number of improvements on the spatial and spectral qualities of the Ti:Sapphire lasers were identified. Currently, the laser system is being upgraded with suggested improvements, including individually pumping each Ti:Sapphire laser. We expect to receive the improved laser system by the end of this year. The new Ti:Sapphire laser system will be installed in a cially designed laser room for IRIS2, while the hot-cavity ion source will be mounted on the target and ion source platform. The Ti:Sapphire laser beams will be carefully collimated and transported into the target and ion source room, and then focused into the hot-cavity ion source through the viewport on the mass analysis magnet. The total laser beam path is more than 20 m. This RIB production technique cannot be implemented at IRIS1 due to space constraints.

Another beam purification technique developed at HRIBF is selective photo-detachment of negative ions. Because the HRIBF tandem post-accelerator requires negatively charged ions as input, a highly efficient method for selectively suppressing the isobaric contaminant in certain negative RIBs by photo-detachment has been developed [16]. In this method, a laser beam having the appropriate photon energy is used to selectively neutralize the contaminant in a gas-filled radio frequency quadrupole (RFQ) ion cooler, if the electron affinity (EA) of the contaminant is smaller than that of the RIB of interest. In off-line experiments with ion beams of stable isotopes, more than 99.9% suppression of Co^-, S^-, and O^- has been demonstrated, while under identical conditions little or no reduction in the corresponding desired beams of Ni^-, Cl^-, and F^- ions was observed [17]. This highly efficient photo-detachment technique will be implemented at IRIS2 for on-line purification of a number of interesting RIBs. A beam of particular interest is ^{56}Ni. The ^{56}Ni beam produced at HRIBF is dominated by about an order of magnitude more abundant ^{56}Co contaminant. A compact Nd:YAG laser at 1064 nm has been purchased and will be used to purify the ^{56}Ni beam. The energy of the 1064 nm photon (1.165 eV) is much larger than the EA of Co (0.661 eV), but only slightly above the EA of Ni (1.156 eV). Our recent off-line studies [18] have shown that with less than 3 W 1064 nm laser power, 99.99% of the Co^- ions can be depleted while under the same conditions, only about 20% reduction in the Ni^- ions is expected. The Nd:YAG laser will also be located in the IRIS2 laser room, while the RF quadrupole ion cooler will be in the target and ion source room, mounted after the mass analysis magnet.

SUMMARY

HRIBF continues to make important contributions to the international RIB science community in beam development and purification techniques, as well as the experimental physics program, and the recent additions of the HPTL and IRIS2 have enhanced facility capabilities and performance. Implementation of laser systems at IRIS2 will lead to the availability of new, high-purity beams for the experimental program, and the proposed replacement of ORIC with a high-power commercial cyclotron would increase facility reliability and RIB yields.

ACKNOWLEDGMENTS

Research sponsored by the Office of Nuclear Physics, U.S. Department of Energy.

REFERENCES

1. B. A. Tatum, D. T. Dowling, J. R. Beene, in "Proceedings of the 16th International Conference on Cyclotrons and their Applications 2001," East, Lansing, MI, AIP Conference Proceedings 600, 1996, pp. 148.
2. D. T. Dowling et al., in "IEEE Conference Proceedings of the 1995 Particle Accelerator Conference," Dallas, TX, Volume 3, 1996, pp. 1897.
3. M. J. Meigs, D. L. Haynes, C. M. Jones, and R. C. Juras, *Nucl. Instr. and Meth.* A **382**, 1996, pp. 51 (1996).
4. C. J. Gross et al., *Nucl. Instr. and Meth.* A **450**, 2000, pp. 12.
5. A. N. James et al., *Nucl. Instr. and Meth.* A **267**, 1988, pp. 144.
6. Tables of available RIB intensities at HRIBF website: phy.ornl.gov/hribf/beams.
7. B. A. Tatum and J. R. Beene, in "IEEE Conference Proceedings of the 2005 Particle Accelerator Conference," Knoxville, TN, p. 3641 (2005).
8. D. W. Stracener et al., *Nucl. Instr. and Meth.* A **521**, 2004, pp. 126.
9. B. A. Tatum, Nucl. Instr. and Meth. B **241**, 2005, pp. 926.
10. K. P. Rykaczewski, Feb. 2009. HRIBF Newsltr. Ed. 17, No. 1 at www.phy.ornl.gov/hribf/news/ feb-09/leribss.shtml.
11. W. T. Diamond, *Nucl. Instr. and Meth.* A **432**, 1999, pp. 471.
12. Report of the HRIBF Users Workshop, found at www.phy.ornl.gov/workshops/users09.
13. J. R. Beene, Feb. 2009 HRIBF Newsletter, Ed. 17, No. 1 at www.phy.ornl.gov/hribf/news/feb-09/feb09.shtml.
14. Y. Liu, C. Baktash, et al., *Nucl. Instr. and Meth.* B **243**, 2006, pp. 442.
15. Y. Liu, J. R. Beene, T. Gottwald, C. C. Havener, C. Mattolat, J. Lassen, K. Wendt, and C. R. Vane, AIP Conference Proceedings, **1099**, 2009, pp. 141.
16. Y. Liu, J. R. Beene, C. C. Havener and J. F. Liang, *Appl. Phys. Letters* **87**, 2005, pp. 113504.
17. Y. Liu, C. C. Havener, T. L. Lewis, A. Galindo-Uribarri, J. R. Beene, AIP Conf. Proc. **1097**, 2009, pp. 431.
18. P. Andersson, A. O. Lindahl, D. Hanstorp, C. C. Havener, Yun Liu, Yuan Liu, J. Appl. Phys. **107**, 2010, pp. 026102.

Neutron Resonance Spin Determination Using Multi-Segmented Detector DANCE

B.Baramsai [a], G. E. Mitchell [a], U. Agvaanluvsan [c], F. Becvar [d], T. A. Bredeweg [b], A. Couture [b], A. Chyzh [a], D. Dashdorj [a], R. C. Haight [b], M. Jandel [b], A. L. Keksis [b], M. Krticka [d], J. M. O'Donnell [b], R. S. Rundberg [b], J. L. Ullmann [b], D. J. Vieira [b], C. Walker [a], J. M. Wouters [b]

[a] *North Carolina State University, Raleigh, NC 27695; TUNL, Durham, NC 27708*
[b] *Los Alamos National Laboratory, PO Box 1663, Los Alamos, NM 87545*
[c] *Stanford University, Palo Alto, CA 94305*
[d] *Charles University in Prague, CZ-180 00 Prague 8, Czech Republic*

Abstract. A sensitive method to determine the spin of neutron resonances is introduced based on the statistical pattern recognition technique. The new method was used to assign the spins of *s*-wave resonances in ^{155}Gd. The experimental neutron capture data for these nuclei were measured with the DANCE (Detector for Advanced Neutron Capture Experiment) calorimeter at the Los Alamos Neutron Science Center. The highly segmented calorimeter provided detailed multiplicity distributions of the capture γ-rays. Using this information, the spins of the neutron capture resonances were determined. With these new spin assignments, level spacings are determined separately for *s*-wave resonances with $J^{\pi} = 1^{-}$ and 2^{-}.

Keywords: neutron capture, resonance spin, γ-ray multiplicity, multiplicity distribution, level density
PACS: 25.40.Ny, 24.30.-v, 21.10.Hw, 21.10.Ma

INTRODUCTION

The γ-ray multiplicity distributions have been used for many years to determine the spins of resonances observed in the (n,γ) reaction. In favorable cases the average multiplicity for different spins is sufficient to determine the spin of the capturing state. (We initially limit the discussion to *s*-wave capture on targets with spin, i.e., to the two-spin problem.) For the DANCE array [1] this is illustrated by work by Sheets *et al.* [2]. If one assumes that resonances with the same spin have identical multiplicity distributions, then by normalizing the cross sections for any two multiplicities (using a strong resonance with a known spin for the normalization), then the difference in the cross sections will remove the resonances with the same spin. There are many multiplicity combinations and this method works reasonably well in practice [3]. This approach is a simplified version of a method in the same spirit that was developed earlier by Koehler [4]. Another approach formulated by Becvar et al. [5] involves starting from prototype resonances with known spin and decomposing the cross section into separate cross sections for the two spins.

All of these methods work reasonably well, but are limited in practice to the two-spin case and do not provide any quantitative measure of the confidence in the spin assignment. We therefore sought to develop a method that would remove these limitations. After a very brief description of the experimental system, we describe below an analysis method that resolves these issues.

DESCRIPTION OF THE METHOD

The DANCE array consists of 160 BaF_2 scintillation crystals surrounding a target with ~4π solid angle. A γ-ray emitted from the compound nucleus partially deposits its energy in several neighboring crystals; we define these as a "cluster". In a typical capture reaction, the excited compound nucleus reaches its ground state after several successive g-decays and creates several clusters; the number of clusters is called the "cluster multiplicity".

Application of Accelerators in Research and Industry
AIP Conf. Proc. 1336, 581-585 (2011); doi: 10.1063/1.3586169

As shown by Monte-Carlo simulation [6], the "cluster multiplicity" is a good representation of the true "γ-ray multiplicity".

The capture events registered by the DANCE array can be sorted by their cluster multiplicity. We denote the experimental yield with cluster multiplicity m at neutron energy E as $Y_m(E)$. The "multiplicity distribution" can be derived by normalizing the $Y_m(E)$ to the total number of counts:

$$y(E) = [y_1(E)\, y_2(E) ..., y_{max}(E)] \qquad (1)$$

where $y_m(E) = Y_m(E)/Y_{total}(E)$ is a normalized yield, $Y_{total}(E) = \sum_{m=1}^{max} Y_m(E)$ is the total counts. Note that the "multiplicity" in the following text will always be the "cluster multiplicity".

The ratio $y_m(E) = Y_m(E)/Y_{total}(E)$ is approximately equal to the probability of emission of m γ-rays in one cascade. Each γ-ray in the cascade carries part of the excitation energy and the total angular momentum. Obviously the total energy of the γ-rays is equal to the initial excitation energy and the vector sum of the angular momenta is equal to the spin difference of the initial and final states. The emission probability of a given γ-ray is dependent on the energy and transition type: mainly electric dipole (E1) and magnetic dipole (M1). The m-step cascade probability, $y_m(E)$, depends on the spin and parity of the initial and final states. Therefore, the multiplicity distribution contains information about the spin of the capturing state; we shall assume that there are different multiplicity distributions for the cascades initiated from states with different spin, but that the distributions are similar for cascades initiated from states with the same spin.

First, we consider two spin states excited by *s*-wave neutrons captured by odd-A targets.

As an example, Fig. 1 shows the multiplicity distributions of ^{155}Gd resonances with spin $J^\pi = 1^-$ and 2^-. One can choose the multiplicity distribution for a resonance with a spin that is well determined from previous experiments as a prototype (Fig. 1) and compare other resonances with the prototype resonance in order to assign their spin.

For resonances with two possible spins, the Eq. (1) can be decomposed as

$$\begin{bmatrix} y_1(E) \\ y_2(E) \\ y_3(E) \\ y_{\max}(E) \end{bmatrix} = \begin{bmatrix} \omega_1^{(1)} & \omega_1^{(2)} \\ \omega_2^{(1)} & \omega_2^{(2)} \\ \dots & \dots \\ \omega_{\max}^{(1)} & \omega_{\max}^{(2)} \end{bmatrix} \begin{bmatrix} \alpha_1(E) \\ \alpha_2(E) \end{bmatrix} \qquad (2)$$

where $\omega_m^{(J)}$ are the prototype multiplicity distributions and $\alpha_1(E)$ and $\alpha_2(E)$ are the weights for the contributions of spin J_1 and spin J_2, respectively, at neutron energy bin E.

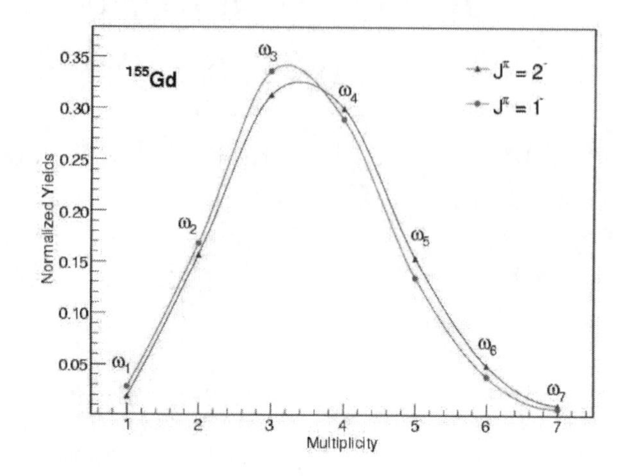

FIGURE 1. Multiplicity distributions of the spin $J^\pi = 1^-$ and 2^- resonances at neutron energy E_n = 21.03 eV and E_n = 14.51 eV in ^{155}Gd.

Equation (2) can be solved analytically, determining the set of quantities $\omega_m^{(J)}$ from the prototype resonances. The equations would give conclusive results, if the multiplicity distributions were exactly the same for all resonances with the same spin.

However, the multiplicity distributions are not the same for the resonances with same spin. Experimental errors and backgrounds distort the multiplicity distributions. Even if the experiment was perfect, the multiplicity distributions will not be identical due to Porter Thomas fluctuations. Consequently, the normalized yields, $y_m(E)$, are distributed around the mean values $\omega_m^{(J)}$. There are two centroids, corresponding to the two spin states. Based on the central limit theorem, we may assume that the distribution function is a Gaussian. Since there are two spin groups, the Probability Density Function (PDF) is a mixture of two Gaussians. The centroids of the Gaussians $\omega_m^{(1)}$ and $\omega_m^{(2)}$ represent the mth component of the multiplicity distribution for the corresponding spins. The widths of the Gaussians are the sum of the experimental uncertainty and the PT fluctuations $\sigma^2 = \sigma_{exp}^2 + \sigma_{PT}^2$.

Generalizing this behavior, the PDF is multivariate normal in multi-dimensional multiplicity space. The joint PDF, $g(y|\Omega)$, will be a mixture of two multivariate normal distributions corresponding to the two spin groups:

$$g(y|\Omega) = \beta_1 f_1(y|\Omega_1,\Sigma_1) + \beta_2 f_2(y|\Omega_2,\Sigma_2), \quad (3)$$

where Ω forms a parameter space: $\Omega = (\Omega_1,\Omega_2,\Sigma_1,\Sigma_2,\beta_1,\beta_2)$. The parameters β_1 and β_2 are the mixing weights that represent the number of spin J_1 and J_2 resonances in the sample. The ratio of the mixing weights is *a priori* probability for any resonance to have spin J_1 or J_2. They satisfy the condition: $\beta_1 \geq 0, \beta_2 \geq 0$ and $\beta_1 + \beta_2 = 1$.

The multivariate normal density functions are described as follows:

$$f_k\left(y|\Omega_{(k)}, \Sigma_{(k)}\right) = \frac{1}{(2\pi)^{N/2}\left|\Sigma_{(k)}\right|^{1/2}} \times$$

$$\times \exp\left(-\frac{1}{2}\left(y-\Omega_{(k)}\right)^T \Sigma_{(k)}^{-1}\left(y-\Omega_{(k)}\right)\right) \quad (4)$$

where $\Omega_{(k)} = \left[\omega_1^{(k)}, \omega_2^{(k)}, \ldots, \omega_{max}^{(k)}\right]$ with $k = 1,2$ is a column vector with $p = max$ components, $\Sigma_{(k)}$ is a covariance matrix that is a real, $p \times p$ dimensional positive definite matrix, N denotes the sample size or bin number in the resonance region, the normalized yields are described as $y = [y(1)\, y(2) \ldots, y(i) \ldots, y(N)]$ and each data point is a p-dimensional vector $y(i) = [y_1(i)\, y_2(i) \ldots, y_{max}(i)]$. Note, the normalized yields are a function of neutron energy, $y_m(E)$, in Eq. (1). The normalized yield samples in Eq. (4) are collected from each neutron energy bin, i, which is equivalent to $y_m(i) = y_m(E)$.

The goal of the analysis is to determine to which class or spin group a given resonance belongs. In order to accomplish this purpose, we need to set a "decision boundary" to separate the distributions into two regions. The boundary is also called the *discriminant function*, and is a function of the parameters of the distribution. The simplest form of the function is called the *Linear Classifier* and the *Bayes classifier* is the best classifier which minimizes the classification error [7].

To illustrate the process, Fig. 2 shows the two dimensional scatter plot with a *linear discriminant function*, where small circle points depict the location of the sample, $y(i) = [y_5(i)\, y_6(i)]$, and the solid elliptical lines are the contour lines of the PDF.

A decision rule is simply based on the probabilities and can be written as follows:

$$\begin{aligned} p_1(y) > p_2(y) \rightarrow J_1 \\ p_1(y) < p_2(y) \rightarrow J_2 \end{aligned} \quad (5)$$

where $p_k(y)$ is *a posteriori* probability determining whether y belongs to J_1 or J_2.

Using Bayes theorem, the *posteriori* probability $p_k(y)$ is calculated from *a priori* probability β_k and the conditional density function $f_k(y)$,

$$p_k(y) = \frac{\beta_k f_k\left(y|\Omega_{(k)}, \Sigma_{(k)}\right)}{g(y|\Omega)}, \quad (6)$$

where the density functions $f_k\left(y|\Omega_{(k)}, \Sigma_{(k)}\right)$ and $g(y|\Omega)$ are given in Eqs. (3) and (4).

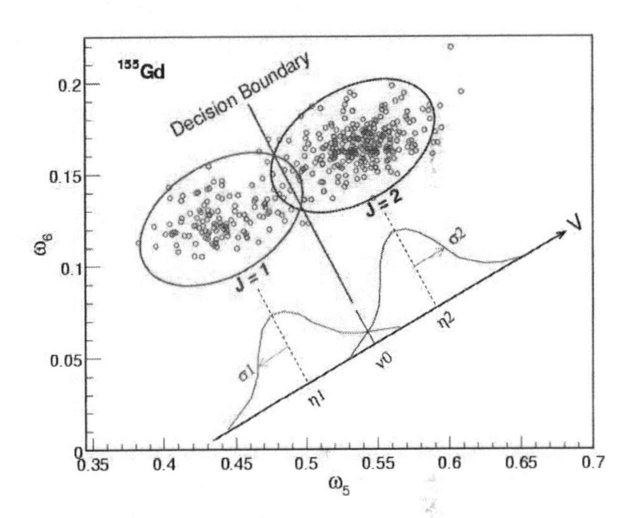

FIGURE 2. Two-dimensional example of the multivariate normal distributions. The optimal choice of the projection axis V maximizes separation between the clusters and the threshold $v0$ minimizes the error.

The unknown parameters, $\Omega = (\Omega_1,\Omega_2,\Sigma_1,\Sigma_2,\beta)$, will be estimated by an iteration method known as the Expectation-Maximization (EM) algorithm [8]. Each EM iteration consists of two steps, an E-step and an M-step. Given an estimate of the population means $\Omega_{(k)}$, covariance matrices $\Sigma_{(k)}$, and mixing proportions β, the E-step computes *a posteriori* probability (Eq. 6) that data y_i belongs to the spin group 1 or 2. In the M-step, parameters are estimated from the data given the conditional probabilities $f_k\left(y|\Omega_{(k)}, \Sigma_{(k)}\right)$.

$$\beta_k = \frac{1}{N}\sum_{i=1}^{N} f_k\left(y_i|\Omega_{(k)}^{old}, \Sigma_{(k)}^{old}\right) \quad (7)$$

$$\Omega_{(k)} = \frac{\sum_{i=1}^{N} y_i f_k\left(y_i \middle| \Omega_{(k)}^{old}, \Sigma_{(k)}^{old}\right)}{\sum_{i=1}^{N} f_k\left(y_i \middle| \Omega_{(k)}^{old}, \Sigma_{(k)}^{old}\right)} \tag{8}$$

$$\Sigma_{(k)} = \frac{\sum_{i=1}^{N} f_k\left(y_i \middle| \Omega_{(k)}^{old}, \Sigma_{(k)}^{old}\right)(y_i - \Omega_{(k)}^{old})(y_i - \Omega_{(k)}^{old})^T}{\sum_{i=1}^{N} f_k\left(y_i \middle| \Omega_{(k)}^{old}, \Sigma_{(k)}^{old}\right)} \tag{9}$$

where the superscript "*old*" represents the last iteration for which the values remained constant in the present iteration.

APPLICATION OF THE METHOD TO ^{155}Gd *S*-WAVE RESONANCES

Using the new technique, we have determined the spins for almost all of the *s*-wave resonances in ^{155}Gd. We assume that only *s*-wave neutron resonances are experimentally observed in ^{155}Gd. The *p*-wave and higher *l* resonances are negligible for ^{155}Gd due to the orbital angular momentum potential barrier and the minimum of the *p*-wave strength function in this region. The ground-state spin of ^{155}Gd is 3/2$^-$. Capturing an s-wave neutron leads to $J^\pi = 1^-$ and 2^- resonances in ^{156}Gd. Since there are two spin states that are experimentally observable, Eqs. (6)–(9) can be use directly to determine spin of the resonances.

In DANCE experiments, the backgrounds for low multiplicity events are large compared to real capture events. From Fig. 1 it is clear that the contribution of multiplicity $m = 1$ and $m \geq 7$ events are less than 5% for both spins; ignoring these components will not greatly affect the total counts. We also ignored the multiplicity 2 components because of the relatively high background. Therefore, the multiplicity space is confined to $3 \leq m \leq 6$ which means $p = 4$ dimensional.

A priori probabilities are initialized from the $2J+1$ level density law and remain constant during the iteration. The covariance matrix initially was set as a 4 dimensional identity matrix. The E-step and M-step are iterated until a maximum number of iterations is reached. Convergence is usually fast for ^{155}Gd data and about 15 iterations will result in solutions that are accurate to about 1 percent.

As an example, Fig. 3 shows a graphical illustration of the results in the neutron energy range between 1 eV and 25 eV.

The last iteration of the EM algorithm gives *a posteriori* probabilities $p_1(y)$ and $p_2(y)$ for each neutron energy bin. The relative yields for spin $J^\pi = 1^-$ and 2^- shown in Fig. 3 were calculated by multiplying the probabilities times the experimental yield. For example, $p_1(y) \sim 1$ and $p_2(y) \sim 0$ for the bins of spin J =1 resonance at 21.03 eV and multiplying them with Y_{exp} gives nearly zero yields for spin J =2 (blue line in Fig. 3) and $Y_1 \approx Y_{\text{exp}}$ (red line in Fig. 3). As we can see, for most cases, the relative yields for spin $J^\pi = 1^-$ or 2^- resonances agree with the experimental yields, indicating that the probabilities are nearly 100 percent. For some resonances the relative yields from both spin groups are noticeable (see for example resonance at 17.77 eV). These cases suggest spin doublets. However, this effect could also be observed for weak resonances due to background contamination. Therefore we only assigned spin doublets for strong resonances.

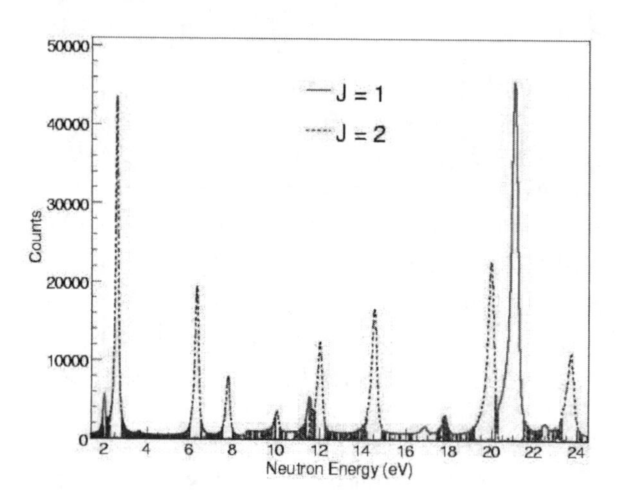

FIGURE 3. Spin assignments of the neutron resonances in ^{155}Gd.

Based on these results, we plot the cumulative number of resonances as a function of resonance energy. From the reciprocal of the slope of the staircase plot, the average level spacing of the s-wave neutron resonances is calculated as $D_0 = 1.76 \pm 0.09$ eV. The error of the average spacings was calculated by the formula given in [9]. The average level spacings were determined as $D_{0,1} = 4.6 \pm 0.4$ eV and $D_{0,2} = 2.9 \pm 0.2$ eV, respectively, for $J^\pi = 1^-$ and 2^- resonances. This is in agreement with the expected $2J+1$ level spacing.

We have also extended this analysis method to p-wave resonances both on spin-zero targets and targets with spin. In these cases one must rely on other experimental evidence to determine the parity of the resonances. Once this is determined, then the pattern recognition technique works rather well, even for e.g., the four-spin p-wave problem.

In summary the pattern recognition technique has been demonstrated to provide an effective method of determining the spins of capture states observed with the DANCE array.

ACKNOWLEDGMENTS

This work was supported in part by U.S. DOE grants No. DE-FG52-09NA29460 and DE-FG02-97ER41402 and performed under the auspices of the U.S. DOE under contracts Nos. DE-AC52-07NA27344 and DE-AC52-06NA25396.

REFERENCES

1. M. Heil, R. Reifarth, M. M. Fowler, R. C. Haight, F. Kappeler, R. S. Rundberg, E. H. Seabury, J. L. Ullmann, J. B. Wilhelmy, and K. Wisshak, Nucl. Instrum. Methods Phys. Res. A 459 (2001), 229.
2. S. A. Sheets, et al., Phys. Rev. C 76 (2007) 064317.
3. A. Chyzh, Ph.D. thesis, NC State Unversity (2009).
4. P. E. Koehler, J. L. Ullmann, T. A. Bredeweg, J. M. O'Donnell, R. Reifarth, R. S. Rundberg, D. J. Vieira, and J. M. Wouters, Phys. Rev. C 76 (2007), 025804.
5. F. Becvar, P. E. Koehler, M. Krticka, G. E. Mitchell, and J. L. Ullmann (2010), An Optimized g-Multiplicity Based Spin Assignment of Neutron Resonances (to be published).
6. R. Reifarth, T. et al., Tech. Rep. LA-UR-03-5560, LANL (2003).
7. K. Fukunaga, Introduction to Statistical Pattern Recognition (Second Edition), Academic Press, Inc., San Diego (1990).

Digital Electronics For The Versatile Array Of Neutron Detectors At Low Energies

M. Madurga[a], S. Paulauskas[a], R. Grzywacz[a], S.W. Padgett[a], D.W. Bardayan[b], J.C. Batchelder[c], J.C. Blackmon[d], J.A. Cizewski[e], R.E. Goans[c,a], S.N. Liddick[a], P. O'Malley[e], C. Matei[c], W.A. Peters[c], C. Rasco[c], F.Raiola[f] and F. Sarazin[f]

[a] *Department of Physics and Astronomy, University of Tennessee, Knoxville, TN 37996 USA*
[b] *Physics Division, Oak Ridge National Laboratory, Oak Ridge, TN 37831 USA*
[c] *Oak Ridge Associated Universities, Oak Ridge, TN 37831 USA*
[d] *Department of Physics and Astronomy, Louisiana State University, Baton Rouge, LA 708034*
[e] *Department of Physics and Astronomy, Rutgers University, New Brunswick, NJ 08903 US*
[f] *Department of Physics, Colorado School of Mines, Golden, CO 80401 USA*

Abstract. A χ^2 minimization algorithm has been developed to extract sub-sampling-time information from digitized waveforms, to be used to instrument the future Versatile Array of Neutron Detectors at Low energies. The algorithm performance has been characterized with a fast Arbitrary Function Generator, obtaining time resolution better than 1 ns for signals of amplitudes between 50 mV and 1V, with negligible walk in the whole range. The proof-of-principle measurement of the beta-delayed neutron emission from ^{89}Br indicates a resolution of 1 ns can be achieved in realistic experimental conditions.

Keywords: Computer data analysis, Data acquisition and sorting
PACS: 29.85.-c, 29.85.Ca

THE VERSATILE ARRAY OF NEUTRON DETECTORS AT LOW ENERGY

The Versatile Array of Neutron Detectors at Low Energy (VANDLE) is under development at the Center for Radioactive Ion Beam Studies for Stewardship Science. The array will consist of 100 3x3x60 cm^3 small bars sensitive to low neutron energies as well as 128 5x5x200 cm^3 large bars for large angular coverage and high overall neutron detection efficiency [1]. The small and long bars will be combined in different geometries for a modular design that can be optimized for the requirements of each experiment. The unique characteristics of VANDLE, a low neutron energy detection threshold and high efficiency, make it an ideal detector for the proposed experimental program. The VANDLE detector array will be used to study the nuclear structure of exotic neutron-rich and proton-deficient nuclei using (d,n) reactions and the study of the beta-delayed neutron emission of neutron-rich fission fragments. Both measurements are not only important to advance the understanding of nuclear structure, but are also necessary to refine the calculations of r-process nucleosynthesis [2].

The large number of channels and low detection threshold present a challenge for the design of an acquisition system for VANDLE. These requirements make a solution based on digital acquisition electronics ideal for VANDLE. Digital acquisition electronics are compact, "one cable" systems, as the detector is directly connected to the digitizer module. Moreover, since the pulse information is extracted directly from the waveform and sophisticated algorithms can be used, digital electronics can have an excellent signal-to-noise ratio and are capable of operating at low thresholds. However, using digitizers with fast scintillators is challenging due to the very fast scintillator signal, with pulses of 10 ns width or shorter. Because of this, the analysis of scintillator signals would require a digitizer with a sampling time of less than 10ns. However, these systems are prohibitively expensive to be used for systems with large number of channels.

Application of Accelerators in Research and Industry
AIP Conf. Proc. 1336, 586-589 (2011); doi: 10.1063/1.3586170

In this paper we report the development of a χ^2 minimization algorithm capable of extracting timing information in the sub-nanosecond range from waveforms acquired using a 10 ns sampling interval. A series of tests using a fast pulser were completed to characterize the performance of the algorithm applied to a wide range of waveform characteristics. Finally we present the result of the proof-of-principle measurement of the ^{89}Br beta-delayed neutron emission.

TIMING ANALYSIS

The work presented in this paper was performed using XIA 100 MSPS, i.e. 10 ns sampling interval, Pixie16 digitizers. Two factors make extracting time information in the sub-sampling-time range possible with Pixie16. First, Pixie16 is capable of analyzing signals of 10 ns width or shorter because it has a 50 MHz high frequency cutoff (Nyquist) filter [3] before the Analog to Digital Conversion (ADC) stage. In a typical example in our work, a plastic scintillator pulse of ~10 ns width will be "spread" to a 40 ns wide pulse after the Nyquist filter. It is important to note that even if the Nyquist filter removes the high frequency components off the waveform, it is immune to phase transformations. Thus, time translations of intervals shorter than 10 ns are preserved. Second, after the pulse is digitized, the information of a time transformation smaller than the sampling time is preserved in the pulse shape. This can be easily demonstrated. Let us assume that the waveform whose leading edge can be described by a function f and that the ADC digitizes the waveform at a time $t= t_d$, thus obtaining the value $f(t_d)$. The next waveform is delayed a time $t'=t+\phi$, where $\phi<10$ ns, obtaining $f'(t')=f(t+\phi)$. This implies that for $f'(t'=t_d)$

$$f'(t'=t_d) = f(t_d - \varphi) \tag{1}$$

Thus, the change of the waveform shape for time transformations below the sampling time is, using eq. 1.

$$\Delta f = \frac{f(t_d) - f(t_d - \phi)}{\phi} \tag{2}$$

After a time translation, the leading edge of the digitized pulses shapes change by a factor close to its first derivative (when $\phi \rightarrow 0$). The result shown in eq. 2 has profound implications for other analysis techniques based on using a constant fraction discriminator combined with linear interpolation of the digitized waveform [4]. The digital Constant Fraction discrimination (CFD) calculates the time from the zero crossing of a pulse constructed from the waveform and its time shifted amplitude inverse. From eq. 2 we see that only in the rare cases where one can construct a digital CFD waveform of perfect linear slope (or very small slope compared to the time transformation) the calculated time will be linearly dependent on the time translation. Here, in the case of obtaining time from signals from fast plastic scintillators, the leading edge is normally characterized mathematically using non linear functions, such as Gaussians, or even higher order exponentials, making the digital CFD technique challenging.

The relation between a time transformation and the waveform shape shown in eq. 2 suggested to us that in order to extract precise sub-sampling-time information, the waveform slope needs to be explicitly taken into account by the timing algorithm. A simple way to achieve this is to fit the waveforms, using a χ^2 minimization, to a template waveform of fixed widths. We used the function,

$$f(t) = Ae^{\frac{-(t-t_0)}{\lambda}}\left(1 - e^{\frac{-(t-t_0)^2}{\sigma}}\right) \tag{3}$$

where the parameters λ and σ were individually optimized for the particular scintillator and pulser waveforms we studied, and the amplitude A and the intercept time t_0 are obtained from the fit.

TESTS AND CALIBRATION WITH AN ARBITRARY FUNCTION GENERATOR

In order to characterize the fitting time algorithm we first studied its behavior for a wide range of waveform properties using a 100 MHz Tektronix Arbitrary Function Generator (AFG). The two channels of the AFG were set to single burst mode, with an external trigger at 1 kHz. In order for the pulse waveform to be as close as possible to the short pulses coming from plastic scintillators we selected a square shape of 10 ns width. The fitting function parameters were optimized to the AFG pulse by minimizing the time resolution of 1 V amplitude signals, obtaining σ=0.9 ns^2 and λ=0.17 ns.

In order to study the response of our system to time translations, the second channels of the AFG was delayed from 0 to 10 ns in 1 ns intervals. In the whole range the fitting algorithm was able to calculate the proper delay between the two signals and the Full Width at Half Maximum (FWHM) resolution was degraded no more than 5% of the average value.

The whole range of delays was repeated with decreasing signals amplitude, defining the resolution from the average of the 10 different delays. The resolution ranges from the excellent 100 ps for 1 V signals to 840 ps for 50 mV (see Fig. 1a). These results are well below the resolution required for VANDLE. We also measured the algorithm walk by using the 1V measurement as a reference (see Fig. 1b). We see that the walk is small over all the voltage range, with a maximum of 240 ps.

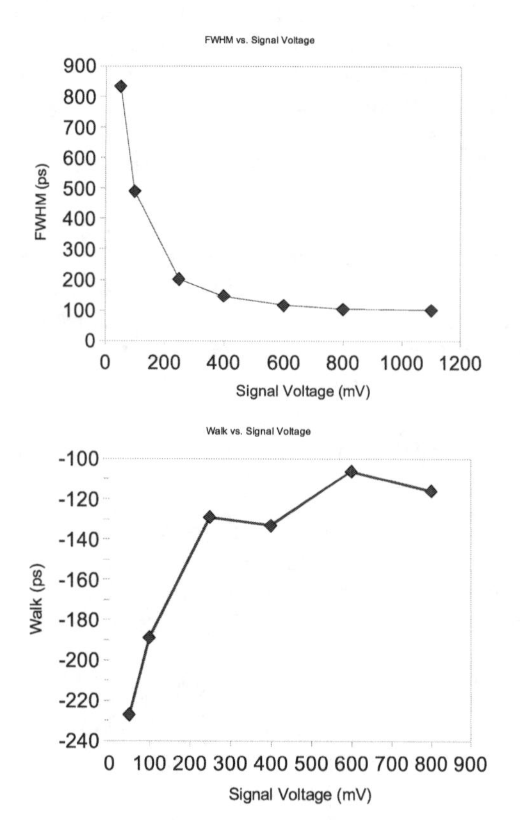

FIGURE 1. Top: time resolution, from the digital fitting algorithm, versus signal amplitude for narrow signals from a Tektronix Arbitrary Function generator. Bottom: Signal walk for the same set of measurements shown in a. The walk was measured with respect of the measurement of 1 V amplitude.

Timing Resolution Measurement Of The VANDLE Plastic Scintillators

The next step consisted in testing the timing algorithm with signals from VANDLE's plastic scintillators and comparing them to tests using analog electronics. A classical timing resolution measurement for plastic scintillators consists of measuring the time resolution from muon events. We arranged two small 5x5x10 cm^3 trigger detectors, made of the same material as VANDLE's detectors, on top and below a 5x5x200 cm^3 VANDLE bar [1]. This ensured that the detected muons passed through a straight path marked by the cross section of the three detectors. Moreover, to reduce natural background contamination we configured the system to record signals only when all detectors fired within a 100 ns coincidence window. The fitting time algorithm parameters were optimized by successive fits of a typical VANDLE pulse, obtaining σ=3.96 10^{-2} ns^2 and λ=0.32 ns. The benchmark result, obtained with analog electronics, was a resolution of FWHM =810 ps for the time of flight between one of the photomultipliers of the VANDLE bar and one of the triggers [5]. We obtain, using digital electronics with the fitting time algorithm, a FWHM=1.14 ns with consistent results for the other photomultiplier and trigger combination.

^{89}Br Proof-Of-Principle Tests

In order to test the performance of the digital acquisition timing algorithm in realistic conditions we measured the beta-delayed neutron emission of ^{89}Br. This is an ideal case to test VANDLE due to its energy distribution, with transitions between 200 keV and 1 MeV [6] . Mass number A=89 fission fragments, produced by bombarding a uranium carbide target with 25 MeV protons from the Holifield tandem accelerator, were separated at the On-line Test Facility at Oak Ridge. The resulting cocktail beam contained up to 1000 ^{89}Br ions per second, with no other beta-delayed neutron emitter produced in significant amounts. The beam was implanted in an aluminized tape to remove long lived mass 89 nuclei from the measuring station. Two plastic scintillators wrapped around the aluminum tube at the implantation point provided start-trigger for the time-of-flight measurement. The stop signal(s) was given by one 3x3x60 cm^3 VANDLE bar was placed 35 cm away from the implantation point and four 5x5x200 cm^3 placed 1.5 m away. All VANDLE photomultipliers were biased with relatively low voltages, optimized for up to 6 MeVee dynamic range for the 60 cm long bar and up to 15 MeVee for the 200 cm long bar. In order to study the low energy response of our system, the detection threshold of the 60 cm long bar was set at around 20 mV. However, thresholds of the 200 cm bars were much higher, around 200 mV, due to the presence of high-rate (~10000 counts/s) low-amplitude signals in these bars. There was no evidence of delayed neutrons in the time-of-flight spectrum of the 200 cm long bars, most likely due to the combination of high threshold and low operating voltage.

Beta-delayed neutrons were identified and individual transitions observed in the time-of-flight spectrum of the 60 cm bar, as seen Fig 2. The inset in Fig. 2 shows a simulation of the ^{89}Br time-of-flight neutron spectrum calculated from neutron energy spectrum in the literature [6]. We see that the neutron peaks at times around 30 ns, corresponding to neutrons

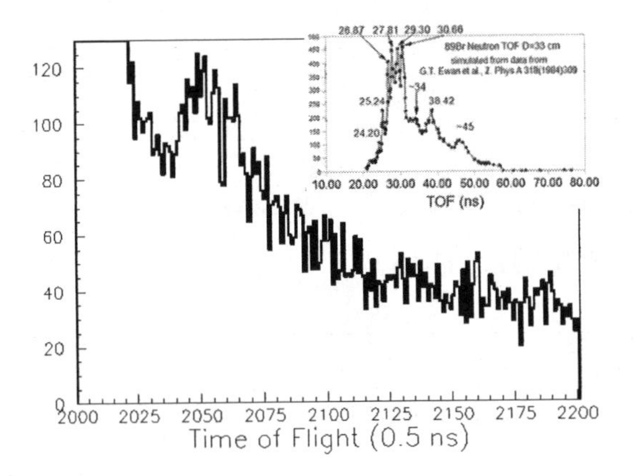

FIGURE 2. Time-of-flight spectrum of mass 89 fission fragments measured at the OnLine Test Facility (OLTF) with a setup consisting of two small plastic scintillators wrapped around the implantation point and a 3x3x60 cm3 VANDLE bar. The only mass 89 beta-delayed neutron precursors that is produced with enough intensity at OLTF is 89Br, thus we can assign the observed neutrons to the decay of 89Br. The inset shows the calculated time-of-flight spectrum using the published neutron energy spectrum [6], the experiment geometry and folded by VANDLE efficiency curve. We see that the experimental distribution follows the calculation closely, thus confirming the assignment to the delayed neutron branch of 89Br.

of energies between 500 and 750 keV, had a resolution at FWHM=1.5 ns. This result is slightly worse than the value expected from the pulser tests. This can be attributed to several factors and it is important to carefully study their individual contribution. First, the VANDLE bars are not curved, meaning the neutrons travel slightly different distances depending on the interaction point in the VANDLE bar. In the final spectrum shown in Fig. 4 the time of flight was recalculated as if the source was at a distance of 35 cm to the VANDLE bar independently of the impinging point. Second, the plastic scintillators currently used as start triggers have a single photomultiplier tube. Because we cannot correct for the scintillation light travel time inside the detector their timing is sensitive to the impinging position and therefore poor for precise timing measurements. Finally, we ran VANDLE's photomultipliers with low bias voltage to achieve a large detection dynamic range. This resulted in signal amplitudes for neutron events consistently lower than 60 mV. From the pulser tests discussed in the section above, we know that the timing algorithm's intrinsic resolution for signals lower than 50 mV will be lower than 1ns. All things considered, the observed resolution of 1.5 ns for neutron transitions indicates that it will be possible to obtain a resolution of 1 ns or better using Pixie16 digital electronics combined with the fit timing algorithm once all this factors are addressed.

CONCLUSIONS

We have developed a timing algorithm capable of extracting sub-sampling-time information from the waveform shape. The study of the dependence of the digitized waveform shape to time transformations smaller than the sampling time indicates that it is driven by the slope of the pulse. In the case that concerns this work, plastic scintillator signals are typically described by combinations of high order exponential functions. Therefore a time transformation will have a highly non-linear behavior. A successful timing algorithm has to include explicitly this dependence to extract precise information. We propose an algorithm based on fitting the digitized waveform to a model function. We performed a series of tests to characterize the behavior of the timing algorithm for different pulse delays and pulse heights using a Arbitrary Function Generator as a pulse source and the 100 MSPS Pixie16 digitizer. The results show that the algorithm is reliable, and capable of resolutions from 100 ps for 1V signals, to 800 ps for 50 mV signals. A time-of-flight resolution of 1.14 ns was obtained with our timing algorithm for cosmic rays measured with a 200 cm long bar compared to 810 ps using analog electronics. We performed a proof of principle measurement of ^{89}Br beta-delayed neutron emission with VANDLE bars at the Online Test Facility, ORNL. Neutrons were detected and their time-of-flight spectrum was calculated using the digital timing algorithm, showing a resolution no better than 1.5 ns. The main sources that contributed to the poor time-of-flight resolution in our experiment were the timing resolution of the start-trigger detectors and the low bias voltage on VANDLE's photomultipliers. By controlling these conditions, the goal of 1 ns resolution for neutrons of energies around 500 keV is possible.

ACKNOWLEDGMENTS

This work was supported by the NNSA through the DOE Cooperative Agreement DE-FG52-08NA28552.

REFERENCES

1. C. Matei et al., PoS(NIC X)138, 1-5 (2008).
2. K. Langanke and M. Wiescher, Rep. Prog. Phys. 64, 1657 (2001).
3. S. K. Mitra, *Digital Signal Processing: A computer based approach,* McGraw-Hill, 2001.
4. A. Fallu-Labruyere, et. al., *Nucl. Inst. and Methods A* 579, 247–251 (2007).
5. I. Spassova and C. Matei, (private communication).
6. G. Ewan *et al., Z. Phys. A* 318, 309–314 (1984).

Using Quasi-Monoenergetic Photon Sources To Probe Photo-Fission Resonances

Micah S. Johnson.[a], James M. Hall[a], Dennis P. McNabb[a], Michael J. Tuffley[b], Mohammed W. Ahmed[c], Sean Stave[c], Henry R. Weller[c], Hugon Karwowski[d], Jeromy Thompkins[d]

[a] *Lawrence Livermore National Laboratory, Livermore CA 94550*
[b] *San Jose State University, San Jose CA 95192*
[c] *Duke University, Durham NC 27708*
[d] *University of North Carolina, Chapel Hill NC 27599*

Abstract. We present preliminary results of photo-fission measurements of uranium isotopes with the quasi-monoenergetic gamma-ray source, HIGS. The measurements were performed to search for photo-fission resonances. We discuss potential applications to use photo-fission resonances to identify special nuclear material in cargo containers. We discuss the importance of quasi-monoenergetic gamma-ray sources for this kind of application.

Keywords: photo-fission, fission, fission neutrons, national security, cross-sections
PACS: 24.30.Gd, 25.40.Ny, 25.60.Dz, 25.85.Jg

INTRODUCTION

Quasi-monoenergetic gamma-ray sources are being developed to provide higher fluxes and lower bandwidths. One application of this type of source is to penetrate containers and probe for special nuclear material (SNM). These sources may be game-changers in the national security sector to detect SNM in various containers. Many efforts are underway to study the efficacy of using these sources with processes such as nuclear resonance fluorescence (NRF) to detect, map, and assay containers for isotopes of interest. LLNL is part of a large collaboration to study the efficacy of NRF for detection and assay [1]. LLNL is also part of a collaboration to look at photo-fission signatures with very intense, low-bandwidth sources to detect SNM. One of the major benefits of using low bandwidth sources is background reduction.

Actinides have a relatively low fission threshold compared to other nuclei. In fact, very few nuclei have neutron production thresholds (fission neutrons, photo-disintegration neutrons, etc…) that are comparable to actinides. Deuterium is an example of a nucleus that has a low neutron production threshold (2.2 MeV). Deuterium is expected to be a large background component for neutron detection with photon sources because of its ubiquity in nature. Quasi-monoenergetic gamma-ray sources can mitigate this background by focusing most of its flux on photo-fission channels. Bremsstrahlung sources integrate the photo-fission channels of the isotope of interest but will also integrate unnecessarily over the photo-neutron production channels of deuterium (and other present isotopes

Photo-fission may offer a fast method to pass or fail a container when looking for fissile material in comparison to other techniques, such as NRF, because the integrated cross section over all beam energies are larger for photo-fission than for NRF. Furthermore, both neutrons and photons are emitted in photo-fission. The presence of two signatures can be a benefit. Photons are attenuated depending on material density and Z. Neutrons, however, have different moderation and attenuation characteristics in different materials. Detailed information regarding the energy distribution and angular anisotropy of neutrons from fission resonances using polarized photon sources and correlations with gamma-rays could be used to help reject backgrounds or perhaps even to isotopically identify fissile materials.

Bohr first suggested [2] that near the saddle point of fission, a nucleus will be relatively cold and states at the saddle point should resemble spectra of its

Application of Accelerators in Research and Industry
AIP Conf. Proc. 1336, 590-593 (2011); doi: 10.1063/1.3586171

normal deformed shape. The incident photon is absorbed, exciting a complicated, normally-deformed state. The excited nucleus has some probability of increasing its deformation and undergoing fission. Near the fission barrier, the possible modes of fission are limited. The deforming nucleus can transition to fission through a lower-energy state near the saddle point and approach the fission barrier with more "kinetic" energy, or can transition through a state of energy similar to the initial state, but with little "kinetic" energy. These states offer a discrete set of resonances near the fission barrier. The search for these structures is an important aspect of our effort.

Detailed information about the structures can provide guidance for the bandwidth, energies, and flux requirements for the next generation of quasi-monoenergetic gamma-ray sources.

EXPERIMENTAL METHOD

Photo-fission measurements were performed using the High Intensity Gamma-ray Source (HIGS) at Duke University's Free Electron Laser Laboratory. The energy of the photon source was stepped up in 25 keV steps from 5.0 MeV to 6.5 MeV. The energy range was chosen to look for resonances, which were expected to be near the fission barrier (approx. 5.2 MeV in uranium isotopes). We used ^{238}U as a feasibility measurement to test the method for ^{235}U. Thin and thick targets of ^{238}U (in the form of depleted uranium) were used in the feasibility test.

Twelve scintillating neutron detectors were used to measure the neutrons from the fission process. Pulse shape discrimination was used to separate gamma-ray events from neutrons. Time-of-flight was used to determine the neutron energies. The neutron separation energy in ^{238}U is 6.153 MeV. This implies that for the energy region that was measured there are no photo-neutrons above 2 MeV. The efficiency of the neutron detectors are such that the thresholds are slightly less than 1 MeV. Since fission neutrons extend above 2 MeV and the signals below 1 MeV are associated with photo-neutrons or are below thresholds, we have gated on neutrons above 2.0 MeV.

The detectors were placed at polar angles of 75-, 90-, and 126-degrees to measure the angular distributions. Three detectors were placed above the beam and three more below the beam, which is the plane perpendicular to the polarization plane. The remainder of the detectors was placed in a similar geometry parallel to the polarization plane (beam-left and beam-right). The placement of the detectors in the different planes was done to measure the anisotropy of neutrons.

The photon flux was measured in-beam with a series of thin plastic paddle scintillators. The paddle detectors were flux-calibrated with a high-efficiency NaI detector and/or an HPGe detector that measured the attenuated flux, on axis, before and after each run.

RESULTS AND DISCUSSION

The results of the photo-fission measurements of ^{238}U are shown in figure 1 (only the range from 5.0- to 6.5-MeV are shown in the interest of clarity). Our cross-section results were calculated from the measured data by averaging over the angles (blue data in Fig. 1) and from the so-called "A0" coefficient from second-order Legendre fit (red data in Fig. 1). We have also compared our data to previous measurements by Dickey and Axel [3] and the results are consistent (see Fig. 1). Deviations from the average cross-sections and the A0 results indicate a lower contribution from the P0 term of the Legendre polynomials. Indeed the A2 coefficients are less than -1.0 in the range from 5.0 MeV to 5.5 MeV. The green curve represents the shape of the beam and is superimposed to give the reader a sense of scale.

It is important to point out that there are two large structures that we have observed and are consistent with Dickey and Axel, centered at 5.6 MeV and 6.2 MeV. These structures are much wider than the beam and could represent cusps or pygmy-like features. From this data it is inconclusive whether or not any resonances exist. Narrow resonances should be at the lowest energies (nearest the barrier). With a broad beam, such as the one from HIGS (approximately 160 keV) any narrow structures should appear as a hump or shoulder on the overall cross-section. The widths of these shoulders should be no greater than the width of the beam. This is challenging with a broad beam unless the resonance(s) were large. We can use the data for ^{238}U to set a limit on the strength of a single photo-fission resonance at the lowest energy, about 3 MeV. Assuming that the green curve is the flux we can integrate the cross-section. We thus set an upper limit on the integrated cross-section at 5.3 MeV to be 95(6) eV barns. This implies that the height, H, of any hidden resonance is determined simply: H=3.545/w, where, w, is the width of the structure, and 3.545 comes from the integrated cross-section.

Recent measurements of ^{235}U (too recent to include here) have shown preliminary indications that there may be underlying structure (resonances) to the above results for ^{238}U. The data collected for ^{235}U were taken with 25-keV steps. With careful deconvolution of the data, it may be possible to gain a higher resolution (25-keV) picture of the underlying structures. If the structures are lower in width than what the analysis is capable of extracting, then it is imperative that these structures be re-measured with much narrower beams. This will enable exact determination of the size and shape of the resonance structure (i.e. measure w and H). Determining the strength and shape of any structures will help guide the models for detection protocols with photo-fission probes.

The anisotropy results for ^{238}U show a ratio (perpendicular to parallel) of 2.0(5) at 90-degrees (averaged for neutron energies between 2.0- and 12.0-MeV). This is in contrast to other nuclei including ^{235}U, which shows a preliminary result of 1.0(7) at 90-degrees. This may provide an additional signature to detect HEU over natural uranium or DU. The angular distributions show a marked difference, in ^{238}U the A2 coefficient is 1.3(3) and in ^{235}U the A2 coefficient is 0.03(4). This may be a result of the even-even versus even-odd excitations where there are many projections in the odd mass case causing a more isotropic distribution.

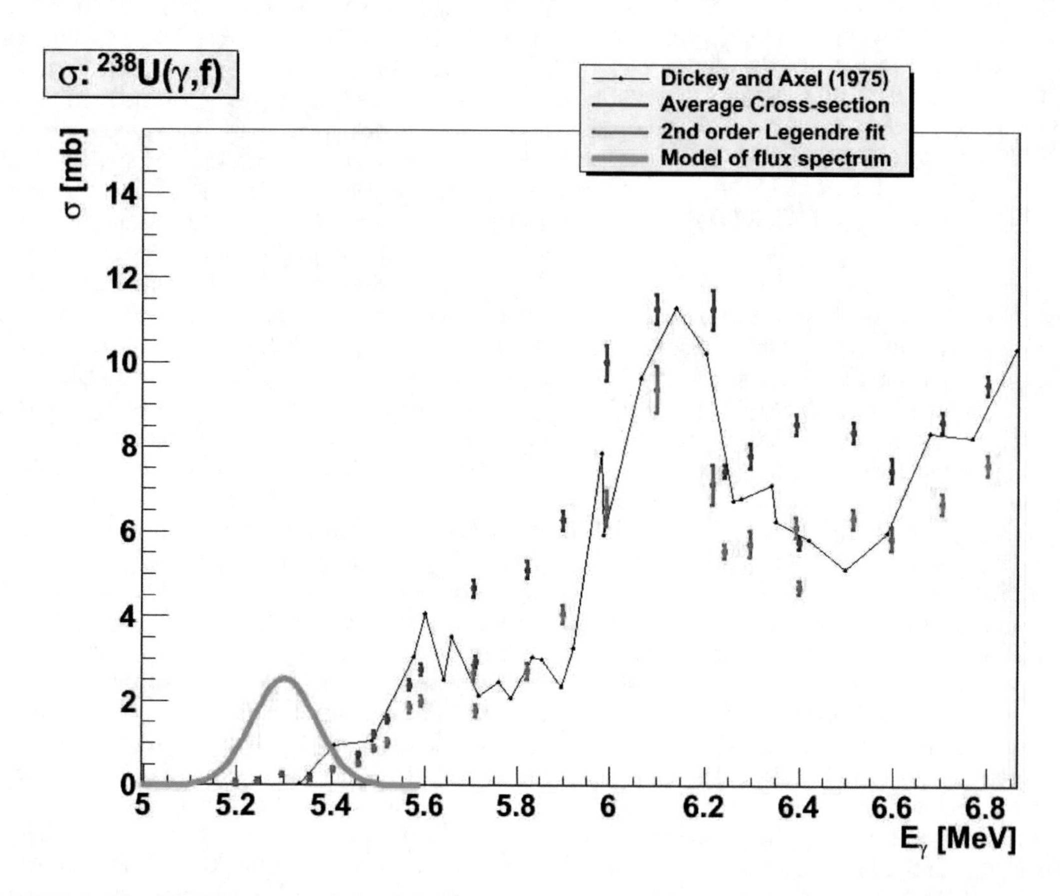

FIGURE 1. Photo-fission cross-section versus incident photon energy. Blue data represents an angle-averaged cross-section, and the red data represents a second-order Legendre fit to the data. The black line represents previous data [3] for comparison. The green curve is a likeness to the beam shape.

The importance of narrow structures for SNM detection is that it allows operators to tune quasi-monoenergetic sources on and off of resonances. By measuring the variations in the emitted photo-neutron count rate, one can determine which isotope (providing no overlaps) is present. Isotopic sensitivity with photo-fission probes would be a game-changer in SNM detection. This is because the integrated cross-section (given above, 95(5) eV barns) is more than an order of magnitude greater than typical NRF strengths, typically 1 to 10 eV barns. With higher efficiency of detecting neutrons than gamma-rays (from NRF) the counting rates for photo-fission will be much greater (than NRF).

SUMMARY AND OUTLOOK

We have performed photo-fission measurements of uranium isotopes with a linearly polarized, quasi-monoenergetic gamma-ray source. Our analysis of ^{238}U shows good agreement with prior results. The results indicate narrow structures or the possibility of photo-fission resonances. Data analysis is ongoing and the refined results will be published later for ^{238}U and ^{235}U. Detailed information about the resonances can provide guidance for the bandwidth, energies, and flux requirements for the next generation of quasi-monoenergetic gamma-ray sources.

We have also analyzed the data for angular distribution effects and anisotropies in the neutron spectrum and have shown differences between the uranium isotopes. This information may be useful as a signature for detecting SNM in cargo containers.

ACKNOWLEDGMENTS

This work performed under the auspices of the U.S. DOE by LLNL under Contract DE-AC52-07NA27344. This effort is supported by DHS/DNDO.

REFERENCES

1. J. Pruet et al., J. Appl. Phys. 99, 123102 (2006)
2. A. Bohr, Proc. UN Int. Conf. PUAE, **2** 151 (1956)
3. P.A. Dickey and P. Axel, PRL, **35** 501 (1975)

Use Of The BigSol Time Of Flight Spectrometer In The Study Of Superheavy Element Production

M. Barbui[a], K. Hagel[a], J.B. Natowitz[a], R. Wada[a], P.K. Sahu[a], T. Materna[a], Z. Chen[a], L. Quin[a], G. A. Souliotis[a,g], G. Chubaryan[a], A. Bonasera[a,h], D. Fabris[b], M. Lunardon[b], M. Morando[b], S. Moretto[b], G. Nebbia[b], S. Pesente[b], G. Viesti[b], F. Bocci[c], M. Cinausero[c], V. Rizzi[c], G. Prete[c], S. Kowalski[d], Z. Majka[e], A. Wieloch[e], F.D. Becchetti[f], T.W. O'Donnel[f] and H.Griffin[f]

[a] *Cyclotron Institute Texas A&M University, College Station, Texas, 77843*
[b] *Dipartimento di Fisica dell'Università di Padova and INFN Sezione di Padova, Italy*
[c] *INFN Laboratori Nazionali di Legnaro, Italy*
[d] *Institute of Physics, Silesian University, Katowice, Poland*
[e] *Smoluchowski Institute of Physics, Jagiellonian University, Krakow, Poland*
[f] *University of Michigan, Ann Arbor, Michigan*
[g] *Laboratory of Physical Chemistry Department of Chemistry, University of Athens, 15771 Athens, Greece*
[h] *INFN Laboratori Nazionali del Sud, Catania, Italy*

Abstract. A time-of-flight spectrometer with the BigSol superconducting solenoid at Texas A&M was used to investigate the possibility to produce heavy and superheavy nuclei by using two body collisions involving heavy projectiles and targets. The reaction $^{197}Au + ^{232}Th$ at 7.5 AMeV is studied in this work. Preliminary results for the yields of heavy nuclei are presented.

Keywords: Time of flight, Z identification, heavy elements
PACS: 25.70.Hi 25.70.Jj

INTRODUCTION

The synthesis of superheavy elements (SHE) has been an important topic in both theoretical and experimental nuclear science for many decades. Two standard experimental techniques are currently used to produce SHE: the "cold" and "hot" fusion reactions. "Cold fusion" reactions with Pb or Bi targets [1] produce SHE which are neutron deficient and have short half-lives, whereas "hot fusion" reactions, with actinide targets and ^{48}Ca beams [2], produce SHE with more neutrons, but still very far away from the predicted "island of stability" for SHE.

Two body reactions involving very heavy nuclei can be considered as an alternative method to produce SHE along the "island of stability". Deep inelastic reactions involving two ^{238}U nuclei at energies around the Coulomb barrier have been studied over the years by several authors from the theoretical [3-5] and the experimental [6-9] point of view. In such experiments the reaction products emitted around the grazing angle were detected. Though the SHE were not detected the results indicate that the observed fragments originate from the binary break up of a rather long living and highly deformed di-nuclear system (DNS). In this work we investigate the reaction $^{197}Au + ^{232}Th$ at 7.5 AMeV. The experiment was performed using the BigSol time of flight spectrometer of Texas A&M. The energy and the charge distribution of the reaction products were measured. Experimentally, five events with atomic number about Z=100 were detected, however further improvements to the experimental setup are needed in order to increase the detection sensitivity.

EXPERIMENTAL SETUP

The experiment was performed using the BigSol superconducting solenoid time-of-flight spectrometer at the Texas A&M University. A complete description of this device can be found in ref [10]. Fig. 1 shows

Application of Accelerators in Research and Industry
AIP Conf. Proc. 1336, 594-597 (2011); doi: 10.1063/1.3586172

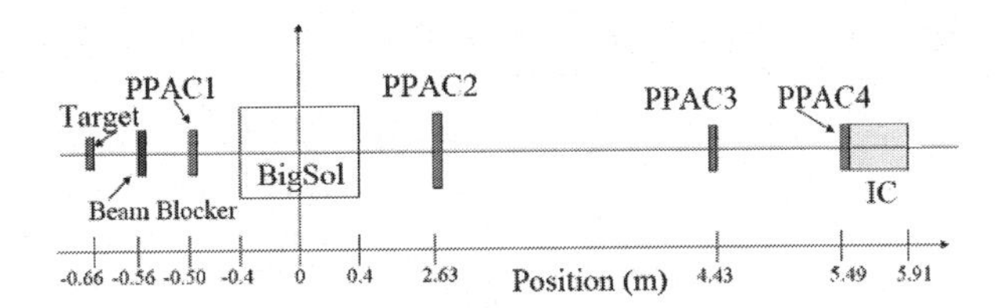

FIGURE 1. Experimental Setup. See the text for details.

the lay-out of the beam line as used in this work. The 7.5 AMeV ^{197}Au beam delivered by the K500 superconducting Cyclotron impinged on a ^{232}Th target of thickness 8.9 mg/cm^2, located near the entrance of the BigSol solenoid. We note that the Coulomb barrier for spherical nuclei in the exit channel is about 680 MeV. Since the ^{232}Th target is very thick the total center of mass energy of the entrance channel drops from 799 MeV to 629 MeV as the beam passes trough the target.

Fragments emitted in the acceptance angle of the solenoid are focused at the focal plane position, approximately 6-meters downstream. A beam blocker and a sixteen-segment annular Parallel Plate Avalanche Counter (PPAC) (PPAC1) placed after the target define the acceptance of the spectrometer in the range 6°-16°. The BigSol magnetic field was tuned in order to optimize the transmission of low energy heavy fragments. The transmission efficiency of the spectrometer for different reaction products was estimated by simulating ion trajectories taking into account the velocity and the average charge state of the ions.

Three position-sensitive PPAC detectors, positioned at 3.3 m (PPAC2), 5.1 m (PPAC3), 6.15 m (PPAC4) from the target, are used to reconstruct fragment trajectories and measure the ions time of flight. The focal plane detector includes PPAC4 and a multi-anode Ionization Chamber (IC), used either as a stopping detector for slow heavy reaction products or as a transmission detector for faster fragments. The IC has an active area of 6.5x6.5 cm^2 and is equipped with 8 parallel anodes, each having a width of 4.65 cm along the beam direction. Several experimental techniques were used in order to eliminate spurious events and pileup. Independent measurements of the time difference between PPAC3 and PPAC4, PPAC2 and PPAC3, PPAC2 and PPAC4 and between PPAC1 and PPAC4 were performed and all time measurements are required to be consistent in order to accept an event. We also implemented a pileup rejection scheme in which we require the measured time between two events in PPAC4 to be more than 8 μs or the event is rejected. Moreover we compared the energy loss in each of the IC anodes measured using both a peak sensing ADC after signal shaping as well as sampling the raw signal using a flash-ADC. These two measurements are also required to agree within 5% in order to accept an event. The analysis of the flash-ADC signals itself also provided a powerful method to reject spurious events.

Time and energy calibrations of the system were obtained by using different heavy ion beams (^{40}Ar, ^{84}Kr, ^{131}Xe, ^{197}Au, ^{238}U) at different energies (7.5 and 15 AMeV) going directly into the detectors with the beam blocker removed. Calibration runs with the direct ^{197}Au beam were repeated at regular time intervals during the measurements. The energy losses in the detectors were calculated using a new effective charge parameterization [11] developed to calculate the stopping powers and the energy loss of the reaction products. This parameterization is based on the existing energy loss data [12, 13] of several heavy ions with 18<Z<92 in the energy range 0.1-15 AMeV, in the detector materials (mylar and isobutane) and can be used to extrapolate the stopping powers for very heavy elements.

The atomic number identification of the reaction products is obtained with a recursive procedure by iterating over the possible Z from 1 to 130 and comparing the stopping powers measured in each anode as a function of the energy per nucleon with the values predicted by our parameterization. The reconstructed Z is the number that minimizes the average of the absolute values of the differences between the measured stopping power and the predicted one. Fig. 2 shows the Z distribution obtained for the direct ^{197}Au beam at 7.5 AMeV; the resolution ΔZ/Z is in this case about 3.5% (FWHM).

In a first approximation, a provisional mass of the fragments is assigned simply by requiring that the A/Z ratio of the entrance channel is conserved. For the ions stopped in the IC, the mass is also reconstructed by dividing the total energy measured in the IC and PPAC4 by the energy per nucleon measured from the time of flight between PPAC3 and PPAC4. Due to the large uncertainty in the measurement of the energy

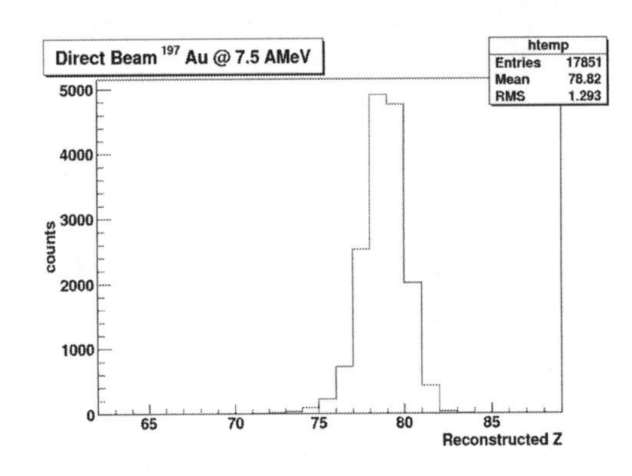

FIGURE 2. Reconstructed Z plot for the direct beam ^{197}Au at 7.5 AMeV.

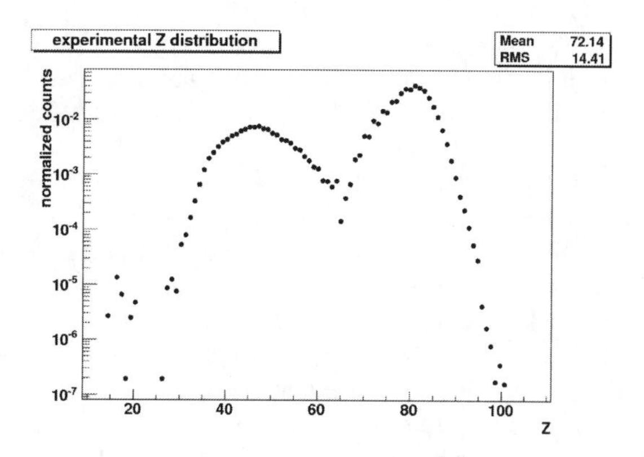

FIGURE 3. Z distribution of the reaction products. The counts are normalized by the total number of events and corrected by the estimated transmission efficiency of the spectrometer.

loss in the PPAC4, the mass resolution ΔA/A is worse than the Z resolution. Therefore in the following analysis we used the reconstructed Z.

The initial energy of the fragments is calculated by correcting the energy measured by the time of flight by the energy losses in the PPACs and in half the target thickness.

EXPERIMENTAL RESULTS

The atomic number distribution of the reaction products emitted in the angular range (6-16°)lab and detected at the BigSol focal plane is shown in Fig.3. By selecting the products emitted in this angular range we intend to filter out relatively central collisions characterized by longer survival time of the DNS and larger energy dissipation compared to peripheral deep inelastic collisions. Different regions in Fig.3 correspond to different reaction mechanisms.

The events with 60<Z<100 result mainly from the break-up of an initial DNS in two fragments. They show a broad peak around the Z=80 corresponding to projectile-like fragments produced in the deep inelastic reaction. The tail of this distribution between Z=60 and Z=70 probably results from charged particle evaporation from the primary excited deep inelastic fragments or from the fission of the very heavy systems. The events with Z<60 are fission fragments from the excited projectile-like or target–like nuclei formed in the deep-inelastic reaction.

A few events (5) with high atomic number 97<Z<102 survived the pileup-rejection filters. The energy loss profile of those events is shown in Fig.4. The energy of those events at the entrance of the IC is about or below 1 AMeV. At this energy the Z

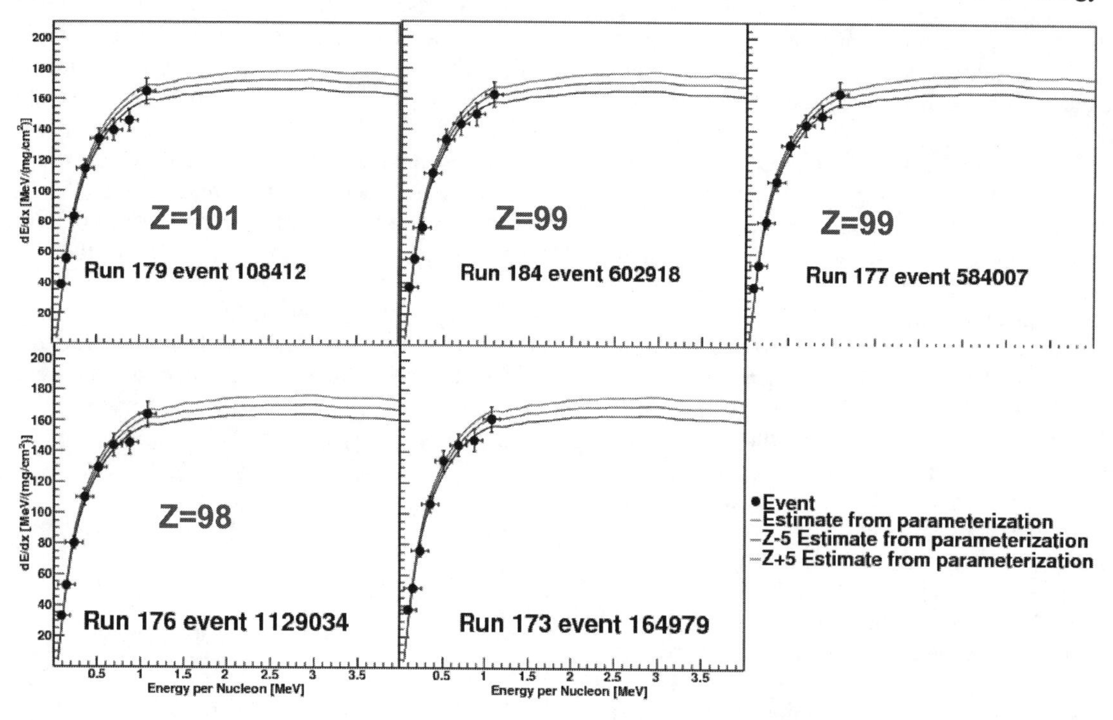

FIGURE 4. Energy loss profile of the five events with Z>97. The red, purple and blue lines show the stopping powers estimated using our parameterization for the identified Z, identified Z+5 and identified Z-5 respectively.

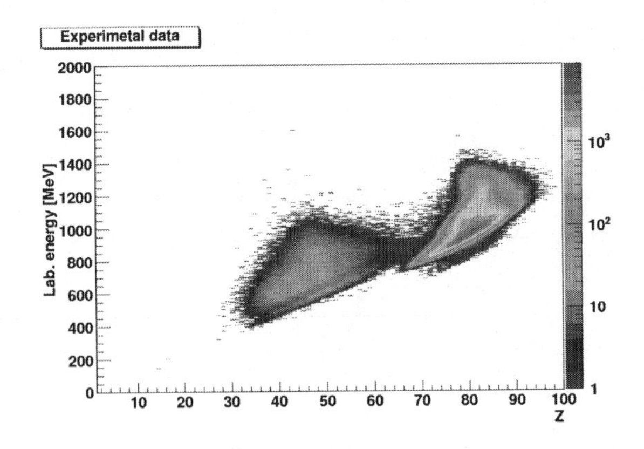

FIGURE 5. Energy of the detected fragment as a function of the atomic number.

resolution of our detector is relatively poor and more accurate measurements are required in order to improve the result. A rough estimate of the reaction cross section for these very heavy elements gives an upper limit of 11 nb/event.

In deep inelastic reactions the kinetic energy of the entrance channel is transformed into internal excitation of different degrees of freedom of the DNS. The analysis of the kinetic energy of the detected fragments as a function of the atomic number can be useful to understand the origin of those events. Unfortunately, with our experimental setup we can detect only one of the reaction products and therefore we cannot obtain a direct measure of the total kinetic energy loss. Fig. 5 shows the energy of the detected fragments as a function of the atomic number. The elastically scattered gold particles are cut by the magnet settings. The low kinetic energy of the products in the range $70<Z<85$ suggests that the detected fragments might originate from relatively long lived systems. The data analysis is still in progress.

CONCLUSIONS

The reaction ^{197}Au+^{232}Th at 7.5 AMeV has been investigated by using the BigSol time of flight spectrometer at Texas A&M. Five nuclei with atomic number larger than 97 were detected during the experiment. The total cross section for those events is about 50 nb. The Z resolution of the system for those events is larger than 5% because their energy is very much degraded before reaching the IC. Therefore a precise identification of the nuclei is not possible. Further improvements of the experimental setup are needed in order to reduce the energy loss of the ions in the detectors. A higher granularity of the IC is also necessary in order to improve the efficiency of the pileup rejection. However the results obtained so far suggest the possibility of producing nuclei with Z about 100.

ACKNOWLEDGMENTS

This work was supported by the US Department of Energy and The Robert A. Welch Foundation (Grant A330) and by the Istituto Nazionale di Fisica Nucleare (project N2P).

REFERENCES

1 S. Hoffmann and G. Münzenberg, *Rev. Mod. Phys.* **72**, 733 (2000).
2. Yu. Ts. Oganessian et al., *Phys. Rev.* C **70**, 064609 (2004).
3. T. Maruyama, A. Bonasera, M. Papa and S. Chiba, *Eur. Phys. J.* A 14, 191 (2002).
4. Wu Xi-Zhen et al., *Chinese Physics* C **33**, 30 (2009).
5. C. Golabek and C. Simenel, *Phys. Rev. Lett.* **103**, 042701 (2009).
6. H. Freiesleben et al., *Z. Phys.* A **292**, 189 (1979).
7. H. Gaggeler et al., *Phys. Rev. Lett.* **45**, 1824 (1980).
8. J.V. Kratz et al., *Phys. Rev.* C **33**, 504 (1986).
9. C. Golabek et al., *Eur. Phys. J.* A **43**, 251 (2010).
10. M. Barbui et al., *Nucl. Instr. and Meth.* B **265**, 605 (2007).
11. M. Barbui et al., *Nucl. Instr. and Meth.* B **268**, 2377 (2010).
12. M. Barbui et al., *Nucl. Instr. and Meth.* B **268**, 20 (2010).
13. A collection of experimental data is available at the H. Paul web page (http://www.exphys.uni-linz.ac.at/stopping/).

Actinide Neutron-Induced Fission Cross Section Measurements At LANSCE

F. Tovesson[a], A.B. Laptev[a] and T.S. Hill[b]

[a]*Los Alamos National Laboratory, Los Alamos NM 87545, United States*
[b]*Idaho National Laboratory, Idaho Falls ID 83415, United States*

Abstract. Fission cross sections of a range of actinides have been measured at the Los Alamos Neutron Science Center (LANSCE) in support of nuclear energy applications in a wide energy range from sub thermal energies up to 200 MeV. Parallel-plate ionization chambers are used to measure fission cross sections ratios relative to the ^{235}U standard while incident neutron energies are determined using the time-of-flight method. Recent measurements include the ^{233,238}U, $^{239-242}$Pu and ^{243}Am neutron-induced fission cross sections. Obtained data are presented in comparison with existing evaluations and previous data.

Keywords: Fission, cross section, actinides, fast neutrons, time-of-flight
PACS: 25.85.Ec, 27.90.+b, 29.30.Hs, 29.40.Cs

INTRODUCTION

To fulfill present and future world energy demand and decrease world economy dependence on fossil fuel the use of nuclear energy should be widely expanded because it is a reliable source of "carbon-free" electricity production. The DOE Fuel Cycle R&D program is supporting research in the technologies foreseen for the next generation of nuclear reactors.

Sensitivity studies carried out as part of the FC R&D program have demonstrated the need to improve the accuracy of some of the nuclear data used for predicting fuel properties and behavior. A program to provide fission cross sections with the required accuracies has been ongoing at the Los Alamos Neutron Science Center (LANSCE) [1] for the last few years. Below we present a general overview of the experimental program and the technical approach used, as well as new result for the ^{233}U and ^{238}U fission cross sections and preliminary result for recent measurements of the ^{243}Am.

EXPERIMENTAL METHOD

The fission cross section measurements at LANSCE are done at a white neutron source, and the time-of-flight (TOF) technique is used to determine the incident neutron energies. The cross sections are measured relative to ^{235}U(*n,f*) using ionization chambers. The cross section of ^{235}U(*n,f*) reaction is a standard at thermal energies and in the range from 0.15 to 200 MeV.

The LANSCE Neutron Source

An 800-MeV linear proton accelerator drives the spallation targets at LANSCE [1]. General lay-out of LANSCE is presented in Fig. 1. The Weapons Neutron Research (WNR) facility gets proton pulses from the accelerator delivered to target 4, which is a bare tungsten target. The pulse repetition rate is 40 Hz, and each macro-pulse has micro-pulse structure of typically 1.8 μs spaced, 150 ps wide proton pulses. The 90L flight path at WNR is used for fission cross section measurement, and is the closest flight path to the spallation target. The fission chambers are at a nominal distance of 10 meters from the spallation target. When running 1.8 μs spacing the lowest accessible neutron energy is 0.15 MeV, and the highest accessible neutron energy is around 200 MeV.

The Lujan Center uses LANSCE target 1, which is surrounded by different moderators. The proton beam repetition rate is 20 Hz, and each pulse is about 250 ns wide. Flight path 5 at the Lujan Center is used for the fission cross section measurements, and the fission chambers are located about 8 meters from target 1 on this flight path. A water moderator shapes the neutron spectrum on this flight path, and the usable neutron energy range is from sub thermal energies, around 1 meV, to 200 keV. The lower limit is set by limited

Application of Accelerators in Research and Industry
AIP Conf. Proc. 1336, 598-602 (2011); doi: 10.1063/1.3586173

statistics, and the higher limit is set by the gamma shower from spallation and the ability to resolve neutron energies. More information on the neutron spectrum for flight path 5 is found in Ref. [2].

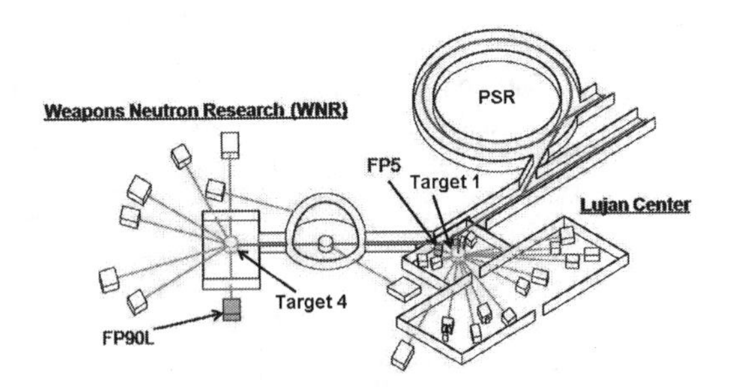

FIGURE 1. Drawing of the Lujan Center and WNR at LANSCE. The proton beam enters from top right. LANSCE targets 1 and 4 and the two flight paths used for the measurements are indicated with arrows. The proton storage ring (PSR) compresses the micro structure of the proton beam into one pulse to be used at the Target 1 of the Lujan Center.

Fission Chambers

The same type of parallel plate ionization chambers commonly used for neutron flux monitoring at WNR is employed for the fission cross section measurements, and the detailed description of the detectors are found in Ref. [3]. The chambers can hold up to four samples, and each sample is mounted in identical geometry in order to perform relative cross section measurements. The chambers typically provide about 1.2 ns (FWHM) timing for fission fragments, and measure the energy loss of particles in the active volume. The gas gap between electrodes is about 14 mm, so only part of total kinetic energy of the fission fragments and alpha-particles is measured for most tracks. Spectra of the measured energy loss are shown in Fig. 2 and are used to identify fission and reject alpha-decay and other backgrounds.

Samples

The samples used for fission cross sections are by necessity very thin, typically about 200 μg/cm^2 and below, so that the fission fragment emitted can escape the material and generate ionization in the active volume of the detector. The efficiency for fission detection depends on the thickness of the target, and at the nominal thickness the detection efficiency is about 97-98%.

The ^{243}Am sample was produced at Idaho Nation Laboratory using electro deposition, and has a total mass of approximately 200 μg (thickness is about 19 μg/cm^2). The sample still needs to be characterized using alpha-counting. The sample preparation procedure is described in Ref. [4]. The ^{233}U and ^{235}U samples were also prepared from electro deposition and had total masses (thicknesses) of 18.7 and 15.8 mg (238 and 201 μg/cm^2), respectively.

Data Acquisition System

The chambers are read out with fast preamplifiers (from RIS Corp.), and NIM electronics is used for signal treatment. A CAMAC system with FERA bus readout is used to digitize the data which is then transferred to a computer.

ANALYSIS

The online data is stored by the MIDAS software package, and is converted to ROOT trees in the post-analysis.

Fission Identification

Fission is identified using software threshold on the fission chamber pulse height. The fission fragments are generally well separated from other events as seen in Fig. 2. The pulse height spectrum from decay radiation is typically seen as a low energy peak in the ADC spectra, and is easy to measure in long beam-off counting. The correction for this type of background is therefore straight forward and associated with relatively small uncertainties. Another, similar type of background is neutron-induced charged particle emission form the backing foils. The response to these events is investigated using data from empty backings, and has been shown to no extend significantly into the ADC peaks produced by fission.

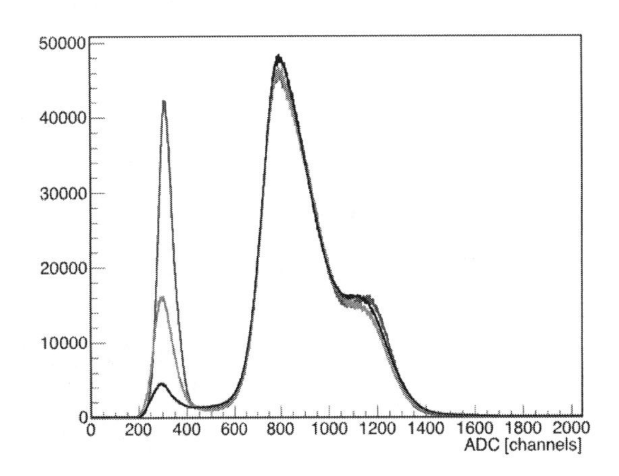

FIGURE 2. Pulse height spectra of signal from fission ionization chamber for fissile targets with different levels of α activity.

Dead Time

The data acquisition system (DAQ) has a fixed dead time of about 17 μs after each event, which is mainly driven by the conversion time of the analogue-to-digital (ADC) conversion. Each foil in the fission chambers is treated by a separate electronic chain, such that the dead times for the different foils are un-coupled.

The dead time correction needed in each measurement is determined by using hardware scalers in the DAQ system. The total number of events is scaled, as well as the number of events accepted by the DAQ, which is used to determine the integral dead time. Typically dead time correction was on the order of 0.5-5%. Details on dead time correction are given in Ref. [5].

Neutron Background

In addition to the neutrons that reach the fission chambers with a one-to-one correlation between time-of-flight and kinetic energy, there are background neutrons with unambiguous relation between TOF and energy. This gives rise to background events in the observed fission spectrum, which needs to be corrected for.

One source of neutron background is "frame overlap", due to the short spacing between proton pulses at WNR. With only 1.8 μs between pulses, some fast neutrons will arrive at the same time as slow neutron from a preceding pulse which leads to ambiguity in the TOF-energy relation. At Lujan Center the pulse spacing is 50 ms, so this problem is eliminated.

Another type of background that only affect WNR experiments is "dark current", which is neutrons produced by proton leakage between pulses. This produces a background of white neutrons that can in some situations make up 1% of the total neutrons.

Neutron scattering around the flight path cave, or "room scattering", affects measurement both at WNR and Lujan Center.

The events from neutron background are carefully corrected for. The "dark current" events are observed in non-fission targets below the fission threshold, and can thus be quantified and subtracted. The "frame overlap" events are modeled and fitted after last micro-pulse in each macro-pulse, as described in Ref. [6]. The room background was investigated by measuring event rates with the fission chambers outside of the neutron beam.

Cross Section Calculation

The normalized ratio of events is used to calculate the measured fission cross section as a ratio to ^{235}U(*n,f*). The ratio is then converted to a cross section using the standard evaluation of ^{235}U(*n,f*) [7], which extends up to 200 MeV.

RESULTS AND CONCLUSIONS

The ^{233}U fission cross section measured at WNR is shown in Fig. 3. The same target was also measured at Lujan Center, so the complete data sets extend down to thermal energies. The low energy data is still being analyzed. The ENDF/B-VII and JENDL-3.3 evaluations are fairly consistent for this reaction, as expected from the large volume of experimental data available. The more interesting part of the current dataset is above 30 MeV, where only two previous data sets are available in EXFOR; the measurement by Lisowski et al. [8] and Shcherbakov et al. [9]. The current results are in very close agreement with those of Shcherbakov et al. at the high energy end, while the Lisowski results are slightly higher.

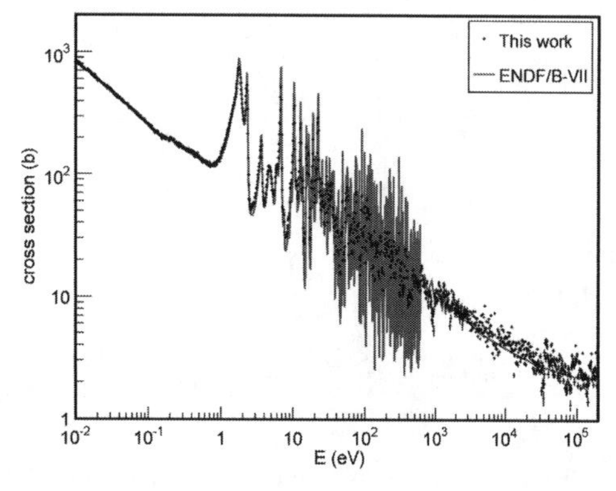

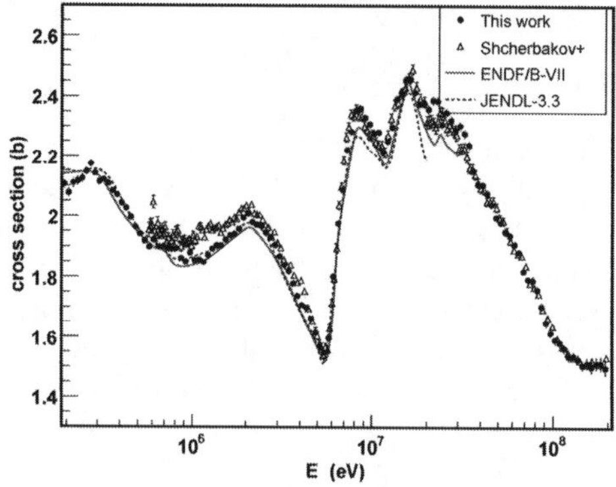

FIGURE 3. Measured fission cross section of ^{233}U from 10 meV to 200 keV (left plot) and from 200 keV to 200 MeV (right plot).

The ^{238}U results above threshold are shown in Fig. 4. The current results are generally higher than the ENDF/B-VII evaluation by about 2%, which is within the normalization uncertainty of the measurement. Above 20 MeV the evaluation is based on the data from Lisowski et al., and the current results are in good agreement with the evaluation. However, the cross sections measured by Shcherbakov et al. [9] are higher by about 8% in this energy region.

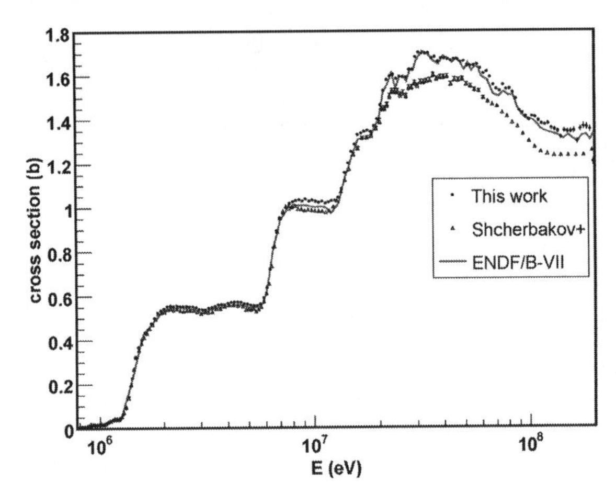

FIGURE 4. Measured fission cross section of ^{238}U from threshold to 200 MeV.

The preliminary result of the ^{243}Am fission cross section measurement is shown in Fig. 5. This is only a shape measurement at this point, since the sample needs to be better characterized. The cross section has been arbitrarily normalized to the average ENDF/B-VII value for first chance fission. For this reaction the absolute value of the cross section for first-chance fission is actually of interest, since values reported in literature have large discrepancies. Recently several measurements of the ^{243}Am fission cross section were done, for example, Laptev et al. [10] and Aiche et al. [11], but absolute normalization still needs to be proved. The current measurement will hopefully be useful in resolving existing discrepancy in worldwide data once the sample mass have been determined.

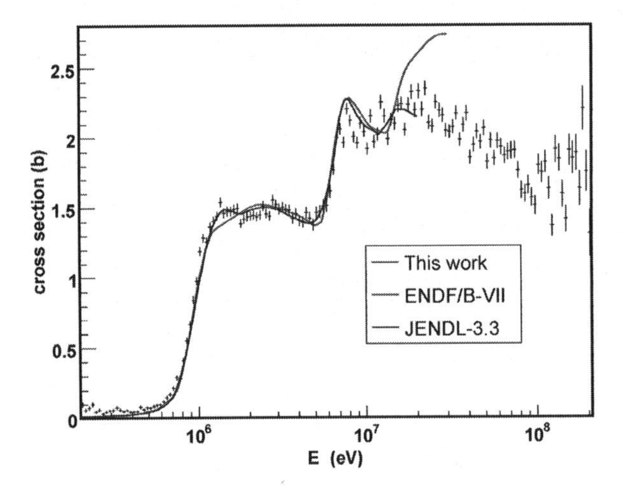

FIGURE 5. Preliminary result for fission cross section of ^{243}Am from 0.2 to 200 MeV.

The results presented here are important a variety of nuclear applications and uncertainties are generally sufficiently low to meet the requirements for those applications. However, in some cases uncertainties below what is attainable with the present techniques are required. An ongoing activity is therefore to develop better detectors for fission measurements that would help reduce systematic uncertainties. A Time Projection Chamber is currently being developed in collaboration between 3 national laboratories and 6 universities [12] for fission measurements, and is planned for use in cross section measurements at LANSCE.

ACKNOWLEDGMENTS

This work has benefited from the use of the Los Alamos Neutron Science Center at the Los Alamos National Laboratory. This facility is funded by the US Department of Energy and operated by Los Alamos National Security, LLC under contract DE-AC52-06NA25396.

REFERENCES

1. P. W. Lisowski and K. F. Schoenberg, *Nucl. Instr. and Meth. A* **562**, 910-914 (2006).
2. M. Mocko, G. Muhrer, and F. Tovesson, *Nucl. Instr. and Meth. A* **589**, 455-464 (2008).
3. S. A. Wender, et al., *Nucl. Instr. and Meth. A* **336**, 226-231 (1993).
4. J. D. Baker, C. A. McGrath, T. S. Hill, R. Reifarth, and F. Tovesson, *J. Radioanalytical Nucl. Chem.* **276**, 555-560 (2008).
5. F. Tovesson and T. S. Hill, *Phys. Rev. C* **75**, 034610 (2007).
6. F. Tovesson and T. S. Hill, *Nucl. Sci. and Eng.* **159**, 94-101 (2008).
7. A. D. Carlson, et al., *Nucl. Data Sheets* **110**, 3215-3324 (2009).
8. P. W. Lisowski, A. Gavron, W. E. Parker, S. J. Balestrini, A. D. Carlson, O. A. Wasson and N. W. Hill, "Fission cross sections ratios for $^{233,234,236}U$ relative to ^{235}U from 0.5 to 400 MeV," in *Nucl. Data for Sci. and Technol., Proc. Int. Conf., Juelich, Germany, 13-17 May 1991*, edited by S.M. Qaim, Berlin: Springer-Verlag, 1992, pp. 732-213.
9. O. Shcherbakov, et al., *J. Nucl. Sci. Technol.* **Supplement 2**, 230-233 (2002).
10. A. B. Laptev, O. A. Shcherbakov, A. S. Vorobyev, R. C. Haight and A. D. Carlson, "Fast neutron-induced fission of some actinides and sub-actinides," in *Fission and Properties of Neutron-Rich Nuclei, Proc. Fourth Int. Conf., Sanibel Island, USA, 11-17 Nov. 2007*, edited by J. H. Hamilton, A. V. Ramayya, and H. K. Carter, New Jersey: World Scientific, 2008, pp. 462-468.
11. M. Aiche, G. Kessedjian, G.Barreau *et al.*, "Quasi-absolute neutron-induced fission cross section of 243Am," in *Proc. Int. Conf. on Nucl. Data for Sci. and Technol., April 22-27, 2007, Nice, France*, edited by O. Bersillon, F. Gunsing, E. Bauge, R. Jacqmin, and S. Leray, Paris: EDP Sciences, 2008, pp. 483-485.
12. M. Heffner, "The Fission Time Projection Chamber Project," in *ND2010, Int. Conf. on Nucl. Data for Sci. and Technol., April 26-30, 2010, Jeju Island, Korea. Abstract Book*, edited by H. I. Kim and Y.-O. Lee, Daejeon: Korea Atomic Energy Research Institute, 2010, p. 143.

Absolute *np* and *pp* Cross Section Determinations Aimed At Improving The Standard For Cross Section Measurements

A.B. Laptev[a], R.C. Haight[a], F. Tovesson[a], R.A. Arndt[b,†], W.J. Briscoe[b], M.W. Paris[b], I.I. Strakovsky[b] and R.L. Workman[b]

[a]*Los Alamos National Laboratory, Los Alamos NM 87545, United States*
[b]*The George Washington University, Washington DC 20052, United States*

Abstract. Purpose of present research is a keeping improvement of the standard for cross section measurements of neutron-induced reactions. The cross sections for *np* and *pp* scattering below 1 GeV are determined based on partial-wave analyses (PWAs) of nucleon-nucleon scattering data. These cross sections are compared with the most recent ENDF/B-VII.0 and JENDL-4.0 data files, and the Nijmegen PWA. Also a comparison of evaluated data with recent experimental data was made to check a quality of evaluation. Excellent agreement was found between the new experimental data and our PWA predictions.

Keywords: Partial-wave analysis, cross section, *np* and *pp* interactions
PACS: 25.40.Dn, 27.10.+h, 28.20.Cz, 29.85.Fj, 29.87.+g

INTRODUCTION

Nucleon-nucleon scattering is the simplest two-body reaction that allows an examination of different nuclear interaction models. Progress in the development of nuclear models is linked to the availability of high-quality data. The *np* scattering is also used as a *primary* standard in measurements of neutron-induced nuclear reactions [1]. Its cross section is used in determining the flux of incoming neutrons.

This information is important in many applications, such as astrophysics, the transmutation of nuclear waste, energy generation, and the conceptual design of an innovative nuclear reactor being carried out in the course of the Generation IV initiative [2]. The increasing quality of neutron-induced nuclear reaction measurements requires a high-quality standard for *np* cross sections, reproducing total *np* cross sections with an accuracy of 1% or better for energies below 20 MeV [1] and [3]. The need for neutron data above 20 MeV up to hundreds of mega-electron-volts with accuracy better than 10% [3] leads to the requirement of cross section data for the *np reference* reaction with uncertainties at the few percent level.

An extensive database exists for nucleon-nucleon scattering, with measurements from laboratories worldwide. These data sets, from the various laboratories, have different statistical and systematic uncertainties that must be taken into account when combined into a single fit. At present, there are several evaluations of the *np* cross sections below 20 MeV. Perhaps most widely known are the ENDF/B [4] and JENDL nuclear data files. The JENDL database has low energy evaluation, below 20 MeV, and high energy data file JENDL/HE [5], up to 3 GeV. Recently the new version of low energy database, JENDL-4.0, was released [6]. For case of *np* cross section data the JENDL-4.0 accepted evaluation from ENDF/B-VII.0 [4]. An R-matrix analysis of the nucleon-nucleon system [7] was used in the course of the ENDF/B-VII.0 (JENDL-4.0) evaluation of *np* cross sections, whereas in the JENDL/HE *np* cross section evaluation, a method based on phase-shift data [8,9] was used.

Here, we will concentrate on total *np* and *pp* cross sections determined on the basis of recent energy dependent (global) fits SP07 [10], LE08 [11] and associated single- energy solutions (SESs) [11] from the George Washington (GW) Data Analysis Center [12]. Precise measurements collected over many years have helped to isolate discrepancies between different experiments and have contributed to a good

† Deceased

Application of Accelerators in Research and Industry
AIP Conf. Proc. 1336, 603-607 (2011); doi: 10.1063/1.3586174

description of nucleon-nucleon scattering at the level of both observables and amplitudes.

DATABASE

The GW fit to nucleon-nucleon elastic scattering data covers an energy range up to 1.3 GeV (for *np* data) and 3 GeV (for *pp* data). The *np* analysis was restricted to 1.3 GeV because of a lack of high-energy data. The full database includes all available unpolarized and polarized measurements. A number of fits, from the GW group and others, are available through the online SAID facility [12].

The latest evolution of the SAID database is summarized in [10]. At present it contains 12,693 (24,916) data points for *np* (*pp*) interactions.

Not all of the available data have been used in each fit. Some data with very large χ^2 contributions have been excluded. Redundant data are also excluded. Polarized measurements with uncertainties >0.2 are not included as they have little influence on GW fits. A complete description of the database, and those data not included in GW fits, is available from the authors [12].

PARTIAL-WAVE ANALYSIS

Simultaneous fits to the full database are possible within the formalism used and described in seven previous GW analyses, which are regularly updated online; the full bibliography can be found in [11]. The observables are represented in terms of partial-wave amplitudes, using a Chew-Mandelstam K-matrix approach, which incorporates the effect of an $N\Delta$ channel on the nucleon-nucleon scattering process. By parameterizing the K-matrix elements as functions of energy, data up to 3 GeV can be fitted simultaneously (both *pp* and *np*, with a 1.3 GeV limit for *np*). In general, GW PWAs have attempted to remain as model-independent as possible.

In fitting the data, systematic uncertainty has been used as an overall normalization factor for angular distributions. With each angular distribution, we associate the pair (X,ε_X): a normalization constant (X) and its uncertainty (ε_X). The quantity ε_X is generally associated with the systematic uncertainty (if known). The modified χ^2 function, to be minimized, is then given by

$$\chi^2 = \sum_i \left(\frac{X\theta_i - \theta_i^{\exp}}{\varepsilon_i} \right)^2 + \left(\frac{X-1}{\varepsilon_X} \right)^2. \quad (1)$$

Here the subscript i labels data points within the distribution, θ_i^{exp} is an individual measurement, θ_i is the calculated value, and ε_i is the statistical uncertainty. For total cross sections and excitation data, we have combined statistical and systematic uncertainties in quadrature. Renormalization freedom significantly improves GW best-fit results. In the global solution SP07 [10] χ^2/data is 21,496/12,693 for *np* and 44,463/24,916 for *pp*.

Starting from this global fit, we have also generated a series of SES results. Each SES is based on a "bin" of scattering data spanning a narrow energy range. A total of 43 SESs have been generated, with central energy values ranging from 5 MeV to 2.83 GeV and bin widths varying from 2 to 75 MeV. In generating the SES, a linearized energy dependence is taken over the energy range, reducing the number of searched parameters. A systematic deviation between the SES and global fits can be an indication of missing structure in the global fit (or possibly problems with a particular data set). An error matrix is generated in the SES fits, which can be used to estimate the overall uncertainty in the global fit. Further details on the global and SES fit results are given in [10] and [11].

TOTAL *np* AND *pp* CROSS SECTIONS

Isovector and isoscalar partial-wave amplitudes, determined through the PWA, have been used to generate total *np* and *pp* cross sections. Also generated was a low-energy fit to 25 MeV, LE08, which searches 19 parameters, scattering length a, and effective range r for three S waves and 13 leading parameters for S, P, and D waves. LE08 results in a χ^2/data = 696/391 for *pp* and 627/631 for *np*. The numerical data for LE08 can be found in [10] or retrieved from the SAID [12]. Errors for LE08 have been generated from the error matrix and require some comments.

In the region below 25 MeV, there are numerous total cross section measurements for *np* but not for *pp*, which is hindered by large Coulomb effects. As a result, the *np* error estimates are more reasonable. Those quoted for *pp* are far too small (lower limits) in the threshold region.

For the region above 25 MeV, the SES errors give a more accurate estimate of the uncertainty in our cross sections. The amplitudes found in GW fits to 1 GeV have remained stable for many years against the addition of new measurements. Sufficient observables exist for a direct amplitude reconstruction at many energies, and we have compared GW amplitudes to those found in this way in [10].

As cross sections change rapidly with energy, we have chosen to display the agreement between various fits in terms of ratios. This gives a clear picture of the overall consistency and reveals cases where systematic deviations are present. The ratios of SES values to the global fit SP07, below 20 MeV, are displayed in Fig. 1a. Also plotted is a band showing the ratio of LE08 to SP07 determinations of the *np* cross section. As expected, this band more closely reproduces the *np* SES, plotted as single points with error bars, than the 3 GeV fit SP07. Deviations are within 1% for the *np* determinations and within 2% for *pp*.

In Fig. 1b, we plot ratios of SP07 and SES, for both *np* and *pp* cases, to the Nijmegen PWA predictions [13]. The low-energy Nijmegen total *pp* cross sections are systematically above SP07 (~2% or less) while *np* cross sections agree with SP07 at better than the 0.3% level.

In Fig. 2, we plot ratios of the GW *np* fits with the ENDF/B-VII.0 [4] and JENDL/HE [5] nuclear data files. A slightly better agreement is found with JENDL/HE than with ENDF/B-VII.0, though the wiggles seen in Fig. 2b reflect a lack of smoothness in JENDL/HE (SP07 and LE08 are a smooth function of energy). Apparently those wiggles resulted from linear interpolation between some reference points in the JENDL/HE evaluation. The ENDF/B-VII.0 result is systematically below SP07 and the Nijmegen fit [13], but the maximal deviation is only 1%. SP07 and JENDL/HE agree at the 0.5% level over most of the region below 20 MeV.

At higher energies (up to 1 GeV), ratios of the grid of SES to SP07 differ from unity by <3% (Fig. 3a). Above 180 MeV, SAID *np* cross sections are larger than JENDL/HE [5] by up to 5% (Fig. 3b).

COMPARISON WITH RECENT DATA

To evaluate a quality of data predictions of obtained PWA solutions, a comparison was done with new data, which were not included in the analysed database to the moment of the last analysis. Fig. 4a and 4c present our global solution SP07 [10] and SES predictions for the *np* total cross section in the energy range from 8 to 500 MeV along with experimental data measured at the LANL [14]. Fig. 4b and 4d display a ratio of experimental cross section [14] to those from the SP07 solution. One can see an excellent agreement of PWA prediction and experimental observable. Maximal discrepancy is nearly 1%. The χ^2 per data point calculated including experimental systematic errors is about 0.4 for dataset [14].

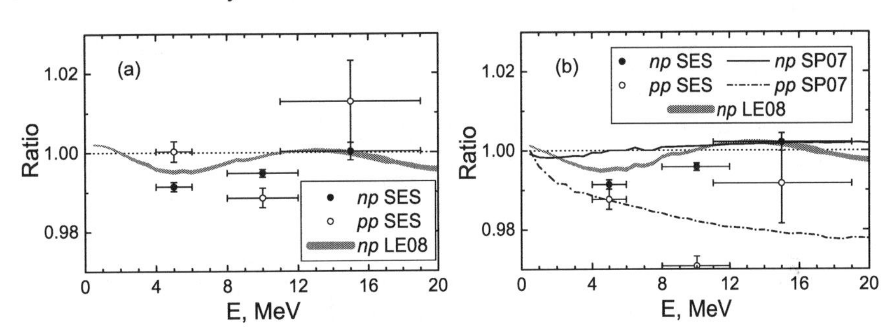

FIGURE 1. Ratios of total *np* and *pp* cross sections below 20 MeV. Horizontal bars give the energy binning of SES. (a) Single-energy to energy-dependent SP07 [10] ratios are plotted. The band represents the ratio of LE08 [11] to SP07 for the *np* case. (b) SES [11] and SP07 divided by Nijmegen PWA predictions [13] are plotted. The band represents the ratio of LE08 to Nijmegen PWA for the *np* case.

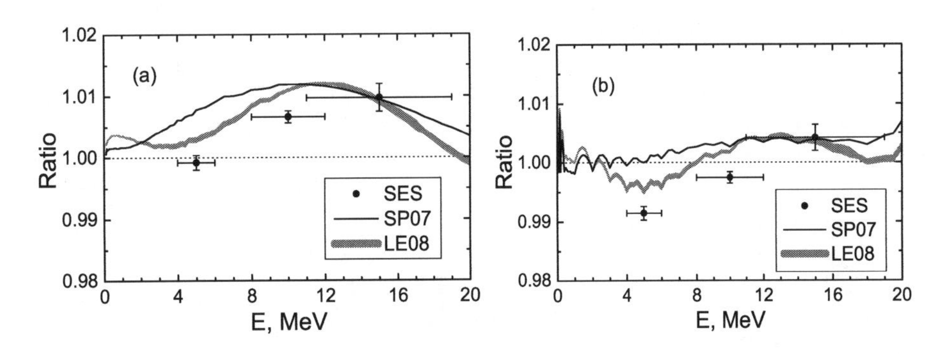

FIGURE 2. Ratio of total *np* cross sections below 20 MeV. Horizontal bars give the energy binning of SES. (a) Single-energy and SP07 [10] fits divided by the ENDF/B-VII.0 [4] results are plotted. The band gives a ratio of LE08 to ENDF/B-VII.0. (b) The same for JENDL/HE [5] evaluated data.

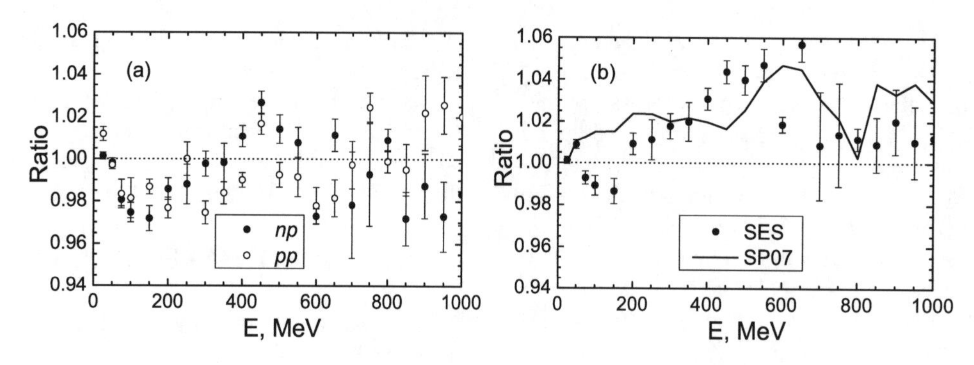

FIGURE 3. Ratios of total *np* and *pp* cross sections between 20 MeV and 1 GeV. (a) SES [11] to SP07 [10] ratios are plotted. (b) *np* SES and SP07 divided by JENDL/HE [5] results are plotted.

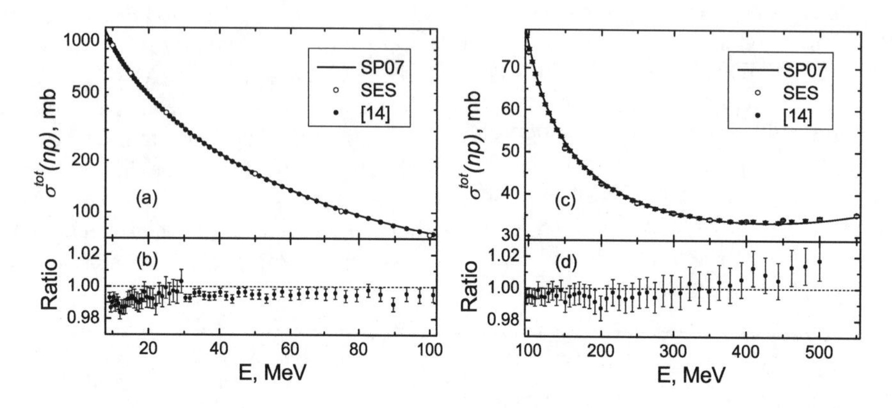

FIGURE 4. Total *np* cross sections in the energy range from 8 to 500 MeV. (a) and (c) SP07 [10] and SES [11] predictions are plotted along with experimental data [14]. (b) and (d) Data [14] divided by the SP07 fit are plotted.

SUMMARY AND CONCLUSIONS

We have generated fits to describe the total *np* and *pp* scattering cross sections below 1 GeV. These fits have been both energy dependent (SP07, LE08) and single energy (analyzing narrow bins of data). The uncertainties associated with our total *np* cross sections below 20 MeV are clearly <1%. The agreement between SP07, JENDL/HE, and the Nijmegen analysis suggests an uncertainty of 0.5% or less. A comparison with ENDF/B-VII.0 shows deviations of 1% or less. Errors on the LE08 solution, while obtained using a well-defined method, are lower bounds as they do not account for systematics effects.

For the *pp* cross sections, uncertainties are larger (e.g. Fig. 1), and systematic disagreements are evident in comparisons with the Nijmegen PWA. The main problem stems from a lack of relevant *pp* data at low energies. Here also, at low energies, the various determinations agree at the few-percent level.

The PWA prediction capabilities checked by comparison with recent experimental data, which were not used in our analysis. Excellent agreement was found.

The advantage of the GW parameterization is its smooth energy dependence and coverage from threshold to high energies. We also have the capability to modify the GW fits to either generate SES centered on a particular energy or produce lower-energy fits when a specific energy region is of interest. We will continue to update both GW energy-dependent solutions and SESs as the new measurements become available.

ACKNOWLEDGMENTS

This work was supported in part by the U.S. Department of Energy under grant DE–FG02–99ER41110.

REFERENCES

1. A. D. Carlson *et al.*, *Nucl. Data Sheets* **110**, 3215-3324 (2009).
2. U.S. Department of Energy Nuclear Energy Research Advisory Committee/Generation IV International Forum "A Technology Roadmap for Generation IV Nuclear Energy Systems", GIF–002–00 (2002).
3. G. Aliberti *et al.*, *Ann. Nucl. Energy* **33**, 700-733 (2006).
4. M. B. Chadwick *et al.*, *Nucl. Data Sheets* **107**, 2931-3060 (2006).
5. Y. Watanabe *et al.*, "Status of JENDL High Energy File," in *ND2010, Int. Conf. on Nucl. Data for Sci. and Technol., April 26-30, 2010, Jeju Island, Korea. Abstract Book*, edited by H. I. Kim and Y.-O. Lee, Daejeon: Korea Atomic Energy Research Institute, 2010, p. 48.
6. K. Shibata *et al.*, "JENDL-4.0: A New Library for Innovative Nuclear Energy Systems," in *ND2010, Int. Conf. on Nucl. Data for Sci. and Technol., April 26-30, 2010, Jeju Island, Korea. Abstract Book*, edited by H. I. Kim and Y.-O. Lee, Daejeon: Korea Atomic Energy Research Institute, 2010, p. 32.
7. G. M. Hale and A. S. Johnson, "Results for n + p Capture from an R-Matrix Analysis of N-N Scattering," in *Few-Body 17, Proc. 17th Int. Union of Pure and Applied Physics Conf. Few-Body Problems in Physics, Durham, North Carolina, USA, June 5-10, 2003*, edited by W. Gloeckle and W. Tornow, Amsterdam: Elsevier B.V., 2004, pp. S120-S122.
8. R. A. Arndt, J. S. Hyslop, and L. D. Roper, *Phys. Rev. D* **35**, 128-144 (1987).
9. R. A. Arndt, L. D. Roper, R. A. Bryan, R. B. Clark, B. J. VerWest, and P. Signell, *Phys. Rev. D* **28**, 97-122 (1983).
10. R. A. Arndt, W. J. Briscoe, I. I. Strakovsky, and R. L. Workman, *Phys. Rev. C* **76**, 025209 (2007).
11. R. A. Arndt, W. J. Briscoe, A. B. Laptev, I. I. Strakovsky, and R. L. Workman, *Nucl. Sci. Eng.* **162**, 312-318 (2009).
12. R. A. Arndt, W. J. Briscoe, M. W. Paris, I. I. Strakovsky, and R. L. Workman (unpublished). The full SAID database and numerous PWAs can be accessed at the website http://gwdac.phys.gwu.edu.
13. V. G. J. Stoks, R. A. M. Klomp, M. C. M. Rentmeester, and J. J. de Swart, *Phys. Rev. C* **48**, 792-815 (1993).
14. W. P. Abfalterer, F. B. Bateman, F. S. Dietrich, R. W. Finlay, R. C. Haight, and G. L. Morgan, *Phys. Rev. C* **63**, 044608 (2001).

Multi-Channel Integrated Circuits For Use In Research With Radioactive Ion Beams

G. L. Engel[a], V. Vangapally[a], N. Duggireddi[a] ,L. G. Sobotka[b], J. M. Elson[b] and R. J. Charity[b]

[a]*Department of Electrical and Computer Engineering, IC Design Research Laboratory, Southern Illinois University Edwardsville, Illinois, USA, 62026*
[b]*Departments of Chemistry and Physics, Washington University, Saint Louis, Missouri, USA, 63130*

Abstract. The Integrated Circuits Design Research Laboratory at Southern Illinois University Edwardsville (SIUE) has been collaborating over the past several years with the Nuclear Reactions Group at Washington University (WU) on the development of a family of custom, multi-channel Integrated Circuits (ICs). To date, the collaboration has successfully produced two micro-chips. The first was an analog shaped and peak sensing chip known as HINP16C (Heavy Ion Nuclear Physics – 16 Channel). The second chip, christened PSD8C (Pulse Shape Discrimination – 8 Channel), was designed to logically complement (in terms of detector types) the HINP16C chip. The HINP16C chip, for use with solid-state detectors, produces sparsified analog pulse trains for both linear (pulse height) and timing (relative to an external reference) signals. A shaper and peak detector are implemented in the linear branch, and a pseudo Constant-Fraction Discriminator (CFD) and Time-to-Voltage Converter (TVC) are implemented in the logic/timing branch. The internal and external gain options make the chip suitable for use in a wide variety of applications. PSD8C greatly simplifies the pulse-processing electronics needed for large arrays of scintillation detectors. As the PSD8C employs user-controlled, multi-region charge integration, particle identification is inherent in the design. Each of the three pulse-height sub-channels consists of a gated integrator with eight programmable charging rates and an externally programmable gate generator that defines the start (with 4 time ranges) and the width (with 4 time ranges) of the gate, relative to an external discriminator signal. Moreover, each channel on the chip also contains a TVC. This paper describes the design, capabilities, and features of the HINP16C and PSD8C ICs along with a description of how the chips are being used. It also discusses modifications and enhancements which have been made to both chips and their associated prototypical systems in order to improve ease of use and increase performance.

Keywords: integrated circuit, pulse shape discrimination, analog signal processing, Si strip detectors
PACS: 85.40-e, 07.50.Qx, 07.05.Hd

INTRODUCTION

The Nuclear Reactions Group at WU and the Integrated Circuits Design Research Laboratory at SIUE have been working over the past decade on a family of multi-channel custom integrated circuits (ICs) for use in research with radioactive ion beams. The group's first success was an analog shaped and peak sensing chip. This IC, known as HINP16C (Heavy-Ion Nuclear Physics – 16 Channel) [1] is designed for use with arrays of Si-strip detectors with a few hundred to a few thousand channels. The second chip, christened PSD8C (Pulse Shape Discrimination – 8 Channel) [2], performs Pulse-Shape Discrimination (PSD), and thus particle identification, if the time dependence of the light output of the scintillator depends on particle type. Moreover, both chips use almost all of the same supporting hardware and both ICs were fabricated in the ON-Semiconductor (formerly AMI) 0.5 μm n-well process (C5N), available through MOSIS (see www.mosis.com).

ANALOG SHAPED AND PEAK SENSING IC

The architecture of the HINP16C chip is illustrated in Fig. 1. The basic architecture has remained virtually unchanged throughout the course of several design revisions. The first generation HINP16C chip is described in detail in [1]. HINP16C produces sparsified pulse trains for both linear (pulse height) and timing (relative to an external reference). A shaper and peak detector are implemented in the linear branch, and a Constant-Fraction Discriminator (CFD)

Application of Accelerators in Research and Industry
AIP Conf. Proc. 1336, 608-613 (2011); doi: 10.1063/1.3586175

and Time-to-Voltage Converter (TVC) is implemented in the logic/timing branch. Two internal Charge Sensitive Amplifier (CSA) gain options and an external gain option along with the preparation of both pulse height and timing pulse trains with synchronized addresses (suitable for pipelined ADCs) make the chip suitable for a wide variety of applications. The pulse-height and the relative time are all stored on capacitors and are either reset (by the auto-reset circuits depicted in Fig. 1), after a user controlled time, or sequentially read out if acquisition of the event is desired.

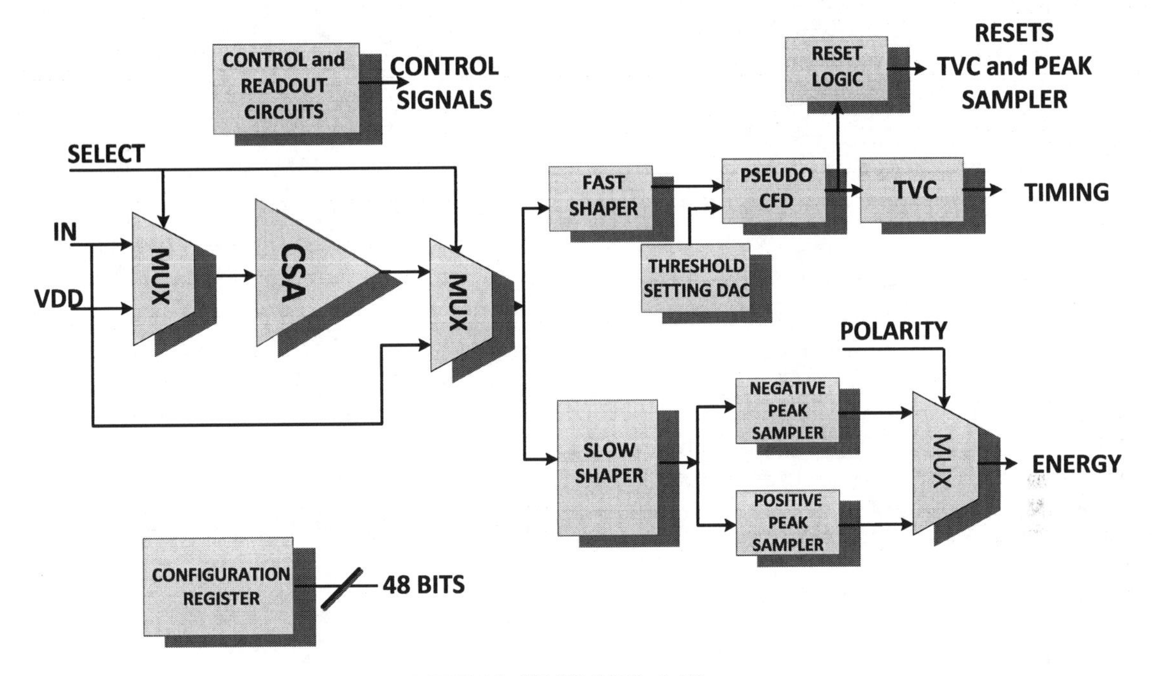

FIGURE 1. HINP16C Block Diagram.

The current version of HINP16C is 4 mm x 6.4 mm and is packaged in a 14 x 14 mm, 128 lead, thin quad flat pack. Power consumption is approximately 800 mW, and the cost per channel of the ASIC varies from \$15 - \$25 depending on the quantity ordered. Important features of HINP16C are summarized in Table 1.

Unfortunately, some changes made in order to improve the IC adversely impacted the performance of the aforementioned auto-reset circuit. The newest version of HINP16C is expected to be ready for fabrication in late 2010. In this upcoming version of HINP16C, problems with the auto-reset circuit will be fixed. Moreover, we plan to modify the existing CSA circuit (see Fig. 2) so as to include a single CSA gain (*high-gain*, 0.5 mV/fC), but the circuit may be programmed to include a diode-connected FET in the feedback path which will extend the dynamic range of this mode to approximately 400 MeV. This extension of range is needed for heavy-ion break-up experiments, conducted at hundreds of MeV per nucleon, when heavy ion residues, as well as light ions, hit the same detector. (The heavy-ions can drop 100's of MeV in a 300 μm Si while protons will drop less than a MeV). Simulation results demonstrating the CSA's extended dynamic range (electron collection) are presented in Fig. 3. Similar results were obtained when holes were collected.

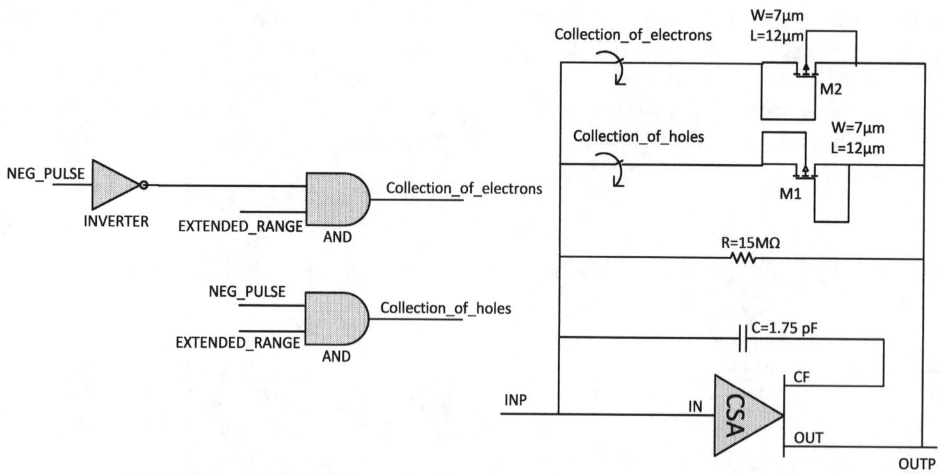

FIGURE 2. HINP16C (Rev. 4) CSA circuit with extended dynamic range.

PSD-ENABLING IC

Our PSD8C chip greatly simplifies the pulse-processing electronics needed for large arrays of scintillation detectors. The structure of a single PSD8C sub-channel is illustrated in Fig. 4. Each of the chip's 8 channels possesses 3 sub-channels. The sub-channels are referred to as A, B, and C. A sub-channel consists of an integrator and a gate generator. External control voltages (D_X, W_X where x = A, B, C) determine the gate delay and gate width. Because PSD8C employs (user-controlled) multi-region charge integration, particle identification is incorporated into the basic design. Each channel on the chip also contains a TVC which provides relative time information. The pulse-height integrals and the relative time are all stored on capacitors and are either reset, after a user controlled time, or sequentially read out if acquisition of the event is desired in a manner similar to that of HINP16C.

PSD8C is described in detail in [2]. The IC is 2.25 mm x 5.7 mm and is packaged in a 14 x 14 mm, 128 lead thin quad flat pack. Power consumption is 60 mW (low-bias mode) or 135 mW (high-bias mode). The cost per channel is $25. Key features of the design are summarized in Table 2.

A second version of PSD8C was submitted for fabrication in May 2010. This version attempts to correct several minor problems identified (over the past year or so) during testing of the PSD8C IC. First, it was found that the TVC circuits could inadvertently be re-started. This logic error has been fixed.

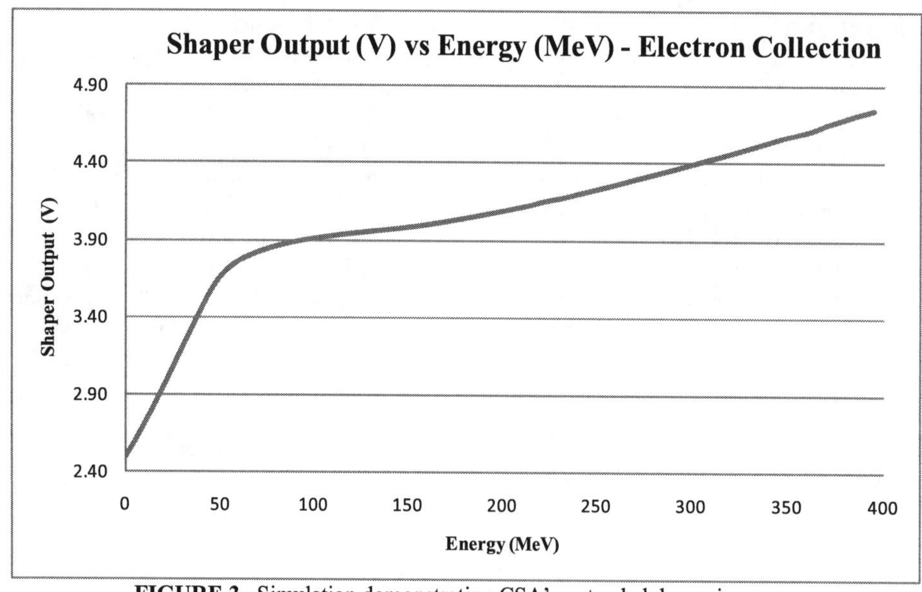

FIGURE 3. Simulation demonstrating CSA's extended dynamic range.

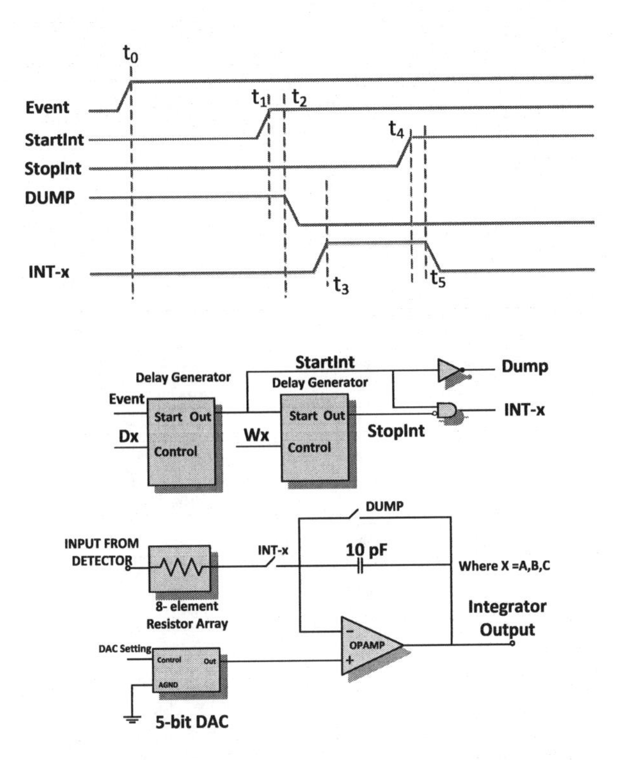

FIGURE 4. Each channel contains 3 sub-channels referred to in the text as sub-channels A, B and C. Here we show a representative sub-channel, X, where X is A, B or C

Second, undesirable temperature dependence (1 ns / C) in the TVC circuit was noted and traced to the local channel buffer. The buffer was re-designed, and the TVC temperature sensitivity has been greatly reduced (5 ps/C in the 500 ns mode, 40 ps/C in the 2 μs mode). Third, some TVC crosstalk issues were remedied.

Finally, at the system level, the chip-boards (printed-circuit boards) are being redesigned to include on-board ADCs (one for each of the chip's four analog output pulse trains). In the future, data will be transmitted in a digital format to the motherboard. This change will potentially reduce the time required for readout (since data from all chips in the system would be read out in parallel), make the IC easier to use, and improve performance since digital data is less susceptible to environmental noise than are low-level analog signals.

WHO IS USING THE ICs?

Nuclear Physics groups at Washington University (WU), Michigan State University (MSU), Indiana University (IU), Texas A&M University (TAMU), Oak Ridge National Laboratory (ORNL), Louisiana State University (LSU), and Florida State University (FSU) are either using HINP16C or will be doing so by summer of 2010.

The most productive use of the original HINP16C chip to date has been for the study of high-order multi-particle correlations in the study of the continuum structure of light nuclei. These works include the study of ^{6}Be [3, 4], ^{10}C [5, 6, 7], ^{12}Be and other light neutron-rich nuclei [8, 9]. However it has also been used for the study of nucleon knock-out from secondary beams [10] and for transfer reactions on both light and heavy Ar isotopes, also with secondary beams [11]. The current version of the HINP16C chip was used for a 5-particle correlation study of the decay of ^{8}C (into an alpha and 4 protons) [12] and for a set of transfer reactions on heavier systems than in the initial transfer work.

A group at Los Alamos National Laboratory (LANL) has been helping to characterize the PSD8C's performance using several different scintillators [13]. Ongoing work at LANL is using this chip for both the very bright Lathinum-Halide scintillators, (e.g. $LaBr_3$(Ce)) and the new class of CLYC scintillators (e.g. Cs_2LiYCl_6(Ce)) [14]. In the former case, it is just the simplification offered by the chip in running large arrays which is utilized while in the second case, the pulse-shape discrimination offered by the chip is used (along with pulse height) to differentiate neutrons from gamma rays.

At Washington University In St. Louis, the group led by D. G. Sarantites is planning on using this chip to provide the signal processing for a large array of scintillators using the fast plastic – CsI(Tl) phoswich concept employed by the ``Dwarf Ball and Wall'' device [15]. Light charged particles (p,d,t,^{3}He, ^{4}He, ^{6}He) are identified using the time dependence of the light output of the CsI(Tl) while the atomic number of elements with Z<8 are identified from the light partition between the fast plastic and the CsI(Tl). The latter concept is similar to that employed in standard dE-E telescopes. This application makes use of all three integration windows: one (prompt and narrow) for the fast plastic and two (delayed – to different extents) for the CsI(Tl). The device being built, unlike the Dwarf project, makes use of multi-element PMT's, have a position resolution of close to 1 mm, and will be designed to study heavy-ion transfer reactions using secondary beams.

SUMMARY

Since 2001, our university-based group has been working on the development of a "toolbox" of IC circuits useful for researchers working with radioactive ion beams. The circuits which we have designed can be composed in different ways to meet the researchers' evolving needs and desires. To date, the group has produced two micro-chips: one analog shaped and peak sensing (HINP16C) while the other multi-sampling and PSD-enabling (PSD8C). The ICs are currently being used by a growing number of researchers across the country and the world.

TABLE 1. HINP16C parameters.

Parameter	Value
Number of channels	16
Programmable	via 48-bit register
Data sparsification	yes
CSA gain modes	0.5 mV/fC (high-gain) 0.1mV/fC (low-gain) external preamp
CSA decay time constant	25 μs
Full scale range	100 MeV (high-gain mode) 500 MeV (low-gain mode)
Energy resolution (C_{det} = 75 pF)	high-gain mode: 28 keV low-gain mode: 42 keV
Polarities supported	holes and electron
Slow shaper peaking time	1.2 μs (fixed)
Slow shaper return to baseline	< 20 μs
Fast shaper time constants	250 ns 620 ns
Automatic channel reset	set by external voltage
Discriminators on chip	Nowlin CFDs
Threshold setting DAC	6 bits
Range of CFD threshold setting DAC	1 MeV (step size 35 keV)
TVC ranges (full-scale)	500 ns and 2 μsec
Walk through CFD	< 1 ns over range of 40 dB
Jitter	1 ns
Diagnostic modes	yes
Power consumption	800 mW
Supply voltage	5 VDC
Cost per channel	$15 -$25
Die area	4 mm x 6.4 mm
Package	128 lead, TQFP
Technology	0.5 micron CMOS

TABLE 2. PSD8C parameters.

Parameter	Value
Number of channels	8
Programmable	via 48 bit register
Data sparsification	yes
Polarities supported	+/-
Input/Output voltage range	+/- 2V rel. AGND
Automatic channel reset	yes
TVC ranges (full scale)	500 ns and 2 μsec
Triggering modes	3
Integration regions	3
Delay and integration width ranges	20 ns – 70 ns 50 ns - 300 ns 200 ns - 1.5 μs 1 μs - 10 μs
Integrator programmable charging rates (8)	500 Ω - 100 kΩ
Integrator offset canceling DAC	5 bits, +/- 25 mV FS
Diagnostic modes	yes
Power modes	2 (slow/fast scintillators)
Power consumption	60 mW (low-power) 135 mW (high-power)
Supply voltage	5 VDC
Cost per channel	$25
Die area	2.25 mm x 5.7 mm
Package	128 lead, TQFP
Technology	0.5 micron CMOS

ACKNOWLEDGMENTS

Early work on HINP16C was supported in part by an NSF MRI grant to build the High Resolution Si Array (HiRA) and the U.S. Department of Energy under Grant No. DE-FG02-87ER-40316. The support for the PSD8C chip development was from NSF Grant #06118996 while the implementation support came from the U. S. Department of Energy, Division of Nuclear Physics under grant # DOE-FG02-87ER-40316. Currently, work on PSD8C is sponsored by a grant from LANL. For the latter we are indebted to Dr. Mark Wallace.

REFERENCES

1. G. Engel, M. Sadasivam, M. Nethi, J. M. Elson, L.G. Sobotka and R. J. Charity, *Multi-Channel Integrated Circuit for Use in Low and Intermediate Energy Nuclear Physics - HINP16C*, Nucl. Instru. Meth. A 573, 418-426, (2007).
2. G.L. Engel, M.J. Hall, J.M. Proctor, J.M. Elson, L.G. Sobotka, R. Shane, R.J. Charity, *Design and Performance of a Multi-Channel, Multi-Sampling*, PSD-Enabling Integrated Circuit, Nucl. Instru. Meth. A, 612, 161-170 (2009).
3. L.V. Grigorenko, et al., *Three-body decay of* ^{6}Be, Phys. Rev. C 80, 034602 (2009).
4. L.V. Grigorenko, et al., *Complete correlation studies of two-proton decays:* ^{6}Be and ^{45}Fe, Phys. Lett. B 677, 30 (2009).
5. R. J. Charity, K. Mercurio, L. G. Sobotka, J. M. Elson, M. Famiano, A. Banu, C. Fu, L. Trache, and R. E. Tribble, *Decay of* ^{10}C *excited states above the* $2p+2\alpha$ *threshold and the contribution from 'democratic' two-proton emission*, Phys. Rev. C 75, 051304(R) (2007).
6. K. Mercurio, et al., *Correlated two-proton decay from* ^{10}C, , Phys. Rev. C 78, 031602(R) (2008).
7. R. J. Charity, T. D. Wiser, K. Mercurio, R. Shane, L. G. Sobotka, A. H. Wuosmaa, A. Banu, L. Trache, and R. E. Tribble, *Continuum spectroscopy with a* ^{10}C *beam; Cluster structure and three-body decay*, Phys. Rev. C 80, 024306 (2009).
8. R. J. Charity, et al., *Particle decay of* ^{12}Be *excited states*, Phys. Rev. C 76, 064313 (2007).
9. R. J. Charity, et al.,*Investigation of particle-unbound excited states in light nuclei with resonance-decay spectroscopy using a* ^{12}Be *beam*, Phys. Rev. C 78, 054307 (2008).
10. D. Bazin, et al., *Mechanisms in knockout reactions*, Phys. Rev. Lett. 102, 232501 (2009).
11. Jenny Lee, et al., *Neutron Spectroscopic Factors of* ^{34}Ar *and* ^{46}Ar *– Study of asymmetry dependence of reduction factors in transfer reactions*, Phys. Rev. Lett. 104, 112701 (2010).
12. R. J. Charity, et al., ^{8}C *caught in the act of decaying: sequential steps of simultaneous two-proton emission*, submitted for publication 2010.
13. Dongwon Lee, Ed McKigney, Laura Stonehill, Shawn Tornga, Mark Wallace, Lee G. Sobotka, George L. Engel, Jon M. Elson, *Pulse shape discrimination analysis of various scintillators using PSD-capable ASIC readout*, SORMA Conference, (2010).
14. Jarek Glodo, William M. Higgins, Edgar V. D. van Loef, and Kanai S. Shah, *Scintillation Properties of 1 Inch* $Cs_2LiYCl_6(Ce)$ *Crystals*, IEEE Trans. on Nucl. Sci. Vol. 55, No. 3, June 2008.
15. D.W. Stracener, D.G. Sarantites, L.G. Sobotka, J. Elson, J.T. Hood, Z. Majka, V. Abenante, A. Chbihi, and D.C. Hensley, *Dwarf Ball and Dwarf Wall: Design, Instrumentation, Response Characterisics of a* 4π *CsI(TI) Plastics Phoswich Multidetector System for Light Charged Particle and Intermediate Mass Fragment Spectrometer*, Nucl. Instru. Meth. A 294, 485 (1990).

Design Features Of K=100 Cyclotron Magnet For ISOL RIB Production

[a]JIN AH PARK, [b]Kh.M.M.Gad, [a,b]Jong-Seo Chai*

[a]*Lab. of Accelerator and Medical Engineering, School of Information and Communication, Engineering, SungKyunKwan University, 300 Cheoncheon–dong, Suwon Gyeonggido 440-746, Korea*
[b]*Department of Energy Science, SungKyunKwan University, 300 Cheoncheon–dong, Suwon Gyeonggi-do 440-330*
**corresponding author: jschai@skku.edu*

Abstract. K=100 Separated Sector Cyclotron was designed in conceptual for the ISOL driver. It has 4 separated sector magnets. Two SF cyclotrons will be used as the injectors for separated sector cyclotron. RF frequency is 70 MHz, 4th harmonics. We have designed sector magnet without trim and harmonic coils. Minimum radius of the magnet is 55 cm and maximum radius is 1.8 m. Designed magnets were calculated and simulated by OPERA 3D (TOSCA) code. Ion beam dynamics calculations have been done using particle studio code to prove the focusing properties of the designed magnets.

Keywords: Sector Separated cyclotron
PACS: 29.20.dg

INTRODUCTION

Korean National Project, KoRIA, was started on April, 2010 for rare isotope production using radioactive beam by both the Isotope Separator On-line (ISOL) and in flight fragmentation. KoRIA facility will contribute to the various basic research fields such as nuclear physcis, atomic physics, material science, biology and medical science. [1] Fig.1 shows the layout of the KoRIA. For the projectile in-flight fragmentation, a superconducting cavity linear accelerator will be used and for the ISOL SSC (Sector Separated Cyclotron) was selected with high beam current about 1mA. The whole structure of the main cyclotron is shown in Fig. 3. The maximum radius is 1.8m and the minimum radius is 0.55m. Parameters of the main cyclotrons are shown in Table 1. Magnet design is the most important part of the cyclotron and requires an iterative process. This paper mainly introduces an optimization of the K=100 cyclotron magnet design which will have to satisfy the isochronism and transversal focusing to avoid dangerous resonances. This process is supported by POISSON [2] and OPERA-3D TOSCA [3].

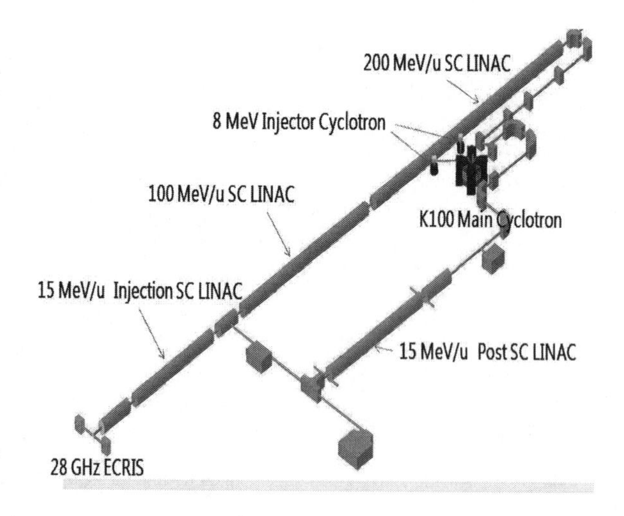

FIGURE 1. The layout of the KoRIA Project

Application of Accelerators in Research and Industry
AIP Conf. Proc. 1336, 614-617 (2011); doi: 10.1063/1.3586176

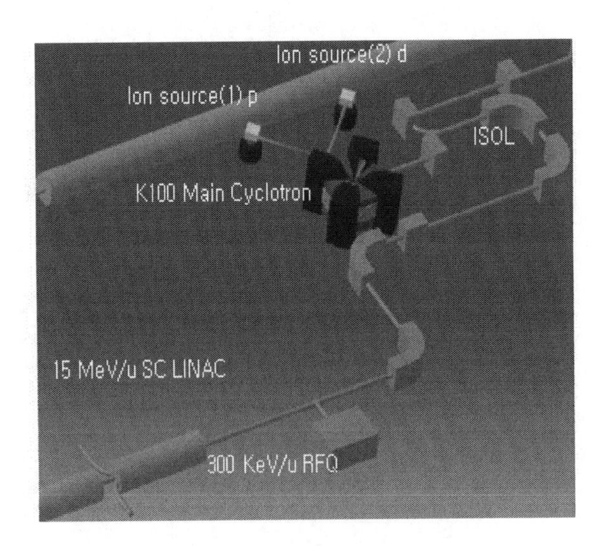

FIGURE 2. Block 1 layout of the KoRIA Project

FIGURE 3. The whole structure of the main cyclotron

TABLE 1. Parameters of K=100 cyclotron magnet

Injection Energy	8MeV
Extraction Energy	70MeV
Estimated maximum beam intensity	1mA
RF – frequency	70MHz
Harmonic numbers	4
Pole radii – r_{min}/r_{max}	0.55/1.8m
Extraction radius	1.2m
Number of magnet sectors	4
Magnetic gap	3cm
Sector angle on injection radius	14 deg
Number of RF cavities	4
Peak voltage	Delta-2gaps
Peak radial gain/turn	150/4500 kVolts
Estimated total power required for 3Ma	50 / 20 mm
Estimated total accelerator efficiency (phase 1)	~25%

DESIGN CONSIDERATION

Acceleration of intense beams requires a very efficient focusing and extraction process free of beam loss. The main magnet design parameters should satisfy the following criteria:

Single turn extraction: A large radial gain per turn is requested, i.e. a high energy gain per turn in order to get an effective turn separation on the extraction radius.

Powerful vertical and radial focusing: the problem of space charge effect is not fully understandable because of the very complicated nonlinearity of it. However many systems that have been designed were very successful for overcoming this problem. A deep valley sector focused cyclotron have been designed to be the injector for the final stage separated sector cyclotron which is the most appropriate choice for our project because the powerful focusing of sectors is a solution for the space charge problem.

Magnet Design And Field Computation

This is the paragraph spacing that occurs when you use the Enter key. There exist several criteria for magnet design: field isochronisms during acceleration, beam focusing both radial and axial directions, and avoiding passing of the beam through dangerous resonances zones. To achieve these goals, hand calculations were first performed for determining the basic geometry of the magnet, the isochronous magnetic field was calculated by the equation [4].

$$B_{iso}(r) = \gamma \cdot B_c = (1 + T / E_0) \cdot B_C$$

Relativistic factor γ is calculated from the equation below as a function of radius. [4]

$$B(r) / B_0 = \gamma (R) = [\, 1 - (qB_0 / m_i c)^2 \,]^{-1/2}$$

There are 3 main methods for achieving the isochronisms:

1- Decreasing the gap between magnet pole,
2- Increasing the hill--to--valley ratio,
3- Trim coils.

In this work we are going to attain the isochronous field by increasing the ratio of hill--to--valley along the radius. Increasing the angle of hill is optimized for hydrogen. However for deuterium we will use some simple trim coils.

2D simulation had been carried out for determining the basic magnet dimensions. Poisson 2D code had been used and the advantage of 2D simulation is the ability to use small mesh size for areas with high magnetic field gradient. Magnetic field lines of the center cross-section of the hill are shown in Fig.4.

A CATIA V5 [5] was used for drawing the whole magnet in 3D perspective followed with a 3D FEM (TOSCA) code for the final calculations of the model. A repeated iteration process was carried out. Fig. 5 shows the process steps of the magnet design. The required isochronous field within the tolerances was obtained after several iterations by shimming the angular width of the hill as a function of radius. Fig. 6 shows the required isochronous magnetic field and the simulated one. The average errors in simulations are about 20 ~ 40 Gauss.

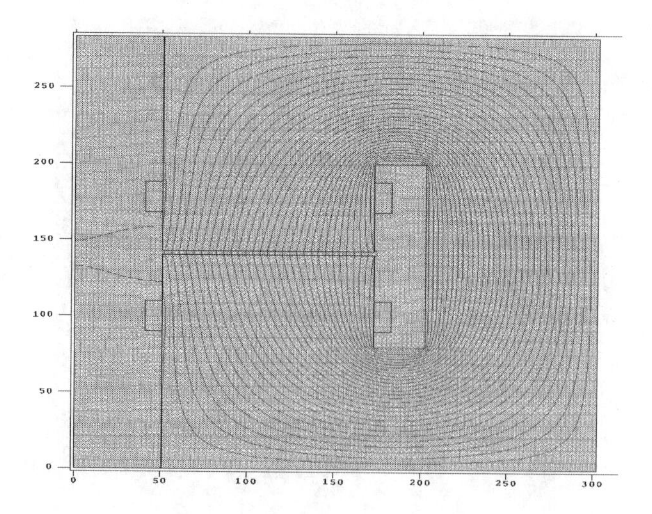

FIGURE 4. Magnetic field lines of the center cross-section of the hill calculated by POISSION

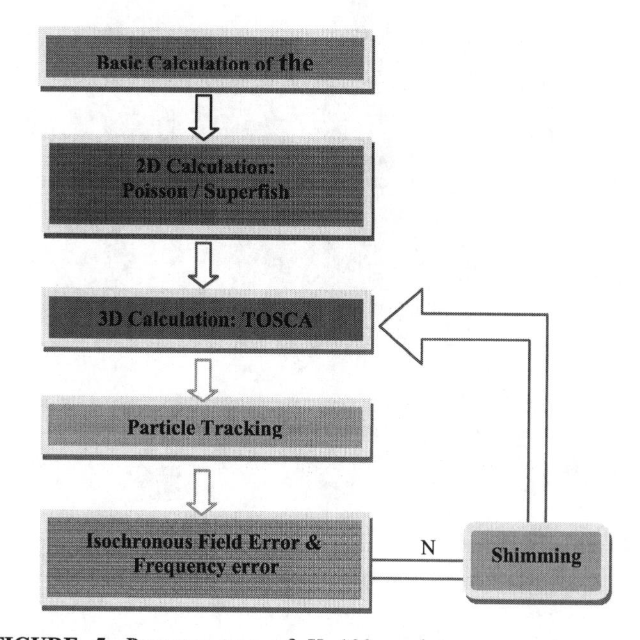

FIGURE 5. Process steps of K=100 cyclotron magnet design

Beam Properties And Beam Trajectory

The properties of the equilibrium orbit i.e., the radial and axial tunes, frequency error, and integrated phase shifts were optimized after several iterations. Fig. 7 shows radial and vertical betatron frequency υ_r, υ_z with many resonance lines. We confirmed that the working points are far away from dangerous resonances zones.

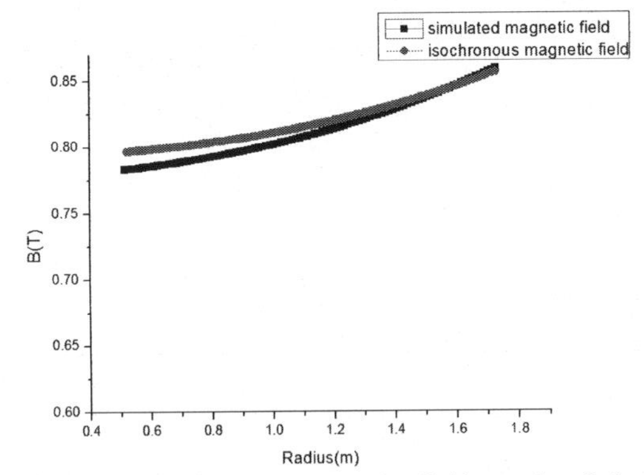

FIGURE 6. Average magnetic field graph of designed magnet with ideal isochronous filed graph

The beam tracking simulation had been done by OPERA-3D Post-Processor in Fig.8 In this simulation, space charge effects were not considered and the beam trajectory was plotted. The RF dee voltage is about 150 kV.

The values of each radius from 8MeV to 100MeV are shown in Fig.9. Particles gain 600kV per turn and about 140 steps are needed for achieving 100 MeV energy. All trajectories and radii were calculated by the OPERA-3D post-processor and that data was used in plotting.

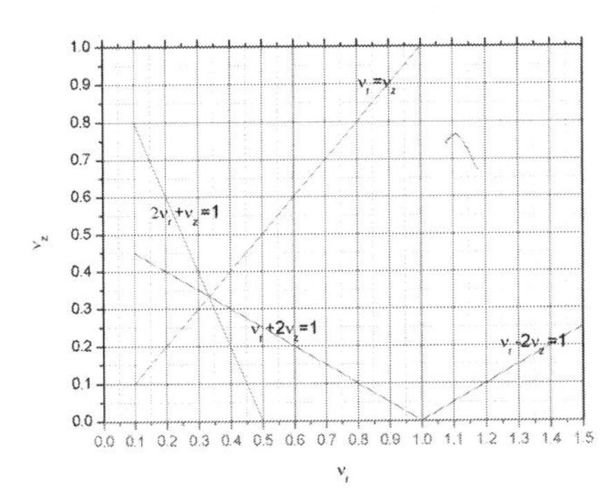

FIGURE 7. Tune Diagram

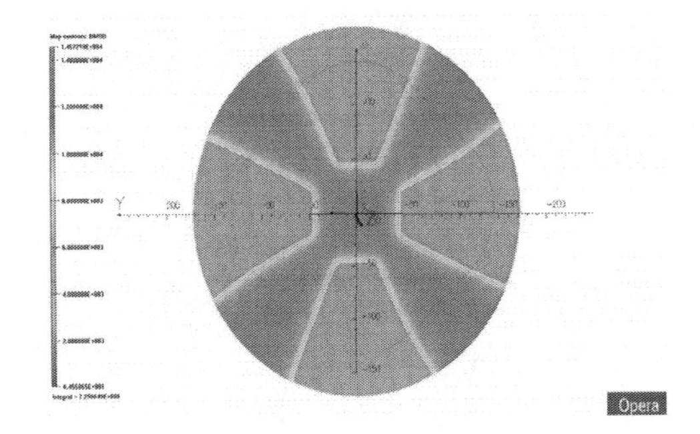

FIGURE 8. Beam trajectories at energy 70 MeV

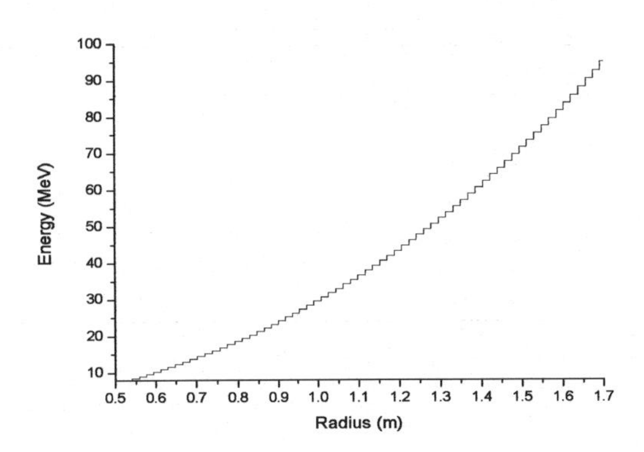

FIGURE 9. Beam trajectories on the mid-gap of the hill from 8MeV to 100MeV

CONCLUSION

The design of a K=100 cyclotron magnet has been carried out without trim and harmonic coils. Sector magnet design was optimized. We confirmed the isochronous field of the magnet with high precision. This magnet is optimized for proton beam extraction.

ACKNOWLEDGMENTS

This work was supported by National R&D Program through the National Research Foundation of Korea (NRF), funded by the Ministry of Education, Science and Technology & Ministry of Knowledge Economy (2009-0094278). Also by the Department of Energy Science, and School of Information and Communication Engineering of SungKyunKwan University supported this project.

REFERENCES

1. J.S. CHAI etc, Facility of KoRIA project, Review Journal of Korean Physical Society, June 2010
2. User's Guide for the POISSON/SUPERFISH Group of Codes, LA-UR-87-115, Los Alamos Accelerator Code Group
3. Cobham, Vector Fields Ltd, UK
4. Stanley Humphries, Jr., "Principles of Charged Particle Accelerator", JOHN WILEY & SONS.
5. CATIA V5, IBA

SECTION XI – RADIATION EFFECTS

In-Situ RBS Channelling Studies Of Ion Implanted Semiconductors And Insulators

E. Wendler

Friedrich-Schiller-Universität Jena, Institut für Festkörperphysik,
Max-Wien-Platz 1, 07743 Jena, GERMANY

Abstract. The experimental set-up at the ion beam facility in Jena allows the performance of Rutherford backscattering spectrometry (RBS) in channeling configuration at any temperature between 15 K and room temperature without changing the environment or the temperature of the sample. Doing RBS channeling studies at 15 K increases the sensitivity to defects, because the influence of lattice vibrations is reduced. Thus, the very early processes of ion induced damage formation can be studied and the cross section of damage formation per ion in virgin material, *P*, can be determined. At 15 K ion-beam induced damage formation itself can be investigated, because the occurrence of thermal effects can be widely excluded. In AlAs, GaN, and ZnO the cross section *P* measured at 15 K can be used to estimate the displacement energy for the heavier component, which is in reasonable agreement with other experiments or theoretical calculations. For a given ion species (here Ar ions) the measured cross section *P* exhibits a quadratic dependence $P \propto P_{SRIM}^2$ with P_{SRIM} being the value calculated with SRIM using established displacement energies from other sources. From these results the displacement energy of AlN can be estimated to about 40 eV. Applying the computer code DICADA to calculate the depth distribution of displaced lattice atoms from the channeling spectra, indirect information about the type of defects produced during ion implantation at 15 K can be obtained. In some materials like GaN or ZnO the results indicate the formation of extended defects most probably dislocation loops and thus suggest an athermal mobility of defect at 15 K.

Keywords: ion implantation, radiation damage, Rutherford backscattering spectrometry, low temperatures
PACS: 61.85.+p;61.80.-x;61.72.uf;61.72.uj

INTRODUCTION

Ion implantation is a powerful technique used extensively in the semiconductor industry. The use of ion beam modification is continuously broadening to less common wide gap semiconductors and optoelectronic materials. However, a successful application of this technique in industry requires a certain amount of knowledge about the primary processes of ion-solid interaction taking place in the material of interest, in order to ensure a reproducible and reliable production of devices. Furthermore, there is also a continuous scientific interest to study the defect production in ion implanted materials and to understand the different susceptibility of materials to ion beam induced damage formation. Depending on the temperature of the target and the material studied, ion beam induced effects may be strongly superimposed by thermal effects. Therefore very low irradiation temperatures and in situ techniques are the matter of choice in order to minimize the influence of thermal effects and to study purely ion beam induced damage formation. In this paper results are presented applying Rutherford backscattering spectrometry (RBS).

EXPERIMENTAL TECHNIQUE

At the ion beam facility in Jena a two-beam chamber exists which allows subsequent ion implantation and ion beam analysis without changing the temperature and the environment of the target [1, 2]. Two beam lines from a 3 MeV Tandetron and a 400 kV implanter, respectively, enter the chamber at an angle of 30 degrees. The sample holder is mounted on a goniometer to position the sample 7° off-axis towards the ion beam for implantation and to align it to the high-energetic ions for RBS. By performing subsequent stepwise implantation and analysis a quasi in-situ procedure is realized. The cooling of the target holder is obtained with a closed-cycle refrigerator. The connection to the sample holder with copper flexes results in a minimum temperature of 15 K and gives flexibility to tilt the holder. Additionally the sample

Application of Accelerators in Research and Industry
AIP Conf. Proc. 1336, 621-625 (2011); doi: 10.1063/1.3586177

holder contains a heating cable so that any temperature between 15 K and room temperature can be reached.

The experimental data presented throughout this paper are for implantation and subsequent measurement at 15 K. For all materials Ar ions were used and the energies were chosen to obtain a layer thickness of 0.2 to 0.3 μm in all materials studied.

To measure the defect concentration after each implantation step, RBS is applied in channeling condition using 1.4 or 2 MeV He ions and a backscattering angle of 170°. It should be mentioned that during the procedure described the analyzing ion beam itself may damage the sample or anneal defects. This has to be checked beforehand. To overcome such problems one can use a fresh spot for each measurement by translation of the sample or study the effect and incorporate it in the data evaluation [3]. The described facility can be also used to study defect annealing between 15 K and room temperature by cooling the sample back to 15 K for measurement after each annealing step.

The relative concentration of displaced lattice atoms versus depth, $n_{da}(z)$ (referred to as defect concentration or defect profile), is calculated from the RBS spectra with the help of the computer code DICADA [4] (exploiting only the part of the spectrum of the heavier element in case of compounds). This code is based on the discontinuous model of dechanneling [5]. The dechanneling background is calculated analytically for a given type of defects. In our evaluation of the channeling spectra the existence of uncorrelated displaced lattice atoms is assumed, which are randomly distributed within the lattice cell. Uncorrelated displaced lattice atoms result from point defects, complexes of point defects or amorphous zones. For comparison the number of displacements per ion and unit length versus depth, $N_{displ}^*(z)$, is calculated with the computer code SRIM [6] using version 2003.26. The resulting cross section of damage formation per ion, P_{SRIM}, is given by $P_{SRIM}= N_{displ}^*(z) /N_0$ (with $N_{displ}^*(z)$ taken at the maximum of the distribution and N_0 atomic density).

EXAMPLES FOR APPLICATION

Mechanisms Of Damage Formation - Information About Type Of Defects

Figure 1 shows a set of channeling spectra measured on Ar implanted GaP at 15 K. From these spectra the defect profiles $n_{da}(z)$ were calculated as described in Sect. 2 and plotted in Fig. 2 (a). Further defect profiles for Ar implanted GaN are shown in Fig. 2 (b) to (d) [7]. Measurement at low temperatures in combination with the use of the computer code DICADA enhances the sensitivity of RBS significantly thus allowing the determination of defect profiles for rather low defect concentrations (see Fig. 2 (b)). A defect concentration of n_{da}=1 is commonly taken as an indication for amorphisation. For GaP the defect concentration reaches zero behind the implanted layer which indicates that the assumption of uncorrelated lattice atoms used for the calculation of the defect profiles with DICADA is correct. Therefore, it can be concluded that this implanted layer contains point defects, complexes of point defects and amorphous zones.

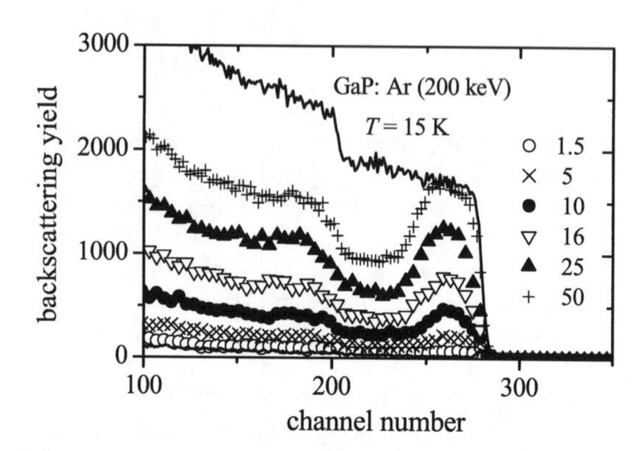

FIGURE 1. RBS spectra of 1.4 MeV He ions backscattered on 200 keV Ar implanted GaP. Implantation and measurement were performed at 15 K. The ion fluence N_I are given in 10^{12} cm^{-2}. The solid lines are the random and aligned spectrum of the perfect crystal.

Information about the mechanisms of damage formation is obtained from the analysis of the fluence dependence of the defect concentration [8, 9]. This is taken at the maximum of the measured distribution. Figure 3 shows examples of dependencies $n_{da}(N_I)$ which were found for Ar ion implantation at 15 K. In the case of GaP (Fig. 3 (a)) a straight forward amorphization of the implanted layer is obtained by accumulation and stimulated growth of heavily damaged and/or amorphous clusters produced within a single impact in the crystalline material. This coincides with the fact that the shape of the measured defect profiles is in good agreement with that calculated using SRIM (see Fig. 2 (a)). In AlAs (Fig. 3 (b)) a broad plateau of the defect concentration is observed over a wide range of ion fluences at a rather low value of n_{da}≈0.2, which can be attributed to a balance between defect formation and the recombination of defects from collision cascades of different ions [10]. A similar effect occurs for GaN (Fig. 3 (c)) and ZnO (Fig. 3 (d)) with a saturation value of n_{da}≈0.05 and n_{da}≈0.02, respectively [7, 11]. This suggests that in these materials within a single ion impact in crystalline material mainly point defects or complexes of them are produced. In AlAs amorphous seeds nucleate when the concentration of implanted ions exceed a critical value [10]. These nuclei grow rapidly during further implantation thus rendering the implanted layer amorphous. In GaN and ZnO a second intermediate plateau is found at a defect concentration well below amorphization. This can be modeled assuming the formation of non-recombinable defect clusters with a saturation concentration of $n_{da}\approx0.7$ for GaN and $n_{da}\approx0.4$ for ZnO [7, 11]. GaN is finally amorphized within a small interval of the ion fluence which - as in the case of AlAs - can be explained by the nucleation and growth of amorphous seeds. However, the mechanism for nucleation of amorphous seeds is not clear [7]. In ZnO no indication of amorphization is found up to the highest ion fluence applied of 5×10^{16} cm^{-2}.

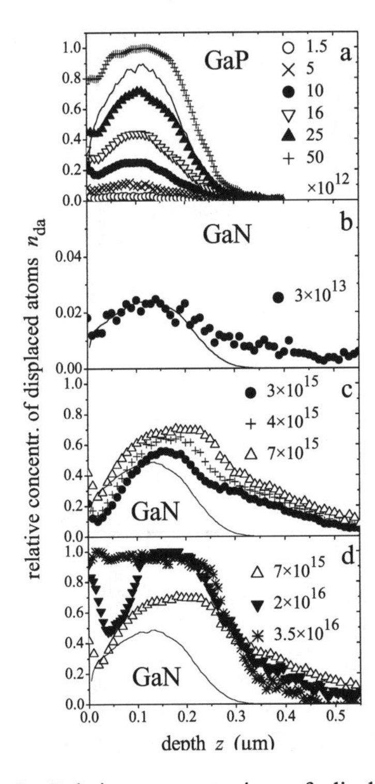

FIGURE 2. Relative concentration of displaced lattice atoms, n_{da}, versus depth z for 200 keV Ar implanted GaP (a) and 300 keV Ar implanted GaN (b) to (d). The ion fluences N_I are given in cm^{-2}. The distributions of the primary displacements, $N_{displ}^{*}(z)$, calculated by SRIM (thin solid lines) are included in arbitrary units.

To obtain more information about the kind of defects for ion fluences within the second plateau in GaN and ZnO, it is useful to analyse the defect profiles in more detail. For GaN examples are given in Fig. 2 (c). The profiles of Ar implanted ZnO have the same structure in the corresponding range of ion fluences

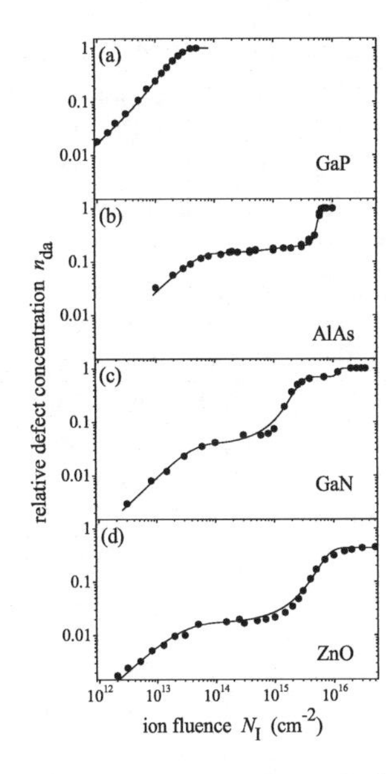

FIGURE 3. Relative concentration of displaced lattice atoms, n_{da}, taken at the maximum of the measured distribution versus ion fluence N_I for Ar (200 keV) implanted GaP (a), Ar (200 keV) implanted AlAs (b), Ar (300 keV) implanted GaN (c) and Ar (200 keV) implanted ZnO (d) at 15 K. The lines are model fits to the experimental data (for details see text).

[12]. The profiles exhibit a long artificial tail which extends far behind the implanted layer. Once the implanted layer gets amorphized, this tail disappears (see Fig. 2 (d)), confirming that this tail is not real but caused by the evaluation of the spectra. Calculations with the code DICADA show that such tails occur when the assumption of the defect type made to evaluate the spectra is not correct [12]. Because uncorrelated displaced lattice atoms were assumed for the calculation of the profiles in Fig. 2, the occurrence of the tails indicates that these implanted layers do not solely contain uncorrelated displaced lattice atoms but a mixture with correlated displaced lattice atoms. The occurrence of correlated displaced lattice atoms points to the existence of extended defects which are most probably dislocation loops. The formation of these loops requires some kind of defect mobility. Implantation and analysis were performed at 15 K at which thermal processes are widely suppressed. Therefore, the results suggest an athermal defect mobility which may be connected with the electronic energy loss of the ions.

Cross Section Of Damage Formation

As mentioned above, the curves in Fig. 3 are fitted to the experimental data applying simple models [8, 9]. One main parameter is the cross section of damage formation per individual ion, P, which follows from the almost linear increase of the defect concentration at very low ion fluences with $dn_{da}/dN_I \cong P$. Such a linear increase always occurs provided the defects remaining after relaxation of the primary collision cascade do not anneal during further implantation. In Fig. 3 one can recognize a linear increase for all examples and P is obtained by fitting the model curves to the experimental data. The cross section P obtained from the modeling of the curves n_{da} (N_I) can be compared to the corresponding value P_{SRIM} (see Sect. 2). The displacement energies E_d of the target atoms is varied until P_{SRIM} is in agreement with the measured value P (not shown). This yields E_d=(17±3)eV for the As atoms in case of AlAs which is in good agreement with the value of E_d=20eV from MD calculations [13]. This procedure works similarly well for GaN [14] and ZnO [11] but not for other materials. The reason for that becomes obvious in Fig. 4 which plots the cross section P determined from the RBS date as a function of P_{SRIM} using well established values of E_d from literature [see e.g. 9, 15]. Despite two misfits, all data are close to a line with $P \propto P_{SRIM}^2$. The data of the materials for which a reasonable displacement energy could be estimated, are found around the intersection with $P=P_{SRIM}$. This explains the reasonable determination of the displacement energies as described above. Believing the experimental dependence $P \propto P_{SRIM}^2$ at least for the III-V compounds, the displacement energies of AlN can be estimated from $P=4\times10^{-16}$ cm^2 [18]. Assuming both sublattices to be similarly damaged we obtain E_d(Al)=(42±4)eV and E_d(N)=(38±4)eV. So far no other data exist for this material.

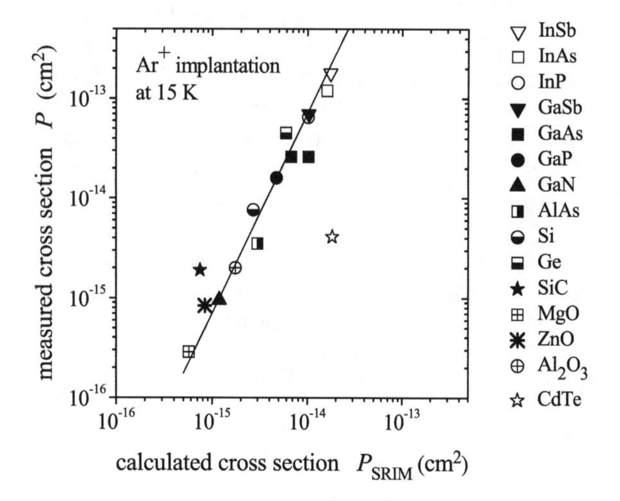

FIGURE 4. Cross section of damage formation P versus the cross section calculated with SRIM, P_{SRIM}, for Ar ions implanted into various materials at 15 K. The data of P for Si and Ge are taken from Ref. 16 and 17, for further details see text.

The dependence $P \propto P_{SRIM}^2$ is a clear proof of non-linear processes occurring within the primary collision cascades. The results in Fig. 4 suggest that for low-density cascades, i.e. for low values of P_{SRIM}, significant defect recombination occurs within the lifetime of the cascade, resulting in $P<P_{SRIM}$. In the case of MgO for instance a defect survival rate of 15...50% is found [15, 19]. Contrary, for high-density cascades non linear processes force the damage production leading to $P>P_{SRIM}$. One possible explanation would be that the bonds are softened once the collision cascade has started which may reduce the displacement energy in comparison to that found for an atom in perfect crystalline surrounding.

The two misfits in Fig. 4 are SiC and CdTe [20]. SiC is exceptional because it forms very many stable polytypes. It might be possible that this ability prevents defect recombination and stabilizes damage, thus resulting in a relatively large value of P. In the case of CdTe, P is strikingly low. More date for other materials of the group of tellurides and selenides have to be measured at 15 K to understand this behavior.

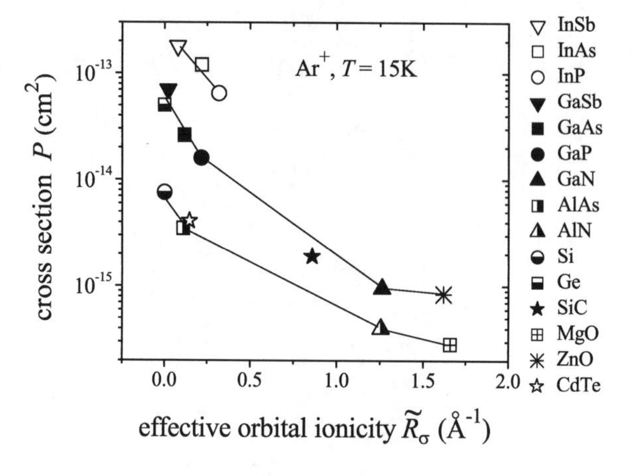

FIGURE 5. Cross section of damage formation P versus R_s representing an effective orbital ionicity. The lines indicate the anion series for the cations of one period (for details see text).

The susceptibility of materials to ion-beam induced damage formation must be related to the chemical bonds and the bond strength of the materials. In this context one of the most commonly used ordering parameters is the ionicity. We found the quantity R_σ to be useful which is quoted as to be an effective orbital ionicity [21]. This is demonstrated in Fig. 5 which plots P as a function of R_σ. It is found that anion series form for the cations of one period. As in Fig. 4, the data for SiC and CdTe appear as misfits. Within each anion series the cross section decreases with rising ionicity. This coincides with the well-known picture with respect to the amorphizability of materials under ion irradiation. Already in a very early paper it was observed that the resistance to ion beam induced amorphization increases with the bond ionicity [22]. In a more recent review about understanding the resistance to amorphization by ion irradiation it is stated that a material is amorphizable by radiation damage if its chemistry allows it to form a covalent network [23]. Here we do not regard the amorphizability of materials but try to understand the trend of the cross section of damage formation per individual ion in the virgin (undamaged) materials. However, it is obvious that the mechanisms discussed above are also important for this subject, because they pertain to the effects within the primary collision cascades. The occurrence of anion series being different for the various cations was found in other context too (see e.g. [24]).

SUMMARY AND CONCLUSION

Using in-situ RBS allows the study of damage formation during ion implantation at a given temperature without change of the environment. When using the computer code DICADA to evaluate the channeling spectra measured at very low temperatures, the sensitivity is significantly enhanced and the early stage of damage formation can be investigated. For the example of Ar ion implantation in various materials at 15 K it is shown that the mechanisms of damage formation can be identified from extended fluence dependencies analyzed on the base of simple models. Because the computer code DICADA uses an analytical description of the dechanneling of the analyzing ions as a function of defect type and target temperature, indirect information about the type of defects in the implanted layers can be obtained. It is shown that even at 15 K in some materials extended defects form during the implantation, which suggests an athermal mobility of defects.

The cross section of damage formation P per Ar ion implanted at 15 K in a wide range of materials depends on the cross section P_{SRIM} calculated with SRIM according to $P \propto P_{SRIM}^2$. This quadratic dependence proofs the occurrence of non-linear processes within the primary collision cascades. Furthermore, it can be used to estimate a displacement energy for AlN, for which no data exist so far, of about 40 eV for both atomic species. The cross section P is closely related to the ionicity of the materials. However, no uniform dependence but separate anion series for the cations of one period are found.

ACKNOWLEDGMENTS

The author acknowledges the technical support of U. Barth and G. Lenk during the measurements.

REFERENCES

1. B. Breeger, E. Wendler, W. Trippensee, Ch. Schubert, W. Wesch, *Nucl. Instr. and Meth. B* **174**, 199-204 (2001).
2. R. Lauck, E. Wendler, W. Wesch, *Nucl. Instr. and Meth. B* **242**, 484-486 (2006).
3. C.S. Schnohr, E. Wendler, K. Gärtner, W. Wesch, *J. Appl. Phys.* **99**, 123511-10 p. (2006).
4. K. Gärtner, *Nucl. Instr. and Meth. B* **132**, 147-158 (1997).
5. K. Gärtner, K. Hehl, G. Schlotzhauer, *Nucl. Instr. and Meth.* **216**, 275-286 (1983); *B* **4** 55-62 and 63-71 (1984).
6. J.P. Biersack and J.F. Ziegler, *The Stopping and Ranges of Ions in Matter* **1**, Pergamon Press Oxford, 1985; see also www.srim.org.
7. E. Wendler, A. Kamarou, E. Alves, K. Gärtner, W. Wesch, *Nucl. Instr. and Meth. B* **206**, 1028-1032 (2003).
8. W.J. Weber, *Nucl. Instr. and Meth. B* **166-167**, 98-106 (2000).
9. E. Wendler, *Nucl. Instr. and Meth. B* **267**, 2680-2689 (2009).
10. E. Wendler, in: *Proceedings of the 17th International Conference on the Application of Accelerators in Research and Industry 2002*, edited by J.L. Duggan and I.L. Morgan, AIP Conference Proceedings 680, American Institute of Physics, Melville, NY, 2003, p. 670-674.
11. K. Lorenz, E. Alves, E. Wendler, O. Bilani, W. Wesch, M. Hayes, *Appl. Phys. Lett.* **87**, 191904-3 p. (2005).
12. E. Wendler, O. Bilani, K. Gärtner, W. Wesch, M. Hayes, F.D. Auret, K. Lorenz, E. Alves, *Nucl. Instr. and Meth. B* **267**, 2708–2711 (2009).
13. K. Gärtner, *Nucl. Instr. and Meth. B* **252**, 190–196 (2006).
14. E. Wendler, W. Wesch, E. Alves, A. Kamarou, *Nucl. Instr. and Meth. B* **218**, 36–41 (2004).
15. S.J. Zinkle, C. Kinoshita, *J. Nucl. Mater.* **251**, 200-217 (1997).
16. A. Schroeter and E. Wendler (private communication).
17. M. Hayes, A. Schroeter, E. Wendler, W. Wesch, F.D. Auret, J.M. Nel, *Physica B-Condensed Matter* **404**, 4382-4385 (2009).
18. E. Wendler, W. Wesch, *Nucl. Instr. and Meth. B* **242**, 562-564 (2006).
19. E. Wendler, K. Gärtner, W. Wesch, *Nucl. Instr. and Meth. B* **266**, 2872-2876 (2008).
20. C.W. Rischau, C.S. Schnohr, E. Wendler, W. Wesch, to be presented at the IBMM 2010.
21. C.-Y. Yeh, Z.L. Lu, S. Froyen, A. Zunger, *Phys. Rev. B* **46**, 10086-10097 (1992).
22. H.M. Naguib, R. Kelly, *Radiat. Eff.* 25, 1-12 (1975).
23. K. Trachenko, *J. Phys.: Condens. Matter* **16**, R1491-R1515 (2004).
24. J.J. Gilman, *J. Phys. D: Appl. Phys.* **41**, 074020-5 p. (2008).

Plasma-Surface Interaction Research At The Cambridge Laboratory Of Accelerator Studies Of Surfaces

G.M Wright, H.S. Barnard, Z.S. Hartwig, P.W. Stahle, R.M. Sullivan, K.B. Woller and D.G. Whyte

Plasma Science and Fusion Center, Massachusettes Institute of Technology, 77 Massachusetts Ave. Cambridge, MA, 02139

Abstract. The material requirements for plasma-facing components in a nuclear fusion reactor are some of the strictest and most challenging facing us today. These materials are simultaneously exposed to extreme heat loads (20 MW/m^2 steady-state, 1 GW/m^2 in millisecond transients) and particle fluxes (>10^{24} m^{-2}s^{-1}) while also undergoing high neutron irradiation (10^{18} neutrons/m^{2}s). At the Cambridge Laboratory of Accelerator Studies of Surfaces (CLASS), many of the most important issues in plasma-surface interaction research, such as plasma-driven material erosion and deposition, material transport and irradiation and hydrogenic retention are investigated with the use of a 1.7 MV tandem ion accelerator. Ion-Beam Analysis (IBA) is used to investigate and quantify changes in materials due to plasma exposure and ion irradiation is used as a proxy for neutron irradiation to investigate plasma-surface interactions for irradiated materials. This report will outline the capabilities and current research activities at CLASS.

Keywords: Plas ma-Surface Interactions, Nuclear Fusion, PIXE, ERD
PACS: 29.20.Ej, 28.52.Fa, 52.40.Hf

INTRODUCTION

In the pursuit of a commercially viable nuclear fusion reactor, it is quickly becoming apparent that one of the greatest challenges is the selection of the material used for plasma-facing components (PFC) and understanding the complex interactions between the edge plasma in the reactor and the material walls [1]. The bombardment of the PFC with high fluxes of low energy (<1 keV) hydrogen/deuterium/tritium (H/D/T) ions is known to have a wide range of detrimental effects to both the material and the plasma, such as PFC erosion, material transport, deposition and mixing inside the reactor, and hydrogenic implantation and retention. To further complicate matters, PFC in a fusion reactor will experience high rates of neutron irradiation (10^{18} neutrons/m^{2}s) and operate over a wide range of temperatures (e.g. 300-1200 K). These detrimental effects can compromise component life-time (e.g. PFC erosion), reduce plasma performance (e.g. impurity injection due to PFC erosion, loss of fuel due to H/D/T retention in the wall), modify or compromise material properties (e.g. neutron irradiation, material mixing/alloying), and even raise serious safety concerns (e.g. radioactive T trapped in the walls, thick deposits flaking off to produce activated dust).

The goal of plasma-surface interaction (PSI) research is to eliminate or minimize these effects through appropriate selection or creation of PFC materials, proper material preparation and controlling and manipulating edge-plasma conditions. In such a complex environment, experimental data is invaluable, thus appropriate diagnostics are of critical importance. For materials research, ion-beam analysis (IBA) is one of the most powerful and flexible diagnostics available, and thus plays a large role in PSI research in determining how plasma exposure has modified PFC. As a further benefit, high-energy (~MeV) ions can be used as a proxy for neutron irradiation, such that PSI research can be performed on highly irradiated and damaged materials as we look toward PSI in a neutron-producing fusion reactor. In this report, we will present the capabilities and recent PSI research at the Cambridge Laboratory of Accelerator Studies of Surfaces (CLASS), where IBA and material irradiation play a prominent role.

Application of Accelerators in Research and Industry
AIP Conf. Proc. 1336, 626-630 (2011); doi: 10.1063/1.3586178

THE CLASS FACILITY

The CLASS facility is built around a 1.7 MV tandem ion accelerator. There are two negative ion sources for the accelerator. One is a cathode cesium-sputter source, which provides ion beams of a wide range of atomic species. Access to high-Z ion beams provides a high number of target atom displacements per incident ion during target irradiations so that high levels of displacements per atom (dpa) can be achieved without excessively long irradiation times. High-Z ion beams also allow for more options using elastic recoil detection analysis (ERDA), such as depth profiling of helium profiles in the near-surface region. The second ion source is a RF charge exchange negative ion source designed for ^{4}He and ^{3}He ion beams. The ^{4}He beams are valuable for Rutherford backscattering (RBS) studies of the layers deposited on substrates and the ^{3}He ion beams are of critical importance to PSI research due to the ability of d(^{3}He,p)^{4}He nuclear reaction analysis (NRA) to quantify and resolve implanted deuterium in the surface of PFC.

On the high-energy side of the accelerator, there are two beam lines currently in use. On one is the PSI-dedicated Dynamics of ION Implantation and Sputtering Of Surfaces (DIONISOS) experiment [2]. In the DIONISOS experiment, an IBA chamber is combined with helicon RF plasma exposure chamber such that the target material can be simultaneously irradiated with ~MeV ions from the accelerator and high-flux ($\leq 10^{22}$ m^{-2}s^{-1}), low energy (<500 eV) plasma ions (see Figure 1). The DIONISOS facility is designed to provide abundant diagnostic and physical access to the specimen surface and to induce a large range of surface conditions for plasma exposures. The target is rotatable through 360° such that the various geometries for the different IBA techniques (RBS, ERDA, NRA, etc.) are available and can be optimized. The use of different target holders allows the target temperature to be varied from 300-1273 K. Electrically biasing the target (0-350 V) allows for the plasma ion energy to be controlled while having no discernible effect on the ions from accelerator. Solid-state detectors in DIONISOS are protected by water-cooled housings to ensure detectors do no overheat.

With simultaneous bombardment of a target with high-energy ions and plasma ions, DIONISOS can investigate the dynamics of PSI with time-resolved IBA during a plasma exposure. As an example, the evolution of hydrogenic retention can be dynamically monitored if a target is exposed to deuterium plasma ions while simultaneously bombarded with ^{3}He ions for d(^{3}He,p)^{4}He NRA. The ion beam can also be used

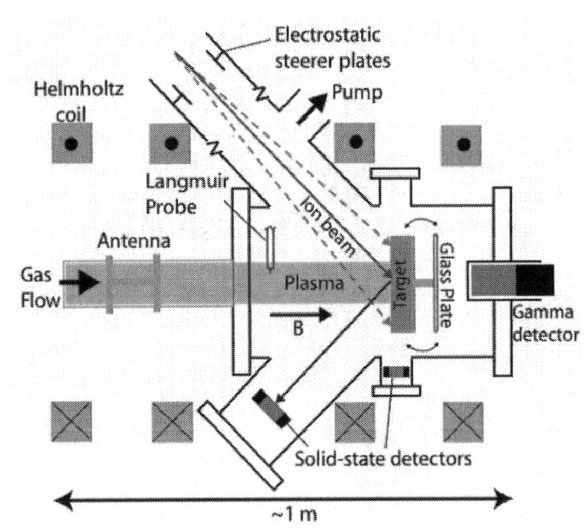

FIGURE 1. A schematic drawing of the DIONISOS experiment showing the geometry of the simultaneous bombardment of the target by ion beam and plasma.

to cause irradiation damage in the target, approximating neutron damage. The ability to study PSI in a simultaneously *in situ* irradiating environment is unique to DIONISOS [3] where other studies are limited to investigations with sequential ex-situ irradiations and plasma exposures [4,5].

On the second beam line, there is a vacuum vessel dedicated for IBA. This chamber is equipped with a rotatable target holder with vertical translation of up to 25 mm and several options for detector mounting allowing for a wide range of beam-target-detector geometries. This chamber is also equipped with a port and aperture for an external ion beam set-up. The external beam is only applicable for proton beams. The proton beam is brought out into air through a 3.5 mm aperture and a 7.5 μm kapton window (see Figure 2a). The external beam is exclusively used for particle-induced X-ray/gamma emission (PIXE/PIGE) since emitted X-rays and gammas will not be significantly affected by travelling through a short distance of atmosphere. The advantage of the external beam is that it greatly simplifies the analysis of large or odd-shaped targets that are difficult to mount in vacuum. It can also greatly accelerate the analysis of targets since no vacuum break is necessary to re-position, remove or install new targets.

PLASMA-SURFACE INTERACTION RESEARCH AT THE CLASS FACILITY

The CLASS facility studies a wide range of PSI issues through internal and external collaborations, as well as on-site dedicated experiments, such as DIONISOS. There is also an effort to develop new ion-accelerator applications and techniques for PSI

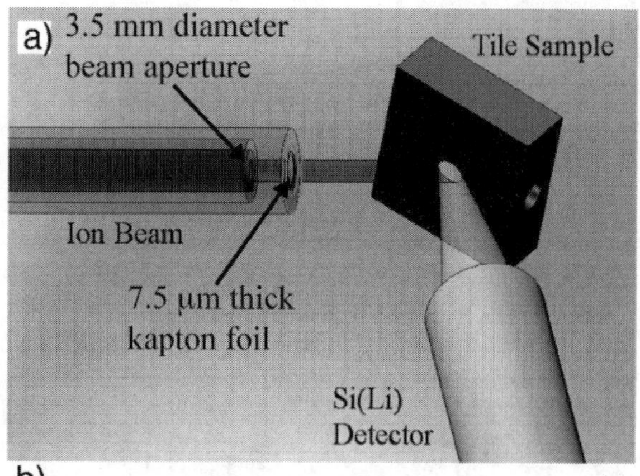

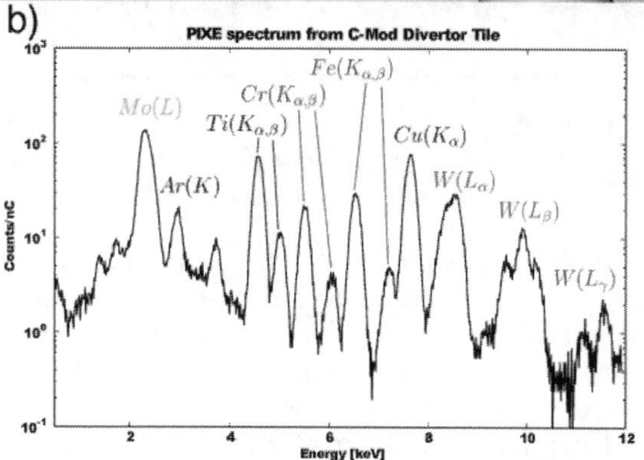

FIGURE 2. a) A schematic of the external beam set-up at CLASS and b) the PIXE spectra obtained from an Alcator C-Mod Mo tile after extended exposure to plasma in the tokamak.research

at CLASS to provide better diagnostic access to PFC in tokamaks and to investigate new fields of PSI research.

A recent focus at CLASS is the understanding of tungsten transport in the edge plasma of the Alcator C-Mod tokamak. Alcator C-Mod uses molybdenum as a PFC material but installed a row of tungsten tiles in the outer divertor, a region of net erosion in the tokamak. At CLASS, we can monitor the locations of deposition of W in the tokamak and, since the location of the source of W is known, this yields information on the W transport in Alcator C-Mod. W deposition on the Mo tiles is characterized using PIXE analysis. For these experiments, the external beamline was used to allow for easy positioning of the C-Mod tiles with respect to the ion beam (Figure 2a). The incident ion beam was 1-3 MeV (varied to probe to various depths) protons at the surface, which allows for the detection of a wide range of high-Z elements, including W. In the X-ray energy spectrum from 0-12 keV, peaks corresponding to W ($W(L_\alpha)$, $W(L_\beta)$, and $W(L_\gamma)$) and Mo (Mo(L)) can be indentified as well as peaks from other components found in the tokamak (e.g. stainless steel, copper, titanium) as can be seen in Figure 2b. The W peaks can then be quantified into an areal density of deposited W by comparing the magnitude of the measured W peaks to thick-target x-ray yields from a block of pure tungsten.

This has allowed for the tracking and inventory of transported W and yielded insights into the transport mechanisms taking place in the edge plasma. A large amount of the tungsten is deposited on tiles poloidally adjacent to the W tiles indicating a prompt ionization and re-deposition of many of the eroded W atoms, resulting in a material transport on the scale of millimeters. However, W deposition was also measured on Mo tiles in distant areas of the machine, such as the inner wall and inner divertor, indicating material transport on the order of centimeters or even meters is also taking place. This is indicative of neutral transport of the eroded W atoms (i.e. neutral atoms can freely cross the confining magnetic field lines) and ion transport through the edge plasma and even through the core plasma where small amounts of W can be seen spectroscopically during plasma discharges [6]. By defining both the location and magnitude of W deposition, the PIXE analysis has given us a better understanding of high-Z transport in the Alcator C-Mod tokamak.

In a fusion reactor burning D and T, the products are a high energy neutron and He. Therefore a fraction of the plasma ions bombarding the reactor wall will be He ions. Thus, the effects of He plasma bombardment on PFC materials must also be understood. Recent work has shown that bombarding a high temperature W surface ($T_W > 900$ K) with low energy (<500 eV) He ions induces the extrusion of W nano-filaments, or W fuzz, from the surface of the target [7]. At CLASS, we have investigated this process by measuring the depth profiles of the implanted He in tungsten targets at various stages of nano-filament growth. For this analysis, a 7 MeV O^{4+} ion beam was used to perform heavy-ion ERDA for He detection. The W fuzz has only 5-10 % density of bulk W [7], meaning the stopping power is also reduced by a factor of 10-20 so that the grazing-angle O^{4+} ions can penetrate to a depth of ~950 nm in the W fuzz.

Depth profiles of He concentration in a thick W fuzz target, a W surface in the very early stages of fuzz growth, and a W surface before fuzz growth begins can be seen in Figure 3. Two important observations can be made from these depth profiles. First, the He concentration is evenly distributed at a concentration of ~1 at.% through the fuzz layer as is shown by the flat He profile in the thick W fuzz target (Figure 3a). Since the implantation depth of these low energy He ions in bulk W is ≤10 nm, the even distribution of He through the W fuzz layer suggests that He can still efficiently diffuse through this W fuzz layer. Second, from the pre-fuzz target (Figure 3c), we see that the He

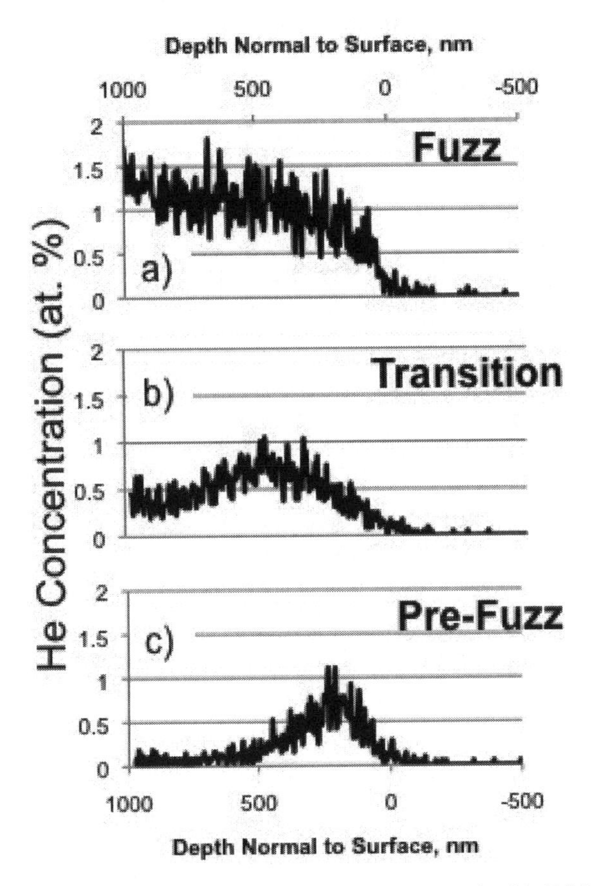

FIGURE 3. Helium-depth profiles measured with 7 MeV O^{4+} ERDA of tungsten targets with a) a thick layer of W fuzz, b) very early W fuzz growth, and c) before W fuzz growth has begun.

profile is peaked at the surface at a concentration below the saturation level measured in the thick W fuzz target. This may be indicating that the saturation concentration must be reached before the W nano-filament extrusion is initiated. It should be noted that the depth scale (which is calibrated to penetration depth in W fuzz) is overestimating the depth of the He profile in the pre-fuzz W target, which can be assumed to be bulk W. The transition target (Figure 3b) in the early stages of nano-filament growth, demonstrates how the initiation of nano-filament growth initially disperses the He across a larger physical depth (although not necessarily in areal density). Further investigations are on-going but the heavy-ion ERDA of these W fuzz targets have yielded additional clues towards the physical mechanism that is driving this W nano-filament growth.

While IBA techniques are a powerful tool for PSI research, it has limitations in terms of integration directly into fusion experiments. IBA of plasma-exposed targets is typically done in an *ex situ* fashion,

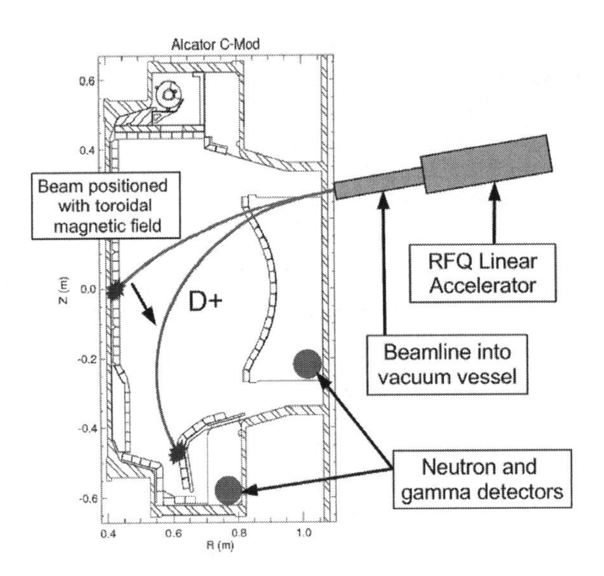

FIGURE 4. A schematic for the installation of an RFQ accelerator for *in situ* IBA in the Alcator C-Mod tokamak.

and IBA for targets exposed in tokamaks are investigating changes in the surface induced over weeks, months, or even years of plasma exposure. At CLASS, the DIONISOS experiment allows for *in situ* IBA of plasma-exposed targets to monitor surface and material changes in the target under controlled plasma exposures. While DIONISOS brings the plasma to the ion accelerator, there is an on-going effort at CLASS to also integrate an ion accelerator into a tokamak diagnostic suite.

The key aspects to integrate IBA into a tokamak is to use the abilities of a tokamak to your advantage. Due to limited physical access to tokamaks, a compact radio-frequency quadrupole (RFQ) accelerator is connected to one of the upper-radial ports in Alcator C-Mod. So, the RFQ runs a 1 mA, 0.9 MeV D^+ ion beam, which can induce many nuclear reactions with various low-Z elements resulting in neutron (e.g. D + D → ^{3}He + n) and gamma emission (e.g. D + ^{A}X → p + γ + $^{A+1}$X). A tokamak is a high heat flux environment with a large surface area, so using IBA techniques with solid state detectors (i.e. RBS, ERDA, etc) is not feasible. While there are neutron and gamma detectors around Alcator C-Mod for safety reasons, none of these meet the requirements for these IBA techniques. New, custom scintillation detectors (LaBr3:Ce inorganic scintillator for gamma detection and BC501A organic liquid scintillator for fast neutron detection) will be integrated into Alcator C-Mod for this system. Another concern is diagnostic access to surfaces in a tokamak. The inside wall of a tokamak is a large area exposed to very different conditions, so it is desirable to have diagnostic access to a wide variety of locations inside the tokamak. In this respect, the toroidal magnets on the tokamak can be used as steering magnets to bend the 0.9 MeV D^+ beam to a wide range of surfaces in the tokamak (see Figure 4). ACRONYM, a synthetic diagnostic code, has shown this accelerator-based diagnostic is able to effectively monitor *in situ* boron deposition due to frequent boronizations of the Mo-tiles in Alcator-Mod and the subsequent erosion of these deposited boron layers after subsequent plasma discharges [8].

SUMMARY

The CLASS facility is based around a 1.7 MV tandem ion accelerator equipped with a negative ion cesium sputter source and a RF plasma charge exchange negative ion source. At the beam ends, the facility is equipped with the DIONISOS experiment and a vacuum chamber for IBA as well as an external beam set-up to bring the ion beam out into air. The DIONISOS experiment allows for *in situ*, simultaneous bombardment of targets by high-energy ions from the accelerator and low-energy plasma ions. The IBA chamber and external beam set-up allow for analysis of targets of any shape or size.

The accelerator has recently been used for IBA of targets obtained from internal (Alcator C-Mod) and external collaborations. PIXE analysis of W deposition on Mo tiles from Alcator C-Mod has helped better understand high-Z transport in the edge plasma in the tokamak. ERDA of He depth profiles in layers of W nano-filaments, or W fuzz, grown by the bombardment of high-temperature W by low-energy He ions has yielded new insights into the potential growth mechanisms for these W nano-filaments. Aside from the PSI research taking place at CLASS, there is also an effort to develop new applications of accelerator applications in PSI research. DIONISOS is a unique experiment that allows for *in situ*, time-resolved IBA of plasma-exposed targets allowing for the dynamics of many of these PSI processes to be investigated. Work is also being done to incorporate an RFQ into the diagnostic suite of Alcator C-Mod to use a 0.9 MeV D^+ beam for *in situ* IBA of a variety of plasma-facing surfaces for D and low-Z isotope measurement.

ACKNOWLEDGMENTS

This work is support by US DoE Award DE-SC00-02060, DoE Grant DE-FG02-94ER54235, and DoE Cooperative Agreement DE-FC02-99ER54512.

REFERENCES

1. V. Philipps, *Phys. Scr.* **T123**, 24-32 (2006).
2. G.M. Wright, D.G. Whyte, B. Lipschultz, R.P. Doerner, J.G. Kulpin, *J. Nucl. Mater.* **363-365,** 977-980 (2007).
3. G.M. Wright, D.G. Whyte, B. Lipschultz, *J. Nucl. Mater.* **390-391,** 544-549 (2009).
4. B. Tyburska, V. Kh. Alimov, O. V. Ogorodnikova, K. Schmid, K. Ertl, *J. Nucl. Mater.* **395** 150-156 (2009).
5. W.R. Wampler, R.P. Doerner, *Nucl. Fusion* **49** 115023 (2009).
6. H.S. Barnard, D.G. Whyte, B. Lipschultz, in: Proc. 19th Int. Conf. on Plasma-Surface Interactions, San Diego, USA 2010, *submitted to J. Nucl. Mater.*
7. M.J. Baldwin, R.P. Doerner, *Nucl. Fusion* **48** 035001 (2008).
8. Z.S. Hartwig, D.G. Whyte, *Rev. Sci. Instrum.* **81** 10E106 (2010).

Heavy Ion Radiation Effects Studies With Ion Photon Emission Microscopy

J. V. Branson[1], K. Hattar[1], G. Vizkelethy[1], C. J. Powell[1], P. Rossi[2], B. L. Doyle[1]

[1]*Sandia National Laboratories, P.O. Box 5800 MS 1056, Albuquerque, NM, USA, 87185-1056*
[2]*Department of Physics of the University of Padua and INFN, via Marzolo 8, 35131 Padua, Italy*

Abstract. The development of a new radiation effects microscopy (REM) technique is crucial as emerging semiconductor technologies demonstrate smaller feature sizes and thicker back end of line (BEOL) layers. To penetrate these materials and still deposit sufficient energy into the device to induce single event effects, high energy heavy ions are required. Ion photon emission microscopy (IPEM) is a technique that utilizes coincident photons, which are emitted from the location of each ion impact to map out regions of radiation sensitivity in integrated circuits and devices, circumventing the obstacle of focusing high-energy heavy ions. Several versions of the IPEM have been developed and implemented at Sandia National Laboratories (SNL). One such instrument has been utilized on the microbeam line of the 6 MV tandem accelerator at SNL. Another IPEM was designed for ex-vacu use at the 88" cyclotron at Lawrence Berkeley National Laboratory (LBNL). Extensive engineering is involved in the development of these IPEM systems, including resolving issues with electronics, event timing, optics, phosphor selection, and mechanics. The various versions of the IPEM and the obstacles, as well as benefits associated with each will be presented. In addition, the current stage of IPEM development as a user instrument will be discussed in the context of recent results.

Keywords: radiation effects, cyclotron, emission microscopy, microprobe
PACS: 78.45.th, 81.05.Ea, 85.40.Qx

INTRODUCTION

Testing the reliability of microelectronic components that will be exposed to radiation environments is critical before their application. Space radiation can easily lead to failure in critical components, such that being able to study single event effects in these devices will allow one to pinpoint radiation sensitivity and work to engineer the component for radiation hardness. The most common means for performing these studies is to use high-energy broad beam testing. These experiments are typically carried out at cyclotrons where the entire device can be irradiated at once in air with high fluencies, providing high linear energy transfer (LET), and thus a high probability of causing device failure. The one major problem with this technique is that it does not provide position-specific information. Therefore, it is known if the device fails or not, but not precisely where the failure occurred. With this in mind, tests are also performed at nuclear microprobe facilities. With this technique, lower energy ion accelerators on which the ion beam can easily be focused to micron-size are used. The beam is focused and scanned across the sample, such that the position information is obtained from the scanners and the electronic signal can be read-out from the device. This technique allows one to pinpoint radiation-sensitive regions and therefore engineer out this sensitivity before using the device in practice [1,2,3].

Microbeam testing has proven to be a useful method for several decades, and has led to some important findings for devices requiring radiation tolerance. However, progressing technology in the integrated circuit (IC) industry is introducing very thick back-end-of-line (BEOL) layers on top of the active silicon region (Figure 1). There are metal, dielectric, and passivation layers, adding up to tens of microns of material in current technologies. With traditional nuclear microscopy, high enough energies cannot be achieved to penetrate through these layers and still deposit sufficient energy in the active region to cause single event effects. In order to penetrate through that material and still obtain the desired radiation-sensitivity information, it is necessary to use higher energy beams, such as those produced with a cyclotron. The ideal technique would therefore use

Application of Accelerators in Research and Industry
AIP Conf. Proc. 1336, 631-636 (2011); doi: 10.1063/1.3586179

cyclotron beams to irradiate devices in air and still obtain spatial resolution information [4].

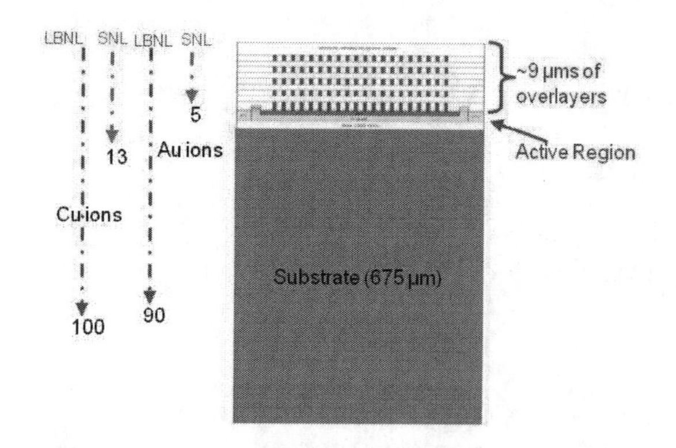

FIGURE 1. Schematic of overlayers present on top of active silicon region in modern ICs. Arrows represent the penetration depth possible with gold and copper beams from SNL's tandem accelerator and LBNL's cyclotron.

At Sandia National Laboratories (SNL), two nuclear emission microscopy techniques have been developed to satisfy these requirements. With this method, a broad beam is used, and spatial information is collected from secondary particles emitted from the point of ion impact. These secondary particles can be either electrons or photons. More specifically, ion photon emission microscopy (IPEM) utilizes photons emitted from a luminescent film, which has been placed on top of the device, to determine the location of each individual ion strike [3,5,6]. This technique is still in the development stage, although several versions have been created and installed with various ion sources [3,4]. The components, electronics, results obtained, and obstacles of each of these systems will be described in this work.

INSTRUMENTATION

Tabletop IPEM

The tabletop IPEM was designed as an initial proof-of-concept system to demonstrate that the idea behind this technique would be feasible. This system was fairly small and easily fit on a benchtop in the lab. A Po-210 alpha-particle source was used to induce charge in the devices, and the luminescent film used was either an organic scintillator (e.g. Bicron BC400) or a GaN thin film. An Olympus BX40 microscope was used to collect the light produced from the film, with the alpha source being positioned directly underneath the objective lens. In this orientation, light could still be collected with the lens and the alpha source was located directly above the sample to avoid non-uniform energy loss through the air. With the working distance of the objective lens being 8 mm, the 3.5 MeV alpha particles had about 2.7 MeV of energy remaining by the time they reached the device [3-6].

Once the light was collected with the microscope, it was focused onto the sensitive area of a Quantar Technology single photon position sensitive detector (PSD). The PSD is a Mepsicron-II™ high-resolution imaging photomultiplier tube (PMT) with a 24.5 mm active area. The PSD used on this system has a bialkali photocathode, which is blue sensitive and has the lowest dark count rate of the available photocathodes. The detector is powered by a low noise high-voltage power supply with a customized voltage divider network optimized for imaging PMTs, and SHV connections. This system also contains an overcount trigger module to disable the voltage if a preset count rate limit is exceeded. The signals from the detector are readout with a 2401B position analyzer, which outputs 0-5V analog pulses whose amplitude are linearly proportional to the event position on the x and y axes for each processed photon [7].

The electronics for this system consisted of two main lines, one for the photon detection, and the other for the operation and readout of the device. Most of the experiments on the tabletop IPEM used a silicon PIN diode as the sample, with a copper TEM grid and thin luminescent film on top. The PIN diodes were biased with 40V. The ion beam induced charge (IBIC) signal from the diode first went through an Ortec 142 pre-amplifier, followed by an Ortec 671 amplifier. The amplified signal was then fed through an analog-to-digital converter (ADC) before entering the FastComTec MPA-3 data acquisition system. The MPA-3 system feeds the information into the computer where it can be processed as necessary, and coincidence mapping can be set up. On the photon side, the signals from the position analyzer are sent through ADCs and into the same MPA-3 data acquisition system. Using the software, the electronic signals from the device can be put in coincidence with the photon signal from the detector, with a selected coincidence time window. A block diagram of the electronics for the tabletop IPEM can be seen in Figure 2.

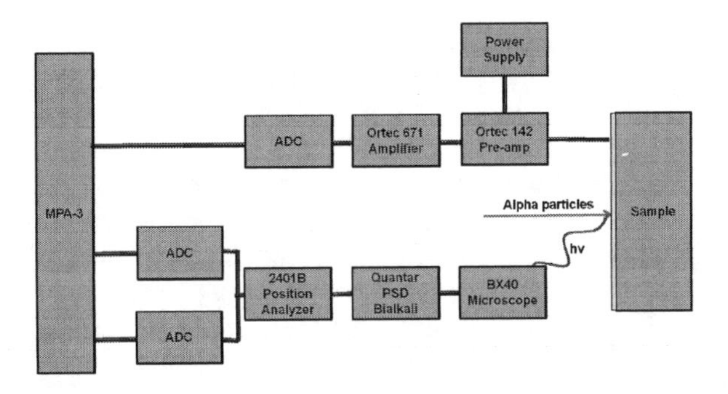

FIGURE 2. Electronics diagram for the tabletop IPEM.

μ-ONE IPEM

To create an IPEM system that most closely reflects what will be eventually be used on a cyclotron, the second major IPEM was built and installed on the nuclear microprobe line (μ-ONE) of Sandia's 6 MV tandem accelerator. With the beam from this accelerator, single event effects can be observed in select devices that upset at low LETs. One major advantage of installing it on μ-ONE is the ability to directly compare and contrast the results obtained using traditional nuclear microscopy with those from IPEM [4]. This system consists of a focused, scanned beam of various ions and energies. The beam spot can be focused to less than 1 μm^2 and scanned across a 100 μm X 100 μm area. The end station consists of a Raith vacuum chamber with laser interferometric stages achieving ~ 50 nm resolution. A JEOL OM-40 microscope is inside the chamber, with a prism having a hole bored through the middle to allow the beam to pass directly through the microscope. In this way, the microscope looks directly at the location where the beam will strike the sample.

There exist a few major differences between the μ-ONE IPEM and the tabletop version. The most significant difference is the availability of ions and energies with the accelerator. Experiments could be carried out with anything from low-energy protons to 50 MeV copper ions. The system being in vacuum also played a major role, limiting the effect of air luminescence, which will be described in detail later. In addition, the light was collected with the OM-40 microscope instead of the BX-40. Therefore there was no viewing port, and the optics line instead led to a 90° mirror, which could be pointed towards a CCD for real-time imaging or to the PSD for IPEM experiments. No changes were made to the electronics or data acquisition system.

Cyclotron IPEM

Once realizing the μ-ONE IPEM and collecting results with that system, the next step was to design a similar setup that could easily be installed on one of the beamlines at the 88" cyclotron at Lawrence Berkeley National Laboratory (LBNL) for performing SEE experiments [4]. The first version of the cyclotron IPEM was installed on an external beamline in Cave 4A at LBNL. Several things differentiated this from the previous IPEMs used. It was decided to use an external beam from the cyclotron and do the experiments in air for ease-of-use as well as to allow industries to continue testing their components in similar environments as what they are used to. Instead of having one microscope that sent light to a rotating mirror, two separate microscopes were used; one which illuminated the sample for real-time viewing with a CCD, and the other which focused light onto the PSD (Figure 3). A Newport 3-axis stage held the samples and could be remotely controlled with the computer system. A translation vector was determined which moved the sample from the focal point of the CCD to that of the PSD, with the thought that specific locations could be located on the sample and easily translated to be imaged with the PSD.

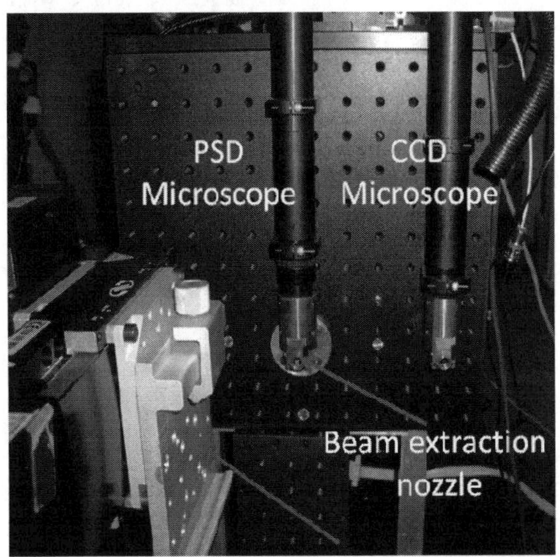

FIGURE 3. Image of the optics for the cyclotron IPEM, demonstrating the two separate OM-40 microscopes used for real-time imaging and IPEM.

Several obstacles were encountered with the cyclotron IPEM system. One of the first problems realized was associated with the translation vector. Once the position of the sample was varied, even by a small step, the translation vector was no longer reliable. Therefore for each position, the focus had to be adjusted without real-time capabilities. This significantly increased the time necessary for each experiment. A second major problem, which was alluded to earlier, is that of air luminescence. The high-energy heavy ion beams from the cyclotron tend to ionize molecules in the air, which luminesce, producing a significant amount of light. The light from the air is very fast and well matched to the response of the bialkali photocathode on the PSD (Figure 4). Therefore after a signal was received in the data processing system from the electronics, the first photon signal to be detected by the detector was often from air luminescence and not from the luminescent film on the sample. This led to a high background and accidental coincidences, hindering the resolution of the system. A third problem was the timing of the photon and device signals. Often, the photons were being detected before the IBIC signal from the device. This is a problem since the first photon to be produced after an ion strike contains the most accurate position information. Ignoring that photon because it comes before the device signal limits the resolution of the system.

Cyclotron IPEM Mark II

Considering the obstacles described above with the original cyclotron IPEM, a myriad of modifications were made to the mark II system. To overcome the unreliability of the translation vector, a new optics line was designed with only one OM-40 microscope. A 92/8 beamsplitter was installed after the zoom/focus lens, which sends 8% of the light to the CCD camera for real-time imaging, which the other 92% can be used for IPEM imaging with the PSD. This eliminates all moving objects, making the focusing process easier and more reliable. Three steps were taken to avoid the detection of the air luminescence. First, a long-pass filter was installed just after the microscope which only allows wavelengths longer than ~500 nm to pass. Second, a new PSD was purchased which is equipped with an S25 photocathode. The response curve for this photocathode can been seen in Figure 4, and demonstrates fairly low sensitivity in the wavelength range where the air emits. The third step was to alter the luminescent materials research to focus instead on longer-wavelength emitting materials that could match the sensitivity of the new photocathode. A fairly simple resolution was found for the signal timing. Ortec 427A delay amplifiers were installed on the photon signal lines to make sure they are read into the data acquisition system only after the device signal comes through. The use of the S25 photocathode, however, introduced a new problem. This type of detector has an inherently higher dark count rate, on the order of 500 counts per second when cooled to -25°C. To minimize the effect of the dark counts, a Hamamatsu photomultiplier tube (PMT) is being installed in front of the sample, with the same longpass filter used on the microscope. The signal from the PMT can then be put in coincidence with the photon signals from the detector, such that a signal will only be processed if it comes from the film luminescence. A schematic of the new optics and the electronic system for the mark II cyclotron IPEM can be seen in Figures 5 and 6 respectively.

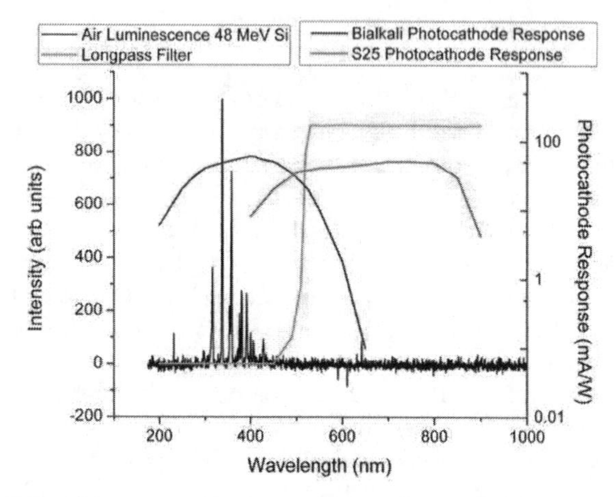

FIGURE 4. Graph demonstrating the spectrum for air luminescence from a 48 MeV silicon beam, overlayed with the response curves for the bialkali and S25 photocathodes. The transmittance of the longpass filter shows its effectiveness at eliminating the air luminescence.

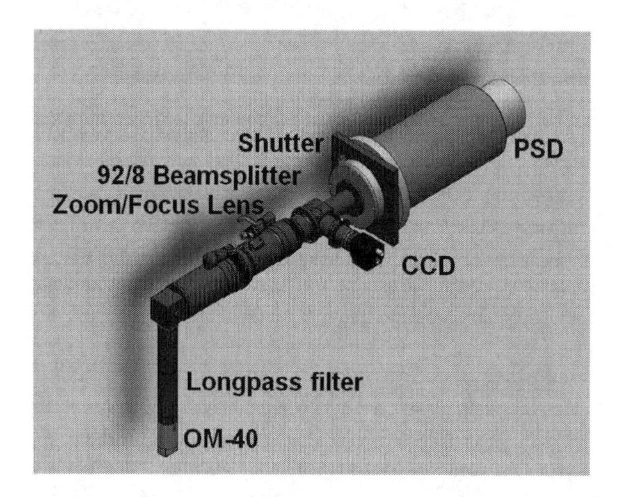

FIGURE 5. Schematic of the mark II cyclotron IPEM optics.

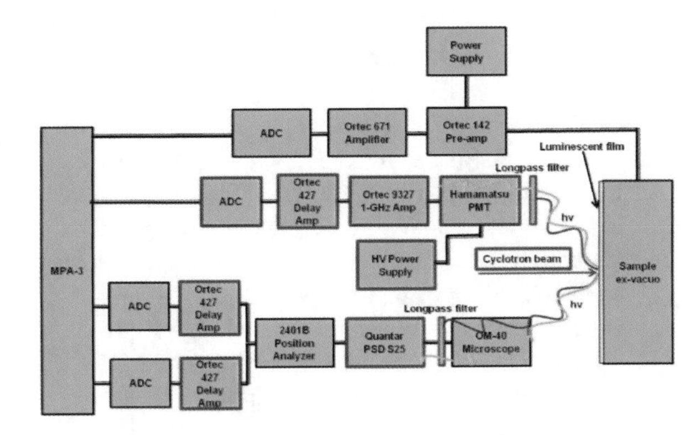

FIGURE 6. Schematic of the mark II cyclotron IPEM electronics. The variations from the previous systems are highlighted.

RESULTS

Experiments have been carried out and results collected on all the IPEM systems described here, except for the mark II cyclotron IPEM. Figure 7 shows IBIC images obtained of a Sandia TA788 SRAM device, using the μ-ONE system. Images were simultaneously collected using traditional nuclear microscopy (scanning, focused beam) and using IPEM. The luminescent film used in these experiments was 5 μm-thick n-type GaN. It is clear from the results that imaging can be performed with IPEM, however the resolution (~ 2 μm) is not yet comparable to that possible with traditional radiation effects microscopy. Figure 8 is the first IPEM image collected with the cyclotron IPEM, using a 10 MeV/amu neon beam. The sample here is also the TA788 device with a 5 μm GaN luminescent film. The resolution achieved is poor (~ 5 μm), which is thought to be a result of the intense air luminescence. The mark II cyclotron IPEM has been designed, developed, and it currently being installed at LBNL, as a permanent endstation for one of their beamlines. Once installation is complete, initial experiments will be run to determine the effectiveness of the modifications.

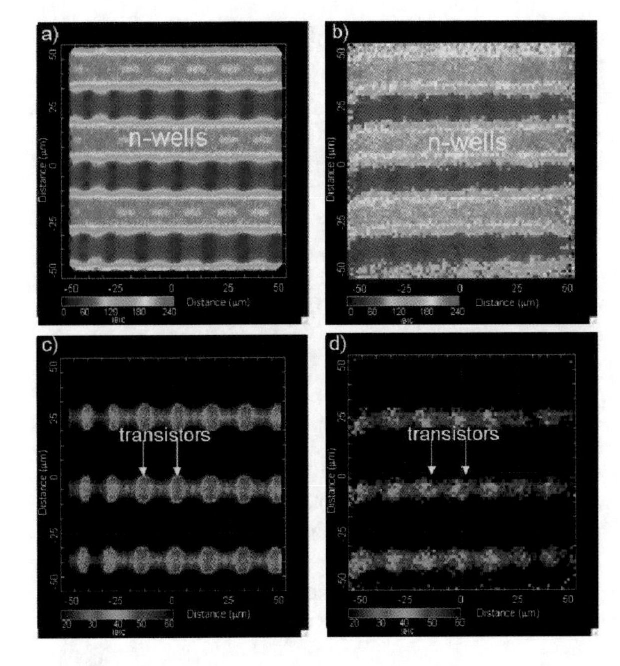

FIGURE 7. a) IBIC image of Sandia's TA788 device b) Simultaneous IPEM-IBIC image of identical region as 3a c) IBIC image in same region as 3a and 3b with modified intensity range to show details in the p substrate d) Simultaneous IPEM-IBIC image of identical region as 3c.

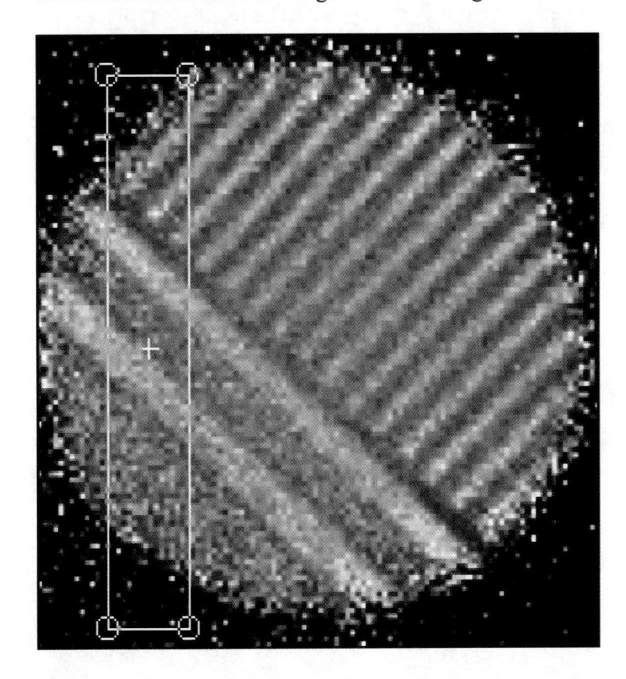

FIGURE 8. IPEM image of Sandia's TA788 device taken with the cyclotron IPEM. This image demonstrates approximately 5 μm resolution for the first experiments on this system.

CONCLUSIONS

Ion photon emission microscopy is a technique developed at Sandia that will allows the radiation effects community to continue to study progressing IC technology with high LET beams and still obtain crucial spatial resolution for radiation sensitivity mapping. Several versions of the IPEM have been designed and developed at Sandia and at LBNL, with promising results. The obstacles encountered with the ex-vacuo cyclotron IPEM have been identified and modifications have been made to overcome these problems with the next generation system. Initial results from the μ-ONE and cyclotron IPEM have demonstrated its ability to be used as a radiation effects microscope. Having the ability to use GeV-energy ions while achieving the desired resolution will allow for the continued use of REM into the future as feature sizes shrink, and the thicknesses of overlayers continue to increase.

ACKNOWLEDGEMENTS

Sandia National Laboratories is a multi-program laboratory operated by Sandia Corporation, a wholly-owned subsidiary of Lockheed Martin company, for the U.S. Department of Energy's National Nuclear Security Administration under contract DE-AC04-94AL85000. Work supported in part by the Italian "Instituto Nazionale di Fisica Nucleare" (INFN). Work at LBNL was supported by the Director, Office of Basic Energy Sciences, of the US Department of Energy under Contract No. DE-AC03-76SF00098.

REFERENCES

1. G. Vizkelethy, B.L. Doyle, D.K. Brice, P.E. Dodd, M.R. Shaneyfelt, J.R. Schwank, Nucl. Instr. And Meth. B 231 (2005) 467
2. B.L. Doyle, D.S. Walsh, S.N. Renfrow, G. Vizkelethy, T. Schenkel, A.V. Hamza, Nucl. Instr. and Meth. B 181 (2001) 199
3. B.L. Doyle, D.S. Walsh, G. Vizkelethy, P. Rossi, F.D. McDaniel, T. Schenkel, J. McDonald, A.V. Hamza, Current Appl. Phys. 3 (2003) 31
4. J.V. Branson, B.L. Doyle, G. Vizkelethy, P. Rossi, J.A. Knapp, M.A. McMahan, Nucl. Instr. And Meth. B 267 (2009) 2085
5. P. Rossi, B.L. Doyle, J.C. Banks, A. Battistella, G. Gennaro, F.D. McDaniel, M. Mellon, E. Vittone, G. Vizkelethy, N.D. Wing, Nucl. Instr. And Meth. B 210 (2003) 123
6. P. Rossi, B.L. Doyle, V. Auzelyte, F.D. McDaniel, M. Mellon, Nucl. Instr. And Meth. B 249 (2006) 242
7. Quantar Technology, "Series 2600 Imaging Photon Detector Systems" http://www.quantar.com/pages/QTI/optical.htm

LET Spectrum Measurements In CR-39 PNTD With AFM

C. E. Johnson[a], J. M. DeWitt[b], E. R. Benton[b], N. Yasuda[c] and E. V. Benton[d]

[a]*Los Alamos National Laboratory, Los Alamos, NM 87545 USA*
[b]*Dept. of Physics, Oklahoma State University, Stillwater, OK 74078 USA*
[c]*National Institute of Radiological Sciences, Chiba, Japan*
[d]*Dept. of Physics, University of San Francisco, San Francisco, CA 94117 USA*

Abstract. Energetic protons, neutrons, and heavy ions undergoing collisions with target nuclei of varying Z can produce residual heavy recoil fragments via intra-nuclear cascade/evaporation reactions. The particles produced in these non-elastic collisions generally have such extremely short range (~<10 μm) that they cannot be directly observed by conventional detection methods including CR-39 plastic nuclear track detector (PNTD) that has been chemically etched for analysis by standard visible light microscopy. However, high-LET recoil fragments having range on the order of several cell diameters can be produced in tissue during radiotherapy using proton and carbon beams. We have developed a method to analyze short-range, high-LET tracks in CR-39 plastic nuclear track detector (PNTD) using short duration chemical etching (~<1 μm) following by automated atomic force microscope (AFM) scanning. The post-scan data processing used in this work was based on semi-automated matrix analysis opposed to traditional grey-scale image analysis. This method takes advantage of the 3-D data obtained via AFM to achieve robust discrimination of nuclear tracks from other features inherently present in the post-etch detector surface. Through automation of AFM scanning, sufficient AFM scan frames were obtained to attain an LET spectrum spanning the LET range from 200-1500 keV/μm. In addition to our experiments, simulations were carried out with the Monte Carlo transport code, FLUKA. To demonstrate this method, CR-39 PNTD was exposed to the proton therapy beam at Loma Linda University Medical Center (LLUMC) at 60 and 230 MeV. Additionally, detectors were exposed to 1 GeV protons at the NASA Space Radiation Laboratory (NSRL) at Brookhaven National Laboratory (BNL). For these exposures CR-39 PNTD, Al and Cu target foils were used between detector layers.

Keywords: CR-39, AFM, short-range, high-LET
PACS: 07.05.Kf, 25.40.Ep, 29.40.Wk

INTRODUCTION

Energetic protons and neutrons, such as those encountered in earth's orbit and those used to treat cancer patients undergoing proton radiotherapy, are capable of undergoing non-elastic collisions with the nuclei present within the body of the astronaut or patient respectively. Non-elastic reactions can be described by a two step reaction model[10,14] also known as an intra-nuclear cascade/evaporation reaction. In the first step of the reaction, immediately following the collision, individual nucleons may be ejected along with light charged particles (LCP's), which include nuclei with $Z < 10$. In the second step of the reaction, equilibrium is attained by ejecting additional nucleons. To conserve momentum, the remaining nucleus recoils in an appropriate direction. This heavy nuclear recoil fragment (HNR) typically has a range < 10-15 μm in low Z materials. In this manner a source of low LET primary particles will readily generate secondary particles with much higher LET through non-elastic cascade/evaporation reactions.

The dose deposited by secondary particles formed from non-elastic reactions is difficult to measure directly due to their extremely short range. Short-range, high-LET particles cannot enter the sensitive volume of active detectors due to their short range. Passive detectors, such as CR-39 plastic nuclear track detector (PNTD) prepared for optical analysis using a ≈ 40 μm bulk etch (B) also fail to detect short-range, high-LET particle tracks because these tracks are completely removed by the etching process. A reduced value of B can be used that preserves the tracks formed by short-range, high-LET particles but an alternative, non-optical, technology must be employed to complete the detector analysis. It has been shown previously that CR-39 PNTD used with a short etch ($B \leq 1.0$ μm) followed by analysis using AFM can be used to measure the high-LET dose contribution for both cancer treatment patients and for astronaut dose

Application of Accelerators in Research and Industry
AIP Conf. Proc. 1336, 637-642 (2011); doi: 10.1063/1.3586180

determination[2,3]. In this paper we present our recent efforts to simultaneously improve and automate the AFM/CR-39 PNTD method.

Atomic force microscopy (AFM) has been applied to the study of both etched and un-etched nuclear track detectors and has proven effective in the analysis of CR-39 PNTD[4,11,12]. Unlike standard optical analysis, which has an inherent fluence limit of $\sim 10^5$ cm^{-2} as a result of the difficulty in analyzing overlapped tracks, AFM analysis of CR-39 PNTD permits fluences in excess of $\sim 10^8$ cm^{-2}. CR-39 PNTD exposed to high primary particle fluences, such as on the exterior of spacecraft (for long duration and under low shielding) cannot be analyzed by optical microscopy methods, but are amenable to AFM analysis.

METHODS

The experimental approach used with CR-39 PNTD intended for AFM analysis is similar to that used for detectors destined for optical analysis with the exceptions of the etch parameters employed, the etch measurement technique used, and the methods used for detector analysis. Each set of detectors are placed in a proton beam in a stack configuration with one detector in front and one in back (downstream and upstream with respect to the beam). Detector sets were exposed to 10 Gy doses of 60 and 230 MeV protons at the Loma Linda University Medical Center Proton Therapy Facility and a 2 Gy dose of 1 GeV protons at the Brookhaven National Laboratory. For the detectors analyzed in this work, a 2 h etch in 50 °C, 6.25 *N* NaOH was used. This etch procedure was determined to provide a bulk etch of ≈ 0.5 μm. The step method[16] of bulk etch determination must be used since *B* is on the order of measurement error for traditional methods (such as Henke[7]). It is also noteworthy that the 50 °C etch solution has a slower rate of attack during etch that leaves a smoother post-etch surface, which is beneficial for AFM analysis.

AFM Parameters For High-Volume CR-39 Scanning

The AFM used in this work is a Veeco, Dimension 3100 with Nanoscope IIIa controller. This AFM is equipped with a motorized stage and is mounted on an air pressurized vibration damping table. By using a 'Programmed Move' command, which is included in the Veeco, Inc. AFM control software, it is possible to complete many AFM scans sequentially. In this mode a surface scan is acquired, the motorized stage repositions (in x and/or y), and scanning is resumed in the new stage position. A 250-500 μm separation spacing (pitch) between AFM scans was used to avoid scanning the same area repeatedly. The Programmed Move command can buffer up to 99 consecutive scan positions which will be automatically scanned without operator involvement. During automated scanning, it is important to utilize 'automatic drive feedback' (also included in the Veeco control software). This setting allows the scan head to maintain tracking for the duration of the Programmed Move command, which takes approximately two days to complete.

The remaining AFM scan parameters used are typical settings with the possible exception of the AFM scan rate. In this work, a scan rate speed limit of 40 – 50 μm/s was found to provide artifact free scans using scan sizes up to 50 μm × 50 μm. The presence of artifacts in an AFM scan complicates track analysis in later steps.

In AFM applications, a 50 μm × 50 μm scan size is fairly large, but since we are interested in the comparatively rare short-range, high-LET tracks a large scan area is a necessity to achieve statistical significance. The selection of a large scan area has direct consequences on the lower threshold of detection in particle LET. We have found that for this scan size; information is lost on tracks with LET < 100-150 keV/μm. Loss of information on low LET tracks is related to the distribution of height measurements (digital sampling resolution) across the surface as opposed to an inherent feature of CR-39 PNTD.

Data Analysis Methods

A novel analysis method was developed for this research[8,9] which differs from traditional image processing in that the 3-D nature of AFM data is leveraged to extract topographic features. In gray-scale processing with optical microscope images there exist multiple features types that can give rise to a particular shade of gray. With AFM, a series of physical measurements are made in the x-y plane of a detector's surface. These deterministic measurements correspond to actual physical height with respect to x and y. AFM measurement accuracy is typically ~ 1 nm in x and y and < 1 nm in z depending on the AFM tip used. The algorithms developed for this work have been designed to analyze topographic features, rejecting artifacts and/or noise, and extract individual track parameters for actual nuclear tracks.

The first step in processing AFM data consists of flattening any surface curvature and removing vertical offset, if present. This process is crucial since track parameter measurements in subsequent steps require a surface level at zero height. A plane is selected beneath the surface level of the AFM data and individual matrices corresponding to the points inside individual tracks cones are populated. An example AFM scan shown in the contour analysis approach is shown in Figure 1. Each matrix of points is considered a track candidate (TC). An ellipse-specific least squares ellipse fitting[6] is applied to each TC matrix. This is followed by a gradient weighted estimation of fitting[13] (EOF) to test the 'goodness' of the ellipse fit. The EOF is used to certify that a TC is in fact a nuclear track using an empirically determined threshold. Additionally, the EOF is useful for the separation of overlapped tracks. If the EOF is found to be high, the TC matrix can be separated into two new matrices which are then fitted with new ellipses. If the new EOF values are improved, then the two new tracks are kept. The fitted ellipses are used to determine the semi-major axis, a, and the semi-minor axis, b. The reduced etch rate ratio, V_R, is then calculated using,

$$V_R = \sqrt{1 + \frac{4\left(\frac{a}{B}\right)^2}{\left[1-\left(\frac{b}{B}\right)^2\right]^2}}\,. \quad (1)$$

For undercut nuclear track cones, it is common for the AFM tip to miss the bottom of the cone. This has been dubbed the AFM tip artifact. An example is shown in Figure 2. It is important to select an analysis plane below the level of surface roughness, but high enough to avoid the AFM tip artifact. This level was empirically determined to be ~5-10 μm.

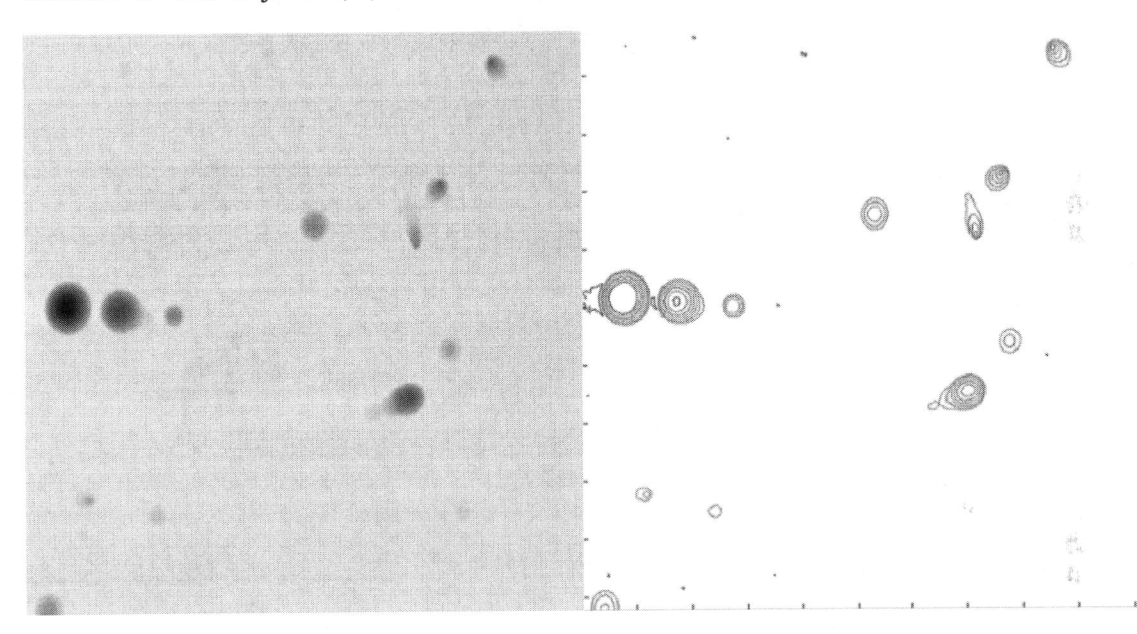

FIGURE 1. An AFM scan is displayed in gray scale (left image) and as a contour map (right image). The contour display places emphasis on the details of track interiors as well as reducing the visual impact of surface irregularities.

FLUKA Models

Target fragmentation simulations were performed using the Monte Carlo radiation transport code FLUKA[1,5]. The simulations involved the irradiation of target-detector stacks with 60, 230, and 1000 MeV proton beams using targets of aluminum, copper, and CR-39 PNTD. Each target-detector stack consists of a sandwich arrangement of CR-39 and a target material of interest. The proton beam was simulated to irradiate a cross sectional area of 10 × 10 cm while passing through a 200 cm column of air (0.24 g/cm^2) before arriving at the target-detector stack measuring 2 × 4 cm in cross sectional area. The thickness of the CR-39 was specified to be 0.05 cm, with the target foils having a thickness of 0.0025 cm. The CR-39 was specified as the scoring volume of the simulation to obtain LET_∞CR-39 spectra of protons, alpha particles, heavy ions, and all charged particles. Additionally, ^{24}Na recoil nuclei were scored, in the case of the aluminum target, as well as ^{28}Mg recoil nuclei in the case of the copper target[15]. No evidence of these recoils was found in the simulations described here.

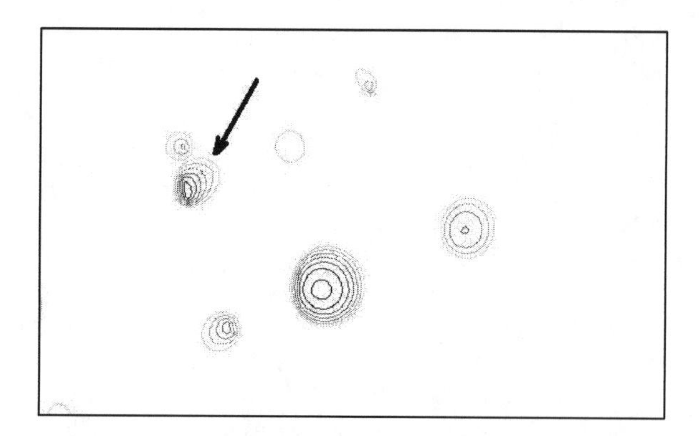

FIGURE 2. The nuclear track on the left side (arrow) illustrates an example of the AFM tip artifact. The left edge of this track appears to be square with the scan frame. This is not a real feature and must be carefully avoided during analysis by extracting track parameters at a depth above the onset of the artifact.

RESULTS AND DISCUSSION

In Figure 3, the results obtained using a CR-39 PNTD target layer are shown. The FLUKA simulations clearly show the presence of light charged particles ejected in the first step of non-elastic nuclear reactions. These particles consist of knock-out protons, alphas, and Li nuclei but can range higher for higher Z nuclear targets. The AFM/CR-39 PNTD results lose accuracy below 100-150 keV/μm due to the distribution of AFM height measurements as mentioned previously. Other types of experiments, such as CR-39 PNTD prepared for optical analysis can be used in this low LET region if desired. In the LET region between roughly 100 and 250 keV/μm there is a reasonable agreement between FLUKA and AFM/CR-39 PNTD results.

In the high-LET region above 200-250 keV/μm, FLUKA differs markedly from our experimental results. The models completed in this work used 1×10^9 primary protons to complete the simulation. It is possible that an increased number of primaries would provide more data at higher LET's in the model. Alternatively, it may be the case that FLUKA does not propagate HNR particles along their complete trajectories although the FLUKA dose estimations may be accurate.

A comparison between AFM results and FLUKA for Cu and Al target layers is shown in Figure 4. A similar range of agreement was found in Cu and Al, approximately between 100-250 keV/μm. It is noted that the FLUKA output predicts sharp peaks in the LET spectra which coincide with the LET maxima of protons, He, Li, and higher Z nuclei. These peaks are not observed in our AFM results. This is likely due to the fact that the actual value of *B* varies slightly across the surface of a detector. This variation in *B* broadens out any sharp peaks in LET. Additionally, the experimental data are affected by other factors, such as a very small air gap between target and detector.

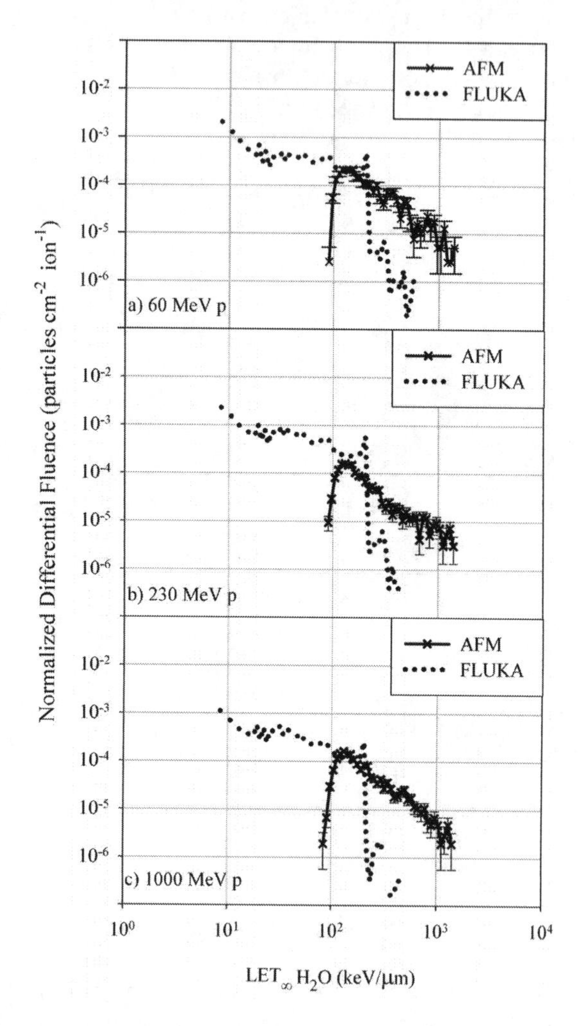

FIGURE 3. LET spectra comparisons between AFM results and FLUKA model for a) 60 MeV p, b) 230 MeV p, and c) 1000 MeV p on a CR-39 PNTD target.

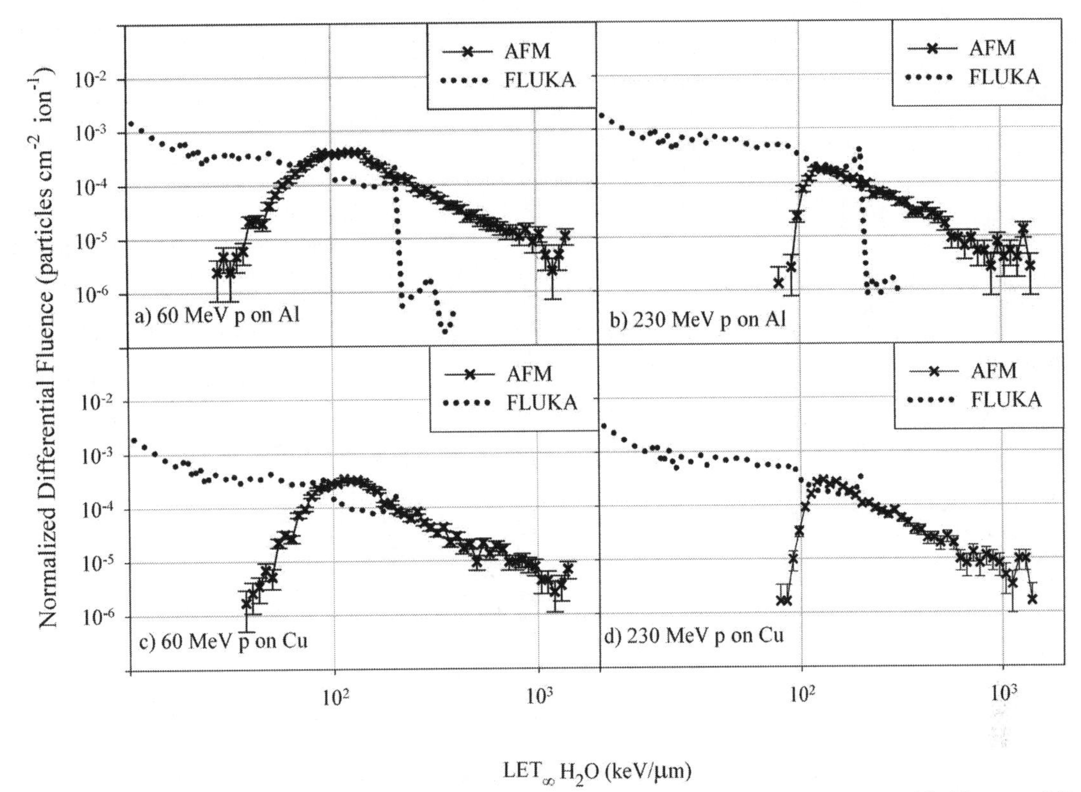

FIGURE 4. LET spectra comparisons between AFM results and FLUKA model for a) 60 MeV p with Al target foil, b) 230 MeV p with Al target foil, c) 60 MeV p with Cu target foil, and d) 230 MeV p with Cu target foil.

CONCLUSIONS

The AFM/CR-39 PNTD method was used successfully to measure the LET spectra of secondary particles formed as products of proton induced non-elastic nuclear collisions in Al, Cu, and CR-39 PNTD targets exposed to 60, 230, and 1000 MeV protons. The short-range, high-LET HNR particle tracks measured in this work are typically removed by the long duration etch used for optical analysis. A short etch can be used to gain additional information from exposed CR-39 PNTD, such as those exposed to high fluences or for detectors in which short-range tracks are of interest, provided the additional analysis time requirements associated with AFM scanning are justifiable. For example, the CR-39 PNTD dosimeter used by an astronaut on a ~1.8 yr mission to mars could not be analyzed by optical methods but it could be analyzed using the AFM/CR-39 PNTD methods presented here.

ACKNOWLEDGMENTS

We greatly appreciate the assistance of the staff of Loma Linda University Medical Center, for both proton beam exposures and the loan of an atomic force microscope. We are also grateful to Koichi Ogura for many helpful discussions.

REFERENCES

1. G. Battistoni, F. Cerutti, A. Fasso, A. Ferrari, S. Muraro, J. Ranft, S. Roesler and P. R. Sala, (2007). Proceedings of the Hadronic Shower Simulation Workshop 2006, Fermilab 6-8 September 2006, M. Albrow and R. Raja, (Eds.), AIP Conference Proceeding **896**, 31-49.
2. E. R. Benton, E. V. Benton and A. L. Frank, Rad. Meas. **35** (5), 457-471 (2002).
3. E. R. Benton, E. V. Benton and A. L. Frank, Internal Technical Report No. ERI-02021, 2002.
4. M. Drndic, Y. D. He, P. B. Price, D. P. Snowden-Ifft and A. J. Westphal, Nucl. Instr. and Meth. B **93** (1), 52-56 (1994).
5. A. Ferrari, P. R. Sala, A. Fasso and J. Ranft, Report No. CERN-2005-10 (2005), INFN/TC_05/11, SLAC-R-773, 2005.
6. A. Fitzgibbon, M. Pilu and R. B. Fisher, Pattern Analysis and Machine Intelligence, IEEE Transactions on **21** (5), 476-480 (1999).
7. R. P. Henke, K. Ogura and E. V. Benton, Nucl. Tracks and Rad. Meas. **12** (1-6), 307-310 (1986).
8. C. E. Johnson, E. R. Benton, N. Yasuda and E. V. Benton, Rad. Meas. **44** (9-10), 742-745 (2009).
9. C. E. Johnson, "Atomic Force Microscopy Methods for the Analysis of High-LET Tracks in CR-39 Plastic Nuclear Track Detector", Ph.D. Thesis, Oklahoma State University, 2009.
10. W. G. Lynch, Annual Review of Nuclear and Particle Science **37** (1), 493-535 (1987).
11. D. Nikezic, J. P. Y. Ho, C. W. Y. Yip, V. S. Y. Koo and K. N. Yu, Nucl. Instr. and Meth. B **197** (3-4), 293-300 (2002).
12. P. B. Price, Nucl. Tracks and Rad. Meas. **22** (1-4), 9-21 (1993).
13. P. L. Rosin, Pattern Recognition Letters **17** (14), 1461-1470 (1996).
14. R. Serber, Phys. Rev. **72** (11), 1114 (1947).
15. L. Winsberg, Phys. Rev. C **22**, 2123-2134 (1980).
16. N. Yasuda, M. Yamamoto, N. Miyahara, N. Ishigure, T. Kanai and K. Ogura, Nucl. Instr. and Meth. B **142** (1-2), 111-116 (1998).

Applications Of Monte Carlo Radiation Transport Simulation Techniques For Predicting Single Event Effects In Microelectronics

Kevin Warren, Robert Reed, Robert Weller, Marcus Mendenhall, Brian Sierawski, Ronald Schrimpf

ISDE, Vanderbilt University, 1025 16th Avenue South, Nashville, TN 37212, United States

Abstract. MRED (Monte Carlo Radiative Energy Deposition) is Vanderbilt University's Geant4 application for simulating radiation events in semiconductors. Geant4 is comprised of the best available computational physics models for the transport of radiation through matter. In addition to basic radiation transport physics contained in the Geant4 core, MRED has the capability to track energy loss in tetrahedral geometric objects, includes a cross section biasing and track weighting technique for variance reduction, and additional features relevant to semiconductor device applications. The crucial element of predicting Single Event Upset (SEU) parameters using radiation transport software is the creation of a dosimetry model that accurately approximates the net collected charge at transistor contacts as a function of deposited energy. The dosimetry technique described here is the multiple sensitive volume (MSV) model. It is shown to be a reasonable approximation of the charge collection process and its parameters can be calibrated to experimental measurements of SEU cross sections. The MSV model, within the framework of MRED, is examined for heavy ion and high-energy proton SEU measurements of a static random access memory.

Keywords: Radiation effects, single event upset, SEU, single event effects, SEE, geant4, MRED, sensitive volume models
PACS: 61.82.Fk

INTRODUCTION

Ionizing radiation generates free carriers within semiconductor devices [1]. The motion and collection of the liberated carriers causes currents to flow within the circuit. In sequential logic, the magnitude and duration of the current may be sufficient to corrupt the internal state(s) of the storage element. The ultimate corruption of the state is called a single event upset (SEU) and represents a reliability concern for electronics in space, atmospheric, and terrestrial environments.

The problem of predicting SEU rates, for the purpose of this paper, will be divided into three categories: determination of the magnitude and location of energy deposition within the solid model of the transistor or circuit, quantification of the node or contact currents or the total collected charge from the event, and evaluation of the resultant circuit level response (e.g., SEU). Monte-Carlo physics-based radiation transport tools, such as those derived from the Geant4 class libraries [2], provide a powerful capability for determining energy deposition from typical environmental radiation species (e.g., alpha particles, galactic cosmic ray neutrons, thermal neutrons, protons, etc.) [3]. At the circuit level, Spice, has long been used for estimating the total quantity of charge required to induce SEU [4]. In this work, a method for bridging energy loss calculations to circuit response is presented. The proposed model, called the multiple sensitive volume (MSV) model, is integrated into the machinery of the radiation transport software.

MULTIPLE SENSITIVE VOLUME MODEL

The concept of using a single cuboidal volume to approximate energy loss, charge generation, and circuit level SEU has been central to the problem of predicting SEU rates in space environments [5]. The general method, known as the rectangular parallelepiped (RPP) model, assumes that the deposited energy within a volume associated with a particular node is related to the collected charge at that node. A detailed treatment of the RPP and alternative

Application of Accelerators in Research and Industry
AIP Conf. Proc. 1336, 643-648 (2011); doi: 10.1063/1.3586181

methods to SEU rate prediction are given in [7]. The mathematical expression is given by (1), where E_{dep} is the deposited energy, q is the electron charge and ε is the charge-energy conversion factor (3.6 eV/carrier Si) [6]. If a particle event generates more charge within the sensitive volume than the critical charge of the circuit (Q_{crit}), a SEU occurs. Q_{crit} is the salient circuit parameter (as determined by Spice) that defines the quantity of charge required to change the logic state of the storage element.

$$Q_{coll} = \frac{qE_{dep}}{\varepsilon} \tag{1}$$

A single dosimetry region, such as that given by the RPP method, predicts a step function response in heavy ion SEU cross-section measurements as a function of ion linear energy transfer (LET), which is inconsistent with typically measured SEU cross-section curves [7]. Given that there is a spatial variability in collected charge with respect to ion strike locations, an intuitive alternative is to define the total collected charge as a combination of weighted deposited energies, where the weight represents the efficiency of the charge collection process shown in (2) in which α_i is the efficiency and E_i is the energy deposited with the ith volume [8]. A conceptual drawing of a typical MSV arrangement is shown in Figure 1.

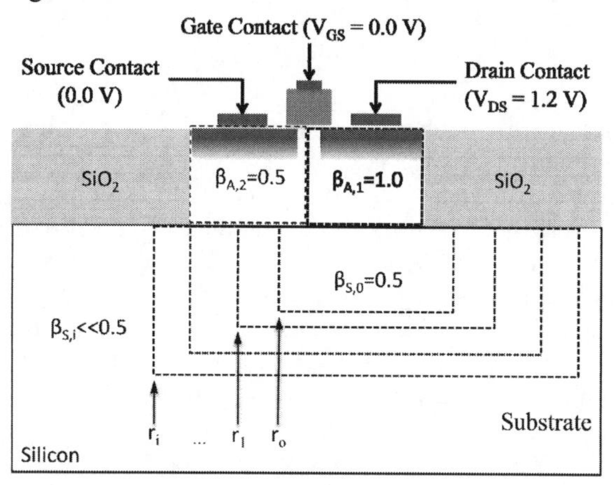

FIGURE 1. Conceptual drawing of a MSV configuration for a bulk NMOS transistor, biased in the off-state. The highest region of efficiency is near the drain junction (β_{A1}). Charge is also collected at the drain contact from charge generated in proximity to the source contact (region A2). Collection efficiency decreases with distance from the active opening in the nested substrate sensitive volumes out to a minimum value at the ith volume.

$$Q_{coll} = \frac{q}{\varepsilon}\sum_{i=1}^{N}\alpha_i E_i \tag{2}$$

Barring parasitic coupling to adjacent transistors and structures, charge collection at a transistor contact from the substrate, with respect to the location of ionization, can be viewed as a centrally weighted mechanism. The farther the ionization occurs from the active silicon opening and the transistor junctions, the less efficiently it is collected at the transistor contacts. Consequently, it is reasonable to arrange the substrate sensitive volumes into a nested configuration (Figure 1). While mathematically the independent terms of (2) are valid, conceptually, the differential volume is better described by a cumulative efficiency factor, β. The relationship between α and β is shown in (3), where each volume index, i, represents a cuboid contained entirely within the volumes i+1 to i=N_s, the total number of substrate volumes.

$$\alpha_i = \begin{cases} \beta_i - \beta_{i+1} & 1 \le i < N_S \\ \beta_i & i = N_S \end{cases} \tag{3}$$

The central nature of the charge collection process lends itself to parameterization of the MSV group. For example, (4) will be demonstrated as a suitable parametric representation of the cumulative collection efficiency as a function of distance from the active silicon opening. In (4), r_i-r_o is the distance of the ith volume from the innermost volume edge (r_o), β_o is the innermost volume's collection efficiency, and m and s are adjustable fitting parameters. The fitting parameters m and s effectively drive the efficiencies to the appropriate values during the calibration process.

$$\beta_i(r_i) = \beta_o e^{-\left(\frac{r_i - r_o}{m}\right)^s} \tag{4}$$

MODEL VALIDATION

To demonstrate the validity of the MSV model in the nested configuration is sufficient to describe the process of energy loss and charge collection, a charge transport simulation, using the Synopsys device simulator, was performed of a singly diffused n-type contact region in a p-type substrate as shown in Figure 2. The active area opening is 0.40 x 0.40 μm^2 and the active silicon thickness is 0.35 μm. Junction depths and doping concentrations are consistent with a 90 nm bulk complimentary metal oxide semiconductor (CMOS) process. The total collected charge for each single event simulation was determined by integrating the contact current from pre-strike to post-strike steady state conditions. The total collected charge in the MSV group was calculated by using chord-length calculations (not physics based transport).

A base MSV model composed of 30 sensitive volumes, parameterized according to (4), was constructed and fit to the charge collection database. Error reduction was achieved by randomly sampling values of s, m, and β_o and comparing the MSV collected charge to charge transport predictions until subsequent cumulative errors for the entire dataset changed by no more than 1% between logged improvements in the error. The sensitive volumes (SV) were spaced equally in the x,y dimensions and the depth (z) was varied as a free parameter in addition to the fitting parameters, s, m, and β_0. The results of the normally incident and 90° simulations are shown in Figure 3 and Figure 4, respectively. The SEU cross section curve for the single-node example is shown in Figure 5 for a series of Q_{crit}. In the limiting case of small Q_{crit}, the curve more closely resembles a step function with an inflection point at an LET of zero. The cross section rapidly saturates to the outermost substrate SV areas. For larger values of critical charge, the shape of the curve follows the increasing area of the substrate SV series, as each contributes to the cumulative cross section according to the SV weight.

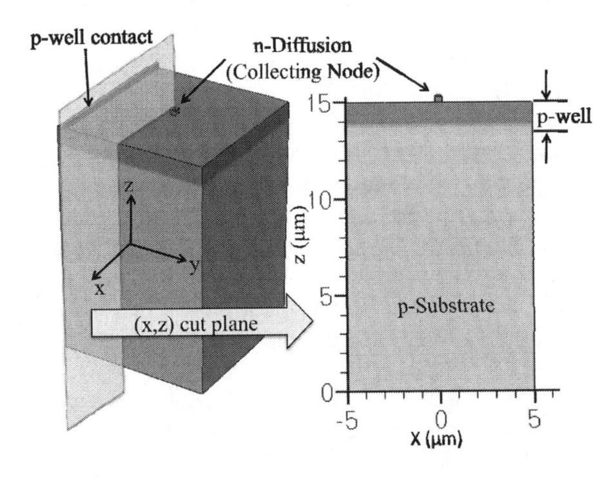

FIGURE 2. TCAD model used in charge collection simulations. The collecting node is a 0.4 x 0.4 x 0.35 µm active silicon region on silicon substrate. The collecting node was biased at 1.2 V with respect to the p-well and substrate contacts.

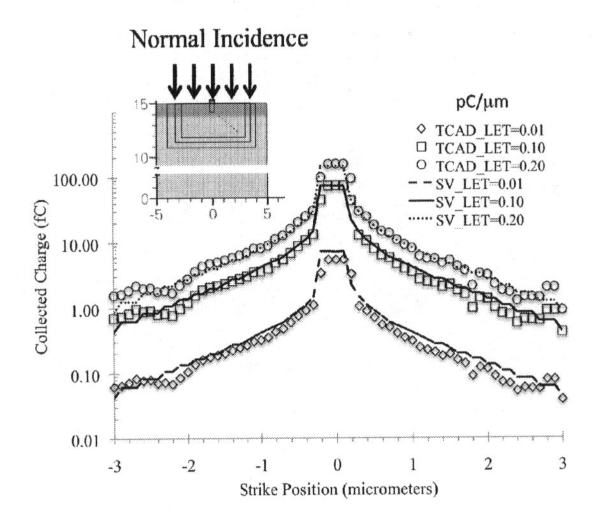

FIGURE 3. Charge collection simulation results (TCAD) of collected charge as a function of ion strike locations across the device surface at normal incidence and MSV model predictions of the collected charge following calibration (30 volumes, s=0.48, m=0.06, β_o=0.44.

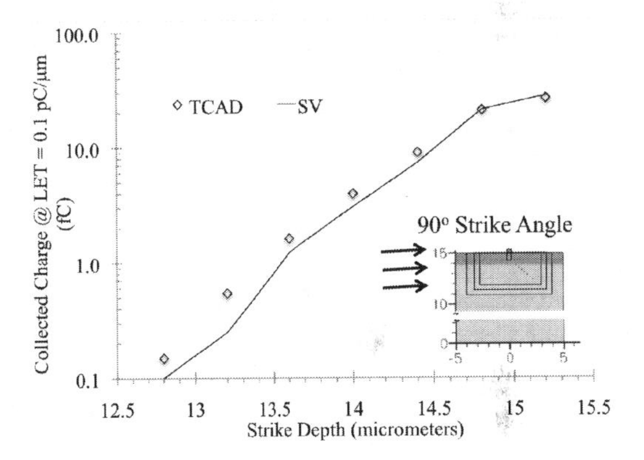

FIGURE 4. Charge collection simulation results (TCAD) of collected charge as a function of ion strike locations across the device surface at grazing incidence and MSV model predictions of the collected charge following calibration (30 volumes, s=0.48, m=0.06, β_o=0.44).

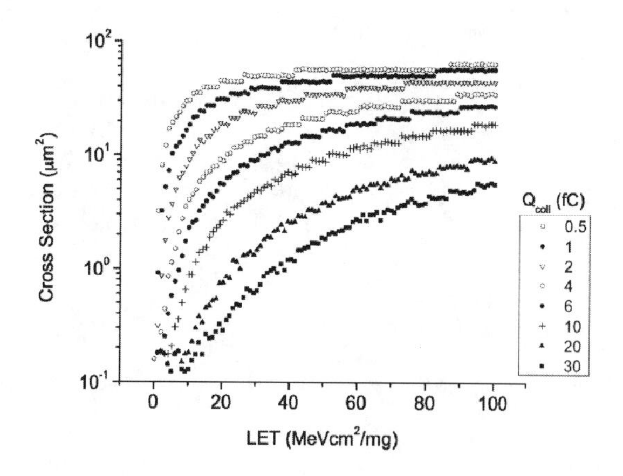

FIGURE 5. Monte-Carlo generated SEU cross-sections curves based on collected charge estimates of the MSV model. The shape of the cross section curve is a function of the critical charge requirements and the sensitive volume parameters.

PHYSICS-BASED MONTE CARLO SIMULATION

The previous section demonstrated the viability of the MSV model to represent the process of charge collection based upon constant LET chord length calculations. However, the chord length methodology is an overly restrictive approximation to all but the inter-planetary space environment where high energy particles are highly penetrating and whose LET does not vary appreciably over short distances. In reality, detailed interaction of the primary particle with the material comprising the electrical devices, such as simple slowing in back end of the line (BEOL) layers (e.g., as would occur for alpha particles) and scattered particles from nuclear interactions (neutrons and protons) dominate the SEU rate in terrestrial and some space environments. Integration of the MSV model into Monte-Carlo radiation transport software is a natural union of the capabilities of the high-energy physics community and the problem of predicting SEU rates in the electronics community.

The Monte-Carlo Radiative Energy Deposition (MRED) tool, developed at Vanderbilt University, is specifically engineered for the purpose of analyzing radiation effects on electronics [9]. MRED is essentially a framework for computing single event effects in semiconductors in which the component tool describing radiation interactions and transport in matter is a built-in, Monte Carlo, binary-collision code. Geant4 provides almost all of the basic radiation physics of MRED. Machinery for the tracking of energy loss in solids of arbitrary type and complexity and material types, user specific dosimetry models (including the MSV model), sampling algorithms suitable for radiation spectra of arbitrary environments, as well as variance reduction enhancements are part of the MRED tool.

To illustrate the application of MRED with the MSV model we consider the case of a 4 MBit static random access memory (SRAM) fabricated on a 250 nm process line. Heavy ion and proton data SEU data are presented with a detailed description of the SRAM in [8] and [10]. Technology computer aided design (TCAD) simulations of the full SRAM (shown in Figure 6) were used to identify the sensitive transistors within the device, which are the off-state N- and P-MOSFET transistors. TCAD, in this study, was used to identify the sensitive regions rather than explicitly determine charge collection profiles. The TCAD results established a guideline for the placement of the MSV within the solid model. SPICE circuit analysis was used to determine the critical charge of the circuit, which was found to be 4.7 and 7.0 fC for strikes to the NMOS and PMOS transistors, respectively. Note that this was accomplished by sourcing or sinking current on the circuit node with a double exponential ideal current source.

A geometrically and compositionally accurate solid model of an individual cell was created using the Synopsys *Procem* process emulation software for the purpose of radiation transport, and is shown in Figure 6. Within the active silicon and substrate regions a separate MSV model was constructed for both the NMOS and PMOS transistors. These are shown in the lower left-hand side of Figure 6. Unlike the previous example, the sizes of the substrate SVs were determined by calibration to experimental heavy ion SEU cross sections, shown in Figure 7. Note that the cross sections are normalized to the SRAM bit count. The efficiencies and sizes of the SVs were adjusted until a best-fit (least error) was achieved between the MRED MSV model and the experimental results.

At SEU onset (LET ~ 1.5 MeVcm^2/mg), the relatively extreme sensitivity of the SEU cross section to LET and model parameters results in larger relative errors than at LET values closer to the saturation cross section. Given the average experimental cross section at the smallest LET values is approximately 0.001 to 0.01 μm^2/bit, it is unlikely that a physically identifiable region can be defined that directly accounts for the measured value. Potential sources of error in the low LET regime are believed to be a result of inter-cell variability in SEU sensitivity [4], which can arise from process variability as well as layout driven sensitivities, such as cell-to-contact spacing [11], or variability in the energy deposition that results from the stochastic nature of energy loss, especially over small volumes [12].

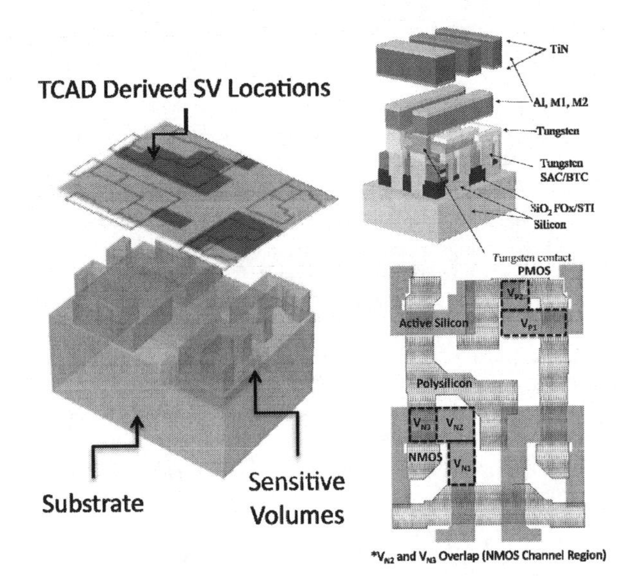

FIGURE 6. Placement of MSV with respect to transistor location in process stack (left). Sensitive transistors were identified by TCAD analysis. Active silicon SVs were located as shown on right, corresponding to the drain/body region of the devices.

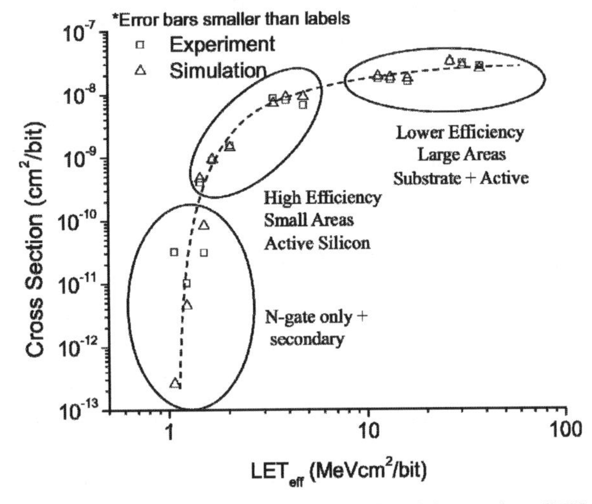

FIGURE 7. Experimental and simulated heavy ion SEU cross sections following calibration.

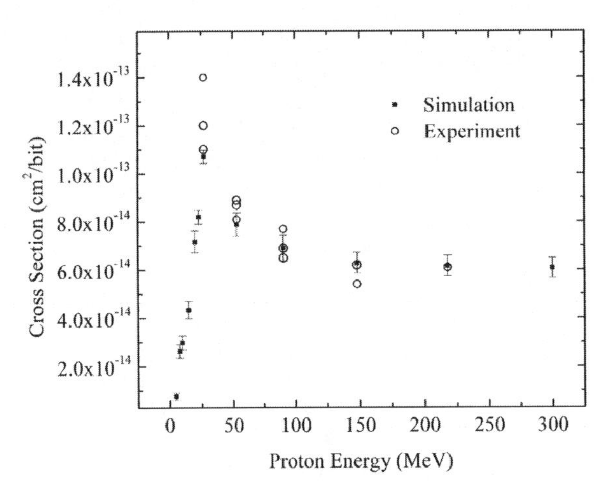

FIGURE 8. Experimental and MRED derived proton SEU cross sections. Note that the lowest proton energy data point was 26 MeV. Experimental errors were less than 10% of the mean and are not shown for clarity.

The heavy-ion calibrated MSV model (including the Q_{crit} parameters and the solid model) was applied to the problem of predicting proton SEU cross section, which are of particular importance for calculating terrestrial SEU rates for galactic cosmic ray neutrons [13]. With the addition of nuclear physics models (Screened Coulombic and the Binary Cascade [9]), the calibrated model is shown to agree well with the experimental results given in Figure 8.

CONCLUSIONS

It has been demonstrated that the MSV model is a valid approximation to charge collection, assuming a suitable method for determining the magnitude and spatial profiles of energy loss and carrier generation can be calculated. However, transistor level charge transport simulation may not always be possible, in which case sensitive regions of the circuit can be identified by laser [14] or micro-beam probing [15]. In the absence experimental and simulation data beyond basic heavy ion SEU cross sections, a basic understanding of the circuit topology (e.g., number of sensitive transistors) is useful for determining the number of MSV (and Q_{crit}) groups that are needed.

Broad-beam heavy ion experiments irradiate the entire silicon surface, but the SEU cross section is a sampling of only the total sensitive area of the circuit, which is also a function of the ion's LET. One may derive the lateral dimensions of the SVs from the experimental heavy ion SEU cross section alone. In the nested configuration, this has the effect of activating progressively larger SV of lesser efficiency, with increasing LET. The investigator must have a precise calculation of the LET of the ion near the electrically active silicon regions and assume that it does not change appreciably over the anticipated depths of the sensitive volumes. Angular heavy ion broad-beam cross-section measurements, with respect to the surface of the device, are further used to calibrate SV depths and efficiencies.

REFERENCES

1. G. C. Messenger and M. S. Ash, Single Event Phenomena. Chapman and Hall Publishing, 1997.
2. S. Agostinelli, et al., *Geant4-a simulation toolkit*, Nuclear Instruments and Methods in Physics Research A, Vol. 506, Jul. 2003. pp. 250–303.
3. J. F. Ziegler and W. A. Lanford, *Effect of Cosmic Rays on Computer Memories*, Science, Vol. 206, No. 4420, 1979. pp. 776–788.
4. L.W. Massengill, M.L. Alles, S.E. Kerns, K.L. Jones, *Effects of process parameter distributions and ion strike locations on SEU cross-section data [CMOS SRAMs]*, IEEE Trans. on Nucl. Sci., Vol. 40, No. 6, Dec. 1993. pp. 1804-1811.
5. E. L. Petersen, J. C. Pickel, E. C. Smith, P. J. Rudeck, and J. R. Letaw, *Geometrical factors in see rate calculations*, IEEE Trans. Nucl. Sci., Vol. 40, No. 6, Dec. 1993. pp. 1888–1909.
6. R. C. Alig and S. Bloom, *Electron-hole-pair creation energies in semiconductors*, Phys. Rev. Lett., Vol. 35, No. 22, Dec. 1975. pp. 1522–1525.
7. E. L. Petersen, *Single-event analysis and prediction*, in 2008 IEEE NSREC Short Course, Section III, 2008.
8. K. M. Warren, et al., *Application of RADSAFE to Model Single Event Upset Response of a 0.25 um CMOS SRAM*, IEEE Trans. Nucl. Sci., Vol. 54, No. 4, Aug. 2007, pp 898-903.
9. R. A. Weller, M. H. Mendenhall, R. A. Reed, R. D. Schrimpf, K. M. Warren, B. D. Sierawski, and L. W. Massengill, *Monte Carlo Simulation of Single Event Effects*, IEEE Trans. Nucl. Sci., Vol. 57, No. 4, Aug. 2010 (in print).
10. K.M. Warren, B.D. Sierawski, R. A. Reed, M. E. Porter, M. H. Mendenhall, R. D. Schrimpf, L. W. Massengill, *Predicting Neutron Induced Soft Error Rates: Evaluation of Accelerated Ground Based Test Methods*, Proceedings of the 2008 International Reliability and Physics Symposium. Apr. 2008.
11. D. Radaelli, H. Puchner, S. Wong, and S. Daniel, *Investigation of multi-bit upsets in a 150 nm technology sram device*, IEEE Trans. Nucl. Sci., Vol. 52, No. 6, Dec. 2005. pp. 2433–2437.
12. M. A. Xapsos, E. A. Burke, P. Shapiro, and G. P. Summers, *Probability Distributions of Energy Deposition and Ionization in Sub-Micrometer Sites of Condensed Media*, Radiation Measurements, vol. 26, no. 1, Jan. 1996. pp. 1-9.
13. JEDEC and Standards, *Measurement and reporting of alpha particles and terres- trial cosmic ray-induced soft errors in semiconductor devices*, JEDEC Solid State Technology Association, Aug 2001.
14. D. McMorrow, et al., *Single-Event Upset in Flip- Chip SRAM induced by Through-Wafer, Two-Photon Absorption*. IEEE Trans. Nucl. Sci, Vol. 52, No. 6, Dec. 2007. pp. 2421-2425.
15. R. A. Reed, et al., *Applications of heavy ion microprobe for single event effects analysis*, Nuclear Instruments and Methods in Physics Research Section B: Beam Interactions with Materials and Atoms, Vol. 261, No. 1-2, Aug. 2007, pp. 443-446.

Radiation Effects In Space

Ram K. Tripathi

NASA Langley Research Center, MS – 188 E, Hampton VA 23681, United States

Abstract. Protecting space missions from severe exposures from radiation, in general, and long duration/ deep space human missions, in particular, is a critical design driver, and could be a limiting factor. The space radiation environment consists of galactic cosmic rays (GCR), solar particle events (SPE), trapped radiation, and includes ions of all the known elements over a very broad energy range. These ions penetrate spacecraft materials producing nuclear fragments and secondary particles that damage biological tissues and microelectronic devices. One is required to know how every element (and all isotopes of each element) in the periodic table interacts and fragments on every other element in the same table as a function of kinetic energy ranging over many decades. In addition, the accuracy of the input information and database, in general and nuclear data in particular, impacts radiation exposure health assessments and payload penalty. After a brief review of effects of space radiation on materials and electronics, human space missions to Mars is discussed.

Keywords: Space Radiation, Cosmic Rays, Solar Particle Events, Radiation Effects on Space Missions, astronauts, materials and Electronics
PACS: 85.30.-z, 87.53-j, 89.90. +n, 24, 25

SPACE ENVIRONMENT

In order to quantify radiation exposure in space, it is required that the external ambient ionizing radiation environment be specified in terms of individual constituents and their respective energy fluxes. A great quantity of observational space environmental data from instrumented space platforms has been amassed in recent decades and used in developing computer models serving to define, as well as possible, the composition and temporal behavior of the space environment [1-3]. From the standpoint of radiation protection for humans and electronics in interplanetary space, the heavy ions (atomic nuclei with all electrons removed) of the galactic cosmic rays (GCR) and the sporadic production of energetic protons from large solar particle events (SPE) must be dealt with. The GCR environmental model used herein is based on a current version in which ion spectra are modulated between solar maxima and minima according to terrestrial neutron monitor data assuming the radial dependent diffusion model of Badhwar et al. [2], as described in reference [4]. The modeled spectra for solar minimum in 1977 and Solar Maximum in 1990 as given by Badhwar are shown in figure 1.

The environment near a large celestial body is modified by interaction with local materials producing an induced environment and shielding within the subtended angle of such a large body. The surface exposure on a lunar plain is shielded below the horizon but experiences an induced environment (mainly but not exclusively neutrons) produced in the local surface. The lunar surface GCR environment is shown in figure 1 at the 1977 Solar Minimum and the 1990 Solar Maximum. In addition to the GCR ions streaming from overhead, large numbers of neutrons are produced in the lunar surface materials and diffuse from below the surface as shown in the figure. Similar results are obtained [5] on the surface of Mars. The main difference is the presence of the Martian atmosphere that attenuates the incident ions and produces additional GCR fragments and more energetic neutrons in the atmosphere overhead.

In addition to the galactic cosmic rays able to penetrate the geomagnetic field to LEO, there are occasional solar particle events able to penetrate the geomagnetic field. The solar particle source is mainly composed of protons of similar quality as the trapped protons. Large SPE have only been observed to occur during times of increased solar activity conditions, and very large energetic events of grave important to human and electronics protection occur only infrequently (avg. 1 or 2 per cycle) and only outside of two years of solar minimum. Among the large events, the largest observed ground level event of the last 60 years of observation is that of February 23, 1956

Application of Accelerators in Research and Industry
AIP Conf. Proc. 1336, 649-654 (2011); doi: 10.1063/1.3586182

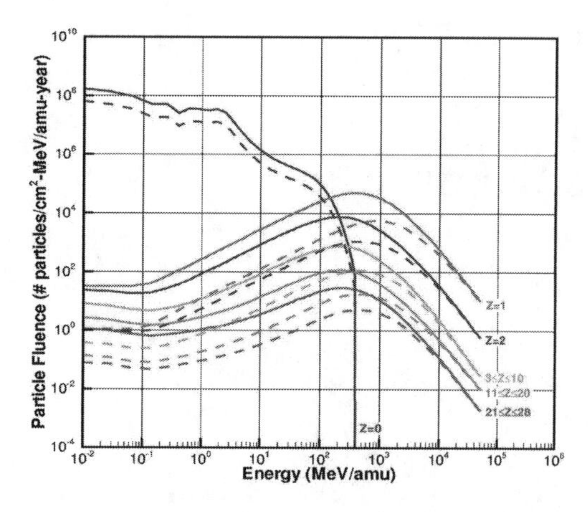

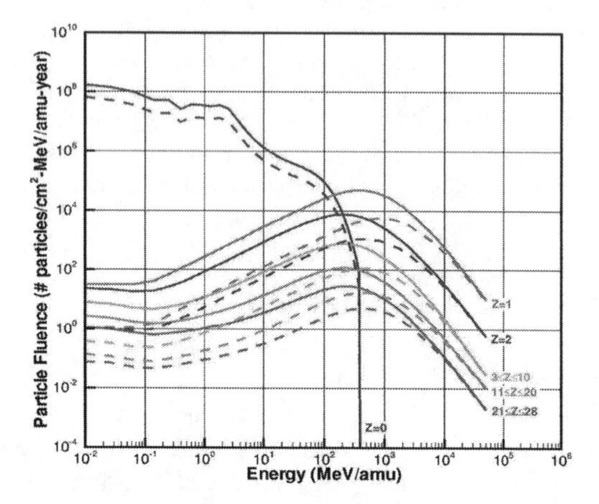

FIGURE 1. Galactic cosmic ray spectra at the 1997 Solar Minimum (full lines) and 1990 Solar Maximum (dashed lines) according to Badhwar et al., upper pane free space, lower pane lunar surface.

which produced a 3600 percent increase in neutron monitor levels on the terrestrial surface. The next largest event observed is the September 29, 1989 event with ground level increases of 400 percent or an order of magnitude smaller than that of Feb. 1956 event. Numerous other ground level events of smaller magnitude have occurred but are a factor of four and more lower in magnitude than the Sept. 1989 event. It is known that large SPEs are potentially mission threatening and astronauts in deep space must have access to adequate shelter from such an occurrence. The selection of electronics needs to account for such events. The SPE particle energy spectrum, usually, is used derived from the event which took place on September 29, 1989. To provide a baseline worst-case scenario we assume, for design purpose, an event of the order of four times larger than the September 29, 1989 event as an event comparable to the August 4, 1972 event from the point of view of space exposure. The September 1989 SPE spectrum is shown in figure 2. If we meet 30-day dose rate constraints for human exposure on an event four times larger than the September 1989 event then it is unlikely that an added factor of two or so larger events (like that of Feb. 23, 1956) would have serious protection potential.

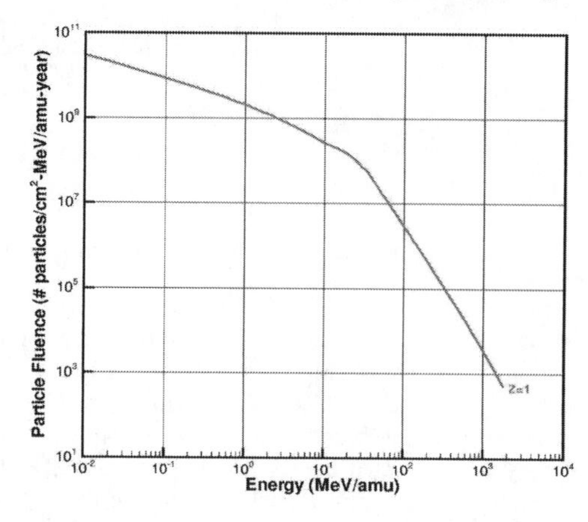

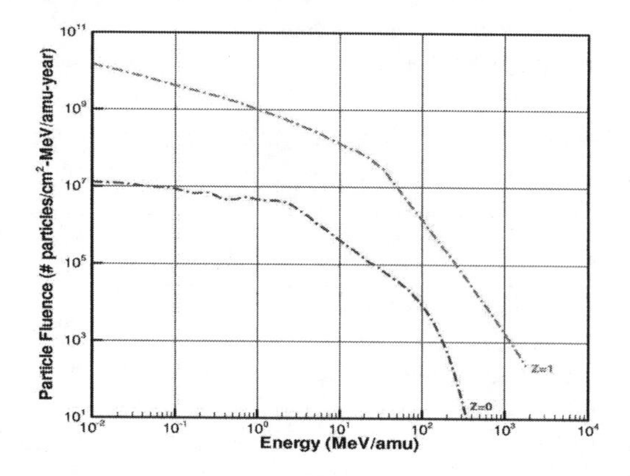

FIGURE 2. Solar particle event September 1989, upper pane free space and lower pane lunar surface.

The SPE are likewise altered by the presence of a large body similar to the GCR. The corresponding lunar surface environment is shown in figure 2. The role of the neutrons on the lunar surface is less effective in causing exposure relative to the protons streaming from overhead. Note that is in contrast to the more energetic GCR wherein large numbers of neutrons are produced in the lunar surface materials (see figure 1). Neutrons play a relatively more important role on the Martian surface where added neutrons are produced in the overhead atmosphere and the SPE protons are greatly attenuated [5].

The effectiveness of a given shield material is characterized by the transport of energetic particles within the shield, which is in turn defined by the interactions of the local environmental particles (and in most cases, their secondaries) with the constituent atoms and nuclei of the shield material. These interactions vary greatly with different material types. For space radiation shields, materials with high hydrogen content generally have greater shielding effectiveness, but often do not possess qualities that lend themselves to the required structural integrity of the space vehicle or habitat. Organic polymers are the exception. The design of properly-shielded spacecraft and habitats for long-duration human presence in interplanetary space will thus require an approach tending toward optimization of a compromise between protective shielding and various other functional aspects of the onboard materials.

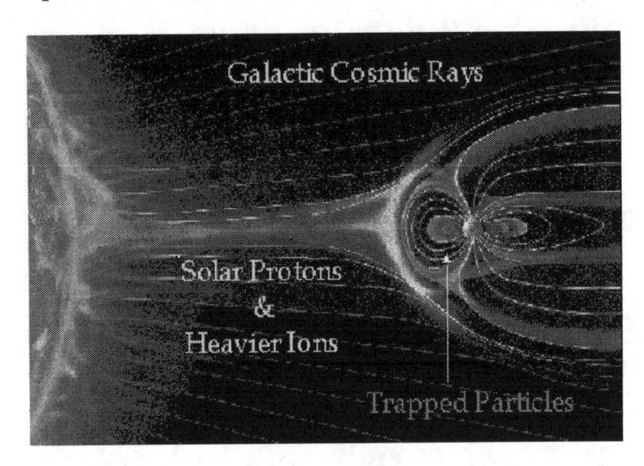

FIGURE 3. Schematic Picture of Space Radiation Environment

Salient Attributes Of Space Radiation

Figure 3 illustrates some of the salient features of space radiation with following characteristics. The Sun is on the left and earth is on right side of the figure, where magnetic field lines are displayed that trap the charge particles from the solar winds giving rise to trapped radiation.

Trapped radiation has Geo-trapped electrons, from few keV to several MeV. A few penetrate bumper/pressure vessel, and are of concern for spacesuits and electronics damage, in particular, for low earth orbit missions (LEO).

Solar Particle Events Are Infrequent But A Potentially Serious Hazard

Solar particle events mostly consist of protons and few 4-He (alpha particle) from a few keV to GeV. Protons are main penetrating component. Alpha particles are effectively shielded by geomagnetic field or thick materials. There are some (not many) heavy charged ions; these are shielded by thin layers of materials and are stopped by spacesuit garment. Figure 2 mainly displays the spectrum of proton distribution for the September 1989 SPE.

Galactic Cosmic Rays Are The Main Long-Term Exposure Hazard

Figure 1 displays the actual spectral distribution of charge ions for solar particle minimum (1977) and maximum (1990). We note that protons and 4-He energies typically run from tens MeV to many GeV and are always present in GCR. The protons are ~ 85 percent of GCR and alpha particles are ~ 14 percent. High charge and energy (HZE) ions' energies typically run from tens MeV to many GeV and, usually, are small fraction (1 -2%) of GCR but are very damaging for biological tissues due to their high charge numbers. The lower energy heavy ions are shielded by geomagnetic field or thick materials, but higher energies are difficult to shield against and are very damaging both for biological tissues and electronics.

HUMAN SPACE MISSIONS TO MARS

An enabling technology for the exploration, the development, and the commercialization of space is a cost-effective means of reducing the health risks from exposures to galactic cosmic rays (GCR) and a possible solar particle event (SPE). This has been a well-recognized challenge and a critical enabling technology for exploration in which astronaut health effects are of principal concern. Even more so with the development of space infrastructure and the eventual commercialization of space as new materials and other space products are identified and as larger numbers of civilians become involved in space based careers. Shield mass can be a high cost factor in system designs for the long-term operations. Optimization methods in the design process will be critical to cost- effective progress in space development. Limiting the time of transfer to duty station or the mission time within the solar cycle as well as the choice of materials used in construction can reduce the shield mass required on specific missions.

To demonstrate the vital role of radiation on space missions, we consider a 655-day Mars/Venus swing-by mission [5]. We assume mixed crew of 6 astronauts, four males and two females. The optimum volume of living space is taken as 114 m^3 and crew age is considered in the trade study. It is assumed that the living space is a right circular cylinder 2.2 m high. We also assume that the crew quarter /SPE shelter is a right circular cylinder of 2 m height with Tolerable living requirements. Shield optimization was investigated for a variety of materials: Aluminum (ALM), polyethermide (PEI), polysulfone (PSF), polyethylene (PET), lithium hydride (LIH), liquid methane (LME), hydrogenated nanofiber (HGNF), and liquid hydrogen (LH2). The reason for the choice of the materials is that there is increasing hydrogen content in the materials as we go down the list from aluminum to liquid hydrogen. We have established that hydrogen is a better space shielding material. As a result, the more the hydrogen content the better the material is expected to perform for space radiation shielding. There is a sensitivity of optimization on age.

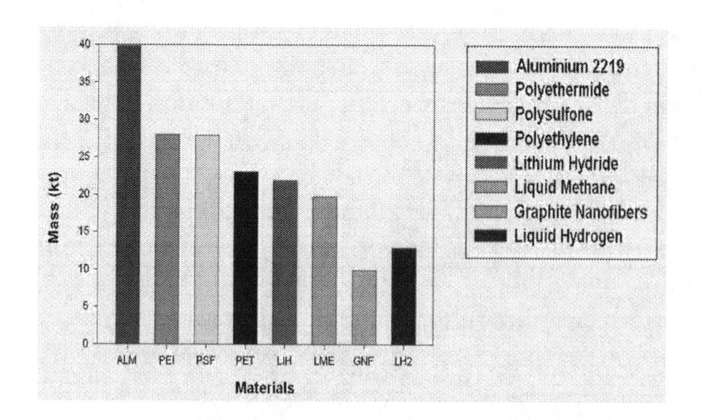

FIGURE 4. Optimized mass for mars mission various materials for age 50 (the minimum age of optimization for Al 2219).

As an example, for the Mars mission, optimization has been studied for all materials with varying age of 55 and below until optimization is possible. For a given material type, at each age, the shield and shelter thickness are varied between 1 and 100 g/cm^2 till an optimized profile that satisfies the constraints is found. Optimization is not practical when shield and shelter thickness between 1 and 100 g/cm^2 cannot be found that satisfies the dose exposure constraints set by the National Council on Radiation Protection and Measurements (NCRP). For Al 2219 the minimum age for optimized profile is 50 years. Figure 4 shows the optimized mass for different materials at age 50. There is an interesting feature to the optimization process in finding the optimum thickness. At higher ages, e.g. at 55, shield thickness is 1 g/cm2 and the optimization is achieved by varying the shelter thickness. As the age is lowered progressively shield thickness increases and the shelter thickness decreases till at certain age optimization is not possible. The minimum age down to which optimization is possible depends on the material considered (see Fig. 4).

Figure 4 shows that liquid hydrogen and HGNF out performs other materials. Although this is not the exact geometry and only the shield wall is represented, we see a large impact on the optimized mass for the Mars mission. It is clear that few material options are available for future deep space mission explorations and other methods such as biological countermeasures will play a pivotal role. At the very minimum from an engineering design point of view, reliability based methods must be implemented to accurately portray the shielding component of risk mitigation in mission design.

EFFECTS OF RADIATION ON ELECTRONICS

The first generation of satellites has revolutionized human development and exploration of space, communications, entertainment, and information industries and the way we live here on earth today. This progress has occurred despite the need for satellites to function in the harsh radiation environment of space. Electronics is a very rapidly growing and changing technology. With time the size of the components is decreasing and their functionality is dramatically increasing, and the speed is exponentially increasing. These developments put ever growing and changing demand on designing electronic components/devices which will provide maximum failure tolerance from space radiation effects. About 33 percent of anomalies in space are attributed to electronics failure as shown in Fig. 5.

EFFECTS OF RADIATION ON DEVICES

Heavy ions of various energies cause single event effects (SEE). A convenient way to express the transient charge generated by these heavy ions or charged particles is in charge per unit length, e.g. pC/micron. However, a more frequently used term (but less intuitive) to express the same thing is called, 'Linear Energy Transfer or LET', which in electronics literature is expressed in MeV.cm^2/mg.

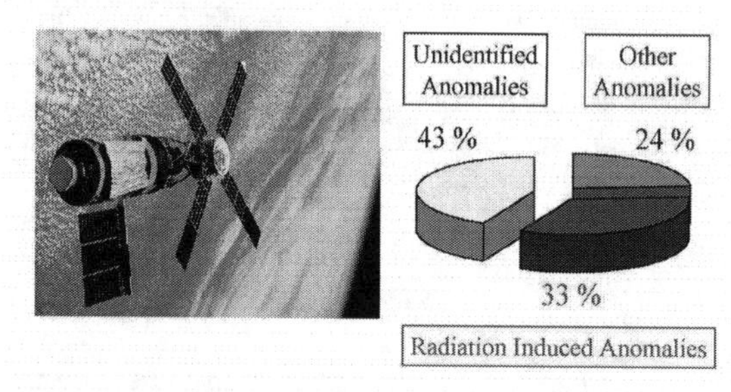

FIGURE 5. Role of electronics in space anomalies.

In bit-storage devices the high energy heavy-ions cause bits to change, and are expressed in terms of the bit error rates or SEU 'Single Event Upset' Error Probability. The SEU Error Probability is a number generated by computer from three data input: (a) the expected distribution of particles vs. LET, (b) the device cross-section for upset or latch-up as a function of LET, usually obtained from laboratory measurements, and (c) a calculation of expected error rate that combines the first two relationships. With a calculation of the effect of the omni-directional particle flux on charge produced in the device by the incident particles. Computer programs are available that perform this calculation. The net result is a fixed number for the upset or latch-up probability.

The total-dose hardness of conventional commercial MOS or bipolar electronics ranges from less than 1.0 krad (SiO_2) to greater than 100 krad (SiO_2). Most standard commercial devices fail somewhere in the range of 3-30 krad (SiO_2) [6]. However, higher dose rad-hard electronics can be fabricated depending upon the specific needs at higher cost due to their very limited market.

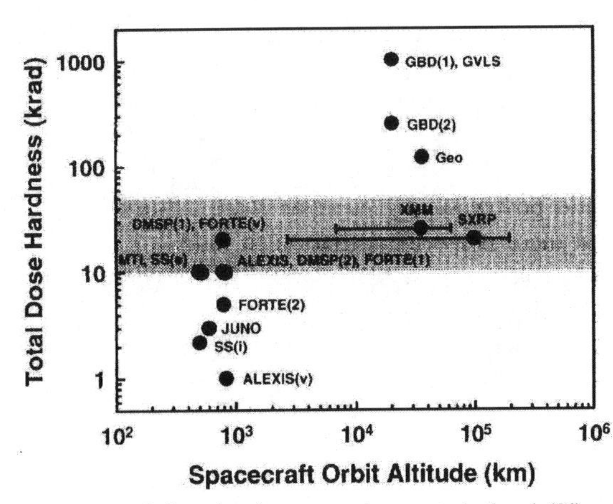

FIGURE 6. Total-dose hardness requirements in krad (Si) for several satellite systems as a function of orbit altitude (After Ref. 6)

Using rad-soft component does not significantly reduce cost, but greatly increases risk. There are no components that are ideal for all parameters. Integrated circuits (IC) design requires many tradeoffs in performance (Cost included as performance parameter). Commercially components are useful only for commercial applications, where low cost, latest technology (even if it is immature), and high speed takes precedence over extreme temperature and voltage ranges. Shielding these devices in Space Applications is a futile effort, especially for Single Event Effects such as SEU and Single Event Latch-up (SEL).

Figure 6 shows total dose rad-hard needs with altitude for some missions. The rad-hard requirement gets increasingly demanding as the altitude and duration of the mission increases. It is important to note that although a device may pass total-dose testing, it may still fail for other radiation damage effects like single-event effects (SEE), upset (SEU), latch-up (SEL), and burnout (SEB), gate rupture (SEGR) and other latch-up destructive or non destructive effects.

TEST METHODS

The most successful space systems begin their parts selection and/or testing program before system designs are finalized. One simply cannot decide that a particular part will be a key element of a stem without confidence the device will function as intended in the environment. Moreover, if the device is a commercial part, prior test data may only provide a general guideline for these decisions, as commercial manufacturers periodically make changes to process flows to improve yield and/or performance that can often negatively impact radiation response. In addition, some types of commercial devices exhibit a large amount of variability in their radiation response, especially in total dose effects. NASA and NASA Langley Research Center has governing procedures for parts selection guidelines like – GSFC EEE-INST-002, and uses other governing procedures – OP-5515, NHB-5300.x, NPD 8730.2C, 8739.x, Practice No. PD-ED-1258. NASA is AS 9100 compliant. The process followed is that of selection, review, performance testing, definition of environment, requirements, approval and shielding and mitigation approaches. In-house testing is carried out at the NASA Langley Research Center (LaRC), Goddard Space Flight Center (GSFC), and at the Jet Propulsion Laboratory (JPL). There are a number of accelerator facilities where the testing is also done e.g. Thomas Jefferson National Accelerator Facility (TJNAF), Los Alamos National Laboratory (LANL), Brookhaven National Laboratory (BNL), Indiana University Cyclotron Facility (IUCF), Michigan State University (MSU), Texas A&M University Cyclotron (TAMU), and Lawrence Berkeley Laboratory (LBL). There is some collaborative effort with European Space Agency (ESA). There are several tools and activities that are leveraged in this effort. Some of them are as follows:

- Web based information
- GSFC parts data base
- JPL Data base
- JSC (Johnson Space Center) info on parts heritage from station and shuttle
- PC based modeling
- Other (ERRIC-Electronics Radiation Response Information Center (DoD); ESA, IEEE, REDW-Radiation Effects Data Workshop, a part of the Nuclear and Space Radiation Effect Conference (NSREC); DoD, Test Beds i.e. Living with Star).

CONCLUSION

An overview of effects of radiation has been presented. We have discussed some of the state-of-the-art technologies for human exploration and development of space with a focus on human space flights and challenging ever-changing electronics needs for space exploration. The impact of space radiation on space missions has been shown by the assessment of dose exposure on mission to Mars behind a number of materials with increasing hydrogen contents known to be a better radiation shielding material. For career astronauts who are exposed to longer duration space radiation over the period of time the choice of material plays a very critical role. It is found that improved shield materials will be required to enable a Mars mission in which middle-aged astronauts can participate. If the age of the astronauts is allowed to be 55 and older then more options are available. The analysis provides enabling technology for protecting astronauts and missions for long duration and deep space missions. Current technology is adequate for a single lunar mission. For career astronauts, exposed to long duration space radiation revolutionary technology needs to be developed. For the success and safety of space missions, electronics plays a very critical role. Copping up with the ever changing electronics market and the need for space missions is a very challenging task. Effects of radiation in space is a very exciting area of research, development and technology, where several disciplines of sciences merge, and lead to game changing technologies those amongst other things benefit life here on earth.

ACKNOWLEDGEMENTS

Information contained in this document represents the opinion of the author(s) only, and does not represent the views or policies of NASA.

REFERENCES

1. Badhwar, G.D. and O'Neill, P.M., Improved model of galactic cosmic radiation for space exploration mission, Nucl. Tracks & Radiat., 20, 403-410, 1992
2. Badhwar G.D., et. al, Inter-comparison of radiation measurements on STS-63; Radiat. Meas. 26; 147-158; 1997
3. O'Neill, P.M. Badhwar, G.D., Galactic Cosmic Ray Model update based on advanced composition explorer (ACE) energy spectra from 19977 to present. Adv. Space Res. 37, 1227-1733, 2006.
4. Wilson, J.W., Kim M-H, Shinn, J.L., Tai, H., Cucinotta, F.A., Badhwar, G.D., Badavi, F.F., Atwell, W.: Solar Cycle Variation and Application and Applications to the Space Radiation Environment, NASA/TP-1999-209369
5. R.K. Tripathi, J.W. Wilson, F.A. Cucinotta, J.E. Nealy, M.S Clowdsley, M.Y. Kim, Deep space mission shielding optimization, SAE 2001-01-2326 (2001).
6. M.R. Shaneyfelt, P.S. Winikur, T.L. Meisenheimer, F.W. Sexton, S.B. Roeske, and M.G. Knoll, Hardness variability in commercial technologies, IEEE Trans. Nucl. Sci. 41, 2536 (1994).

Recoil-Implantation Of Multiple Radioisotopes Towards Wear Rate Measurements And Particle Tracing In Prosthetic Joints

Jacob A. Warner[a], Paul N. Smith[b], Jennifer M. Scarvell[b], Laura Gladkis[a,b], Heiko Timmers[a]

[a] *School of Physical, Environmental and Mathematical Sciences, University of New South Wales at ADFA, Canberra, ACT 2600, Australia*
[b] *Trauma and Orthopaedic Research Unit, The Canberra Hospital, PO BOX 11, Woden, ACT 2606, Australia*

Abstract. This study demonstrates a new method of radioisotope labeling of ultra-high molecular weight polyethylene inserts in prosthetic joints for wear studies. The radioisotopes ^{97}Ru, ^{100}Pd, ^{100}Rh, and ^{101m}Rh are produced in fusion evaporation reactions induced by ^{12}C ions in a ^{92}Zr target foil. The fusion products recoil-implant into ultra-high molecular weight polyethylene plugs, machined to fit into the surface of the inserts. During laboratory simulations of the joint motion, a wear rate of the labeled polyethylene may be measured and the pathways of wear debris particles can be traced by detecting characteristic gamma-rays. The concentration profiles of the radioisotopes extend effectively uniformly from the polyethylene surface to a depth of about 4 μm. The multiplicity of labeling and the use of several gamma-ray lines aids with avoiding systematic measurement uncertainties. Two polyethylene plugs were labeled and one was fitted into the surface of the tibial insert of a knee prosthesis, which had been worn in. Actuation over close to 100,000 cycles with a 900 N axial load and a 24° flexion angle removed (14 ± 1) % of the gamma-ray activity from the plug. Most of this activity dispersed into the serum lubricant identifying this as the important debris pathway. Less than 1 % activity was transferred to the femoral component of the prosthesis and the measured activity on the tibial tray was insignificant. Assuming uniform wear across the superior surface of the insert, a wear rate of (12 ± 3) mm^3/Megacycle was determined. This is consistent with wear rate measurements under similar conditions using other techniques.

Keywords: radioisotope tracing, UHMWPE, ion implantation, thin layer activation
PACS: 25.70.Jj, 29.20.-c, 29.30.Kv

INTRODUCTION

Worldwide, 600,000 knee replacement surgeries are performed every year [1]. The types of prosthetic materials vary from ceramic-on-ceramic, to metal-on-metal, to metal-on-polymer bearing combinations [2-4]. Currently the preferred bearing combination is the metal-on-polymer system, specifically CoCr-on-UHMWPE (ultra-high molecular weight polyethylene) because of UHMWPE's excellent wear characteristics [5]. UHMWPE debris particles, however, can lead to harmful immunological responses, eventually causing osteolytic bone resorption and necessitating revision surgery [6, 7].

The measurement of wear rates has become a useful method of evaluating prosthetic designs *in vitro*. Presently, the most common method of wear rate measurements is gravimetric analysis, which involves calculating the mass of the tibial insert both before and after the *in vitro* wear process occurs [8, 9]. This method is insensitive to small amounts of wear loss due to the subtraction of two relatively large numbers. The absorption of lubricant into the insert is also problematic [1].

A more sensitive method termed 'thin layer activation' involves the implantation of radioisotopes into the near-surface of the UHMWPE tibial component, followed by wear studies in a simulator [10]. This method commonly uses ^{7}Be as a label. The ^{7}Be isotope may be implanted over a depth of 0-30 μm (using appropriate stopper foils), with a flat implantation profile for the first 5 μm [11, 12]. Typical ^{7}Be fluences are on the order of 1×10^{10} ions·cm^{-2} [13]. Significant changes to the surface properties of UHMWPE have not been found[14].

The current study explores a new method of thin layer activation for wear rate measurements and particle tracing. Employing fusion evaporation

Application of Accelerators in Research and Industry
AIP Conf. Proc. 1336, 655-659 (2011); doi: 10.1063/1.3586183

reactions induced by ^{12}C in a ^{92}Zr target, multiple tracers can be introduced in a single implantation. The efficacy of this method is demonstrated in this study with measurements for a knee prosthesis with a labeled UHMWPE plug in the tibial insert using a constant force prosthesis actuator.

EXPERIMENTAL DETAILS

A 2.5 μm-thick ^{92}Zr production foil was bombarded with 70 MeV ^{12}C projectiles. The radioisotopes ^{97}Ru, ^{100}Pd, ^{100}Rh, and ^{101m}Rh were produced by the fusion evaporation reactions: ^{92}Zr(^{12}C,α3n)^{97}Ru, ^{92}Zr(^{12}C,p3n)^{100}Rh, ^{92}Zr(^{12}C,4n)^{100}Pd, and ^{92}Zr(^{12}C,p2n)^{101}Rh, respectively. The radioisotopes recoil-implanted into 1.2 mm thick cylindrical UHMWPE plugs with a diameter of 10 mm. A fraction of the ^{100}Rh in the samples originated from the decay of ^{100}Pd. The beam current was typically 1.0 – 1.5 μA. The total implanted recoil fluence was estimated to be 8×10^{11} ions·cm^{-2}, and the total implanted fluence of scattered ^{12}C projectiles was estimated to be 1×10^{14} ions·cm^{-2}. Along with unrelated samples, the UHMWPE plugs (see Figure 1), were placed 30 mm behind the production foil, at an average scattering angle of θ = 26°. The depth profile of the implanted ions extended relatively uniformly from the UHMWPE surface to a depth of (4.4 ± 0.3) μm according to TRIM [15, 16].

An LCS Complete Rotating-Platform (Depuy Orthopaedics, Inc., Warsaw, Indiana, US) tibial insert, with a matching CoCr component, was worn-in over 500,000 gait cycles. Following wear-in, a cylindrical recess, measuring 10 mm in diameter and 1.2 mm deep, was milled from the topside posterior of the tibial insert using a CNC-milling machine (Deckel Maho, Kempten, Germany). The recess for the flat plug was positioned at one of the two lowest points of the superior surface of the tibal insert, where the curvature is low and approximately symmetric (Figure 1 b and c).

A constant force prosthesis actuator was used at 1 cycle per second. The constant axial force was set to 900 N, and a flexion angle of 24° was chosen. A detailed description of the constant force actuator is provided by Gladkis et al. [17]. It should be noted that this actuator is not a state-of-the-art knee motion simulator and does not conform to American Society for Testing and Materials (ASTM) standards. Fetal Bovine Serum (Sigma-Aldrich, St. Louis, Missouri, US) diluted to 25 % with deionised water, was used as a lubricant. Sodium azide (0.1 % w/v) and EDTA (4 % w/v) were added as an anti-bacterial agent and protein accumulation inhibitor, respectively[3].

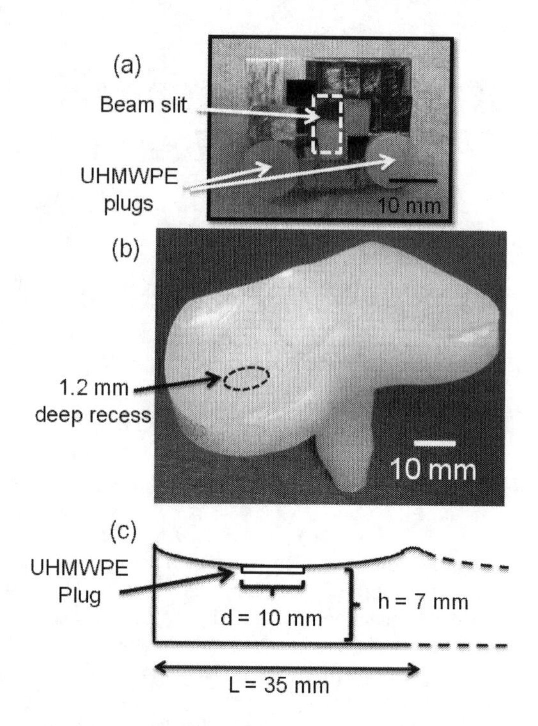

FIGURE 1. Mass-specific activities measured in this work with a resolution-deteriorated Germanium detector without thin entrance window (filled symbols). The data are compared with independent measurements by a commercial laboratory,

The actuator was halted after 50, 500, 1500, 4500, 14500, 39500, and 99500 cycles. The tibial insert, femoral component, tibial tray, and lubricant were separated and gamma-ray spectrum were recorded using a high-purity Germanium gamma-ray detector. The components were each placed at a distance of 1 cm from the detector which provided high statistics, yet a deadtime < 2% and approximately equal detection efficiencies. A 20-minute spectrum was taken which allowed a quantification of the number of detected counts in photopeaks associated with the tracer isotopes. Since debris particles tend to float on the surface of the lubricant, gamma-ray absorption by the lubricant has not been considered. A Gaussian fit was applied to each peak to determine the total number of counts (see Figure 2). The typical statistical uncertainty for the activity on the tibial insert was less than 1 %.

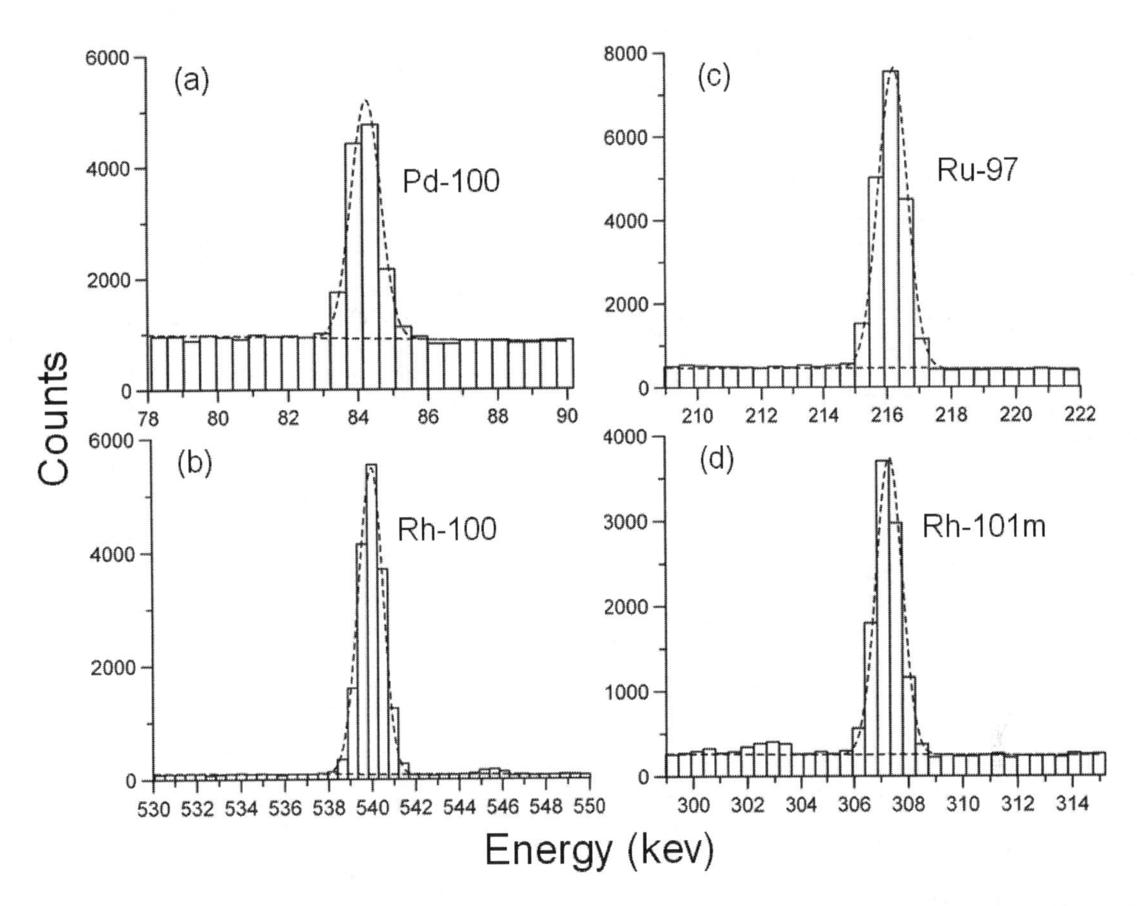

FIGURE 2. Mass-specific activities measured in this work with a resolution-deteriorated Germanium detector without thin entrance window (filled symbols). The data are compared with independent measurements by a commercial laboratory,

The number of detected counts for the insert before commencement of actuation was considered the original activity and used to normalize the detected number of counts from all measurements. Activity changes due to radioactive decay have been corrected using the appropriate half-lives and the growth of the ^{100}Rh activity due to the decay of ^{100}Pd has been considered.

RESULTS AND DISCUSSION

The fraction of activity removed from the labeled plug in the UHMWPE insert is compared in Figures 3a and b with the activity measured for the debris floating on the lubricant. The data show that after each halt of the actuator the amount of activity removed from the plug is effectively equal to the activity measured for debris in the lubricant. As can be seen clearer in Figure 3b, the removal of activated UHMWPE from the tibial insert is linear with the number of gait cycles. Such linear removal of debris has also been shown in previous gravimetric wear studies using knee prostheses [4, 10, 20].

The new method provides for the labeling of different locations on the tibial insert. Using different tracers may also allow the simultaneous tracing of debris from different departure points. With appropriate implantation chambers, tracer ions may be directly implanted into tibial inserts, avoiding the fitting of plugs.

An approximation of the wear surface has been calculated to be (2033 ± 10) mm^2 by using a 2D projection of the superior surface. The activated volume of the plug is 0.346 mm^3. By extrapolating the labeled debris from the plug uniformly over the entire superior surface of the insert, a wear rate of (13 ± 2) mm^3·Megacycles^{-1} can be derived. It should be noted though that in general wear rates depend strongly on location on the superior surface of the tibial insert. However, previous gravimetric wear studies with the same actuator have determined a wear rate of approximately 20 mm^3·Megacycles^{-1} [17] for similar loading conditions and using de-ionized water as a lubricant. Realistic knee simulator results with bovine serum typically range from 2 mm^3·Megacycles^{-1} to 49 mm^3·Megacycles^{-1} [18, 19].

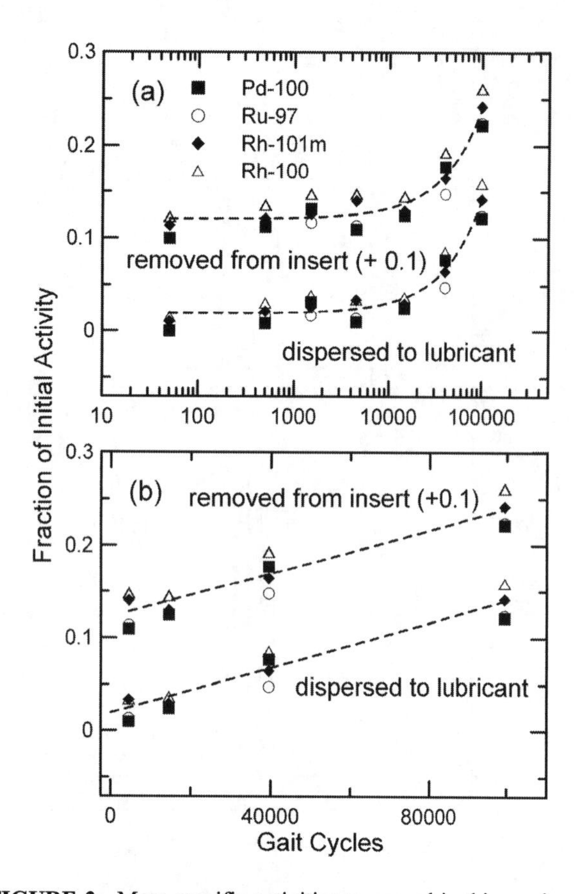

FIGURE 3. Mass-specific activities measured in this work with a resolution-deteriorated Germanium detector without thin entrance window (filled symbols). The data are compared with independent measurements by a commercial laboratory,

While wear debris removal can be quantified with the new method, the pathways of the UHMWPE wear debris particles can be additionally mapped. It was found that after 99,500 cycles no detectable amount of activity was found on the tibial tray. Thus a debris transport pathway from the superior surface of the tibial insert to the tray underneath the insert is insignificant. Also, no labeled debris was detectable at locations on the superior surface of the insert, other than that of the plug, indicating that no significant amount of debris particles is reintegrated into the UHMWPE. In contrast, a detectable amount, however, less than 1 % of the labeled polyethylene was transferred from the insert to the femoral component at the end of the study after 99,500 cycles. Thus such a pathway exists, however, it is of minor importance. This may be expected with bovine serum as the lubricant, which prevents the formation of an adhesive transfer layer of UHMWPE forming on the femoral component [20].

The main debris transport pathway from the superior surface of the tibial insert is directly to the lubricant. Figures 3a and 3b show that at the end of the study the total activity dispersed to the bovine serum was (13 ± 1) % of the original activity.

CONCLUSIONS

The recoil implantation of ^{97}Ru, ^{100}Pd, ^{100}Rh, and ^{101m}Rh radioisotopes into UHMWPE has been shown to be a viable alternative method of measuring wear rates of inserts in joint prostheses and tracing wear debris. The implantation into small plugs, which are then press-fit into the tibial inserts, allows the isolation of specific locations on the insert surface to be investigated. This labeling process gives a uniform and flat activity profile within the first 4.4 μm of the material, which allows a direct correlation with the volumetric wear rate. Furthermore, the new method allows the simultaneous placing of four radioisotopes into the UHMWPE as a result of one ion implantation, which each give an independent measurement of the wear rate.

A constant load actuator was used to simulate wear in a knee prosthesis. Of the initially labeled debris, (13 ± 1) % was dispersed into the bovine serum and less than 1 % was transferred to the femoral component. Assuming uniform wear, the measured wear rate of the labeled plug on the superior lateral side was extrapolated to give a wear rate for the tibial insert of (13 ± 2) $mm^3 \cdot Megacycles^{-1}$. Both the lack of significant debris transfer to the femoral component and the measured wear rate agree well with previous studies despite the limitations of the wear actuator. This demonstrates that the new alternative method is effective and that radioisotopes can be used to sensitively trace UHMWPE debris pathways.

Future applications of the new method may include the comparison of wear rates at different locations on the superior surface of tibial inserts as well as their backside. The method can be combined with a realistic, state-of-the-art knee simulator.

ACKNOWLEDGMENTS

The authors would like to thank Johnson & Johnson for the donation of the LCS-RP UHMWPE-CoCr prosthetic.

REFERENCES

1. T. J. Joyce and A. Unsworth, *Journal of App. Biomaterials and Biomechanics* **2**, 29 (2004).
2. I. C. Clarke and G. Willmann, *Bone Implant Interface*, 203 (1994).
3. A. A. J. Goldsmith, D. Dowson, G. H. Isaac, and J. G. Lancaster, *Proc. Instn. Mech. E.* **214**, 39 (2000).
4. O. K. Muratoglu, C. R. Bragdon, M. Jasty, D. O. O'Connor, A. S. V. Knoch, and W. H. Harris, *Journal of Arthroplasty* **19**, 887 (2004).
5. S. M. Kurtz, *The UHMWPE Handbook* (2004).
6. E. Ingham and J. Fisher, *Proc. Instn. Mech. E.* **214**, 21 (2000).
7. T. Nizegorodcew, G. Gasparini, G. Maccauro, A. Todesca, and E. D. Santis, *International Orthopaedics (SICOT)* **21**, 14 (1997).
8. D. W. V. Citters, F. E. Kennedy, and J. P. Collier, *Wear* **263**, 1087 (2007).
9. G. A. Engh, K. A. Dwyer, and C. K. Hanes, *Journal of Bone and Joint Surgery* **74B**, 9 (1992).
10. M. F. Stroosnijder, M. Hoffmann, T. Sauvage, G. Blondiaux, and L. Vincent, *Nuclear Instruments and Methods in Physics Research B* **227**, 597 (2005).
11. C. M. Wall, D. C. Eberle, M. B. Treuhaft, and J. H. Arps, *Wear* **259**, 964 (2005).
12. G. Dearnaley, J. Asher, A. T. Peacock, S. J. Allen, and R. E. J. Watkins, *Surface and Coatings Technology* **201**, 8070 (2007).
13. T. Sauvage, O. Lacroix, G. Blondiaux, and P. Boyer, *Nucl. Inst. and Methods in Physics Research B* **143**, 397 (1998).
14. J. D. Carlson, J. E. Bares, A. M. Guzman, and P. P. Pronko, *Nuclear Instruments and Methods in Physics Research B* **7**, 507 (1985).
15. J. F. Ziegler and J. P. Biersack, in *SRIM-2003*, 2003).
16. J. F. Ziegler, J. P. Biersack, and U. Littmark, *The Stopping and Range of Ions in Matter* (Pergamon, New York, 1985).
17. L. G. Gladkis, R. W. Li, J. M. Scarvell, P. N. Smith, and H. Timmers, *Wear* **267**, 632 (2009).
18. R. W. Treharrne, R. W. Young, and S. R. Young, *Engineering and Medicine* **10**, 137 (1981).
19. M. Endo, J. L. Tipper, D. C. Barton, M. H. Stone, E. Ingham, and J. Fisher, *Proc. Instn. Mech. E.* **216**, 111 (2002).
20. J. R. Cooper, D. Dowson, and J. Fisher, *Clinical Materials* **14**, 295 (1993).

Refreshable Decrease In Peak Height Of Ion Beam Induced Transient Current From Silicon Carbide Metal-Oxide-Semiconductor Capacitors

T. Ohshima[1], N. Iwamoto[1,2], S. Onoda[1], T. Makino[1], M. Deki[1,3] and S. Nozaki[2]

[1]*Japan Atomic Energy Agency, 1233 Watanuki, Takasaki, Gunma 370-1292, Japan*
[2]*The University of Electro-Communications, 1-5-1, Chofugaoka, Chofu, Tokyo 182-8585, Japan*
[3]*The University of Tokushima, 2-24 Shinkura-cho, Takushima 770-8501, Japan*

Abstract. MOS capacitors were fabricated on an n-type 6H-SiC epitaxial layer, and charge induced in the nMOS capacitors by 15 MeV oxygen ion microbeams was measured using Transient Ion Beam Induced Current (TIBIC). The peak amplitude of TIBIC signals decreases and the fall time increases with increasing number of incident ions. The TIBIC signal peak value can be refreshed to its original value by applying a forward bias of + 1V to the oxide electrode. Since the values of the effective bias applied to gate oxide are changed by the charge state of deep hole traps, the decrease and refreshment of TIBIC peak can be interpreted in terms of change in the effective gate bias due to charge trapping and recombination by deep hole traps, respectively.

Keywords: Silicon Carbide (SiC), Metal-Oxide-Semiconductor (MOS), Transient Ion Beam Induced Current (TIBIC)
PACS: 73.40.Qv, 61.82.Fk, 61.80.Jh, 72.20.Jv

INTRODUCTION

Silicon carbide (SiC) is a promising wideband gap semiconductor suitable for high-power and high-frequency applications due to its excellent physical and chemical properties [1-3]. Furthermore, SiC electronic devices are likely to be used in high-radiation environments, such as in space [4-7]. For the application of SiC devices, it is very important to understand their performance reliability. In previous studies [8, 9], the failure of silicon (Si) power devices caused by single event effects (SEEs) with destruction due to cosmic rays (high energy neutrons from space) was reported, and the reliability of Si devices against SEEs with destruction is required to be greater than 10 FIT (1FIT = 1 failure in 109 hours). The failure rate due to SEEs with destruction in Si power devices increases with increasing electric field [8, 9]. Since SiC devices are expected to be operating under even higher electric fields than Si devices, it is very important to understand the SEEs in SiC device. In earlier works, we have reported that SiC Metal-Oxide-Semiconductor (MOS) Field Effect Transistors (FET) exhibited higher radiation resistance than Si MOSFETs [4, 5]. SiC Static Induction Transistors (SITs) [7] and Metal-Semiconductor FETs (MESFETs) [6] were found to be able to operate up to 10 MGy without significant degradation. These results demonstrated that SiC has superior radiation tolerance comparing to ordinary Si devices from the point of view of Total Ionizing Dose effects (TIDs). As for SEEs, Soelkner et al. [10] reported the heavy-ion radiation-induced effects in SiC Schottky diodes and Junction Field Effect Transistors (JFETs). Also from the point of view of charged particle detectors, the charge collection in SiC diodes by incidence ions has been reported [11-13]. There are very few studies of SEE in SiC Metal Oxide Semiconductor (MOS) [14, 15] and the mechanisms governing the SEEs in SiC MOS devices have not yet been understood. Therefore, it is necessary to elucidate and clarify the role of SEEs in SiC MOS devices.

In this study, we detected transient current collected from SiC MOS capacitors under applying reverse biases by a MeV oxygen (O) ion using a Transient Ion Beam Induced Current (TIBIC) measurement system, and investigated the change in transient current with increasing number of incident ionsTIBIC signal peaks decreased with increasing number of incident ions, and the TIBIC peak was recovered to the initial value by applying forward biases, suggesting that charge trapping and

Application of Accelerators in Research and Industry
AIP Conf. Proc. 1336, 660-664 (2011); doi: 10.1063/1.3586184

recombination by deep hole traps affect ion induced current for the SiC MOS capacitors.

EXPERIMENTS

The samples used in this study were MOS capacitors of 200 μm diameter fabricated on an n-type 6H-SiC epitaxial layer grown on n-type 6H-SiC substrate (3.5°off, Si face, CREE). The thickness of the epitaxial layer and its donor concentration are 4.9 μm and $5x10^{15}$ /cm^3, respectively. Firstly, the field oxide at a thickness of 100 nm was grown on the sample surface by pyrogenic oxidation ($H_2 : O_2 = 1:1$) at 1100 °C, and following the field oxide with a diameter of 200 μm was removed using standard photolithography technique and wet etch ($HF:NH_4F$ = 1:5). Then, the gate oxide was grown on the SiC surface by dry oxidation at 1180 °C for 120 min followed by re-oxidation under pyrogenic condition ($H_2 : O_2 = 1:1$) at 800 °C for 15 min. During cooling down to 800 from 1180 °C, which took 1 hour, the samples were kept under argon ambient. The thickness of the gate oxide was estimated to be 39 nm from capacitance - voltage (*C-V*) measurements. Circular electrodes of 200 μm diameter were formed on the gate oxide by Al evaporation (300 nm thick) and lift-off technique. The leakage current through the gate oxide for the samples was below 1.5 pA at a reverse bias of -15V.

Ion-induced currents through gate oxide on semiconductors are known as displacement currents. In this study, TIBIC signals induced through gate oxide on SiC by 15 MeV oxygen ion microbeams were measured using a single ion hitting system in conjunction with the 3 MV tandem accelerator at JAEA, Takasaki [16]. During the TIBIC measurements, the reverse biases up to -15V were applied to the gate oxide. Since the system consists of a 3 GHz Tektronix TDS694C oscilloscope and a fast beam switch, the transient current-induced in MOS capacitors by a single ion incidence can be observed. To minimize damage formation due to ion incidence, the microbeams were scanned across the sample during the TIBIC measurements (typical scanning size was 200 x 200 μm^2, and 30 x 30 points of TIBIC signals were collected in the area of interest). Details of the TIBIC system are described in Ref. [17].

RESULTS AND DISCUSSION

Figure 1 shows the TIBIC signals for a 6H-SiC nMOS capacitor induced by 15 MeV-O ions. A reverse bias of -15 V was applied to the MOS capacitor during the TIBIC measurements. The average of the TIBIC signals obtained from the first 50 measurement points is depicted as a solid line in Fig. 1. The peak amplitude of TIBIC signal is 0.10 mA and the fall time, which is defined as the decay time from 90 % to 10 % of the peak value, is 1.75 ns. With

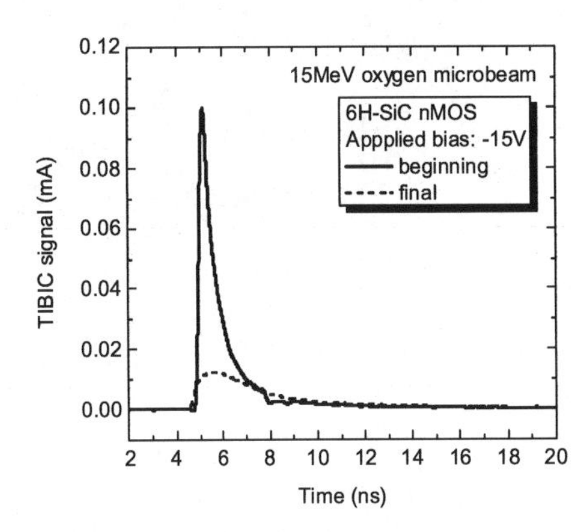

FIGURE 1. TIBIC signals of a 6H-SiC n MOS capacitor induced by 15 MeV-O ions. A reverse bias of -15 V was applied to the MOS capacitor during the TIBIC measurements. The solid and broken lines represent the results observed from the first 50 transients and from measurements between 1000th and 3000th, respectively.

increasing number of incident ions, the TIBIC peak signal decreases and the fall time value increases. The broken line in Fig. 1 indicates the averaged TIBIC signal between 1000 and 3000th measurements. The peak value and the fall time of the TIBIC signal are 0.013mA and 6.83ns, respectively. This result suggests that the internal electric field in the MOS capacitor becomes weaker with increasing incident ions.

Next, the TIBIC peak and the collected charge as a function of number of incident ions are plotted in Fig. 2. The values of collected charge are computed by integration of the TIBIC signals. As shown in Fig. 2 (a), the peak amplitude decreases with increasing number of incident ions up to 1000 ions, and eventually the TIBIC signal peaks show a constant value (0.013 mA) after 1000 ions. On the 3150th ion incident, a positive bias of +1 V was applied to the sample for 1 min, and following a negative bias of -15 V was reapplied to the sample. It is noteworthy to point out that the brief duration of positive gate bias helps the peak amplitude of the TIBIC signal to recover to 0.1 mA. This cycle of the peak refreshes to its initial value upon application of +1 V to the gate electrode and subsequent reduction of the peak value to a saturation value of 0.013 mA is repeatable. Similarly, the collected charge decreases with increasing number of incident ions and, finally, becomes a constant value of ≈ 0.5 pC after 1000th ion incidence. By applying a positive bias of 1V after the 3150th ion incidence, the collected charge also recovers to its initial value (shown in Fig. 2 (b)).

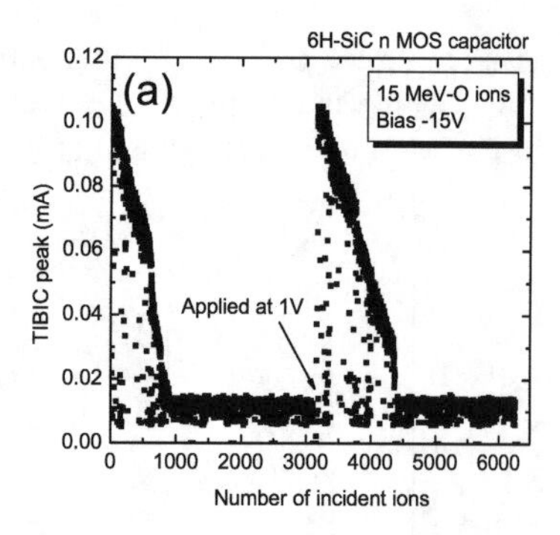

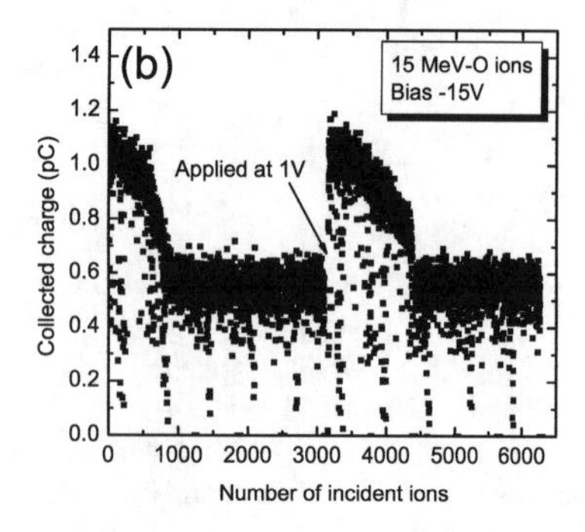

FIGURE 2. (a) peak amplitude of TIBIC and (b) collected charge for a SiC nMOS capacitor as a function of number of incident ions. During the TIBIC measurements, a reverse bias of -15 V was applied to the MOS capacitor.

Figures 3 (a) and (b) show the reverse bias dependence of peak amplitude of TIBIC signal and collected charge for the MOS capacitor, respectively.

Square and circle symbols represent the averaged values obtained from the incidence of 50 ions after refreshing by applying a positive bias of +1 V and the saturated values after incidence of more than 1000 ions, respectively. The peak amplitude of TIBIC signal for both initial and saturated regime increases with increasing the reversed bias. This can be explained in terms of increasing electric field in SiC due to increasing applied bias to the gate. The difference of peak amplitude of TIBIC signal between the initial and saturated ones becomes larger with increasing bias voltage as shown in Fig. 3 (a). For example, the saturated value at a reverse bias of -8 V (7.29×10^{-6} A) is 18.2 % of the initial value (4.01×10^{-5} A). In the case of applied bias at -15 V, the saturated value (1.30×10^{-5} A) is only 12.4 % of the initial value (1.05×10^{-4} A). For the charge collection, the total charge collected tends to saturate around -14 V although it increases with increasing applied bias at applied biases below -12 V (especially in the case of the saturated one).

The difference between the initial and the saturated ones for collected charge is not as large as the difference observed in the case of TIBIC peak. For example, the saturate values for the collected charge at a reverse bias of -8 V and -15 V are 49 and 46 % of the initial values, respectively. The smaller decrease in the charge collection as compared to the peak amplitude is attributed to a significant broadening of the full-wave half maximum of the saturated current transients, resulting in a not so drastic difference in the charge collected despite that there is a large decrease in the peak amplitude of TIBIC.

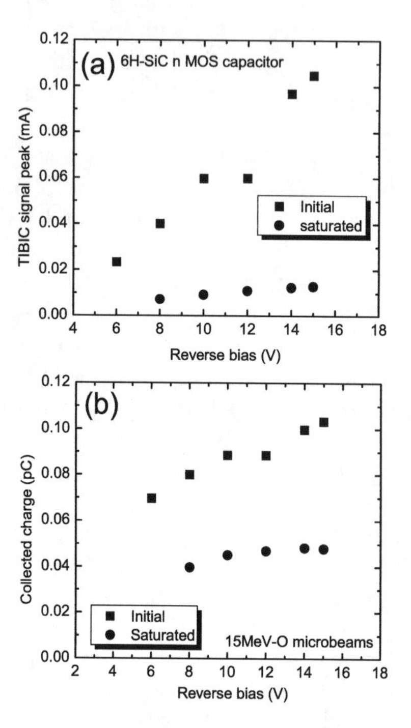

FIGURE 3. Reverse bias dependence of (a) peak amplitude of TIBIC signal and (b) collected charge obtained for a 6H-SiC nMOS capacitor, respectively. Square and circle symbols represent the averaged values obtained from the incidence of 50 ions after refreshing by applying a positive bias of +1 V and saturated values after incidence of more than 1000 ions, respectively.

Since the reduction in peak amplitude of TIBIC and the collected charge during measurements can be refreshed by applying a positive bias to the 6H-SiC n MOS capacitors, this behavior cannot be explained in terms of damage creation (displacement damage effects) in SiC crystal due to ion incidence. Instead, we believe the intrinsic defects within the oxide which can act as charge trapping might be the origin of this behavior. In the case of Si, it is reported that strained Si-Si bonds near the interface act as hole trapping when the Si-Si bonds were broken (deep hole trapping near Si/SiO_2 interface), and the charge state of this

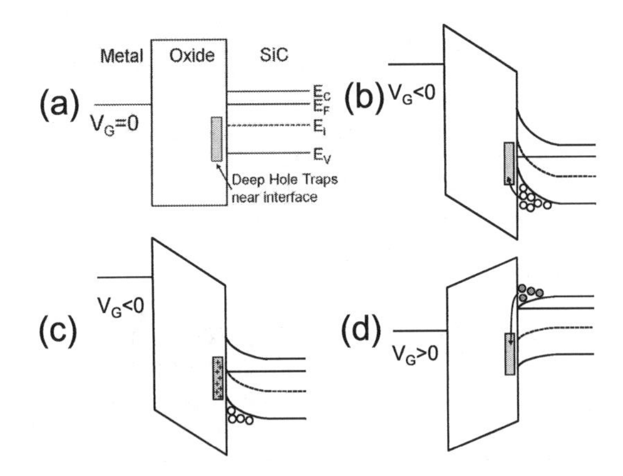

FIGURE 4. Schematic views of the change in MOS band bending during the TIBIC measurements. (a) VG = 0 (before TIBIC measurements), (b) VG < 0 (initial), (b) VG < 0 (saturated), and (d) VG > 0 (for refreshment).

trapping can be changed from/to neutral to/from positive by supplying holes or electrons [18]. The imperfection of chemical bonding between atoms, such as Si-Si bonds might also exist near the interface between SiO_2 and SiC, as well as in the SiO_2 - Si system, since oxidation of SiC was performed by the standard thermal oxidation in this study. Figure 4 shows the schematics of the change in MOS band diagram during TIBIC measurements. The "deep hole traps" exist near the interface and their charge state is neutral at the gate bias of 0 V (Fig. 4 (a)). When an oxygen ion incidences on the sample, a high density of electron-hole pairs is generated in both SiC and oxide.

When a reverse gate bias was applied to the MOS capacitor during TIBIC measurements, the holes generated by ion incidence in SiC drift toward the interface due to the internal electric field, and following some holes tunnel into the oxide and are captured by deep traps located near the interface (Fig. 4 (b)). Upon trapping holes, the deep hole traps are positively charged. These positively charged traps contribute to the weakening of the electric field within the oxide (Fig. 4 (c)) and as a result the surface potential of SiC is smaller and the band bending of SiC is reduced. Also as a consequence of increasing trapping of positive charge and the reduced band bending in the SiC, the peak height of TIBIC signals and collected charge decrease with increasing number of incident ions. Finally, saturation of those values is observed when all the deep hole traps are positively charged. In the case, when a forward bias is applied to the sample, electrons are injected into the oxide and recombination of the positively charged hole traps occurs (Fig. 4 (d)).

This explains the refreshment of TIBIC signals to its original value. The above-mentioned model can plausible explain the current result of TIBIC signals in 6H-SiC nMOS capacitors, however, further investigations are necessary to clarify the origin of the deep hole traps in SiC MOS devices.

SUMMARY

In order to investigate ion-induced charge observed through oxide on SiC, MOS capacitors were fabricated on an n-type 6H-SiC epitaxial layer by thermal oxidation. Charge induced in the MOS capacitors was measured by a single ion hitting TIBIC measurement system using 15 MeV-O microbeams. The peak of TIBIC signals decreases and the fall time increases with increasing number of incident ions. The peak amplitude of TIBIC signal saturated after an incidence of 1000 ions and it can recovered to its initial value after applying a positive bias of + 1V for 1min. Collected charge which is calculated by the integration of TIBIC signals also shows similar dependence on the number of incident ions. These phenomena are attributed to charge trapping and recombination at deep hole traps located in the oxide but near the interface of SiO_2 and SiC. Since the fabrication of SiO_2/SiC system without deep levels is very difficult so far, the decrease and refresh of TIBIC peak might be observed from almost all SiC MOS devices as well. Thus, the results obtained in this study indicate that it is important to consider the charge states (charge trapping or recombination) of hole traps when we understand mechanism of SEEs for SiC MOS devices. The details of hole traps such as structures, energy levels are open question at present, and to clarify those, further investigations are necessary.

ACKNOWLEDGMENTS

Authors are very thankful to Dr. Kin Kiong Lee of JAEA (JSPS fellow) for fruitful discussion of this work. This study was partially supported by the Ministry of Education, Science, Sports and Culture, Grant-in-Aid for Scientific Research (B), 2009, 21360471.

REFERENCES

1. W. E. Nelson, F. A. Halden and A. Rosengreen, J. Appl. Phys. 37, 333-336 (1966).
2. D. K. Ferry: Phys. Rev. B 12, 2361-2369 (1975).
3. P. Friedrichs, T. Kimoto, L. Ley, G. Pensl, "Silicon Carbide" Weinheim, WILEY-VCH Verlag GmbH & Co. KGaA, 2010.
4. T. Ohshima, H. Itoh, M. Yoshikawa, J. Appl. Phys. 90, 3038-3041 (2001).
5. K. K. Lee, T. Ohshima, H. Itoh, IEEE. Trans.Nucl. Sci. 50, 194-200 (2003).
6. S. Onoda, N. Iwamoto, S. Ono, S. Katakami, M. Arai, K. Kawano and T. Ohshima, IEEE Trans. Nucl. Sci. 56, 3218-3222 (2009).
7. Y. Tanaka, S. Onoda, A. Takatsuka, T. Ohshima and T. Yatsuo, Mater. Sci. Forum 645-648, 941-944 (2010).
8. H. R. Zeller, Solid-State Electronics 38, 2041-2046 (1995).
9. C. Findeisen, E. Herr, M. Schenkel, R. Schlegel and H. R. Zeller, Microelectronics Reliability 38, 1335-1339 (1998).
10. G. Soelkner, W. Kaindl, M. Treu and D. Peters, Mater. Sci. Forum 556-557, 851-856 (2007).
11. F. Nava, E. Vittone, P. Vanni, G. Verzellesi, P.G. Fuochi, C. Lanzieri, M. Glaser, Nucl. Instr. and Meth. A 505, 645-655 (2003).
12. T. Ohshima, N. Iwamoto, S. Onoda, T. Kamiya, K. Kawano, Nucl. Instrum. Meth. B 267, 2189-2192 (2009).
13. S. Onoda, N. Iwamoto, M. Murakami, T. Ohshima, T. Hirao, K. Kojima, K. Kawano and I. Nakano, Mater. Sci. Forum 615-617, 861-864 (2009).
14. N. Iwamoto, S. Onoda, T. Ohshima, K. Kojima and K. Kawano, Mater. Sci. Forum 615-617, 517-520 (2009).
15. K. K. Lee, J. S. Laird, T. Ohshima, S. Onoda, T. Hirao, and H. Itoh, Mater. Sci. Forum 645-648, 1013-1016 (2010).
16. JAEA Takasaki, http://www.taka.jaea.go.jp/index_e.html
17. J. S. Laird, T. Hirao, H. Mori, S. Onoda, T. Kamiya, H. Itoh, Nucl. Instr. and Meth. B 181, 87-94 (2001).

18 T. R. Oldham and F. B. McLean, IEEE. Trans.Nucl. Sci. 50, 483-499 (2003).

High-Reliability Computing For The Smarter Planet

Heather Quinn[1], Andrea Manuzzato[1], Paul Graham[1], Andre' DeHon[2], Nicholas Carter[3]

[1]ISR3, Space Data Systems, LANL, PO Box 1663, MSD440, Los Alamos NM 87545
[2]Electrical and Systems Engineering, University of Pennsylvania, 200 S. 33rd St, Philadelphia, PA 19104
[3]Intel Corporation, 2200 Mission College Blvd, RNB6-61, Santa Clara, CA 95054

Abstract. As computer automation continues to increase in our society, the need for greater radiation reliability is necessary. Already critical infrastructure is failing too frequently. In this paper, we will introduce the Cross-Layer Reliability concept for designing more reliable computer systems.

Keywords: Reliability, Computing, Radiation Effects in Semiconductors
PACS: 89.20.Ff, 89.20.Kk

INTRODUCTION

The geometric rate of improvement of transistor size and integrated circuit performance, known as Moore's Law, has been an engine of growth for our economy that has enabled new products and services, created new value and wealth, increased safety, and removed menial tasks from our daily lives. Affordable, highly integrated components have enabled both life-saving technologies and rich entertainment applications. Anti-lock brakes, insulin monitors, and GPS-enabled emergency response systems save lives. Cell phones, internet appliances, video games, and mp3 players enrich our lives and connect us together. Over the past 40 years of silicon scaling, the increasing capabilities of inexpensive computation have transformed our society through automation and ubiquitous communications. In this paper, we will present the concept of the smarter planet, how reliability failures affect current systems, and methods that can be used to increase the reliable adoption of new automation in the future.

THE SMARTER PLANET

Recently IBM has been presenting the idea of a "smarter planet." In smarter planet documents IBM states, "With computational power now being put into things we wouldn't recognize as computers, any person, any object, any process or service and any organization — large or small — can become digitally aware, connected and smart. With so much technology and networking available at such low cost, what wouldn't you enhance?[1]" In these documents, IBM discusses increased computer automation of roadways, banking, healthcare, and infrastructure, as automation could create more efficient systems that could reduce traffic jams, increase transparency in the financial institutions, increase information connectivity between doctors and hospitals, and reduce infrastructure strain. Increasing computation, though, necessitates that we increase reliability, as unmitigated failures in automobiles, infrastructure, medical-implantable devices and avionic systems could be dangerous.

In particular, we have found that with extreme-scale systems or systems in extreme environments, the effect of radiation-induced failures can dominate other reliability problems. In extreme-scale systems, such as supercomputing or infrastructure computing, the vast number of components in the system contribute to increase the likelihood of radiation-induced failures. In extreme-environment systems, such as avionics, the heightened radiation environment increases the likelihood of radiation-induced upsets. To illustrate both the exteme-scale systems and the extreme environment systems, we will present examples from automotive and avionic systems.

Application of Accelerators in Research and Industry
AIP Conf. Proc. 1336, 665-668 (2011); doi: 10.1063/1.3586185

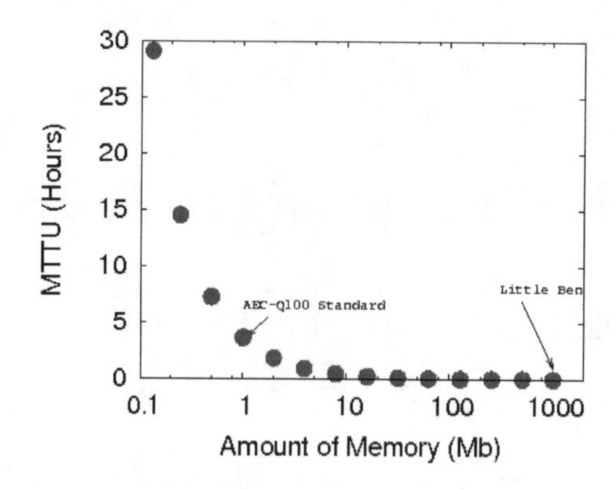

Figure 1: Worldwide mean-time-to-upset in hours for different car memory sizes.

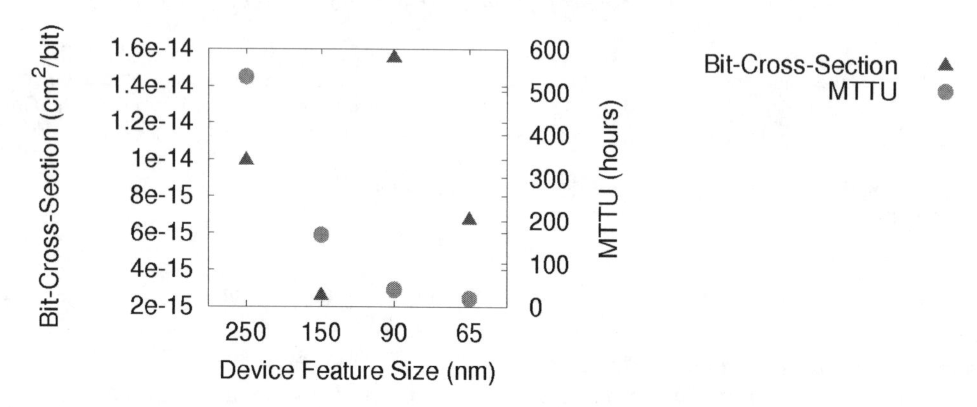

Figure 2: Mean-Time-To-Upset (MTTU) decreases as feature sizes shrink [shown at a flux of23,082 neutrons-cm2/hour corresponding to the flux seen by an airplane flying at 60,000 feet over the North Pole where the magnetic field lines converge]; sensitivity per bit (Bit-Cross-Section) does not show direct correlation with feature size.

AUTOMOTIVE SYSTEMS

One area of recent concern has been the reliability of cars. The most recent car reliability standard, "Failure Mechanism Based Stress Test Qualification for Integrated Circuits (AEC-Q100-Rev-G)," states that "Soft Error Rate (SER)] testing is needed for devices with large numbers of SRAM or DRAM cells (≥ 1 Mbit). For example: Since the SER rates for a 130 nm technology are typically near 1000 FIT/Mbit, a device with only 1,000 SRAM cells will result in an SER contribution of 1 FIT." In Figure 1 the mean-time-to-upset (MTTU) in hours is shown for the worldwide population of cars as a function of the memory capacity in each car. These calculations take into account that approximately 250 million [2] cars are on the road every day with an average time on the road of three hours. Of the 250 million cars, we assume that the new cars are more likely to have more memory, so the MTTU is further rated to indicate that 60% of all cars on the road are manufactured in the last ten years. From this graph we see that if all of these cars had only 1Mb of memory, a single-event upset will occur approximately every 3.6 hours. For the computer-driven Grand and Urban Challenge [3] cars, such as Little Ben [4], which can have up to 128MB of memory, the MTTU is 12.6 seconds. These autonomous vehicles are indicative of the level of electronics we may see in cars over the next 10–20 years as automobile electronics continues to assist with more of the driving functions, such as automated parking and distance sensing cruise control.

While error-correcting codes will likely correct some of these errors in both of these scenarios, the increasingly likelihood of multiple-bit upsets, where a single ionizing particle creates multiple memory bit flips, can make correcting the errors expensive. Depending on the layout of the memory devices, multiple-bit upsets can be as high as 90% [5] of all events. Furthermore, single-event functional interrupts that destroy entire pages of memory are now as common as single-event upsets in dynamic random access memories (DRAMs). While there are many error-correcting codes that can correct multiple faults simultaneously, such as Reed-Solomon, often times they require a particular memory usage pattern that will make random accesses to memory very expensive.

AVIONIC SYSTEMS

In avionic systems we have seen an increase in the use of Field-Programmable Gate Arrays (FPGAs). FPGAs are programmable devices where the user's circuit is implemented in programmable logic and routing. Due to the flexible architecture, these devices have always been popular for image processing. Recently, we have seen real-time image processing systems being incorporated in unmanned aerial vehicles (UAVs). These systems can collect real-time images from an aerial vantage point and can be helpful guiding wild fire firefighters to and from the hottest parts of the fire or supporting in-theater warfighters.

In the past we have used accelerated radiation experiments to measure the sensitivity of the device to radiation, called the cross-section. As these devices leverage memory cells, the sensitivity to radiation-induced failures is similar to traditional memory, such as static random access memory (SRAM). Figure 2 shows that while the per-bit cross-section has not changed much over several generations of FPGAs, the device's MTTU at airplane altitudes continues to decrease exponentially. As shown in Figure 3, this correlation is due to the increasing amount of memory used in each generation of FPGAs. Therefore, even if the sensitivity to radiation decreases on a per-bit basis, the increasing number of bits in the component negates these effects. Unlike the automotive example, the use of error-correcting codes is not useful for FPGAs. In these cases, triple-modular redundancy is used to mask the effect of radiation-induced failures until the system can be reset.

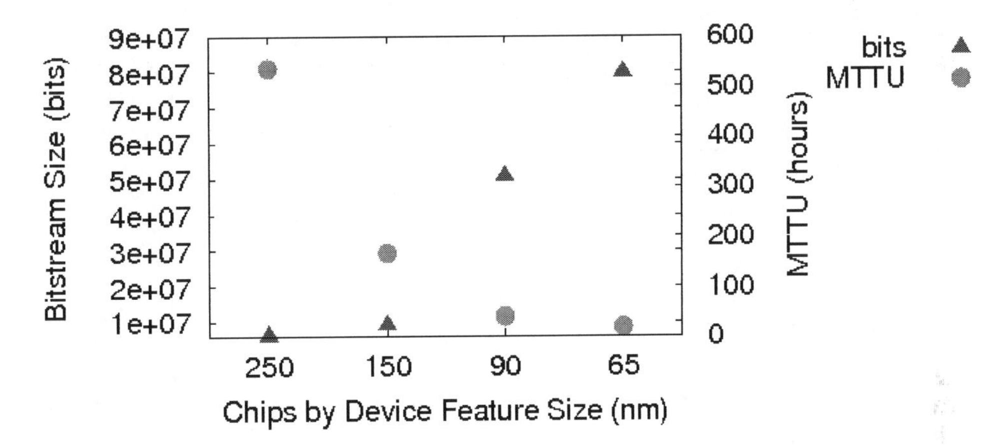

Figure 3: Chip capacity continues to increase as feature sizes shrink; capacity increase correlates well with decreasing MTTU.

CROSS-LAYER RELIABILITY

In both of these previous examples, it is necessary to mitigate radiation-induced failures in electronics to ensure reliable system execution. Unfortunately, radiation is only one source of reliability problem with modern electronics. In many current systems all of these reliability problems are mitigated at the hardware level, so that errors are not observable in the operating system or application. We are finding, though, that the need to mitigate all of these problems in the hardware is expensive in terms of area and power increases tohardware devices. To combat these problems NSF funded a Computing Community Consortium Visioning program in Cross-Layer Reliable Computing [6] to study new methods of reliability engineering. This program brought reliability researchers from many different fields together to discuss current and future reliability problems. After discussing these problems with the areospace, consumer, infrastructure, large-scale, and life-critical industries we found several key points these researchers converged on:

- Reliability problems in electronic systems must be addressed but the cost of mitigating reliability problems is expensive,
- Information regarding the reliability of electronics is hard to determine and often comes too late in the design cycle to address it,
- No single layer of the computing stack has enough information to mitigate reliability problems efficiently, and
- The onus for system reliability should be shared across the computing stack.

Because of these problems, system reliability is often designed as shown in Figure 4. In this scenario, the hardware layers focus on preventing reliability failures and the software layers are blind to reliability failures. The solution to these problems is to build systems that are designed for repair, can filter errors, can detect and correct errors, can adapt to changing scenarios and environments, and degrade gracefully as the system comes to the end of its usable life.

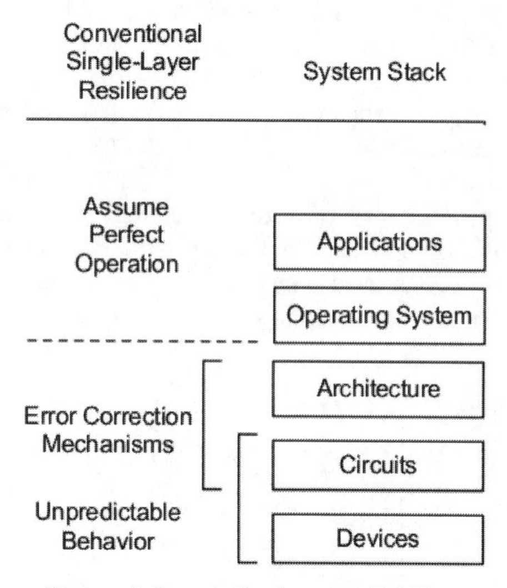

Figure 4: System Stack and Reliability

An illustration of cross-layer system design is shown in Figure 5. In this illustration the application layer determines a bound on the value x. Unlike conventional systems, the cross-layer version shares this information with lower layers. The architecture layer can use this bound to detect misbehaving hardware. However, the architecture layer does not need to capture and correct the problem. When it does notice misbehavior, it can signal the the middleware and OS layers to contain and correct the error before it becomes visible to the application. We are beginning to see scattered solutions with this flavor. Nonetheless, this approach demands a wholesale paradigm shift in the way we design and engineer computer systems. Multi-level solutions are not without their precedent in today's computer systems. They are regularly employed to protect storage and communications. While multi-level solutions have been useful in protecting bulk storage, such as DRAMs and persistent storage, similar solutions for computation currently do not exist. In part, the heterogeneous design of computing systems and their ability to transform data makes posing simple solutions that do not rely on brute-force replication difficult.

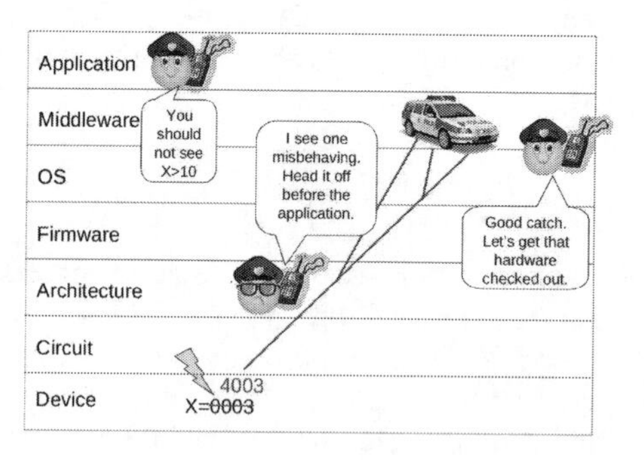

Figure 5: Cartoon Illustration of Cross-Layer Cooperation

Moore's Law feature-size scaling has been an economic and capability engine fueling wealth creation and quality-of-life improvements for the past 40 years. We are now moving into a qualitatively new regime where device energy will be the dominant limitation for the exploitation of additional capacity and where device are inherently unpredictable and unreliable. In this new regime our old solutions to reliability no longer make sense and will lead to an early end to the benefits of scaling. However, by exploiting information rather than energy to tolerate errors at higher levels in our system stack, we can productively exploit these smaller technologies to continue reducing energy while ensuring high system-level safety.

REFERENCES

1. "Conversations for a Smarter Planet," found on the web at http://www.ibm.com/smarterplanet, last accessed on 8/5/2010.
2. U.S. Bureau of Transporation Statistics. Number of u.s. aircraft, vehicles, vessels, and other conveyances. on web at http://www.bts.gov/publications/national_transportation_statistics/html/table_01_11.html, July 2010.
3. U.S. Defense Advanced Research Projects Agency. Darpa grand challenge and darpa urban challenge. on web at http://www.darpa.mil/grandchallenge/index.asp, 2007.
4. Jonathan Bohren, Tully Foote, Jim Keller, Alex Kushleyev, Daniel Lee, Alex Stewart, Paul Vernaza, Jason Derenick, John Spletzer, and Brian Satterfield. Little ben: The ben franklin racing team's entry in the 2007 darpa urban challenge. Journal of Field Robotics, 25(9):598–614, 2008.
5. Gasiot et al. Multiple cell upsets as the key contribution to the total SER of 65 nm CMOS SRAMs and its dependence on well engineering. IEEE Transactions on Nuclear Science, 54(6):2468–2473, Dec 2007.
6. "Cross-Layer Reliable Computing" found on the web at http://relxlayer.org, last accessed on 8/5/2010

Swift Ion Implanted Optical Waveguide In Chalcogenide Glass

Feng Qiu, Tadashi Narusawa

Kochi University of Technology, Tosayamada-cho, Kochi, 782-8502, Japan

Abstract. This work demonstrates that a holmium doped chalcogenide glass waveguide can be fabricated by ultralow dose swift ion implantation method. More than three modes exist in the waveguide at the wavelength of 632.8nm and the refractive index profile is reconstructed. The near-field pattern of the transmitted light is obtained and the propagation loss of the waveguide is about 4.0dB/cm.

Keywords: Swift ion implantation; Optical waveguides
PACS: 42.70.-a

INTRODUCTION

Optical waveguide is one of the basic components in integrated optics. It is characterized by a region of high refractive index surrounded by regions with lower refractive index. The small size of waveguide structures possesses several desirable features. It offers high light densities produced by even very low powers, consequently, the nonlinearities or laser actions in waveguides may be more efficient than those in bulk materials [1]. Furthermore, active and passive components can be combined on a single substrate, resulting in reduction in device size and cost, and the waveguide components may be connected together with optical fibers quite easily due to the compatible dimensions. There have been some techniques for fabricating waveguides, such as diffusion of metal ions, ion exchange, sol gel, pulsed laser deposition and ion implantation [2]. Among these techniques, the advantages of ion implantation are its wide applicability to most optical materials (more than 100 optical materials), precise control of the refractive index profiles by adjusting the implantation parameters and geometries and the ability of combination with other techniques. Conventional ion implantation has been extensively investigated [2, 3]. Recently, swift ion implantation has been paid much attention because it is more effective than conventional ion implantation [4].

Chalcogenide glasses are excellent candidates for active applications due to their wide optical transmission band. Among chalcogenide glass, gallium lanthanum sulfide (GLS) glass has greatly improved thermal and mechanical properties and excellent rare-earth solubility, which makes it more suitable for waveguide lasers. Ho^{3+}-doped GLS glass is of great interest for its infrared emission at 3.9 and 4.9μm [5]. The fluorescence band at 3.9μm agrees with the atmospheric transmission window and the fluorescence band at 4.9μm overlaps with the fundamental absorption of carbon monoxide, making the glass a potential waveguide laser source for remote sensing and gas sensing applications. So far, there are not any reports on the fabrication of Ho^{3+}-doped GLS glass waveguides. In this work, we use 60MeV Ar^{4+} at the dose of 2×10^{12} ions/cm^2 to fabricate an optical waveguide in Ho^{3+}-doped GLS glass. The relevant optical properties are characterized.

EXPERIMENTAL

The glass with composition of $70Ga_2S_3$:$29La_2S_3$:$1Ho_2S_3$ (molar ratio) was cut into $10 \times 10 \times 2$ mm^3. After optical cleaning and polishing, the sample was implanted with 60MeV Ar^{4+} at the dose of 2×10^{12} ions/cm^2 using the tandem accelerator at Tokai Research and Development Center, Japan Atomic Energy Agency. The ion beam was focused on the sample surface and then scanned over the area of 10×10 mm^2 to achieve the dose uniformity. The beam current density was maintained at about 10nA/cm^2 to minimize charging and heating effect during irradiation.

The prism-coupling method was used to observe the waveguide dark modes. The effective refractive indices of all the dark modes were measured by using

Application of Accelerators in Research and Industry
AIP Conf. Proc. 1336, 669-671 (2011); doi: 10.1063/1.3586186

Model 2010 prism coupler (Metricon Corp., USA). After polishing the end-faces of the implanted glass, the near-field pattern of transmitted light through the waveguide was measured by end-fire coupling. The light of 632.8nm was coupled into the sample by a calibrated 25× lens. The movement of the waveguide was controlled by a six-dimensional optical stage where the sample was mounted. The output light was collected by passing a 25× lens and then focused onto a CCD camera to measure the near-field pattern. The optical loss was measured by the back-reflection method [6].

Results And Discussions

Figure 1 shows the TE dark mode spectrum of the waveguide, in which one sharp dip corresponds to one optically allowed mode. There are more than three modes existing for the waveguide, and the effective refractive index of the first mode (2.4567) is higher than the refractive index of the substrate (2.4560), which means that the light can propagate in the waveguide without tunneling loss. This non-leaky mode strongly suggests a possibility of channel waveguide fabrication by simply adding standard photolithographic procedures.

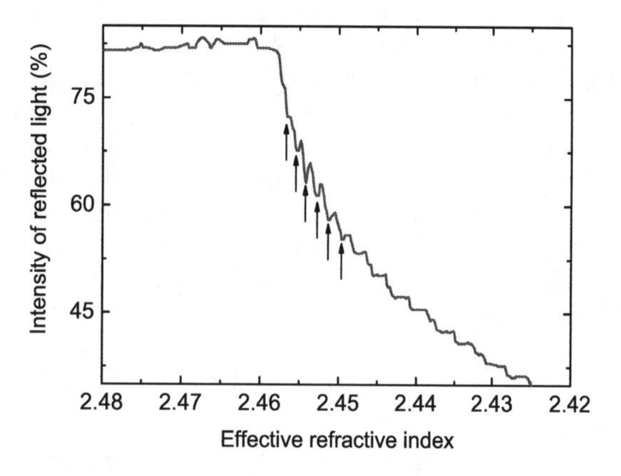

Fig. 1 The results of prism coupling at the wavelength of 632.8nm.

Figure 2 shows the reconstructed refractive index profile of the waveguide by means of RCM [7] and the refractive index of the substrate (dotted line). As we can see, a "well" with a refractive index increase of 0.2% and a width of ~12.8μm is generated beneath the surface of the sample. A negative index change happens at the end of the profile, which is usually called a "barrier".

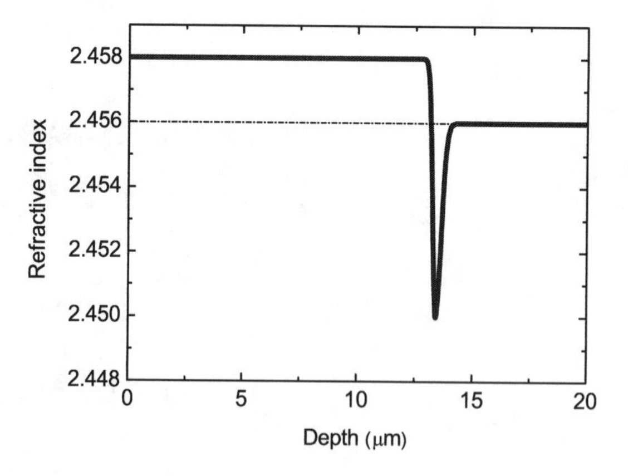

Fig. 2 The reconstructed refractive index profile of the implanted waveguide.

To estimate the energy loss of the implanted Ar^{4+} ions, SRIM' 2008 (http://www.srim.org) was used, and the obtained results are shown in Fig. 3. By comparing Fig.2 with Fig.3, it can be seen clearly that the enhanced refractive index well is mainly located in the electronic damage region, whilst the barrier is formed in the nuclear damage region. In our previous research [4], the swift ion implantation induced a darkening phenomenon in chalcogenide glass which was very similar to photodarkening and thus the mechanism of enhanced index well formation was supposed that the change in local structure order of the substrate caused the "well" formation. Moreover, the barrier formation is usually attributed to the nuclear damage [3]. As a result, the mechanism of waveguide formation in this work can be summarized as follow. The violent collisions between the incident ions and target nuclei mainly occur at the end of the ion range. The nuclear damage caused by these collisions should be the main reason for the reduction of the refractive index of glass to generate an "optical barrier", although the physics behind this is not clear yet. Within the region between this barrier and the surface, the electronic effect is dominant, and the ionization processes caused by electronic energy deposition induce the change in local structure order of the substrate and the "well" formation. Thus, the region between air and the optical barrier is sandwiched by low-refractive index regions, which acts as a waveguide.

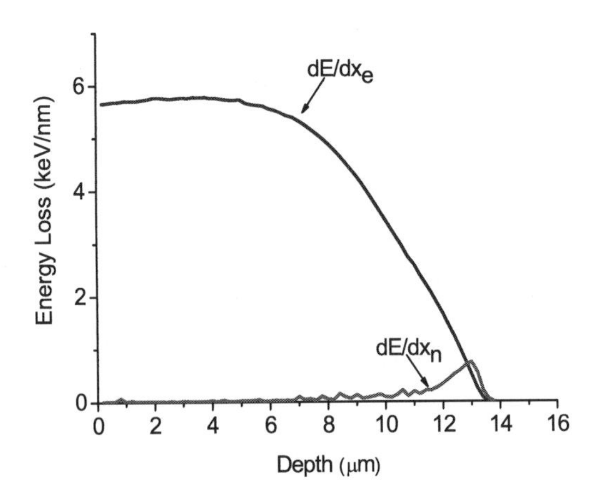

Fig. 3 The electronic energy loss (dE/dx_e) and nuclear energy loss (dE/dx_n) of the implanted Ar^{4+} ions.

The near field pattern profile of the waveguide is shown in Fig. 4. We can see clearly that the waveguide can confine light well and no light is tunneled into the substrate, which is consistent with our expectation described above. The propagation loss of the waveguide is about 4.0dB/cm measured by back-reflection method. We suppose that the main reason of the loss is the absorption of color centers [8] caused by the implantation. In future work, the waveguide will be annealed at appropriate temperatures to decrease the propagation loss.

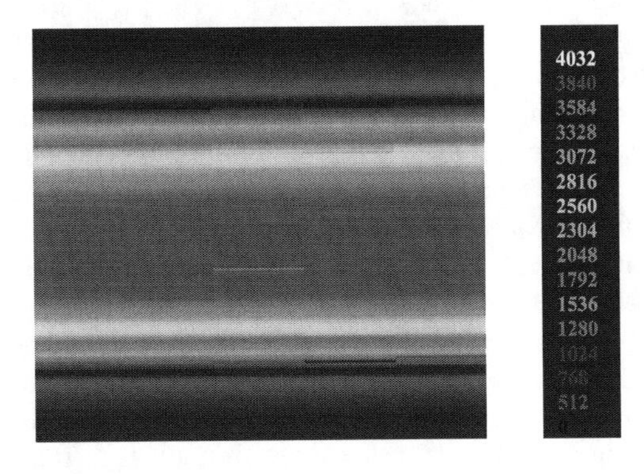

Fig. 4 The near field pattern of the waveguide at the wavelength of 632.8nm.

CONCLUSIONS

We demonstrate the fabrication and characterization of a planar waveguide in Ho^{3+}-doped GLS glass by 60MeV Ar^{4+} at the dose of 2×10^{12} ions/cm^2. According to the results of prism coupling, one non-leaky mode exists in the waveguide, which means that the light can propagate without tunneling loss. The reconstructed refractive index profile has a "well + barrier" type distribution, and the refractive index increases by 0.2% in the "well" region. The near-field pattern also exhibits no light tunneled into the substrate. The propagation loss of the waveguide is 4.0dB/cm which may be decreased by thermal annealing of the sample at appropriate temperatures. Our results show that swift ion implantation is a potential method to fabricate optical waveguides.

ACKNOWLEDGMENTS

The authors would like to thank Prof. A. Osa and Prof. M. Sataka at Japan Atomic Energy Agency (JAEA) for the swift ion implantation.

REFERENCES

1. 1.J. I. Mackenzie, IEEE J. Sel. Top. Quantum Electron. 13, 626 (2007).
2. F. Chen, X. L. Wang, andK. M. Wang, Opt. Mater. 29, 1523 (2007).
3. P. D. Townsend, P. J. Chandler, and L. Zhang, Optical Effects of Ion Implantation (Cambridge University Press, Cambridge, 1994).
4. F. Qiu and T. Narusawa, Opt. Mater. (2010) submitted.
5. T. Schweizer, B. N. Samsom, J. R. Hector, W. S. Brocklesby, D. W. Hewak, and D. N. Payne, Infrared Phys. Technol. 40, 329 (1999).
6. R. Ramponi, R. Osellame, and M. Marangoni: Rev. Sci. Instrum. 73, 1117 (2002).
7. P. J. Chandler and F. L. Lama: Opt. Acta. 33, 127(1986).
8. F. Qiu and T. Narusawa, Jpn. J. Appl. Phys. 49, 092503 (2010).

Simulation Of Ion Implantation Into Nuclear Materials And Comparison With Experiment

Z. Insepov[a], D. Yun[a, b], B. Ye[b], J. Rest[a], S. Starikov[a, c], A. M. Yacout[a]

[a]*Argonne National Laboratory, 9700 S. Cass Ave, Lemont, IL 60439*
[b]*University of Illinois at Urbana-Champaign, 104 S. Wright Street, Urbana, IL 61801*
[c]*Joint Institute for High Temperatures RAS, 13 Izhorskaya , Moscow, 125412*

Abstract. A new many-body potential is proposed for pure molybdenum that consists of using ab initio and atomistic MD simulation methods verified against existing surface erosion experimental data. Mo is an important material for metallic U-Mo alloys for using them in low-enriched fuels. Several new Xe-Mo potentials were also parameterized by comparing the calculated sputtering yield of a Mo-surface bombarded with Xe ions with experimental data. Calculated results were also compared with defect distributions in CeO2 crystals obtained from experiments by 500 keV Xe implantation at the doses of 1×1017 ions/cm2 at several temperatures.

Keywords: Radiation defects, Mo, CeO_2, TEM, Xe ion, MD and MC simulations
PACS: 61.72.uf, 61.72.uj, 61.72.up, 61.72.J-, 61.43.Bn, 61.43.Bn

INTRODUCTION

The evolution of defects and gas bubbles in U-Mo and CeO_2 are qualitatively similar. By studying both materials using realistic interatomic potentials the phenomena that are common and dissimilar can be elucidated. One aim of this work is to develop a multiscale methodology that is applicable to a wide range of candidate nuclear materials. The identification of phenomenon that are common across a range of different materials types provides leverage not only for the understanding of the underlying physics, but for the development of a tool kit that is applicable for the study of a variety of nuclear materials.

The interaction of energetic heavy Xe ions with nuclear fuel related materials such as Mo, U-Mo, U-Zr, and UO_2 is an important stage in the modeling of nuclear fuel behavior where generation and accumulation of radiation induced defects can ultimately lead to fuel failure. Interaction of heavy particles with both oxide and metal materials has many similarities. For example, formation and growth of fission gas bubbles by ion beam irradiations have been observed in both oxide and metallic fuels.

Cerium dioxide is a very attractive material for catalysts and is a natural surrogate material for UO_2. However, computer simulations on an atomistic scale cannot well reproduce experimentally measurable data due to inherent difficulties to simulate oxides. For example, the oxygen diffusion coefficient in CeO_2 obtained with different potentials of a Born-Mayer-Huggins type can vary up to 33% [1-3]. There are no simulation studies of radiation induced defect formation, fission gas accumulation on cerium dioxide. This is mainly due to the lack of accurate interatomic potentials that describe both the behaviors of point defects, defect clusters and high energy collision cascade during irradiation.

Therefore, a qualitative comparison of the experimental defect and fission gas properties can be made with an accurate simulation data obtained for simple materials. In this paper, we have developed a new interatomic potential and new simulation models for such processes as gas bubble formation, diffusion of radiation-induced defects, defect depth profile calculation, and lattice swelling. To verify new approaches, we initiated ion beam irradiation experiments for comparison to simulation data. Since gas bubble formation has many similarities in metal (alloy) and oxide fuels, we qualitatively compared our simulation results with obtained experimental data.

Fuel behavior codes are sensitive to materials parameters, many of which have large uncertainties, or have not been measured. Atomistic simulation is a powerful method for tracking defect accumulation during ion implantation and for estimating the values of critical materials properties and parameters used in kinetic fuel-behavior models. Whereas first principle

Application of Accelerators in Research and Industry
AIP Conf. Proc. 1336, 672-676 (2011); doi: 10.1063/1.3586187

simulations are limited to a few hundred atoms at most, classical molecular dynamic (MD) calculations with many millions of atoms are routinely performed. However, the reliability and predictive power of classical MD depend crucially on the quality of the effective potential used. For elementary solids, such potentials are usually obtained by adjusting a few potential parameters to optimally reproduce a set of reference data, which typically includes a number of experimental values such as lattice constants, cohesive energies, or elastic constants, sometimes supplemented with ab-initio cohesive energies and stresses. However, for more complex systems with a large variety of local environments and many potential parameters to be determined, such an approach cannot help. A new method of deriving realistic interatomic potentials is presented in this work.

SIMULATION

Empirical interaction potentials were widely used in classical molecular dynamics studies for studying ion–solid interactions and surface erosion effects. However, the empirical potentials failed to reproduce fine defect structures and correct activation energies and therefore failed to predict correct basic defect structures and diffusion coefficient in molybdenum.

Table 1: Comparison of simulation results (with new potential) with experimental data for pure Mo

	Cohesive energy (eV)	Lattice parameters (Å)	C11 (GPa)	C12 (GPa)	Melting temperature (K)
Simulation	6.91	3.1469	560	225	2630
Experiment	6.82	3.15	464	163	2890

Table 2: Comparison of simulation results (with new potential) with VASP calculation for Mo defects

	Formation energy of crowdion <111> (eV)	Formation energy of dumbbell <111> (eV)	Formation energy of dumbbell<110>(eV)	Formation energy of vacancy (eV)
Simulation	6.42	6.43	6.67	2.79
VASP	6.89	6.88	7.02	2.40

The force matching method (FMM) provides a way to construct physically justified potentials even under such circumstances as absence of experimental data. This method provides new interatomic potentials obtained on the basis of synthesis of quantum and statistical mechanics. The idea is to compute force and energies from first principles for a suitable selection of small reference system and to adjust the parameters of the interatomic potential to optimally reproduce them [4, 5]. The method allows creating correct potentials for simulation of various processes such as phase transitions, deformation at different temperature, and fracturing.

With the use of FMM, we developed an interatomic potential for Mo-Xe systems. The reference data was calculated with the VASP[1] code [6]. The following parameters were used: the electron orbitals represented using plane-waves, with a cut-off energy of 400 eV; the generalized gradient approximation (GGA) for pseudopotential; 2×2×2-points in k-space. We used 81 various configurations with total number of atoms of 10746. These configurations included different Mo-Xe systems: 39 states with pure Mo (liquid and bcc solid states at different densities; solid states with SIAs and/or vacancies and/or surface), 20 states with pure Xe (liquid and solid states) and 22 states with Mo-Xe (including a single Xe atom in pure Mo). Generally the matching was carried out with three value types: energy (only 1 in every configuration), stress tensor (6 in every configuration), and forces (fx, fy and fz per every atom in configurations). Only the energy at matching was used in the configurations with SIA and vacancies in Mo. Table 1 shows comparison of the Mo parameters calculated using the new many-body Mo potential with experimental data.

Subsequently, the force matching procedure was implemented. The potential is realized in EAM (embedded atom method) form with seven independent functions. In our investigation the potential functions were set by splines with 10 independent parameters. The search algorithm for potential parameters was obtained with the *potfit*-code [5]. A search was performed for a minimum of the target function

[1] Vienna Ab-initio Simulation Package

$$Z = \sum_{i}^{N} \sum_{\alpha=x,y,z} \frac{\left(f_{i\alpha} - f_{0,i\alpha}\right)^2}{f_{0,ix}^2 + f_{0,ix}^2 + f_{0,ix}^2 + \varepsilon}, \quad (1)$$

where $f_{0,\alpha i}$ and $f_{\alpha i}$ are value of force components of atom i calculated by VASP. Summation is carried out over atoms and configurations in the reference data.

The derived potential reproduces the reference data with good precision. For instance, energy difference equals 0.01 eV (precision about 0.1 %). The new interatomic potential simulates experimental equations of states of pure Mo and pure Xe. Simulation results with the new EAM potential were compared with some experimental data performed for verification in

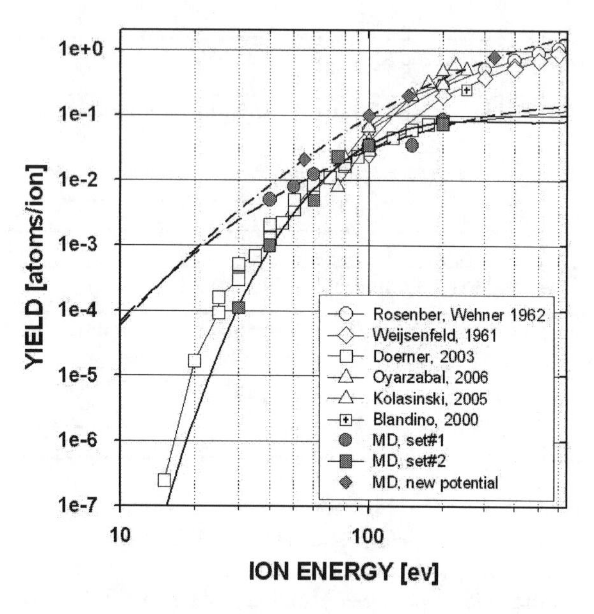

FIGURE 1. Comparison of the sputtering yield of a Mo (100) surface bombarded by accelerated Xe+-ions interacting with Mo atoms via a Morse potential with the experimental data from Refs. [7-11].

Table 1. Table 2 shows comparison of the defect properties calculated with the new EAM-potential with the VASP data.

Note that a difference between formation energies of defects is more important than total energies. Two configurations of SIA with minimal energy (Crowdion and Dumbbell <111>) have almost equal values. This fact explains the high one-dimensional mobility of SIA in pure Mo at low temperatures.

In addition the new interatomic potential reproduces experimental results on scattering Mo atoms by Xe atoms (see Fig. 1)

A combination of *in situ* TEM (transmission electron microscopy) and *ex situ* TEM experiments on Mo were used to study the evolution of defect clusters during implantation of Xe and Kr ions at energies of 150-700 keV, depending on experimental conditions. The simulation and irradiation were performed on thin–film, single crystal materials. Figure 1 compares the sputtering yields for Xe^+ ion bombardment of a Mo (100) surface calculated in the present paper with the experimental data obtained from the literature [7–11]. We used three relatively new interatomic potentials to describe the Mo-Xe interaction. The parameters for the pair potential function for set #1 yield data close to experiment at higher energies, namely, ~100 eV.

However, the calculated yields are much higher than the data at energies lower than 60 eV. Set #2 gives calculated yields close to experiment for both high and low energy regions. In addition, a new many-body interatomic potential for Mo-Xe system was also developed that gave close sputtering yields compared with experimental data in Fig. 1.

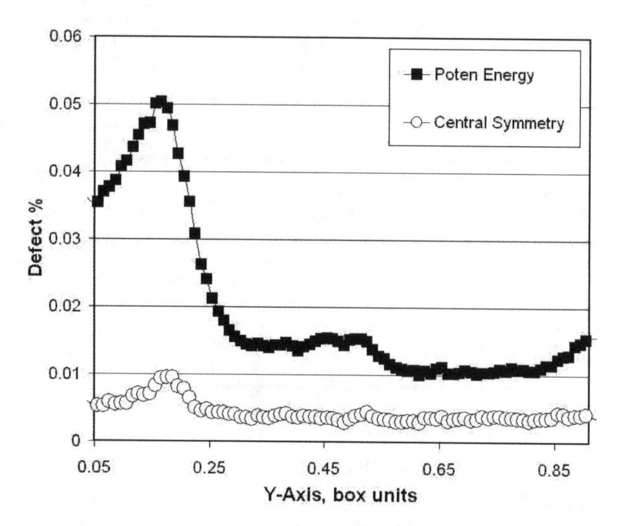

FIGURE 2. Distribution of vacancies in molybdenum calculated by molecular dynamics for a Xe+ ions, with energy of 25 keV. The defects were identified by their potential energy and the central symmetry number.

In what follows, set #2 based on pair potential and the new many-body Mo-Xe potentials were applied for studying Xe-bubble properties in Mo. The Mo-Mo EAM potential presented in this work reproduces the cold curve in agreement with the experimental data up to approximately 600 GPa (corresponding compression $V/V_0 \sim 0.5$). In addition, the description of thermal expansion is replicated well up to the melting point. The most stable configurations of interstitial defects in Mo are <111> dumbbell and <111> crowdion with very small differences in formation energies. This configuration provides for one-dimensional migration of self-interstitial atoms at very low temperatures [13] in agreement with the resistivity recovery measurements following electron irradiation [14]. With increasing temperature the

<111>-<110> dumbbell transitions are activated, providing a rotation of the axis of migrating crowdions, and hence a basis for a viable mechanism for three-dimensional diffusion.

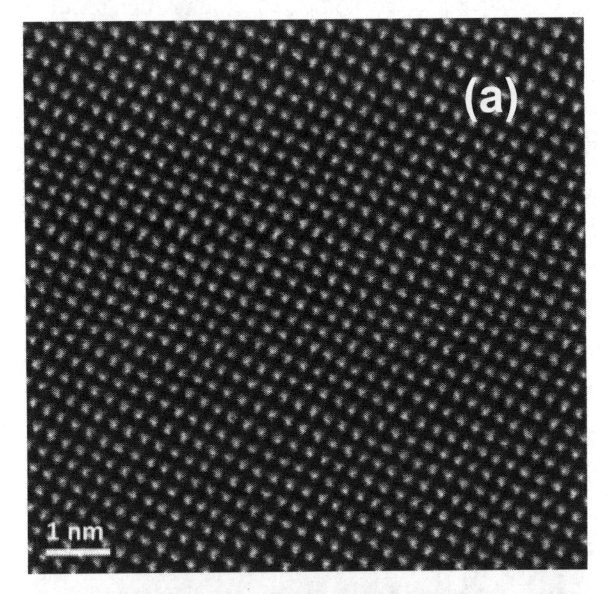

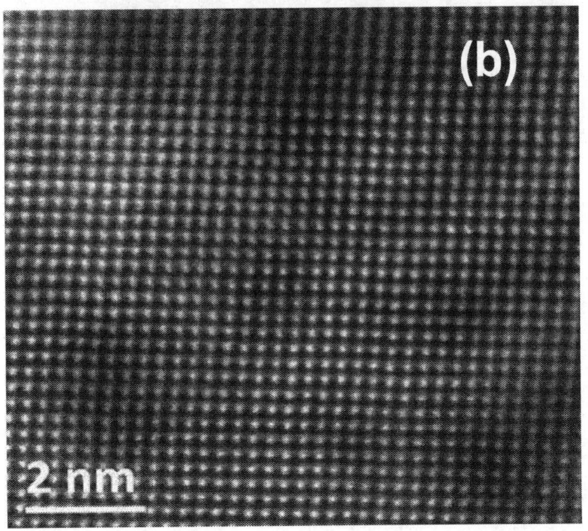

FIGURE 3. Scanning TEM micrograph of single crystal CeO_2 (a) before irradiation (b) after irradiated with 500 keV Xe to a dose of 2×10^{16} ions/cm^2 at 600°C. Electron beam direction is along <001>.

Fast parallel calculations were carried out on a Blue Gene supercomputer by using a Lammps MD simulation package [15].

A typical MD simulation of a system containing 22,000 Mo atoms in 10 ns was completed in six hours of computing time. Figure 2 shows the distribution of defects along the ion track in pure Mo irradiated with energy of 500 keV. The length of the sample was chosen to be 100 nanometers, comparable to an experimental thickness 160 nm of the CeO2 samples used in irradiation experiments..

EXPERIMENT

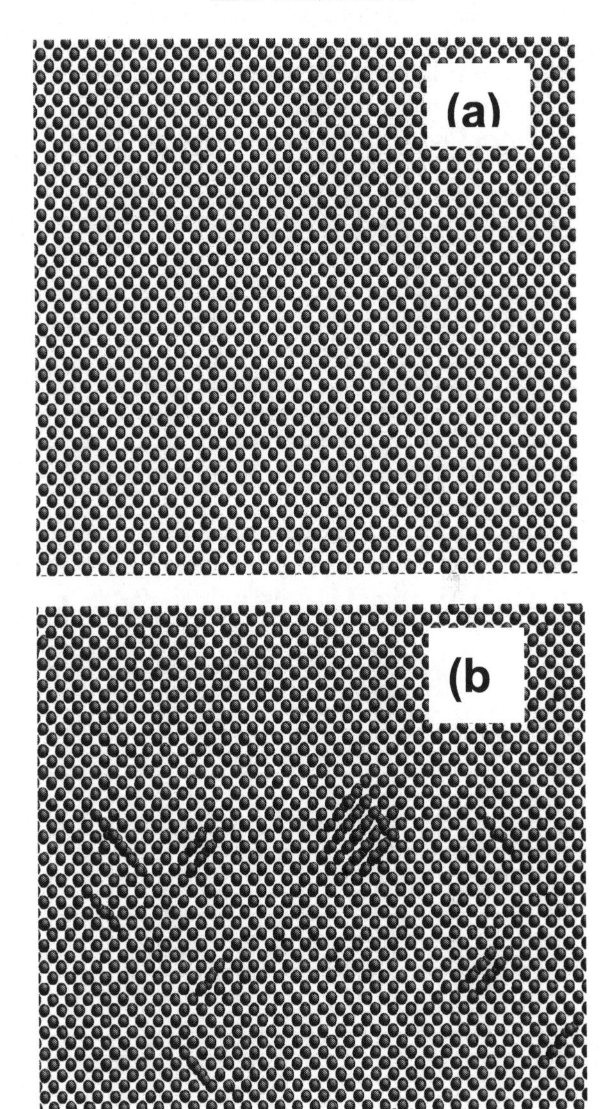

FIGURE 4. MD simulation of radiation damage of single crystal Mo (a) before irradiation and (b) after irradiation with 50 keV Xe to a dose of 1×10^{12} ions/cm^2 at 300°C. Dark spots show cluster of interstitial atoms formed by association of SIAs within 0.5 ns of simulation.

Ion irradiation was done in the IVEM-Tandem facility at Argonne National Laboratory, and the scanning transmission electron microscopy (STEM) was carried out on JEOL 2200Fs TEM at the University of Illinois. 500 keV Xe ions were implanted into single crystal CeO_2 TEM thin foil specimen at 600°C to an accumulated dose of $2x10^{16}$ ions/cm^2. The specimen were mechanically thinned first followed by ion polishing with 5keV Ar^+ beam at a shallow angle of 7 degrees until perforation in a precision ion polishing system (PIPS). Figure 3 shows the atomic level crystal structures of the specimen before and after irradiation obtained by STEM. The specimen was in a perfect crystal structure before irradiation (Fig. 3 (a), distortions in the lattice image are due to a rather slow scan rate used to produce the image) and remains crystalline after bombardment (Fig. 3(b)). The areas with darker contrast in Fig.3 (b) suggest formation of defect clusters, which are in a size range of 1 – 3 nm in diameter.

However, the nature of these defect clusters is not clear yet. Therefore, any direct comparison of these TEM experiments to MD simulations is not yet possible since these two methods provide information on very different time scales.

Figure 4 qualitatively compares our MD simulation results before and after radiation. The dark spots in Fig. 4 (b) are clusters of interstitial atoms formed by an association mechanism of single dumbbells and crowdions representing the main type of radiation defects in pure irradiated Mo.

Since the irradiation dose in simulation was much lower (1×10^{12} ions/cm^2), the sizes of the clusters are smaller. The detailed analysis shows that they can be formed by two or three interstitial dislocation loops with the diameters of 10-15 Å.

SUMMARY

Kinetic mesoscale models are directly comparable to reactor experiments. Our new concept is based on kinetic rate-equations for radiation damage, energetics and kinetics of defects, and swelling of fuels as a function of temperature and burnup. Quantum and classical atomistic simulation methods are applied to deepen our understanding of defect formation and growth processes during radiation damage and to calculate the probabilities of elemental processes and reactions applicable to irradiated nuclear materials. Since the interaction potentials are critical for the new concept, they were developed based on a force-matching method data from ab initio calculations or were fitted to existing experimental data.

A new many-body potential is proposed for pure molybdenum. The potential is benchmarked against existing surface erosion experimental data using ab initio and atomistic MD simulation methods. Several new Xe-Mo potentials were also parameterized by comparing the calculated sputtering yield of a Mo-surface bombarded with Xe ions with experimental data.

The interaction of energetic heavy Xe ions with Mo was studied as an important stage in the modeling of nuclear fuel behavior where generation and accumulation of radiation induced defects can ultimately lead to fuel failure. Formation and growth of fission gas bubbles in nuclear fuel related materials by ion beam irradiations have been studied in oxide fuels and qualitatively compared to simulation results.

ACKNOWLEDGMENTS

We thank Dr. James Belak for his valuable suggestions that helped build the Xe-Mo potential.

Argonne National Laboratory's work was supported under U.S. Department of Energy contract DE-AC02-06CH11357.

REFERENCES

1. C. Ciccotti, G. Jacucci and I.R. McDonald, Phys. Rev. A13 (1976) 426.
2. S. Gruenhut, D.R. Macfarlane, J. of Non-Cryst. Sol. 184 (1995) 356-362.
3. H. Hayashi, R. Sagawa, H. Inaba, K. Kawamura, Solid State Ionics 131 (2000) 281-290.
4. F. Ercolessi and J. B. Adams, Europhys. Lett. V. 26(8), P. 583 (1994).
5. P. Brommer and F. Gahler, Modelling Simulation Mater. Sci. Eng. V 15, P. 295 (2007).
6. G. Kresse and J. Furthmuller, Phys. Rev. B V. 54, P. 11169 (1996).
7. C.H. Weijsenfeld, A. Hoogendoorn, M. Koedam, Physica 27 (1961) 763.
8. D. Rosenberg, G.K. Wehner, J. Appl. Phys. 33 (1962) 1842.
9. J.J. Blandino, D.G. Goodwin, C.E. Garner, Diamond and Rel. Mat. 9 (2000) 1992-2001.
10. R.P. Doerner, D.G. Whyte, D.M. Goebel, J. Appl. Phys. 93 (2003) 5816.
11. R. Kolasinski, AIAA Paper No. 2005-3526, 2005.
12. E. Oyarzabal, J. H. Yu, R.P. Doerner, G.R. Tynan, K. Schmid, J. Appl. Phys. 100 (2006) 063301.
13. S.P. Fitzgerald, D. Nguyen-Manh, Phys. Rev. Lett. 101 (2008) 115504.
14. P. Ehrhart, P. Jung, H. Shultz, H. Ullmaier, Atomic Defects in Metals, ed. by H. Ullmaier Landolt-Bornstein New Series, Group III, Vol. 25 (Springer-Verlag, Berlin, 1991).
15. S. J. Plimpton, J. Comp. Phys. V. 117, P. 1 (1995).

Radiation Effect On Gas Electron Multiplier Detector Performance

Kwang June Park[a], Edwin Baldeloma[b], Seongtae Park[b], Andrew P. White[b] and Jaehoon Yu[b]

[a]*Korean Atomic Energy Research Institute, Daejeon, 305-600, Korea*
[b]*Department of Physics, University of Texas at Arlington, Arlington TX 76019, United States*

Abstract. Gas Electron Multiplier (GEM) detector is a gas device with high gain and high efficiency. These detectors use chemically perforated 65μm thick copper clad Kapton polyimide foils. Given its potential for detecting X-rays and other radiations, GEM detectors may be used in an environment with high radioactivity. The Kapton foils manufacturer, Du Pont Inc., claims that the foils are radioactive resistant. To verify whether the GEM detector performance is affected by the exposure to radiation, several GEM foils were irradiated to a ^{60}Co source at the gamma-ray irradiation facility at Sterigenics, Tustin, CA. Four sets of GEM foils were exposed to the level of 10kGy, 100kGy, 1,000kGy and 10,000kGy. The output signal from the GEM detectors with irradiated GEM foils were measured and compared to the detector with no irradiation. We observed that the shapes of the peaks from 5.9KeV ^{55}Fe X-ray were distorted and that the detector gain increased compared to that of the un-irradiated detector. In particular, the detector with 10,000kGy irradiation appeared to have the biggest peak distortion and increased gain. It was also found from that additional electrons from radiation-induced free radicals in the Kapton film contribute to output signal of the irradiated GEM detectors. Further studies are needed to explain the mechanism of these detector performance changes.

Keywords: Gas Electron Multiplier, GEM foil, Kapton polyimide film, Gamma irradiation, Electronic signal gain, Peak distortion
PACS: Replace .89.+b, 07.57.Kp, 42.88.+h, 82.50.-m

INTRODUCTION

It has been well-known that the polymer materials can be damaged by the exposure to radiation [1]. When they are used in a device in a radioactive environment, effects of radiation to poly materials must be well understood for a safe operation of the system. The polymers used in radiation facilities are exposed to ionizing radiation, leading to a radiation-induced degradation. Many studies on the radiation degradation of the polymer materials have focused on the radiation-induced oxidation phenomena, because most polymers in radiation devices are used in air [2]. The radiation-induced free radicals in a polymer react with the oxygen in air, and the degradation of the polymer proceeds through the similar process in other harsh environments, such as in a high temperature environment, in the ultra-violet light, under the mechanical stress, and through the chemical initiation involving radicals, etc. It is known that oxidation induced by free radicals progress as follows; 1) the creation of free radicals by absorbing radiation energy, 2) the propagation and chain branching of free radicals leading to a scission and cross-linking and 3) the termination stage at which all the free radicals are completely consumed [3,4].

In this study, radiation effects on GEM foils and the resulting detector performance were evaluated for the detectors built with double layers of GEM foils (D-GEM) as shown in Fig. 1. The GEM foils in these detectors are made of 60μm thick polyimide material Kapton® for the electrical insulation between the copper electrodes clad on either side of the foil. To establish the benchmark, the five D-GEM detectors were characterized through the signal shape measurements, gain dependence to high voltage and dependence to atmospheric pressure and others, were carried out. Once the benchmark was established, the eight GEM foils from four of the five detectors were extracted irradiated at the level of 10kGy, 10kGy, 1,000kGy and 10,000kGy together with the same number of bare Kapton polyimide film in the gamma-ray irradiation facility at Sterigenics, Tustin, CA. The shapes of the electronic output signal of the

Application of Accelerators in Research and Industry
AIP Conf. Proc. 1336, 677-680 (2011); doi: 10.1063/1.3586188

reassembled D-GEM detectors after the irradiation of the GEM foils were measured and analyzed periodically for time dependent variation. Finally, the radiation effects were evaluated by comparing the gains of the irradiated D-GEM detectors to that of the un-irradiated D-GEM detector.

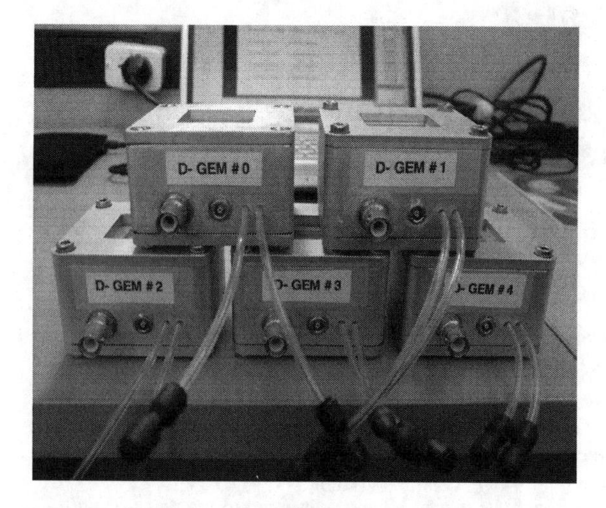

FIGURE 1. Photograph of the five D-GEM detectors used in this study.

EXPERIMENT

D-GEM Detector Construction And Characterization

Five small D-GEM detectors were constructed using two 50mm x 50mm GEM foils each. The detectors have 5mm drift gap and two 2mm induction and transfer gaps. The operational high voltage across the entire chamber structure was maintained at -1950V to provide 395V potential across each GEM foil. The signal is then collected on single 50mm x 150mm pad and read out through a single channel house-developed electronics with peak holding circuitry. LabView software is used for data acquisition and recording. The chamber body had a unibody structure made of solid Aluminum block to minimize external noise pickup. Gas used in the detectors was Ar/CO2 at the mixing proportion of 80/20. Once the detectors were constructed, each of the detectors was characterized for signal output shape, gain and efficiencies using radioactive source, in particular 5.9 keV ^{55}Fe X-rays. In addition, the high voltage dependence and the ambient pressure dependence of the chamber gains were determined. Pressure dependence of the chamber responses have been corrected for the final comparisons.

Gamma Irradiation Of GEM Foils

Disassembly Of Four Specimen D-GEM Detectors

Once the benchmarks were established, four of the five D-GEM detectors were disassembled to extract the eight GEM foils. They were carefully marked to associate them to the detectors they were taken from, were packed with a set of bare Kapton polyimide films and were transferred to Sterigenics's irradiation facility in Tustin, CA [6]. Since GEM foil has copper-clad electrodes on either side of Kapton polyimide film and had to be reintegrated into the detectors for performance study, the bare Kapton samples were used for the destructive material property examinations. The bare Kapton foils are identical samples obtained from the GEM foil manufacturer by chemically stripping off the copper layers.

Gamma Irradiation

The GEM foils and bare Kapton samples were irradiated to the dosage of 10kGy, 100kGy, 1,000kGy and 10,000kGy at Sterigenics irradiation facility. D-GEM#1, D-GEM#2, D-GEM#3 and D-GEM#4 detectors correspond to the dose level of 10kGy, 100kGy, 1,000kGy and 10,000kGy, respectively.

Post-Irradiation Detector Characterization

Once the irradiated GEM foils were returned, four D-GEM detectors were reassembled with eight irradiated foils in the HEP laboratory of UTA. High voltage dependence studies of the gains of the irradiated GEM detectors and the un-irradiated chamber were performed following the identical procedure as before the irradiation to minimize any systematic bias. This measurement was repeated periodically for possible time dependence of the post-irradiation effects.

RESULTS AND DISCUSSIONS

Tests With ^{55}Fe Radiation Source

We observe that the shapes of the ^{55}Fe 5.9keV peaks were distorted, and the gains of the chamber increased as a function of the irradiation level in the detectors with irradiated GEM foils compared to that of the detector with no exposure to radiation. Figure 2 compares the peak shapes and the gains of D-GEM#3 detector before and after the irradiation. As it is apparent, the gain of the detector with irradiated foils is increased significantly compared to that before the irradiation. It can also be seen that while the two clear peaks for the 4.1keV and 5.9 keV are seen distinctively in the un-irradiated detector, three peaks are seen in the irradiated detector, though the peak for the 4.1keV is not distinctive as before irradiation. The prominent peak at the low signal amplitude implies additional sources of electrons contributing to the detector signal in addition to the electrons produced in the ionization of $Ar\text{-}CO_2$ gas in the detector. While statistical fluctuation and external electronic could broaden the width of the Gaussian, affecting the resolution [7], these effects do not change the shape of the pulse or the location of the peak thereby the gain as seen in Fig. 2.

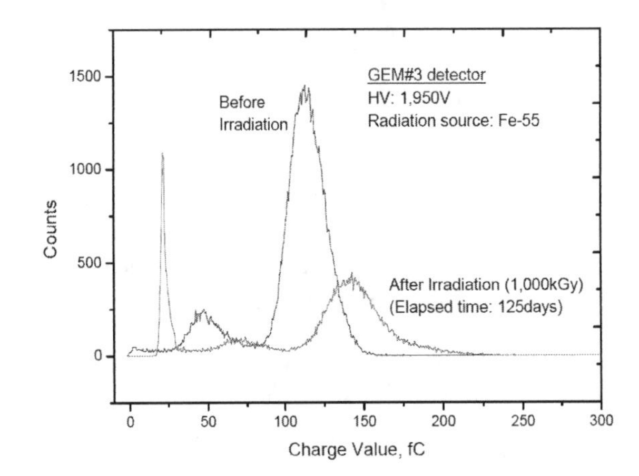

FIGURE 2. ^{55}Fe Spectra of small D-GEM detector before (black line) and after (red line) irradiation.

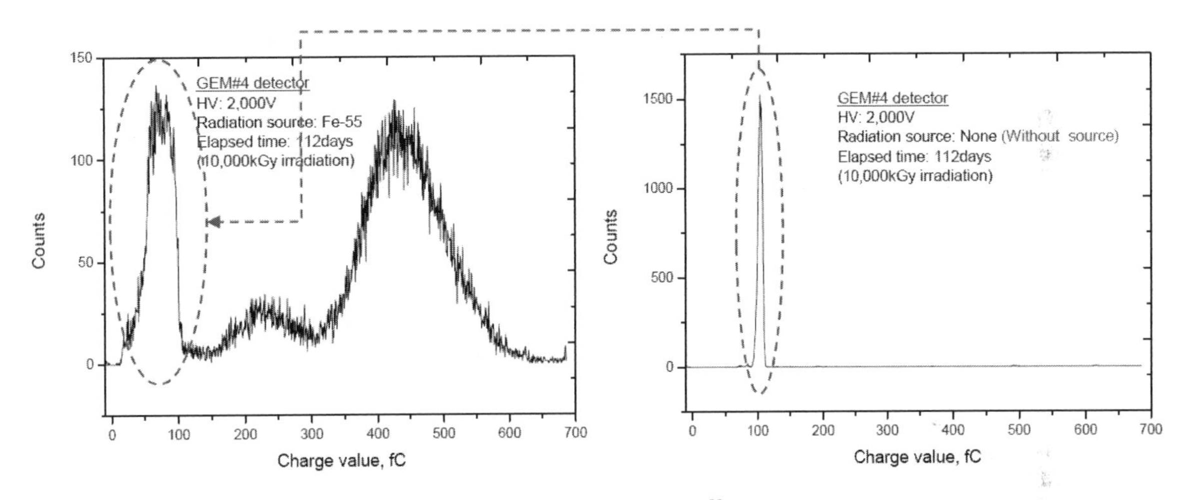

FIGURE 3. Spectra of irradiated GEM#4-foiled detector measured with ^{55}Fe source (left) and without the radioactive source (right).

Tests Without Radiation Source

A set of tests were carried out to understand the cause of the peak distortion and gain variation in the irradiated GEM detectors. To determine where the additional electrons contributing to the detector signal come from, the output of GEM#4 was measured with and without the ^{55}Fe radioactive source. We observed that while the distorted two characteristic peaks disappear when ^{55}Fe source was removed, a sharp peak at 100fC still remained as shown in the plots in Fig. 3. Given the fact that the peak at 100fC was observed without the radiation source, we conclude that the electrons emitted from the irradiated GEM foils themselves are accelerated and cause avalanche as the normal ionization electrons. .These processes generate the underlying constant flow of additional ionization electrons and result in the peak distortion and gain variation. Generally, it has been reported that the radiation-induced free radicals are produced when polymer materials are exposed to a gamma radiation. In this study, it is confirmed experimentally that the additional electrons contributing to the electronic signals of the irradiated GEM detectors come from the free radicals in the Kapton films of irradiated GEM foils. Further studies in radiation effect to the material properties and more detailed studies on understanding the exact mechanism for radiation damage will be carried in the future.

CONCLUSIONS

To verify the effect of radiation exposure of the GEM foil to the detector performance, GEM foils were exposed to a gamma-ray source to the level of 10kGy-10,000kGy. The change of output signal shape of the irradiated GEM detectors were observed and evaluated experimentally. We observe that the signal peak shapes were distorted and the detector gains increased before and after the irradiation. From a test of chamber responses with and without ^{55}Fe radioactive source, we determined that the shape distortion and the gain increase were caused by the possible flow of constant electron emissions from GEM foils contributing to the signal from actual particles incident to the detector. More details of the mechanism on the peak distortion and gain variation will be examined through further investigation and material property analyses.

ACKNOWLEDGMENTS

We would like to acknowledge the support from the U.S. Department of Energy (DOE-FG02-96ER40943, LCRD), and from the University of Texas at Arlington.

REFERENCES

1. James H. O'Donnell, *Chemistry of Radiation Degradation of Polymers*, Radiation Effects on Polymers, ACS Symposium Series **475**, American Chemical Society, Washington DC, 1991, pp. 402-413.
2. Roger L. Clough and Shalaby W. Shalaby, *Degradation and Stabilization of Irradiated Polymers*, Radiation Effects on Polymers, ACS Symposium Series **475**, American Chemical Society, Washington DC, 1991, pp. 401.
3. J.W.T. Spinks and R.J. Woods, *An Introduction to Radiation Chemistry*, **3rd Ed.**, 1990, pp.187.
4. Jacqueline I. Kroschwitz et al., *Encyclopedia of Polymer Science and Engineering*, Vol.**13**, John Wiley & Sons, 1988, pp. 667-707.
5. F. Sauli, *GEM: A New Concept for Eelectron Amplification in Gas Detectors*, Nuclear Instruments and Methods in Physics Research **A 386** (1997) pp.531-534.
6. S. Goetz (private communication).
7. George Chabot, *Answer to Question #5855 Submitted to "Ask the Experts", Health Physics Society(HPS), 2006.*

SECTION XII – SAFETY, SECURITY, AND CONTRABAND DETECTION

Lessons Learned In Developing The VACIS™ Products

Victor J. Orphan

Science Applications International Corporation
7455 W. Washington Avenue, Suite 290, Las Vegas, NV 89428

Abstract. SAIC's development of VACIS™ provides useful "lessons learned" in bridging the gap from an idea to a security or contraband detection product. From a gamma densitometer idea for solving a specific Customs Service (CS) requirement (detection of drugs in near-empty tanker trucks) in mid-1990's, SAIC developed a broad line of vehicle and cargo inspections systems (over 500 systems deployed to date) based on a gamma-ray radiographic imaging technique. This paper analyzes the reasons for the successful development of VACIS and attempts to identify "lessons learned" useful for future security and contraband detection product developments.

Keywords: Radiography Systems, Cargo Inspection, Gamma-ray Inspection, Contraband Detection

INTRODUCTION

SAIC's development of VACIS™ provides useful "lessons learned" in bridging the gap from an idea to a security or contraband detection product. From a gamma densitometer idea for solving a specific Customs Service (CS) requirement (detection of drugs in near-empty tanker trucks) in the mid-1990's, SAIC developed a broad line of vehicle and cargo inspections systems (over 500 systems deployed to date) based on a gamma-ray radiographic imaging technique. This paper analyzes the reasons for the successful development of VACIS and attempts to identify "lessons learned" useful for future security and contraband detection product developments.

BACKGROUND ON VACIS™ DEVELOPMENT

SAIC initially developed its Vehicle and Cargo Inspection Systems (VACIS) technology in 1994 in response to an urgent requirement from the U.S. Customs Service (USCS) for a low cost, non-intrusive means of inspecting propane tanker trucks entering the U.S. from Mexico. Using a seized propane tanker truck in which 8,000 pounds of cocaine had been hidden, SAIC demonstrated the effectiveness of a gamma-ray transmission radiography technique suggested by Dr. Victor Verbinski. Based on this successful proof-of-concept study which showed that a drug stimulant only a few inches thick could be rapidly detected inside the tanker, the Office of National Drug Control Policy (ONDCP) funded the development of a prototype track-and-trolley based system, referred to as VACIS I with rather crude image resolution (2 inch pixels).

VACIS°I was evaluated at several southwest border port of entries. Even though it was designed for a limited test of a few months, the USCS inspectors found VACIS I so useful that the unit was in daily operation at the Santa Teresa port of entry until it was decommissioned early in 2002 and replaced with an improved (VACIS II) system. This early fielding of the first VACIS prototype provided valuable feedback which was quite valuable. For instance, CS suggested the image resolution be enhanced to allow broader use than tanker truck inspection. We reduced the size of the NaI detectors in the imaging array to provide an ~0.5 inch pixel size. CS ordered 30 VACIS II systems, also called Relocatable VACIS and deployed them at Ports of Entry (POEs) principally at land border crossings.

For seaport POEs CS required a mobile inspection capability since the space at seaports is limited. When it became clear that government funding of a mobile system was not available in a time frame consistent with the urgent requirement of CS, SAIC funded the development of Mobile VACIS in a 6 month effort and was rewarded with a CS order for 11 Mobile VACIS systems in 2000. Mobile VACIS has proven to be the most popular of the six VACIS systems (over 200 systems deployed). We recommended to CS that the Cs-137 source used in VACIS II be replaced with a Co-60 source in Mobile VACIS for improved

AIP Conf. Proc. 1336, 683-685 (2011); doi: 10.1063/1.3586189

penetration. We obtained CS concurrence by performing a side-by-side evaluation of contraband detection capabilities of each source using realistic cargo and CS inspectors to interpret the images.

The Department of Defense also bought Mobile VACIS systems for force protection. In some theater locations the military needed a more robust vehicle than the commercial truck used for Mobile VACIS. A Military Mobile VACIS (MMV) prototype was developed in which the detector array was mounted on a HMMWV and the source was on a small "out-vehicle" designed to track the movement of the detector array on the HMMWV. To address a concern that the VACIS source holder might be breached if the truck being inspected exploded, we subjected the source holder to an explosion of 10,000 pounds of TNT which demonstrated that the source holder was able to contain the source. Successful evaluation of the MMV prototype in Kosovo and later in Afghanistan led to the Army ordering 75 MMV systems.

The simplicity, reliability, and low cost of our gamma-transmission radiography technology enabled extremely rapid development of custom systems tailored to other specific inspection requirements (see Figure 1). Railroad VACIS was designed to address CS need for a means to perform 100% inspection of cargo railcars entering the US from Mexico and Canada without impeding commerce. Railroad VACIS scans at speeds up to 10 mph and has a detector array adequate for inspecting double deck railcars. Images from a performance demonstration of a Railroad VACIS prototype showed that, as CS expected, illegal drug contraband but, surprisingly, a large number of people. In fact, the smuggling operation of people was so well organized that it appeared to offer two classes of service: first class in SUVs being imported and economy class in hopper cars.

Portal VACIS was developed to satisfy need for an inspection system with a standard traffic lane footprint. It featured two sets of source-detector array pairs; one for imaging the upper part of the container or truck and the other the lower section.

The Pallet VACIS addressed CS need for a means of inspecting higher density cargo pallets which could not be penetrated by conventional pallet x-ray systems. This VACIS, which required no additional shielding, used a horizontal detector array which scanned the pallet in the vertical direction and minimized the footprint of the system. The system featured higher resolution (3/8 inch pixels) to accommodate CS input during our development that smugglers had started to use smaller packages of drugs hidden in cargo pallets.

In response to a requirement by DHS/DNDO for a means of detecting high-Z materials (indicative of nuclear materials) in cargo at high scanning speeds, SAIC developed a dual high energy x-ray system using a unique interlaced 6 and 9 MeV Linac. The technique was successfully demonstrated with a breadboard system, VACIS-Z, and a prototype was recently completed under the CAARS program. In support of the VACIS-Z development, Dr. Rex Richardson developed an innovative Cherenkov detector array which provided significant advantages over scintillation detectors: Inherent threshold energy which reduces scattered x-rays enhancing contrast sensitivity, more affordable detector array and wider dynamic range. SAIC's CAARS system fully meet DNDO's ambitious requirements such as scanning at 33 inches/sec, contrast sensitivity of 1%, spatial resolution of 0.28 inches, penetration of 16 inches of steel, and automated detection of 100 cc of high Z-material shielded by 10 inches of steel.

LESSONS LEARNED

Lessons learned during the development of the family of VACIS products are as follows:

1. Ensure that the idea addresses an important real-world requirement. In the case of VACIS, more cost-effective means of detecting contraband in cargo was a high priority which provided "pull" for the VACIS technology
2. Invest in understanding customer requirements and user CONOPS (Concept of Operations), especially constraints on CONOPS of implementing proposed system. For example, CS was initially quite concerned about the safety of implementing a system with a 1 Ci Cs-137 source in a port environment. We addressed radiation safety concerns by implementing safety features such as a fast acting shutter on the source which would close if the railcar stopped during a scan by Railroad VACIS and by radiation safety training.
3. Make sure that the proposed product offers significant advantages over conventional inspection approaches—VACIS provided inherent enhanced reliability compared to x-ray systems (availability from 95 to 99%), rapid setup, faster scanning and no need for expensive infrastructure.
4. Performance metrics should be user-oriented. For instance, we arranged experiment allowing CS inspectors to evaluate ability to detect contraband using Cs-137 compared to Co-60 using realistic cargo.
5. Obtain user feedback as early and as often as possible during the development. VACIS I prototype use at several POEs on the southern border led to design of VACIS II with improvements (such as higher spatial resolution) desired by CS.

6. Work with user to identify related applications. With VACIS we explored its application to a wide-range of CS inspection requirements leading to a family of VACIS systems for trucks, cargo containers, railcars, and pallets.
7. Invest in developing the infrastructure (manufacturing capability, training and maintenance) required to support the product. Even a good product will ultimately fail if it is not effectively manufactured and maintained.
8. Identify as broad a range of customers as possible and invest in product modifications, when necessary, to address identified new applications. Modification of Mobile VACIS to suit military Force Protection requirements resulted in the successful MMV product.
9. Pursue government research and development funding; but be prepared to invest private funds when government funding is not available or is slow to materialize. SAIC funded the development of Mobile VACIS which became the most successful of the VACIS products. Results of government-sponsored R&D can be leveraged into new or improved products. When the government funds the development it tends to "buy in" to the technical solution. Some private funding can also help ensure intellectual property protection.

Combined Photoneutron And X Ray Interrogation Of Containers For Nuclear Materials

Tsahi Gozani, Timothy Shaw, Michael J. King, John Stevenson, Mashal Elsalim, Craig Brown and Cathie Condron

Rapiscan Laboratories, Inc., 520 Almanor Ave., Sunnyvale, CA 94085, USA

Abstract. Effective cargo inspection systems for nuclear material detection require good penetration by the interrogating radiation, generation of a sufficient number of fissions, and strong and penetrating detection signatures. Inspection systems need also to be sensitive over a wide range of cargo types and densities encountered in daily commerce. Thus they need to be effective with highly hydrogenous cargo, where neutron attenuation is a major limitation, as well as with dense metallic cargo, where x-ray penetration is low. A system that interrogates cargo with both neutrons and x-rays can, in principle, achieve high performance over the widest range of cargos. Moreover, utilizing strong prompt-neutron (~3 per fission) and delayed-gamma ray (~7 per fission) signatures further strengthens the detection sensitivity across all cargo types. The complementary nature of x-rays and neutrons, used as both probing radiation and detection signatures, alleviates the need to employ exceedingly strong sources, which would otherwise be required to achieve adequate performance across all cargo types, if only one type of radiation probe were employed. A system based on the above principles, employing a commercially-available 9 MV linac was developed and designed. Neutrons are produced simultaneously with x-rays by the photonuclear interaction of the x-ray beam with a suitable converter. A total neutron yield on the order of 10^{11} n/s is achieved with an average electron beam current of 100 μA. If fissionable material is present, fissions are produced both by the high-energy x-ray beam and by the photoneutrons. Photofission and neutron fission dominate in hydrogenous and metallic cargos, respectively. Neutron-capture gamma rays provide information on the cargo composition. The prompt neutrons resulting from fission are detected by two independent detector systems: by very efficient Differential Die Away Analysis (DDAA) detectors, and by direct detection of neutrons with energies higher than 3 MeV using a recently developed fluorine-based threshold activation detector (TAD). The delayed gamma-ray signals are measured with high efficiency with the same TAD and with additional lower-cost plastic scintillators.

Keywords: Nuclear Material Detection; Photofission; Neutron Fission, Photoneutrons; Prompt Neutrons; Threshold Activation Detection; Delayed Gamma Rays; Differential Die Away Analysis
PACS: 25.20.-x, 25.85.Ec, 25.85.Jg, 29.25.Dz

INTRODUCTION

The detection of illicit fissile materials, such as ^{235}U and ^{239}Pu that could be smuggled in various conveyances is critical to national and international security. Great efforts are being made to develop reliable means to detect these materials. Paramount among them is active interrogation of conveyances using fast neutrons or high energy x-rays. These particles stimulate fission in all nuclear materials. The fission process is rich in detectable signatures [1]. The primary signatures are prompt neutrons and gamma rays and delayed neutrons and gamma rays (see Table 1). The delayed gamma-ray signature and the much smaller delayed neutron signature are the most commonly used fission indicators. The signatures employed in the current system are indicated in the last column of Table 1.

The prompt neutron and gamma-ray signatures are strong, but they are created during the time the much stronger stimulating x-ray and neutron radiations are present. This fact makes their detection quite difficult. In fact the detection of prompt gamma rays is currently not feasible.

Application of Accelerators in Research and Industry
AIP Conf. Proc. 1336, 686-692 (2011); doi: 10.1063/1.3586190

TABLE 1. Approximate strength of fission signatures (in units of particles per fission).

Signature Name	Isotope Signature	^{235}U	^{239}Pu	^{238}U	Used in Current System
Prompt neutrons	ν_p from (n,f) $E_n \leq 2.5$ MeV or (γ,f) E_γ<9MeV	2.8	3.2	2.9	-
	ν_p with $E_n \geq 3$ MeV	0.6	0.7	0..6	yes
Prompt gamma rays	γ_p	6.7	6.7	7.2	no
Delayed neutrons	n_d from (n,f)	0.015	0.0061	0.044	yes
	n_d from (γ,f)	0.01	0.004	0.028	yes
Delayed gamma rays	γ_d(total)	6.7	6.7	7.2	-
	γ_d w/ $E\gamma \geq 2$ MeV	0.5	0.3	0.5	yes

Effective cargo inspection systems for nuclear material detection require good penetration of the interrogating radiation, generation of a sufficient number of fissions, and strong and penetrating detection signatures. Furthermore, inspection systems need to be sensitive over a wide range of cargo types and densities encountered in daily commerce. Thus they need to be effective in highly-hydrogenous cargo, where neutron attenuation is a major limitation, as well as in dense metallic cargo where x-ray penetration is low.

A system that interrogates cargo simultaneously with neutrons and x-rays can, in principle, achieve high performance over the widest range of cargo types. Moreover, utilizing the strong prompt neutron (~3 neutrons per fission) and delayed-gamma ray (~7 gamma rays per fission) signatures further strengthens the detection sensitivity across all cargo types. The complementary nature of x-rays and neutrons, used as both probing radiation and detection signatures, alleviates the need to employ exceedingly strong sources, which are otherwise required to provide adequate performance when only one probing particle type is used with the more difficult-to-penetrate cargo type.

PHYSICS PRINCIPLES OF THE SYSTEM

The detection sensitivity of the system, for a given source intensity, is determined by the level of cargo penetration achieved by the interrogating radiation, the fission cross section of the probing radiation, the penetration capability of the resulting fission signatures and the detection efficiency of the various detectors. Each one of these factors was simulated and experimentally validated in a small scale laboratory system.

Neutron And Photon Penetration

Neutrons and photons have very different attenuation characteristics. Neutrons are highly attenuated by low-Z materials, especially hydrogen. X-rays, on the other hand, are absorbed more by dense metallic cargoes, which have a higher Z than hydrogenous materials. As an example, the total attenuation coefficients (in units of cm^2/g) of photons and neutrons versus their energy are plotted in Figure 1 for water and iron.

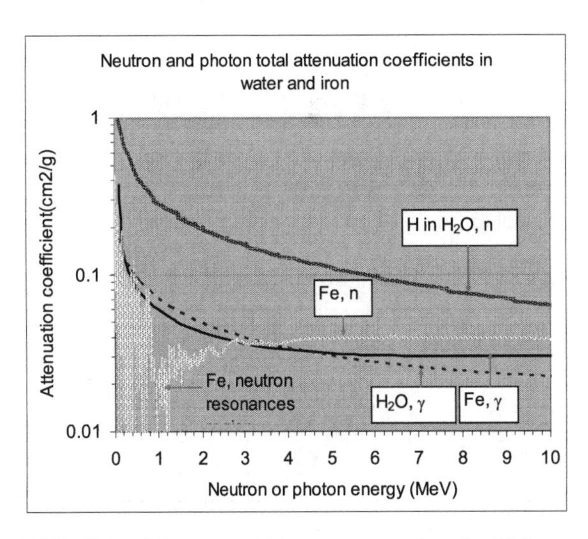

FIGURE 1. Neutron and Photon Attenuation Coefficients in Water and Iron vs. Energy.

The attenuation coefficient for photons in water (designated as "H_2O,γ") is significantly lower, by more than a factor of 3, than for neutrons. Furthermore, in the case of elastic scattering in hydrogen the neutron can lose all its energy. On average it loses half of its energy and then is prone to be absorbed in the cargo. The curve designated as "H in H_2O,n" is actually the attenuation coefficient for the hydrogen content of water. Scattering off oxygen, which is important for higher energies, further increases the attenuation of neutrons in water. The photon and neutron attenuation in iron ("Fe,γ" and "Fe,n", respectively) have seemingly similar behaviors, especially for particle energies above 2 MeV. Iron has many neutron scattering resonances in the epithermal region (~10 - 100 keV). But there are large fundamental differences between photon and neutron interactions with iron. The neutron, while changing direction, loses only a small fraction of its energy in each collision with iron nuclei and is otherwise available for interaction with fissile material, if present. However when a high-energy photon (near the photofission threshold) Compton-scatters even by a modest angle, (e.g., 10 degrees) it loses a substantial amount of energy, considerably reducing its ability to induce fission in the nuclear material. At the low- and high-energy ends, iron attenuates x-rays more than hydrogenous materials because of the photoelectric and pair-production effects, respectively.

Neutron And Photon Fission Cross Sections

The probing particles that penetrate the cargo to reach the nuclear material induce fissions in accordance with the corresponding fission cross sections. The fission cross sections relevant to our system (Fig. 2) span six orders of magnitude, from about 600 barns for thermal neutron fission (detected by the DDAA detectors) to epi-thermal and fast neutron fission, from 300 to 1 barn cross section, respectively, to the tens of a millibarn cross section for photofission. The signatures of the non-thermal neutron and photon fissions are efficiently detected by the TAD and plastic scintillators. It should be noted that the low photofission cross section is generally compensated by the copious amount of x-rays produced by a 9 MV linac. Three energy ranges are highlighted in Fig. 2: thermal, epi-thermal and fast. The DDAA works in the thermal region, whereas the direct fission detection used in the current system is applicable over all neutron and photon energies where fission is possible.

PHOTO-NEUTRON/X-RAY INSPECTION SYSTEM

System Overview

We have studied, developed, designed, and are building a system based on the above principles of dual cargo interrogation by high-energy x-rays and integrally produced photoneutrons. The system (see Fig. 3) uses a commercially-available 9 MV linac. Neutrons are produced simultaneously with x-rays by the photonuclear interaction of the x-ray beam with a suitable D_2O converter built around the x-ray source, which uses a tungsten target.

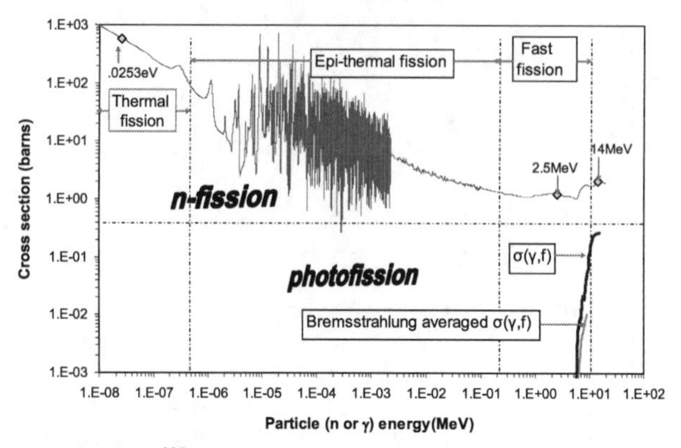

FIGURE 2. ^{235}U Fission Cross Section of Neutrons and Photons Employed in the System. The cross sections of ^{239}Pu have a similar shape but are generally higher.

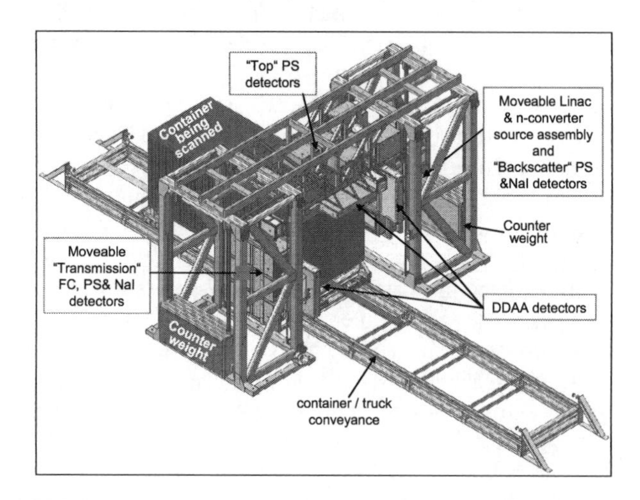

FIGURE 3. Scaled schematic Diagram of Combined Photo-neutron/ X-ray Inspection System.

A cargo container (or truck) is moved through a portal. The x-ray source is pulsed typically at 30 to 300 pulses per second and produces high-energy x-rays. A photoneutron beam is generated by the same x-rays impinging on a heavy water converter. If fissionable material is present in the cargo, fissions are produced by the high-energy x-ray beam as well as by the photoneutrons. Typically the photofission process dominates in hydrogenous cargos while the neutron fission process dominates more in dense metallic cargos.

Detectors

Cylindrical NaI (15.2 cm diameter x 15.2 cm long) scintillators detect, between beam pulses, the thermal-neutron-capture gamma rays which are characteristic to the cargo elemental composition. They provide information on the average cargo composition [2], which in turn allows for an automatic adjustment of various operational parameters, including the scanning time. The fission prompt neutrons are detected by two independent detectors: high efficiency Differential Die Away Analysis (DDAA) [3] detectors (see also Fig. 4a), and by the direct detection of neutrons with energies higher than 3 MeV using the fluorine-based threshold activation detector (TAD) [4]. The relatively abundant delayed gamma rays are measured with the same high efficiency TAD and with additional lower-cost plastic scintillators of the same size (Fig. 4b). All of these measurements are taken during the time interval (10-30 ms) between the linac pulses in order to avoid the radiation produced by the source. Typical pulse-height distributions of these detectors due to photofissions and photoneutron fissions, measured in the laboratory, are shown in Fig. 5.

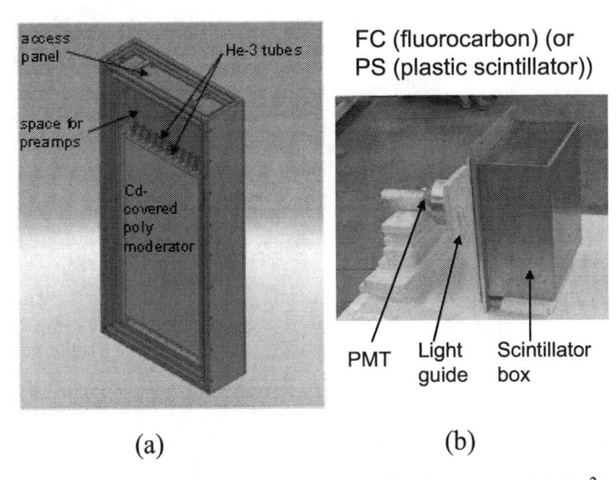

(a) (b)

FIGURE 4. (a) DDAA (with an effective area of $0.5m^2$) and (b) FC or PS Detectors ($40X40X20cm^3$, with a 12.7cm diameter PMT).

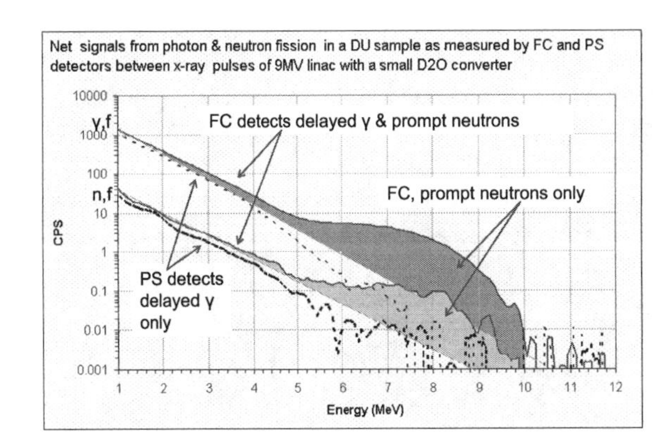

FIGURE 5. Pulse Height Distributions of FC and PS Detectors due to Fissions in a Deleted Uranium (DU) Sample in Air (see also [4]). The solid lines represent the total signal measured in the FC detector. The shaded area is the net signal attributed to the contribution of the prompt neutrons via the $^{19}F(n,\alpha)^{16}N$ reaction. The contribution of the delayed gamma ray signal to the high energy spectrum, above 5.5 MeV is very small in the FC detector. The broken lines are the spectra measured in the PS detector due to the delayed gamma rays alone (after subtraction of the prompt neutron (n,α) contribution). Natural background is subtracted from all spectra.

Interrogation Sources

The interrogation source employed in the system consists of a commercial 9 MeV linac [Varian model L2000] with a built-in heavy water converter, embedded within a compact composite shielding. The role of the shielding is to reduce the dose rate in all directions, except in the irradiation zone, and to allow operation of the detectors (see Fig. 3) in proximity of the source. The photoneutron source, described in another paper presented at this conference [5], was optimized to provide the maximum yield of the desired spectrum while minimizing the weight and size of the source assembly as much as possible. A schematic diagram of the source assembly is shown in Fig. 6.

Expected System Performances

Extensive laboratory experiments, using an available 9 MV linac and various types and sizes of detectors, were conducted to establish the validity of the various techniques to be employed in the combined photoneutron and x-ray cargo interrogation system. The experimental configurations were simulated using MCNP [6,7]. Good agreement with the various net signals was obtained for delayed γ rays and prompt neutrons from both fast-neutron and x-ray stimulated fissions. The prompt neutron signal from thermal neutron fission i.e., DDAA signal strength, was validated earlier in our SNMDS program [3]. The validated simulation approach was then used to guide the design and predict the net signal count rates in the final system as designed.

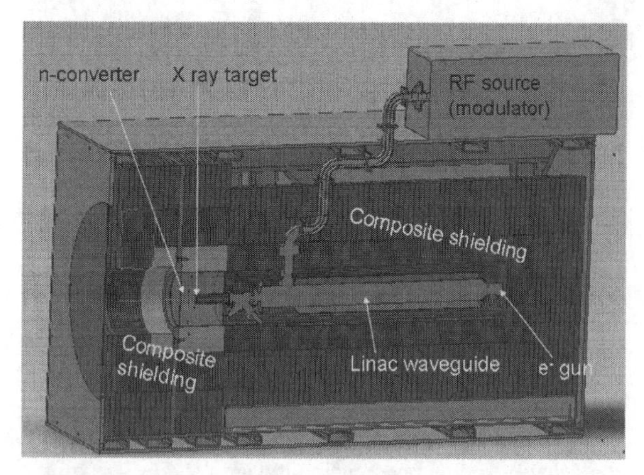

FIGURE 6. Combined X-ray/Photoneutron Source Assembly. (the external linac source box, excluding the RF source, has the following dimensions: 137 cm x 137 cm x 241 cm).

The beneficial complementarity between neutron fission and photofission is exemplified by the various fission rates (normalized to unity in air) stimulated in a ^{235}U sample at the center of homogenous cargos made of highly hydrogenous (water) or highly metallic (iron) cargoes, vs. the cargo density (Fig. 7). The cargo is assumed to fully fill a standard 20′ container. The densities covered in the simulation range from air to liquid water (up to 240 g/cm^2) and iron (up to ≈550 g/cm2). This range arguably exceeds the densities likely to be encountered in real cargo, so the design is likely to be conservative.

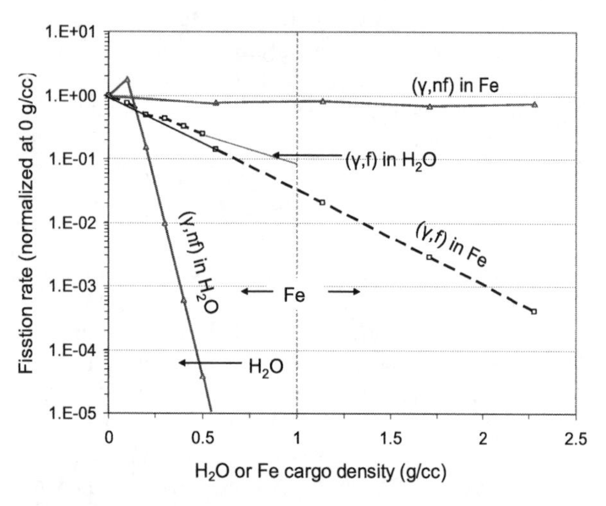

FIGURE 7. Normalized Fission Rate vs. Density of Homogenous Water and Iron Cargos.

The photofission (γ,f) rate decreases modestly with the increasing density. It is almost independent of the cargo material, consistent with the dominance of Compton scattering for most of the contributing x-rays. The decrease in water is slightly smaller than in iron, consistent with the relatively weak dependence of the x-ray attenuation coefficient on the atomic number Z between water and iron. The fission rate due to the photoneutron (γn,f) is dramatically different. It precipitously decreases in water beyond a density of 0.15 - 0.2 g/cc, while it hardly changes over the entire density range of iron.

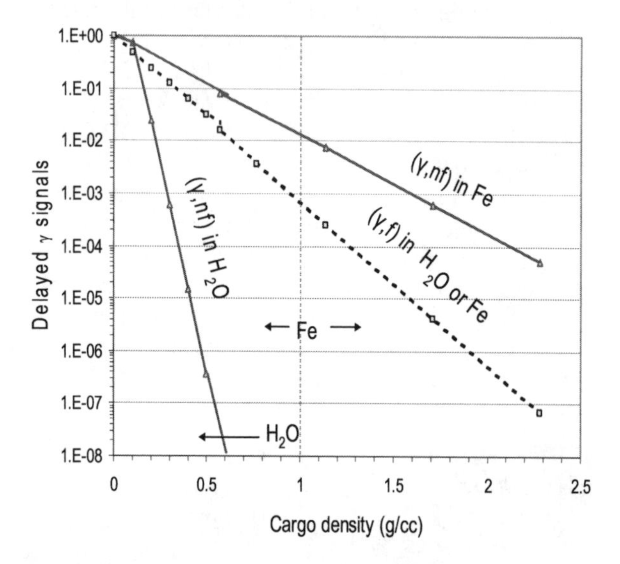

FIGURE 8(a). Normalized Signals of Delayed Gamma Rays vs. Density of Homogenous Water and Iron.

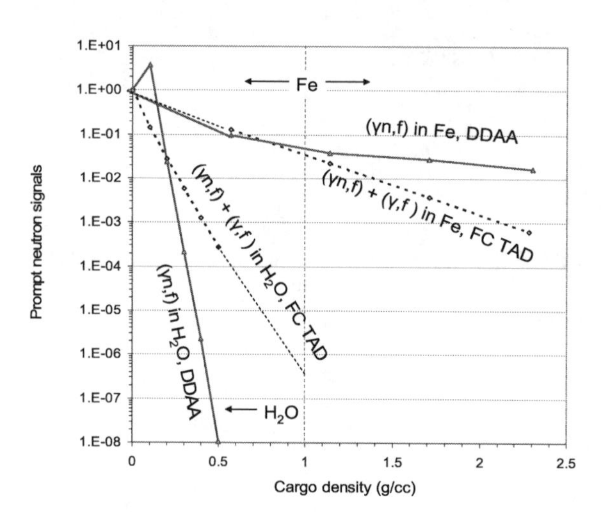

FIGURE 8(b). Prompt Neutrons vs. Density of Homogenous Water and Iron Cargos.

TABLE 2. Count rate (cps/detector/g ^{235}U) averaged over the plane of inspection for 100μA 9 MV linac.

Fission Signature	Delayed Gamma Rays Measured with PS+FC	Prompt Neutrons Measured Via (n, a) in FC TAD	Prompt Neutrons Measured with DDAA
CPS per detector per gram ^{235}U averaged over container inspection plane	3400	310	160

The normalized fission signals recorded by the various detectors surrounding the container (see Fig. 3) as a function of density are shown in Fig. 8. The curves represent the convolution of the fission rates and the probability of the specific fission signature to penetrate the cargo and deposit directly, or via secondary nuclear interaction, enough energy to be counted. The dominance of neutron fission in iron and photofission in water is amply manifested by the delayed γ-ray signal. The same can be said for the prompt neutron signal due to both types of fission, but here the DDAA signal dominates in the denser (>1 g/cc) iron cargos. The absolute signal averaged over the plane of inspection per detector per gram of a small ^{235}U sample in air is given in Table 2 for a 9-MV linac operating at a 100 μA average current. The x-ray dose rate at this current is 2260 rad/min at one meter away from the x ray source. The dose from the neutrons is about 3 orders of magnitude lower. The "small sample" designation means that no neutron multiplication nor x-ray or any signature self shielding in the sample are necessary. However, for larger samples these factors are important and need to be accounted for. The absolute values of the signal per gram and the large difference between them, seen in the table greatly decline, as the sample becomes larger and is embedded in cargos of different type and densities, as seen in Figs 7 and 8. All signals, i.e., prompt neutrons and delayed gamma rays contribute to the success of the concept over wide range of cargos.

The final system performance obviously depends not only on the signal strength but also on the magnitude of the background. Several sources of backgrounds can be identified: thermal neutron capture gamma rays, cargo material activation (with seconds' decay time) and the natural background. The first one decays fast relative to the signal and can be eliminated. The second one is generally low because the neutron spectrum is relatively soft, eliminating the high energy neutron activation, which typically generates the high energy gamma rays. The natural background, mostly from cosmic rays depositing more than 2.5 MeV energy, is the ultimate source of background. Based on the calculated signal count rates, the combined system as described here, employing multiple detectors of the types described, can rapidly determine if a threat of a much smaller mass than a nuclear device is present in the cargo.

CONCLUSION

The dual-species photoneutron/x-ray based interrogation system efficiently combines the penetrability of probing photons in hydrogenous materials with that of neutrons in metallic cargos. The photofission rate in highly-hydrogenous (e.g., food) cargo is very high using a commercial 9 MV linac. It is indeed higher than what available intense (d,T) or various energy (d,D) neutron sources can provide.

The system beneficially combines the high penetrability of delayed γ-ray fission signature in hydrogenous cargo with the high penetrability of the high-energy fission prompt neutron signature in metallic cargos, yielding strong signals from a standard commercial 9 MV linac. These signatures are amongst the most abundant in the fission process. The prompt neutrons are detected by two methods: high efficiency DDAA and the novel high energy threshold FC detector. Delayed γ rays are detected by the latter detectors as well as by low cost plastic scintillators of the same size.

The same, relatively compact source (linac) is employed to generate both the probing neutrons and x rays. This source is less expensive than that which would otherwise be required if it were a neutron source alone, such as a high energy accelerator generating neutrons via the prolific reactions of (d,D) or (d,Be).

The ability of the source to scan vertically (see Fig. 3) assures that the path length in the cargo of the probing radiations and induced signatures are minimized, maximizing the detection sensitivity regardless of the threat position.

The result is that high detection sensitivity is attained in all types of cargo, higher than what can be achieved by systems that are based on neutron-only (independent of their energy) or photon-only probes.

ACKNOWLEDGMENTS

Research was funded under contract HSHQDC-08-C-00125 from the U.S. Department of Homeland Security, Transformational and Applied Research Directorate (TARD) Domestic Nuclear Detection Office (DNDO).

REFERENCES

1. T. Gozani, "Fission Signatures for Nuclear Material Detection," *IEEE Trans NS,* **56 3**, 726 (June 2009).
2. R. M. Yee, Timothy Shaw and Tsahi Gozani, "Thermal Neutron Die–Away Studies in a 14 Mev Neutron Based Active Interrogation System," *IEEE Trans. Nuclear Science*, **56 3** (June 2009).
3. K. A. Jordan and T. Gozani, "Pulsed Neutron Differential Die Away Analysis for Detection of Nuclear Materials," *NIM B,* **261,** 365-368 (2007).
4. T. Gozani, J. Stevenson, M. J. King, "Neutron threshold activation detectors (TAD) for the detection of fissions," NIM A, SORMA Conference Proceedings 2010, Ann Arbor, MI, to be published.
5. T. Shaw, M. J. King and T. Gozani, "Intense Photoneutron Sources for Nuclear Material Detection," American Institute of Physics Conference Proceedings Series 2010, CAARI 2010 Proceedings, Dallas Ft. Worth, TX.
6. M. King, T. Shaw, J. Stevenson, M. Elsalim, C. Brown, C. Condron and T. Gozani, "Simulation Methods for Photoneutron-Based Active Interrogation Systems," Nucl. Instr. and Meth. A, Conference Proceedings, SORMA 2010 Proceedings, Ann Arbor, MI.
7. M. King, T. Gozani, J. Stevenson and T. Shaw, "Simulation of a Photofission-Based Cargo Interrogation System," American Institute of Physics Conference Proceedings Series 2010, CAARI 2010 Proceedings, Dallas Ft. Worth, TX.

The Field Use Of The EURITRACK Tagged Neutron Inspection System

G. Viesti [a,b], D. Fabris [b], M. Lunardon [a,b], S. Moretto [a,b], G. Nebbia [b], S. Pesente [a,b]

[a] *Dipartimento di Fisica dell', Università di Padova, Via Marzolo 8, I-35131 Padova, Italy*
[b] *INFN Sezione di Padova, Via Marzolo 8, I-35131 Padova, Italy*

Abstract. The EURITRACK portal makes use of the tagged neutron technique to inspect suspect regions of cargo containers identified in preliminary x-ray inspections. The portal has been in operation at the Rijeka seaport during the last three years. The performances of sub-systems, with particular interest on the sealed tube neutron generator, is presented.

Keywords: EURITRACK, associated particle technique, fast neutron inspection
PACS: 28.20.-V, 29.30.Kv, 89.20.Bb, 89.40.-Cc

INTRODUCTION

Non-intrusive inspection of cargo containers has become a key issue in the Homeland Security programs during the last decade, with a continuous search of new technologies to improve speed and security of Customs operations. Gamma-rays following fast neutron induced reactions are used to characterize the presence in the sample of nuclei such as carbon, oxygen and nitrogen, which are the major components of explosives or narcotics. Moreover, the use of a tagged neutron generator allows one to inspect specific regions of the container [1].

The Tagged Neutron Inspection System (TNIS) concept has been successfully adopted in the EURITRACK project [2] to develop a portal for cargo containers to be used after X-ray scanners. The X-ray image determines the suspect region(s) inside the cargo container to be probed with neutrons, thus optimizing the use of such technique. The EURITRACK project started in 2004 and the portal was installed in 2007 at the Rijeka seaport in Croatia for a first demonstration. Since then, the portal has been operated (non continuously) during the last three years, building a database of more than 150 inspected containers [3,4].

From the hardware point of view, the EURITRACK portal is based on few major components: the EADS SODERN sealed-tube neutron generator, which includes a 8×8 matrix of YAP:Ce alpha particle detectors coupled to a multi-anode photomultiplier, a set of 22 high-efficiency 5"×5" and 5"×5"×10" NaI(Tl) detectors which are located around the cargo container and a dedicated CAEN VME front-end electronics to build coincidences between any alpha and gamma-ray detectors and measure their relative time delays [2,5].

The partners of the EURITRACK project developed and tested the TNIS hardware components during 2005. Since then, components were used in laboratory tests (Zagreb,Croatia), in the integration studies (Saclay, France) and then in the demonstration campaigns in Rijeka.

We present the hardware performance during the field measurements. Such performance study is aimed at determining the optimal use of the technique in future second generation inspection systems.

EXPERIMENTAL RESULTS FROM THE FIELD

During the operations at the Rijeka seaport, the majority (about 60%) of the inspected container were filled with organic materials having an average density (defined as the net load divided by the nominal volume) of about 0.13 g/cm^3 . Such figures are well in agreement with a recent statistical analysis of container manifests imported through US ports [6].

The SODERN neutron generator was used for about 720 hours. 390 hours were spent during the laboratory and integration tests with neutron fluxes up to the maximum allowed (2×10^8 neutron/s) to verify the count rate capability of the hardware components. During the field use in the Rijeka seaport (about 330

Application of Accelerators in Research and Industry
AIP Conf. Proc. 1336, 693-695 (2011); doi: 10.1063/1.3586191

hours), the neutron generator was operated at the nominal flux of 1-1.5 $\times 10^7$ neutron/s, this value being determined by the radiation hazard outside the available exclusion area. At this neutron flux, the standard measure time was 10 minutes for each suspect point. The flux of 1×10^7 neutron/s corresponds to a rate of 76 kHz of the associated alpha particle detector.

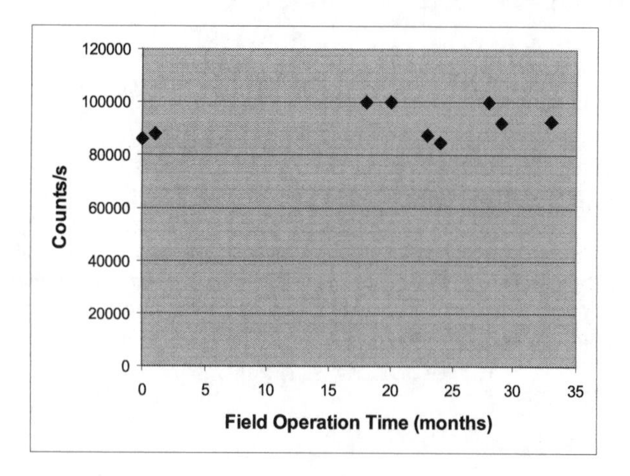

FIGURE 1. Average rate of the alpha particle detector during the field operation.

In Fig.1 the average alpha particle rate registered during the different demonstration campaigns is reported as a function of time. The results demonstrate that the tagged neutron source exhibited a good stability over the 3 years of field running. A more quantitative analysis based on the data of the 2008-9 campaigns yields the average value of 77 $\pm$ 5 kHz for the nominal flux of 1×10^7 neutron/s, very close to the design value.

However, as described in [7], the present alpha particle detector suffers from some cross talk between the 64 pixels because of the light dispersion inside the glass flange placed between the scintillation array inside the neutron generator and the PMT placed on air. The signal due to the alpha particle hitting a given pixel is normally higher than the cross talk light. Consequently, a careful tuning of the threshold of each alpha pixel discriminator is needed to cut the cross talk and minimize the possibility of multiple alpha events. Such events are indeed discarded in the data analysis since the alpha particle pulse height is not recorded, making impossible to discriminate off-line the cross talk signals and then to reconstruct the neutron path. The fraction of events with the alpha double hits is normally about 3% of the total, out of which 1/3 is due to random coincidences between the pixels and 2/3 to the cross talk. Such situation was not stable during the field operation, with some runs with a significantly larger number of rejected events due to multiple alpha hits.

In any case, it is worth noting that each YaP:Ce crystal was irradiated with about 3×10^9 alpha particles (corresponding to about 10^{10} particles/cm^2) without showing signs of radiation damage. Similarly, the HAMAMATSU H8500 is not showing degradation after processing about 2×10^{11} pulses.

The array of the 22 gamma-ray detector was characterized by an average total rate of 70 $\pm$ 17 kHz during the field operations, with values ranging from about 50 kHz to 120 kHz depending on the low energy thresholds (1.3 or 0.6 MeV) of the gamma-ray discriminators and the type of container. Taking into account that the counting rate of the individual detectors changes as a function of position relative to the neutron source, such values do not rise particular problems for NaI(Tl) scintillators equipped with fast PMTs.

The rate of coincidences between alpha particles and gamma rays was on average about 1.5 kHz, with peak values of about 4 kHz. The corresponding data flow is relatively low compared with the possibility of the VME front-end, for which a dead time of about 15% at 35 kHz was measured [5].

Looking at the field performance of individual sub-systems, the possibility of running a tagged neutron inspection system at significantly higher neutron flux (i.e. 10^8 neutron/s) seem to be feasible, once the radiation hazard problems are solved.

One distinctive feature of the TNIS is the reconstruction of the source position for the gamma rays produced inside the cargo, from the identified neutron flight path and the measured delay between the alpha particle and the gamma ray. In the EURITRACK software, the gamma source position is plotted in a range of about 7 m that includes the 14 MeV neutrons travel distance from the neutron generator to the far wall of the container and an additional region used to evaluate the random coincidence contribution. The latter information is needed to derive the net gamma-ray energy spectra by subtracting the random events. In such task, it is of primary importance to define the signal to noise ratio (S/N) for each time interval, were S are the true coincidences and N is the random background. Such (S/N) ratio is determined by the counting rate of the alpha and gamma-ray detectors and by the channel width in the time spectrum [8]. Operating the system at 1×10^7 neutron/s, a typical value is about S/N=0.8 for the total spectrum, which improves to S/N=1.6 for the window corresponding to the cargo container. This S/N ratio strongly deteriorates when increasing the neutron flux and therefore the detector count rate. This fact sets limitations that need to be carefully explored in developing new inspection systems.

TOWARDS A SECOND GENERATION TNIS

The experience obtained so far with the EURITRACK hardware components has been extremely positive taking into account that the system was operated well below the limiting performance of the subsystems.

On the other side, despite the low average density of the cargo container inspected, the quantitative analysis of the inspected material is generally obtained only in the first half of the container depth. This is due to scattering of the primary neutrons that makes impossible the determination of the correct neutron path [9]. Consequently the gamma ray detectors close to the far half of the container are not contributing in an efficient way to the global statistics.

In this respect a possible design for a second generation system might simply consist in maintaining the portal structure adding a second neutron generator on the far side were the so called transmission box is located. This solution has been studied in details during the Eritr@C project.

A second solution is represented by implementing for the container case the well known reflection geometry used in other neutron based inspection systems [10]. This solution is currently being studied within an Italian national program. In the latter case, a substantial change in the front-end electronics is foreseen, by using digital signal processing, to solve also the problem of the cross talk between alpha particle pixels when optical flanges are used inside the neutron generator.

ACKNOWLEDGMENTS

Tanks are due to all participants to EURITRACK (FP6-2003-IST-2 Proposal/Contract 511471). and Eritr@C (JLS/2007/ISEC/55030-CE-0179232/00-41) projects.

REFERENCES

1. G. Viesti et al., Nucl. Instr. Meth. B **241** 748-752 (2005).
2. B. Perot et al., Nucl. Instr. Meth. B **261** 295-298 (2007).
3. C. Carasco et al., Nucl. Instr. Meth. A **588** 397-405 (2008).
4. J. Obhodas et al., Nucl. Instr. Meth. A **619** 460-466 (2010).
5. M. Lunardon et al., Nucl. Instr. Meth. B **261** 391-395 (2007).
6. M.A. Descalle, D. Manatt, D. Slaughter, Nucl. Sci. Symp. Conf. Rec. IEEE 275–280 (2006).
7. S. Pesente et al., Nucl. Instr. Meth. B **261** 268-271 (2007).
8. H. Jeremie, Nucl. Instr. Meth. A **244** 587-588 (1986).
9. S. Pesente et al., Applied Radiation and Isotopes **65** 1322-1329 (2007).
10. T. Gozani and D. Strellis, Nucl. Instr. Meth. B **261** 311-315 (2007).

Intense Photoneutron Sources For Nuclear Material Detection

Tsahi Gozani, Timothy Shaw and Michael King

Rapiscan Laboratories, Inc., 520 Almanor Ave., Sunnyvale, CA 94085

Abstract. Intense neutron sources are essential for cargo inspection for a broad range of threats from explosives, to contraband, to nuclear materials and especially SNM (Special Nuclear Materials). To be effective over a wide range of cargo materials, in particular for hydrogenous cargo such as food, and to offer practical inspection times, the neutron source must be very strong, typically $>10^{10}$ neutrons per second. Unfortunately there are currently no reasonably compact and economical neutron generators with the required intensities. The insufficiency and inadequacy of intense neutron sources are especially conspicuous in the $\leq$2.5 MeV range (low voltage (d,D) generator). This energy range is needed if the strong signature of prompt fission neutrons ($\approx$3 per fission) is to be detected and discerned from the numerous source neutrons. The photonuclear reactions of x-rays from commercial linacs in appropriate converters can provide ample intensities of neutrons. These converters have a very low (γ,n) energy threshold: 1.67 MeV for beryllium and 2.23 MeV for deuterium. The intense x-ray beams provided by commercial x-ray systems, more than compensate for the relatively low (γ,n) cross-sections which are in the milli-barn range. The choice of converter material, the geometrical shape, dimensions and location relative to the x-ray source, determine the efficiency of the neutron conversion. For electron accelerators with less than 10 MeV, the preferred converters, Be and D_2O, are also very good neutron moderators. Thus, while increasing the converters' thickness leads to an increase in the overall neutron yield, this causes the softening of the neutron spectrum, which reduces the neutron penetration especially in hydrogenous cargos. Photoneutron sources can be optimized to meet specific needs such as maximum fission signals in various cargo materials of interest. Efficient photoneutron sources with different energy spectra were investigated. Conversion efficiency of more than 10-4 neutron per 9 MeV electron yielding neutron intensities of more than 10^{11} n/s for commercial 100 μAmp electron accelerators has been modeled and designed. The simulation was validated in laboratory experiments using small Be and D_2O converters.

Keywords: Photoneutron Production, Linacs, Neutron Sources, Nuclear Material Detection, Neutron Fission, Prompt Neutrons, Delayed Gamma Rays, Differential Die Away Analysis
PACS: 25.85.Ec, 29.20.Ej, 29.25.Dz

INTRODUCTION

Active interrogation methods for detecting nuclear materials in large objects such as truck and containers are essential components to the mission of intercepting attempts to smuggle nuclear materials across borders or into populated areas.

Inspection systems based on neutrons offer the ability not only to detect nuclear materials but also to other threats and contraband such as conventional military and improvised explosives, narcotics, dangerous chemicals, and to classify and characterize cargo materials. While fast neutrons are highly penetrating especially in metallic and most manufactured materials, they are highly attenuated in dense hydrogenous cargo such as food. To overcome this limitation and be effective over a wide range of commercial cargos, strong sources are required. Neutron intensities well above 10^{10} neutrons per second are highly desirable.

Sources with this intensity are available, in principle, by using high voltage (such as 4 to 6 MeV) accelerators of protons and deuterons. Employing the prolific nuclear reactions of (p,Li), (d,D) and (d,Be) intense sources with outputs between 10^9 to 10^{11} n/s with different energy spectra can be generated. These accelerators, some commercially available, are relatively large and expensive, typically \$1.5 to \$2.5 M.

The insufficiency and inadequacy of intense neutron sources are especially conspicuous in the $\leq$2.5 MeV neutron energy range provided by low voltage (d,D) generators. This energy range is needed mainly if the strong signature of prompt fission neutrons, where ~3 neutrons are emitted per fission, is to be

Application of Accelerators in Research and Industry
AIP Conf. Proc. 1336, 696-699 (2011); doi: 10.1063/1.3586192

detected and discerned from the numerous source neutrons. The lower energy of the interrogating neutrons is also desirable to reduce the cargo activation, which can constitute a background for the other very important fission signature, the delayed gamma rays.

The generation of neutrons through the photonuclear reactions of x-rays with an appropriate photonuclear converter has been used in the past for nuclear material safeguards [1, 2, 3]. It is shown here to be a viable and cost effective alternative to intense neutron source generators, emitting a $\leq$ 2.5 MeV energy region for modern cargo inspection applications.

BREMSSTRAHLUNG BASED PHOTONEUTRON PRODUCTION

The most common commercially available and reliable x-ray sources are electron linear accelerators (linacs) with electron energy at or below 9 MeV. In this energy range, materials that can efficiently generate a prolific neutron output through the (γ,n) reactions are solid beryllium and deuterium, in the form of heavy water. They have the lowest threshold energy among all stable isotopes, 1.76 MeV for beryllium and 2.23 MeV for deuterium. Their (γ,n) cross sections and the Bremsstrahlung spectra from electrons of 6 MeV and 9 MeV linacs impinging on a 1.4 mm tungsten target are shown in Fig.1.

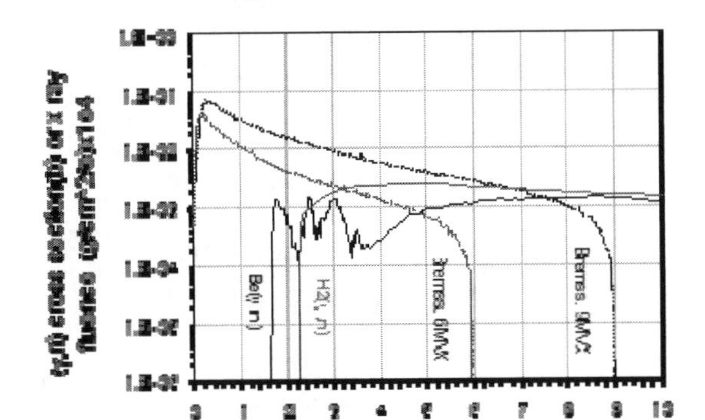

FIGURE 1. Photonuclear Cross Sections for Be and D, and 6 MV and 9 MV Bremsstrahlung Spectra.

The x-ray spectra show that a significant part of the spectra are above the (γ,n) thresholds of the Be and D_2O converter. The copious amount of x rays generated by commercial linacs more than compensates for the low, millibarn level, (γ,n) reaction cross sections. The maximum energy of the created neutrons is about equal to the amount of energy the x-ray exceeds the threshold of the photonuclear reaction of Be. In the case of the deuteron (in the D_2O), the neutron energy is half of the excess above the threshold, as the proton and the neutrons resulting from the (γ,n) reaction, equally share the energy. The ideal neutron spectra from thin Be and D converters are shown in Fig. 2.

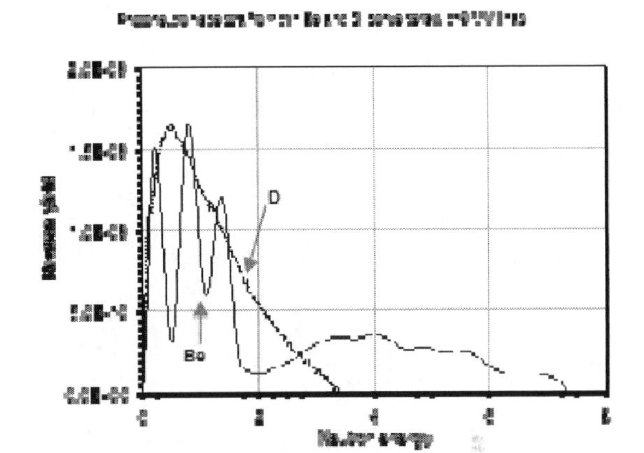

FIGURE 2. Thin Converter Neutron Spectra of 9 MV X Ray in Be and D.

Both D_2O and Be are very good neutron moderators. Thus the hardiest energy spectrum without neutron moderation and slowing down can be obtained only when the converters are very thin. The bulk of the neutrons are below 2 MeV with both converters. The main difference are the higher energy tail of the neutron spectrum from the Be , and to a lesser extent, the structure below 2 MeV. In some applications, such as those requiring the detection of prompt fission neutrons, the high energy tail of Be may be very deleterious, confusing fission neutrons from source neutrons. On the other hand, better penetration can be achieved with this type of hardier spectrum.

The low x-ray attenuation of Be and D_2O allows the use of a thick converters in order to increase the neutron yield generation. While the yield may increase, thicker Be and D_2O converters will soften the energy spectrum. The trade-off between neutron intensity and converter dimensions and thicknesses in particular, affords the optimization of the photoneutron source to a range of applications.

9 MEV LINAC BASED PHOTONEUTRON SOURCE

When the main objective of the source is to stimulate the detection of nuclear materials, it is desirable to maximize the fission rate at the center of a large cargo being inspected. We have selected a wide range of iron cargos with densities ranging from 140 to 400 g/cm^2 and investigated the induced fission rate, at the center of the cargo, as a function of converter thickness. The optimum thickness for both converters is about 8 to10 cm. The schematic design of such a compact photoneutron source, built around a Varian L2000 9 MeV linac, is shown in Fig. 3. A short, small-diameter evacuated drift tube, which terminates with a tungsten target, is appended to the linac to allow the maximum interception of the x rays with the converter material (D_2O, in this case). To reduce the radiation footprint of the source, the entire linac and converter are embedded in x-ray and neutron shielding, leaving the front open.

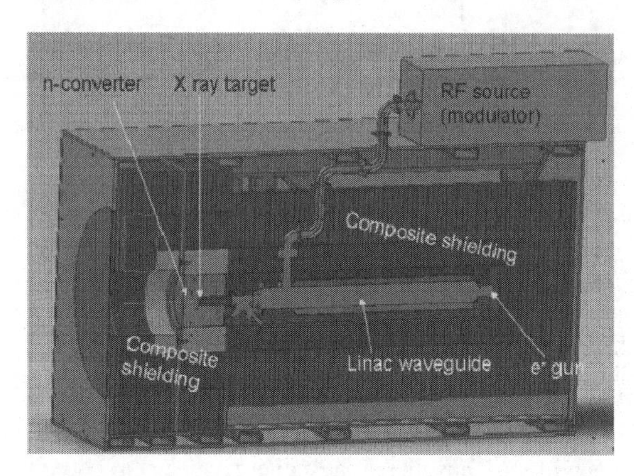

FIGURE 3. Compact Linac L2000 Based Photonuclear Neutron Source Assembly.

The calculated total neutron yield, which is integrated over all angles, is about 2x10^{-4} neutron per electron. Thus a linac operating at 9 MeV with an average current of 100 μA, is capable of generating in excess of 10^{11} neutrons per second. The spectra of neutrons generated by Be and D_2O converters and contained within cone of ±22° are shown in Fig. 4.

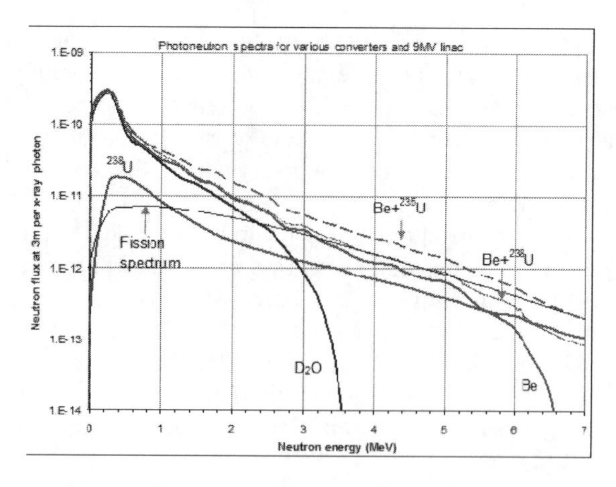

FIGURE 4. Photoneutron Spectra for 9 MV Bremsstrahlung, Be and D2O Converters and Spectrum Modifiers (^{235}U and ^{238}U).

In the energy region below 1 MeV the two spectra are similar. The main difference is in the high-energy region, above 3 MeV, where there are very few neutrons generated by the D_2O converter. This is advantageous, as it allows for the detection of fission neutrons, 20% of which have energies above 3 MeV. If one needs to increase the fraction of the high energy neutrons, a combined converter made of 1 cm ^{238}U followed by 8 cm of Be will yield about 35% more neutrons above 3 MeV (see Be/^{238}U curve in Fig. 4). An overall more efficient source is Be (or D_2O) encapsulated with a very thin (1 mm) of ^{235}U. The latter converts most of the thermal and low epi-thermal neutrons from the Be (or D_2O) to fission neutrons, increasing the total number of neutrons by about 30% and neutrons above 3 MeV by more than 200% (see Be/^{235}U curve in Fig. 4). A pure ^{238}U converter generates far less neutrons (see ^{238}U curve, Fig. 4). The spectrum of prompt fission neutrons is also shown in Fig. 4 for comparison to other spectra.

6 MEV LINAC BASED PHOTONEUTRON SOURCE

There are currently many more radiography systems based on the lower cost 6 MeV linacs. These lower energy linac fitted with Be or D_2O converters could deliver sizable neutron intensity, though with a much softer spectrum. Fig. 5 shows the total and above-1 MeV neutron intensities and the full energy spectra with the two converters. The soft neutron spectrum could be converted, at least in part, to a fission spectrum by encapsulating a large (γ,n) converter by a very thin (<1 mm) ^{235}U layer, which would be fissioned by all thermal and most of the low epithermal neutrons, similar to the Be/^{235}U curve in

Fig. 4. Because of the softness of the neutron spectrum here, the ^{235}U effect will be much stronger.

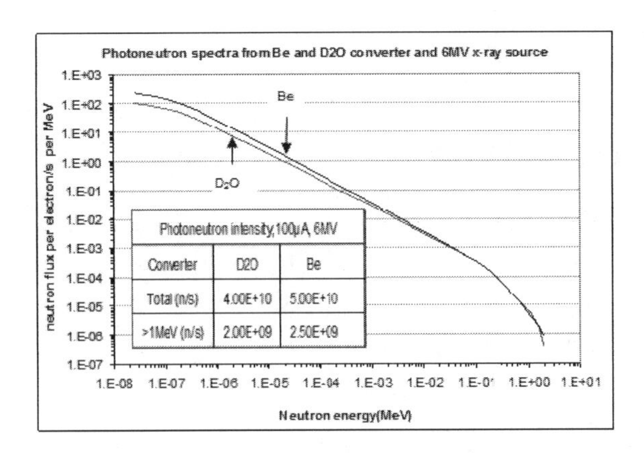

FIGURE 5. Photoneutron Spectrum and Total Yield of 6 MV Bremsstrahlung, and Be and D_2O converters.

CONCLUSION

An electron linac in the energy range of <10 MeV in conjunction with an appropriate (γ,n) converter, typically beryllium or heavy water, by themselves, or in combination with ^{235}U or ^{238}U provide an intense, fast neutron source with desirable spectral features. The intense x-ray beams provided by the linacs more than compensate for the relatively low millibarn level (γ,n) cross-sections. The choice of converter, size, shape, location relative to the x-ray target, as well as the use of spectral modifiers such as ^{235}U, determine the efficiency of neutron conversion and the spectrum.

Efficient photoneutron sources with different spectra were designed. Conversion efficiency in excess of 10^{-4} neutrons per 9 MeV electron, yielding neutron intensities of more than 10^{11} n/s for a commercial 100 μA linac, has been modeled and designed. The linac-based photoneutron source, properly designed for inducing fission of ^{235}U in a large cargo container, generates a much higher fission rate than can be achieved with the best currently available neutron sources (Fig. 6).

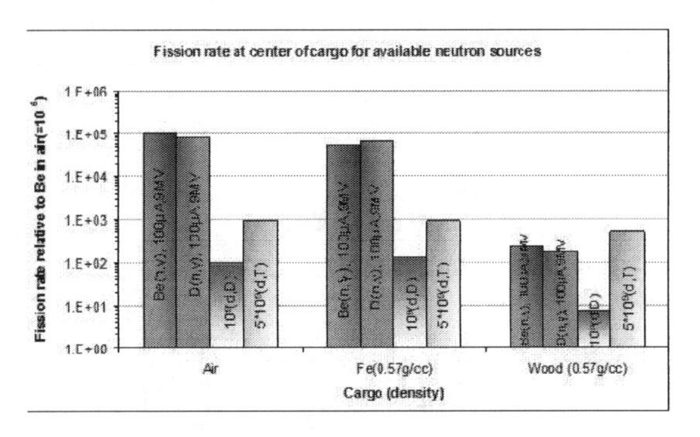

FIGURE 6. Neutron-stimulated fission rate at center of various uniform cargo for the proposed photoneutron sources and the strongest available compact neutron generators.

ACKNOWLEDGMENTS

The research was funded under contract HSHQDC-08-C-00125 from the U.S. Department of Homeland Security, Transformational and Applied Research Directorate (TARD) Domestic Nuclear Detection Office (DNDO).

REFERENCES

1. T. Gozani, D. E.Runquist, R. O. Ginaven and R. L Bramblett, "New Development in Nuclear Material Assay Using Photonuclear Reactions," IAEA Symposium on Nuclear Material Safeguards, "Safeguards Techniques," Vol. II, 143-163, 1970.
2. T. Gozani, "Neutron Production by Electron Accelerators," American Nuclear Society Transactions 19, 63 (1974).
3. T. Gozani, "Active Nondestructive Assay of Nuclear materials-Principles and Applications," U.S. Nuclear Regulatory Commission Book, NUREG/CR-0602, Chapter 4 (1981).
4. T. Gozani, T. Shaw, M. J. King, J. Stevenson, Ma. Elsalim, C. Brown and C. Condron, "Combined Photoneutrons and X-rays Detection of Nuclear Materials," American Institute of Physics Conference Proceedings Series 2010, CAARI 2010 Proceedings, Dallas Ft. Worth, TX.

Simulation Of A Photofission-Based Cargo Interrogation System

Michael King, Tsahi Gozani, John Stevenson and Timothy Shaw

Rapiscan Laboratories, Inc., 520 Almanor Avenue, Sunnyvale, CA 94085

Abstract. A comprehensive model has been developed to characterize and optimize the detection of Bremsstrahlung x-ray induced fission signatures from nuclear materials hidden in cargo containers. An effective active interrogation system should not only induce a large number of fission events but also efficiently detect their signatures. The proposed scanning system utilizes a 9-MV commercially available linear accelerator and the detection of strong fission signals i.e. delayed gamma rays and prompt neutrons. Because the scanning system is complex and the cargo containers are large and often highly attenuating, the simulation method segments the model into several physical steps, representing each change of radiation particle. Each approximation is carried-out separately, resulting in a major reduction in computational time and a significant improvement in tally statistics. The model investigates the effect on the fission rate and detection rate by various cargo types, densities and distributions. Hydrogenous and metallic cargos, homogeneous and heterogeneous, as well as various locations of the nuclear material inside the cargo container were studied. We will show that for the photofission-based interrogation system simulation, the final results are not only in good agreement with a full, single-step simulation but also with experimental results, further validating the full-system simulation.

Keywords: MCNPX, MCNP, Photofission, Active Interrogation
PACS: 07.05.Tp, 29.50.+v

INTRODUCTION

Active interrogation techniques have been developed to address the current deficiencies in air and seafaring cargo container inspection for hidden nuclear materials. These non-intrusive inspection systems utilize highly-penetrating probing radiations to induce fission and sensitive detectors to detect characteristic fission signatures. One approach that is capable of fulfilling the inspection needs is an x-ray interrogation system. An intense Bremsstrahlung photon beam from a high-energy linear accelerator (linac) has the ability to penetrate dense cargo and induce photofission. For optimal performance, the development of the final design must consider many issues including extrapolation of experimental data, system sensitivity versus source strength, number, type and distribution of detectors, radiation shielding and cost, all of which require extensive simulation and modeling.

A full simulation of a high energy x-ray inspection system is described in this paper. The system utilizes a 9-MV Bremsstrahlung source to inspect a cargo container for the presence of U-235. Plastic and fluorocarbon scintillators are placed around the cargo to detect the delayed gamma and prompt neutron fission signatures. [1, 2] The Bremsstrahlung-based cargo inspection system will be segmented for simulation into multiple, manageable steps, each utilizing only common desktop quad-core computers. It is shown that these segments minimize computational time as well as demonstrate the underlying physical events of each step, which allows for efficient performance optimization of the system.

COMPUTATIONAL APPROACH

The Monte Carlo N-Particle eXtended (MCNPX) simulation code is chosen for this task because it provides the user with the powerful capability to model radiation transport in all nuclear-related applications. The code encompasses an extensive nuclear cross-section database, validated physics models and a strong user community. Our simulation effort involves three steps: approximating an electron accelerator with a Bremsstrahlung source, detailing photofission inside the U-235 sphere and approximating the emission and energy spectrum of fission signatures.

Application of Accelerators in Research and Industry
AIP Conf. Proc. 1336, 700-704 (2011); doi: 10.1063/1.3586193

MCNPX SIMPLIFICATION METHODOLOGY

The simulation code, MCNPX, contains the ability to perform complex, full-scale simulations of any radiation inspection system through the individual monitoring of radiation particles. A photon-based active interrogation system detects nuclear materials when x-ray induced fission signatures are identified by the detectors. Figure 1 shows all the key radiation conversion steps that appear in the scanning system. The system utilizes an electron accelerator to generate a Bremsstrahlung beam. The Bremsstrahlung photons then induce photofission in the nuclear material (such as U-235 and U-238), if present, in the cargo. Finally, the fission signatures traverse the cargo contents and are detected. Although MCNPX can effortlessly model and simulate each of these physical processes, the results would be inadequate unless the statistics were reliable and the simulation was completed in a reasonable amount of time. An MCNPX simulation of $1*10^9$ electron particles may take one hour to generate a useful fission rate. However, it would take 71 years to simulate what a 100 μA electron linac or $6.25*10^{14}$ electrons/sec generates in one second!

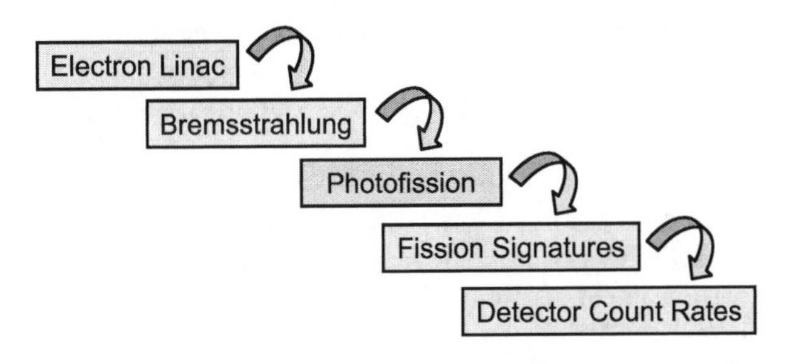

FIGURE 1. The five main steps in a Bremsstrahlung based inspection system.

For all of the steps shown in Figure 1, it is highly desirable to develop reliable simplifications and approximations, which will speed up the overall computation and provide better statistical accuracy to the Monte Carlo results. This is described in the next few sections.

Bremsstrahlung Photons

Bremsstrahlung photons are generated when high-energy electrons decelerate in a high-Z target, usually tungsten. The intensity and energy spectra of the photons that are emitted are highly angular dependent; the photons have higher energy and intensity in the forward direction. An MCNPX photon approximation, taking into account the angular dependency, can be made to eliminate the need to start the simulation with an electron source. Starting the calculation with an effective source of Bremsstrahlung photons instead of electrons increases, in effect, the photofission conversion efficiency where more photofissions are created per electron.

Tallies in MCNPX can profile an angular and intensity dependent Bremsstrahlung spectrum, which can then be used to define the approximate Bremsstrahlung source. The spectrum was generated with a 0.14 cm thick tungsten target coupled to a 0.25 cm thick copper cooling plate. A spherical surface tally surrounded the electron beam target. The tally recorded the outgoing photons in 0.1 MeV energy and 5° angular bins. Because particle current tallies in MCNPX defines the angle of the outgoing particle as normal to the surface, a spherical surface would incorrectly bin the angle of the photon; a backward (180°) photon would be tallied in the same angular bin as a forward directed (0°) photon. To eliminate this effect, a transformation vector was used to replace the original vector normal to the tally surface as the reference direction. With one simple approximation, the need for the time consuming electron beam transport was eliminated from all future simulations. Figure 2 shows the intensity and energy spectra of 18 (out of 36) five degree angular bins. The source input now contains photons that are energy and intensity dependent on angle.

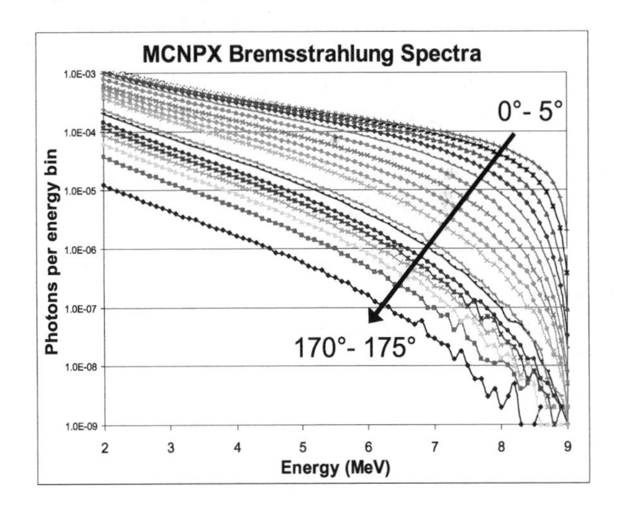

FIGURE 2. The MCNPX generated 9-MV Bremsstrahlung spectra in 5° angular bins. Each curve represents the intensity and energy spectra for a particular angular bin.

Photofission

One of the main goals of the simulations is to accurately model and predict the photon fission rate in nuclear material, which can be hidden at various locations in cargo of different types and contents. In MCNPX, a point tally can be used to model the photofission rate at any position inside the cargo container. The flux tally, multiplied by the photofission cross-section and atom density, provides the photofission density at the location of the U-235 sphere. Multiplying the photofission density by volume provides the number of fissions in the U-235 sphere. However, the photofission rate from the point tally is inaccurate because the rate depends not only on the volume but also on the shape of the nuclear material. In the case of the sphere geometry, the photons are highly attenuated and induce most of the fissions near the front spherical surface. To verify this affect, the sphere was segmented in half, perpendicular to the Bremsstrahlung source, as well as in radius in order to depict the positional dependence of the photofission in more detail. The photofission tally confirmed that 85% of the photofission occurred in the hemisphere closest to the source. The simulated emission of the various fission signatures is dependent on the location of the where the photofission occurred.

Fission Signatures

Tallying the detection of fission signatures is the final simulation step as seen in Figure 1. The simulations of detector count rates are separated from the simulations of photon fission rates, even though, MCNPX is capable of simulating both the fission and detector count rates within the same simulation. A new feature, tally tagging, allows the separation of scoring particles of the same type. The tally could be set so that, for example a photon induced prompt fission gamma ray is binned separately from the source photon. Although a one-step calculation is simpler, tally tagging is statistically imprecise, unless run for long time, as very few fission gamma rays and neutrons make it to the detector. By propagating the fission signatures to their respective detectors in a separate simulation, detection efficiency increases enhancing the number of detected particles per particle simulated.

Delayed gamma-ray emission is a new feature in MCNPX but currently, the sampling of individual gamma ray lines greatly slows the computation. The delayed gamma-ray spectrum is generally a featureless continuum, especially when detected with medium resolution detectors (e.g., NaI, BGO and CsI). Thus the simulated delayed gamma-ray continuum can be employed instead of sampling individual gamma ray lines. An approximate delayed gamma-ray spectrum, suitable for medium and low resolution detector is shown to be well represented by:

$$E_\gamma(E) = e^{-1.064 \bullet E} \quad (1)$$

For prompt neutrons, the Watt fission spectrum is used to describe the neutron emission spectrum from fissioned nuclear material. The constants used are for thermal neutron fission of U-235 where a = 0.988 and b = 2.249.

$$f(E) = e^{\left(\frac{-E}{a}\right)} \sinh\left[(b \bullet E)^{\frac{1}{2}}\right] \quad (2)$$

RESULTS

Since the final design of a photon-based active interrogation system relies heavily on the computer-based modeling of fission rates and detector count rates, the segmented simulated results should be judged by how close they are to the full-particle simulation results. The ratio of the photofission rates of utilizing the Bremsstrahlung approximation to starting the simulation with electrons was 0.84 for air, water (0.57 g/cm3) and iron (0.57 g/cm3) uniform density cargo. The constant ratio, which is reasonably close to unity, demonstrates the adequacy of the approximations. Although the fission rates do not exactly match, the computational time saved greatly outweighs the disagreement. By comparing the runs performed in vacuum, the time saved was on the order of ~500.

TABLE 1. The simulation time is compared for two cases that determine the detection count rate with < 1% relative error of prompt fission neutrons from photofission.
*The time to complete the non-segmented calculation was extrapolated due to the length of the calculation.

Case Type	Description of Simulation	Computational Time for Neutron Detection Count Rate (min)
Non-Segmented	1) Bremsstrahlung Photon → Photofission → Neutron Detection	1,000,000*
Segmented	1) Bremsstrahlung Photon → Photofission	100
	2) Prompt Neutron → Neutron Detection	1000
	Time Saved	~1000

The gain in the amount of computational time saved by utilizing segmented steps for detection count rates is shown in Table 2. The computational time required to generate a < 1% relative error in the neutron detection rate was compared for two cases. The first case was non-segmented; the simulation's starting particle was the Bremsstrahlung source. The source photons first generated prompt fission neutrons and then the neutrons propagated to the detector, all in one simulation. The second case utilized segmentation where the simulation starting particle was prompt neutrons. In order to directly compare the computational times for both cases, the same detection units of neutron counts per Bremsstrahlung source photon must be used. So for the segmented case, the additional computational time it took to determine the photofission rate with the Bremsstrahlung photons was taken into account. The overall time savings was tremendous, on the order of ~1000 and it would be even greater for the delayed gamma ray detection rate because multi-group delayed-gamma emission can require 4-5 times longer execution times in MCNPX.

Experimental Corroboration

The modeling simplification technique was also applied to the modeling of small-scale photofission experiments. Benchmarking the simulation results against experimental results provides confidence in the results of the full-scale system simulation. The experiment consisted of a well-collimated 9-MV Bremsstrahlung beam inducing photofission in a depleted uranium (DU) slab. Plastic and fluorocarbon scintillators, detecting delayed gamma rays, were placed next to the wood and iron cargos that surrounded the DU slab. In Figure 3, the MCNPX simulated delayed gamma-ray spectrum is plotted against the experimental spectrum recorded from the plastic scintillator illustrating good agreement. The insert shows a 3-dimensional model of the experimental setup. The main transparent box is the wood cargo and inside the box is a depleted uranium slab. The fluorocarbon and plastic scintillators are positioned perpendicularly on each side.

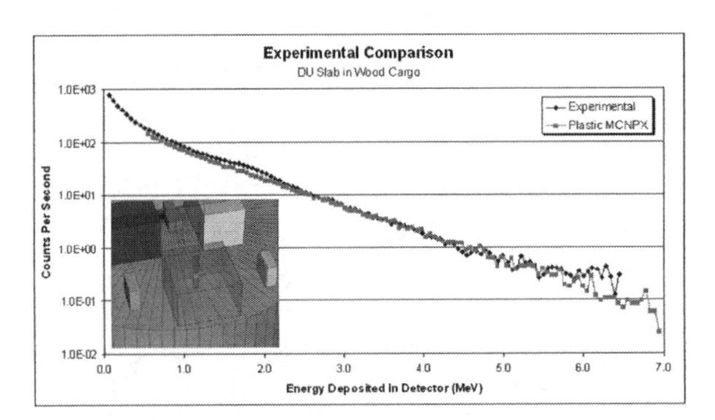

FIGURE 3. The simulated delayed gamma count rate is plotted against the experimental spectrum. The illustration insert shows a 3-D model of the experiment.

Table 2 compares the ratio of experimental to simulated delayed gamma counts, integrated between 3-5 MeV, in both the plastic and fluorocarbon scintillators. The results show that the calculated results are in good agreement with the experimental results for both the wood and iron cargos.

TABLE 2. Experiment to calculation ratio of delayed gamma ray counts from DU between 3-5 MeV in plastic and fluorocarbon scintillators.

Simulated Cargo	Plastic Scintillator	Fluorocarbon Scintillator
	(Exp / MCNPX)	(Exp / MCNPX)
Wood	1.1	0.9
Iron	1.1	1.0

System Detection Efficiency

The size and cost of fabricating a large-scale interrogation system is high. Utilizing simulations to predict the detection efficiency and optimize the detector geometry of the large-scale system can alleviate the risks associated with such a large and costly undertaking. Figure 4 shows the detection and photofission results from the model. Photofission as well as delayed gamma-ray and prompt neutron detection rates were calculated for water cargos with densities that range from 0 to 0.5 g/cm^3, wood with density of 0.57 g/cm^3 and iron with densities that range from 0.57 to 2.28 g/cm^3.

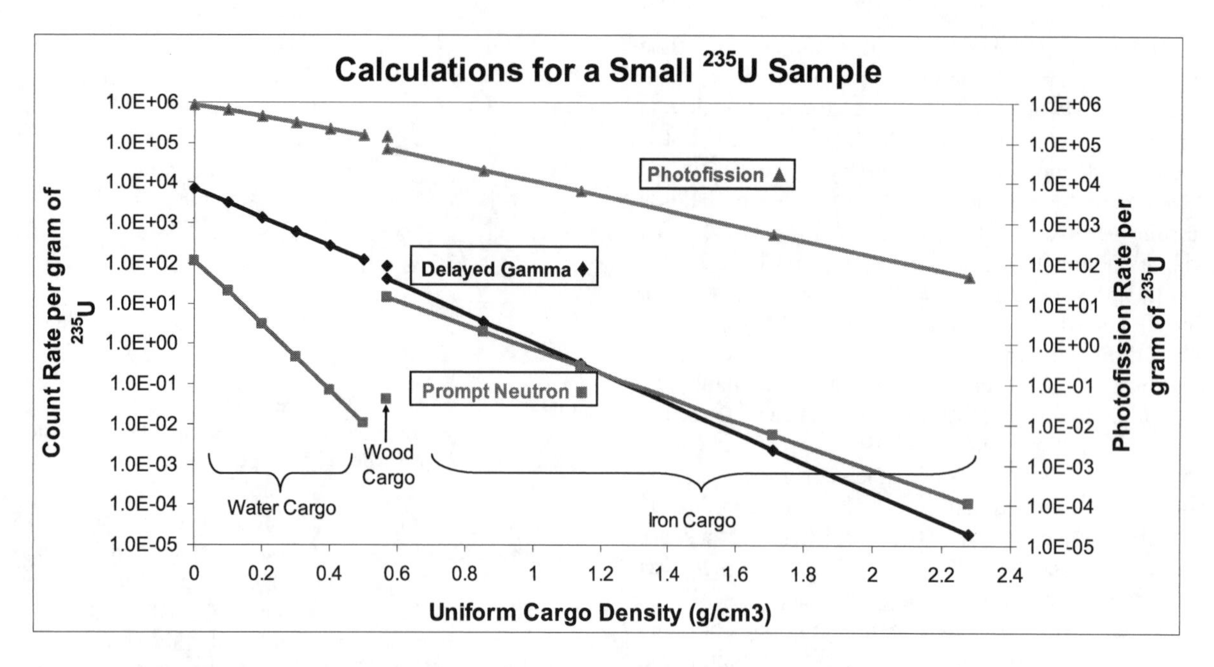

FIGURE 4. Photofission and detector count rates averaged over the inspection plane for a small U-235 sample are plotted for different cargos and densities.

The plot shows the results for a U-235 per gram sample averaged over the entire scanned area or slice of the cargo container. In iron cargo, the prompt neutron detection count rates are similar to that of the delayed gamma rays. The former provides a distinct and independent confirmation for the presence of nuclear material beyond that of the fission delayed gamma rays. This signature greatly enhances the overall sensitivity and robustness of the inspection system. It should be noted that the higher average iron density cargos, where both signals are relatively low, represent, in general, unlikely cargo contents, e.g. a solid iron sphere of 70 cm in diameter.

DISCUSSION

Simulations are a powerful tool which provides us the ability to model a complete experimental test stand or a full inspection system, corroborating experimental data as well as provide important physics insight such as the photofission distribution inside the U-235 sphere. The simple Bremsstrahlung approximation provided accurate enough photofission results while avoiding exceedingly long computational times. By separating photofission from the emission of fission signatures and approximating the emission energy spectra, sampling of detector count rates significantly improves. Ultimately, the main advantage of utilizing simulations is the ability to generate accurate results in the place of or in support of experimental data. Very often carrying out a full-sized experiment is difficult to execute due to size, time or economical constraints. Utilizing MCNPX prior to or in parallel with experiments not only provides feasibility analyses but also valuable forecasts and data.

ACKNOWLEDGMENTS

This work is supported by the Department of Homeland Security under the contracts HSHQDC-08-C-00125 and HSHQDC-08-C-00132. The first author, Michael King, would like to thank Dr. Tak Pui Lou from the Lawrence Berkeley National Laboratory (LBNL) and Dr. John Hendricks from Los Alamos National Laboratory (LANL) for many fruitful discussions.

REFERENCES

1. J. Stevenson, C. Brown, C. Condron, M. Elsalim, T. Gozani, "Linac Based Photofission Inspection System Employing Novel Detection Concepts", Nucl. Instr. And Meth. A Conference Proceedings, to be published
2. T. Gozani, J. Stevenson and M.J. King," Neutron Threshold Activation Detectors (TAD) for the Detection of Fissions", Nucl. Instr. And Meth. A Conference Proceedings, to be published

Performance Characteristics Of An Intensity Modulated Advanced X-Ray Source (IMAXS) For Homeland Security Applications

Dr. Willem G.J. Langeveld[a], Dr. Craig Brown[a], Phil. A. Christensen[b], Dr. Cathie Condron[a], Dr. Michael Hernandez[c], Mike Ingle[a], William A. Johnson[b], Roger D. Owen[b], Randy Ross[d], and Russell G. Schonberg[†c]

[a]Rapiscan Laboratories, Inc., 520 Almanor Ave., Sunnyvale, CA 94085
[b]HESCO/PTSE Inc., 2501 Monarch St., Alameda, CA 94501
[c]XScell Corp., 2134 Old Middlefield Way, Mountain View, CA 94043
[d]Stangenes Industries, Inc., 1052 East Meadow Circle, Palo Alto, CA 94303

Abstract. X-ray cargo inspection systems for the detection and verification of threats and contraband must address stringent, competitive performance requirements. High x-ray intensity is needed to penetrate dense cargo, while low intensity is desirable to minimize the radiation footprint, i.e. the size of the controlled area, required shielding and the dose to personnel. In a collaborative effort between HESCO/PTSE Inc., XScell Corp., Stangenes Industries, Inc. and Rapiscan Laboratories, Inc., an Intensity Modulated Advanced X-ray Source (IMAXS) was designed and produced. Cargo inspection systems utilizing such a source have been projected to achieve up to 2 inches steel-equivalent greater penetration capability, while on average producing the same or smaller radiation footprint as present fixed-intensity sources. Alternatively, the design can be used to obtain the same penetration capability as with conventional sources, but reducing the radiation footprint by about a factor of three. The key idea is to anticipate the needed intensity for each x-ray pulse by evaluating signal strength in the cargo inspection system detector array for the previous pulse. The IMAXS is therefore capable of changing intensity from one pulse to the next by an electronic signal provided by electronics inside the cargo inspection system detector array, which determine the required source intensity for the next pulse. We report on the completion of a 9 MV S-band (2998 MHz) IMAXS source and comment on its performance.

Keywords: Accelerator control systems, electron linear accelerators, intensity modulation, cargo inspection, accelerator shielding, homeland security
PACS: 52.38.Ph, 52.59.Px, 87.59.Bh

INTRODUCTION

Cargo inspection systems typically use linear-accelerator (linac)-based pulsed high-energy x-ray sources. Existing sources operate at a fixed x-ray intensity, which we shall call the Output Set Point (OSP). For much of a typical image, the OSP is higher than needed, and performance may be limited when the OSP is chosen to be less than the maximum rated output of the source to reduce the radiation footprint. In previous papers[1,2], we have shown that an Intensity Modulated Advanced X-ray Source (IMAXS) can be used to significantly enhance the imaging penetration performance of cargo inspection systems while maintaining the same average radiation footprint in terms of size of exclusion zone and radiation dose to cargo and personnel. This is accomplished by varying, on a pulse-to-pulse basis, the x-ray intensity of the source, based on the signal strengths measured in the inspection system's detector array during the previous pulse. The average dose reduction made possible through the use of an IMAXS (at 6 MV) was estimated to be a factor of three[1], while maintaining the same imaging penetration. With the same radiation footprint as a conventional source, the imaging penetration of a system using an IMAXS is up to two inches[1] better than the state of the art,

† Deceased

Application of Accelerators in Research and Industry
AIP Conf. Proc. 1336, 705-710 (2011); doi: 10.1063/1.3586194

which, at 6 MV source energy, is roughly 16 inches of steel.

Recently, a 9 MV IMAXS was completed and tested. In the following sections we report on the physical characteristics of this source and discuss experiments demonstrating the IMAXS operation. We conclude with a summary.

FIGURE 1. XScell beam centerline of the 9 MV IMAXS.

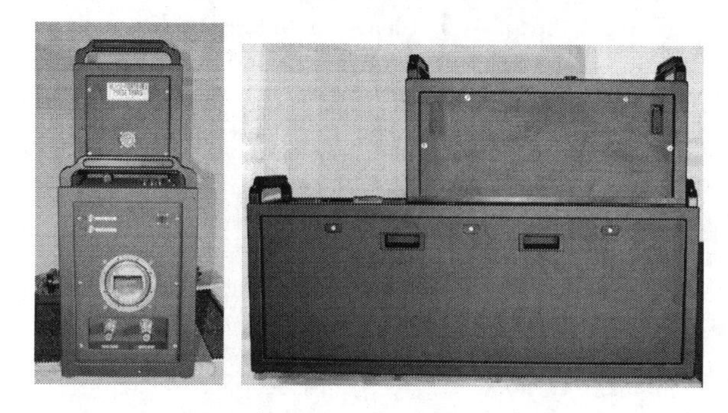

FIGURE 2. HESCO RF Head.

FIGURE 3. HESCO control console.

PHYSICAL CHARACTERISTICS

The IMAXS project was a collaboration of four companies: HESCO/PTSE Inc., XScell Corp., Stangenes Industries, Inc. (SI), and Rapiscan Systems, Inc. The x-ray source reported on in this paper consists of a number of components. The beam centerline was manufactured by XScell, the RF head, control console and PLC touch screen (Figure 3) were produced by HESCO, the solid-state modulator by Stangenes, and the IMAXS steering electronics by Rapiscan. A commercially available chiller was used. The system was integrated by HESCO and tested at their facilities in Alameda, CA.

The linear accelerator for the IMAXS system is a standing wave side-cavity coupled type operating at 2.998 GHz. The electron gun used to perform the IMAXS operation is a standard M592 gridded gun and the electron target is isolated. One of the requirements for this particular linear accelerator was that its beam centerline (BCL) was to occupy a smaller longitudinal footprint than a standard industrial 9 MeV linac, in order to be integrated into an inspection system under development under a separate DNDO program. Hence, the BCL for this IMAXS (see Figure 1) is about 27 inches long, compared to standard 44 inch long 9 MeV industrial accelerators. Including the gun and target, the length is about 30 inches (75 cm). This length requirement necessitated designing the cavities of the accelerator with higher shunt impedance for greater efficiency. Increasing the shunt impedance can present consequences such as arcing and prolonged RF conditioning. The cavities used in the IMAXS system have been successfully employed in a smaller 6 MeV system, run with a 2.6 MW E2V MG5193 magnetron. This system produced 1100R/min @ 1m. The linac RF conditioned in less than one day and has operated for hours at full drive with no arcing or vacuum issues. The 9 MeV IMAXS linac was initially designed to operate with >3 MW of input RF power and with this power it is projected to produce 3000R/min @ 1m at 9 MeV energy, based on data from 2.6 MW operation.

The RF head (Figure 2) houses a high-voltage pulse transformer, a magnetron, RF circulator, water load, the solid-state gun controller and all the support electronics. It is designed for portable operation and is fully self-contained. The system is controlled using an off-the-shelf Programmable Logic Controller (PLC, see Figure 3). All general control interlocks, system warm-up, and other operational functions are displayed and controlled using a touch screen interface. To control the intensity on a pulse-to-pulse basis, a custom Master Pulse Controller (MPC) was designed using a RISC-based microcontroller that calculates and delivers in real time a varying pulse width to a solid-state gun grid controller. The MPC has 16 distinct, customizable, pulse width settings, which translate to intensity levels. The pulse widths can vary between 0.1 and 4.7 microseconds. The IMAXS steering electronics (discussed below) send requests to the MPC to either increase or decrease power based on current needs.

The SI model SSM-6-7-M1 (Figure 4) is a magnetron modulator system compatible with the E2V MG5193 S-band magnetron. This modulator system supports 0 to 50 kV output pulses with beam currents up to 130 amps peak. The pulse flat tops are adjustable from 1 to 4 microseconds continuously. Pulse rise time (10%-90%) is less than 500 nanoseconds.

FIGURE 4. Stangenes Industries model SSM-6-7-M1 modulator.

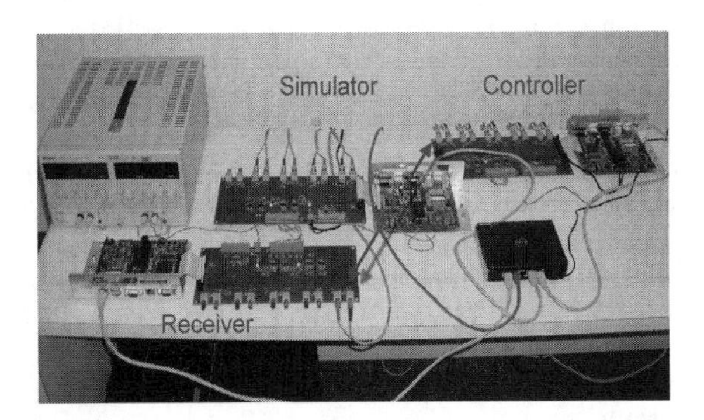

FIGURE 5. IMAXS electronics, produced by Rapiscan; source intensity controller and source intensity control receiver are shown, together with the system simulator.

Pulse repetition rates are 0 to 300 Hz. Amplitude jitter of the pulse is less than 0.5%. The modulator consists of three chassis, mounted in a standard 19" rack (measuring 26.5 inches by 24 inches by 35 inches tall) and a pulse transformer in a separate enclosure. The three chassis contain, respectively, a mini digital control chassis, a 10 kW high-frequency, high-voltage capacitor-charging power supply, and a variable pulse width Solid State Marx fast pulse modulator. The oil-filled pulse transformer tank measures 11 inches wide by 10.5 inches tall by 22 inches long. The 19" rack and its three chassis are connected to the pulse tank by 3 cables: two RG-8 pulse cables and 1 control cable. Insulated Gate Bipolar Transistors (IGBTs) are used as the solid state switches for the Solid State Marx Chassis in order to provide improved reliability compared to thermionic devices. Combining Solid State Marx with pulse transformer topology enables flexible pulse width capability (which is, however, not required for the IMAXS mode of operation) and supports highly repeatable operation to within 0.1%, while incorporating fault power limiting. A digital control system coordinates system operation, provides unified fault management, filament current setback for high power operation, automatic warm up and customer interface. The weight of the modulator is 300 lbs, and that of the pulse transformer 115 lbs. The system has a universal power input and can operate from 3 phase 200 to 400 VAC 50/60 Hz with neutral for electrical power. It requires 5 GPM of cooling water in the range of 20° to 40° C at the pulse transformer, and is designed for operation in interior environments with air temperatures from 5° to 40° C.

The IMAXS steering electronics (Figure 5) consist of a Source Intensity Controller (SIC) and a Source Intensity Control Receiver (SICR). The SIC, which is a Technologic Systems TS-7300 embedded computer running LINUX [3] with a user-programmable Altera Cyclone II FPGA, interfaces to the cargo inspection system's detector array by means of a custom I/O card. The FPGA is programmed to read out the detector signals after an x-ray pulse and then determines the lowest signal level. Using two preprogrammed threshold levels A and B < A, it determines whether the lowest signal is above A, between A and B, or below B. It then sends a signal to the SICR to decrease, maintain, or increase the source intensity, respectively, followed by a PULSE command (see Ref. 1 and 2 for details). The PULSE timing can be controlled either by an internal timer, or by an external encoder. Another feature of the SIC is the capability of streaming the detector data by 10/100 MBit Ethernet to the cargo inspection system's data acquisition computer. The SIC is logistically part of the inspection system detector array, the SICR part of the x-ray source. The SICR, also comprised of a TS-7300 with an I/O card, receives the SIC signal over an optical connection, and translates the instructions into voltage levels which are then interpreted by the MPC. For test purposes, a Simulator was also constructed, again based on a TS-7300 with an I/O card. The Simulator can be used to simulate the presence of an entire detector array, and also partially simulates the response of the x-ray source to the instructions generated by the SIC and received by the SICR. Incidentally, all functionality for the three I/O cards was integrated into a single circuit board, which is why the SIC, SICR and Simulator look identical. The only differences are in FPGA firmware and cabling.

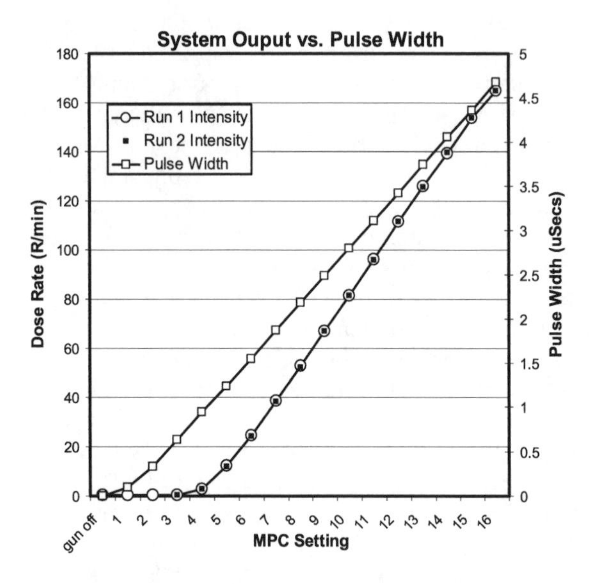

FIGURE 6. Repeatability demonstration of the intensity settings. Shown are the grid pulse width (open squares), as well as the recorded source intensities during two runs (circles and black squares).

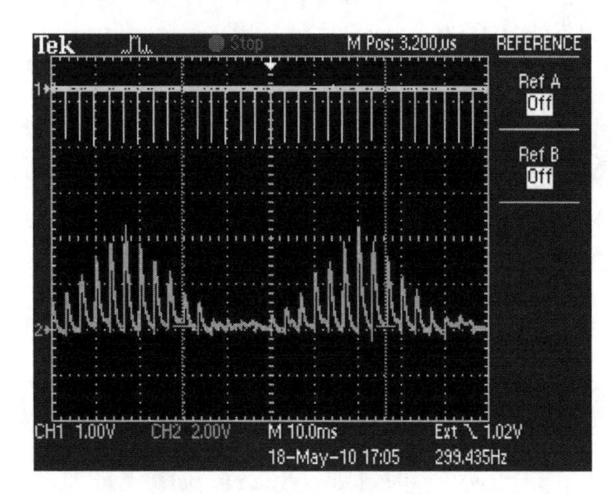

FIGURE 7. Oscilloscope trace showing pulse-to-pulse IMAXS intensity modulation at 300 pps.

IMAXS TESTS AND OPERATION

A number of tests were performed to demonstrate the system capabilities and operation. Some tests were performed using an alternative modulator, while the SI modulator was under development.

One test was designed to gauge the system's ability to consistently deliver distinct intensity levels. For this test, the 16 available pulse width settings were programmed into the MPC in a linearly increasing fashion. The test was performed using two calibrated RadCal ion-chamber dosimeters located 1 meter from the target. The machine was run at 100 pulses/second (pps) at each of 16 preprogrammed intensity levels. Measurements were taken and averaged from both dosimeters to determine the actual system output in Rads/min @ 1 m. This measurement was then repeated to determine consistency. Figure 6 shows that the measured pulse widths (open squares) indeed follow a linear curve, and, except at small pulse widths, so does the measured dose (open circles and black squares). Since the dose also depends non-linearly on the electron beam energy, one can conclude that the electron beam energy does not change appreciably over the useful range of pulse widths, which is, of course, the reason that pulse-width modulation was employed in the IMAXS [1, 2]. The open circles show the x-ray intensities for the first measurement and the black squares those for the second measurement, demonstrating that the system produces repeatably the same output to within 0.75 R/min, at any given intensity level. The output was shown to be consistently adjustable from 100% down to 1.8% by varying the gun pulse width.

Another test was performed by programming the SICR to send a repeated saw-tooth test pattern of eight "increase" commands, followed by eight "decrease" commands to the MPC. The pulse width settings were, again, programmed in a linearly increasing fashion. The system was run at 300 pps and an ion chamber was placed in the beam path, and the direct output of the ion chamber was measured with an oscilloscope. Figure 7 shows a screen capture from the oscilloscope, showing that the actual measured intensity of the source follows the saw-tooth pattern as desired. Note, that in an actual inspection system, the pulse width settings would probably be programmed to provide factors of two changes in intensity, and only 6 or 7 different intensity values would be used, e.g., 100%, 50%, 25%, 12.5%, 6.3%, 3.2% and 1.6% (or the minimum value allowed by the hardware). For the demonstration, a linear sequence was easier to display visually.

In the absence of a full cargo inspection system with integrated IMAXS source, separate tests were performed to establish that the SIC/SICR system functions as designed. The SIC was tested with available inspection system hardware, to verify that the hardware protocols match. The SIC/SICR system was tested using a loop including the Simulator: the Simulator sent detector data (based on an actual radiography image from an existing system) to the SIC, the SIC determined the appropriate source intensity change and sent the appropriate command to the SICR, and the SICR sent the proper source adjustment back to the Simulator. The Simulator then used this information to apply the appropriate intensity change to the next column of the image, and sent the corresponding data to the SIC. This sequence could be timed using either an internal timer or an external encoder. The IMAXS-encoded image arriving in the SIC was sent, by the Ethernet connection, to an external computer, and using an also simulated reference detector, the IMAXS-encoded image was restored to its original form and compared successfully with the input image. In this way, the full system design was tested and verified.

As stated before, pulse-width variation is employed in the IMAXS in order to minimize changes in electron beam energy. If one varies the peak pulse current, significant energy changes result due to beam loading (unless some form of energy compensation is used). Linac energy measurements were performed as a function of peak pulse current by precisely measuring the Bremsstrahlung x-ray spectrum produced by the source using a well-shielded and collimated, calibrated, NaI detector positioned and laser-aligned at a 55 ft distance from the source, which was itself well shielded and collimated. Using an extrapolation technique, the Bremsstrahlung spectrum end point was measured with good statistical precision, determining the electron beam energy with much greater accuracy than the standard half-value-layer technique. A 52-inch-long graphite attenuator was used in order to suppress the low-energy x-rays and reduce the NaI detector count rate. The accelerator settings were optimized and the gun current was systematically adjusted until the maximum electron energy was found. In these measurements, the accelerator had an effective beam-on-target fixed pulse width of about 2.85 μsec and ran at 270 pulses per second and was operated with a 2.6 MW magnetron.

Using these settings, the source can achieve bremsstrahlung endpoint energies ranging from about 6 MeV at 800 mA (peak) injected current, to about 8 MeV at 80 mA injected current (see Figure 8). The maximum dose rate was measured to be 1680 R/min @ 1m, at an energy of 7.1 MeV and 480 mA injected current, with an estimated beam capture efficiency at this setting of about 30%.

Based on these results, we expect that the accelerator will be able to achieve 9 MeV with a 3.1 MW magnetron.

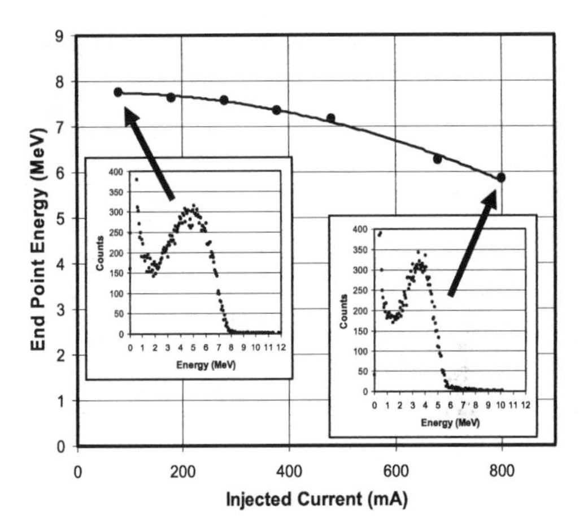

FIGURE 8. Measured energies achieved by the IMAXS as a function of injected current. The inserts show example NaI spectra for 80 and 800 mA injected current.

SUMMARY

In this paper, we have demonstrated, for the first time, a high-energy IMAXS accelerator system capable of switching intensities under computer control within 3 msec. The system supports up to 16 different intensity settings, using grid pulse width modulation. Also in other respects, the system described here sports several innovations. The length of 27 inches is considerably shorter than other 9 MeV S-band accelerators, and the Solid State Modulator is also a novel development. Overall, the system performance is as expected.

DEDICATION

This paper is dedicated to one of the authors, Russell Schonberg, a true pioneer in the accelerator community. While the IMAXS project turned out to be his last, he demonstrated once more his life-long knack for innovation. He was actively involved in the project until two weeks before his death. He will be missed.

ACKNOWLEDGEMENT

This work was supported in part by the U.S. Department of Homeland Security, Domestic Nuclear Detection Office, under Contract HSHQDC-08-C00023.

REFERENCES

1. W.G.J. Langeveld, W.A. Johnson, R.D. Owen and R.G. Schonberg, "Intensity Modulated Advanced X-ray Source (IMAXS) for Homeland Security Applications(I)", Proceedings of the Symposium on Radiation Measurements and Applications (SORMA) 2008, IEEE Trans. Nucl. Sc., 56 (2009) 1288.
2. W.G.J. Langeveld, W.A. Johnson, R.D. Owen and R.G. Schonberg, "Intensity Modulated Advanced X-ray Source (IMAXS) for Homeland Security Applications(II)", Proceedings of the Twentieth International Conference on Applications of Accelerators in Research and Industry (CAARI) 2008, AIP Conference Proceedings 1099 (2009) 628.
3. See https://www.embeddedarm.com.

Nuclear-Reaction-Based Radiation Source For Explosives- And SNM-Detection In Massive Cargo

Michal Brandis[a,c], Volker Dangendorf[b], Christian Piel[d], David Vartsky[a], Benjamin Bromberger[b], Doron Bar[a], Eliahu Friedman[c], Israel Mardor[a], Ilan Mor[a], Kai Tittelmeier[b], Mark B. Goldberg[a,b]

[a]*Nuclear Physics Division, Soreq NRC, Yavne 81800, Israel*
[b]*Physikalisch-Technische Bundesanstalt (PTB), Braunschweig 38116, Germany*
[c]*Racah Institute of Physics, The Hebrew University of Jerusalem, Jerusalem 91904, Israel*
[d]*RI Research-Instruments GmbH, Bergisch-Gladbach 51429, Germany*

Abstract. An automatic, nuclear-reaction-based, few-view transmission radiography method and system concept is presented, that will simultaneously detect small, operationally-relevant quantities of chemical explosives and special nuclear materials (SNM) in objects up to the size of LD-3 aviation containers. Detection of all threat materials is performed via the ^{11}B(d,n+γ) reaction on thick, isotopically-enriched targets; SNM are primarily detected via Dual Discrete-Energy Radiography (DDER), using 15.11 MeV and 4.43 MeV ^{12}C γ-rays, whereas explosives are primarily detected via Fast Neutron Resonance Radiography (FNRR), employing the broad-energy neutron spectra produced in a thick ^{11}B-target. To achieve a reasonable throughput of ~20 containers per hour, ns-pulsed deuteron beam of the order of 0.5 mA intensity at energies of 5 – 7 MeV is required. As a first step towards optimizing parameters and sensitivities of an operational system, the 0° spectra and yields of both γ-rays and neutrons in this reaction have been measured up to E_d=6.65 MeV.

Keywords: SNM detection, explosives detection, nuclear-reaction-based source, pulsed beams, FNRR, DDER
PACS: 29.25.-t, 29.20.Ej, 25.45.-z, 29.30.Hs, 29.25.Dz

INTRODUCTION

The threat to civil aviation, international trade and homeland security posed by illicitly-transported explosives and fissile materials (SNM) is of ever-increasing concern. Moreover, contemporary terror organizations have become highly skilled in devising bombs that are smaller, more potent and, no less significantly, progressively harder to detect with the screening techniques currently deployed or approaching maturity.

In response, the U.S. Government passed a Bill named "Implementing Recommendations of the 9/11 Commission Act" (Public Law 110-53, Aug. 3, 2007), which mandates that within five years, 100% of all maritime cargo and aviation cargo loaded onto passenger aircraft bound for the U.S. be screened for the above threat materials in foreign ports prior to loading.

This requirement implies that a very large number of screening systems with capabilities well beyond the current state-of-the-art will be needed in the immediate future, both in the U.S. and other countries menaced by terrorist activity, at a total hardware outlay estimated (ibid) to exceed 5000 M$ worldwide.

Such systems should be able to detect automatically small amounts of nuclear materials (HEU, ^{239}Pu), as well as standard and improvised explosives in aviation cargo, trucks and rolling stock. Inspection times must be short (< 5 min/container) and contraband detected with high reliability, at low false alarm rates.

The method presented in this paper combines few-view high-spatial-resolution Fast Neutron Resonance Radiography (FNRR) [1-5] and Dual Discrete-Energy γ-Radiography (DDER) [6, 7]. FNRR is a neutron imaging method that exploits the characteristic cross-section structure (resonances) of C, O, & N in the En=1–10 MeV range. Its pulsed variant is known as Pulsed Fast Neutron Transmission Spectroscopy (PFNTS) [8-11]. The transmission spectrum carries information about the elemental composition of the cargo contents. Thus, a pre-requisite for FNRR is precise knowledge of the neutron energy. This is

Application of Accelerators in Research and Industry
AIP Conf. Proc. 1336, 711-716 (2011); doi: 10.1063/1.3586195

achieved by Time-Of-Flight (TOF) spectroscopy, requiring ns-pulsed beams. TOF also helps to discriminate the neutrons from the γ-rays employed in DDER.

DDER makes use of the nuclear reaction $^{11}B(d,n+\gamma)$. Its γ-spectrum is dominated by two discrete, well separated γ-rays at 15.11 and 4.43 MeV. These give rise to an image contrast sensitivity (CS) for high-Z materials which is about a factor of 5 higher than that achievable with Dual-Energy Bremsstrahlung at electron energies of 5 and 9 MeV [6].

Unlike conventional X-ray radiographic systems, FNRR and DDER do not rely on the skills of a human operator to identify the threat object by its shape and density. Instead, they are based on automatic identification of the concealed contraband by its elemental composition. FNRR can reliably detect standard and improvised explosives in solid or liquid form. Moreover, it possesses sufficient granularity to permit detection of thin threat objects such as sheet-explosives. The role of DDER is to counter the threat of improvised nuclear devices being assembled from small quantities of SNM that were smuggled in piecemeal, to avoid detection. An operational system combining these two state-of-the-art inspection techniques will thus permit reliable and automatic detection of small quantities of explosives and SNM, respectively.

MEASURMENTS OF $^{11}B(,n+\gamma)\ ^{12}C$ REACTION THICK-TARGET YIELD

Experimental Setup

The yields were measured using the PTB Van-De-Graaff accelerator for $E_d < 3.5$ MeV and the cyclotron for higher beam energies. Both accelerators deliver pulsed deuteron beams (τ~1.5 ns fwhm, T<500 ns) suitable for TOF measurement. Beam intensities are several μA PC averaged at the Van-De-Graaff and ~10 μA PC averaged at the cyclotron.

An NE213 liquid scintillator spectrometer was positioned at 0°, at a distance of 7 m from the thick ^{11}B target. Neutrons were distinguished from γ-rays via Pulse-Shape-Discrimination and Time-Of-Flight.

Target Configuration

The target consisted of a 3 mm thick layer of compressed ^{11}B granules, contained between a water-cooled Cu backing and a 5 μm thick Molybdenum foil. The isotopically-enriched ^{11}B contained less than 5% ^{10}B. Another Molybdenum foil of the same thickness served as safety valve between target assembly and accelerator beam line. The 5 mm gap between the two Mo windows was filled with He gas, to avoid parasitic nuclear reactions with O, N and also to prevent in-beam degradation of the Mo foils by atmospheric oxygen. The deuteron energies indicated in the abscissa of Figs. 1-2 below are those incident on the ^{11}B target surface.

Gamma-Ray Yields

The measured γ-ray yields are presented in Fig. 1. They increase with deuteron energy, but at different rates for the two lines. Preferred beam energies for our application are in the 5 – 7 MeV range, in which the 15.1 MeV yield is comparable to that of 4.43 MeV.

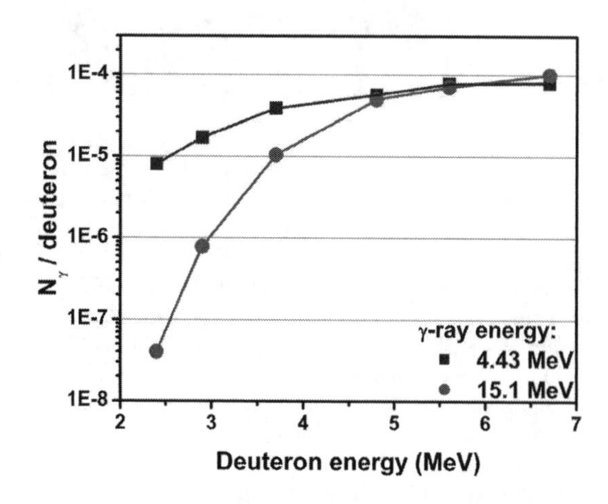

FIGURE 1. Measured γ-ray Yields as Function of Deuteron Energy.

Neutron Yields

Fig. 2 presents the thick-target neutron yields measured at several deuteron energies in the range of E_d= 2.9 - 6.7 MeV. The highest energy neutron groups leave ^{12}C in its ground state and first excited state at 4.43 MeV, respectively. Most of the continuum at energies < 7 MeV is due to neutrons leaving ^{12}C in higher excited states, along with deuterium-ion breakup neutrons [12] at energies below half the beam energy. As deuteron energy increases, the spectral shape is qualitatively maintained, but all neutron groups progressively extend to higher energies.

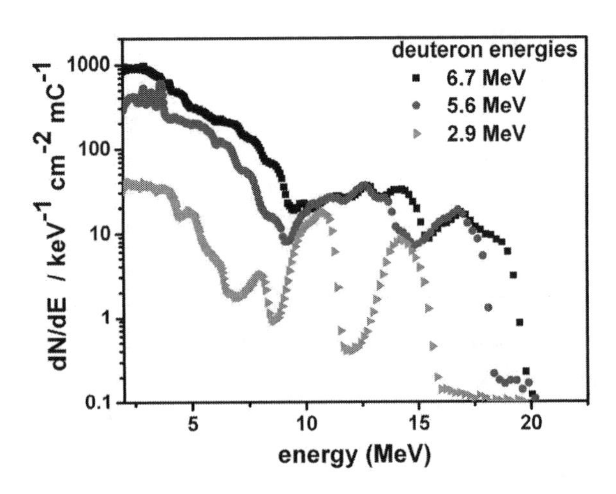

FIGURE 2. Neutron Yield Spectra as Function of Deuteron Energy.

COMBINED FNRR/DDER INSPECTION SYSTEM: REQUIREMENTS ON RADIATION SOURCE

The requirements for the combined FNRR/DDER radiation source are:

- Broad-energy 1 - 10 MeV neutron spectrum.
- Required neutron flux on detector face: $>5x10^5$ n/s/cm^2 integrated over the entire spectrum.
- Two intense γ-rays: one ~4 MeV and one in the energy region of more than 10 MeV
- Focal spot of radiation source: < 5 mm (for detection of thin sheet explosives).
- TOF distances (target to detector) of ~6 m.
- Beam time-structure: pulse-width ~1 ns at a repetition rate of ~4 MHz (for TOF neutron spectroscopy).
- Time averaged beam current: ~0.5 mA.

Accelerator and Beam Pulsing

For RF ion-acceleration, the beam must be in the stable phase regime of the accelerating field (-45° to -15°, see Fig. 3). Meeting the above-mentioned beam requirements thus determines an accelerator frequency of ~83 MHz. For CW operation, the instantaneous beam-current (within the individual pulse) that the accelerator delivers will then need to be ~10 mA.

Several accelerator types can deliver such deuteron beams:

- DC accelerators
- Cyclotrons
- Dielectric wall accelerators
- RFQ's, optionally followed by an ion-energy booster, such as an IH-structure

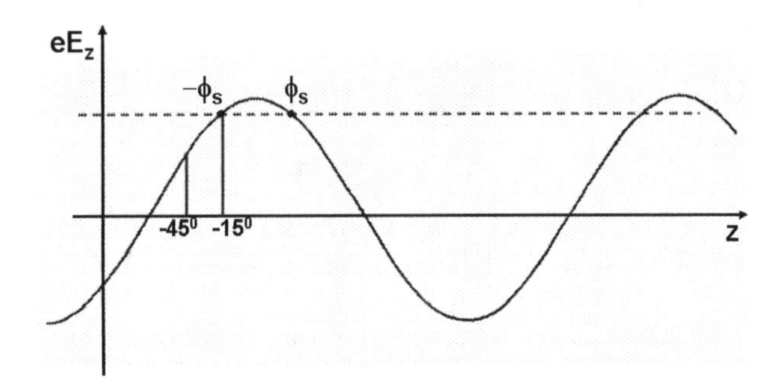

FIGURE 3. Stable Acceleration of an Ion Beam Requires Longitudinal Focusing by the RF Field.

Although highly efficient DC accelerators have been in operation for decades, they are gradually being replaced these days by RFQ's, which generally possess a smaller footprint and permit reliable, long-term operation.

Compact Cyclotrons employing superconducting magnets are also extensively used, but their weak focusing cause space charge limitations on beam intensity. Calculations on the beam charge limit for compact cyclotrons have been performed by TRIUMF [13] exhibit a space charge limit of several mA, well in line with measurements performed on existing systems. With Sector Cyclotrons [14], this limitation can be overcome.

Dielectric wall accelerators constitute a recent development that promises compact and efficient structures, but they are limited to repetition rates of several 10Hz and are therefore unsuitable for this application.

RFQ's have been operated with gradients of 2MV/m in pulsed and continuous mode, at currents of up to 100mA. They are thus good candidates to be the base accelerator of choice. With quality factors of about 4000 the filling time of an RFQ is in the μs range. At repetition rates of ~4 MHz, they will need to operate in CW mode.

Such an accelerator (Fig. 4) would therefore consist of 4 principal components: A DC beam is generated by an ion source at 40-90 kV, then passing a pulse selector to ensure that only every 9th beam bucket is filled. Subsequently, the beam will be accelerated by an RFQ to 1-2 MeV/u. At ~2MeV/u the acceleration efficiency of a RFQ becomes poor, so that other structures such as IH or DTL (not shown) might take over to provide the ultimately desired energy. This is mandatory as e.g. LEDA [15] required 1.2 MW rf power to reach 6.7 MeV. It is expected that at a frequency of 83 MHz and with a judicious choice of accelerating structure for the high energy end, the power requirements of the accelerator can be reduced to below 150 kW.

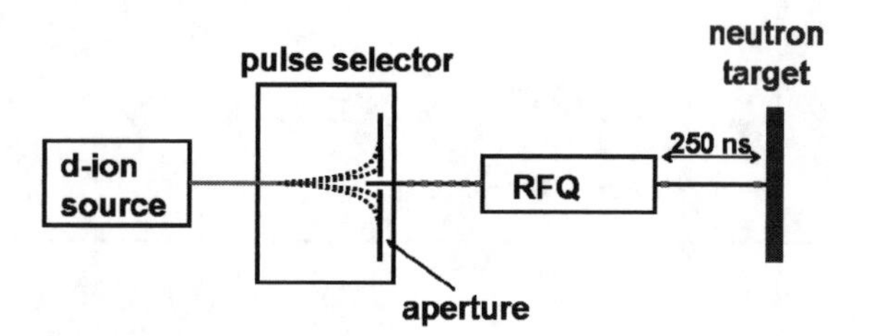

FIGURE 4. Schematic of Accelerator and Beam Pulsing Modules.

The low-energy pulsing system will need to be designed and optimized for the accelerator of choice. This task is foreseen within the framework of the ACCIS project, proposed by the Institutes participating in the present work and recently approved by GIST, a new bi-national German/Israel Fund for Civil Security Research.

FNRR/DDER: SALIENT FEATURES OF METHODOLOGY

The Radiation Source

Soreq and PTB have experimentally studied accelerator-based nuclear reactions that produce broad-energy neutron and dual discrete energy γ-ray beams. The $^{11}B(d,n+\gamma)$ reaction is, to the best of our knowledge, the only radiation source that fulfils these requirements and permits simultaneous, automatic detection of SNM and explosives [6] via the combination of FNRR and DDER. For FNRR, the yields and spectral ranges are poorer by a factor of ~10 than with $^{9}Be(d,n)$, but the γ-ray spectra and yields with a ^{11}B target, that are dominated by the 4.43 MeV and 15.1 MeV lines, are much better adapted for DDER than those obtained with a ^{9}Be target. It should however be borne in mind that there is "conflict of interest" between these two techniques as to the optimal bombarding energy: whereas for FNRR the neutron yields increase steadily up to E_d = 12 MeV [16], DDER requires E_d that is no higher than 7 MeV, because other reaction channels open and the γ-ray spectra becomes cluttered with intermediate-energy lines that spoil the contrast sensitivity for detecting high-Z materials.

Sensitivities To The Threat Materials

DDER

From the Z and E_γ behaviour of the photon mass attenuation coefficients μ, it can easily be shown that the Contrast Sensitivity (CS) for detecting high-Z materials

$$CS = \Delta\mu = \mu(E_{high}) - \mu(E_{low}) \qquad (1)$$

is, to a very good approximation, simply proportional to the energy difference between the low-E and high-E γ-rays. Thus, the low-E line should be at an energy around the global minimum of μ for all absorbers (which is in the 1-4 MeV range) and the high-E line should be at an energy that is as high as practically possible (in terms of reaction yield and spectrum cleanliness). A comprehensive literature study [6] of the spectroscopy of light-ion-induced low-E nuclear reactions and the excitation curves of the present work have demonstrated that the thick target γ-spectra from $^{11}B(d,n)$ at energies below E_d = ~7 MeV are optimal in this respect.

By the same considerations, it can also be shown that CS for the 15.1 and 4.43 MeV γ-rays employed here will exceed that of dual-, continuous-E Bremsstrahlung (BS) sources at electron-beam energies of 5 & 9 MeV by a factor as large as ~5 [6]. This is because the 2 BS-spectra differ only by a small high energy tail, even when they are heavily filtered. In terms of an overall figure-of-merit for detecting SNM, this factor more than offsets the higher BS yield.

The high CS achieved by DDER is also conducive to detecting small quantities of SNM with high sensitivity and reliability. This relates to the methodology of determining the density of items within an inspected object via spatial reconstruction from a small number of radiographic views (typically, 2-4). Thereby, false alarms related to the presence of innocuous high-Z materials (Pb, Bi, etc.) are also heavily suppressed (Fig. 5). However, few-view reconstruction can only be performed when the cargo images exhibit a low degree of object clutter [17]. Intuitively, such clutter is a very sharp function of contrast sensitivity, which underscores the advantages of the variant of DDER and related inspection methodology advocated here.

Fig. 6 is a schematic of an FNRR/DDER cargo inspection system, showing how few-view spatial reconstruction can be achieved in a single pass, by means of several detector arrays.

FNRR

In recent years, the Australian Commonwealth Scientific and Industrial Research Organization (CSIRO) integrated, for the first time, fast neutron radiography using 14 MeV neutrons and Bremsstrahlung X-ray radiography into a single system [18]. This combination enhances the information on high-Z objects contained within the cargo. Further performance enhancement can be obtained by using FNRR with neutron beams in the 1-10 MeV energy range, where the transmission spectrum carries stoichiometric information about low-Z items. Thus, it permits distinguishing a broad spectrum of explosives from innocuous organic materials.

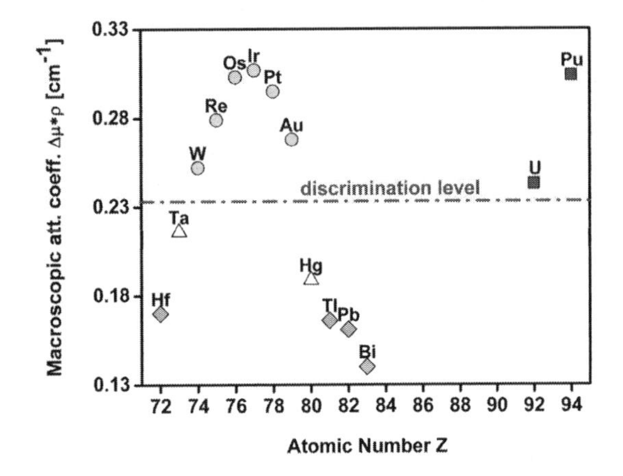

FIGURE 5. Difference of macroscopic attenuation coefficients ($\Delta\mu*\rho$) as function of Z for heavy elements, calculated at $E_{\gamma,low}$=4.43 MeV and $E_{\gamma,high}$=15.1 MeV. Note that this quantity can only be determined reliably from a few-view spatial reconstruction procedure. At the discrimination level indicated, only SNM and noble metals will alert the DDER system, thereby drastically reducing the false alarm rate.

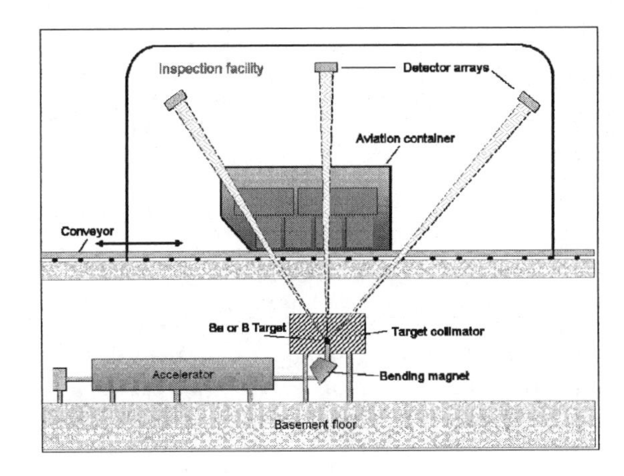

FIGURE 6. Schematic of Combined FNRR/DDER System, showing how few-view inspection can be achieved in a single pass by means of several detector arrays.

FNRR/DDER

Clearly, if each method (FNRR, DDER) is close to optimal by itself, as suggested above, then this is a-fortiori true of their simultaneous combination (Fig. 6), which addresses the detection of both threats, SNM and explosives.

REFERENCES

1. D. Vartsky, *Prospects of Fast-Neutron Resonance Radiography and its Requirements for Instrumentation*, Proc. of International Workshop on Fast Neutron Detectors and Applications (Capetown, 2006).
2. G. Chen, R.C. Lanza, *Fast Neutron Resonance Radiography for Elemental Imaging: Theory and Applications*, IEEE Trans. Nucl. Sci., 49 1919 (2002).
3. Dangendorf V., Kersten C., Laczko G., Jagutzky O., Spillman U., *Fast Neutron Resonance Radiography in a Pulsed Neutron Beam*, (arxiv).
4. Dangendorf V., Kersten C.et al, *Detectors for Energy Resolved Fast Neutron Imaging*, Nucl. Instr. and Meth. A 535. 93-97 (2004).
5. I. Mor, D. Vartsky, et al., *High spatial resolution fast-neutron imaging detectors for Pulsed Fast-Neutron Transmission Spectroscopy,* Journal of Instrumentation JINST 4 (2009) P05016.
6. M.B. Goldberg, Soreq NRC, U.S. Patent No. 7,381,962 (2008).

7. M.B. Goldberg, D. Bar, et al *Detecting Small Amounts of SNM in Cargo-Containers/Vehicles via Dual-Discrete-E γ-Ray Radiography following* $^{11}B(d,n)^{12}C$, 9th Intl. Conference on Applications of Nuclear Techniques, (Crete, June 2008).
8. J.C. Overley, *Element-sensitive computed tomography with fast neutrons*, IEEE Trans. Nucl. Sci. 30 (1983) 1677.
9. J.C. Overley et al., *Results of tests for explosives in luggage from fast-neutron time-of-flight transmission measurements,* Proc. SPIE Vol. 2867 (1997) 219.
10. T.G. Miller, P.K. van Staagen, B.C. Gibson, R.A. Krauss, *Contraband identification in sealed containers using neutron transmission*, Proc. SPIE Vol. 2867 (1997) 215.
11. *The practicality of Pulsed Fast Neutron Transmission Spectrometry for aviation safety*, National Research Council, publication NMAB-482-6 (1999) National Academy Press, Washington DC.
12. M.A. McMahan, L. Ahle et al., *Neutron beams from deuteron breakup at the 88-Inch cyclotron at Lawrence Berkeley National Laboratory*, Proceedings of Nuclear Data for Science and Technology (France, 2007)
13. L.M. Onischenkov et al., *Current Limit in the Compact Cyclotron with External Injection*, Proc. of Cyclotron and Their Applications conf. (Italy, 2007).
14. F. Marti, H. Blosser et al. *A Separated Sector Cyclotron for the Production of High Intensity Protons*, Particle accelerator conf. Proc. (Vancouver 1997)
15. L.M Young et al, *High Power Conditioning Of The LEDA RFQ*, Proceedings of the Particle Accelerator Conference (1999).
16. P.R Almond, J.R. Risser, *Differential cross sections at 0° and angular distributions of resolved neutrons from* $^{11}B+d$, Nucl. Phys. Vol 72 (1965) 436-448.
17. G. Feldman, D. Vartsky, et al *Analysis of Gamma ray nuclear resonant absorption (NRA) images for automatic explosive detection*, Int. Conf. on Image processing and its applications, IEEE Conf. Pub. 456, 789-793, July 1999.
18. Y. Liu, B.D. Sowerby, J.R. Tickner, *Comparison of neutron and high-energy X-ray dual-beam radiography for air cargo inspection*,Appl. Rad. & Isot. 66 (2008) 463.

A Neutron Based Interrogation System For SNM In Cargo

Steven Z. Kane and David S. Koltick

Purdue University, Physics Department, 525 Northwestern Avenue, W. Lafayette, IN 47907, USA

Abstract. A complete system has been simulated using experimentally obtained input parameters for the detection of special nuclear materials (SNM). A variation of the associated particle imaging (API) technique, referred to as reverse associated particle imaging detection (RAPID), has been developed in the context of detecting 5-kg spherical samples of U-235 in cargo. The RAPID technique allows for the interrogation of containers at neutron production rates between ~$1x10^8$ neutrons/s and ~$3x10^8$ neutrons/s. The merit of performance for the system is the time to detect the threat material with 95% probability of detection and 10^{-4} false positive rate per interrogated voxel of cargo. Detection times of 5 minutes were found for a maximally loaded cargo container uniformly filled with iron and as low as 1 second in containers loaded to ¼ of full capacity with either iron or wood. The worse case system performance, 30 minutes interrogation time, occurs for a maximally loaded container containing wood at 0.4 g/cm^3.

Keywords: SNM, API, prompt gamma-rays
PACS: 28.20.-v, 25.85.Ec

INTRODUCTION

In 2008, cargo container traffic bound for the United States was estimated at over 39 million TEU (twenty-foot equivalent unit) [1]. In addition, hundreds of metric tons of highly enriched uranium and plutonium are currently in existence, the United States and Russia having the largest stockpiles [2]. To combat the transportation of illicit special nuclear materials (SNM), the concept for a cargo inspection system, based on neutron-induced fission followed by the detection of multi-gamma and multi-detector coincidences, has been studied using Monte Carlo simulation code MCNP-PoliMi ver. 1.0 [3]. In contrast to systems that operate at neutron production rates on the order of 10^{11-12} neutrons/s to utilize SNM signatures based on delayed fission products [4], the proposed system operates between ~$1x10^8$ neutrons/s and ~$3x10^8$ neutrons/s, depending on cargo material and density. Lower neutron production rates are possible because the system utilizes the prompt gamma-ray signal. On average, ~7 prompt gamma-rays are emitted per fission [5] compared to ~0.127 delayed gamma-rays above the 3-MeV natural gamma-ray background [4] and ~0.015 delayed neutrons [4]. The reduced neutron production rates limit radiation safety hazards for operators and possible stowaways.

GAMMA-RAY JET-LIKE STRUCTURE

MCNP simulations were performed to study the angular distribution of prompt fission gamma-rays that exit a 5-kg spherical sample of U-235. These gamma-rays were found to exhibit an asymmetry in angular distribution similar to a "jet-like" structure [6]. The term "jet" is used strictly to denote gamma-rays that have exited the sample. The nuclear fission process itself produces gamma-rays uniformly in space. The jet-like structure is due to self-shielding of gamma-rays which are emitted toward the center of the uranium sample. These gamma-rays have a significant probability of being attenuated because the mean free path of 1-MeV gamma-rays in uranium is ~1 cm in contrast to the diameter of a 5-kg sample of U-235, which is ~8 cm. The jet-like structure is a surface effect. Due to a large total cross section of ~700 b [11], thermal neutrons only penetrate millimeters of the U-235 sample while the sample is nearly transparent to 14-MeV neutrons having a relatively low total cross section of ~6 b [11]. For thermal neutron-induced fission, 60% of the time, all gamma-rays were found to exit the sample in a solid angle of 2π [6]. In contrast, this occurred with a frequency of ~46% for 14-MeV neutron-induced fission [6]. Benign materials of comparable mass, such as iron and copper, were also found to produce gamma-ray jets through nuclei cascade de-excitations but with

Application of Accelerators in Research and Industry
AIP Conf. Proc. 1336, 717-722 (2011); doi: 10.1063/1.3586196

approximately half the jet rate [6]. Tungsten was found to have a slightly higher jet production rate than uranium but while producing on average fewer gamma-rays per jet [6]. By demanding the gamma-ray jet-like structure in a coincidence of detector panels surrounding a cargo container, background reduction is achieved.

THE SNM SCANNER CONCEPT

In Figure 1, the SNM scanner is depicted with an associated particle neutron generator, such as model A-920 from Thermo Scientific, capable of producing 10^9 neutrons/s at 14-MeV. The detector panels have dimension 122 cm by 305 cm (4 ft by 10 ft) and a depth of 30 cm, designed to detect gamma-rays with ~94% detection efficiency at 1-MeV. The goal was to cover ~1/2 Ω to detect fission events produced in the geometric center plane of the scanner. The motivation for this choice is to have excellent acceptance for multi-gamma events consisting of four or more gamma-rays. To first order, the probability of detecting an event is proportional to the fourth power of the product of the detector efficiency times the solid angle coverage. The design includes lead shielding to reduce Compton scattering of gamma-rays between detector panels.

The ideal gamma-ray detectors would have sub-nanosecond time resolution, σ_{time} ~250 ps, and corresponding position sensitivity, $\sigma_x = \sigma_y = \sigma_z$ ~7.5 cm. In addition, the detectors require high-rate capability, ~10 MHz/panel, to optimize throughput, and moderate energy resolution, σ_{energy} ~0.25 MeV, to provide additional background reduction. Liquid noble scintillators match this application well having high stopping power, high light output, radiation hardness, and sub-nanosecond time resolution [7]. Liquid xenon has the best scintillation properties but costs thousands of dollars per kilogram [8]. However, liquid krypton is relatively cheap at ~$200 per kilogram [8] and it has been demonstrated that a mixture of ~2% liquid xenon and liquid krypton yields practically the same scintillation properties as pure liquid xenon [7] [9]. In addition, using a depth of ~30 cm, MCNP simulations shown that liquid krypton achieves 94% detection efficiency for 1-MeV gamma-rays [6]. To obtain position sensitivity and to sustain singles rate up to 10 MHz, the detector panels would be comprised of cells of liquid krypton detectors.

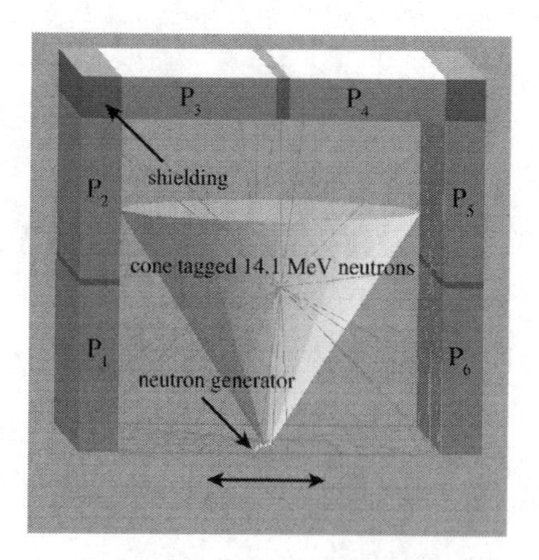

FIGURE 1. SNM scanner concept (neutron generator not drawn to scale, the container is shown transparent to reveal the cone of tagged 14-MeV neutrons).

MODES OF DETECTION

To optimally interrogate high-Z and low-Z cargo, the scanner simultaneously detects multi-gamma, multi-detector coincidence events produced from the interaction of 14-MeV neutrons and moderated neutrons. This is achieved by observing gamma-ray jets with and without alpha-gating, respectively. The detection of a gamma-ray starts a 10-ns logic gate, which is set to match the maximum gamma-ray time-of-flight across a container. The scanner can trigger at the gamma-ray detector level in two ways. The first condition is for a minimum of four gamma-ray logic gates to overlap in time such that the gamma-rays are distributed between three adjacent detector panels. These events, depicted in Figure 2, are labeled as three-fold panel coincidences. Following a three-fold panel trigger, a post-trigger processor determines whether or not the event is correlated to an alpha particle. The second condition is for the gamma-rays to be distributed between four adjacent detector panels. These four-fold panel coincidences are detected in the ungated mode of detection. The readout signal from the alpha detector is segmented into pixels using a multi-anode photomultiplier tube. The detection of an alpha particle starts a 60-ns logic gate equal to the time-of-flight for an unscattered 14-MeV neutron to cross to the farthest corner of a container.

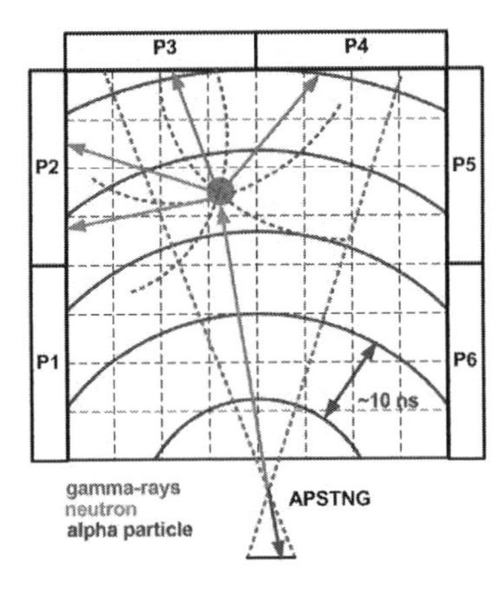

FIGURE 2. Depiction of an alpha-gated event: a 14-MeV neutron induces fission in SNM to produce a multi-gamma jet resulting in a three-fold panel coincidence between adjacent detector panels P_2, P_3, and P_4, in coincidence with the correlated alpha particle (not drawn to scale).

THE REVERSE ASSOCIATED PARTICLE IMAGING DETECTION (RAPID) TECHNIQUE

First examine how a three-fold panel coincidence would be processed using the traditional approach to associated particle imaging (API) [10]. In this scenario, the neutron generator is operated at $\sim 10^7$ neutrons/s so that the excess alpha pixel detector occupancy is near zero and a single detected alpha particle is uniquely able to generate a trigger. Consequently, as depicted in Figure 3, any three-fold panel coincidence overlapping in time with the alpha logic gate constitutes a true trigger. The neutron time-of-flight is estimated from the time difference between the detection of the gamma-rays and the start of the alpha logic gate. The directionality of the neutron is determined from the location of the D-T beam spot size on the generator target and the triggered alpha detector pixel. Using the neutron time-of-flight and direction, the neutron interaction point is imaged in 3-D. However, as neutron production rate increases and the sum of all pixel occupancies approaches one, "ghost" events appear due to the multiplicity of alpha-gamma coincidence associations. The traditional approach of API cannot suppress occurring random triggers and falsely associated alpha-gamma coincidences.

The approach taken is to exploit the fine gamma-ray detection time information and coordinate location information to perform a geometric reconstruction [6] of the source location of multiple coincident gamma-rays. If the gamma-ray information for a given trigger is uncorrelated to a given alpha particle, as is the case of random triggers or falsely associated alpha-gamma coincidences, the location of the reconstruction origin for the gamma-ray jet is statistically likely to disagree with the location obtained using the traditional approach to API. Because of these disagreements, random events and false alpha-gamma coincidence associations can be eliminated to some extent. The proposed technique allows higher neutron production rates to be used in the interrogation process. Consequently, the cargo image develops more rapidly.

The total alpha-gamma coincidence trigger, $R_{alpha\text{-}gamma}$, is the sum of the random background trigger due to random gamma-rays and gamma-rays from cascade de-excitations, $R_{uncorrelated\ background}$, and the alpha-correlated background trigger, $R_{correlated\ background}$. It is expressed as

$$R_{alpha-gamma} \approx 2R_{\alpha}R_{uncorrelated\ background}\tau + R_{correlated\ background}, \quad (1)$$

where τ is the alpha logic gate width. To first order, alpha-correlated background events reconstruct to a voxel matching a pixel with the detected correlated alpha particle. Therefore, the alpha-correlated background is not suppressed. However, random background triggers are suppressed. To first order, the random background suppression factor is

$$f \approx \frac{G(F)+1}{N_{Pixels}-1} \times \frac{1}{N_{Time\ Bins}}, \quad (2)$$

where $G(F)$ is the average number of ghost events created in a given length of time at a given neutron production rate F, N_{pixels} is the number of alpha pixels, and $N_{Time\ Bins}$ is the number of time bins in the radial direction from the neutron generator. The background reduction factor is the probability that a random background trigger coincidentally reconstructs into a voxel matching the gamma-ray jet location predicted using the traditional approach to API.

For the ungated mode of detection, the same geometric reconstruction is performed but no random background suppression is achieved using the alpha detector.

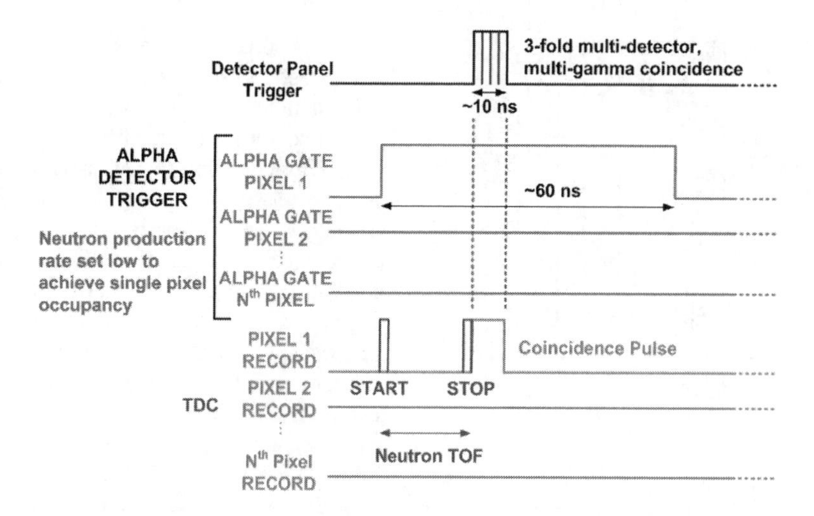

FIGURE 3. Detection of a three-fold panel coincidence using the traditional approach of API.

DECISION HYPERSPACE

The 2-D decision hyperspace plots the ratio of ungated signal to alpha-gated signal in a voxel, $R_{U/G}$, versus the average energy deposited in the detectors from that voxel, $<E>$. As an example, the decision hyperspace for a container uniformly filled with wood at 0.1 g/cm^3 is shown in Figure 4. Benign materials are found to cluster in a ball centered at mean signal ratio R_{Benign} and mean energy E_{benign}. Similar separations were observed at higher cargo densities [6]. A 5-kg sample of U-235 separates from 5-kg samples of U-238, tungsten, iron, and copper in wood and iron cargo [6]. In Figure 6, the statistical errors for the signal ratios and mean energies are negligible because a large number of neutrons were transported in the MCNP simulations to reduce statistical errors. However, δR_{Ball} and δE_{Ball} are non-negligible systematic errors, which are representative of the spread in benign material signatures. The decision parameter has the form:

$$\theta = \sqrt{\left(\frac{R - R_{Benign}}{\delta R}\right)^2 + \left(\frac{E - E_{Benign}}{\delta E}\right)^2}. \tag{3}$$

Note in Figure 4 that the signal ratio is enough to separate U-235 from benign materials. Thus, the signal ratio term in the decision parameter θ dominates the energy term.

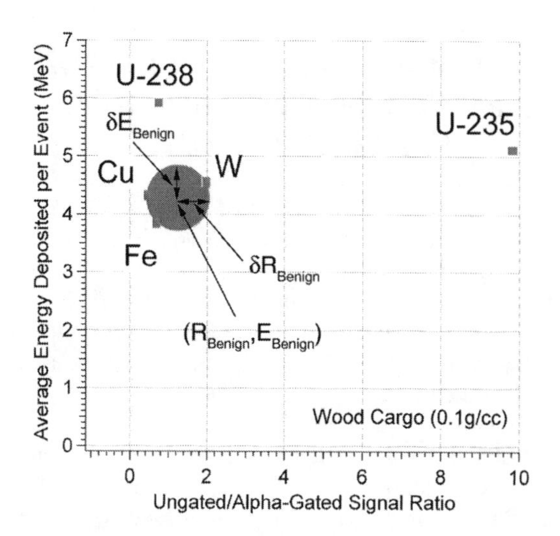

FIGURE 4. Decision hyperspace for a container uniformly filled with wood at 0.1g/cm^3 for a four detector panel scanner.

EXPECTED DETECTION TIMES

Because the largest "distance" in the hyperspace is the signal ratio information, for simplification and clarity, detection times were calculated without using the energy information. However, this information would improve the results. The ability to separate materials depends on the measuring error in $R_{U/G}$, given by

$$\delta R_{U/G} = \frac{S_U}{S_G}\sqrt{\left(\frac{\delta S_U}{S_U}\right)^2 + \left(\frac{\delta S_G}{S_G}\right)^2}. \tag{4}$$

However, for all cargo configurations studied, S_G, was found to be significantly smaller than the ungated signal, S_U [6]. Thus, the error in $R_{U/G}$ is dominated by the error in the alpha-gated signal, δS_G. Consequently, the error in $R_{U/G}$ is approximately

$$\delta R_{U/G} \approx \frac{S_U}{S_G}\sqrt{\left(\frac{\delta S_G}{S_G}\right)^2} = R_{U/G} \times \frac{1}{\theta_G}, \tag{5}$$

where θ_G is a decision parameter for detecting the alpha-gated signal. In terms of relative (percent) error this equation is expressed as

$$\frac{\delta R_{U/G}}{R_{U/G}} \approx \frac{1}{\theta_G}. \tag{6}$$

The decision parameter θ_G is similar to a statistical significance and is determined as

$$\theta_G \equiv \frac{C_G - B_G}{\sqrt{C_G}} = \frac{S_G^r}{\sqrt{S_G^r + B_G^r}}\sqrt{t}, \tag{7}$$

where C_G is the total number of counts observed in the alpha-gated mode of detection in a given voxel, and B_G is the expected mean background in the voxel. This equation assumes C_G and B_G obey Gaussian statistics. The equation can also be expressed in terms of the signal rate, mean background rate, and scan time.

For two materials in the decision hyperspace, the measuring error is used as the unit of distance separating one material from the threat group and one material from the benign group. A signal ratio cutoff, R_{cut}, can be set along the $R_{U/G}$ axis such that benign materials are below the cutoff and U-235 is above the cutoff. Using the true signal ratio value, R_{true}, and the measuring error in the signal ratio, δR_{true}, the probability, $P_{R,pass}$, that a material passes a chosen cut is given by

$$P_{\mathrm{R,pass}} = \frac{1}{2}\,\mathrm{erfc}\left[\frac{R_{cut} - R_{true}}{\delta R_{true}\sqrt{2}}\right]. \tag{8}$$

Substituting equation 6 into equation 9, $P_{R,pass}$ becomes

$$P_{\mathrm{R,pass}} = \frac{1}{2}\,\mathrm{erfc}\left[\frac{R_{cut} - R_{true}}{R_{true}\sqrt{2}}\,\theta_G\right]. \tag{9}$$

By inserting for R_{true} the signal ratio for a threat material, R_T, and benign material, R_B, two equations are obtained. For a given set of true positive, P_{TP}, and false positive, P_{FP}, rates of detection, these equations may be solved to find the R_{cut} value and mean decision parameter required for the signal ratio decision. It follows that the mean decision parameter required for the signal ratio decision is

$$\theta_G = \frac{\sqrt{2}}{R_T - R_B}\left[R_B\mathrm{erfc}^{-1}\left(2P_{FP}\right) - R_T\mathrm{erfc}^{-1}\left(2P_{TP}\right)\right]. \tag{10}$$

The time of detection is calculated from equation 7 as

$$t = \frac{S_G^r(F) + B_G^r(F)}{S_G^r(F)^2}\theta_G^2(F), \tag{11}$$

where θ_G is obtained from equation 10, and the signal and background rates are expressed as a function of neutron production rate.

A program was written to populate a container with reconstructed signal and background data from MCNP-PoliMi simulations. The random background trigger rates were reduced by the expected amount of background reduction predicted by equation 2. To use equation 2, a separate Monte Carlo simulation described in [6] was performed to estimate the number of ghosts produced at a given neutron production rate. For each mode of detection, the mean signal in the sample voxel was obtained and the mean background was estimated using the average counts in neighboring voxels. Plots of search time versus neutron production rate were obtained for the detection of 5-kg samples of U-235 in the center of containers uniformly filled with wood (low-Z) and iron (high-Z) cargo. From these plots, the shortest time of detection was determined. Expected detection times are shown in Table 1 for a scanner with time resolution of 250 ps.

CONCLUSION

Timing resolution of 250 ps produced detection times of 5 minutes in a maximally loaded cargo container uniformly filled with iron and as low as 1 second in containers loaded to ¼ of full capacity with either iron or wood. The worse case system performance, 30 minutes interrogation time, occurs for a maximally loaded container containing wood [12] at 0.4 g/cm^3. For general cargo, these detection times would suggest that the RAPID system would perform well as a confirmation scanner for SNM, especially with its ability to distinguish U-238 from U-235. The RAPID technique may be best suited for the interrogation of containers the size of airfreight cargo or similar sized vehicles, in this case, the worse case expected detection times for SNM would be a few

TABLE 1. Expected detection times for 5-kg of U-235 in center of container for scanner with time resolution of 250 ps (false positive rate set to 10^{-4}).

Cargo Material	Optimal Neutron Production Rate	Minimal Detection Time (to interrogate a voxel)
Iron at 0.1 g/cm^3	~$2.4x10^8$ neutrons/s	~1s for T.P. = 0.95
Iron at 0.4 g/cm^3	~$1.4x10^8$ neutrons/s	~5 min for T.P. = 0.90
Wood at 0.1 g/cm^3	~$2.5x10^8$ neutrons/s	~1 s for T.P. = 0.95
Wood at 0.4 g/cm^3	~$1.1x10^8$ neutrons/s	~ 30 min for T.P. = 0.90

minutes. Note that the Monte Carlo simulations performed in this study assumed an ideal alpha detector. However, the alpha detection efficiency was experimentally measured to be ~85% [13], which worsens actual detection times. More work, including a series of experiments, is necessary to validate the model and method.

REFERENCES

1. http://www.bts.gov/publications/americas_container_ports/2009/html/trends_in_container_throughput.html, accessed July 2010.
2. "Global Fissile Material Report 2009: A Path to Nuclear Disarmament, International Panel on Fissile Materials", IPFM, September 2009.
3. E. Padovani and S.A. Pozzi, MCNP-PoliMi ver. 1.0 User's Manual, Department of Nuclear Engineering, Polytechnic of Milan, Italy (2002).
4. D. Slaughter *et al.*, "Detection of special nuclear material in cargo containers using neutron interrogation", DOE, Lawrence Livermore National Laboratory, UCRL-ID-155315, August 2003.
5. T.E. Valentine, Annals of Nuclear Energy 28 (2001), 191-201.
6. S.Z. Kane, Detection of Special Nuclear Materials Using Prompt Gamma-Rays from Fast and Slow Neutron-Induced Fission, Ph.D. Thesis, Purdue University, 2010.
7. E. Aprile, A.I. Bolozdynya, A.E. Bolotnikov, T. Doke, "Noble Gas Detectors", December 2006.
8. R.S. Chandrasekharan, G. Davatz, A. Rubbia, "Noble Gas Scintillation-Based Radition Portal Monitors and Active Interrogation Systems", IEEE Nuclear Science Symposium Conference Record (2006).
9. D.Yu. Akimov *et al.*, IEEE Transactions on Nuclear Science, Vol. 42, No. 6, December 1995.
10. E. Rhodes, Dickerman, "Associated-Particle Sealed-Tube Neutron Probe for Non-Intrusive Inspection", Proc. of the 14th International Conference on the App. Of Accel. in Res. and Indust., Denton, TX, November 6-9, 1996.
11. National Nuclear Data Center, 92-U-235 ENDF/B-VII.0.
12. Wood definition in weight fraction: $H_{0.057}C_{0.48}O_{0.45}$, Compendium of Material Composition Data for Radiation Transport Modeling, Pacific Northwest National Laboratory, April 2006.
13. D.S. Koltick, S.Z. Kane, M. Lvovsky, E.K. Mace, S.M. McConchie, J.T. Mihalczo, IEEE Transactions on Nuclear Science 56, 3 PN Part 2 (2009) 1301-1305.

Monte Carlo Simulations Of The Response Of Shielded SNM To A Pulsed Neutron Source

E.H. Seabury and D.L. Chichester

Idaho National Laboratory, 2525 N. Fremont Ave., Idaho Falls, ID 83415

Abstract. Active neutron interrogation has been used as a technique for the detection and identification of special nuclear material (SNM) for both proposed and field-tested systems. Idaho National Laboratory (INL) has been studying this technique for systems ranging from small systems employing portable electronic neutron generators to larger systems employing linear accelerators as high-energy photon sources for assessment of vehicles and cargo. In order to assess the feasibility of new systems, INL has undertaken a campaign of Monte Carlo simulations of the response of a variety of masses of SNM in multiple shielding configurations to a pulsed neutron source using the MCNPX code, with emphasis on the neutron and photon response of the system as a function of time after the initial neutron pulse. We present here some preliminary results from these calculations.

Keywords: MCNP,MCNPX, Active Interrogation, Electronic Neutron Generator
PACS: 25.85Ec , 28.20Gd

INTRODUCTION

Neutron active interrogation has been used in a number of systems designed to detect special nuclear material (SNM). Systems based on this technique have been designed to inspect cargo[1], vehicles, and smaller items[2] to determine if a nuclear weapon or other SNM-containing device is being moved along with legitimate cargo. These systems are generally designed to examine only one or two signatures of SNM, e.g. neutron die-away or neutron multiplicity, and ignore some other signatures such as buildup of fission product activities or specific fission product gamma line emission. At Idaho National Laboratory (INL) we are interested in examining the complete space of signatures expected from neutron irradiation of shielded SNM and developing technologies for detection. As part of that effort, we have undertaken an extensive campaign of modeling Active Interrogation scenarios using the MCNPX[3] Monte Carlo radiation transport code developed at Los Alamos National Laboratory..

Active Interrogation Signatures

The expected response of SNM to a pulsed neutron source is shown schematically in Figure 1 below. As can be seen in the figure, there are a number of signatures that vary as a function of time after the initial neutron burst. There are prompt gammas and neutrons after the neutron burst, neutron die away in between neutron pulses, beta-delayed neutron and gamma emission in the interpulse region, and beta-delayed gamma and neutron emission in the post-irradiation time period. All of these signatures can be used to detect SNM, in particular highly-enriched uranium (HEU) which is difficult to detect passively if it is well shielded. In order to examine these signatures, appropriate tallies have been incorporated into the simulations that have been undertaken.

MCNPX MODELS

The geometry of the models is relatively simple. A sphere of uranium is surrounded by spheres of near-shielding material, and this assembly is embedded in a 3 meter by 3 meter cube of bulk shielding material. A neutron source is placed 20 cm from the face of the bulk shield and an array of ^{3}He detectors is placed near the neutron source and bulk shield. The geometry is shown schematically in Figure 2. The composition of the near and bulk shielding materials, masses of SNM and location within the bulk shield of the SNM are shown in Table 1. For all the simulations, the SNM was composed of 90% ^{235}U and 10% ^{238}U.

Application of Accelerators in Research and Industry
AIP Conf. Proc. 1336, 723-726 (2011); doi: 10.1063/1.3586197

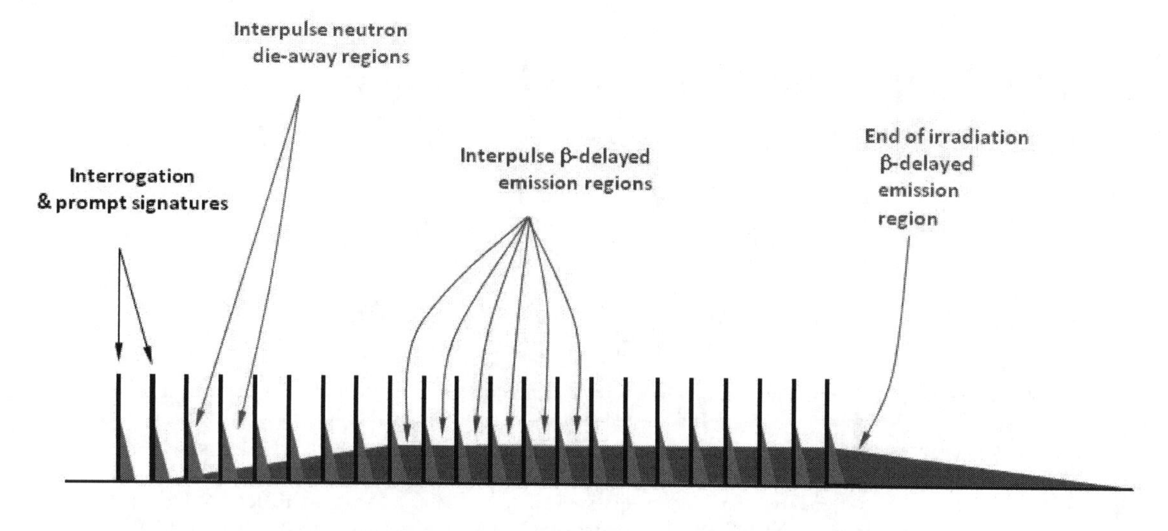

FIGURE 1. Response of SNM to a pulsed neutron source.

The neutron source used in the calculations was monoenergetic with either 2.5 MeV (deuterium-deuterium fusion) or 14.1 MeV energy (deuterium-tritium fusion). In all the simulations, the bulk shield is surrounded by air at sea-level pressure. The ^{3}He detectors are filled with 4 atmospheres of ^{3}He and are not shielded from the incident neutron source.

There are a number of tallies in the simulations. In addition to the ^{3}He detectors, there are tally locations above, behind, and to the sides of the bulk shield. The neutron and photon fluences are energy-binned and tallied as functions of time after the neutron burst at a number of locations around the bulk shield. Neutron captures in the ^{3}He gas of the detectors are tallied, fissions are tallied in the HEU sphere, and neutrons and photons are energy-binned and tallied as functions of time both into and out of the HEU sphere. In addition to the tallies described above, some mesh calculations of neutron fluence through the bulk shielding material were also performed.

Example Case: Concrete Bulk Shield

The entire set of calculations summarized in Table 1 comprises over 2500 individual runs. Each combination of model parameters in the table constitutes a run, e.g. there are runs for 0-20 kg HEU, at distances of 0.2-2.5 m, with all near-shielding materials in the concrete bulk shield. The first subset of calculations has been performed at INL, namely all the models using the concrete bulk shield. The concrete bulk shield consisted of standard materials, at a density of 2.35 g/cm3. One of the first items to be calculated was the neutron fluence mesh throughout the bulk shield with a bare 20 kg sphere of HEU placed within the bulk shield, 0.5 meter from the shield face. A contour plot of this simulation for the 2.5 MeV (DD) source case is shown in Figure 3. As can be seen in the figure, the HEU sphere does multiply some of the incident neutron flux. The plot also shows that air scatter around the bulk shield provides much of the neutron flux around the edges of the bulk shield, with the center of the shield having very low neutron flux as would be expected.

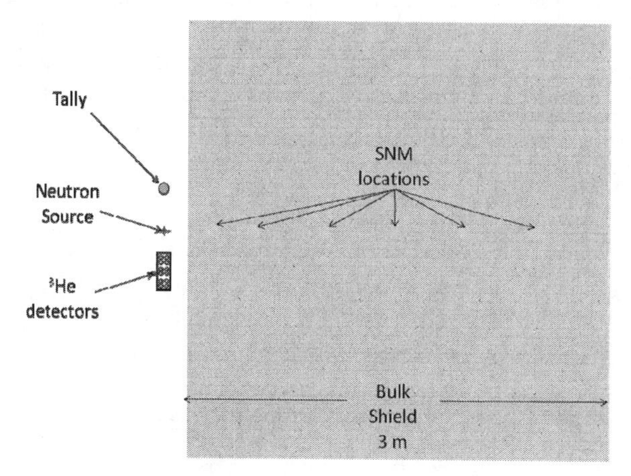

FIGURE 2. Model Geometry.

One of the more interesting results from these calculations is examining the effect of the near shield around the HEU on various tallies. For example, Figure 4 shows the effect of the near shield material on the total photon flux for the 20 kg HEU sphere 0.2 meter into the bulk shield and with a 14.1 MeV (DT) source. It is noteworthy that with the HEU encased in a steel shield, there is less total photon flux outside the bulk shield than in the case where there is no HEU or steel present at all. The effect of the near shield on the counts in the ^{3}He detectors can be seen in Figure 5.

TABLE 1. Model parameters for MCNPX simulations

HEU Mass (kg)	Near Shield	Bulk Shield	Hidden Depth	Neutron Source (MeV)
0	None	Air	0.2 m	14.1
1	5 cm Pb	Concrete	0.5 m	2.5
2	5 cm Steel	Plywood	1.0 m	
5	1mm Cd	Steel (0.6 g/cc)	1.5 m	
10	Pb + Cd	Aluminum (0.6 g/cc)	2.0 m	
20		Polyethylene Borated poly.	2.5 m	

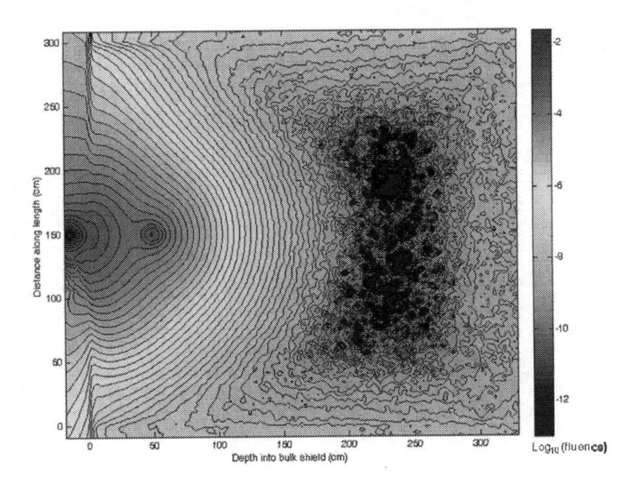

FIGURE 3. Mesh plot of neutron fluence in the concrete bulk shield for the 2.5 MeV neutron source and 20 kg HEU placed 0.5 meter from edge of bulk shield.

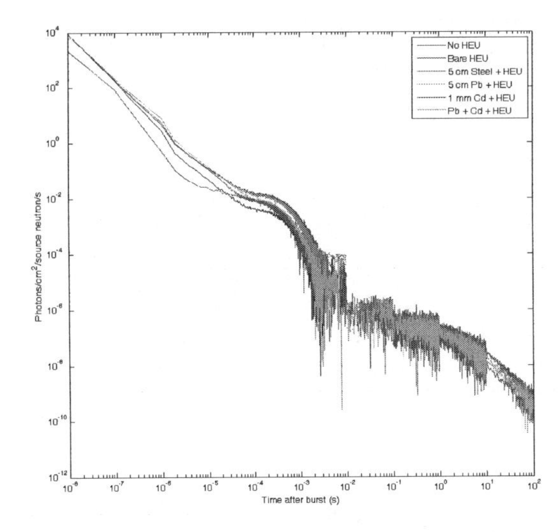

FIGURE 4. Total photon fluence outside the bulk shield as a function of time after the incident DT neutron burst for the 20 kg HEU 0.2 meter into the bulk shield case.

In this case, the HEU sphere had a mass of 10 kg and was placed 0.5 meter into the bulk shield. As can be seen in the Figure, 5 cm of lead shielding does not have a large effect on the counts in the detector. In contrast, the steel shield does have a significant effect, and the 1 mm Cd shield indeed gives rise to the most significant drop in count rates, as would be expected.

The effect of the depth within the bulk shield of the HEU can be seen in Figures 6 and 7. Figure 6 shows the total neutron fluence outside the bulk shield for the bare 20-kg sphere and DT neutron source case.

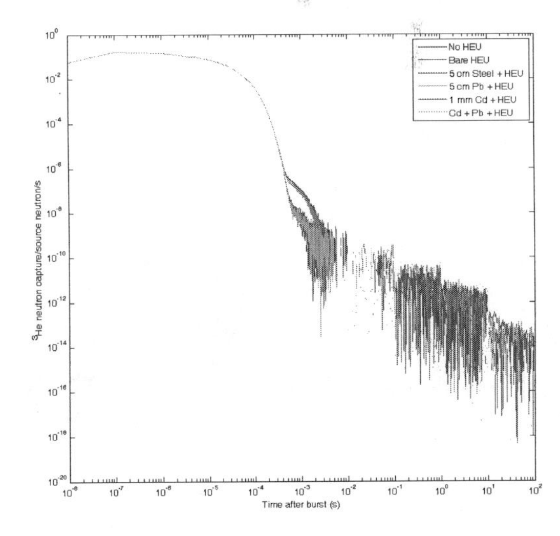

FIGURE 5. Effect of the near-shield material on neutron capture in all the ^{3}He tubes for the 10 kg, 0.5 meter, and DD neutron source case.

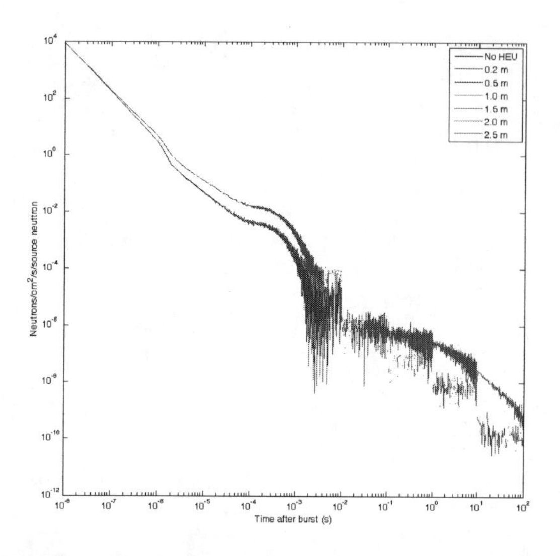

FIGURE 6. Total neutron fluence outside the concrete bulk shield as a function of hidden depth for the bare 20 kg, DT neutron source case.

As would be expected, the neutron counts seen in the tally drop as one places the HEU further and further into the bulk shield. The effect of the sphere placement on the number of fissions within the HEU can be are shown in Figure 7. In this case, the bare sphere contains 5 kg of HEU and is irradiated with a DD neutron source. With the sphere placed directly in the middle of the concrete bulk shield, the number of fissions observed drops by three orders of magnitude when compared to the sphere placed at 0.2 meter from the edge.

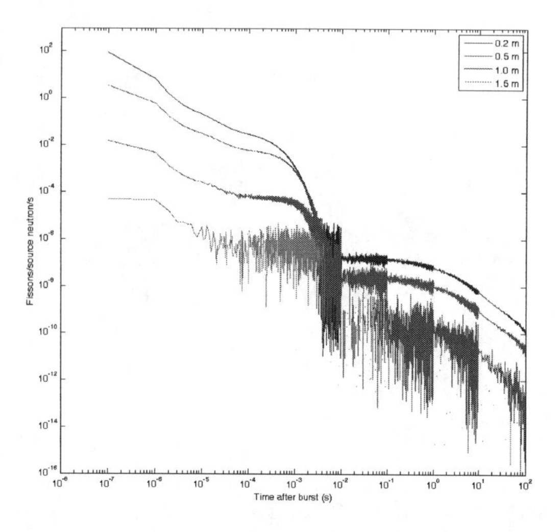

FIGURE 7. Total number of fissions in the bare 5 kg DD neutron source case as a function of hidden depth.

CONCLUSIONS AND FUTURE WORK

The value of these calculations lies in their use for the design of future Active Interrogation systems and their associated neutron sources. Knowing the expected response of shielded HEU as a function of time will allow us to concentrate on signatures that can be most easily observed for a given shielding scenario. Knowing the expected count rates of neutrons and photons outside the bulk shield as functions of energy will allow us to design detectors of sufficient and appropriate efficiency to determine the presence of SNM.

The remaining calculations for the other bulk shield materials will continue and provide a complete set of results for a wide selection of shielding scenarios.

ACKNOWLEDGEMENTS

The authors would like to thank the administrators of INL's High Performance Computing environment, whose help ensured that this large set of computer simulations could be performed in a timely manner. This work was supported by the U.S. Department of Energy National Nuclear Security Administration Office of Nonproliferation Research and Development (NA-22). Idaho National Laboratory is operated for the U.S. Department of Energy by Battelle Energy Alliance under DOE contract DE-AC07-05-ID14517.

REFERENCES

1. D. R. Slaughter *et al.*, "The nuclear car wash: A system to detect nuclear weapons in commercial cargo shipments," NIM-A, **579** , (2007), pp 349-352.
2. D.L. Chichester and E.H. Seabury, "Using Electronic Neutron Generators in Active Interrogation to Detect Shielded Nuclear Material," IEEE Trans. On Nuc. Sci, **56** (2009), pp 441-447.
3. D.B. Pelowitz, "MCNPX^{TM} User's Manual version 2.6.0," Los Alamos National Laboratory Report LA-CP-07-1473 (2008).

SECTION XIII – TEACHING WITH ACCELERATORS

Nuclear Physics Experiments Below The Coulomb Barrier

J. M. Sanders[a], J. R. Morales Cifuentes[a,b] and R. K. Clark[a]

[a]*Department of Physics, ILB 115, University of South Alabama, Mobile, AL, 36608, USA*
[b]*Department of Physics, University of Maryland, College Park, MD, 20742, USA*

Abstract. In 1932, Cockcroft and Walton showed that (p,α) reactions with lithium were possible at energies near 100 keV. We report an undergraduate laboratory experiment with 90 keV protons colliding with a thick lithium target. The experiment allows students to observe the products of two reactions, to determine the product masses, and to learn techniques for deconvolving experimental spectra profiles.

Keywords: Education, Nuclear Physics
PACS: 01.50 Pa, 25.40 Hs

INTRODUCTION

With a low-energy (150 kV maximum) accelerator, choices for nuclear physics experiments are limited. Although d-t and d-d reactions are possible, indeed they were the intended use of our accelerator, both deuterium and tritium are expensive, and tritium and the fast neutrons produced by the reactions raise radiation safety issues. We are fortunate that the ^{7}Li(p,α)^{4}He and ^{6}Li(p,α)^{3}He reactions are feasible at very low energy. These reactions were the first to be induced by accelerated particles when J.D. Cockcroft and E.T.S. Walton bombarded lithium with 125 to 400 keV protons in 1932 [1,2]. We describe introductory and advanced experiments using these reactions.

Lithium has two naturally occurring isotopes: ^{7}Li with an abundance of 92.4% and ^{6}Li with an abundance of 7.6%. When a low energy proton collides with either of these nuclides, a non-resonant reaction may occur leading to the emission of a high energy alpha particle. The two reactions are:

$$^{1}\mathrm{H} + {}^{6}\mathrm{Li} \rightarrow {}^{3}\mathrm{He} + {}^{4}\mathrm{He} \qquad (1)$$

$$^{1}\mathrm{H} + {}^{7}\mathrm{Li} \rightarrow {}^{4}\mathrm{He} + {}^{4}\mathrm{He} \qquad (2)$$

At the low energies of this experiment, the reaction can be viewed as an example of quantum tunneling. The distance of closest approach for a 90 keV proton to the lithium nucleus is

$$r_c = \frac{1}{4\pi\epsilon_0}\frac{Ze^2}{E_p} = 48\ \mathrm{fm}, \qquad (3)$$

while the nuclear radius for ^{7}Li is

$$r_7 = (1.25\ \mathrm{fm})A^{1/3} = 2.4\ \mathrm{fm}. \qquad (4)$$

Thus the distance of closest approach is twenty times the nuclear radius. For the projectile to interact with the nucleus, quantum tunneling is required.

For kinematic analysis, the Q value of the reaction is easily computed:

$$Q = (M_p + M_t - M_A - M_B)c^2, \qquad (5)$$

where M_p is the mass of the incident projectile, M_t is the mass of the target nucleus, and M_A and M_B are the masses of the products. Since the energy of the incident proton and also the resulting Q values are small compared to the rest energies of the reactants and products, non-relativistic expressions for momentum and kinetic energy may be used in further analysis.

Application of Accelerators in Research and Industry
AIP Conf. Proc. 1336, 729-733 (2011); doi: 10.1063/1.3586198

Elementary Kinematic Analysis

If the lithium (p,α) reaction is to be used in an introductory laboratory, the very low projectile energy in this experiment allows a simple calculation of the kinetic energies of the products of the reaction to be made. Since the initial momentum of the incident proton is small compared to the momenta of the emitted reaction products, we may treat the reaction as the breakup of a compound nucleus at rest. In this simple analysis, the products would be emitted in opposite directions and isotropically in the lab frame. These considerations lead by straightforward algebra to the expected kinetic energy of product A:

$$E_A = \frac{M_B}{M_A + M_B} Q , \tag{6}$$

and a similar expression is obtained for E_B by exchanging subscripts A and B.

If the energies of the particles E_A and E_B are measured, they may be compared with the expected energies. Alternatively, the masses M_A and M_B may be calculated from the energies assuming that the total mass of the products is not greatly different from the total mass of the reactants, viz. $A=M_A+M_B \approx M_p+M_t$. Then the mass of product A is given by

$$M_A = A \frac{E_B}{E_A + E_B} , \tag{7}$$

and the mass of product B is given by a similar expression obtained by interchanging the subscripts in Eq. 7.

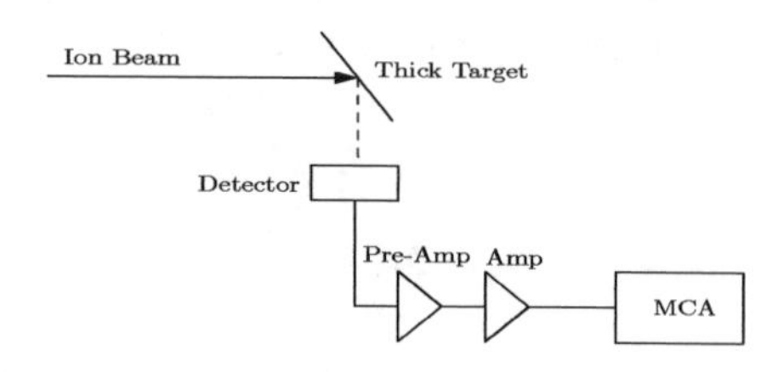

FIGURE 1. Schematic diagram of the experimental apparatus.

Advanced Kinematic Analysis

If the lithium (p,α) reaction is to be studied in an advanced laboratory, a full non-relativistic treatment of the kinematics of the reaction can be done using conservation of momentum along with the Q value to calculate the expected kinetic energies of the products. The algebra is somewhat involved and the result depends on the angles at which the products are detected. For product M_A detected at angle θ, the energy is given by

$$E_A = \left(\sqrt{H+G} - \sqrt{H}\right)^2 , \tag{8}$$

where

$$H = \frac{M_A M_p}{(M_B + M_A)^2} E_p \cos^2 \theta , \tag{9}$$

and

$$G = \frac{M_B}{M_A + M_B} Q + \frac{M_B - M_p}{M_A + M_B} E_p . \tag{10}$$

It may be noted that the exact result, Eq. (8), reduces to the result of the elementary analysis, Eq. (6), if E_p is set to zero.

If the observation angle is 90°, as it was in our case, the term H is zero, and the expression for the energy of the detected particle is much simpler

$$E_A = G = \frac{M_B}{M_A + M_B} Q + \frac{M_B - M_p}{M_A + M_B} E_p . \tag{11}$$

The first term in Eq. (11) is the same as in the elementary analysis, so the second term may be thought of as a correction to Eq. (6) that accounts for the non-zero energy of the incoming proton.

If we assume that the product masses are unknown, we may use the measured energies to compute the product masses. Taking again $A \approx M_p+M_t$, the expression for the mass of product A is

$$M_A = \frac{AE_B - M_p E_p}{E_A + E_B + 2\frac{M_p E_p}{A}} , \tag{12}$$

and the mass of product B is given by an expression obtained by exchanging the subscripts A and B in Eq. (12). If the projectile energy, E_p, is set to zero, Eq. (12) reduces to the approximate expression found earlier, Eq. (7).

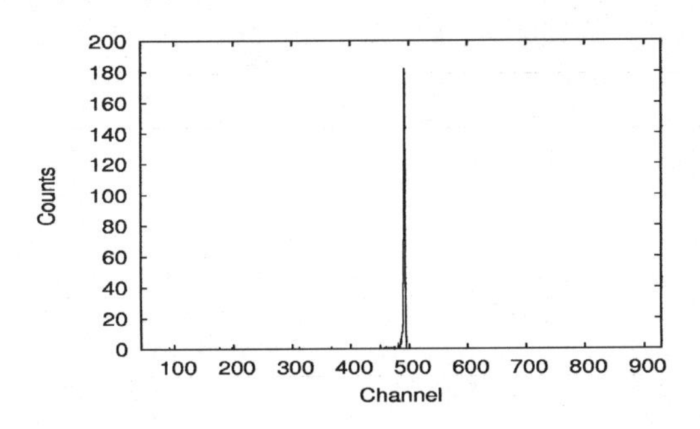

FIGURE 2. Energy calibration spectrum using alpha particles from ^{210}Po.

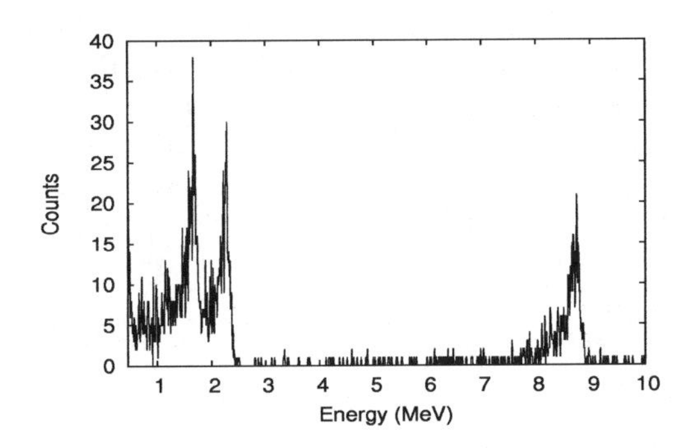

FIGURE 3. Alpha particle spectrum resulting from 90 keV proton bombardment of natural lithium fluoride.

EXPERIMENT

A schematic diagram of the experimental apparatus is provided in Fig. 1. A proton beam from the accelerator was directed onto a thick LiF target on a brass substrate. The target was oriented at 45° to the beam direction, and the beam current on target was monitored with an electrometer. Reaction products emitted from the target were detected by a surface barrier detector placed at 90° to the beam direction.

Signals from the surface barrier detector were amplified and sent to a multichannel analyzer (MCA) for pulse-height analysis. A one-point energy calibration of the electronics system was performed by using 5.3 MeV alpha particles emitted from a thin ^{210}Po source placed directly in front of the detector. This one point calibration assumes that the amplifier and MCA are completely linear and that zero energy corresponds to the zero channel on the MCA. These assumptions were made in the interest of making the calibration quickly in a short lab period. A better, but more time-consuming, calibration could be done by using other alpha emitters or by using the pulser method to find the zero-energy channel on the MCA [3].

The ^{210}Po calibration spectrum is shown in Fig. 2. The spectrum from proton bombardment of lithium is shown in Fig. 3. This spectrum was accumulated in approximately forty minutes with 60–70 nA of unsuppressed proton current on target. In the lithium spectrum, the lower two peaks arise from ^{6}Li and the upper peak from ^{7}Li. The ^{210}Po source is a thin one, so the peak in Fig. 2 is very narrow (25 keV FWHM). In contrast, the peaks from lithium disintegration in Fig. 3 are emitted from a thick target and are considerably broader (approximately 300 keV FWHM) due to energy loss of the alpha particles as they exit the target.

ANALYSIS

The broad, asymmetrical peaks in the alpha particle spectrum can be analyzed in a simple way that ignores their distorted shape or in a more advanced way that models their shape. In the sections below, both types of analysis will be discussed and an example of results from each will be presented.

Simple Analysis

An analysis of the data suitable for an introductory course can be done by simply taking the alpha particle energy to be the energy at the maximum of each peak. Since the peaks are asymmetrically skewed to lower energies, such a procedure will result in the alpha particle energies being underestimated. In Table 1, the results of the simple procedure are given. As can be seen, the results of the simple analysis are nevertheless consistent with the expected energies, and the calculated masses agree with the actual masses of the particles within one standard deviation.

Advanced Analysis

In an advanced laboratory, students can be introduced to deconvolution of spectral lines through the analysis of the alpha particle spectrum arising in this experiment. The intrinsic line shape of the spectrum of emitted alpha particles is Gaussian as can be observed in the ^{210}Po alpha spectrum shown in Fig. 4.

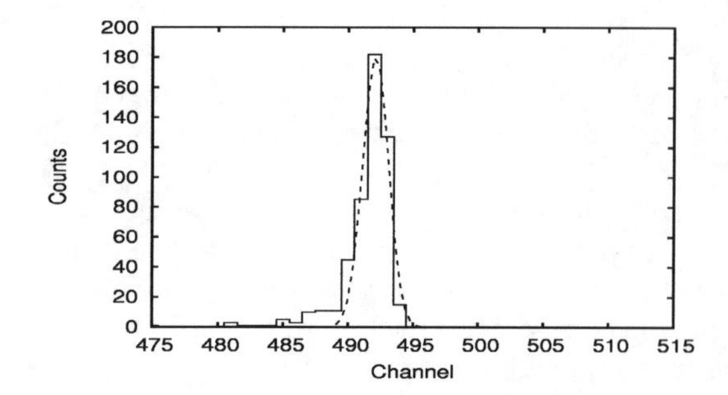

FIGURE 4. Gaussian fit to the ^{210}Po calibration spectrum. The solid line is the experimental data, and the dotted curve is the Gaussian fit. This spectrum is from a thin alpha source.

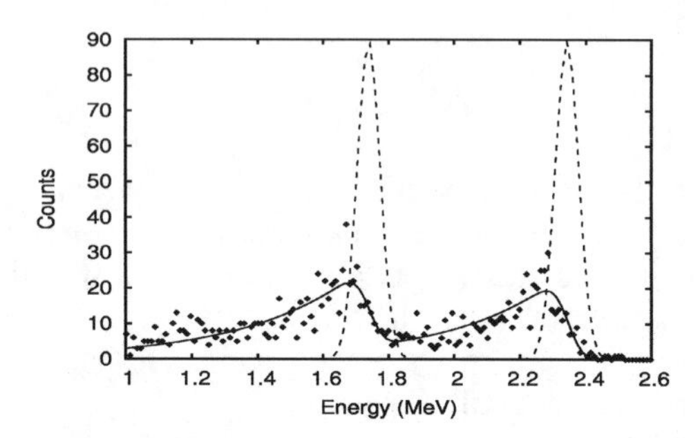

FIGURE 5. Fit of Eq. (16) to the peaks arising from the disintegration of ^{6}Li. The dashed curves are the undistorted Gaussian portions of the profiles.

So we take the line shape to be given by

$$G(E) = \frac{A}{\sigma\sqrt{2\pi}} e^{-\frac{1}{2}\left(\frac{E-\mu}{\sigma}\right)^2}. \tag{13}$$

We model the distortion distortion due to energy loss as a left exponential distribution, so that the probability of an alpha particle emitted with energy E' is detected with energy E is given by

$$F(E'-E) = \frac{1}{\lambda} e^{-\left(\frac{E'-E}{\lambda}\right)} \quad \text{for} \quad E \leq E' \leq \infty. \tag{14}$$

Here E is the energy, A is the number of alpha particles in the peak, σ is the Gaussian width, and μ is the mean alpha particle energy.where E is the detected energy, E' is the emitted energy, and λ is an adjustable parameter controlling the amount of the distortion. The number of alpha particles detected at energy E will be the convolution of the line shape and the resolution function

$$n(E) = \int_0^\infty G(E')F(E'-E)dE'. \tag{15}$$

TABLE 1. Reaction product energies and masses. The theoretical energies were obtained from Eq. (6). The measured energies result from identifying the energy of the maximum of each peak. The masses were computed using Eq. (7).

Product	Theoretical Energy (MeV)	Measured Energy (MeV)	Calculated Mass (u)
^{4}He (^{6}Li)	1.727	1.7 ± 0.2	4.0 ± 0.2
^{3}He (^{6}Li)	2.292	2.3 ± 0.2	3.0 ± 0.2
^{4}He (^{7}Li)	8.673	8.7 ± 0.2	4.00 ± 0.13

TABLE 2. Reaction product energies and masses. The theoretical energies were obtained from Eq. (11). The measured energies result from fitting Eq. (16) to the data. The masses were computed using Eq. (12).

Product	Theoretical Energy (MeV)	Measured Energy (MeV)	Calculated Mass (u)	Area (counts)
^{4}He (^{6}Li)	1.727	1.7 ± 0.2	4.0 ± 0.2	730 ± 30
^{3}He (^{6}Li)	2.292	2.3 ± 0.2	3.0 ± 0.2	730 ± 30
^{4}He (^{7}Li)	8.673	8.7 ± 0.2	4.00 ± 0.13	460 ± 20

The integral in Eq. (15) can be performed analytically and we find the model profile

$$n(E) = \frac{A}{2\lambda} e^{\sigma^2/2\lambda^2} \operatorname{erfc}\left(\frac{E}{\sigma\sqrt{2}} + C \right) \tag{16}$$

where

$$C = \frac{\sigma^2 - \mu\lambda}{\sigma\lambda\sqrt{2}}. \tag{17}$$

A fit to the peaks resulting from disintegration of ^{6}Li is shown in Fig. 5. Also shown in the figure is the undistorted Gaussian peak is determined from the fit. In Table 2, the results of the profile fits are given. In this case, there is a statistically significant discrepancy between the theoretical energy and the measured energy for the ^{4}He from ^{7}Li disintegration; this is most likely due to the crudity of the one-point energy calibration. In addition, the table also lists the reaction product masses based on their detected energies and the known masses of the proton and lithium isotopes.

Also shown in Table 2 are the integrated areas of the alpha peaks. The fit to the 6Li peaks was performed subject to the constraint that the areas of the ^{4}He and ^{3}He peaks were equal. From the integrated areas and the isotopic abundances, we obtained the ratio of the (p,α) cross section for ^{6}Li to that for ^{7}Li. We find the ratio of the cross sections at 90 keV is

$$\frac{\sigma_6}{\sigma_7} = 39 \pm 2.$$

This calculation assumes that the detection efficiency is the same for all of the detected particles.

CONCLUSION

Even though 90 keV is well below any resonances in the p + Li reaction, a usable amount of data can be accumulated in 40 minutes with 70 nA on target. That the reaction is seen at all at this low energy is an indication of quantum tunneling, since the distance of closest approach at this energy is much larger than the nuclear radius. The low energy of this experiment also allows for a very simple two-body break-up kinematic analysis to be done. The simple kinematic analysis allows this experiment to be considered even for students in their second semester introductory physics course.

For the elementary data analysis, where conceptual understanding is emphasized over experimental precision, agreement is good both with expected energies and masses. For the advanced analysis, agreement within two standard deviations was generally obtainable for both energies and masses. The most significant discrepancy was between the theoretical and experimental energy for ^{4}He emitted from ^{7}Li disintegration. We hypothesize that this discrepancy may be attributed to the limitations of the one-point energy calibration used in this experiment.

REFERENCES

1. J. Cockcroft, and E. Walton, *Nature* **129**, 649 (1932).
2. B. Cathcart, *The fly in the cathedral*, New York: Farrar, Strauss and Giroux, 2004.
3. J. L. Duggan, *AN34: Experiments in Nuclear Science*, Oak Ridge, TN: EG&G ORTEC,1984.

Teaching And Training Tools For The Undergraduate: Experience With A Rebuilt AN-400 Accelerator

Andrew D. Roberts

Department of Physics and Astronomy, Minnesota State University, Trafton Science Center N14,1 Mankato MN 56001, United States

Abstract. There is an increasingly recognized need for people trained in a broad range of applied nuclear science techniques, indicated by reports from the American Physical Society and elsewhere. Anecdotal evidence suggests that opportunities for hands-on training with small particle accelerators have diminished in the US, as development programs established in the 1960's and 1970's have been decommissioned over recent decades. Despite the reduced interest in the use of low energy accelerators in fundamental research, these machines can offer a powerful platform for bringing unique training opportunities to the undergraduate curriculum in nuclear physics, engineering and technology. We report here on the new MSU Applied Nuclear Science Lab, centered around the rebuild of an AN400 electrostatic accelerator. This machine is run entirely by undergraduate students under faculty supervision, allowing a great deal of freedom in its use without restrictions from graduate or external project demands.

Keywords: accelerator, nuclear, education
PACS: 01.50Pa

INTRODUCTION

It has long been recognized that small charged particle accelerators (<few MV) provide an excellent platform for the teaching and training of students at all levels in experimental techniques, with associated research planning, performance, analysis and presentation. Despite the increasing demand for staff skilled in nuclear techniques [1], anecdotal evidence suggests the use of small accelerators for use in the training of undergraduate students in particular has declined over recent decades. There is continued growth in the numbers of particle accelerators in use [2], but these machines are typically unavailable for training activities. For example, while there has been a boom of small cyclotrons (<16 MeV for protons) installed worldwide for medical radioisotope production, and an order of magnitude more machines for radiotherapy, the complete use and investigation by undergraduates is typically severely limited given the demands for their uninterrupted use and application based regulatory limitations.

In this light, we have established the Applied Nuclear Physics Lab at Minnesota State University, Mankato (MSU), an extension of the nuclear physics lab facilities established in the early 1970's [3,4,5]. The accelerator is the High Voltage Engineering AN-400, a 400 kV versatile multi-particle machine housed in a shielded vault (figure 1). The modest site requirements and simple components of this small accelerator make it ideally suited to an undergraduate teaching lab. While the machine was reliably used for two decades, its basic operating parameters were beginning to show some deterioration toward the end. An identical machine, decommissioned at St. Cloud State University, MN, was acquired to serve as spare parts to rebuild the MSU accelerator. While numerous experiments have been performed with undergraduates using this revitalized facility, we report here on one example of the types of investigations performed by our students.

Application of Accelerators in Research and Industry
AIP Conf. Proc. 1336, 734-737 (2011); doi: 10.1063/1.3586199

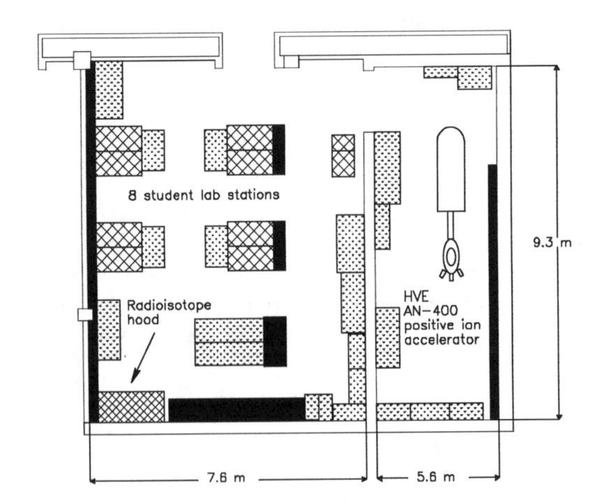

FIGURE 1. Schematic floorplan for the MSU Applied Nuclear Science Lab.

SAMPLE EXPERIMENTS

Several factors come into play when devising effective experiments for teaching undergraduates with particle accelerators. While undergraduates are often enthusiastic and eager, they are also inexperienced and pushed for available time. It is important for motivation to keep some big picture aims in mind, but to recognize that achieving these research aims should be reasonably parsed into smaller projects, achievable in a limited time frame. There are excellent references for particular experiments devised for this accelerator which can and should be consulted [6,7].

One example involves the production of ^{13}N, both for radiotracer studies in plants with [^{13}N]ammonia, and to investigate the atomic composition of the surfaces of target materials (an introduction to ion beam analysis). While these are not novel technologies in their own right, and typically exceed the time constraints of a typical undergraduate project, the ownership of such studies does provide ample opportunity for a series of training exercises.

Operation

In order to fully take advantage of the use of an accelerator for undergraduate training it is critical that the students have an in-depth understanding of the basic engineering of the machine. While there is relatively little physics involved in this step, spending the time to do a full maintenance (or repair, as required) and basic operations test on the machine with the students right away goes far toward establishing the confidence needed to approach further experiments.

Particle Identification

Another often overlooked exercise is to answer the basics question, what is the ion beam composed of? One simple test which requires little specialized detection equipment is to look at the steering of the beam with a suitable dipole magnet. The MSU AN-400 accelerator is equipped with a 1.2 kG dipole magnet with 0° and 15° beam ports, suitable for proton and deuteron extraction to 400 keV. Measurement of the magnetic field and range with a simple handheld gauss probe along with the physical dimensions of the field extent and target geometry are sufficient to provide a calculation of the expected steering conditions for the mass 1 and 2 beams vs. particle speed. Figure 2 shows a sample measure for the AN-400 particle beam at a nominal 300 kV steered through the 15° port vs. dipole magnet current, using a (presumed) H_2 gas supply on the ion source.

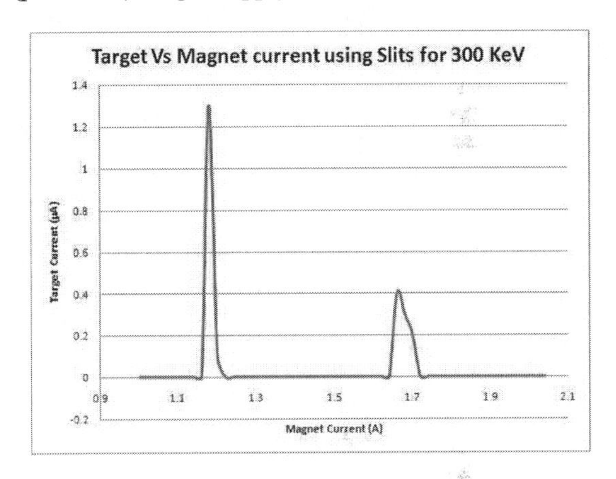

FIGURE 2. Sample beam current measurements vs. dipole magnet current for a nominal 300kV beam, showing the location and separation of the mass 1 (left) and mass 2 (right) peaks.

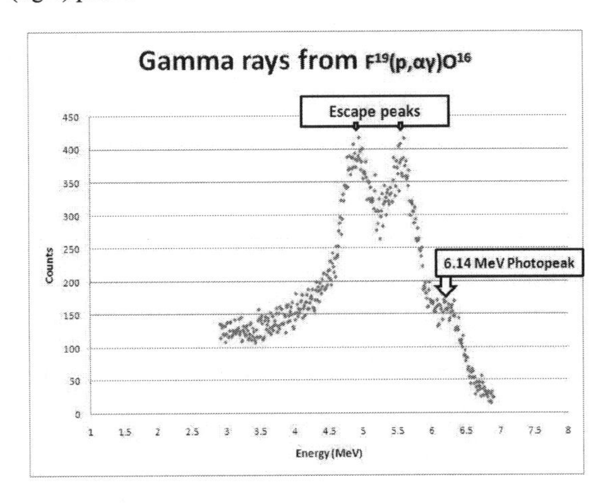

FIGURE 3. Sample gamma energy spectrum showing the 6.14 MeV gammas following the $^{19}F(p,\alpha)^{16}O$ reaction for E>340 keV.

Energy Calibration

Although the particle identification project does provide a crude measure of beam energy, it is limited by the accuracy of the magnetic field and target/switch magnet geometry measurements. True validation of the beam energy and species is best determined through measurement of a well established nuclear reaction. One such reaction suitable for a low energy proton beam is $^{19}F(p,\alpha)^{16}O$, which has a narrow resonance at 340 keV [6,8]. The signature for this reaction is the prompt 6.1 MeV gamma emitted by the decay of the excited ^{16}O nucleus after production. A sweep of the reaction across some reference for beam energy (such as the measured acceleration column current or the generating voltmeter reading) then gives a good single energy reference calibration. This procedure is particularly amenable to student application, since the high energy signature gamma is easily detected by a single 2"x2" NaI detector, placed external to the target vacuum chamber, with few stringent requirements on detector placement or geometry. The ^{19}F target is easily manufactured by short exposure of standard aluminum foil to HF vapor, then placed on our copper target rod. Figure 3 shows a typical energy spectrum for the 6 MeV gamma emitted with beam energy in excess of the 340 keV resonance. Normalizing for beam current and irradiation time gives the measured data rate vs. beam energy (column current in our case), shown for our machine in figure 4.

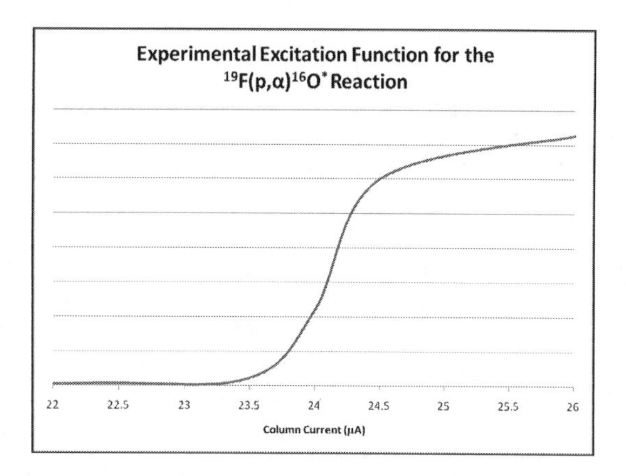

FIGURE 4. Measured excitation function for the narrow $^{19}F(p,\alpha)^{16}O$ reaction resonance at 340 keV with a thick ^{19}F target on aluminum foil. The relative rates are plotted against accelerator column current as an uncalibrated measure of beam energy.

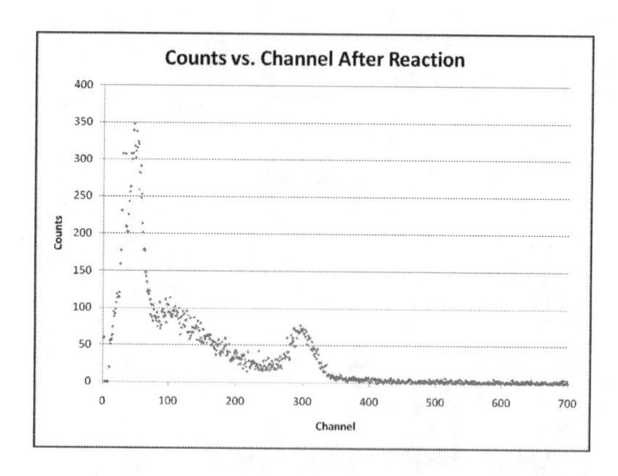

FIGURE 5. Sample gamma energy spectrum with a single 2"x2" NaI detector for the 511 keV photons from positron decay of ^{13}N after production by the $^{12}C(d,n)^{13}N$. The 511 keV peak sits approximately at channel 300 on this spectrum.

Deuteron Reaction

The next experiment is to investigate the production of ^{13}N by the $^{12}C(d,n)^{13}N$ reaction, which has a Q-value of -281 keV for a reaction threshold of 328 keV. For this measurement the target is a thick slab of graphite, mounted to the same copper target post used in previous measurements. The signature for the reaction is the 511 keV peak from positron decay of the ^{13}N nucleus. While this lower energy gamma does suffer from some absorption in the target chamber walls and beam pipe, the effect is minimal and the same external NaI detector is used. Coincident detection of the paired gammas from the target can be used, however we found the signal sufficiently clean to accurately measure the decay half life of 10.0 min using the singles rates from one detector (figure 5). While this reaction doesn't exhibit the narrow resonance used in the $^{19}F(p,\alpha)^{16}O$ reaction, the reaction rate vs. beam energy does show the expected behavior consistent with the known threshold and cross sections (figure 6).

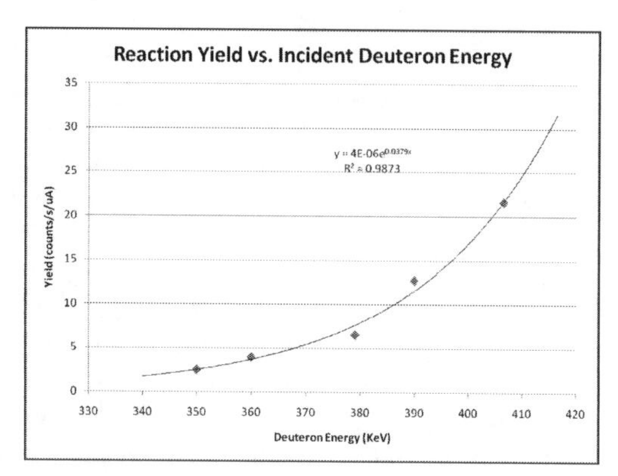

FIGURE 6. Measured excitation function for the $^{12}C(d,n)^{13}N$ reaction on a thick graphite target.

SUMMARY

The accelerator projects listed above are well within the skill levels of typical undergraduates at the level of introductory/modern physics training. While much more would need to be done to approach the main goal of applied ^{13}N use as suggested above, these experiments do offer a solid experimental training base for physics students, with only minimal requirements for specific calculations in nuclear physics. Furthermore, the series is relatively simple, given access to the accelerator, and can, with modest planning be successfully performed in a 1 or 2 semester sequence.

ACKNOWLEDGMENTS

Undergraduate research at the Applied Accelerator Facility was funded in part through internal grants through the MSU Undergraduate Research Conference, and facilities funding through the MSU College of Science, Engineering and Technology. Special recognition should be made to the following MSU undergraduate students for their efforts: C. Anderson, S. Clarke, J. Clymer, N. Compton, H. Dam, A. Hanson, O. Oliver, C. Prokop, A. Timilsina, and R. Wickland.

REFERENCES

1. American Physical Society, Panel on Public Affairs, R.H. Socolow chair, 2008 report, Readiness of the U.S. Nuclear Workforce for 21st Century Challenges.
2. W. Maciszewski and W. Scharf, Particle Accelerators for Radiotherapy. Int. J. Radiation Oncology, (2004), 76.
3. Udeh, N.V.; Chmielewski, J.J.; Flinner, J.L.; Inman, F.W.; Sollenberger, B. (1981) "Tantalum L-X-ray-production cross sections for 280-400-keV proton bombardment," *Physical Review* **A** (General Physics), Oct. 1981, vol.24, no.4, pp. 1746-9.
4. Chmielewski, J.J.; Flinner, J.L.; Inman, F.W.; Sollenberger, B.; Udeh, N.V. (1981) "Study of the L-shell X-ray production cross section of cadmium by proton bombardment," *Physical Review* **A** (General Physics), July 1981, vol.24, no.1, pp. 29-32.
5. Fast, S.; Flinner, J.L.; Glick, A.; Inman, F.W.; Oolman, L.; Pearson, C.; Wickelgren, D. L (1982) "X-ray-production cross sections for (120-400)-keV proton bombardment of indium", *Physical Review* **A** (General Physics), Nov. 1982, vol.26, no.5, pp. 2417-20.
6. Accelerator Nuclear Physics, W. Bygrave, P. Treado and J. Lambert, High Voltage Engineering Corp., 1970
7. *Experiments in Nuclear Science*, J. L. Duggan, AN34 Laboratory Manual, ORTEC, http://www.ortec-online.com/Library/an34.aspx
8. J.B. Marion, Rev. Mod. Phys. 38, 660 (1966) 660-668.

Undergraduate Measurements For Fission Reactor Applications

S. F. Hicks[1] L. J. Kersting[1], C. J. Lueck[1], P. McDonough[1], B. P. Crider[2], M. T. McEllistrem[2], E. E. Peters[2] and J. R. Vanhoy[3]

[(1)]Department of Physics, University of Dallas, Irving TX 75019, USA
[(2)]Department of Physics and Astronomy, University of Kentucky, Lexington, KY 40506, USA
[(3)]Department of Physics, United States Naval Academy, Annapolis MD 21402, USA

Abstract. Undergraduate students at the University of Dallas (UD) have investigated elastic and inelastic neutron scattering cross sections on structural materials important for criticality considerations in nuclear fission processes. Neutrons scattered off of ^{23}Na and ^{Nat}Fe were detected using neutron time-of-flight techniques at the University of Kentucky Low-Energy Nuclear Accelerator Facility. These measurements are part of an effort to increase the efficiency of power generation from existing fission reactors in the US and in the design of new fission systems. Students have learned the basics of how to operate the Model CN Van de Graaff generator at the laboratory, setup detectors and electronics, use data acquisition systems, and they are currently analyzing the angular dependence of the scattered neutrons for incident neutron energies of 3.57 and 3.80 MeV. Most students participating in the project will use the research experience as the material for their undergraduate research thesis required for all Bachelor of Science students at the University of Dallas. The first student projects on this topic were completed during the summer of 2010; an overview of student participation in this investigation and their preliminary results will be presented.

Keywords: Undergraduate research, neutron spectroscopy, elastic and inelastic neutron scattering differential cross sections
PACS: 25.40.Dn, 25.40.Fq, 25.55.Ci, 27.30.+t, 27.40.+z, 28.20.Cz, 28.41.-I, 28.41.Qb, 28.52.Fa, 29.20.Ba

INTRODUCTION

Undergraduate students at the University of Dallas have participated in an experimental program to investigate the scattering of neutrons off of materials such as ^{23}Na and ^{Nat}Fe that are important in fission reactor technology. Nuclear energy currently accounts for about 20% of the total electricity production in the U.S and 16% globally. Because of global population, environmental protection, and energy concerns, there is a need to increase substantially the nuclear reactors available for electricity production in the U.S. and around the world[1].

Both existing and new reactors depend heavily on nuclear data for their design and for determining safe operating parameters. This is done primarily through computer modeling which depends heavily on experimental data and data uncertainties for model input. In addition to knowing how much energy is generated by the fission reactions, one must also understand coolant properties, heat transfer, and structural material properties in an environment filled with neutrons and γ rays. Cross sections, reaction rates and production rates are some of the important model input, and much of the needed data are either missing or were taken years ago and have large uncertainties. Inelastic and elastic neutron scattering cross sections and γ-ray production cross sections on the structural materials of reactors, particularly Fe and Na, and have all been identified as areas where new data are needed[2-5].

The University of Dallas (UD) is a liberal arts institution with limited in-house research laboratories, but a collaboration between scientists and students at UD, the United States Naval Academy (USNA), and the University of Kentucky (UK) has been ongoing for more than 20 years and has focused on the study of nuclei with neutrons as the probe for both scattering and structure measurements. Because of this long history of neutron physics research and, of equal importance, the history of including undergraduates in this research, a project to measure Fe and Na elastic and inelastic neutron scattering cross sections was

Application of Accelerators in Research and Industry
AIP Conf. Proc. 1336, 738-741 (2011); doi: 10.1063/1.3586200

funded by the Department of Energy's Nuclear Energy Universities Program (NEUP) to begin in 2010.

Students at UD receive an education with a strong emphasis on liberal learning at the undergraduate level. Physics majors complete an extensive core curriculum in humanities while carrying out the requirements of a rigorous major in physics, which for a B.S. student includes an undergraduate research project, a research thesis, and the presentation of their work at a professional meeting. Student projects are typically completed during a 10-week summer research period. Two students worked on the neutron scattering measurements during the summer of 2010 and completed projects which included deducing differential angular distributions of scattered elastic and inelastic neutrons from ^{Nat}Fe and ^{23}Na, and modifying a computer program for fitting neutron TOF spectra so that multiple kinematic groups could be fit simultaneously. The focus of the research presented here is the role these two UD undergraduate students have played in the measurement of elastic and inelastic neutron scattering cross sections from ^{Nat}Fe at E_n=3.57 and 3.80 MeV. These two incident neutron energies were chosen because they are in an energy region sparse of such data and needed for calculations of neutron processes in fission reactors.

EXPERIMENTAL METHODS

The experiments discussed were all completed at the UK Low-Energy Accelerator Facility using the neutron production and detection facilities located there. The accelerator is a modified 7 MV Model CN Van de Graaff manufactured by High Voltage Engineering Corporation and is capable of producing the pulsed proton beams necessary for neutron production and time-of-flight (TOF) measurements. Undergraduates participating in these measurements are introduced to the operation of the accelerator, neutron detectors and shielding, electronics, and data acquisition systems while at the laboratory in a very hands-on way.

The ^{3}H(p,n)^{3}He reaction was used as a neutron source. The scattering samples were 59.223 g of natural Fe in cylindrical shape with a radius of 1.90 cm and height of 2.54 cm. Neutrons were detected using TOF techniques with a 2.54 cm by 11.4 cm hexafluorbenzene liquid scintillator detector located 3.22 m and 3.17 m from the scattering sample at 3.57 and 3.80 MeV, respectively. The main detector apparatus is shown in Fig. 1. A second NE218 liquid scintillation detector was used as a monitor detector and kept at a fixed location in the laboratory for all measurements; this detector looked at the middle of the ^{3}H gas cell.

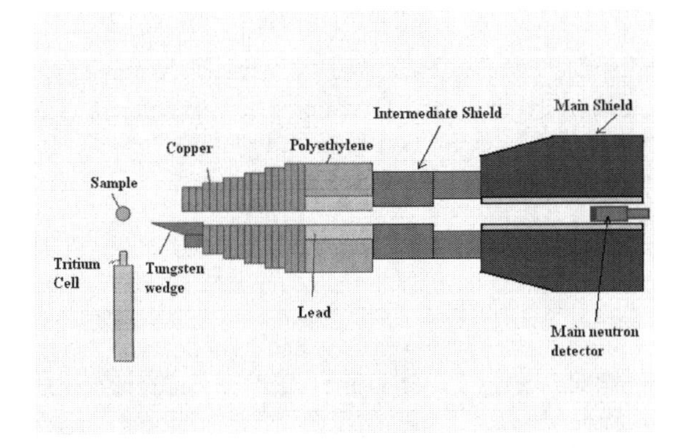

FIGURE 1. Experimental setup for the main detector in relation to the scattering sample and neutron source. The detector is located 3.22 m from the sample and shielded from the neutron source by the apparatus shown.

Further details of the experimental techniques used can be found in Ref. 6. An example neutron TOF spectrum is shown in Fig. 2.

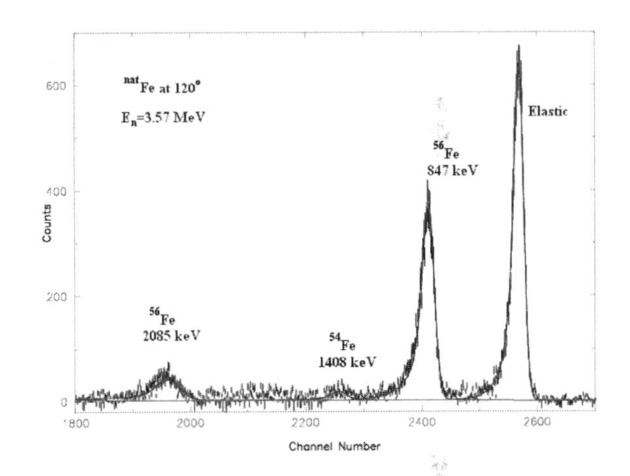

FIGURE 2. ^{Nat}Fe TOF spectrum at E_n=3.57 MeV. Peaks are labeled in the spectrum from elastic and inelastic neutron scattering from ^{54}Fe and ^{56}Fe.

Neutron Detector Efficiency

The first analysis completed by students was the relative efficiency of the neutron detector. To complete this part of the project, students had to fit main detector and monitor detector spectra with a neutron peak fitting program developed at UK. Scattered neutron energies vary for a given incident neutron energy because of kinematic shifts as a function of detector angle and due to inelastic neutron scattering. The relative efficiency of the detector as a function of neutron energy was determined by measuring the angular distribution of neutrons produced by the ^{3}H(p,n)^{3}He reaction. Experimentally this required placing the pivot point of the rotatable carriage under the center of the gas cell and detecting the neutrons produced from about 5-135 degrees. The relative efficiency was determined by

$$\varepsilon(E_n) = \frac{Y_{main}(\theta)}{Y_{mon}(\theta)\sigma(E_n,\theta)}, \qquad (1)$$

where $Y_{main}(\theta)$ is the yield of neutrons detected in the main detector, $Y_{mon}(\theta)$ is the yield of neutrons in the monitor detector, and $\sigma(E_n,\theta)$ is the known source reaction cross section. The relative detector efficiency for the Fe experimental run is shown in Fig. 3; this analysis was completed by UD student Luke Kersting.

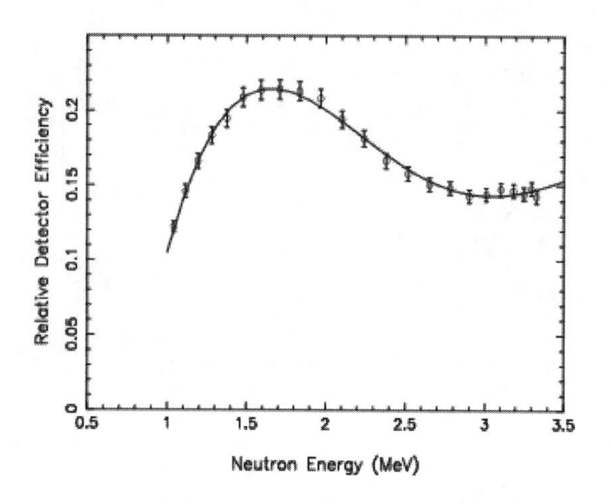

FIGURE 3. Neutron detector efficiency for the main detector. The solid line is a polynomial fit to the data.

Angular Distributions Of Scattered Neutrons

The primary goal of the project is to measure the angular distributions of scattered elastic and inelastic scattered neutrons as a function of scattering angle. These measurements are completed by moving the pivot point of the rotatable carriage under the scattering sample and detecting the neutrons scattered off of the sample. At E_n= 3.57 MeV, both elastic and inelastic scattering peaks are observed in the resulting spectrum, as shown in Fig. 1. Peak areas were extracted, normalized to the yield from the neutron monitor spectrum, and corrected for neutron detector efficiency by UD students to determine the relative angular distribution as a function of scattering angle, which are defined at each angle as

$$W(\theta) = \frac{Y_{main}(\theta)}{Y_{mon}(\theta)\varepsilon(E_n)}, \qquad (2)$$

where each of these quantities is defined above. The relative angular distributions for both elastic scattering from ^{Nat}Fe and from the inelastic level at 847 keV in ^{56}Fe are shown in Fig. 4 and Fig. 5, respectively. Once these data are corrected for multiple scattering and normalized to the well known *n-p* scattering cross sections, which were also measured at each incident neutron energy, the final absolute cross sections required for the NEUP project will be complete at this energy. University of Dallas student Collin Lueck has completed a programming project to analyze the *n-p* scattering data; that analysis is currently in progress.

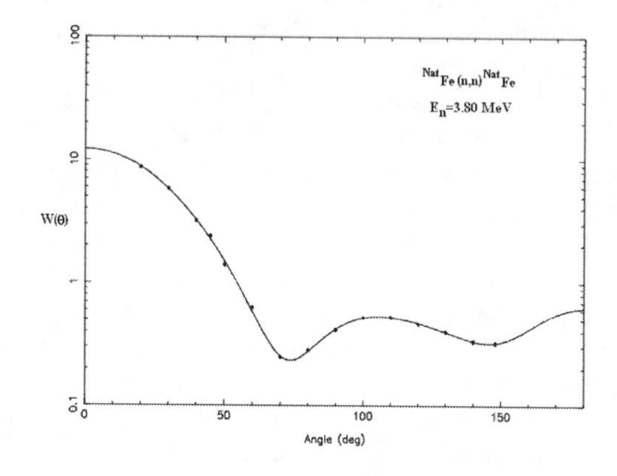

FIGURE 4. Relative elastic neutron scattering angular distribution at E_n=3.8 MeV from ^{Nat}Fe. These data have been corrected for the detector efficiency but not multiple scattering. Statistical uncertainities are less than 3%. The solid line is a Legendre polynomial fit to the data.

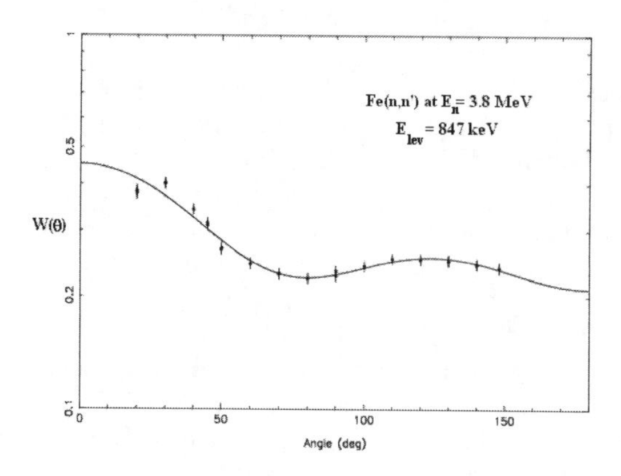

FIGURE 5. Relative inelastic neutron scattering from the 847 keV first excited state of ^{56}Fe. These data have been corrected for detector efficiency but not multiple scattering. Statistical uncertainites are less than 3.8%.

RESULTS

University of Dallas students have participated in a series of neutron scattering measurements on ^{Nat}Fe and ^{23}Na. They have analyzed monitor and main detector spectra to determine the neutron detector efficiency and the relative scattering probabilities for elastic and inelastic scattering from a ^{Nat}Fe sample at E_n=3.57 and 3.80 MeV. While participating in these measurements and analyzing data, they have learned much about experimental techniques, data analysis, and nuclear physics. Furthermore, their work has contributed to the national need for high quality neutron scattering data important for fission energy research. The data will be included in the National Nuclear Data Center Compilation as soon as all corrections have been completed, and the students involved will submit a thesis to the UD Physics Department as part of their B.S. requirements.

ACKNOWLEDGMENTS

This work was supported in part by the Department of Energy through the NEUP program and by the Cowan Physics Fund at the University of Dallas. We acknowledge with appreciation the contributions of Harvey Baber the UK accelerator engineer and UK Professor Steven W. Yates for his many contributions to this project.

REFERENCES

1. *A Technological Roadmap for Generation IV Nuclear Energy Systems Executive Summary*, issued by the U.S. Doe Nuclear Energy Research Advisory Committee and the Generation IV International Forum, http://nuclear.energy.gov/genIV/documents/Roadmap_Executive_Summary_4-15-03.pdf.
2. Nuclear Energy Agency, http://www.nea.fr/html/dbdata/welcome.html.
3. Nuclear Data Center, Japan Atomic Energy Agency, http://www.ndc.tokai.sc.jacea.go.jp/jendl/jendl.html.
4. R. Juzaitis, Los Alamos Science **28**, 32 (2003).
5. G. Aliberti, G.Palmiotti, M.Slvatores, T.K.Kim, T.A.Taiwo, M.Anitescu, I.Kodeli, E.Sartori, J.C.Bosq, J.Tommasi, Annals of Nuclear Energy **33**, 700 (2006).
6. Sally F. Hicks, S. E. Hicks, G.R. Shen, and M. T. McEllistrem, Phys. Rev. C**40**, 2560 (1990).

Educational Activities At The Nuclear Engineering Teaching Laboratory

Tracy N. Tipping

The University of Texas at Austin, Nuclear Engineering Teaching Laboratory, 1 University Station, R9000, Austin, Texas, USA 78712-0742

Abstract. The Nuclear Engineering Teaching Laboratory (NETL) at the University of Texas at Austin performs a wide variety of educational activities for students at various levels. Regular on-site courses in the areas of health physics, radiochemistry, and reactor operations are offered for university credit. Along with on-site courses, access to the reactor facility via a remote console connection allows students in an off-site classroom to conduct experiments via a "virtual" control console. In addition to the regularly scheduled courses, other programs, such as the Nuclear Regulatory Commission Summer Nuclear Engineering Institute and Office of Naval Research partnerships with Historically Black Colleges and Universities, provide access to the facility for students from other universities both domestic and foreign. And NETL hosts professional development programs such as training programs for Nuclear Regulatory Commission personnel and International Atomic Energy Agency fellowships.

Keywords: Education
PACS: 01.40.-d

INTRODUCTION

The Nuclear Engineering Teaching Laboratory (NETL) at The University of Texas at Austin (UT-Austin) would appear to be a typical university research facility. This facility houses a TRIGA Mark II research reactor with associated experimental beamlines, a 14 MeV neutron generator, laboratory space, classroom space, office space, and conference room. The facility has a very strong research program as well as providing service (primarily in the form of nuclear analytical services) to a variety of internal and external customers. However, as the name implies, a large part of the function of the facility is education. From the NETL mission statement, the primary mission is to "educate the next generation of leaders in nuclear science and engineering." Education at NETL is performed via a variety of methods.

Through the Nuclear and Radiation Engineering Program, an undergraduate nuclear certificate is offered and graduate degrees are offered within the Mechanical Engineering Department. Of the courses required for these degrees, approximately a dozen courses (see Table 1.) utilize the NETL facilities.

Some of the courses require direct access to the facility for hands-on education in reactor physics and nuclear science. As the NETL is located on the J.J. Pickle Research Campus (approximately ten miles north of the main academic campus of The University of Texas at Austin), it is convenient to have remote access to the facility for some activities. Thus, some classes access the facility via a "virtual console" which allows students to observe reactor parameters and participate in experiments from remote locations such as the main academic campus.

Not wanting to limit the teaching mission to just students at UT-Austin, NETL hosts the Summer Nuclear Engineering Institute to further the education mission. This one-month institute is primarily for sophomores and juniors majoring in technical fields outside of nuclear engineering and typically from universities other than UT-Austin. Students study fundamental health physics and nuclear engineering concepts, and gain hands-on experience at the reactor and other experimental facilities with credit hours available to the students via the Division of Continuing Education. Additionally, NETL partners with Historically Black Colleges and Universities to allow access to students from participating institutions.

Application of Accelerators in Research and Industry
AIP Conf. Proc. 1336, 742-743 (2011); doi: 10.1063/1.3586201

TABLE 1. Nuclear and Radiation Engineering Program courses utilizing the Nuclear Engineering Teaching Laboratory.

Course Number	Course Name
ME 136N	Concepts in Nuclear and Radiological Engineering
ME 337C	Introduction to Nuclear Power Systems
ME337F	Nuclear Environmental Protection
ME 361E	Nuclear Operations and Reactor Engineering
ME 361F	Radiation and Radiation protection Laboratory
ME 388C	Nuclear Power Engineering
ME 388N	Design of Nuclear Systems I
ME 389C	Nuclear Environmental Protection
ME 390F	Nuclear Analysis Techniques
ME 390G	Nuclear Engineering Laboratory
ME390T	Nuclear- and Radiochemistry

Professional development is another component of the education mission at NETL with various seminars and short courses offered to nuclear professionals. One such offering is a special training program for U.S. Nuclear Regulatory Commission personnel without prior reactor experience to give them hands-on understanding of areas that they may be regulating. Through its strong relationship with the International Atomic Energy Agency (IAEA), NETL conducts training for IAEA Fellows and visiting scientist in the form of on-site, on-the-job training in subjects such as reactor operations, radiochemistry, health physics, and related topics. At times NETL personnel travel to other IAEA institutions to provide training as well. Additionally, NETL offers a training program for individuals to become licensed by the NRC as reactor operators. And NETL has a number of part-time undergraduate research assistant positions where students receive on-the-job training in various laboratory techniques.

Although not a traditional classroom activity, public education is also included in the education mission of the NETL. NETL personnel routinely conduct such activities as seminars for special interest groups (such as Boy Scouts earning the Nuclear Science Merit Badge), tours for public school classes, and specialized training for emergency responders. These activities help educate the public in relevant topics in nuclear science to increase their awareness of nuclear science issues relating to everyday life.

Along with its ability to conduct leading edge research, the Nuclear Engineering Teaching Laboratory is truly an educational facility.

ACKNOWLEDGMENTS

The author would like to thank Drs. Steven Biegalski and Sheldon Landsberger for their valuable discussions during the preparation of this manuscript.

Enhancing The Undergraduate Experience: Measuring Film Thicknesses Using A Helium Ion Beam

Rahul Mehta[a], Stephen R. Addison[a], Jerome L. Duggan[b]

[a]Physics and Astronomy, University of Central Arkansas, 201 Donaghey Avenue, Lewis Science Center 171, Conway AR 72035, United States
[b]Physics, Ion Beam and Modification Analysis Laboratory, University of North Texas, P.O. Box 311427, Denton TX 76203-1427, United States

Abstract. Students performed Rutherford back scattering (RBS) experiments using 1.5 MeV helium ion (+1) beam from a Van de Graaff accelerator. The ion beam was incident normally on target samples located in a scattering chamber. The samples were single and multilayered thin films on substrate, prepared using a vacuum evaporator. The scattered ions were measured using a particle detector located at 150° from incident beam direction. The elastically scattered particle energies were predicted from kinematical scattering factor (derived using conservation of energy and momentum) and experimentally verified. Using samples of known elemental thicknesses, the data was normalized and thicknesses of layers in unknown samples determined. For multilayer samples, the scattered particle spectra were analyzed for particle yield per ion, energy shift in the centroid and energy at the leading edge for the elemental peak. The Rutherford scattering formula and the kinematical scattering factor together with the energy loss of the helium ion through the various layers were used to identify elemental layers, their positions and thicknesses. The multilayer films were identified as gold-tungsten-tin (Au-W-Sn). The normalization procedure allowed the thicknesses to be determined with uncertainties of a few percent.

Keywords: accelerator, helium, kinematics, RBS, thickness, particle detector
PACS: 1.50.Qb

INTRODUCTION

Junior and senior physics majors used ion beam [1] to study ion-atom collision. Several experiments [2,3] were performed and this report shows the comprehensive nature of the training the students got in this capstone activity. In another paper [4] the details of this type of educational trip to an accelerator facility including details of the experiments are summarized. Here we are specifically interested in using ion beam to measure thicknesses of multilayered films. The aim is to teach students to combine various aspects of ion-atom collision measurement- mainly kinematics of the scattering, energy loss of the scattered ion in traversing various layers and relating the particle detector spectra to the multiple processes taking place in the journey of the ion.

Experiments, Results And Discussion

1.50 MeV He^+ ions from the Van de Graaff accelerator at the IBMAL facility at University of North Texas were incident on sample placed in the scattering chamber. The Rutherford back scattered (RBS) ions at 150° were measured by a silicon surface barrier particle detector which subtended a solid angle ($\Delta\Omega$) of 1.10 mSr. The scattering was treated as an elastic collision thereby Rutherford formula for scattering cross section σ_R can be used.

$$\sigma_R = \frac{d\sigma}{d\Omega} = \left[\frac{Z_1 Z_2 e^2}{4E_1}\right]^2 \left[\frac{1}{\sin^4\left(\theta/2\right)} - 2\left(\frac{M_1}{M_2}\right)^2\right]$$

for $M_1 \ll M_2$ ....(1)

Application of Accelerators in Research and Industry
AIP Conf. Proc. 1336, 744-747 (2011); doi: 10.1063/1.3586202

Here E_1 is the initial energy of the beam and θ is the scattering angle. M_1 and M_2 are the masses of the ion and the target atom, respectively, while Z_1 and Z_2 represent the atomic numbers of incident ion (helium) and that of the target atoms. The scattered ion particle net yield (Σ − β), together with the number of incident ions N_0 and the thickness of the target atom s N_1 relate to above calculated σ_R as shown in equation below.

$$N_1 = \frac{\Sigma - \beta}{\sigma_R N_0 \Delta\Omega} \quad \text{...... (2)}$$

The assumption of perfectly elastic collision during scattering meant one can use the energy and momentum conservation to relate the velocity of incident v_1 and scattered ions v_2 and hence derive the standard kinematical scattering factor K (ratio of incident (E_0) and scattered (E_1) kinetic energies) equation

$$K = \left[\frac{\sqrt{M_2^2 - M_1^2 \sin^2 \vartheta} + M_1 \cos\theta}{M_1 + M_2}\right]^2 \quad \text{.... (3)}$$

K values allowed one to determine energy of the scattered ions at different depth within the sample provided one knows the incident energy at that depth. The analysis of the spectra together with the knowledge about how helium ion loses energy with thickness in different elements provides this incident energy. Table 1 provides the K for several elements. These elements (selected from gold to oxygen) were used as samples (some as thickness standard) and scattered energy measured. Table 1 compares the theoretical prediction of eq. (3) for scattered energy with the measured energy. The last column in Table 1 compares well with the theoretical prediction for gold, copper and chromium. For elements of tungsten, tin, and silicon, the measured value is much lower than the prediction as should be for deeper layers in the sample. For each of these, the elemental layer was not the top layer and so the helium ion beam suffered energy loss in the upper layers before it scattered from these elements. Once the scattered ion energy was determined, the thickness of different layers could be established by using the energy loss tables [5].

Students leaned the technique of making samples by first making thin single elemental layer samples followed by multilayer samples. The samples were made using a vacuum evaporator. Each of the samples was evaluated using RBS.

Figure 1 shows the RBS spectra for a standard gold sample on a silicon substrate. The kinematical scattering factor K(see Table 1) confirmed that the centroid of the peak came from helium ion scattering from the gold. The leading edge of the gold corresponds to the gold atoms on the surface that the helium ion first encounters. The thick silicon of the substrate appears as a plateau with small oxygen layer. The K for silicon, oxygen and other elements is in Table 1 together with scattered energy values.

TABLE 1. For selected elements, Kinematical Scattering factor K, scattered energy E1 for E0 = 1.50 MeV for helium (He+) incident ion.

Element	**K factor**	**E1 (Theoretical) =K*E0**	**E1 (Measured)**
Au (Z=79)	0.927	1.390	1.39
W(Z=74)	0.922	1.383	1.33
Sn(Z=50)	0.882	1.323	1.21
Cu(Z=29)	0.790	1.185	1.18
Cr(Z=24)	0.750	1.125	1.12
Si(Z=14)	0.585	0.878	0.806
O(Z=8)	0.385	0.577	-

TABLE 2. Stopping powers from Northcliff and Shillings table [4] at selected E (MeV) per mass (u) for atomic numbers Z=47, 63, 73 and 79. Others have been extrapolated with a linear fit. For scattered energy E(MeV) for 4He+ ions, dE/dx (in MeV/mg/cm2)

E (MeV)	**Z=47**	**Z=50**	**Z=63**	**Z=73**	**Z=74**	**Z=79**
1.28	0.585	0.554	0.419	0.372	0.368	0.349
1.6	0.555	0.527	0.404	0.359	0.356	0.338
1.5	0.564	0.535	0.409	0.363	0.359	0.341
1.46	0.568	0.539	0.411	0.365	0.361	0.343
1.44	0.570	0.540	0.412	0.366	0.362	0.344
1.35	0.578	0.548	0.416	0.369	0.365	0.347
1.2	0.592	0.561	0.423	0.375	0.371	0.352

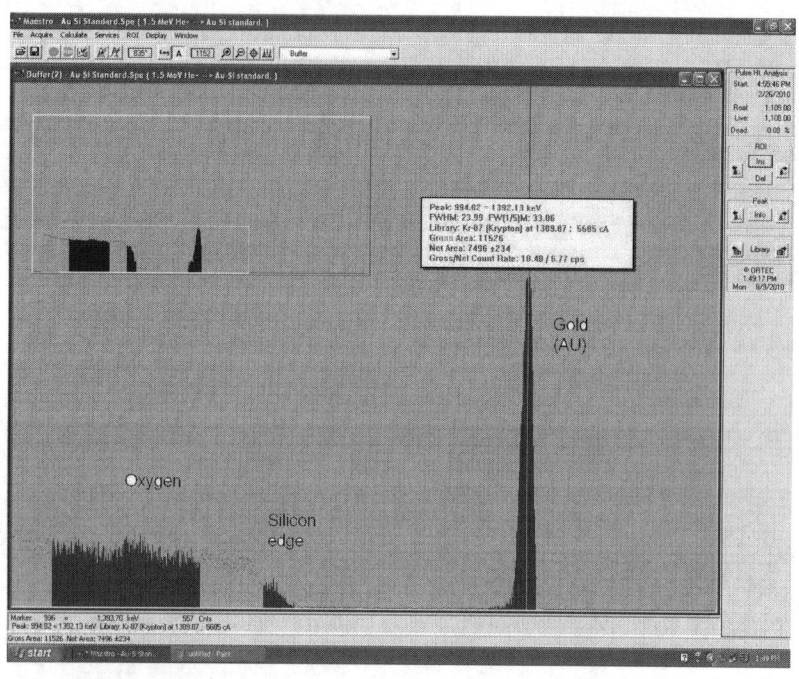

Figure 1. The scattered particle spectrum from standard sample (gold-silicon). Vertical scale shows count on a linear scale (insert shows the spectrum on a log scale). The energy of the scattered ion (in MeV) is on the x-axis. The gold centroid is at 1.39 MeV and the leading edge of Silicon at 0.806 MeV (shifted as gold is on top of the silicon substrate). A 50 mm^2 silicon surface barrier detector was used at 150° from the incident beam direction.

Leading edge of silicon provides the energy loss (ΔE) of the incident helium ion in traversing the gold layer. This shift is an indication of the thickness of the gold layer. The energy loss tables of Northcliff and Shillings [5], provides the energy loss (dE/dx) in MeV per mg per cm^2 (or keV/μg/cm^2). The thickness of the gold layer is determined (as a first iteration) from $\Delta t = (dE/dx)\ \Delta E$. As Table 2 indicates, the dE/dx varies with incident energy and therefore the dE/dx values have to be interpolated as the ion energy keeps changing in its path to the detector. The calculation gets more complex when multilayer targets are used and one is trying to determine inside layers thicknesses.

For single elemental samples, Eq. (1) and Eq. (2) were used to evaluate the thickness of the elemental layer (ref [4] describes this). The thickness of the gold layer determined this way was within 5% of the value determined using dE/dx technique described earlier. The standard gold sample was used to normalize gold layer in multi layer sample. Figure 2 is a particle spectrum for a multilayered target of gold-tungsten-tin (Au-W-Sn) with gold the top layer. The thickness of the gold, tungsten and tin layers were determined as follows. First the incident 1.50 MeV He^+ ion scattered from gold with scattered energy of 1.39 MeV. The ion loses energy in the gold layer as shown by a full-width at half maximum (FWHM) of $\Delta E(FWHM)$ = 36.8 keV for the gold peak. One can show that thickness of the gold layer that helium ion traversed in a normal direction is ΔE divided by ($\sqrt{K}\ (dE/dx)_{in} + (dE/dx)_{out}/\cos 30°$). As a first iteration one can simplify this by using an average value for dE/dx. The gold layer was found to be 51 μg/cm^2. To find the thickness of the next layers (of tungsten followed by tin), measured value of ΔE (FWHM) = 24.6 keV and ΔE(FWHM) = 26.0 keV was employed. Taking into account the energy losses, the helium ion when it first scatters from tungsten, it has an energy of 1.463 MeV and 1.438 MeV when it scatters from atoms near the end of tungsten layer. The scattered energy of helium ion is 1.35 MeV (front of tungsten) and 1.326 MeV (from the back part). The scattered helium ion has now to traverse the layers of tungsten and gold (in a direction 150 ° from the incident direction) and undergoes energy loss in each of the layers before reaching the detector. The same procedure has to be applied in determining the tin layer where the helium ion loses energy in gold and tungsten layer before being scattered from the tin atoms. The scattered helium ion has now 3 layers to traverse before reaching the detector. For tin layer, the helium ion of 1.438 MeV scatters from tin (front) at 1.245 MeV and helium ion of 1.156 MeV scatters from tin (back) at 1.098 MeV.

The thickness of the gold, tungsten and tin can also be determined using Eqs. (1) and (2) but one has to take into account the actual energy of the helium ion at the layer location and absorption in the layers. One simplification is to compare the yields (from the net peak area for W and Sn) with the yields for gold taking into account the only factors that will be different in the Rutherford scattering cross section equation (energy and the atomic numbers) and also include the atomic mass differences when converting thickness from atoms per cm^2 to $\mu g/cm^2$. Using these various methods the students were able to determine the thicknesses of various layers and found the different methods to agree within measurement uncertainty.

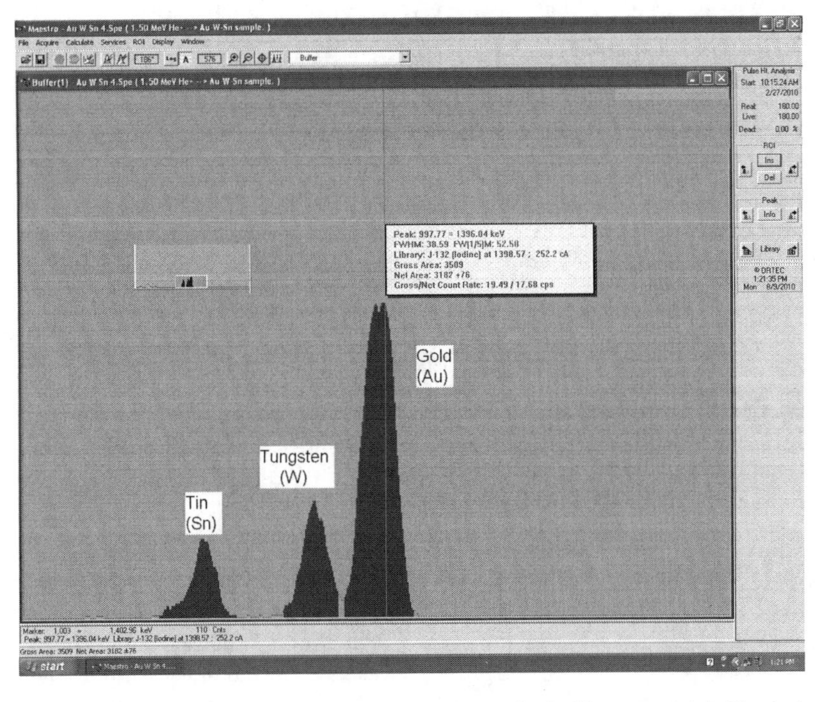

FIGURE 2. The scattered particle spectrum from the multilayered sample (gold-tungsten-tin). Vertical scale is counts and horizontal scale is energy in MeV. The gold is the top (first) layer of the sample.

CONCLUSION

Students successfully applied RBS to determine thicknesses of layers for single and multilayered samples. The thickness determination involved using kinematical scattering factor, Rutherford scattering equation and energy loss of incident helium ion though various layers in the sample. This experiment coalesced data measurement and analysis techniques learned before in analyzing a more complex sample.

ACKNOWLEDGMENTS

Authors wish to acknowledge the support of the Department of Physics and astronomy, the College of Natural Sciences and Mathematics at the University of Central Arkansas, and the Department of Physics at the University of North Texas.

REFERENCES

1. W. Scharf, *Particle Accelerators and their uses*, Amsterdam: Hardwood Academic Publishers, 1986, pp 679-697
2. J.L. Duggan , *Laboratory Investigations in Nuclear Science*, Oak Ridge: Nucleus and Tennelec, 1988, pp 43-83
3. *Experiments in Nuclear Science* AN34 Laboratory Manual, 3rd edition, EG&G ORTEC 1984 pp 15-52.
4. R.Mehta, S.R. Addison, J.L. Duggan, AIP Conference Proceedings 1099 (2009) pp 234-239.
5. C. Northcliffe, R.F. Schilling, Nucl. Data Tables A7, 233 (1970).

Using PIXE To Teach Materials Analysis At Union College

Scott M. LaBrake, Michael F. Vineyard, Maria V. Battaglia, Katie J. Schuff, Colin L. Gleason, Charles I. Harrington, Shivani Pathak, Colin F. Turley, Robert D. Moore

Department of Physics and Astronomy, Union College, Schenectady, NY 12308

Abstract. Particle Induced X-ray Emission (*PIXE*) spectroscopy, an ion-beam analysis technique, is a powerful tool for studying the environment. The Union College Ion-Beam Analysis Laboratory (*UC – IBAL*) provides a small group of students opportunities to explore experimental physics and complement their classroom training, by studying atmospheric aerosols and liquid precipitation from upstate New York using a tandem electrostatic particle accelerator. The students learn how to operate the accelerator and gain valuable laboratory skills in sample collection and preparation procedures and data acquisition and analysis techniques, as well as presentation skills.

Keywords: *PIXE*, environmental pollution, atmospheric aerosols, liquid precipitation, ion-beam analysis, undergraduate research, accelerator based teaching and research
PACS: 29.20.Ba, 92.60.Mt, 92.60.Sz, *92.40.kc, *92.40.qc, 41.75.Ak, 01.50.Pa

INTRODUCTION

Union College, located in Schenectady New York, is primarily an undergraduate institution with approximately 2200 full time undergraduate students. There are 10 full-time faculty members in the department of Physics and Astronomy and the department graduates 10 - 12 majors and minors per year. The Union College Ion-Beam Analysis Laboratory (*UC – IBAL)* is operated by two physics faculty with generally 4 - 5 undergraduate students, most of whom are declared physics majors or minors, but occasionally we get students with different backgrounds and majors, for instance biology, chemistry, geology, mathematics, and engineering.

Involving undergraduates in research can be a rewarding experience for both faculty and student. At the *UC – IBAL* the aim is to blend the experiences the student has in the classroom with a unique educational experience in the laboratory. *PIXE* is an easy way to foster the connections between the classroom and the, perhaps to the student, intimidating and daunting challenge of completing a successful research experience in the laboratory. Our students often marvel at the fact that we are going to teach them how to operate a sophisticated instrument and further how the accelerator will be used to undertake a research project in materials analysis that is complementary to those analyses that might be found in other fields. In addition the student has the opportunity to contribute to a larger scale scientific project by identifying and determining the creation, transport, transformation, deposition and uptake of pollution in the environment.

UC – IBAL operates a 1.0-*MV* tandem electrostatic Pelletron particle accelerator purchased from the National Electrostatics Corporation in 1991. Throughout the accelerator's now almost 20-year history, it has been used primarily as a teaching tool in our undergraduate curriculum and to perform undergraduate research projects in environmental materials analysis. In the past few years we have begun expanding our undergraduate research program in the analysis of environmental materials and thus *UC - IBAL* was born in 2009. The complete systematic study of environmental materials relies upon the simultaneous ion-beam analysis techniques of proton induced x-ray and gamma-ray emission spectroscopies (*PIXE*[1] & *PIGE*), Rutherford backscattering spectroscopy (*RBS*) and proton elastic scattering analysis (*PESA*). *PIXE* has been the backbone of our research efforts and *PIGE*, *RBS* and *PESA* are currently being investigated and should be fully integrated into the study of environmental materials in the coming year.

The study of environmental pollution in New York State involves our undergraduate research students in every aspect. Our students first receive instruction on the basics of *PIXE* and on the workings

Application of Accelerators in Research and Industry
AIP Conf. Proc. 1336, 748-752 (2011); doi: 10.1063/1.3586203

of the accelerator, mostly a review from what they have been taught in the classroom. They are then trained to operate the accelerator and after training, they start preparing thin targets of liquid precipitation and atmospheric aerosol samples for *PIXE* analysis. Our undergraduates do the site selection and sample collection within a geographic radius of about 50 miles surrounding Union College. Data collection for *PIXE* is performed using an Amptek[2] silicon drift detector and the trace elemental composition and concentrations for elements in the target are determined using *GUPIX*[3].

Environmental pollutants, in particular those that are discharged into or found to precipitate into our water supply and with those that are contained in airborne particulate matter are continually causes for concern. Pollution in aerosols and liquid precipitation has been extensively investigated using *PIXE*[4-7]. Particulate matter in the air plays a role in climate change[8,9] and the smaller the particulate matter the more harmful the effects to the human body. Accelerator-based ion-beam analysis studies, especially *PIXE*, provide a powerful, quick, non-destructive means of quantifying the extent of environmental pollution in liquid and airborne forms and provide meaningful research projects for undergraduates. Elemental composition and concentrations are easily determined from a *PIXE* analysis. However, *PIXE* is insensitive to very light elements from hydrogen to sodium and the *PIXE* analysis technique is not able to provide chemical state information.

PIXE ANALYSIS OF ATMOSPHERIC AEROSOLS

Ion-beam analysis of atmospheric aerosols begins with site selection. Previous studies have suggested that emissions from crematoria may contain trace amounts of mercury[10,11] and other toxic metals. To investigate possible environmental contamination from mercury, a crematorium, which routinely performs 90 plus cremations per month, located in Schenectady, NY, was chosen as our first sampling site along with the Union College Boathouse, located approximately one-mile from the crematorium. Aerosol samples were collected using a 9-stage PIXE International cascade impactor[12] during winter and summer of 2010 with sample collection times of approximately 50 hours at a flow rate of about 1.0 L/min. The impaction foils are thin Kapton and after impaction the deposited spots were measured on a stereomicroscope and the average visible deposit sizes ranged from 0.01 – 0.05 cm^2.

PIXE is the go-to standard technique in ion-beam analysis and is the most utilized technique in our laboratory for the analysis of environmental materials. Elemental sensitivity for *PIXE* ranges from sodium to uranium. Our *PIXE* setup uses beams of protons with energies around 2.0-*MeV* to eject inner shell electrons from the target material. Electrons from higher energy states de-excite to fill the vacancy created by the incident proton accompanied by an emission of an x-ray photon characteristic of the element-proton interaction. Typical *PIXE* spectra of aerosol samples are shown in Figure 1 with corresponding elemental concentrations (in ng/cm^2 as determined using *GUPIX*) in Figure 2. *GUPIX* determines the concentration (and uncertainties) C_Z of a particular element Z according to

$$C_Z = \frac{Y_Z}{Y_t H Q \varepsilon T}. \qquad (1)$$

Y_Z is the intensity of the principle x-ray line for element Z, Y_t is the theoretical intensity (per microCoulomb of charge), H is an experimental constant determined by taking data on a set of standards, Q is the measured beam charge incident on the sample, ε is the intrinsic detector efficiency, and T is the transmission coefficient through any filters between the detector and the target.

The crematorium aerosol sample in Figure 1 represents particles in the aerodynamic size range of 2.0 – 4.0-μm and Figure 1 shows a variety of elements ranging from aluminum to lead. The highest concentrations are due to chlorine and calcium from the crematorium emission while sulfur and lead dominate the boathouse sample. The high concentrations of chlorine could be due to body preparation before cremation, lead from the combustion of general aviation fuel from a neighboring airport, and the sulfur from acid rain or the combustion of diesel fuel from nearby railroad activity. Interestingly, there is no obvious mercury peak from the crematorium sample, however, initial studies do indicate a possible weak mercury signature on some of the impacted foils and more investigation is needed.

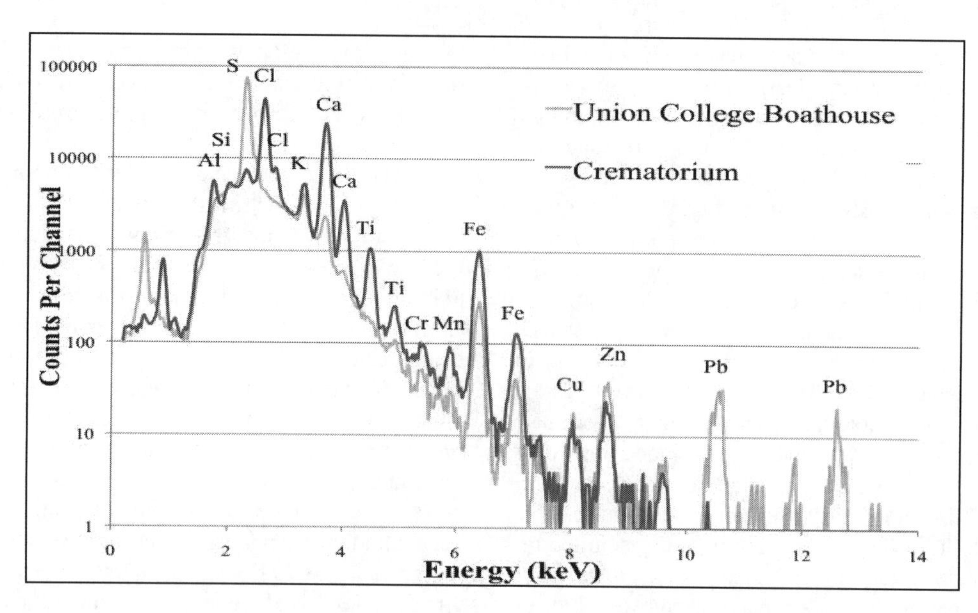

FIGURE 1. *PIXE* spectra for impacted aerosol samples for particle sizes in the range of 2.0 – 4.0-μm taken at the Union College Boathouse and a crematorium. The beam energy was 2.0-*MeV* and 15-μC of charge were collected for each sample.

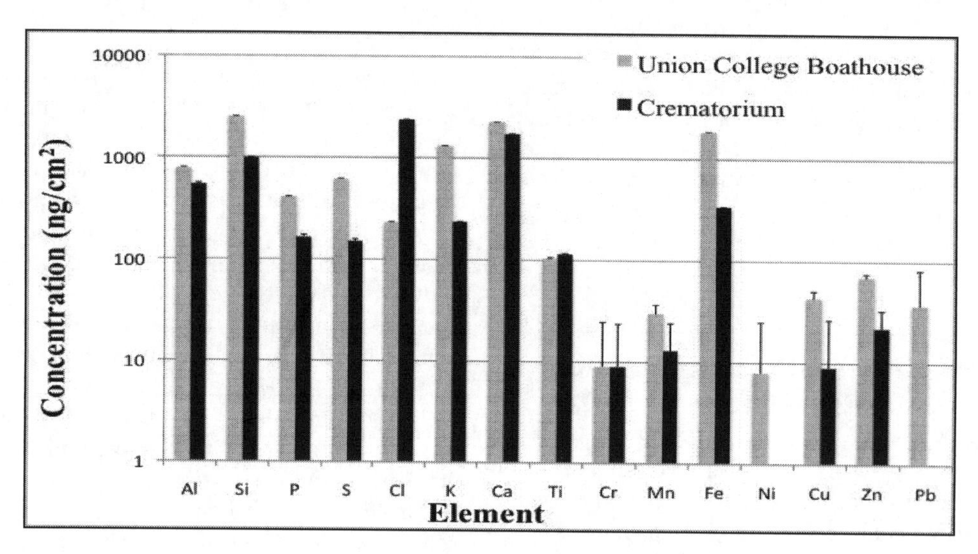

FIGURE 2. Concentrations with uncertainties (in ng/cm^2) of the elements found in the impacted aerosol samples as determined by *GUPIX*.

PIXE ANALYSIS OF LIQUID PRECIPITATION

Clifton Park, NY, located 10 miles northeast of Union College, was selected as the site for studying environmental pollution in liquid precipitation samples. Approximately 200-mL of rain and snow samples were collected in plastic containers each season during the period from June 2009 to June 2010. The precipitation was prepared following the method of Ghorai[13] by slowly evaporating 10-mL of the liquid in acid washed glassware at an approximately constant temperature of 80°C until 5-mL of concentrated rain or snowfall remained. 50-μL of a concentrated liquid sample were pipetted onto 12-μm aluminized Mylar backings and further dried under vacuum for 20 hours.

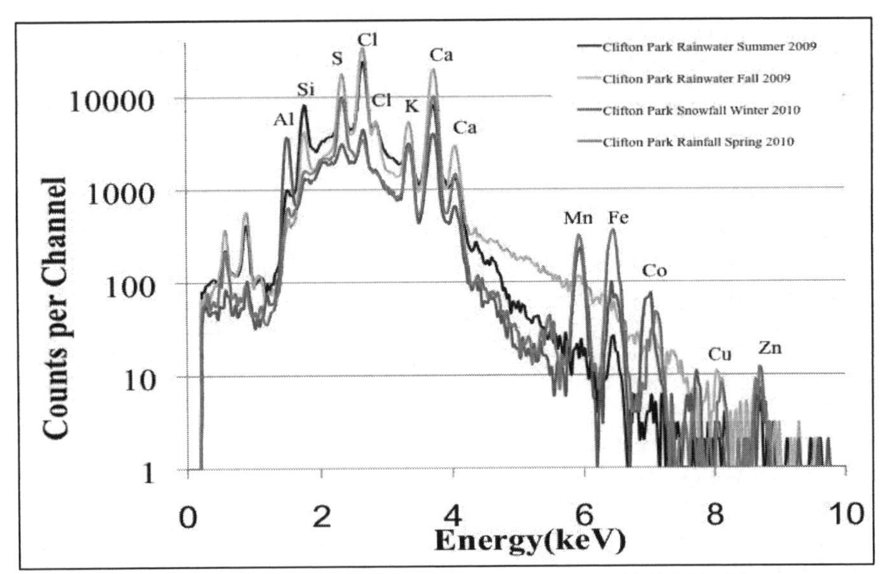

FIGURE 3. *PIXE* spectra for precipitation samples collected in the four seasons of 2009-10. The spectra were acquired using a 1.8-*MeV* proton beam and 10μC of charge was accumulated for each sample.

Four-season precipitation spectra are shown in Figure 3 with corresponding concentrations (in ng/cm^2) in Figure 4. Comparing Figures 1 and 3 we see that Figure 1 shows a greater elemental distribution. Figure 3 shows a constant elemental distribution over the four seasons and a varying elemental concentration with the largest due to sulfur (again maybe a signature of acid rain), chlorine and calcium for which we do not have a cause. We do note the presence of cobalt in the winter and spring samples for Clifton Park and a future study for our research students will be to collect and analyze more liquid precipitation samples, determine average essential concentrations and then formulate a probable cause for these higher concentrations. Finally, Figure 4 suggests a possible seasonal variation in the concentrations and undergraduate researchers are currently investigating this. The analysis of these two environmental samples has produced two projects that have involved our undergraduate team and the excitement among the students, and faculty, is certainly palpable. To complete the analysis and determine the causes for high elemental concentrations, the three ion-beam techniques of *PIGE*, *RBS* and *PESA* will be incorporated.

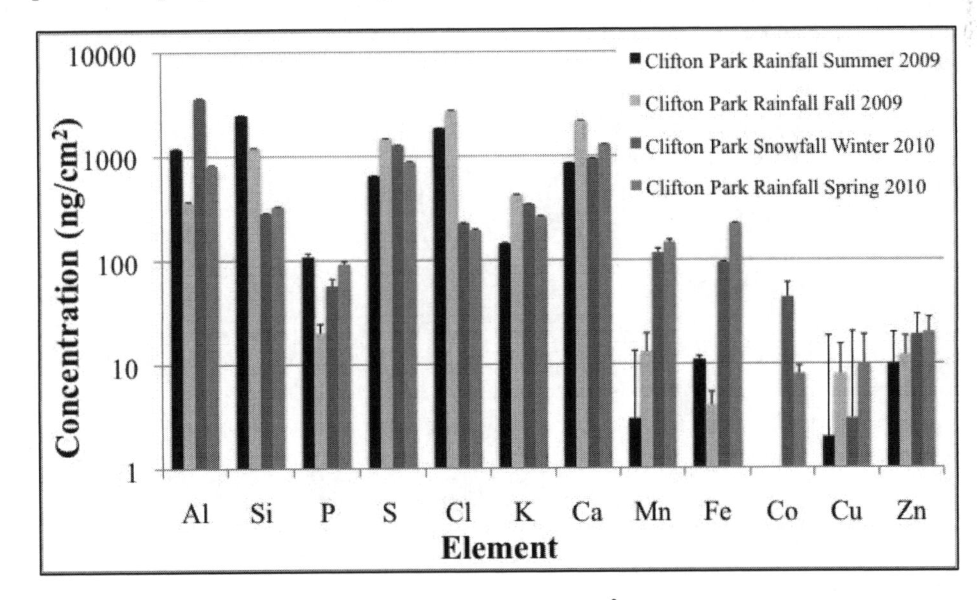

FIGURE 4. A graph of the concentrations with uncertainties (in ng/cm^2) of the elements found in each liquid precipitation sample as determined by *GUPIX*.

SUMMARY AND CONCLUSIONS

The ion-beam analysis technique of *PIXE* is well suited to the analysis of environmental pollution for elements in the range of sodium to uranium, especially those found in atmospheric aerosols and liquid precipitation samples. In addition, *UC - IBAL* involves undergraduate students in all aspects of research projects on the analysis of environmental materials. An all-undergraduate team of researchers performed data collection and subsequent data analysis using a 1.0-*MV* tandem electrostatic particle accelerator and *GUPIX*. The *UC – IBAL* provides an exciting experience for undergraduates to do research in physics and in particular the analysis of environmental materials. The undergraduate researchers are provided with a first-hand, hands-on introduction to research in experimental physics that fully complements their classroom experiences and provides a basis for an on-going basic research project that they can really tackle and own. The *UC – IBAL* provides an avenue that takes our students from the classroom to the laboratory and we value our undergraduates and the experiences that our undergraduates have in the laboratory.

ACKNOWLEDGMENTS

The author wishes to thank Mike Vineyard for all of this patience and guidance in the completion of this project and for reading and editing many versions of this manuscript, as well as Union College, and in particular the Department of Physics & Astronomy, the Dean of Academic Departments and The Dead of Studies for their financial support. Of course to my wife Sandra and my kids Alexandra and Samantha for their understanding with me being away from home so much. And it goes without saying that this research project and paper would not have been possible with out the dedication and tireless effort put forth by of all of the research students involved. You are truly valued.

REFERENCES

1. Particle Induced X-ray Emission Spectroscopy. Johannson, S. Campbell, J. Malmgvisit, K. Wiley, NY 1995.
2. Amptek Inc.: http://www.amptek.com/drift.html.
3. GUPIX Analysis software, University of Guelph: http://pixe.physics.uoguelph.ca/gupix/main/.
4. K. S. Johnson, A. Laskin, J. L. Jimenez, V. Shutthanandan, L. T. Molina, D. Salcedo, K. Dzepina, and M. J. Molina, *Environmental Science & Technology*, Vol. 42, No. 17 (2008).
5. S. M. Almeida, M. C. Freitas, M. A. Reis, C. A. Pio, M. A. Trancoso, *Nuclear Instruments and Methods in Physics Research A*, 564 (2006) 752 – 760.
6. D. D. Cohen, E. Stelcer, O. Hawas, and D. Garton, *Nuclear Instruments and Methods in Physics Research B*, 219 – 220 (2004) 145 – 152, and references therein.
7. G. Ghermandi, R. Cecchi, F. Costa, and R. Zonta, *Nuclear Instruments and Methods in Physics Research B*, 56 - 57 (1991) 677 – 682.
8. Intergovernmental Panel on Climate Change, Climate Change 2007: The Physical Science Basis. Contribution of Working Group I to the Fourth Assessment Report of the Intergovernmental Panel on Climate Change (2007).
9. D. Shindell and G. Faluvegi, "Climate response to regional radiative forcing during the twentieth century," Nature Geoscience 2, 294 (2009); http://www.nasa.gov/topics/earth/features/warming_aerosols.html.
10. D. Arenholt-Bindslev, *European Journal of Oral Sciences*, Vol 106, No 2p2 (2003) 713-720.
11. L. Hylander, A. Lindvall, and L. Gahnberg, *Science of the Total Environment*, Vol. 362, No. 1 – 3, (2006), 74 – 84.
12. PIXE International Corporation: http://pixeintl.com/Impactor.asp.
13. Ghorai, S.K. Tekyi-Mensah, O. Sims, J.F. Williams, J.R. Alford, W.L. *Nuclear Instruments and Methods in Physics Research B*, 62, (1991), 139 – 142.

The Scanning Electron Microscope As An Accelerator For The Undergraduate Advanced Physics Laboratory

Randolph S. Peterson[1,2], Karl K. Berggren[2] and Mark Mondol[2]

[(1)] *Department of Physics and Astronomy, The University of the South, 735 University Avenue, Sewanee TN 37383, United States*
[(2)] *Department of Electrical Engineering and Computer Science, Massachusetts Institute of Technology, 50 Vassar Street, Cambridge MA 01239, United States*

Abstract. Few universities or colleges have an accelerator for use with advanced physics laboratories, but many of these institutions have a scanning electron microscope (SEM) on site, often in the biology department. As an accelerator for the undergraduate, advanced physics laboratory, the SEM is an excellent substitute for an ion accelerator. Although there are no nuclear physics experiments that can be performed with a typical 30 kV SEM, there is an opportunity for experimental work on accelerator physics, atomic physics, electron-solid interactions, and the basics of modern e-beam lithography.

Keywords: SEM, secondary electron, x-ray emission, lithography, beam optics
PACS: 68.37.Hk, 79.20.Hx, 78.70.En, 85.40.Hp, 41.85.-p

INTRODUCTION

An accelerator is a modern symbol of physics research and a glance through any modern physics textbook will yield a chapter or major section on accelerators and their use in fundamental research and for applications in medicine. In the 2009 conference on advanced undergraduate laboratories held at the University of Michigan[1], the use of accelerator experiments in advanced laboratories was mentioned in only a few of the more than fifty university programs discussed. Only a fraction of physics departments have local access to a traditional accelerator, yet many, if not most departments have access to a scanning electron microscope (SEM).

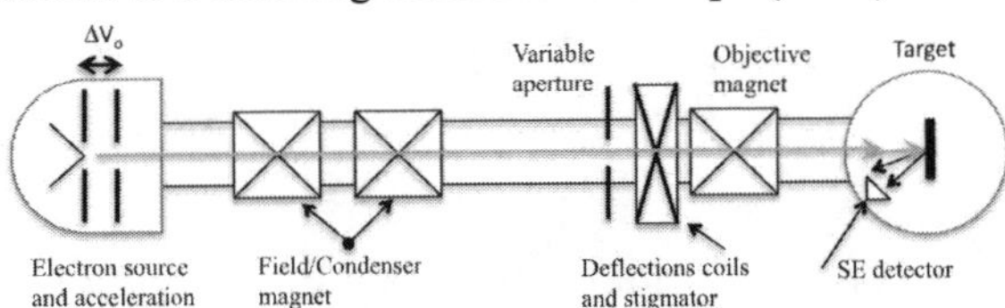

FIGURE 1. This is a schematic for a scanning electron microscope, but it could also represent a single-end accelerator if the first magnet is a bending magnet and the others are quadrupole-focusing magnets, instead of the dipole magnets of a SEM. The acceleration region of a 0-30 kV SEM is much smaller and less complex than higher energy accelerators.

Although accelerator experiments usually bring to mind nuclear and elementary particle themes, this accelerator can be integrated into the advanced laboratory course to study some very basic accelerator physics and technology. It can be used for a wide variety of advanced, non-nuclear experiments that can be open-ended, giving students an opportunity to use an instrument that they have invested time learning to use. Students can learn to operate a SEM in a couple of lab periods.

ACCELERATOR PHYSICS

A simple schematic drawing of a typical electron microscope, shown in Figure 1, looks similar to a single-ended accelerator of the 50 kV – 3 MV range of accelerating potentials. Many SEM accelerators are limited to a 30 kV maximum accelerating potential. With such a small potential there is no need for the complex accelerating column used in higher potential accelerators. A bending magnet to separate masses and charge states is not necessary on a SEM, so all magnets with a SEM are dipole (solenoidal) magnets for focusing the electron beam. Focusing magnets on higher potential accelerators are usually quadrupole magnets.

Application of Accelerators in Research and Industry
AIP Conf. Proc. 1336, 753-757 (2011); doi: 10.1063/1.3586204

Accelerators, including SEMs, can be thought of as a source for producing and accelerating a beam of particles through beam-shaping magnets with instrumentation to study the beam-target interaction. There is also a Faraday cup to measure the beam current at the target. The beam of a SEM is scanned to produce an image, one pixel in the image for each position of the beam.

An analogy between an optical system, such as a microscope, and an accelerator is often used to help students understand the operation of the accelerator system. However, students rarely have more than cursory experience with optical systems and usually a shallow or even incorrect understanding of their operational principles. It is better to compare similar types of accelerators and mention the electromagnetic behaviors imbedded in the accelerator systems than to overuse the optical analogy.

Control labels on a SEM, such as contrast, focus, brightness, and magnification, shown in Figure 2, refer to imaging and not directly to the accelerator system operation. Relabeling a few of the controls on the SEM to reflect the devices they control, rather than the imaging effects that result, can help physics students make the connection between the operation of the SEM's accelerator system and the electromagnetism that the students have studied. An example of SEM controls is given in Figure 2 with some suggested name changes.

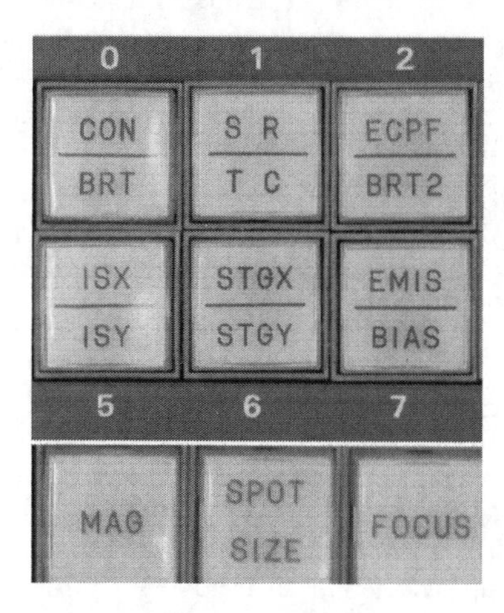

FIGURE 2. This is a selection of a few of the controls found on a typical SEM that uses a heated tungsten filament source. Terminology is often related to the control of the image rather than to the device in the accelerator being controlled. Changing some of the control names can help a student relate the control to the actual device and its operation in the accelerator system. For example, "Contrast & Brightness" (CON/BRT) can be "Detector Range & Gain"; "Magnification" (MAG) can be "Scanning Magnet Control"; "FOCUS" can be "Objective Magnet Control"; and "SPOT SIZE" can be "Condenser Magnet Control".

The magnetic fields of the SEM's field and objective magnets will always act as (positive) focusing lenses for a beam of electrons. This is not obvious to many students, but simple models can be used to explain the magnetic focusing[2] and can be used to predict an observable effect, the direction of the image rotation in a changing magnetic field. The objective lens (the final magnet as the electrons leave the SEM) is adjusted to bring the beam to a very small spot (between 0.5 nm and 10 μm, depending upon the SEM design). As an image is formed by deflecting (with scanning coils) the electron beam across an object, the image is rotated from the expected position by the interaction of the beam with the focusing magnetic field ($-ev \times B$) of the objective magnet. The dependence of the rotation angle upon the magnetic field strength is shown in Figure 3. The two images of the silicon wafer in Figure 3 were made at constant beam energy but different magnetic field strengths, causing the electron beam to rotate through different angles for the two cases, and producing a rotated image. The two images were produced by physically moving the object closer to the objective magnet while simultaneously increasing the objective magnet's field strength. This effect can also be observed dynamically as the SEM beam energy is changed.

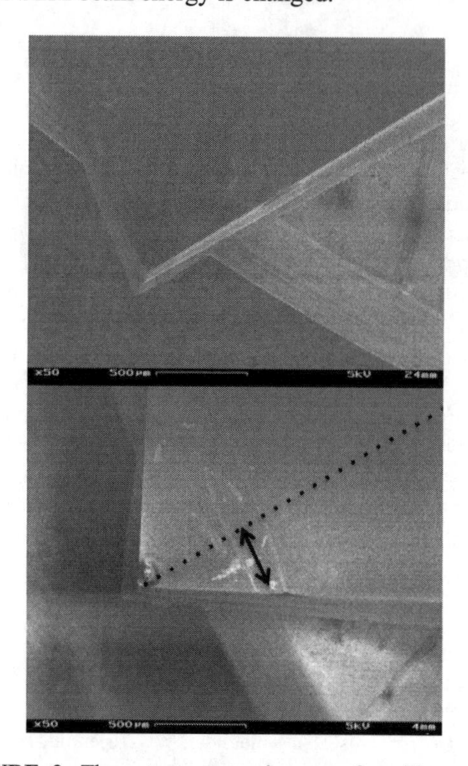

FIGURE 3. These two corner images of a silicon wafer were made at the same beam energy but with different objective magnetic field strengths. The lower image is rotated by about 25° from the upper image, and is an example of the spiral motion of the electrons due to the Lorentz force on them.

Each accelerator has a source with beam-shaping electric or magnetic fields, plus an acceleration region. Apertures are used to collimate and spatially filter the beam. A good image from the SEM requires that the cross sectional area of the beam is adjusted to be nearly circular, and good resolution requires a small spot size. The image resolution achieved by a SEM depends not only upon the beam diameter, but also on the precision of the electron beam scanning, and the interaction of the beam with the target. The beam diameter at the electron source, the chromatic aberrations, and the spherical aberrations of the focusing magnets are usually much more important contributions to the beam diameter than the increase from aperture diffraction of the electrons due to their deBroglie wavelength. These features and requirements are not usually found on a typical particle accelerator, and they add to the richness of using the SEM as an accelerator.

A biased Faraday cup is often used to measure the beam current in an accelerator system, and Faraday cups are also used on a SEM, usually without biasing[3]. In a SEM secondary electrons escaping the target are the source of electrons most often used for imaging in a SEM. By magnifying the image of the Faraday cup until only the inside of the hole is "visible" as a black image, few if any electrons are escaping and a good measurement of the beam current can be made. Drill bits from #65 to #115 can be used to make the holes in a metal target to use as the SEM Faraday cup (20:1 depth to diameter ratio makes for an effective Faraday cup). Beam currents usually range from picoamperes up to a few nanoamperes, depending upon the aperture size selected.

BEAM-TARGET INTERACTIONS

The beam electrons interact with the atoms of the target solid by scattering. Electrons come from the target solid as backscattered electrons, secondary electrons, or Auger electrons. Backscattered and Auger electrons provide information about the atomic composition of the solid, especially near the surface. Secondary electrons (SE) provide information about the surface topography. More secondary electrons are released from vertical edges than from level surfaces providing great image contrast. These electrons are accelerated to a scintillation detector biased positively to the target.

Photons with wavelengths ranging from x rays (characteristic and Bremsstrahlung) to the visible and infrared are emitted and provide information about the atomic composition and atomic structure of the solid. Any of these emissions can be used to create an image of the target. As an example in Figure 4, a SE image (4c) and two x-ray images (4a and 4b) are shown, along with the x-ray spectrum (4d). Each image provides a different perspective on the target. The secondary electron images require only a few seconds to acquire, while the x-ray images can require several hours, depending upon detector type, dead-time counting rates, and detector efficiency. The identification of elements from an x-ray spectrum produced with the SEM is often more convenient than using x-ray fluorescence or PIXE techniques, and is just as interesting.[4]

Only a few special types of ion-beam accelerators are capable of performing lithography, but all SEMs

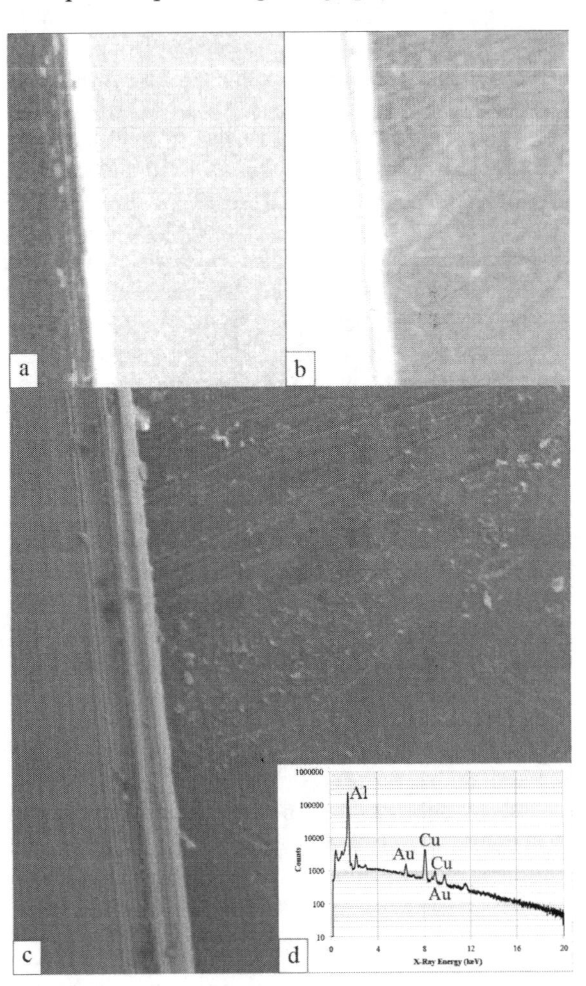

FIGURE 4. A piece of copper tape on a sanded aluminum stub, previously coated with a thin layer of gold, was imaged at 24 keV and 105x magnification. Images created with characteristic x rays from the copper and the aluminum are shown in (a) and (b) respectively, and the total x-ray spectrum for the region that was imaged is shown in (d). The darker areas represent the copper or the aluminum. The image created with secondary electrons is shown in (c). All three images cover the same area and are only displayed at different sizes in this Figure.

can be used for some elementary lithography, while others have features to perform more advanced lithography. Lithography with a SEM is very similar to optical photography. The "film" for SEM lithography is called the resist, which can be made of different polymers. Some polymers are very sensitive to electron (beam) interactions and their chain bonds are easily broken. A piece of a silicon wafer can be spin-coated[5] with a thin layer of the resist PMMA, poly(methyl methacrylate), a common polymer (Plexiglas) but one of the most commonly used for electron-beam lithography. A short exposure time with the electron beam scanning the field of view will decrease the average molecular weight of the PMMA. These molecules are easily removed by briefly immersing the wafer in a common solvent at room temperature. This will expose the underlying silicon and leave the high atomic-weight molecules of the PMMA nearly intact. Electrons at 10 keV and a magnification of 620x (100 μm x 100 μm field of view) were used to expose the pattern shown in Figure 5a.

Depending upon the size of the field of view (or the magnification used for the exposure), the exposed areas with some or all the PMMA removed during development can be viewed with an optical microscope (as shown in Figure 5) or viewed with the SEM. For the lithography shown in Figure 5a, the reflectivity of the exposed rectangular areas increases as remaining PMMA thins, until all the PMMA has been removed, exposing the highly reflective silicon substrate. An exposure-developing curve (bottom of Figure 5) is measured from the average observed brightness for each exposed area and the best dose density and developing time can be determined from this graph (Figure 5b). The exposure curve in Figure 5b is for 40 nm thick PMMA on silicon, developed in a 2:1 concentration of isopropyl alcohol and methyl isobutyl ketone for 90 seconds, rinsed in IPA for 60 seconds, and dried with N_2. Other developing techniques can also be used[6]. If a metal layer is deposited on the sample after developing the PMMA resist, all areas will be coated with the metal. Washing the wafer with acetone removes the remaining (metal-covered) PMMA. The silicon is left with a metalized pattern, that is, metal covering all the exposed regions.

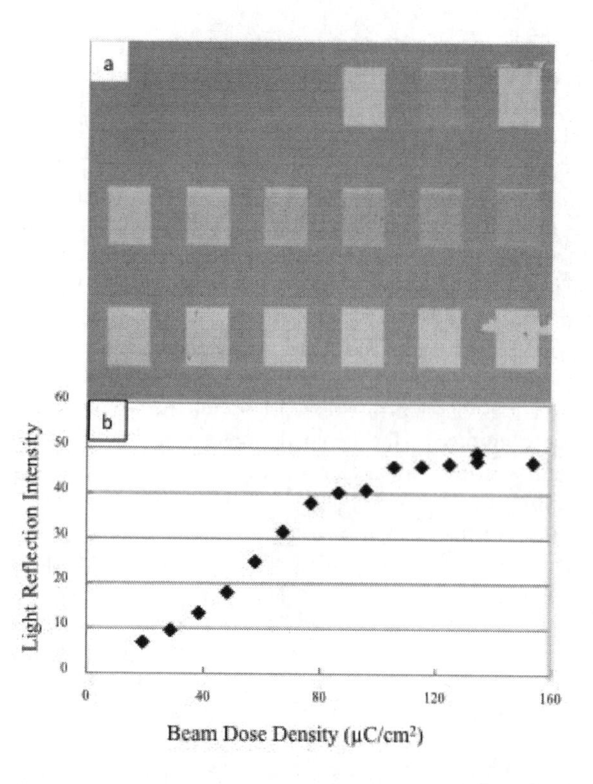

FIGURE 5. The upper image is an optical microscope image of rectangular regions created by electron-beam lithography. Each rectangle is the field-of-view of the SEM at a magnification of 620x, approximately 100 μm x 120 μm. Exposure times varied from 10 - 80 seconds for the 10 keV electrons incident upon the PMMA resist on a silicon wafer. After the resist was developed, the reflectivity of each exposed rectangle was measured using ImageJ software. The reflectivity of the silicon is much higher than the PMMA covered silicon and the graph of the reflectivity as a function of the beam dose density can be interpreted as doses above 100 $\mu C/cm^2$ will result in complete removal of the exposed PMMA.

Depending upon the SEM and its control software, complex lithography can be done by students at the micro-scale or the nanoscale. Without any special SEM features, rectangular exposures like those in Figure 5a can be made end-to-end to produce a single reflecting slit that can be tested optically. More complex arrays of slits can be made, though it is a time-consuming but rewarding endeavor. With better software, etch-a-sketch type lithographic images can be created and written into the PMMA resist with electron beams. Students who develop some skill with the basic lithography may continue to exploit the properties of the SEM to make lithographic patterns to create and test mechanical devices, electrical devices, and their combinations (MEMs and NEMs).

CONCLUSIONS

The SEM can be used as a tool in the advanced laboratory to provide a rich, new variety of experiments. Exploring the operation of the SEM as an accelerator provides students with a hands-on opportunity to operate an accelerator and to test the standard operating characteristics of an accelerator. Although nuclear experiments cannot be done with a SEM, the variety of photon and electron detectors that are standard with most SEMs opens up many other experiments, especially related to characterizing solids. In addition, simple lithography can be performed with any SEM for the creation of micron-sized objects.

ACKNOWLEDGMENTS

We would like to thank the National Science Foundation for support under NSF grant EECS-0823778. RSP would like to thank the Massachusetts Institute of Technology and The University of the South for their support for this sabbatical work.

REFERENCES

1. Topical Conference on Advanced Laboratories (2009), http://advlabs.aapt.org/events/event.cfm?ID=2
2. V. Kumar, Am. J. Phys. **77**, 737 (2009).
3. J. A. Simpson, Rev. Sci. Instr. **32**, 1283 (1961).
4. C. W. S. Conover and John Dudek, Am. J.Phys. **64**, 335 (1996).
5. J. D. Musgraves, B. T. Close, and D. M. Tanenbaum, Am. J. Phys. **73**, 980 (2005).
6. J-S. Wi, H-S. Lee, and K-B. Kim, Elect. Mat. Lett. **3**, 1 (2007).

An Apparatus For Student Projects Using External-Beam PIXE And PIGE

Francis D. Correll, Douglas W. Edsall, Katherine A. DePooter, Nicholas D. Maskell and Jeffrey R. Vanhoy

Physics Department, United States Naval Academy, 572C Holloway Road, Annapolis, MD 21402, USA

Abstract. We recently installed a simple endstation at the Naval Academy Tandem Accelerator Laboratory to support student projects using external-beam PIXE and PIGE. It consists of a short, graphite-lined beamline extension with a thin window, an interlocked box that surrounds the target, detectors for x- and gamma rays, provision for flooding the target with helium gas, easily changed x-ray absorbers, and a compact video camera for monitoring the position of the beam spot. We used this system to measure the elemental composition of colonial-era architectural materials, principally bricks and mortar, from James Madison's Montpelier, the reconstructed Virginia estate of the fourth President of the United States. We describe the design and construction of the system, relate some of our experiences using it, and present some preliminary data from our investigations.

Keywords: student projects; external-beam; PIXE; PIGE; archaeometry; cultural heritage
PACS: 01.50.Pa, 29.30.Kv, 81.70.Jb, 82.80.Ej

INTRODUCTION

The Naval Academy Tandem Accelerator Laboratory was established about 20 years ago to support upper-division undergraduate courses in atomic and nuclear physics and to provide opportunities for student research. Experiments we perform in the courses include Rutherford and non-Rutherford scattering, Coulomb excitation, and the polarization of gamma rays from nuclear reactions [1]. Over the years, we have found that external-beam PIXE and PIGE provide rich and varied opportunities for student research. Our students have studied 1000-year-old rhyolite projectile points, oyster shells, oyster tissue, river-bottom sediments, 17th-Century clay smoking pipes, U.S. coins, nails and fasteners, piston rods, meteorites, and many other materials.

Recently, following a relocation of the laboratory, we built a new external-beam apparatus. In this paper, we describe its design and a current application.

Advantages Of External-Beam Analysis

External-beam analysis has several well-known advantages over in-vacuum analysis, especially for environmental or cultural heritage samples like the ones that our students often study. Sample mounting and positioning are simpler in an external-beam apparatus. Large, porous and even moist samples, which cannot be placed in a vacuum chamber, can be studied with an external beam. Sample heating and charging are reduced. The distance and the material between the sample and the detectors can be minimized.

Hazards Of External-Beam Analysis

Although they have advantages, external-beam systems do pose significant hazards. The greatest is direct exposure of the user to an energetic particle beam, which can deliver a very high radiation dose. Additional hazards can arise from exposure to scattered ions or reaction products from the sample and the atmosphere [2].

A simple calculation illustrates the hazard posed by direct exposure to an external beam. The biological effect of radiation is expressed in terms of the dose equivalent, which is the absorbed dose multiplied by a quality factor. The SI unit of absorbed dose is the gray (1 Gy = 1 J/kg = 100 rad), and that of dose equivalent is the sievert (1 Sv = 100 rem). We estimate the dose equivalent rate for a 10-nA beam of 3-MeV protons, with 2-mm diameter and range of 140 μm in tissue [3].

Application of Accelerators in Research and Industry
AIP Conf. Proc. 1336, 758-762 (2011); doi: 10.1063/1.3586205

The beam deposits energy at the rate of 3.0×10^{-2} J/s in a cylinder of tissue whose volume and mass are 4.5×10^{-10} m^3 and about 4.5×10^{-7} kg, giving an absorbed dose rate of 6.6×10^4 Gy/s. The quality factor for high-energy protons is generally taken to be 10 [4], so the dose equivalent rate is 6.6×10^5 Sv/s. This exceeds every regulatory limit by many orders of magnitude. For example, the Navy limits Radiation Workers to less than 0.50 Sv/year to the skin or extremities [5].

Because the range of the beam is comparable to the thickness of the epidermis, the damage to underlying tissue from such a beam may be small, but damage to living tissue exposed by cuts, or to the cornea of the eye, could be very serious. Exposure to the external beam must therefore be avoided.

DESCRIPTION OF THE LABORATORY

Major equipment in our laboratory includes the accelerator, two ion sources, two beamlines with one target chamber each, and the external-beam apparatus. Figure 1 shows a view of the external-beam apparatus with the accelerator in the background.

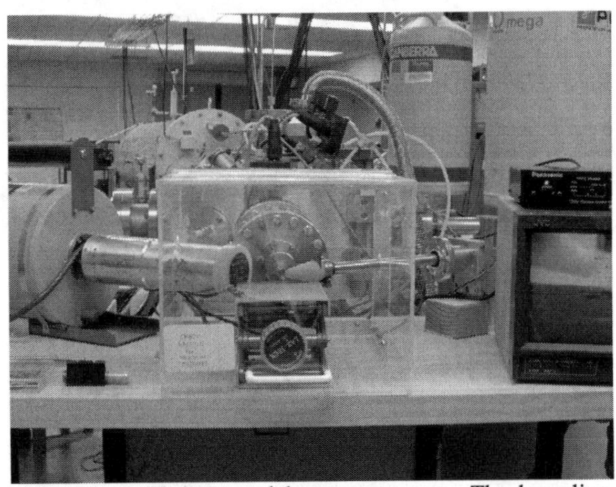

FIGURE 1. The external beam apparatus. The beamline extension is at the center of the acrylic box. Samples are placed on the lab jack and viewed by detectors on either side.

The accelerator is a National Electrostatics Corporation (NEC) model 5SDH tandem Pelletron. Its maximum terminal potential is 1.7 MV and it can supply beam currents of several hundred nA. The Alphatross source produces negative helium and hydrogen ions, whereas the SNICS source produces a wide variety of negative ions. One beamline leads to a large-volume target chamber for nuclear scattering and reaction experiments. The other leads to a modified NEC model RC43 RBS chamber, and the external-beam apparatus is mounted just downstream of the RC43 chamber.

The external-beam apparatus consists of a short beamline extension that ends in a thin window, detectors for x- and gamma rays, a device for positioning the sample, a camera for observing the position of the beam on the sample, and a safety enclosure that surrounds the beamline, sample and detectors.

The beamline extension is machined from a 6.4-cm-long aluminum rod that is 2.5 cm in diameter where it connects to the main beamline by a compression fitting and 1.3 cm in diameter at the end nearest the sample. It is lined with graphite to prevent the beam from striking the aluminum. The inner diameter of the graphite liner is 3 mm. We can focus beams of currents greater than 30 nA through this assembly.

The beam enters the atmosphere through a 7.5-μm-thick kapton window that is attached to the end of the beamline extension with epoxy. A 3.0-MeV proton beam loses about 115 keV as it passes through the window [6]. As a result, the window eventually weakens and breaks. The rising pressure closes an interlocked beamline valve to protect the accelerator. The lifetime of the kapton window varies with beam current but we rarely replace one more often than once per semester. When this happens, we can install a spare, restore the vacuum, and resume measurements in half an hour.

We measure x-rays with a Princeton Gamma Tech model 0S26-9118 Si(Li) detector and gamma rays with an Ortec model GMX25P4 hyperpure Ge detector. They are placed on opposite sides of the beam at 120° with respect to the beam direction. A tapered x-ray collimator, machined from nylon rod, has a 4.7-mm diameter bore and a slot to hold thin foils that absorb low-energy x-rays and reduce pileup. A flexible tube entering the side of the collimator allows us to flood the sample with helium gas to reduce the production of characteristic x-rays from atmospheric argon and the absorption of low-energy x-rays by the air.

We place the sample on a lab jack that allows us to raise and lower it to expose the desired portion to the beam. A lead from the current integrator is connected to the jack. A video camera is mounted below the beamline extension and pointing at the sample. Before a measurement, we observe the position of the beam spot on a slice of CsI scintillator at the same location as the sample. We mark the position of the beam spot on a video monitor so that we know where the beam will strike the sample.

Because the external beam is a radiation hazard, we enclose the sample in an acrylic box. The range of 3-MeV protons in air is about 18 cm [6] and the box is larger than this. The box has a hinged front to allow us to change samples. When it is opened, a magnetic interlock switch keeps a beamline valve closed so that the beam cannot pass into the atmosphere. In addition, we require that a Faraday cup is inserted in the beamline before opening the box, and we check irradiated samples for residual activity before we handle them.

Although the design of this system is simple and useful, it can certainly be improved. We plan to incorporate a beam-position monitor and an electrostatic quadrupole lens just before the external-beam extension to increase transmission and reduce background. We also plan to study charge collection with different sample types and geometries so that normalizing the results of different runs will become more straightforward.

APPLICATION: COLONIAL BRICKS AND MORTARS

Since 2001, researchers with the National Trust for Historic Preservation have been restoring Montpelier, the Virginia estate of former President James Madison and his wife, Dolley. As part of the project, they have returned the house to its appearance circa 1820, removing over a century of extensive renovations by later owners.

During any undertaking of this kind, many questions arise concerning the materials found in the building and the surrounding site. Were they part of the original construction or added later? Were they locally made or imported? Did the restorers correctly understand their original functions and did they properly place them during the reconstruction? Sometimes measurements of the elemental compositions of the materials can help to answer these questions.

We obtained samples of bricks, mortar, and limestone from Montpelier. The samples come from different portions of the home and surrounding yards. We used external-beam PIXE and PIGE to see if we could extract reliable information about the elemental compositions of the bricks. If this proved possible, we thought that further measurements might help clarify the origin and history of the bricks.

PIXE Analysis Of Bricks

In one study, Midshipman 1/C Katie DePooter (Class of 2009) used PIXE to analyze eight bricks, six from Montpelier and two from Historic St. Mary's City (HSMC) Maryland. She also analyzed SRM679, a National Bureau of Standards Brick Clay reference material. We compared the Montpelier bricks to those from HSMC because they were from a different location and a different time. The idea was that, if we were able to detect different compositions at all, we should be able to do so easily with bricks of such different origins.

Of the six Montpelier Bricks, four were from the house. They include one from a hearth in a 1760's-era cellar, one from a basement window in the Northeast wing known as Dolley's Wing, one from a window header in the Southwest wing known as Nelly's Wing, and one from a chimney constructed around 1797. Two bricks were from 2008 excavations of the South yard. The bricks from Historic St. Mary's City include one from a window mullion in the Great Brick Chapel of 1667 and one yellow brick believed to be Dutch.

Bricks are inhomogeneous and their surfaces are usually rough, potentially complicating PIXE analyses. To reduce these complications, we prepared homogeneous, smooth pellets by removing a 1-cm^3 sample from the interior of each brick, grinding it to a fine powder in a mortar, and pressing for several minutes at 3×10^7 Pa. The resulting pellets were 13 mm in diameter and about 4 mm thick.

We bombarded each pellet with a nominal 3-MeV proton beam and flooded it with helium gas. Because of energy loss in the kapton foil and a 2-cm path length of helium gas between the kapton and the sample, the proton energy at the sample surface was about 2.8 MeV [6]. The beam current was between 1 and 5 nA, and the integrated charge on each sample was between 1.2 and 6.0 μC. We placed 60-μm-thick beryllium foils between the sample and the detector to absorb some of the low-energy x-rays and reduce pileup. With the foils, x-ray count rates were about 2000 counts per second. Figure 2 shows a typical PIXE spectrum obtained from a brick pellet. The energy range extends from 1 to 20 keV, corresponding to elements from aluminum to zirconium.

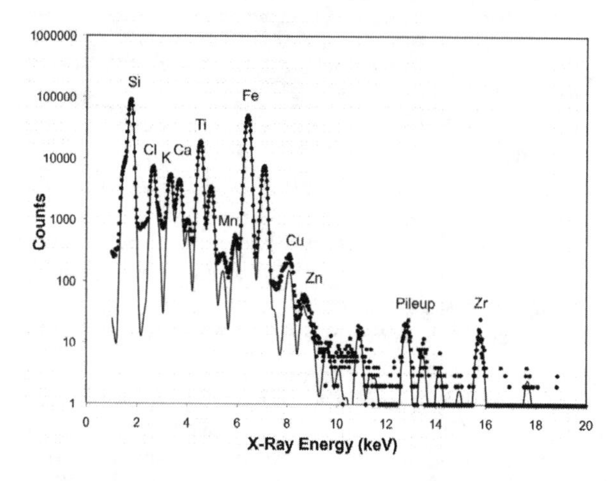

FIGURE 2. PIXE spectrum for a pellet made from the Dolley's Window brick. Dots are data and the line is a GUPIX fit. K-α x-ray peaks for some elements are labeled.

We used Windows GUPIX 2007 [7] in the thick target mode to calculate elemental concentrations. Table 1 lists the deduced concentrations for each sample. GUPIX estimates a lower limit of detectability for each element based, in part, on the spectral background. When the best-fit concentration for an element was zero or less than the limit of detectability, the table entry was set to "<".

TABLE 1. Measured compositions of brick samples from Montpelier, VA, and Historic St. Mary's City, MD. Uncertainties include contributions from counting statistics, absorber thickness, and fitting. The symbol "<" indicates that the best-fit concentration was either zero or less than the detectability limit.

Major Constituents (Weight per cent)

Element	1760's Cellar	Dolley's Window	Nelly's Wing	1797 Chimney	South Yard Red	South Yard Dark	HSMC Chapel	HSMC Dutch
Al	6.80 ± 0.45	7.14 ± 0.48	8.86 ± 0.55	8.24 ± 0.55	10.49 ± 0.69	4.81 ± 0.35	7.22 ± 0.48	6.20 ± 0.59
Si	32.75 ± 0.68	32.16 ± 0.61	29.49 ± 0.60	30.07 ± 0.59	24.69 ± 0.57	37.62 ± 0.72	36.38 ± 0.71	30.22 ± 0.65
K	0.63 ± 0.01	0.69 ± 0.01	0.86 ± 0.01	0.78 ± 0.01	0.69 ± 0.01	0.77 ± 0.01	1.57 ± 0.02	2.41 ± 0.03
Ca	0.36 ± 0.01	0.47 ± 0.01	0.43 ± 0.01	1.21 ± 0.01	0.34 ± 0.01	0.22 ± 0.01	0.46 ± 0.01	9.99 ± 0.06
Ti	3.42 ± 0.02	2.30 ± 0.01	2.01 ± 0.01	1.84 ± 0.01	3.05 ± 0.01	1.85 ± 0.01	0.85 ± 0.01	0.50 ± 0.01
Fe	7.65 ± 0.04	8.32 ± 0.02	11.50 ± 0.07	11.04 ± 0.02	16.11 ± 0.03	4.55 ± 0.01	3.44 ± 0.01	3.94 ± 0.02

Minor Constituents (Weight parts per million)

Element	1760's Cellar	Dolley's Window	Nelly's Wing	1797 Chimney	South Yard Red	South Yard Dark	HSMC Chapel	HSMC Dutch
S	<	<	<	<	<	451 ± 103	<	383 ± 212
Cl	122 ± 62	14235 ± 212	162 ± 65	335 ± 59	254 ± 68	<	<	5712 ± 214
V	517 ± 35	364 ± 32	479 ± 32	368 ± 33	437 ± 44	251 ± 39	141 ± 31	<
Cr	<	104 ± 11	80 ± 12	133 ± 14	<	143 ± 14	91 ± 13	272 ± 30
Mn	572 ± 17	539 ± 18	1489 ± 27	505 ± 20	579 ± 24	398 ± 20	107 ± 16	505 ± 38
Ni	49 ± 18	72 ± 18	112 ± 23	79 ± 23	162 ± 31	28 ± 16	<	73 ± 24
Cu	406 ± 25	324 ± 24	388 ± 28	293 ± 28	307 ± 34	107 ± 23	39 ± 21	<
Zn	212 ± 14	139 ± 14	208 ± 17	174 ± 16	194 ± 21	26 ± 11	43 ± 12	36 ± 18
Rb	66 ± 28	<	<	<	<	<	129 ± 45	<
Sr	48 ± 21	94 ± 25	43 ± 26	48 ± 31	<	<	105 ± 46	487 ± 118
Y	<	61 ± 25	<	83 ± 30	65 ± 38	<	<	<

We detected 12 to 18 elements, from aluminum to zirconium, in all of the pellets. Compositions have been published for the Brick Clay Standard [8] and for HSMC bricks similar to the ones we have [9]. Our results are reasonably consistent with the published ones, giving us confidence in our procedures.

As expected, the Montpelier bricks are distinguishable from the HSMC bricks: both HSMC bricks contain less iron and titanium than any Montpelier brick, but much more potassium. The Yellow Dutch brick contains much more calcium than any Montpelier brick and more chlorine than any except the one from Dolley's Window. There are also differences among the Montpelier bricks. Considering major elements, the brick from Dolley's Window contains more silicon, titanium, and chlorine than the one from Nelly's wing, but less aluminum, iron, and manganese. The South Yard Red Brick contains more aluminum, titanium, and iron than the South Yard Dark Brick, but less silicon. The 1797 Chimney brick contains more calcium than any of the other Montpelier bricks. Considering minor and trace elements, the 1797 Chimney brick contains much less zirconium than the other Montpelier bricks. The South Yard Dark brick contains much less zinc, copper, and nickel than other Montpelier samples, and is the only one to contain a detectable quantity of sulfur. The brick from Dolley's Window contains about twice as much strontium than the other Montpelier bricks, but less than the bricks from HSMC.

PIGE Analysis Of Limestone And Mortars

In a second study, Midshipman 3/C Nick Maskell (Class of 2012) acquired PIGE spectra for the same samples plus several additional ones, including two different types of mortar and limestone from which the mortars may have been made. PIGE complements PIXE because it can detect light elements that are difficult or impossible to detect with external-beam PIXE.

We made pellets from the new samples using the same procedure as before and bombarded them with 2.6 MeV protons (after energy losses in kapton and air). The beam current was about 15 nA and the integrated charge for each run was 20 μC. Figure 3 shows a PIGE spectrum for white mortar used in construction of a section of the mansion dated to 1765. Gamma rays that are commonly used [10] for quantifying light elements are labeled (^{19}F, 110 and 197 keV; ^{23}Na, 440 keV; ^{25}Mg, 585 keV; ^{27}Al, 844 and 1014 keV; ^{28}Si, 1779 keV; ^{29}Si, 1273 keV; and ^{35}Cl, 1219 keV).

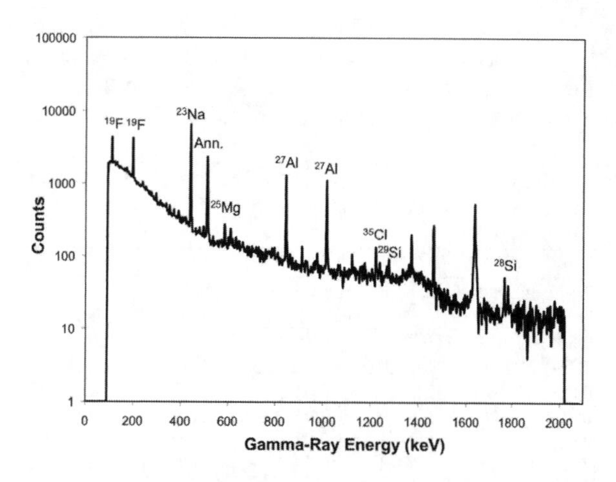

FIGURE 3. PIGE spectrum for a pellet made from the 1765 white mortar. Peaks used for quantifying the concentrations of several light elements are labeled.

Analysis of the PIGE results has just begun. However, we note that the composition of the limestone found in the underpinning of a wing built in 1810 is very similar to that of the 1765 white mortar, and that the composition of a 1765 tan mortar suggests that it is a mixture of the white mortar together with clay whose composition is much like that of the bricks. We also note that the bricks generally contain less fluorine, sodium and chlorine than the limestone and mortars, but much more aluminum and silicon. These results are not surprising, but they give us confidence that PIGE will be a useful technique for studying the compositions of these building materials.

SUMMARY

We have described an external-beam apparatus that we recently installed to support student research using PIXE and PIGE. In addition, we have presented some preliminary results from an ongoing study of colonial bricks and mortars from James Madison's Montpelier. More measurements and analysis are needed before we can draw any firm conclusions about the origins of and relationships among the materials.

ACKNOWLEDGMENTS

We wish to thank our technician, David Moore, who very ably maintains and runs our accelerator, and our machinist, Jeff Walbert, who expertly fabricated much of the apparatus. We also express our appreciation to Dr. Matthew Reeves, the Director of Archaeology at James Madison's Montpelier, who provided the samples.

REFERENCES

1. J. R. Vanhoy, D. J. Hartley, F. D. Correll, D. M. Moore and J. R. Huddle, "Accelerator-Based Laboratory Activities at USNA" in Application of Accelerators in Research and Industry: 20th International Conference, edited by F. D. McDaniel and B. L. Doyle, AIP Conference Proceedings 1099, American Institute of Physics, Melville, NY, 2009, pp. 214 – 219.
2. B. L. Doyle, D. S. Walsh and S. R. Lee, Nucl. Instr. and Meth. B 54, 244-257 (1991).
3. J. F. Ziegler, SRIM, http://www.srim.org.
4. Code of Federal Regulations, www.nrc.gov/reading-rm/doc-collections/cfr/part020/part020-1004.html.
5. NAVMED P-5055, Chapter 4, Table III, page 4-1, www.med.navy.mil/directives/Pub/5055.pdf.
6. National Institute of Standards and Technology, PSTAR, http://physics.nist.gov/PhysRefData/Star/Text/PSTAR.html.
7. Guelph PIXE Group, "Gupix and Gupixwin," http://pixe.physics.uoguelph.ca/gupix/.
8. NBS Certificate of Analysis, SRM 679, https://www-s.nist.gov/srmors/view_detail.cfm?srm=679.
9. R. A. Armitage, L. Minc, D. V. Hill and S. D. Hurry, J. Archaeological Sci. 33, 615-627 (2006).
10. A. Savidou, X. Aslanoglou, T. Paradellis and M. Pilakouta, Nucl. Instr. and Meth. B 152, 12-18 (1999).

Pyroelectric Crystal Accelerator In The Department Of Physics And Nuclear Engineering At West Point

Don Gillich[a], Mike Shannon[b], Andrew Kovanen[a], Tom Anderson[a], Kevin Bright[a], Ronald Edwards[a], Yaron Danon[c], Brian Moretti[a], and Jeffrey Musk[b]

a Department of Physics and Nuclear Engineering, United States Military Academy, West Point, NY 10996
b Nuclear Science and Engineering Research Center, Defense Threat Reduction Agency, West Point, NY 10996
c Department of Mechanical, Aerospace, and Nuclear Engineering, Rensselaer Polytechnic Institute, Troy, NY 12180.

Abstract. The Nuclear Science and Engineering Research Center (NSERC), a Defense Threat Reduction Agency (DTRA) office located at the United States Military Academy (USMA), sponsors and manages cadet and faculty research in support of DTRA objectives. The NSERC has created an experimental pyroelectric crystal accelerator program to enhance undergraduate education at USMA in the Department of Physics and Nuclear Engineering. This program provides cadets with hands-on experience in designing their own experiments using an inexpensive tabletop accelerator. This device uses pyroelectric crystals to ionize and accelerate gas ions to energies of ~100 keV. Within the next year, cadets and faculty at USMA will use this device to create neutrons through the deuterium-deuterium (D-D) fusion process, effectively creating a compact, portable neutron generator. The double crystal pyroelectric accelerator will also be used by students to investigate neutron, x-ray, and ion spectroscopy.

Keywords: Pyroelectric Crystal Accelerator, Educational Accelerator, Nuclear Fusion, Neutrons
PACS: 01, 24

INTRODUCTION

The Nuclear Science and Engineering Research Center (NSERC), a Defense Threat Reduction Agency (DTRA) office located at the United States Military Academy (USMA) at West Point, sponsors and manages cadet and faculty research in support of DTRA objectives. The NSERC mission is to increase USMA faculty and cadet research in support of DTRA objectives; enhance the professional development of USMA faculty; and contribute to the education of cadets, especially those majoring in nuclear engineering or physics.

The cadets at USMA represent the next generation of Army leaders and as such must be given the education and experience necessary for them to be successful junior officers. The NSERC-USMA link facilitates cadet work on projects to support DTRA scientific objectives, provides cadets with valuable educational opportunities, and directly addresses important scientific questions of the Department of Defense and other national agencies.

From the time the NSERC opened in 2007 to the present, the focus of cadet projects has primarily been theoretical and computational in nature. In many of these DTRA-sponsored projects, cadets used the Monte Carlo N-Particle (MCNP) transport code, to answer scientific questions related to safeguarding America and its allies from weapons of mass destruction. To further expand the educational opportunities provided to cadets, the NSERC has now established a number of experimental projects. These projects both provide the cadets hands-on experimental experience, and assist DTRA in developing compact and portable systems for homeland security and defense applications.

NSERC RESEARCH STRATEGY

Cadets majoring in nuclear engineering at USMA are required to complete a two-semester capstone design project. The topic and scope of these projects are provided by the NSERC, which works with DTRA to identify research areas appropriate for undergraduates. The NSERC then sponsors and funds summer internships for cadets providing the cadets with a "head start" on their project and extending the two semester project to a full year design experience.

Application of Accelerators in Research and Industry
AIP Conf. Proc. 1336, 763-766 (2011); doi: 10.1063/1.3586206

In addition, the NSERC funds faculty summer research in support of these projects.

Cadets and faculty conduct this NSERC-funded research at DTRA or other Department of Defense and Department of Energy locations during the summer and at USMA during the academic year. Project sponsors at DTRA (i.e. the "client") are actively engaged in these projects via video teleconference, periodic travel to USMA to work with cadets, and by attending USMA Projects Day in April. The NSERC also arranges for cadets and faculty to travel to DTRA to present their findings to the clients and DTRA leadership. In addition, the NSERC often funds cadets to attend professional conferences, such as the Hardened Electronics and Radiation Technology (HEART) Conference and conferences run by the American Physical Society and the American Nuclear Society. Presenting at these conferences provides cadets both verbal professional experience and the opportunity to get results published.

PYROELECTRIC ACCELERATOR

The pyroelectric accelerator project is an ideal example of NSERC-sponsored research and provides a valuable experimental experience for faculty and undergraduates. The project is relatively inexpensive (crystals cost less than $1000) and the vacuum system and radiation detection equipment were already on hand in the Department of Physics and Nuclear Engineering. In total, this project cost less than $10,000 to begin.

In addition to its relatively low cost, another important advantage to a pyroelectric accelerator is that the basic physics of pyroelectric acceleration is taught during introductory- level electricity and magnetism (E&M) physics courses. At USMA, cadets are required to take two semesters of calculus-based physics, which includes E&M. As such, cadets already have the basic knowledge to understand the underlying principles of a pyroelectric accelerator. Detecting the x-rays and neutrons generated by a pyroelectric crystal accelerator also provides a significant educational opportunity for cadets.

Nominal Experimental Equipment

A pyroelectric accelerator experimental apparatus requires a pyroelectric crystal (lithium tantalate (LiTaO3) is used here), a heating/cooling source (a thermoelectric heater/cooler), a direct current power supply, and a vacuum chamber. An x-ray detector is needed to detect the Bremsstrahlung radiation resulting from energetic electrons interacting with surrounding vacuum chamber materials. In addition, deuterium gas, a deuterated target, and a neutron detector are needed to conduct D-D nuclear fusion experiments. For this project, we assume that the measured x-ray end-point energy equals the maximum acceleration field potential influencing the charged particles. At USMA, a two crystal system is used which pairs a +z face opposing a –z face to allow for double the acceleration potential between the two crystals (superposition).

The Physics Of Pyroelectric Acceleration

Each pyroelectric crystal unit cell has an intrinsic dipole moment with the bulk crystal material exhibiting spontaneous polarization when heated or cooled. Spontaneous polarization, P_s, is defined as the dipole moment per unit volume of the material and is non-zero for pyroelectric materials[1]. Pyroelectric crystals are generally cut such that the dipole moments of the unit cells are arranged perpendicularly to the flat surfaces (named –z and +z faces).

During equilibrium conditions (constant temperature), free charged particles assemble at the faces of the crystal and screen the spontaneous polarization charge. When the crystal is heated (or cooled), however, the spontaneous polarization decreases (or increases) and charge moves to the crystal face to counteract the change. A thermal cycle is heating the crystal from equilibrium conditions (usually room temperature) to approximately 120 °C, holding the crystal at that temperature for a time to ensure thermal equilibrium across the crystal and then cooling the crystal back down to room temperature.

Experiments are usually done under vacuum at low ambient pressure (1 to 10 mTorr) to reduce the crystal discharge rate and to allow the electric potential to build by reducing the amount of free charge available at the crystal surface. The charge at the crystal surface creates an electrostatic potential which is used to ionize and accelerate residual gas particles. These accelerated ions (and electrons) interact with the surrounding environment (usually a vacuum chamber) to create Bremsstrahlung x-ray radiation.

The potential (V) generated by the pyroelectric effect can be calculated using the charge (Q) on the crystal surface divided by the equivalent capacitance (C_{eq}) of the system:

$$V = \frac{Q}{C_{eq}} \tag{1}$$

Typically a single crystal-target system is modeled as two parallel-plate capacitors added in parallel with one capacitor being the crystal itself and the other being the gap between the crystal and the target. The capacitance (C) of a parallel-plate capacitor is given by:

$$C = \frac{k\varepsilon_0 A}{d} \quad (2)$$

where k is the dielectric constant, ε_0 is the permittivity of free space, A is the plate area, and d is the distance between the two plates. Adding the two capacitors in parallel yields:

$$V = \frac{Q}{C_{eq}} = \frac{Q}{C_{cr} + C_{gap}} \quad (3)$$

where C_{cr} is the capacitance of the crystal and C_{gap} is the capacitance of the gap between the crystal and target.

The amount of surface charge on each face of the crystal is equal to the change in spontaneous polarization, ΔPs. This change can be calculated by multiplying the change in temperature by the pyroelectric coefficient, γ, such that:

$$\Delta P_s = Q = \gamma \Delta T \quad (4)$$

where γ is typically given in μC per m^2-K and ΔT is in K. For the $LiTaO_3$ crystals used in this project in the temperature range of interest (293 to 313 K), the value[2] for γ is 190 $\mu C/m^2$-K and ΔT is approximately 100.

POTENTIAL CADET PROJECTS

There are several aspects of the cadet pyroelectric crystal project that directly relate to DTRA's mission of safeguarding America and its allies from weapons of mass destruction. These include development of a portable, battery operated neutron source, a portable mass spectrometer for aerosol chemical detection, and a compact x-ray generator. All three of these are outlined in this paper. In addition to these specific applications, cadets also gain experience designing their own experiments to investigate a wide variety of other physical phenomena using a pyroelectric crystal accelerator.

Portable Battery Operated Neutron Source

Pyroelectric crystal accelerators have been used to achieve D-D nuclear fusion.[3,4] One of the products of D-D nuclear fusion is a highly penetrating 2.45 MeV neutron. The ultimate goal of this project is to develop an inexpensive, hand-carried neutron source for active interrogation of suspect packages at close range. Further work is required to develop pyroelectric acceleration technology to improve neutron yield per thermal cycle.

In the experimental setup presented here, a copper disk is used to collect the charge from the crystal surface and transfer it to a sharp tip. This tip is used to locally enhance the electric field thereby increasing ionization of deuterium gas introduced into the vacuum chamber. This same electric field accelerates the deuterium ions to high enough energies (100 – 300 keV) to achieve D-D fusion and produce 2.45 MeV neutrons as the ions impact a deuterated target. Figure 1 is a graphical depiction of a two-crystal pyroelectric crystal fusion experimental setup.

Potential student projects relating to neutron generation include, to name a few: designing, constructing, and testing a portable prototype neutron generator; novel pyroelectric crystal heating and cooling schemes; designing a thermal management system to precisely control the thermal cycles of the crystals; and using more than 2 pyroelectric crystals to achieve pyroelectric crystal fusion.

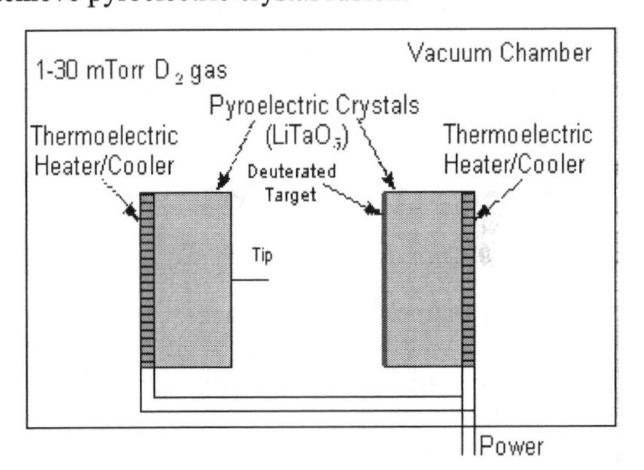

Figure 1. Experimental setup for a pyroelectric crystal accelerator neutron source.

Portable, Robust, Mass Spectrometer For Aerosol Chemical Detection

Previous work[5] has shown that pyroelectric crystal ionization can accelerate gases through a magnetic field. As such, a novel mass spectroscopy system using pyroelectric crystals as the ion source can be created using a pyroelectric-generated electric field to ionize the gas, accelerate ions through a magnetic field, and measure their deflection. More experimental work needs to be done to determine if different gas species can be detected and resolved. Additionally, field-enhancing tips or an array of nanorods, nanotubes, or nanowires may improve ionization of the various gases. Figure 2 graphically depicts an experimental setup for these types of experiments.

Additional materials necessary to conduct these types of experiments include: an ion detector (previous work used a low-light camera and a ZnS screen which illuminated when charge particles impacted it), various gases and gas mixtures, and various field-enhancing tips. Potential student projects in this area include: designing, constructing, and testing a mass spectrometer using pyroelectric crystals; conducting magnetic deflection simulations and experiments to determine an optimal magnetic field configuration; conducting experiments using various nanostructures to enhance the electric field; designing, constructing and testing an ion detection system; and using an air sampling system in conjunction with the mass spectrometer.

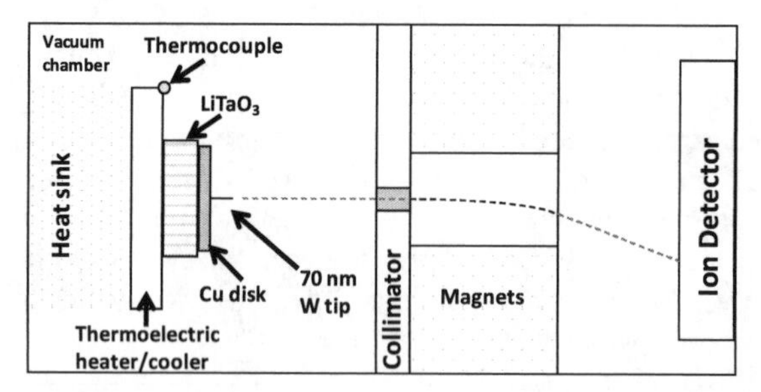

FIGURE 2. Experimental setup for a pyroelectric crystal accelerator Mass Spectrometer. The dotted line represents the path an ion would take through the magnetic field.

Compact X-Ray Generator

Building an inexpensive, hand-carried, energetic x-ray source for imaging of suspect packages is also a potential student project. Previous work[2,5] has shown that a one-crystal system, using a10-mm thick, 20-mm diameter cylindrical LiTaO3 crystal, will yield ~100 keV x-rays and a two-crystal system will yield ~200 keV x-rays. The x-rays generated can be used to interrogate suspect materials to determine elemental composition. Ultimately, this new technology may provide a highly compact, battery operated, and inexpensive x-ray source. A pocket-sized x-ray generator using pyroelectric crystals is commercially available with the Amptek Cool-X[6] which provides approximately 35 keV x-rays. The experimental project proposed here would use larger crystals to achieve higher energies and more penetrating x-rays. Figure 3 provides a sketch of the x-ray generator apparatus.

Potential student projects here include: designing, constructing and testing a portable, battery operated, and energetic x-ray source; conducting experiments using various target materials to determine which materials provide the highest x-ray yield; conducting experiments with an array of crystals (more than two) or larger crystals to achieve more energetic x-rays; designing and conducting x-ray fluorescence experiments for classroom use and for elemental analysis of unknown objects; and conducting imaging experiments to determine the feasibility of a portable x-ray machine that can be used for homeland security operations.

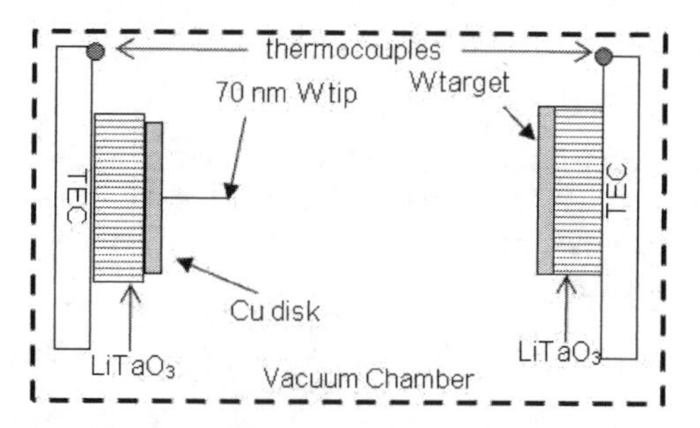

FIGURE 3. Experimental setup for a pyroelectric crystal accelerator x-ray generator. The tungsten (W) target is only an example, other types of target materials may be used to facilitate generation of x-rays.

CONCLUSION

The NSERC, through DTRA, is providing cadets at USMA with a hands-on educational experience using a pyroelectric crystal accelerator. The pyroelectric crystal accelerator at USMA has already been constructed and preliminary results are reported in Reference 7. The many student projects outlined in this paper are presented as examples of experiments that may be conducted once the experimental accelerator is fully constructed. This pyroelectric crystal accelerator is a relatively inexpensive and effective tool developed to give undergraduate students the opportunity to design and conduct their own experiments.

Disclaimer: The views expressed herein are those of the author and do not reflect the position of the United States Military Academy, the Department of the Army, the Defense Threat Reduction Agency, or the Department of Defense.

REFERENCES

1. Lang, S. B., "Pyroelectricity: From ancient curiosity to modern imaging tool", *Physics Today*, Aug 2005, pp. 31 – 36.
2. Geuther, J. A., Danon, Y., "High-Energy X Ray Production with Pyroelectric Crystals", *J. Appl. Phys.* 97, 104916 (2005).
3. Naranjo, B., Gimzewski, J., and Putterman, S., "Observation of nuclear fusion driven by a pyroelectric crystal," *Nature*, 434, (2005) pp. 1115-1117.
4. Geuther, J., Danon, Y., and Saglime, F., "Nuclear reactions induced by a pyroelectric accelerator," *Phys. Rev. Lett.* 96, 054803 (2006).
5. Gillich, D. J., "Particle Acceleration with Pyroelectric Crystals", Ph.D. Thesis, Rensselaer Polytechnic Institute, April 2009.
6. Amptek Website, http://www.amptek.com/coolx.html.
7. Anderson, T., et al., "Preliminary Results Using a Pyroelectric Crystal Accelerator at West Point," Proceedings of the 21st International Conference on the Application of Accelerators in Research and Industry (CAARI 2010), Ft Worth TX, August 8-13, (2010).

Preliminary Results from Pyroelectric Crystal Accelerator

Tom Anderson[1], Ronald Edwards[1], Kevin Bright[1], Andrew Kovanen[1], Yaron Danon[3], Brian Moretti[1], Jeffrey Musk[2], Mike Shannon[2], Don Gillich[1]

[1]*Department of Physics and Nuclear Engineering, United States Military Academy, West Point, NY 10996*
[2]*Nuclear Science and Engineering Research Center, Defense Threat Reduction Agency, West Point, NY 10996*
[3]*Dept. of Mechanical, Aerospace, and Nuclear Engineering, Rensselaer Polytechnic Institute, Troy, NY 12180*

Abstract. The Nuclear Science and Engineering Research Center (NSERC), a Defense Threat Reduction Agency (DTRA) office located at the United States Military Academy (USMA), sponsors and manages cadet and faculty research in support of DTRA objectives. Cadets in the Department of Physics and Nuclear Engineering at USMA are using pyroelectric crystals to ionize and accelerate residual gas trapped inside a vacuum system. A system using two lithium tantalate crystals with associated diagnostics was designed and is now operational. X-ray energies of approximately 150 keV have been achieved. Future work will focus on developing a portable neutron generator using the D-D nuclear fusion process.

Keywords: Pyroelectric Crystal Accelerator, Educational Accelerator, Nuclear Fusion, Neutrons
PACS: 24.90+d

INTRODUCTION

Pyroelectric crystals have been used to achieve deuterium-deuterium (D-D) nuclear fusion by ionizing and accelerating deuterium gas particles into a deuterated target.[1,2] Cadets at the United States Military Academy are conducting pyroelectric crystal research to ionize and accelerate residual gas particles, generate Bremsstrahlung x-rays, and ultimately achieve D-D nuclear fusion. The maximum end-point energy of these x-rays gauges the acceleration potential of the system and an end-point acceleration potential energy of the two crystal system of approximately 150 keV has been achieved.

THE PHYSICS OF PYROELECTRIC ACCELERATION

The driving force behind the ionization and acceleration of the residual gas particles is the electrostatic force generated by the pyroelectric effect. These crystal unit cells have intrinsic dipole moments that exhibit spontaneous polarization during heating or cooling past their equilibrium temperature. Spontaneous polarization is the dipole moment per unit volume of material and is non-zero for a heated or cooled pyroelectric crystal.[3] Typically, pyroelectric crystals are cut to align the dipoles along the flat surfaces, thereby creating +z and –z faces. Spontaneous polarization of the crystal decreases or increases during heating and cooling and a charge is generated on the flat crystal surfaces to compensate for these polarization changes. For example, a +z face during cooling will result in a net change in the polarization and the face becomes less positive. As a result, an electric field is generated and attracts ions towards and electrons away from the flat crystal surface. During cooling, the electric field will attract electrons towards and accelerate positive ions away from the crystal face. The negative face of the crystal, typically grounded for the crystal with the field-enhancing tip, displays the same properties for charges of the opposite sign.[4]

The electric potential generated by a pyroelectric crystal can be calculated by:

$$V = \frac{Q}{C_{eq}} = \frac{\gamma * \Delta T \cdot A}{C_{eq}} \qquad (1)$$

where γ, the pyroelectric coefficient, is normally given in terms of $\mu C/m^2$-K and ΔT is approximately 100 K for the experiments presented in this paper. C_{eq} represents the equivalent capacitance of the two-crystal system and A is the surface area of the crystal face. For $LiTaO_3$ crystals in the temperature range of interest, the value[5] for γ is 190 $\mu C\ m^{-2}\ K^{-1}$.

Application of Accelerators in Research and Industry
AIP Conf. Proc. 1336, 767-769 (2011); doi: 10.1063/1.3586207

Charged particles that are accelerated by the crystals interact with the chamber walls and other materials inside the chamber. The change in kinetic energy of these charged particles produces a Bremsstrahlung x-ray spectrum. The energy scale and maximum end-point energy of this spectrum is an indication of the acceleration potential of the system. In fusion experiments, it is assumed that deuterium ions are influenced by the same end-point energy of the acceleration potential between the two crystals.

EXPERIMENTAL SET-UP

The vacuum chamber was pumped down to an ambient pressure of 1 x 10^{-5} Torr to prevent electrostatic discharges from the crystals. Such discharges would necessitate a new thermal cycle to recharge the crystal faces. The vacuum was measured through pressure gauges. In fusion experiments, a gas leak valve will control the flow of D_2 gas into the vacuum chamber. In the experiments presented in this paper, the valve remained closed and only residual gas particles were ionized and accelerated to measure the Bremsstrahlung end-point energy and yield. Figure 1 shows a picture of the experimental setup.

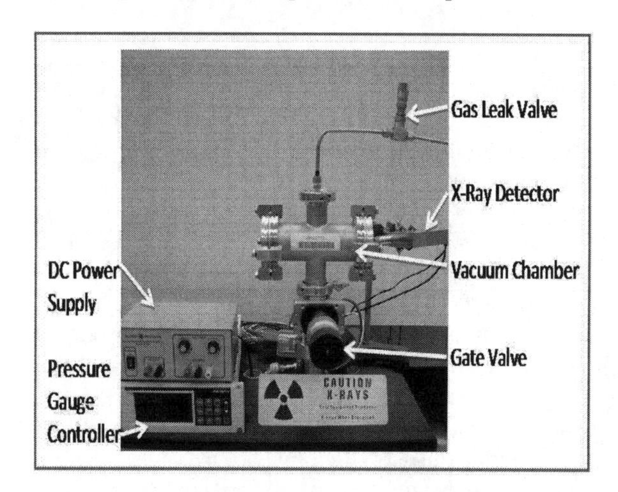

FIGURE 1. A picture of the experimental setup with associated diagnostics and power supply. The gas leak valve is not used until D_2 gas is added to the system. In the experiments presented here, Bremsstrahlung x-rays were generated by the ionization and acceleration of residual gas particles in the vacuum system.

Using thermally and electrically conductive epoxy, the 20 mm diameter, 10 mm thick crystals were glued to thermoelectric heater/coolers (TECs) which provided energy for the thermal cycles of the crystals. On the +z crystal face of one of the crystals, a copper disk collected the charge from the crystal surface and a 70 nm radius Tungsten tip was used to locally enhance the electric field and thereby, the ability to ionize residual gas particles. Both crystals had one of its faces grounded. The crystals are positioned on opposite ends of the cylindrical chamber such that the two electric fields generated by the crystals added by superposition to effectively double the potential between the crystals. A thermocouple was used to measure the TEC temperature and was assumed to be the same temperature at the grounded crystal surfaces. Figure 2 graphically depicts the experimental setup inside the vacuum chamber.

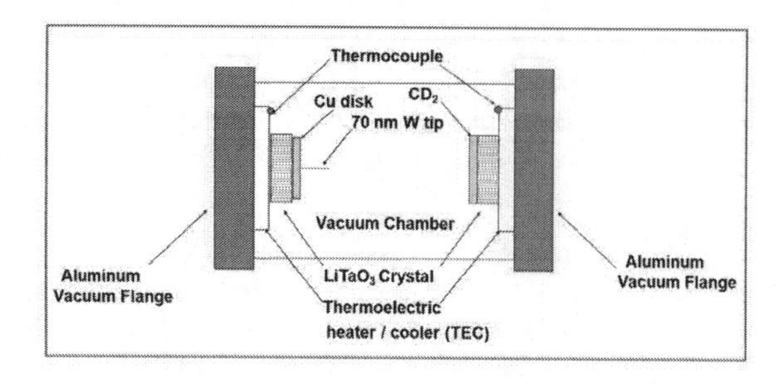

FIGURE 2. A diagram illustrating the experimental setup inside the vacuum chamber for the two-crystal pyroelectric accelerator at West Point.

The thermal cycle consisted of heating the crystal via the TEC for 5 minutes. The crystal was then set at a constant temperature for 5 minutes to allow for a more uniform heat distribution throughout the crystal. The TEC was then turned off and the crystal cooled down via conduction through the flange upon which it was mounted. The data was collected for 300 seconds.

EXPERIMENTAL RESULTS

Figure 3 shows the x-ray spectrum generated by a two-crystal pyroelectric system during the cooling phase of the thermal cycle as gathered by a sodium Iodide (NaI) detector. The Bremmstrahlung spectrum is an effect of the acceleration of positive residual gas ions away from the positive crystal face into the opposing target crystal and surrounding vacuum chamber walls. The maximum end-point energy of the x-rays is approximately 150 keV and it is assumed that residual gas ions experienced the same electric potential field. Previous results[1] have shown that this acceleration potential is adequate to achieve D-D nuclear fusion using a one crystal system. The Bremmstrahung spectrum appears cut off at lower energies. This is most likely due to absorption of lower energy x-rays in the aluminum flanges and the stainless steel chamber walls. To increase the probability of achieving fusion, USMA will replicate the two crystal system used at Rensselaer Polytechnic Institute.[2] Additional experimental work is planned to achieve 200 keV ion energies at USMA. This work will include synchronizing the thermal cycles of the two crystals with a feedback control loop. The goal will be to match the electric potentials from both crystals such that superposition of the two potentials will result in an increased electric potential between the crystals.

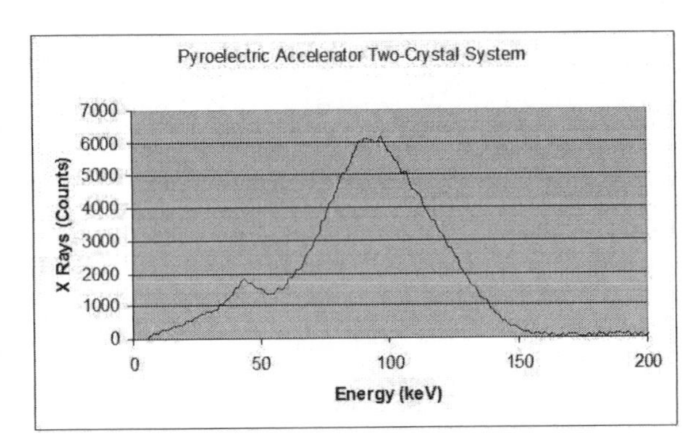

FIGURE 3. A graph of a Bremsstahlung x-ray spectrum generated by the field ionization and acceleration of residual gas particles from a two-crystal system inside a vacuum chamber over a 5 minute counting period. The end-point energy indicates an acceleration potential of approximately 150 keV. The lower energy x-rays may have been absorbed or scattered by the stainless steel vacuum chamber wall. A 5 minute background radiation spectrum was stripped out of these results.

FUTURE WORK

The ultimate goal of this project is to facilitate D-D fusion by leaking deuterium gas into the system and coating the negative crystal face (the target crystal) with deuterated polyethylene. Unlike fission, the probability of fusion is increased with high kinetic energy ions. To achieve higher energies, the acceleration potential must be optimized.

One method to optimize the acceleration potential is to precisely control the thermal cycle of the crystals through a proportional thermal controller.[4] This controller will utilize a LabVIEW program to read the temperature of the thermocouple, analyze the desired temperature at that time in the cycle, and adjust the TEC current accordingly. Once the acceleration potential is raised to 180- 200 keV, deuterium gas will be introduced into the system to achieve D-D nuclear fusion

Disclaimer: The views expressed herein are those of the author and do not reflect the position of the United States Military Academy, the Department of the Army, the Defense Threat Reduction Agency, or the Department of Defense.

REFERENCES

1. Naranjo, B., Gimzewski, J., and Putterman, S., Nature, 434, pp. 1115-1117 (2005).
2. Geuther, J., Danon, Y., and Saglime, F., Phys. Rev. Lett. 96, 054803 (2006).
1. Lang, S. B., “Pyroelectricity: From ancient curiosity to modern imaging tool”, Physics Today, pp. 31 - 36, Aug. 2005.
4. Gillich, D. J., “Particle Acceleration with Pyroelectric Crystals”, (PhD. Thesis, Rensselaer Polytechnic Institute, April 2009).
5. Geuther, J. A., Danon, Y., “High-Energy X Ray Production with Pyroelectric Crystals”, J. Appl. Phys. 97, 104916 (2005).

AUTHOR INDEX

C

D

E

F

G

H

I

J

K

L

M

N

O

P

Q

R

S

T

X

Y

Z